REMOTE SENSING HANDBOOK
VOLUME I

REMOTELY SENSED DATA CHARACTERIZATION, CLASSIFICATION, AND ACCURACIES

Remote Sensing Handbook

Remotely Sensed Data Characterization, Classification, and Accuracies

Land Resources Monitoring, Modeling, and Mapping with Remote Sensing

Remote Sensing of Water Resources, Disasters, and Urban Studies

REMOTE SENSING HANDBOOK
VOLUME I

REMOTELY SENSED DATA CHARACTERIZATION, CLASSIFICATION, AND ACCURACIES

Edited by
Prasad S. Thenkabail, PhD
United States Geological Survey (USGS)

CRC Press
Taylor & Francis Group
Boca Raton London New York

CRC Press is an imprint of the
Taylor & Francis Group, an **informa** business

CRC Press
Taylor & Francis Group
6000 Broken Sound Parkway NW, Suite 300
Boca Raton, FL 33487-2742

First issued in paperback 2021

Version Date: 20150513

ISBN 13: 978-0-367-86895-6 (pbk)
ISBN 13: 978-1-4822-1786-5 (hbk)

Library of Congress Cataloging-in-Publication Data

Remotely sensed data characterization, classification, and accuracies / editor, Prasad S. Thenkabail.
pages cm
Includes bibliographical references and indexes.
ISBN 978-1-4822-1786-5 (alk. paper)
1. Remote sensing--Data processing. I. Thenkabail, Prasad Srinivasa, 1958- editor of compilation.

G70.4.R49 2015
621.36'780285--dc23 2014050332

Visit the Taylor & Francis Web site at
http://www.taylorandfrancis.com

and the CRC Press Web site at
http://www.crcpress.com

I dedicate this work to my revered parents whose sacrifices gave me an education, as well as to all those teachers from whom I learned remote sensing over the years.

Contents

SECTION IV Vegetation Index Standardization and Cross-Calibration of Data from Multiple Sensors

SECTION V Image Processing Methods and Approaches

SECTION VI Change Detection

SECTION VII Integrating Geographic Information Systems (GIS) and Remote Sensing in Spatial Modeling Framework for Decision Support

Foreword: Satellite Remote Sensing Beyond 2015

Satellite remote sensing has progressed tremendously since Landsat 1 was launched on June 23, 1972. Since the 1970s, satellite remote sensing and associated airborne and in situ measurements have resulted in vital and indispensible observations for understanding our planet through time. These observations have also led to dramatic improvements in numerical simulation models of the coupled atmosphere–land–ocean systems at increasing accuracies and predictive capabilities. The same observations document the Earth's climate and are driving the consensus that *Homo sapiens* is changing our climate through greenhouse gas (GHG) emissions.

These accomplishments are the combined work of many scientists from many countries and a dedicated cadre of engineers who build the instruments and satellites that collect Earth observation (EO) data from satellites, all working toward the goal of improving our understanding of the Earth. This edition of the Remote Sensing Handbook (*Remotely Sensed Data Characterization, Classification, and Accuracies*; *Land Resources Monitoring, Modeling, and Mapping with Remote Sensing*; and *Remote Sensing of Water Resources, Disasters, and Urban Studies*) is a compendium of information for many research areas of our planet that have contributed to our substantial progress since the 1970s. The remote sensing community is now using multiple sources of satellite and in situ data to advance our studies, whatever they may be. In the following paragraphs, I will illustrate how valuable and pivotal satellite remote sensing has been in climate system study over the last five decades. The chapters in the handbook provide many other specific studies on land, water, and other applications using EO data of the last five decades.

The Landsat system of Earth-observing satellites has led the way in pioneering sustained observations of our planet. From 1972 to the present, at least one and sometimes two Landsat satellites have been in operation (Irons et al. 2012). Starting with the launch of the first NOAA–NASA Polar Orbiting Environmental Satellites NOAA-6 in 1978, improved imaging of land, clouds, and oceans and atmospheric soundings of temperature was accomplished. The NOAA system of polar-orbiting meteorological satellites has continued uninterrupted since that time, providing vital observations for numerical weather prediction. These same satellites are also responsible for the remarkable records of sea surface temperature and land vegetation index from the advanced very-high-resolution radiometers (AVHRRs) that now span more than 33 years, although no one anticipated these valuable climate records from this instrument before the launch of NOAA-7 in 1981 (Cracknell 1997).

The success of data from the AVHRR led to the design of the moderate-resolution imaging spectroradiometer (MODIS) instruments on NASA's Earth-Observing System (EOS) of satellite platforms that improved substantially upon the AVHRR. The first of the EOS platforms, Terra, was launched in 2000; and the second of these platforms, Aqua, was launched in 2002. Both of these platforms are nearing their operational life, and many of the climate data records from MODIS will be continued with the visible infrared imaging radiometer suite (VIIRS) instrument on the polar orbiting meteorological satellites of NOAA. The first of these missions, the NPOES Preparation Project (NPP), was launched in 2012 with the first VIIRS instrument that is operating currently among several other instruments on this satellite. Continuity of observations is crucial for advancing our understanding of the Earth's climate system. Many scientists feel that the crucial climate observations provided by remote sensing satellites are among the most important satellite measurements because they contribute to documenting the current state of our climate and how it is evolving. These key satellite observations of our climate are second in importance only to the polar orbiting and geostationary satellites needed for numerical weather prediction.

The current state of the art for remote sensing is to combine different satellite observations in a complementary fashion for what is being studied. Let us review climate change as an excellent example of using disparate observations from multiple satellite and in situ sources to observe climate change, verify that it is occurring, and understand the various component processes:

1. *Warming of the planet, quantified by radar altimetry from space*: Remotely sensed climate observations provide the data to understand our planet and what forces our climate. The primary climate observation comes from radar altimetry that started in late 1992 with Topex/Poseidon and has been continued by Jason-1 and Jason-2 to provide an uninterrupted record of global sea level. Changes in global sea level provide unequivocal evidence if our planet is warming,

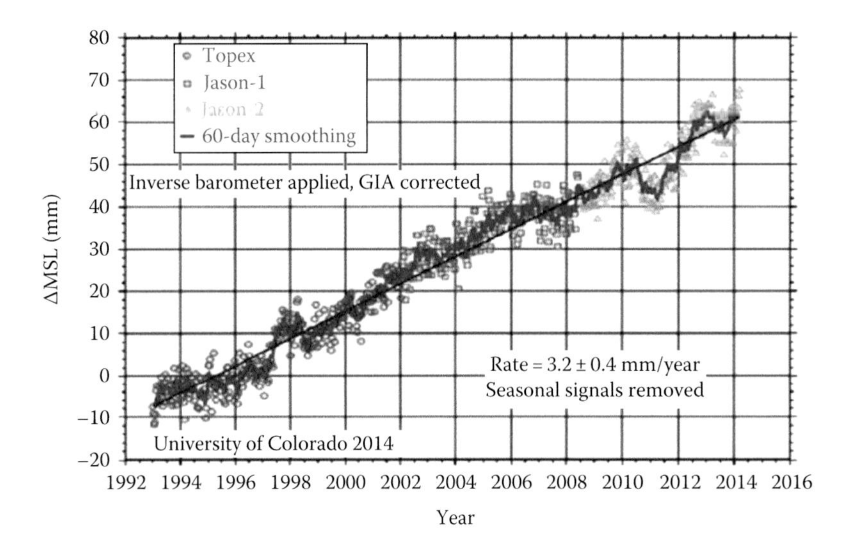

FIGURE F.1 Warming of the planet quantified by radar altimetry from space. Sea level determined from three radar altimeters from late 1992 to the present shows global sea level increases of ~3 mm/year. Sea level is the unequivocal indicator of the Earth's climate—when sea level rises, the planet is warming; when sea level falls, the planet is cooling. (From Gregory, J.M. et al., *J. Climate*, 26(13), 4476, 2013.)

cooling, or staying at the same temperature. Radar altimetry from 1992 to date has shown global sea level increases of ~3 mm/year, and hence, our planet is warming (Figure F.1). Sea level rise has two components, thermal expansion and ice melting in the ice sheets of Greenland and Antarctica, and to a much lesser extent, in glaciers.

2. The Sun is not to blame for global warming, based on solar irradiance data from satellites. Next, we consider two very different satellite observations and one in situ observing system that enable us to understand the causes of sea level variations: total solar irradiance, variations in the Earth's gravity field, and the Argo floats that record ocean temperature and salinity with depth, respectively.

Observations of total solar irradiance have been made from satellites since 1979 and show total solar irradiance has varied only ±1 part in 500 over the past 35 years, establishing that our Sun is not to blame for global warming (Figure F.2). Thus, we must look to other remotely sensed climate observations to explain and confirm sea level rise.

3. Sea level rise of 60% is explained by a mass balance of melting of ice measured by GRACE satellites. Since 2002, we have measured gravity anomalies from the Gravity Recovery and Climate Experiment Satellite (GRACE) dual satellite system. GRACE data quantify ice mass changes from the Antarctic and Greenland ice sheets (AIS and GIS) and concentrations of glaciers, such as in the Gulf of Alaska (GOA) (Luthcke et al. 2013). GRACE data are truly remarkable—their retrieval of variations in the Earth's gravity field is quantitatively and directly linked to mass variations. With GRACE data, we are able to determine for the first time the mass balance with time of the AIS and GIS and concentrations of glaciers on land. GRACE data show sea level rise of 60% explained by ice loss from

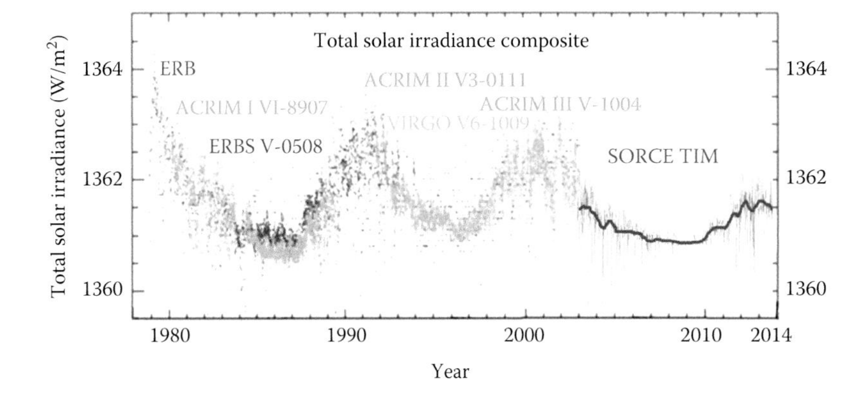

FIGURE F.2 The Sun is not to blame for global warming, based on solar irradiance data from satellites. Total solar irradiance reconstructed from multiple instruments dates back to 1979. The luminosity of our Sun varies only 0.1% over the course of the 11-year solar cycle. (From Froehlich, C., *Space Sci. Rev.*, 176(1–4), 237, 2013.)

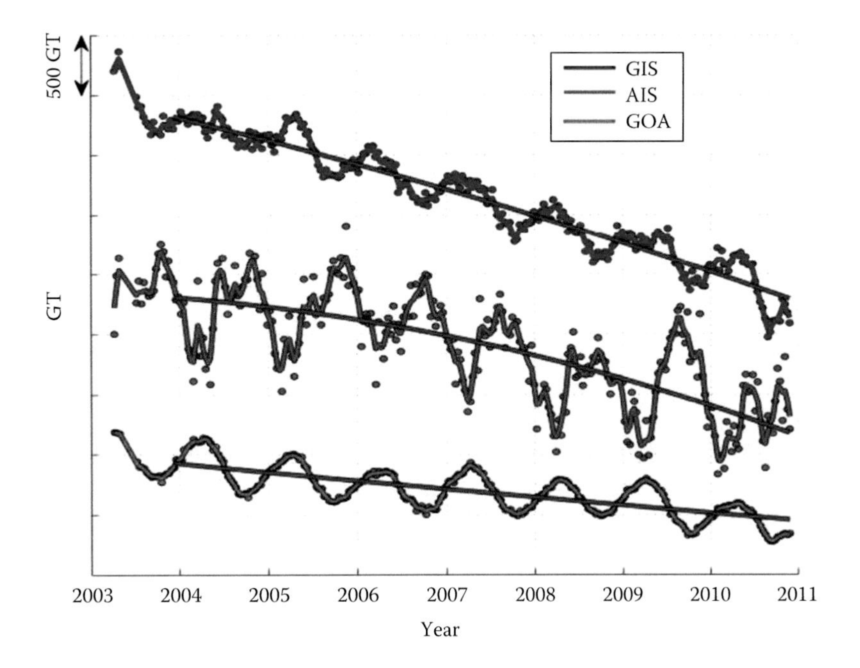

FIGURE F.3 Sea level rise of 60% explained by mass balance of melting of ice measured by GRACE satellites. Ice mass variations from 2003 to 2010 for the Antarctic ice sheets (AIS), Greenland ice sheets (GIS) and the Gulf of Alaska (GOA) glaciers using GRACE gravity data. (From Luthcke, S.B. et al., *J. Glaciol.*, 59(216), 613, 2013.)

land (Figure F.3). GRACE data have many other uses, such as indicating changes in groundwater storage, and readers are directed to the GRACE project's website if interested (http://www.csr.utexas.edu/grace/).

4. Sea level rise of 40% is explained by thermal expansion in the planet's oceans measured by in situ ~3700 drifting floats. The other contributor to sea level rise is thermal expansion in the planet's oceans. This necessitates using diving and drifting floats in the Argo network to record temperature with depth (Roemmich et al. 2009 and Figure F.4). Argo floats are deployed from ships; they then submerge and descend slowly to 1000 m depth, recording temperature, pressure, and salinity as they descend. At 1000 m depth, they drift for 10 days continuing their measurements of temperature and salinity. After 10 days, they slowly descend to 3000 m and then ascend to the surface, all the time recording their measurements. At the surface, each float transmits all the data collected on the most recent excursion to a geostationary satellite and then descends again to repeat this process.

Argo temperature data show that 40% of sea level rise results from the warming and thermal expansion of our oceans. Combining radar altimeter data, GRACE data, and Argo data provides a confirmation of sea level rise and shows what is

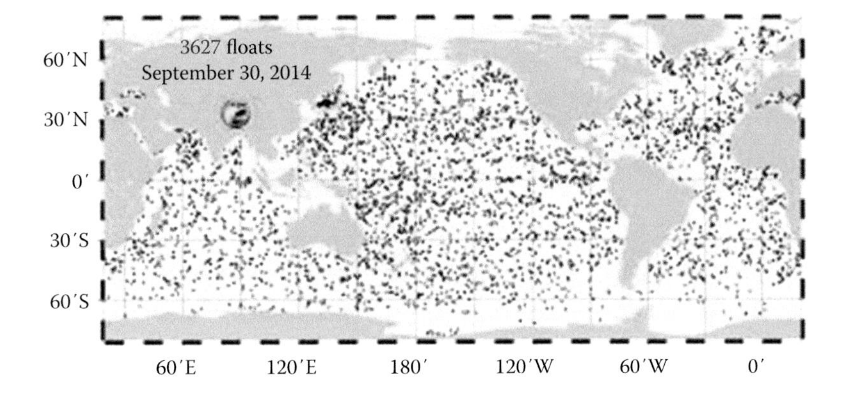

FIGURE F.4 Sea level rise of 40% explained by thermal expansion in the planet's oceans measured by in situ ~3700 drifting floats. This is the latest picture of the 3627 Argo floats that were in operation on September 30, 2014. These floats provide the data needed to document thermal expansion of the oceans. (From http://www.argo.ucsd.edu/.)

responsible for it and in what proportions. With total solar irradiance being near constant, what is driving global warming can be determined. The analysis of surface in situ air temperature coupled with lower tropospheric air temperature and stratospheric temperature data from remote sensing infrared and microwave sounders shows that the surface and near surface are warming while the stratosphere is cooling. This is an unequivocal confirmation that greenhouse gases are warming the planet.

Many scientists are actively working to study the Earth's carbon cycle, and there are several chapters in the handbook that deal with the components of this undertaking. Much like simultaneous observations of sea level, total solar irradiance, the gravity field, ocean temperature, surface temperature, and atmospheric temperatures were required to determine if the Earth is warming and what is responsible; the carbon cycle (Figure F.5) will require several complementary satellite and in situ observations (Cias et al. 2014).

Carbon cycles through reservoirs on the Earth's surface in plants and soils exist in the atmosphere as gases, such as carbon dioxide (CO_2) and methane (CH_4), and in ocean water in phytoplankton and marine sediments. CO_2 and CH_4 are released into the atmosphere by the combustion of fossil fuels, land cover changes on the Earth's surface, respiration of green plants, and decomposition of carbon in dead vegetation and in soils, including carbon in permafrost. The atmospheric concentrations of CO_2 and CH_4 control atmospheric and oceanic temperatures through their absorption of outgoing long-wave radiation and thus also indirectly control sea level via the regulation of planetary ice volumes.

Satellite-borne sensors provide simultaneous global carbon cycle observations needed for quantifying carbon cycle processes, that is, to measure atmospheric CO_2 concentrations and emission sources, to measure land and ocean photosynthesis, to measure the reservoir of carbon in plants on land and its change, to measure the extent of biomass burning of vegetation on land, and to measure soil respiration and decomposition, including decomposing carbon in permafrost. In addition to the required satellite observations, in situ observations are needed to confirm satellite-measured CO_2 concentrations and determine soil and vegetation carbon quantities. Understanding the carbon cycle requires a full court press of satellite and in situ observations because all of these observations must be made at the same time. Many of these measurements have been made over the past 30–40 years, but new measurements are needed to quantify carbon storage in vegetation, atmospheric measurements are needed to quantify CH_4 and CO_2 sources and sinks, better measurements are needed to quantify land respiration, and more explicit numerical carbon models need to be developed.

Similar work needs to be performed for the role of clouds and aerosols in climate because these are fundamental to understanding our radiation budget. We also need to improve our understanding of the global hydrological cycle.

The remote sensing community has made tremendous progress over the last five decades as discussed in this edition of the handbook. Chapters on aerosols in climate, because these are fundamental, provide comprehensive understanding of land and water studies through detailed methods, approaches, algorithms, synthesis, and key references. Every type of remote

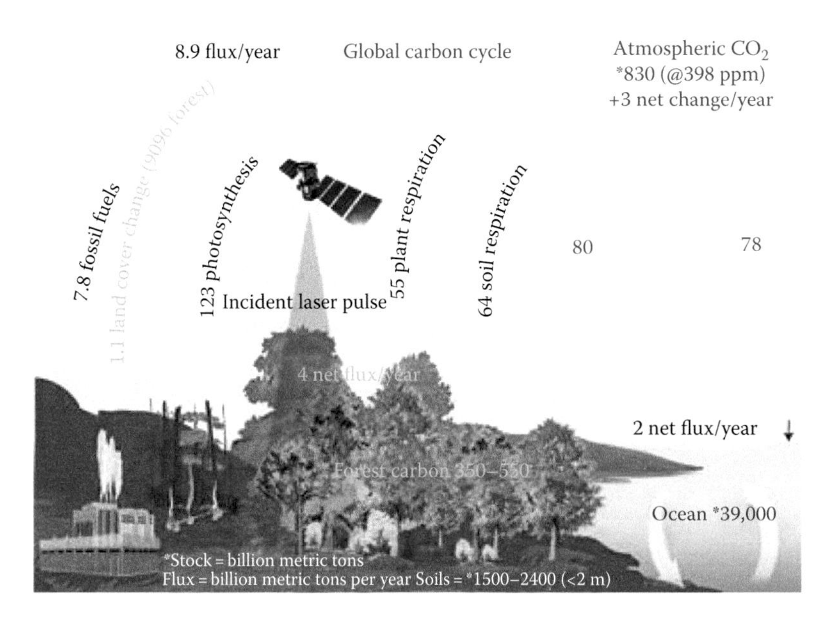

FIGURE F.5 Global carbon cycle measurements from a multitude of satellite sensors. A representation of the global carbon cycle showing our best estimates of carbon fluxes and carbon reservoirs. A series of satellite observations are needed simultaneously to understand the carbon cycle and its role in the Earth's climate system. (From Cias, P. et al., *Biogeosciences*, 11(13), 3547, 2014.)

sensing data obtained from systems such as optical, radar, light detection and ranging (LiDAR), hyperspectral, and hyperspatial is presented and discussed in different chapters. *Remotely Sensed Data Characterization, Classification, and Accuracies* sets the stage with chapters in this book addressing remote sensing data characteristics, within and between sensor calibrations, classification methods, and accuracies taking a wide array of remote sensing data from a wide array of platforms over the last five decades. *Remotely Sensed Data Characterization, Classification, and Accuracies* also brings in technologies closely linked with remote sensing such as global positioning system (GPS), global navigation satellite system (GNSS), crowdsourcing, cloud computing, and remote sensing law. In all, the 82 chapters in the 3 volumes of the handbook are written by leading and well-accomplished remote sensing scientists of the world and competently edited by Dr. Prasad S. Thenkabail, Research Geographer-15, at the United States Geological Survey (USGS).

We can look forward in the next 10–20 years to improving our quantitative understanding of the global carbon cycle, understanding the interaction of clouds and aerosols in our radiation budget, and understanding the global hydrological cycle. There is much work to do. Existing key climate observations must be continued and new satellite observations will be needed (e.g., the recently launched NASA's Orbiting Carbon Observatory-2 for atmospheric CO_2 measurements), and we have many well-trained scientists to undertake this work and continue the legacy of the past five decades.

References

Ciais, P. et al. 2014. Current systematic carbon-cycle observations and the need for implementing a policy-relevant carbon observing system. *Biogeosciences* 11(13): 3547–3602.

Cracknell, A. P. 1997. *The Advanced Very High Resolution Radiometer (AVHRR).* Taylor & Francis, U.K., 534pp.

Froehlich, C. 2013. Total solar irradiance: What have we learned from the last three cycles and the recent minimum? *Space Science Reviews* 176(1–4): 237–252.

Gregory, J. M. et al. 2013. Twentieth-century global-mean sea level rise: Is the whole greater than the sum of the parts? *Journal of Climate* 26(13): 4476–4499.

Irons, J. R., Dwyer, J. L., and Barsi, J. A. 2012. The next Landsat satellite: The Landsat data continuity mission. *Remote Sensing of Environment* 122: 11–21. doi:10.1016/j.rse.2011.08.026.

Luthcke, S. B., Sabaka, T. J., Loomis, B. D. Arendt, A. A., McCarthy, J. J., and Camp, J. 2013. Antarctica, Greenland, and Gulf of Alaska land-ice evolution from an iterated GRACE global mascon solution. *Journal of Glaciology* 59(216): 613–631.

Roemmich, D. and the Argo Steering Team, 2009. Argo—The challenge of continuing 10 years of progress. *Oceanography* 22(3): 46–55.

Compton J. Tucker
Earth Science Division
Goddard Space Flight Center
National Aeronautics and Space Administration
Greenbelt, Maryland

Preface: Remote Sensing Advances of the Last 50 Years and a Vision for the Future

The overarching goal of the Remote Sensing Handbook (*Remotely Sensed Data Characterization, Classification, and Accuracies*; *Land Resources Monitoring, Modeling, and Mapping with Remote Sensing*; and *Remote Sensing of Water Resources, Disasters, and Urban Studies*), with 82 chapters and about 2500 pages, was to capture and provide the most comprehensive state of the art of remote sensing science and technology development and advancement in the last 50 years, by clearly demonstrating the (1) scientific advances, (2) methodological advances, and (3) societal benefits achieved during this period, as well as to provide a vision of what is to come in the years ahead. The three books are, to date and to the best of my knowledge, the most comprehensive documentation of the scientific and methodological advances that have taken place in understanding remote sensing data, methods, and a wide array of land and water applications. Written by 300+ leading global experts in the area, each chapter (1) focuses on a specific topic (e.g., data, methods, and applications), (2) reviews the existing state-of-the-art knowledge, (3) highlights the advances made, and (4) provides guidance for areas requiring future development. Chapters in the books cover a wide array of subject matter of remote sensing applications. The Remote Sensing Handbook is planned as a reference material for remote sensing scientists, land and water resource practitioners, natural and environmental practitioners, professors, students, and decision makers. The special features of the Remote Sensing Handbook include the following:

1. Participation of an outstanding group of remote sensing experts, an unparalleled team of writers for such a book project
2. Exhaustive coverage of a wide array of remote sensing science: data, methods, and applications
3. Each chapter being led by a luminary and most chapters written by teams who further enriched the chapters
4. Broadening the scope of the book to make it ideal for expert practitioners as well as students
5. Global team of writers, global geographic coverage of study areas, and a wide array of satellites and sensors
6. Plenty of color illustrations

Chapters in the books cover the following aspects of remote sensing:

State of the art
Methods and techniques
Wide array of land and water applications
Scientific achievements and advancements over the last 50 years
Societal benefits
Knowledge gaps
Future possibilities in the twenty-first century

Great advances have taken place over the last 50 years using remote sensing in the study of the planet Earth, especially using data gathered from a multitude of Earth observation (EO) satellites launched by various governments as well as private entities. A large part of the initial remote sensing technology was developed and tested during the two world wars. In the 1950s, remote sensing slowly began its foray into civilian applications. During the years of the Cold War, remote sensing applications, both civilian and military, increased swiftly. But it was also an age when remote sensing was the domain of a very few top experts and major national institutes, having multiple skills in engineering, science, and computer technology. From the 1960s onward, there have been many governmental agencies that have initiated civilian remote sensing. The National Aeronautics and Space Administration (NASA) and the United States Geological Survey (USGS) have been in the forefront of many of these efforts. Others who have provided leadership in civilian remote sensing include, but are not limited to, the European Space Agency (ESA) of the European Union, the Indian Space Research Organization (ISRO), the Centre National d'Études Spatiales (CNES) of France, the Canadian Space Agency (CSA), the Japan Aerospace Exploration Agency (JAXA), the German Aerospace Center (DLR), the China National Space Administration (CNSA), the United Kingdom Space Agency (UKSA), and the Instituto Nacional de Pesquisas Espaciais (INPE) of Brazil. Many private entities have launched and operated satellites. These government and private agencies and enterprises launched and operated a wide array of satellites and sensors that captured the data of the planet Earth in various regions of the electromagnetic spectrum and in various spatial, radiometric, and temporal resolutions, routinely and repeatedly. However, the real thrust for remote sensing advancement came during the last decade of the twentieth century and the beginning of the twenty-first century. These initiatives included a launch of a series of new-generation EO satellites to gather data more frequently and routinely, release of pathfinder datasets, web enabling the data for free by many agencies (e.g., USGS release of the entire Landsat archives as well as real-time acquisitions of the world for free dissemination by web-enabling), and providing processed data ready to users (e.g., surface reflectance products of moderate-resolution imaging spectroradiometer [MODIS]). Other efforts like Google Earth made remote sensing more popular and brought in a new platform for easy visualization and navigation of remote sensing data. Advances in computer hardware and software made it possible to handle Big Data. Crowdsourcing, web access, cloud computing, and mobile platforms added a new dimension to how remote sensing data are used. Integration with global positioning systems (GPS) and global navigation satellite systems (GNSS) and inclusion of digital secondary data (e.g., digital elevation, precipitation, temperature) in analysis have made remote sensing much more powerful. Collectively, these initiatives provided a new vision in making remote sensing data more popular, widely understood, and increasingly used for diverse applications, hitherto considered difficult. The free availability of archival data when combined with more recent acquisitions has also enabled quantitative studies of change over space and time. The Remote Sensing Handbook is targeted to capture these vast advances in data, methods, and applications, so a remote sensing student, scientist, or a professional practitioner will have the most comprehensive, all-encompassing reference material in one place.

Modern-day remote sensing technology, science, and applications are growing exponentially. This growth is a result of a combination of factors that include (1) advances and innovations in data capture, access, and delivery (e.g., web enabling, cloud computing, crowdsourcing); (2) an increasing number of satellites and sensors gathering data of the planet, repeatedly and routinely, in various portions of the electromagnetic spectrum as well as in an array of spatial, radiometric, and temporal resolutions; (3) efforts at integrating data from multiple satellites and sensors (e.g., sentinels with Landsat); (4) advances in data normalization, standardization, and harmonization (e.g., delivery of data in surface reflectance, intersensor calibration); (5) methods and techniques for handling very large data volumes (e.g., global mosaics); (6) quantum leap in computer hardware and software capabilities (e.g., ability to process several terabytes of data); (7) innovation in methods, approaches, and techniques leading to sophisticated algorithms (e.g., spectral matching techniques, and automated cropland classification algorithms); and (8) development of new spectral indices to quantify and study specific land and water parameters (e.g., hyperspectral vegetation indices or HVIs). As a result of these all-around developments, remote sensing science is today very mature and is widely used in virtually every discipline of the earth sciences for quantifying, mapping, modeling, and monitoring our planet Earth. Such rapid advances are captured in a number of remote sensing and earth science journals. However, students, scientists, and practitioners of remote sensing science and applications have significant difficulty gathering a complete understanding of the various developments and advances that have taken place as a result of their vastness spread across the last 50 years. Therefore, the chapters in the Remote Sensing Handbook are designed to give a whole picture of scientific and technological advances of the last 50 years.

Today, the science, art, and technology of remote sensing are truly ubiquitous and increasingly part of everyone's everyday life, often without the user knowing it. Whether looking at your own home or farm (e.g., see the following figure), helping you navigate when you drive, visualizing a phenomenon occurring in a distant part of the world (e.g., see the following figure), monitoring events such as droughts and floods, reporting weather, detecting and monitoring troop movements or nuclear sites, studying deforestation, assessing biomass carbon, addressing disasters such as earthquakes or tsunamis, and a host of other applications (e.g., precision farming, crop productivity, water productivity, deforestation, desertification, water resources management), remote sensing plays a pivotal role. Already, many new innovations are taking place. Companies such as the Planet Labs and Skybox are planning to capture very-high-spatial-resolution imagery (typically, sub-meter to 5 meters), even videos from space using a large number of microsatellite constellations. There are others planning to launch a constellation of hyperspectral or other sensors. Just as the smartphone and social media connected the world, remote sensing is making the world our backyard. No place goes unobserved and no event gets reported without a satellite or other kinds of remote sensing images or their derivatives. This is how true liberation for any technology and science occurs.

Google Earth can be used to seamlessly navigate and precisely locate any place on Earth, often with very-high-spatial-resolution data (VHRI; submeters to 5 m) from satellites such as IKONOS, QuickBird, and GeoEye (Note: the image below is from one of the VHRI). Here, the editor-in-chief (EiC) of this handbook located his village home (Thenkabail) and surroundings that have land covers such as secondary rainforests, lowland paddy farms, areca nut plantations, coconut plantations, minor roads, walking routes, open grazing lands, and minor streams (typically, first and second order) (note: land cover detailed is based on the ground knowledge of the EiC). The first primary school attended by him is located precisely. Precise coordinates (13 degree 45 minutes 39.22 seconds northern latitude, 75 degrees 06 minutes 56.03 seconds eastern longitude) of Thenkabail's village house on the planet and the date

of image acquisition (March 1, 2014) are noted. Google Earth images are used for visualization as well as for numerous science applications such as accuracy assessment, reconnaissance, determining land cover, and establishing land use for various ground surveys. It is widely used by lay people who often have no idea on how it all comes together but understand the information provided intuitively. This is already happening. These developments make it clear that we not only need to understand the state of the art but also have a vision of where the future of remote sensing is headed. Therefore, in a nutshell, the goal of this handbook is to cover the developments and advancement of six distinct eras in terms of data characterization and processing as well as myriad land and water applications:

1. *Pre-civilian remote sensing era of the pre-1950s*: World War I and II when remote sensing was a military tool
2. *Technology demonstration era of the 1950s and 1960s*: Sputnik-I and NOAA AVHRR era of the 1950s and 1960s
3. *Landsat era of the 1970s*: when the first truly operational land remote sensing satellite (Earth Resources Technology Satellite or ERTS, later renamed Landsat) was launched and operated in the 1970s and early 1980s by United States
4. *Earth observation era of the 1980s and 1990s*: when a number of space agencies began launching and operating satellites (e.g., Landsat 4,5 by the United States; SPOT-1,2 by France; IRS-1a, 1b by India) from the middle to late 1980s onward till the middle of 1990s
5. *Earth observation and the first decade of the New Millennium era of the 2000s*: when data dissemination to users became as important as launching, operating, and capturing data (e.g., MODIS Terra\Aqua, Landsat-8, Resourcesat) in the late 1990 and the first decade of the 2000s
6. *Second decade of the New Millennium era starting in the 2010s*: when new-generation micro-\nanosatellites (e.g., PlanetLabs, Skybox) are added to the increasing constellation of multiagency sensors (e.g., Sentinels, and the next generation of satellites such as SMAP, hyperspectral satellites like NASA's HyspIRI and others from private industry)

Motivation for the Remote Sensing Handbook started with a simple conversation with Irma Shagla-Britton, acquisitior editor for remote sensing and GIS books of Taylor & Francis Group/CRC Press, way back in early 2013. Irma was informally getting my advice about "doing a new and unique book" on remote sensing. Neither the specific subject nor the editor was identified. What was clear to me though was that I certainly did not want to lead the effort. I was nearing the end of my third year of recovery from colon cancer, and the last thing I wanted to do was to take any book project, forget a multivolume remote sensing magnum opus, as it ultimately turned out. However, mostly out of courtesy for Irma, I did some preliminary research. I tried to identify a specific topic within remote sensing where there was a sufficient need for a full-fledged book. My research showed that there was not a single book that would provide a complete and comprehensive coverage of the entire subject of remote sensing starting from data capture, to data preprocessing, to data analysis, to myriad land and water applications. There are, of course, numerous excellent books on remote sensing, each covering a specific subject matter. However, if a student, scientist, or practitioner of remote sensing wanted a standard reference on the subject, he or she would have to look for numerous books or journal articles and often a coherence of these topics would still be left uncovered or difficult to comprehend for students and even for many experts with less experience. Guidance on how to approach the study of remote sensing and capture its state of the art and advances remained hazy and often required referring to a multitude of references that may or may not be immediately available, and if available, how to go about it was still hazy to most. During this process, I asked myself, several times, what remote sensing book will be most interesting, productive, and useful to a broad audience? The answer, each time, was very clear: "A complete and comprehensive coverage of the state-of-the-art remote sensing, capturing the advances that have taken place over the last 50 years, which will set the stage for a vision for the future." When this became clear, I started putting together the needed topics to achieve such a goal. Soon I realized that the only way to achieve this goal was through a multivolume book on remote sensing. Because the number of chapters was more than 80, this appeared to be too daunting, too overwhelming, and too big a project to accomplish. Yet I sent the initial idea to Irma, who I thought would say "forget it" and ask me to focus on a single-volume book. But to my surprise, Irma not only encouraged the idea but also had a number of useful suggestions. So what started as intellectual curiosity turned into this full-fledged multivolume Remote Sensing Handbook.

However, what worried me greatly was the virtual impossibility (my thought at that time) of gathering the best authors. What was also crystal clear to me was that unless the very best were attracted to the book project, it was simply not worth the effort. I had made up my mind to give up the book project, unless I got the full support of a large number of the finest practitioners of remote sensing from around the world. So, I spent a few weeks researching the best authors to lead each chapter and wrote to them to participate in the Remote Sensing Handbook project. What really surprised me was that almost all the authors I contacted agreed to lead and write a chapter. This was truly surreal. These are extremely busy people of great scientific reputation and achievements. For them to spend the time, intellect, and energy to write an in-depth and insightful book chapter spread across a year or more is truly amazing. Most also agreed to put together a writing team, as I had requested, to ensure greater perspective for each chapter. In the end, we had 300+ authors writing 82 chapters.

At this stage, I was somewhat drawn into the project as if by destiny and felt compelled to go ahead. One of the authors who agreed to lead the chapter mentioned ".....whether it was even possible." This is exactly what I felt, too. But I had reached the stage of no return, and I took on the book project with all the seriousness it deserved. It required some real changes to my lifestyle: professional and personal. Travel was reduced to bare minimum during most of the book project. Most weekends were spent editing, writing, and organizing, and other social activities were reduced. Accomplishing such complex work requires the highest levels of discipline, planning, and strategy. But, above all, I felt blessed with good health. By the time the book is published, I will have completed about 5 years from my colon cancer surgery and chemotherapy. So I am as happy to see this book released as I am with the miracle of cancer cure (I feel confident to say so).

But it is the chapter authors who made it all feasible. They amazed me throughout the book project. First, the quality and content of each of the chapters were of the highest standards. Second, with very few exceptions, chapters were delivered on time. Third, edited chapters were revised thoroughly and returned on time. Fourth, all my requests on various formatting and quality enhancements were addressed. This is what made the three-volume Remote Sensing Handbook possible and if I may say so, a true *magnum opus* on the subject. My heartfelt gratitude to these great authors for their dedication. It has been my great honor to work with these dedicated legends. Indeed, I call them my *heroes* in a true sense.

Overall, the preparation of the Remote Sensing Handbook took two and a half years, from the time book chapters and authors were being identified to its final publication. The three books are designed in such a way that a reader can have all three books as a standard reference or have individual books to study specific subject areas. The three books of Remote Sensing Handbook are

Remotely Sensed Data Characterization, Classification, and Accuracies: 31 Chapters
Land Resources Monitoring, Modeling, and Mapping with Remote Sensing: 28 Chapters
Remote Sensing of Water Resources, Disasters, and Urban Studies: 27 Chapters

There are about 2500 pages in the 3 volumes.

The wide array of topics covered is very comprehensive. The topics covered in *Remotely Sensed Data Characterization, Classification, and Accuracies* include (1) satellites and sensors; (2) remote sensing fundamentals; (3) data normalization, harmonization, and standardization; (4) vegetation indices and their within- and across-sensor calibration; (5) image classification methods and approaches; (6) change detection; (7) integrating remote sensing with other spatial data; (8) GNSS; (9) crowdsourcing; (10) cloud computing; (11) Google Earth remote sensing; (12) accuracy assessments; and (13) remote sensing law.

The topics covered in *Land Resources Monitoring, Modeling, and Mapping with Remote Sensing* include (1) vegetation and biomass, (2) agricultural croplands, (3) rangelands, (4) phenology and food security, (5) forests, (6) biodiversity, (7) ecology, (8) land use/land cover, (9) carbon, and (10) soils.

The topics covered in *Remote Sensing of Water Resources, Disasters, and Urban Studies* include (1) hydrology and water resources; (2) water use and water productivity; (3) floods; (4) wetlands; (5) snow and ice; (6) glaciers, permafrost, and ice; (7) geomorphology; (8) droughts and drylands; (9) disasters; (10) volcanoes; (11) fire; (12) urban areas; and (13) nightlights.

There are many ways to use the Remote Sensing Handbook. A lot of thought went into organizing the books and chapters. So you will see a *flow* from chapter to chapter and book to book. As you read through the chapters, you will see how they are interconnected and how reading all of them provides you with greater in-depth understanding. Some of you may be more interested in a particular volume. Often, having all three books as reference material is ideal for most remote sensing experts, practitioners, or students; however, you can also refer to individual books based on your interest. We have also made attempts to ensure the chapters are self-contained. That way you can focus on a chapter and read it through, without having to be overly dependent on other chapters. Taking this perspective, there is a slight (~5%–10%) material that may be repeated in some of the chapters. This is done deliberately. For example, when you are reading a chapter on LiDAR or radar, you don't want to go all the way back to another chapter (e.g., Chapter 1, *Remotely Sensed Data Characterization, Classification, and Accuracies*) to understand the characteristics of these sensors.

Similarly, certain indices (e.g., vegetation condition index [VCI], temperature condition index [TCI]) that are defined in one chapter (e.g., on drought) may be repeated in another chapter (also on drought). Such minor overlaps are helpful to the reader to avoid going back to another chapter to understand a phenomenon or an index or a characteristic of a sensor. However, if you want a lot of details on these sensors or indices or phenomena or if you are someone who has yet to gain sufficient expertise in the field of remote sensing, then you will have to read the appropriate chapter where there is in-depth coverage of the topic.

Each book has a summary chapter (the last chapter of each book). The summary chapter can be read two ways: (1) either as a last chapter to recapture the main points of each of the previous chapters or (2) as an initial overview to get a feeling for what is in the book. I suggest the readers do it both ways: Read it first before going into the details and then read it at the end to recollect what was said in the chapters.

It has been a great honor as well as a humbling experience to edit the Remote Sensing Handbook (*Remotely Sensed Data Characterization, Classification, and Accuracies*; *Land Resources Monitoring, Modeling, and Mapping with Remote Sensing*; and *Remote Sensing of Water Resources, Disasters, and Urban Studies*). I truly enjoyed the effort. What an honor to work with luminaries in this field of expertise. I learned a lot from them and am very grateful for their support, encouragement, and deep insights. Also, it has been a pleasure working with outstanding professionals of Taylor & Francis Group/CRC Press. There is no joy greater than being immersed in pursuit of excellence, knowledge gain, and knowledge capture. At the same time, I am happy it is over. The biggest lesson I learned during this project was that if you set yourself to a task with dedication, sincerity, persistence, and belief, you will have the job accomplished, no matter how daunting.

I expect the books to be standard references of immense value to any student, scientist, professional, and practical practitioner of remote sensing.

Prasad S. Thenkabail, PhD
Editor-in-Chief

Acknowledgments

The Remote Sensing Handbook (*Remotely Sensed Data Characterization, Classification, and Accuracies*; *Land Resources Monitoring, Modeling, and Mapping with Remote Sensing*; and *Remote Sensing of Water Resources, Disasters, and Urban Studies*) brought together a galaxy of remote sensing legends. The lead authors and coauthors of each chapter are internationally recognized experts of the highest merit on the subject about which they have written. The lead authors were chosen carefully by me after much thought and discussions, who then chose their coauthors. The overwhelming numbers of chapters were written over a period of one year. All chapters were edited and revised over the subsequent year and a half.

Gathering such a galaxy of authors was the biggest challenge. These are all extremely busy people, and committing to a book project that requires a substantial work load is never easy. However, almost all those whom I asked agreed to write the chapter, and only had to convince a few. The quality of the chapters should convince readers why these authors are such highly rated professionals and why they are so successful and accomplished in their field of expertise. They not only wrote very high quality chapters but delivered on time, addressed any editorial comments timely without complaints, and were extremely humble and helpful. What was also most impressive was the commitment of these authors for quality science. Three lead authors had serious health issues and yet they delivered very high quality chapters in the end, and there were few others who had unexpected situations (e.g., family health issues) and yet delivered the chapters on time. Even when I offered them the option to drop out, almost all of them wanted to stay. They only asked for a few extra weeks or months but in the end honored their commitment. I am truly honored to have worked with such great professionals.

In the following list are the names of everyone who contributed and made possible the Remote Sensing Handbook. In the end, we had 82 chapters, a little over 2500 pages, and a little over 300 authors.

My gratitude to the following authors of chapters in *Remotely Sensed Data Characterization, Classification, and Accuracies*. The authors are listed in chapter order starting with the lead author.

- Chapter 1, Drs. Sudhanshu S. Panda, Mahesh Rao, Prasad S. Thenkabail, and James P. Fitzerald
- Chapter 2, Natascha Oppelt, Rolf Scheiber, Peter Gege, Martin Wegmann, Hannes Taubenboeck, and Michael Berger
- Chapter 3, Philippe M. Teillet
- Chapter 4, Philippe M. Teillet and Gyanesh Chander
- Chapter 5, Rudiger Gens and Jordi Cristóbal Rosselló
- Chapter 6, Dongdong Wang
- Chapter 7, Tomoaki Miura, Kenta Obata, Javzandulam T. Azuma, Alfredo Huete, and Hiroki Yoshioka
- Chapter 8, Michael D. Steven, Timothy Malthus, and Frédéric Baret
- Chapter 9, Sunil Narumalani and Paul Merani
- Chapter 10, Soe W. Myint, Victor Mesev, Dale Quattrochi, and Elizabeth A. Wentz
- Chapter 11, Mutlu Ozdogan
- Chapter 12, Jun Li and Antonio Plaza
- Chapter 13, Claudia Kuenzer, Jianzhong Zhang, and Stefan Dech
- Chapter 14, Thomas Blaschke, Maggi Kelly, and Helena Merschdorf
- Chapter 15, Stefan Lang and Dirk Tiede
- Chapter 16, James C. Tilton, Selim Aksoy, and Yuliya Tarabalka
- Chapter 17, Shih-Hong Chio, Tzu-Yi Chuang, Pai-Hui Hsu, Jen-Jer Jaw, Shih-Yuan Lin, Yu-Ching Lin, Tee-Ann Teo, Fuan Tsai, Yi-Hsing Tseng, Cheng-Kai Wang, Chi-Kuei Wang, Miao Wang, and Ming-Der Yang
- Chapter 18, Daniela Anjos, Dengsheng Lu, Luciano Dutra, and Sidnei Sant'Anna
- Chapter 19, Jason A. Tullis, Jackson D. Cothren, David P. Lanter, Xuan Shi, W. Fredrick Limp, Rachel F. Linck, Sean G. Young, and Tareefa S. Alsumaiti
- Chapter 20, Gaurav Sinha, Barry J. Kronenfeld, and Jeffrey C. Brunskill
- Chapter 21, May Yuan
- Chapter 22, Stefan Lang, Stefan Kienberger, Michael Hagenlocher, and Lena Pernkopf
- Chapter 23, Mohinder S. Grewal
- Chapter 24, Kegen Yu, Chris Rizos, and Andrew Dempster
- Chapter 25, D. Myszor, O. Antemijczuk, M. Grygierek, M. Wierzchanowski, and K.A. Cyran
- Chapter 26, Fabio Dell'Acqua
- Chapter 27, Ramanathan Sugumaran, James W. Hegeman, Vivek B. Sardeshmukh, and Marc P. Armstrong
- Chapter 28, John Bailey
- Chapter 29, Russell G. Congalton

- Chapter 30, P.J. Blount
- Chapter 31, Prasad S. Thenkabail

My gratitude to the following authors of chapters in *Land Resources Monitoring, Modeling, and Mapping with Remote Sensing*. The authors are listed in chapter order starting with the lead author.

- Chapter 1, Alfredo Huete, Guillermo Ponce-Campos, Yongguang Zhang, Natalia Restrepo-Coupe, Xuanlong Ma, and Mary-Susan Moran
- Chapter 2, Frédéric Baret
- Chapter 3, Wenge Ni-Meister
- Chapter 4, Clement Atzberger, Francesco Vuolo, Anja Klisch, Felix Rembold, Michele Meroni, Marcio Pupin Mello, and Antonio Formaggio
- Chapter 5, Agnès Bégué, Damien Arvor, Camille Lelong, Elodie Vintrou, and Margareth Simoes
- Chapter 6, Pardhasaradhi Teluguntla, Prasad S. Thenkabail, Jun Xiong, Murali Krishna Gumma, Chandra Giri, Cristina Milesi, Mutlu Ozdogan, Russell G. Congalton, James Tilton, Temuulen Tsagaan Sankey, Richard Massey, Aparna Phalke, and Kamini Yadav
- Chapter 7, David J. Mulla, and Yuxin Miao
- Chapter 8, Baojuan Zheng, James B. Campbell, Guy Serbin, Craig S.T. Daughtry, Heather McNairn, and Anna Pacheco
- Chapter 9, Prasad S. Thenkabail, Pardhasaradhi Teluguntla, Murali Krishna Gumma, and Venkateswarlu Dheeravath
- Chapter 10, Matthew Clark Reeves, Robert A. Washington-Allen, Jay Angerer, E. Raymond Hunt, Jr., Ranjani Wasantha Kulawardhana, Lalit Kumar, Tatiana Loboda, Thomas Loveland, Graciela Metternicht, and R. Douglas Ramsey
- Chapter 11, E. Raymond Hunt, Jr., Cuizhen Wang, D. Terrance Booth, Samuel E. Cox, Lalit Kumar, and Matthew C. Reeves
- Chapter 12, Lalit Kumar, Priyakant Sinha, Jesslyn F. Brown, R. Douglas Ramsey, Matthew Rigge, Carson A. Stam, Alexander J. Hernandez, E. Raymond Hunt, Jr., and Matt Reeves
- Chapter 13, Molly E. Brown, Kirsten M. de Beurs, and Kathryn Grace
- Chapter 14, E.H. Helmer, Nicholas R. Goodwin, Valéry Gond, Carlos M. Souza, Jr., and Gregory P. Asner
- Chapter 15, Juha Hyyppä, Mika Karjalainen, Xinlian Liang, Anttoni Jaakkola, Xiaowei Yu, Mike Wulder, Markus Hollaus, Joanne C. White, Mikko Vastaranta, Kirsi Karila, Harri Kaartinen, Matti Vaaja, Ville Kankare, Antero Kukko, Markus Holopainen, Hannu Hyyppä, and Masato Katoh
- Chapter 16, Gregory P. Asner, Susan L. Ustin, Philip A. Townsend, Roberta E. Martin, and K. Dana Chadwick
- Chapter 17, Sylvie Durrieu, Cédric Véga, Marc Bouvier, Frédéric Gosselin, and Jean-Pierre Renaud Laurent Saint-André
- Chapter 18, Thomas W. Gillespie, Andrew Fricker, Chelsea Robinson, and Duccio Rocchini
- Chapter 19, Stefan Lang, Christina Corbane, Palma Blonda, Kyle Pipkins, and Michael Förster
- Chapter 20, Conghe Song, Jing Ming Chen, Taehee Hwang, Alemu Gonsamo, Holly Croft, Quanfa Zhang, Matthew Dannenberg, Yulong Zhang, Christopher Hakkenberg, Juxiang Li
- Chapter 21, John Rogan and Nathan Mietkiewicz
- Chapter 22, Zhixin Qi, Anthony Gar-On Yeh, and Xia Li
- Chapter 23, Richard A. Houghton
- Chapter 24, José A.M. Demattê, Cristine L.S. Morgan, Sabine Chabrillat, Rodnei Rizzo, Marston H.D. Franceschini, Fabrício da S. Terra, Gustavo M. Vasques, and Johanna Wetterlind
- Chapter 25, E. Ben-Dor and José A.M. Demattê
- Chapter 26, Prasad S. Thenkabail

My gratitude to the following authors of chapters in *Remote Sensing of Water Resources, Disasters, and Urban Studies*. The authors are listed in chapter order starting with the lead author.

- Chapter 1, Sadiq I. Khan, Ni-Bin Chang, Yang Hong, Xianwu Xue, and Yu Zhang
- Chapter 2, Santhosh Kumar Seelan
- Chapter 3, Trent W. Biggs, George P. Petropoulos, Naga Manohar Velpuri, Michael Marshall, Edward P. Glenn, Pamela Nagler, and Alex Messina
- Chapter 4, Antônio de C. Teixeira, Fernando B. T. Hernandez, Morris Scherer-Warren, Ricardo G. Andrade, Janice F. Leivas, Daniel C. Victoria, Edson L. Bolfe, Prasad S. Thenkabail, and Renato A. M. Franco
- Chapter 5, Allan S. Arnesen, Frederico T. Genofre, Marcelo P. Curtarelli, and Matheus Z. Francisco
- Chapter 6, Sandro Martinis, Claudia Kuenzer, and André Twele
- Chapter 7, Chandra Giri
- Chapter 8, D. R. Mishra, Shuvankar Ghosh, C. Hladik, Jessica L. O'Connell, and H. J. Cho
- Chapter 9, Murali Krishna Gumma, Prasad S. Thenkabail, Irshad A. Mohammed, Pardhasaradhi Teluguntla, and Venkateswarlu Dheeravath
- Chapter 10, Hongjie Xie, Tiangang Liang, Xianwei Wang, and Guoqing Zhang
- Chapter 11, Qingling Zhang, Noam Levin, Christos Chalkias, and Husi Letu
- Chapter 12, James B. Campbell and Lynn M. Resler
- Chapter 13, Felix Kogan and Wei Guo
- Chapter 14, Felix Rembold, Michele Meroni, Oscar Rojas, Clement Atzberger, Frederic Ham, and Erwann Fillol
- Chapter 15, Brian Wardlow, Martha Anderson, Tsegaye Tadesse, Chris Hain, Wade T. Crow, and Matt Rodell
- Chapter 16, Jinyoung Rhee, Jungho Im, and Seonyoung Park
- Chapter 17, Marion Stellmes, Ruth Sonnenschein, Achim Röder, Thomas Udelhoven, Stefan Sommer, and Joachim Hill

- Chapter 18, Norman Kerle
- Chapter 19, Stefan Lang, Petra Füreder, Olaf Kranz, Brittany Card, Shadrock Roberts, and Andreas Papp
- Chapter 20, Robert Wright
- Chapter 21, Krishna Prasad Vadrevu and Kristofer Lasko
- Chapter 22, Anupma Prakash and Claudia Kuenzer
- Chapter 23, Hasi Bagan and Yoshiki Yamagata
- Chapter 24, Yoshiki Yamagata, Daisuke Murakami, and Hajime Seya
- Chapter 25, Prasad S. Thenkabail

These authors are "who is who" in remote sensing and come from premier institutions of the world. For author affiliations, please see "Contributors" list provided a few pages after this. My deepest apologies if I have missed any name. But, I am sure those names are properly credited and acknowledged in individual chapters.

The authors not only delivered excellent chapters, they provided valuable insights and inputs for me in many ways throughout the book project.

I was delighted when Dr. Compton J. Tucker, senior Earth scientist, Earth Sciences Division, Science and Exploration Directorate, NASA Goddard Space Flight Center (GSFC), agreed to write the foreword for the book. For anyone practicing remote sensing, Dr. Tucker needs no introduction. He has been a *godfather* of remote sensing and has inspired a generation of scientists. I have been a student of his without ever really being one. I mean, I have not been his student in a classroom but have followed his legendary work throughout my career. I remember reading his highly cited paper (now with citations nearing 4000!):

- Tucker, C.J. (1979) Red and photographic infrared linear combinations for monitoring vegetation, *Remote Sensing of Environment*, **8(2)**,127–150.

That was in 1986 when I had just joined the National Remote Sensing Agency (NRSA; now NRSC), Indian Space Research Organization (ISRO). After earning his PhD from the Colorado State University in 1975, Dr. Tucker joined NASA GSFC as a postdoctoral fellow and became a full-time NASA employee in 1977. Since then, he has conducted path-finding research. He has used NOAA AVHRR, MODIS, SPOT Vegetation, and Landsat satellite data for studying deforestation, habitat fragmentation, desert boundary determination, ecologically coupled diseases, terrestrial primary production, glacier extent, and how climate affects global vegetation. He has authored or coauthored more than 170 journal articles that have been cited more than 20,000 times, is an adjunct professor at the University of Maryland, is a consulting scholar at the University of Pennsylvania's Museum of Archaeology and Anthropology, and has appeared on more than twenty radio and TV programs. He is a fellow of the American Geophysical Union and has been awarded several medals and honors, including NASA's Exceptional Scientific Achievement Medal, the Pecora Award from the U.S. Geological Survey (USGS), the National Air and Space Museum Trophy, the Henry Shaw Medal from the Missouri Botanical Garden, the Galathea Medal from the Royal Danish Geographical Society, and the Vega Medal from the Swedish Society of Anthropology and Geography. He was the NASA representative to the U.S. Global Change Research Program from 2006 to 2009. He was instrumental in releasing the AVHRR 32-year (1982–2013) Global Inventory Monitoring and Modeling Studies (GIMMS) data. I strongly recommend that everyone read his excellent foreword before reading the book. In the foreword, Dr. Tucker demonstrates the importance of data from EO sensors from orbiting satellites to maintaining a reliable and consistent climate record. Dr. Tucker further highlights the importance of continued measurements of these variables of our planet in the new millennium through new, improved, and innovative EO sensors from Sun-synchronous and/or geostationary satellites.

I am very thankful to my USGS colleagues for their encouragement and support. In particular, I mention Edwin Pfeifer, Dr. Susan Benjamin, Dr. Dennis Dye, Larry Gaffney, Miguel Velasco, Dr. Chandra Giri, Dr. Terrance Slonecker, Dr. Jonathan Smith, and Dr. Thomas Loveland. There are many other colleagues who made my job at USGS that much easier. My thanks to them all.

I am very thankful to Irma Shagla-Britton, acquisition editor for remote sensing and GIS books at Taylor & Francis Group/CRC Press. Without her initial nudge, this book would never have even been completed. Thank you, Irma. You are doing a great job.

I am very grateful to my wife (Sharmila Prasad) and daughter (Spandana Thenkabail) for their usual unconditional love, understanding, and support. They are always the pillars of my life. I learned the values of hard work and dedication from my revered parents. This work wouldn't have come about without their sacrifices to educate their children and their silent blessings. I am ever grateful to my former professors at The Ohio State University, Columbus, Ohio, United States: Prof. John G. Lyon, Dr. Andrew D. Ward, Prof. (Late) Carolyn Merry, Dr. Duane Marble, and Dr. Michael Demers. They have taught, encouraged, inspired, and given me opportunities at the right time. The opportunity to work for six years at the Center for Earth Observation of Yale University (YCEO) was incredibly important. I am thankful to Prof. Ronald G. Smith, director of YCEO, for his kindness. At YCEO, I learned and advanced myself as a remote sensing scientist. The opportunities I got from working for the International Institute of Tropical Agriculture (IITA), Africa and International Water Management Institute (IWMI) that had a global mandate for water were very important, especially from the point of view of understanding the real issues on the ground. I learned my basics of remote sensing mainly working with Dr. Thiruvengadachari of the National Remote Sensing Agency/Center (NRSA/NRSC), Indian Space Research Organization (ISRO), India, where I started my remote sensing career as a young scientist. I was just 25 years old then and had joined NRSA after earning my masters of engineering (hydraulics and water resources) and bachelors of engineering (civil engineering). During my first day in the office, Dr. Thiruvengadachari asked me how much remote sensing did I know. I said, "zero" and instantly thought that I would be thrown out of the room. But he said "very good" and gave me a manual on remote sensing from the Laboratory for Applications of Remote Sensing (LARS), Purdue. Those were the days where there was no formal training in remote sensing in any Indian universities. So my remote sensing lessons began working practically on projects and one of our first projects was "drought monitoring for India using NOAA AVHRR data." This was an intense period of learning remote sensing by actually practicing it on a daily basis. Data came on 9 mm tapes; data were read on massive computing systems; image processing was done, mostly working on night shifts by booking time on centralized computing; field work was conducted using false color composite outputs and topographic maps (not the days of global positioning systems); geographic information system was in its infancy; and a lot of calculations were done using calculators. So when I decided to resign my NRSA job and go to the United States to do my PhD, Dr. Thiruvengadachari told me, "Prasad, I am losing my right hand, but you can't miss opportunity." Those initial wonderful days of learning from Dr. Thiruvengadachari will remain etched in my memory. Prof. G. Ranganna of the Karnataka Regional Engineering College (KREC; now National Institute of Technology), Karnataka, India, was/is one of my most revered gurus. I have learned a lot observing him, professionally and personally, and he has always been an inspiration. Prof. E.J. James, former director of the Center for Water Resources Development and Management (CWRDM), was another original guru from whom I have learned the values of a true professional. I am also thankful to my good old friend Shri C. J. Jagadeesha, who is still working for ISRO as a senior scientist. He was my colleague at NRSA/NRSC, ISRO, and encouraged me to grow as a scientist. This Remote Sensing Handbook is a blessing from the most special ones dear to me. Of course, there are many, many others to thank especially many of my dedicated students over the years, but they are too many to mention here. I thank the truly outstanding editing work performed by Arunkumar Aranganathan and his team at SPi Global.

It has been my deep honor and great privilege to have edited the Remote Sensing Handbook. I am sure that I won't be taking on any such huge endeavors in the future. I will need time for myself, to look inside, understand, and grow. So thank you all, for making this possible.

Prasad S. Thenkabail, PhD
Editor-in-Chief

Editor

Prasad S. Thenkabail, PhD, is currently working as a research geographer-15 with the U.S. Geological Survey (USGS), United States. Currently, at USGS, Prasad leads a multi-institutional NASA MEaSUREs (Making Earth System Data Records for Use in Research Environments) project, funded through NASA ROSES solicitation. The project is entitled Global Food Security-Support Analysis Data at 30 m (GFSAD30) (http://geography.wr.usgs.gov/science/croplands/index.html also see https://www.croplands.org/). He is also an adjunct professor at three U.S. universities: (1) Department of Soil, Water, and Environmental Science (SWES), University of Arizona (UoA); (2) Department of Space Studies, University of North Dakota (UND); and (3) School of Earth Sciences and Environmental Sustainability (SESES), Northern Arizona University (NAU), Flagstaff, Arizona.

Dr. Thenkabail has conducted pioneering scientific research work in two major areas:

1. Hyperspectral remote sensing of vegetation
2. Global irrigated and rainfed cropland mapping using spaceborne remote sensing

His research papers on these topics are widely quoted. His hyperspectral work also led to his working on the scientific advisory board of Rapideye (2001), a German private industry satellite. Prasad was consulted on the design of spectral wavebands.

In hyperspectral research, Prasad pioneered in the following:

1. The design of optimal hyperspectral narrowbands (HNBs) and hyperspectral vegetation indices (HVIs) for agriculture and vegetation studies.
2. Certain hyperspectral data mining and data reduction techniques such as now widely used concepts of lambda by lambda plots.
3. Certain hyperspectral data classification methods. This included the use of a series of methods (e.g., discriminant model, Wilk's lambda, Pillai trace) that demonstrate significant increases in classification accuracies of land cover and vegetation classes as determined using HNBs as opposed to multispectral broadbands.

In global croplands, Prasad conducted seminal research that led to the first global map of irrigated and rainfed cropland areas using multitemporal, multisensor remote sensing, one book, and a series of more than ten novel peer-reviewed papers. In 2008, for one of these papers, Prasad (lead author) and coauthors (Pardhasaradhi Teluguntala, Trent Biggs, Murali Krishna Gumma, and Hugh Turral) were the second-place recipients of the 2008 John I. Davidson American Society of Photogrammetry and Remote Sensing (ASPRS) President's Award for practical papers. The paper proposed a novel spectral matching technique (SMT) for cropland classification. Earlier, Prasad (lead author) and coauthors (Andy Ward, John Lyon, and Carolyn Merry), won the 1994 Autometric Award for outstanding paper on remote sensing of agriculture from ASPRS. Recently, Prasad (seccond author) with Michael Marshall (lead author), won the ASPRS ERDAS award for best scientific paper on remote sensing for their hyperspectral remote sensing work.

Earlier to this **path-breaking Remote Sensing Handbook**, Prasad has published two seminal books (both published by Taylor & Francis Group/CRC Press) related to hyperspectral remote sensing and global croplands:

- Thenkabail, P.S., Lyon, G.J., and Huete, A. 2011. *Hyperspectral Remote Sensing of Vegetation*. CRC Press/Taylor & Francis Group, Boca Raton, FL, 781pp.

Reviews of this book:

- http://www.crcpress.com/product/isbn/9781439845370.
- Thenkabail, P., Lyon, G.J., Turral, H., and Biradar, C.M. 2009. *Remote Sensing of Global Croplands for Food Security*. CRC Press/Taylor & Francis Group, Boca Raton, FL, 556pp (48 pages in color).

Reviews of this book:

- http://www.crcpress.com/product/isbn/9781420090093.
- http://gfmt.blogspot.com/2011/05/review-remote-sensing-of-global.html.

He has guest edited two special issues for the American Society of Photogrammetry and Remote Sensing (PE&RS):

- Thenkabail, P.S. 2014. Guest editor of special issue on "Hyperspectral remote sensing of vegetation and agricultural crops." *Photogrammetric Engineering and Remote Sensing* 80(4).
- Thenkabail, P.S. 2012. Guest editor for Global croplands special issue. *Photogrammetric Engineering and Remote Sensing* 78(8).

He has also guest edited a special issue on global croplands for the *Remote Sensing Open Access Journal* (ISSN 2072-4292):

- Thenkabail, P.S. 2010. Guest editor: Special issue on "Global croplands" for the MDPI remote sensing open access journal. Total: 22 papers. http://www.mdpi.com/journal/remotesensing/special_issues/croplands/.

Prasad is, currently editor-in-chief, *Remote Sensing Open Access Journal,* an on-line journal, published by MDPI; editorial board member, *Remote Sensing of Environment;* editorial advisory board member, *ISPRS Journal of Photogrammetry and Remote Sensing.*

Prior to joining USGS in October 2008, Dr. Thenkabail was a leader of the remote sensing programs of leading institutes International Water Management Institute (IWMI), 2003–2008; International Center for Integrated Mountain Development (ICIMOD), 1995–1997; International Institute of Tropical Agriculture (IITA), 1992–1995.

He also worked as a key remote sensing scientist for Yale Center for Earth Observation (YCEO), 1997–2003; Ohio State University (OSU), 1988–1992; National Remote Sensing Agency (NRSA) (now NRSC), Indian Space Research organization (ISRO), 1986–1988.

Over the years, he has been a principal investigator (PI) of NASA, USGS, IEEE, and other funded projects such as inland valley wetland mapping of African nations, characterization of eco-regions of Africa (CERA), which involved both African savannas and rainforests, global cropland water use for food security in the twenty-first century, automated cropland classification algorithm (ACCA) within WaterSMART (Sustain and Manage America's Resources for Tomorrow) project, water productivity mapping in the irrigated croplands of California and Uzbekistan using multisensor remote sensing, IEEE Water for the World Project, and drought monitoring in India, Pakistan, and Afghanistan.

The USGS and NASA selected Dr. Thenkabail to be on the Landsat Science Team (2007–2011) for a period of five years (http://landsat.gsfc.nasa.gov/news/news-archive/pol_0005.html; http://ldcm.usgs.gov/intro.php). In June 2007, his team was recognized by the Environmental System Research Institute (ESRI) for "special achievement in GIS" (SAG award) for their tsunami-related work (tsdc.iwmi.org) and for their innovative spatial data portals (http://waterdata.iwmi.org/dtViewCommon.php; earlier http://www.iwmidsp.org). Currently, he is also a global coordinator for the Agriculture Societal Beneficial Area (SBA) of the Committee for Earth Observation

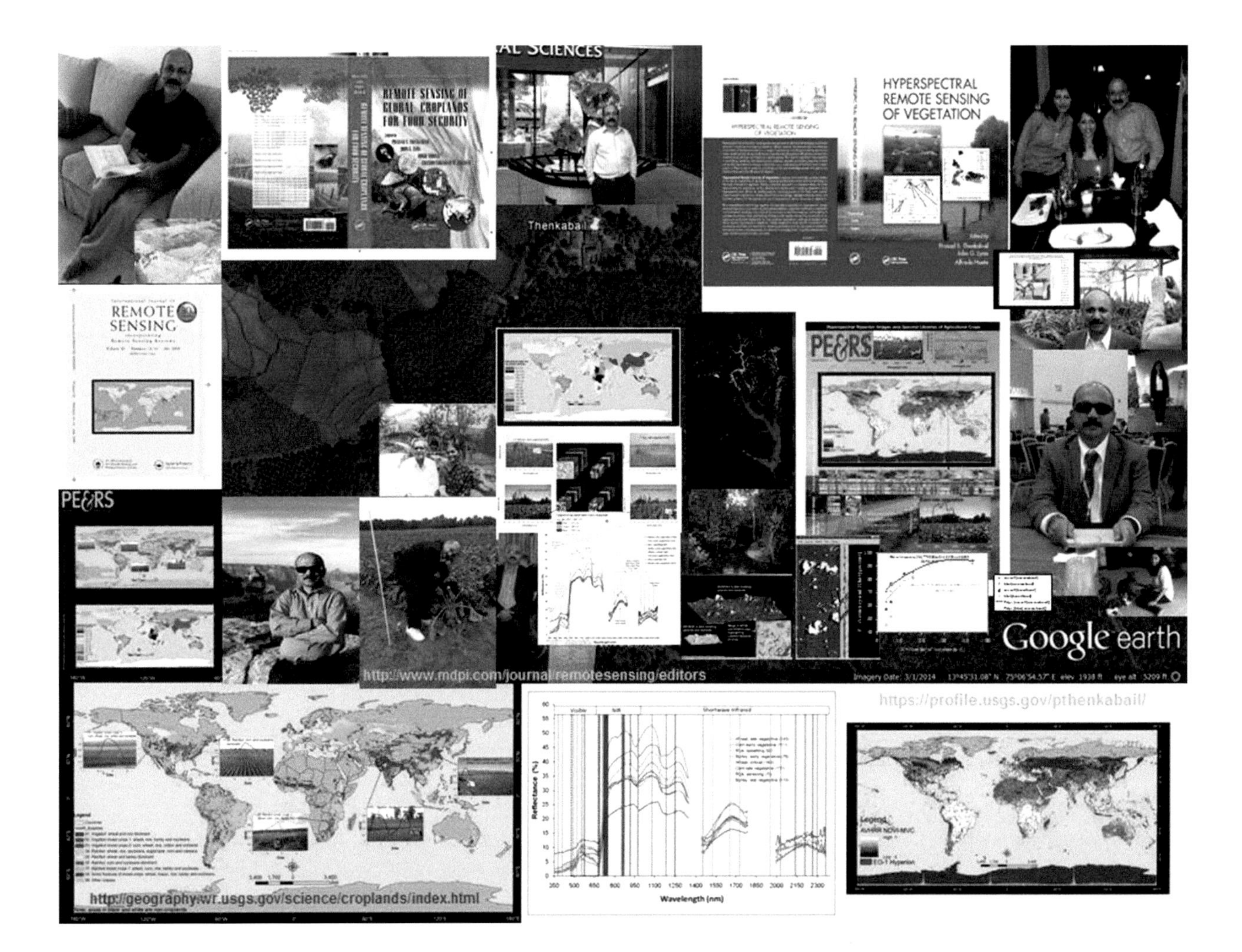

Systems (CEOS). He is active in the Group on Earth Observation (GEO) agriculture and water efforts through Earth observation. He was a co-lead of the Water for the World Project (IEEE effort). He is the current chair of the International Society of Photogrammetry and Remote Sensing (ISPRS) Working Group WG VIII/7: "Land Cover and Its Dynamics, including Agricultural & Urban Land Use" for the period 2013–2016. Thenkabail earned his PhD from The Ohio State University (1992). His master's degree in hydraulics and water resources engineering (1984) and bachelor's degree in civil engineering (1981) were from India. He began his professional career as a lecturer in hydrology, water resources, hydraulics, and open channel in India. He has 100+ publications, mostly peer-reviewed research papers in major international remote sensing journals: http://scholar.google.com/citations?user=9IO5Y7YAAAAJ&hl=en. Prasad has about 30 years' experience working as a well-recognized international expert in remote sensing and geographic information systems (RS/GIS) and their application to agriculture, wetlands, natural resource management, water resources, forests, sustainable development, and environmental studies. His work experience spans over 25+ countries spread across West and Central Africa, Southern Africa, South Asia, Southeast Asia, the Middle East, East Asia, Central Asia, North America, South America, and the Pacific.

Contributors

Selim Aksoy
Department of Computer Engineering
Bilkent University
Ankara, Turkey

Tareefa S. Alsumaiti
University of Arkansas
Fayetteville, Arkansas

Daniela Anjos
Image Processing Division
National Institute for Space Research
São José dos Campos, Brazil

O. Antemijczuk
Faculty of Automatic Control,
Electronics and Computer Science
Institute of Computer Science
Silesian University of Technology
Gliwice, Poland

Marc P. Armstrong
Department of Geographical and
Sustainability Sciences
The University of Iowa
Iowa City, Iowa

Javzandulam T. Azuma
Department of Natural Resources and
Environmental Management
University of Hawaii at Manoa
Honolulu, Hawaii

John E. Bailey
Google Inc.
Mountain View, California

Frédéric Baret
National Institute of Agronomic
Research
Avignon, France

Michael Berger
European Space Agency
ESA secondment at European Commission
Directorate General Research and
Innovation
European Commission
Brussels, Belgium

Thomas Blaschke
Department of Geoinformatics-Z_GIS
University of Salzburg
Salzburg, Austria

P.J. Blount
School of Law
University of Mississippi
Oxford, Mississippi

Jeffrey C. Brunskill
Department of Environmental,
Geographical, and Geological
Sciences
University of Pennsylvania
Bloomsburg, Pennsylvania

Gyanesh Chander
Science Data Systems Branch
Goddard Space Flight Center
National Aeronautics and Space
Administration
Greenbelt, Maryland

Shih-Hong Chio
Department of Land Economics
National Chengchi University
Taipei, Taiwan

Tzu-Yi Chuang
Department of Civil Engineering
National Taiwan University
Taipei, Taiwan

Russell G. Congalton
Department of Natural Resources and
the Environment
University of New Hampshire
Durham, New Hampshire

Jackson D. Cothren
Department of Geosciences
University of Arkansas
Fayetteville, Arkansas

K.A. Cyran
Faculty of Automatic Control,
Electronics and Computer Science
Institute of Computer Science
Silesian University of Technology
Gliwice, Poland

Stefan Dech
German Aerospace Center (DLR)
and
Earth Observation Center (EOC)
and
German Remote Sensing Data Center,
DFD of DLR
Oberpfaffenhofen, Germany

and

Department for Geography and Geology
University of Wuerzburg
Wuerzburg, Germany

Fabio Dell'Acqua
Department of Electrical, Computer and
Biomedical Engineering
University of Pavia
Pavia, Italy

Andrew Dempster
School of Electrical Engineering &
Telecommunications
University of New South Wales
Sydney, New South Wales, Australia

Luciano Dutra
Image Processing Division
National Institute for Space Research
São José dos Campos, Brazil

James E. Fitzerald
Cobb County, Georgia

Peter Gege
Earth Observation Center
and
Remote Sensing Technology Institute
and
German Aerospace Centre (DLR)
Oberpfaffenhofen, Germany

Rudiger Gens
Geophysical Institute
University of Alaska Fairbanks
Fairbanks, Alaska

Mohinder S. Grewal
Electrical Engineering Department
California State University
Fullerton, California

M. Grygierek
LG Nexera Business Solutions AG
Vienna, Austria

Michael Hagenlocher
Department of Geoinformatics–Z_GIS
Paris-Lodron University of Salzburg
Salzburg, Austria

James W. Hegeman
Department of Computer Science
The University of Iowa
Iowa City, Iowa

Pai-Hui Hsu
National Taiwan University
Taipei, Taiwan

Alfredo Huete
Plant Functional Biology and Climate Change Cluster
University of Technology, Sydney
Ultimo, New South Wales, Australia

Jen-Jer Jaw
Department of Civil Engineering
National Taiwan University
Taipei, Taiwan

Maggi Kelly
Department of Environmental Science, Policy, and Management
University of California
Berkeley, California

Stefan Kienberger
Department of Geoinformatics–Z_GIS
Paris-Lodron University of Salzburg
Salzburg, Austria

Barry J. Kronenfeld
Department of Geology and Geography
Eastern Illinois University
Charleston, Illinois

Claudia Kuenzer
German Aerospace Center (DLR)
and
Earth Observation Center (EOC)
and
German Remote Sensing Data Center, DFD of DLR
Oberpfaffenhofen, Germany

Stefan Lang
Department of Geoinformatics–Z_GIS
Paris-Lodron University of Salzburg
Salzburg, Austria

David P. Lanter
CDM Smith
Philadelphia, Pennsylvania

Jun Li
School of Geography and Planning
Sun Yat-Sen University
Guangzhou, People's Republic of China

W. Fredrick Limp
Department of Geosciences
University of Arkansas
Fayetteville, Arkansas

Shih-Yuan Lin
Department of Land Economics
National Chengchi University
Taipei, Taiwan

Yu-Ching Lin
Department of Environmental Information and Engineering
Chung Cheng Institute of Technology
National Defense University
Taoyuan, Taiwan

Rachel F. Linck
Department of Geosciences
University of Arkansas
Fayetteville, Arkansas

Dengsheng Lu
Center for Global Change and Earth Observations
Michigan State University
East Lansing, Michigan

Timothy J. Malthus
Commonwealth Scientific and Industrial Research Organisation
Ecosciences Precinct
Dutton Park, Queensland, Australia

Paul Merani
Department of Earth and Atmospheric Sciences
University of Nebraska–Lincoln
Lincoln, Nebraska

Helena Merschdorf
Interfaculty Department of Geoinformatics–Z_GIS
Paris-Lodron University of Salzburg
Salzburg, Austria

Victor Mesev
Department of Geography
Florida State University
Tallahassee, Florida

Tomoaki Miura
Department of Natural Resources and Environmental Management
University of Hawaii at Manoa
Honolulu, Hawaii

Soe W. Myint
School of Geographical Sciences and Urban Planning
Arizona State University
Tempe, Arizona

D. Myszor
Faculty of Automatic Control, Electronics and Computer Science
Institute of Computer Science
Silesian University of Technology
Gliwice, Poland

and

LG Nexera Business Solutions AG
Vienna, Austria

Sunil Narumalani
Department of Earth and Atmospheric Sciences
University of Nebraska–Lincoln
Lincoln, Nebraska

Kenta Obata
Institute of Geology and Geoinformation
National Institute of Advanced Industrial Science and Technology
Tsukuba, Japan

Natascha Oppelt
Department of Geography
Christian Albrechts Universität zu Kiel
Kiel, Germany

Mutlu Özdoğan
Nelson Institute for Environmental Studies
and
Department of Forest and Wildlife Ecology
University of Wisconsin–Madison
Madison, Wisconsin

Sudhanshu S. Panda
Institute of Environmental Spatial Analysis
University of North Georgia
Oakwood, Georgia

Lena Pernkopf
Department of Geoinformatics–Z_GIS
Paris-Lodron University of Salzburg
Salzburg, Austria

Antonio Plaza
Hyperspectral Computing Laboratory
University of Extremadura
Cáceres, Spain

Dale A. Quattrochi
NASA Marshall Space Flight Center
National Aeronautics and Space Administration
Huntsville, Alabama

Mahesh N. Rao
Department of Forestry and Wildlife Resources
Humboldt State University
Arcata, California

Chris Rizos
School of Civil and Environmental Engineering
University of New South Wales
Sydney, New South Wales, Australia

Jordi Cristóbal Rosselló
Geophysical Institute
University of Alaska–Fairbanks
Fairbanks, Alaska

Sidnei Sant'Anna
Image Processing Division
National Institute for Space Research
São José dos Campos, Brazil

Vivek B. Sardeshmukh
Department of Computer Science
The University of Iowa
Iowa City, Iowa

Rolf Scheiber
Microwaves and Radar Institute
German Aerospace Center (DLR)
Oberpfaffenhofen, Germany

Xuan Shi
Department of Geosciences
University of Arkansas
Fayetteville, Arkansas

Gaurav Sinha
Department of Geography
Ohio University
Athens, Ohio

Michael D. Steven
School of Geography
University of Nottingham
Nottingham, United Kingdom

Ramanathan Sugumaran
Department of Geographical and Sustainability Sciences
The University of Iowa
Iowa City, Iowa

and

John Deere
Moline, Illinois

Yuliya Tarabalka
Sophia-Antipolis Méditerranée
Sophia-Antipolis, France

Hannes Taubenboeck
German Remote Sensing Data Center
German Aerospace Centre (DLR)
Oberpfaffenhofen, Germany

Philippe M. Teillet
Department of Physics and Astronomy
University of Lethbridge
Lethbridge, Alberta, Canada

Tee-Ann Teo
National Chiao Tung University
Hsinchu City, Taiwan

Prasad S. Thenkabail
Geography, Flagstaff Science Center
and
Western Geographic Science Center
and
United States Geological Survey
Flagstaff, Arizona

Dirk Tiede
Department of Geoinformatics–Z_GIS
Paris-Lodron University of Salzburg
Salzburg, Austria

James C. Tilton
Goddard Space Flight Center
National Aeronautics and Space Administration
Greenbelt, Maryland

Fuan Tsai
Center for Space and Remote Sensing Research
National Central University
Zhongli, Taiwan

Yi-Hsing Tseng
National Cheng Kung University
Tainan City, Taiwan

Jason A. Tullis
Department of Geosciences
University of Arkansas
Fayetteville, Arkansas

Cheng-Kai Wang
Department of Geomatics
National Cheng Kung University
Tainan City, Taiwan

Chi-Kuei Wang
Department of Geomatics
National Cheng Kung University
Tainan City, Taiwan

Dongdong Wang
Department of Geographical Sciences
University of Maryland
College Park, Maryland

Miao Wang
Department of Geomatics
National Cheng Kung University
Tainan City, Taiwan

Martin Wegmann
Department for Geography and Geology
Ludwig-Maximilians-Universität Würzburg
Würzburg, Germany

Elizabeth A. Wentz
School of Geographical Sciences and Urban Planning
Arizona State University
Tempe, Arizona

M. Wierzchanowski
Faculty of Automatic Control, Electronics and Computer Science
Institute of Computer Science
Silesian University of Technology
Gliwice, Poland

and

LG Nexera Business Solutions AG
Vienna, Austria

Ming-Der Yang
Department of Geomatics
National Chung Hsing University
Taichung, Taiwan

Hiroki Yoshioka
Department of Information Science and Technology
Aichi Prefectural University
Nagakute, Japan

Sean G. Young
College of Liberal Arts and Sciences
University of Iowa
Iowa City, Iowa

Kegen Yu
School of Geodesy and Geomatics
Wuhan University
Wuhan, People's Republic of China

May Yuan
School of Economic, Political, and Policy Sciences
University of Texas at Dallas
Richardson, Texas

Jianzhong Zhang
Freelance Remote Sensing Consultant
Kaufering, Germany

I

Satellites and Sensors from Different Eras and Their Characteristics

1

Remote Sensing Systems—Platforms and Sensors: Aerial, Satellite, UAV, Optical, Radar, and LiDAR

Sudhanshu S. Panda
University of North Georgia

Mahesh N. Rao
Humboldt State University

Prasad S. Thenkabail
United States Geological Survey

James E. Fitzerald
Department of Land Management

Acronyms and Definitions

ACRIM	Active cavity radiometer irradiance monitor
ADEOS	Advanced Earth observing satellite
AEP	Architecture evolution plan
AIRS	Atmospheric infrared sounder
AIRSAR	Airborne synthetic aperture radar
ALADIN	Atmospheric laser doppler instrument
ALI	Advanced land imager
ALOS	Advanced land observing satellite (from JAXA)
AMS	Airborne multispectral scanner
AOI	Area of interest
ASAR	Advanced synthetic aperture radar (from ESA)
ASFA	Aquatic sciences and fisheries abstracts (FAO)
ASTER	Advanced spaceborne thermal emission and reflection radiometer
ATM	Airborne thematic mapper
ATSR	Along-track scanning radiometer
AUV	Autonomous underwater vehicle
AVHRR	Advanced very-high-resolution radiometer
AVIRIS	Airborne visible/infrared imaging spectrometer
AVNIR	Advanced visible and near-infrared radiometer
AWiFS	Advanced wide field sensor (IRS)
BDRF	Bidirectional reflectance function
BV	Brightness value
CASI	Compact airborne spectrographic imager
CBERS-2	China–Brazil Earth Resources Satellite
C-CAP	Coastal Change Analysis Program
CCD	Charge-coupled device
CCRS	Canadian Center for Remote Sensing
CCT	Computer-compatible tape
CEO	Centre for Earth Observation
CEOS	Committee on Earth Observation Satellites
CERES	Clouds and the Earth's Radiant Energy System
CIR	Color infrared
CMOS	Complementary metal-oxide semiconductor
CNES	Centre National d'Etudes Spatiales (French space agency)
DEM	Digital elevation model
DGPS	Differential global positioning

DIDSON	Dual-frequency identification sonar
DLR	German Aerospace Research Establishment
DN	Digital number (pixel value)
DOQQ	Digital orthophoto (quarter) quad
DTM	Digital terrain model
EDC	EROS Data Center
EME	Electromagnetic energy
EMR	Electromagnetic radiation
EMS	Electromagnetic spectrum
ENVISAT	Polar-orbiting environmental satellite by ESA
EOS	Earth Observing System
EOSAT	Earth observation satellite
EOSDIS	Earth Observing System Data and Information System
EOSP	Earth Observing Scanning Polarimeter
EROS	Earth Resources Observation Systems
ERS	European Remote Sensing Satellite
ERS-1	Earth Remote Sensing Satellite
ERTS	Earth Resources Technology Satellite
ESA	European Space Agency
ESRI	Environmental Systems Research Institute
ETM+	Enhanced Thematic Mapper Plus
EUMETSAT	European Organisation for the Exploitation of Meteorological Satellites
FAS	Foreign Agricultural Service
FGDC	Federal Geographic Data Committee
FIR	Far infrared
FLIR	Forward-looking infrared
FORMOSAT	Taiwanese satellite operated by the National Space Organization of Taiwan (NSPO)
FOV	Field of view
FSA	Farm Service Agency (USDA)
G3OS	Global Observing Systems of GCOS, GOOS, and GTOS
GAP	National Gap Analysis Program
GCP	Ground control point
GEMS	Global Environment Monitoring System
GEO	Group on Earth Observations
GEOSS	Global Earth Observation System of Systems
GIS	Geographic information system
GLAS	Geoscience Laser Altimeter System
GNSS	Global Navigation Satellite System
GOES	Geosynchronous Orbiting Environmental Satellite
GOES	Geostationary Operational Environmental Satellite
GOMOS	Global Ozone Monitoring by Occultation of Stars
GOOS	Global Ocean Observing System
GOS	Global Observing System (of the WMO)
GOMS	Geostationary Operational Meteorological Satellite
GPM	Global Precipitation Measurement
GPS	Global Positioning System
GSD	Ground sample distance
GSFC	Goddard Space Flight Center
GSTC	Geospatial Service and Technology Center
HIRS	High-resolution infrared sounder
Hyperion	First spaceborne hyperspectral sensor onboard Earth observing-1 (EO-1)
IFOV	Instantaneous field of view
IKONOS	High-resolution satellite operated by GeoEye
IPCC	Intergovernmental Panel on Climate Change
IRS	Indian Remote Sensing Satellite
IVOS	Infrared and visible optical sensor
Jason	The satellite series planned as follow on to TOPEX/Poseidon (Jason-1 is the first of these)
JAXA	Japan Aerospace Exploration Agency
JERS	Japanese Earth Resources Satellite
JPSS	Joint Polar Satellite System
KOMFOSAT	Korean Multipurpose Satellite (data marketed by SPOT Image)
LANDSAT	Land Remote Sensing Satellite
LASER	Light Amplification by Stimulated Emission of Radiation
LCCP	Land cover characterization program
LCT	Laser communication terminal
LIDAR	Laser Image Detection and Ranging
LISS	Linear Imaging Self-Scanning Sensor (IRS)
LRA	Laser reflectometry array
LWIR	Longwave infrared
MERIS	Medium-resolution imaging spectrometer
METEOSAT	Meteorological satellite by ESA
MIR	Mid-infrared
MIRAS	Microwave Imaging Radiometer with Aperture Synthesis
MISR	Multi-angle Imaging SpectroRadiometer
MLS	Microwave Limb Scanner
MODIS	Moderate-Resolution Imaging Spectroradiometer
MOS	Marine Observation Satellite
MRLC	Multi-Resolution Land Characteristics Consortium
MSS	Multispectral scanner
MTF	Modular transfer function
MTSAT	Multipurpose transport satellite by JAXA
NAIP	National Agriculture Imagery Program
NALC	North American Landscape Characterization
NAOS	North American Atmospheric Observing System
NCAR	National Center for Atmospheric Research (United States)
NASA	National Aeronautics and Space Administration
NAPP	National Aerial Photography Program
NAWQA	National Water Quality Assessment Program (a USGS program)
NCDC	National Climatic Data Center
nDSM	Normalized digital surface model
NDVI	Normalized difference vegetation index
NED	National Elevation Dataset
NEXRAD	Next-Generation Radar

NHAP	National High Altitude Program
NIMA	National Image and Mapping Agency
NIR	Near infrared
NLCD	National Land Cover Data
NOAA	National Oceanic and Atmospheric Administration
NPOESS	National Polar-Orbiting Environmental Satellite System
NRCS	National Resources Conservation Service (a USDI agency)
NRI	National Resources Inventory
NRSC	National Remote Sensing Centre (United Kingdom)
NSSDC	National Space Science Data Center
NWS	National Weather Service
OCTS	Ocean Color and Temperature Scanner
OCX	Operational Control System
OSCAR	Ocean Surface Current Analysis Real Time
OSTM	Ocean Surface Topography Mission
PALSAR	Phased Array–Type L-Band Synthetic Aperture Radar
PBS	Public Broadcasting Service
QuickBird	Satellite from DigitalGlobe, a private company in United States
RADAR	Radio detection and ranging
RADARSAT	Canadian radar satellite
RapidEye	Satellite constellation from RapidEye, a German company
RESOURCESAT	Satellite launched in India
RGB	Red, green, blue
RKA	Russian Federal Space Agency
S/N	Signal-to-noise ratio
SAR	Synthetic aperture radar
SeaWiFS	Sea-Viewing Wide Field-of-View Sensor
SEVIRI	Spinning Enhanced Visible and Infrared Imager
SIRS	Satellite infrared spectrometer
SLAR	Side-looking airborne radar
SODAR	Sonic detection and ranging
SONAR	Sound navigation and ranging
SPOT	Systeme Probatoire D'Observation De La Terre (France)
SSM	Special sensor microwave
SSM/I	Special sensor microwave/imager
SSM/T	Special sensor microwave/temperature
SSM/T-2	Special sensor microwave/water vapor
SSS	Sea surface salinity
SWIR	Shortwave infrared sensor
SST	Sea surface temperature
SWAT	Soil and Water Assessment Tool
TES	Tropospheric Emission Spectrometer
TIMS	Thermal infrared multispectral scanner
TIR	Thermal infrared
TM	Thematic Mapper
TOMS	Total Ozone Mapping Spectrometer
TOPEX	Ocean Topography Experiment
TOPSAR	Topographic Synthetic Aperture Radar
TRMM	Tropical Rainfall Measuring Mission
UAS	Unmanned aircraft system
UAV	Unmanned aerial vehicle
UHF	Ultrahigh frequency
USGS	United States Geological Survey
UV	Ultraviolet
VIS	Visible spectrum
VMS	Vessel monitoring system
VNIR	Visible near-infrared sensor
WFOV	Wide field of view
WRS	Worldwide Reference System
X-SAR	X-band synthetic aperture radar

1.1 Introduction

1.1.1 Remote Sensing Definition and History

The American Society of Photogrammetry and Remote Sensing defined remote sensing as the measurement or acquisition of information of some property of an object or phenomenon, by a recording device that is not in physical or intimate contact with the object or phenomenon under study (Colwell et al., 1983). Environmental Systems Research Institute (ESRI) in its geographic information system (GIS) dictionary defines remote sensing as "collecting and interpreting information about the environment and the surface of the earth from a distance, primarily by sensing radiation that is naturally emitted or reflected by the earth's surface or from the atmosphere, or by sending signals transmitted from a device and reflected back to it (ESRI, 2014)." The usual source of passive remote sensing data is the measurement of reflected or transmitted electromagnetic radiation (EMR) from the sun across the electromagnetic spectrum (EMS); this can also include acoustic or sound energy, gravity, or the magnetic field from or of the objects under consideration. In this context, the simple act of reading this text is considered remote sensing. In this case, the eye acts as a sensor and senses the light reflected from the object to obtain information about the object. It is the same technology used by a handheld camera to take a photograph of a person or a distant scenic view. Active remote sensing, however, involves sending a pulse of energy and then measuring the returned energy through a sensor (e.g., Radio Detection and Ranging [RADAR], Light Detection and Ranging [LiDAR]). Thermal sensors measure emitted energy by different objects. Thus, in general, passive remote sensing involves the measurement of solar energy reflected from the Earth's surface, while active remote sensing involves synthetic (man-made) energy pulsed at the environment and the return signals are measured and recorded.

Remote sensing functions in harmony with other *spatial* data collection techniques or tools of the *mapping sciences*, including cartography and GIS (Jensen, 2009). Remote sensing is a tool or technique similar to mathematics where each sensor is used to measure from a distance the amount of EMR exiting an object, or geographic area, then extracting valuable information from the data using mathematical and statistical algorithms (Jensen, 2009). In fact, the science and art of obtaining reliable measurements using photographs is called photogrammetry (American Society of Photogrammetry, 1952, 1966). According to Aronoff (2004), remote sensing is the technology, science, and art of obtaining information about objects from a distance over regions too costly, dangerous, or remote for human observers to directly access. This includes target areas beyond the limits of human ability. Avery and Berlin (1992) simply defined remote sensing as the "technique of reconnaissance from a distance" in which information about objects are obtained through the analysis of data collected by special instruments that are not in physical contact with the objects of investigation. Therefore, remote sensing is different from in situ observation that involves use of sensing instruments in direct contact with the objects under consideration but not proximal in situ sensing, where data are collected from the in situ objects with controlled spectra. In situ observations of the Earth's land cover and other physical phenomena are often time-consuming, costly, and impossible to have a full spatial coverage. In contrast, remote sensing acquires information about an object or phenomenon on the Earth's surface and even beneath the surface without making physical contact with the object by use of propagated signals, for example, EMR (Lillesand et al., 2004).

The very first successful photographic image of nature, considered as a permanent photograph or remotely sensed image, was captured by a French native Joseph Nicephore Niepce (Gernsheim and Gernsheim, 1952). Modern remote sensing started in 1858 at Paris, France, with Gaspard-Felix Tournachon first taking aerial photographs of the city of Paris from a hot air balloon (Briney, 2014). Remote sensing continued to grow from there but at a slow rate. During the U.S. Civil War (1861–1865), messenger pigeons, kites, and unmanned balloons were flown over enemy territory with cameras attached to them, which is considered as one of the very first planned usages of remote sensing technology (Trenear-Harvey, 2009; Briney, 2014). During World War I and II, the first government-organized air photography or photogrammetry missions were developed for military surveillance and later adopted by many other countries for other applications, including land surveying (Briney, 2014). After World War II, during the Cold War era, aerial photo reconnaissance became much more prominent and prevalent between the United States and the Soviet Union to collect information about each other. In December 1954, U.S. President Dwight Eisenhower approved the U-2 reconnaissance program (Brugioni and Doyle, 1997) for World War II reconnaissance from space. In recent years, tremendous advancements in technology have resulted in a rapid growth in the remote sensing industry (Jensen, 2009). During the last few decades, growth in the civilian sector has far surpassed the defense and military applications. However, the recent years have seen new applications of miniature remote sensors or camera systems that are mounted on both manned and unmanned aerial platforms, which are used by law enforcement and military sectors for surveillance purposes (Briney, 2014). Unmanned aerial vehicles (UAVs) are one of the most important advancements in aerial remote sensing in recent times and used even for fighting stealth wars (Russo et al., 2006). UAVs are now controlled from home base through onboard Global Positioning System (GPS) and forward-looking infrared (FLIR) and/or videography (FLIR) technology (Jensen, 2009). Other advances in remote sensing technology include RADAR, LiDAR, sound navigation and ranging (SONAR), sonic detection and ranging (SODAR), microwave synthetic aperture radar (SAR), infrared sensors, hyperspectral imaging, spectrometry, Doppler radar, and space probe sensors, in addition to improvements in conventional aerial photography, hyperspectral imaging, and imaging spectroradiometer.

1.1.2 Data Collection by Remote Sensing and Usage

Remote sensing in comparison to other methods of data collection is much more advantageous as it provides an overview of the Earth's phenomena that allow users to discern patterns and relationships not apparent from the ground (Aronoff, 2004). EMR (usually solar energy or any form of energy) is reflected or emitted from the object and is detected by the devices called remote sensors or simply *sensors* (similar to cameras or scanners) fitted on platforms such as aircrafts, satellites, or ships. Traditional remote sensing involves two basic processes: data acquisition and data analysis. Figure 1.1 illustrates the data collection and processing involved in a typical remote sensing operation.

According to Lillesand et al. (2004), the elements of remote sensing data acquisition involve (1) energy sources such as solar energy, self-produced heat/light energy, or sound energy; (2) propagation of energy through the atmosphere; (3) energy interaction with the Earth's surface features; (4) retransmission of energy through the atmosphere; (5) sensing platforms like airborne, spaceborne, and onboard ship sensors; and (6) resultant data in the form of pictorial or digital format. There are numerous factors that affect the remote sensing data collection process such as (1) solar position, (2) atmospheric condition, (3) weather and meteorology, (4) season of data collection, (5) ground condition, (6) sensor characteristics, and (7) sensor position. As one can realize, most of the aforementioned factors directly relate to the intensity of the solar energy in terms of enhancing or attenuating it. On the other hand, remote sensing data analysis involves (1) data interpretation using various interpreting/viewing devices like computers and software, (2) analyzing the data in collaboration with other geospatial data, and (3) applying the interpreted and analyzed data for Earth management decision support.

Basically, in terms of the energy source and data collection mode, there are two types of remote sensing: (1) passive remote sensing, when reflected or emitted energy from an object or phenomenon is recorded by sensors mounted on airborne or spaceborne platforms and (2) active remote sensing,

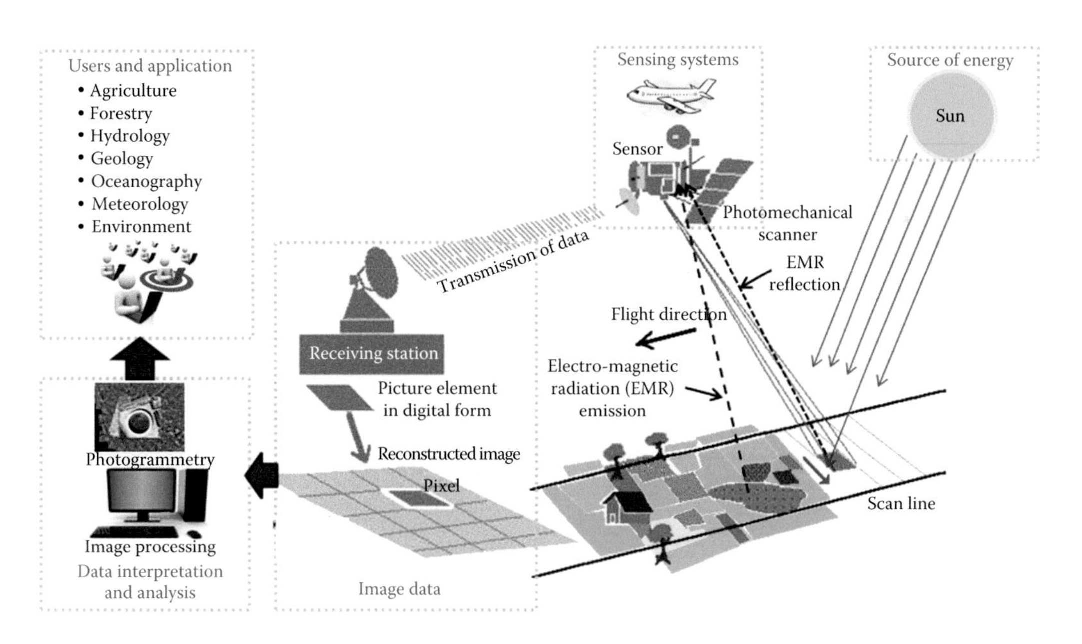

FIGURE 1.1 Illustration of remote sensing data collection for user application. Sensors onboard aircraft or satellite platforms collect and record reflected energy from target features. The reflected data are collected across the electromagnetic spectrum and are used for various studies such as agriculture and forestry.

when reflectance of synthetic light (nonsolar) that is actively pulsed or emitted from an aircraft, satellite, or any other energy-producing/recording platform is recorded (Schott, 2007; Schowengerdt, 2007). Passive remote sensing systems such as film photography have become obsolete due to several disadvantages, including low exposure time, low spectral range (i.e., usually below 700 nm only), low dynamic range (saturation/minimum limit) of only around 100, and other image characteristic issues, including linearity and photometric accuracy. Currently, most systems use charge-coupled device (CCD) and complementary metal-oxide semiconductor (CMOS)—very common in digital cameras and camcorders that convert light energy into electrons, which is then measured and converted into radiometric intensity values (McGlone, 2004). However, the active remote sensing systems such as RADAR, LiDAR, SONAR, and GPS detect not only the backscattered or energy reflecting off of Earth objects but also record time lag and intensity (Schott, 2007; Schowengerdt, 2007). Thus, unlike passive remote sensing, which only detects the location of an object, active remote sensing uses the time delay between transmission and reception of energy pulse to establish the location, height, depth, speed, and direction of an object. Figure 1.2 illustrates the difference between active and passive remote sensing with examples of different platforms.

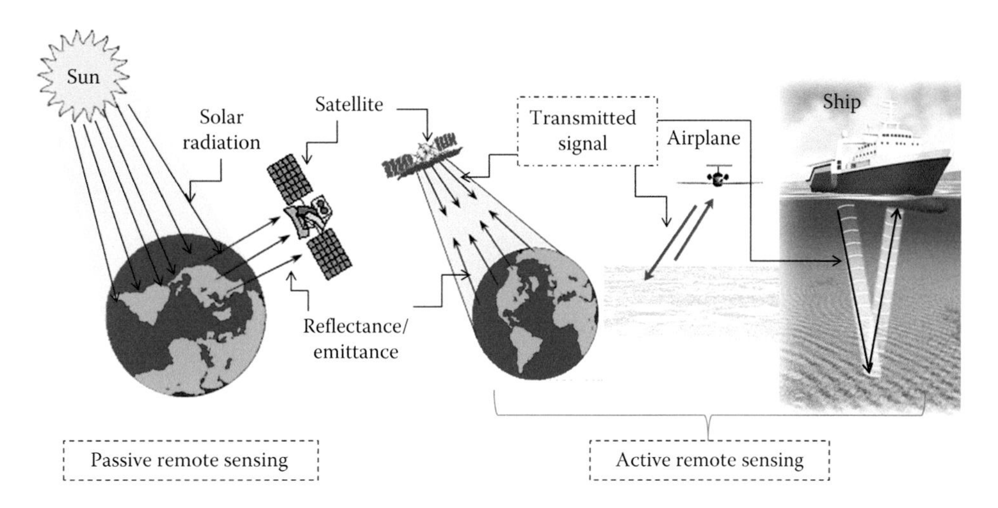

FIGURE 1.2 Illustration of passive vs. active remote sensing with different platforms. In passive remote sensing, the sun acts as the source of energy. Sensors then measure the reflected solar energy off the targets. Active remote sensing sends a pulse of synthetic energy and measures the energy reflected off the targets.

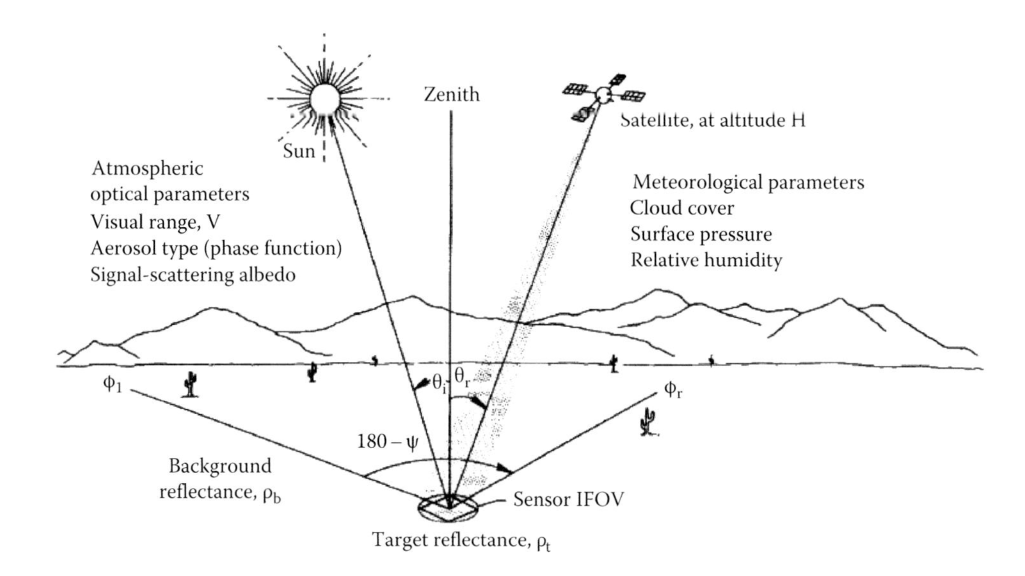

FIGURE 1.3 Illustration of factors affecting apparent reflectance in passive remote sensing. (Adapted and modified from Bowker, D.E. et al., Spectral reflectances of natural targets for use in remote sensing studies, NASA reference publication # 1139, Hampton, VA, 1985.) Note that the solar energy passes through the atmosphere and hits the target (any object on the planet). Then, depending on the type of the object, certain portion of the energy reflects off the target, again passes back through the atmosphere, and is captured by the sensor. Such reflected energy can be captured in various portions of the electromagnetic spectrum (EMS) (e.g., visible, near infrared, shortwave infrared). Also, there can be several wave bands within each portion of the EMS.

National Aeronautics and Space Administration (NASA) reference publication # 1139 by Bowker et al. (1985) list different factors affecting apparent reflectance, which subsequently determines the object or phenomena on the Earth or in the atmosphere. These factors are (1) viewing geometry (solar zenith angle, viewing angle, azimuthal angle, relative azimuthal angle, and altitude of the sensor), (2) meteorological parameters (relative humidity, cloud cover, and surface pressure), (3) atmospheric optical parameters (aerosol optical thickness, atmospheric visual range, aerosol types, and single-/multiple-scattering albedo), and (4) target and background parameters (target size, target reflectance, background reflectance, instantaneous field of view [IFOV], and bidirectional reflectance function [BDRF]). Figure 1.3 extracted from Bowker et al. (1985) pictorially depicts these factors affecting target reflectance measured by a passive remote sensing system.

1.1.3 Principles of Electromagnetic Spectrum in Remote Sensing

Light or radiant flux in the form of electromagnetic energy (EME), which includes visible light, infrared, radio waves, heat, ultraviolet (UV) rays, and x-rays, is the primary form of energy used in remote sensing (Lillesand et al., 2004). EMR is a carrier of EME by transmitting the oscillation of the electromagnetic field through space or matter based on Maxwell's equation (Murai, 1993).

EMR has characteristics of both wave and particle motion. EMR essentially follows the basic wave theory ($c = \nu\lambda$), which describes that EME travels in a harmonic or sinusoidal fashion (Figure 1.4) at the velocity of light ($c = 2.998 \times 10^8$ m/s); the distance from one wave peak to another is wavelength, λ, and the number of peaks passing a fixed point in space per unit time is wave frequency, ν (Lillesand et al., 2004).

According to the Murai (1993), EMR consists of four elements such as (1) frequency/wavelength, (2) transmission direction, (3) amplitude, and (4) polarization. These four features are important for remote sensing as each corresponds to different features. Wavelength/frequency corresponds to the color of an object in the visible region, which is represented by a unique characteristic curve, aka reflectance curve correlating wavelength and the radiant energy from the object (Figure 1.5). Transmission direction and amplitude, corresponding to the direction of propagation and magnitude of the waves, are influenced by the spatial location and shape of the objects. Finally, the plane of polarization, that is,

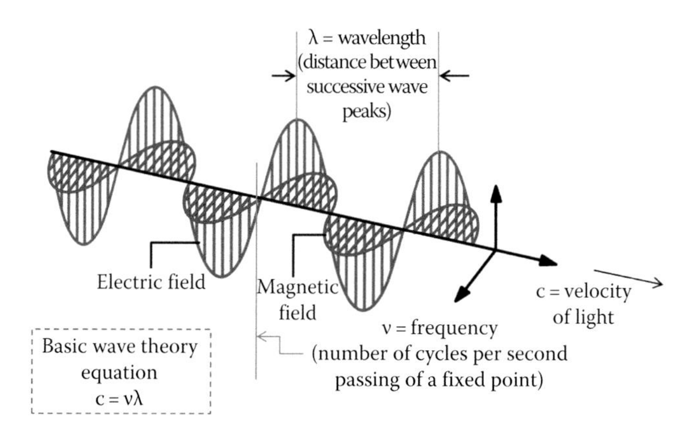

FIGURE 1.4 Illustration of an electromagnetic wave.

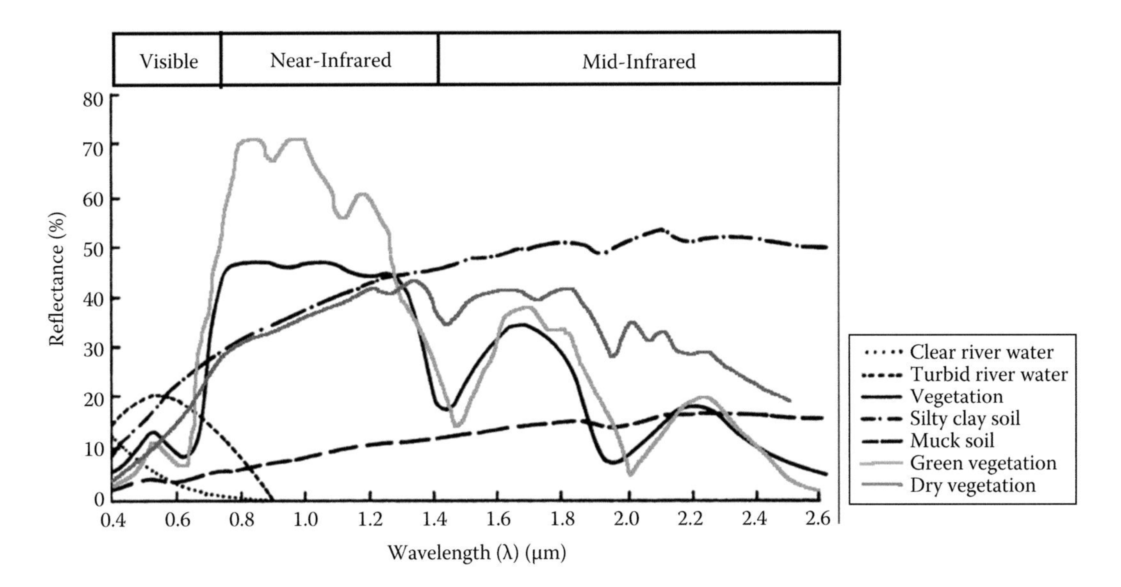

FIGURE 1.5 Typical reflectance curve of different objects generated with electromagnetic radiation frequency characteristics that are useful to distinguish/interpret objects. (Adapted and modified from Murai, S., *Remote Sensing Note*, JARS, Tokyo, Japan, 1993.) Note how energy is reflected or absorbed in various portions of the spectrum that forms the basis of remote sensing. For example, healthy vegetation absorbs heavily in the red band (especially around 0.68 μm) but reflects heavily in near-infrared 0.76–0.90 μm.

orientation of the electric field of the radiant energy, is influenced by the geometric shape of the objects under investigation (Murai, 1993). Therefore, EMR provides detailed information about the object(s) under investigation based on the spectral characteristics of the objects.

NASA reference publication # 1139 by Bowker et al. (1985) is one of the most comprehensive articles on spectral reflectance curves and curve development processes of natural targets for use in remote sensing studies. Most investigations based on remote sensing data are models that are developed between the amount of electromagnetic or light energy reflected, emitted, transmitted, or backscattered from the object at different frequencies of the EMS and the biophysical/chemical properties of the object or phenomena under investigation (Jensen, 2009). The EMS is divided into several wavelength (frequency) regions, such as gamma rays (10^{-6}–10^{-5} μm), x-ray (10^{-5}–10^{-2} μm), UV (0.1–0.4 μm), visible (0.4–0.7 μm), infrared (0.7–10^{3} μm), microwaves (10^{3}–10^{6} μm), and radio waves (>10^{6} μm). Panchromatic (i.e., grayscale) and color (i.e., red, green, blue [RGB]) imaging systems have dominated electro-optical sensing in the visible region of the EMS (Figure 1.6), which describe the efficacy of a remotely sensed imagery (Shaw and Burke, 2003). Figure 1.7 depicts the bandwidth (wavelength) of different bands associated with the spectral regions, along with the different remote sensing systems that acquire data in these bands.

1.2 Remote Sensing Platforms and Sensor Characteristics

Traditional remote sensing has its roots in aircraft-based platforms where photographic instruments capture Earth features as either single images or dual overlapping stereo images. The landscape images provided the basis for some of the early photogrammetric and stereo analysis of images, which are still considered as foundations for aerial surveys. The aircrafts used for airborne remote sensing are usually single- or twin-engine propeller or turboprop platforms. However, small jets (Learjet) are used for high-altitude reconnaissance or mapping surveys. These aerial platforms usually fly at about 1,200–12,000 ft above the terrain with flying speeds of about 100 knots. Aerial platforms provide the main advantage of responding to the need of the user application. For example, in an emergency forest fire and flooding event, an aerial survey can be quickly planned and implemented. Thus, an aerial platform is a quick response system that one can easily adapt to changes in weather conditions. Moreover, airborne platforms enable near-real-time review of acquired data and provide great control of data quality. In contrast to aerial platforms, satellite platforms are enormously expensive and complex, and the development and deployment of sensors as payload on satellites can take 5–10 years. Unlike aerial platforms, satellite platforms due to orbital characteristics have limitations of revisiting the same land area at user-needed time interval. This revisit cycle that relates to the time taken by the satellite to take a subsequent image acquisition of the same land area might not be practical. However, once a satellite-based imaging sensor becomes operational, the spaceborne image data are usually very consistent and allow end users to develop robust applications such as land cover change analysis, particularly for large tracts of the Earth's surface. Hence, depending on the resolutions provided by the onboard sensors, satellite platforms offer a wide array of scale-based mapping. Over the past few decades, numerous satellites have been launched by various private

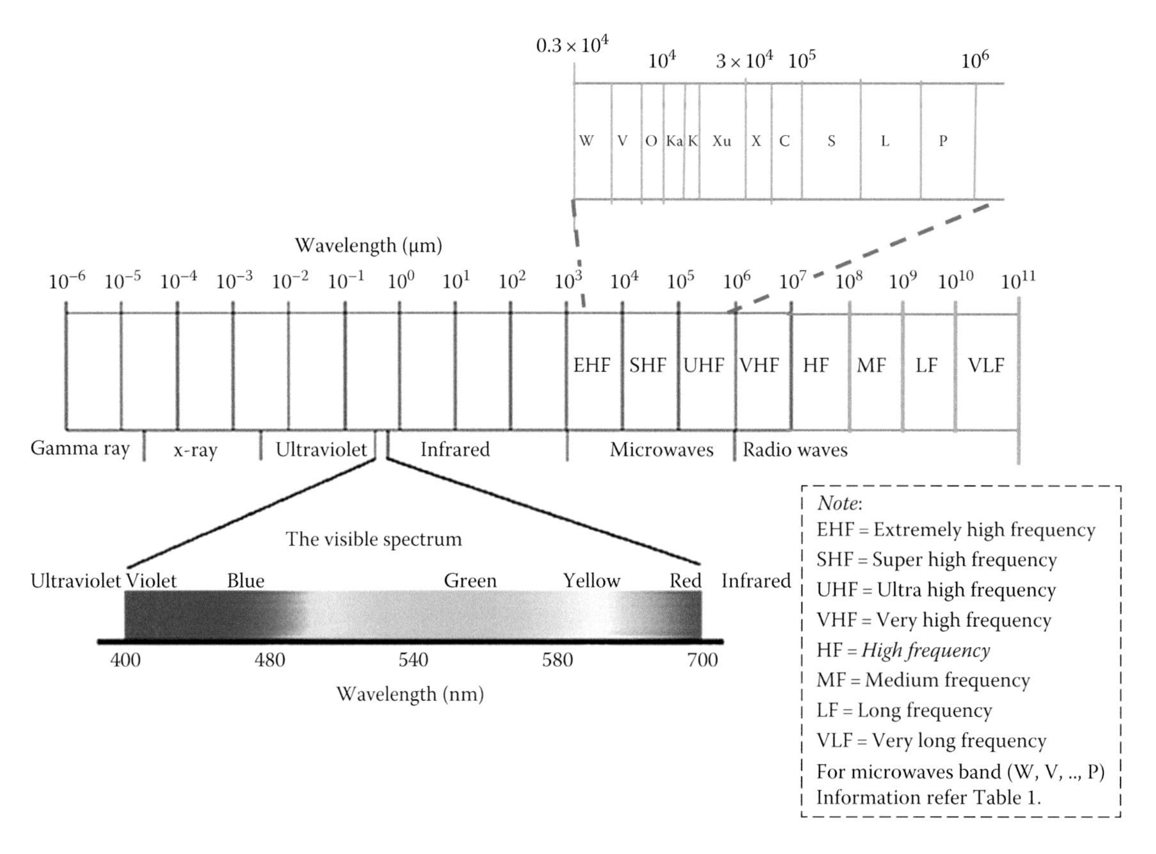

FIGURE 1.6 Electromagnetic spectrum on which remote sensing systems are based. For example, visible portion of electromagnetic radiation is in 400–700 nm. The data can be captured in narrowbands (e.g., 1 nm) or over broad wave bands (e.g., a single band over 630–690 nm).

enterprises and government organizations to acquire remote sensing data. Remote sensors are energy-sensitive devices mounted on the particular satellite to view and take images of the Earth in different bands of the EMS (Avery and Berlin, 1992). In addition to the two types of sensors (discussed in Section 1.1.2) used in remote sensing—passive and active sensors—remote sensors are classified into various types based on their scanning and imaging mechanism.

One of the main advantages of using active systems relate to the characteristics of the active energy (RADAR, LiDAR, etc.) used, which is solar energy independent and least affected by atmospheric constituents. This provides huge advantages for remote sensing over regions such as tropical environments where clouds and rain are frequent weather events that interfere with traditional optical/passive systems. Furthermore, active sensors do not require solar energy and hence can operate during night. Additionally, active systems have better capability of sensing vegetation and soil attributes that are dependent on moisture content, thus having immense value in various applications, including hydrology, geology, glaciology, forestry, and agriculture. The disadvantages include lower spectral characteristics, complicated data processing, massive data volume, and higher cost.

Passive and active sensors are divided into nonscanning and scanning types, which in turn are divided into imaging and nonimaging sensors (Murai, 1993). Microwave radiometer, magnetic sensor, gravimeter, Fourier spectrometer, microwave altimeter, laser water depth meter, laser distance meter, etc., are examples of nonimaging scanners (Murai, 1993). Figure 1.8 graphically depicts the sensor types.

According to NASA's Earth Observation System Data and Information System (EOSDIS), accelerometer, radiometer, imaging radiometer, spectrometer, spectroradiometer, hyperspectral radiometer, and sounders are examples of passive remote sensors. Linear Imaging Self-Scanning Sensor (LISS), advanced very-high-resolution radiometer (AVHRR), Coastal Zone Color Scanner (CZCS), Sea-Viewing Wide Field-of-View Sensor (SeaWiFS), Moderate-Resolution Imaging Spectroradiometer (MODIS), Active Cavity Radiometer Irradiance Monitor (ACRIM II and III), Advanced Spaceborne Thermal Emission and Reflection Radiometer (ASTER), Imaging Infrared Radiometer (IIR), Clouds and the Earth's Radiant Energy System (CERES), special sensor microwave radiometer (SMMR), airborne visible/infrared imaging spectrometer (AVIRIS), Polarization and Directionality of the Earth's Reflectance (POLDER), and Atmospheric Infrared Sounders (AIRSs) are some of the prominent passive sensors used in remote sensing data collection in the present day (EOSDIS, 2014). As discussed in the previous section, RADAR, ranging instruments, scatterometer, LiDAR, laser altimeter,

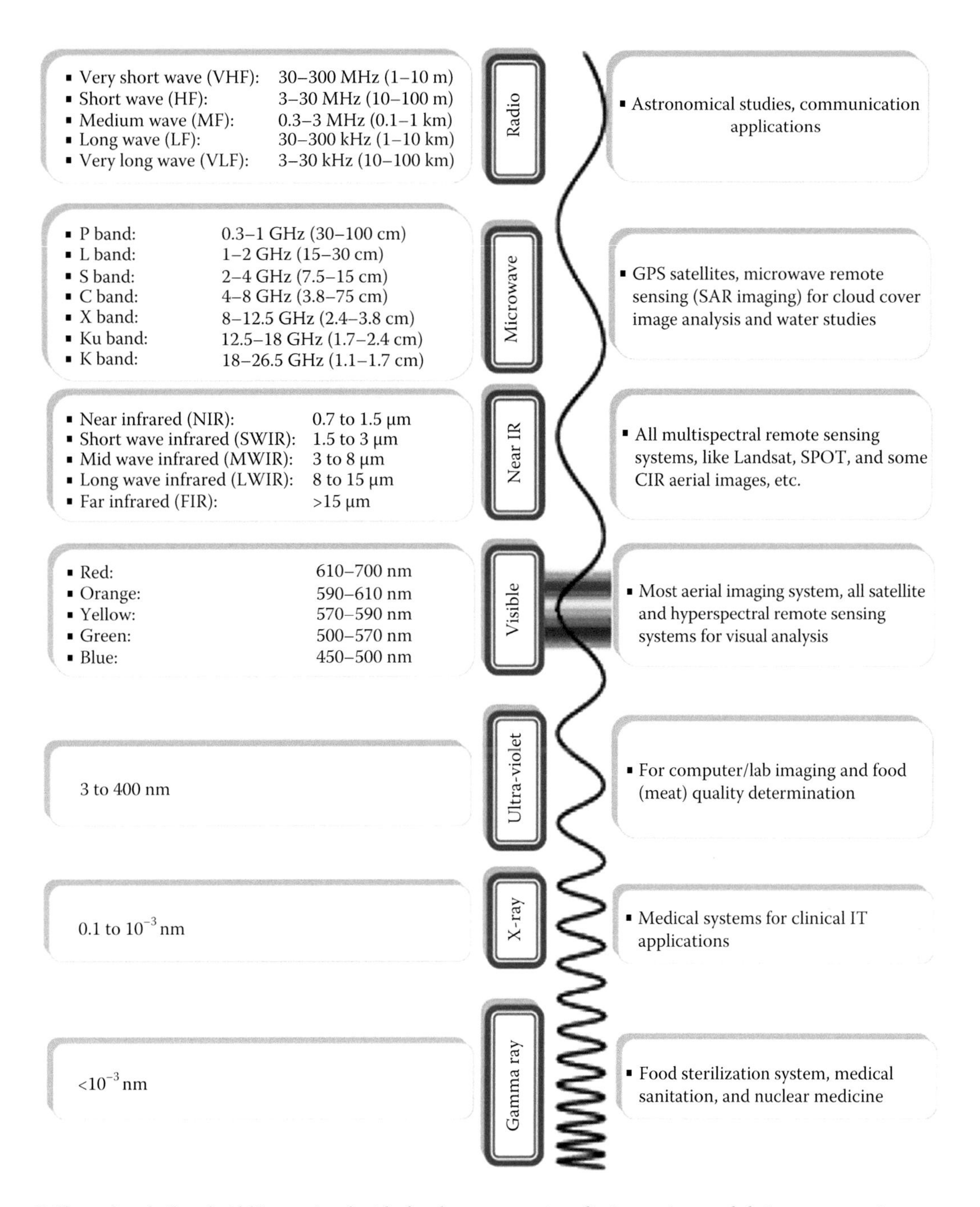

FIGURE 1.7 Different bands (bandwidth) associated with the electromagnetic radiation regions and their remote sensing systems. Gathering data in various spectral ranges is extremely important for characterizing Earth systems. For example, radar penetrates the clouds and is important to gather data during cloudy days when optical remote sensing is infeasible.

and sounders are examples of active sensors (EOSDIS, 2014). Table 1.1 provides a detailed list of remote sensors with the instrument name, type of sensor, platform, data center, and other descriptions about the sensors.

Remote sensing optical sensors are characterized by spectral, radiometric, and geometric performance (Murai, 1993). Observation range of the electromagnetic wave, center wavelength of a band, changes at both ends of a band, sensitivity of a band, polarization sensitivity, and ratio of sensitivity difference between different bands are some of the spectral characteristics that define the types of optical sensors (Murai, 1993). Detection accuracy, signal-to-noise ratio (S/N), dynamic range, quantization level, sensitivity difference between pixels, linearity of sensitivity, and noise equivalent power are a few radiometric characteristics of optical sensors (Murai, 1993). Field of view (FOV), IFOV, registration between different spectral bands, modular transfer function (MTF), and optical distortions are examples of

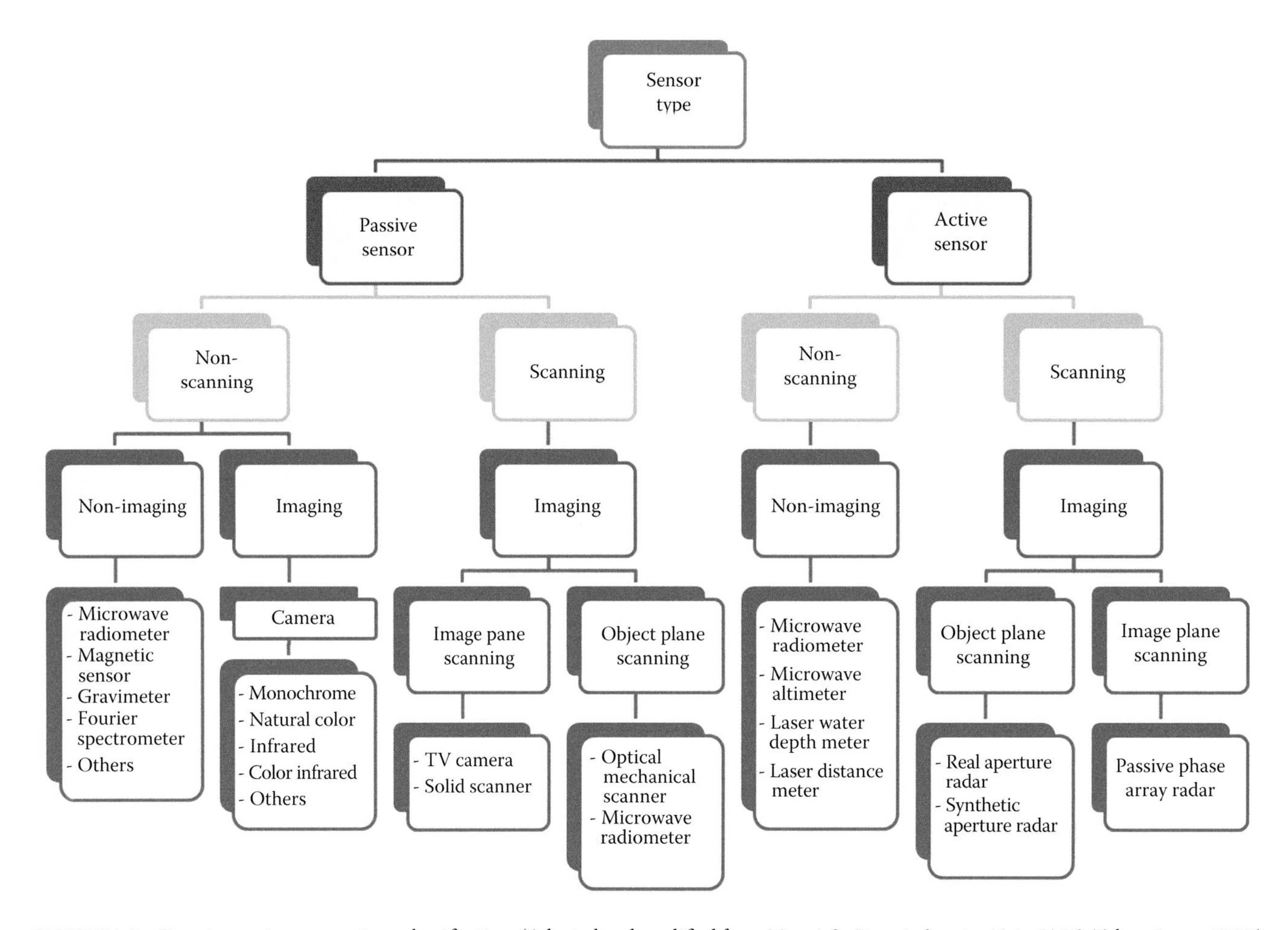

FIGURE 1.8 Remote sensing sensor–type classification. (Adapted and modified from Murai, S., *Remote Sensing Note*, JARS, Tokyo, Japan, 1993.)

geometric characteristics of optical sensors that classify the sensors (Murai, 1993). These sensor characteristics determine the spatial, spectral, radiometric, and temporal resolutions of remote sensing data.

1.2.1 Image Characteristics

Different sensors continuously scan the Earth's surface to produce imagery. Each image is a matrix of pixels or picture elements defined by columns and rows as shown in Figure 1.9. Each pixel records a numeric value representative of the brightness or intensity level of the reflected energy as discussed in Section 1.2. The image data produced by different sensors on satellite systems have unique characteristics that relate to the sensor's resolutions—spatial, spectral, radiometric, and temporal.

Radiometric resolution refers to the data depth indicative of the sensitivity of the sensor to incoming energy. About 8-bit data have higher contrast (0–255 digital number [DN] range) and higher radiometric resolution than a 6-bit sensor that provides data in a lower range (0–63 DN range) and thus lower contrast. For example, Landsat 7 ETM+ sensor provides 8-bit data for individual bands, while the latest version of Landsat satellite series, Landsat 8, provides 12-bit data with DN values ranging from 0 to 4095. Temporal resolution refers to the revisit frequency/time of the sensor to a specific location on the Earth's surface. For example, Landsat revisit time is 16 days over a specific geographic location on Earth. Table 1.2 provides a list of selected remote sensing systems (sensors) with detailed data characteristics including the spatial, spectral (band ranges), and frequency of revisit over a location on Earth.

1.2.1.1 Spatial Resolution

Spatial resolution implies the unit ground area for which the sensor records the reflected energy providing a unique DN or brightness value (BV) (refer to Figure 1.9). For example, a Landsat sensor senses a ground space that is 30 m × 30 m. Usually, the image spatial resolution is equal to the ground sample distance (GSD), which is the smallest discernible detail in an image (Gonzalez and Woods, 2002). The GSD could be smaller than the spatial resolution of an image acquired by a remote sensing system, and they vary from a fraction of a meter to tens of meters (Shaw and Burke, 2003). The image spatial resolution is recognized primarily by the sensor aperture and platform altitude. Platform altitude is loosely constrained by the class of

TABLE 1.1 List of Selected Active and Passive Sensors and Their Types

Instrument	Type	Platform	Data Center	Description
Passive sensors				
Power radiometers and imagers				
ACRIM II	Total power radiometer	UARS	LaRC ASDC	Measures total solar irradiance.
ACRIM III	Total power radiometer	ACRIMSAT	LaRC ASDC	Measures total solar irradiance.
TIM	Total power radiometer	SORCE	GES DISC	Measures total solar irradiance.
LIS	Imager	TRMM	GHRC DAAC	Detects intracloud and cloud-to-ground lightning, day and night.
WFC	Wide Field Camera	CALIPSO	LaRC ASDC	Fixed, nadir-viewing imager with a single spectral channel covering the 620–670 nm region.
Multispectral instruments				
AMPR	Microwave radiometer	ER-2 and DC-8	GHRC DAAC	Cross track scanning total power microwave radiometer with four channels centered at 10.7, 19.35, 37.1, and 85.5 GHz (FIRE ACE, Teflon-B, TRMM-LBA, CAMEX-4. TCSP, TC4 Projects).
AMSR-E	Multichannel microwave radiometer	Aqua	NSIDC DAAC GHRC DAAC	Measures precipitation, oceanic water vapor, cloud water, near-surface wind speed, sea and land surface temperature, soil moisture, snow cover, and sea ice. Provides spatial resolutions of 5.4, 12, 21, 25, 38, and 56 km and a 0.25° resolution.
ASTER	Multispectral radiometer	Terra	LP DAAC ORNL DAAC	Measures surface radiance, reflectance, emissivity, and temperature. Provides spatial resolutions of 15, 30, and 90 m.
AVHRR	Multispectral radiometer	NOAA/POES	GES DISC NSIDC DAAC ORNL DAAC PO.DAAC	Four or six bands, depending on platform. Telemetry resolutions are 1.1 km (HRPT data), 4 km (Pathfinder V5 and GAC data), 5, and 25 km spatial resolution.
CERES	Broadband scanning radiometer	Aqua Terra TRMM NPP	LaRC ASDC	Four to six channels (shortwave, longwave, total). Measures atmospheric and surface energy fluxes. Provides 20 km resolution at nadir.
IIR	IIR	CALIPSO	LaRC ASDC	Nadir-viewing, nonscanning imager having a 64 km swath with a pixel size of 1 km. Provides measurements at three channels in the TIR window region at 8.7, 10.5, and 12.0 mm.
MAS	Imaging spectrometer	NASA ER-2 aircraft	GES DISC GHRC DAAC LaRC ASDC ORNL DAAC	Fifty spectral bands that provide spatial resolution of 50 m at typical flight altitudes.
MISR	Imaging spectrometer	Terra	LaRC ASDC ORNL DAAC	Obtains precisely calibrated images in four spectral bands, at nine different angles, to provide aerosol, cloud, and land surface data. Provides spatial resolution of 250 m to 1.1 km.
MODIS	Imaging spectroradiometer	Aqua Terra	GES DISC GHRC DAAC LP DAAC MODAPS NSIDC DAAC OBPG ORNL DAAC PO.DAAC	Measures many environmental parameters (ocean and land surface temperatures, fire products, snow and ice cover, vegetation properties and dynamics, surface reflectance and emissivity, cloud and aerosol properties, atmospheric temperature and water vapor, ocean color and pigments, and ocean biological properties). Provides moderate spatial resolutions of 250 m (bands 1 and 2), 500 m (bands 3–7), and 1 km (bands 8–36).
SSM/I	Multispectral microwave radiometer	DMSP	GHRC DAAC LaRC ASDC NSIDC DAAC PO.DAAC ORNL DAAC	Has seven channels and four frequencies. Measures atmospheric, ocean, and terrain microwave brightness temperatures, which are used to derive ocean near-surface wind speed, atmospheric integrated water vapor, and cloud/rain liquid water content and sea ice extent and concentration.
SMMR	Multispectral microwave radiometer	NIMBUS-7	GES DISC LaRC ASDC NSIDC DAAC PO.DAAC	Ten channels. Measures SSTs, ocean near-surface winds, water vapor and cloud liquid water content, sea ice extent, sea ice concentration, snow cover, snow moisture, rainfall rates, and differential of ice types.
TMI	Multispectral microwave radiometer	TRMM	GES DISC GHRC DAAC	TMI measures the intensity of radiation at five separate frequencies: 10.7, 19.4, 21.3, 37, and 85.5 GHz. TMI measures microwave brightness temperatures, water vapor, cloud water, and rainfall intensity.

(*Continued*)

TABLE 1.1 (*Continued*) List of Selected Active and Passive Sensors and Their Types

Instrument	Type	Platform	Data Center	Description
Hyperspectral instruments				
AVIRIS	Imaging spectrometer	Aircraft	ORNL DAAC	224 contiguous channels, approximately 10 nm wide. Measurements are used to derive water vapor, ocean color, vegetation classification, mineral mapping, and snow and ice cover (BOREAS Project).
SOLSTICE	Spectrometer	SORCE	GES DISC	Measures the solar spectral irradiance of the total solar disk in the UV wavelengths from 115 to 430 nm.
Polarimetric instruments				
POLDER	Polarimeter	Aircraft	ORNL DAAC	Measures the polarization and the directional and spectral characteristics of the solar light reflected by aerosols, clouds, and the Earth's surface (BOREAS Project).
PSR	Microwave polarimeter	Aircraft	GHRC DAAC	Measures wind speed and direction (CAMEX-3 Project).
Ranging and sounding instruments				
ACC	Accelerometer	GRACE	PO.DAAC	The Onera SuperSTAR Accelerometer measures the nongravitational forces acting on the GRACE satellites.
AIRS	Sounder	Aqua	GES DISC	Measures air temperature, humidity, clouds, and surface temperature. Provides spatial resolution of ~13.5 km in the IR channels and ~2.3 km in the visible. Swath retrieval products are at 50 km resolution.
AMSU	Sounder	Aqua	GES DISC GHRC DAAC	Has 15 channels. Measures temperature profiles in the upper atmosphere. Has a cloud filtering capability for tropospheric temperature observations. Provides spatial resolution of 40 km at nadir.
HAMSR	Sounder	DC-8	GHRC DAAC	Measures vertical profiles of temperature and water vapor, from the surface to 100 mb in 2–4 km layers (CAMEX-4, NAMMA Projects).
HIRDLS	Sounder	Aura	GES DISC	Measures infrared emissions at the Earth's limb in 21 channels to obtain profiles of temperature, ozone, CFCs, various other gases affecting ozone chemistry, and aerosols at 1 km vertical resolution. In addition, HIRDLS measures the location of polar stratospheric clouds.
MLS	Sounder	Aura	GES DISC	Five broadband radiometers and 28 spectrometers measure microwave thermal emission from the Earth's atmosphere to derive profiles of ozone, SO_2, N_2O, OH group, and other atmospheric gases, temperature, pressure, and cloud ice.
MOPITT	Sounder	Terra	LaRC ASDC ORNL DAAC	Measures carbon monoxide and methane in the troposphere. Is able to collect data under cloud-free conditions. Provides horizontal resolution of ~22 km and vertical resolution of ~4 km.
OMI	Multispectral radiometer	Aura	GES DISC	740 wavelength bands in the visible and UV. Measures total ozone and profiles of ozone, N_2O, SO_2, and several other chemical species.
TES	Imaging Spectrometer	Aura	LaRC ASDC	High-resolution imaging infrared Fourier transform spectrometer that operates in both nadir and limb-sounding modes. Provides profile measurements of ozone, water vapor, carbon monoxide, methane, nitric oxide, nitrogen dioxide, nitric acid carbon dioxide, and ammonia.
Active sensors				
Radar and laser (LiDAR)				
ALT-A, -B	Radar altimeter	TOPEX/Poseidon	PO.DAAC	Dual-frequency altimeter that measures height of the satellite above the sea (satellite range), wind speed, wave height, and ionospheric correction.
CALIOP	Cloud and aerosol LiDAR	CALIPSO	LaRC ASDC	Two-wavelength polarization-sensitive LiDAR that provides high-resolution vertical profiles of aerosols and clouds.
Altimeters: radar and laser (LiDAR) GLAS	Laser altimeter	ICESat	NSIDC DAAC	The main objective is to measure ice sheet elevations and changes in elevation through time. Secondary objectives include measurement of cloud and aerosol height profiles, land elevation and vegetation cover, and sea ice thickness.
Poseidon-1	Radar altimeter	TOPEX/Poseidon	PO.DAAC	Single-frequency altimeter that measures height of the satellite above the sea (satellite range), wind speed, and wave height.

(*Continued*)

TABLE 1.1 (*Continued*) List of Selected Active and Passive Sensors and Their Types

Instrument	Type	Platform	Data Center	Description
Poseidon-2	Radar altimeter	Jason-1	PO.DAAC	Measures sea level, wave height, wind speed, and ionospheric correction.
SAR	SAR	ERS-1 ERS-2 JERS-1 RADARSAT-1 PALSAR UAVSAR	ASF SDC NSIDC DAAC ORNL DAAC	Provides high-resolution surface imagery at 7–240 m. Multiple polarizations are utilized by some SAR instruments.
KBR	Ranging instrument	GRACE	PO.DACC	The dual-frequency KBR instrument measures the range between the GRACE satellites to extremely high precision.
Scatterometers				
NSCAT	Radar scatterometer	ADEOS-I	PO.DAAC	Dual Fan-Beam Ku Band that measures ocean vector winds at a nominal grid resolution of 25 km.
SASS	Radar scatterometer	Seasat	PO.DAAC	Dual Fan-Beam Ku Band that measures ocean vector winds at a nominal grid resolution of 25 km.
Seawinds	Radar scatterometer	QuikSCAT ADEOS-II	PO.DAAC	Dual Pencil-Beam Ku Band that measures ocean vector winds at a nominal grid resolution of 25 km.
Sounding instruments				
CLS	LiDAR	ER-2	LaRC ASDC	Determines vertical cloud structure (FIRE Project).
LASE	LiDAR	DC-8	GHRC DAAC	Measures water vapor, aerosols, and clouds throughout the troposphere (CAMEX-4, TCSP, NAMMA Projects).
PR	Phased array radar	TRMM	GES DISC ORNL DAAC	Measures 3D distribution of rain and ice. Provides horizontal resolution of 250 m and vertical resolution of 5 km.
VIL	LiDAR	Ground	LaRC ASDC ORNL DAAC	Determines vertical cloud structure (FIFE, FIRE, and BOREAS Projects).

Source: Adapted and modified from Earth Observing System Data and Information System (EOSDIS), Earth Data, 2014, https://earthdata.nasa.gov/, accessed on July 5, 2014.

sensor platform, that is, either spaceborne or airborne (Shaw and Burke, 2003). Sensor aperture size, particularly for spaceborne systems, determines the cost of the remote sensing system; many times to cut cost, the aperture size is made bigger, thus providing low-spatial-resolution image. According to Shaw and Burke (2003), the best detection performance is expected when the angular resolution of the sensor, specified in terms of the GSD, is commensurate with the footprint of the targets of interest. For example, if the application requires observation of global land cover, then low-resolution imagery is suitable. On the other hand, if the application requires local land cover information such as water stress in a corn field, then high-resolution imagery is ideal. According to Belward and Skoien (2014), land cover datasets at 1 km resolution (National Oceanic and Atmospheric

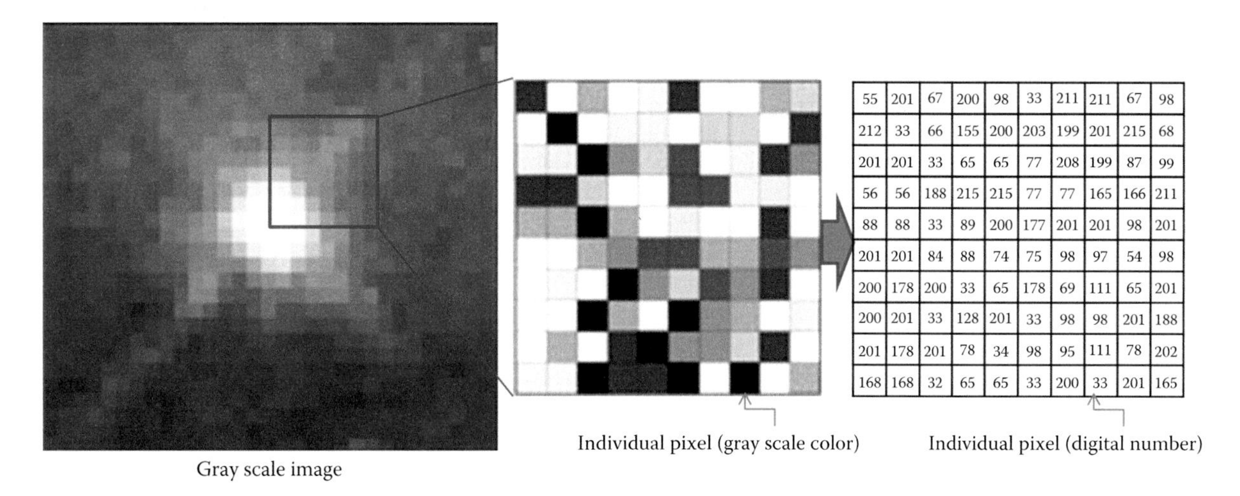

FIGURE 1.9 Basic matrix configuration of an image showing a pixel (picture element). Each pixel denotes an intensity value (e.g., reflected energy, radar backscatter). Each pixel can have data values between 0 and 256 for 8-bit (quantization) data, 0 and 4096 for 12-bit data, and so on.

TABLE 1.2 Data Characteristics of Selective Satellite Sensors

Sensor	Image Characteristics				Band Range (μm)
	Spatial (m)	Spectral (#)	Radiometric (bit)	Temporal (Days)	
Coarse-resolution sensors					
AVHRR	1000	4	11	Daily	0.58–0.68
					0.725–1.1
					5.55–3.95
					10.30–10.95
					10.95–11.65
MODIS	250; 500; 1000	36	12	Daily	0.62–0.67
					0.84–0.876
					0.459–0.479
					0.545–0.565
					1.23–1.25
					1.63–1.65
					2.11–2.16
Multispectral sensors					
Landsat 1, 2, 3 MSS	56 × 79	4	6	16	0.5–0.6
					0.6–0.7
					0.7–0.8
					0.8–1.1
Landsat 4, 5 TM	30	7	8	16	0.45–0.52
					0.52–0.60
					0.63–0.69
					0.76–0.90
					1.55–1.74
					10.4–12.5
					2.08–2.35
Landsat 7 ETM+	30	8	8	16	0.45–0.52
					0.52–0.60
					0.63–0.69
					0.50–0.75
					0.75–0.90
					10.4–12.5
					1.55–1.75
					0.52–0.90 (p)
Landsat 8	15; 30; 100	11	12	16	0.43–0.45
					0.45–0.51
					0.53–0.59
					0.64–0.67
					0.85–0.88
					1.57–2.29
					0.50–0.68 (p)
					1.36–1.38
					10.60–11.19
					11.50–12.51
ASTER	15; 30; 90	15	8	16	0.52–0.63
					0.63–0.69
					0.76–0.86
					1.60–1.70
					2.145–2.185

(Continued)

TABLE 1.2 (*Continued*) Data Characteristics of Selective Satellite Sensors

Sensor	Image Characteristics				Band Range (μm)
	Spatial (m)	Spectral (#)	Radiometric (bit)	Temporal (Days)	
					2.185–2.225
					2.235–2.285
					2.295–2.365
					2.360–2.430
					8.125–8.475
					8.475–8.825
					8.925–9.275
					10.25–10.95
					10.95–11.65
ALI	30	10	12	16	0.48–0.69 (p)
					0.433–0.453
					0.450–0.515
					0.425–0.605
					0.633–0.690
					0.775–0.805
					0.845–0.890
					1.200–1.300
					1.550–1.750
					2.080–2.350
SOPT-5	2.5–20	5	8	2–3	0.48–0.71 (p)
					0.5–0.59
					0.61–0.68
					0.78–0.89
					1.58–1.75
IRS-1A, IRS-1B	36; 73	4	8	22	0.45–0.52
					0.52–0.68
					0.62–0.68
					0.77–0.86
IRS-1C	5.8 (p); 23.5; 70.5 (B5)	5	8	24	0.50–0.75 (p)
					0.52–0.59
					0.62–0.68
					0.77–0.86
					1.55–1.70
IRS-P6 AWiFS	56	4	10	16	0.52–0.59
					0.62–0.68
					0.77–0.86
					1.55–1.70
JERS-1	18	8	8	44	0.52–0.60
					0.63–0.69
					0.76–0.86 (2 bands)
					1.60–1.71
					2.01–2.12
					2.13–2.25
					2.27–2.40
CBERS-2	20 (p)	5	8	26	0.51–0.73 (p)
CBERS-3B	20 (MS)				0.45–0.52
CBERS-3	5 (p)				0.52–0.59
CBERS-4	20 (MS)				0.63–0.69
					0.77–0.89

(*Continued*)

TABLE 1.2 (*Continued*) Data Characteristics of Selective Satellite Sensors

	Image Characteristics				
Sensor	Spatial (m)	Spectral (#)	Radiometric (bit)	Temporal (Days)	Band Range (μm)
Hyperspectral sensor					
Hyperion	30	196	16	16	196 effective calibrated bands
					VNIR (bands 8–57) 0.427–0.925
					SWIR (band 79–224) 0.93–2.4
Hyperspatial sensor					
IKONOS	1–4	4	11	5	0.445–0.516
					0.506–0.595
					0.632–0.698
					0.757–0.853
QuickBird	0.61–2.44	4	11	5	0.45–0.52
					0.52–0.60
					0.63–0.69
					0.76–0.89
RESOURCESAT	5.8	3	10	24	0.52–0.59
					0.62–0.68
					0.77–0.86
RapidEye A–E	6.5	5	12	1–2	0.44–0.51
					0.52–0.59
					0.63–0.68
					0.69–0.73
					0.77–0.89
WorldView	0.55	1	11	1.7–5.9	0.45–0.51
FORMOSAT-2	2–8	5	11	Daily	0.45–0.52
					0.52–0.60
					0.63–0.69
					0.76–0.9
					0.45–0.9 (p)
KOMPSAT-2	1–4	5	10	3–28	0.45–0.52
					0.52–0.60
					0.63–0.69
					0.76–0.9
					0.5–0.9 (p)

Source: Adopted from Melesse et al. (2007).
Note: (p) is the panchromatic band.
(MS) is multispectral bands

Administration [NOAA] AVHRR) are *low*-resolution data when compared with a 30 m Landsat data product but become acceptable when land cover information is gathered for a global assessment perspective or when it is used in a climate model running with a 100 km cell size. In addition to sensor optical characteristics, the flight height of the satellite or the airplane determines the spatial resolution. The higher the location of the satellite in space, the lower will be the image resolution.

Sensor specifications are usually provided by satellite operator and/or sponsoring space agency and/or manufacturer, reproduced by three agencies (National Space Science Data Center [NSSDC], Committee on Earth Observation Satellites [CEOS], and Ocean Surface Current Analysis Real Time [OSCAR]) (Belward and Skoien, 2014). Some remote sensors provide data at two or more spatial resolutions with panchromatic spatial resolution being higher than the multispectral scanner (MSS) sensors. MODIS, Earth Observing System (EOS) Terra, Landsat, and Systeme Probatoire D'Observation De La Terre (SPOT) are the examples of dual spatial resolution–based image acquisition. Some SAR missions, such as Canada's RADARSAT, also provide multiple-resolution images at different band frequencies (Belward and Skoien, 2014). In general, spatial resolution can be categorized into five broad classes: 0.5–4.9 m (very high resolution), 5.0–9.9 m (high resolution), 10.0–39.9 m (medium resolution), 40.0–249.9 m (moderate resolution), and 250 m–1.5 km (low resolution). Any such grouping is somewhat arbitrary; the low-resolution class acknowledges the threshold of 250 m established for global monitoring of land transformations

(Townshend and Justice, 1988), and the moderate-/medium-resolution classes include imagery available through *free-and-open* data policies. The high- and very-high-resolution classes reference commercial distinctions. The upper limit of 50 cm is set because the U.S. Government licensing limits unrestricted distribution of spatial data at this resolution. Table 1.2 provides the spatial resolution information of a selected list of remote sensing systems (sensors).

Generally, panchromatic bands of satellites are of very high spatial resolution. For example, SPOT 1 acquired the first 10 m resolution images in 1986; then in 1995, Indian Remote Sensing Satellite (IRS-1C) image exceeded the 6 m spatial resolution. In 1999, IKONOS satellite provided imagery with 1 m spatial resolution, the 0.5 m spatial resolution imagery was obtained by WorldView-1 after that, and finally, the highest resolution imagery of 41 cm was obtained with GeoEye in 2007. All these high-spatial-resolution satellite images are acquired with panchromatic bands.

1.2.1.2 Spectral Resolution

Satellite sensors measure EMR in different portions of the EMS. Spectral resolution is the ability of the sensor to resolve spectral features and *bands* into their separate components in the EMS. It also describes the ability of a sensor to distinguish between wavelength intervals in EMS. Thus, in general, spectral resolution determines the number of bands a satellite sensor can sense. And so, higher-spectral-resolution imagery will provide more spectral information when compared to lower-spectral-resolution imagery. The spectral information is particularly useful in applications dealing with mapping and modeling biophysical properties of objects such as water quality, plant vigor, and soil nutrients. As shown in Figure 1.10, U.S. Landsat has higher spectral resolution than the French SPOT sensor.

MODIS sensor has a high spectral resolution because it senses in 36 wavelength regions of the EMS, in comparison to Landsat TM that senses in only 7 EMR regions.

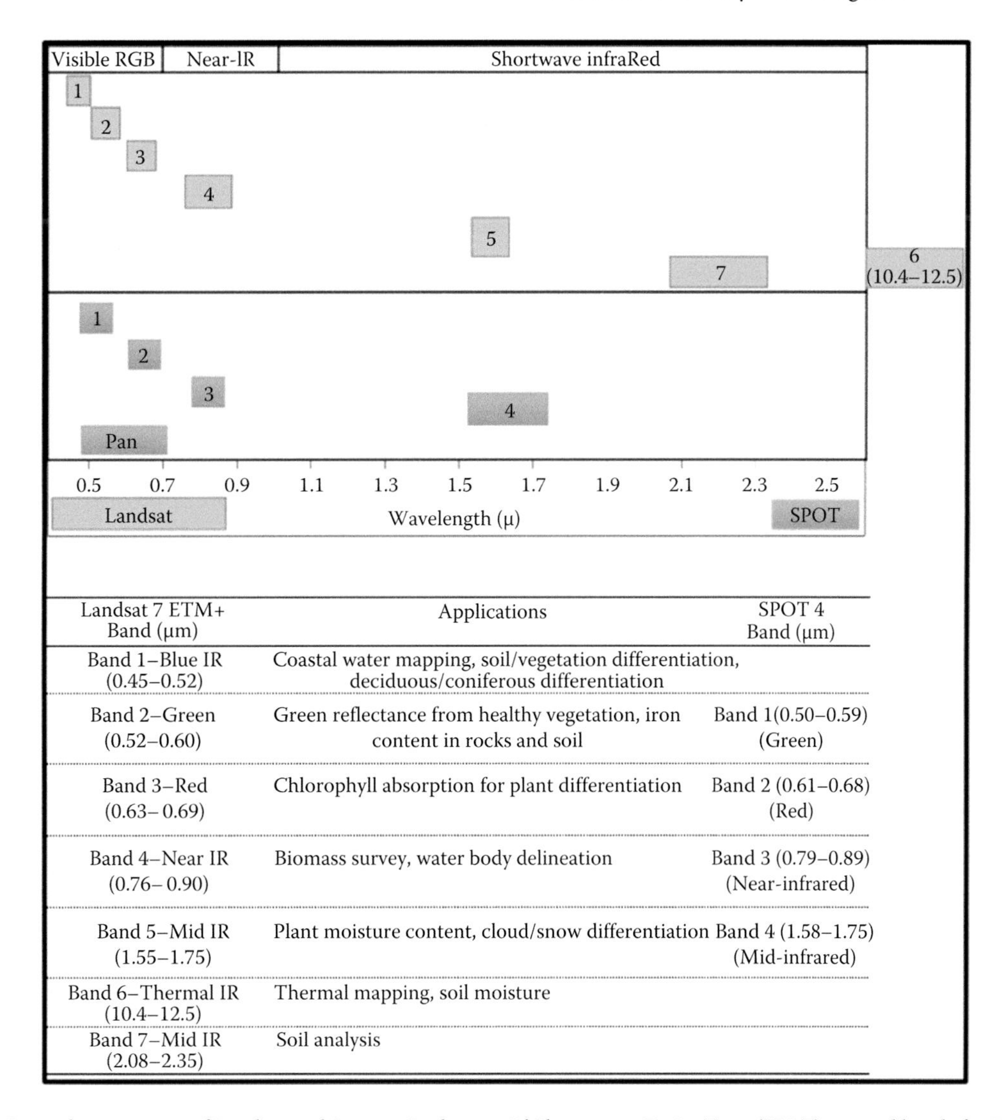

Landsat 7 ETM+ Band (μm)	Applications	SPOT 4 Band (μm)
Band 1–Blue IR (0.45–0.52)	Coastal water mapping, soil/vegetation differentiation, deciduous/coniferous differentiation	
Band 2–Green (0.52–0.60)	Green reflectance from healthy vegetation, iron content in rocks and soil	Band 1(0.50–0.59) (Green)
Band 3–Red (0.63–0.69)	Chlorophyll absorption for plant differentiation	Band 2 (0.61–0.68) (Red)
Band 4–Near IR (0.76–0.90)	Biomass survey, water body delineation	Band 3 (0.79–0.89) (Near-infrared)
Band 5–Mid IR (1.55–1.75)	Plant moisture content, cloud/snow differentiation	Band 4 (1.58–1.75) (Mid-infrared)
Band 6–Thermal IR (10.4–12.5)	Thermal mapping, soil moisture	
Band 7–Mid IR (2.08–2.35)	Soil analysis	

FIGURE 1.10 General comparison of Landsat and Systeme Probatoire D'Observation De La Terre (SPOT) spectral bands for Earth observation application.

Similarly, Hyperion sensor provides hyperspectral capabilities of resolving 220 spectral bands in the 0.4–2.6 μm range. The position, number, and width of spectral bands in an image determine the degree to which individual targets can be discriminated. Multispectral imagery has a higher degree of individual target discrimination power than a panchromatic single-band image. Table 1.2 provides a list of spectral bands available for selective remote sensing sensors.

Spectral reflectance curves (Figure 1.5) and discussion about EMS in Section 1.2 explain the advantages of discerning Earth objects or phenomena from the images using the spectral band information. One of the important advantages of separately sensing different spectral bands relates to the ability of combining the bands in various ways to enhance the visual information from the image. These combinations are easily implemented using software routines that stack three image bands and assign the reflected data values to three fundamental colors (red, green, and blue) to create image composites. Thus, when we have higher spectral resolution in our satellite data, more Earth's surface features can be discerned using different band combinations resulting in different color composites. In addition to visual enhancements, higher spectral resolution provides higher capabilities to combine band information using *band algebra* to extract additional biophysical information about the Earth's surface such as plant vigor and soil moisture status. These are implemented as spectral indices that mathematically combine band data into derivate, for example, normalized difference vegetation index (NDVI). Figure 1.10 shows some of the important applications of individual bands in satellite images, for example, Landsat 7 ETM+ and SPOT 4. Thus, higher-spectral-resolution images are better for feature extraction and decision support.

1.2.1.3 Radiometric Resolution

As satellite sensors capture reflected energy (range of spectral region) from the Earth's surface (ground area unit) and the energy is quantified into a DN, where the value depends on the intensity of the energy sensed. ESRI GIS Dictionary (2014) defines radiometric resolution as "the sensitivity of a sensor to incoming reflectance. Radiometric resolution refers to the number of divisions of bit depth (for example, 255 for 8-bit, 65,536 for 16-bit, and so on) in data collected by a sensor." While the pixel arrangement describes the spatial structure of an image, the radiometric characteristics describe the actual information content in a remotely sensed image. When an image is acquired on film or by a sensor, its sensitivity to the magnitude of the EMR determines the radiometric resolution (CCRS, 2014). The radiometric resolution of an imaging system describes its ability to differentiate slight differences in energy. The finer the radiometric resolution of a sensor, the more sensitive it is in detecting small differences in reflected or emitted energy (CCRS, 2014). As can be seen in Figure 1.11, a sensor with low radiometric resolution (1 bit) can resolve the incoming reflectance into two levels of contrast, while sensors with higher radiometric resolution provide higher levels of energy quantization. These results not only provide better contrast to the display but also facilitate the numerical analysis of the subtle differences in the radiometry. The downside to this is the high data volumes and storage issues associated with high-radiometric-resolution imagery. Table 1.2 provides the radiometric resolution of many selective remote sensing sensors.

1.2.1.4 Temporal Resolution

Satellite revisit period refers to the length of time it takes for the satellite to complete one entire orbit cycle around the Earth, for example, Landsat satellite takes 16 days. The revisit period of a satellite sensor is usually several days

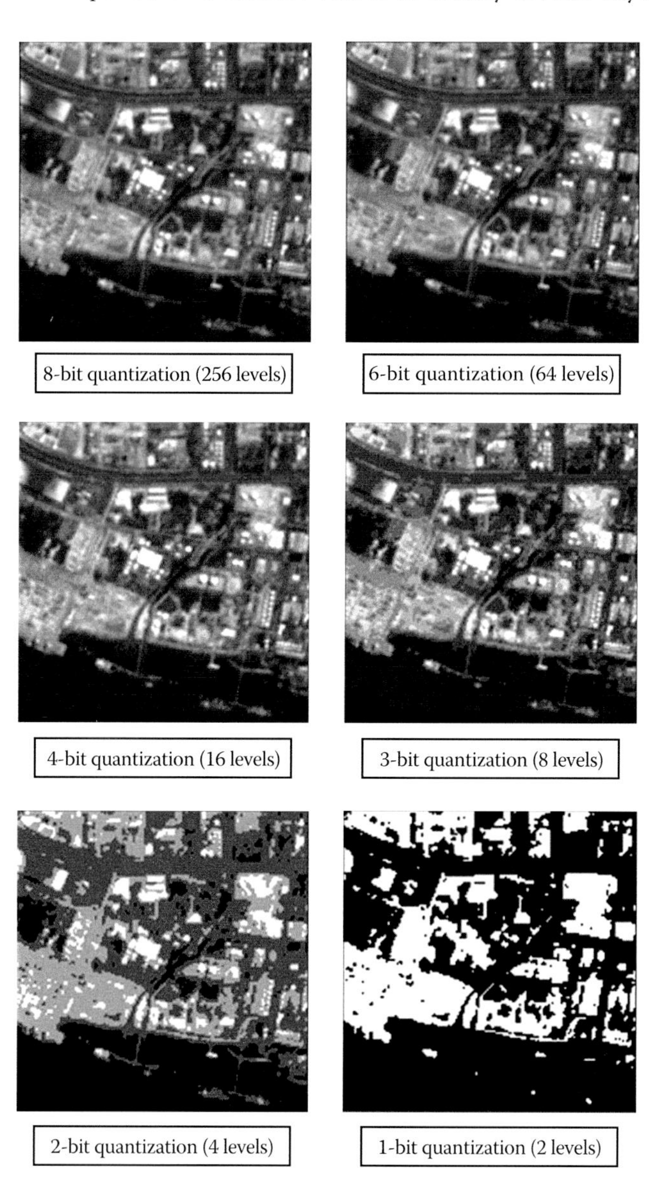

FIGURE 1.11 Image captured at different levels of energy quantization to illustrate concept of radiometric resolution. Center for Remote Imaging, Sensing, and Processing (CRISP) (2014). (From http://www.crisp.nus.edu.sg/~research/tutorial/image.htm.) The more the quantization, the greater is the detail of the information captured, thus enabling greater separability between features.

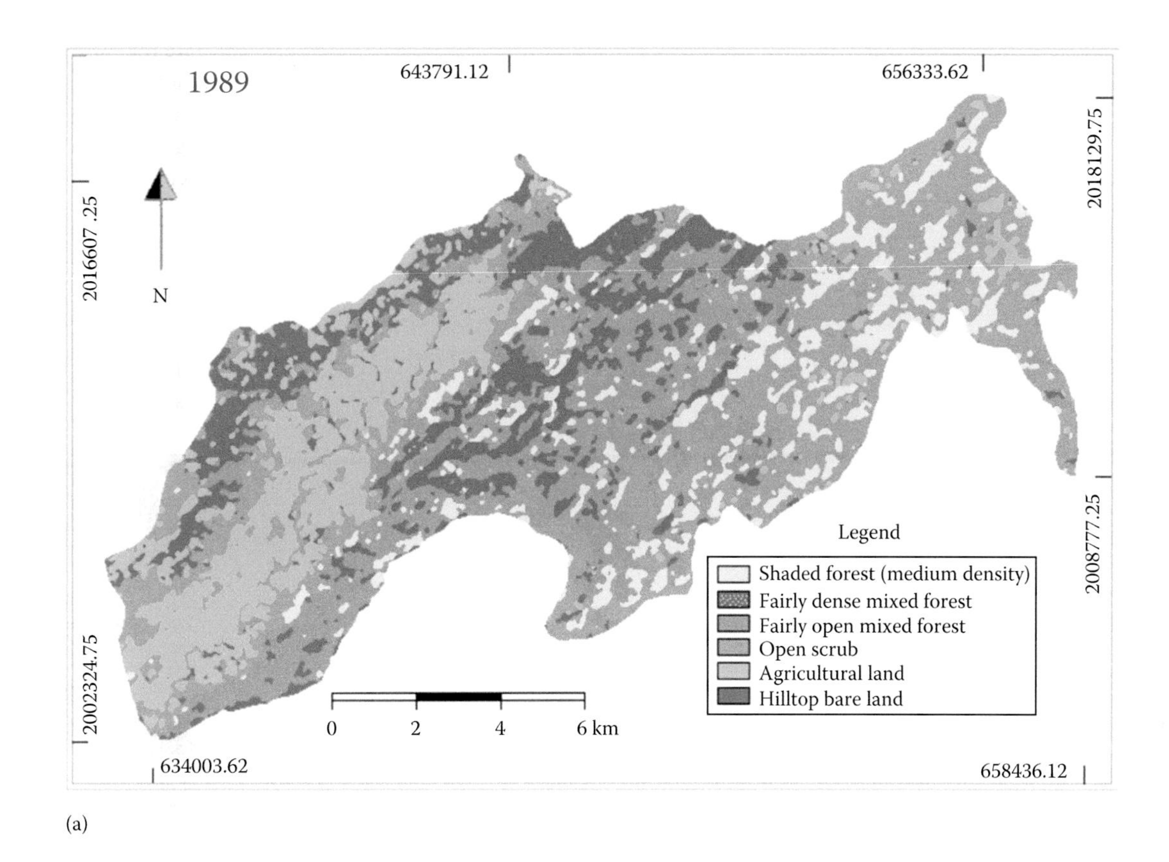

(a)

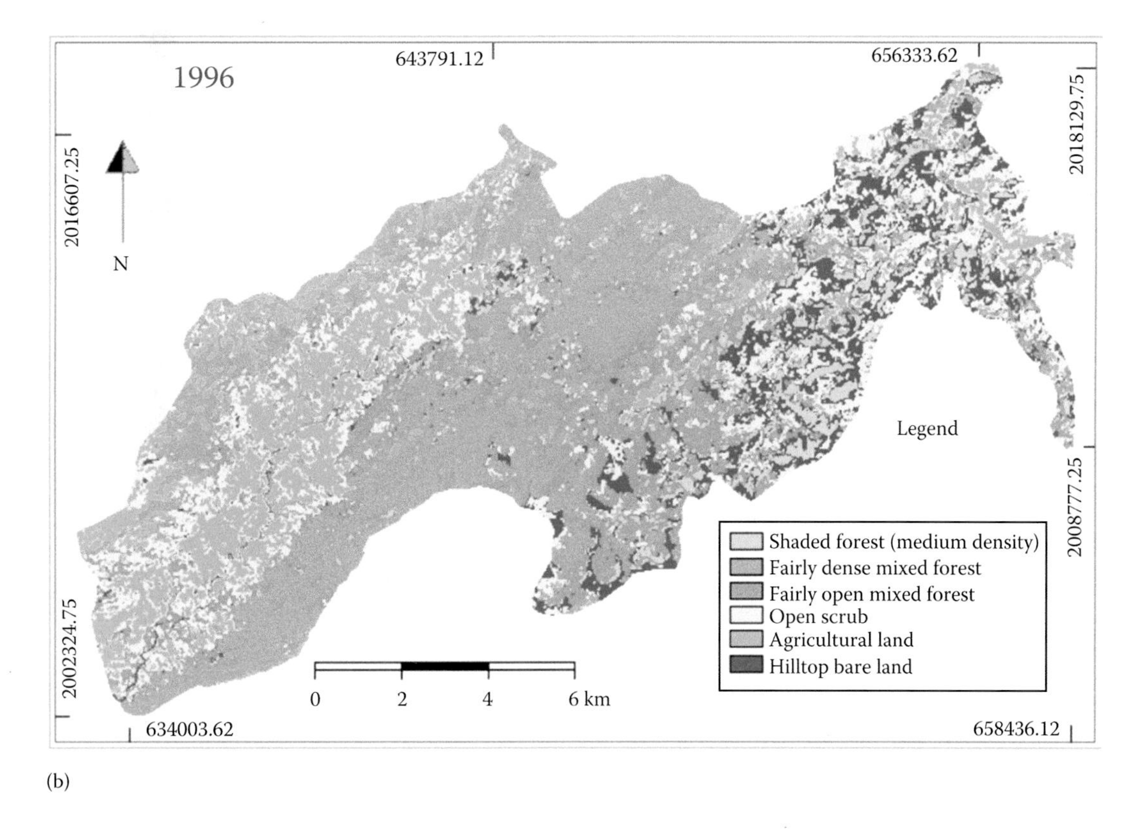

(b)

FIGURE 1.12 Land use change analysis within a watershed in Orissa (India) using Indian Remote Sensing Satellite 1 Linear Imaging Self-Scanning Sensor A satellite imagery: (a) 1989 classified land use map with six classes (From IRS-1A-LISS2 satellite image); (b) 1996 classified land use map, also with 6 classes (From IRS-1B-LISS2 satellite image).

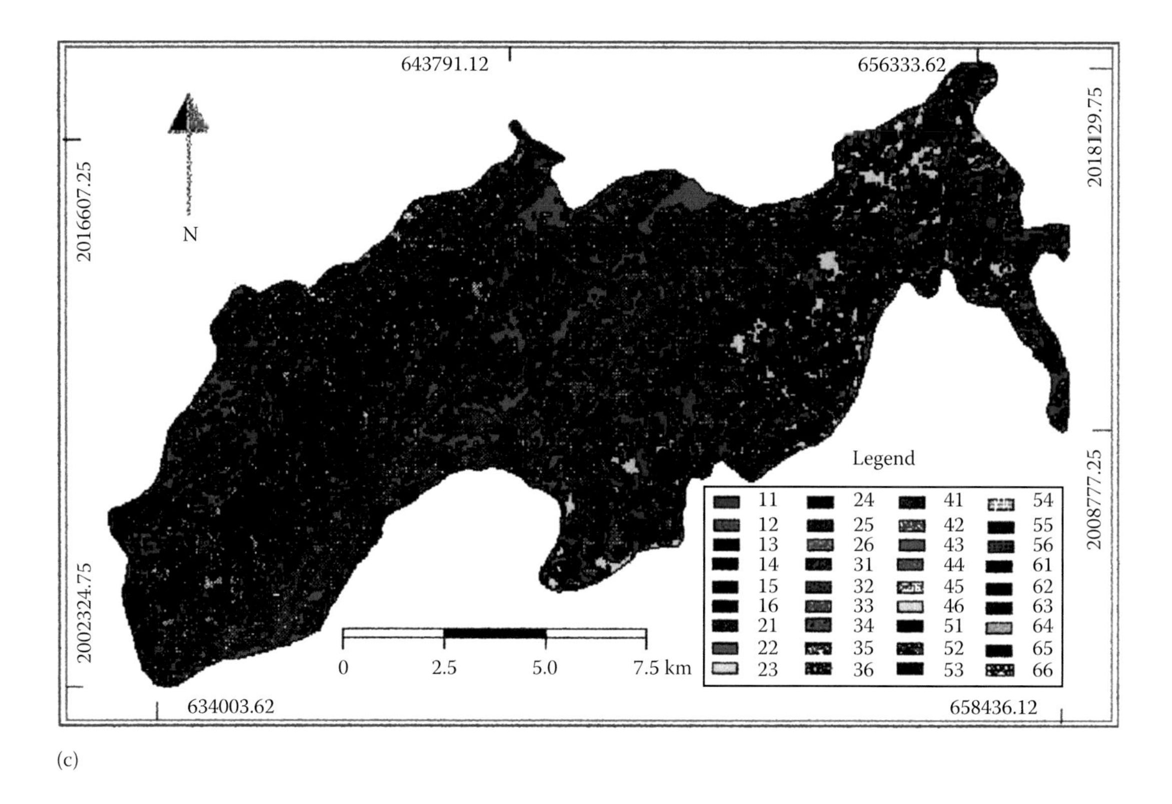

(c)

FIGURE 1.12 Land use change analysis within a watershed in Orissa (India) using Indian Remote Sensing Satellite 1 Linear Imaging Self-Scanning Sensor A satellite imagery: (c) land use change matrix analysis showing 36 classes (6 classes of 1989 × 6 classes of 1996) in the map—Class 11 means Class 1 (medium density shaded forest) in year 1989 remains same Class 1 in the year 1996 and so on (From GIS analysis of two date satellite images).

and also known as the temporal resolution of the sensor. Table 1.2 provides the temporal resolution of many selective remote sensing sensors. With ever-changing global environment due to events such as urban sprawl, deforestation, and natural disasters, including drought, flood, landslides, wildfire, avalanches, and earthquake, it is extremely essential for end users to conduct temporal analysis. There is an increase in studies pertaining to land use forecasting more so now than earlier due to increased awareness and implications of rapid land cover–land use changes occurring in the landscape. Temporal image analysis is highly essential for these studies. Better temporal resolution of images such as RapidEye, WorldView, MODIS, and even SPOT 5 can provide rapid land cover change information in scenarios such as tracking the spread of wildfire or damages caused by any natural disaster. Aerial imaging can be conducted daily over a study area and even more than once in a day. Therefore, aerial imaging does not have a specific temporal resolution. The advantages of temporal image resolution are shown in Figure 1.12a and b, in which IRS-1 LISS-A satellite imageries of two different years, 1989 and 1996, were used to classify the land use of a remote watershed in Orissa (India) and perform land use change analysis to determine a positive environmental effect (increase in forest cover and increase in changes from low-density forest to high-density forest cover) over 7 years using social awareness in forest preservation (forming village-level forest preservation committee and implementing with earnest) through the state department of forestry and village committees of the watershed (Panda et al., 2004).

1.2.2 Categorization of Satellites by Country of Ownership

It is a matter of national pride to have an Earth-observing program, and the acquired satellite data are very important for applications pertaining to resource inventory and management. Based on the study of Baker et al. (2001), the civilian remote sensing systems are government owned, the commercial systems are government licensed but privately owned, and the military/intelligence gathering platforms are government owned but with highly restricted access to the systems and products. According to Martin (2012), South Africa's scientific and engineering capacities were enhanced by the experience of designing, building, and launching SumbandilaSat; at the same time, the country obtained a great deal of social benefit due to the Earth-observing program. Belward (2012) opined that national space programs and their success not only enhance the nation's self-esteem but also reinforce national identity and sense of purpose for the country. This is why many countries in the world are in the forefront of the Earth observation missions. Table 1.3 presents a comprehensive list of countries that have Earth-observing programs and the associated operational satellite series in the program.

TABLE 1.3 Selected List of Earth-Observing Programs of Different Countries in the World

Country	Agency Name	Prominent Satellite Programs
France	Centre National d'Études Spatiales (CNES)	• SPOT • Pléiades • TOPEX/Poseidon
United States	National Aeronautics and Space Administration (NASA)	• Thermosphere, Ionosphere Mesosphere, Energetics and Dynamics (TIMED) • TOPEX/Poseidon • Upper Atmosphere Research Satellite • Landsat • NOAA–AVHRR • SRTM • Vanguard
Russia	Russian Federal Space Agency (RKA; Roscosmos)	• Elektro-L • Monitor-E • Resurs-DK1 • Resurs-P No. 1
Japan	Japan Aerospace Exploration Agency (JAXA) (former NASDA)	• MOS-1 (Momo-1) • MOS-1b (Momo-1b) • JERS-1 (Fuyo-1) • ADEOS (Midori) • ADEOS-II (Midori II) • GOSAT (Ibuki) • ALOS (Daichi) PALSAR, AVNIR-2, and PRISM
India	Indian Space Research Organisation (ISRO)	• Oceansat-2, September 23, 2009 • IMS-1, April 28, 2008 • Cartosat-2A, April 28, 2008 • Cartosat-2, January 10, 2007 • IRS-P5 (Cartosat-1), May 5, 2005 • IRS-P6 (RESOURCESAT-1), October 17, 2003 • IRS-P4, May 27, 1999 • IRS-P3, March 21, 1996 • IRS-P2, October 15, 1994 • IRS-1D, September 29, 1997 • IRS-1C, December 28, 1995 • IRS-1B, August 29, 1991 • IRS-1A, March 17, 1988
Canada	Canadian Space Agency	• RADARSAT
China	Chinese Space Agency	• Yogan-21
Europe	European Space Agency (ESA)	• ENVISAT • ERS (1 and 2) • CryoSat-2
Brazil	Brazilian Space Agency (AEB)	• CBERS-1 • CBERS-2 • CBERS-2B
Argentina	Argentina Space Agency (CONAE)	• SAC-A • SAC-B • SAC-C • SAC-D • SAOCOM
Indonesia	Lembaga Penerbangan dan Antariksa Nasional (LAPAN, Indonesia)	• Lapan (TUBsat)
Belarus	National Academy of Sciences of the Republic of Belarus	• BelKA
Pakistan	SUPARCO (Space and Upper Atmosphere Research Commission)	• Badr-1 • Badr-B • Paksat-1R • Pakistan Remote Sensing Satellite
Sweden	Swedish National Space Board	• Munin
Bolivia	Bolivarian Agency for Space Activities	• VRSS-1
South Korea		• KOMPSAT-2
Thailand		• Thaichote
Turkey		• Göktürk-2 (2012) (IMINT), Turkish Armed Forces, Intelligence [10] • RASAT (2011), mapping • BILSAT-1 (2003–2006), part of the Disaster Monitoring Constellation Project • Göktürk-1 (2013) (IMINT), Turkish Ministry of National Defense, Intelligence

1.2.3 Types of Sensing Technologies Based on Applications

The NASA NSSDC's master catalog (http://nssdc.gsfc.nasa.gov/nmc/spacecraftSearch.do) lists 7262 spacecraft that have been launched between October 4, 1957, and March 31, 2015 (NSSDC, 2014). They were listed in the following categories based on their application: astronomy (316), earth science (904), planetary sciences (313), solar physics (202), space physics (650), human crew (324), life science (97), microgravity (71), communication (2054), engineering (419), navigation and global positioning (448), resupply/refurbishment/repair (204), and surveillance and other military (2280), and technology applications (201). The number in parentheses depicts the number of spacecraft launched with application mission objective. These numbers are getting updated on the site frequently. Among these, 919 spacecraft were launched for earth science application. As discussed earlier, earth science as a broad discipline encompasses land use–land cover issues including surface (biophysical or chemical features) or underground (minerals) feature identification, but clearly, not all 890 missions can be used for global land cover mapping. All of these near-polar-orbiting Earth science application–oriented satellites are searchable in the NSSDC website. For example, upon clicking on the satellite Aryabhata in the search result, the following description shows:

> This spacecraft, named after the famous Indian astronomer, was India's first satellite and was completely designed and fabricated in India. It was launched by a Soviet rocket from a Soviet cosmodrome. The spacecraft was quasispherical in shape containing 26 sides and contained three experiments for the measurement of cosmic X-rays, solar neutrons, and Gamma rays, and an ionospheric electron trap along with a UV sensor. The spacecraft weighed 360 kg, used solar panels on 24 sides to provide 46 watts of power, used a passive thermal control system, contained batteries, and a spin-up gas jet system to provide a spin rate of not more than 90 rpm. There was a set of altitude sensors comprised of a triaxial magnetometer, a digital elevation solar sensor, and four azimuth solar sensors. The data system included a tape recorder at 256 b/s with playback at 10 times that rate. The PCM-FM-PM telemetry system operated at 137.44 MHz. The necessary ground telemetry and telecommand stations were established at Shar Centre in Sriharikota, Andhra Pradesh.

The link page to the satellite also includes an alternate name of the spacecraft, other brief facts, funding agencies names, and discipline names for which the spacecraft was launched. The reader is directed to the NASA website (http://nssdc.gsfc.nasa.gov/nmc) for additional metadata information about the Earth-observing satellites.

On another note, among several types, the three most important types of satellite orbits are (1) near-polar orbits, (2) sun synchronous orbits, and (3) geosynchronous orbits. In near-polar orbits, the satellites' inclination to the Earth is nearly 90°. Thus, the satellite can virtually see every part of the Earth as the Earth rotates beneath the satellite. It takes approximately 90 min for the near-polar-orbiting satellites to complete one orbit and is very useful for atmospheric (stratosphere) measurements like greenhouse gas concentration, ozone concentration, temperature, and water vapor (Montenbruck and Gill, 2000). Sun synchronous satellites like Landsat can pass over a section of the Earth at the same time of a day due to these orbits (Montenbruck and Gill, 2000). These satellites, in general, orbit the Earth at an altitude of approximately 700–800 km. As some of these satellites take pictures of the Earth, they work best with bright sunlight, that is, pass over a section of Earth during late morning to until 2-3 PM or early afternoon. When sun synchronous satellites measure longwave radiation, they would work best in complete darkness and hence pass over the Earth section at night. In geosynchronous orbits, also known as geostationary orbits, satellites circle the Earth at the same rate as the Earth spins (Montenbruck and Gill, 2000). These orbits make the satellites stay over a location of the Earth constantly throughout and observe almost the full hemisphere of the Earth.

According to Jensen (2009), remote sensing systems can directly measure the fundamental biological and/or physical characteristics of Earth features without using other ancillary data. The following are a few example of how remotely sensed data (MSS, hyperspectral, LiDAR, etc.) can help find water and nutrient stress in agricultural crop fields and support farmers to schedule irrigation and fertilizer applications for increased crop yields (Panda et al., 2011a,b): finding the growth stages of blueberry crops and conducting site-specific crop management to enhance the crop yield (Panda et al., 2009, 2010a,b, 2011a,b; Panda and Hoogenboom, 2013), finding and assessing the drainage characteristics of low-gradient coastal watershed for watershed management decision support (Amatya et al., 2013), and estimating the deciduous forest structure for forest management decision support (Defibaugh et al., 2013). Engman (1995) shows the advantage of microwave remote sensing to accurately measure soil moisture amounts. Panda et al. (2012a,b) and Amatya et al. (2011) used Landsat data for estimating soil moisture, stomatal conductance, leaf area index (LAI), and canopy temperature of pine forest. Hale et al. (2012), Phillips et al. (2012), Rylee et al. (2012), and Ertberger et al. (2012) used remote sensing and other watershed geospatial data for urban and rural watershed development decision support, while Cash and Panda (2012) used high-resolution National Agriculture Imagery Program (NAIP) imagery, SSURGO soil data, and LiDAR data to develop a hydrologic model using Soil and Water Assessment Tool (SWAT). Numerous studies have already been conducted with the use of remote sensing to detect, interpret, and analyze the Earth's biophysical characteristics for management decision support. More studies for environmental and Earth resource management using remote sensing are provided later in individual sections. Jensen (2009) has developed a database showing the potential remote sensing systems for detection, interpretation, and analysis of various biophysical variables (Table 1.4).

TABLE 1.4 Potential Remote Sensing Systems for Detection, Interpretation, and Analysis of Biophysical Variables

Biophysical Variables	Potential Remote Sensing Systems
• Geodetic control • Location from orthocorrected imagery	• GPS • Analog and digital stereoscopic aerial photography, space imaging • IKONOS, DigitalGlobe, QuickBird, OrbView-3, SPOT, Landsat (Thematic Mapper, Enhanced TM+), Indian IRS-ICD, European ERS-1 and ERS-2 microwave and ENVISAT MERIS, MODIS, LiDAR, Canadian RADARSAT-1 and -2, etc.
Topography/bathymetry	
• DEM • Digital Bathymetric Model	• GPS, stereoscopic aerial photography, and other location identification satellite systems • SONAR, bathymetric LiDAR, stereoscopic aerial photography
Vegetation	
• Chlorophyll a and b • Canopy structure • Biomass • LAI • Absorbed photosynthetically active radiation • Evapotranspiration	• Color aerial photography, Landsat ETM+, IKONOS, QuickBird, and most other MSS sensor–based satellites • Stereoscopic aerial photography, LiDAR, RADARSAT, IFSAR • CIR aerial photography, most other MSS (Landsat, QuickBird, etc.) and hyperspectral systems (AVIRIS, HyMap, CASI), AVHRR, multiple imaging spectrometer (MISR)
Surface temperature	
• Land, water, and atmosphere	• ASTER, AVHRR, GOES, Hyperian, MISR, MODIS, SeaWiFS, Airborne TIR
Soil and rocks	
• Moisture • Mineral composition • Taxonomy • Hydrothermal alteration	• ASTER, passive microwave (SSM/1), RADARSAT, MISR, ALMAZ, Landsat, ERS-1 and ERS-2, Intermap STAR-3i • ASTER, MODIS, hyperspectral systems (AVIRIS, HyMap) • High-resolution color and CIR aerial photography, airborne hyperspectral systems • Landsat, ASTER, MODIS, airborne hyperspectral systems
Surface roughness	• Aerial photography, ALMAZ, ERS-1 and ERS-2, RADARSAT, Intermap STAR-3i, IKONOS, QuickBird, ASTER, etc.
Atmosphere	
• Aerosols (optical depth) • Clouds • Precipitation • Water vapor (precipitable) • Ozone	• MISR, GOES, AVHRR, MODIS, CERES, MOPITT, MERIS • GOES, AVHRR, MODIS, CERES, MOPITT, MERIS, UARS • TRMM, GOES, AVHRR, SSM/1, MERIS • GOES, MODIS, MERIS • MODIS
Water	
• Color • Surface hydrology, suspended minerals • Chlorophyll • Dissolved organic matter	• CIR aerial photography, most other MSS (Landsat, QuickBird, etc.) and hyperspectral systems (AVIRIS, HyMap, CASI), bathymetric LiDAR, MISR, Hyperion, TOPEX/Poseidon, MERIS, AVHRR, CERES, etc.
Snow and sea ice	
• Extent and characteristics	• CIR aerial photography, most other MSS (Landsat, QuickBird, etc.) and hyperspectral systems (AVIRIS, HyMap, CASI)
• Bidirectional reflectance distribution function	• MISR, MODIS, CERES

Source: Adapted and modified from Jensen, J.R., *Remote Sensing of the Environment: An Earth Resource Perspective 2/e*, Pearson Education India, Delhi, India, 2009, pp. 11–12.

1.3 Remote Sensing Platforms

Depending on the airplane or the Earth-observing satellite platform, there are hundreds of different sensors available for specific usage. These sensors can be sorted by the sensor type and are categorized into the following: (1) acoustic, sound, and vibration; (2) automotive and transportation; (3) chemical; (4) electric current, electric potential, magnetic, and radio; (5) environment, weather, moisture, and humidity; (6) fluid flow; (7) ionizing radiation and subatomic particles; (8) navigation instruments; (9) position, angle, displacement, distance, speed, and acceleration; (10) optical, light, imaging, and photon; (11) pressure; (12) force, density, and level; (13) thermal, heat, and temperature; (14) proximity and presence; (15) sensor technology; and (16) other sensors. However, this chapter is focused on Earth observation remote sensing, covering both active and passive remote sensing technologies as discussed in Section 1.1.2. These Earth-observing satellites and aerial imaging systems are grouped into several types based on their operational principles within the EMS and means of obtaining images. They are optical, radar, microwave, hyperspectral, sonar, etc.

Typical remote sensing systems operate in the visible (0.4–0.7 μm) and IR (0.7–1000 μm) portion of the EMS, known as optical remote sensing (Aronoff, 2004). Other sensors such as microwave sensors operate in the microwave region of the EMS (0.3 mm–1 m). RADAR uses radio waves (10^4–10^{11} μm) to determine the range, altitude, direction, and speed of an object or phenomena on Earth. On the other hand, hyperspectral imaging sensors, although operate in the visible/NIR ranges of the EMS, are marked by very high spectral resolutions resulting in many numbers of spectral bands (refer to Figure 1.6).

Another active sensor LiDAR is a remote sensing technology in which the signal (return) distances are measured based on the lag time of the pulsed signal. Based on the different returns recorded, it is possible to accurately measure Earth objects such as canopy height and ground surface elevation. SONAR is another active remote sensing technology platform. It uses sound propagation usually in water for depth mapping, object detection, and navigation underwater. These different platform-based remote sensing technologies have their individual applications and advantages in real-world problem solving and decision support.

1.3.1 Aerial Imaging

Aerial photography is a remote sensing system in which photographs of the Earth's surface are taken from an elevated position, mostly by airplanes flying within a 10 mile height from the ground. Platforms for aerial photography include fixed-wing aircraft, helicopters, multirotor unmanned aircraft systems (UASs), balloons, blimps and dirigibles, rockets, kites, parachutes, stand-alone telescoping, and vehicle-mounted poles and in recent times military operated drones. Recently, radio-controlled model aircrafts–drones, for example, are very popular in taking images of the Earth's surface from a much lower height. These aerial images are of high resolution and have a wide application in environmental and natural resource management fields.

The U.S. Geological Survey (USGS) initiated novel programs called the National High Altitude Program (NHAP) and the National Aerial Photography Program (NAPP) in the 1980s to acquire aerial photograph for the conterminous 48 states to support operational and mapping needs of the local, state, and federal agencies (USGS, 1992a,b). The black-and-white and color-infrared (CIR) products from these programs provide an invaluable resource for historical assessment of land cover/use at map scales ranging from 1:58,000 to 1:40,000.

Building on the efforts of the USGS, the USDA Farm Service Agency (FSA) launched and administered the NAIP aimed at acquiring high-resolution color and near-infrared (NIR) imagery during the agricultural growing seasons in the conterminous United States. These annual image data products are orthorectified and GIS-ready, and are primarily used by USDA to maintain the Common Land Unit database. NAIP imagery is one of the major resources for the geospatial community working on vegetation mapping and precision agriculture due to its high resolution (1 and 2 m). More information on the NAIP imagery can be obtained from http://www.fsa.usda.gov/FSA website. Use of aerial images in urban management (Hodgson et al., 2003; Cleve et al., 2008), site-specific crop management (Jhang et al., 2002; Panda et al., 2009, 2010a,b, 2011a,b), forest and ecological resource management (Suarez et al., 2005; Morgan et al., 2010; Rao, 2013), water resource management (Ritchie et al., 2003; Jha et al., 2007), and many other natural resource management is abundant. As discussed previously, photogrammetry or aerial stereo photography was the earliest form of remote sensing. Even today, image and terrain analysts use stereographic pairs of aerial photographs to make topographic maps by image and terrain analysts for digital elevation–based land management decision support such as road construction, watershed delineation, stream flow direction and flow accumulation mapping, and traffic ability applications. According to Dornaika and Hammoudi (2009) and Bulstrode et al. (1986), stereoscopy is a photogrammetric technique for creating or enhancing the illusion of depth in an image by means of stereopsis for binocularl vision called Stereogram. Adjacent but overlapping aerial photos are called stereopairs and are needed to determine parallax and stereo/3D viewing. Advances in the hardware and software sector simulate such procedures in the digitial and visualization domains, called soft-copy photogrammetry.

1.3.2 Optical Remote Sensing

As discussed earlier, optical remote sensing uses visible, near infrared and short-wave infrared sensors to acquire images of the earth's surface (Figure 1.13). Optical remote sensing systems are classified into several types, depending on the number of spectral bands used in the imaging process, such as (1) panchromatic (PAN) imaging system, (2) multispectral (MSS) imaging system, (3) superspectral imaging system, and (4) hyperspectral imaging system. In panchromatic imaging system, the sensor is a single-channel detector sensitive to radiation within a broad wavelength. For example, IKONOS PAN image covering a broad bandwidth of 0.45–0.9 μm, which is the range of multispectral B, G, R, and IR bands. Similarly, the SPOT PAN image has a band range of 0.51–0.73 μm, which covers most of the multispectral region of B and G bands. PAN image spatial resolution is mostly better than the multispectral images of the same sensor, for example, IKONOS PAN and SPOT PAN spatial resolutions are 1 and 10 m, respectively, whereas the spatial resolutions of their MSS bands are 4 and 20 m. In multispectral imaging system, the sensor uses a multichannel detector with a few spectral bands like B, G, R, and IR within a narrow wavelength band (refer to Table 1.2). Table 1.2 provides a detailed list of MSS and PAN bands of some select optical sensors. In contrast to MSS imaging system, in a superspectral imaging system, there are many more spectral channels (typically >10). The bands have narrower bandwidths than MSS imaging system, enabling the finer spectral characteristics of the targets to be captured by the sensor. MODIS and medium-resolution imaging spectrometer (MERIS) are a few examples of the superspectral systems. MODIS superspectral imaging system has 36 bands ranging from 0.405 to 14.385 μm with each band having narrower bandwidths; MODIS band 11 and band 12 have bandwidths of 0.526–0.536 and 0.546–0.556 μm, respectively, whereas the bandwidth of Landsat MSS imaging system has little wider bandwidths. For instance, bands 4 and 5 of Landsat MSS have bandwidths of 0.76–0.90 and 1.55–1.75 μm, respectively. However, hyperspectral imaging systems acquire images in about 100 or more contiguous spectral bands. Hyperspectral imaging system is discussed in a later section.

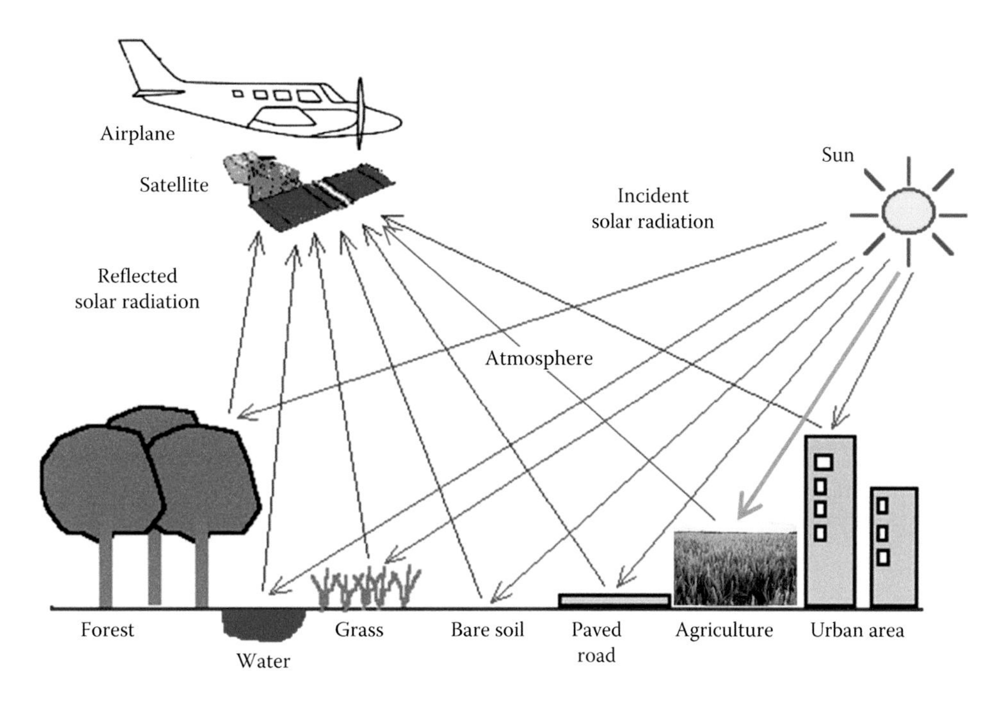

FIGURE 1.13 Basic image acquisition process of an optical imaging system.

According to Avery and Berlin (1992), unlike photographic cameras that record radiation reflected from ground scene directly onto films, electro-optical sensors use nonfilm detectors, which convert the reflected and/or emitted radiation from the object or phenomena on ground/earth to proportional electrical signals that are used to construct 2D images for conventional viewing. Avery and Berlin (1992) categorized electro-optical sensors into video camera, vidicon camera, across-track scanner, and along-track scanner. According to Murai (1993), optical sensors are characterized by spectral (spectral band, bandwidth, central wavelength, response sensitivity at the edge of band, spectral sensitivity at outer wavelength, and sensitivity of polarization), radiometric (radiometry of the sensor, sensitivity in noise equivalent power, dynamic range, S/N, and other quantification noise), and geometric (FOV, IFOV, band-to-band registration, MTF, geometric distortions, and alignment of optical elements) performance. Table 1.5 lists optical sensor–based remote sensing satellites with their spectral band information. The table also contains satellites based on other different sensing platforms as discussed in this section. The description column of the table explains about the sensor platform of the satellite. In addition to the mechanics of image acquisition, Figures 1.14a and b illustrate a comprehensive working principle of optical remote sensing, which includes across-track/whiskbroom, pushbroom, and multispectral scanning processes. As the names suggest, in a wishbroom system, the sensor detectors scan and record the reflected energy using oscillation mirrors that move across the satellite track, while the pushbroom system employ an array of detectors that are perpendicular to the satellite track. Landsat sensor systems until Landsat 8 OLI use the wishbroom system. Sensors that use the push broom design include SPOT, IRS, QuickBird, OrbView, and IKONOS.

1.3.3 Hyperspectral Remote Sensing

Hyperspectral imaging system is an optical imaging system but is known as an *imaging spectrometer* as the system acquires an image from a satellite or airplane in about 100 or more contiguous spectral bands. Imaging spectrometry is discussed in a later section. The targets on Earth are better identified and characterized by a hyperspectral imaging system than other optical imaging systems because of the system's precise spectral information. Hyperspectral imaging or field spectrometry has potential applications in precision agriculture or site-specific crop management (e.g., monitoring the health and growth of crops and measuring water stress, nutrient stress, or stress from crop diseases in crop fields), intense forest management (e.g., forest species differentiation, timber readiness for harvesting determination and analysis, wildfire susceptibility determination and assessment), and coastal management (e.g., monitoring of water quality parameters such as phytoplankton, organic matter, sediment pollution, and bathymetry changes).

Hyperspectral imaging produces an image where each pixel has full spectral information with imaging narrow spectral bands over a contiguous spectral range of the EMS. Hyperspectral imaging is a very-fast-growing area in remote sensing that expands and improves the efficiency/capability of multispectral satellite imaging (Lees and Ritman, 1991; Goetz, 2009). Hyperspectral imagers are used in various applications since the late 1980s including mineralogy, biology (food safety), defense, and environmental measurements (Clark et al., 1990; Kim et al., 2001; Chang, 2003; Lu, 2003; Van Wagtendonk et al., 2004; Asner et al., 2007; Gowen et al., 2007). Hyperion, AVIRIS, HyMap, and compact airborne spectrographic imager (CASI) are examples of a high-spatial-resolution hyperspectral system.

TABLE 1.5 List of Selected Optical Remote Sensing Satellites along with Their Spectral Band Information

Name of Sensor	Description	Mission	No. of Bands
ACE-FTS	Atmospheric Chemistry Experiment Fourier Transform Spectrometer	SCISAT	
ACS	Atmospheric Correction Sensor	RESOURCESAT-3	
ADCS/SARSAT	Advanced Data Collection System/Search and Rescue Satellite Aided Tracking	JPSS-1 (NPOESS)	4
AEISS	Advanced Earth Imaging Sensor System	KOMPSAT-3 (Arirang 3)	9
ALI	Advanced Land Imager	EO-1	3
AlSat	Standard DMC sensor	AlSat-1	
APS	Aerosol Polarimetry Sensor	GLORY	
ARGOS		Metop-A, Metop-B, SARAL/AltiKa	4
AWFI	Wide Field Imaging Camera	CBERS-3, CBERS-4	4
AWiFS	Advanced Wide Field Sensor	RESOURCESAT-1, RESOURCESAT-2	3
BEIJING-1-MS	Standard DMC sensor	BEIJING-1	4
BILSAT-MS		BILSAT	1
BILSAT-PAN		BILSAT	
Camera	Multispectral camera with black-and-white and color imaging	Svea	4
CCD	Multispectral camera	CBERS, CBERS-2, CBERS-3, CBERS-4	
CCD/TDI	Selectable-multispectral possibility	EROS C	1
CCD/TDI	Selectable	EROS B	
CCSP	Cloud Camera Sensor Package	GLORY	3
CERES	Clouds and Earth's Radiant Energy System	Aqua, JPSS-1 (NPOESS), Suomi NPP (SNPP), Terra, TRMM	1
Chinese Mapping Telescope		BEIJING-1	1
CIRC	Compact infrared camera	ALOS-2	9
COBAN		BILSAT	6
CZCS	Coastal Zone Color Scanner	Nimbus-7	3
DMC2	Standard DMC sensor	UK-DMC 2	2
DORIS	Doppler Orbitography and Radiopositioning Integrated by Satellite	CryoSat-2, ENVISAT, Jason-1, Jason-2 (OSTM), SPOT 2, SPOT 3, SPOT 4, SPOT 5, TOPEX/Poseidon	1
EOC	Electro-optical camera	KOMPSAT-1	
E-OP1	Panchromatic and 5 band color imager	NigeriaSat-2	
E-OP2	Wide area coverage imager	NigeriaSat-2	13
ERB	Earth Radiation Budget	Nimbus-7	4
ERBE	Earth Radiation Budget Experiment	NOAA-10	1
EROS A		EROS A	8
ETM	Enhanced Thematic Mapper	Landsat 7	
FRS	Strip map mode	RISAT-1	2
GERB	Geostationary Earth Radiation Budget	MSG-1, MSG-2, MSG-3	4
GIS-MS	Visible and NIR	GeoEye-1 (OrbView-5)	1
GIS-PAN	Panchromatic	GeoEye-1 (OrbView-5)	36
GLI	Global Imager	ADEOS-II	5
GOES Imager	Multichannel instrument to sense radiant and solar-reflected energy	GOES-14	6
GOME	Global Ozone Monitoring Experiment	ERS-2	3
GOMOS	Global Ozone Monitoring by Occultation of Stars	ENVISAT	
GUVI	Global Ultraviolet Imager	TIMED	1
HCS	High-sensitivity camera	SAC-C	7
HEPD	High-energy particle detector	KOMPSAT-1	
HiRI	High-resolution optical imager	Pleiades HR 1A, Pleiades HR 1B	
HPO	High-performance optical sensor	ASNARO-1	1
HRC	High-resolution camera	PROBA-1	4

(*Continued*)

TABLE 1.5 (*Continued*) List of Selected Optical Remote Sensing Satellites along with Their Spectral Band Information

Name of Sensor	Description	Mission	No. of Bands
HRG	High-resolution geometric	SPOT 5	1
HRS	High-resolution stereoscopic	SPOT 5	1
HRTC	High-resolution technological camera	SAC-C	4
HRV	High-resolution visible	SPOT 1, SPOT 2, SPOT 3	4
HRVIR	High-resolution visible and infrared	SPOT 4	2
HSRS	Hot Spot Recognition Sensor System	BIRD	64
HySI	Hyper Spectral Camera with 64 fixed bands	IMS 1	
IDEE	Instrument for Detection of High-Energy Electrons	TARANIS	3
IKAR-DELTA		PRIRODA-MIR	5
IKAR-N		PRIRODA-MIR	2
IKAR-P		PRIRODA-MIR	
IMA	Multispectral Imager operating in 5 bands	Argo (RapidEye 6)	5
Imager		GOES-10, GOES-11, GOES-13 = GOES-East, GOES-8, GOES-9	5
Imager-M		GOES-12 = GOES-South America	1
IMG	Interferometer Monitor for Greenhouse Gases	ADEOS	
IMM	Instrument for magnetic measurements	TARANIS	1
IMS	Ionosphere Measurement Sensor	KOMPSAT-1	4
IPEI	Ionosphere Plasma and Electrodynamics Instrument	FORMOSAT-1	4
IRMSS	Medium-resolution scanner	CBERS-3, CBERS-4	4
IR-MSS	Infrared multispectral scanner	CBERS, CBERS-2	
ISUAL	Imager of Sprites and Upper Atmospheric Lightning	FORMOSAT-2	5
JAMI	Japanese Advanced Meteorological Imager	MTSAR-1R	2
KBR	K-Band Ranging System	GRACE	4
KOMPSAT-MSC	B&W panchromatic, MSS and merged 1 m resolution images	KOMPSAT-2	1
KVR-1000		Cosmos	256
LAC	LEISA Atmospheric Corrector, corrects high-spatial-resolution multispectral data for atmospheric effects on surface reflectance	EO-1	
Laser reflector		CryoSat-2	6
LIMS	Limb infrared monitor of the atmosphere	Nimbus-7	1
LIS	Lightning Imaging Sensor	TRMM	4
LISS-1	Linear Imagine Self-Scanning System	IRS-1A, IRS-1B	4
LISS-2	Linear Imagine Self-Scanning System	IRS-1A, IRS-1B, IRS-P2	4
LISS-3	Linear Imagine Self-Scanning System	IRS-1C, IRS-1D	4
LISS-3*	Linear Imagine Self-Scanning System	RESOURCESAT-1	3
LISS-4	Linear Imagine Self-Scanning System	RESOURCESAT-1	4
LISS-3	Linear Imagine Self-Scanning System	RESOURCESAT-2	3
LISS-4	Linear Imagine Self-Scanning System	RESOURCESAT-2	4
LISS-III-WS	Wide-swath sensor	RESOURCESAT-3	4
MAC	MS	RazakSAT (MACSAT), RazakSAT (MACSAT)	
MAESTRO	Measurements of Aerosol Extinction in the Stratosphere and Troposphere Retrieved by Occultation	SCISAT	
MCP	Micro Camera and Photometer	TARANIS	15
MERIS	Medium-resolution imaging spectrometer	ENVISAT	
MEXIC	Multiexperimental Interface Controller	TARANIS	5
MMRS	Multispectral Medium-Resolution Scanner	SAC-C	
MODI	Moderate-Radiation Visible and NIR Imager	FengYun-3A	4
MOMS-2P	Modular Optoelectronic Multispectral Scanner	PRIRODA-MIR	64

(*Continued*)

TABLE 1.5 (*Continued*) List of Selected Optical Remote Sensing Satellites along with Their Spectral Band Information

Name of Sensor	Description	Mission	No. of Bands
MOPITT	Measurement of Pollution in the Troposphere	Terra	4
MS camera	Derived from SPOT camera, has thermoelastic stability	Deimos 1, Kanopus-Vulkan, THEOS	12
MSC	Multispectral	VENUS	13
MSI	Multispectral instrument	EarthCARE, Sentinel-2	4
MSS	Multispectral scanner	AlSat-2A, Landsat 4, Landsat 5	5
MSS (LS 1–3)	Multispectral scanner Landsat 1, 2, 3	Landsat 1, Landsat 2, Landsat 3	3
MSU-E1		Resurs-O1	3
MSU-E2		PRIRODA-MIR	5
MSU-GS	Multispectral Scanner Geostationary	Electro-L/GOMS 2	5
MSU-SK		PRIRODA-MIR	5
MSU-SK (Resurs)		Resurs-O1-3	5
MSU-SK1		Resurs-O1	16
MTI	Multispectral Thermal Imager	MTI	3
MUX	Multispectral CCD Camera	CBERS-3, CBERS-4	4
Mx	Four band multispectral CCD camera	IMS 1	5
NAOMI	New AstroSat Optical Modular Instrument	SPOT 6, SPOT 7, VNREDSat-1A	3
NigeriaSat-1	Standard DMC sensor	NigeriaSat-1	1
NSCAT	NASA Scatterometer	ADEOS	7
OCI	Ocean Color Imager	FORMOSAT-1	8
OCM	Ocean Color Monitor	IRS-P4 (Oceansat), Oceansat-2	12
OCTS	Ocean Color and Temperature Scanner	ADEOS	3
OIS	Optical Imaging System	RASAT	
OLCI	Ocean and Land Color Instrument	Sentinel-3	9
OLI	Operational Land Imager	Landsat 8 (LDCM)	3
OLS	Operational Linescan System	DMSP-16	3
OMI	Ozone Monitoring Instrument	AURA	
OMPS	Ozone Mapping and Profiler Suite	JPSS-1 (NPOESS), Suomi NPP (SNPP)	8
OPS	Optical Sensor	JERS-1	6
Optical Imaging (GOKTURK-2)		GOKTURK-2	5
OrbView-3		OrbView-3	5
OSA	Optical Sensor Assembly	IKONOS	
OSIRIS	Optical Spectrograph and Infrared Imaging System	ODIN	6
OSMI	Ocean Scanning Multispectral Imager	KOMPSAT-1	1
OTD	Optical Transient Detector	OrbView-1	4
OZON-M		PRIRODA-MIR	4
PAMELA		Resurs-DK1	1
PAN		Cartosat-2B, CBERS-3, CBERS-4, IRS-1C, IRS-1D	1
PAN Camera	Panchromatic camera	Cartosat-2 (IRS-P7), Kanopus-Vulkan	1
PAN Telescope		THEOS	1
PAN-A	Panchromatic aft pointing	Cartosat-1	1
PANC	Panchromatic camera	Cartosat-2A	1
PAN-F	Panchromatic forward pointing	Cartosat-1	9
POAM-II	Polar Ozone and Aerosol Measurement	SPOT 3	9
POAM-III	Polar Ozone and Aerosol Measurement	SPOT 4	15
POLDER	Polarization and Directionality of the Earth's Reflectance	ADEOS, ADEOS-II, Parasol	2
Poseidon 2 altimeter		Jason-1	
POSEIDON-3	Altimeter	Jason-2 (OSTM)	1
PRISM	Three panchromatic sensors for stereomapping	ALOS	4

(*Continued*)

TABLE 1.5 (*Continued*) List of Selected Optical Remote Sensing Satellites along with Their Spectral Band Information

Name of Sensor	Description	Mission	No. of Bands
QuickBird	High-resolution PAN, 61 cm (nadir) to 72 cm (25° off nadir); MS, 2.44–2.88 m	QuickBird	1
R-400		PRIRODA-MIR	1
Radar altimeter		GFO	
Radar sensor		SMAP	
Radiometer	L-band	SMAP	4
RBV	Return Beam Vidicon Camera	Landsat 1	5
REIS	Records in 5 spectral bands on VIR and NIR	RapidEye	1
RIS	Reflector in space	ADEOS	5
ROCSAT-2		FORMOSAT-2	4
RSI	Remote sensing instrument	FORMOSAT-5	1
SAM II	Stratospheric Aerosol Measurement	Nimbus-7	1
SASS	Seasat-A Satellite Scatterometer	Seasat	12
SBUV/2	Solar Backscatter Ultraviolet Instrument	NOAA-11, NOAA-12, NOAA-14, NOAA-15, NOAA-16, NOAA-17, NOAA-18 (NOAA-N), NOAA-19 (NOAA-N Prime)	6
SBUV/TOMS	Solar Backscatter Ultraviolet/Total Ozone Mapping—failed in 1993	Nimbus-7	
SCARAB	Scanning Radiative Budget Instrument	Megha-Tropiques	
Scatterometer		Oceansat-2	8
SeaWiFS	Sea-Viewing Wide Field-of-View Sensor	OrbView-2	1
Seawinds		ADEOS-II, QuikSCAT	
SEE	Solar Extreme Ultraviolet Experiment	TIMED	
SEM	Space Environment Monitor	GOES-14	7
SEM	To measure solar radiation in the x-ray and extreme ultraviolet (EUV) region	GOES-13 = GOES-East	12
SEVIRI	Spinning Enhanced Visible and Infrared Imager	MSG-1, MSG-2, MSG-3	35
SGLI	Second-Generation Global Imager	GCOM-C1	1
SIM	Spectral Irradiance Monitor	SORCE	1
SIR-C		SRTM	3
SLIM6	Surrey Linear Imager Multispectral 6 channels—but 3 spectral bands	NigeriaSat-X	1
SOLSTICE A and B	Solar–Stellar Irradiance Comparison Experiment	SORCE	19
Sounder		GOES-10, GOES-11, GOES-12 = GOES-South America, GOES-13 = GOES-East, GOES-8, GOES-9	
SSULI	Ultraviolet Limb Imager	DMSP-16	
SXI	Solar X-ray Imager	GOES-14	
SXI Solar X-Ray Imager	To monitor the sun's x-rays	GOES-13 = GOES-East	4
TANSO-CAI	Thermal and NIR Sensor for Carbon Observation, Cloud and Aerosol Imager	GOSAT (Ibuki)	1
TDI	Panchromatic	AlSat-2A	
TIDI	TIMED Doppler Interferometer	TIMED	1
TIM	Total Irradiance Monitor	SORCE	
TIP	Tiny Ionosphere Photometer	FORMOSAT-3 (COSMIC)	2
TIRS	Thermal Infrared Sensor	Landsat 8 (LDCM)	1
TK-350		Cosmos	7
TM	Thematic Mapper	Landsat 4, Landsat 5	1
TopSat		TopSat	
TOR Package	Tracking, Occupation, and Ranging	TanDEM-X	2
Travers SAR		PRIRODA-MIR	1
TV Camera		PRIRODA-MIR	3
UK-DMC	Standard DMC sensor	UK-DMC	4

(*Continued*)

TABLE 1.5 (*Continued*) List of Selected Optical Remote Sensing Satellites along with Their Spectral Band Information

Name of Sensor	Description	Mission	No. of Bands
VEGETATION		SPOT 4, SPOT 5	4
VGT-P	Vegetation instrument on PROBA-V	PROBA-V	5
VIRS	Visible and Infrared scanner	TRMM	2
WAOSS-B	Wide-Angle Optoelectronic Stereo Scanner	BIRD	2
Water Vapor Radiometer		GFO	2
WFI	Wide Field Imager	CBERS, CBERS-2	3
WiFS	Wide Field Sensor	IRS-1C, IRS-1D, IRS-P3	5
WindSat		Coriolis	1
WorldView-1	Provides highly detailed imagery for precise map creation and in-depth image analysis	WorldView-1	8
WV 3 MSS	WorldView-3 multispectral sensor	WorldView-3	1
WV 3 PAN	WorldView-3 panchromatic sensor	WorldView-3	8
WV 3 SW	WorldView-3 shortwave infrared sensor	WorldView-3	8
WV110	Standard 4 colors + new 4 colors	WorldView-2	1
WV60	PAN band for WorldView-2	WorldView-2	
XGRE	X-ray and Gamma Relativistic Electron Detector	TARANIS	1
XPS	XUV photometer	SORCE	

Source: Adapted and modified from ITC, ITC's database of satellites and sensors, 2014, http://www.itc.nl/research/products/sensordb/allsensors.aspx, accessed on May 2, 2014.

Hyperion high-resolution (30 m) hyperspectral imaging system acquires images in 220 contiguous spectral bands with very high radiometric accuracy within a bandwidth range of 0.4–2.5 μm. With these large numbers (220) of bands, complex land characteristics can be identified. Kruse et al. (2003) used Hyperion imagery to map the minerals in and around Cuprite, NV, with high accuracy.

Figure 1.15 is an example of AVIRIS hyperspectral imaging system data taken over Yellowstone National Park acquired in 224 continuous bands having a 10 nm band pass over the spectral wavelength range of 350–2500 nm (from visible light to NIR). AVIRIS collects 20 m wide pixels at approximately 14 m spacing. The sensor swath width is approximately 10.5 km. Kokaly et al. (2003) used AVIRIS data from Yellowstone National Park to map the vegetation types in the park with very high accuracy. Kokaly et al. (1998) used AVIRIS imaging spectrometry data to characterize and map the biology and mineralogy of Yellowstone National Park. Above all, commercial companies also acquire hyperspectral imageries on demand. Table 1.6 contains the selective list of hyperspectral imaging system satellites with pertinent specifications on number of bands, spatial resolution, and applications.

Over the past years, the demands for better mapping and characterization of Earth objects and related phenomena have increased, resulting in a higher need to better understand the biophysical interactions of radiation, atmospheric effects, and albedo properties of surface materials. Ground-based remote sensing using spectroradiometers provide an innovative approach to measuring surface reflectance of materials while removing the interfering effects of atmospheric path radiance, absorption, and scattering effects (Clark et al., 2002). Thus, spectroscopy serves as a tool to map specific material and mineral for environmental assessments including vegetation health studies under field or laboratory conditions (Yang et al., 2005; Swayze et al., 2014). Hyperspectral imaging and imaging spectroscopy are the technologies used to acquire a spectrally resolved image of an object or scene (Butler and Laqua, 1996). Also, spectrometry is used as lab/computer imaging technology for food quality—meat deterioration (Panigrahi et al., 2003; Savenije et al., 2006), fruit sweetness analysis (Guthrie and Walsh, 1997; Nicolai et al., 2007), leaf chlorophyll measurement (Shibata, 1957), etc.—and thus images acquired through space imaging spectrometry platforms would analyze the Earth objects and phenomena in the most qualitative means.

1.3.4 RADAR and SODAR Remote Sensing

The earliest development of radar technology was in 1886 by a German physicist named Heinrich Hertz. A Russian physicist, Alexander Popov in 1985 showed an application of radar technology in detecting far lightning strikes. While the early applications of RADAR remote sensing were focused on defense and military applications primarily for reconnaissance surveys of tropical environments, recent applications in the past few decades have seen more diverse applications. Conventional radar such as Doppler radar is mostly associated with weather forecasting, aerial traffic control, and subsequent early warning. Doppler radar is used by local law enforcements' monitoring of speed limits and in enhanced meteorological data collection such as wind speed, direction, precipitation location, and intensity. Interferometric SAR is used to produce precise digital elevation models (DEMs) of large-scale terrain using RADARSAT, TerraSAR-X, Magellan, etc. Refer to Table

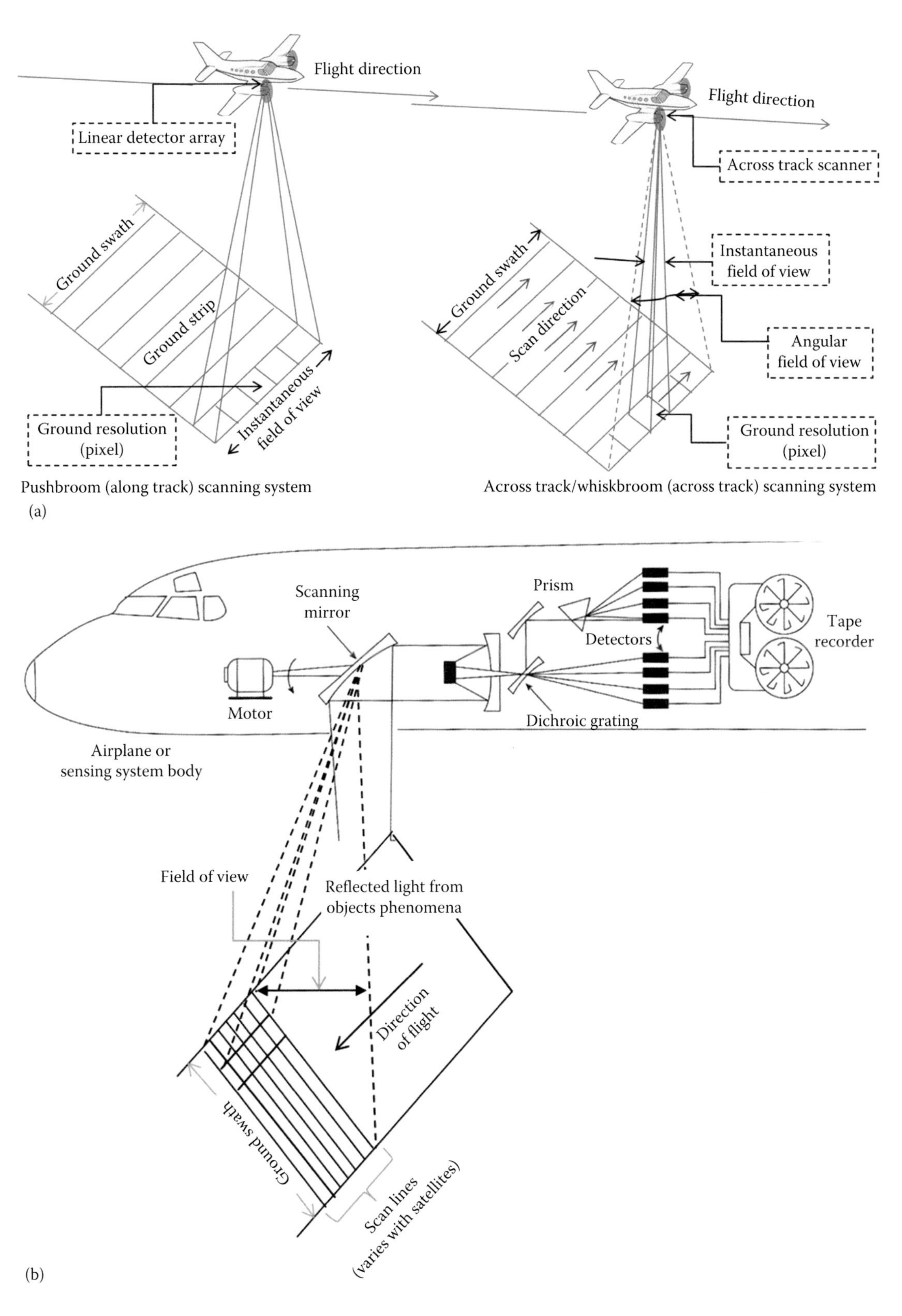

FIGURE 1.14 Basic operating system of an optical sensor: (a) across-track or whiskbroom scanner, and (b) multispectral scanner, onboard an airplane or satellite acquiring images. (Adapted and modified from Avery, T.E. and Berlin, G.L., *Fundamentals of Remote Sensing and Airphoto Interpretation*, Prentice Hall, Upper Saddle River, NJ, 1992.)

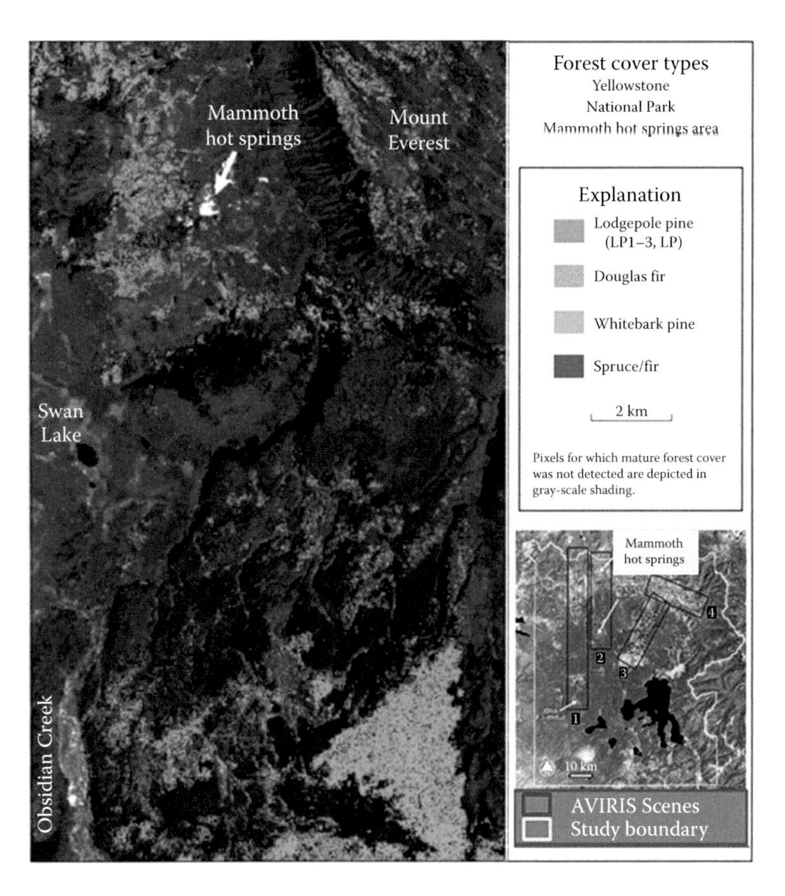

FIGURE 1.15 AVIRIS image coverage for Yellowstone National Park collected on August 7, 1996, overlaid on Landsat TM imagery. (From http://speclab.cr.usgs.gov/national.parks/Yellowstone/ynppaper.html.)

1.3 for specifications about RADAR data. In the United States, the National Weather Service (NWS) uses Next-Generation Radar (NEXRAD) satellites to detect atmospheric events such as precipitation and wind movement and thus becomes useful in tracking tornados, thunderstorms, and other weather-related hazards. Figure 1.16 is an example of a freely accessible NEXRAD data analysis by NWS to detect probable tornado sighting. NASA's next-generation dual-frequency precipitation radar whose mission was Global Precipitation Measurement (GPM) can accurately measure rain and snow worldwide every 3 h. The European Space Agency (ESA) SAR imagery is the most popular microwave radar imagery used for land cover mapping in cloudy, humid, and wet areas similar to coastal regions. Microwaves can penetrate the water vapor in the atmosphere and collect discernable images for analysis. SAR imageries are very useful when conventional multispectral satellite imaging cannot acquire cloud-free images. Figure 1.17 is an example of a SAR image acquired to distinguish the land inundation in China due to the Three Gorges Dam. Table 1.7 shows a list of radar-based satellite information.

In addition to applications in weather and climate analysis, RADAR data are used in various applications such as precision agriculture (Brisco et al., 1989), mapping soil water distribution with ground-penetrating RADAR (Dobson and Ulaby, 1998; Huisman et al., 2003; D'urso and Minacapilli, 2006), forestry application (Leckie and Ranson, 1998), ecosystem studies (Waring et al., 1995; Kasischke et al., 1997), geomorphic and hydrologic applications (Lewis and Henderson, 1998; Glenn and Carr, 2004), snow and ice mapping and analysis (Hall et al., 2012), urban remote sensing (Dong et al., 1997; Chen et al., 2003), and land use and land cover mapping (Haack and Bechdol, 2000).

The multiple look direction (ascending and descending) characteristic of RADARSAT is especially helpful in distinguishing differently oriented linear features such as the traces of fracture and faults (Riopel, 2000). Some of the fracture traces that might be hidden in one look direction due to a shadowing effect become visible from the opposite look direction. Natural corner reflection is another special property of radar. If any lineament and fracture along a small fault is present, the backscatter energy from the corner will be highlighted, revealing faults or other lineaments that would be otherwise hidden (Riopel, 2000).

1.3.5 LASER and RADAR Altimetry Imaging

Laser and radar altimeters on satellites have provided a wide range of data on detecting the bulges of water caused by gravity in the ocean. It helps map features on the seafloor to a

TABLE 1.6 List of Selected Hyperspectral Sensor–Based Satellites along with Detailed Specification

Name of Sensor	Description	Mission	Platform (Airborne/Spaceborne)		No. of Bands	Spatial Resolution	Revisit Period (Days)	Operation Since	Country/Agency
			Air	Space					
AVIRIS	Airborne Visible/Infrared Imaging Spectrometer	JPL Earth Remote Sensing	X		224	20 m and less	At wish	1987	NASA JPL, United States
HSI	Hyperspectral imager	EnMAP		X	244	30 m	4	2008	Germany
ACE-FTS	Atmospheric Chemistry (Ozone) Experiment Fourier Transform Spectrometer	SCISAT		X		n/a	15 times/day	2003	Canada
CHRIS	Compact High-Resolution Imaging Spectrometer	PROBA-1		X	153 (5 modes)	17 m/34 m		2002	ESA, Europe
EPTOMS	Earth Probe Total Ozone Mapping Spectrometer	Earth Probe		X	6	n/a	n/a	1996	NASA, United States
Hyperion	High-resolution hyperspectral imager with 220 spectral bands (from 0.4 to 2.5 μm)	EO-1		X	220	30		2001	USGS, United States
ILAS	Improved Limb Atmospheric Spectrometer	ADEOS		X	2	2,000–13,000 m (IFOV)		2002	JAXA, Japan
ILAS-II	Improved Limb Atmospheric Spectrometer-II	ADEOS-II		X	4	1,000 m (IFOV)		2002	JAXA, Japan
ISTOK-1	Infrared Spectrometer	PRIRODA-MIR		X	64	750 m		1996	RKA, Russia
MISR	Multi-angle Imaging SpectroRadiometer	Terra		X	4	275 m	9	1999	JPL, United States
MODIS	Moderate-Resolution Imaging Spectroradiometer (PFM on Terra, FM1 on Aqua)	Aqua, Terra		X	36	250, 500, 1,000 m	2	1999	NASA, United States
MOS	Modular Optoelectronic Scanning Spectrometer	IRS-P3		X	18	500 m	3 (approx.)	1996	DLR, Germany
MOS-A	Modular Optoelectronic Scanning Spectrometer	PRIRODA-MIR		X	4	2.87 km	3 (approx.)	1996	DLR, Germany
MOS-B	Modular Optoelectronic Scanning Spectrometer	PRIRODA-MIR		X	13	0.7 × 0.65 km	3 (approx.)	1996	DLR, Germany
SCIAMACHY	Scanning Imaging Absorption Spectrometer for Atmospheric Chartography	ENVISAT		X	8	30 × 27–30 × 240 km (IFOV)	3	2002	ESA, Europe
SSJ/5	Precipitating Particle Spectrometer	DMSP-16		X				2003	U.S. Air Force, United States
SSUSI	Special Sensor Ultraviolet Spectrographic Imager	DMSP-16		X				2003	U.S. Air Force, United States
TANSO-FTS	Thermal and NIR Sensor for Carbon Observation, Fourier transform spectrometer	Greenhouse Gas Observing Satellite (GOSAT) (Ibuki)		X	4	0.5 km	3	2009	JAXA, Japan
TES	Tropospheric Emission Spectrometer	AURA		X	12	0.5 km	16	2004	NASA, United States
TOMS	Total Ozone Mapping Spectrometer (see also SBUV/TOMS)	ADEOS		X	6	47 km × 3.1 km (IFOV)	41	1996–1997 (out of service)	ILRS, NASA

Source: Adapted and modified from ITC, ITC's database of satellites and sensors, 2014, http://www.itc.nl/research/products/sensordb/allsensors.aspx, accessed on May 2, 2014.

resolution of approximately 1 mile. The Light Amplification by Stimulated Emission of Radiation (LASER) and RADAR altimeters measure the height and wavelength of ocean waves, wind speeds, wind direction, surface ocean currents and their directions, etc. (Brenner et al., 2007; Giles et al., 2007; Connor et al., 2009). Connor et al. (2009) used satellite microwave altimeters boarded on ENVISAT/RA-2 to measure the Arctic Sea ice depth and height accurately. ESA-operated CryoSat-2 interferometric radar altimeter acquires accurate measurements of the thickness of floating sea ice for annual variations and surveys the surface of ice sheets accurately enough to detect small changes. This supports the study of global warming and climate change effect on present-day global land cover changes and especially diminishing snow and ice cover. Figure 1.18 shows a graphical depiction of how satellite altimetry works with Jason-2 Ocean Surface Topography Mission (OSTM). Jason-2 OSTM measures

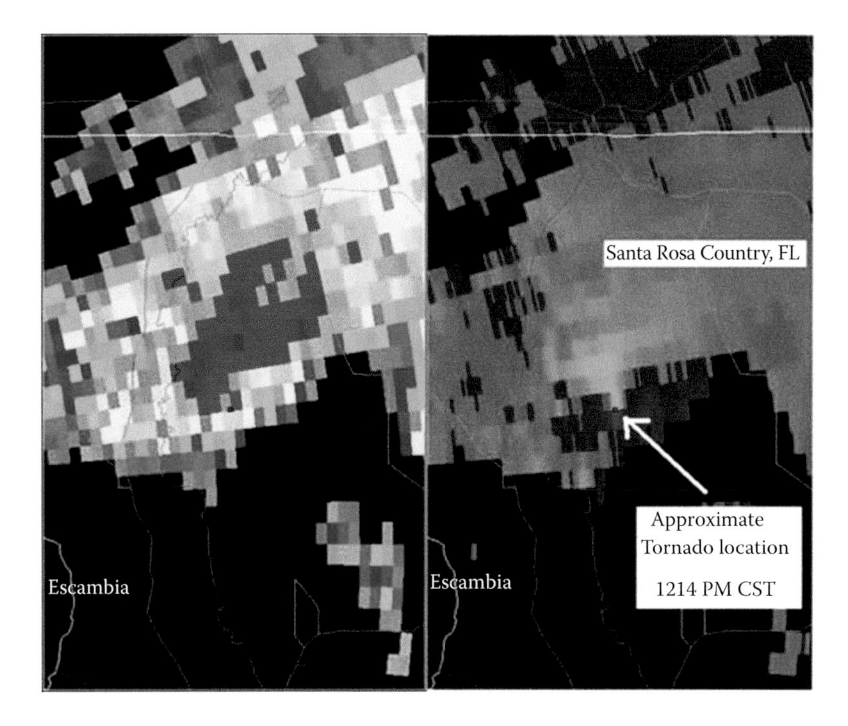

FIGURE 1.16 Sample of a Next-Generation Radar imagery from a tornado event.

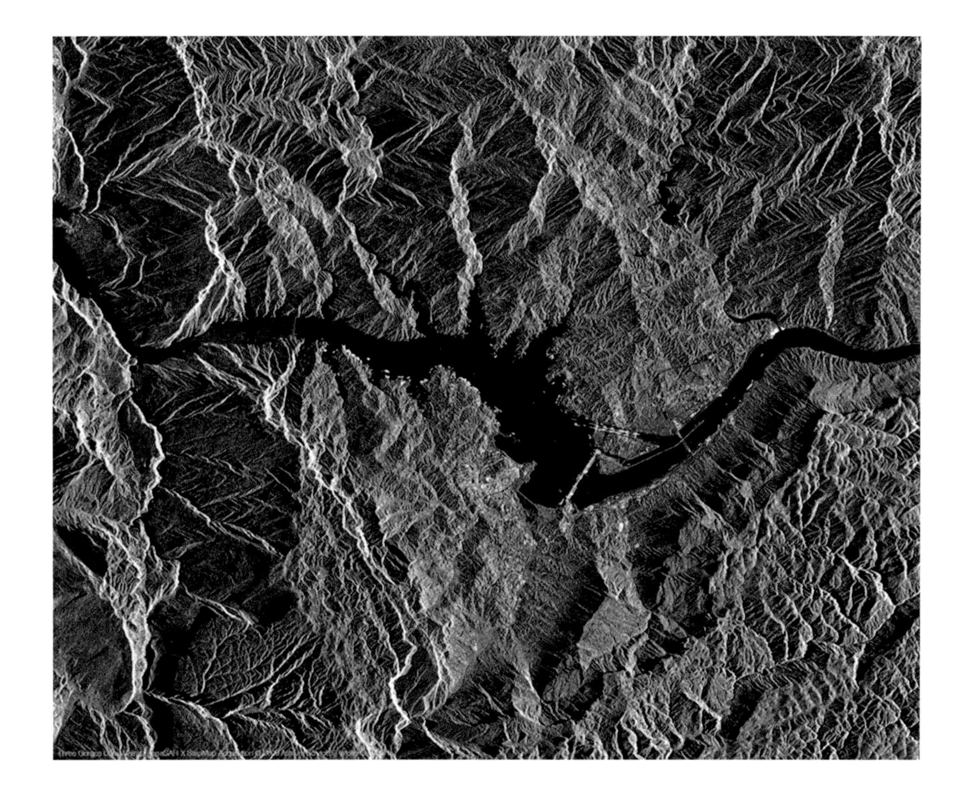

FIGURE 1.17 On October 21, 2009, a scene of the Three Gorges Dam, China, was captured using the TerraSAR-X sensor. It is an X-band radar sensor, operating in different modes and recording images with different swath widths, resolutions, and polarizations. (From www.geo-airbusds.com.)

TABLE 1.7 List of Selected RADAR Sensor-Based Satellites along with Detailed Specification

Name of Sensor	Description	Mission	Platform (Airborne/Spaceborne)		No. of Bands	Spatial Resolution/IFOV	Revisit Period (Days)	Operation Since	Agency/Country
			Air	Space					
ALMAZ-1	SAR	ALMAZ-1		X	1	15 m		1991–1992 a Out of Service	NPO Mashinostroyenia, Russia
Active Phased Array	SAR-X band	TanDEM-X		X	1 (SAR-X)		11	2010	DLR, Germany
ALT	Dual-Frequency Radar Altimeter	TOPEX/Poseidon		X	2	6 km	10	1992–2005 b Proposed	NASA, United States
ASAR	Advanced SAR	ENVISAT		X	1 (SAR-C)	30 (150) m	35	2002–2012 (out of service)	ESA, Europe
COSI	Corea SAR Instrument	KOMPSAT-5 (Arirang 5)		X	1	1, 3, 20 m	28	2013	KARI, South Korea
CPR	Cloud Profiling RADAR	EarthCARE		X	1		25	2015 (proposed)	ESA, Europe
DPR	Dual-Frequency Precipitation Radar	GPM Core		X	2	5 km		2014	NASA, United States
GMI	Microwave Radar Instrument	GPM Core		X	13	6–26 km (IFOV)		2014	NASA, United States
HQSAR	High-Quality SAR	TerraSAR-X		X	1 SAR-X)	1 m	11	2007	DLR, Germany
LBI	L-band SAR	SAOCOM-1A		X	1 (SAR-L)	10 m	16	2015 (proposed)	CONAE, Argentina
MMSAR	Multimode SAR	TecSAR		X	1 SAR-X)	0.1 m		2008	IAI, Israel
MRS, CRS	Scan SAR mode	RISAT-1		X	1 (SAR-C)	25 m	25	2012	ISRO, India
PALSAR	L-band SAR	ALOS		X	1 (SAR-L)	10 m	46	2006–2011 (out of service)	JAXA, Japan
PALSAR-2	L-band SAR-2	ALOS-2		X	1 (SAR-L)	3 m	14	2014	JAXA, Japan
PR	Precipitation radar	TRMM		X	2	25 km	0.5	1997	NASA, United States
RADARSAT-2	Radar	RADARSAT-2		X	1 (SAR-C)	3–100 m	24	2007	MDA Geospatial Services Inc.
SAR (JERS-1)	SAR	Japanese Earth Resource satellite (JERS-1)		X	1 (SAR-L)	18 m	44	1992–1998 (out of service)	JAXA, Japan
SAR (RADARSAT-1)	SAR	RADARSAT-1		X	1	8.4–100 m	24	1995–2007 (out of service)	MDA Geospatial Services Inc.
SAR (Seasat)	SAR	Seasat		X	1	25 m		1978–1978 (out of service)	NASA, United States
SAR (Sentinel-1A)	C-band SAR on Sentinel-1A	Sentinel-1A		X	1 (SAR-C)	5 m	12	2014	ESA, Europe
SAR2000	RADAR	COSMO-SkyMed		X	1	1 m	16	2007	ASI, Italy
SAR-L	L-band SAR	MapSAR		X		3 m		2011	INPE, Brazil
SIRAL	Interferometric Radar Altimeter	CryoSat-2		X		250 m	369 with 30 days subcycle	2010	ESA, Europe
SRAL	SAR Radar Altimeter	Sentinel-2		X	13	10, 20, 60 m	5	2007 (proposed)	ESA, Europe
SRAL	SAR Radar Altimeter	Sentinel-3		X		0.5–1 km	27	2017 (proposed)	ESA, Europe
X-SAR	SAR	SRTM		X	1	25–90 m		2000–2000 (out of service)	NASA, United States

Source: Adapted and modified from ITC, ITC's database of satellites and sensors, 2014, http://www.itc.nl/research/products/sensordb/allsensors.aspx, accessed on May 2, 2014.

sea surface height through imaging that ultimately determines ocean circulation, climate change, and sea-level rise. Table 1.8 contains a selected list of LASER and RADAR altimetry imaging sensor platform with their ancillary information.

1.3.6 LiDAR Remote Sensing

LiDAR technology, a type of active remote sensing, was developed in the early 1960s following the invention of LASER and was initially used to measure distance by illuminating a target

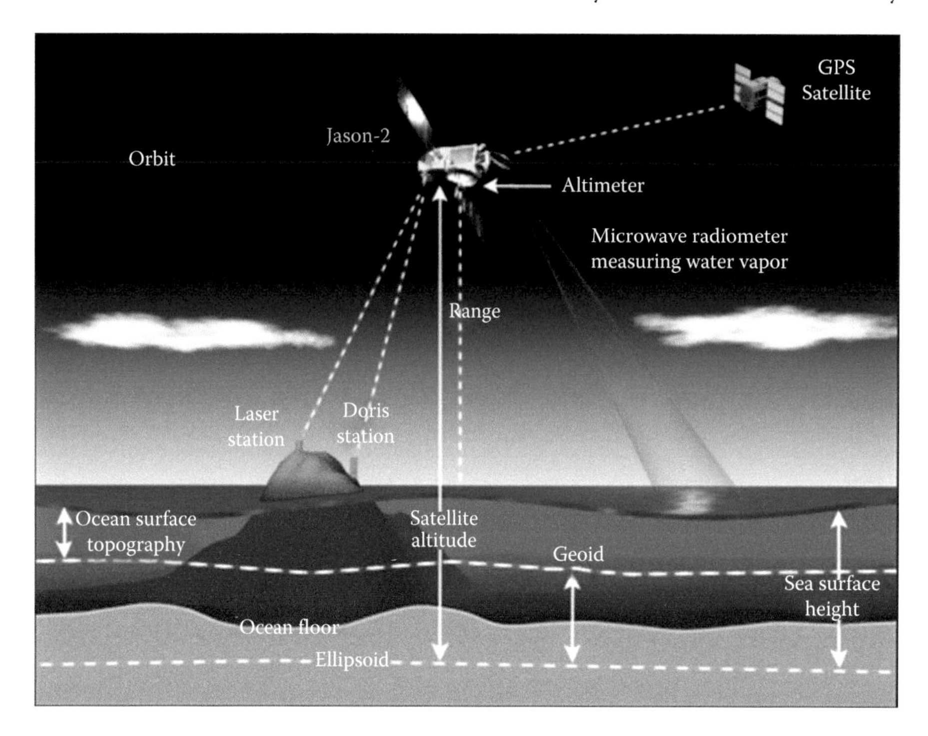

FIGURE 1.18 Graphical depiction of how Jason-2 satellite altimetry works to monitor the ocean. (From http://www.ppi.noaa.gov/bom_chapter3_satellite_management/National Oceanic and Atmospheric Administration - Office of Program Planning and Integration. 2015. Chapter 3 - NOAA Operations - Satellite Management. Retrieved from http://www.ppi.noaa.gov/bom_chapter3_satellite_management/ on 3/30/15) Please read the foreword to gather the importance of radar altimetry in recording climate impacts such as sea-level rise.

TABLE 1.8 List of Selected LASER and RADAR Altimetry Sensor–Based Satellites along with Detailed Specification

Name of Sensor	Description	Mission	Platform (Airborne/Spaceborne)		No. of Bands	Spatial Resolution	Revisit Period (Days)	Operation Since	Agency/Country
			Air	Space					
ALT	Dual-Frequency Radar Altimeter	TOPEX/Poseidon		X	2	6 km	10	1992–2005 (out of service)	NASA, United States
ALT (Seasat)	Altimeter	Seasat		X	1 (SAR-K)	1.6 km		1978–1978 (out of service)	NASA, United States
AltiKa	High-resolution altimeter including bifrequency radiometric function	SARAL/AltiKa		X	2	2 km	35	2013	CNES, France
GLAS	GLAS	ICESat		X	2	100 m–1 km	Monthly/seasonal	2003–2010 (out of service)	NASA, United States
RA	Radio altimeter	ERS-1		X	2	16 km	35	1991–2000 (out of service)	ESA, Europe
RA	Radio altimeter	ERS-2		X	2	16 km	35	1995–2011 (out of service)	ESA, Europe
SIRAL	SAR Interferometric Radar Altimeter	CryoSat-2		X	1 (Radar altimeter)		369 with 30 day subcycle	2010	ESA, Europe
SRAL	SAR Radar Altimeter	Sentinel-3		X	1 (Radar altimeter)	0.5–1 km	27	2017 (proposed)	ESA, Europe

Source: Adapted and modified from ITC, ITC's database of satellites and sensors, 2014, http://www.itc.nl/research/products/sensordb/allsensors.aspx, accessed on May 2, 2014.

with LASER. The first LiDAR application was in the field of meteorology; the National Center for Atmospheric Research (NCAR) used LiDAR to measure the cloud distance from ground surface (Goyer and Watson, 1963). The 1971 use of LiDAR technology by the Apollo mission to map the surface of the moon made the technology well known among the general public. Soon LiDAR became a very common tool for geospatial technology researchers and users in the present century when LiDAR was first used by airplanes to map the land topography.

Satellite-based LiDAR data acquisition is spearheaded by NASA, ESA, Russian Federal Space Agency (RKA), and German Aerospace Research Establishment (DLR) with sensors such as laser reflectometry array (LRA) on Jason-2 (OSTM) and Geoscience Laser Altimeter System (GLAS) for Ice, Cloud, and land Elevation Satellite (ICESat) mission, atmospheric laser Doppler instrument (ALADIN) and ATLID EarthCARE, Priroda-Mir mission, and laser communication terminal (LCT) for TanDEM-X mission, respectively (Table 1.9). Table 1.9 also provides the satellite specifications on selected LiDAR remote sensing satellites.

Present-day aerial LiDAR data are collected from airplanes flying around 2–5 miles above the ground surface and provide a minimum of 30 cm resolution data. The data are collected by focusing and scanning a LASER beam from an airplane to the objects on the ground surface, which then acquires the data as different returns (Panda et al., 2012a,b; Amatya et al., 2013; Panda and Hoogenboom, 2013). Table 1.9 includes a list of LiDAR-based satellite information. LiDAR technology is becoming popular since the start of the millennium due to its advantage in mapping the Earth topography along with object heights on the Earth's surface, thus supporting image classification process tremendously (Dubayah et al., 2000). The NOAA mission of collecting LiDAR data of the entire U.S. coast and some ecologically sensitive areas in the United States made LiDAR a well-known technology. NOAA Coastal Services Center's Digital Coast Data Access Viewer (http://www.csc.noaa.gov/dataviewer/#) provides free access to LiDAR data, including some recent acquisitions of 2009 and later, for a major portion of the United States.

LiDAR remote sensing instrument provides point cloud data. The crude point cloud data are processed and each laser shot is converted to a position in a 3D frame of reference with spatially coherent cloud of points. In this processing stage, some LiDAR data provide texture or color information for each point Renslow 2012, Rao, 2013). The processed 3D spatial and spectral information contained in the dataset allows great flexibility to perform manipulations to extract the required information from the point cloud data Renslow 2012, Rao, 2013). Thereafter, visualization, segmentation, classification, filtering, transformations, gridding, and mathematical operations are conducted on the data to obtain required information of Earth objects or phenomena as discussed.

The first return of the LiDAR data is generally from the tallest features, that is, tallest tree canopy or top of high-rise buildings, the intermediate returns are from the canopy of the small trees and shrubs, and the final return is from the ground surface. These individual return data are processed to get height information of the features discussed earlier. Figure 1.19 displays the point cloud LiDAR data showing different returns colorized based on height as visualized using LiDAR Data Viewer of the USGS Fusion software (http://forsys.cfr.washington.edu/fusion/fusionlatest.html). Using interpolation and smothing algorithms, the point cloud is rendered as a grid surface, which can be easily manipulated in a GIS using map algebra operations to produce canopy height, ground elevations, etc. The normalized digital surface model (nDSM) is developed in the LiDAR processing software Quick Terrain Modeler (Applied Imagery, Chevy Chase, MD) from the ground return data to produce the accurate DEM (topography) data of the ground surface (Panda et al., 2012a,b). Panda et al. (2012a,b) with their wetland change and cause recognition study of Georgia, United States, coast found a major discrepancy with the 10-m DEM data available through USGS's National Elevation Dataset (NED). The problem was found while comparing with the DEM created with 30-cm resolution LiDAR for Camden County, GA. They found an average change of ground elevation of 0.9 m with a range of −0.17 to 1.5 m when comparing the elevation at permanent locations. Figure 1.20 shows the DEM developed for Camden County, GA, using the 2010 LiDAR data, the 10 m DEM of the county created in 2000 (based on the raster metadata), the comparison table of

TABLE 1.9 Selected LiDAR Sensor-Based Satellites along with Detailed Specification

Name of Sensor	Description	Mission	Platform (Spaceborne)	No. of Bands	Spatial Resolution	Revisit Period (Days)	Operation Since	Agency/Country
ALADIN	ALADIN	ADM-Aeolus	X			7	2015 (planned)	ESA, Europe
ALISSA	L'atmosphere par Lidar Sur Saliout	PRIRODA-MIR	X	1	150 m		1996	RKA, Russia
ATLID	High-spectral-resolution LiDAR	EarthCARE	X	2 (S and X)		25	2016 (planned)	ESA, Europe, and JAXA, Japan
GLAS	GLAS	ICESat	X	2	70 m and 3 cm vertical	91	2003–2010 (out of service)	NASA, United States
LCT	LCT	TanDEM-X	X			11	2010	DLR, GFZ, Infoterra, Germany
LRA (Jason)	Laser Retroreflector Array	Jason-2 (OSTM)	X			10	2008	NASA, United States

Source: Adapted and modified from ITC, ITC's database of satellites and sensors, 2014, http://www.itc.nl/research/products/sensordb/allsensors.aspx, accessed on May 2, 2014.

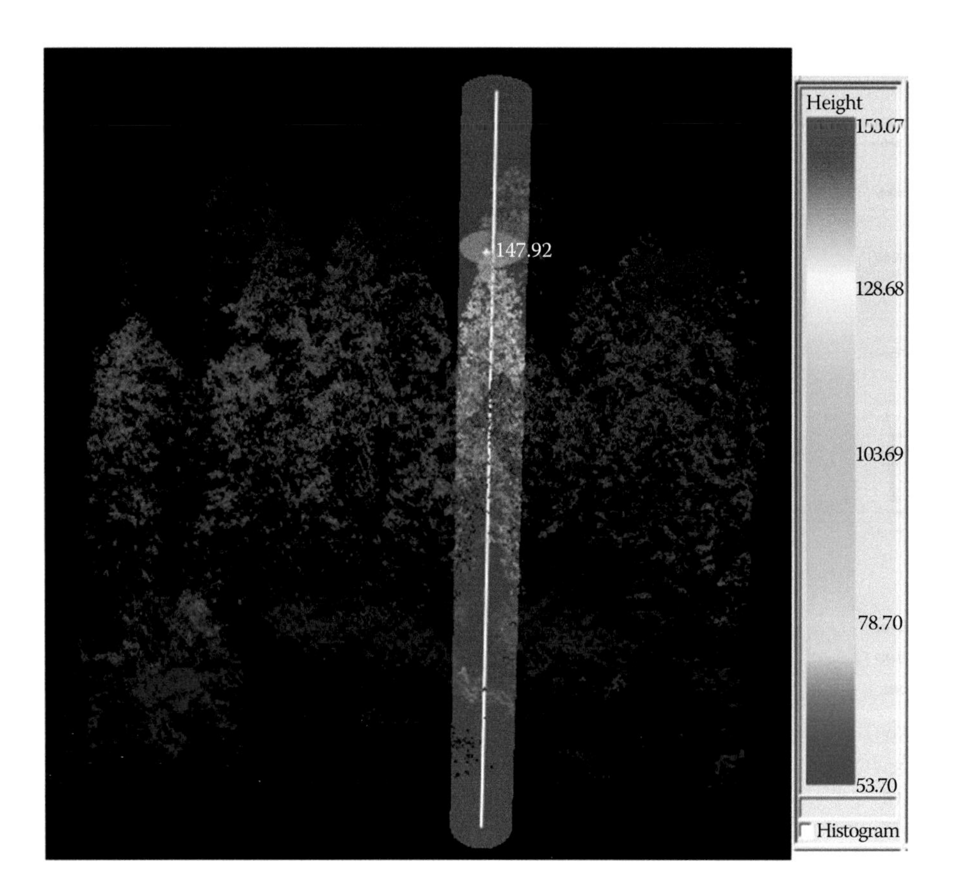

FIGURE 1.19 A 3D display of point cloud light detection and ranging (LiDAR) data showing different returns (colorized based on height as visualized using the LiDAR Data Viewer of the USFS Fusion software). Also, note the interactive measuring cylinder with marker can be moved about the virtual space for measuring height.

elevation differences, and the algorithm developed to modify the 10 m DEM of the study area.

Panda and Hoogenboom (2013) used 30 cm LiDAR from a blueberry orchard in Pierce County, GA, to obtain the height of pine trees, existing structures, and blueberry plants with ground elevation to be used in the object-based image analysis (OBIA)–based image segmentation with eCognition software for blueberry orchard SSCM. Yu et al. (2010) in their study of urban building density determination using airborne LiDAR data explained the development of nDSM and subsequent DEM from the LiDAR data for decision support. Kelly et al. (2014) completed a study of coastal marsh management decision support by using 30 cm LiDAR tiles. The DEM of South Carolina coast was developed to delineate the areas of saltwater intrusion (with the highest possible tide heights) calculated with the latest climate change analysis. Additionally, the researchers used nDSM from LiDAR imagery to classify the NAIP imagery using eCongnition's OBIA image segmentation procedure to produce an output of the coastal forest plant species. These applications show the potential of LiDAR in environmental (forestry, ecology, agriculture, urban, and others) management decision support systems.

Similarly, Rao (2012) demonstrated applications of LiDAR-based mapping and height estimates for snags and coarse woody debris using threshold filtering of the point cloud data. Furthermore, LiDAR data, owing to its properties to penetrate canopy, have high application in improving wildland fire mapping through better estimates of canopy bulk density and base height (Riano et al., 2004; Lee and Lucas, 2007; Erdody and Moskal, 2010), canopy cover (Lefsky et al., 2002; Hall et al., 2005), shrub height (Riano et al., 2007), and aboveground biomass (Edson and Wing, 2011; Vuong and Rao, 2013) and carbon stocks (Asner et al., 2010; Goetz and Dubayah, 2011). Other studies have documented better mapping and classification outputs using an integrated approach where multispectral data are fused with LiDAR data (Garcia et al., 2011; Rao and Miller, 2012). Innovative approaches to map natural resources at global scales have been initiated in the recent past such as using the GLAS (see Figure 1.21) aboard the ICESat in conjunction with MODIS percent tree cover product (MOD44B) to map forest canopy height globally with spaceborne LiDAR (Simard et al., 2011). LiDAR is also used to detect and measure the concentration of various chemicals in the atmosphere including the use of camera LiDAR instrument to accurately measure the particulates (Barnes et al., 2003). These and many other studies show the potential of LiDAR data in environmental (forestry, ecology, agriculture, urban, and others) management and decision support systems, particularly in context of climate change issues at regional and global levels.

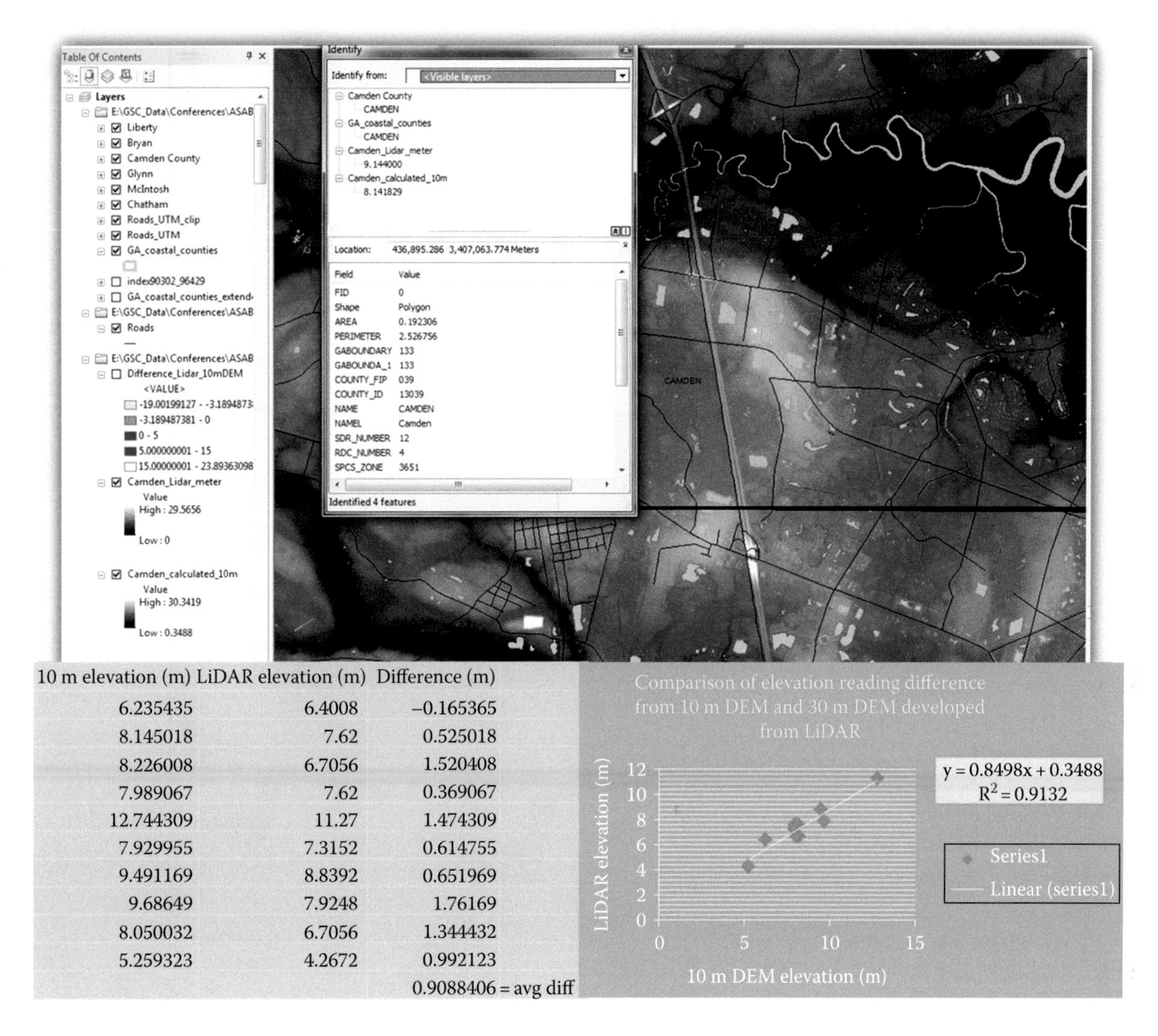

10 m elevation (m)	LiDAR elevation (m)	Difference (m)
6.235435	6.4008	−0.165365
8.145018	7.62	0.525018
8.226008	6.7056	1.520408
7.989067	7.62	0.369067
12.744309	11.27	1.474309
7.929955	7.3152	0.614755
9.491169	8.8392	0.651969
9.68649	7.9248	1.76169
8.050032	6.7056	1.344432
5.259323	4.2672	0.992123
		0.9088406 = avg diff

FIGURE 1.20 Comparison of 10 m digital elevation model (DEM) of National Elevation Dataset and DEM developed with airborne LiDAR of Camden County, GA. (From Panda, S.S. et al., Wetland change and cause recognition in Georgia coastal plain, in: Presented in the *International American Society of Agricultural and Biological Engineers (ASABE) Conference 2012*, July 29–August 2, 2012, Dallas, TX, Paper # 1338205, 2012a; Panda, S.S. et al., Stomatal conductance and leaf area index estimation using remotely sensed information and forest speciation, in: Presented in the *Third International Conference on Forests and Water in Changing Environment 2012*, September 18–20, 2012, Fukuoka, Japan, 2012b.)

Currently, LiDAR is also extensively used by law enforcement for tracking of vehicle speed military for weapon ranging, mine detection for countermine warfare, and laser-illuminated homing of projectiles; in mining industry to calculate the ore volume (3Dlasermapping, 2014); in physics and astronomy by measuring the distance to the reflectors placed on the moon and detecting snow on the Mars atmosphere (NASA-Phoenix Mars Lander, 2014); in the field of robotics for the perception of the environment as well as object classification and safe landing of robots and manned vehicles (Amzajerdian et al., 2011); in spaceflight, surveying using mobile LiDAR instruments (Figure 1.22); in adaptive cruise control; and in many other high-end energy-producing sources like wind farms and solar farms for wind velocity and turbulence measurement and optimizing solar photovoltaic systems by determining shading losses.

1.3.7 Microwave Remote Sensing

Unlike LiDAR remote sensing, microwave sensing encompasses both active and passive forms of remote sensing. Atmospheric scattering affects shorter optical wavelengths but longer wavelengths of EMS are not affected by atmospheric scattering (Aggarwal, 2003). Therefore, longer-wavelength microwave radiation can penetrate through cloud cover, haze, dust, and all but the heaviest rainfall, and thus microwave remote sensing can be useful for geospatial community as data can be collected with microwave remote sensing systems in almost all weather and environmental conditions. RADAR remote sensing is the example of active microwave remote sensing. The radar units transmit short pulses or bursts of microwave radiation when sunlight is unavailable and then obtain reflectance signal from the Earth-based objects (Avery and Berlin, 1992). Therefore,

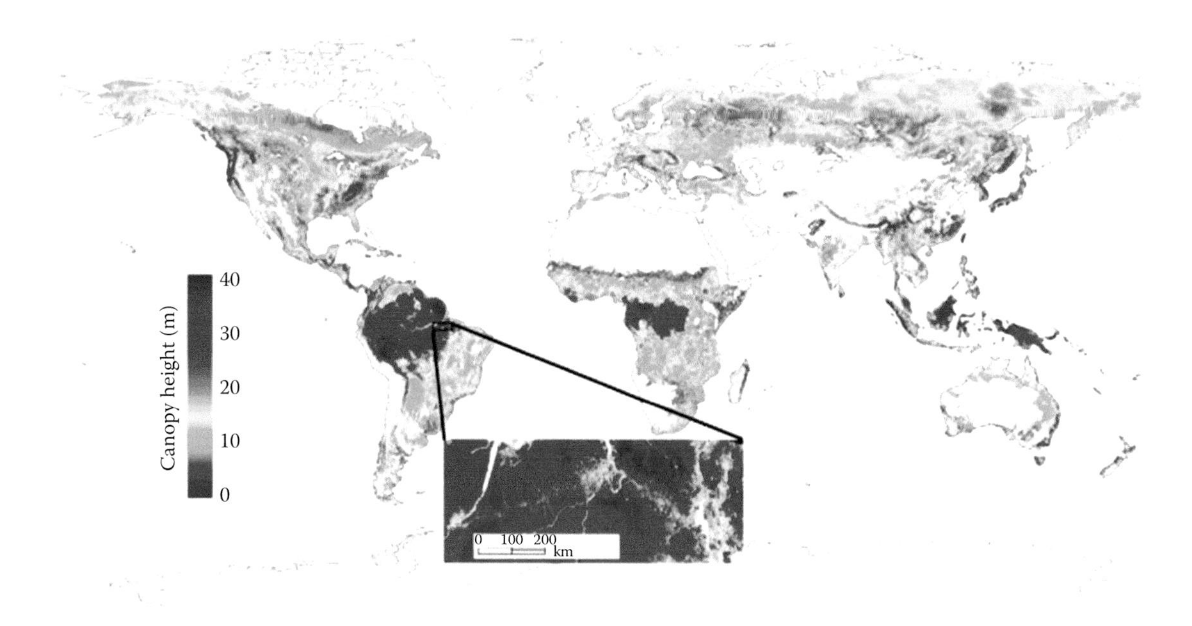

FIGURE 1.21 Wall-to-wall mapping of canopy height implemented using the Geoscience Laser Altimeter System (GLAS) instrument on the Ice, Cloud, and land Elevation Satellite. GLAS data model. The inset shows the sensitivity and capability of the model to disturbance gradient in the Amazon. The GLAS laser transmits short pulses (4 ns) of infrared light at 1064 nm and visible green light at 532 nm 40 times/s. (From Simard, M. et al., *J. Geophys. Res.*, 116, G04021, 2011, doi: 10.1029/2011JG001708.)

FIGURE 1.22 Mobile light detection and ranging for land surveying. (From Photo Science Geospatial Solutions, Lexington, KY, http://www.photoscience.com/services/mobile-mapping.)

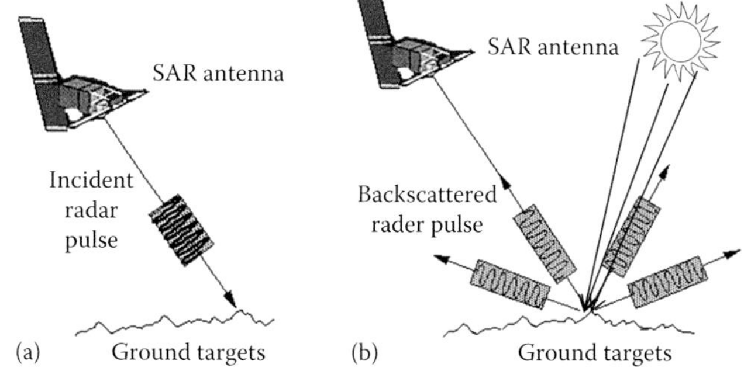

FIGURE 1.23 (a) Active microwave remote sensing: Microwave pulses are incident upon objects from the synthetic aperture radar (SAR) satellite in no-sunlight condition. (b) Passive microwave remote sensing: Microwave radiations are reflected from the objects on ground water receiving light energy from sun.

microwave remote sensing system can be operable in the night. When the microwave remote sensors collect naturally emitted microwave energy, though small, from the objects without sending pulse from the system, it is passive microwave sensing, which is similar in concept to thermal remote sensing (Avery and Berlin, 1992). The naturally emitted microwave energy is related to the temperature and moisture properties of the emitting object or surface. Figure 1.23 shows the image acquisition principle of both active and passive remote sensing systems.

Microwave remote sensors efficiently monitor targets such as soil moisture content (Engman and Chauhan, 1995; Wagner et al., 2007), salinity (Wilson et al., 2001; Matternicht and Zinck, 2003), temperature of the estuary and sea surface temperature (SST) (Blume et al., 1978; Klemas, 2011), and rain, snow, ice, and sea surface condition (Fung and Chen, 2010), water vapor, cloud, oil slick, and different gaseous presence in the atmosphere like CO_x, ozone, and NO_x. Microwave scatterometer, microwave altimeter, and imaging radar are the examples of active microwave sensing, while microwave, radiometers, and scanners are all examples of passive microwave sensing technology. Table 1.10 shows the typical microwave sensor–based satellites and their specifications. Table 1.11 summarizes passive and active sensing–based microwave sensors and their target areas of applications for Earth resources measurement.

TABLE 1.10 List of Selected Microwave Sensor-Based Satellites along with Detailed Specification

Name of Sensor	Description	Mission	Platform (Airborne/Spaceborne) Air	Space	No. of Bands	Spatial Resolution	Revisit Period (Days)	Operation Since	Agency/Country
AMI	Active Microwave	ERS-1		X	1	30 m	35	1991–2000 (out of service)	ESA, Europe
AMI	Active Microwave	ERS-2		X	1	30 m	35	1995–2011 (out of service)	ESA, Europe
AMR	Advanced Microwave Radiometer	Jason-2 (OSTM)		X		3.3 cm sea-level measurement accuracy	10	2001	JPL, NASA, United States
AMSR	Advanced Microwave Scanning Radiometer	ADEOS-II		X	8	5, 10 × 1.6 km (IFOV)	4	2002–2003 (out of service)	JAXA, Japan
AMSR2	Advanced Microwave Scanning Radiometer	GCOM-W1 (SHIZUKU)		X	7	0.15, 0.35, 0.75, 0.65, 1.2, 1.8 km	2	2002	NASA, United States
AMSR-E	Advanced Microwave Scanning Radiometer for EOS	Aqua		X	6	5, 10 × 1.4 km (IFOV)	4	2002	NASA, United States
AMSU-A	Advanced Microwave Sounding (temperature)	Aqua, Metop-A, Metop-B, NOAA-15, NOAA-16, NOAA-17, NOAA-18 (NOAA-N), NOAA-19 (NOAA-N Prime)		X	15	5 × 2.343 km (IFOV)	29 (Metop) and 11 (NOAA)	2002	NOAA, United States
AMSU-B	Advanced Microwave Sounding	Aqua, NOAA-15, NOAA-16, NOAA-17		X	5	15 × 2.343 km (IFOV)	11	1998	NOAA, United States
CrIMSS	Cross track Infrared and Advanced Technology Microwave Sounder	JPSS-1 (NPOESS)		X	3	0.025, 0.0125, 0.0625 m		2017 (proposed)	NOAA, United States
CrIMSS	Cross track Infrared and Advanced Technology Microwave Sounder	Suomi NPP (SNPP)		X	3	0.025, 0.0125, 0.0625 m		2011	NASA, United States
GMI	Microwave Radar Instrument	GPM Core		X	13			2014 (proposed)	NOAA, United States
JMR	Jason-1 Microwave Radiometer	Jason-1		X	3		10	2001–2013 (out of service)	NASA, United States
MADRAS	Microwave Imager	Megha-Tropiques		X	9	6 × 40 km	1–6 times a day	2011	CNES, France
MHS	Microwave Humidity Sounder	Metop-A, Metop-B, NOAA-18 (NOAA-N), NOAA-19 (NOAA-N Prime)		X	5	16.3 × 1.078 km (IFOV)	29 (Metop) and 11 (NOAA)	2006, 2012, 2005, and 2009, respectively	ESA, Europe
MIRAS	Microwave Imaging Radiometer Aperture Synthesis	SMOS		X	1 (SAR-L)	35 km	3	2009	ESA, Europe
MIS	Microwave Imager/Sounder	JPSS-1 (NPOESS)		X	1 (SAR-L)			2017 (proposed)	NOAA, United States

(Continued)

TABLE 1.10 (*Continued*) List of Selected Microwave Sensor–Based Satellites along with Detailed Specification

Name of Sensor	Description	Mission	Platform (Airborne/ Spaceborne) Air	Space	No. of Bands	Spatial Resolution	Revisit Period (Days)	Operation Since	Agency/ Country
MIS	Microwave Imager/ Sounder	Suomi NPP (SNPP)		X	1 (SAR-L)			2011	NOAA, United States
MLS	Microwave Limb Sounder	AURA		X	5	3 km	16	2004	NASA, United States
MSMR	Multifrequency Scanning Microwave Radiometer	IRS-P4 (Oceansat)		X	8	360 × 236 m	2	1999–2010 (out of service)	ISRO, India
MSR	Microwave Scanning Radiometer	MOS-1		X	2	32 km	17	1987–1995 (out of service)	JAXA, Japan
MSR	Microwave Scanning Radiometer	MOS-1b		X	2	32 km	17	1990–1996 (out of service)	JAXA, Japan
MSU	Microwave Sounding Unit	NOAA-10, NOAA-11, NOAA-12, NOAA-14		X	4		11	1986–2001 (all out of service)	NOAA, United States
MSU-E	Microwave Sounding Unit	RESURS-O1-3		X	3	34 m	18	1994	SRC Planeta, Russia
MWR	Microwave Radiometer	ENVISAT		X	2	1040 × 1200 m for marine and 2630 × 300 m for land and coastal application	35	2002–2012 (out of service)	ESA, Europe
MWR	Microwave Radiometer	Sentinel-3		X	2	0.3 km	27	2015 (proposed)	ESA, Europe
MWRI	Microwave Radiation Imager	FengYun-3A		X		1 km, 250 m		2008	NSMC, China
SAPHIR	Microwave Sounding Instrument	Megha-Tropiques		X		10 km at NADIR	1–6 times a day	2011	CNES, France
SMMR	Scanning Multichannel Microwave Radiometer	Nimbus-7		X	5	27–149 km	6	1978–1994 (out of service)	NASA, United States
SMMR	Scanning Multichannel Microwave Radiometer	Seasat		X	5	27–149 km	6	1978–1978 (out of service)	NASA, United States
TMI	TRMM Microwave Imager	TRMM		X	5	4.4 km	0.5	1997	NASA, United States
TMR	TOPEX Microwave Radiometer	TOPEX/ Poseidon		X	3	23.5–44.6 km	10	1992–2005 (out of service)	NASA, United States

Source: Adapted and modified from ITC, ITC's database of satellites and sensors, 2014, http://www.itc.nl/research/products/sensordb/allsensors.aspx, accessed on May 2, 2014.

1.3.8 SONAR and SODAR Remote Sensing

SONAR is a technique that uses sound propagation to navigate and communicate with or detect objects on or under the surface of the water (Figure 1.24). Two types of technology share the name *sonar*: *passive* sonar is essentially listening for the sound made by vessels; *active* sonar is emitting pulses of sound energy and listening for the echoes. Blakinton et al. (1983) used SONAR technology to map the seafloor with the development of SeaMARC II. Kidd et al. (1985) used long-range side-scan SONAR to map sediment distributions over wide expansions of the ocean floor. Fairfield and Wettergreen (2008) recently used multibeam SONAR to map the ocean floor. As sound can propagate through water, the sensing system uses it to know the objects underneath water. Sonar may also be used in air for robot navigation. Skarda et al. (2011) used SONAR technology–based fish-finder instrument to map the bottom of two retention ponds to study the advantages of retention pond in soil conservation. Similarly, Dual-Frequency Identification Sonar (DIDSON) uses sound to produce images using high- and low-frequency modes in the range of about 15–40 m and have demonstrated benefit in fish inventorying (Belcher et al., 2001; Moursund et al., 2003; Tiffan et al., 2004).

SODAR (Sonic detection and ranging) is upward-looking, in-air sonar used for atmospheric investigations. SODAR remote sensing systems are like a LiDAR/RADAR system in which sound waves are propagated instead of light or radio waves in order to detect the object of interest (Bailey, 2000). Doppler SODAR (Figure 1.25) is a widely used remote sensing system for weather forecasting (Goel and Srivastava, 1990; Beyrich, 1997). Wind farms are using fulcrum compact-beam multiple-axis 3D SODAR to increase efficiency of wind power production. Satellite

TABLE 1.11 Microwave Remote Sensors and Their Application Areas

Sensor		Target
Passive sensor		
	Microwave radiometer	Near-sea-surface wind SST Sea condition Salinity and sea ice condition Water vapor, cloud water content Precipitation intensity Air temperature Ozone, aerosol, NO_x, other atmospheric constituents
Active sensor		
	Microwave scatterometer	Soil moisture content Surface roughness Lake and ocean ice distribution Snow distribution Biomass SST Sea condition Salinity and sea ice condition Water vapor Precipitation intensity Wind direction and velocity
	Microwave altimeter	Sea surface topography, geoid Ocean wave height Change of ocean current Mesoscale eddy, tide, etc. Wind velocity
	Imaging radar	Image of ground surface Ocean wave Near-sea-surface wind Topography and geology Submarine topography Ice monitoring

platform-based SONAR systems are operated by the NOAA, NASA, U.S. Department of Defense, ESA, and Space Agency of France (Centre National d'Etudes Spatiales [CNES]). Table 1.12 provides selected list of these SONAR satellite systems with their specifications. These SONAR systems are mostly used for ocean phenomena study, atmospheric measurement, and ice and snow studies.

1.3.9 Global Positioning System

The GPS, now known as Global Navigation Satellite System (GNSS), is a space-based satellite navigation system that provides location (coordinates and elevation) and time information in all weather conditions, anywhere on or near the Earth, where there is an unobstructed line of sight to four or more GPS satellites (Van Diggelen, 2009). GPS satellites use the trilateration technique to know the location of a receiver by coordinating signals received from four satellites. It is an application of active remote sensing system as unique signals (pseudorandom code) in longwaves, L1 = 1575.42 MHz (19 cm wavelength) and L2 = 1227.6 MHz (24.4 cm wavelength), are transmitted to the receivers along with accurate time stamps and other satellite information. These signals are intercepted by the GPS receivers where the timing code is translated to interpret the precise time it took the signal to reach the receiver. This precise time is used with the signal speed to calculate distance or range. The intersection of the three ranges from three satellites is used to accurately determine the location of the GPS receiver.

There are three segments involved in GPS data collection: (1) space segment, (2) control segment, and (3) user segment. Figure 1.26 shows the GPS geosynchronous satellite constellations and their orbit. The control segment consists of five ground monitoring stations that continuously monitor the satellite trajectory and atmospheric conditions and help develop an error model for the atmospheric interference. The error correction based on the model is uploaded back to the GPS satellites via the master control facility at Colorado Springs, Colorado, United States. The receiver segment is the most important part, because

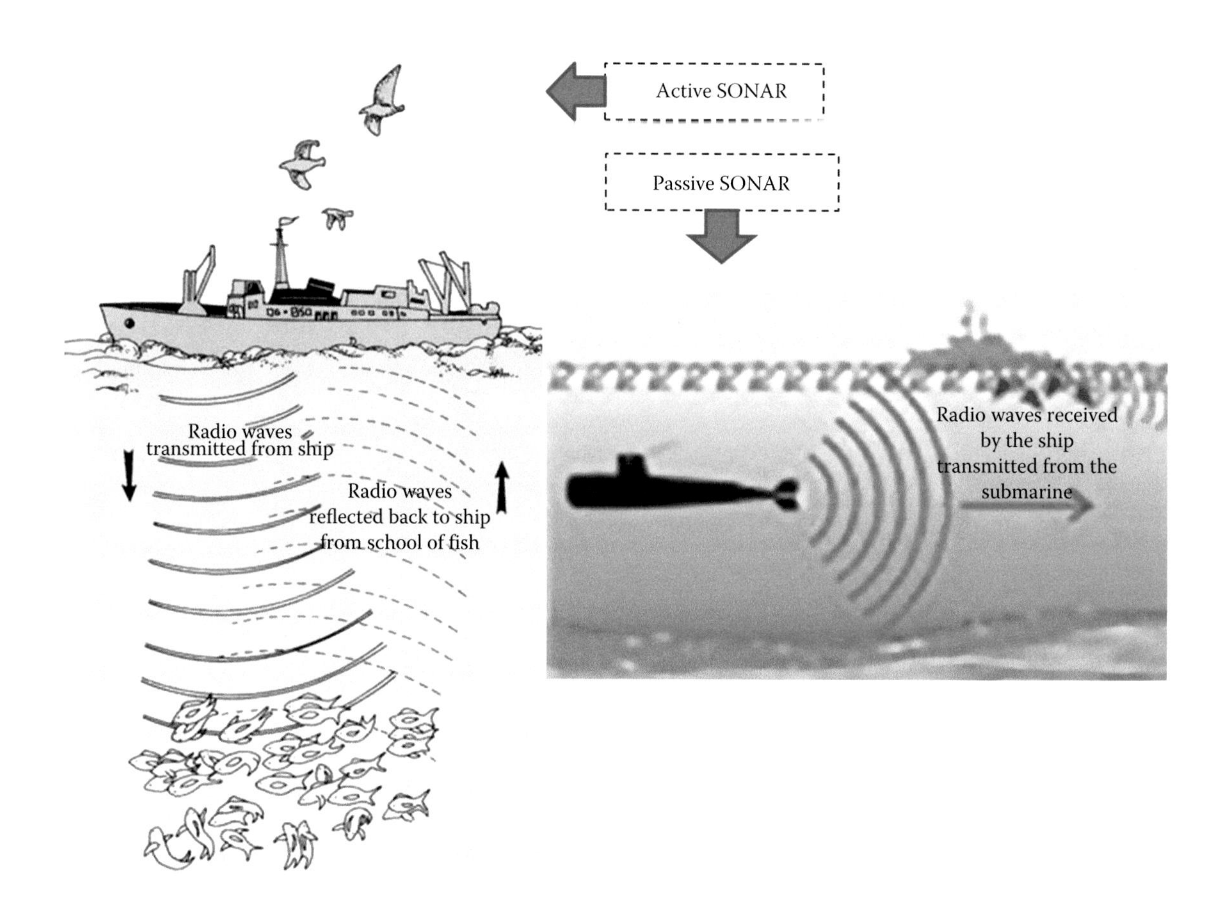

FIGURE 1.24 Example of active and passive sound navigation and ranging (SONAR) detection system emitting ultrasound and radio waves to find the location of fish in the ocean and receiving radio waves from undersea submarine and detecting its location. SONAR is used in navigation, communication, and detection of objects on surface and underwater.

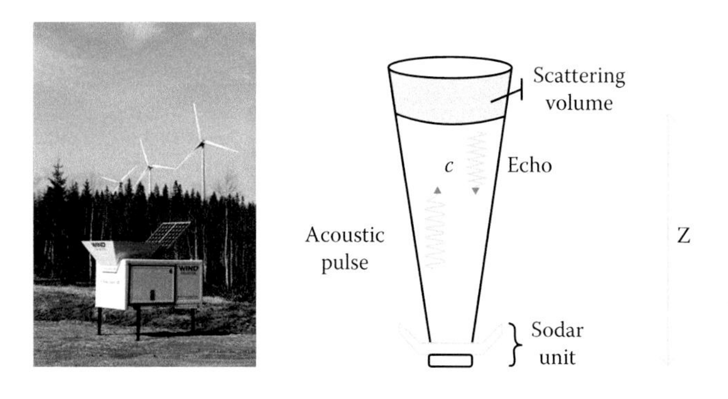

FIGURE 1.25 Wind measurement with a phased array SODAR SFAS from Scintec Corporation (Louisville, CO) and the working principle of SODAR sensing system.

it involves the end user application needs. With the advancement of technology, we have centimeter accuracy GPS receivers that are able to process signal correction information (kinematic) on the fly, thereby facilitating accurate location data collection. Follow this (http://www.gps.gov/) site for more information about GPS. GPS is described in detail in another chapter of this book.

While the history of the GPS is rooted in military applications, over the past several decades, the civilian applications of GPS have far exceeded the military applications. The literature is replete with numerous applications of GPS data that include telecommunications, commerce and retailing industry, and recreation, in addition to the traditional applications in natural resource management and planning. The GPS.gov—the official U.S. government information portal about GPS and related topics—provides a good treatise about the diverse applications of GPS in everyday life limited only by the human imagination. More importantly, the website provides an important repository of information pertaining to the GPS modernization program that includes new series of satellite acquisitions (GPS IIR, GPS IIF, and GPS III). The improvements bring about new features such as the Architecture Evolution Plan (AEP) and the Next-Generation Operational Control System (OCX) that are geared to deliver enormous improvements in the robustness and accuracy of the entire GPS system in the coming years (Figure 1.27). These infrastructure and signal developments are designed not only to increase operational capacity but also to provide better geometry in *shadow* environments of urban canyons and mountainous terrain.

TABLE 1.12 List of Selected SONAR Technology-Based Satellites along with Detailed Specification

Name of Sensor	Description	Mission	Platform (Spaceborne)	No. of Bands	Spatial Resolution	Revisit Period (Days)	Operation Since	Agency/ Country
AIRS	AIS	Aqua	X	7	13.5 km	0.5	2002	NASA, United States
AMSU-A	AIS	Aqua	X	15	50 km		2002	NASA, United States
AMSU-A	AIS	Metop-A	X	15	50 km	29	2006	ESA, Europe
AMSU-A	AIS	Metop-B	X	15	50 km	29	2012	ESA, Europe
AMSU-A	AIS	NOAA-15	X	15	50 km	11	1998	NOAA, United States
AMSU-A	AIS	NOAA-16	X	15	50 km	11	2000	NOAA, United States
AMSU-A	AIS	NOAA-17	X	15	50 km	11	2002	NOAA, United States
AMSU-A	AIS	NOAA-18 (NOAA-N)	X	15	50 km	11	2005	NOAA, United States
AMSU-A	Advanced Microwave Sounding	NOAA-19 (NOAA-N Prime)	X	15	50 km	11	2009	NOAA, United States
HSB	Humidity Sounder for Brazil	Aqua	X	4	13.5 km	0.5	2002	NASA, United States
MHS	Microwave Humidity Sounder	Metop-A	X	5	50 km	29	2006	ESA, Europe
MHS	Microwave Humidity Sounder	Metop-B	X	5	50 km	29	2012	ESA, Europe
MHS	Microwave Humidity Sounder	NOAA-18 (NOAA-N)	X	5	50 km	11	2005	NOAA, United States
MHS	Microwave Humidity Sounder	NOAA-19 (NOAA-N Prime)	X	5	50 km	11	2009	NOAA, United States
MLS	Microwave Limb Sounder	AURA	X	5	3 km	16	2004	NASA, United States
MSU	Microwave Sounding Unit	NOAA-10	X	4	105 km	11	1986–2001 (out of service)	NOAA, United States
MSU	Microwave Sounding Unit	NOAA-11	X	4	105 km	11	1988–2004 (out of service)	NOAA, United States
MSU	Microwave Sounding Unit	NOAA-12	X	4	105 km	11	1991–2007 (out of service)	NOAA, United States
MSU	Microwave Sounding Unit	NOAA-14	X	4	105 km	11	1994–2007 (out of service)	NOAA, United States
SAPHIR	Microwave Sounding Instrument	Megha-Tropiques	X		1°	1–6 times/day	2011	CNES, France
SSMIS	Special sensor microwave imager/ sounder	DMSP-16	X	4	13–43 m		2003	USDD, United States
SSU	Stratospheric Sounder Unit	NOAA-11	X	3	147.3 km	11	1988–2004 (out of service)	NOAA, United States
SSU	Stratospheric Sounder Unit	NOAA-14	X	3	147.3 km	11	1997–2007 (out of service)	NOAA, United States

Source: Adapted and modified from ITC, ITC's database of satellites and sensors, 2014, http://www.itc.nl/research/products/sensordb/allsensors.aspx, accessed on May 2, 2014.

The obvious benefits of location-based information collected through GPS are immense and beyond the scope of this chapter to discuss. While these benefits are invaluable for applications involving field surveys, navigation, and basic GIS mapping of what features are located where, there are additional benefits that directly relate to remote sensing applications. Some of these include geometric rectification and registration of image data by way of ground control points (GCPs) collected through high-accuracy GPS units. Similarly, location information collected through ground-truthing surveys using high-accuracy GPS units plays an important role in image classification and accuracy assessment. Using the location information collected using

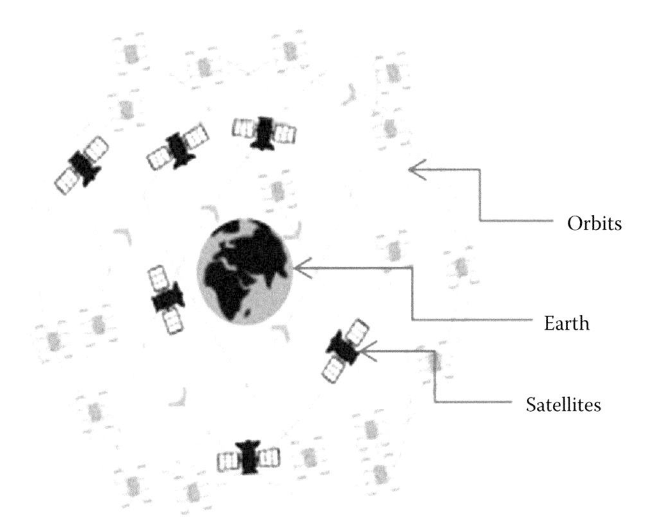

FIGURE 1.26 Global Positioning System space segment showing 24 satellites orbiting the Earth at an altitude of 20, 200 km in 6 orbits with 4 satellites in each. (From http://www.gps.gov/multimedia/images/constellation.jpg.)

GPS, better *training* sites could be used for algorithm *learning* phases during image classification procedures. Furthermore, as a postclassification routine, the field-collected GPS data are useful as a *reference* information about the landscape feature or biophysical phenomenon that could be used in the accuracy assessment exercise of comparing and validating outputs generated from image classification algorithms (Powell and Matzke, 2004; Congalton and Green, 2008).

1.4 Upcoming New Satellite Sensor Platforms

According to NSSDC Master Catalog (2014), since the beginning of 2014, 126 new satellites have been sent to space for Earth observation and all areas of application. Some of the prominent among them are as follows: (1) ALOS 2 is a RADAR imaging satellite system operated by the Japan Aerospace Exploration Agency (JAXA). The primary functions are for land resource studies, disaster monitoring, and environmental research. (2) GPM Core Observatory satellite is a core mission of NASA and JAXA to provide real-time accurate information of rain and snow every 3 h, along with soil moisture, carbon cycle, winds, and aerosol estimation. (3) Hodoyoshi 3 and 4 is an experimental Earth-observing microsatellite built by the University of Tokyo. (4) IRNSS 1B is a satellite-based navigation system developed in India to be compatible with GPS and Galileo (European navigation system). (5) Sentinel-1A is an ESA's two-satellite constellation with the prime objective of land and ocean monitoring with C-band SAR continuity. (6) SkySat 2 is a commercial Earth-observing satellite by Skybox Imaging to acquire very-high-resolution images. (7) TSAT is a 2U CubeSat microsatellite communication network from Taylor University, Upland, IN, used to research about a nanosat network to observe space weather with a plasma probe, three-axis magnetometer, and three UV photodiodes and is part of the NASA ELaNa 5 CubeSat initiative. (8) UKube is the U.K. Space Agency's pilot program with the miniature cube-shaped satellite that allows the United Kingdom to test cutting-edge new technologies in space. (9) Velox 1 is a small satellite from Singapore's Nanyang Technological University with a high-resolution camera for Earth imaging and later split into two smaller satellites—N-sat and P-sat—to conduct communications experiments. (10) WorldView-3 is a 0.5 m panchromatic and 1.8 m multispectral

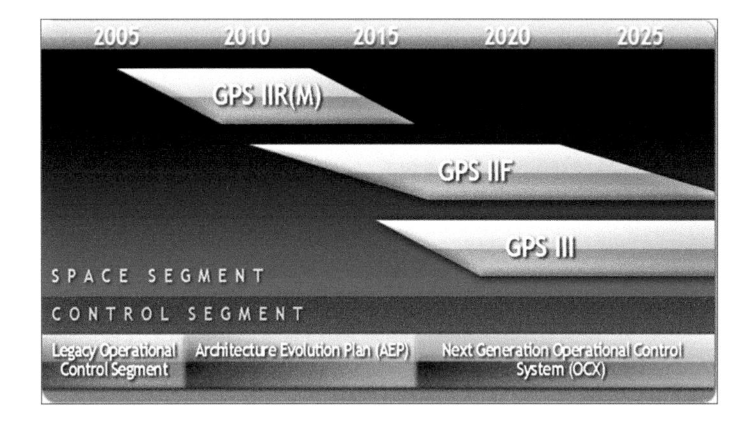

FIGURE 1.27 Global Positioning System modernization program schedule. The graphic provides links to the important information about developments in the space segment (http://www.gps.gov/systems/gps/space/#IIRM) and control segment (http://www.gps.gov/systems/gps/control/), including Architecture Evolution Plan (http://www.gps.gov/systems/gps/control/#AEP) and Operational Control System (http://www.gps.gov/systems/gps/control/#OCX). (From GPS.gov.)

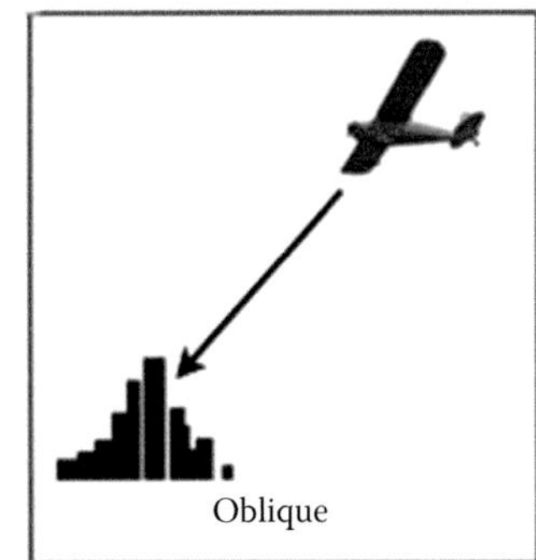

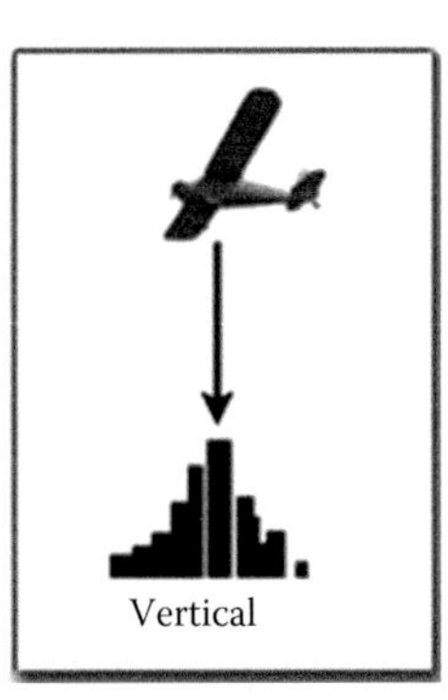

FIGURE 1.28 Examples of an oblique aerial image and graphics of the image acquisition processes.

resolution Earth observation imaging satellite by DigitalGlobe. 11. NASA's SMAP (Soil Moisture Active Passsive) mission is slated to provide global soil moisture measurement and soil freeze/thaw state using a combination of active radar and passive radiometer. From this present trend of latest satellite launches, it is presumed that the upcoming satellite launches will be with the research-related nanosatellite platforms, satellites doing dual duties after expiration of successful first job assigned to satellites, and satellites launched with very-high-resolution imaging capabilities.

Newer satellites from older programs like NASA, CNES, JAXA, and ESA are being programmed for future space launches with minimum two new improvements, that is, collecting data with larger swath so that revisit time of a satellite will be faster and including more bands into the system to help collect data for better and improved Earth observation. For example, according to Astrium Inc. (http://www.astrium-geo.com/en/147-spot-6-7-satellite-imagery), France's CNES is expanding SPOT imaging satellite series to 2024, and the latest series of SPOT 6 and SPOT 7 emphasize on collecting data by covering larger areas, that is, up to 6 million km^2 area/day, an area larger than the entire European Union. SPOT 6 and 7 satellites also include 4-day weather forecasting to the satellite's tasking. NASA's Landsat 8 satellite includes two new bands to help in analyzing more Earth resources like coastal resources with more cloud cover. Joint Polar Satellite System (JPSS) is the next-generation satellite system of the United States and a collaborative program between NOAA and its acquisition agent NASA, and its probable launch date is scheduled to be 2017. It will help weather and natural hazard forecasting at a newer level.

Oblique aerial photograph is the latest development in aerial imaging area, which has high potential growth in future years. Figure 1.28 shows an example of oblique aerial photograph. Oblique aerial photographs or a combination of oblique and vertical photographs, which are widely used for high-density urban land use mapping, are reviewed by Petrie (2009), and according to him, oblique aerial photography have been used since the 1960s, but its advantages in environmental management and other spatial analyses are observed more recently. More images will be acquired with oblique imaging platform in the next years.

1.5 Future of Remote Sensing and Evolving Microsatellites

According to a study by U.S.-based Forecast International, in order to address the need for better imaging data driven by aerospace and defense requirements, more satellites will be launched from high-resolution imagery acquiring platforms (Anonymous, 2012). For this purpose, UAS and UAV sensor platforms would serve the purpose. As discussed in Section 1.4, the present trends in satellite launches are from very high-resolution, research-oriented nanosatellite platforms; the UAS and UAV platforms would suffice the need. According to Everaerts (2008), the ISPRS Congress in Istanbul passed a resolution that UAVs provide a new controllable platform for remote sensing and permit data acquisition in inaccessible and dangerous environments. UAS or UAVs are the prominent part of the entire system of flying an aircraft and acquiring ultraspatial-resolution imagery. The aircraft is controlled from the ground control station with reliable communication network through air traffic control. Over the last four years, the number of UAV systems increased by leaps and bounds in the area of remote sensing and mapping (Everaerts, 2008). Everaerts (2008) also state that much of the work in the use of UAVs in remote sensing is in research stage presently and the future of remote sensing in essence depends on the growth of UAS- and UAV-based remote sensing. As in most countries aviation regulations are adapted to include UAS and UAV systems into the general airspace, these microsatellite systems will become the preferred platforms of future remote sensing. The following is an example of the utility of UAVs for Earth observation and analysis in most advanced manner.

One of the popular fully autonomous UAV eBee (Figure 1.29a) manufactured by a Swiss sensor manufacturer senseFly (senseFly

SA, Cheseaux-Lausanne, Switzerland) can acquire 1.5 cm spatial resolution aerial images (photos) that can be transformed into 2D orthomosaics and 3D models. The eBee can cover up to 12 km^2 (4.6 $miles^2$) in a single flight. When it flies over smaller areas at lower altitudes, it can acquire high resolution imagery at 1.5 cm/pixel with overlap. (senseFly, 2014). It uses a built-in 12-megapixel cameras, one for visible spectrum (VIS) (R-, G-, and B-bands) and another for NIR-band image acquisition to create aerial maps for supporting like wildlife, crops, and traffic management to name a few. Thermal infrared (TIR) cameras can be mounted in the UAV for TIR image acquisition. The intuitive eMotion software of the senseFly makes it easy to plan and simulate your mapping mission, which also generates a full flight path by calculating the UAV's required altitude and projected trajectory (senseFly, 2014). Sharma and Husley (2014) of University of North Georgia used eBee as a research and teaching tool. Figure 1.29b shows the simulated flight paths of the UAV created by Sharma and Husley (2014) to acquire aerial images at 1 inch spatial resolution for their study. On the site, it is launched by hand, after which it autonomously follows a flight path along with the simulated image acquisition schedule mapped out in advance by the user, via the included eMotion 2 software. Users can take control of the UAV at any point and pilot it remotely in real time (senseFly, 2014). Upon completion of its flight, it can land itself. The eBee website (https://www.sensefly.com/drones/ebee.html) contains other important specifications (weight, wingspan, material, propulsion, battery, camera, maximum flight time, nominal cruise speed, radio link range, maximum coverage, wind resistance, GSD, relative orthomosaic/3D model accuracy, absolute vertical/horizontal accuracy [with or without GCPs], multidrone operation, automatic 3D flight planning, linear landing accuracy, etc.) about the UAV that can be useful for readers to make decisions to own one and use for Earth observation and analysis in very advanced manner.

The eBee's acquired data are processed by senseFly's postflight Terra 3D software to produce 2D orthomosaics, 3D point clouds, triangle models, and DEMs. Figure 1.29c is the example of all the aerial photos acquired by Sharma and Husley (2014) over the university campus, and they mosaicked the acquired images using in-house software of the UAV to produce the 2D (Figure 1.29d), stereoscopic analysis–based 3D point cloud (Figure 1.29e) and subsequent highly accurate DEM. Sharma and Husley (2015) compared the DEM with the DEM created by LiDAR data acquired over the campus (Figure 1.29f). Triangulation between the points together with the onboard GPS camera helped in the image development with high accuracy (1.5–4.5 cm).

The future trend in remote sensing is to have full free-and-open data access, even for higher-resolution data (Belward and Skoien, 2014). NASA Landsat mission paved the way in providing free imagery to the public via the Internet (http://earthexplorer.usgs.gov/), and others with commercially driven programs such as Planet Labs and Skybox are following to make their data available

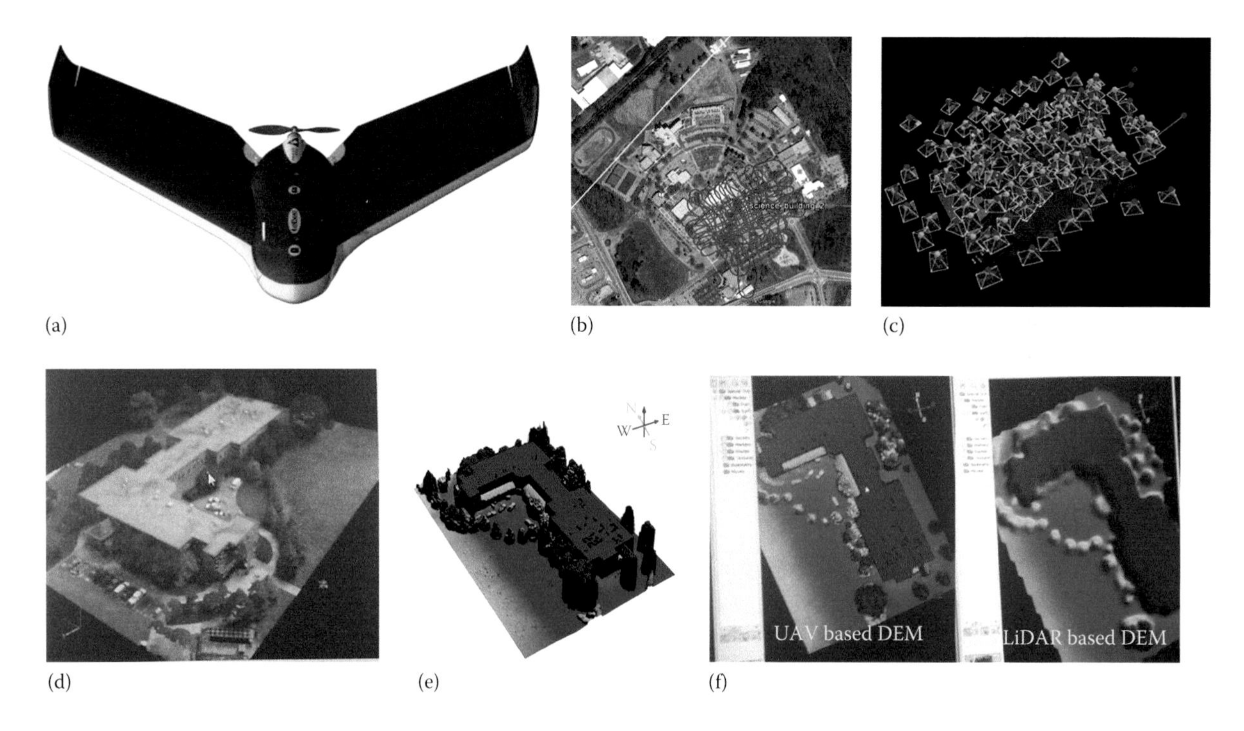

FIGURE 1.29 Example a unmanned aerial vehicle (UAV) (senseFly's eBee) used for ultrahigh-resolution orthophoto collection and ancillary derivative data development: (a) eBee UAV, (b) flight path simulated by the Institute of Environmental Spatial Analysis (IESA), (c) all acquired aerial orthophotos by the UAV over UNG Gainesville Campus, (d) mosaicked 2D orthophoto of the campus (partial), (e) stereoscopic analysis–based 3D point cloud, and (f) comparison of digital elevation models developed by imaging through UAV (eBee) and LiDAR technology. (From https://www.sensefly.com/drones/ebee.html and Sharma and Husley, 2014.

free to academics and nongovernment organizations, and more would follow in the future (Butler, 2014). This approach of free data sharing is a core element of the Group on Earth Observation's Global Earth Observing System of System's goal of data sharing and data management (Withee et al., 2004; GEO, 2012).

1.6 Discussions

In addition to the rich visual information in remote sensing data that can be directly interpreted to obtain the geographic information about landscape features (e.g., the boundary extent of forest cover), remote sensing data also provide additional biophysical information about the object (e.g., forest health, biomass). The EMR reflected, emitted, or backscattered from an object or geographic area is used as a *surrogate* for the actual property under investigation (Jensen, 2009). The EME measurements must be calibrated and turned into information using visual and/or digital image processing techniques. According to Belward and Skoien (2014), to meet the food, fuel, freshwater, and fiber requirements of the Earth's seven-plus billion humans, obtaining correct land cover information is vital and its importance is enhanced when the world is facing severe consequences from global warming and climate change in the form of land degradation such as desertification, wildfire, flooding, landslides, coastal erosion, and other hazards.

A major portion of our Earth-orbiting satellites provide a unique vantage point from which to map, measure, and monitor how, when, and where land resources are changing across the globe (Townshend et al., 2008). The importance of remote sensing has increased abundantly in the present century due to growing and shifting human population, and therefore, the resulting land use pattern has changed with urbanization and resource use for food, fiber, and fuel production (De Castro et al., 2013). Additionally, with growing impacts of climate change, newer approaches to resource conservation aimed at sustainability such as biodiversity offsetting (McKenney and Kiesecker, 2010) and carbon markets (Mollicone et al., 2007) are increasingly using remote sensing data to study and analyze the spatial distribution of land use. Land cover mapping using remote sensing can be implemented at different scales ranging from global (Loveland et al., 1999; Friedl et al., 2002; Bartholome and Belward, 2005; Arino et al., 2007; Gong et al., 2013) to more regional and localized levels (Cihlar, 2000; Panda et al., 2010a,b).

To better plan and balance resource management issues, global land cover mapping with remote sensing supports decision makers and policy analysts at national and international scales (Belward and Skoien, 2014). The success of land use management plans depends to a large extent on reliable information concerning how, when, and where about resource use and its changing pattern. A great deal of the reliable information leading to land use characterization for ecological habitat assessment and planning can be accomplished using geospatial technologies such as remote sensing and GIS (Osborne et al., 2001; Rao et al., 2005; Rao et al., 2007; Valavanis et al., 2008; Vierling et al., 2008).

1.7 Summary and Conclusion

This chapter provides an exhaustive overview of the remote sensing satellites, sensors, and their characteristics. *First*, the chapter provides definition and an understanding of remote sensing from best known sources. This common understanding is required for any student and practitioner of remote sensing. *Second*, the principles of EMS are enumerated. *Third*, the spatial, spectral, radiometric, and temporal resolution are explained. *Fourth*, over 7000 satellites have been launched and operated by various governments and private enterprises from the mid-1950s to the present. The characteristics of many of the important satellite and sensor systems have been described. *Fifth*, remote sensing is both active and passive and is gathered in various EMSs such as the visible, NIR, shortwave infrared, mid-infrared (MIR), TIR, microwave, and radio waves (high-frequency waves). Radar, SAR, LiDAR, SONAR, and SODAR technologies operate with microwave and radio wave region of the EMS, while optical and hyperspectral imaging operates with visible to TIR region of EMS. The importance of gathering these data across such wide range of EMS has been highlighted. *Sixth*, remote sensing data gathering is done in various modes such as nadir, off nadir, hyperspatial, hyperspectral, and multispectral. All of these have various implications on what we study and the level of these accuracies as a result of such acquisitions. This aspect has been implied throughout the chapter. *Seventh*, remote sensing data are gathered in various platforms: ground based, platform mounted, airborne, spaceborne, undersea, and UAV. The chapter focuses on spaceborne but does discuss several other platforms. Reading through these sections, it becomes clear that remote sensing has truly evolved in both sensor design and acquisition platforms. *Eighth*, the chapter provides a window into what to expect in near future through upcoming newer satellites and sensors. *Ninth*, the evolution of microsatellites has been highlighted. Like miniaturization of computing technology, satellites and sensors are undergoing revolutionary technological innovations, which make them provide smaller, cheaper, and yet better-quality data. *Tenth*, the chapter captures in a nutshell key developments in remote sensing from 1858 to the present and provides a glimpse on where the future of remote sensing satellites, sensors, and data are headed. Finally, the authors recommend readers to watch the Public Broadcasting Station (PBS) NOVA documentary, "Earth from Space" (http://www.pbs.org/wgbh/nova/earth/earth-from-space.html) to understand the amazing ability of remote sensing satellites to monitor, measure, and analyze the Earth's (oceans, land, and atmosphere) resources for prudent decision support to ensure a sustainable life on it.

References

3Dlasermapping. 2014. Volume measuring. http://3dlasermapping.com/index.php/mining-monitoring-applications/volume-measuring. Accessed on July 2, 2014.

Aggarwal, S. 2003. Principles of remote sensing. In: *Satellite Remote Sensing and GIS Applications in Agricultural Meteorology*, p. 23.

Amatya, D., Panda, S. S., Cheschair, G., Nettles, J., Appleboom, T., and Skaggs, W. 2011. Evaluating evapotranspiration and stomatal conductance of matured pine using geospatial technology. In: *American Geophysical Union Conference 2011*, San Fransisco, CA, December 5–9.

Amatya, D., Trettin, C., Panda, S. S., and Ssegane, H. 2013. Application of LiDAR data for hydrologic assessments of low-gradient coastal watershed drainage characteristics. *Journal of Geographic Information System*, 5(2), 175–191. doi: 10.4236/jgis.2013.52017.

American Society of Photogrammetry and Remote Sensing. 1952, 1966. *Manuals of Photogrammetry*. Bethesda, MD: ASPRS.

Amzajerdian, F., Pierrottet, D., Petway, L., Hines, G., and Roback, V. 2011. LiDAR systems for precision navigation and safe landing on planetary bodies. In: *International Symposium on Photoelectronic Detection and Imaging 2011*, Beijing, China. International Society for Optics and Photonics, p. 819202.

Anonymous. 2012. Imaging to drive remote sensing satellite market. http://optics.org/news/3/5/43. Accessed on September 1, 2014.

Arino, O., Gross, D., Ranera, F., Bourg, L., Leroy, M., Bicheron, P., Latham, J. et al. 2007. GlobCover: ESA service for global land cover from MERIS. In: *Proceedings of the IEEE International Geoscience and Remote Sensing Symposium, 2007 (IGARSS'07)*, Barcelona, Spain, pp. 2412–2415. http://dx.doi.org/10.1109/IGARSS.2007.4423328.

Aronoff, S. 2004. *Remote Sensing for GIS Managers*. Redlands, CA: ESRI Press.

Asner, G. P., Boardman, J., Field, C. B., Knapp, D. E., Kennedy-Bowdoin, T., Jones, M. O., and Martin, R. E. 2007. Carnegie airborne observatory: In-flight fusion of hyperspectral imaging and waveform light detection and ranging for three-dimensional studies of ecosystems. *Journal of Applied Remote Sensing*, 1(1), 013536–013536.

Asner, G. P., Powell, G. V., Mascaro, J., Knapp, D. E., Clark, J. K., Jacobson, J., Kennedy-Bowdoin, T. et al. 2010. High-resolution forest carbon stocks and emissions in the Amazon. *Proceedings of the National Academy of Sciences*, 107(38), 16738–16742.

Avery, T. E. and Berlin, G. L. 1992. *Fundamentals of Remote Sensing and Airphoto Interpretation*. Upper Saddle River, NJ: Prentice Hall.

Bailey, D. T., 2000. *Meteorological Monitoring Guidance for Regulatory Modeling Applications*. Environmental Protection Agency, Office of Air Quality Planning. EPA document EPA-454/R-99-005.

Baker, J. C., O'Connell, K. M., and Williamson, R., eds. 2001. *Commercial Observation Satellites*. Rand Corporation, Santa Monica, CA, p. 668.

Bartholome, E. M. and Belward, A. S. 2005. GLC.2000. A new approach to global land cover mapping from earth observation data. *International Journal of Remote Sensing*, 26(9), 1959–1977.

Belcher, E. O., Matsuyama, B., and Trimble, G. M. 2001. Object identification with acoustic lenses. In: *Proceedings of Oceans 2001 Conference*, November 5–8, Honolulu, HI: Marine Technical Society/IEEE, pp. 6–11.

Belward, A. S. 2012. Europe's relations with the wider world—A unique view from space. In: Venet, C. and Baranes, B., eds. *European Identity through Space; Space Activities and Programmes as a Tool to Reinvigorate the European Identity*. Vienna, Austria: Springer, p. 318.

Belward, A. S. and Skoien, J. O. 2014. Who launched what, when and why; trends in global land-cover observation capacity from civilian earth observation satellites. *ISPRS Journal of Photogrammetry and Remote Sensing*.

Beyrich, F. 1997. Mixing height estimation from SODAR data—A critical discussion. *Atmospheric Environment*, 31(23), 3941–3953.

Blackinton, J. G., Hussong, D. M., and Kosalos, J. G. 1983. First results from a combination side-scan sonar and seafloor mapping system (SeaMARC II). In: *Offshore Technology Conference*, Huston, TX.

Blume, H. J. C., Kendall, B. M., and Fedors, J. C. 1978. Measurement of ocean temperature and salinity via microwave radiometry. *Boundary-Layer Meteorology*, 13(1–4), 295–308.

Bowker, D. E., Davis, R. E., Myrick, D. L., Stacy, K., and Jones, W. T. 1985. Spectral reflectances of natural targets for use in remote sensing studies. NASA reference publication # 1139, Hampton, VA.

Brenner, A. C., DiMarzio, J. P., and Zwally, H. J. 2007. Precision and accuracy of satellite radar and laser altimeter data over the continental ice sheets. *IEEE Transactions on Geoscience and Remote Sensing*, 45(2), 321–331.

Briney, A. 2014. An overview of remote sensing. http://geography.about.com/od/geographictechnology/a/remotesensing.htm. Accessed on May 2, 2014.

Brisco, B., Brown, R. J., and Manore, M. J. 1989. Early season crop discrimination with combined SAR and TM data. *Canadian Journal of Remote Sensing*, 15(1): 44–54.

Brugioni, D. A. and Doyle, F. J. 1997. Arthur C. Lundahl: Founder of the image exploitation discipline. In: *Corona between the Sun and the Earth: The First NRO Reconnaissance Eye in Space*, Bethesda, MD: American Society for Photogrammetry and Remote Sensing. pp. 159–166.

Bulstrode, C. J. K., Goode, A. W., and Scott, P. J. 1986. Stereophotogrammetry for measuring rates of cutaneous healing: A comparison with conventional techniques. *Clinical Science*, 71(4), 437–443.

Butler, D. 2014. Many eyes on earth. *Nature*, 505, 143–144.

Butler, L. R. P. and Laqua, K. 1996. Nomenclature, symbols, units and their usage in spectrochemical analysis—IX. Instrumentation for the spectral dispersion and isolation of optical radiation. *Spectrochimica Acta Part B: Atomic Spectroscopy*, 51(7), 645–664.

Canadian Center for Remote Sensing (CCRS). 2014. Fundamentals of remote sensing. https://www.nrcan.gc.ca/sites/www.nrcan.gc.ca/files/earthsciences/pdf/resource/tutor/fundam/pdf/fundamentals_e.pdf. Accessed on July 5, 2014.

Cash, M. and Panda, S. S. 2012. Urban stream water quality control with green island provision. In: Presented in *Southeast Lake and Watershed Management Conference*, May 13–15, 2012, Columbus, GA.

Center for Remote Imaging, Sensing, and Processing (CRISP). 2014. Electromagnetic waves. http://www.crisp.nus.edu.sg/~research/tutorial/em.htm. Accessed on May 2, 2014.

Chang, C. I. (Ed.). 2003. *Hyperspectral imaging: Techniques for spectral detection and classification* (Vol. 1). Springer Science & Business Media: New York, NY.

Chen, C. M., Hepner, G. F., and Forster, R. R. 2003. Fusion of hyperspectral and radar data using the IHS transformation to enhance urban surface features. *ISPRS Journal of Photogrammetry and Remote Sensing*, 58(1), 19–30.

Cihlar, J. 2000. Land cover mapping of large areas from satellites: Status and research priorities. *International Journal of Remote Sensing*, 21 (6–7), 1093–1114.

Clark, R. N., King, T. V. V., Klejwa, M., and Swayze, G. A. 1990. High spectral resolution spectroscopy of minerals. *Journal of Geophysical Research*, 95(B8), 12653–12680.

Clark, R. N., Swayze, G. A., Livo, K. E., Kokaly, R. F., King, T. V. V., Dalton, J. B., Vance, J. S., Rockwell, B. W., Hoefen, T., and McDougal, R. R. 2002. Surface reflectance calibration of terrestrial imaging spectroscopy data: A tutorial using AVIRIS. In: *Proceedings of the 10th Airborne Earth Science Workshop*, Pasadena, CA: JPL Publication 02-1.

Cleve, C., Kelly, M., Kearns, F. R., and Moritz, M. 2008. Classification of the wildland–urban interface: A comparison of pixel-and object-based classifications using high-resolution aerial photography. *Computers, Environment and Urban Systems*, 32(4), 317–326.

Colwell, R. N., Ulaby, F. T., Simonett, D. S., Estes, J. E., and Thorley, G. A. 1983. *Manual of Remote Sensing. Interpretation and Applications*, Vol. 2. American Society of Photogrammetry, Falls Church, VA. 2143pp.

Congalton, R. G. and Green, K. 2008. *Assessing the Accuracy of Remotely Sensed Data: Principles and Practices*. Boca Raton, FL: CRC Press.

Connor, L. N., Laxon, S. W., Ridout, A. L., Krabill, W. B., and McAdoo, D. C. 2009. Comparison of ENVISAT radar and airborne laser altimeter measurements over Arctic sea ice. *Remote Sensing of Environment*, 113(3), 563–570.

D'urso, G. and Minacapilli, M. 2006. A semi-empirical approach for surface soil water content estimation from radar data without a-priori information on surface roughness. *Journal of Hydrology*, 321(1), 297–310.

De Castro, P., Adinolfi, F., Capitanio, F., Di Falco, S., Di Mambro, A., eds. 2013. *The Politics of Land and Food Scarcity*. Abingdon, U.K.: Routledge, p. 154.

Defibaugh, Y., Chávez, J., and Tullis, J. A. 2013. Deciduous forest structure estimated with LiDAR-optimized spectral remote sensing. *Remote Sensing*, 5(1), 155–182.

Dobson, M. C. and Ulaby, F. T. 1998. Mapping soil moisture distribution with imaging radar. In: *Principles and Applications of Imaging Radar. Manual of Remote Sensing*, Vol. 2, New York, NY: John Wiley, pp. 407–433.

Dong, Y., Forster, B., and Ticehurst, C. 1997. Radar backscatter analysis for urban environments. *International Journal of Remote Sensing*, 18(6), 1351–1364.

Dornaika, F. and Hammoudi, K. 2009. Extracting 3D polyhedral building models from aerial images using a featureless and direct approach. In: *MVA*, IAPR Conference on Machine Vision Applications, May 20–22, 2009, Yokohama, JAPAN, pp. 378–381.

Dubayah, R. O., Knox, R. G., Hofton, M. A., Blair, J. B., and Drake, J. B. 2000. Land surface characterization using lidar remote sensing. In: M. Hill, and R. Aspinall (Eds.), *Spatial information for land use management*, Singapore: International Publishers Direct, pp. 25–38.

Earth Observing System Data and Information System (EOSDIS). 2014. Earth Data. https://earthdata.nasa.gov/. Accessed on July 5, 2014.

Edson, C. and Wing, M. G. 2011. Airborne light detection and ranging (LiDAR) for individual tree stem location, height, and biomass measurements. *Remote Sensing*, 3, 2494–2528. doi: 10.3390/rs3112494.

Engman, E. T. and Chauhan, N. 1995. Status of microwave soil moisture measurements with remote sensing. *Remote Sensing of Environment*, 51(1), 189–198.

Erdody, T. L. and Moskal, L. M. 2010. Fusion of LiDAR and imagery for estimating forest canopy fuels. *Remote Sensing of Environment*, 114, 725–737.

Ertberger, C., Jaume, A., and Panda, S. S. 2012. Estimation and evaluation of bacterial loadings of the Upper Chattahoochee watershed. In: Presented in *Southeast Lake and Watershed Management Conference*, May 13–15, 2012, Columbus, GA.

ESRI. 2014. GIS Dictionary. http://support.esri.com/en/knowledgebase/GISDictionary/term/remote%20sensing. Accessed on July 5, 2014.

Everaerts, J. 2008. The use of unmanned aerial vehicles (UAVs) for remote sensing and mapping. *The International Archives of the Photogrammetry, Remote Sensing and Spatial Information Sciences*, 37, 1187–1192.

Fairfield, N. and Wettergreen, D. 2008. Active localization on the ocean floor with multibeam sonar. In: *Proceedings of the IEEE/MTS OCEANS Conference and Exhibition*, pp. 1–10.

Friedl, M. A., McIver, D. K., Hodges, J. C. F., Zhang, X. Y., Muchoney, D., Strahler, A. H., Woodcock, C. E et al. 2002. Global land cover mapping from MODIS: Algorithms and early results. *Remote Sensing of Environ*ment, 83(1), 287–302.

Fung, A. K. and Chen, K. S. 2010. *Microwave Scattering and Emission Models for Users*. Artech House, Norwood, MA.

García, M., Riaño, D., Chuvieco, E., Salas, J., and Danson, F. M. 2011. Multispectral and LiDAR data fusion for fuel type mapping using Support Vector Machine and decision rules. *Remote Sensing of Environment*, 115(6), 1369–1379.

GEO. 2012. *The Group on Earth Observation 2012–2015 Work Plan*, GEO Secretariat. Geneva, Switzerland: WMO, p. 79. http://www.earthobservations.org/. Accessed on July 2, 2014.

Gernsheim, H. and Gernsheim, A. 1952. Re-discovery of the world's first photograph. *Photographic Journal*, 118.

Giles, K. A., Laxon, S. W., Wingham, D. J., Wallis, D. W., Krabill, W. B., Leuschen, C. J., McAdoo, D. et al. 2007. Combined airborne laser and radar altimeter measurements over the Fram Strait in May 2002. *Remote Sensing of Environment*, 111(2), 182–194.

Glenn, N. F. and Carr, J. R. 2004. The effects of soil moisture on synthetic aperture radar delineation of geomorphic surfaces in the Great Basin, Nevada, USA. *Journal of Arid Environments*, 56(4), 643–657.

Goel, M. and Srivastava, H. N. 1990. Monsoon trough boundary layer experiment (MONTBLEX). *Bulletin of the American Meteorological Society*, 71(11), 1594–1600.

Goetz, A. F. 2009. Three decades of hyperspectral remote sensing of the Earth: A personal view. *Remote Sensing of Environment*, 113, S5–S16.

Goetz, S. and Dubayah, R. 2011. Advances in remote sensing technology and implications for measuring and monitoring forest carbon stocks and change. *Carbon Management*, 2(3), 231–244.

Gong, P., Wang, J., Yu, L., Zhao, Y., Zhao, Y., Liang, L., Niu, Z. et al. 2013. Finer resolution observation and monitoring of global land cover: First mapping results with Landsat TM and ETM+ data. *International Journal of Remote Sensing*, 34(7), 2607–2654.

Gonzalez, R. C. and Woods, R. E. 2002. *Digital Image Processing*. Delhi, India: Pearson Education (Singapore) Ltd.

Gowen, A. A., O'Donnell, C., Cullen, P. J., Downey, G., and Frias, J. M. 2007. Hyperspectral imaging–an emerging process analytical tool for food quality and safety control. *Trends in Food Science and Technology*, 18(12), 590–598.

Goyer, G. G. and Watson, R. 1963. The laser and its application to meteorology. *Bulletin of the American Meteorological Society*, 44(9), 564.

Guthrie, J. and Walsh, K. 1997. Non-invasive assessment of pineapple and mango fruit quality using near infra-red spectroscopy. *Animal Production Science*, 37(2), 253–263.

Haack, B. and Bechdol, M. 2000. Integrating multisensor data and RADAR texture measures for land cover mapping. *Computers and Geosciences*, 26(4), 411–421.

Hale, J. D., Tamblyn, C., Cash, M., and Panda, S. S. 2012. Fecal coliform and stream health analysis of Flat Creek with geospatial model development. In: Presented in *Southeast Lake and Watershed Management Conference*, May 13–15, 2012, Columbus, GA.

Hall, D. K., Fagre, D. B., Klasner, F., Linebaugh, G., and Liston, G. E. 2012. Analysis of ERS 1 synthetic aperture radar data of frozen lakes in northern Montana and implications for climate studies. *Journal of Geophysical Research: Oceans (1978–2012)*, 99(C11), 22473–22482.

Hall, S. A., Burke, I. C., Box, D. O., Kaufmann, M. R., and Stoker, J. M. 2005. Estimating stand structure using discrete-return LiDAR: An example from low density, fire prone ponderosa pine forests. *Forest Ecology and Management*, 208, 189–209.

Hodgson, M. E., Jensen, J. R., Tullis, J. A., Riordan, K. D., and Archer, C. M. 2003. Synergistic use of LiDAR and color aerial photography for mapping urban parcel imperviousness. *Photogrammetric Engineering and Remote Sensing*, 69(9), 973–980.

Huisman, J. A., Hubbard, S. S., Redman, J. D., and Annan, A. P. 2003. Measuring soil water content with ground penetrating radar. *Vadose Zone Journal*, 2(4), 476–491.

ITC. 2014. ITC's database of satellites and sensors. http://www.itc.nl/research/products/sensordb/allsensors.aspx. Accessed on May 2, 2014.

Jensen, J. R. 2009. *Remote Sensing of the Environment: An Earth Resource Perspective 2/e*. Pearson Education India, Delhi, India, pp. 11–12.

Jha, M. K., Chowdhury, A., Chowdary, V. M., and Peiffer, S. 2007. Groundwater management and development by integrated remote sensing and geographic information systems: Prospects and constraints. *Water Resources Management*, 21(2), 427–467.

Jhang, J., Panigrahi, S., Panda, S. S., and Borhan, M. S. 2002. Techniques for yield prediction from corn aerial images—A neural network approach. *International Journal of Agricultural and Biosystems Engineering*, 3(1), 18–28.

Kasischke, E. S., Melack, J. M., and Craig Dobson, M. 1997. The use of imaging radars for ecological applications—A review. *Remote Sensing of Environment*, 59(2), 141–156.

Kelly, B., Panda, S., Trettin, C., and Amatya, D. 2014. Assessment of the reach and ecological condition of freshwater tidal creeks in the lower coastal plain, Charleston County, South Carolina with advanced geospatial technology application. In: Poster Presented in *South Carolina Water Resources Conference*, October 15–16, 2014, Columbia, SC.

Kidd, R. B., Simm, R. W., and Searle, R. C. 1985. Sonar acoustic facies and sediment distribution on an area of the deep ocean floor. *Marine and Petroleum Geology*, 2(3), 210–221.

Kim, M. S., Chen, Y. R., and Mehl, P. M. 2001. Hyperspectral reflectance and fluorescence imaging system for food quality and safety. *Transactions of the American Society of Agricultural Engineers*, 44(3), 721–730.

Klemas, V. 2011. Remote sensing of coastal plumes and ocean fronts: Overview and case study. *Journal of Coastal Research*, 28(1A), 1–7.

Kokaly, R. F., Clark, R. N., and Livo, K. E. 1998. Mapping the biology and mineralogy of Yellowstone National Park using imaging spectroscopy. In: *JPL Airborne Earth Science Workshop*, Vol. 7, Pasadena, CA, pp. 97–21.

Kokaly, R. F., Despain, D. G., Clark, R. N., and Livo, K. E. 2003. Mapping vegetation in Yellowstone National Park using spectral feature analysis of AVIRIS data. *Remote Sensing of Environment*, 84(3), 437–456.

Kruse, F. A., Boardman, J. W., and Huntington, J. F. 2003. Comparison of airborne hyperspectral data and EO-1 Hyperion for mineral mapping. *IEEE Transactions on Geoscience and Remote Sensing*, 41(6), 1388–1400.

Leckie, D. G. and Ranson, K. J. 1998. Forestry applications using imaging radar. In: *Principles and Applications of Imaging Radar*, John Wiley, New York, NY, Vol. 2, pp. 435–509.

Lee, A. C. and Lucas, R. M. 2007. A LiDAR-derived canopy density model for tree stem and crown mapping in Australian forests. *Remote Sensing of Environment*, 111, 493–518. doi: 10.1016/j.rse.2007.04.018.

Lees, B. G. and Ritman, K. 1991. Decision-tree and rule-induction approach to integration of remotely sensed and GIS data in mapping vegetation in disturbed or hilly environments. *Environmental Management*, 15(6), 823–831.

Lefsky, M. A., Cohen, W. B., Parker, G. G., and Harding, D. J. 2002. LiDAR Remote Sensing for Ecosystem Studies LiDAR, an emerging remote sensing technology that directly measures the three-dimensional distribution of plant canopies, can accurately estimate vegetation structural attributes and should be of particular interest to forest, landscape, and global ecologists. *BioScience*, 52(1), 19–30.

Lewis, A. J. and Henderson, F. M. 1998. Geomorphic and hydrologic applications of active microwave remote sensing. In: *Principles and Applications of Imaging Radar. Manual of Remote Sensing*, John Wiley, New York, NY, pp. 567–629.

Lillesand, T. M., Kiefer, R. W., and Chipman, J. W. 2004. *Remote Sensing and Image Interpretation*, 5th edn. John Wiley & Sons Ltd., New York, NY.

Loveland, T. R., Zhu, Z., Ohlen, D. O., Brown, J. F., Reed, B. C., and Yang, L. 1999. An analysis of the IGBP global land-cover characterization process. *Photogrammetric Engineering & Remote Sensing*, 65(9), 1021–1032.

Lu, R. 2003. Detection of bruises on apples using near-infrared hyperspectral imaging. *Transactions of the American Society of Agricultural Engineers*, 46(2), 523–530.

Martin, G. 2012. Sumbandilasat beyond Repair, African Defense and Security News Portal Defence web, January 25, 2012. http://www.defenceweb.co.za/. Accessed May 3, 2014.

McGlone, J. C. 2004. *Manual of Photogrammetry*, 5th edn. Bethesda, MD: ASPRS.

McKenney, B. A. and Kiesecker, J. M. 2010. Policy development for biodiversity offsets: A review of offset frameworks. *Environmental Management*, 45(1), 165–176.

Melesse, A. M., Weng, Q., Thenkabail, P., and Senay, G. 2007. Remote Sensing Sensors and Applications in Environmental Resources Mapping and Modelling. *Special Issue of Remote Sensing of Natural Resources and the Environment. Sensors Journal*, 7, 3209–3241.

Metternicht, G. I. and Zinck, J. A. 2003. Remote sensing of soil salinity: Potentials and constraints. *Remote Sensing of Environment*, 85(1), 1–20.

Mollicone, D., Achard, F., Federici, S., Eva, H. D., Grassi, G., Belward, A., Raes, F. et al. 2007. An incentive mechanism for reducing emissions from conversion of intact and non-intact forests. *Climatic Change*, 83(4), 477–493.

Montenbruck, O. and Gill, E. 2000. *Satellite Orbits*. Springer New York, NY.

Morgan, J. L., Gergel, S. E., and Coops, N. C. 2010. Aerial photography: A rapidly evolving tool for ecological management. *BioScience*, 60(1), 47–59.

Moursund, R. A., Carlson, T. J., and Peters, R. D. 2003. A fisheries application of a dual-frequency, identification sonar, acoustic camera. *ICES Journal of Marine Science*, 60, 678–683.

Murai, S. 1993. *Remote Sensing Note*. Tokyo, Japan: JARS.

NASA-Phoenix Mars Lander. 2014. http://www.nasa.gov/mission_pages/phoenix/news/phoenix-20080929.html. Accessed on July 2, 2014.

Nicolaï, B. M., Beullens, K., Bobelyn, E., Peirs, A., Saeys, W., Theron, K. I., and Lammertyn, J. 2007. Nondestructive measurement of fruit and vegetable quality by means of NIR spectroscopy: A review. *Postharvest Biology and Technology*, 46(2), 99–118.

NSSDC. 2014. The NASA Master Directory held at the NASA Space Science Data Center. http://nssdc.gsfc.nasa.gov/nmc/SpacecraftQuery.jsp. Accessed on May 2, 2014.

Osborne, P. E., Alonso, J. C., and Bryant, R. G. 2001. Modelling landscape-scale habitat use using GIS and remote sensing: A case study with great bustards. *Journal of Applied Ecology*, 38(2), 458–471.

Panda, S. S., Ames, D. P., and Panigrahi, S. 2010a. Application of vegetation indices for agricultural crop yield prediction using neural network. *Remote Sensing*, 2(3), 673–696.

Panda, S. S., Andrianasolo, H., Murty, V. V. N., and Nualchawee, K. 2004. Forest management planning for soil conservation using satellite images, GIS mapping, and soil erosion modeling. *Journal of Environmental Hydrology*, 12(13), 1–16.

Panda, S. S., Burry, K., and Tamblyn, C. 2012a. Wetland change and cause recognition in Georgia coastal plain. In: Presented in the *International American Society of Agricultural and Biological Engineers (ASABE) Conference 2012*, July 29–August 2, 2012, Dallas, TX. Paper # 1338205.

Panda, S. S. and Hoogenboom, G. 2013. Blueberry orchard site specific crop management with geospatial based yield modeling. In: Presented in the *International American Society of Agricultural and Biological Engineers (ASABE) Conference 2012*, July 21–24, 2013, Kansas City, MO. Paper # 1620890.

Panda, S. S., Hoogenboom, G., and Paz, J. 2009. Distinguishing blueberry bushes from mixed vegetation land-use using high resolution satellite imagery and geospatial techniques. *Computers and Electronics in Agriculture*, 67(1–2), 51–59.

Panda, S. S., Hoogenboom, G., and Paz, J. 2010b. Remote sensing and geospatial technological applications for site-specific management of fruit and nut crops: A review. *Remote Sensing*, 2(8), 1973–1997.

Panda, S. S., Martin, J., and Hoogenboom, G. 2011a. Blueberry crop growth analysis using climatologic factors and multi-temporal remotely sensed imageries. In: Carroll, D., ed. Published in the Peer Reviewed *Proceedings of 2011 Georgia Water Resources Conference*, Athens, GA, April 11–13, 2011. http://www.gawrc.org/2011proceedings.html.

Panda, S. S., Nolan, J., Amatya, D., Dalton, K., Jackson, R. M., Ssegane, H., and Chescheir, G. 2012b. Stomatal conductance and leaf area index estimation using remotely sensed information and forest speciation. In: Presented in the *Third International Conference on Forests and Water in Changing Environment 2012*, September 18–20, 2012, Fukuoka, Japan.

Panda, S. S., Steele, D. D., Panigrahi, S., and Ames, D. P. 2011b. Precision water management in corn using automated crop yield modeling and remotely sensed data. *International Journal of Remote Sensing Applications*, 1(1), 11–21.

Panigrahi, S., Gautam, R., Gu, H., Panda, S. S., Venugopal, M., and Kizil, U. 2003. Fluorescence imaging for quality assessment of meat. ASAE Paper No. RRV03-0025, St. Joseph, MI.

Petrie, G. 2009. Systematic oblique aerial photography using multiple digital cameras. *Photogrammetric Engineering & Remote Sensing*, 75(2), 102–107.

Phillips, J., Tamblyn, C., Smith, A., and Panda, S. S. 2012. Impact of urbanization and point source on changes in water quality in upstream of Upper Chattahoochee River. In: Presented in *Southeast Lake and Watershed Management Conference*, May 13–15, 2012, Columbus, GA.

Powell, R. L. and Matzke, N. 2004. Sources of error in accuracy assessment of thematic land-cover maps in the Brazilian Amazon. *Remote Sensing of Environment*, 90(2), 221–234.

Rao, M., Awawdeh, M., and Dicks, M. 2005. Spatial allocation and environmental benefits: The impacts of the conservation reserve program in Texas County Oklahoma. In: Allen, A. W. and Vandever, M. W., eds. *The Conservation Reserve Program—Planting for the Future: Proceedings of a National Conference*, June 8–9, 2004, Fort Collins, CO: U.S. Geological Survey, Biological Resources Discipline, Scientific Investigations Report 2005-5145, pp. 174–182.

Rao, M. and Miller, E. 2012. Mapping of serpentine soils in Lassen and Plumas National Forests. *CSU Geospatial Review*, 10, 6.

Rao, M. N. 2012. Mapping snag locations in the blacks mountain experimental forest using LiDAR data. Technical Report Submitted to USFS-PSW Research Station, CA, April 19, 2012.

Rao, M. N. 2013. Mapping of serpentine soils in the Lassen and Plumas National Forest using integrated multispectral and LiDAR data. Technical Report Submitted to USFS-Mt. Hough Ranger District, CA, March 13, 2013.

Rao, M. N., Fan, G., Thomas, J., Cherian, G., Chudiwale, V., and Awawdeh, M. 2007. A web-based GIS decision support system for managing and planning USDA's Conservation Reserve Program (CRP). *Environmental Modelling and Software*, 22, 1270–1280.

Renslow, M.S. 2012. Manual of Airborne Topographic LiDAR. ASPRS publication. ISBN 1-57083-097-5 Stock # 4587.

Riaño, D., Chuvieco, E., Condes, S., Gonzalez-Matesanz, J., and Ustin, S. L. 2004. Generation of crown bulk density for *Pinus sylvestris* L. from LiDAR. *Remote Sensing of Environment*, 92, 345–352.

Riaño, D., Chuvieco, E., Ustin, S. L., Salas, J., Rodríguez-Pérez, J. R., Ribeiro, L. M., Viegas, D. X., Moreno, J. M., and Helena, F. 2007. Estimation of shrub height for fuel-type mapping combining airborne LiDAR and simultaneous color infrared ortho imaging. *International Journal of Wildland Fire*, 16(3), 341–348.

Riopel, S. 2000. *The Use of RADARSAT-1 Imagery for Lithological and Structural Mapping in the Canadian High Arctic*. Ottawa, Ontario, Canada: University of Ottawa.

Ritchie, J. C., Zimba, P. V., and Everitt, J. H. 2003. Remote sensing techniques to assess water quality. *Photogrammetric Engineering and Remote Sensing*, 69(6), 695–704.

Russo, J. C., Amduka, M., Gelfand, B., Pedersen, K., Lethin, R., and Springer, J. 2006. Enabling cognitive architectures for UAV mission planning. http://www.atl.external.lmco.com/papers/1396.pdf. Accessed on July 5, 2014.

Rylee, J., Panda, S. S., Fitzgerald, J., and Hohnhorst, D. 2012. Geospatial technology based suitability analysis for new additional reservoirs in Hall County, GA. In: Presented in *Southeast Lake and Watershed Management Conference*, May 13–15, 2012, Columbus, GA.

Savenije, B., Geesink, G. H., Van der Palen, J. G. P., and Hemke, G. 2006. Prediction of pork quality using visible/near-infrared reflectance spectroscopy. *Meat Science*, 73(1), 181–184.

Schott, J. R. 2007. *Remote Sensing: The Image Chain Approach*, 2nd edn. Oxford University Press, New York, NY, p. 1.

Schowengerdt, R. A. 2007. *Remote Sensing: Models and Methods for Image Processing*, 3rd edn. Academic Press, Chicago, IL, p. 2.

senseFly. 2014. eBee: The professional mapping drone. https://www.sensefly.com/drones/ebee.html. Accessed on October 10, 2014.

Sharma, J. B. and Hushley, D. 2014. Integrating the UAS in Undergraduate Teaching and Research-Opportunities and Challenges at the University of North Georgia. Proceedings of the Joint ASPRS Pecora 2014 Conference and the ISPRS Technical Commision 1 and IAG Commision 4 Meeting in Denver, CO, Nov 2014.

Shaw, G. A. and Burke, H. H. K. 2003. Spectral imaging for remote sensing. *Lincoln Laboratory Journal*, 14(1), 3–28.

Shibata, K. 1957. Spectroscopic studies on chlorophyll formation in intact leaves. *Journal of Biochemistry*, 44(3), 147–173.

Simard, M., Pinto, N., Fisher, J. B., and Baccini, A. 2011. Mapping forest canopy height globally with spaceborne LiDAR. *Journal of Geophysical Research*, 116, G04021. doi: 10.1029/2011JG001708.

Skarda, R. J., Panda, S. S., and Sharma, J. B. 2011. An assessment of the impact of retention ponds for sediment trapping in the Ada Creek and Longwood Cove using remotely sensed data and GIS analysis. In: Published in the *Proceedings of International Symposium on Erosion and Landscape Evolution Hilton Anchorage Hotel*, September 18–21, 2011, Anchorage, AK. ISELE Paper Number 11025.

Suárez, J. C., Ontiveros, C., Smith, S., and Snape, S. 2005. Use of airborne LiDAR and aerial photography in the estimation of individual tree heights in forestry. *Computers and Geosciences*, 31(2), 253–262.

Swayze, G. A., Clark, R. N., Goetz, A. F. H., Livo, K. E., Breit, G. N., Kruse, F. A., Stutley, S. J. et al. 2014. Mapping advanced argillic alteration at Cuprite, Nevada using imaging spectroscopy. *Economic Geology*, 109(5), 1179–1221. doi: 10.2113/econgeo.109.5.1179.

Tiffan, K. F., Rondorf, D. W., and Skalicky, J. J. 2004. Imagining fall Chinook salmon redds in the Columbia River with a dual frequency identification sonar. *North American Journal of Fisheries Management*, 24, 1421–1426.

Townshend, J. R. and Justice, C. O. 1988. Selecting the spatial resolution of satellite sensors required for global monitoring of land transformations. *International Journal of Remote Sensing*, 9(2), 187–236.

Townshend, J. R., Latham, J., Arino, O., Balstad, R., Belward, A., Conant, R., Elvidge, C. et al. 2008. Integrated global observation of the land: An IGOS-P theme, IGOL Report No. 8, GTOS 54, FAO, Rome, Italy, p. 74.

Trenear-Harvey, G. S. 2009. *Historical Dictionary of Air Intelligence*. Scarecrow Press.

U.S. Geological Survey. 1992a. The National Aerial Photography Program (NAPP), factsheet: Reston, VA, U.S. Geological Survey, p. 1.

U.S. Geological Survey. 1992b. NHAP and NAPP photographic enlargements, factsheet: Reston, VA, U.S. Geological Survey, p. 1.

Valavanis, V. D., Pierce, G. J., Zuur, A. F., Palialexis, A., Saveliev, A., Katara, I., and Wang, J. 2008. Modelling of essential fish habitat based on remote sensing, spatial analysis and GIS. *Hydrobiologia*, 612(1), 5–20.

Van Diggelen, F. S. T. 2009. *A-GPS: Assisted GPS, GNSS, and SBAS*. Artech House Norwood, MA.

Van Wagtendonk, J. W., Root, R. R., and Key, C. H. 2004. Comparison of AVIRIS and Landsat ETM+ detection capabilities for burn severity. *Remote Sensing of Environment*, 92(3), 397–408.

Vierling, K. T., Vierling, L. A., Gould, W. A., Martinuzzi, S., and Clawges, R. M. 2008. LiDAR: Shedding new light on habitat characterization and modeling. *Frontiers in Ecology and the Environment*, 6(2), 90–98.

Vuong, H. and Rao, M. 2013. Using LiDAR to estimate total aboveground biomass of redwood stands in the Jackson Demonstration State Forest, Mendocino, California, B23B-0557. In: *Proceedings of the 2013 Fall Meeting*, December 9–13, American Geophysical Union (AGU), San Francisco, CA.

Wagner, W., Bloschl, G., Pampaloni, P., Calvet, J. C., Bizzarri, B., Wigneron, J. P., and Kerr, Y. 2007. Operational readiness of microwave remote sensing of soil moisture for hydrologic applications. *Nordic Hydrology*, 38(1), 1–20.

Waring, R. H., Way, J., Hunt, E. R., Morrissey, L., Ranson, K. J., Weishampel, J. F., and Franklin, S. E. 1995. Imaging radar for ecosystem studies. *BioScience*, 45, 715–723.

Wilson, W. J., Yueh, S. H., Dinardo, S. J., Chazanoff, S. L., Kitiyakara, A., Li, F. K., and Rahmat-Samii, Y. 2001. Passive active L-and S-band (PALS) microwave sensor for ocean salinity and soil moisture measurements. *IEEE Transactions on Geoscience and Remote Sensing*, 39(5), 1039–1048.

Withee, G. W., Smith, D. B., and Hales, M. B. 2004. Progress in multilateral Earth observation cooperation: CEOS, IGOS and the ad hoc group on earth observations. *Space Policy*, 20(1), 37–43.

Yang, Z., Rao, M., Elliott, N., Kindler, S., and Popham, T. W. 2005. Using ground-based multispectral radiometry to detect stress in wheat caused by Greenbug (Homoptera: Aphididae) infestation. *Computers and Electronics in Agriculture*, 47, 121–135.

Yu, B., Liu, H., Wu, J., Hu, Y., and Zhang, L. 2010. Automated derivation of urban building density information using airborne LiDAR data and object-based method. *Landscape and Urban Planning*, 98(3), 210–219.

Fundamentals of Remote Sensing: Evolution, State of the Art, and Future Possibilities

2

Fundamentals of Remote Sensing for Terrestrial Applications: Evolution, Current State of the Art, and Future Possibilities

Natascha Oppelt
Christian-Albrechts-Universität zu Kiel

Rolf Scheiber
German Aerospace Center (DLR)

Peter Gege
German Aerospace Center (DLR)

Martin Wegmann
Ludwig-Maximilians-Universität Würzburg

Hannes Taubenboeck
German Aerospace Center (DLR)

Michael Berger
ESA secondment at European Commission

Acronyms and Definitions

ADEOS	Advanced Earth Observing Satellite
ALS	Airborne laser scanning
ALOS	Advanced Land Observing Satellite
ASAR	Advanced SAR
ASTER	Advanced spaceborne thermal emission and reflection radiometer
AVHRR	Advanced very-high-resolution radiometer
CBERS	China/Brazil Earth Resources Satellite
CEOS	Committee on Earth Observation Satellites
CHRIS	Compact high-resolution imaging spectrometer
CIR	Color infrared
CNES	Centre National d'Etudes Spatiales
CORINE	CoORdination of INformation on the Environment
DA	U.S. Departments of Agriculture
DEM	Digital elevation model
D-InSAR	Differential interferometry synthetic aperture radar
DLR	German Aerospace Center
DMSP	Defense Meteorological Satellite Program
DOI	US Departments of Interior
DSM	Digital surface model
DTED	Digital terrain elevation data
DTM	Digital terrain model
EEA	European Environment Agency
EnMAP	Environmental Mapping and Analysis Program, Germany
EO	Earth observation
ERS	European Remote Sensing
ERTS	Earth Resources Technology Satellite
ESA	European Space Agency
EU	European Union
EUMETSAT	European Organisation for the Exploitation of Meteorological Satellites

fAPAR	Fraction of absorbed photosynthetically active radiation
G8	Group of Eight
GCOS	Global Climate Observation System
GDEM	Global digital elevation model
GEO	Group on Earth Observation
GEO-CAPE	Geostationary Coastal and Air Pollution Events
GEOSS	Global Earth Observation System of Systems
GFW	Global Forest Watch
GIS	Geographic information system
GMES	Global Monitoring for Environment and Security
GOCI	Geostationary ocean color imager
HICO	Hyperspectral imager for the coastal ocean
HISUI	Hyperspectral imager suite, Japan
HRCCD	High Resolution Charge Coupled Device
HSE RESURS-P	Hyperspectral equipment on the Russian Resurs-P platform
HyspIRI	Hyperspectral infrared imager
InSAR	Airborne interferometric synthetic aperture radar
IOCCG	International Ocean Colour Coordinating Group
IRMSS	InfraRed MultiSpectral Scanner
IRS	Indian Remote Sensing
JPL	Jet Propulsion Laboratory
LAI	Leaf area index
LULC	Land use and land cover
MERIS	Medium-resolution imaging spectrometer
METI	Ministry of Economy, Trade, and Industry of Japan
MIR	Mid-infrared
MODIS	Moderate-resolution imaging spectrometer
MSI	Multispectral imager
MSS	Multispectral scanner system
NASA	National Aeronautics and Space Administration
NDVI	Normalized difference vegetation index
NIR	Near-infrared
NOAA	National Oceanic and Atmospheric Administration
OCM	Ocean color monitor
OLCI	Ocean and land color instrument
OLS	Operational linescan system
PALSAR	Phased array L-band synthetic aperture radar
Pol-InSAR	Polarimetric SAR interferometry
PRISMA	Hyperspectral Precursor and Application Mission, Italy
SAOCOM	Satellites for Observation and Communications, Argentina
SAR	Synthetic aperture radar
SeaWiFSS	Sea viewing wide field-of-view sensor
SMAP	Soil Moisture Active Passive
SPOT	Satellite Pour l'Observation de la Terre
SRTM	Shuttle Radar Topography Mission
SWIR	Short-wave infrared
TIR	Thermal or far infrared
TIROS	Television Infrared Observation Satellite
UN	United Nations
US	United States
USDA	United States Department of Agriculture
USDOI	United States Department of Interior
USGS	US Geological Survey
VHRR	Very-high-resolution radiometer
VIS	Visible light

2.1 Introduction

Remote sensing is the acquisition of information about objects from a distance. In general, information is acquired by detecting and measuring an electromagnetic field emitted or reflected by the objects, but it may also be an acoustic field or gravity/magnetic potential.

Today remote sensing is applied in various fields such as astronomy, medicine, industry, and environmental science; the latter includes atmospheric, oceanic, and terrestrial applications—the subject of our attention. In general, terrestrial remote sensing is related to gravity, magnetic fields, and electromagnetic radiation; the techniques of the latter cover the electromagnetic spectrum from the visible (VIS) to the microwave region. Table 2.1 indicates the wavelength ranges of different spectral regions of the electromagnetic spectrum. There are, however, no clear-cut divisions of the spectrum; the partitions have rather developed from different application fields and may vary (Campbell and Wynne, 2011).

Passive remote sensing in all of these regions includes sensors measuring radiation naturally reflected or emitted from the Earth surface, atmosphere, or clouds. The VIS, near-infrared (NIR), and mid-infrared (MIR) regions correspond to the reflective spectral range, because the radiation is essentially solar radiation reflected from the Earth. In the literature, the MIR region may also be referred to as short-wave infrared (SWIR). The thermal or far-infrared (TIR) consists of wavelengths beyond the MIR and extends into regions that border the microwave region and cover radiation that is emitted by the Earth; this kind of radiation is often referred to as thermal energy. This self-emitted radiation can be sensed even in the microwave region using passive systems. At this end of the electromagnetic spectrum, active remote sensing is, employing an artificial source of radiation, used to analyze the scattering

TABLE 2.1 Primary Spectral Regions Used in Terrestrial Remote Sensing

Name	Wavelength Range
VIS	0.38–0.70 μm
NIR	0.70–1.30 μm
MIR, SWIR	1.30–3 μm
TIR	3–14 μm
Microwave, radar	1 mm–1 m

TABLE 2.2 Subdivisions of Active Microwave Sensory

Band Name	Wavelength Range (cm)	Band Name	Wavelength Range (cm)
Ka	0.75–1.18	C	3.75–7.50
Ku	1.18–1.67	S	7.50–15.0
K	1.67–2.40	L	15.0–30.0
X	2.40–3.75	P	77–107

properties of the surface. Imaging radars operate within small ranges of wavelengths within the broad range indicated in Table 2.1; Table 2.2 lists the subdivisions of active microwave regions (Campbell and Wynne, 2011).

The immediate next few chapters discuss the history of terrestrial applications and image processing since the advent of the term *remote sensing* in the 1960s. Section 2.2 gives an overview of present terrestrial applications; it is, however, beyond the scope of this chapter to provide details for all fields of terrestrial remote sensing, but the reader is encouraged to look through other relevant books and chapters of the Remote Sensing Handbook. Therefore, we picked some relevant and representative applications, which will be discussed briefly. These applications provide good understanding of how remote sensing data gathered from different portions of the electromagnetic spectrum is used. The selected examples represent a broad range of studies at different scales (Section 2.3.1), and applications of certain specific disciplines (Section 2.3.2). These applications are characteristic for vegetation and Earth sciences, the hydrosphere, land use and land cover (LULC); moreover, they cover a variety state of art sensors, including passive reflective multispectral and hyperspectral (imaging spectrometers), emissive techniques, and active instruments. Multispectral refers to sensors, which collect the radiation reflected by the Earth's surface in a few, predefined and rather broad spectral bands. Hyperspectral or imaging spectroscopy sensors collect the reflected radiation contiguously across the electromagnetic spectrum in several narrow wavebands (Campbell and Wynne, 2011). In practice, however, imaging spectroscopy is more a conceptual term used for sensors with more than a few spectral bands—often, few 10s or 100s but contiguously over a range (e.g., 0.4–2.5 μm) of the electromagnetic spectrum. After discussing thematic applications, data policy strategies are introduced (Section 2.3.4). Data availability and comparability are discussed in Section 2.3.5. Section 2.4 outlines future possibilities including enhanced technologies (Section 2.4.1), data availability, and continuity issues (Section 2.4.2) before main conclusions are discussed in the last section of this chapter.

2.2 Evolution of Terrestrial Applications

2.2.1 From Qualitative Description and Visual Interpretation of the Earth Surface to Digital Data Processing

Aerial photography is the original form of remote sensing; throughout its history, proponents of aerial photography have sought improved technology to collect information about Earth's resources. During the 1950s and 1960s, most analyses were visual interpretations of prints or transparencies of aerial images (Fischer et al., 1976). Traditionally, black-and-white (panchromatic, 400–700 nm) aerial photography has been common. With panchromatic images, different patterns of the land surface appeared as a series of grey values ranging from black to white. In land applications, the development of color infrared (CIR) photographs (covering green, red, and NIR wavelengths), sometimes referred to as false color, may be attributed among the most significant developments. Colwell (1956) was one of the first who successfully used CIR images to address crop diseases. Figure 2.1 demonstrates that besides analysis of shapes, sizes, pattern, texture, shadow, or spatial relationships, CIR images improved interpretation of vegetation due to the coverage of the NIR plateau of healthy vegetation (for vegetation spectra, see Figure 2.4 later in the chapter). In CIR imagery healthy, dense vegetation appears in bright red. Lighter tones of red generally represent vegetation with a less distinct NIR plateau such as mature stands of evergreens. Agricultural fields approaching the end of growing season, and dead or unhealthy plants often appear in less intense reds or green. Soils appear in white, blue-green, or tan, while water bodies are dark blue to black (USGS, 2001). Image interpretation, however, was limited to visual analysis and therefore remained qualitative.

In 1957, the former Soviet Union launched the first artificial satellite into space: Sputnik, carrying four radio antennas to broadcast radio pulses, triggered the space race and therefore was the starting signal for the space age (Siddiqi, 2003). NASA's (National Aeronautics and Space Administration) Television Infrared Observation Satellite, TIROS-1, was launched in 1960 and provided the first systematic images of the Earth from space (Allison and Neil, 1962; see also Table 2.3). A single television camera pointed at the Earth surface for a limited time each orbit, and mainly collected images of North America. However, TIROS-1 was the first of a series of experimental weather satellites, which through TIROS-X contained television cameras; four of them also included IR sensors (Hastings and Emery, 1992).

By the late 1960s, scientists in the U.S. Departments of Agriculture (DA) and Interior (DOI) had proposed a space mission dedicated to acquire synoptic, multispectral images of the Earth's surface. These data could be used in a wide range of applications such as agriculture, forestry, mineral exploration, land resource evaluation, land use planning, water resources, mapping and charting, and environmental protection. The mission was launched in 1972 as the Earth Resources Technology Satellite (ERTS), and later renamed Landsat. The launch of the multispectral scanner system (MSS) onboard the ERTS/Landsat represents the beginning of spaceborne terrestrial remote sensing. For the first time an Earth-orbiting system provided systematic, repetitive observations of the Earth's land areas. Landsat MSS depicted large areas of the Earth surface and, although only with four spectral bands and a pixel size of 80 m, provided routinely available data (Lillesand et al., 2008).

Also in the 1960s, National Oceanic and Atmospheric Administration (NOAA) launched the very-high-resolution radiometer (VHRR), which was based on developments around the TIROS sensors on board of the NOAA-4 satellite.

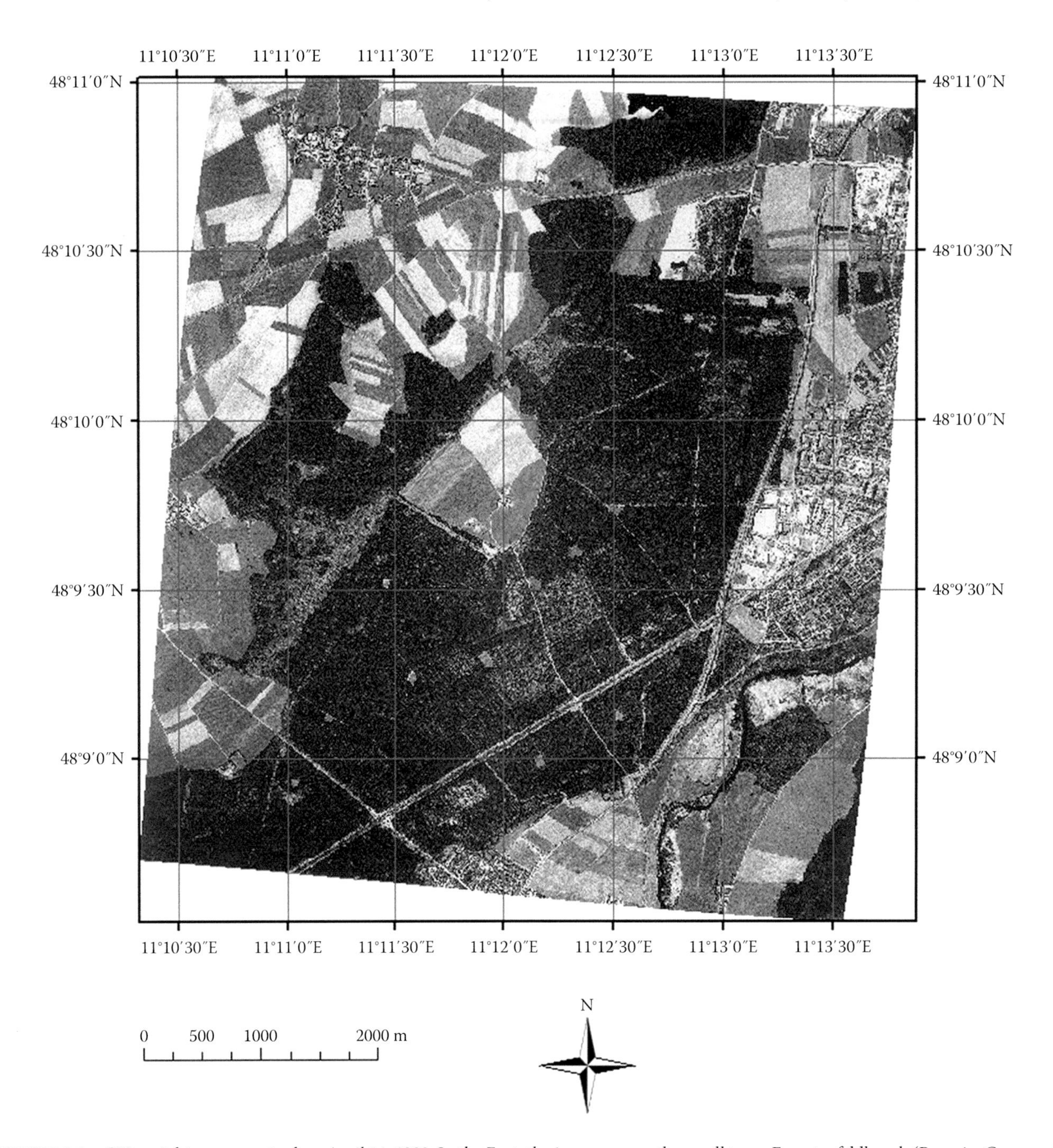

FIGURE 2.1 CIR aerial image acquired on April 14, 1983. In the East, the image covers the small town Fuerstenfeldbruck (Bavaria, Germany), a mixed forest stand crosses the image in SW–NE direction, surrounded by agricultural fields (bare soil, winter cereals, and pasture). (Geospatial data © Bavarian Surveying and Mapping Authority, Bavaria, Germany, 2014.)

NOAA also introduced the direct reception of digital VHRR data free of charge to ground stations installed by users already in 1972 (Hastings and Emery, 1992). Being available also in digital form, VHRR and Landsat MSS marked the beginning of the era of digital image analysis (Campbell and Wynne, 2011), which, however, was limited to specialized research institutions. The technical standard of personal computers and limited availability of image analysis software prohibited the analysis of data that now is regarded as commonplace (Jones and Vaughan, 2010).

The availability of digital data, however, was a seeding point for the development of image analysis software and computer-based processing techniques. In 1978, the advanced very-high-resolution radiometer (AVHRR) was on board of TIROS-N (TIROS-NOAA) and the seventh generation of NOAA, which up to then also carried the VHRR instrument (Hastings and Emery, 1992). Initially, the NOAA satellites were designed to observe the Earth's weather. Nevertheless, subsequent sensors were specifically designed to measure other phenomena including terrestrial vegetation, where the AVHRR-derived

TABLE 2.3 Advent of Satellite-Borne Instruments Used for Terrestrial Applications

Sensor/Satellite	Launch	Band Setting	Spatial Resolution (m)[a]	Revisit Time at Equator (days)[b]
TIROS-1	1960	4 bands VIS	8000	—
VHRR/NOAA	1966	1 band VIS, 1 band TIR	900	1
MSS/ETRS (Landsat-1)	1972	2 bands VIS, 3 bands NIR/MIR	68 × 83	18
AVHRR/TIROS-NOAA	1978	3 bands VIS, 2 bands TIR	1100	3
HRV/SPOT	1986	1 band PAN, 2 bands VIS, 1 band NIR	10 (PAN), 20 (VIS/NIR)	26 (1 with pointing)
LISS/IRS	1988	3 bands VIS, 1 band NIR	72.5 (LISS-I), 36 (LISS-II)	22
ERS-1/2	1991/1995	C-band (5.3 GHz)	30 × 30	35/1
Radarsat-1/2	1995/2007	C-band (5.4 GHz)	3–100	24
AVNIR/ADEOS	1996	1 band PAN, 3 bands VIS, 1 band NIR	8 (PAN), 16 (VIS–NIR)	41
HRCCD/CBERS	1999	1 band PAN, 3 bands VIS, 1 band NIR	20	26
IRMSS/CBERS	1999	1 band PAN, 2 bands MIR, 1 band TIR	80 (PAN, MIR), 160 (TIR)	26
ASTER/Terra	1999	14 band (VIS–TIR)	15 (VIS)–90 (TIR)	16
MODIS/Terra, Aqua	1999/2002	36 bands (VIS–TIR)	250–1000	1–2
Hyperion/EO-1	1999	36 bands (VIS–TIR)	250–1000	16
ALI/EO-1	2000	1 band PAN, 4 bands VIS, 5 bands NIR/MIR	10 (PAN), 30 (VIS–MIR)	16
CHRIS/Proba	2001	up to 63 bands (VIS–NIR)	17	16 (2 with pointing)
MERIS/Envisat	2002	15 selectable bands (VIS–NIR)	260 × 300 (land and coast) 1040 × 1200 (ocean)	3
ASAR/Envisat	2002	C-band (5.3 GHz)	30 (image mode)–150 (wide swath)	35
ALOS	2007	L-band (1.3 GHz)	10–100	46
TerraSAR-X/TanDEM-X	2007/2010	X-band (9.6 GHz)	1–40	11
CosmosSkyMed Constellation	2007/2010	X-band (9.6 GHz)	1–100	Up to 1
Sentinel-1a	2014	C-band (5.4 GHz)	5 × 5–5 × 20	12

[a] With passive, reflective sensors the spatial resolution is provided at nadir.
[b] For SAR sensors the orbit repeat cycles are indicated instead of revisit time.

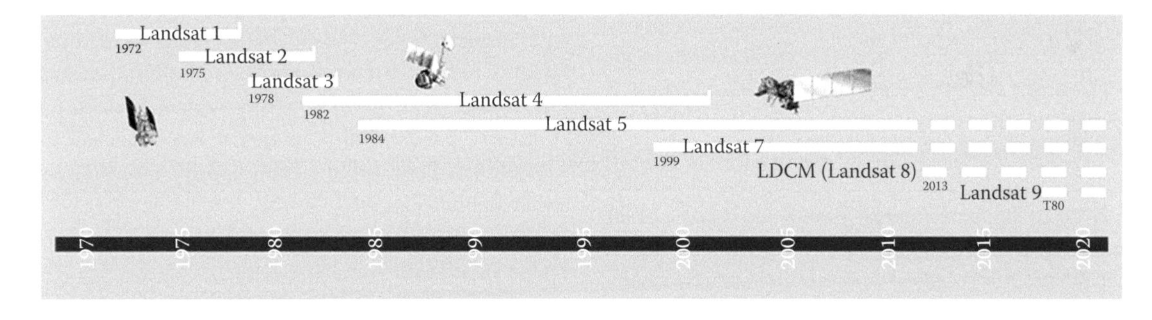

FIGURE 2.2 Landsat mission timeline. (Courtesy of NASA Goddard Space Flight Center and U.S. Geological Survey, Greenbelt, MD.)

regional and global indices have proven to be a very robust and useful quantity (Gutman, 1991). Also in the 1970s, the next two Landsat satellites have been launched (see Figure 2.2). During this time period, other nations decided to operate own sensors, which led to a rapid increase of data availability; for example, in the 1980s, SPOT-1 (Satellite Pour l'Observation de la Terre), the first of a series of multispectral satellites, was launched by the French Centre National d'Etudes Spatiales (CNES) in cooperation with Belgium and Sweden (Chevrel et al., 1981), and India launched the first sensor from its Indian Remote Sensing (IRS) satellite program (Misty, 1998). In the 1990s, other nations launched their own Earth Observation (EO) satellites such as China/Brazil (China/Brazil Earth Resources Satellite [CBERS]; De Oliveira et al., 2000) or Japan (Advanced Earth Observing Satellite [ADEOS]; Shimoda, 1999). This development pushed forward further developments in data processing; dedicated software and the rapid advances in computer technology enabled the users to apply image processing such as geometric and radiometric correction, image enhancement, data classification, or data merging (Lillesand et al., 2008).

The launch of the first civil synthetic aperture radar (SAR) in space started a similar rapid development of applications as with optical imagery. Since that time, a number of satellites have been deployed that carry microwave radiometers and imaging radars (Table 2.3; Elachi, 1988; Kuschel and O'Hagan, 2010; Sullivan, 2000). In spite of data complexity, the parallel advances in computer techniques have escalated the number of radar data for terrestrial applications (Jones and Vaughan, 2010). Prominent examples were the European Remote Sensing (ERS) satellites, ERS-1 and ERS-2, which were launched in the 1990s and acquired tandem C-band SAR data for many applications for nearly two decades (see also Section 2.2.3).

These data offered the very first opportunity for global coverage every 3–22 days; with two Landsat or ERS sensors in orbit, the coverage time could be reduced. Nevertheless, true global coverage was limited by a host of issues such as cloud cover, data downlink limitations, and absence of sufficient ground stations to receive data (Goward et al., 2001). In the 1990s, however, a series of EO satellites came through with a wide range of spatial, spectral, temporal coverage capabilities, which allowed for true global coverage. In 1999, Terra was launched, a joint spaceborne mission of the Ministry of Economy, Trade, and Industry (METI) of Japan and NASA (Running et al., 1998). Terra was the first satellite of a system specifically designed to acquire global coverage for monitoring the Earth's ecosystems. Terra carries several sensors such as the advanced spaceborne thermal emission and reflection radiometer (ASTER) and the moderate-resolution imaging spectrometer (MODIS) (Xiong et al., 2011). For global applications, MODIS was a substantial improvement over the AVHRR sensor in spatial resolution, band setting, with on-board calibration and enhanced radiometric accuracy (Running et al., 1994). In 2002, the European Space Agency (ESA) launched ENVISAT—with 10 instruments, the largest civil EO mission. With the advanced SAR (ASAR) aboard, ENVISAT ensured the data continuity of ERS, but new instruments supplemented the mission such as two atmospheric sensors to monitor trace gases and the medium-resolution imaging spectrometer (MERIS) that operated with 15 bands in the VIS/NIR spectral domain and a spatial resolution of 260 m × 300 m. These systems were designed to support broad-scale Earth science research and allowed monitoring spatial patterns of environmental and climate changes (Louet and Bruzzi, 1999).

In contrast to the common approach of distributing radiance or reflectance images, the broad-scale sensors also opened a new era of image-processing techniques, which targeted toward the generation of standard data products. A new goal was the processing of consistent, satellite-based records of physical environmental parameters (e.g., chlorophyll, leaf area index [LAI] or fraction of absorbed photosynthetically active radiation [fAPAR]). These products are focused on monitoring temporal dynamics, and have been successfully integrated into Earth system models to improve the understanding of connections between driver and responsive variables (Pielke, 2005; Sellers et al., 1997).

During the first decade of the twenty-first century, internet began to influence public access to remote sensing imagery (Goodchild, 2007; Gould et al., 2008; see also Chapters 9 through 11). Nowadays, many image-based products and radiance or reflectance data are web-enabled and delivered free of cost such as the MODIS (Woodcock et al., 2008) or MERIS (Bontemps et al., 2011) products.

2.2.2 Development of Indices and Quantitative Assessment of Environmental Parameters

Computer-based processing of digital data directly led to the development of quantitative assessment of environmental parameters, at which the development of indices was an important step. Spectral indices do aim at deriving information about the land surface by using specific spectral responses. Rouse observed that vegetation shows a typical reflectance behavior in the VIS and NIR (see also Figure 2.5 later in the chapter) and introduced the normalized difference vegetation index (NDVI) for MSS data already in 1973 (Rouse, 1974):

$$\text{NDVI} = \frac{(\text{NIR} - \text{RED})}{(\text{NIR} + \text{RED})} \tag{2.1}$$

where

NIR reflectance in the NIR (MSS, 0.8–1.1 μm)
RED reflectance in the VIS (MSS, 0.6–0.7 μm)

Foremost, the NDVI was used to analyze any terrestrial green vegetation status and spatiotemporal dynamics of herbaceous vegetation or forests (Sellers, 1985). The normalization proved to be an important advantage for the success of the NDVI as a descriptor of vegetation variations in spite of atmospheric effects (Holben et al., 1990). Accordingly, it was used in numerous regional and global applications for studying vegetation with various sensors (e.g., Hame et al., 1997; Tucker and Sellers, 1986). Although the NDVI suffers from specific deficiencies (e.g., soil background influence), it is probably still the most well-known and common index. Nowadays, there are numerous alternative forms of indices that correct for specific deficiencies of the NDVI or are adapted to specific sensors such as MODIS (Huete et al., 2002; see also Section 2.3.1.2). For hyperspectral instruments, a broad variety of narrow band indices have been developed to assess parameters such as LAI, fractional cover, fAPAR, or to assess and quantify biophysical and biochemical plant parameters such as chlorophyll, accessory pigments, or plant water content (Thenkabail et al., 2012). Hyperspectral indices are also commonly used for mineral identification (Sabins, 1999; van der Meer et al., 2012), monitoring of inland water constituents (Odermatt et al., 2012), and assessment of soil conditions (Pettorelli, 2013).

2.2.3 SAR Imaging from 2D to 3D and Beyond

Fostered by the availability of regular SAR observations from space, using satellite sensors like ENVISAT ASAR (C-band, ESA), Radarsat1 and 2 (C-band, Canada), and ALOS-PALSAR (L-band, Japan), interferometric techniques have been developed for several terrestrial applications, which until then were not supported by remote sensing data (Rosen et al., 2000). Primarily dedicated to the generation of digital elevation models (DEMs), culminating in the Shuttle Radar Topography Mission (SRTM), the application range quickly extended toward the measurement of displacements with centimeter accuracy using differential interferometry SAR (D-InSAR; Figure 2.3) techniques. Thus, mapping

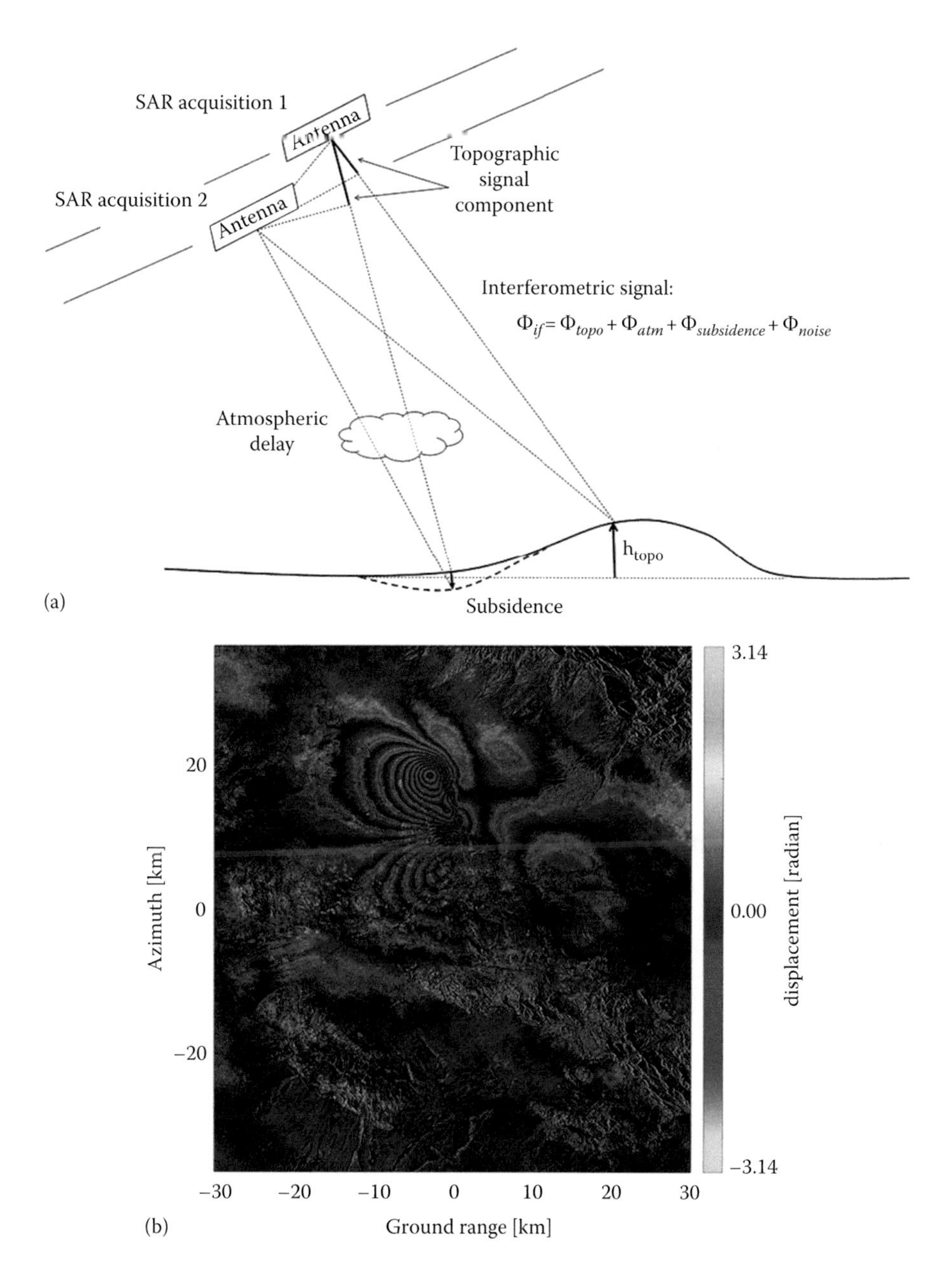

FIGURE 2.3 Geometry for differential interferometric SAR (D-InSAR) data acquisition (a). Differential interferometric phase signature of coseismic displacement of Bam earthquake, December 26, 2003, Iran as derived from ASAR data on ENVISAT superimposed on radar reflectivity (b). One phase/color cycle corresponds to 2.8 cm displacement in radar line of sight. Maximum uplift amounts to 31 cm at [-2, 20] km (geographic location ~29.05°N, 58.36°E), maximum subsidence is 17 cm at [-1,5] km.(© Data provided by ESA and processed by DLR.)

of co- and postseismic displacements, volcano inflation, and glacier surface velocities became possible over large areas supporting traditional point-wise GPS measurements. To counteract the disturbing influence of atmosphere and temporal changes of the scattering in between acquisitions, D-InSAR approaches quickly developed toward the evaluation of time series of SAR images.

The common primary objective of all these D-InSAR stacking approaches is to remove the disturbing signal component of the atmosphere (correlated in space, but random in time), and of DEM errors (using a geometric model) from the displacement signal (correlated in time and spatially limited). In this way, D-InSAR measurements became more trustworthy and accurate; for example, the measurement of seasonal thermal dilation of buildings with accuracies down to few millimeters was reported. With the availability of spaceborne SAR data with resolutions in the order of 1 m (presently TerraSAR-X [Germany] and CosmoSkyMed [Italy]), the identification of scattering centers in layover became feasible, allowing the monitoring of individual building infrastructures (Zhu and Bamler, 2014). Further applications of the D-InSAR technique are for the detection of landslides, measurements of subsidence due to ground water extraction, mining activities, gas and petrol prospection, as well as control of inflation due to carbon capture sequestration (Moreira et al., 2013).

2.3 State of the Art

At present, a large diversity of data sets is available, ranging from very high spatial resolution optical data such as WorldView, Ikonos, GeoEye, and Quickbird (commercial sensors operated by Digital Globe) to Aster, Landsat, SPOT toward medium-resolution MODIS and coarse-scale NOAA AVHRR data; furthermore, a broad variety of satellite-borne SAR data from sensors like ALOS, RADARSAT, TerraSAR-X, or Sentinel-1 are available. According to their spectral and spatial characteristics, they support various terrestrial applications (Figure 2.4).

2.3.1 Terrestrial Applications Using Multisensor Data

2.3.1.1 Precision Farming: Local Level

Precision farming, or information-based management of agricultural production systems, emerged in the mid-1980s as a way to manage within and between field spatial and temporal variabilities associated with all aspects of agricultural production for the purpose of improving crop performance and environmental quality (Pierce and Nowak, 1999). Apart from field crop production, precision farming technologies have been applied successfully in viticulture and horticulture, including orchards, and in livestock production, as well as pasture and turf management (Gebbers and Adamchuk, 2010). Due to the need for high spatial resolution, satellite-based remote sensing applications have seen limited use for quite a long time. Nowadays, remote sensing is a key component of precision farming and is used by an increasing number of scientists, engineers, and large-scale farmers and focuses on a wide range of applications; these include crop yield, leaf area, and biomass (Doraiswamy et al., 2003; Shanahan et al., 2001; Yang et al., 2000), crop nutrient and water status (Oppelt and Mauser, 2004; Tilling et al., 2007), weed infestation (Lamb and Brown, 2001; Thorp and Tian, 2004), plant diseases (Mahlein et al., 2012), and soil properties such as organic matter, moisture, and pH (Christy, 2008) or salinity (Corwin and Lesch, 2003).

To cover the wide range of applications, a variety of sensors and techniques are used. Multispectral and hyperspectral sensors are applied as well as thermal imagery. Using multispectral and hyperspectral imageries, vegetation analysis generally bases on the dependence of vegetation spectra on plant pigments and vegetation water content (Blackburn, 2007), mostly by applying spectral indices (Mulla, 2013). Figure 2.5 depicts how plant reflectance changes with varying pigment and plant water content as modeled with Prospect 5 (Feret et al., 2008).

The most appropriate spatial and spectral resolution for precision farming depends on crop management, farm equipment, and dominant field sizes. The requirements regarding spatial resolution also differ for the respective application and vary between 1 m for weed sensing and 10 m for variable rate application of fertilizer (Mulla, 2013).

Satellite instruments such as Ikonos, GeoEye, Quickbird, and WorldView offer high spatial resolution and high revisit times; combination with (relatively) high geolocation accuracies (Table 2.4) led to a broad range of applications, especially for WorldView and GeoEye data. These instruments offer spectral bands most suitable for calculating multiple spectral indices.

Until present, imaging spectroscopy is often confined to airborne imaging spectrometers; satellite-based hyperspectral instruments

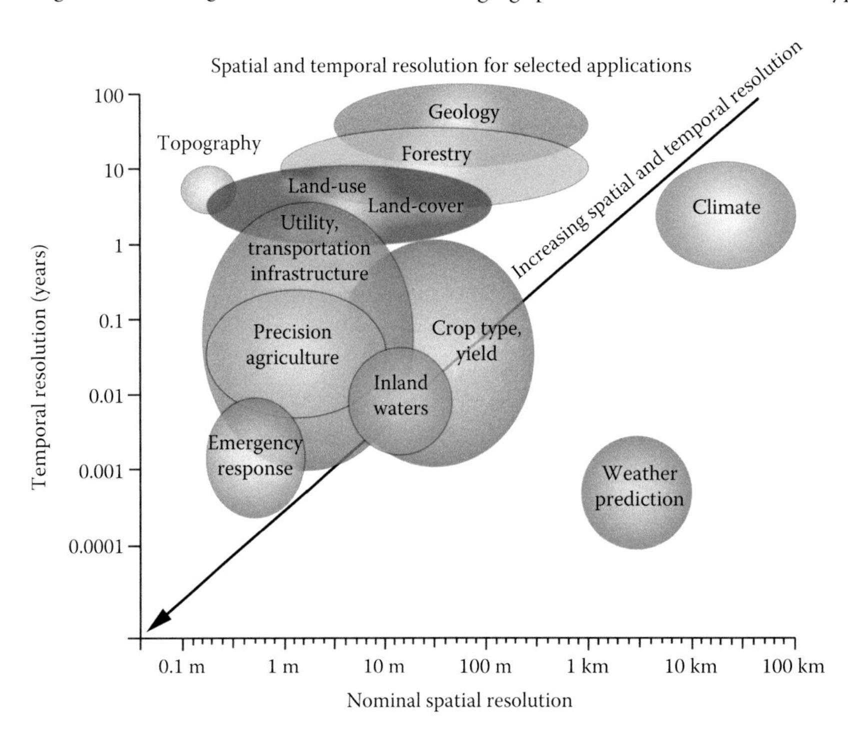

FIGURE 2.4 Terrestrial remote sensing applications. (Modified from a figure originally published by the American Society for Photogrammetry and Remote Sensing, Bethesda, MD, www.asprs.org; Davis, F. et al., *Photogramm. Eng. Remote Sens.*, 57(6), 689, 1991.)

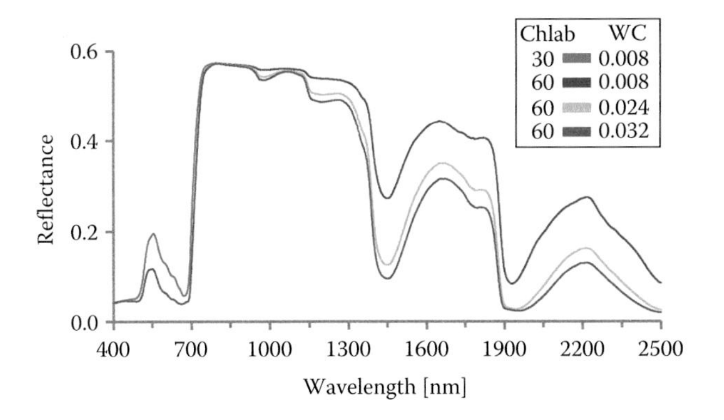

FIGURE 2.5 Reflectance signatures of a clover leaf (*Trifolium pratense*) with varying contents of chlorophyll a + b (Chlab, in μg/cm²) and plant water (WC, in g/m²).

are rare. An example may be the compact high-resolution imaging spectrometer (CHRIS), launched in 2001 by ESA. CHRIS covers the VIS and NIR with up to 63 spectral bands and a spatial resolution of 17 m at nadir (Barnsley et al., 2004). Using CHRIS, various studies have been conducted for precision farming applications (Delegido et al., 2008; Kneubühler et al., 2006; Oppelt, 2010).

In addition to reflectance, plant leaves can emit energy by fluorescence or thermal emission. The former is used to investigate the photosynthetic activity by solar-induced chlorophyll fluorescence (Meroni et al., 2009). The latter has been shown to be useful for the delineation of management zones within fields—based on the transpiration of crop canopies related to plant biomass—as well as for the assessment of plant vigor (Stenzel et al., 2007) and pathogens (Hillnhütter et al., 2011); however, airborne applications dominate fluorescence measurements and thermography. Besides, a new trend toward the use of robust, low-cost, and preferably, real-time remote sensing systems—unmanned airborne sensors—also can be observed (Zhang and Kovacs, 2012).

In recent years also, polarimetric SAR data started to be used for investigating the temporal evolution/phenology of crops as well as to extract soil moisture information. Being in a preoperational research stadium, reported results are primarily based on airborne SAR data. However, being independent from daylight conditions, operational use starts to emerge based on the polarimetric capabilities of Radarsat-2, ALOS (Japan), and Risat-1 (India) sensors. Related investigations make use of polarimetric decomposition techniques to distinguish reflections from bare soil, a vegetation layer or man-made structures. Supervised and unsupervised classifications of polarimetric SAR data enable, for example, distinguishing between different crop types (Moreira et al., 2013).

2.3.1.2 From Space to Species: Assessment of Plant Species and Monitoring of Biodiversity and Habitat Dynamics: Regional Level

Understanding the global distribution of biodiversity has been advanced in the past also by means of remote sensing (Gaston, 2000). Biodiversity, the variety of species, is increasingly affected by the ongoing global change, and its decrease since decades has been identified internationally (Balmford et al., 2005; Collen et al., 2013). Monitoring the status and its changes is a high priority for conservation (Myers et al., 2000; Pereira et al., 2013) and can be achieved using remote sensing data (Horning et al., 2010; Kerr and Ostrovsky, 2003) combined with in situ measurements and statistical models (Figure 2.6).

AVHRR time series data has already been used in the mid-1990s for analyzing biodiversity hotspots (Fjeldså et al., 1997) and has been advanced to broad application in environmental management (Duro et al., 2007; Pettorelli et al., 2005; Strand et al., 2007). Especially the widely applied NDVI has proven to be valuable to contribute to an explanation of spatial ecological patterns (Pettorelli et al., 2005). Recent developments of remote sensing for biodiversity managed to show statistical correlation of multisensor approaches with plant species richness and provide spatial prediction models (Buermann et al., 2008; Saatchi et al., 2008). Further examples by Vierling et al. (2008) and Nagendra et al. (2013) show the potential for a variety of environmental applications.

2.3.1.3 Changing Land Use and Forest Monitoring: Global Level

LULC, and alterations of LULC, play a major role in global-scale patterns of the climate and biogeochemistry of the Earth system (Friedl et al., 2002); therefore, monitoring LULC and its changes is one of the most important agents of environmental change and has significant implications for ecosystem health, water quality, and sustainable land management (Friedl et al., 2010). An important goal in developing these products is to meet the needs of the modeling community and to attempt to better understand the role of human impacts on Earth systems through LULC conversions (Hansen et al., 2000). Therefore, many LULC classifications are globally available such as the global land cover

TABLE 2.4 Characteristics of Satellite Instruments Used for Precision Farming,[a] Geolocation Accuracy is Specified According to the CE90 Standard at Nadir over Flat Terrain

Name	Launch	Band Setting	Spatial Resolution (m)	Revisit Time (days)	Geolocation Accuracy (m)
IKONOS	1999	PAN, B, G, R, NIR	0.8 (PAN)–4 (MS)	5	15
Quickbird	2001	PAN, B, G, R, NIR	0.6 (PAN)–2.4 (MS)	1–3.5 (dep. on latitude)	23
GeoEye	2008	PAN, B, G, R, NIR	0.4 (PAN)–1.6 (MS)	<3	5
RapidEye	2008	B, G, R, NIR	5.0	1	23.6
WorldView2	2009	PAN, COASTAL, B, G, Y, R, RED EDGE, NIR	0.4 (PAN)–3.0 (MS)	1	5

Source: Digital Globe. 2014. Digital Globe satellite information. https://www.digitalglobe.com/resources/satellite-information (accessed March 24, 2014).
[a] Band setting codes: PAN, panchromatic; MS, multispectral; B, blue; G, green; Y, yellow; R, red; NIR, near-infrared COASTAL, violet/blue; RED EDGE, near infrared.

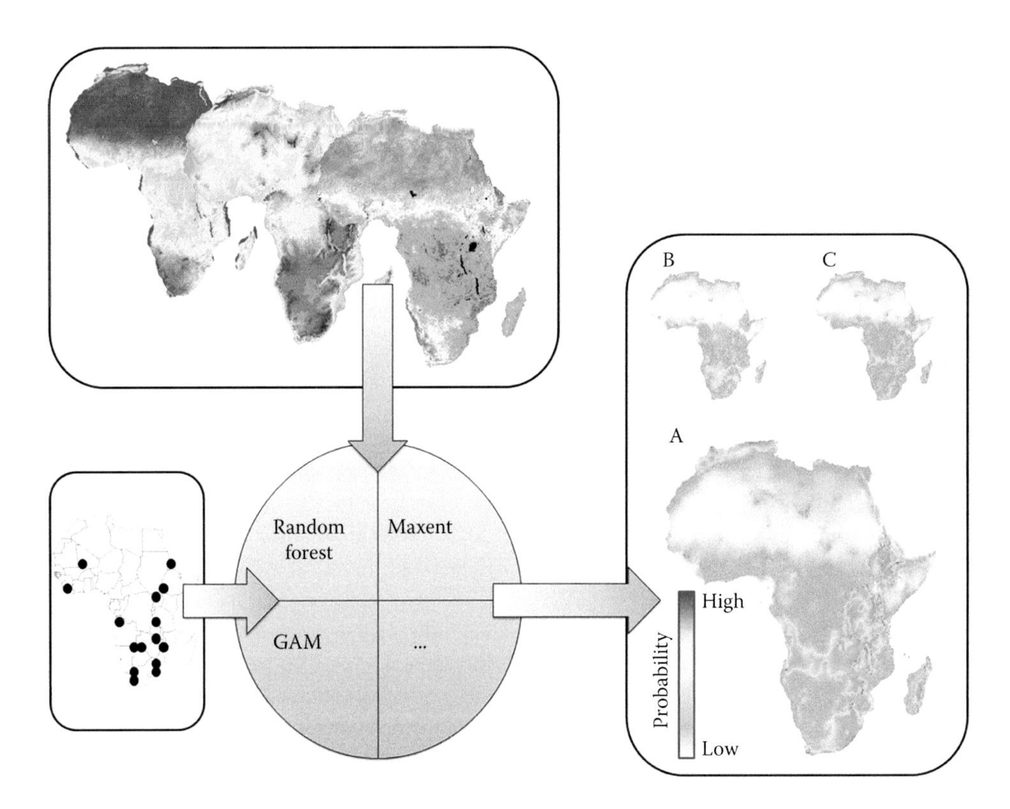

FIGURE 2.6 Workflow of the spatial prediction modeling of species by combining species occurrence points and environmental data sets with statistical models. Resulting prediction maps (A, B, C) highly depend on chosen input variables and occurrence points as well as selected model. All models do not consider the biogeography, and hence, areas such as islands might have a high probability of occurrence just based on their environmental conditions.

products derived from AVHRR data (DeFries et al., 1995, 1998). As newer, medium-resolution data sources have emerged, substantial effort has been made on developing improved characterizations of global land cover. The current generation of global land cover products includes the Global Land Cover 2000 product, produced from SPOT VEGETATION (4 spectral bands in the VIS/NIR/MIR and a spatial resolution of 1 km; Bartholomé and Belward, 2005), the MODIS Land Cover Product (Friedl et al., 2010), and the GlobCover product produced using data from MERIS (Arino et al., 2008). Mora et al. (2014) discusses present LULC products and future trends in detail.

Since LULC maps became increasingly important sources of baseline information for a wide variety of applications, validation also became increasingly important. LULC maps are commonly validated using higher-quality reference data, such as independent validation data sets or regional maps, or they are cross validated against field data (Friedl et al., 2002; Hansen et al., 2013). Global Land Cover 2000, MODIS Land Cover, or MERIS GlobCover are examples for such independently validated LULC, which rely on validation based on visual interpretation of satellite images, regional maps, and geotagged photographs (Tsendbazar et al., 2014). As a consequence, the accuracies of the resulting LULC maps vary between 67% and 81% (Tsendbazar et al., 2014); in some regions of the world, accuracies range between 10% and 50% (Mora et al., 2014). Being validated independently, there also exist critical differences (spatial disagreements) between LULC products. For GlobCover and MODIS, for example, Fritz et al. (2011) reported a discrepancy of approximately 20% of the global cropland area. Differences in classification methodology, training and ground reference data, the type of satellite sensors used, and georeferencing errors may cause these disagreements. A small portion of the differences may also be attributed to differences in the date of the satellite data acquisition used in creating land cover maps (Fritz et al., 2011).

Accuracy assessment and generation of reference data sets for accuracy assessments therefore remain difficult. To address these problems, Oloffson et al. (2013) claimed definition of accuracy standards apart from established quantities such as overall accuracy and kappa. They stated the high value of confusion matrices, which provide valuable information on the magnitude of the classification errors as well as on area estimates and provision of confidence intervals for each class. To move toward a more efficient use of reference data, Oloffson et al. (2012) proposed generation of a new reference data set that is created without the use of remote sensing data. An improvement in training data coverage or data sharing may also help to improvefuture global land cover data sets (Fritz et al., 2011). Detailed information of methods and reporting of results may enhance the use of accuracy assessment data beyond the present common practice; moreover, it will lead to better estimates of area, accuracy, and the uncertainty associated with these estimates (Oloffson et al., 2013).

Within the LULC mapping, particular attention has been paid to the global observation of forest cover with a special emphasis on forest loss. Since the highest losses in forest cover have been documented in tropical and subtropical regions (Hansen et al., 2013), microwave applications play an important role in forest mapping. Not being affected by cloud cover or changing illumination conditions, spaceborne SAR data are best suited for mapping forested areas on a global scale. ERS-1/2 data were used to monitor deforestation especially in tropical regions and to classify on a large-scale boreal forest (Wagner et al., 2003). Identification of the response of forested areas to changing environmental conditions, as, for example, water level changes in mangroves, was demonstrated with L-band data of the PALSAR sensor of the Japanese ALOS (Lucas et al., 2009). Present day interest is particularly focused on the estimation of above ground forest biomass, due to its impact on climate change. Different evaluation methods may be applied for obtaining estimates of biomass and its temporal evolution. First, there is a direct relationship of radar backscatter to above ground biomass. Second, forest height may also be estimated from the interferometric coherence values. This technique can also be applied to small wavelength SAR data, as, for example, the ones of the presently ongoing German TanDEM-X mission (Krieger et al., 2007), which offers the advantage of not being affected by temporal decorrelation, which otherwise may lead to height bias. Having an estimate of forest height, the biomass may be inferred through allometry relations, which are specific for different tree species. Finally, polarimetric SAR interferometry (Pol-InSAR) and SAR tomography offer an increased observational space, allowing for additional information on the forest structure and thus improved classification possibilities For these reasons ESA has selected in 2013 BIOMASS, a P-band radar, to be implemented as the 7-th Earth Explorer mission. (ESA, 2012; Moreira et al., 2013).

There exists a variety of initiatives which point towards the importance of remote sensing based forest monitoring such as the United Nations Collaborative Programme on Reducing the Emissions from Deforestation and Forest Degradation in Developing Countries (REDD+, Angelsen et al. 2012), the Global Forest Resources Assessment (FRA; FAO, 2001) or the Global Forest Observation Initiative (GFOI; Mitchell and Hoekman, 2013).

In the context of global forest monitoring using passive sensors, a new initiative was announced in 2014: Global Forest Watch (GFW). The World Resources Institute initiated the project together with 40 other partners along with Google to set up a near-real-time monitoring system to provide information for an enhanced forest management (Showstack, 2014). GFW represents the latest iteration of an initiative, which began in 1997, and at that time was mainly based on reports, static maps, and other materials not updated frequently (WRI, 2002). The new online tool is

FIGURE 2.7 Global tree cover loss (2002–2012). (Courtesy of Global Forest Watch.)

based on satellite data, open access to data, cloud computing, and crowdsourcing (Showstack, 2014). In support of this activity, the space agencies of the Committee on Earth Observation Satellites (CEOS) established a Space Data Coordination Group (representing 12 space agencies) to coordinate the acquisition and provision of satellite data (Baltuck et al., 2013). GFW provides standardized products not only of forest cover, but also of change (gain or loss, Figure 2.7), use (e.g., logging, oil or wood fiber plantations), and protected areas online and free of charge (GFW, 2014).

2.3.2 Thematic Applications

EO allows monitoring of patterns, structures, morphology, and the relationships of the built environment. In the past decades, the expansion of satellite-derived data products expanded incorporating Earth science products into the mainstream environmental, meteorological, and other user communities. This includes movement of essentially science-produced research products derived from missions into operational observations (Brown et al., 2013). This results in a vast range of applications where remote sensing plays a significant role. Land applications involve mapping, change detection, monitoring, and modeling, and other observations of phenomena at the Earth's surface (see other contributions in *Remotely Sensed Data Characterization, Classification, and Accuracies*). A majority of applications is integrated in a geographic information system (GIS) environment (see Chapter 7), mostly developed to provide thematic maps, which may then be used by specialized groups of scientists, stakeholder, or policy makers (Barrett and Curtis, 2013). Following are examples for applications that already are established or mainstream (Section 2.3.2.1) or represent an emerging field of applications (Section 2.3.2.2). Furthermore, the Remote Sensing Handbook covers a detailed view on a broad range of land (*Land Resources Monitoring, Modeling, and Mapping with Remote Sensing*) and water (*Remote Sensing of Water Resources, Disasters, and Urban Studies*) applications.

2.3.2.1 Urban Applications

Urban applications are a dynamically growing research field in remote sensing (for more detail see *Remote Sensing of Water Resources, Disasters, and Urban Studies*, Chapter 12). Measurements all along the electromagnetic spectrum (optical, radar, thermal, and lidar) enable detection of objects, patterns, and conditions of the Earth's surface as well as the condition of the atmosphere. With spatial resolutions capable to capture the complex patterns and small objects of urban areas "[...], the question has become 'Now that we have the technology, how do we use it?'" (Weng and Quattrochi, 2006).

Naturally, the classification of the built environment at various scales—from global to local scales—plays a central role in urban remote sensing. Global urban maps relying on optical satellite data (e.g., MODIS, DMSP/OLS [Defense Meteorological Satellite Program/operational linescan system]) are available at comparatively coarse resolution (Potere and Schneider, 2009) with a maximum spatial resolution of roughly 300 m (e.g., Bartholomé and Belward, 2005). Currently, new EO initiatives aim at improving the geometric level for global urban mapping based on optical (Miyazaki et al., 2012; Pesaresi et al., 2013) or radar data (Esch et al., 2012, Gamba and Lisini, 2013) to spatial resolution better than 100 m; however, to date, a multitemporal global component is still absent. Many projects such as the CORINE Land Cover (EEA, 2014) produce consistent and thus comparable data with an urban class at continental scale. Another initiative is the European Urban Atlas (Seifert, 2009) that delivers a consistent and thematically higher-resolution classification of more than 300 cities across Europe. Multitemporal data sets (e.g., Landsat, SPOT, and DMSP/OLS) have also been applied for macro- to medium-scale monitoring of urban growth with geometric resolution up to 10 m (Frolking et al., 2013). Taubenböck et al. (2012) classified and monitored spatial variations of urbanization from 1970s to present, taking 27 megacities across the world (Figure 2.8a and b).

On a local scale, active and passive sensor systems are used for thematically and geometrically highly resolved land cover classifications in 2D (e.g., Blaschke, 2010; Figure 2.8c) and 3D (e.g., Rottensteiner et al., 2014). Especially multisensorial approaches combining optical satellite data and digital surface models allow derivation of highly detailed 3D city models for characterizing urban morphology (Sirmacek et al., 2012; Figure 2.8d).

The multiscale capabilities of capturing urban patterns, dynamics, and morphologies have driven researchers toward many different geographic fields: Most obvious, studies analyzing the physical dimension of cities have developed: evaluating and comparing spatial urban pattern configurations and their temporal evolution (Angel et al., 2005) or spatial characteristics have been suggested to turn qualitative urban concepts such as megaregions into spatial ones (Taubenböck et al., 2014). Characterizing urban morphology based on EO data has been presented for biotope types or particular urban structure types such as Central Business Districts (Taubenböck et al., 2013; Figure 2.8d), slums (Baud et al., 2010) or arrangements of green spaces (Zhou and Wang, 2011). Multitemporal approaches also allow for postdisaster damage assessments of buildings (Wegscheider et al., 2013). Beyond the capabilities of capturing urban patterns, dynamics, and morphologies, thermal EO data also have been applied for mapping urban climates and gradients of urban heat islands (Weng, 2009) and its effects toward the rural surroundings (Voogt and Oke, 2003).

2.3.2.2 Inland Water

Inland water ecosystems play an essential role for all human life. They serve for water supply, energy production, transport, recreation and tourism; maintenance of the hydrological balance; retention of sediments and nutrients; and provision of habitats for various fauna and flora. Since they are often extensively modified by humans and are among the most threatened ecosystem types of all, sustainable management of freshwater resources has gained importance at regional and global scales (UN, 2002, 2005), and a number of directives regulate their monitoring, for example, EU (2000). Although it may appear obvious to use

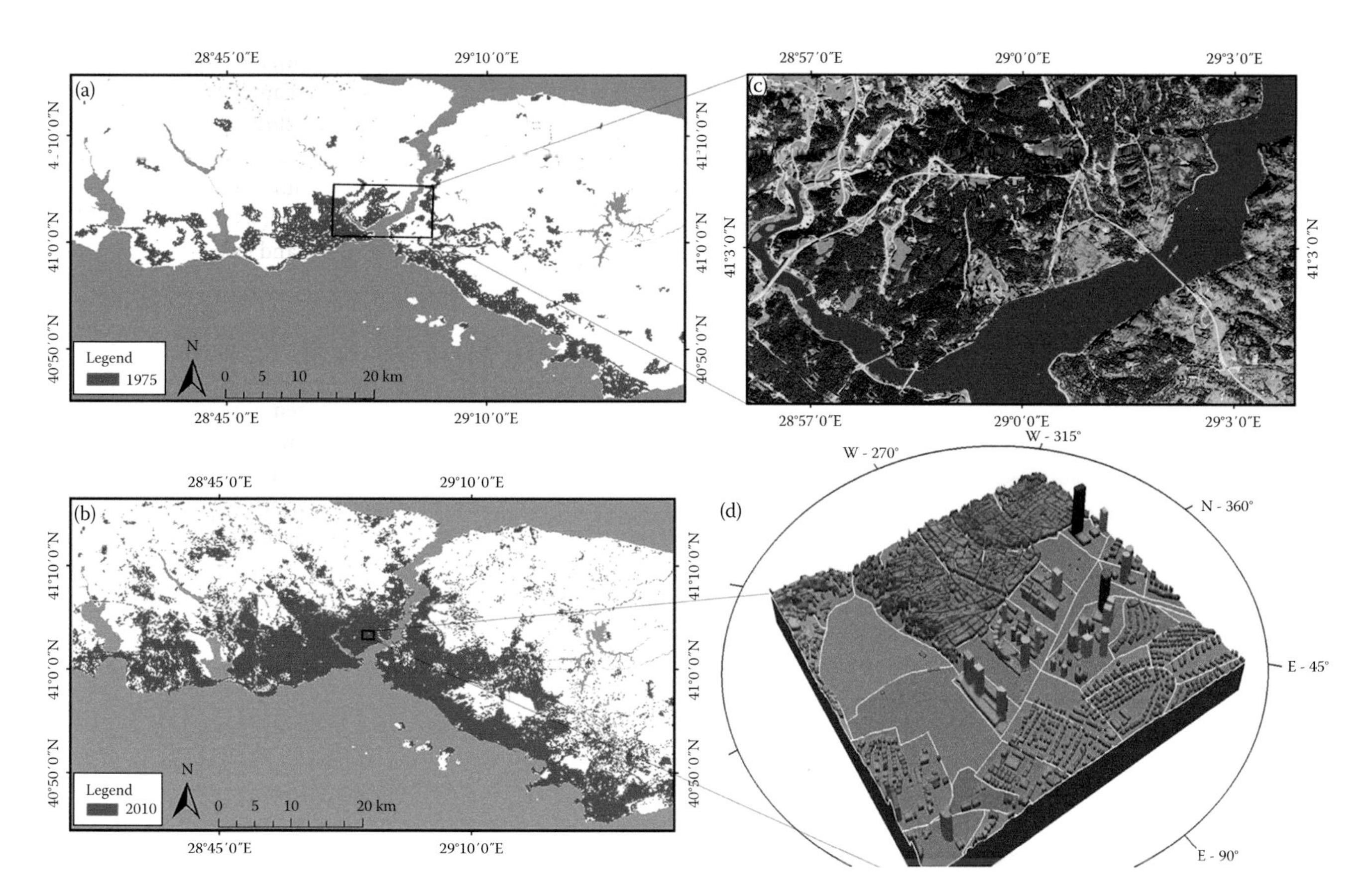

FIGURE 2.8 (a) Urban footprint classification of megacity Istanbul for 1975 based on Landsat MSS data; (b) Urban footprint classification of megacity Istanbul for 2010 based on TerraSAR-X data visualizing spatial urban growth; (c) 2D classification of buildings (red), vegetation (green), streets (yellow) of urban patterns based on Ikonos data; (d) 3D city model based on a Cartosat-1 digital surface model—the Central Business District of Levent, Istanbul.

satellites or airplanes to get a synoptic picture of entire ecosystems, routine monitoring of inland waters is still based on traditional water sampling and does not make use of remote sensing. Many studies demonstrated the potential of remote sensing (overviews in Dekker et al., 1995; Lindell et al., 1999; Papoutsa and Hadjimitsis, 2013), but operational services such as known for marine and coastal sectors are still missing.

Remote sensing of lake water quality from satellite began with Landsat MSS (Lindell et al., 1999). The first water quality maps were based on very general concepts (trophic class), but chlorophyll-a was already mapped in 1974 (Strong, 1974). The high spatial resolution of multispectral sensors designed for land applications allows mapping of many inland water systems including shallow areas. For this reason, Landsat MSS, which had a pixel size of 60 m, and its successors like Landsat TM and ETM+, but also SPOT, IRS, Ikonos, Quickbird, and the German commercial RapidEye sensors, with pixel sizes down to 2.5 m, were used for inland water studies (Dekker et al., 1995; Lindell et al., 1999). However, their coarse spectral resolution and low radiometric dynamics poses strong limitations to water applications (Lindell et al., 1999). The lack of spectral and radiometric details restricts data analysis to empirical approaches (Matthews, 2011), which require calibration against field data and cannot be transferred to other regions and seasons. This impracticality for automated data analysis makes them unattractive for routine monitoring tasks. Better suited are multispectral sensors designed for ocean color and imaging spectrometers. A number of lake-related studies were made with both sensor types, but past and current ocean color sensors like SeaWiFSS (Sea Viewing Wide Field-of-View Sensor, NASA), OCM (Ocean Color Monitor, Japan), MODIS, MERIS, or GOCI (Geostationary Ocean Color Imager, Korea) (IOCCG, 2010) provide spatial resolutions (0.3–1 km) too low for the majority of inland waters, while the current spaceborne imaging spectrometers with pixel sizes below 100 m (Hyperion (USGS), CHRIS (ESA), HJ-1A (China), HICO (hyperspectral imager for the Coastal Ocean, NASA), HSE RESURS-P (hyperspectral equipment on the Russian Resurs-P platform) (Staenz et al., 2013) are experimental instruments and do not provide data on a regular basis.

The parameters detectable in inland waters include concentrations of phytoplankton as an indicator of the trophic state, dissolved organic components and suspended matter as tracers for inflow of pollutants, optical properties like turbidity, attenuation, Secchi depth and euphotic depth, and the surface temperature (Lindell et al., 1999). In shallow areas, additionally, the bottom type can be classified and water depth determined. With the exception of chlorophyll-a, for which empirical algorithms using two or three bands in the red and NIR provide robust estimates in turbid waters (Gitelson et al., 2011), reliable data analysis requires algorithms that are based on radiative transfer models.

For operational ocean color sensors, such algorithms have been developed and are used to produce maps of water constituents on a routine basis. However, since they are designed to process global oceanic data sets, these generic procedures often fail in coastal and inland waters (Cui et al., 2010; Huth and Gege, 2004).

The only way to determine water quality parameters reliably, that is, with known uncertainties, is to model the underlying physical processes using the optical properties of all involved components. A number of models and inversion techniques have been developed over the years, but no one can be considered optimal for all situations (Odermatt et al., 2012). Major problems are the ambiguity of the reflectance spectrum (Defoin-Platel and Chami, 2007); for example, different combinations of water constituents can lead to the same spectrum, and the large variability of type and composition of water constituents and bottom substrates. Because of this, accurate results require site-specific information. Although many algorithms were developed in the last decades (Dekker et al., 2011), the first software tools that can be tuned by the user were published only recently (Gege, 2014; Giardino et al., 2012).

So far, not only the complex optics of inland waters prohibits their routine quality control from satellite, but also the problems of atmosphere and sun glint correction are not yet solved. Compared to the atmosphere, the water body reflects little radiation and therefore appears dark; from satellite, only 5%–10% of the measured signal describes the water body (IOCCG, 2010). An inland water-specific problem is the adjacency effect, where scattering in the atmosphere contaminates water pixels with radiation from the surrounding land. The range of this effect can be in the order of kilometers, and reliable correction is not yet possible. For inland waters also, the standard algorithm for correcting the unavoidable reflections at the water surface (Cox and Munk, 1954) is not applicable, because the underlying statistical relationship requires undisturbed wind fields for many kilometers.

Due to the described challenges, remote sensing of inland waters is still in its infancy. However, the upcoming new generation of hyperspectral satellite sensors with pixel sizes of 30 m (e.g., HyspIRI (hyperspectral infrared imager, NASA), PRISMA (Hyperspectral Precursor and Application Mission, Italy), HISUI (hyperspectral imager suite, Japan), EnMAP (Environmental Mapping and Analysis Program, Germany) (Staenz et al., 2013), and modern multispectral sensors like Sentinel-2 and -3 (ESA) will foster the development of robust algorithms. Synergetic usage of sensors with complementary properties, for example, with high spectral and geometric resolution, bears the potential to solve some of the problems and to extend the range of applications.

2.3.3 Digital Elevation Models

DEMs are of fundamental importance for a broad range of commercial and scientific applications in areas like hydrology, glaciology, forestry, geology, and urban planning. Nowadays, high-quality DEMs are being derived mostly by means of remote sensing from a variety of data acquired by air- and spaceborne sensors using laser scanning and radar (interferometric SAR) principles, providing users with a multitude of products with different horizontal and vertical resolutions, accuracies, and spatial coverage. The following terms are distinguished:

- Digital surface model (DSM): Height information of the scenery envelope including man-made features and vegetation. Usually, all remote sensing systems (laser and radar) provide this type of data.
- Digital terrain model (DTM): Height information of the bare Earth surface without man-made features and vegetation. In case of InSAR-derived DSMs, this product is usually computed using specific value adding techniques. For airborne laser scanning (ALS) DSMs and DTMs are generated simultaneously from the set of raw data.

Regional scale DEMs mostly are products of lidar and airborne InSAR measurements. ALS is based on lidar principles (light detection and ranging). The travel distance of short laser pulses to objects on ground is measured and evaluated. Different scanning principles are possible, and the aircraft position attitude is precisely considered to infer the 3D location of the reflection from ground. Due to their exceptional resolution in the decimeter range, the applications of these DEMs are in areas of flood and coastline protection, mapping of open cast mining activities, archeology, and line routing planning, usually being limited to regional scales.

In airborne interferometric synthetic aperture radar (InSAR) systems, the reflected radar signal is received in a side-looking geometry. The data are afterward combined interferometrically to derive the topography information providing swath width of several kilometers, thus being more economic during data acquisition compared to ALS. In consequence, during the last decade, companies like Intermap (United States/Germany) and OrbiSAT (Brazil) were able to generate countrywide DEM products, with vertical accuracies in the 1–2 m range and horizontal posting of approximately 5 m. Besides these commercial service providers, airborne InSAR-derived DEMs are being generated occasionally on request by several national research organizations using their own sensors, especially if the application requires up-to-date topography information.

There has also been a tremendous development for generating global scale DEMs, however, at reduced height resolution and posting. On a global scale, the most prominent DEMs originate from SRTM, ASTER-GDEM, and the TanDEM-X mission. The SRTM mission, a joint initiative of American and German space agencies NASA and the German Aerospace Center (DLR), was the first system fulfilling the requirements for a global, homogeneous, and reliable DEM with the DTED-2 specification (digital terrain elevation data, level 2 with 30 m posting, 12 m height accuracy) at least for latitudes up to approx. 60° north and 58° south, with polar regions not being accessible due to the shuttle orbit. While NASA used a C-band radar to record the entire accessible land surface of the Earth with an height precision of ±10 m, DLR used an X-band radar to cover a smaller swath, but with a precision of ±6 m. For more than one decade, researchers worldwide use these data sets, as they were made accessible (at least partially) free of cost.

A second type of global DEM has been computed from ASTER stereopair images. The improved global digital elevation model (GDEM V2) was released in 2011 (Fujisada et al., 2012); its coverage spans from 83° north latitude to 83° south, encompassing 99% of Earth's landmass including polar regions. The spatial resolution is similar to one of the SRTM data set. Although version 2 shows significant improvements over the previous release, users are advised that the data contains anomalies and artifacts that will impede effectiveness for use in certain applications. An attempt for improving the quality of this free-of-charge data set has been made by Intermap that offers the NextMap World30™ product based on the height values of ASTER GDEM V2, but with reduced artifacts.

A new worldwide DEM, from pole to pole, with an accuracy of a new dimension, has been acquired from 2010 to 2013 by the TanDEM-X mission (Krieger et al., 2007). Flying only few hundred meters apart, the two twin satellites, TerraSAR-X and TanDEM-X (TerraSAR-X add-on for Digital Elevation Measurement), form the first configurable SAR interferometer in space. The derived DSMs are characterized by a 12 m posting and vertical accuracy of 2 (relative) and 10 m (absolute, see also Figure 2.9 and Table 2.5).

For scientific use, DLR is offering preliminary sample data products (intermediate DEMs) within announcements of opportunity, while Airbus Defence and Space holds the exclusive commercial marketing rights for the WorldDEM™, which is the adaptation of the TanDEM-X elevation model to the needs of commercial users worldwide.

2.3.4 Data Policy

Remote sensing data policy underwent an alternating history. The availability of remote sensing data strongly depends on national distribution policies and pricing. In general there coexist three models of data distribution policy:

1. Web-enabled, free: Organizations that provide data and associated products freely available for every interested user (e.g., NASA's/USGS's Landsat data or MODIS products, Brazil's CBERS data);
2. Commercial, at a cost: Commercial remote sensing industry, which mainly is located in the very high spatial resolution domain, began with the launch of Ikonos and Quickbird (see also Section 2.3.1.1). Most commercial systems, however, operate with meter or submeter resolutions (e.g., WorldView, GeoEye, Pleiades [France]) and allow the detection of small-scale objects, such as elements of residential housing, commercial buildings, transportation systems, and other utilities, which can assist industrial geospatial applications (Navulur et al., 2013), but may also be purchased by noncommercial users or scientist.

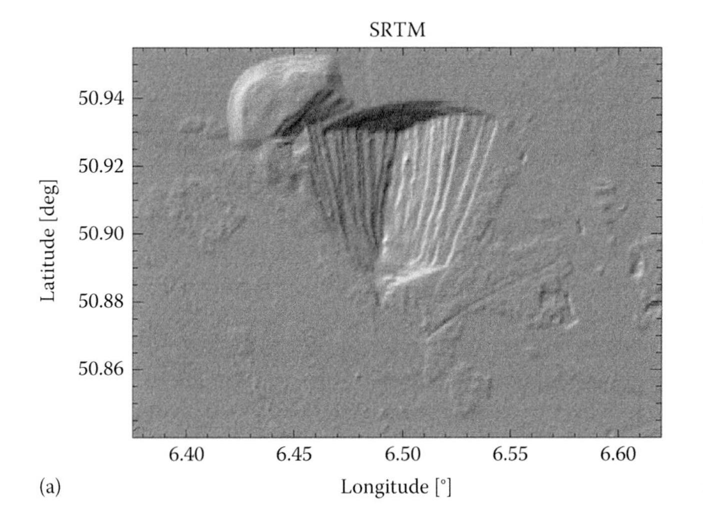

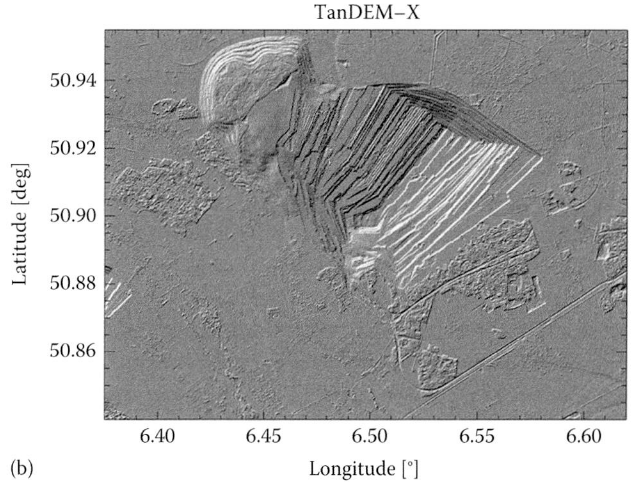

FIGURE 2.9 Example for DEM quality and timeliness for an open cast coal mine in Germany. Shaded relief representation of (a) SRTM on 30 m grid, 6 m height accuracy, recorded in 2000; (b) TanDEM-X on 12 m grid, 2 m height accuracy, recorded in 2011. (© DLR.)

TABLE 2.5 Available DEMs, Corresponding Resolutions and Accuracies; Accuracies Are Indicated According to the EC90 Standard

DEM	Spatial Resolution	Absolute Vertical Accuracy	Relative, Point-To-Point Vertical Accuracy
ALS	<1 m × 1 m	<1 m	<10 cm
Airborne InSAR DEMs (e.g., NextMap)	5 m × 5 m	<2 m	<1 m
ASTER GDEM V2 and World30	30 m × 30 m	<20 m	<17 m
SRTM DTED-1	90 m × 90 m (3 arcsec)	<30 m	<20 m
SRTM DTED-2	30 m × 30 m (1 arcsec)	<18 m	<12 m
TanDEM-X DEM; WorldDEM	12 m × 12 m (0.4 arcsec)	<10 m	<2 m

3. Databuy: Public–private partnerships support the scientific community with cost-free data, but also seek to establish commercial EO markets, that is, assist in developing remote sensing business in a way that follow-ups may be financed by industry using the profit from selling data or products to nonscientific clients. Examples for public–private partnerships are the RapidEye fleet as well as the TerraSAR and TanDEM-X missions. The SPOT sensors are a successful example for a public–private partnership, which led to a commercial funding and operation of SPOT-6 by Airbus (launch in 2014).

2.3.5 Data Availability and Comparability

In the past two decades, the availability of remote sensing data has been growing enormously, while the real costs decreased. In parallel, the costs and efforts to gather/download, process, and analyze remote sensing data have decreased. The development of personal computers and image-processing software was tremendous, and at present, sophisticated processing is possible on single computers or laptops. The use of remote sensing data, however, not only depends on its availability, but also on its prize. Since Landsat data are freely available, scientific investigations and applications have spurred a rapid increase (Wulder et al., 2012); similar experiences have been made for MODIS (Masuoka et al., 2011) and MERIS (Curran and Steele, 2005) products.

At present, a diversity of data sets is available, ranging from very high to coarse-scale resolution data. Converting these data into valuable products for terrestrial application is a challenge and has already been aimed at by Bartholomé and Belward (2005). However, the definition of products for terrestrial applications is challenging and needs to be improved. Prominent examples are global products such as LULC classifications, which provide a good global representation of general land cover classes but are neither easily comparable nor congruent to other land cover classifications (Herold et al., 2008; McCallum et al., 2006; Section 2.3.1.3). Questions of integrity and reusability of global LULC maps therefore will have to be further discussed. In response to this need, international bodies such as Group on Earth Observation (GEO), Global Climate Observation System (GCOS), and EO communities in Europe have been involved to establish an operational and continuous LULC observing system that includes integration, harmonization, and validation of data sets (Mora et al., 2014). Existing global LULC and forest initiatives serve as pioneers for future initiatives providing cloud computing and crowd funding. The commitment of these many agents including space agencies and industrial partners to supporting an ongoing program such as GFW was unprecedented (Baltuck et al., 2013).

GEO also promotes free and open data within its scope to build a Global Earth Observation System of Systems (GEOSS). GEO has been launched as a response to the World Summit on Sustainable Development and the Group of Eight (G8) leading industrialized countries, which called for action concerning international collaboration for exploiting the potential of EO (Nativi et al., 2013). In 2014, it includes 97 national governments including the European Commission as members; the numbers of 87 partizipating organizations includes intergovernmental, international, and regional organizations such as space agencies and United nation bodies (GEO, 2015; Nativi et al., 2013). GEOSS is conceived as a global network of product providers aiming provision of decision-support tools to a great variety of stakeholder. Focused community portals have developed to meet the particular needs of individual communities and to develop solutions targeted to these specialized users (GEO, 2010). As GEO approaches the end of its first 10-year (2005-2015) mandate, the ministers from GEO member governments and leaders from GEO participating organizations decided to extend their political commitment to the GEO vision through 2025. Discussions on the implementation plan for GEOSS are ongoing (GEO, 2015). This kind of coordinated development of methodologies, processing, and validation may lead to a new era of generating standardized data products.

SAR data of the last two decades are available within the archives of several space agencies. With respect to radar, the ESA-owned ERS-1/2 (1991–2000/1995–2011) as well ENVISAT (2002–2012) data, which are available on request and usually delivered free of charge for scientific applications, whereas for ALOS PALSAR (2006–2011) and Radarsat-2 data (since 2007) commercial distribution is favored. Free-of-charge data for supporting scientific applications of the worldwide EO community are further available within announcements of opportunity of the high-resolution SAR satellites (e.g., TerraSAR-X and TanDEM-X, Radarsat-2, and CosmoSkyMed).

2.4 Future Possibilities

2.4.1 Enhanced Technology for New Applications

Various new sensors including new technologies have been recently launched or are planned: the U.S. National Space Policy of 2010 stated that a primary goal is the improvement of spaceborne EO to conduct science, monitor climate and global change, manage natural resources, and to support disaster response and recovery; moreover, it included the need of long-term observations (USA, 2010). The latest generation of NASA's flagship, Landsat 8, has been launched in early 2013 and provides the very valuable continuity of the Landsat missions. Other missions such as the Soil Moisture Active Passive (SMAP, United States) or the FLuorescence EXplorer (FLEX, ESA) are also highly relevant for environmental monitoring. The SMAP has been launched in early 2015 and carries a SAR (L-band, 1.2-1.4 GHz) and a radiometer, aiming at soil moisture measurement for water and energy cycling analysis. The antenna covers a 1000 km swath, providing global coverage within 3 days at the equator and 2 days at boreal latitude (NASA EOSPO, 2014).

FLEX is a candidate for ESA's eighth Earth Explorer and aims to provide global maps of vegetation fluorescence, specifically designed for the estimation of vegetation fluorescence on a global scale. The FLORIS (Fluorescence Imaging Spectrometer) instrument onboard FLEX will measure the radiance between 500 and 800 nm with a bandwidth between 0.1 nm and 2 nm, providing images with a 150 km swath and 300 m pixel size. This information will allow a detailed monitoring of vegetation dynamics, by improving the methods for the estimation of classical biophysical parameters (Moreno and Moran, 2014).

Another future U.S. mission suitable for land applications is the coarse-scale hyperspectral GEO-CAPE (Geostationary Coastal and Air Pollution Events), which is planned to be in orbit in 2020. GEO-CAPE may be an improved MODIS successor and will provide data for a variety of land applications such as water resources, agriculture, disasters monitoring, and ecological forecasting (NASA, 2014). Another hyperspectral, but medium-scale (60 m at nadir) instrument may be the hyperspectral infrared imager (HyspIRI), which at present is at the study stage. HyspIRI includes two instruments: an imaging spectrometer measuring from the VIS to TIR with 10 nm contiguous bands and a multispectral imager measuring from 3 to 12 μm in SWIR and TIR; its data might be used for a wide variety of studies primarily in the carbon cycle and ecosystem and earth surface (JPL, 2014).

To support the highly needed high-resolution environmental monitoring, the European Union (EU) recently implemented the Copernicus program (former, Global Monitoring for Environment and Security [GMES]). Copernicus is the European flagship for EO, tailored to the monitoring of the environment and security relevant information; therefore, it will provide global information in an operational manner tailored to the needs of a broad range of applications. The Copernicus comprises two types of EO missions: the Sentinel families (Sentinel-1 to 6) and contributing missions. There are around 30 existing or planned contributing missions from ESA or their member states, EUMETSAT, and other third party (European and international) mission operators, which make part of their data available for Copernicus (Berger and Aschbacher, 2012). The EO capabilities provided through the Copernicus system will cause a paradigm shift in EO with the robustness and reliability of a long-term operational infrastructure fostering research and innovation in the field of EO following a similar development as introduced for meteorological mission and services some years back.

With the launch of Sentinel-1A in 2014, new opportunities arise in the microwave domain, offering repeated global acquisitions with a repeat cycle of 12 days, which will be reduced to 6 days with the launch of Sentinel-1B. Due to the free and open data policy, these data are expected to boost the application range of microwave techniques such as D-InSAR. Further SAR data for these applications are provided in L-band by ALOS-2 (Japan, launched in 2014) and will be supplemented by the two upcoming SAOCOM (Satellites for Observation and Communications, Argentina, with expected launches in 2016 and 2017), followed by the Radarsat Constellation Mission (C-band, Canada, launch in 2018). Since the commercial market is demanding high-resolution data, being usually acquired at X-band, new satellites are likely to be launched as private or national initiatives within the next decade for replacing the present generation of TerraSAR-X and CosmoSkyMed sensors.

The pair of Sentinel-2 satellites (2015/2016) will routinely provide medium spatial resolution optical images over land surfaces globally. Sentinel-2 will ensure continuity of SPOT type observation with improvements in terms of spectral resolution and temporal coverage. The multispectral imager (MSI) instrument aboard Sentinel-2 covers the VIS, NIR, and SWIR in 13 spectral bands with a spatial resolution of 10 m × 10 m (continuity of SPOT observation), 6 bands of 20 m × 20 m, and with 3 bands of 60 m × 60 m dedicated to atmospheric corrections and cloud screening. The increased swath width of about 290 km together with having two satellites in orbit at the same time will enable a revisit time of 5 days at equator and 2–3 days at midlatitudes. This will allow the monitoring of rapid changes such as vegetation characteristics during the growing season and improved change detection techniques.

Sentinel-3 is primarily a mission to support services relating to the marine environment, and the first Sentinel-3 satellite is expected to launch in 2015/2016. The system, however, will also carry the *ocean and land color instrument* (OLCI), which is based on heritage from the MERIS instrument but with improved wavelength bands (21 compared to 15 on MERIS) with a spatial resolution of 300 m × 300 m (full resolution mode).

National space missions will trump international space programs; for example, the German EnMAP (launch in 2018) will set a milestone for spaceborne imaging spectroscopy with Landsat-like spatial resolution. With 244 bands in the VIS and SWIR and a, by pointing increased, revisit time of 4 days, EnMAP will provide highly resolved time-series to decipher the response of ecosystems on natural and man-made pressure and to push forward multidisciplinary applications on climate change, LULC changes, hazards, and environmental pollution (Kaufmann et al., 2012).

2.4.2 Data Availability and Continuity

As already depicted, data continuity and availability are two important decisions of international data policies, which will improve EO data for a wide range of applications (Lynch et al., 2013). To contribute to standardized measures especially for global monitoring, continuous information of our planet and its availability are of high importance (Pereira et al., 2013). In this context, the Landsat (see also Figure 2.2), SPOT as well as ESA's C-band SAR missions already have set standards.

Copernicus has a free and open data policy, following the GEO data principle, which is expected to further boost research and innovation. To further increase the use of remote sensing data for new applications, the capability to develop standardized products based on automatic thematic processors for consistent knowledge generation and provision may be a prerequisite. Furthermore, the foreseeable increase in data volume will induce new challenges in terms of data storage, handling, mining, processing techniques, and accuracy requirements. Investment in processing is still comparatively high due to mostly not full automation of information generation procedures due to, for example, different atmospheric conditions, land cover types, different user's requirements, or the algorithms still being in experimental status. Therefore, robust methodologies are needed.

As it becomes more and more understood and accepted that single disciplines are decreasingly able to progress individually, this paradigm shift may be observed increasingly. Multidisciplinary approaches using EO-based data in combination with other data sources are developing with the idea for geographic value-adding to these spatial data beyond the sole physical measurements. Besides multidisciplinary approaches, multisensor applications will form a further spreading technique to enhance existing approaches, which requires temporally overlapping and well-calibrated sensors. One example may be the synergistic use of Landsat-8 and Sentinel-2 data, which paves the way to capture near-daily data (Wulder et al., 2012) and further ensures continuity of data if one sensor fails. International cooperation and data sharing will also reduce instrument and launch costs and mission redundancy (Durrieu and Nelson, 2013). Joint commitments between national and international space agencies may therefore be a powerful means of stabilizing new space initiatives, and hence data availability and continuity (Morel, 2013).

2.5 Conclusions

From the early beginning of satellite-based remote sensing in the 1970s, the use of remote sensing for terrestrial applications has increased tremendously. Manifold applications from field to global scales have been established since then. The needs and uses of the terrestrial EO community, however, are divers, with various user communities, sensor types, and data needs. Different systems are therefore needed, which depend on the required information. Redundant developments and the replication of existing EO programs have often been perceived as a waste of resources, but may also be seen as potential reservoir for innovations. The synergy between missions and the potential to capitalize upon sensors with different focus and output is high, as could have been experienced with Landsat, MODIS, and MERIS. Moreover, multidisciplinary studies that use the increased data density to investigate temporal dynamics on various scales point toward future development of well-calibrated, global products.

Besides affordability and sustainability, long-term scenarios may aim to satisfy the priorities of continuity, frequency of observations, (inter-)calibration of sensors, and evolution of the system: continuity of operational observations drives the nature of the observations and the type of missions; the frequency of observations drives the number of simultaneous sensors in orbit for each class of observations; (inter-)calibration of sensors drives data continuity, and the evolution of the sensors based on user needs, which accounts for the demand for new services and products, demand for new observations and technology improvements.

References

Allison, L.J. and Neil, E.A. 1962. Final report on the TIROS 1 meteorological satellite system. NASA technical report R-131, Goddard Space Flight Center, Greenbelt, MD. 20771.

Angel, S., Sheppard, S., Civco, D. L., and Buckley, R. 2005. The dynamics of global urban expansion. Department of Transport and Urban Development, The World Bank: Washington, DC.

Angelsen, A., Brockhaus, M., Sunderlin, W.D. and Verchot, L.V. (eds.). 2012. Analysing REDD+: Challenges and choices. CIFOR, Bogor (Indonesia).

Arino, O., Bicheron, P., Achard, F., Latham, J., and Witt, R. 2008. The most detailed portrait of Earth. *ESA Bulletin*, 136, 25–31.

Balmford, A., Bennun, L., ten Brink, B., Cooper, D., Côté, I. M., Crane, P., Dobson, A. et al. 2005. The convention on biological diversity's 2010 target. *Science*, 307, 212–213.

Baltuck, M., Briggs, S., Loyche-Wilkie, M., McGee, A., Muchoney, D., and Skøvseth, P. E. 2013. The Global Forest Observation Initiative: Fostering the use of satellite data in forest management, reporting and verification. *Carbon Management*, 4(1), 17–21.

Barnsley, M. J., Settle, J. J., Cutter, M. A., Lobb, D. R., and Teston, F. 2004. The CHRIS/Proba Misison: A low cost Smallsat for hyperspectral, multi-angle, observations of the earth surface and atmosphere. *IEEE Transactions on Geoscience and Remote Sensing*, 42(7), 1512–1520.

Barrett, E. C. and Curtis, L. F. 2013. *Introduction to Environmental Remote Sensing*. New York: Taylor & Francis.

Bartholomé, E. and Belward, A. S. 2005. GLC2000: A new approach to global land cover mapping from Earth observation data. *International Journal of Remote Sensing*, 26(9), 1959–1977.

Baud, I., Kuffer, M., Pfeffer, K., Sliuzas, R., and Karuppannan, S. 2010. Understanding heterogeneity in metropolitan India: The added value of remote sensing data for analyzing sub-standard residential areas. *International Journal of Applied Earth Observation and Geoinformation*, 12(5), 359–374.

Berger, M. and Aschbacher, J. 2012. The sentinel missions—New opportunities for science. *Remote Sensing of Environment*, 120, 1–276.

Blackburn, A. 2007. Hyperspectral remote sensing of plant pigments. *Journal of Experimental Botany*, 58(4), 855–867.

Blaschke, T. 2010. Object based image analysis for remote sensing. *ISPRS Journal of Photogrammetry and Remote Sensing*, 65(1), 2–16.

Bontemps, S., Defourny, P., Bogaert, E. V., Arino, O., Kalogirou, V., and Perez, J. R. 2011. GLOBCOVER 2009. Products description and validation report, UCV Louvain and ESA, Louvain, Belgium.

Brown, M. E., Escobar, V. M., Aschbacher, J., Milagro-Perez, M. P., Doorn, B., Macauley, M. K., and Friedl, L. 2013. Policy for robust space-based earth science, technology and applications. *Space Politics*, 29(1), 76–82.

Buermann, W., Saatchi, T., Smith, T. B., Zutta, B. R., Chaves, J. A., Milá, B., and Graham, C. H. 2008. Predicting species distributions across the Amazonian and Andean regions using remote sensing data. *Journal of Biogeography*, 35(7), 1160–1176.

Campbell, J. B., and Wynne, R. H. 2011. *Introduction to Remote Sensing*. New York: The Guilford Press.

Chevrel, M., Courtois, M., and Weill, G. 1981. The SPOT satellite remote sensing mission. *Photogrammetric Engineering & Remote Sensing*, 47, 1163–1171.

Christy, C. D. 2008. Real-time measurement of soil attributes using on-the-go near infrared reflectance spectroscopy. *Computers and Electronics in Agriculture*, 61(1), 10–19.

Collen, B., Pettorelli, N., Baillie, J. E. M., and Durant, S. 2013. *Biodiversity Monitoring & Conservation: Bridging the Gap between Global Commitment and Local Action*. Cambridge, U.K.: Wiley-Blackwell.

Colwell, R. 1956. Determining the prevalence of certain cereal crop diseases by means of aerial photography. *Hilgardia*, 26, 223–286.

Corwin, D. L. and Lesch, A. M. 2003. Application of soil conductivity to precision agriculture. *Agronomy Journal*, 95(3), 455–471.

Cox, C. and Munk, W. 1954. Statistics of the sea surface derived from sun glitter. *Journal of Marine Research*, 13, 198–227.

Cui, T., Zhang, J., Groom, S., Sun, L., Smyth, T., and Sathyendranath, S. 2010. Validation of MERIS ocean-color products in the Bohai Sea: A case study for turbid coastal waters. *Remote Sensing of Environment*, 114, 2326–2336.

Curran, P. J. and Steele, C. M. 2005. MERIS: The re-branding of an ocean sensor. *International Journal of Remote Sensing*, 26(9), 1781–1798.

Davis, F., Quattrochi, D., Ridd, M., Lam, N. Walsh, S., and Michaelson, J. 1991. Environmental analysis using integrated GIS and remotely sensed data: Some research needs and priorities. *Photogrammetric Engineering & Remote Sensing*, 57(6), 689–697.

Defoin-Platel, M. and Chami, M. 2007. How ambiguous is the inverse problem of ocean color in coastal waters? *Journal of Geophysical Research*, 112, C03004, doi:10.1029/2006JC003847.

DeFries, R., Hansen, M., and Townshend, J. 1995. Global discrimination of land cover types from metrics derived from AVHRR pathfinder data. *Remote Sensing of Environment*, 54(3), 209–222.

DeFries, R., Hansen, M., and Townshend, J. 1998. Global land cover classifications at 8 km spatial resolution: The use of training data derived from Landsat imagery in decision tree classes. *International Journal of Remote Sensing*, 19(16), 3141–4168.

Dekker, A. G., Malthus, T. J., and Hoogenboom, H. J. 1995. The remote sensing of inland water quality. In Danson, F. M. and Plummer, S. E. (eds.). *Advances in Environmental Remote Sensing*. New York: Wiley & Sons Ltd.

Dekker, A. G., Phinn, S. R., Anstee, J., Bissett, P., Brando, V. E., Casey, B., Fearns, P. et al. 2011. Intercomparison of shallow water bathymetry, hydro-optics, and benthos mapping techniques in Australian and Caribbean coastal environments. *Limnology and Oceanography: Methods*, 9, 396–425.

Delegido, J., Fernandez, G., Gandla, S., and Moreno, J. 2008. Retrieval of chlorophyll and LAI of crops using hyperspectral techniques: Application to CHRIS/PROBA data. *International Journal of Remote Sensing*, 29(24), 7107–7127.

De Oliveira, L. C., Lima, M. G. R., and Hubscher, G. L. 2000. CBERS—An international space cooperation program. *Acta Astronautica*, 47, 559–564.

Digital Globe. 2014. Digital Globe satellite information. https://www.digitalglobe.com/resources/satellite-information (accessed March 24, 2014).

Doraiswamy, P. C., Moulin, S., Cook, P. W., and Stern, A. 2003. Crop yield assessment from remote sensing. *Photogrammetric Engineering & Remote Sensing*, 69(6), 665–674.

Duro, D., Coops, N. C., Wulder, M. A., and Han, T. 2007. Development of a large area biodiversity monitoring driven by remote sensing. *Progress in Physical Geography*, 31, 235–260.

Durrieu, S. and Nelson, R. F. 2013. Earth observation from space—The issue of environmental sustainability. *Space Politics*, 29(4), 238–250.

EEA (European Environment Agency). 2014. CORINE Land Cover. http://www.eea.europa.eu/publications/COR0-land cover (accessed March 27, 2014).

Elachi, C. 1988. *Spaceborne Radar Remote Sensing: Applications and Techniques*. New York: IEEE Press.

ESA (European Space Agency). 2012. *Report for Mission Selection: Biomass*. ESA SP-1324/1 (3). Noordwijk, the Netherlands: ESA.

Esch, T., Taubenböck, H., Roth, A., Heldens, M., Felbier, A., and Thiel, M. 2012. TanDEM-X mission: New perspectives for the inventory and monitoring of global settlement patterns. *International Journal of Applied Earth Observation and Geoinformation*, 6(1), 061702, doi: 10.1117/1.JRS.6.061702.

EU (European Union). 2000. European Water Framework Directive (Directive 2000.60/EC) of the European parliament and of the Council establishing a framework for the Community action in the field of water policy. Official Journal L327 of the EU. http://eur-lex.europa.eu/legal-content/EN/TXT/?uri=CELEX:02000L0060-20090625 (accessed March 3, 2014).

FAO (Food and agriculture organization of the United nations). 2001. Global Forest Resources Assessment 2000: Main report. FAO, Rome.

Feret, J. B., Francois, C., Asner, G. P., Gitelsen, A. A., Martin, R. E., Bidel, L. P., Ustin, S. L., le Maire, G., and Jaquemoud, S. 2008. Prospect 4 and 5: Advances in leaf optical properties model separating photosynthetic pigments. *Remote Sensing of Environment*, 112, 3030–3043.

Fischer, W. A., Hemphill, W. R., and Kover, A. 1976. Progress in remote sensing. *Photogrammetria*, 32, 33–72.

Fjeldså, J., Ehrlich, D., Lambin, E., and Prins, E. 1997. Are biodiversity "hotspots" correlated with current ecoclimatic stability? A pilot study using the NOAA-AVHRR remote sensing data. *Biodiversity and Conservation*, 6, 401–422.

Friedl, M. A., McIver, D. K., Hodges, J. C., Zang, X. Y., Muchoney, D., Strahler, A. H., Woodcock, C. E. et al. 2002. Global land cover mapping from MODIS: Algorithms and early results. *Remote Sensing of Environment*, 83(1–2), 287–302.

Friedl, M. A., Sulla-Menashe, D., Tan, B., and Schneider, A. 2010. MODIS Collection 5 global land cover: Algorithms refinements and characterization of new data sets. *Remote Sensing of Environment*, 114(1), 168–182.

Fritz, S., See, L., McCallum, I., Schill, C., Obersteiner, M., van der Velde, M., and Achard, F. 2011. Highlighting continued uncertainty in global land cover maps for the user community. *Environmental Research Letters*, 6(4), 044005, doi: 10.1088/1748-9326/6/4/044005.

Frolking, S., Milliman, T., Seto, K., and Friedl, M. A. 2013. A global fingerprint of macro-scale changes in urban structure from 1999 to 2009. *Environmental Research Letters*, 8, 024004, doi: 10.1088/1748-9326/8/2/024004.

Fujisada, H., Urai, M., and Iwasaki, A. 2012. Technical methodology for ASTER global DEM. *IEEE Transactions on Geosciences and Remote Sensing*, 50(1), 3725–3736.

Gamba, P. and Lisini, G. 2013. Fast and efficient urban extent extraction using ASAR wide swath mode data. *IEEE Journal of Selected Topics in Applied Earth Observations and Remote Sensing*, 6(5), 2184–2195.

Gaston, K. J. 2000. Global patterns in biodiversity. *Nature*, 405, 220–227.

Gebbers, R. and Adamchuk, V. L. 2010. Precision agriculture for food security. *Science*, 327(5967), 828–831.

Gege, P. 2014. WASI-2D: A software tool for regionally optimized analysis of imaging spectrometer data from deep and shallow waters. *Computers & Geosciences*, 62, 08–215.

GEO (Group on Earth Observations). 2010. Report on Progress Beijing Ministerial Summit—Observe, Share, Inform. Geneva, Switzerland: GEO Secretariat.

GEO (Group on Earth Observations). 2015. The GEO Website. https://www.earthobservations.org (accessed March 30, 2015).

GFW (Global Forest Watch). 2014. The Global Forest Watch Website. http://www.globalforestwatch.org/ (accessed April 7, 2014).

Giardino, C., Candiani, G., Bresciani, M., Lee, Z. P., Gagliano, S., and Pepe, M. 2012. BOMBER: A tool for estimating water quality and bottom properties from remote sensing images. *Computers & Geosciences*, 45, 313–318.

Gitelson, A. A., Gurlin, D., Moses, W. J., and Yacobi, Y. Z. 2011. Remote estimation of chlorophyll-a concentration in inland, estuarine, and coastal waters. In Weng, Q. (ed.). *Advances in Environmental Remote Sensing: Sensors, Algorithms and Applications*. Boca Raton, FL: CRC Press.

Goodchild, M. F. 2007. Citizens as sensors: The world of volunteered geography. *GeoJournal*, 69(4), 211–221.

Gould, M., Craglia, M., Goodchild, M. F., Annoni, A., Camara, G., Kuhn, K., Mark, D., Masser, I., Liang, S., and Parsons, E. 2008. Next Generation Digital Earth: A position paper from the Vespucci initiative for the advancement of geographic information science. *International Journal of Spatial Data Infrastructures Research*, 3, 146–167.

Goward, S. N., Masek, J. G., Williams, D. L., Irons, J. R., and Thompson, R. J. 2001. The Landsat 7 mission: Terrestrial research and applications for the 21st century. *Remote Sensing of Environment*, 78, 3–12.

Gutman, G. G. 1991. Vegetation indices from AVHRR: An upgrade and future prospects. *Remote Sensing of Environment*, 35, 121–136.

Hame, T., Salli, K., Andersson, K., and Lohi, A. 1997. A new methodology for the estimation of biomass of conifer-dominated boreal forest using NOAA AVHRR data. *International Journal of Remote Sensing*, 18(15), 3211–3243.

Hansen, M. C., de Fries, R. S., and Townshed, J. R. 2000. Global land cover classification at 1 km spatial resolution using a classification tree approach. *International Journal of Remote Sensing*, 21(6–7), 1331–1364.

Hansen, M. C., Potapov, P. V., Moore, R., Hancher, M., Turubanova, S. A., Tyukavina, A., Thau, S. et al. 2013. High-resolution global maps of 21st century forest cover change. *Science*, 342, 850–853.

Hastings, D. A. and Emery, W. 1992. The advanced very high resolution radiometer (AVHRR)—A brief reference guide. *Photogrammetric Engineering & Remote Sensing*, 58(8), 1183–1188.

Herold, M., Mayaux, P., Woodcock, C. E., Baccini, A., and Schmullius, C. 2008. Some challenges in global land cover mapping: An assessment of agreement and accuracy in existing 1 km datasets. *Remote Sensing of Environment*, 112(5), 2538–2556.

Hillnhütter, C., Mahlein, A.-K., Sikora, R. A., and Oerke E. C. 2011. Remote sensing to detect plant stress induced by *Heterodera schachtii* and *Rhizoctonia solani* in sugar beet fields. *Field Crop Research*, 122, 70–77.

Holben, B. N., Kaufman, Y. J., and Kenall, J. D. 1990. NOAA-11 AVHRR visible and near-IR inflight calibration. *International Journal of Remote Sensing*, 11(8), 1511–1519.

Horning, N., Robinson, J. A., Sterling, E. J., Turner, W., and Spector, S. 2010. *Remote Sensing for Ecology and Conservation*. New York: Oxford University Press.

Huete, A., Didan, D., Miura, T., Rodriguez, E. P., and Gao, X. 2002. Overview of the radiometric and biophysical performance of MODIS vegetation indices. *Remote Sensing of Environment*, 83(1–2), 195–213.

Huth, J. and Gege, P. 2004. Suspended matter dynamics in Lake Constance in 2003 derived from MERIS and MODIS satellite data. *Proceedings of ENVISAT & ERS Symposium*, Salzburg, Austria, September 6–10. ESA Special publication, SP-572 (CD-ROM).

IOCCG (International Ocean Colour Coordinating Group). 2010. Atmospheric correction for remotely-sensed ocean-colour products. In Wang, M. (ed.). *Reports of the IOCCG*, No. 10. Dartmouth, Nova Scotia, Canada: IOCCG.

Jones, H. G. and Vaughan, R. A. 2010. *Remote Sensing of Vegetation: Principles, Techniques, and Applications*. New York: Oxford University Press.

JPL (Jet Propulsion Laboratory, California Institute of Technology). 2014. HyspIRI mission study http://hyspiri.jpl.nasa.gov/ (accessed April 4, 2014).

Kaufmann, H., Förster, S., Wulf, H., Segl, K., Guanter, L., Bochow, M., Heiden, U. et al. 2012. *Science Plan of the Environmental Mapping and Analysis Program (EnMAP)*. Potsdam, Germany: German Research Center for Geosciences, Scientific Technical Report.

Kerr, J. T. and Ostrovsky, M. 2003. From space to species: Ecological applications for remote sensing. *Trends in Ecology & Evolution*, 18(6), 299–305.

Kneubühler, M., Koetz, B., Huber, S., Schöpfer, J., Richter, R., and Itten, K. 2006. Monitoring vegetation growth using multitemporal CHRIS/PROBA data. *IEEE Geoscience and Remote Sensing Symposium* (IGARSS 2006), IEEE, Denver, CO, pp. 2677–2680.

Krieger, G., Moreira, A., Fiedler, H., Hajnsek, I., Werner, M., Younis, M., and Zink, M. 2007. TanDEM-X: A satellite formation for high resolution SAR interferometry. *IEEE Transactions on Geosciences and Remote Sensing*, 45(11), 3317–3341.

Kuschel, H. and O'Hagan, D. 2010. Passive radar from history to future. *Proceedings of the 11th International Radar Symposium*, June 16–18, Vilnuis, Lithuania, pp. 1–4.

Lamb, D. W. and Brown, R. B. 2001. Precision agriculture: Remote sensing and mapping of weeds in crops. *Journal of Agricultural Engineering Research*, 78, 117–125.

Lillesand, T., Kiefer, R. W., and Chipma, J. 2008. *Remote Sensing and Image Interpretation*. New York: Wiley & Sons.

Lindell, T., Pierson, D., Premazzi, G., and Zilioli, E. (eds.). 1999. Manual for Monitoring European Lakes Using Remote Sensing Techniques. EUR 18665 EN, Joint Research Centre, Ispra, Italy.

Louet, J. and Bruzzi, S. 1999. ENVISAT mission and system. *Geoscience and Remote Sensing Symposium*. IGARSS'99 Proc 3, pp. 1680–1682.

Lucas, R. M., Bunting, P., Clewley, D., Proisy, C., Filho, P. W. M., Viergever, K., Woodhouse, I. et al. 2009. Characterisation and monitoring of mangroves using ALOS PALSAR data. JAXA Kyoto & Carbon Initiative, Phase 1 report. JAXA/Meti, Japan.

Lynch, M., Maslin, H., Balzter, H., and Sweeting, M. 2013. Sustainability: Choose satellites to monitor deforestation. *Nature*, 496, 293–294.

Mahlein, A. K., Oerke, E. C., Steiner, U., and Dehne, W. 2012. Recent advances in sensing plant diseases for precision crop protection. *European Journal of Plant Pathology*, 133, 197–209.

Masuoka, E., Roy, D., Wolfe, R., Morisette, J., Sinno, S., Teague, M., Saleous, N., Devadiga, S., Justice, C. O., and Nickeson, J. 2011. MODIS Land data products: Generation, quality assurance and validation. In Ramachandran et al. (eds.). *Land Remote Sensing and Global Environmental Change*, Remote Sensing and Digital Image Processing Series, Vol. 11. Springer, New York.

Matthews, M. W. 2011. A current review of empirical procedures of remote sensing in inland and near-coastal transitional waters. *International Journal of Remote Sensing*, 32, 6855–6899.

McCallum, I., Obersteiner, M., Nilsson, S., and Shvidenko, A 2006. A spatial comparison of four satellite derived 1 km global land cover datasets. *International Journal of Applied Earth Observation and Geoinformation*, 8(4), 246–255.

Meroni, M., Rossini, M., Guanter, L., Alonso, L., Rascher, U., Colombo, R., and Moreno, J. 2009. Remote sensing of solar-induced chlorophyll fluorescence: Review of methods and applications. *Remote Sensing of Environment*, 113, 2037–2051.

Misty, D. 1998. India's emerging space program. *Pacific Affairs*, 71, 151–174.

Mitchell, A. and Hoekman, D.H. Global Forest Observation Initiative (GFOI) - Review of priority research and development topics. GEO, Geneva (Switzerland), 159 p.

Miyazaki, H., Shao, X., Iwao, K., and Shibasaki, R. 2012. An automated method for global urban area mapping by integrating ASTER satellite images and GIS data. *IEEE Journal of Selected Topics in Applied Earth Observations and Remote Sensing*, 6, 1004–1019.

Mora, B., Tsendbazar, N. E., Herold, M., and Arino, O. 2014. Global land cover mapping: Current status and future trends. *Land Use Land Cover: Remote Sensing and Digital Image Processing*, 18, 11–30.

Moreira, A., Prats-Iraola, P., Yonis, M., Krieger, G., Hajnsek, I., and Papathanassiou, K. P. 2013. A tutorial on synthetic aperture radar. *IEEE Geoscience and Remote Sensing Magazine*, 43pp.

Morel, P. 2013. Advancing Earth observation from space: A global challenge. *Space Policy*, 29, 175–180.

Moreno, J. and Moran, S. 2014. Vegetation stress from soil moisture and chlorophyll fluorescence: Synergy between SMAP and FLEX approaches. *Proceedings of the EGU General Assembly*, 27 April - 2 May, 2014, Vienna, Austria.

Mulla, D. J. 2013. Twenty five years of remote sensing in precision agriculture: Key advances and remaining knowledge gaps. *Biosystems Engineering*, 114, 385–371.

Myers, N., Mittermeier, R. A., Mittermeier, C. G., da Fonseca, G. A. B., and Kent, J. 2000. Biodiversity hotspots for conservation priorities. *Nature*, 403, 853–858.

Nagendra, H., Lucas, R., Honrado, J. P., Jongman, R. H. G., Tarantino, C., Adamo, M., and Mairota, P. 2013. Remote sensing for conservation monitoring: Assessing protected areas, habitat extent, habitat condition, species diversity, and threats. *Ecological Indicators*, 33, 45–59.

NASA (National Aeronautics and Space Administration). 2014. GEO-Cape website. http://geo-cape.larc.nasa.gov/ocean-instrumentdesign.html (accessed April 4, 2014).

NASA EOSPSO (NASA's Earth Observing System Project Science Office). 2014. SMAP mission. http://eospso.nasa.gov/missions/soil-moisture-active-passive (accessed April 4, 2014).

Nativi, S., Mazzetti, P., Craglia, M., and Pirrone, N. 2013. The GEOSS solution for enabling data interoperability and integrative research. *Environmental Science and Pollution Research*, 21, 4177–4192.

Navulur, K., Pacifici, F., and Baugh, B. 2013. Trends in optical commercial remote sensing industry. *IEEE Geoscience and Remote Sensing Magazine*, December, 57–64.

Odermatt, D., Gitelson, A., Brando, V. E., and Schaepman, M. 2012. Review of constituent retrieval in optically deep and complex waters from satellite imagery. *Remote Sensing of Environment*, 118, 116–126.

Oloffson, P., Foody, G. M., Stehman, S. V., and Woodcock, C. E. 2013. Making better use of accuracy data in land change studies: Estimating accuracy and area and quantifying uncertainty using stratified estimation. *Remote Sensing of Environment*, 129, 122–131.

Oloffson, P., Stehman, S. V., Woodcock, C. E., Sulla-Menashe, D., Sibley, A. M., Newell, J. D., Friedl, M., and Herold, M. 2012. A global land-cover validation data set, part I: Fundamental design principles. *International Journal of Remote Sensing*, 33, 5768–5788.

Oppelt, N. 2010. Monitoring of the biophysical status of vegetation using multi-angular, hyperspectral remote sensing for the optimization of a physically-based SVAT model. In Kieler Geogr Schriften 121:123f. Kiel, Germany: Kiel University.

Oppelt, N. and Mauser, W. 2004. Hyperspectral monitoring of physiological parameters of wheat during a vegetation period using AVIS data. *International Journal of Remote Sensing*, 25(1), 145–160.

Papoutsa, C. and Hadjimitsis, D. G. 2013. Remote sensing for water quality surveillance in inland waters: The case study of Asprokremmos dam in Cyprus. In Hadjimitsis, D. G. (ed.). *Remote Sensing of Environment—Integrated Approaches*, InTech, Croatia, 211pp.

Pereira, M., Ferrier, S., Walters, M., Geller, G. N., Jongman, R. H. G., Scholes, R. J., Bruford, M. W. et al. 2013. Essential biodiversity variables. *Science*, 339, 277–278.

Pesaresi, M., Guo, H., Blaes, X., Ehrlich, D., Ferri, S., Gueguen, L., Halkia, M. et al. 2013. A Global Human Settlement Layer from optical HR/VHR RS data: Concept and first results. *IEEE Journal of Selected Topics in Applied Earth Observations and Remote Sensing*, 6, 2102–2131.

Pettorelli, N. 2013. *The Normalized Difference Vegetation Index*. New York: Oxford University Press.

Pettorelli, N., Vik, J. O., Mysterud, A., Gaillard, J. M., Tucker, C. J., and Stenseth, N. C. 2005. Using the satellite-derived Normalized Difference Vegetation Index (NDVI) to assess ecological effects of environmental change. *Trends in Ecology & Evolution*, 20, 503–510.

Pielke, R. A. 2005. Land use and climate change. *Science*, 310, 1625–1626.

Pierce, F. J. and Nowak, P. 1999. Aspects of precision agriculture. *Advances in Agronomy*, 67, 1–85.

Potere, D. and Schneider, A. 2009. Comparison of global urban maps. In. Gamba, P. and M. Herold (eds.). *Global Mapping of Human Settlements: Experiences, Data Sets, and Prospects*. CRC Press, Boca Raton, FL, pp. 269–308.

Rosen, P., Hensley, S., Joughin, I. R., Li, F. K., Madsen, S. N., Rodriguez, E., and Goldstein, R. M. 2000. Synthetic aperture radar interferometry. *Proceedings of the IEEE*, 88, 333–382.

Rottensteiner, F., Sohn, G., Gerke, M., Wegner, J. D., Breitkopf, U., and Jung, J. 2014. Results of the ISPRS benchmark on urban object detection and 3D building reconstruction. *ISPRS Journal of Photogrammetry and Remote Sensing*, 93, 256–271.

Rouse, J. W. 1974. Monitoring the vernal advancement and retrogradation of natural vegetation. NASA/GSFCT Type II Report, Greenbelt, MD.

Running, S. W., Collatz, G. J., Diner, D. J., Kahle, A. B., and Salomonson, V. V. (eds.). 1998. Special issue on EOS AM-1 platform, instruments, and scientific data. *IEEE Transactions on Geoscience and Remote Sensing*, 36, 1056–1349.

Running, S. W., Justice, C. O., Salomonson, V., Hall, D., Barker, J., Kaufmann, Y. J., Strah-ler, A. H. et al. 1994. Terrestrial remote sensing science and algorithms planned for EOS/MODIS. *International Journal of Remote Sensing*, 15, 3587–3620.

Saatchi, S., Buermann, W., ter Steege, H., Mori, S., and Smith, T. B. 2008. Modeling distribution of Amazonian tree species and diversity using remote sensing measurements. *Remote Sensing of Environment*, 112, 2000–2017.

Sabins, F. F. 1999. Remote sensing for mineral exploitation. *Ore Geology Reviews*, 14, 157–183.

Seifert, F. M. 2009. Improving urban monitoring toward a European Urban Atlas. In: Gamba, P. and Herold, M. (eds.). *Global Mapping of Human Settlements: Experiences, Data Sets, and Prospects*. CRC Press, Boca Raton, FL, pp. 231–249.

Sellers, P. J. 1985. Canopy reflectance, photosynthesis and transpiration. *International Journal of Remote Sensing*, 6, 1335–1372.

Sellers, P. J., Dickinson, R. E., Randall, D. A., Betts, A. K., Hall, F. G., Berry, J. A., Collatz, G. J. et al. 1997. Modeling the exchanges of energy, water, and carbon between continents and the atmosphere. *Science*, 275, 502–509.

Shanahan, J. F., Schepers, J. S., Francis, D. D., Varvel, G. E., Wilhelm, W. W., Tringe, J. M., Schlemmer, M. R., and Major, D. J. 2001. Use of remote sensing imagery to estimate corn grain yield. *Agronomy Journal*, 93, 583–589.

Shimoda, H. 1999. ADEOS overview. *IEEE Transactions on Geoscience and Remote Sensing*, 37, 1465–1471.

Showstack, R. 2014. Global Forest Watch Initiative provides opportunity for worldwide monitoring. *Eos, Transactions, American Geophysical Union*, 95, 77–79.

Siddiqi, A. A. 2003. *Sputnik and the Soviet Space Challenge*. Gainesville, FL: University of Florida Press.

Sirmacek, B., Taubenboeck, H., and Reinartz, P. 2012. Performance evaluation for 3-D city model generation of six different

DSMs from air- and spaceborne sensors. *IEEE Selected Topics in Applied Earth Observations and Remote Sensing*, 5, 59–70.

Staenz, K., Mueller, A., and Heiden, U. 2013. Overview of terrestrial imaging spectroscopy missions. *Proceedings of the International Geoscience and Remote Sensing Symposium* July 21-26, Melbourne (AUS) *IGARSS 2013*, pp. 3502–3505.

Stenzel, I., Steiner, U., Dehne, H.-W., and Oerke, E. C. 2007. Occurrence of fungal leaf pathogens in sugar beet fields monitored with digital infrared thermography. In Stafford, J. V. (ed.). Precis Agric'07. *Proceedings of the Sixth European Conference on Precision Agriculture*, October 10-12, Bonn (Germany) Wageningen Academic Publisher, pp. 529–535.

Strand, H., Höft, R., Strittholt, J., Miles, L., Horning, N., Fosnight, E., and Turner, W. 2007. Sourcebook on remote sensing and biodiversity indicators. Montreal: Secretariat of the Convention on Biological Diversity, Technical Series no. 32, 203pp. http://cce.nasa.gov/pdfs/cbd-ts-32_sourcebook.pdf

Strong, A. E. 1974. Remote sensing of algal blooms by aircraft and satellite in Lake Erie and Utah Lake. *Remote Sensing of Environment*, 3, 99–107.

Sullivan, R. J. 2000. *Microwave Radar: Imaging and Advanced Concepts*. Norwood, MA: Artech House Inc.

Taubenböck, H., Esch, T., Felbier, A., Wiesner, M., Roth, A., and Dech, S. 2012. Monitoring of mega cities from space. *Remote Sensing of Environment*, 117, 162–176.

Taubenböck, H., Klotz, M., Wurm, M., Schmieder, J., Wagner, B., Wooster, M., and Esch, T. 2013. Central Business Districts: Delineation in mega city regions using remotely sensed data. *Remote Sensing of Environment*, 136, 386–401.

Taubenböck, H., Wiesner, M., Felbier, A., Marconcini, M., Esch, T., and Dech, S. 2014. New dimensions of urban landscapes: The spatio-temporal evolution from a polynuclei area to a mega-region based on remote sensing data. *Applied Geography*, 47, 137–153.

Thenkabail, P. S., Lyon, J. G., and Huete, A. 2012. *Hyperspectral Remote Sensing of Vegetation*. Boca Raton, FL: CRC Press.

Thorp, K. R. and Tian, L. F. 2004. A review on remote sensing of weeds in agriculture. *Precision Agriculture*, 5, 477–508.

Tilling, S. K., O'Leary, G. J., Ferwerda, J. G., Jones, S. D., Fitzgerald, G. J., and Rodriguez, D. 2007. Remote sensing of nitrogen and water stress in wheat. *Field Crops Research*, 104, 77–85.

Tsendbazar, N. E., de Bruin, S., and Herold, M. 2014. Assessing global land cover reference datasets for different user communities. *ISPRS Journal of Photogrammetry and Remote Sensing*, available online March 17, 2014. org/10.1016/j.isprsjprs.2014.02.008.

Tucker, C. J. and Sellers, P. J. 1986. Satellite remote sensing of primary production. *International Journal of Remote Sensing*, 7, 1395–1416.

USA. 2010. *National Space Policy of the United States of America*. Washington, DC: United States Federal Government.

USGS (US Geological Survey). 2001. Understanding Color-Infrared Photography. USGS Fact Sheet, 129(1).

Van der Meer, F. D., van der Werff, H., vanRuitenbeek, F. J., Hecker, C. A., Bakker, W. H., Noomen, M. F., van der Meijde, M., Carranza, E. J. M., de Smeth, J. B., and Woldai, T. 2012. Multi-and hyperspectral geologic remote sensing: A review. *International Journal of Applied Earth Observations and Remote Sensing*, 14, 112–128.

Vierling, K. T., Vierling, L. A., Gould, W. A., Martinuzzi, S., and Clawges, R. M. 2008. Lidar: Shedding new light on habitat characterization and modeling. *Frontiers in Ecology and the Environment*, 6, 90–98.

Voogt, J. A. and Oke, T. R. 2003. Thermal remote sensing of urban climates. *Remote Sensing of Environment*, 86, 370–384.

Wagner, W., Luckman, A., Vietmeier, J., Tansey, K., Balzter, H., Schmullius, C., Davidsone, M. Et al. 2003. Large-scale mapping of boreal forest in SIBERIA using ERS tandem coherence and JERS backscatter data. *Remote Sensing of Environment*, 85, 125–144.

Wegscheider, S., Schneiderhan, T., Mager, A., Zwenzner, H., Post, J., and Strunz, G. 2013. Rapid mapping in support of emergency response after earthquake events. *Natural Hazards*, 68(1), 181–195.

Weng, Q. 2009. Thermal infrared remote sensing for urban climate and environmental studies: Methods, applications, and trends. *ISPRS Journal of Photogrammetry and Remote Sensing*, 64(4), 335–344.

Weng, Q. and Quattrochi, A. 2006. *Urban Remote Sensing*. Boca Raton, FL: CRC Press Taylor & Francis.

Woodcock, C. E., Allen, R., Anderson, M., Belward, A., Bindschadler, R., Cohen, W., Gao, F. et al. 2008. Free access to Landsat imagery. *Science*, 320(5879), 1011. doi: 10.1126/science.320.5879.1011a.

WRI (World Resources Institute). 2002. Global Forest Watch. *WRI Annual Review*, 2002, 20–21.

Wulder, M. A., Masek, J. G., Cohen, W. B., Loveland, T. R., and Woodcock, C. E. 2012. Opening the archive: How free Landsat data has enables the science and monitoring promise of Landsat. *Remote Sensing of Environment*, 122, 2–10.

Xiong, X., Wenny, B. N., and Barnes, W. L. 2011. Overview of NASA Earth Observing Systems Terra and Aqua moderate resolution spectroradiometer instrument calibration algorithms and on-orbit performance. *Journal of Applied Remote Sensing*, 3(1), 032501. doi: 10.1117/1.3180864.

Yang, G., Everitt, J. H., Bradford, J. M., and Escobar, D. E. 2000. Mapping grain sorghum growth and yield variations using airborne multispectral digital imagery. *Transactions of the ASAE*, 43, 1927–1938.

Zhang, C. and Kovacs, J. M. 2012. The application of small unmanned aerial systems for precision agriculture: A review. *Precision Agriculture*, 13, 693–712.

Zhou, X. and Wang, Y.-C. 2011. Spatial–temporal dynamics of urban green space in response to rapid urbanization and greening policies. *Landscape and Urban Planning*, 100, 268–277.

Zhu, X. and Bamler, R. 2014. Super-resolving SAR tomography for multi-dimensional imaging of urban areas. *IEEE Signal Processing Magazine*, 1053–5888.

Remote Sensing Data Normalization, Harmonization, and Intersensor Calibration

3

Overview of Satellite Image Radiometry in the Solar-Reflective Optical Domain

Philippe M. Teillet
University of Lethbridge

Acronyms and Definitions

AVHRR	Advanced very-high-resolution radiometer
BRDF	Bidirectional reflectance distribution function
CEOS	Committee on Earth Observation Satellites
CLARREO	Climate Absolute Radiance and Refractivity Observatory
DEM	Digital elevation model
DTM	Digital terrain model
ETM+	Enhanced Thematic Mapper Plus
GEO	Group on Earth Observations
GEOSS	Global Earth Observation System of Systems
GIFOV	Ground instantaneous field of view
GIS	Geographic information system
IFOV	Instantaneous field of view
IVOS	Infrared Visible Optical Subgroup
LAI	Leaf area index
MODIS	Moderate resolution Imaging spectroradiometer
MODTRAN	Moderate resolution atmospheric transmission
NASA	National Aeronautics and Space Administration
NIST	National Institute of Standards and Technology
NOAA	National Oceanic and Atmospheric Administration
NOMAD	Networked online mapping of atmospheric data
NRC	National Research Council
PSF	Point spread function
QA4EO	Quality Assurance Framework for Earth Observation
SI	International system of units
SRBC	Solar-radiation-based calibration
TM	Thematic Mapper
TOA	Top of atmosphere
TRUTHS	Traceable Radiometry Underpinning Terrestrial and Helio Studies
UTM	Universal Transverse Mercator
WGCV	Working Group on Calibration and Validation

3.1 Introduction

Modern-day remote sensing satellite systems yield high-quality digital images that provide both synoptic and detailed observations of the Earth from space. The steps that have led to this unprecedented geospatial technology are many, and even a summary of that development is beyond the scope of this chapter. Details of the key historical elements of Earth observation remote sensing, such as aviation, rockets, space travel, orbiting satellites, imaging, and lunar and planetary exploration, can be found in numerous books (e.g., Burrows, 1999; Kramer, 2001; Jensen, 2006). Nonetheless, it is worth noting that it has long been known that seeing our planet from above brings significant advantages.*

Countless applications of satellite imagery have been and continue to be developed, the vast majority for qualitative, everyday uses that benefit nevertheless from the laboratory geometric and radiometric quality of the digital images available. That said, there is also tremendous interest in the quantitative use of satellite images to retrieve and monitor information about the current and changing states of geophysical and biophysical variables, particularly those involved in climate, Earth resources, and environment. Applications include, but are not limited to, vegetation analysis (e.g., agriculture, forestry, precision farming), disaster monitoring, environmental monitoring, watershed management, urban growth analysis, bathymetry, geological mapping, mineral exploration, and intelligence data gathering. In that light, this chapter outlines the key considerations that need to be addressed to ensure that terrestrial variables derived from satellite sensor systems operating in the solar-reflective optical domain are calibrated radiometrically to a common physical scale.

Around 30 space agencies worldwide are operating currently, or planning to launch over the coming 15 years, 258 Earth observation satellite missions carrying 767 sensors (383 different sensors, some being repeats in a series).† The data quality varies significantly between sensor systems due to differences in mission requirements, sensor characteristics, and the different calibration methodologies utilized.

In this chapter, emphasis is placed on the postlaunch methodologies used to convert digital image data in the solar reflective domain to radiometrically calibrated products for the user community. For most targets of interest, the retrieval of land, atmosphere, and ocean data and information from satellite image data requires sensor radiometric calibration, retrieval of surface reflectance (involving correction for atmospheric effects), allowance for geometric effects on image radiometry, and an understanding of scene spectral reflectance behavior. The primary terrestrial variable that serves as an example and a common thread in this short treatment is that of reflectance, an Earth surface parameter of wide interest.

* "Man must rise above the atmosphere and beyond to understand fully the world in which he lives." Socrates, 700 BC.

† These numbers are according to *The Earth Observation Handbook* (2014) (http://www.eohandbook.com/) of the international Committee on Earth Observation Satellites (CEOS), which consists of the majority of the space agencies worldwide. The numbers exclude missions by military agencies and commercial companies.

3.2 Need for Data Standardization

Environmental monitoring requirements and responsibilities spanning from local communities to planetary scales continue to multiply. Accordingly, building to a significant extent on military technologies, government agencies in many countries have developed geostrategic technologies and applications that make use of space-based observations of the Earth. Much of the focus of Earth sciences that make use of these satellite sensor systems is on improving predictions of Earth system changes, both short term and long term, primarily with respect to climate, population growth, land transformation, pollution, and biodiversity. As such, remote sensing systems help to provide more solid underpinnings for decisions that impact us all in terms of quality of life and economic consequences.

The extent to which new Earth observation technologies can contribute information of significant economic, social, environmental, strategic, and political value depends critically on developments in data standardization and quality assurance (MacDonald, 1997; Teillet et al., 1997a). Thus, remote sensing calibration and validation are essential aspects of Earth observation measurements and methods used to estimate terrestrial variables to ensure that they are not compromised by sensor effects and data processing artifacts. Accordingly, research progress in remote sensing calibration has been featured periodically in review articles (e.g., Slater, 1984, 1985, 1988a; Duggin, 1986, 1987; Fraser and Kaufman, 1986; Teillet, 1986; Price, 1987a; Ahern et al., 1989; Duggin and Robinove, 1990; Che and Price, 1992; Nithianandan et al., 1993; Slater et al., 1995, 2001; Teillet, 1997a; Teillet et al., 1997a; Secker et al., 2001; Teillet, 2005; Chander et al., 2013a), special journal issues and conference proceedings (e.g., Markham and Barker, 1985; Price, 1987b; Slater, 1988b; Jackson, 1990; Guenther, 1991; Connolly and Tolar, 1994; Bruegge and Butler, 1996; Teillet, 1997b; Chander and Teillet, 2010; Chander et al., 2013b), as well as book chapters and reports (e.g., Muench, 1981; Malila and Anderson, 1986; Teillet et al., 2004a; Butler et al., 2005). The challenge is to ensure that measurements and methods yield self-consistent and accurate geophysical and biophysical data, even though the measurements are made with a variety of different satellite sensors under different observational conditions and the parameter retrieval methodologies vary.

The most stringent satellite data requirements are driven by the need for long-term monitoring of terrestrial variables to determine condition and detect change, as well as to provide inputs to regional and global carbon, energy, and water process models. These requirements have been well documented and include climate and land surface variables such as surface temperature, albedo, fraction of absorbed photosynthetically active radiation, and net primary productivity (e.g., Guenther et al., 1996, 1997; Ohring et al., 2007). Similar user requirements have been documented for oceans (e.g., sea surface temperature and ocean color) (e.g., Gordon, 1987) and atmospheres (e.g., temperature, water vapor, precipitation, ozone, clouds and aerosols, radiation, and trace species) (e.g., Guenther et al., 1996, 1997; Edwards et al., 2004).

Calibration is defined by the international Committee on Earth Observation Satellites (CEOS) Working Group on Calibration and Validation as the process of quantitatively defining the system response to known, controlled signal inputs (www.ceos.org). Generally, *calibration-validation*, or *cal-val*, refers to the entire suite of processing algorithms used to convert raw data into accurate and useful Earth science data that are verified to be self-consistent (Teillet, 1997a). Calibration can include radiometric, geometric, spectral, temporal, and polarimetric aspects. This chapter focuses on radiometric considerations in the solar-reflective optical domain, but it is clear that all these different aspects are important considerations for any measurement device regardless of what part of the electromagnetic spectrum it covers (optical, microwave, or others) if it is to yield useful data and information (Teillet et al., 2004a). Chapters 3 through 6 provide readers with a comprehensive overview of remote sensing calibration.

The Global Earth Observation System of Systems (GEOSS) of the Group on Earth Observations (GEO) aims to deliver timely and comprehensive *knowledge information products* to meet the needs of its nine *societal benefit areas*, of which the most demanding, in terms of accuracy, is climate (Ohring et al., 2005, 2007). This vision builds on the synergistic use of a system of disparate sensing systems that were or are being built for a multitude of applications, and requires the establishment of a worldwide coordinated operational framework to facilitate interoperability and harmonization. CEOS, considered to be the space arm of GEO, has led the development of a Quality Assurance Framework for Earth Observation (QA4EO),* which is based on the adoption of key guidelines. These guidelines have been derived from *best practices* for implementation by the community under the auspices of GEO. The guidelines define the generic processes and activities needed to put in place an operational QA4EO. Their use will facilitate the assignment of quality indicators to the output of every step in Earth observation data-processing chains to demonstrate the level of traceability to internationally agreed reference standards (SI where possible). The QA4EO was endorsed by the CEOS Working Group for Calibration and Validation (WGCV) at its 29th plenary in September 2008 and endorsed by the CEOS Plenary in November 2008.

The inadequacy of proper calibration and validation in diverse Earth observation applications has been identified at many workshops and in a variety of user need and market studies (e.g., Sweet et al., 1992; Sellers et al., 1995; Hall et al., 1991; Horler, 1996; Horler and Teillet, 1996; Teillet, 1997c, 1998; Hegyi, 2004). Raw or uncorrected imagery cannot be used to provide meaningful information for natural resource management, environmental monitoring, and climate studies. As Schowengerdt (2007) puts it: raw sensor digital counts are "simply numbers, without physical units." Thus, quantitative uses of Earth observation rely on conversion of the data to physical units, certifiably traceable to the international system of units (SI) standards, in order to enable the comparison of Earth data from different sensors or from any given sensor over time, whether it be from days to decades (e.g., Fox, 1999; Helder et al., 2012). Moreover, even for qualitative applications, remote sensing remains in some measure untapped† and it will only become a mainstream information technology when it provides reliability of supply, consistent data quality, and plug-and-play capability (Teillet et al., 1997a).

* http://qa4eo.org/index.html.

3.3 Overview

Radiometry and radiation propagation in the remote sensing setting are well described in textbooks (e.g., Slater, 1980; Chen, 1996; Wyatt et al., 1998; Morain and Budge, 2004; Schott, 2007). It is not the role of this chapter to cover fundamental radiometric terms and concepts but rather to provide a brief overview of the key steps involved in converting digital image data from satellite optical systems to calibrated surface quantities, spectral reflectance in particular, in the Earth science context. To begin, it is instructive to look at the paths photons take from their source, the Sun, to the entrance aperture of any given satellite sensor in Earth orbit. Figure 3.1 portrays the processes involved, with five main pathways and associated interactions, from a phenomenological perspective.

1. Photons from the Sun propagate through the Earth's atmosphere, are reflected by the target of interest, and then propagate back through the atmosphere to the satellite sensor (direct sunlight path).
2. Photons from the Sun are scattered by the Earth's atmosphere (thus generating diffuse sky illumination), are reflected by the target of interest, and then propagate back through the atmosphere to the satellite sensor (diffuse skylight path).
3. Photons from the Sun are scattered by the Earth's atmosphere back to the sensor without reaching the surface of the Earth (i.e., path radiance, which can be as much as 10% of the total signal at visible wavelengths on hazy days).
4. Photons from the Sun propagate through the Earth's atmosphere, are reflected from background objects and subsequently from the target of interest, and then propagate back through the atmosphere to the satellite sensor (multiple reflections).
5. Photons from the Sun propagate through the Earth's atmosphere, are reflected from surrounding surfaces, and then scattered into the line of sight of the sensor (the so-called adjacency effect).

Atmospheric radiative transfer codes used for satellite image correction typically take (1–3) into account. Some codes take (5) into account, but those that include (4) are less common.

† A recent estimate indicates that 99.5% of newly created digital data of all kinds are never analyzed (Regalado, 2013).

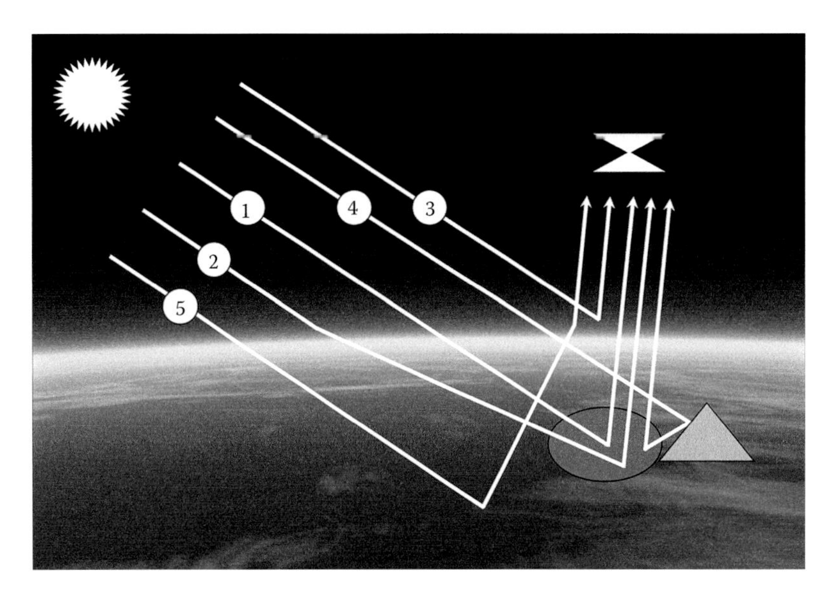

FIGURE 3.1 Schematic depicting photon pathways from the Sun to the entrance aperture of a satellite sensor: (1) direct solar illumination; (2) diffuse sky illumination; (3) atmospheric path radiance; (4) background object reflections; and (5) adjacency effects.

Teillet (2005) lists over 60 considerations and factors that affect the radiometry of Earth observation image data, grouped into seven categories: sensor characteristics, sensor regime, illumination conditions, observation domain, path medium, target characteristics, and product generation domain. While some of the factors are straightforward, such as the variation in Earth–Sun distance, others merit extensive treatments in their own right, such as atmospheric scattering and absorption. This chapter touches on the key postlaunch factors, but it is beyond the scope of the chapter to provide details on the remaining factors and considerations.

Figure 3.2 presents a data flow scheme that emphasizes the main radiometric preprocessing steps prior to information extraction. The main elements portrayed are sensor radiometric calibration, surface reflectance retrieval (i.e., atmospheric correction), radiometric normalization, and information extraction. Information extraction is a separate subject that is not treated in this chapter, and normalization is touched on briefly only. However, the additional dimensions of spectral characterization and georadiometric effects on image radiometry, not easily shown in the data flow in Figure 3.2 because they enter into all of the elements, will be discussed.

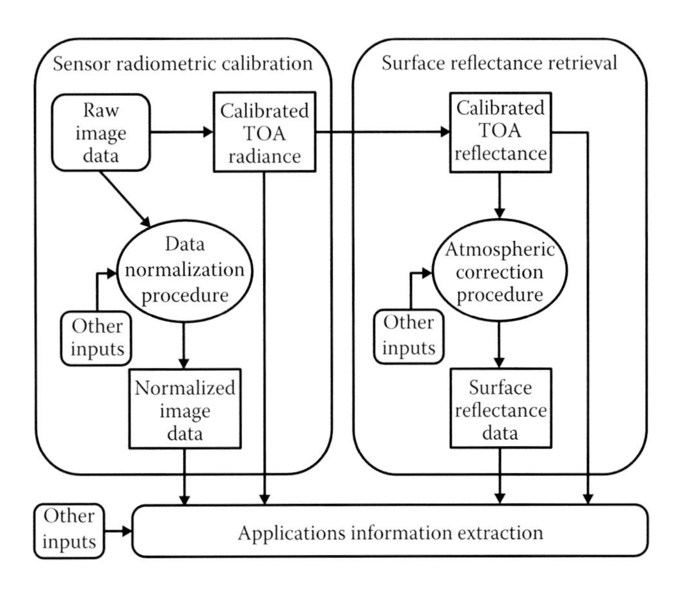

FIGURE 3.2 Radiometric data flow options in preparation for information extraction. The additional dimensions of spectral characterization and georadiometric effects on image radiometry (not shown, but discussed in the text) enter into all of the elements portrayed.

Raw image data and calibrated radiances at the top of the atmosphere (TOA) (Figure 3.2) are referred to as Level 0 and Level 1 data, respectively. Level 2 data consisting of radiometrically calibrated, and geolocated physical variables such as surface reflectance, emittance, and temperature are preferred or required in applications dependent on quantitative analyses. Still higher data product levels consist of information products that are spectral, spatial, or temporal integrations or aggregates of lower-level data. Few remote sensing data product generation systems and services offer Level 2 or higher-level products. Therefore, users must typically undertake this processing work themselves using image analysis software or in collaboration with private industry.

3.4 Sensor Radiometric Calibration

The most fundamental part of the calibration-validation process is sensor radiometric calibration, a broad and complex field that imposes the greatest limitations on quantitative applications of remote sensing (e.g., Teillet, 1997a; Teillet et al., 1997a). The methodologies and instrumentation involved can be grouped into three domains (e.g., Dinguirard and Slater, 1999): (1) on the ground before launch; (2) onboard the spacecraft postlaunch, including reference to lamp sources and/or solar illumination (e.g., Markham et al., 1997); and (3) vicarious approaches using suitable Earth scenes or extraterrestrial targets imaged in-flight (cf., later in this chapter and the next chapter (Teillet and Chander, 2014)).

As for any scientific instrument, extensive and expensive efforts are devoted to building stable optical imaging sensors and to characterizing them to the fullest extent possible before launch (e.g., Guenther et al., 1996). Prelaunch calibrations in a laboratory are easier to control and perform than methods used after launch. Sensor calibration coefficients are determined before launch using radiation sources, whose calibrations are traceable to national laboratory standards.

After launch, sensor radiometric response usually degrades over time such that it must be monitored continually and recalibrated as necessary throughout the life of the system to ensure the maintenance of high data quality. In many instances, considerable time and effort go into the development and use of onboard systems such as standard lamps (blackbody sources for thermal bands), and/or solar diffuser panels to monitor the postlaunch calibration performance of satellite sensors. Nevertheless, even the status of the onboard calibration systems must be verified over time via independent means. Vicarious methods (described in Section 3.4.4) provide these independent data and yield calibration information over the mission lifetime. Further discussion of prelaunch and onboard calibration systems is beyond the scope of this chapter.

Whereas prelaunch methods encompass a vast array of painstaking sensor characterizations in the laboratory (and on very rare occasions outdoors, so-called solar-radiation-based calibration [SRBC] (e.g., Biggar et al., 1993; Mueller et al., 1996; Dinguirard et al., 1997)), postlaunch radiometric calibration based on onboard systems or vicarious methods is devoted primarily to the monitoring of the radiometric responsivities or gain coefficients for each sensor in each spectral band over time. Bias coefficients are determined generally on the basis of deep space or dark-shutter readings. In recent years, unified approaches have been pursued to link the three domains (prelaunch, onboard, and vicarious calibrations), plus the field instruments deployed at the surface in support of vicarious calibration (e.g., Slater et al., 2001; Butler et al., 2005). Error analyses of such approaches indicate that uncertainties in the 2%–3% (1σ) range with respect to exo-atmospheric solar irradiance are attainable in each domain and can be accurately related to national laboratory standards (Dinguirard and Slater, 1999). In all cases, the objective is traceability of radiometric calibration accuracies to the international system of units (SI) for science users (e.g., Pollock et al., 2003) and data products with consistent quality for the broader user community.

As an example, the prelaunch coefficients of the Landsat-5 Thematic Mapper (TM) were revised after launch, because the percent changes in radiometric gain calibration coefficients in the blue, green, and red spectral bands attained −1.0%, −0.85%, and −0.55% per annum, respectively, and the coefficients in the near-infrared and two shortwave infrared spectral bands changed by 0%, +0.1%, and −0.2% per annum, respectively (Teillet et al., 2004a). These changes are significant given the almost 30-year lifetime of the Landsat-5 mission. The impact of these radiometric gain degradations is exemplified in one study by the underestimation of leaf area index (LAI) over the course of a decade by 4% and 1% for rangeland and grassland, respectively, whereas there would be negligible impact on the LAI of black spruce (Teillet et al., 2004a).

3.4.1 Dynamic Range

Radiance describes the flux of energy impinging on a given area from a specified direction in a given spectral band. Conventional units for radiance, used in this document, are Watts per square meter per steradian per micrometer ($W \cdot m^{-2} \cdot sr^{-1} \cdot \mu m^{-1}$), although some data suppliers use slightly different units. The dynamic range in radiance units is set by sensor design and calibration specialists to cover the full radiance range of scenes to be imaged in that satellite mission. For sensor systems with linear response in the radiance range of interest, the dynamic range is specified in terms of the minimum radiance, L^*_{min}, which corresponds to a digital image level of $Q_{cal,\,min}$ (counts), and the maximum radiance, L^*_{max}, which corresponds to a digital image level of $Q_{cal,\,max}$ (counts) (e.g., Markham and Barker, 1987). The asterisk indicates a TOA quantity. The four parameters, $Q_{cal,\,min}$, $Q_{cal,\,max}$, L^*_{min}, and L^*_{max}, establish a known linear relationship between calibrated digital data, Q_{cal}, and radiance at the sensor, L^* (cf., next section).

3.4.2 Converting Digital Counts to At-Sensor Radiance

In each spectral band, the outputs of most optical satellite remote sensing systems are quantized raw image data, Q, which are transmitted to the ground (or recorded onboard and then subsequently transmitted to the ground). After housekeeping corrections and relative radiometric calibration, the raw data are archived on the ground as Level-0 data in units of digital counts. Relative radiometric calibration of satellite imagery primarily involves characterizing and correcting the differences between detector gain and bias levels in sensor systems that utilize multiple detectors in a given spectral band.

To derive radiometrically calibrated (Level 1) data requires knowing the relationship between Q and the TOA at-satellite radiance input to the sensor (absolute radiometric calibration). Rather than computing radiance directly, image product generation systems often transform raw data, Q, to scaled calibrated data, Q_{cal}, taking care of changes in sensor calibration performance in the process. For such Q_{cal} products, the user can then use a set of time-invariant calibration coefficients to convert calibrated digital counts, Q_{cal}, to TOA radiance, L^*. Thus, product generation systems require time-dependent information about the sensor's radiometric calibration performance, a requirement that remains one of the most challenging aspects of the production and quantitative use of satellite image data (e.g., Slater and Biggar, 1996; Slater et al., 2001; Teillet and Chander, 2014).

The linear radiometric calibration equation is

$$L^* = \frac{1}{G}\left[Q_{cal} - Q_o\right]. \tag{3.1}$$

The gain and bias calibration parameters G and Q_o are specified as follows:

$$G = \frac{Q_{\text{cal,max}} - Q_{\text{cal,min}}}{L^*_{\max} - L^*_{\min}} \text{ (counts per unit radiance)} \tag{3.2}$$

and

$$Q_o = Q_{\text{cal,min}} - (Q_{\text{cal,max}} - Q_{\text{cal,min}}) \frac{L^*_{\min}}{L^*_{\max} - L^*_{\min}} \text{ (counts).} \tag{3.3}$$

If a particular application involves the utilization of images from multiple sensors, it is necessary to transform the image data from the different sensors to common physical units such as radiance before they can be compared. If a particular application involves the relative intercomparison of several images from a single sensor, the data can be used in the form of calibrated counts, Q_{cal}. In the latter case, it is not necessary to transform to radiances.

It is worth noting that, while the sensor radiometric calibration equation is as given in Equation 3.1, thus relating the input to the sensor (at-sensor radiance) to the output (digital counts) in proper engineering terms, users are more interested in the inverse computation that converts digital counts in their image data to at-sensor radiance. Unfortunately, the coefficients for this inverse calculation are at times also called *gain* and *offset*, which can lead to confusion. These inverse computation *gain* and *offset* coefficients have units that differ from those of the sensor radiometric calibration gain and offset coefficients defined in Equations 3.2 and 3.3, that is, the inverse computation *gain* and *offset* coefficients are in units of *radiance per unit count* and *radiance*, respectively.

3.4.3 Converting At-Sensor Radiance to At-Sensor Reflectance

A full description of reflectance, whether it be TOA or at the surface, is given by the bidirectional reflectance distribution function (BRDF), which is a function of all incident (i) and reflected (r) zenith angles (θ) and azimuthal angles (ϕ) (Nicodemus, 1965, 1970):

$$\rho_{\text{BRDF}} = \frac{L(\theta_r, \phi_r)}{E(\theta_i, \phi_i)} (\text{sr}^{-1}). \tag{3.4}$$

Units for radiance, L, are as already defined ($\text{W}\cdot\text{m}^{-2}\cdot\text{sr}^{-1}\cdot\mu\text{m}^{-1}$) and units for irradiance, E, are Watts per square meter per micrometer ($\text{W}\cdot\text{m}^{-2}\cdot\mu\text{m}^{-1}$). Note that the units of BRDF are per steradian (sr^{-1}). Since full BRDF characterizations of terrestrial surfaces are nontrivial, remote sensing most often makes use of simplified representations such as the Lambertian reflectance. The radiance from the idealized surface known as a Lambertian surface is the same in all directions, that is, the decreasing intensity with angle follows a cosine law and is exactly compensated by a decrease in projected area. More advanced treatments are needed to deal with surfaces that are, to any significant extent, non-Lambertian and/or sloped with respect to the horizontal.

The TOA or at-sensor reflectance, ρ^*, can be defined as a function of the upwelling radiance, L^* ($\text{W}\cdot\text{m}^{-2}\cdot\text{sr}^{-1}\cdot\mu\text{m}^{-1}$), observed by the satellite sensor divided by the exo-atmospheric solar irradiance, E_0 ($\text{W}\cdot\text{m}^{-2}\cdot\mu\text{m}^{-1}$), as follows (Schott, 2007):

$$\rho^*(\bar{\lambda}) = \frac{\pi d_s^2 L^*(\bar{\lambda})}{E_0(\bar{\lambda})\cos(\theta_z)}, \tag{3.5}$$

where

λ indicates quantities that are wavelength dependent

$\bar{\lambda}$ is used to indicate that these variables are integrated over a given spectral band

As formulated earlier, the reflectance ρ^* is a dimensionless factor between 0 and 1. Equation 3.5 is exact only for a Lambertian surface, that is, one for which the radiance is constant as a function of angle (hence the geometric factor π). E_0 is the exo-atmospheric solar irradiance (Watts per square meter per micrometer), θ_z is the solar zenith angle, and d_s is the Earth–Sun distance in dimensionless astronomical units. Exo-atmospheric solar irradiance is a difficult quantity to measure, and it does vary a few percent over time. Hence, an average model for E_0 is always used and several such models have been developed (e.g., Thuillier et al., 2003). Therefore, care must be taken to ensure that comparisons between different calibration methods and results are made on the basis of the same E_0 data.

The use of TOA reflectance as opposed to TOA radiance corrects for at least some of the sources of variation that affect satellite data, in particular, variations in solar illumination caused by the diurnal cycle and by cyclical changes in the Earth–Sun distance, as well as differences in solar irradiance due to differences between similar spectral bands from sensor to sensor (discussed later in this chapter).

3.4.4 Vicarious Calibration

Earth surfaces with appropriate characteristics have long been used to provide postlaunch updates of the radiometric calibration of satellite sensors, a methodology often referred to as vicarious or ground-look calibration (e.g., Slater et al., 1987, Teillet et al., 1990; Biggar et al., 1994; Thome, 2001). A comprehensive summary of this topic is given in Chapter 4 (Teillet and Chander, 2014). Accordingly, only a few aspects are mentioned here.

Reflectance-based or radiance-based methods use surface and atmospheric measurements to estimate TOA radiance at the entrance aperture of a given satellite sensor in order to monitor and, if needed, provide updates of sensor radiometric calibration. Initially, field measurement campaigns by specialized teams at calibration reference test sites targeted only one sensor per sortie. In subsequent years, with the increase in the number of sensors that passed over a given test site on a given day, it became possible with careful planning to undertake vicarious calibrations for several sensors per sortie (e.g., Thome et al., 1998). Efforts such as these are resource intensive and take time to complete. Hence, it has been of considerable interest to develop less expensive complementary approaches that can provide more frequent calibration updates, even if they are less accurate individually. The use of standard reference test sites to compare or transfer radiometric calibration between satellite sensors, that is, cross-calibration, has also been investigated (e.g., Teillet et al., 2001a; Thome et al., 2003; Chander et al., 2013a; Helder et al., 2013), with and without near-simultaneous or coincident surface measurements. With greater timeliness and reduced costs in mind, new methods are being explored that yield useful results without near-simultaneous or coincident measurements by field teams (e.g., Teillet et al., 2001b; Rao et al., 2003; Thome, 2005; Thome et al., 2008; McCorkel et al., 2013).

Vicarious calibration can also be undertaken with respect to lunar views (e.g., Kieffer and Wildey, 1985, 1996; Godden and McKay, 1997; Kieffer et al., 2002; Barnes et al., 2004; Stone et al., 2005, 2013; Stone, 2008; Xiong et al., 2008) and/or bright stars (e.g., Chang et al., 2012). Although such ultrastable targets are imaged without atmospheric effects, they are relatively low radiance sources and they require high-risk spacecraft platform maneuvers to achieve. Some sensors are designed to acquire space views to provide dark target calibration data.

Other vicarious calibration or cross-calibration methods take advantage of nonland Earth targets, including atmospheric Rayleigh scattering (e.g., Vermote et al., 1992; Kaufman and Holben, 1993; Dilligeard et al., 1997; Meygret et al., 2000), ocean sun-glint (e.g., Kaufman and Holben, 1993; Vermote and Kaufman, 1995; Luderer et al., 2005), snow/ice fields (e.g., Loeb, 1997; Tahnk and Coakley, 2001; Nieke et al., 2003; Six et al., 2004), and clouds (e.g., Vermote and Kaufman, 1995; Iwabuchi, 2003; Doelling et al., 2004, 2010; Hu et al., 2004; Fougnie and Bach, 2009).

Early and ongoing work worth mentioning in the context of vicarious calibration is the case of the series of advanced very-high resolution radiometer (AVHRR) sensors operating on National Oceanic and Atmospheric Administration (NOAA) satellites since 1978. The AVHRR program did not include onboard postlaunch radiometric calibration for the solar-reflective spectral bands. As a result, a wide variety of vicarious methodologies were developed to generate updates of prelaunch radiometric calibrations during the course of the mission on-orbit. AVHRR radiometric gain degradation proved to be significant postlaunch, and the degradations differed considerably from sensor to sensor in the AVHRR series (e.g., Frouin and Gautier, 1987; Smith et al., 1988; Holben et al., 1990; Staylor, 1990; Teillet et al., 1990; Whitlock et al., 1990; Brest and Rossow, 1992; Che and Price, 1992; Mitchell et al., 1992; Teillet, 1992; Vermote et al., 1992; Abel et al., 1993; Kaufman and Holben, 1993; Vermote and Kaufman, 1995; Loeb, 1997; Cabot et al., 2000; O'Brien and Mitchell, 2001; Cao and Heidinger, 2002; Iwabuchi, 2003; Wu et al., 2003; Doelling et al., 2004; Vermote and Saleous, 2006). Quantitative Earth observation applications taking advantage of AVHRR synoptic daily coverage found it difficult to keep up with radiometric calibration changes until various researchers and, eventually, NOAA developed and implemented appropriate approaches and communication tools (e.g., Teillet and Holben, 1994; Cihlar and Teillet, 1995; Rao and Chen, 1995, 1996, 1999; Brest et al., 1997; Tahnk and Coakley, 2001).

Compared to onboard radiometric calibration techniques, the main disadvantages of vicarious calibration are that the methods suffer from lower precision and lower temporal sampling frequency. The main advantage of vicarious calibration is that the sensor typically acquires calibration data in the same modality as it acquires Earth image data, that is, with the same source spectrum, similar illumination conditions, and the same spectral bands.

3.5 Surface Reflectance Retrieval

Surface reflectance in the solar reflective part of the electromagnetic spectrum is a primary geophysical variable of interest in many applications, either for its own sake or to generate other geophysical or biophysical quantities. The Lambertian surface spectral reflectance, ρ_{surf} (dimensionless between 0 and 1), is defined as π times the ratio of the radiance, L ($W{\cdot}m^{-2}{\cdot}sr^{-1}{\cdot}\mu m^{-1}$), upwelling from the target (as measured just above the target) divided by the downwelling irradiance, E ($W{\cdot}m^{-2}{\cdot}\mu m^{-1}$), that illuminates the target (Schott, 2007):

$$\rho_{surf}(\bar{\lambda}) = \frac{\pi L(\bar{\lambda})}{E(\bar{\lambda})}. \quad (3.6)$$

As already noted (Figure 3.1), the downwelling irradiance signal at the surface, E, is made up of several components in addition to the direct solar and diffuse sky contributions, a consideration that must also be taken into account when making ground-based radiance measurements. The sky irradiance distribution is anisotropic, but it can be modeled reasonably well to take into account the increase in diffuse radiation in the circumsolar region and toward the horizon (Dave, 1977; Steven, 1977; Temps and Coulson, 1977; Klucher, 1979; Hooper and Brunger, 1980; Kirchner et al., 1982; Hooper et al., 1987; Brunger and Hooper, 1993). This refinement is especially helpful where topography obscures parts of the sky from a given point on the ground (Hay and McKay, 1985; Hay et al., 1986; Duguay, 1993).

3.5.1 Atmospheric Correction

Scattering and absorption due to aerosols and gases in the atmosphere modify radiation, making its way from outside the atmosphere down to the target, as well as the surface reflectance propagated through the atmosphere to the satellite sensor (e.g., Chahine, 1983; Schott, 2007). In the solar-reflective spectral domain, molecular (Rayleigh) scattering (λ^{-4}) is strongest in the blue (e.g., contributing ~0.07 in reflectance for 1013 mbar), aerosol scattering (λ^{-2} to $\lambda^{+0.6}$) is strongest in the visible (e.g., contributing ~0.04 in reflectance for 10 km visibility), ozone gas absorption is strongest in the green (e.g., ~10% effect on atmospheric transmission for 0.35 cm·atm), and water vapor absorption is strongest in the near-infrared and shortwave infrared (e.g., ~12% effect on transmission for 3 g·cm^{-2}). Interestingly, atmospheric source and loss effects roughly cancel out for Earth surface reflectances around 0.25; hence, this is a good reflectance for vicarious calibration reference sites, since atmospheric effects are minimized.

Given that the optical properties of the Earth's atmosphere are not uniform spatially or temporally, image corrections for these effects in the solar-reflective spectral bands are needed to put satellite data on the same radiometric scale for investigations intended to monitor terrestrial surfaces quantitatively over time and space. A compact form of the relationship between surface and TOA reflectance is given as (Tanré et al., 1990)

$$\rho^{\star} = \tau_g \left(\frac{\tau_s \tau_v \rho_{\text{surf}}}{1 - \bar{\rho} s} + \rho_a \right). \tag{3.7}$$

All quantities are wavelength dependent. τ_g is atmospheric gas transmittance, τ_s is downward scattering transmittance in the solar illumination direction, τ_v is upward scattering transmittance in the sensor view direction, ρ_a is atmospheric reflectance, $\bar{\rho}$ is the average reflectance of surrounding surfaces, and s is the spherical albedo (i.e., the reflectance) of the atmosphere. All reflectances and transmittances in Equation 3.6 are dimensionless factors between 0 and 1. The atmospheric quantities are computed using radiative transfer codes.

The detailed textbook treatment by Schott (2007) is also instructive and worth citing. It explains all of the contributions to the upwelling spectral radiance reaching the satellite sensor, including the thermal energy paths not presented in this chapter. The result is the so-called *big equation* for the TOA spectral radiance (cf., Schott (2007) for details).

The need for and various approaches to atmospheric correction have long been examined in the literature (e.g., Turner and Spencer, 1972; O'Neill et al., 1978; Ahern et al., 1979; Richardson, 1982; Slater and Jackson, 1982; Tanré et al., 1983; Moran et al., 1992; Kaufman et al., 1997), and atmospheric radiative transfer codes are available in modern remote sensing image analysis systems. Some solutions may involve the use of a radiative transfer code, but many make use of precomputed look-up table results to save computation time. A widely accepted atmospheric correction code for research purposes is MODTRAN5 (moderate resolution atmospheric transmission code version 5) (Berk et al., 2006). Most of the predominantly used codes tend to disagree significantly only for very large aerosol optical depths and/or large off-nadir illumination or viewing geometries (>60°). Thus, the choice of code is not an important factor except for the correction of high spectral resolution (hyperspectral) data (e.g., Staenz et al., 1994, 2002). Monochromatic computations should not be used (Teillet, 1989), and bandpass calculations based on relative spectral response profiles with 0.0025 μm grid spacing or finer are recommended. Another caution is that Rayleigh scattering, a long-understood phenomenon, is not always computed accurately in atmospheric correction codes (Teillet, 1990a; He et al., 2006).

Operational atmospheric correction for surface reflectance retrieval depends on ready access to timely and accurate information on atmospheric variables such as aerosol optical depth (e.g., Teillet et al., 1994) and water vapor content. Although the computational tools for image correction are available, the user is still left with the problem of obtaining the necessary atmospheric variables for input to the image correction. The main possibilities in this respect are as follows:

1. Measure the required parameters in the field at the same time as image acquisition.
2. Query an online source of radiosonde and/or sunphotometer network data that may have acquired parameters near the time and location of the image acquisition (e.g., Holben et al., 1998).
3. Assume fixed standard values for the atmospheric parameters for the geographic region of interest (e.g., Fedosejevs et al., 2000).
4. Use climatological values developed over time (e.g., Bokoye et al., 2001).
5. Estimate the required parameters from the image data themselves, using techniques such as the dark target approach (e.g., Chavez, 1988, 1996; Teillet and Fedosejevs, 1995; Liang et al., 1997; Song et al., 2001) and atmospheric absorption feature extraction (e.g., Staenz et al., 2002).
6. Use data assimilation results based on dynamic models driven by analyzed meteorological data (e.g., O'Neill et al., 2002).

The challenge is to make optimum use of available ground-based and satellite-based atmospheric optical measurements and to ensure they are consistent and easily accessible. A prototype atmospheric optical parameter estimation system was developed to provide optical parameters online for any time and place across extended regions such as Canada or North America based on available climate to meteorological scale data and models (Freemantle et al., 2002; O'Neill et al., 2002). Aside from making atmospheric corrections more operational, this concept of networked online mapping of atmospheric data (NOMAD) transfers the responsibility of making available quality atmospheric parameters from the user to the scientific and technical specialists maintaining the atmospheric parameter server. Nevertheless, at the end of the day, the significant uncertainties that can arise from the atmospheric correction process with uncertain inputs (e.g., Kaufman et al., 1997; Teillet et al., 1994), too varied to characterize in a summary way, are such that users often prefer to seek out imagery acquired on relatively haze-free days, or to apply normalization techniques (e.g., Schroeder et al., 2006) rather than undertake atmospheric correction.

3.6 Geometric Effects on Image Radiometry

3.6.1 Pixel

The nature of the remote sensing image pixel has been given some but, arguably, not enough attention (Townshend, 1981; Duggin, 1986; Fisher, 1997; Cracknell, 1998; Townshend et al., 2000). Clearly, a pixel is not a true geographical object, nor does it correspond exactly to the spatial resolution of the imaging sensor. The pixel is a sampling of the nonuniform point spread function (PSF) of a given instantaneous field-of-view (IFOV) of the sensor. Space does not allow elaboration on the subject of PSFs of satellite imaging sensors. A good treatment of this topic can be found in Schowengerdt (2007).

It is easy to forget that a substantial proportion of the signal apparently coming from the surface area represented by a given pixel comes from surrounding areas and from the atmospheric path. A significant proportion of any given satellite image consists of mixed pixels. Moreover, in most imaging systems, pixels at large off-nadir view angles each encompass significantly larger surface areas than do pixels closer to nadir because of the projection effect. Large field-of-view whiskbroom sensors such as the moderate-resolution imaging spectroradiometer (MODIS) are subject to the panoramic *bow tie* effect, yielding scans that partially overlap at off-nadir angles (Souri and Azizi, 2013). Further still, cross-track ground sampling intervals increase with increasing view angle because of Earth curvature; the flat-Earth approximation is good to within 4% for off-nadir view angles less than 23° (Schowengerdt, 2007). These considerations have important ramifications on the per-pixel characterization of regions of interest and on the integration of remote sensing data into geographical information systems (GISs). Thus, contrary to common practice, land cover properties should be reported at spatial resolutions coarser than the individual remote sensing pixels.

The case of the AVHRR sensor series illustrates this point. Nominally, the nadir-view size of AVHRR pixels is 1.1 km by 1.1 km. With the large off-nadir angles that constitute a significant portion of each AVHRR image, the majority of pixels are greater than 1.1 km in size, reaching as much as 10 times the area of nadir pixels at the largest off-nadir angles (55° at ground level). Moreover, it has been estimated that the nadir pixel sampling at 1.1 km represents approximately only 27% of the radiation captured by the ground IFOV (GIFOV) (personal communication, NASA Langley Research Center). It has also been found that, in the case of the Landsat-5 TM, the cross-track GIFOV is on the order of 40–45 m instead of the nominal resolution of 30 m (Schowengerdt et al., 1985).

3.6.2 Illumination and Viewing Geometries through the Atmosphere

Except for large angles, the influence of varying illumination and viewing paths through the atmosphere is generally well handled by most atmospheric correction algorithms. However, image data acquired for high-latitude regions, where sun angles are typically far from the zenith, are more difficult to correct for atmospheric effects. In addition, while some atmospheric correction codes incorporate simple BRDF models to allow for multiple surface-atmosphere scatterings (e.g., the 6S code*: Kotchenova et al., 2006; Vermote et al., 2006; Kotchenova and Vermote, 2007), computationally intensive multiple scattering in the radiative transfer is not always a component of atmospheric correction algorithms and, hence, is not always taken into consideration when deriving reflectance information about a surface from satellite image data. The inclusion of multiple scattering is a greater concern for ocean color parameter retrieval (e.g., Gordon and Wang, 1994) than for land surface reflectance retrieval.

3.6.3 Atmospheric Refraction

For typical air densities, the atmospheric refraction correction for the entire atmosphere down to sea level is less 0.1° for zenith angles less than 80° (Scarpace and Wolf, 1973; McCartney, 1976; Chu, 1983; Egan, 1985). Accordingly, atmospheric refraction only becomes important for satellite sensors designed to study the Earth's atmosphere at near-grazing angles.

3.6.4 Reflectance Anisotropy

Reflectance anisotropy as a function of illumination and viewing geometries is a fundamental property of any terrestrial surface and is best described in terms of BRDF (Equation 3.4). Some terrestrial surfaces have highly anisotropic reflectance properties, and this has an important influence on image radiometric characteristics. Forest canopies are particularly subject to these effects (cf., Figure 3.3). Methodologies to model and deal with BRDF effects are well documented in the literature (e.g., Li and Strahler, 1992; Chen and Cihlar, 1997; White et al., 2002a; Pinty et al., 2004). Nevertheless, although anisotropic reflectance effects have been studied extensively, they remain challenging to deal with in an operational setting and there are many other geometric effects to consider apart from the BRDF.

3.6.5 Adjacency Effects

As already noted, some of the photons reaching the satellite sensor would have propagated through the Earth's atmosphere, reflected from surfaces surrounding any given IFOV, and then scattered into the line of sight of the sensor (Otterman and Fraser, 1979; Dave, 1980; Dana, 1982; Meckler and Kaufman, 1982; Deschamps et al., 1983; Otterman et al., 1983; Kaufman and Fraser, 1984). This adjacency effect can give rise to blurring and contrast reduction at the boundaries of IFOVs. While the preference here is to include it as a georadiometric effect, the adjacency effect is commonly discussed as part of atmospheric correction. However, more often than not, the adjacency effect

* 6S: Second Simulation of the Satellite Signal in the Solar Spectrum.

FIGURE 3.3 Digital imagery showing the same black spruce canopy taken at the indicated angles from a truck-based platform. The effect of reflectance anisotropy is clearly evident (solar illumination is from the right). Because these anisotropy effects are at the scale of trees and their components, they will give rise to sub-pixel variations in satellite sensor data of any spatial resolution.

is ignored or, at best, taken into account in an approximate way (Tanré et al., 1990; Vermote et al., 2006).

3.6.6 Topographic Effects

A number of approaches have been developed to address the influence of surface topography on the radiometric properties of pixels in remotely sensed image data. Terrain elevation variations cause variable atmospheric path lengths that image correction algorithms have to address in the contexts of remote sensing from both airborne platforms in the atmosphere and satellite platforms above the atmosphere (e.g., Teillet and Santer, 1991; Teillet and Staenz, 1992). Terrain elevation variations also cause shadow effects that introduce errors in image understanding in general (Woodham and Lee, 1985; Woodham and Gray, 1987) and in surface reflectance retrieval in particular (Giles, 2001; Adeline et al., 2013).

Image corrections for terrain slope and aspect variations are more difficult to carry out. Some approaches have used normalization techniques (e.g., Holben and Justice, 1981; Allen, 2000), with mixed results (Richter, 1997, 1998). Instead, most radiometric corrections for topographic effects use digital terrain elevation, slope, aspect, and other derivatives (i.e., a digital terrain model (DTM)) to describe explicitly the surface topography. Two main categories of these approaches are: (1) slope-aspect corrections that use some function of the cosine of the incident solar angle (Smith et al., 1980; Teillet et al., 1982; Cavayas, 1987; Meyer et al., 1993; Ekstrand, 1996; Riaño et al., 2003; Soenen et al., 2005, 2008) and (2) model-based corrections that deal with solar radiation interactions with surface targets (Hugli and Frei, 1983; Proy et al., 1989; Gu and Gillespie, 1998; Dymond and Shepherd, 1999; Shepherd and Dymond, 2003; Li et al., 2012) and with atmospheric radiative transfer codes (Richter, 1997, 1998).

In practice, slope-aspect corrections may involve the prior rectification of an image to a map-based coordinate system, typically the Universal Transverse Mercator (UTM) projection, which matches the map coordinate system of the DTM. Although this simplifies image processing and facilitates image classification studies involving various images in a common map projection, there are disadvantages to the image rotation and resampling that occur during the map rectification process. Significant radiometric errors can be introduced as a result of resampling (Forster and Trinder, 1984), even using an optimum parametric cubic convolution algorithm (Schowengerdt et al., 1983). Additionally, the slope-aspect correction procedure has to keep track of the scan direction to specify view angles properly. A better approach is to not apply the map transformation to the image before slope-aspect correction and, instead, make use of the map transformation equations in reverse to apply the slope-aspect correction in image space.

A rule of thumb to keep in mind is that the spatial resolution of the digital elevation model (DEM) used to define the slope and aspect data in the DTM should be twice as fine as the spatial resolution of the image data to be corrected for slope-aspect effects (Hodgson, 1995). For example, terrain elevation data with horizontal spatial resolutions of 15 m by 15 m should be used to generate the slope and aspect angles to be used to correct imagery whose spatial resolution is 30 m by 30 m.

In principle, topographic slope-aspect corrections should also involve atmospheric and surface reflectance models to allow for the proper treatment of the effects due to terrain elevation variations, BRDF, irradiance from adjacent slopes, and diffuse sky illumination distribution (Kimes and Kirchner, 1981; Bruhl and Zdunkowski, 1983; Cavayas et al., 1983; Sjoberg and Horn, 1983; Woodham and Lee, 1985).

3.6.7 Position of the Sun

Uncertainties in knowledge of the position of the Sun can give rise to significant errors in shadow area estimation in rugged terrain (Teillet et al., 1986). The largest error by far is the use of a single set of solar zenith and azimuth angles for the radiometric processing and correction of an entire satellite image. For example, the lack of specification of within-scene location for a Landsat image leads to uncertainties in solar position of 4.5° (zenith angle) by 2.5° (azimuth angle). It is not uncommon to use a single set of solar angles for Landsat images, whereas, for radiometric corrections applied to wide-area coverage satellite imaging sensors such as AVHRR or MODIS, it is routine to specify solar angles on a subscene basis. Other contributions to the uncertainty in solar position include, in decreasing order of importance: finite size of the solar disk (30 min), lack of specification of within-scene time (15 min [azimuth angle] by 8 min [zenith angle]), use of solar ephemeris from an inappropriate epoch, atmospheric refraction effects for large solar zenith angles, and horizontal parallax (target at Earth surface, not Earth center). The potential cumulative uncertainty effect in the Landsat case is 5° (zenith angle) by 3.5° (azimuth angle).

3.7 Spectral Characterization

Although it has received less attention than other aspects of satellite image radiometry, spectral characterization is an important consideration for proper surface reflectance retrieval, regardless of how wide or narrow the spectral bands may be. Spectral bands designed for specific applications and data products are susceptible to postlaunch variations in spectral bandpasses and they usually differ between different sensors and missions. If spectral bands have changed in wavelength position or bandwidth postlaunch, or if there are uncertainties as to their characteristics, there is a direct impact on radiometric and atmospheric processing, as well as on derived information products (Suits et al., 1988; Teillet, 1990b; Teillet and Irons, 1990; Flittner and Slater, 1991; Goetz et al., 1995; Teillet et al., 1997b; Steven et al., 2003; Trishchenko, 2009). In practice, there is little that users can do to take into account postlaunch changes in spectral band performance. Even when spectral bands perform as designed, users should be aware that similar information products derived from different sensors with analogous spectral bands that do not match exactly are not directly comparable (Teillet et al., 2007; Teillet and Ren, 2008). These spectral band difference effects are scene dependent and, hence, not corrected easily.

A limited number of investigations have been undertaken to assess radiometric calibration errors due to differences in spectral response functions between satellite sensors when attempting cross-calibration based on near-simultaneous imaging of common ground look targets in analogous spectral bands (Teillet et al., 2001c, 2004b, 2007; Trishchenko et al., 2002; Rao et al., 2003; Doelling et al., 2012; Chander et al., 2013c; Henry et al., 2013). A specific example is the radiometric cross-calibration of the Landsat-7 Enhanced Thematic Mapper Plus (ETM+) and Landsat-5 Thematic Mapper (TM) sensors based on early-mission tandem-orbit data sets (Teillet et al., 2001c, 2004b), which included adjustments for spectral band differences between the two Landsat sensors. Spectral band difference effects were shown to be significant (an effect that can be 5% or more), despite the close similarity in the spectral filters and response functions of the Landsat sensors. A variety of terrestrial surfaces were assessed regarding their suitability for Landsat radiometric cross-calibration in the absence of surface reflectance spectra. This line of inquiry was extended to Earth observation sensors on several satellite platforms, indicating that large differences can arise (Teillet et al., 2007; Teillet and Ren, 2008).

3.8 Normalization Approaches

Many applications are not dependent on securing radiometric calibration on an absolute scale. Only relative changes between images are of interest such that normalization procedures can be used to bypass steps in Figure 3.2. Spectral band transformations (e.g., band ratios or principal component analyses) have long been used to mitigate atmospheric and/or topographic effects (e.g., Crist, 1985). A variety of empirical radiometric normalization methods have been developed (e.g., Schott et al., 1988; Hall et al., 1991; Yuan and Elvidge, 1996; Schroeder et al., 2006).

3.9 Processing Considerations

This chapter has so far provided an overview of the principal physical effects affecting satellite image radiometry. Additional concerns arise in the domain of image processing that, under certain circumstances, can affect image radiometry significantly.

Nominally, image-processing software will first calibrate digital counts to TOA radiance and, in some cases, convert TOA radiance to TOA reflectance. In any case, the image data are then in physical units such that atmospheric correction to estimate surface reflectance can be done. Adjustments for bidirectional and/or topographic effects should be included in the latter step. It is important to note that these radiometry-related image-processing steps involve equations that are linear or nonlinear. The associated software must be constrained to prevent user application of the various calibration/correction steps in different sequences because linear and nonlinear transformations are not commutative (Teillet 1986). In principle, all radiometric corrections should be completed prior to any geometric processing.

Moreover, it is important to understand how the sequencing of spectral band integration can affect image radiometry, especially in processing for surface reflectance retrieval (i.e., atmospheric correction). Detailed atmospheric computations can be done monochromatically and the results integrated over the spectral band involved, or band-integrated quantities can be used in the atmospheric computations. Depending on the surface reflectance spectrum and the relative spectral response profile of the spectral band involved, the two approaches can yield results that differ significantly, and the proper choice of approach depends mainly on whether the image data are multispectral or hyperspectral in character (Teillet 1989; Teillet and Irons 1990).

3.10 Discussion of Future Trends

Although radiometric calibration is a very specialized aspect of remote sensing, it constitutes an essential component of Earth observation systems and their utilization. It is crucial for the generation of useful long-term data records, as well as for independent data quality control. In that light, this chapter outlines the key postlaunch methodologies used to convert digital image data to calibrated data products. Nevertheless, postlaunch calibration and the associated quality assurance infrastructures, which should be behind-the-scenes activities as with any mature technology, remain semioperational at best. The CEOS QA4EO effort, noted in Section 3.2 in the context of GEOSS, is an important step in the direction of operational best practices for traceability and interoperability. Two other calibration-related initiatives for the future are noted here, both representing challenging space-based undertakings.

The Traceable Radiometry Underpinning Terrestrial and Helio Studies (TRUTHS) mission concept (Fox et al., 2002, 2011) offers a novel approach to the provision of key scientific data with unprecedented radiometric accuracy for Earth observation and solar studies. TRUTHS will calibrate its instrumentation directly to SI on orbit, overcoming the usual uncertainties associated with drifts of sensor calibration by using an electrical rather than an optical standard as the basis of its radiometric calibration. A space-based cryogenic radiometer together with its associated calibration chain to a terrestrial primary standard will provide a space-based standard reference for measurements of both the Sun and the Earth. The TRUTHS mission has the potential to improve the performance and accuracy of Earth observation missions by an order of magnitude (to accuracies on the order of 0.3% in the solar reflective domain). As a result, TRUTHS will provide the necessary advances to ensure that Earth system data sets have sufficient radiometric long-term precision for the reliable detection and evaluation of global change.

One of the missions recommended for earliest possible implementation by the U.S. National Research Council (NRC) Decadal Survey report (NRC, 2007) is the Climate Absolute Radiance and Refractivity Observatory (CLARREO), a joint NASA/NOAA mission. NOAA's contribution is to be sensors for total and spectral solar irradiance measurements and Earth energy budget climate data records. NASA's contribution is to be sensors for the measurement of spectrally resolved thermal infrared and reflected solar radiation with high absolute accuracy. The CLARREO mission objective pertinent to the topic of this chapter is the provision of a space-based high-accuracy calibration standard to enable calibration and intercalibration of Earth observation satellite sensors. To accomplish this, CLARREO will include a Reflected Solar Suite consisting of two push-broom hyperspectral imagers covering 320–2300 nm in a single instrument package. Reflectance will be obtained via ratios of Earth-view data to solar-view data. High accuracy will be achieved by precisely calibrating the instruments via the solar view combined with a sensor model to transfer National Institute of Standards and Technology (NIST) laboratory standards to orbit (Anderson et al., 2008; Sandford et al., 2010; Thome et al., 2010; Lukashin et al., 2013). As for TRUTHS, the CLARREO mission has the potential to improve the performance and accuracy of Earth observation missions by an order of magnitude (on the order of 0.3% in the solar reflective domain).

The success of QA4EO is far from guaranteed, and the TRUTHS and CLARREO missions have yet to be approved. In the early twenty-first century, innovation in Earth observation as well as the subfield of data standardization requires making judicious choices. As the geostrategic technologies of Earth observation seek adoption by mainstream information society, it is hoped that young people pursuing optical engineering, radiometry, metrology, and remote sensing physics will strive to ensure the data quality and product validation needed for Earth observation to underpin sound decision making.

3.11 Concluding Remarks

This chapter has provided a broad overview of the key considerations involved in the radiometric calibration and correction of image data from satellite sensor systems operating in the solar-reflective optical domain. Relevant pioneering research over the past four decades has been featured, with emphasis on sensor radiometric calibration, retrieval of surface reflectance (involving correction for atmospheric effects), allowance for geometric effects on image radiometry, and an understanding of target spectral reflectance behavior. Surface reflectance served as an important Earth surface variable of wide interest. The need for radiometric calibration and correction is well documented in the literature and growing in importance as Earth science data products are being derived increasingly from a multiplicity of different satellite sensor systems, and data and information demands for the purposes of sound decision-making require increasingly greater precision and accuracy.

Sensor radiometric calibration is the most fundamental consideration given that uncertainties in radiometric calibration translate directly into the same amount of radiometric uncertainty in all products derived downstream. Uncertainties within 2%–3% (1σ) with respect to exo-atmospheric solar irradiance are attainable routinely, relative to national laboratory standards, and missions such as TRUTHS and CLARREO promise an order-of-magnitude improvement. Recommendations for future research on vicarious calibration using terrestrial reference standard sites can be found in the next chapter (Teillet and Chander, 2014).

However, the retrieval of terrestrial surface variables is subject to many other considerations, as outlined extensively in this chapter. Significant spatial and temporal variations in the optical properties of the Earth's atmosphere necessitate image corrections for these effects in the solar-reflective spectral bands to put satellite data on the same radiometric scale for studies intended to monitor terrestrial surfaces over time and space. Although the computational tools for image correction for atmospheric effects are available, they require user inputs reliant on ready access to timely and accurate data on atmospheric variables such as aerosol optical depth and water vapor content. The net result is that surface reflectance retrieval from remotely sensed image data is operational for larger programs (e.g., MODIS—Vermote et al., 1997, 2002; Schaaf et al., 2002), but remains laborious for individual user studies. It would be very helpful if space agencies and data suppliers more often took the extra steps to generate and offer surface reflectance and other higher-level products to the user community at reasonable or no cost. Progress can also come about if an agency with the pertinent scientific and technical resources and specialists maintained a dynamic open-access atmospheric parameter server along the lines of NOMAD.

The chapter also described numerous geometric factors and phenomena that give rise to radiometric effects in satellite image data. Most of these geometric considerations are tractable, but a few are more challenging to address, especially anisotropic reflectance properties of surfaces and topographic effects. A progressive perspective on BRDF is to view angular signatures of spectral reflectance as an information source as opposed to something that has to be corrected (e.g., White et al., 2001, 2002b). Indeed, a greater number of satellite sensor systems than can perform angular remote sensing would yield substantial information returns on investment (M. Verstraete, personal communication). Slope-aspect corrections will benefit from the highest-resolution DEMs, which are becoming increasingly available for the entire surface of the Earth.*

Although spectral considerations have received relatively less attention than other aspects of satellite image radiometry, the literature has grown considerable over the past decade. The increased attention is partly because of the advent of hyperspectral remote sensing (beyond the scope of this chapter to treat), which demands numerous preprocessing corrections. Also, efforts devoted to radiometric calibration, a first-order consideration, have been well advanced by the research community such that increased attention is now being devoted, belatedly, to spectral considerations. Proper surface reflectance retrieval depends on a good understanding of the state of sensor spectral band characteristics postlaunch, given that they have a direct impact on radiometric and atmospheric processing, as well as on the spectral properties of the scene (the atmosphere and the surface targets of interest).

Based on best practices and recommended for implementation and use throughout the GEO community, QA4EO is the way forward for sensor radiometric calibration. It is hoped that QA4EO will grow to encompass the other important elements of satellite image radiometry, including atmospheric, geometric, and spectral influences on Earth science data.

Acknowledgments

The author gratefully acknowledges guidance over the years from Philip N. Slater and many substantive discussions on remote sensing radiometry with Kurtis J. Thome, Nigel P. Fox, Robert P. Gauthier, and Gunar Fedosejevs. The author also thanks Gyanesh Chander for taking the time to read and provide detailed comments on drafts of this chapter.

References

Abel, P., B. Guenther, R.N. Galimore, and J.W. Cooper, 1993. Calibration results for NOAA-11 AVHRR channels 1 and 2 from congruent path aircraft observations, *Journal of Atmospheric and Oceanic Technology*, 10(4): 493–508.

Adeline, K., M. Chen, X. Briottet, S. Pang, and N. Paparoditis, 2013. Shadow detection in very high spatial resolution aerial images: A comparative study, *ISPRS Journal of Photogrammetry and Remote Sensing*, 80: 21–38.

Ahern, F.J., R.J. Brown, J. Cihlar, R. Gauthier, J. Murphy, R.A. Neville, and P.M. Teillet, 1989. Radiometric correction of visible and infrared remote sensing data at the Canada centre for remote sensing, In: A. Cracknell and L. Hayes, eds., *Remote Sensing Yearbook 1988/89*, Taylor and Francis, Philadelphia, PA, pp. 101–127.

Ahern, F.J., P.M. Teillet, and D.G. Goodenough, 1979. Transformation of atmospheric and solar illumination conditions on the CCRS image analysis system, in: *Proceedings of the Fifth International Symposium on Machine Processing of Remotely Sensed Data*, West Lafayette, IN, pp. 34–51.

Allen, T.R., 2000. Topographic normalization of Landsat Thematic Mapper data in three mountain environments, *Geocarto International*, 15: 13–19.

Anderson, D., K.W. Jucks, and D.F. Young, 2008. The NRC decadal survey climate absolute radiance and refractivity observatory: NASA implementation, in: *Proceedings of the IEEE International Geoscience and Remote Sensing Symposium (IGARSS)*, Boston, MA, pp. 9–11.

Barnes, R.A., R.E. Eplee, Jr., F.S. Patt, H.H. Kieffer, T.C. Stone, G. Meister, J.J. Butler, and C.R. McClain, 2004. Comparison of SeaWiFS measurements of the moon with the U.S. geological survey lunar model, *Applied Optics*, 43(31): 5838–5854.

Berk, A., G.P. Anderson, P.K. Acharya, L.S. Bernstein, L. Muratov, J. Lee, M. Fox et al., 2006. MODTRAN5: 2006 update, in: *Proceedings of SPIE Conference on Algorithms and Technologies for Multispectral, Hyperspectral, and Ultraspectral Imagery XII*, Vol. 6233, Orlando, FL, p. 62331F.

Biggar, S.F., P.N. Slater, and D.I. Gellman, 1994. Uncertainties in the in-flight calibration of sensors with reference to measured ground sites in the 0.4 to 1.1 μm range, *Remote Sensing of Environment*, 48: 245–252.

Biggar, S.F., P.N. Slater, K.J. Thome, A.W. Holmes, and R.A. Barnes, 1993. Preflight solar-based calibration of SeaWiFS, in: *Proceedings of SPIE Conference 1939*, Orlando, FL, pp. 233–242.

Bokoye, A.I., A. Royer, N.T. O'Neill, P. Cliche, G. Fedosejevs, P.M. Teillet, and L.J.B. McArthur, 2001. Characterization of atmospheric aerosols across Canada from a ground-based sunphotometer network: AEROCAN, *Atmosphere-Ocean*, 39(4): 429–456.

Brest, C.L. and W.B. Rossow, 1992. Radiometric calibration and monitoring of NOAA AVHRR data for ISCCP, *International Journal of Remote Sensing*, 13(2): 235–273.

Brest, C.L., W.B. Rossow, and M.D. Roiter, 1997. Update of radiance calibrations for ISCCP, *Journal of Atmospheric and Oceanic Technology*, 14(5): 1091–1109.

Bruegge, C. and J. Butler, eds., 1996. Special issue on earth observing system calibration, *Journal of Atmospheric and Oceanographic Technology*, 13(2): 273–544.

Bruhl, C. and W. Zdunkowski, 1983. An approximate calculation method for parallel and diffuse irradiances on inclined surfaces in the presence of obstructing mountains or buildings, *Archives for Meteorology, Geophysics, and Bioclimatology (Series B)*, 32(2–3): 111–129.

* For example, WorldDEM: http://www.astrium-geo.com/worlddem/ and OpenTopography: http://www.opentopography.org/index.php.

Brunger, A.P. and F.C. Hooper, 1993. Anisotropic sky radiance model based on narrow field of view measurements of shortwave radiance, *Solar Energy*, 51(1): 53–64, Erratum: 51(6): 523.

Burrows, W.E., 1999. *This New Ocean: The Story of the First Space Age*, Modern Library Paperbacks, 752p.

Butler, J.J., B.C. Johnson, and R.A. Barnes, 2005. The calibration and characterization of Earth remote sensing and environmental monitoring instruments, in: A.C. Parr, R.U. Datla, and J.L. Gardner, eds., *Optical Radiometry, Experimental Methods in the Physical Sciences*, Vol. 41. R. Celotta and T. Lucatorto, eds., *Treatise*, Elsevier/Academic Press, San Diego, CA, pp. 453–534.

Cabot, F., O. Hagolle, and P. Henry, 2000. Relative and multitemporal calibration of AVHRR, SeaWiFS, and VEGETATION using POLDER characterization of desert sites, in: *Proceedings of the International Geoscience and Remote Sensing Symposium 2000*, Honolulu, HI, pp. 2188–2190.

Cao, C. and A.K. Heidinger, 2002. Inter-comparison of the longwave infrared channels of MODIS and AVHRR/NOAA-16 using simultaneous nadir observations at orbit intersections, in: W.L. Barnes, ed., *Proceedings of SPIE Conference on Earth Observing Systems VII*, Vol. 4814, Seattle, WA. SPIE, Bellingham, WA, pp. 306–316.

Cavayas, F., 1987. Modelling and correction of topographic effect using multi-temporal satellite images, *Canadian Journal of Remote Sensing*, 13(2): 49–67.

Cavayas, F., G. Rochon, and P. Teillet, 1983. Estimation des reflectances bidirectionnelles par analyse des images Landsat: Problèmes et possibilités de solutions, in: *Comptes rendus du 8ème Symposium canadien sur la télédétection et 4ème Congres de l'Association québécoise de télédétection*, Montréal, Québec, Canada, p. 645.

Chahine, M.T., 1983. Interaction mechanisms within the atmosphere, in: R.N. Colwell, ed., Chapter 5: *Manual of Remote Sensing*, 2nd edn., American Society of Photogrammetry, Falls Church, VA.

Chander, G., T.J. Hewison, N. Fox, X. Wu, X. Xiong, and W.J. Blackwell, 2013a. Overview of intercalibration of satellite instruments, *IEEE Transactions on Geoscience and Remote Sensing*, 51(3 SI): 1056–1080.

Chander, G., T.J. Hewison, N. Fox, X. Wu, X. Xiong, and W.J. Blackwell, Guest Editors, 2013b. Special issue on intercalibration of satellite instruments, *IEEE Transactions on Geoscience and Remote Sensing*, 51(3 SI): 491p.

Chander, G., N. Mishra, D.L. Helder, D. Aaron, A. Angal, T. Choi, X. Xiong, and D. Doelling, 2013c. Applications of Spectral Band Adjustment Factors (SBAF) for cross-calibration, *IEEE Transactions on Geoscience and Remote Sensing*, 51(3 SI): 1267–1281.

Chander, G. and P.M. Teillet, Guest Editors and N. Coops, Editor-in-Chief, 2010. Special issue on terrestrial reference standard test sites for postlaunch calibration, *Canadian Journal of Remote Sensing*, 36(5): 437–630, doi:10.5589/cjrs3605fi.

Chang, I.L., C. Dean, Z. Li, M. Weinreb, X. Wu, and P.A.V.B. Swamy, 2012. Refined algorithms for star-based monitoring of GOES Imager visible channel responsivities, in: *Proceedings of SPIE Earth Observing Systems XVII*, San Diego, CA, p. 85100R.

Chavez, P.S., 1988. An improved dark-object subtraction technique for atmospheric scattering correction of multispectral data, *Remote Sensing of Environment*, 24: 459–479.

Chavez, P.S., 1996. Image-based atmospheric correction—Revisited and improved, *Photogrammetric Engineering and Remote Sensing*, 62(9): 1025–1036.

Che, N. and J.C. Price, 1992. Survey of radiometric calibration results and methods for visible and near-infrared channels of NOAA-7, NOAA-9, and NOAA-11 AVHRRs, *Remote Sensing of Environment*, 41(1): 19–27.

Chen, H.S., 1996. *Remote Sensing Calibration Systems—An Introduction*, A. Deepak Publishing, Hampton, VA, ISBN: 0-937194-38-7.

Chen, J.M. and J. Cihlar, 1997. A hotspot function in a simple bidirectional reflectance model for satellite applications, *Journal of Geophysical Research—Atmospheres*, 102(D22): 25907–25913.

Chu, W.P., 1983. Calculations of atmospheric refraction for spacecraft remote sensing applications, *Applied Optics*, 22(5): 721–725.

Cihlar, J. and P.M. Teillet, 1995. Forward piecewise linear calibration model for quasi-real-time processing of AVHRR data, *Canadian Journal of Remote Sensing*, 21(1): 22–27.

Connolly, J.I. and B. Tolar, 1994. Standards and calibration, Workshop II, in: *Proceedings of the First International Symposium on Spectral Sensing Research* (*ISSSR*), San Diego, CA, pp. 87–111.

Cracknell, A.P., 1998. Synergy in remote sensing—What's in a pixel? *International Journal of Remote Sensing*, 19(11): 2025–2047.

Crist, E.P., 1985. A TM Tasseled cap equivalent transformation for reflectance factor data, *Remote Sensing of Environment*, 17: 301–306.

Dana, R.W., 1982. Background reflectance effects in Landsat data, *Applied Optics*, 21(22): 4106–4111.

Dave, J.V., 1977. Validity of the isotropic-distribution approximation in solar energy estimations, *Solar Energy*, 19(4): 331–333.

Dave, J.V., 1980. Effect of atmospheric conditions on remote sensing of a surface nonhomogeneity, *Photogrammetric Engineering and Remote Sensing*, 46: 1173–1180.

Deschamps, P.Y., M. Herman, and D. Tanré, 1983. Definitions of atmospheric radiance and transmittances in remote sensing, *Remote Sensing of Environment*, 13(1): 89–92.

Dilligeard, E., X. Briottet, J.L. Deuze, and R.P. Santer, 1997. SPOT calibration of blue and green channels using Rayleigh scattering over clear oceans, in: *Proceedings of SPIE Conference on Advanced Next-Generation Satellites II*, Taormina, Italy, pp. 373–379.

Dinguirard M., J. Mueller, F. Sirou, and T. Tremas, 1997. Comparison of ScaRaB ground calibration in the short wave and long wave domains, *Metrologia*, 35: 597–601 (*Special Issue on NEWRAD'97*, Tucson, AZ).

Dinguirard, M. and P.N. Slater, 1999. Calibration of space-multispectral imaging sensors: A review, *Remote Sensing of Environment*, 68(3): 194–205.

Doelling, D.R., G. Hong, D. Morstad, R. Bhatt, A. Gopalan, and X. Xiong, 2010. The characterization of deep convective cloud albedo as a calibration target using MODIS reflectances, in: X. Xiong, C. Kim, and H. Shimoda, eds., *Proceedings of SPIE Conference on Earth Observing Missions Sensors: Development, Implementation, and Characterization*, Vol. 7862, Incheon, Republic of Korea, p. 78620I.

Doelling, D.R., C. Lukashin, P. Minnis, B. Scarino, and D. Morstad, 2012. Spectral reflectance corrections for satellite intercalibrations using SCHIAMACHY data, *IEEE Geoscience and Remote Sensing Letters*, 9(1): 119–123.

Doelling, D.R., L. Nguyen, and P. Minnis, 2004. On the use of deep convective clouds to calibrate AVHRR data, in: W.L. Barnes and J.J. Butler, eds., *Proceedings of SPIE Conference on Earth Observing Systems IX*, Vol. 5542, Denverr, CO, pp. 281–289.

Duggin, M.J., 1986. Variance in radiance recorded from heterogeneous targets in the optical-reflective, middle-infrared, and thermal-infrared regions, *Applied Optics*, 25(23): 4246–4252.

Duggin, M.J., 1987. Impact of radiance variations on satellite sensor calibration, *Applied Optics*, 26(7): 1264–1271.

Duggin, M.J. and C.J. Robinove, 1990. Assumptions implicit in remote sensing data acquisition and analysis, *International Journal of Remote Sensing*, 11(10): 1669–1694.

Duguay, C.R., 1993. Radiation modeling in mountainous terrain review and status, *Mountain Research and Development*, 13(4): 339–357.

Dymond, J.R. and J.D. Shepherd, 1999. Correction of the topographic effect in remote sensing, *IEEE Transactions on Geoscience and Remote Sensing*, 37: 2618–2620.

Edwards, D.P., L.K. Emmons, D.A. Hauglustaine, D.A. Chu, J.C. Gille, Y.J. Kaufman, G. Petron et al., 2004. Observations of carbon monoxide and aerosols from the Terra satellite: Northern Hemisphere variability, *Journal of Geophysical Research*, 109: D24202, doi:10.1029/2004JD004727.

Egan, W.G., 1985. *Photometry and Polarization in Remote Sensing*, Elsevier, New York, 503p.

Ekstrand, S., 1996. Landsat TM-based forest damage assessment: Correction for topographic effects, *Photogrammetric Engineering and Remote Sensing*, 62(2): 151–161.

Fedosejevs, G., N.T. O'Neill, A. Royer, P.M. Teillet, A.I. Bokoye, and B. McArthur, 2000. Aerosol optical depth for atmospheric correction of AVHRR composite data, *Canadian Journal of Remote Sensing*, 26(4): 273–284.

Fisher, P., 1997. The pixel: A snare and a delusion, *International Journal of Remote Sensing*, 18(3): 679–685.

Flittner, D.E. and P.N. Slater, 1991. Stability of narrow-band filter radiometers in the solar-reflective range, *Photogrammetric Engineering and Remote Sensing*, 57(2), 165–171.

Forster, B.C. and J.C. Trinder, 1984. An examination of the effects of resampling on classification accuracy, in: *Proceedings of the Third Australasian Remote Sensing Conference*, Brisbane, Queensland, Australia, pp. 106–115.

Fougnie, B. and R. Bach, 2009. Monitoring of radiometric sensitivity changes of space sensors using deep convective clouds: Operational application to PARASOL, *IEEE Transactions on Geoscience and Remote Sensing*, 47(3): 851– 861.

Fox, N., J. Aiken, J.J. Barnett, X. Briottet, R. Carvell, C. Frohlich, S.B. Groom et al., 2002. Traceable radiometry underpinning terrestrial- and helio-studies (TRUTHS), in: *Proceedings of SPIE Conference on Sensors, Systems, and Next-Generation Satellites VIII*, Vol. 4881, Heraklion, Crete, Greece, 12p.

Fox, N., A. Kaiser-Weiss, W. Schmutz, K. Thome, D. Young, B. Wielicki, R. Winkler, and E. Woolliams, 2011. Accurate radiometry from space: An essential tool for climate studies, *Philosophical Transactions of the Royal Society A—Mathematical, Physical, and Engineering Sciences*, 369(1953): 4028–4063.

Fox, N.P., 1999. Improving the accuracy and traceability of radiometric measurements to SI for remote sensing instrumentation, in: *Proceedings of the Fourth International Airborne Remote Sensing Conference and Exhibition/21st Canadian Symposium on Remote Sensing*, Vol. I, Ottawa, Ontario, Canada. ERIM International, pp. 304–311.

Fraser, R.S. and Y.J. Kaufman, 1986. Calibration of satellite sensors after launch, *Applied Optics*, 25(7): 1177–1185.

Freemantle, J., N.T. O'Neill, P.M. Teillet, A. Royer, J.-P. Blanchet, M. Aubé, S. Thulasiraman, F. Vachon, S. Gong, and M. Versi, 2002. Using web services for atmospheric correction of remote sensing data, in: *Proceedings of the 2002 International Geoscience and Remote Sensing Symposium (IGARSS'02) and the 24th Canadian Symposium on Remote Sensing*, Vol. V, Toronto, Ontario, Canada, pp. 2939–2941, also on CD-ROM.

Frouin, R. and C. Gautier, 1987. Calibration of NOAA-7 AVHRR, GOES-5, and GOES-6 VISSR/VAS solar channels, *Remote Sensing of Environment*, 22(1): 73–101.

Giles, P., 2001. Remote sensing and cast shadows in mountainous terrain, *Photogrammetric Engineering and Remote Sensing*, 67(7): 833–840.

Godden, G.D. and C.A. McKay, 1997. A strategy for observing the moon to achieve precise radiometric stability monitoring, *Canadian Journal of Remote Sensing*, 23(4): 333–341.

Goetz, A.F.H., K.B. Heidebrecht, and T.G. Chrien, 1995. High accuracy in-flight wavelength calibration of imaging spectrometry data, in: *Summaries of the Fifth Annual JPL Airborne Earth Science Workshop, AVIRIS Workshop*, Vol. 1, Pasadena, CA, pp. 67–69.

Gordon, H.R., 1987. Calibration requirements and methodology for remote sensors viewing the ocean in the visible, *Remote Sensing of Environment*, 22: 103–126.

Gordon, H.R. and M. Wang, 1994. Retrieval of water-leaving radiance and aerosol optical thickness over the oceans with SeaWiFS: A preliminary algorithm, *Applied Optics*, 33(3): 443–452.

Gu, D. and A. Gillespie, 1998. Topographic normalization of Landsat TM images of forest based on sub-pixel sun-canopy-sensor geometry, *Remote Sensing of Environment*, 64: 166–175.

Guenther, B., W. Barnes, E. Knight, J. Barker, J. Harnden, R. Weber, M. Roberto, G. Godden, H. Montgomery, and P. Abel, 1996. MODIS calibration: A brief review of the strategy for the at-launch calibration approach, *Journal of Atmospheric and Oceanographic Technology*, 13(2): 274–285.

Guenther, B.W., 1991. Calibration of passive remote observing optical and microwave instrumentation, in: *Proceedings of SPIE*, Vol. 1493, Orlando, FL. SPIE, The International Society for Optical Engineering, Bellingham, WA, 304p.

Guenther, B.W., J. Butler, and P. Ardanuy, eds., 1997. *Workshop on Strategies for Calibration and Validation of Global Change Measurements*, May 10–12, 1995, NASA Reference Publication 1397, NASA/GSFC, Greenbelt, MD, 125p.

Hall, F.G., D.E. Strebel, J.E. Nickerson, and S.J. Goetz, 1991. Radiometric rectification: Toward a common radiometric response among multidate, multisensor images, *Remote Sensing of Environment*, 35: 11–27.

Hay, J.E. and D.C. McKay, 1985. Estimating solar irradiance on inclined surfaces: A review and assessment of methodologies, *International Journal of Solar Energy*, 3(4–5): 203–240.

Hay, J.E., R. Perez, and D.C. McKay, 1986. Addendum and Errata to the Paper "Estimating solar irradiance on inclined surfaces: A review and assessment of methodologies", *International Journal of Solar Energy*, 4(1): 321–324.

He, X., D. Pan, Y. Bai, and F. Gong, 2006. A general purpose exact Rayleigh scattering look-up table for ocean color remote sensing, *Acta Oceanologica Sinica*, 25(1): 48–56.

Hegyi, F., 2004. Alignment of earth observation calibration and validation activities, Report delivered by Hegyi Geomatics International Inc. (HGI) to the Canada Centre for Remote Sensing, March 2004, Ottawa, Ontario, Canada, 47p.

Helder, D., K. Thome, D. Aaron, L. Leigh, J. Czapla-Myers, N. Leisso, S. Biggar, and N. Anderson, 2012. Recent surface reflectance measurement campaigns with emphasis on best practices, SI traceability and uncertainty estimation, *Metrologia*, 49(2): S21–S28.

Helder, D., K. Thome, N. Mishra, G. Chander, X. Xiong, A. Angal, and T. Choi, 2013. Absolute radiometric calibration of Landsat using a pseudo invariant calibration site, *IEEE Transactions on Geoscience and Remote Sensing*, 51(3 SI): 1360–1369.

Henry, P., G. Chander, B. Fougnie, C. Thomas, and X. Xiong, 2013. Assessment of spectral band impact on inter-calibration over desert sites using simulation based on EO-1 Hyperion data, *IEEE Transactions on Geoscience and Remote Sensing*, 51(3 SI): 1297–1308.

Hodgson, M.E. 1995. What Cell Size Does the Computed Slope/Aspect Angle Represent?, *Photogrammetric Engineering and Remote Sensing*, 61(5): 513–517.

Holben, B.N., T.F. Eck, I. Slutsker, D. Tanré, J.P. Buis, A. Setzer, E. Vermote et al., 1998. AERONET—A federated instrument network and data archive for aerosol characterization, *Remote Sensing of Environment*, 66: 1–16.

Holben, B.N. and C.O. Justice, 1981. An examination of spectral band rationing to reduce the topographic effect on remotely sensed data, *International Journal of Remote Sensing*, 2(2): 115–133.

Holben, B.N., Y.J. Kaufman, and J.D. Kendall, 1990. NOAA-11 AVHRR visible and near-IR inflight calibration, *International Journal of Remote Sensing*, 11(8): 1511–1519.

Hooper, F.C. and A.P. Brunger, 1980. A model for the angular distribution of sky radiance, *Journal of Solar Energy Engineering*, 102: 196–202.

Hooper, F.C., A.P. Brunger, and C.S. Chan, 1987. A clear sky model of diffuse sky radiance, *Journal of Solar Energy Engineering*, 109: 9–14.

Horler, D.N.H., 1996. Framework study on calibration/validation user requirements as part of the ground infrastructure program of the Canadian long-term space plan, Final Report to the Canada Centre for Remote Sensing, Ottawa, Ontario, Canada, by Horler Information Inc. for Contract 23413-5-E178/01-SQ, 53p.

Horler, D.N.H. and P.M. Teillet, 1996. *Workshop on Canadian Earth Observation Calibration and Validation, Workshop Report*, Canada Centre for Remote Sensing, Ottawa, Ontario, Canada, 59p.

Hu, Y., B.A. Wielicki, P. Yang, P.W. Stackhouse, Jr., B. Lin, and D.F. Young, 2004. Application of deep convective cloud albedo observation to satellite-based study of the terrestrial atmosphere: Monitoring the stability of spaceborne measurements and assessing absorption anomaly, *IEEE Transactions on Geoscience and Remote Sensing*, 42(11): 2594–2599.

Hugli, H. and W. Frei, 1983. Understanding anisotropic reflectance in mountainous terrain, *Photogrammetric Engineering and Remote Sensing*, 49(5): 671–683.

Iwabuchi, H., 2003. Calibration of the visible and near-infrared channels of NOAA-11 and -14 AVHRRs by using reflections from molecular atmosphere and stratus cloud, *International Journal of Remote Sensing*, 24(24): 5367–5378.

Jackson, R.D., ed., 1990. Special issue on coincident satellite, aircraft, and field measurements at the Maricopa Agricultural Center (MAC), *Remote Sensing of Environment*, 32(2&3): 77–228.

Jensen, J.R., 2006. *Remote Sensing of the Environment: An Earth Resources Perspective*, 2nd edn., Prentice Hall, Upper Saddle River, NJ, 608p.

Kaufman, Y.J. and R.S. Fraser, 1984. Atmospheric effects on classification of finite fields, *Remote Sensing of Environment*, 15: 95–118.

Kaufman, Y.J. and B.N. Holben, 1993. Calibration of the AVHRR visible and near-IR bands by atmospheric scattering, ocean glint and desert reflection, *International Journal of Remote Sensing*, 14(1): 21–52.

Kaufman, Y.J., D. Tanré, H.R. Gordon, T. Nakajima, J. Lenoble, R. Frouin, H. Grassl, B.M., Herman, M.D. King, and P.M. Teillet, 1997. Passive remote sensing of tropospheric aerosol and atmospheric correction for the aerosol effect, *Journal of Geophysical Research*, 102(D14): 16815–16830.

Kieffer, H.H., T.C. Stone, R.A. Barnes, S. Bender, R.E. Eplee, Jr., J. Mendenhall, and L. Ong, 2002. On-orbit radiometric calibration over time and between spacecraft using the moon, in: *Proceedings of SPIE Conference on Sensors, Systems, and Next-Generation Satellites VI*, Crete, Greece, pp. 287–298.

Kieffer, H.H. and R.L. Wildey, 1985. Absolute calibration of Landsat instruments using the moon, *Photogrammetric Engineering and Remote Sensing*, 51(9): 1391–1393.

Kieffer, H.H. and R.L. Wildey, 1996. Establishing the moon as a spectral radiance standard, *Journal of Atmospheric and Oceanic Technology*, 13(2): 360–375.

Kimes, D.S. and J.A. Kirchner, 1981. Modeling the effects of various radiant transfers in mountainous terrain on sensor response, *IEEE Transactions on Geoscience and Remote Sensing*, 19: 100–108.

Kirchner, J.A., S. Youkhana, and J.A. Smith, 1982. Influence of sky radiance distribution on the ratio technique for estimating bidirectional reflectance, *Photogrammetric Engineering and Remote Sensing*, 48: 955–959.

Klucher, T.M., 1979. Evaluation of models to predict insolation on tilted surfaces, *Solar Energy*, 23(2): 111–114.

Kotchenova, S.Y. and E.F. Vermote, 2007. Validation of a vector version of the 6S radiative transfer code for atmospheric correction of satellite data. Part II: Homogeneous Lambertian and anisotropic surfaces, *Applied Optics*, 46(20): 4455–4464.

Kotchenova, S.Y., E.F. Vermote, R. Matarrese, and F. Klemm, 2006. Validation of a vector version of the 6S radiative transfer code for atmospheric correction of satellite data. Part I: Path radiance, *Applied Optics*, 45: 6762–6774.

Kramer, H.J., 2001. *Observation of the Earth and Its Environment: Survey of Missions and Sensors*, 4th edn., Springer-Verlag, Berlin, Heidelberg, New York, 1510p.

Li, F., D.L.B. Jupp, M. Thankappan, L. Lymburner, N. Mueller, A. Lewis, and A. Held, 2012. A physics-based atmospheric and BRDF correction for Landsat data over mountainous terrain, *Remote Sensing of Environment*, 124: 756–770.

Li, X. and A.H. Strahler, 1992. Geometric-optical bidirectional reflectance modeling of the discrete crown vegetation canopy: Effect of crown shape and mutual shadowing, *IEEE Transactions on Geoscience and Remote Sensing*, 30(2): 276–292.

Liang, S., H. Fallah-Adl, S. Kalluri, J. JaJa, Y.J. Kaufman, and J.R.G. Townshend, 1997. An operational atmospheric correction algorithmfor Landsat Thematic Mapper imagery over the land, *Journal of Geophysical Research*, 102: 17173–17186.

Loeb, N.G., 1997. In-flight calibration of NOAA AVHRR visible and near-IR bands over Greenland and Antarctica, *International Journal of Remote Sensing*, 18(3): 477–490.

Luderer, G., J.A. Coakley, Jr., and W.R. Tahnk, 2005. Using sun glint to check the relative calibration of reflected spectral radiances, *Journal of Atmospheric and Oceanic Technology*, 22(10): 1480–1493.

Lukashin, C., B. Wielicki, D. Young, K. Thome, Z. Jin, and W. Sun, 2013. Uncertainty estimates for imager reference inter-calibration with CLARREO reflected solar spectrometer, *IEEE Transactions on Geoscience and Remote Sensing*, 51(3 SI): 1425–1436.

MacDonald, J.S., 1997. From space data to information, in: G. Konecny, ed., *Proceedings of the ISPRS Joint Workshop on Sensors and Mapping From Space*, Institute of Photogrammetry and Engineering Surveys, University of Hannover, Hannover, Germany, pp. 233–240.

Malila, W.A. and D.M. Anderson, 1986. *Satellite Data Availability and Calibration Documentation for Land Surface Climatology Studies*, Environmental Research Institute of Michigan, Contract NAS5-28715, Report No. 180300-1-F, NASA/GSFC, Greenbelt, MD, 214p.

Markham, B.L. and J.L. Barker, eds., 1985. Special issue on Landsat image data quality assessment, *Photogrammetric Engineering and Remote Sensing*, 51: 1245–1493.

Markham, B.L. and J.L. Barker, 1987. Radiometric properties of U.S. processed Landsat MSS data, *Remote Sensing of Environment*, 22: 39–71.

Markham, B.L., W.C. Boncyk, D.L. Helder, and J.L. Barker, 1997. Landsat-7 enhanced thematic mapper plus radiometric calibration, *Canadian Journal of Remote Sensing*, 23(4): 318–332.

McCartney, E.J., 1976. *Optics of the Atmosphere*, John Wiley, New York, 408p.

McCorkel, J., K. Thome, and R. Lockwood, 2013. Absolute radiometric calibration of narrow-swath imaging sensors with reference to non-coincident wide-swath sensors, *IEEE Transactions on Geoscience and Remote Sensing*, 51(3 SI): 1309–1318.

Meckler, Y. and Y.J. Kaufman, 1982. Contrast reduction by the atmosphere and retrieval of nonuniform surface reflectance, *Applied Optics*, 21: 310–316.

Meyer, P., K.I. Itten, T. Kellenberger, S. Sandmeier, and R. Sandmeier, 1993. Radiometric corrections of topographically induced effects on Landsat TM data in an alpine environment, *ISPRS Journal of Photogrammetry and Remote Sensing*, 48(4): 17–28.

Meygret, A., X. Briottet, P.J. Henry, and O. Hagolle, 2000. Calibration of SPOT4 HRVIR and vegetation cameras over Rayleigh scattering, in: W.L. Barnes, ed., *Proceedings of SPIE Conference on Earth Observing Systems V*, Vol. 4135, San Diego, CA. SPIE, Bellingham, WA, pp. 302–313.

Mitchell, R.M., D.M. O'Brien, and B.W. Forgan, 1992. Calibration of the NOAA AVHRR shortwave channels using split pass imagery: I. Pilot study, *Remote Sensing of Environment*, 40(1): 57–65.

Morain, S.A. and A.M. Budge, eds., 2004. *Postlaunch Calibration of Satellite Sensors*, ISPRS Book Series, Vol. 2, *Proceedings of the International Workshop on Radiometric and Geometric Calibration*, Gulfport, MS. A.A. Balkema Publishers, New York, 193p.

Moran, M.S., R.D. Jackson, P.N. Slater, and P.M. Teillet, 1992. Evaluation of atmospheric correction procedures for visible and near-infrared satellite sensor output, *Remote Sensing of Environment*, 41: 169–184.

Mueller, J., R. Stulhmann, R. Becker, E. Raschke, J.L. Monge, and P. Burkert, 1996. Ground based calibration facility for the scanner for radiation budget instrument in the solar spectral domain, *Metrologia*, 32: 657–660.

Muench, H.S., 1981. Calibration of geosynchronous satellite video sensors, Report No. AFGL-TR-81-0050, Air Force Geophysical Laboratory, Hanscom, MA.

Nicodemus, F.E., 1965. Directional reflectance and emissivity of an opaque surface, *Applied Optics*, 4(7): 767–775.

Nicodemus, F.E., 1970. Reflectance nomenclature and directional reflectance and emissivity, *Applied Optics*, 9(6): 1474–1475.

Nieke, J., T. Aoki, T. Tanikawa, H. Motoyoshi, M. Hori, and Y. Nakajima, 2003. Cross-calibration of satellite sensors over snow fields, in: W.L. Barnes, ed., *Proceedings of SPIE Conference on Earth Observing Systems VIII*, Vol. 5151, San Diego, CA. SPIE, pp. 406–414.

Nithianandam, J., B.W. Guenther, and L.J. Allison, 1993. An anecdotal review of NASA earth observation satellite remote sensors and radiometric calibration methods, *Metrologia*, 30: 207–212.

NRC, 2007. *Earth Science and Applications from Space: National Imperatives for the Next Decade and Beyond*, National Research Council, The National Academies, Washington, DC, 428p.

O'Brien, D.M. and R.M. Mitchell, 2001. An error budget for cross-calibration of AVHRR shortwave channels against ATSR-2, *Remote Sensing of Environment*, 75(2): 216–229.

O'Neill, N.T., J.R. Miller, and F.J. Ahern, 1978. Radiative transfer calculations for remote sensing applications, in: *Proceedings of the Fifth Canadian Symposium on Remote Sensing*, Victoria, British Columbia, Canada, p. 572.

O'Neill, N.T., P.M. Teillet, A. Royer, J.-P. Blanchet, M. Aubé, J. Freemantle, S. Gong, D. Stanley, S. Thulasiraman, and F. Vachon, 2002. Concept of a central optical parameter server for atmospheric corrections of remote sensing data, in: *Proceedings of the 2002 International Geoscience and Remote Sensing Symposium (IGARSS'02) and the 24th Canadian Symposium on Remote Sensing*, Vol. V, Toronto, Ontario, Canada, pp. 2951–2953, also on CD-ROM.

Ohring, G., J. Tansock, W. Emery, J. Butler, L. Flynn, F. Weng, K.S. Germain et al., 2007. Achieving satellite instrument calibration for climate change, *EOS, Transactions American Geophysical Union*, 88(11): 136.

Ohring, G., B. Wielicki, R. Spencer, B. Emery, and R. Datla, 2005. Satellite instrument calibration for measuring global climate change: Report of a workshop, *Bulletin of the American Meteorological Society*, 86: 1303–1313.

Otterman, J., M. Dishon, and S. Rehavi, 1983. Point spread functions in imaging a Lambert surface from zenith through a thin scattering layer, *International Journal of Remote Sensing*, 4: 583.

Otterman, J. and R.S. Fraser, 1979. Adjacency effects on imaging by surface reflection and atmospheric scattering: Cross radiance to zenith, *Applied Optics*, 18: 2852.

Pinty, B., J.-L. Widlowski, M. Taberner, N. Gobron, M.M. Verstraete, M. Disney, F. Gascon et al., 2004. Radiation transfer model intercomparison (RAMI) exercise: Results from the second phase, *Journal of Geophysical Research*, 109: D06.

Pollock, D.B., T.L. Murdock, R.U. Datla, and A. Thompson, 2003. Data uncertainty traced to SI units. Results reported in the International System of Units, *International Journal of Remote Sensing*, 24(2): 225–235.

Price, J.C., 1987a. Radiometric calibration of satellite sensors in the visible and near infrared: History and outlook, *Remote Sensing of Environment*, 22: 3–9.

Price, J.C., ed., 1987b. Special issue on radiometric calibration of satellite data, *Remote Sensing of Environment*, 22(1): 1–158.

Proy, C., D. Tanré, and P.Y. Deschamps, 1989. Evaluation of topographic effects in remotely sensed data, *Remote Sensing of Environment*, 30(1): 21–32.

Rao, C.R.N., C. Cao, and N. Zhang, 2003. Inter-calibration of the moderate-resolution imaging spectroradiometer and the along-track scanning radiometer-2, *International Journal of Remote Sensing*, 24(9): 1913–1924.

Rao, C.R.N. and J. Chen, 1995. Intersatellite calibration linkages for the visible and near-infared channels of the Advanced Very High-Resolution Radiometer on the NOAA-7, NOAA-9, and NOAA-11 spacecraft, *International Journal of Remote Sensing*, 16(11): 1931–1942.

Rao, C.R.N. and J.H. Chen, 1996. Postlaunch calibration of the visible and near-infrared channels of the Advanced Very High-Resolution Radiometer on the NOAA-14 spacecraft, *International Journal of Remote Sensing*, 17(14): 2743–2747.

Rao, C.R.N. and J.H. Chen, 1999. Revised postlaunch calibration of the visible and near-infrared channels of the Advanced Very High-Resolution Radiometer (AVHRR) on the NOAA-14 spacecraft, *International Journal of Remote Sensing*, 20(18): 3485–3491.

Regalado, A., 2013. The data made me do it, *MIT Technology Review*, 116(4): 63–64.

Riaño, D., E. Chuvieco, J. Salas, and I. Aguado, 2003. Assessment of different topographic corrections in Landsat-TM data for mapping vegetation types (2003), *IEEE Transactions on Geoscience and Remote Sensing*, 41(5): 1056–1061.

Richardson, A.J., 1982. Relating Landsat digital count values to ground reflectances for optical thin atmospheric conditions, *Applied Optics*, 21: 1457.

Richter, R., 1997. Correction of atmospheric and topographic effects for high spatial resolution satellite images, *International Journal of Remote Sensing*, 18: 1099–1111.

Richter, R., 1998. Correction of satellite imagery of mountainous terrain, *Applied Optics*, 37(18): 4004–4014.

Sandford, S.P., D.F. Young, J.M. Corliss, B.A. Wielicki, M.J. Gazarik, M.G. Mlynczak, A.D. Little et al., 2010. CLARREO: Cornerstone of the climate observing system measuring decadal change through accurate emitted infrared and reflected solar spectra and radio occultation, in: R. Meynart, S.P. Neeck, and H. Shimoda, eds., *Proceedings of SPIE Conference on Sensors, Systems, and Next-Generation Satellites XIV*, Vol. 7826, Toulouse, France, p. 782611.

Scarpace, F.L. and P.R. Wolf, 1973. Atmospheric refraction, *Photogrammetric Engineering*, 39: 521.

Schaaf, C.B., F. Gao, A.H. Strahler, W. Lucht, X. Li, T. Tsang, N.C. Strugnell et al., 2002. First operational BRDF, albedo nadir reflectance products from MODIS, *Remote Sensing of Environment*, 83(1–2): 135–148.

Schott, J.R., 2007. *Remote Sensing, The Image Chain Approach*, 2nd edn., Oxford University Press, New York.

Schott, J.R., C. Salvaggio, and W.J. Volchok, 1988. Radiometric scene normalization using pseudo-invariant features, *Remote Sensing of Environment*, 26: 1–16.

Schowengerdt, R.A., 2007. *Remote Sensing: Models and Methods for Image Processing*, 3rd edn., Academic Press, San Diego, CA, 515p.

Schowengerdt, R.A., C. Archwamety, and R.C. Wrigley, 1985. Landsat thematic mapper image-derived MTF, *Photogrammetric Engineering and Remote Sensing of Environment*, 51(9): 1395–1406.

Schowengerdt, R.A., S.K. Park, and R.T. Gray, 1983. An optimized cubic interpolator for image resampling, in: *Proceedings of the 17th International Symposium on Remote Sensing of Environment*, Ann Arbor, MI, p. 1291.

Schroeder, T.A., W.B. Cohen, C. Song, M.J. Canty, and Z. Yang, 2006. Radiometric correction of multi-temporal Landsat data for characterization of early successional forest patterns in western Oregon, *Remote Sensing of Environment*, 103: 16–26.

Secker, J., K. Staenz, R.P. Gauthier, and B. Budkewitsch, 2001. Vicarious calibration of hyperspectral sensors in operational environments, *Remote Sensing of Environment*, 26: 81–92.

Sellers, P.J., B.W. Meeson, F.G. Hall, G. Asrar, R.E. Murphy, R.A. Schiffer, F.P. Bretherton et al., 1995. Remote sensing of the land surface for studies of global change: Models—Algorithms—Experiments, *Remote Sensing of Environment*, 51: 3–26.

Shepherd, J.D. and J.R. Dymond, 2003. Correcting satellite imagery for the variance of reflectance and illumination with topography, *International Journal of Remote Sensing*, 24(17): 3503–3514.

Six, D., M. Fily, S. Alvain, P. Henry, and J.P. Benoist, 2004. Surface characterisation of the Dome Concordia area (Antarctica) as a potential satellite calibration site, using SPOT4/VEGETATION instrument, *Remote Sensing of Environment*, 89(1): 83–94.

Sjoberg, R.W. and B.K.P. Horn, 1983. Atmospheric effects in satellite imaging of mountainous terrain, *Applied Optics*, 22: 1702.

Slater, P.N., 1980. *Remote Sensing, Optics and Optical Systems*, Addison-Wesley Publishing Company, Reading, MA.

Slater, P.N., 1984. The importance and attainment of absolute radiometric calibration, in: *Proceedings of SPIE Critical Review of Remote Sensing*, Vol. 475, Arlingrton, VA, pp. 34–40.

Slater, P.N., 1985. Radiometric considerations in remote sensing, *Proceedings of IEEE*, 73(6): 997–1011.

Slater, P.N., 1988a. Review of the calibration of radiometric measurements from satellite to ground level, *International Archives of Photogrammetry and Remote Sensing*, 27(B11): 726–724.

Slater, P.N., 1988b. Recent advances in sensors, radiometry, and data processing for remote sensing, in: *Proceedings of SPIE*, Vol. 924, Orlando, FL. SPIE, The International Society for Optical Engineering, Bellingham, WA, 341p.

Slater, P.N. and S.F. Biggar, 1996. Suggestions for radiometric calibration coefficient generation, *Journal of Atmospheric and Oceanographic Technology*, 13(2): 376–382.

Slater, P.N., S.F. Biggar, R.G. Holm, R.D. Jackson, Y. Mao, M.S. Moran, J.M. Palmer, and B. Yuan, 1987. Reflectance- and radiance-based methods for the in-flight absolute calibration of multispectral sensors, *Remote Sensing of Environment*, 22: 11–37.

Slater, P.N., S.F. Biggar, J.M. Palmer, and K.J. Thome, 1995. Unified approach to pre- and in-flight satellite-sensor absolute radiometric calibration, in: *Proceedings of the SPIE Europto Symposium*, Vol. 2583, SPIE, Paris, France, pp. 130–141.

Slater, P.N., S.F. Biggar, J.M. Palmer, and K.J. Thome, 2001. Unified approach to absolute radiometric calibration in the solar-reflective range, *Remote Sensing of Environment*, 77: 293–303.

Slater, P.N. and R.D. Jackson, 1982. Atmospheric effects on radiation reflected from soil and vegetation as measured by orbital sensors using various scanning directions, *Applied Optics*, 21: 3923.

Smith, G.R., R.H. Levin, P. Abel, and H. Jacobowitz, 1988. Calibration of the solar channels and NOAA-9 AVHRR using high altitude aircraft measurements, *Journal of Atmospheric and Oceanic Technology*, 5: 631–639.

Smith, J.A., T.L. Lin, and K.J. Ranson, 1980. The Lambertian assumption and Landsat data, *Photogrammetric Engineering and Remote Sensing*, 46: 1183–1189.

Soenen, S., D. Peddle, C. Coburn, R. Hall, and F. Hall, 2008. Improved topographic correction of forest image data using a 3-d canopy reflectance model in multiple forward mode. *International Journal of Remote Sensing*, 29(4): 1107–1027.

Soenen, S.A., D.R. Peddle, and C.A. Coburn, 2005. A modified sun-canopy-sensor topographic correction in forested terrain, *IEEE Transactions on Geoscience and Remote Sensing*, 43(9): 2148–2159.

Song, C., C.E. Woodcock, K.C. Seto, M. Pax-Lenney, and S.A. Macomber, 2001. Classification and change detection using Landsat TM data: When and how to correct atmospheric effects, *Remote Sensing of Environment*, 75: 230–244.

Souri, A.H. and A. Azizi, 2013. Removing bowtie phenomenon by correction of panoramic effect in MODIS imagery, *International Journal of Computer Applications*, 68(3): 12–16.

Staenz, K., J. Secker, B.C. Gao, C. Davis, and C. Nadeau, 2002. Radiative transfer codes applied to hyperspectral data for the retrieval of surface reflectance, *ISPRS Journal of Photogrammetry and Remote Sensing*, 57(3): 194–203.

Staenz, K., D.J. Williams, G. Fedosejevs, and P.M. Teillet, 1994. Surface reflectance retrieval from imaging spectrometer data using three atmospheric codes, in: *Proceedings of SPIE EUROPTO'94*, Vol. 2318, Rome, Italy. SPIE, pp. 17–28.

Staylor, W.F., 1990. Degradation rates of the AVHRR visible channel for the NOAA 6, 7, and 9 spacecraft, *Journal of Atmospheric and Oceanic Technology*, 7(3): 411–423.

Steven, M.D., 1977. Standard distribution of clear sky radiance, *Quarterly Journal of the Royal Meteorological Society*, 106: 57.

Steven, M.D., T.J. Malthus, F. Baret, H. Xu, and M.J. Chopping, 2003. Intercalibration of vegetation indices from different sensor systems, *Remote Sensing Environment*, 88(4): 412–422.

Stone, T.C., 2008. Radiometric calibration stability and intercalibration of solar-band instruments in orbit using the moon, in: *Proceedings of SPIE Conference on Earth Observing Systems XIII*, San Diego, CA, 2008, p. 70810X.

Stone, T.C., H.H. Kieffer, and I.F. Grant, 2005. Potential for calibration of geostationary meteorological satellite imagers using the moon, in: *Proceedings of SPIE Conference on Earth Observing Systems X*, San Diego, CA, pp. 1–9.

Stone, T.C., W.B. Rossow, J. Ferrier, and L.M. Hinkelmann, 2013. Evaluation of ISCCP multi-satellite radiance calibration for geostationary imager visible channels using the moon, *IEEE Transactions on Geoscience and Remote Sensing*, 51(3): 1255–1266.

Suits, G.H., W.A. Malila, and T.M. Weller, 1988. The prospects for detecting spectral shifts due to satellite sensor ageing, *Remote Sensing of Environment*, 26: 17–29.

Sweet, R.J.M, J.C. Elliott, and J.R. Beasley, 1992. Research needs to encourage the growth of the earth observation applications market, in: P.A. Cracknell and R.A. Vaughan, eds., *Proceedings of the 18th Annual Conference of the Remote Sensing Society: Remote Sensing from Research to Operation*, September 15–17, University of Dundee, Dundee, U.K., pp. 399–407.

Tahnk, W.R. and J.A. Coakley, 2001. Updated calibration coefficients for NOAA-14 AVHRR channels 1 and 2, *International Journal of Remote Sensing*, 22(15): 3053–3057.

Tanré, D., C. Deroo, P. Duhaut, M. Herman, J.J. Morcrette, and J. Perbos, 1990. Description of a computer code to simulate the satellite signal in the solar spectrum: The 5S code, *International Journal of Remote Sensing*, 11: 659–668.

Tanré, R.C., Herman, M., and Deschamps, P.Y., 1983. Influence of the atmosphere on space measurements of directional properties, *Applied Optics*, 22: 733.

Teillet, P.M., 1986. Image correction for radiometric effects in remote sensing, *International Journal of Remote Sensing*, 7: 1637–1651.

Teillet, P.M., 1989. Surface reflectance retrieval using atmospheric correction algorithms, in: *Proceedings of the 1989 International Geoscience and Remote Sensing Symposium (IGARSS'89) and the 12th Canadian Symposium on Remote Sensing*, Vancouver, British Columbia, Canada, pp. 864–867.

Teillet, P.M., 1990a. Rayleigh optical depth comparisons from various sources, *Applied Optics*, 29: 1897–1900.

Teillet, P.M., 1990b. Effects of spectral shifts on sensor response, in: *Proceedings of the ISPRS Commission VII Symposium*, Victoria, British Columbia, Canada, pp. 59–65.

Teillet, P.M., 1992. An algorithm for the radiometric and atmospheric correction of AVHRR data in the solar reflective channels, *Remote Sensing of Environment*, 41: 185–195.

Teillet, P.M., 1997a. A status overview of earth observation calibration/validation for terrestrial applications, *Canadian Journal of Remote Sensing*, 23(4): 291–298.

Teillet, P.M., ed., 1997b. Special issue on calibration/validation, *Canadian Journal of Remote Sensing*, 23(4): 289–423.

Teillet, P.M., 1997c. *Report on the Second Canadian Workshop on Earth Observation Calibration and Validation*, Canada Centre for Remote Sensing, Ottawa, Ontario, Canada, 66p.

Teillet, P.M., 1998. *Report on the Third Canadian Workshop on Earth Observation Calibration and Validation*, Canada Centre for Remote Sensing, Ottawa, Ontario, Canada, 74p.

Teillet, P.M., September/October 2005. Satellite image radiometry: From photons to calibrated earth science data, *Physics in Canada*, 61(5), 301–310.

Teillet, P.M. and G. Chander, 2014. Post-launch radiometric calibration of satellite-based optical sensors with emphasis on terrestrial reference standard sites, in: P.S. Thenkabail, ed., Chapter 4: *The Remote Sensing Handbook*, Vol. 1, CRC Press, Boca Raton, FL.

Teillet, P.M. and G. Fedosejevs, 1995. On the dark target approach to atmospheric correction of remotely sensed data, *Canadian Journal of Remote Sensing*, 21(4): 374–387.

Teillet, P.M., G. Fedosejevs, F.J. Ahern, and R.P. Gauthier, 1994. Sensitivity of surface reflectance retrieval to uncertainties in aerosol optical properties, *Applied Optics*, 33(18): 3933–3940.

Teillet, P.M., G. Fedosejevs, R.P. Gauthier, N.T. O'Neill, K.J. Thome, S.F. Biggar, H. Ripley, and A. Meygret, 2001a. A generalized approach to the vicarious calibration of multiple earth observation sensors using hyperspectral data, *Remote Sensing of Environment*, 77(3): 304–327.

Teillet, P.M., G. Fedosejevs, R.K. Hawkins, T.I. Lukowski, R.A. Neville, K. Staenz, R. Touzi, J. van der Sanden, and J. Wolfe, 2004a. *Importance of Data Standardization for Generating High Quality Earth Observation Products for Natural Resource Management*, Earth Sciences Sector, Natural Resources Canada, Ottawa, Ontario, Canada, 42p.

Teillet, P.M., G. Fedosejevs, and K.J. Thome, 2004b. Spectral band difference effects on radiometric cross-calibration between multiple satellite sensors in the Landsat solar-reflective spectral domain, in: R. Meynart, S.P. Neeck, and H. Shimoda, eds., *Workshop on Inter-Comparison of Large-Scale Optical and Infrared Sensors, Proceedings of SPIE Conference on Sensors, Systems, and Next-Generation Satellites VIII*, Vol. 5570, Maspalomas, Canary Islands, Spain, pp. 307–316.

Teillet, P.M., G. Fedosejevs, K.J. Thome, and J.L. Barker, 2007. Impacts of spectral band difference effects on radiometric cross-calibration between satellite sensors in the solar-reflective spectral domain, *Remote Sensing of Environment*, 110(3): 393–409.

Teillet, P.M., B. Guindon, and D.G. Goodenough, 1982. On the slope aspect correction of multi-spectral scanner data, *Canadian Journal of Remote Sensing*, 8: 84–106.

Teillet, P.M. and B.N. Holben, 1994. Towards operational radiometric calibration of NOAA AVHRR imagery in the visible and near-infrared channels, *Canadian Journal of Remote Sensing*, 20(1): 1–10.

Teillet, P.M., D.N.H. Horler, and N.T. O'Neill, 1997a. Calibration, validation, and quality assurance in remote sensing: A new paradigm, *Canadian Journal of Remote Sensing*, 23(4): 401–414.

Teillet, P.M. and J.R. Irons, 1990. Spectral variability effects on the atmospheric correction of imaging spectrometer data for surface reflectance retrieval, in: *Proceedings of the ISPRS Commission VII Symposium*, Victoria, British Columbia, Canada, pp. 579–583.

Teillet, P.M., M. Lasserre, and C.G. Vigneault, 1986. An evaluation of sun angle computation algorithms, in: *Proceedings of the 10th Canadian Symposium on Remote Sensing*, Edmonton, Alberta, Canada, pp. 91–100.

Teillet, P.M. and X. Ren, 2008. Spectral band difference effects on vegetation indices derived from multiple satellite sensor data, *Canadian Journal of Remote Sensing*, 34(3): 159–173.

Teillet, P.M. and R.P. Santer, 1991. Terrain elevation and sensor altitude dependence in a semi-analytical atmospheric code, *Canadian Journal of Remote Sensing*, 17: 36–44.

Teillet, P.M., P.N. Slater, Y. Ding, R.P. Santer, R.D. Jackson, and M.S. Moran, 1990. Three methods for the absolute calibration of the NOAA AVHRR sensors in-flight, *Remote Sensing of Environment*, 31: 105–120.

Teillet, P.M. and K. Staenz, 1992. Atmospheric effects due to topography on MODIS vegetation index data simulated from AVIRIS imagery over mountainous terrain, *Canadian Journal of Remote Sensing*, 18(4): 283–291.

Teillet, P.M., K. Staenz, and D.J. Williams, 1997b. Effects of spectral, spatial, and radiometric characteristics on remote sensing vegetation indices for forested regions, *Remote Sensing of Environment*, 61: 139–149.

Teillet, P.M., K.J. Thome, N. Fox, and J.T. Morisette, 2001b. Earth observation sensor calibration using a global instrumented and automated network of test sites (GIANTS), in: H. Fujisada, J.B. Lurie, and K. Weber, eds., *Proceedings of SPIE Conference on Sensors, Systems, and Next-Generation Satellites V*, Vol. 4550, Toulouse, France. SPIE, pp. 246–254.

Temps, R.C. and K.L. Coulson, 1977. Solar radiation incident upon slopes of different orientations, *Solar Energy*, 19: 179.

Thome, K., 2005. Sampling and uncertainty issues in trending reflectance-based vicarious calibration results, in: J.J. Butler, ed., *Proceedings of SPIE Conference on Earth Observing Systems X*, Vol. 5882, San Diego, CA. SPIE, Bellingham, WA, pp. 1–11.

Thome, K., R. Barnes, R. Baize, J. O'Connell, and J. Hair, 2010. Calibration of the reflected solar instrument for the climate absolute radiance and refractivity observatory, in: *Proceedings of the 2010 IEEE International Geoscience and Remote Sensing Symposium (IGARSS)*, Honolulu, HI, pp. 2275–2278.

Thome, K., J. Czapla-Myers, N. Leisso, J. McCorkel, and J. Buchanan, 2008. Intercomparison of imaging sensors using automated ground measurements, in: *Proceedings of IEEE Conference on Remote Sensing: The Next Generation*, Boston, MA. IEEE, Piscataway, NJ, pp. 1332–1335.

Thome, K., S. Schiller, J. Conel, K. Arai, and S. Tsuchida, 1998. Results of the 1996 earth observing system vicarious calibration joint campaign at Lunar Lake Playa, *Metrologia*, 35: 631–638.

Thome, K.J., 2001. Absolute radiometric calibration of Landsat 7 ETM+ using the reflectance-based method, *Remote Sensing of Environment*, 78(1–2): 27–38.

Thome, K.J, S.F. Biggar, and W. Wisniewski, 2003. Cross comparison of EO-1 sensors and other earth resources sensors to Landsat-7 ETM+ using Railroad Valley Playa, *IEEE Transactions on Geoscience and Remote Sensing*, 41: 1180–1188.

Thuillier, G., M. Hersé, D. Labs, T. Foujols, W. Peetermans, D. Gillotay, P.C. Simon, and H. Mandel, 2003. The solar spectral irradiance from 200 to 2400 nm as measured by the SOLSPEC spectrometer from the Atlas and Eureca missions, *Solar Physics*, 214(1): 1–22.

Townshend, J.R.G., 1981. The spatial resolving power of earth resources satellites, *Progress in Physical Geography*, 5(1): 32–55.

Townshend, J.R.G., C. Huang, S.N.V. Kalluri, R.S. Defries, and S. Liang, 2000. Beware of per-pixel characterization of land cover, *International Journal of Remote Sensing*, 21(4): 839–843.

Trishchenko, A.P., 2009. Effects of spectral response function on surface reflectance and NDVI measured with moderate resolution satellite sensors: Extension to AVHRR NOAA-17, 18 and METOP-A, *Remote Sensing of Environment*, 113(2): 335–341.

Trishchenko, A.P., J. Cihlar, and Z. Li, 2002. Effects of spectral response function on surface reflectance and NDVI measured with moderate resolution satellite sensors, *Remote Sensing of Environment*, 81(1): 1–18.

Turner, R.E. and M.M. Spencer, 1972. Atmospheric model for correction of spacecraft data, in: *Proceedings of the Eighth International Symposium on Remote Sensing of Environment*, Ann Arbor, MI, p. 895.

Vermote, E. and Y.J. Kaufman, 1995. Absolute calibration of AVHRR visible and near-infrared channels using ocean and cloud views, *International Journal of Remote Sensing*, 16(13): 2317–2340.

Vermote, E.F., N. El Saleous, C.O. Justice, Y.J. Kaufman, J.L. Privette, L. Remer, J.C. Roger, and D. Tanre, 1997. Atmospheric correction of visible to middle-infrared EOS-MODIS data over land surfaces: Background, operational algorithm and validation, *Journal of Geophysical Research*, 102: 17131–17141.

Vermote, E.F., N.Z. El Saleous, and C.O. Justice, 2002. Atmospheric correction of MODIS data in the visible to middle infrared: First results, *Remote Sensing of Environment*, 83(1–2): 97–111.

Vermote, E.F. and N.Z. Saleous, 2006. Calibration of NOAA16 AVHRR over a desert site using MODIS data, *Remote Sensing of Environment*, 105(3): 214–220.

Vermote, E.F., R. Santer, P.Y. Deschamps, and M. Herman, 1992. In-flight calibration of large field of view sensors at short wavelengths using Rayleigh scattering, *International Journal of Remote Sensing*, 13(18): 3409–3429.

Vermote, E.F., D. Tanré, J.L. Deuzé, M. Herman, J.J. Morcrette, S.Y. Kotchenova, and T. Miura, 2006. Second simulation of the satellite signal in the solar spectrum (6S), 6S user guide version 3, November 2006, http://www.6s.ltdri.org. Accessed May 25, 2015.

White, H.P., J.R. Miller, and J.M. Chen, 2001. Four-scale linear model for anisotropic reflectance (FLAIR) for plant canopies. I: Model description and partial validation, *IEEE Transactions on Geoscience and Remote Sensing*, 39(5): 1072–1083.

White, H.P., J.R. Miller, and J.M. Chen, 2002b. Four-scale linear model for anisotropic reflectance (FLAIR) for plant canopies II: Partial validation and inversion using field measurements, *IEEE Transactions on Geoscience and Remote Sensing*, 40(5): 1038–1046.

White, H.P., L. Sun, K. Staenz, R.A. Fernandes, and C. Champagne, 2002a. Determining the contribution of shaded elements of a canopy to remotely sensed hyperspectral signatures, in: *Proceedings of the First International Symposium on Recent Advances on Quantitative Remote Sensing*, Torrent, Valencia, Spain.

Whitlock, C.H., W.F. Staylor, J.T. Suttles, G. Smith, R. Levin, R. Frouin, C. Gautier et al., 1990. AVHRR and VISSR satellite instrument calibration results for both cirrus and marine stratocumulus IFO periods, in: *Proceedings of FIRE Science Meeting*, Vail, CO. NASA Langley Research Center, Hampton, VA, pp. 141–146.

Woodham, R.J. and M.H. Gray, 1987. Analytic method for radiometric correction of satellite multispectral scanner data, *IEEE Transactions on Geoscience and Remote Sensing*, GE-25: 258–271.

Woodham, R.J. and T.K. Lee, 1985. Photometric method for radiometric correction of multispectral scanner data, *Canadian Journal of Remote Sensing*, 11: 132.

Wu, A., C. Cao, and X. Xiong, 2003. Intercomparison of the 11- and 12-μm bands of Terra and Aqua MODIS using NOAA-17 AVHRR, in: W.L. Barnes, ed., *Proceedings of SPIE Conference on Earth Observing Systems VIII*, Vol. 5151, San Diego, CA. SPIE, Bellingham, WA, pp. 384–394.

Wyatt, C.L., V. Privalsky, and R. Datla, 1998. *Recommended Practice: Symbols, Terms, Units and Uncertainty Analysis for Radiometric Sensor Calibration*, NIST Handbook, Vol. 152, National Institute of Standards and Technology, Gaithersburg, MD, 120p.

Xiong, X., J. Sun, and W. Barnes, 2008. Intercomparison of on-orbit calibration consistency between Terra and Aqua MODIS reflective solar bands using the moon, *IEEE Geoscience and Remote Sensing Letters*, 5(4): 778–782.

Yuan, D. and C.D. Elvidge, 1996. Comparison of relative radiometric normalization techniques, *ISPRS Journal of Photogrammetry and Remote Sensing*, 51: 117–126.

4

Postlaunch Radiometric Calibration of Satellite-Based Optical Sensors with Emphasis on Terrestrial Reference Standard Sites

Philippe M. Teillet
University of Lethbridge

Gyanesh Chander
Goddard Space Flight Center (GSFC)

Acronyms and Definitions

ADEOS	Advanced Earth observing satellite
ALI	Advanced land imager
AVHRR	Advanced very-high-resolution radiometer
CBERS	China–Brazil Earth Resources Satellite
CEOS	Committee on Earth Observation Satellites
CLARREO	Climate Absolute Radiance and Refractivity Observatory
CNES	Centre National d'Etudes Spatiales
DC	Digital count
EO	Earth observation
EO-1	Earth observing-1
EOS	Earth Observing System
EROS	Earth Resources Observation and Science
ESA	European Space Agency
ETM+	Enhanced Thematic Mapper Plus
FASC	Full-aperture solar calibrator
GCOS	Global Climate Observing System
GEO	Group on Earth Observations
GEOSS	Global Earth Observation System of Systems
GIANTS	Global instrumented and automated network of test sites
GLI	Global imager
GSICS	Global Space-Based Inter-Calibration System
HgCdTe	Mercury cadmium telluride
IC	Internal calibrator
InSb	Indium antimonide
IRMSS	Infrared multispectral scanner
ISO	International Organization for Standardization
IVOS	Infrared Visible Optical Sensors
JAXA	Japan Aerospace Exploration Agency
L4	Landsat-4
L5	Landsat-5
L7	Landsat-7
L8	Landsat-8
MOBY	Marine optical BuoY
MODIS	MODerate resolution Imaging Spectroradiometer
MSS	Multispectral scanner
NASA	National Aeronautics and Space Administration
NIST	National Institute of Standards and Technology
NOAA	National Oceanic and Atmospheric Administration
NPL	National Physical Laboratory
OLI	Optical land imager
PASC	Partial-aperture solar calibrator
PICS	Pseudoinvariant calibration site
PIF	Pseudoinvariant feature
QA4EO	Quality Assurance Framework for Earth Observation
QUASAR	Quality assurance and stability reference
SBDE	Spectral band difference effects
SI	International system of units
SPOT	Satellite Pour l'Observation de la Terre
SRCA	Spectro-radiometric calibration assembly
TM	Thematic Mapper

TOA	Top of atmosphere
TRUTHS	Traceable Radiometry Underpinning Terrestrial and Helio Studies
USDA	United States Department of Agriculture
USGS	United States Geological Survey
WGCV	Working Group on Calibration and Validation
WMO	World Meteorological Organization
WSNM	White Sands, New Mexico

4.1 Introduction

Scientists and decision makers addressing local, regional, and/or global issues rely increasingly on operational use of data and information obtained from a multiplicity of satellite-based Earth observation (EO) sensor systems. It is imperative that they be able to rely on the accuracy of EO data and information products (e.g., Ohring et al., 2007). Accordingly, the characterization and calibration of these sensors are critical elements of EO programs.

Chapter 3 provided an overview of satellite image radiometry in the solar-reflective domain (Teillet, 2014). This chapter delves more deeply into postlaunch sensor radiometric calibration methodologies, encompassing optical sensors operating in the visible, near-infrared, and shortwave infrared spectral regions, with particular emphasis on the use of terrestrial reference standard sites. Although a lot of time and effort are devoted to prelaunch radiometric calibration, postlaunch changes to satellite sensors and their radiometric calibration performance necessitate ongoing calibration monitoring and maintenance (e.g., Slater et al., 2001; Butler et al., 2005).

Calibration of satellite sensor radiometric response is generally classified into relative and absolute calibration (e.g., Kastner and Slater, 1982; Slater, 1984; Teillet, 1986). Relative calibration mainly deals with compensation for detector-to-detector differences. Ideally, detectors constructed from the same materials should respond identically to the same incident energy. However, typically, detectors do not respond identically, resulting in differences in detector gain and bias levels that lead to *striping* in the image data. This striping can be corrected by selecting a stable reference detector, then scaling and shifting the other detector responses to identical targets to the reference detector's gain and bias. This process is called detector-to-detector relative radiometric calibration. Absolute radiometric calibration enables the conversion of image digital counts (DCs) to physical units in the International System of Units (SI) such as at-sensor spectral radiance (W m^{-2} sr^{-1} μm^{-1}). Because DCs from one sensor bear no relation to DCs from a different sensor, conversion to at-sensor spectral radiance is a fundamental step that enables the comparison of similar products from different sensors. The additional step of converting the at-sensor radiance to top-of-atmosphere (TOA) reflectance is also important, as outlined in Chapter 3 (Teillet, 2014). Absolute radiometric calibration procedures (Slater, 1985) include (1) prelaunch calibration, with only focal plane calibration after launch; (2) postlaunch reference to onboard standard sources or to the Sun via a diffuser, usually not illuminating the full aperture; and (3) postlaunch reference to an Earth target of known reflectance, to the Moon, or to bright stars. Procedure category (3) is usually called vicarious calibration. Vicarious calibration can also be used for relative calibration to compensate for temporal trends in sensor response.

Onboard radiometric calibration systems can provide good temporal sampling with high precision that allows trending of system responses. A disadvantage of onboard calibration approaches is that they add significantly to the complexity and cost of satellite missions. Vicarious calibration techniques involving Earth targets provide independent, full-aperture calibrations with relatively high accuracy. While reference to an Earth target requires an accurate determination of atmospheric optical properties (mainly aerosol optical depth) at the time of satellite overpass, it has the important advantage of replicating actual conditions of image acquisition, with a full irradiation of the entrance aperture. The sensor acquires data in the same modality as it acquires Earth image data, that is, in the same spectral bands, with the same source spectrum and under typical illumination conditions. The main disadvantages of vicarious calibration are that the methods yield lower precision and have lower temporal sampling frequencies compared to onboard radiometric calibration approaches.

Terrestrial surfaces with suitable characteristics have long served as benchmarks or reference standard sites, via either vicarious calibration or cross-calibration, to assess the postlaunch radiometric calibration performance of satellite optical sensors (e.g., Teillet and Chander, 2010; Chander et al., 2013a,b). The use of such sites is the only practical way of evaluating radiometric calibration biases between sensors, and it also provides a means of bridging gaps in measurement continuity. Accordingly, after a brief review of onboard calibration and other vicarious calibration methodologies, this chapter focuses on the use of terrestrial reference standard sites for postlaunch sensor radiometric calibration of optical sensors operating in the visible, near-infrared, and shortwave infrared spectral regions.

4.2 Postlaunch Sensor Radiometric Calibration Methodologies

As noted in Chapter 3 (Teillet, 2014), the establishment of worldwide coordinated and operational calibration efforts is critical to achieving the goals of quantitative EO programs in general, and the goals of endeavors such as the Global Earth Observation System of Systems (GEOSS)* of the Group on Earth Observations (GEO), as well as the Global Space-Based Inter-Calibration System (GSICS).† Accordingly, the international

* http://www.earthobservations.org/geoss.shtml.

† http://gsics.wmo.int/ GSICS is part of the WMO. It is an international collaboration to monitor, improve and harmonize data quality from operational environmental satellites for climate monitoring and weather forecasting.

Committee on Earth Observation Satellites (CEOS), considered to be the space arm of GEO, has led the development of a Quality Assurance Framework for Earth Observation (QA4EO),* which is based on the adoption of key guidelines to help address difficult issues of traceability and interoperability, especially in the postlaunch environment.

Whereas prelaunch calibration methodologies typically include radiometric, geometric, spectral, and polarization characterizations, postlaunch calibration typically involves primarily radiometric and geometric calibration. One exception is National Aeronautics and Space Administration (NASA)'s Terra and Aqua MODerate-resolution Imaging Spectroradiometers (MODIS), which include an onboard spectro-radiometric calibration assembly (SRCA) designed to monitor the visible, near-infrared, and short-wave infrared spectral bands on MODIS (e.g., Xiong et al., 2006a, 2010a; Choi et al., 2013). Postlaunch radiometric calibration can be categorized into onboard calibration and vicarious calibration, the topics of the following sections.

4.2.1 Onboard Radiometric Calibration

Some satellite sensor systems utilize onboard devices to help monitor and, if necessary, update radiometric calibration coefficients. Partial-aperture calibrators include standard lamps and/or solar diffuser panels. If done well, such approaches can capture sensor drift, instability, and sensor and electronic ageing problems. Shutters or deep space views provide dark readings.

Although there are numerous satellite sensor systems that make use of onboard radiometric calibration systems, the Landsat series of whisk-broom sensors is a notable example. Tremendous efforts have been dedicated to understanding and using the onboard calibration lamps (Barker, 1985; Markham and Barker, 1985) and partial-aperture or full-aperture diffuse reflectance panels (Markham et al., 2003). These efforts are a measure of the importance of the Landsat series of sensors that have provided a readily accessible, global, and seasonal archive of relatively high spatial resolution data spanning over four decades (Goward and Masek, 2001; Goward et al., 2006; Markham and Helder, 2012; Wulder and Masek, 2012). It should also be noted that the Landsat radiometric calibration experience was a critical contribution to the challenging radiometric calibration of the Earth Observing System (EOS) MODIS whisk-broom sensors.

Key characteristics of the multispectral scanners (MSS) sensors on Landsat missions 1–5 are documented mainly in the so-called gray literature (e.g., Bartolucci and Davis, 1983). For the purposes of this chapter, the main MSS onboard calibration systems were internal calibrators (ICs) consisting of a shutter wheel with a mirror and a neutral density filter, and two redundant tungsten-filament lamps (e.g., Ahern and Murphy, 1979; Murphy, 1984, 1986; Markham and Barker, 1985; Thome et al., 1997b). The mirror reflected light from the IC lamps through the neutral density filter and onto the focal plane. The neutral density filter was wedge shaped, yielding a variable attenuation as it rotated with the shutter wheel. The IC did not test the full optical path of the MSS sensors.

The Thematic Mapper (TM) sensors on Landsat-4 (L4), Landsat-5 (L5), and the Landsat-7 (L7) Enhanced TM Plus (ETM+) have been well documented (e.g., Markham et al., 1998, 2012). Briefly, the TMs and the ETM+ have an uncooled primary focal plane containing 16 silicon (Si) detectors per band for the four visible and near-infrared bands. The L7 ETM+ also has a panchromatic band with 32 detectors. The TMs and the ETM+ have a cold focal plane containing 16 indium antimonide (InSb) detectors for each of the two short-wave infrared bands, plus 4 mercury cadmium telluride (HgCdTe) detectors (8 for the ETM+) for the thermal emissive band.

The Landsat TM and ETM+ sensors also incorporate IC systems for onboard radiometric calibration monitoring for the solar-reflective spectral bands (Mika, 1997).† These IC devices have a shutter arm that oscillates back and forth directly in front of the primary focal plane. On the TMs, the IC features three lamps that cycle through an eight-lamp state sequence over the course of a 24 s *scene* of data, whereas the ETM+ has two lamps (four lamp states). A complex, synchronized system involving the shutter, a scan mirror, and photodiodes records IC lamp signals, dark shutter signals, and Earth-reflected light for each scan line. Algorithms to use the IC data for radiometric calibration were developed and implemented in operational Landsat product generation systems around the world. Techniques were developed to analyze the IC data, mainly for the long lifetime of the L5 TM (Metzler and Malila, 1985; Helder, 1996; Helder et al., 1996, 1997, 1998a,b; Markham et al., 1998), but also for the L4 TM (Chander et al., 2007a; Markham and Helder, 2012; Helder et al., 2013). Over time, these trend analyses led to a detailed understanding of several image artifacts introduced by various TM sensor characteristics. Within-scene relative calibration algorithms were developed and implemented in product generation systems to remove most of these artifacts, which include striping, scan-correlated shift, memory effect, coherent noise, and oscillations due to cold focal plane icing (e.g., Helder and Micijevic, 2004; Helder and Ruggles, 2004; Teillet et al., 2004a). Landsat processing systems make corrections for all artifacts except the coherent noise effect, which is typically on the order of 0.15 DCs or less.

Historically, the TM radiometric calibration procedure used the instrument's response to the IC lamps on a scene-by-scene basis to determine the gain and offset of each detector. Since May 2003, the L5 TM data processed and distributed by the U.S. Geological Survey (USGS) Earth Resources Observation and Science (EROS) Center were updated to use a lifetime gain look-up-table (LUT) model to radiometrically calibrate the L5 TM data instead of using scene-specific IC lamp–based gains (Chander and Markham, 2003). The new procedure for

* http://qa4eo.org/index.html.

† An assessment of the Landsat-5 thermal band calibration has been carried out by Barsi et al. (2003, 2007) and Padula et al. (2010).

the reflective bands (1–5, 7) is based on a lifetime radiometric calibration curve for the instrument derived from the instrument's IC, cross-calibration with the L7 ETM+, and vicarious measurements. This change has improved absolute calibration accuracy, consistency over time, and consistency with L7 ETM+ data (Chander et al., 2004a). With this change, the radiometric scaling coefficients were updated as well (Chander et al., 2007b, 2009; Helder et al., 2008). Users need to apply the new L5 TM rescaling factors to convert calibrated DCs to TOA at-sensor spectral radiance.

The L7 ETM+ has a full aperture solar calibrator (FASC). This capability features a near-Lambertian YB71 panel of known reflectance that is deployed periodically (approximately monthly) in front of the ETM+ aperture and diffusely reflects solar radiation into the full aperture of the sensor. With knowledge of the solar irradiance and geometric conditions, the FASC serves as an independent, full-aperture calibrator. The ETM+ also has a partial-aperture solar calibrator (PASC) that consists of a small passive device with a set of optics that allow the ETM+ to image the Sun (daily) through small holes. On-orbit performance of the ETM+ calibrator systems can be found in Markham et al. (2003, 2004a,b, 2012).

While the IC data from the onboard systems played a central role in the radiometric calibration of each Landsat mission, the vicarious calibration techniques described in the following sections made it possible to establish and validate radiometric calibration across the full family of Landsat sensors going back to 1972, including the MSS sensors (Helder et al., 2012b). The overall effort was undertaken over the past 15 years, approximately, and is summarized by Markham and Helder (2012). The key result from the tremendous amount of work involved is a consistently calibrated Landsat data archive that spans more than four decades with total uncertainties on the order of 10% or less for most sensors and bands.

In particular (Markham and Helder, 2012), the L1–L5 MSS sensors are within 10% absolute radiometric accuracy in all but near-infrared spectral band 4, and with the other exceptions of L2 red spectral band 3 (11%) and L1 visible spectral bands 1–3 (11%, 11%, and 12%, respectively). MSS band 4 is very broad spectrally and includes large atmospheric absorption features. Thus, the uncertainty for this band in the MSS series grows much more rapidly to the point where the absolute radiometric calibration uncertainty for L1 MSS band 4 approaches 25%. For the solar reflective spectral bands of the L4 TM, L5 TM, and L7 ETM+ sensors, Markham and Helder (2012) report absolute radiometric calibration uncertainties of 9%, 7%, and 5%, respectively.

The optical land imager (OLI) on Landsat-8 (L8) uses push-broom scanning technology. As a result, OLI radiometric calibration benefits from the long history of calibration experience with the push-broom sensors on the French Satellite Pour l'Observation de la Terre (SPOT) satellites (e.g., Valorge et al., 2004), as well as from the advanced land imager (ALI) technology demonstration sensor flown on NASA's Earth Observing-1 (EO-1) satellite (e.g., Mendenhall et al., 2003, 2005). The spectral bands of the L8 OLI sensor* include the L7 ETM+ sensor bands augmented by a deep blue spectral band designed for water resources and coastal zone investigation, and an infrared spectral band at 1.37 μm for cirrus cloud detection. A quality assurance band is also included with each data product to indicate the presence of clouds, water, and/or snow.

The L8 OLI radiometric calibration system† consists of two solar diffusers and a shutter, and two stimulation lamp assemblies (Markham et al., 2010; Reuter et al., 2011). The diffusers provide a full-system, full-aperture calibration when the shutter is open and a dark reference when the shutter is closed. Each lamp assembly contains three tungsten lamps, operated at constant current and monitored by a silicon photodiode, which illuminate the OLI detectors through the full optical system. One *working* lamp is used daily for intraorbit calibration, a *reference* lamp is used approximately monthly, and a *pristine* lamp is used approximately twice a year. L8 sensor characterization and calibration performance is reviewed in a special journal issue edited by Markham et al. (2015).

4.2.2 Vicarious Calibration Methodologies

The predominant vicarious calibration approach is the measurement, on satellite overpass days, of pertinent surface and atmospheric optical properties at terrestrial sites with suitable characteristics to estimate at-sensor radiance or TOA reflectance. Comparisons of these estimates with image-based values computed using the canonical calibration coefficients provide postlaunch monitoring of the radiometric calibration of satellite sensors, either individually or by cross-calibration between sensors (e.g., Slater et al., 1987, Teillet et al., 1990; Biggar et al., 1994; Thome, 2001; Teillet, 2014). Well-understood or instrumented sites serve as reference standard sites, whereas other terrestrial targets that change very little in surface reflectance serve as pseudoinvariant features (PIFs), also known as pseudo-invariant calibration sites (PICS) in the present context. A comprehensive review of terrestrial reference standard sites used for postlaunch radiometric calibration is given in the next section of this chapter.

Error analyses of vicarious calibration approaches indicate that uncertainties in the 2%–3% (1 σ) range with respect to exo-atmospheric solar irradiance (W m^{-2} μm^{-1}) are attainable and can be accurately related to national laboratory standards (Dinguirard and Slater, 1999). The objective is traceability of radiometric calibration accuracies to SI units for science users (e.g., Pollock et al., 2003) and data products with consistent quality for the broader user community.

Although different satellite sensors image the Earth in analogous spectral bands, the spectral bands almost never match exactly such that data acquired over identical targets and derived information products are not directly comparable

* http://landsat.usgs.gov/band_designations_landsat_satellites.php.
† https://directory.eoportal.org/web/eoportal/satellite-missions/l/landsat-8-ldcm.

(Teillet et al., 2007a; Teillet and Ren, 2008). Relatively few investigations have been undertaken to assess radiometric calibration errors due to differences in spectral band response functions between satellite sensors when attempting cross-calibration based on near-simultaneous imaging of common ground look targets in analogous spectral bands (Teillet et al., 2001a, 2004b, 2007a; Trishchenko et al., 2002; Rao et al., 2003; Teillet and Ren, 2008; Doelling et al., 2012; Chander et al., 2013c; Henry et al., 2013). Such spectral band difference effects (SBDEs) can be significant. For instance, Teillet et al. (2007a) simulated SBDEs affecting vicarious-calibration-based TOA reflectance comparisons between many satellite sensors. The following summary examples are for Railroad Valley Playa in Nevada, a frequently used vicarious calibration test site, and for an atmospheric aerosol optical depth of 0.05 and a solar zenith angle of 60°. Comparisons in visible spectral bands were generally within 3%, some were within 5%, and worst cases were around 7% between Terra-MISR* and various other sensors. Comparisons in near-infrared spectral bands were generally within 3%, some were within 10%, and worst cases were around 21% between advanced very-high-resolution radiometer (AVHRR) and various other sensors. Comparisons in shortwave infrared spectral bands around 1.65 μm were generally within 3%, and worst cases were around 8% between TM or China–Brazil Earth Resources Satellite (CBERS) infrared multispectral scanner (IRMSS)† and various other sensors. Comparisons in shortwave infrared spectral bands around 2.2 μm were generally within 6%, and worst cases were around 21% between MODIS or advanced Earth observing satellite (ADEOS)-global imager (GLI)‡ and various other sensors.

4.3 Vicarious Calibration via Terrestrial Reference Standard Sites

Publications in the form of research articles, special journal issues, reports, and books have periodically provided overviews or reviews of satellite sensor radiometric calibration with some mention of vicarious calibration (Slater, 1980, 1984, 1985, 2001; Markham and Barker, 1985; Malila and Anderson, 1986; Price, 1987a,b; Ahern et al., 1988; Jackson, 1990; Nithianandam et al., 1993; Bruegge and Butler, 1996; Chen, 1996; Slater and Biggar, 1996; Teillet, 1997a,b; Dinguirard and Slater, 1999; Morain and Budge, 2004; Valorge et al., 2004; Butler et al., 2005; Chander et al., 2013a,b). This section provides an overview of the use of terrestrial reference standard sites for postlaunch sensor radiometric calibration from historical, current, and future perspectives. Emphasis is placed on optical sensors operating in the visible, near-infrared, and shortwave infrared spectral regions.

* Terra-MISR: Multi-angle Imaging Spectro-Radiometer on NASA's Terra satellite.

† CBERS-IRMSS: China–Brazil Earth Resources Satellite Infrared Multi-Spectral Scanner.

‡ ADEOS-GLI: Japan's Advanced Earth Observing Satellite Global Imager.

4.3.1 Historical Perspective

Reflectance and other properties of land surfaces and atmospheric phenomena now used by many for vicarious calibration were first examined decades ago using surface and airborne measurements in the contexts of the micrometeorology of arid zones (Ashburn and Weldon, 1956), investigations of natural landing areas for aircraft (Molineux et al., 1971), and other analyses (Davis and Cox, 1982; Smith, 1984; Bowker and Davis, 1992; Warren et al., 1998). In scene understanding studies in support of the first meteorological satellite observations, Salomonson and Marlatt (1968) examined the directional reflectance properties of cloud tops, snow, and white gypsum sand (hydrous calcium sulfate) at *White Sands*, New Mexico (WSNM) in the United States, which are three surface target types that have since been used extensively for vicarious calibration of satellite imaging sensors.

The potential use of suitable terrestrial targets on or near the Earth's surface for the postlaunch radiometric calibration of satellite sensors was first examined by a number of investigators for various satellite sensor systems and for several surface types, including WSNM, deserts, playas, salt flats, ocean, cloud tops, snow or ice fields, PIFs, and uniformly vegetated areas (Coulson and Jacobowitz, 1972; Muench, 1981; Kastner and Slater, 1982; Koepke, 1982; Hovis et al., 1985; Mueller, 1985; Fraser and Kaufman, 1986; Staylor, 1986; Duggin, 1987; Schott et al., 1988; Biggar et al., 1994; Teillet et al., 1998a. 2010; Le Marshall et al., 1999; Smith et al., 2002; Kamstrup and Hansen, 2003; Nieke et al., 2004; Martiny et al., 2005; Anderson and Milton, 2006; Wu et al., 2008a,c). Criteria for terrestrial reference site selection have been well documented (e.g., Scott et al., 1996):

- High spatial uniformity over a large area (within 3%)
- Surface reflectance greater than 0.3 across solar-reflective wavelengths
- Flat spectral reflectance across solar-reflective wavelengths
- Temporally invariant surface properties (within 2%)
- Horizontal surface with nearly Lambertian reflectance
- High altitude, far from ocean, urban, and industrial areas
- Arid region with low probability of cloud cover or airborne particles

It is also worth noting that the characterization of any given calibration reference site usually involves *nested* sites: the specific surface measurement site and the overall calibration reference site to be imaged by satellite sensors, which must be well represented by the measurement site's spectral reflectance properties.

Early research on surface measurement methodologies for land surfaces focused to a considerable extent on the alkali flats region of WSNM (cf. Table 4.1 for selected literature references). These vicarious calibration methodologies were used to provide postlaunch radiometric calibration updates for satellite sensor systems such as the L4 and L5 TM sensors (Slater, 1986; Slater et al., 1987). A notable focus of attention was the provision of postlaunch radiometric calibration for the AVHRR sensors on the series of National Oceanic and Atmospheric Administration

TABLE 4.1 Selected Literature for Vicarious Calibration Targets Mentioned in the Text

Test Site	Literature References
White Sands, New Mexico	Castle et al. (1984), Smith and Levin (1985), Smith et al. (1985), Whitlock et al. (1990a,b), Wheeler et al. (1994)
The Moon	Kieffer and Wildey (1985, 1996), Godden and McKay(1997), Kieffer et al. (2002), Barnes et al. (2004), Stone et al. (2005, 2013), Stone (2008), Xiong et al. (2008)
La Crau, France	Xing-Fa et al. (1990), Santer et al. (1992, 1997), Gu et al. (1992), Moran et al. (1995), Richter (1997), Rondeaux et al. (1998), Six (2002)
Saharan and Arabian Deserts, North Africa	Henry et al. (1993), Cosnefroy et al. (1994, 1996, 1997), Cabot (1997), Cabot et al. (1998, 1999, 2000), Miesch et al. (2003), Rao et al. (2003), Markham et al. (2006), Vermote and Saleous (2006), Wu et al. (2008b), Gamet et al. (2011), Lachérade et al. (2013)
Atmospheric molecular (Rayleigh) scattering	Vermote et al. (1992), Kaufman and Holben (1993), Dilligeard et al. (1997), Meygret et al. (2000)
Saharan and Arabian Deserts, North Africa	Henry et al. (1993), Cosnefroy et al. (1994, 1996, 1997), Cabot (1997), Cabot et al. (1998, 1999, 2000), Miesch et al. (2003), Rao et al. (2003), Markham et al. (2006), Vermote and Saleous (2006), Wu et al. (2008b), Gamet et al. (2011), Lachérade et al. (2013)
Ocean sun-glint	Kaufman and Holben (1993), Vermote and Kaufman (1995), Hagolle et al. (2004), Luderer et al. (2005)
Clouds	Desormeaux et al. (1993), Vermote and Kaufman (1995), Iwabuchi (2003), Doelling et al. (2004, 2010, 2013), Hu et al. (2004), Fougnie and Bach (2009)
Sites in Australia	Graetz et al. (1994), Prata et al. (1996), Mitchell et al. (1997), O'Brien and Mitchell (2001), De Vries et al. (2007)
Dunhuang, China	Min and Zhu (1995), Wu et al. (1994, 1997), Xiao et al. (2001), Hu et al. (2001, 2009, 2010), Liu et al. (2004), Zhang et al. (2001, 2004, 2008, 2009), Li et al. (2009)
Railroad Valley Playa, USA	Scott et al. (1996), Snyder et al. (1997), Thome (2001, 2005), Matsunaga et al. (2001), Thome et al. (2000, 2001, 2003a,b, 2004b, 2006, 2007, 2008), Biggar et al. (2003), Chander et al. (2004a, 2007a), McCorkel et al. (2006), D'Amico et al. (2006), Czapla-Myers et al. (2007, 2008, 2010), Angal et al. (2008), Kerola and Bruegge (2009)
Ivanpah Playa	Thome et al. (1996, 2004b), Villa-Aleman et al. (2003a,b), Thome (2005), Rodger et al. (2005)
Snow/ice fields	Loeb (1997), Tahnk and Coakley (2001), Nieke et al. (2003), Six et al. (2004)
Negev Desert, Israel	Bushlin et al. (1997), Gilead and Karnieli (2004), Faiman et al. (2004), Gilead (2005)
Lunar Lake Playa, USA	Thome et al. (1998), Bannari et al. (2005)
Dome-C, Antarctica	Six et al. (2004, 2005), Tomasi et al. (2008), Wu et al. (2008b), Wenny and Xiong (2008), Wenny et al. (2009), Xiong et al. (2009a,b), Bouvet and Ramoino (2010), Cao et al. (2010), Potts et al. (2013)
Frenchman Flat, USA	Gross et al. (2007), Helmlinger et al. (2007), Kerola et al. (2009)
Tuz Golu, Turkey	Gurol et al. (2008, 2010)
Sonoran Desert, Mexico, aka Yuma, Arizona	Angal et al. (2010a,b, 2011), Kim and Lee (2013)

(NOAA) satellites, because of the widespread utilization of AVHRR image data and because AVHRRs do not have onboard calibration systems (Duggin, 1987; Frouin and Gautier, 1987; Smith et al., 1988; Holben et al., 1990; Staylor, 1990; Teillet et al., 1990; Brest and Rossow, 1992; Che and Price, 1992; Mitchell et al., 1992; Abel et al., 1993; Kaufman and Holben, 1993; Teillet and Holben, 1994; Cihlar and Teillet, 1995; Rao and Chen, 1995, 1996, 1999; Brest et al., 1997; Tahnk and Coakley, 2001; Cao and Heidinger, 2002; Wu et al., 2003, 2008a,c, 2010; Cao et al., 2005).

Other deserts and playas used as reference standard sites include (cf. Table 4.1 for selected literature references) *La Crau* in France, several *sites in Australia*, *Dunhuang* in China, *Railroad Valley Playa* in the United States, *Ivanpah Playa* in the United States, the *Negev Desert* in Israel, *Lunar Lake Playa* in the United States, *Frenchman Flat* in the United States, the *Sonoran Desert* in Mexico (also known as Yuma, Arizona, United States), and *Tuz Golu* in Turkey. About 20 large areas in the *Saharan and Arabian Deserts* of North Africa have been used without surface observations to enable cross-calibration and/or uniformity calibration for a variety of sensor systems for several decades. These large uniform sites are important for the calibration of pushbroom sensor detector arrays, such as those on the SPOT satellite systems, given that it is difficult to avoid detector-to-detector differences if the whole array cannot be exposed easily to a uniform radiance field. Because significant episodes of airborne particles can occur in these regions, only the clearest image acquisitions are used for radiometric calibration purposes.

Additionally, other vicarious calibration or cross-calibration methods take advantage of nonland Earth targets, including (cf. Table 4.1 for selected literature references) atmospheric molecular (Rayleigh) scattering, ocean sun-glint, snow/ice fields, and clouds. A snowfield of particular interest is the *Dome-C* site in Antarctica, where image data from many overpasses are available during the summer months. A key facility for the vicarious calibration of ocean color satellite sensors is the Marine Optical BuoY (MOBY), an autonomous optically instrumented buoy moored off the island of Lanai, Hawaii, United States (Clark et al., 1997; Brown et al., 2007).

The Remote Sensing Group at the University of Arizona pioneered vicarious calibration methodologies that involved not only WSNM but also the playas at Edwards Air Force Base, Railroad Valley, Lunar Lake, and Ivanpah. Many of the key surface measurement methodologies central to vicarious calibration were developed as a result of close collaborations between the University of Arizona and the U.S.

Department of Agriculture (USDA) (Phoenix), focused to a considerable extent on field campaigns at the Maricopa Agricultural Center (e.g., Jackson, 1990), but also involving many field campaigns at WSNM and elsewhere. These pioneering efforts, together with leading-edge work by NASA scientists and collaborations with scientists at the National Institute for Standards Technology (NIST), played an important role in the calibration programs of NASA's EOS (e.g., Bruegge and Butler, 1996; Slater et al., 1996; Xiong et al., 2010b). Similar methodologies and efforts have also been key elements of other EO systems (e.g., Hill and Aifadopoulou, 1990; Guenther et al., 1996; Markham et al., 1997; Thome et al., 1997a,b, 2003a,b; Schroeder et al., 2001; Secker et al., 2001; Teillet et al., 2001a, 2006, 2007a; Biggar et al., 2003; Black et al., 2003; Schiller, 2003; Chander et al., 2004b, 2007a,c, 2013b; Xiong et al., 2006b; Henry et al., 2013). There have also been systematic and sustained vicarious calibration efforts over many years by the Centre National d'Etudes Spatiales (CNES), the European Space Agency (ESA), the UK's National Physical Laboratory (NPL), the Japan Aerospace Exploration Agency (JAXA), and the USGS EROS Center.

Vicarious calibration can also be undertaken by imaging the Moon, a stable target whose dynamic illumination variations can be computed (cf. Table 4.1 for selected literature references), and/or by imaging bright stars (e.g., Chang et al., 2012). Although ultrastable stellar targets are imaged without atmospheric effects, they are relatively low radiance sources and, as with imaging the Moon, they require high-risk spacecraft platform maneuvers to achieve. Satellite sensors often use data acquisitions of deep space views without spacecraft platform maneuvers to provide dark target calibration data.

As the importance of remote sensing calibration was increasingly acknowledged and as more terrestrial sites were investigated for use as reference standard test sites for postlaunch sensor calibration, specialists in the international community endeavored to put together databases of worldwide calibration facilities, including test sites and instruments. Early pilot efforts were those of the international CEOS in 1993 (spearheaded by Barton, Guenther, and others) and NASA's EOS in 1995 (spearheaded by Butler, Starr, Reber, and others). Calls for contributions led to editions of such databases (e.g., Butler et al., 2001), but these early databases were never fully populated and they no longer appear to be active. Current efforts in this regard by CEOS, ESA, and the USGS are outlined in the next section.

During the 1990s, the role of Earth-monitoring systems of all kinds was increasingly cast in the context of climate observations. For example, the Global Climate Observing System (GCOS) was established in 1992 to focus on satellite and in situ observations for climate in the atmospheric, oceanic, and terrestrial domains.* As part of the process of identifying climate observation requirements in 1999, GCOS established a set of climate-monitoring principles† that were adopted in 2003 by the World Meteorological Organization (WMO) and CEOS. The GCOS climate-monitoring principles refer specifically to the critical role of calibrated satellite sensor observations.

A generalized approach to the vicarious calibration of multiple EO sensors using hyperspectral data was demonstrated by Teillet et al. (1998a,b, 2001b) in a project on quality assurance and stability reference (QUASAR) monitoring. The approach uses spatially extensive airborne or satellite hyperspectral imagery of a terrestrial reference standard site as spectroradiometric reference data to carry out vicarious radiometric calibrations for all optical satellite sensors that image the site on the same day.

The establishment of a global instrumented and automated network of test sites (GIANTS) for postlaunch radiometric calibration of EO sensors was outlined by Teillet et al. (2001c). The GIANTS concept proposes that a small number of well-characterized benchmark test sites and datasets be supported for the calibration of all space-based optical sensors imaging the Earth. A core set of surface sensors, measurements, and protocols should be standardized across all participating test sites and measurement datasets should undergo identical processing at a central secretariat. GIANTS is intended to supplement calibration information already available from other calibration systems and efforts, reduce the resources required by individual agencies, and provide greater consistency for terrestrial monitoring studies based on multiple sensor systems. The network approach is also intended to explore the use of automation, communication, co-ordination, visibility, and education, all of which can be facilitated by greater use of advanced ground-based sensor and telecommunication technologies.

4.3.2 Current Developments

More recent initiatives in the domain of reference standard sites include ongoing methodology research (e.g., McCorkel et al., 2013; Wang et al., 2011; Helder et al., 2012a, 2013; Thome, 2012; Thome and Fox, 2011; Anderson et al., 2013; Govaerts et al., 2013; Thome et al., 2013), as well as the GEOSS, CEOS, and QA4EO efforts mentioned in the Introduction. Notably and significantly, the CEOS Working Group on Calibration and Validation (WGCV) subgroup on Infrared Visible Optical Sensors (IVOSs) has worked in recent years with collaborators around the world to establish a core set of CEOS-endorsed, globally distributed, reference standard test sites for the postlaunch calibration of space-based optical imaging sensors. There are now eight CEOS reference instrumented sites (Table 4.2; Figure 4.1) and six CEOS reference PICS (Table 4.3; Figure 4.2). The instrumented sites are mainly used for field campaigns to obtain or update radiometric gain coefficients, and they serve as a focus for international efforts, facilitating traceability and cross-comparison to evaluate biases between current and future sensors and methods in a harmonized manner. The pseudoinvariant reference standard test sites are desert areas characterized by high surface reflectance, some sand dunes, little or no vegetation, and low atmospheric aerosol loadings. Such sites can be used to evaluate the long-term stability of a given sensor and to facilitate cross-comparisons between multiple sensors (e.g., Helder et al., 2010; Chander et al., 2013b). They can also be used for the validation

* http://www.wmo.int/pages/prog/gcos/index.php.

† http://www.wmo.int/pages/prog/gcos/index.php?name=ClimateMonitoringPrinciples.

TABLE 4.2 Core Set of CEOS-Endorsed Instrumented Reference Standard Test Sites

#	Site	WRS-2 Path/Row	Centre Latitude (Degrees), Longitude (Degrees), and Altitude ASL (m)	Point of Contact
1	Dome C, Antarctica	88-89-90/113	−74.50, +123.00, 3215	Stephen Warren, University of Washington, USA
2	Dunhuang, China	137/32	+40.13, +94.34, 1220	Xiuqing Hu, National Satellite Meteorological Center, China
3	Frenchman Flat, USA	40/34	+36.81, −115.93, 940	Carol J. Bruegge, NASA Jet Propulsion Laboratory, USA
4	Ivanpah Playa, USA	39/35	+35.5692, −115.3976, 813	Kurtis J. Thome, NASA Goddard Space Flight Center, USA
5	La Crau, France	196/30	+43.47, +4.97, 28	Patrice Henry, CNES, France
6	Negev Desert, Israel	174/39	+30.11, +35.01, 334	Arnon Karnieli, Ben Gurion University, Israel
7	Railroad Valley Playa, USA	40/33	+38.50, −115.69, 1435	Kurtis J. Thome, NASA Goddard Space Flight Center, USA
8	Tuz Golu, Turkey	177/33	+38.83, +33.33, 905	Selime Gurol, Tubitak Uzay (Space Technologies Research Institute), Turkey

Notes: WRS, Landsat Worldwide Reference System; ASL, above sea level. Other characteristics of some of these test sites are given in Teillet et al. (2007b).

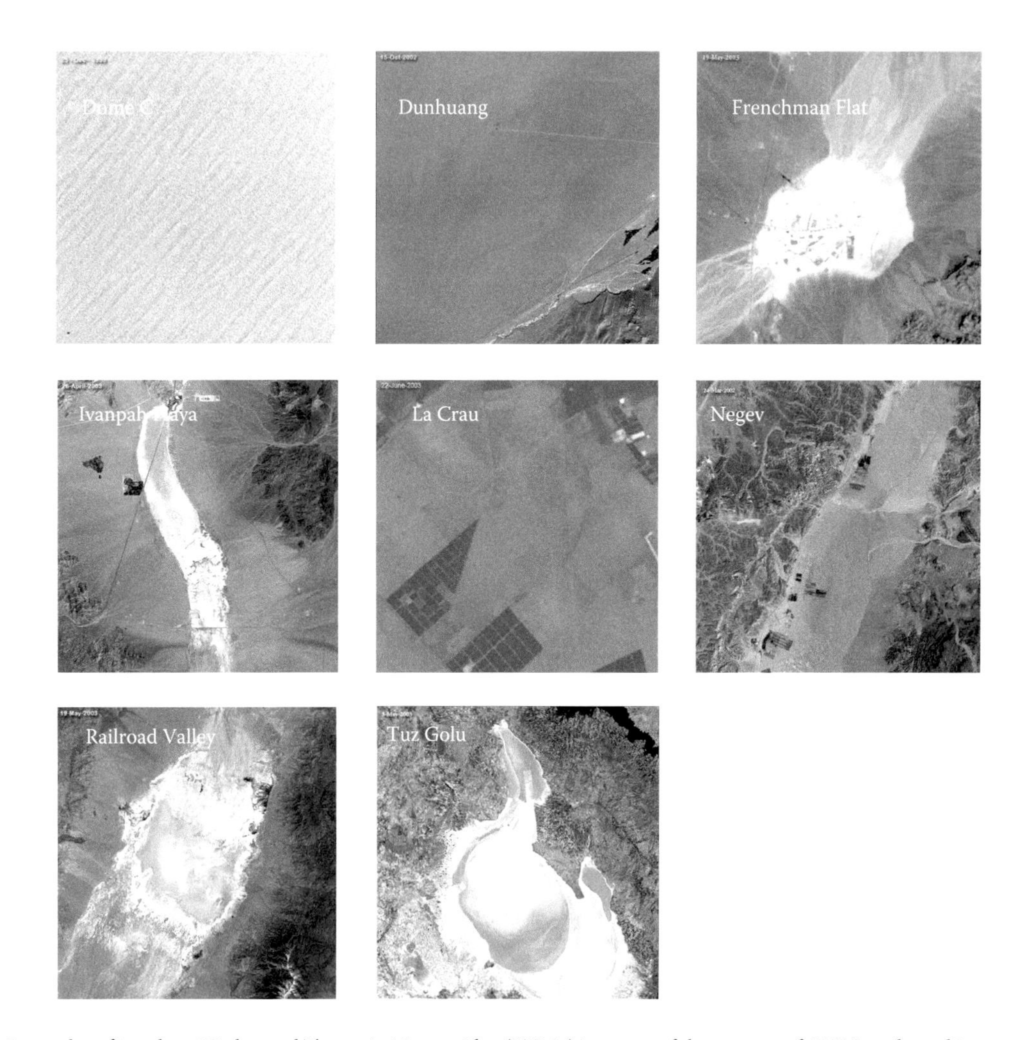

FIGURE 4.1 Examples of Landsat-7 Enhanced Thematic Mapper Plus (ETM+) imagery of the core set of CEOS-endorsed instrumented reference standard test sites (N.B., scales differ, north is up in all cases, and site coordinates can be found in Table 4.2).

TABLE 4.3 Core Set of CEOS-Endorsed Pseudoinvariant Reference Standard Test Sites

#	Site	WRS-2 Path/Row	Centre Latitude (Degrees), Longitude (Degrees), and Altitude ASL (m)
1	Libya 4	181/40	+28.55, +23.39, 118
2	Mauritania 1	201/47	+19.40, −9.30, 392
3	Mauritania 2	201/46	+20.85, −8.78, 384
4	Algeria 3	192/39	+30.32, +7.66, 245
5	Libya 1	187/43	+24.42, +13.35, 648
6	Algeria 5	195/39	+31.02, +2.23, 530

Notes: WRS, Landsat Worldwide Reference System; ASL, above sea level. Note that Mauritania 1 and 2 are considered to be one site. The point of contact for these sites is Patrice Henry, CNES, France. Other characteristics of these test sites are given in Teillet et al. (2007b).

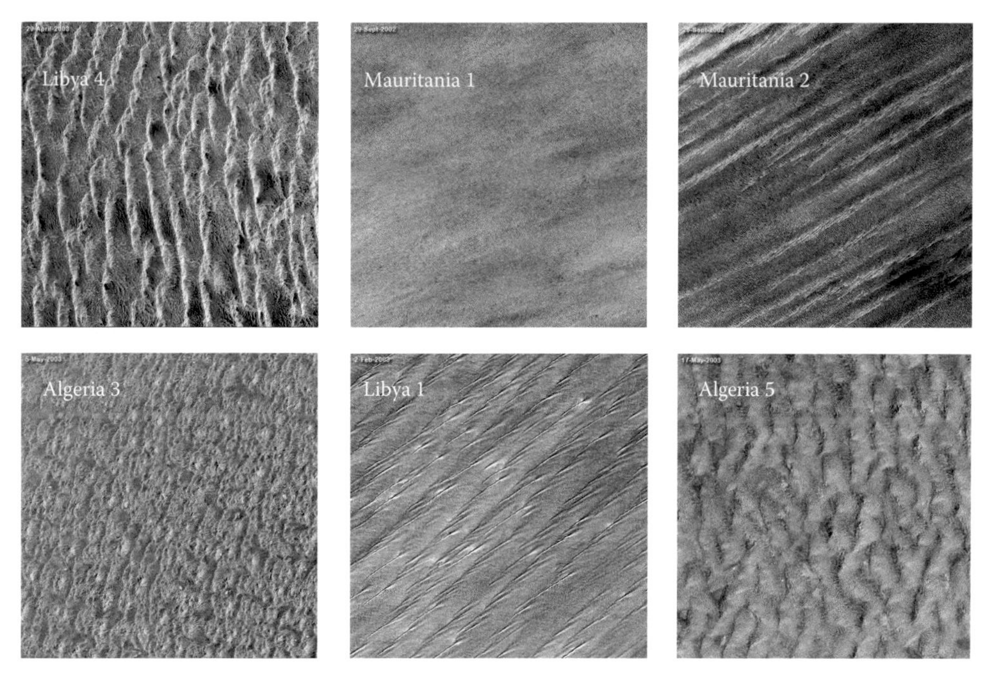

FIGURE 4.2 Examples of Landsat-7 Enhanced Thematic Mapper Plus (ETM+) imagery of the core set of CEOS-endorsed pseudo-invariant reference standard test sites (N.B., scales differ, north is up in all cases, and site coordinates can be found in Table 4.3).

of selected higher-order climate variable products. As a precursor to the selection of the CEOS-endorsed core sites, a list of potential candidates was developed (Teillet et al., 2007b) and an online catalog of worldwide test sites used for sensor characterization was implemented by the USGS (Chander et al., 2007d).* The catalog provides ready access to this vital information by the global community.

CEOS/WGCV/IVOS is also working toward the establishment of optimum methodologies for the characterization and use of the endorsed reference instrumented sites. The principal criterion in selecting these sites, other than spatial uniformity and brightness, was that they were all fully and regularly calibrated by ground-based instrumentation. In some cases, instrumentation is permanently deployed; in others, it is transported to the site for specific characterization campaigns. However, in all cases, the basis for assigning a value to the surface reflectance and its subsequent propagation to the TOA as radiance (for comparison to satellite imager radiances) is derived from measurements made by ground survey teams. Cost is a significant issue for the instrumentation (and associated performance) needed for test sites for establishing strategies for the long-term maintenance of CEOS test sites. In addition, having a limited number of sites provides an opportunity for the owners/operators of Earth-viewing optical imagers (including commercial operators) to routinely collect data over the CEOS-endorsed reference sites.

In principle, radiometric calibration coefficients should not depend on the geographic locations of reference standard sites. However, the methodologies involved in vicarious calibration via instrumented sites involve characterization of surface properties and atmospheric conditions by different teams and, pending the adoption of standardized approaches, different instrumentation. Similarly, vicarious calibration via PICS involves methods for atmospheric compensation. These diversities are deemed beneficial in that they can be used to yield an improved understanding of the role of atmospheric effects and different modeling approaches on calibration results.

* http://calval.cr.usgs.gov/sites_catalog_map.php.

In any given year, there are typically several field campaigns undertaken by various agencies to carry out vicarious calibrations for a number of satellite optical sensor systems. Thus, a logical next step is to formalize and test the GIANTS concept via the concatenation of CEOS-endorsed core sites (Teillet and Fox, 2009). The concatenation concept proposes the deployment of as many field campaigns at as many of the core sites as possible during a given time period (e.g., 1 month) to generate updates for as many satellite optical sensors as possible. The pilot project would make a concrete initial test of the GIANTS approach (at this stage excluding an *automated* aspect in most if not all instances).

The trade-off between automation and the accuracy demands of radiometric calibration is an important consideration that has yet to be investigated to a significant extent. Several organizations have begun to explore automated radiometric systems for vicarious calibration (Thome et al., 2004a; Czapla-Myers et al., 2007, 2008; Gross et al., 2007; Helmlinger et al., 2007; Kerola et al., 2009; Meygret et al., 2011). Another topic that has received insufficient attention by the calibration research community is the use of data at wavelengths outside the solar-reflective spectrum to help characterize reference standard sites used for optical sensor calibration (e.g., Teillet et al., 1995; Blumberg and Freilikher, 2001; Floury et al., 2002; Brogioni et al., 2006).

The U.K.'s NPL has contributed significantly in a number of ways to new methods and systems, as well as to international collaboration, for radiometric calibration of satellite sensors (Fox, 1999), including the development of bold new space mission concepts for advanced cryogenic-based radiometric calibration capabilities (Fox et al., 2002). NPL has also coordinated several CEOS campaigns in the context of GEO tasks to compare satellite calibration approaches, recently using the Dome-C site and also leading major comparison exercises focused on the Tuz Golu site in Turkey.*

CEOS is providing data and information for calibration and validation (cal/val) of EO data through its Cal/Val Portal.† The original Cal/Val Portal started in 2007 and a new version is now moving to a completely operational system providing new tools, datasets, results, and upcoming campaign information on a regular basis. Among other initiatives, the enhanced Cal/Val Portal supports measurement and comparison campaigns by means of a prototype database with data and information on reference standard sites.

4.3.3 Next Steps

Advocacy of the use of CEOS-endorsed reference standard sites and the establishment and use of some sort of certification process as part of *best practice* methodologies should lead to improved measurement consistency between satellite sensors, reduce costs, and help underpin the accurate monitoring of planetary change.

With respect to the CEOS-endorsed reference standard instrumented sites, the following steps are recommended for the near term: (1) Increase the number of sites from eight to ten. (2) Gather more complete site characterization data and information, especially with respect to their reflectance anisotropy properties and their temporal stabilities. (3) Define a recommended standard set of core measurements. (4) Formalize the set of sites into an operational network. (5) Organize local, regional, national, and international field campaigns. (6) Acquire and archive imagery of all of the sites on an ongoing basis. (7) Develop online calibration data access infrastructures. (8) Continue to improve vicarious calibration methodologies. The current set of endorsed sites is distributed unevenly across the globe, which should be kept in mind to the extent possible when selecting additional instrumented sites. However, the adoption and use of new sites will depend not only on the suitability of site characteristics, but also on the interests, capabilities, and funding of research groups, as well as logistical and political accessibilities.

Still with respect to the CEOS-endorsed reference standard instrumented sites, the following steps are recommended for the longer term: (1) Continue to acquire and archive imagery on an ongoing basis for all of the sites. (2) Continue to improve vicarious calibration methodologies. (3) Establish traceability chains for site measurement data. (4) Develop recommended guidelines and best practices for using the network of sites. (5) Endorse and advocate compliance with recommended guidelines and best practices. ISO standards for radiometric calibration of remote sensing data have been proposed (Di, 2004), making a high-level start toward certification.

4.4 Concluding Remarks

This chapter has provided a broad overview of the postlaunch radiometric calibration of image data from satellite sensor systems operating in the solar-reflective optical domain. Even though satellite sensors are very well characterized radiometrically prior to launch, postlaunch changes to satellite sensors and their radiometric calibration performance necessitate ongoing calibration monitoring and adjustment.

Onboard radiometric calibration systems have been implemented since the early days of satellite imaging sensors. This chapter highlighted the rich history of Landsat onboard sensor calibration and the significant efforts that were required by the absence of such capability for the AVHRR sensors, which provided data for countless broad-scale studies of the Earth (e.g., Townshend, 1994; Townshend et al., 1994; Teillet et al., 2000). The chapter focused extensively on vicarious calibration methodologies, important complements to the use of complex and expensive onboard calibration systems that seldom test the full optics of sensors. Terrestrial reference standard sites for vicarious calibration were featured, and the relevant pioneering research over many decades was documented.

Today's mainstream information society routinely makes extensive qualitative use of satellite imagery. With respect to quantitative use, the significant improvement in the knowledge of the radiometric performance of satellite sensor systems, as described in this chapter, is indicative of the advancement in the state of

* http://calvalportal.ceos.org/cvp/web/guest/tuz-golu-campaign.

† http://calvalportal.ceos.org.

the art of postlaunch radiometric calibration methods. It is also indicative of the need for persistent calibration of satellite sensors over mission lifetimes, as well as the need for a variety of calibration methodologies in order to assess the true radiometric performance of any given sensor as accurately as possible. Accordingly, the use of terrestrial reference standard sites for postlaunch monitoring of satellite sensor radiometric performance increasingly constitutes a key component of current and future satellite sensor calibration strategies. With continued international cooperation on test site calibration methodologies among remote sensing agencies and organizations, remotely sensed EO data can be made more accurate and useful to help develop an improved and sound understanding of our planet and its climate.

The success of climate monitoring missions will depend critically on significant advances in postlaunch calibration. The planned Climate Absolute Radiance and Refractivity Observatory (CLARREO) (Anderson et al., 2008; Sandford et al., 2010; Thome et al., 2010; Lukashin et al., 2013) and the proposed Traceable Radiometry Underpinning Terrestrial and Helio Studies (TRUTHS) mission (Fox et al., 2002, 2011) offer novel approaches to the provision of key scientific data with unprecedented radiometric accuracy for terrestrial and solar studies. Missions such as CLARREO and TRUTHS will establish well-calibrated reference targets and standards to support other EO missions as well as solar and lunar observations. Once launched, these sensor systems would in effect provide the de facto primary reference standards for the global ensemble of EO systems. CLARREO and TRUTHS promise to improve the performance and accuracy of EO missions by an order of magnitude (to accuracies on the order of 0.3% in the solar reflective domain). Postlaunch radiometric calibration based on metrology-quality observatories in space lies at the heart of these proposed missions, which, as a result, will be integral and indispensable tools to help maximize the benefits of all EO systems.

Acknowledgments

The postlaunch radiometric calibration methodologies documented in this chapter are the culmination of decades of effort on the part of many individuals and agencies. The authors thank CEOS/WGCV/IVOS members in particular for their stewardship and contributions toward progress in EO calibration and validation.

References

Abel, P., Guenther, B., Galimore, R.N., and Cooper, J.W. 1993. Calibration results for NOAA-11 AVHRR channels 1 and 2 from congruent path aircraft observations. *Journal of Atmospheric and Oceanic Technology*, 10(4): 493–508.

Ahern, F.J., Brown, R.J., Cihlar, J., Gauthier, R., Murphy, J., Neville, R.A., and Teillet, P.M. 1988. Radiometric correction of visible and infrared remote sensing data at the Canada Centre for Remote Sensing. In *Remote Sensing Yearbook 1988*, A. Cracknell and L. Hayes, eds. Taylor & Francis Group, Philadelphia, PA.

Ahern, F.J. and Murphy, J. 1979. Radiometric calibration and correction of LANDSAT 1, 2, and 3 MSS Data. Canada Centre for Remote Sensing Research Report 78-4, Canada Centre for Remote Sensing, Ottawa, Ontario, Canada.

Anderson, D., Jucks, K.W., and Young, D.F. 2008. The NRC decadal survey climate absolute radiance and refractivity observatory: NASA implementation. *Proceedings of the IEEE International Geoscience and Remote Sensing Symposium (IGARSS)*, Boston, MA, pp. 9–11.

Anderson, K. and Milton, E.J. 2006. On the temporal stability of ground calibration targets: Implications for the reproducibility of remote sensing methodologies. *International Journal of Remote Sensing*, 27(16): 3365–3374.

Anderson, N., Czapla-Myers, J., Leisso, N., Biggar, S., Burkhart, C., Kingston, R., and Thome, K. 2013, Design and calibration of field deployable ground-viewing radiometers. *Applied Optics*, 52(2): 231–240.

Angal, A., Chander, G., Choi, T., Wu, A., and Xiong, X. 2010a. The use of the Sonoran Desert as a pseudo-invariant site for optical sensor cross-calibration and long-term stability monitoring. *Proceedings of the 2010 International Geoscience and Remote Sensing Symposium (IGARSS)*, Honolulu, HI, pp. 1656–1659.

Angal, A., Chander, G., Xiong, X., Choi, T., and Wu, A. 2011. Characterization of the Sonoran desert as a radiometric calibration target for Earth observing sensors. *Journal of Applied Remote Sensing*, 5(1): 059502.

Angal, A., Choi, T.J., Chander, G., and Xiong, X. 2008. Monitoring on-orbit stability of Terra MODIS and Landsat 7 ETM+ reflective solar bands using Railroad Valley Playa, Nevada (RVPN) test site. *Proceedings of IEEE Conference on Remote Sensing: The Next Generation*, Vol. IV, Boston, MA, pp. 1364–1367. IEEE, Piscataway, NJ.

Angal, A., Xiong, X., Choi, T.-Y., Chander, G., and Wu, A. 2010b. Using the Sonoran and Libyan desert test sites to monitor the temporal stability of reflective solar bands for Landsat 7 Enhanced Thematic Mapper Plus and Terra MODerate-resolution imaging spectroradiometer sensors. *Journal of Applied Remote Sensing*, 4(1): 043525-12.

Ashburn, E.V. and Weldon, R.G. 1956. Spectral diffuse reflectance of desert surfaces. *Journal of the Optical Society of America*, 46(8): 583–586.

Bannari, A., Omari, K., Teillet, P.M., and Fedosejevs, G. 2005. Potential of Getis statistics to characterize the radiometric uniformity and stability of test sites used for the calibration of Earth observation sensors. *IEEE Transactions on Geoscience and Remote Sensing*, 43(12): 2918–2926.

Barker, J.L., ed. 1985. *Proceedings of the Landsat-4 Science Characterization Early Results Symposium*, February 22–24, 1983, Greenbelt, MD, NASA Conference Publication 2355.

Barnes, R.A., Eplee, R.E. Jr., Patt, F.S., Kieffer, H.H., Stone, T.C., Meister, G., Butler, J.J., and McClain, C.R. 2004. Comparison of SeaWiFS measurements of the moon with the U.S. geological survey lunar model. *Applied Optics*, 43(31): 5838–5854.

Barsi, J.A., Hook, S.J., Schott, J.R., Raqueno, N.G., and Markham, B.L. 2007. Landsat-5 Thematic Mapper thermal band calibration update. *Geoscience and Remote Sensing Letters*, 4(4): 552–555.

Barsi, J.A., Schott, J.R., Palluconi, F.D., Helder, D.L., Hook, S.J., Markham, B.L., Chander, G., and O'Donnell, E.M. 2003. Landsat TM and ETM+ thermal band calibration. *Canadian Journal of Remote Sensing*, 29(2): 141–153.

Bartolucci, L.A. and Davis, S.M. 1983. The calibration of Landsat MSS data as an analysis tool. LARS Technical Reports Paper 71, LARS Technical Report 062283. *Proceedings of the 1983 Machine Processing of Remotely Sensed Data Symposium*, West Lafayette, IN, pp. 279–287.

Biggar, S.F., Slater, P.N., and Gellman, D.I. 1994. Uncertainties in the in-flight calibration of sensors with reference to measured ground sites in the 0.4 to 1.1 μm range. *Remote Sensing of Environment*, 48: 245–252.

Biggar, S.F., Thome, K.J., and Wisniewski, W. 2003. Vicarious radiometric calibration of EO-1 sensors by reference to high-reflectance ground targets. *IEEE Transactions on Geoscience and Remote Sensing*, 41(6): 1174–1179.

Black, S.E., Helder, D.L., and Schiller, S.J. 2003. Irradiance-based cross-calibration of Landsat-5 and Landsat-7 Thematic Mapper sensors. *International Journal of Remote Sensing*, 24(2): 287–304.

Blumberg, D.G. and Freilikher, V. 2001. Soil water-content and surface roughness retrieval using ERS-2 SAR data in the Negev Desert, Israel. *Journal of Arid Environments*, 49(3): 449–464.

Bouvet, M. and Ramoino, F. 2010. Radiometric intercomparison of AATSR, MERIS, and Aqua MODIS over Dome Concordia (Antarctica). *Canadian Journal of Remote Sensing*, 36(5): 464–473.

Bowker, D.E. and Davis, R.E. 1992. Influence of atmospheric aerosols and desert reflectance properties on satellite radiance measurements. *International Journal of Remote Sensing*, 13(16): 3105–3126.

Brest, C.L. and Rossow, W.B. 1992. Radiometric calibration and monitoring of NOAA AVHRR data for ISCCP. *International Journal of Remote Sensing*, 13(2): 235–273.

Brest, C.L., Rossow, W.B., and Roiter, M.D. 1997. Update of radiance calibrations for ISCCP. *Journal of Atmospheric and Oceanic Technology*, 14(5): 1091–1109.

Brogioni, M., Macelloni, G., and Pampaloni, P. 2006. Temporal and spatial variability of multi-frequency microwave emission from the east Antarctic plateau. *Proceedings of IEEE Conference on Remote Sensing: A Natural Global Partnership*, Denver, CO, pp. 3820–3823. IEEE, Piscataway, NJ.

Brown, S.W., Flora, S.J., Feinholz, M.E., Yarbrough, M.A., Houlihan, T., Peters, D., Kim, Y.S., Mueller, J., Johnson, B.C., and Clark, D.K. 2007. The Marine Optical BuoY (MOBY) radiometric calibration and uncertainty budget for ocean color satellite sensor vicarious calibration. *Proceedings of SPIE Conference 6744 on Sensors, Systems, and Next-Generation Satellites XI*, Florence, Italy, R. Meynart, S.P. Neeck, H. Shimoda, and S. Habib, eds. SPIE, Bellingham, WA, pp. 67441M.

Bruegge, C. and Butler, J., eds. 1996. Special issue on Earth observing system calibration. *Journal of Atmospheric and Oceanographic Technology*, 13(2): 273–544.

Bushlin, Y., Ben-Shalom, A., Sheffer, D., Steinman, A., Dimmeler, A., Clement, D., and Strobel, R. 1997. Background properties in arid climates: Measurements and analysis. *Proceedings of SPIE Conference 3062 on Targets and Backgrounds: Characterization and Representation III*, Orlando, FL, W.R. Watkins and D. Clement, eds., pp. 311–321. SPIE, Bellingham, WA.

Butler, J.J., Johnson., B.C., and Barnes, R.A. 2005. The calibration and characterization of Earth remote sensing and environmental monitoring instruments. In *Optical Radiometry, Vol 41: Experimental Methods in the Physical Sciences*, A.C. Parr, R.U. Datla, and J.L. Gardner, eds., R. Celotta and T. Lucatorto, treatise eds. Elsevier/Academic Press, Waltham, MA, pp. 453–534.

Butler, J.J., Wanchoo, L., and Truong, L. 2001. CEOS database of worldwide calibration facilities and validation test sites. *Proceedings of SPIE Conference 4169 on Sensors, Systems, and Next-Generation Satellites IV*, Barcelona, Spain, H. Fujisada, J.B. Lurie, A. Ropertz, and K. Weber, eds., pp. 202–208. SPIE, Bellingham, WA.

Cabot, F. 1997. Proposal for the development of a repository for in-flight calibration of optical sensors over terrestrial targets. *Proceedings of SPIE Conference 3117 on Earth Observing Systems II*, San Diego, CA, W.L. Barnes, ed., pp. 148–155. SPIE, Bellingham, WA.

Cabot, F., Hagolle, O., Cosnefroy, H., and Briottet, X. 1998. Intercalibration using desertic sites as a reference target. *Proceedings of International Geoscience and Remote Sensing Symposium*, Seattle, WA, Vol. 5, pp. 2713–2715.

Cabot, F., Hagolle, O., and Henry, P. 2000. Relative and multitemporal calibration of AVHRR, SeaWiFS, and VEGETATION using POLDER characterization of desert sites. *Proceedings of International Geoscience and Remote Sensing Symposium*, Honolulu, HI, pp. 2188–2190.

Cabot, F., Hagolle, O., Ruffel, C., and Henry, P.J. 1999. Remote sensing data repository for in-flight calibration of optical sensors over terrestrial targets. *Proceedings of SPIE Conference 3750 on Earth Observing Systems IV*, Denver, CO, pp. 514–523, W.L. Barnes, ed. SPIE, Bellingham, WA.

Cao, C. and Heidinger, A.K. 2002. Inter-comparison of the longwave infrared channels of MODIS and AVHRR/NOAA-16 using simultaneous nadir observations at orbit intersections. *Proceedings of SPIE Conference 4814 on Earth Observing Systems VII*, Seattle, WA, W.L. Barnes, ed., pp. 306–316. SPIE, Bellingham, WA.

Cao, C., Uprety, S., Xiong, X., Wu, A., Jing, P., Smith, D., Chander, G., Fox, N., and Ungar, S. 2010. Establishing the Antarctic Dome C community reference standard site towards consistent measurements from Earth observation satellites. *Canadian Journal of Remote Sensing*, 36(5): 498–513.

Cao, C., Xu, H., Sullivan, J., McMillin, L., Ciren, P., and Hou, Y. 2005. Intersatellite radiance biases for the high resolution infrared radiation sounders (HIRS) onboard NOAA-15, -16, and -17 from simultaneous nadir observations. *Journal of Atmospheric and Oceanic Technology*, 22(4): 381–395.

Castle, K.R., Holm, R.G., Kastner, C.J., Palmer, J.M., Slater, P.N., Dinguirard, M., Ezra, C.E., Jackson, R.D., and Savage, R.K. 1984. In-flight absolute radiometric calibration of the Thematic Mapper. *IEEE Transactions on Geoscience and Remote Sensing*, 22(3): 251–255.

Chander, G., Angal, A., Choi, T.J., Meyer, D.J., Xiong, X.J., and Teillet, P.M. 2007c. Cross-calibration of the Terra MODIS, Landsat 7 ETM+ and EO-1 ALI sensors using near-simultaneous surface observation over the Railroad Valley Playa, Nevada, test site. *Proceedings of SPIE Conference 6677 on Earth Observing Systems XII*, San Diego, CA, J.J. Butler and J. Xiong, eds., pp. 66770Y. SPIE, Bellingham, WA.

Chander, G., Christopherson, J.B., Stensaas, G.L., and Teillet, P.M. 2007d. Online catalogue of worldwide test sites for the post-launch characterization and calibration of optical sensors. *Proceedings of the 58th International Astronautical Congress*, Hyderabad, India. International Astronautical Federation, pp. 2043–2051.

Chander, G., Helder, D.L., Malla, R., Micijevic, E., and Mettler, C.J. 2007a. Consistency of L4 TM absolute calibration with respect to the L5 TM sensor based on near-simultaneous image acquisition. *Proceedings of SPIE Conference 6677 on Earth Observing Systems XII*, San Diego, CA, J.J. Butler and J. Xiong, eds., pp. 66770F. SPIE, Bellingham, WA.

Chander, G., Helder, D.L., Markham, B.L., Dewald, J.D., Kaita, E., Thome, K.J., Micijevic, E., and Ruggles, T.A. 2004a. Landsat-5 TM reflective-band absolute radiometric calibration. *IEEE Transactions on Geoscience and Remote Sensing*, 42: 2747–2760.

Chander, G., Hewison, T.J., Fox, N., Wu, X., Xiong, X., and Blackwell, W.J., Guest eds. 2013a. Special issue on Intercalibration of satellite instruments. *IEEE Transactions on Geoscience and Remote Sensing*, 51(3 SI): 491 pp.

Chander, G., Hewison, T.J., Fox, N., Wu, X., Xiong, X., and Blackwell, W.J. 2013b. Overview of intercalibration of satellite instruments. *IEEE Transactions on Geoscience and Remote Sensing*, 51(3 SI): 1056–1080.

Chander, G. and Markham, B.L. 2003. Revised Landsat-5 TM radiometric calibration procedures and postcalibration dynamic ranges. *IEEE Transactions on Geoscience and Remote Sensing*, 41: 2674–2677.

Chander, G., Markham, B.L., and Barsi, J.A. 2007b. Revised Landsat-5 Thematic Mapper radiometric calibration. *IEEE Geoscience and Remote Sensing Letters*, 4: 490–494.

Chander, G., Markham, B.L., and Helder, D.L. 2009. Summary of current radiometric calibration coefficients for Landsat MSS, TM, ETM+, and EO-1 ALI sensors. *Remote Sensing of Environment*, 113: 893–903.

Chander, G., Meyer, D.J., and Helder, D.L. 2004b. Cross calibration of the Landsat-7 ETM+ and EO-1 ALI sensors. *IEEE Transactions on Geoscience and Remote Sensing*, 42(12): 2821–2831.

Chander, G., Mishra, N., Helder, D.L., Aaron, D., Angal, A., Choi, T., Xiong, X., and Doelling, D. 2013c. Applications of spectral band adjustment factors (SBAF) for cross-calibration. *IEEE Transactions on Geoscience and Remote Sensing*, 51(3 SI): 1267–1281.

Chang, I.L., Dean, C., Li, Z., Weinreb, M., Wu, X., and Swamy, P.A.V.B. 2012. Refined algorithms for star-based monitoring of GOES Imager visible channel responsivities. *Proceedings of SPIE Earth Observing Systems XVII*, San Diego, CA, pp. 851 00R.

Che, N. and Price, J.C. 1992. Survey of radiometric calibration results and methods for visible and near-infrared channels of NOAA-7, NOAA-9, and NOAA-11 AVHRRs. *Remote Sensing of Environment*, 41(1): 19–27.

Chen, H.S. 1996. *Remote Sensing Calibration Systems – An Introduction*. A. Deepak Publishing, Hampton, VA, ISBN 0-937194-38-7.

Choi, T., Xiong, X., Wang, Z., and Link, D. 2013. Terra and aqua MODIS on-orbit spectral characterization for reflective solar bands. *Proceedings of SPIE Conference 8724, Ocean Sensing and Monitoring V*, 87240Y, doi:10.1117/12.2016640.

Cihlar, J. and P.M. Teillet. 1995. Forward piecewise linear calibration model for quasi-real-time processing of AVHRR data. *Canadian Journal of Remote Sensing*, 21(1): 22–27.

Clark, D., Gordon, H., Voss, K., Ge, Y., Broenkow, W., and Trees, C. 1997. Validation of atmospheric correction over the oceans. *Journal of Geophysical Research*, 102(D14): 17209–17217.

Cosnefroy, H., Briottet, X., Leroy, M., Lecomte, P., and Santer, R. 1994. In field characterization of Saharan sites reflectances for the calibration of optical satellite sensors. *Proceedings of IEEE Conference on Surface and Atmospheric Remote Sensing: Technologies, Data Analysis and Interpretation*, Pasadena, CA, pp. 1500–1502. IEEE, Piscataway, NJ.

Cosnefroy, H., Briottet, X., Leroy, M., Lecomte, P. and Santer, R. 1997. A field experiment in Saharan Algeria for the calibration of optical satellite sensors. *International Journal of Remote Sensing*, 18(16): 3337–3359.

Cosnefroy, H., Leroy, M., and Briottet, X. 1996. Selection and characterization of Saharan and Arabian desert sites for the calibration of optical satellite sensors. *Remote Sensing of Environment*, 58(1): 101–114.

Coulson, K.L. and Jacobowitz, H. 1972. Proposed calibration target for the visible channel of a satellite radiometer. NOAA Technical Report NESS 62, UDC 551.507.362.2:551.508.21, National Oceanic and Atmospheric Administration, Washington, DC, 31 pp.

Czapla-Myers, J.S., Thome, K.J., and Buchanan, J.H. 2007. Implication of spatial uniformity on vicarious calibration using automated test sites. *Proceedings of SPIE Conference 6677 on Earth Observing Systems XII*, San Diego, CA, J.J. Butler and J. Xiong, eds., pp. 66770U. SPIE, Bellingham, WA.

Czapla-Myers, J.S., Thome, K.J., Cocilovo, B.R., McCorkel, J.T., and Buchanan, J.H. 2008. Temporal, spectral, and spatial study of the automated vicarious calibration test site at Railroad Valley, Nevada. *Proceedings of SPIE Conference 7081 on Earth Observing Systems XIII*, San Diego, CA, J.J. Butler and J. Xiong, eds., pp. 70810I. SPIE, Bellingham, WA.

Czapla-Myers, J.S., Thome, K.J., and Leisso, N.P. 2010. Radiometric calibration of earth-observing sensors using an automated test site at Railroad Valley, Nevada. *Canadian Journal of Remote Sensing*, 36(5): 474–487.

D'Amico, J., Thome, K., and Czapla-Myers, J. 2006. Validation of large-footprint reflectance-based calibration using coincident MODIS and ASTER data. *Proceedings of SPIE Conference 6296 on Earth Observing Systems XI*, San Diego, CA, J.J. Butler and J. Xiong, eds., pp. 629612–629618. SPIE, Bellingham, WA.

Davis, J.M. and Cox, S.K. 1982. Reflected solar radiances from regional scale scenes. *Journal of Applied Meteorology*, 21(11): 1698–1712.

De Vries, C., Danaher, T., Denham, R., Scarth, P., and Phinn, S. 2007. An operational radiometric calibration procedure for the Landsat sensors based on pseudo-invariant target sites. *Remote Sensing of Environment*, 107(3): 414–429.

Desormeaux, Y., Rossow, W.B., Brest, C.L., and Campbell, G.G. 1993. Normalization and calibration of geostationary satellite radiances for the International Satellite Cloud Climatology Project. *Journal of Atmospheric and Oceanic Technology*, 10(3): 304–325.

Di, L. 2004. A proposed ISO/TC 211 standards project on radiometric calibration of remote sensing data. *Proceedings of the International Workshop on Radiometric and Geometric Calibration, ISPRS Book Series Volume 2: Postlaunch Calibration of Satellite Sensors*, Gulfport, MS, S.A. Morain and A.M. Budge, eds., pp. 53–56. Taylor & Francis Group, London, U.K.

Dilligeard, E., Briottet, X., Deuze, J.L., and Santer, R.P. 1997. SPOT calibration of blue and green channels using Rayleigh scattering over clear oceans. *Proceedings of SPIE Conference on Advanced Next-Generation Satellites II*, Taormina, Italy, pp. 373–379.

Dinguirard, M. and Slater, P.N. 1999. Calibration of space-multispectral imaging sensors: A review. *Remote Sensing of Environment*, 68(3): 194–205.

Doelling, D.R., Hong, G., Morstad, D., Bhatt, R., Gopalan, A., and Xiong, X. 2010. The characterization of deep convective cloud albedo as a calibration target using MODIS reflectances. *Proceedings of SPIE Conference 7862 on Earth Observing Missions Sensors: Development, Implementation, and Characterization*, Incheon, Republic of Korea, X. Xiong, C. Kim, and H. Shimoda, eds., 78620I.

Doelling, D.R., Lukashin, C., Minnis, P., Scarino, B., and Morstad, D. 2012. Spectral reflectance corrections for satellite intercalibrations using SCHIAMACHY data. *IEEE Geoscience and Remote Sensing Letters*, 9(1): 119–123.

Doelling, D.R., Morstad, D., Scarino, B.R., Bhatt, R., and Gopalan, A. 2013. The characterization of deep convective clouds as an invariant calibration target and as a visible calibration technique. *IEEE Transactions on Geoscience and Remote Sensing*, 51(3 SI): 1147–1159.

Doelling, D.R., Nguyen, L., and Minnis, P. 2004. On the use of deep convective clouds to calibrate AVHRR data. *Proceedings of SPIE Conference 5542 on Earth Observing Systems IX*, Denver, CO, Barnes, W.L. and Butler, J.J., eds., pp. 281–289.

Duggin, M.J. 1987. Impact of radiance variations on satellite sensor calibration. *Applied Optics*, 26(7): 1264–1271.

Faiman, D., Feuermann, D., Ibbetson, P., Medwed, B., Zemel, A., Ianetz, A., Liubansky, V., Setter, I., and Suraqui, S. 2004. The Negev radiation survey. *Journal of Solar Energy Engineering-Transactions of the ASME*, 126(3): 906–914.

Floury, N., Drinkwater, M., and Witasse, O. 2002. L-band brightness temperature of ice sheets in Antarctica: Emission modelling, ionospheric contribution and temporal stability. *Proceedings of IEEE Conference on Remote Sensing: Integrating Our View of the Planet*, Toronto, Ontario, Canada, pp. 2103–2105. IEEE, Piscataway, NJ.

Fougnie, B. and Bach, R. 2009. Monitoring of radiometric sensitivity changes of space sensors using deep convective clouds—Operational application to PARASOL. *IEEE Transactions on Geoscience and Remote Sensing*, 47(3): 851–861.

Fox, N., Kaiser-Weiss, A., Schmutz, W., Thome, K., Young, D., Wielicki, B., Winkler, R., and Woolliams, E. 2011. Accurate radiometry from space: An essential tool for climate studies. *Philosophical Transactions of the Royal Society A-Mathematical, Physical & Engineering Sciences*, 369(1953): 4028–4063.

Fox, N.P. 1999. Improving the accuracy and traceability of radiometric measurements to SI for remote sensing instrumentation. *Proceedings of the 4th International Airborne Remote Sensing Conference and Exhibition/21st Canadian Symposium on Remote Sensing*, Ottawa, Ontario, Canada, Vol. I, pp. 304–311. ERIM International, Ann Arbor, MI.

Fox, N.P., Aiken, J., Barnett, J.J., Briottet, X., Carvell, R., Froehlich, C., Groom, S.B., Hagolle, O., Haigh, J.D., Kieffer, H.H., Lean, J., Pollock, D.B., Quinn, T.J., Sandford, M.C.W., Schaepman, M.E., Shine, K.P., Schmutz, W.K., Teillet, P.M., Thome, K.J., Verstraete, M.M., and Zalewski, E.F. 2002. Traceable radiometry underpinning terrestrial- and helio-studies (TRUTHS). *Advances in Space Research*, 32(11): 2253–2261.

Fraser, R.S. and Kaufman, Y.J. 1986. Calibration of satellite sensors after launch. *Applied Optics*, 25(7): 1177–1185.

Frouin, R. and Gautier, C. 1987. Calibration of NOAA-7 AVHRR, GOES-5, and GOES-6 VISSR/VAS solar channels. *Remote Sensing of Environment*, 22(1): 73–101.

Gamet, P., Lachérade, S., Fougnie, B., and Thomas, C. 2011. Calibration of VIS/NIR sensors over desert sites: New results for cross- and multitemporal calibration. *Proceedings of the Annual CALCON Conference on Characterization and Radiometric Calibration for Remote Sensing*, Logan, UT, pp. 1–10.

Gilead, U. 2005. *Locating and Examining Potential Sites for Vicarious Radiometric Calibration of Space Multi-Spectral Imaging Sensors in the Negev Desert*. Ben Gurion University, Beer Sheva, Israel, 105 pp.

Gilead, U. and Karnieli, A. 2004. Locating potential vicarious calibration sites for high-spectral resolution sensors in the Israeli Negev Desert by GIS analysis. In *Postlaunch Calibration of Satellite Sensors, ISPRS Book Series Volume 2, Proceedings of the International Workshop on Radiometric and Geometric Calibration*, Gulfport, MS, S.A. Morain and A.M. Budge, eds., pp. 181–187. A.A. Balkema Publishers, Leiden, the Netherlands.

Godden, G.D. and McKay, C.A. 1997. A strategy for observing the moon to achieve precise radiometric stability monitoring. *Canadian Journal of Remote Sensing*, 23(4): 333–341.

Govaerts Y., Sterckx, S., and Adriaensen, S. 2013. Use of simulated reflectances over bright desert target as an absolute calibration reference. *Remote Sensing Letters*, 4(6): 523–531.

Goward, S.N., Arvidson, T., Williams, D.L., Faundeen, J.L., Irons, J.R., and Franks, S. 2006. Historical record of Landsat global coverage-mission operations, NSLRSDA, and international cooperator stations. *Photogrammetric Engineering and Remote Sensing*, 72(10): 1155–1169.

Goward, S.N. and Masek, J.G., eds. 2001. Special Issue on "Landsat 7." *Remote Sensing of Environment*, 78 (1–2): 222 pp.

Graetz, R.D., Wilson, M.A., Prata, A.J., Barton, I.J., and Mitchell, R.M. 1994. A Continental Instrumented Ground Site Network (CIGSN, Australia): A prerequisite for the detection, interpretation and quantification of global change. *Proceedings of 1994 International Geoscience and Remote Sensing Symposium (IGARSS'94)*, Pasadena, CA, pp. 1254–1256. IEEE, Piscataway, NJ.

Gross, H.N., Bruegge, C.J., and Helmlinger, M.C. 2007. Unattended vicarious calibration of a low Earth orbit visible-near infrared sensor. *Proceedings of AIAA Space 2007 Conference*, Long Beach, CA, pp. 786–793. AIAA, Reston, VA.

Gu, X.F., Guyot, G., and Verbrugghe, M. 1992. Evaluation of measurement errors in ground surface reflectance for satellite calibration. *International Journal of Remote Sensing*, 13(14): 2531–2546.

Guenther, B., Barnes, W., Knight, E., Barker, J., Harnden, J., Weber, R., Roberto, M., Godden, G., Montgomery, H., and Abel, P. 1996. MODIS calibration: A brief review of the strategy for the at-launch calibration approach. *Journal of Atmospheric and Oceanic Technology*, 13(2): 274–285.

Gürol, S., Behnert, I., Özen, H., Deadman, A., Fox, N., and Leloğlu, U.M. 2010. Tuz Gölü: New CEOS reference standard test site for infrared visible optical sensors. *Canadian Journal of Remote Sensing*, 36(5): 553–565.

Gurol, S., Ozen, H., Leloglu, U.M., and Tunali, E. 2008. Tuz Golu: New absolute radiometric calibration test site. *Proceedings of the XXI ISPRS Congress: Silk Road for Information from Imagery*, Beijing, China, pp. 35–40.

Hagolle, O., Nicolas, J.M., Fougnie, B., Cabot, F., and Henry, P. 2004. Absolute calibration of VEGETATION derived from an interband method based on sunglint over ocean sites. *IEEE Transactions on Geoscience and Remote Sensing*, 42(7): 1472–1481.

Helder, D., Boncyk, W., and Morfitt, R. 1997. Landsat TM memory effect characterization and correction. *Canadian Journal of Remote Sensing*, 23(4): 299–308.

Helder, D., Thome, K., Aaron, D., Leigh, L., Czapla-Myers, J., Leisso, N., Biggar, S., and Anderson, N. 2012a. Recent surface reflectance measurement campaigns with emphasis on best practices, SI traceability and uncertainty estimation. *Metrologia*, 49(2): S21–S28.

Helder, D., Thome, K., Mishra, N., Chander, G., Xiong, X., Angal, A., and Choi, T. 2013. Absolute radiometric calibration of Landsat using a pseudo invariant calibration site. *IEEE Transactions on Geoscience and Remote Sensing*, 51(3 SI): 1360–1369.

Helder, D.L. 1996. A radiometric calibration archive for Landsat TM. *Proceedings of the SPIE Conference 2758 on Algorithms for Multispectral and Hyperspectral Imagery*, Orlando, FL, pp. 273–284.

Helder, D.L., Barker, J., Boncyk, W., and Markham, B.L. 1996. Short term calibration of Landsat TM: Recent findings and suggested techniques. *Proceedings of 1996 International Geoscience and Remote Sensing Symposium (IGARSS'96)*, Lincoln, NE, pp. 1286–1289.

Helder, D.L., Basnet, B., and Morstad, D.L. 2010. Optimized identification of worldwide radiometric pseudo-invariant calibration sites. *Canadian Journal of Remote Sensing*, 36(5): 527–539.

Helder, D.L., Boncyk, W., and Morfitt, R. 1998a. Absolute calibration of the Landsat Thematic Mapper using the internal calibrator. *Proceedings of 1998 International Geoscience and Remote Sensing Symposium (IGARSS'98)*, Seattle, WA, pp. 2716–2718.

Helder, D.L., Boncyk, W., and Morfitt, R. 1998b. Landsat TM memory effect characterization and correction. *Canadian Journal of Remote Sensing*, 23(4): 299–301.

Helder, D.L., Karki, S., Bhatt, R., Micijevic, E., Aaron, D., and Jasinski, B. 2012b. Radiometric calibration of the Landsat MSS sensor series. *IEEE Transactions on Geoscience and Remote Sensing*, 50(6): 2380–2399.

Helder, D.L., Markham, B.L., Thome, K.J., Barsi, J.A., Chander, G., and Malla, R. 2008. Updated radiometric calibration for the Landsat-5 Thematic Mapper reflective bands. *IEEE Transactions on Geoscience and Remote Sensing*, 46: 3309–3325.

Helder, D.L. and Micijevic, E. 2004. Landsat-5 Thematic Mapper outgassing effects. *IEEE Transactions on Geoscience and Remote Sensing*, 42: 2717–2729.

Helder, D.L. and Ruggles, T.A. 2004. Landsat Thematic Mapper reflective-band radiometric artifacts. *IEEE Transactions on Geoscience and Remote Sensing*, 42: 2704–2716.

Helmlinger, M.C., Bruegge, C.J., Lubka, E.H., and Gross, H.N. 2007. LED spectrometer (LSpec) autonomous vicarious calibration facility. *Proceedings of SPIE Conference 6677 on Earth Observing Systems XII*, San Diego, CA, J.J. Butler and J. Xiong, eds., pp. 66770V. SPIE, Bellingham, WA.

Henry, P., Chander, G., Fougnie, B., Thomas, C., and Xiong, X. 2013. Assessment of spectral band impact on inter-calibration over desert sites using simulation based on EO-1 Hyperion data. *IEEE Transactions on Geoscience and Remote Sensing*, 51(3 SI): 1297–1308.

Henry, P., Dinguirard, M., and Bidilis, M. 1993. SPOT multitemporal calibration over stable desert areas. *Proceedings of SPIE International Symposium on Aerospace Remote Sensing*, Orlando, FL, pp. 67–76.

Hill, J. and Aifadopoulou, D. 1990. Comparative-analysis of Landsat-5 TM and SPOT HRV-1 data for use in multiple sensor approaches. *Remote Sensing of Environment*, 34(1): 55–70.

Holben, B.N., Kaufman, Y.J., and Kendall, J.D. 1990. NOAA-11 AVHRR visible and near-IR inflight calibration. *International Journal of Remote Sensing*, 11(8): 1511–1519.

Hovis, W.A., Knoll, J.S., and Smith, G.R. 1985. Aircraft measurements for calibration of an orbiting spacecraft sensor. *Applied Optics*, 24: 407–410.

Hu, X., Liu, J., Sun, L., Rong, Z., Li, Y., Zhang, Y., Zheng, Z., Wu, R., Zhang, L., and Gu, X. 2010. Characterization of CRCS Dunhuang test site and vicarious calibration utilization for Fengyun (FY) series sensors. *Canadian Journal of Remote Sensing*, 36(5): 566–582.

Hu, X., Zhang, Y., Liu, Z., Zhang, G., Huang, Y., Qiu, K., Wang, Y., Zhang, L., Zhu, X., and Rong, Z. 2001. Optical characteristics of China radiometric calibration site for remote sensing satellite sensors (CRCSRSSS). *Proceeding of SPIE Conference 4151 on Hyperspectral Remote Sensing of the Land and Atmosphere*, Sendai, Japan, W.L. Smith and Y. Yasuoka, eds., pp. 77–86. SPIE, Bellingham, WA.

Hu, X.Q., Liu, J.J., Qiu, K.M., Fan, T.X., Zhang, Y.X., Rong, Z.G., and Zhang, L.J. 2009. New method study of sites vicarious calibration for SZ-3/CMODIS. *Spectroscopy and Spectral Analysis*, 29(5): 1153–1159.

Hu, Y., Wielicki, B.A., Yang, P., Stackhouse, P.W. Jr., Lin, B., and Young, D.F. 2004. Application of deep convective cloud albedo observation to satellite-based study of the terrestrial atmosphere: Monitoring the stability of spaceborne measurements and assessing absorption anomaly. *IEEE Transactions on Geoscience and Remote Sensing*, 42(11): 2594–2599.

Iwabuchi, H. 2003. Calibration of the visible and near-infrared channels of NOAA-11 and-14 AVHRRs by using reflections from molecular atmosphere and stratus cloud. *International Journal of Remote Sensing*, 24(24): 5367–5378.

Jackson, R.D., ed. 1990. Special issue on coincident satellite, aircraft, and field measurements at the Maricopa Agricultural Center (MAC). *Remote Sensing of Environment*, 32(2–3): 77–228.

Kamstrup, N. and Hansen, L.B. 2003. Improved calibration of Landsat-5 TM applicable for high-latitude and dark areas. *International Journal of Remote Sensing*, 24(24): 5345–5365.

Kastner, C.J. and Slater, P.N. 1982. In flight radiometric cali bration of advanced remote sensing systems. *Proceedings of Field Measurement and Calibration Using Electro-Optical Equipment: Issues and Requirements*, San Diego, CA, F.M. Zweibaum and H. Register, eds., pp. 1–8. SPIE, Bellingham, WA.

Kaufman, Y.J. and Holben, B.N. 1993. Calibration of the AVHRR visible and near-IR bands by atmospheric scattering, ocean glint and desert reflection. *International Journal of Remote Sensing*, 14(1): 21–52.

Kerola, D.X. and Bruegge, C.J. 2009. Desert test site uniformity analysis. *Proceedings of SPIE Conference 7452 on Earth Observing Systems XIV*, San Diego, CA, J.J. Butler, X. Xiong, and X. Gu, eds., pp. 74520C-74528. SPIE, Bellingham, WA.

Kerola, D.X., Bruegge, C.J., Gross, H.N., and Helmlinger, M.C. 2009. On-orbit calibration of the EO-1 hyperion and advanced land imager (ALI) sensors using the LED spectrometer (LSpec) automated facility. *IEEE Transactions on Geoscience and Remote Sensing*, 47(4): 1244–1255.

Kieffer, H.H., Stone, T.C., Barnes, R.A., Bender, S., Eplee, R.E. Jr., Mendenhall, J., and Ong, L. 2002. On-orbit radiometric calibration over time and between spacecraft using the moon. *Proceedings of SPIE Conference on Sensors, Systems, and Next-Generation Satellites VI*, Crete, Greece, pp. 287–298.

Kieffer, H.H. and Wildey, R.L. 1985. Absolute calibration of Landsat instruments using the moon. *Photogrammetric Engineering and Remote Sensing*, 51(9): 1391–1393.

Kieffer, H.H. and Wildey, R.L. 1996. Establishing the moon as a spectral radiance standard. *Journal of Atmospheric and Oceanic Technology*, 13(2): 360–375.

Kim, W. and Lee, S. 2013. Study on radiometric variability of the Sonoran Desert for vicarious calibration of satellite sensors. *Korean Journal of Remote Sensing*, 29(2): 209–218.

Koepke, P. 1982. Vicarious satellite calibration in the solar spectral range by means of calculated radiances and its application to Meteosat. *Applied Optics*, 21(15): 2845–2854.

Lachérade, S., Fougnie, B., Henry, P., and Gamet, P. 2013. Cross calibration over desert sites: Description, methodology, and operational implementation. *IEEE Transactions on Geoscience and Remote Sensing*, 51(3 SI): 1098–1113.

Le Marshall, J.F., Simpson, J.J., and Jin, Z.H. 1999. Satellite calibration using a collocated nadir observation technique: Theoretical basis and application to the GMS-5 pathfinder benchmark period. *IEEE Transactions on Geoscience and Remote Sensing*, 37(1): 499–507.

Li, Y., Zhang, Y., Liu, J., Rong, Z.G., and Zhang, L.J. 2009. Calibration of the visible and near-infrared channels of the FY-2C/FY-2D GEO meteorological satellite at radiometric site. *Guangxue Xuebao/Acta Optica Sinica*, 29(1): 41–46.

Liu, J.J., Li, Z., Qiao, Y.L., Liu, Y.J., and Zhang, Y.X. 2004. A new method for cross-calibration of two satellite sensors. *International Journal of Remote Sensing*, 25(23): 5267–5281.

Loeb, N.G. 1997. In-flight calibration of NOAA AVHRR visible and near-IR bands over Greenland and Antarctica. *International Journal of Remote Sensing*, 18(3): 477–490.

Luderer, G., Coakley, J.A. Jr., and Tahnk, W.R. 2005. Using sun glint to check the relative calibration of reflected spectral radiances. *Journal of Atmospheric and Oceanic Technology*, 22(10): 1480–1493.

Lukashin, C., Wielicki, B., Young, D., Thome, K., Jin, Z., and Sun, W. 2013. Uncertainty estimates for imager reference intercalibration with CLARREO reflected solar spectrometer. *IEEE Transactions on Geoscience and Remote Sensing*, 51(3 SI): 1425–1436.

Malila, W.A. and Anderson, D.M. 1986. Satellite data availability and calibration documentation for land surface climatology studies. NASA/GSFC, Greenbelt, MD, Report No. 180300-1-F, 214 pp.

Markham, B., Storey, J., Williams, D., and Irons, J. 2004a. Landsat sensor performance: History and current status. *IEEE Transactions on Geoscience and Remote Sensing*, 42(12): 2691–2694.

Markham, B., Thome, K., Barsi, J., Kaita, E., Helder, D., Barker, J., and Scaramuzza, P. 2004b. Landsat-7 ETM+ on-orbit reflective-band radiometric stability and absolute calibration. *IEEE Transactions on Geoscience and Remote Sensing*, 42(12): 2810–2820.

Markham, B.L. and Barker, J.L., eds. 1985. Special issue on Landsat image data quality assessment (LIDQA). *Photogrammetric Engineering and Remote Sensing*, 51(9): 1245–1493.

Markham, B.L., Barker, J.L., Kaita, E., Seiferth, J., and Morfitt, R. 2003. On-orbit performance of the Landsat-7 ETM+ radiometric calibrators. *International Journal of Remote Sen*sing, 24(2): 265–286.

Markham, B.L., Barsi, J.A., Helder, D.L., Thome, K.J., and Barker, J.L. 2006. Evaluation of the Landsat-5 TM radiometric calibration history using desert test sites. *Proceedings of SPIE Conference 6361 on Sensors, Systems, and Next-Generation Satellites X*, Stockholm, Sweden, R. Meynart, S.P. Neeck, and H. Shimoda, eds., pp. 63610. SPIE, Bellingham, WA.

Markham, B.L., Boncyk, W.C., Helder, D.L., and Barker, J.L. 1997. Landsat-7 enhanced Thematic Mapper Plus radiometric calibration. *Canadian Journal of Remote Sensing*, 23(4): 318–332.

Markham, B.L., Dabney, P.W., Knight, E.J., Kvaran, G., Barsi, J.A., Murphy-Morris, J.E., and Pedelty, J.A. 2010. The Landsat data continuity mission operational land imager (OLI) radiometric calibration. *Proceedings of IEEE International Geoscience and Remote Sensing Symposium (IGARSS) 2010*, July 25–30, Honolulu, HI, 4 pp.

Markham, B.L., Haque, O., Barsi, J.A., Micijevic, E., Helder, D.L., Thome, K., Aaron, D., and Czapla-Myers, J.S. 2012. Landsat-7 ETM+: 12 years on-orbit reflective-band radiometric performance. *IEEE Transactions on Geoscience and Remote Sensing*, 50(5): 2056–2062.

Markham, B.L. and Helder, D.L. 2012. Forty-year calibrated record of earth-reflected radiance from Landsat: A review. *Remote Sensing of Environment*, 122: 30–40.

Markham, B.L., Seiferth, J.C., Smid, J., and Barker, J.L. 1998. Lifetime responsivity behavior of the Landsat-5 Thematic Mapper. *Proceedings of SPIE Conference 3427*, San Diego, CA, pp. 420–431.

Markham, B.L., Storey, J.C., and Morfitt, R. 2015. Landsat-8 sensor characterization and calibration. *Remote Sensing*, 7(3): 2279–2282.

Martiny, N., Santer, R., and Smolskaia, I. 2005. Vicarious calibration of MERIS over dark waters in the near infrared. *Remote Sensing of Environment*, 94(4): 475–490.

Matsunaga, T., Nonaka, T., Sawabe, Y., Moriyama, M., Tonooka, H., and Fukasawa, H. 2001. Vicarious and cross calibration methods for satellite thermal infrared sensors using hot ground targets. *Proceedings of IEEE Conference on Scanning the Present and Resolving the Future*, Sydney, New South Wales, Australia, pp. 1841–1843. IEEE, Piscataway, NJ.

McCorkel, J., Thome, K., Biggar, S., and Kuester, M. 2006. Radiometric calibration of advanced land imager using reflectance-based results between 2001 and 2005. *Proceedings of SPIE Conference 6296 on Earth Observing Systems XI*, San Diego, CA, J.J. Butler and J. Xiong, eds., pp. 62960G. SPIE, Bellingham, WA.

McCorkel, J., Thome, K., and Ong, L. 2013. Vicarious calibration of EO-1 hyperion. *IEEE Journal of Selected Topics in Applied Earth Observations and Remote Sensing*, 6(2 SI): 400–407.

Mendenhall, J.A., Hearn, D.R., Lencioni, D.E., Digenis, C.J., and Ong, L. 2003. Summary of the EO-1 ALI performance for the first 2.5 years on-orbit. *Proceedings of SPIE Conference 5151 on Earth Observing Systems VIII*, San Diego, CA, W.L. Barnes, ed., pp. 574–585.

Mendenhall, J.A., Lencioni, D.E., and Evans, J.B. 2005. Spectral and radiometric calibration of the advanced land imager. *Lincoln Laboratory Journal*, 15(2): 207–224.

Metzler, M.D. and Malila, W.A. 1985. Characterization and comparison of Landsat-4 and Landsat-5 Thematic Mapper data. *Photogrammetric Engineering and Remote Sensing*, 51(9): 1315–1330.

Meygret, A., Briottet, X., Henry, P.J., and Hagolle, O. 2000. Calibration of SPOT4 HRVIR and vegetation cameras over Rayleigh scattering. *Proceedings of SPIE Conference 4135 on Earth Observing Systems V*, San Diego, CA, W.L. Barnes, ed., pp. 302–313. SPIE, Bellingham, WA.

Meygret, A., Santer, R., and Berthelot, B. 2011. ROSAS: A robotic station for atmosphere and surface characterization dedicated to on-orbit calibration. *Proceedings of SPIE Conference 8153 on Earth Observing Systems XVI*, 815311, San Diego, CA, 12 pp.

Miesch, C., Cabot, F., Briottet, X., and Henry, P. 2003. Assimilation method to derive spectral ground reflectance of desert sites from satellite datasets. *Remote Sensing of Environment*, 87(2–3): 359–370.

Mika, A.M. 1997. Three decades of Landsat instruments. *Photogrammetric Engineering and Remote Sensing*, 63(7): 839–852.

Min, X. and Zhu, Y. 1995. Properties of atmospheric aerosol extinction for the satellite radiometric calibration site of Dunhuang, China. *Proceedings of IEEE Conference on Quantitative Remote Sensing for Science and Applications*, Firenze, Italy, pp. 1858–1860. IEEE, Piscataway, NJ.

Mitchell, R.M., O'Brien, D.M., and Forgan, B.W. 1992. Calibration of the NOAA AVHRR shortwave channels using split pass imagery: I. Pilot study. *Remote Sensing of Environment*, 40(1): 57–65.

Mitchell, R.M., O'Brien, D.M., Edwards, M., Elsum, C.C., and Graetz, R.D. 1997. Selection and initial characterization of a bright calibration site in the Strzelecki Desert, South Australia. *Canadian Journal of Remote Sensing*, 23(4): 342–353.

Molineux, C.E., Bliamptis, E.E., and Neal, J.T. 1971. A remote-sensing investigation of four Mojave playas. Environmental Research Papers, No. 352, AFCRL-71-0235, April 16, 1971. Air Force Cambridge Research Laboratories, Hanscom Field, Bedford, MA, 70 pp.

Morain, S.A. and Budge, A.M., eds. 2004. *Postlaunch Calibration of Satellite Sensors, ISPRS Book Series Volume 2: Proceedings of the International Workshop on Radiometric and Geometric Calibration*, Gulfport, MS. A.A. Balkema Publishers, Leiden, the Netherlands, 193 pp.

Moran, M.S., Jackson, R.D., Clarke, T.R., Qi, J., Cabot, F., Thome, K.J., and Markham, B.L. 1995. Reflectance factor retrieval from Landsat TM and SPOT HRV data for bright and dark targets. *Remote Sensing of Environment*, 52(3): 218–230.

Mueller, J.L. 1985. NIMBUS-7 CZCS—Confirmation of its radiometric sensitivity decay-rate through 1982. *Applied Optics*, 24(7): 1043–1047.

Muench, H.S. 1981. Calibration of geosynchronous satellite video sensors. Report No. AFGL-TR-81-0050, Air Force Geophysical Laboratory, Hanscom, MA.

Murphy, J.M. 1984. Radiometric correction of Landsat Thematic Mapper data. CCRS Digital Methods Division Technical Memorandum DMD-TM#84-368, Canada Centre for Remote Sensing, Ottawa, Ontario, Canada.

Murphy, J.M. 1986. Within-scene radiometric correction of Landsat Thematic Mapper data in Canadian production systems. *Proceedings of SPIE Conference 660 on Earth Remote Sensing using Landsat Thematic Mapper and SPOT Sensor Systems*, Innsbruck, Austria, pp. 25–31.

Nieke, J., Aoki, T., Tanikawa, T., Motoyoshi, H., and Hori, M. 2004. A satellite cross-calibration experiment. *IEEE Geoscience and Remote Sensing Letters*, 1(3): 215–219.

Nieke, J., Aoki, T., Tanikawa, T., Motoyoshi, H., Hori, M., and Nakajima, Y. 2003. Cross-calibration of satellite sensors over snow fields. *Proceedings of SPIE Conference 5151 on Earth Observing Systems VIII*, San Diego, CA, W.L. Barnes, ed., Vol. 5151, pp. 406–414.

Nithianandam, J., Guenther, B.W., and Allison, L.J. 1993. An anecdotal review of NASA Earth observing satellite remote sensors and radiometric calibration methods. *Metrologia*, 30(4): 207–212.

O'Brien, D.M. and Mitchell, R.M. 2001. An error budget for cross-calibration of AVHRR shortwave channels against ATSR-2. *Remote Sensing of Environment*, 75(2): 216–229.

Ohring, G., Tansock, J., Emery, W., Butler, J., L, Weng, F., Germain, K.S., Wielicki, B., Cao, C., Goldberg, M., Xiong, J., Fraser, G., Kunkee, D., Winker, D., Miller, L., Ungar, S., Tobin, D., Anderson, J.G., Pollock, D., Shipley, S., Thurgood, A., Kopp, G., Ardanuy, P., and Stone, T. 2007. Achieving satellite instrument calibration for climate change. *EOS, Transactions American Geophysical Union*, 88(11): 136.

Padula, F.P., Schott, J.R., Barsi, J.A., Raqueno, N.G., and Hook, S.J. 2010. Calibration of Landsat 5 thermal infrared channel: Updated calibration history and assessment of the errors associated with the methodology. *Canadian Journal of Remote Sensing*, 36(5): 617–630.

Pollock, D.B., Murdock, T.L., Datla, R.U., and Thompson, A. 2003. Data uncertainty traced to SI units. Results reported in the International System of Units. *International Journal of Remote Sensing*, 24(2): 225–235.

Potts, D.R., Mackin, S., Muller, J.-P., and Fox, N. 2013. Sensor intercalibration over dome C for the ESA GlobAlbedo Project. *IEEE Transactions on Geoscience and Remote Sensing*, 51(3 SI): 1139–1146.

Prata, A.J., Cechet, R.P., Grant, I.F., and Rutter, G.F. 1996. The Australian Continental Integrated Ground-truth Site Network (CIGSN)—Satellite data calibration and validation, and first results. *Proceedings of the Second SEIKEN Symposium on Global Environmental Monitoring from Space*, Tokyo, Japan, pp. 245–261.

Price, J.C. 1987a. Radiometric calibration of satellite sensors in the visible and near-infrared—History and outlook. *Remote Sensing of Environment*, 22(1): 3–9.

Price, J.C., ed. 1987b. Special issue on radiometric calibration of satellite data. *Remote Sensing of Environment*, 22(1): 1–158.

Rao, C.R.N., Cao, C., and Zhang, N. 2003. Inter-calibration of the moderate-resolution imaging spectroradiometer and the along-track scanning radiometer-2. *International Journal of Remote Sensing*, 24(9): 1913–1924.

Rao, C.R.N. and Chen, J. 1995. Intersatellite calibration linkages for the visible and near-infrared channels of the advanced very high-resolution radiometer on the NOAA-7, NOAA-9, and NOAA-11 spacecraft. *International Journal of Remote Sensing*, 16(11): 1931–1942.

Rao, C.R.N. and Chen, J.H. 1996. Postlaunch calibration of the visible and near-infrared channels of the advanced very high-resolution radiometer on the NOAA-14 spacecraft. *International Journal of Remote Sensing*, 17(14): 2743–2747.

Rao, C.R.N. and Chen, J.H. 1999. Revised postlaunch calibration of the visible and near-infrared channels of the advanced very high-resolution radiometer (AVHRR) on the NOAA-14 spacecraft. *International Journal of Remote Sensing*, 20(18): 3485–3491.

Reuter, D., Irons, J., Lunsford, A., Montanaro, M., Pellerano, F., Richardson, C., Smith, R., Tesfaye, Z., and Thome, K. 2011. Operational land imager (OLI) and the thermal infrared sensor (TIRS) on the Landsat data continuity mission (LDCM). *Proceedings of SPIE Conference 8048 on Algorithms and Technologies for Multispectral, Hyperspectral, and Ultraspectral Imagery XVII*, S.S. Shen and P.E. Lewis, eds., 804819.

Richter, R. 1997. On the in-flight absolute calibration of high spatial resolution spaceborne sensors using small ground targets. *International Journal of Remote Sensing*, 18(13): 2827–2833.

Rodger, A.P., Balick, L.K., and Clodius, W.B. 2005. The performance of the multispectral thermal imager (MTI) surface temperature retrieval algorithm at three sites. *IEEE Transactions on Geoscience and Remote Sensing*, 43(3): 658–665.

Rondeaux, G., Steven, M.D., Clark, J.A., and Mackay, G. 1998. La Crau: A European test site for remote sensing validation. *International Journal of Remote Sensing*, 19(14): 2775–2788.

Salomonson, V.V. and Marlatt, W.E. 1968. Anisotropic solar reflectance over white sand, snow, and stratus clouds. *Journal of Applied Meteorology*, 7: 475–483.

Sandford, S.P., Young, D.F., Corliss, J.M., Wielicki, B.A., Gazarik, M.J., Mlynczak, M.G., Little, A.D., Jones, C.D., Speth, P.W., Shick, D.E., Brown, K.E., Thome, K.J., and Hair, J.H. 2010. CLARREO: Cornerstone of the climate observing system measuring decadal change through accurate emitted infrared and reflected solar spectra and radio occultation. *Proceedings of SPIE Conference 7826 on Sensors, Systems, and Next-Generation Satellites XIV*, R. Meynart, S.P. Neeck, and H. Shimoda, eds., Toulouse, France, 782611.

Santer, R., Gu, X.F., Guyot, G., Deuze, J.L., Devaux, C., Vermote, E., and Verbrughe, M. 1992. SPOT calibration at the La-Crau test site (France). *Remote Sensing of Environment*, 41(2–3): 227–237.

Santer, R., Schmectig, C., and Thome, K.J. 1997. BRDF and surface-surround effects on SPOT-HRV vicarious calibration. *Proceedings of SPIE Conference 2957 on Advanced and Next-Generation Satellites II*, Taormina, Italy, H. Fujisada, G. Calamai, and M.N. Sweeting, eds., pp. 344–354. SPIE, Bellingham, WA.

Schiller, S.J. 2003. Technique for estimating uncertainties in top-of-atmosphere radiances derived by vicarious calibration. *Proceedings of SPIE Conference 5151 on Earth Observing Systems VIII*, San Diego, CA, W.L. Barnes, ed., pp. 502–516. SPIE, Bellingham, WA.

Schott, J.R., Salvaggio, C., and Volchok, W.J. 1988. Radiometric scene normalization using pseudo-invariant features. *Remote Sensing of Environment*, 26: 1–16.

Schroeder, M., Poutier, L., Muller, R., Dinguirard, M., Reinartz, P., and Briottet, X. 2001. Intercalibration of optical satellites—A case study with MOMS and SPOT. *Aerospace Science and Technology*, 5(4): 305–315.

Scott, K.P., Thome, K.J., and Brownlee, M.R. 1996. Evaluation of Railroad Valley Playa for use in vicarious calibration. *Proceedings of SPIE Conference 2818 on Multispectral Imaging for Terrestrial Applications*, Denver, CO, B. Huberty, J.B. Lurie, J.A. Caylor, P. Coppin, and P.C. Robert, eds., pp. 158–166. SPIE, Bellingham, WA.

Secker, J., Staenz, K., Gauthier, R.P., and Budkewitsch, P. 2001. Vicarious calibration of airborne hyperspectral sensors in operational environments. *Remote Sensing of Environment*, 76(1): 81–92.

Six, C. 2002. *Automatic Station for the In-Flight Calibration of the Satellite Sensors: Application to the SPOT/HRV on the La Crau Test Site*. Universite du Littoral-Cote d'Opale, Dunkerque, France, 147 pp.

Six, D., Fily, M., Alvain, S., Henry, P., and Benoist, J.P. 2004. Surface characterisation of the Dome Concordia area (Antarctica) as a potential satellite calibration site, using SPOT4/VEGETATION instrument. *Remote Sensing of Environment*, 89(1): 83–94.

Six, D., Fily, M., Blarel, L., and Goloub, P. 2005. First aerosol optical thickness measurements at Dome C (East Antarctica), summer season 2003–2004. *Atmospheric Environment*, 39(28): 5041–5050.

Slater, P.N. 1980. *Remote Sensing, Optics and Optical Systems*. Addison-Wesley Publishing Company, Reading, MA.

Slater, P.N. 1984. The importance and attainment of accurate absolute radiometric calibration. *Proceedings of SPIE Conference 475 on Remote Sensing*, Arlington, VA, P.N. Slater, ed., pp. 34–40. SPIE, Bellingham, WA.

Slater, P.N. 1985. Radiometric considerations in remote-sensing. *Proceedings of the IEEE*, 73(6): 997–1011.

Slater, P.N. 1986. Variations in in-flight absolute radiometric performance. *International Satellite Land-Surface Climatology Project (ISLSCP) Conference*, Rome, Italy, pp. 357–363. European Space Agency, Paris, France.

Slater, P.N. and Biggar, S.F. 1996. Suggestions for radiometric calibration coefficient generation. *Journal of Atmospheric and Oceanic Technology*, 13(2): 376–382.

Slater, P.N., Biggar, S.F., Holm, R.G., Jackson, R.D., Mao, Y., Moran, M.S., Palmer, J.M., and Yuan, B. 1987. Reflectance-based and radiance-based methods for the in-flight absolute calibration of multispectral sensors. *Remote Sensing of Environment*, 22(1): 11–37.

Slater, P.N., Biggar, S.F., Palmer, J.M., and Thome, K.J. 2001. Unified approach to absolute radiometric calibration in the solar-reflective range. *Remote Sensing of Environment*, 77(3): 293–303.

Slater, P.N., Biggar, S.F., Thome, K.J., Gellman, D.I., and Spyak, P.R. 1996. Vicarious radiometric calibration of EOS sensors. *Journal of Atmospheric and Oceanographic Technology*, 13(2): 349–359.

Smith, D.L., Mutlow, C.T., and Rao, C.R.N. 2002. Calibration monitoring of the visible and near-infrared channels of the along-track scanning radiometer-2 by use of stable terrestrial sites. *Applied Optics*, 41(3): 515–523.

Smith, G.R. 1984. Surface soil moisture measurements of the White Sands, New Mexico. NOAA Technical Report NESDIS 7, PB85 135754. National Oceanic and Atmospheric Administration, Washington, DC, 12 pp.

Smith, G.R. and Levin, R.H. 1985. High altitude measured radiance of White Sands, New Mexico, in the 400-2000 nm band using a filter wedge spectrometer. NOAA Technical Report NESDIS 21, PB85-206084, National Oceanic and Atmospheric Administration, Washington, DC, 17 pp.

Smith, G.R., Levin, R.H., Abel, P., and Jacobowitz., H. 1988. Calibration of the solar channels and NOAA-9 AVHRR using high altitude aircraft measurements. *Journal of Atmospheric and Oceanic Technology*, 5: 631–639.

Smith, G.R., Levin, R.H., and Knoll, J.S. 1985. An atlas of high altitude aircraft measured radiance of White Sands, New Mexico, in the 450–1050 nm band. NOAA Technical Report NESDIS 20, PB85-204501, National Oceanic and Atmospheric Administration, Washington, DC, 29 pp.

Snyder, W.C., Wan, Z.M., Zhang, Y.L., and Feng, Y.Z. 1997. Requirements for satellite land surface temperature validation using a silt playa. *Remote Sensing of Environment*, 61(2): 279–289.

Staylor, W.F. 1986. Site selection and directional models of deserts used for ERBE validation targets. NASA Technical Paper 2540, NASA, Washington, DC.

Staylor, W.F. 1990. Degradation rates of the AVHRR visible channel for the NOAA 6, 7, and 9 spacecraft. *Journal of Atmospheric and Oceanic Technology*, 7(3): 411–423.

Stone, T.C. 2008. Radiometric calibration stability and inter-calibration of solar-band instruments in orbit using the moon. *Proceedings of SPIE Conference on Earth Observing Systems XIII*, San Diego, CA, 70810X.

Stone, T.C., Kieffer, H.H., and Grant, I.F. 2005. Potential for calibration of geostationary meteorological satellite imagers using the moon. *Proceedings of SPIE Conference on Earth Observing Systems X*, San Diego, CA, pp. 1–9.

Stone, T.C., Rossow, W.B., Ferrier, J., and Hinkelmann, L.M. 2013. Evaluation of ISCCP multi-satellite radiance calibration for geostationary imager visible channels using the moon. *IEEE Transactions on Geoscience and Remote Sensing*, 51(3): 1255–1266.

Tahnk, W.R. and Coakley, J.A. 2001. Updated calibration coefficients for NOAA-14 AVHRR channels 1 and 2. *International Journal of Remote Sensing*, 22(15): 3053–3057.

Teillet, P.M. 1986. Image correction for radiometric effects in remote sensing. *International Journal of Remote Sensing*, 7: 1637–1651.

Teillet, P.M. 1997a. A status overview of Earth observation calibration/validation for terrestrial applications. *Canadian Journal of Remote Sensing*: Special issue on Calibration/Validation, 23(4): 291–298.

Teillet, P.M., ed. 1997b. Special issue on calibration/validation. *Canadian Journal of Remote Sensing*, 23(4): 289–423.

Teillet, P.M. 2014. Overview of satellite image radiometry in the solar-reflective optical domain. In *The Remote Sensing Handbook*, P.S. Thenkabail, ed., Vol. 1, Chapter 3, CRC Press, Boca Raton, FL.

Teillet, P.M., Barker, J.L., Markham, B.L., Irish, R.R., Fedosejevs, G., and Storey, J.C. 2001a. Radiometric cross-calibration of the Landsat-7 ETM+ and Landsat-5 TM sensors based on tandem data sets. *Remote Sensing of Environment*, 78(1–2): 39–54.

Teillet, P.M., Barsi, J.A., Chander, G., and Thome, K.J. 2007b. Prime candidate Earth targets for the postlaunch radiometric calibration of space-based optical imaging instruments. *Proceedings of the SPIE Conference 6677 on Earth Observing Systems XII*, September 26, San Diego, CA, J.J. Butler and J. Xiong, eds., pp. 66770S1-12. SPIE, Bellingham, WA.

Teillet, P.M. and Chander, G. 2010. Terrestrial reference standard sites for postlaunch sensor calibration. *Canadian Journal of Remote Sensing*, 36(5): 437–450.

Teillet, P.M., El Saleous, N., Hansen, M.C., Eidenshink, J.C., Justice, C.O., and Townshend, J.R.G. 2000. An evaluation of the global 1-km AVHRR land data set. *International Journal of Remote Sensing*, 21(10): 1987–2021.

Teillet, P.M., Fedosejevs, G., Gauthier, D., D'Iorio, M.A., Rivard, B., and Budkewitsch, P. 1995. Initial examination of radar imagery of optical radiometric calibration sites. *Proceedings of SPIE Conference 2583 on Advanced and Next-Generation Satellites*, December 15, Paris, France, H. Fujisada and M.N. Sweeting, eds., pp. 154–165. SPIE, Bellingham, WA.

Teillet, P.M., Fedosejevs, G., and Gauthier, R.P. 1998a. Operational radiometric calibration of broadscale satellite sensors using hyperspectral airborne remote sensing of prairie rangeland: First trials. *Metrologia*, 35(4): 639–641.

Teillet, P.M., Fedosejevs, G., Gauthier, R.P., O'Neill, N.T., Thome, K.J., Biggar, S.F., Ripley, H., and Meygret, A. 2001b. A generalized approach to the vicarious calibration of multiple Earth observation sensors using hyperspectral data. *Remote Sensing of Environment*, 77(3): 304–327.

Teillet, P.M., Fedosejevs, G., Gauthier, R.P., and Schowengerdt, R.A. 1998b. Uniformity characterization of land test sites used for radiometric calibration of Earth observation sensors. *Proceedings of the Twentieth Canadian Symposium on Remote Sensing*, May, Calgary, Alberta, pp. 1–4. Canadian Remote Sensing Society, Ottawa, Canada.

Teillet, P.M., Fedosejevs, G., and Thome, K.J. 2004b. Spectral band difference effects on radiometric cross-calibration between multiple satellite sensors in the Landsat solar-reflective spectral domain. *Workshop on Inter-Comparison of Large-Scale Optical and Infrared Sensors. Proceedings of SPIE Conference 5570 on Sensors, Systems, and Next-Generation Satellites VIII*, Maspalomas, Canary Islands, Spain, R. Meynart, S.P. Neeck, and H. Shimoda, eds., pp. 307–316.

Teillet, P.M., Fedosejevs, G., Thome, K.J., and Barker, J.L. 2007a. Impacts of spectral band difference effects on radiometric cross-calibration between satellite sensors in the solar-reflective spectral domain. *Remote Sensing of Environment*, 110(3): 393–409.

Teillet, P.M. and Fox, N.P. 2009. Concatenation of terrestrial reference standard sites for systematic postlaunch calibration monitoring of multiple space-based imaging sensors. *Proceedings of SPIE Conference 7474 on Sensors, Systems, and Next-Generation Satellites XIII*, August 31–September 3, Berlin, Germany, R. Meynart, S.P. Neeck, and H. Shimoda, eds., pp. 747410. SPIE, Bellingham, WA.

Teillet, P.M., Helder, D.L., Ruggles, T.A., Landry, R., Ahern, F.J., Higgs, N.J., Barsi, J., Chander, G., Markham, B.L., Barker, J.L., Thome, K.J., Schott, J.R., and Palluconi, F.D. 2004a. A definitive calibration record for the Landsat-5 Thematic Mapper anchored to the Landsat-7 radiometric scale. *Canadian Journal of Remote Sensing*, 30(4): 631–643.

Teillet, P.M. and Holben, B.N. 1994. Towards operational radiometric calibration of NOAA AVHRR imagery in the visible and near-infrared channels. *Canadian Journal of Remote Sensing*, 20(1): 1–10.

Teillet, P.M., Markham, B.L., and Irish, R.R. 2006. Landsat cross-calibration based on near simultaneous imaging of common ground targets. *Remote Sensing of Environment*, 102(3–4): 264–270.

Teillet, P.M., and Ren, X. 2008. Spectral band difference effects on vegetation indices derived from multiple satellite sensor data. *Canadian Journal of Remote Sensing*, 34(3): 159–173.

Teillet, P.M., Ren, X., and Smith, A.M. 2010. Suitability of rangeland terrain for satellite remote sensing calibration. *Canadian Journal of Remote Sensing*, 36(5): 451–463.

Teillet, P.M., Slater, P.N., Ding, Y., Santer, R.P., Jackson, R.D., and Moran, M.S. 1990. Three methods for the absolute calibration of the NOAA AVHRR sensors in-flight. *Remote Sensing of Environment*, 31(2): 105–120.

Teillet, P.M., Thome, K.J., Fox, N.P., and Morisette, J.T. 2001c. Earth observation sensor calibration using a global instrumented and automated network of test sites (GIANTS). *Proceedings of SPIE Conference 4550 on Sensors, Systems, and Next-Generation Satellites V*, September 21, Toulouse, France, H. Fujisada, J.B. Lurie, and K. Weber, eds., pp. 246–254. SPIE, Bellingham, WA.

Thome, K. 2005. Sampling and uncertainty issues in trending reflectance-based vicarious calibration results. *Proceedings of SPIE Conference 5882 on Earth Observing Systems X*, August 22, San Diego, CA, J.J. Butler, ed., pp. 1–11. SPIE, Bellingham, WA.

Thome, K. 2012. Characterization approaches to place invariant sites on traceable scales. *Proceedings of IEEE International Geoscience and Remote Sensing Symposium (IGARSS) 2012*, July 22–27, Munich, Germany, pp. 7019–7022.

Thome, K., Barnes, R., Baize, R., O'Connell, J., and Hair, J. 2010. Calibration of the reflected solar instrument for the climate absolute radiance and refractivity observatory. *Proceedings of the 2010 IEEE International Geoscience and Remote Sensing Symposium (IGARSS)*, July 25–30, Honolulu, Hawaii, pp. 2275–2278.

Thome, K., Crowther, B.G., and Biggar, S.F. 1997a. Reflectance- and irradiance-based calibration of Landsat-5 Thematic Mapper. *Canadian Journal of Remote Sensing*, 23(4): 309–317.

Thome, K., Czapla-Myers J., and Biggar, J. 2004a. Ground-monitor radiometer system for vicarious calibration. *Proceedings of SPIE Conference 5546 on Imaging Spectrometry X*, October 15, Denver, CO, S.S. Shen and P.E. Lewis, eds., pp. 223–232. SPIE, Bellingham, WA.

Thome, K., Czapla-Myers, J., and McCorkel, J. 2007. Retrieval of surface BRDF for reflectance-based calibration. *Proceedings of SPIE Conference 6677 on Earth Observing Systems XII*, September 26, San Diego, CA, J.J. Butler and J. Xiong, eds., pp. 66770–66711. SPIE, Bellingham, WA.

Thome, K., Czapla-Myers, J., Leisso, N., McCorkel, J., and Buchanan, J. 2008. Intercomparison of imaging sensors using automated ground measurements. *Proceedings of IEEE Conference on Remote Sensing: The Next Generation*, Boston, MA, pp. 1332–1335. IEEE, Piscataway, NJ.

Thome, K., D'Amico, J., and Hugon, C. 2006. Intercomparison of Terra ASTER, MISR, and MODIS, and Landsat-7 ETM+. *Proceedings of IEEE Conference on Remote Sensing: A Natural Global Partnership*, Denver, CO, pp. 1772–1775. IEEE, Piscataway, NJ.

Thome, K. and Fox, N. 2011. 2010 CEOS field reflectance inter-comparison lessons learned. *Proceedings of IEEE International Geoscience and Remote Sensing Symposium (IGARSS) 2011*, July 24–29, Vancouver, British Columbia, Canada, pp. 3879–3882.

Thome, K., Gustafson-Bold, C., Slater, P.N., and Farrand, W.H. 1996. In-flight radiometric calibration of HYDICE using a reflectance-based approach. *Proceedings of SPIE Conference 2821 on Hyperspectral Remote Sensing and Applications*, August 4, Denver, CO, S.S. Shen, ed., pp. 311–319. SPIE, Bellingham, WA.

Thome, K., Markham, B., Barker, J., Slater, P., and Biggar, S. 1997b. Radiometric calibration of Landsat. *Photogrammetric Engineering and Remote Sensing*, 63(7): 853–858.

Thome, K., McCorkel, J., and Czapla-Myers, J. 2013. In-situ transfer standard and coincident-view intercomparisons for sensor cross-calibration. *IEEE Transactions on Geoscience and Remote Sensing*, 51(3 SI): 1088–1097.

Thome, K., Schiller, S., Conel, J., Arai, K., and Tsuchida, S. 1998. Results of the 1996 Earth Observing System vicarious calibration joint campaign at Lunar Lake Playa, Nevada. *Metrologia*, 35(4): 631–638.

Thome, K., Smith, N., and Scott, K. 2001. Vicarious calibration of MODIS using Railroad Valley Playa. *Proceedings of IEEE Conference on Scanning the Present and Resolving the Future*, July 9–13, Sydney, New South Wales, Australia, pp. 1209–1211. IEEE, Piscataway, NJ.

Thome, K.J. 2001. Absolute radiometric calibration of Landsat 7 ETM+ using the reflectance-based method. *Remote Sensing of Environment*, 78(1–2): 27–38.

Thome, K.J., Biggar, S.F., and Wisniewski, W. 2003a. Cross comparison of EO-1 sensors and other earth resources sensors to Landsat-7 ETM+ using Railroad Valley Playa. *IEEE Transactions on Geoscience and Remote Sensing*, 41(6): 1180–1188.

Thome, K.J., Czapla-Myers, J., and Biggar, S. 2003b. Vicarious calibration of Aqua and Terra MODIS. *Proceedings of SPIE Conference 5151 on Earth Observing Systems VIII*, November 13, San Diego, CA, W.L. Barnes, ed., pp. 395–405. SPIE, Bellingham, WA.

Thome, K.J., Helder, D.L., Aaron, D., and Dewald, J.D. 2004b. Landsat-5 TM and Landsat-7 ETM+ absolute radiometric calibration using the reflectance-based method. *IEEE Transactions on Geoscience and Remote Sensing*, 42(12): 2777–2785.

Thome, K.J., Whittington, E.E., Smith, N., Nandy, P., and Zalewski, E.F. 2000. Ground-reference techniques for the absolute radiometric calibration of MODIS. *Proceedings of SPIE Conference 4135 on Earth Observing Systems V*, November 15, San Diego, CA, W.L. Barnes, ed., pp. 51–59. SPIE, Bellingham, WA.

Tomasi, C., Petkov, B., Benedetti, E., Valenziano, L., Lupi, A., Vitale, V., and Bonafe, U. 2008. A refined calibration procedure of two-channel sun photometers to measure atmospheric precipitable water at various Antarctic sites. *Journal of Atmospheric and Oceanic Technology*, 25(2): 213–229.

Townshend, J.R.G. 1994. Global data sets for land applications from the advanced very high resolution radiometer: An introduction. *International Journal of Remote Sensing*, 15(17): 3319–3332.

Townshend, J.R.G., Justice, C.O., Skole, D., Malingreau, J.-P., Cihlar, J., Teillet, P., Sadowski, F., and Ruttenberg, S. 1994. The 1-km resolution global data set: Needs of the International Geosphere Biosphere Programme. *International Journal of Remote Sensing*, 15(17): 3417–3441.

Trishchenko, A.P., Cihlar, J., and Li, Z. 2002. Effects of spectral response function on surface reflectance and NDVI measured with moderate resolution satellite sensors. *Remote Sensing of Environment*, 81(1): 1–18.

Valorge, C., Meygret, A., Lebègue, L., Henry, P., Bouillon, A., Gachet, R., Breton, E., Léger, D., and Viallefont, F. 2004. 40 Years of experience with SPOT in-flight calibration. *Proceedings of the International Workshop on Radiometric and Geometric Calibration*: ISPRS Book Series Volume 2, *Postlaunch Calibration of Satellite Sensors*, Gulfport, MS, S.A. Morain and A.M. Budge, eds., pp. 119–133. A.A. Balkema Publishers, Leiden, the Netherlands.

Vermote, E. and Kaufman, Y.J. 1995. Absolute calibration of AVHRR visible and near-infrared channels using ocean and cloud views. *International Journal of Remote Sensing*, 16(13): 2317–2340.

Vermote, E.F. and Saleous, N.Z. 2006. Calibration of NOAA16 AVHRR over a desert site using MODIS data. *Remote Sensing of Environment*, 105(3): 214–220.

Vermote, E.F., Santer, R., Deschamps, P.Y., and Herman, M. 1992. In-flight calibration of large field of view sensors at short wavelengths using Rayleigh scattering. *International Journal of Remote Sensing*, 13(18): 3409–3429.

Villa-Aleman, E., Kurzeja, R.J., and Pendergast, M.M. 2003a. Assessment of Ivanpah Playa as a site for thermal vicarious calibration for the MTI satellite. *Proceedings of SPIE Conference 5093 on Algorithms and Technologies for Multispectral, Hyperspectral, and Ultraspectral Imagery IX*, Orlando, FL, S.S. Shen and P.E. Lewis, eds., pp. 331–342. SPIE, Bellingham, WA.

Villa-Aleman, E., Kurzeja, R.J., and Pendergast, M.M. 2003b. Temporal, spatial, and spectral variability at the Ivanpah Playa vicarious calibration site. *Proceedings of SPIE Conference 5093 on Algorithms and Technologies for Multispectral, Hyperspectral, and Ultraspectral Imagery IX*, Orlando, FL, S.S. Shen and P.E. Lewis, eds., pp. 320–330. SPIE, Bellingham, WA.

Wang, Y., Czapla-Myers, J., Lyapustin, A., Thome, K., and Dutton, E. 2011. AERONET-based surface reflectance validation network (ASRVN) data evaluation: Case study for Railroad Valley calibration site. *Remote Sensing of Environment*, 115 (10): 2710–2717.

Warren, S.G., Brandt, R.E., and Hinton, P.O. 1998. Effect of surface roughness on bidirectional reflectance of Antarctic snow. *Journal of Geophysical Research-Planets*, 103(E11): 25789–25807.

Wenny, B.N. and Xiong, X. 2008. Using a cold Earth surface target to characterize long-term stability of the MODIS thermal emissive bands. *IEEE Geoscience and Remote Sensing Letters*, 5(2): 162–165.

Wenny, B.N., Xiong, X., and Dodd, J. 2009. MODIS thermal emissive band calibration stability derived from surface targets. *Proceedings of SPIE Conference 7474 on Sensors, Systems, and Next-Generation Satellites XIII*, Berlin, Germany, R. Meynart, S.P. Neeck, and H. Shimoda, eds., pp. 74740W. SPIE, Bellingham, WA.

Wheeler, R.J., Lecroy, S.R., Whitlock, C.H., Purgold, G.C., and Swanson, J.S. 1994. Surface characteristics for the Alkali Flats and Dunes regions at White-Sands-Missile-Range, New-Mexico. *Remote Sensing of Environment*, 48(2): 181–190.

Whitlock, C.H., Staylor, W.F., Darnell, W.L., Chou, M.-D., Dedieu, G., Deschamps, P.Y., Ellis, J., Gautier, C., Frouin, R., Pinker, R.T., Laslo, I., Rossow, W.B., and Tarpley, D. 1990a. Comparison of surface radiation budget satellite algorithms for downwelled shortwave irradiance with Wisconsin FIRE/SRB surface truth data. *Proceedings of the 7th AMS Conference on Atmospheric Radiation*, San Francisco, CA, pp. 237–242.

Whitlock, C.H., Staylor, W.F., Suttles, J.T., Smith, G., Levin, R., Frouin, R., Gautier, C., Teillet, P.M., Slater, P.N., Kaufman, Y.J., Holben, B.N., Rossow, W.B., and LeCroy, S.R. 1990b. AVHRR and VISSR satellite instrument calibration results for both cirrus and marine stratocumulus IFO periods. *Proceedings of FIRE Science Meeting*, Vail, Colorado. NASA Langley Research Center, Hampton, Virginia, pp. 141–146.

Wu, A., Cao, C., and Xiong, X. 2003. Intercomparison of the 11-and 12- μm bands of Terra and Aqua MODIS using NOAA-17 AVHRR. *Proceedings of SPIE Conference 5151 on Earth Observing Systems VIII*, San Diego, CA, W.L. Barnes, ed., pp. 384–394. SPIE, Bellingham, WA.

Wu, A., Xiong, X., and Cao, C. 2008b. Examination of calibration performance of multiple POS sensors using measurements over the Dome C site in Antarctica. *Proceedings of SPIE Conference 7106 on Sensors, Systems, and Next-Generation Satellites XII*, October 9, Cardiff, Wales, U.K., R. Meynart, S.P. Neeck, H. Shimoda, and S. Habib, eds., pp. 71060W. SPIE, Bellingham, WA.

Wu, A., Xiong, X., and Cao, C. 2008c. Terra and Aqua MODIS intercomparison of three reflective solar bands using AVHRR onboard the NOAA-KLM satellites. *International Journal of Remote Sensing*, 29(7): 1997–2010.

Wu, A., Xiong, X., Cao, C., and Angal, A. 2008a. Monitoring MODIS calibration stability of visible and near-IR bands from observed top-of-atmosphere BRDF-normalized reflectances over Libyan Desert and Antarctic surfaces. *Proceedings of SPIE Conference 7081 on Earth Observing Systems XIII*, San Diego, CA, J.J. Butler and J. Xiong, eds., pp. 708113–708119. SPIE, Bellingham, WA.

Wu, D., Yin, Y., Wang, Z., Gu, X., Verbrugghe, M., and Guyot, G. 1997. Radiometric characterisation of Dunhuang satellite calibration test site (China). *Proceedings of the Seventh International Symposium on Physical Measurements and Signatures in Remote Sensing*, Courchevel, France, G. Guyot and T. Phulpin, eds., pp. 151–160. Taylor & Francis Group, Boca Raton, FL.

Wu, D., Zhu, Y., Wang, Z., Ge, B., and Yin, Y. 1994. The building of radiometric calibration test site for satellite sensors in China. *Proceedings of the Sixth International Symposium on Physical Measurements and Signatures in Remote Sensing*, Val D'Isere, France, pp. 167–171.

Wu, X., Sullivan, J.T., and Heidinger, A.K. 2010. Operational calibration of the advanced very high resolution radiometer (AVHRR) visible and near-infrared channels. *Canadian Journal of Remote Sensing*, 36(5): 602–616.

Wulder, M.A. and Masek, J.G., eds. 2012. Landsat legacy special issue. *Remote Sensing of Environment*, 122, 202 pp.

Xiao, Q., Liu, J., Yu, H., and Zhang, H. 2001. Analysis and evaluation of optical uniformity for Dunhuang calibration site by airborne spectrum survey data. *Proceedings of China Remote Sensing Sensors Radiometric Calibration*, Ocean Press, Beijing, China, pp. 136–142.

Xing-Fa, G., Guyot, G., and Verbrugghe, M. 1990. Evaluation of measurement errors on the reflectance of "La Crau", the French SPOT calibration area. *Proceedings of the 10th EARSeL Symposium on New European Systems, Sensors and Applications*, Toulouse, France, G. Konecny, ed., pp. 121–133. European Association of Remote Sensing Laboratories, Boulogne-Billancourt, France.

Xiong, X., Choi, T., Che, N., Wang, Z., Dodd, J., Xie, Y., and Barnes, W. 2010a. Results and lessons from a decade of Terra MODIS on-orbit spectral characterization. *Proceedings of SPIE Conference 7862, Earth Observing Missions and Sensors: Development, Implementation, and Characterization*, Incheon, Republic of Korea, X. Xiong, C. Kim, and H. Shimoda, eds., 78620M, doi:10.1117/12.868930.

Xiong, X., Che, N., Xie, Y., Moyer, D., Barnes, W., Guenther, B., and Salomonson, V. 2006a. Four-years of on-orbit spectral characterization results for Aqua MODIS reflective solar bands. *Proceedings of SPIE Conference 6361, Sensors, Systems, and Next-Generation Satellites X*, Stockholm, Sweden, R. Meynart, S.P. Neeck, and H. Shimoda, eds., 63610S, doi:10.1117/12.687163.

Xiong, X., Sun, J., and Barnes, W. 2008. Intercomparison of on-orbit calibration consistency between Terra and Aqua MODIS reflective solar bands using the moon. *IEEE Geoscience and Remote Sensing Letters*, 5(4): 778–782.

Xiong, X., Wu, A., Angal, A., and Wenny, B. 2009a. Recent progress on cross-comparison of Terra and Aqua MODIS calibration using Dome C. *Proceedings of SPIE Conference 7474 on Sensors, Systems, and Next-Generation Satellites XIII*, Berlin, Germany, R. Meynart, S.P. Neeck, and H. Shimoda, eds., pp. 747411. SPIE, Bellingham, WA.

Xiong, X., Wu, A., Sun, J., and Wenny, B. 2006b. An overview of intercomparison methodologies for Terra and Aqua MODIS calibration. *Proceedings of SPIE Conference 6296 on Earth Observing Systems XI*, San Diego, CA, J.J. Butler and J. Xiong, eds., pp. 62960C. SPIE, Bellingham, WA.

Xiong, X., Wu, A., Wenny, B., Choi, J., and Angal, A. 2010b. Progress and lessons from MODIS calibration intercomparison using ground test sites. *Canadian Journal of Remote Sensing*, 36(5): 540–552.

Xiong, X.X., Wu, A.S., and Wenny, B.N. 2009b. Using Dome C for moderate resolution imaging spectroradiometer calibration stability and consistency. *Journal of Applied Remote Sensing*, 3(1): 033520.

Zhang, Y., Li, Y., Rong, Z.G., Hu, X.Q., Zhang, L.J., and Liu, J.J. 2009. Field measurement of Gobi surface emissivity spectrum at Dunhuang calibration site of China. *Spectroscopy and Spectral Analysis*, 29(5): 1213–1217.

Zhang, Y., Qiu, K., Hu, X., Rong, Z., and Zhang, L. 2004. Vicarious radiometric calibration of satellite FY-1D sensors at visible and near infrared channels. *Acta Meteorologica Sinica*, 18(4): 505–516.

Zhang, Y., Rong, Z., Hu, X., Liu, J., Zhang, L., Li, Y., and Zhang, X. 2008. Field measurement of Gobi surface emissivity using CE312 and infragold board at Dunhuang calibration site of China. *Proceedings of IEEE Conference on Remote Sensing: The Next Generation*, Boston, MA, pp. 358–360. IEEE, Piscataway, NJ.

Zhang, Y., Zhang, G., Liu, Z., Zhang, L., Zhu, S., Rong, Z., and Qiu, K. 2001. Spectral reflectance measurements at the China radiometric calibration test site for the remote sensing satellite sensor. *Acta Meteorologica Sinica*, 15(3): 377–382.

5

Remote Sensing Data Normalization

Rudiger Gens
University of Alaska, Fairbanks

Jordi Cristóbal Rosselló
University of Alaska, Fairbanks

Acronyms and Definitions

BRDF	Bidirectional reflectance distribution function
CEOS	Committee on Earth Observation Satellites
DN	Digital number
JAXA	Japan Aerospace Exploration Agency
LSE	Land surface emissivity
LST	Land surface temperature
NDVI	Normalized Difference Vegetation Index
SAR	Synthetic aperture radar
TIR	Thermal infrared
TOA	Top of atmosphere

5.1 Introduction

The increasing access to remote sensing data from different platforms, acquired at different spatial, spectral, and temporal resolutions, are continuously widening the scope of applications of these datasets. Parallel advancements in computational science and technology have led to the development of more sophisticated data processing and analysis tools. While early remote sensing studies were centered on detecting a feature or phenomena, the current practice is to carry out multitemporal studies and time series analysis based on multiple data sources, including optical, microwave, and thermal imagery. There are a number of calibration and normalization issues that need to be resolved before these more complex monitoring and change detection studies can be accomplished.

In this chapter, the terms *remote sensing data* and *remote sensing images* are used interchangeably. In order to be able to use data in any combined fashion, each individual dataset needs to have the same reference that allows quantitative comparisons. The most generic term for this is *data normalization* and for remote sensing data, it is known as *radiometric normalization*. Depending on the data source, the processing and necessary corrections differ. Although the terms and definitions used for remote sensing data from different sources might vary, there are two main categories of radiometric normalization (Bao et al., 2012): absolute normalization (methods based on radiative transfer methods that account for atmospheric, illumination, and sensor differences) and relative normalization (techniques that minimize the effects of changing atmospheric and solar conditions in one or a series of images, relative to a standard image). For optical and thermal imagery, *radiometric corrections* need to be applied to account for atmospheric conditions, solar angle, or sensor view angle (Chen et al., 2005; Du et al., 2002), in addition to the sensor prelaunch and postlaunch calibration. These corrections help to convert the raw signal recorded at the sensor to physically meaningful and measurable values, such as ground reflectance or ground temperatures. The quantitative use of synthetic aperture radar (SAR) data requires *calibrated imagery* (Freeman, 1992). Specifically, the SAR processor used for the image generation needs to be calibrated and the calibration parameters then need to be applied to the data to generate calibrated images.

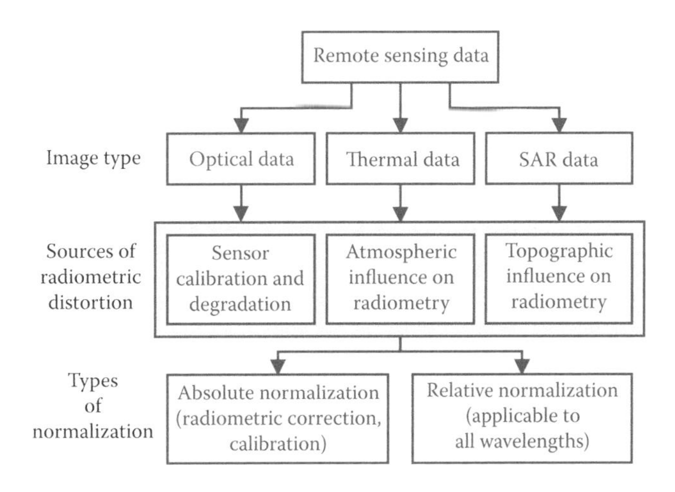

FIGURE 5.1 Remote sensing data normalization overview for various types of imagery.

In combining image datasets, it is assumed that the images are corrected for imaging geometry, that is, optical and thermal imagery have been *orthorectified* (Lillesand et al., 2007) and SAR imagery has been *terrain corrected* (Small, 2011). In the case of SAR data, it should be noted the radiometric terrain correction also corrects the pixel brightness due to geometric distortions (Small, 2011). These geometric corrections are prerequisite for the coregistration of any multitemporal imagery. Image coregistration and radiometric corrections are considered the most important steps in monitoring activities such as change detection (Hussain et al., 2013).

Figure 5.1 summarizes the elements associated with the topic of data normalization covered in this chapter. The chapter focuses on three image types in the optical, thermal, and microwave spectra and the approaches for their absolute and relative normalizations. A number of sources of radiometric distortion such as sensor calibration and degradation as well as the atmospheric and topographic influence on the radiometry are identified that need to be considered. Depending on the wavelength region, absolute normalization may involve radiometric correction or calibration, whereas relative normalization is applicable to all wavelengths.

5.2 Remote Sensing Data

Remote sensing data, as addressed in this chapter, refer to images acquired anywhere from the visible to the microwave region of the electromagnetic spectrum (Figure 5.2).

The visible and infrared regions generally range from 0.4 to 14 μm in wavelength, with visible region occupying the shorter wavelength end from 400 to 700 nm. In remote sensing literature, there is considerable discrepancy and no general consensus on how to classify and name the infrared portion of the spectrum and where to set the boundaries (Gupta, 2003; Quattrochi et al., 2009). The disagreement stems from the fact that the dominant process operating in the shorter wavelength end of the infrared region can either be reflection (at temperatures close to the ambient temperature of Earth), or it can be emission (when the target is at much higher temperatures) with the amount of energy emitted being dependent on the temperature of the target as guided by the Planck's function (Prakash and Gens, 2010; Prakash and Gupta, 1999). For the purpose of data normalization, the logical way to categorize the data is not by the absolute wavelengths, but by the dominant physical processes of reflection and emission. Therefore, we classify the data as optical and thermal, where optical data occupy the shorter wavelength portion and the thermal data occupy the longer wavelength portion of this range.

Optical and thermal remote sensing depend on incoming radiation from the sun and emitted radiation from the Earth's surface, respectively. It does not depend on any external source of energy and, therefore, is also classified as passive remote sensing. Passive remote sensing data can be acquired in broad spectral bands, for example, panchromatic visible imagery, as well as in narrower bands as multispectral (spectral bandwidth in the order of 100 nm) and hyperspectral data (spectral bandwidth in the order of about 10 nm) (Gens, 2009).

SAR data are acquired in the microwave region of the electromagnetic spectrum between 0.75 cm and 1 m in wavelength. As the energy is transmitted as a single frequency from the sensor itself, this technique is classified as active remote sensing. The SAR sensor not only transmits energy, but also records the energy backscattered from the target. SAR data can be acquired day and night and in all weather conditions (Gens, 2009). Unless used for SAR interferometric processing, the imagery is not affected by atmospheric effects and does not need to be corrected for its influence. SAR signals can be transmitted and received with horizontal and vertical polarization. The various combinations of polarizations have different backscatter behaviors and, therefore, provide additional information, complementary to the spectral information. Data

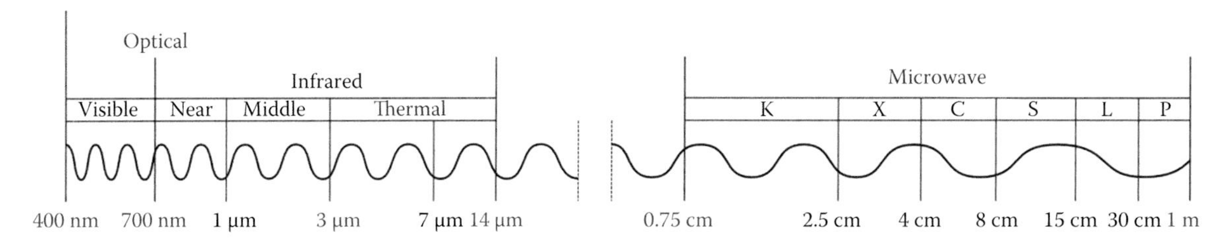

FIGURE 5.2 Optical, thermal, and SAR data and their respective wavelengths in the electromagnetic spectrum. (Adapted from Gens, R., Spectral information content of remote sensing imagery, in Li, D., Shan J., and Gong, J., eds., *Geospatial Technology for Earth Observation*, Springer, New York, 2009.)

acquired in passive mode within the microwave region are not addressed in this chapter. However, readers should refer Chapter 1 to get greater understanding of various sensors and their characteristics.

5.3 Sources of Radiometric Distortion

The main sources of radiometric distortions stem from issues related to sensor calibration and sensor degradation, atmospheric interactions, and influences of topographic variations on image radiometry.

5.3.1 Sensor Calibration and Degradation

The Committee on Earth Observation Satellites (CEOS) defines sensor calibration as the process of quantitatively defining the system response to known, controlled signal inputs. The purpose of calibration activities is to ensure that the user can retrieve as accurate and meaningful quantitative information from the remote sensing images as possible. All sensors and onboard calibration devices undergo a rigorous prelaunch calibration, but need to be routinely recalibrated due to degradation over time (Chander et al., 2009; Wang et al., 2012). Postlaunch calibration activities help to correct the onboard calibrators and to verify that the signal response has not drifted away from the original response for the controlled signal inputs. Any error in sensor calibration will propagate and cause errors in quantitative retrievals from remote sensing data. Excellent reviews of pre- and postlaunch calibration techniques for optical and thermal systems are available in the literature (e.g., Datla et al., 2011; Schott et al., 2012; Xiong et al., 2009)

Typically, the agency responsible for the satellite launch and operation undertakes the calibration tasks and provides calibration parameters, also referred to as calibration constants, as part of the metadata associated with the image data. An end user then applies the correct calibration constants in the respective radiometric correction algorithms to convert the image digital numbers (DNs) to derived physical parameters such as ground reflectances, ground temperatures, etc. For the sake of brevity, a further discussion on optical and sensor calibration is not presented in this chapter. SAR calibration details are presented in Section 5.4.1.2 under absolute radiometric correction.

5.3.2 Atmospheric Influence on Radiometry

The atmosphere plays a huge role in attenuating the signal recorded by the sensor. Atmospheric particles cause selective scattering, absorption, and emission influencing the signal from the target. The atmosphere intervenes twice in optical images: once when the electromagnetic radiations travel from the source (Sun) to the Earth, and the second time when the radiations travel after reflection from the Earth to the sensor. In thermal images, the atmospheric influence comes to play only once as the emitted signal from the Earth travels up to the satellite sensor. The atmosphere is largely transmissive in the microwave region, and for most application purposes, with the exception of applications that rely on SAR interferometric processing; the atmosphere has an insignificant influence on SAR data.

Teasing out the pure signal from the target from a mixed signal response coming from the target and atmosphere then becomes important. Common methods for atmospheric correction in the optical and thermal regions are discussed in Section 5.4.1.1.

5.3.3 Topographic Influence on Radiometry

Surface topographic variations influence the radiometric response of optical, thermal, and SAR data by influencing the source–target–sensor geometry. The influence is most pronounced in high-altitude and high-latitude areas, showing high topographic variations.

5.3.3.1 Effect of Topography on Optical Data

Differences in illumination conditions due to solar position at the moment of image acquisition with respect to surface slope and aspect or elevation can produce similar reflectance responses for similar terrain features (Vanonckelen et al., 2013). Accounting for a topographic correction is then important to calculate surface reflectances accurately, especially in high relief areas (Hantson and Chuvieco, 2011). According to Pons et al. (2014), there are several methodologies that account for topographic effects based on the phenological stage (Hantson and Chuvieco, 2011; Meyer et al., 1993; Riaño et al., 2003; Vincini and Reeder, 2000), the bidirectional reflectance distribution function (BRDF), or the cosine topographic correction model (Teillet et al., 1982). The cosine topographic correction is most convenient for automated radiometric correction procedures. Methodologies accounting for phenological stage require knowledge on the different land cover classes or ground reference information, while BRDF models have limitations, as they at times fail to remove angular effects in spectral bands sensitive to water vapor absorption and caused by large seasonal oscillation (Kim et al., 2012). They also need information that is rarely available and extremely difficult to obtain for regional or long-term studies (Goslee, 2012).

5.3.3.2 Effect of Topography on SAR Data

SAR systems have a side-looking geometry. This viewing geometry leads to geometric distortions, especially in areas with high topography. The information from slopes facing the sensor is compressed, resulting in brighter pixels, while slopes facing away from the sensor in shadow regions, not covered by any signal at all, appear dark.

This geometric distortion can be corrected for using a digital elevation model. While this technique, called terrain correction, shifts the pixels into the correct geolocation, it is not able to completely recover the radiometric information. By calculating the area that was covered by the signal, the correction factors, later multiplied by the geometrically corrected SAR image, can be

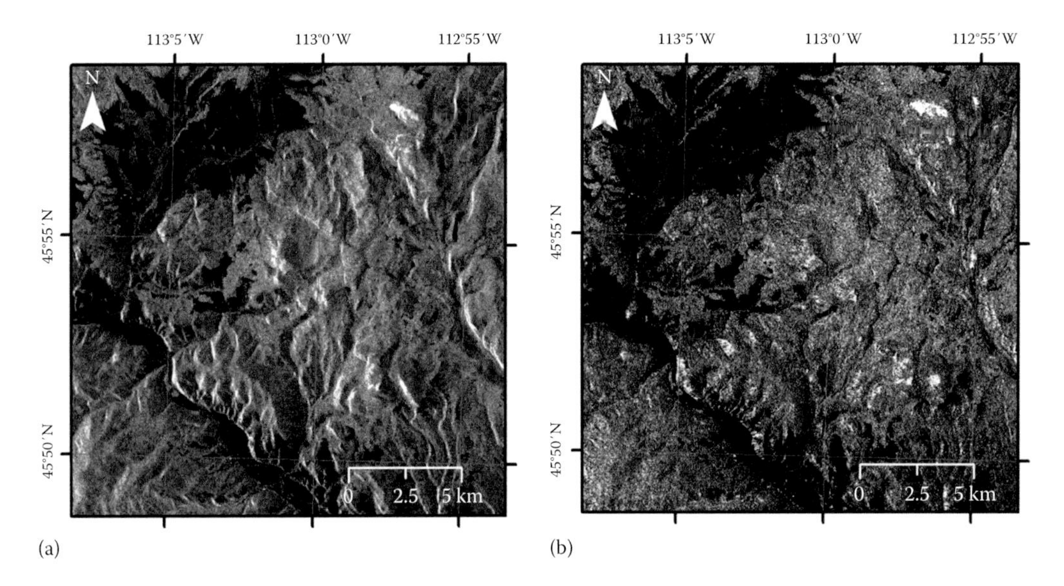

FIGURE 5.3 Example of terrain correction of ALOS PALSAR data in the Rocky Mountains. (a) Uncorrected version, clearly showing the topographic structure of the mountains. In the image on (b), geometric distortions have been removed, so the topography appears flattened. (Imagery copyright Japan Aerospace Exploration Agency (JAXA), Tokyo, Japan, 2006.)

determined (Small, 2011). However, the distribution of the individual scatterers and their contribution to the received backscattered signal within this area remains unknown (Figure 5.3).

5.4 Radiometric Normalization

There are two types of radiometric normalization: absolute and relative. The absolute radiometric normalization, also known as radiometric correction, helps to derive the absolute reflectance of targets at the Earth's surface. The relative radiometric normalization adjusts the radiometric properties of targets within an image to match a reference image (Janzen et al., 2006; Yuan and Elvidge, 1996). Depending on the application needs, an absolute radiometric correction may not be required in all cases. A relative radiometric normalization based on radiometric information in the imagery is sufficient for change detection studies or supervised land cover classifications (Canty et al., 2004). In general, absolute radiometric corrections are required to derive quantitative parameters such as biophysical variables from remote sensing images. For other applications such as studies focused on relative change from one time to another, relative radiometric normalization may be sufficient. The characteristics of the absolute and relative normalizations are summarized in Table 5.1.

5.4.1 Absolute Radiometric Correction

Techniques for absolute radiometric correction use in situ measurements or model data to convert from DN to an absolute scale. Residual effects might still require a relative radiometric normalization.

5.4.1.1 Correction of Optical and Thermal Data

Absolute radiometric correction is a two-step process. In a first step, the sensor-specific calibration parameters (gain and offset), determined usually prior to launch, are applied to convert the DNs (dimensionless) into spectral radiance L_{sat} (in W m^{-2} sr^{-1} μm^{-1}) using the equation (Chen et al., 2005)

$$L_{sat} = \text{DN} \times \text{Gain} + \text{Offset}.$$

L_{sat} can then be converted to top of atmosphere (TOA) reflectances or temperatures. Applying a radiometric correction to

TABLE 5.1 Summary of Absolute and Relative Normalizations for the Various Data Types

Data	Absolute Normalization		Relative Normalization	
	Technique	Effect	Technique	Effect
Optical / Thermal	Radiative transfer methods	Accounts for changes in satellite sensor calibration over time, difference among in-band solar spectral irradiance, solar angle, topography, and atmospheric interferences	Statistical methods using histograms, linear regression, etc.	Minimizes effects of changing atmospheric and solar conditions
SAR	Analysis in homogeneous areas with known backscatter	Adjusts the intensity to reference backscatter level	Statistical methods using information from common area	Preserves dominating scattering mechanism

retrieve surface reflectance or land surface temperature (LST) (kinetic temperature) is needed to be able to analyze long time series of remote sensing data.

Surface reflectance $\rho_{surface}$ can be determined by

$$\rho_{surface} = \frac{\left(L_{sat} - L_{path}\right)\pi}{E\tau},$$

where

L_{path} is the path radiance
E is the exoatmopsheric irradiance on the ground target
τ is the transmission of the atmosphere (Lillesand et al., 2007)

Other methodologies to retrieve surface reflectance can be found in Vicente-Serrano et al. (2008). It is important to note that an absolute method should account for changes in satellite sensor calibration over time, differences among in-band solar spectral irradiance, solar angle, and variability in Earth–Sun distance and atmospheric interferences (de Carvalho et al., 2013). To remove the atmospheric effects, detailed information on atmospheric parameters such as aerosols, water vapor, or ozone is often required. Local atmospheric measurements or reanalysis data are a common source of atmospheric information. However, atmospheric radiosondes are usually not available at the time of satellite pass and a single atmospheric radiosonde might not be representative of the atmospheric conditions of wide-swath satellite images such as the one provided by Landsat, NOAA-AVHRR, or TERRA/AQUA-MODIS sensors, especially in areas with highly variable relief (Cristóbal et al., 2009). AERONET network is another important source of atmospheric data, but as in the case of radiosonde data, ground network distribution might not be wide enough to provide atmospheric parameters over large areas (Themistocleous et al., 2012). Reanalysis data can also provide atmospheric inputs for atmospheric correction; however, its current spatial resolution is still too coarse to be applied to medium or coarse resolution imagery.

An example of atmospheric correction in the thermal infrared (TIR) using a single-channel algorithm proposed by Pons and Solé-Sugrañes (1994) and Pons et al. (2014) using Landsat-5 TM data is shown in Figure 5.4. In remote areas such as Alaska (U.S.), collection of atmospheric data is challenging because of remoteness, winter conditions, and the high costs of maintaining ground-based measurement sensors (Cristóbal et al., 2012). Therefore, there is a real need of methods requiring few atmospheric inputs. The radiometric correction shown in Figure 5.4, based on the dark object subtraction (extracted from the image histogram) and the cosine topographic correction model (computed though a digital elevation model), allows reducing the number of undesired artifacts due to the atmospheric effects or differential illumination that are results of time of day, location on Earth, and relief (zones being more illuminated than others, shadows, etc.), minimizing the effect of these factors on the image data.

In the case of the TIR region, a suitable and popular procedure to retrieve LST is by the inversion of the radiative transfer equation. The following expression can be then applied to a certain sensor channel (or wavelength interval).

$$L_{sensor,\lambda} = \left[\varepsilon_\lambda B_\lambda T_s + \left(1-\varepsilon_\lambda\right) L^{\downarrow}_{atm,\lambda}\right]\tau_\lambda + L^{\uparrow}_{atm,\lambda},$$

where

L_{sensor} is at-sensor radiance (in W m^{-2} sr^{-1} μm^{-1})
ε is the land surface emissivity (dimensionless)
λ is the wavelength (in μm)

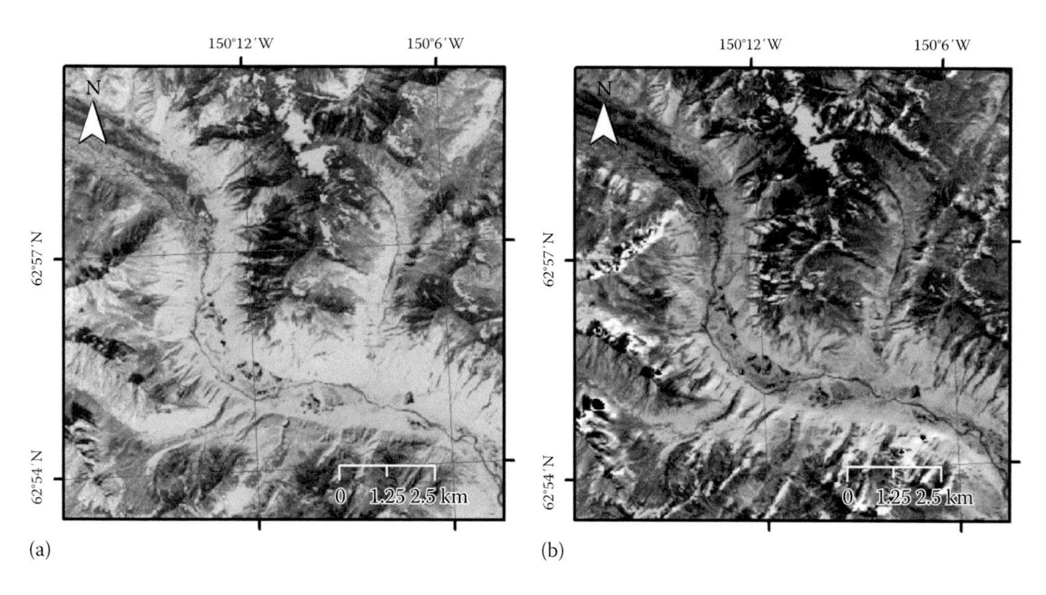

FIGURE 5.4 Example of radiometric correction of optical data in Alaska (US) from a Landsat-5 TM image of 07 July 2009 (standard false color composite). (a) Uncorrected version, clearly showing the topographic structure of the mountains. (b) Radiometrically corrected image where topographic and atmospheric effects have been removed. For example, areas with a similar land cover that occupy different sides of a hilly terrain show self-cast shadows on the left panel but present similar reflectivity in the (b). (Images are courtesy of the U.S. Geological Survey.)

T_s is the LST (in K)

$L_{atm,}\lambda \downarrow$ is the downwelling atmospheric radiance (hemispherical flux divided by pi and in W m^{-2} sr^{-1} μm^{-1})

$L_{atm,}\lambda \uparrow$ is the upwelling atmospheric radiance (path radiance at λ wavelength and in W m^{-2} sr^{-1} μm^{-1})

τ is the atmospheric transmissivity (dimensionless)

B term is Planck's law, expressed as follows:

$$B_\lambda(T_s) = \frac{2\pi hc^2}{\lambda^5\left[\left(\frac{hc}{(k\lambda T_s)}\right) - 1\right]},$$

where

c is the speed of light (2.998 × 10^{8} m s^{-1})

h is the Planck's constant (6.626076 × 10^{-34} Js)

k is the Boltzmann constant (1.3806 × 10^{-23} J K^{-1})

Usually, LST retrieval methods are developed based on the available thermal sensor bands and can be classified in single-channel, split-window or temperature and emissivity separation algorithms. A single-channel algorithm is applied to thermal sensors with only one band in the TIR (Cristóbal et al., 2009; Jiménez-Muñoz and Sobrino, 2003), such as Landsat-5 or Landsat-7 missions. In this case, when two thermal bands are available, such as in NOAA-AVHRR or Landsat-8 TIRS missions, split-window algorithms can be applied (Jimenez-Muñoz et al., 2014; Wan and Dozier, 1996). The temperature and emissivity separation algorithm can be applied when five or more bands are available (Gillespie et al., 1998) as in the ASTER mission. Single-channel and split-window methods also require the knowledge of land surface emissivity as well as water vapor. Land surface emissivity estimates for bare soil and vegetation covers can be retrieved for operational processing using methods based on the Normalized Difference Vegetation Index (NDVI) (Sobrino and Raissouni, 2000; Sobrino et al., 2008; Valor and Caselles, 2005). Water vapor can be obtained by means of a local radiosounding or at regional scale by means of remote sensing data (Sobrino et al., 1999), or via image-based water vapor products such as the TERRA/AQUA MODIS product (MOD/MYD_05).

An example of atmospheric correction in the TIR using a single-channel algorithm proposed by Cristóbal et al. (2009) using Landsat-5 TM data, water vapor estimates from the MODIS water vapor product and emissivity retrieved using Sobrino and Raissouni (2000) are shown in Figure 5.5b (Figure 5.5a shows a radiometric correction of optical bands to assist the interpretation of the LST image). This method works well in areas where only remote sensing estimates of water vapor products and LSE are available as input for LST estimation. The magnitude to which brightness temperature (TOA temperatures not corrected by atmospheric or emissivity effects)

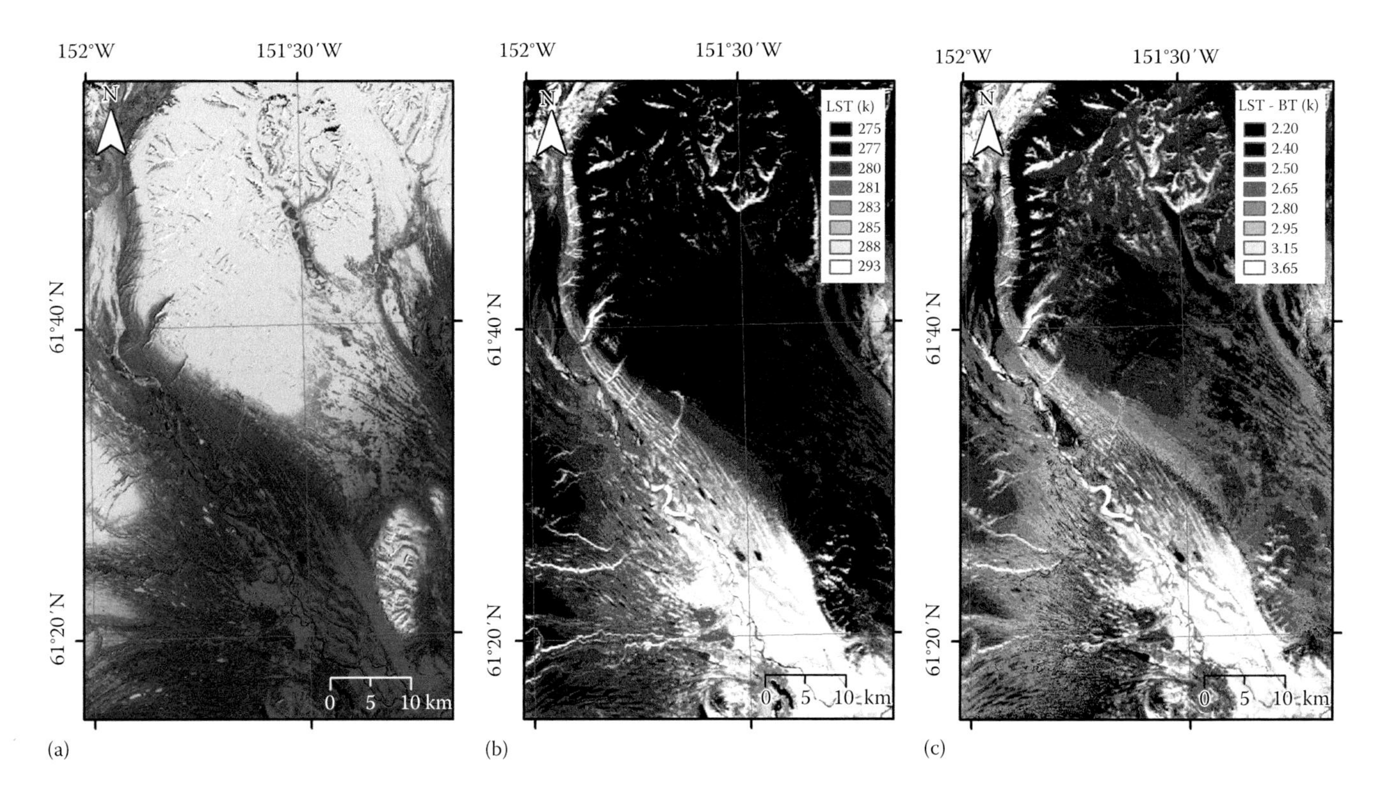

FIGURE 5.5 Example of atmospheric correction of thermal data in Alaska (U.S.) from a Landsat-5 TM image of 04 May 2009 (standard false color composite). (a) Optical image after radiometric correction. (b) LST derived after correcting for atmospheric effects and emissivity. (c) Difference image generated by subtracting the LST corrected for atmospheric effects and emissivity and the brightness temperature, BT (not corrected for emissivity or atmospheric effects). (Images are courtesy of the U.S. Geological Survey, Reston, VA.)

are corrected is shown in Figure 5.5a. This magnitude difference is computed by subtracting the LST corrected by atmospheric and emissivity effects (mid panel) and the brightness temperature. In this case, removing atmospheric and emissivity effects lead to a correction around 2 and 4 K: brightness temperature is clearly underestimating LST, and absolute radiometric correction methodologies are needed for accurate LST retrievals.

5.4.1.2 SAR Calibration

In order to calibrate an SAR image, the SAR processor needs to be calibrated, that is, calibration coefficients have been determined. Applying these coefficients to the SAR data calibrates the SAR image. The DNs of the original image get converted into power scale, a ratio of the power that is backscattered to the power sent within a resolution cell. Calibrated images are often transformed from power scale into the logarithmic dB scale.

In order to derive geophysical parameters, the calibration needs to meet certain requirements. The absolute calibration needs to be ±1 dB, the long-term relative calibration ±0.5 dB, the short-term relative calibration better than 0.5 dB (Freeman, 1992).

Another technique that can provide radiometric calibration uses so-called permanent scatterers. These are natural targets in a stack of SAR images that are stable over time and act as corner reflectors with an unknown radar cross section and with a quality that can be estimated by the repeated observations (D'Aria et al., 2010). These are also exploited in permanent scatterer interferometry that allows long-term time series analysis (Ferretti et al., 2001).

5.4.2 Relative Radiometric Correction

A relative correction transforms the DNs to a common scale, adjusting the radiometric properties of an image to match a reference image (deCarvalho et al., 2013).

5.4.2.1 Correction of Optical and Thermal Data

The goal of image normalization for the relative radiometric correction of optical and thermal data is to reduce the radiometric influence of nonsurface factors, so that the differences in DN between satellite images from different dates will reflect actual changes on the surface of the Earth (Heo and Fitz-Hugh, 2000). Several techniques for relative radiometric correction have been developed in the last decades such as robust and linear regression (El Hajj et al., 2008; Olsson, 1993; Wessman, 1987), histogram matching (Chavez and MacKinnon, 1994; Liang, 2002), the use of invariant areas (Eckhardt et al., 1990; Jensen et al., 1995; Michener and Houhoulis, 1997), the use of pseudoinvariant areas (Bao et al., 2012; Pons et al., 2014; Schott et al., 1988; Zhou et al., 2011), and Gaussian method (Singh, 1989), among others. Linear regression, invariant and pseudoinvariant, methods also require an appropriate selection of stable or quasistable radiometric areas, areas that can be selected using methods such as the multivariate alteration detection (Nielsen et al., 1998; Scheidt et al., 2008). Although there is currently no consensus on what method is more suitable, more objective and automatic methods with less manual intervention and that take advantage of long time series of remote sensing data, such as those based on the pseudo-invariant areas, are preferred for relative radiometric normalization of optical and thermal imagery.

5.4.2.2 Relative Correction of SAR Data

Most often absolute radiometric correction of SAR data is sufficient for analysis. Sometimes a relative radiometric normalization is merited, especially when adjacent scenes have significant seasonal differences. The normalization of SAR polarimetric data is slightly more complicated, as the correction is not supposed to change the scattering mechanism. For relative radiometric normalization, Shimada and Ohtaki (2010) used a polygonal curve approximation to suppress differences in intensity between neighboring image strips. Antropov et al. (2012) extended the relative correlation approach of Shimada and Ohtaki by using the span of the covariance matrix to calculate the corrective gain. There are various ways to apply the radiometric correction in this case. Lee et al. (2004) used only those pixels from the overlapping areas, where a dominating scattering mechanism is preserved. However, for several applications where the purpose is to generate a thematically classified product based on relative clustering of backscatter values within an image, simpler radiometric normalization techniques, such as color balance or histogram matching, used popularly in optical remote sensing can prove to be equally efficient.

Figure 5.6a shows an example mosaic from the northern foothills of Alaska Range generated using two adjacent SAR polarimetric images from different seasons. The images were terrain corrected prior to generating the mosaic. A general brightness contrast is visible across the image boundary (the red dashed line in Figure 5.6b). Note that the boundary is not straight, due to the effect of terrain correction. Relative radiometric normalization using a histogram match of the overlapping area was applied to the terrain corrected image pair. The mosaic generated using these normalized images yielded a superior product (5.6c) for mapping and classification applications.

5.5 Conclusions

In general, absolute radiometric correction methods are required to derive quantitative parameters such as biophysical variables from remote sensing images. However, for other applications such as studies focused on relative change from one time to another, relative radiometric normalization may be sufficient. In the case of absolute radiometric correction methods, removing atmospheric effects often requires detailed information on atmospheric parameters such as aerosols, water vapor, or ozone that are not often available at regional scales. Therefore, absolute radiometric correction methods that minimize input variables are preferred for optical and thermal imagery.

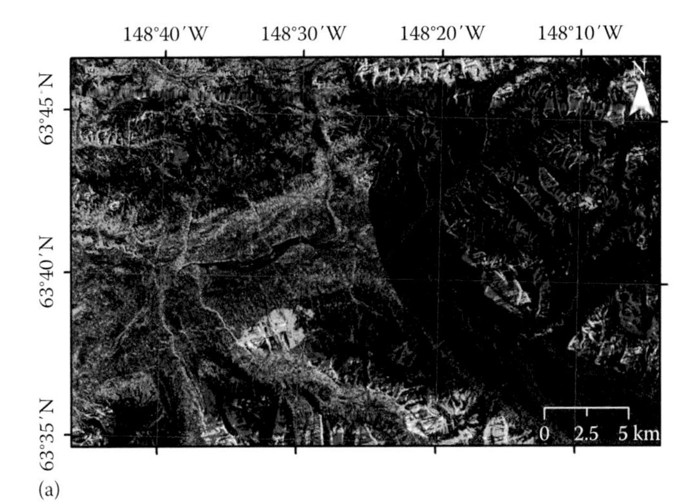

(a)

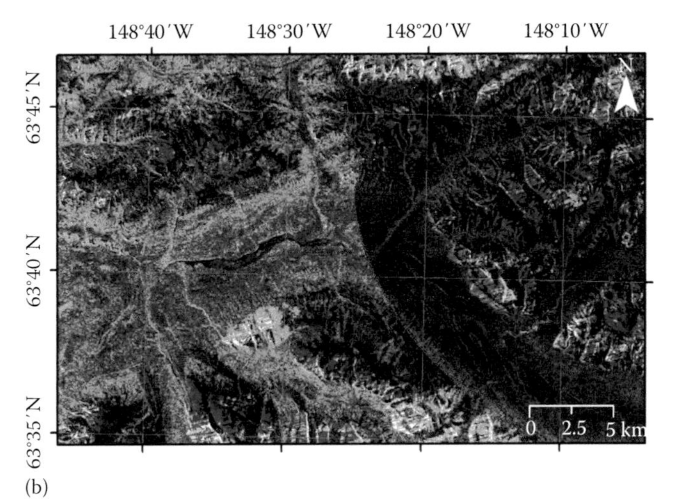

(b)

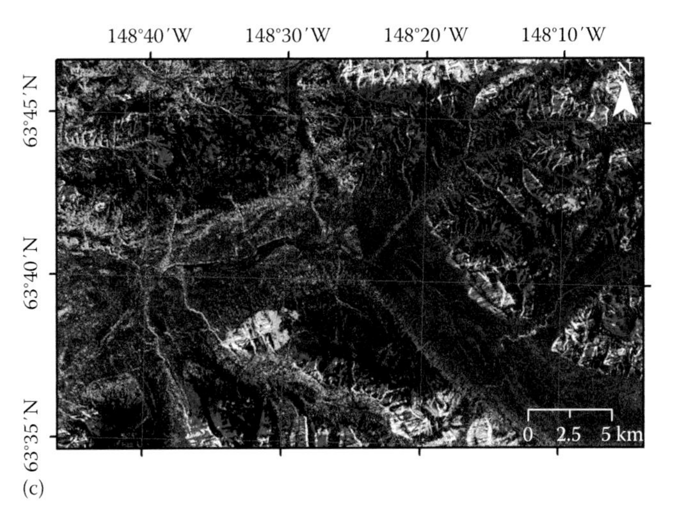

(c)

FIGURE 5.6 Relative correction of adjacent polarimetric ALOS PALSAR images. The Yamaguchi decompositions have been terrain corrected. (a) Uncorrected mosaic. (b) Boundary between the two images that are indicated (not a straight line because of the terrain correction). (c) Relative corrected mosaic. (Imagery copyright Japan Aerospace Exploration Agency (JAXA), Tokyo, Japan, 2007.)

For their study on the radiometric cross-calibration of Landsat sensors, Teillet et al. (2001) concluded that the most limiting factor of their approach is the need to adjust for spectral band differences between the two sensors, introducing a dependency on knowledge mainly about the surface reflectance spectrum of the scene. Particularly, the cross-calibration of the shortwave infrared bands remains an issue without these spectra (Teillet et al., 2001).

References

Antropov, O., Rauste, Y., Lonnqvist, A., and Hame, T. 2012. PolSAR mosaic normalization for improved land-cover mapping. *IEEE Geoscience and Remote Sensing Letters*, 9, 1074–1078.

Bao, N. S., Lechner, A. M., Fletcher, A., Mellor, A., Mulligan, D., and Bai, Z. K. 2012. Comparison of relative radiometric normalization methods using pseudo-invariant features for change detection studies in rural and urban landscapes. *Journal of Applied Remote Sensing*, 6, 063578-063571-063578-063518.

Canty, M. J., Nielsen, A. A., and Schmidt, M. 2004. Automatic radiometric normalization of multitemporal satellite imagery. *Remote Sensing of Environment*, 91, 441–451.

Chander, G., Markham, B. L., and Helder, D. L. 2009. Summary of current radiometric calibration coefficients for Landsat MSS, TM ETM+ and EO-1 ALI sensors. *Remote Sensing of Environment*, 113, 893–903.

Chavez, P. S. and MacKinnon, D. L. 1994. Automatic detection of vegetation changes in the Southwestern United States Using Remotely Sensed Images. *Photogrammetric Engineering and Remote Sensing*, 60, 571–583.

Chen, X. X., Vierling, L., and Deering, D. 2005. A simple and effective radiometric correction method to improve landscape change detection across sensors and across time. *Remote Sensing of Environment*, 98, 63–79.

Cristóbal, J., Jiménez-Muñoz, J. C., Sobrino, J. A., Ninyerola, M., and Pons, X. 2009. Improvements in land surface temperature retrieval from the Landsat series thermal band using water vapor and air temperature. *Journal of Geophysical Research—Atmospheres*, 114, D08103, doi: 10.1029/2008JD010616.

Cristóbal, J., Prakash, A., Starkenburg, D., Fochesatto, J., Anderson, M. A., Kustas, W. P., Alfieri, J. G., Gens, R., and Kane, D. 2012. Energy fluxes retrieval on an Alaskan Arctic and Sub-Arctic vegetation by means MODIS imagery and the DTD method. *AGU Fall Meeting*, San Francisco, CA, December 3–7, 2012.

D'Aria, D., Ferretti, A., Guarnieri, A. M., and Tebaldini, S. 2010. SAR calibration aided by permanent scatterers. *IEEE Transactions on Geoscience and Remote Sensing*, 48, 2076–2086.

Datla, R. U., Rice, J. P., Lykke, K. R., Johnson, B. C., Butler, J. J., and Xiong, X. 2011. Best practice guidelines for pre-launch characterization and calibration of instruments for passive optical remote sensing. *Journal of Research of the National Institute of Standards and Technology*, 116, 621–646.

de Carvalho, O. A., Guimaraes, R. F., Silva, N. C., Gillespie, A. R., Gomes, R. A. T., Silva, C. R., and De Carvalho, A. P. F. 2013. Radiometric normalization of temporal images combining automatic detection of pseudo-invariant features from the distance and similarity spectral measures, density scatterplot analysis, and robust regression. *Remote Sensing*, 5, 2763–2794.

Du, Y., Teillet, P. M., and Cihlar, J. 2002. Radiometric normalization of multitemporal high-resolution satellite images with quality control for land cover change detection. *Remote Sensing of Environment*, 82, 123–134.

Eckhardt, D. W., Verdin, J. P., and Lyford, G. R. 1990. Automated update of an irrigated lands GIS using SPOT HRV imagery. *Photogrammmetric Engineering and Remote Sensing*, 56, 1515–1522.

El Hajj, M., Bégué, A., Lafrance, B., Hagolle, O., Dedieu, G., and Rumeau, M. 2008. Relative radiometric normalization and atmospheric correction of a SPOT 5 time series. *Sensors*, 8, 2774–2791.

Ferretti, A., Prati, C., and Rocca, F. 2001. Permanent scatterers in SAR interferometry. *IEEE Transactions on Geoscience and Remote Sensing*, 39, 8–20.

Freeman, A. 1992. SAR calibration—An overview. *IEEE Transactions on Geoscience and Remote Sensing*, 30, 1107–1121.

Gens, R. 2009. Spectral information content of remote sensing imagery. In: Li, D., Shan, J., and Gong, J. (eds.) *Geospatial Technology for Earth Observation*. Springer, New York.

Gillespie, A., Rokugawa, S., Matsunaga, T., Cothern, J. S., Hook, S., and Kahle, A. B. 1998. A temperature and emissivity separation algorithm for advanced spaceborne thermal emission and reflection radiometer (ASTER) images. *IEEE Transactions on Geoscience and Remote Sensing*, 36, 1113–1126.

Goslee S. C. 2012. Topographic corrections of satellite data for regional monitoring. *Photogrammetric Engineering and Remote Sensing*, 78, 973–981.

Gupta, R. P. 2003. *Remote Sensing Geology*, 2nd edn. Springer, Berlin/Heidelberg, 656 pp., ISBN-13: 978-3540431855.

Hantson, S. and Chuvieco, E. 2011. Evaluation of different topographic correction methods for Landsat imagery. *International Journal of Applied Earth Observation and Geoinformation*, 13, 691–700.

Heo, J. and FitzHugh, T. W. 2000. A standardized radiometric normalization method for change detection using remotely sensed imagery. *Photogrammetric Engineering an Remote Sensing*, 66, 173–181.

Hussain, M., Chen, D. M., Cheng, A., Wei, H., and Stanley, D. 2013. Change detection from remotely sensed images: From pixel-based to object-based approaches. *ISPRS Journal of Photogrammetry and Remote Sensing*, 80, 91–106.

Janzen, D. T., Fredeen, A. L., and Wheate, R. D. 2006. Radiometric correction techniques and accuracy assessment for Landsat TM data in remote forested regions. *Canadian Journal of Remote Sensing*, 32, 330–340.

Jensen, J. R., Rutchey, K., Koch, M. S., and Narumalani, S. 1995. Inland wetland change detection in the Everglades Water Conservation Area 2A using a time series of normalized remotely sensed data. *Photogrammetric Engineering and Remote Sensing*, 61, 199–209.

Jiménez-Muñoz, J. C. and Sobrino, J. A. 2003. A generalized single-channel method for retrieving land surface temperature from remote sensing data. *Journal of Geophysical Research—Atmospheres*, 108, 4688, doi: 10.1029/2003JD003480.

Jimenez-Munoz, J. C., Sobrino, J. A., Skokovic, D., Mattar, C., and Cristobal, J. 2014. Land surface temperature retrieval methods from Landsat-8 thermal infrared sensor data. *IEEE Geoscience and Remote Sensing Letters*, 11, 1840–1843.

Kim, D. S., Pyeon, M. W., Eo, Y. D., Byun, Y. G., and Kim, Y. I. 2012. Automatic pseudo-invariant feature extraction for the relative radiometric normalization of Hyperion hyperspectral images. *Giscience and Remote Sensing*, 49, 755–773.

Lee, J. S., Grunes, M. R., Pottier, E., and Ferro-Famil, L. 2004. Unsupervised terrain classification preserving polarimetric scattering characteristics. *IEEE Transactions on Geoscience and Remote Sensing*, 42, 722–731.

Liang, S. 2002. Estimation of land surface biophysical variables. In: Kong, J. A. (ed.) *Quantitative Remote Sensing of Land Surfaces*. Wiley, Hoboken, NJ, pp. 247–264.

Lillesand, T., Kiefer, R. W., and Chipman, A. 2007. *Remote Sensing and Image Interpretation*. Wiley, Hoboken, NJ.

Meyer, P., Itten, K. I., Kellenbenberger, T., Sandmeier, S., and Sandmeier, R. 1993. Radiometric corrections of topographically induced effects on Landsat TM data in an alpine environment. *ISPRS Journal of Photogrammetry and Remote Sensing*, 48, 17–28.

Michener, W. K. and Houhoulis, P. F. 1997. Detection of vegetation changes associated with extensive flooding in a forested ecosystem. *Photogrammetric Engineering and Remote Sensing*, 63, 173–181.

Nielsen, A. A., Conradsen, K., and Simpson, J. J. 1998. Multivariate alteration detection (MAD) and MAF postprocessing in multispectral, bitemporal image data: New approaches to change detection studies. *Remote Sensing of Environment*, 64, 1–19.

Olsson, H. 1993. Regression functions for multi-temporal relative calibration of thematic mapper data over Boreal forest. *Remote Sensing of Environment*, 46, 89–102.

Pons, X., Pesquer, L., Cristóbal, J., and González-Guerrero, O. 2014. Automatic and improved radiometric correction of Landsat imagery using reference values from MODIS surface reflectance images. *International Journal of Applied Earth Observation and Geoinformation*, 33, 243–254.

Pons, X. and Solé-Sugrañes, L. 1994. A simple radiometric correction model to improve automatic mapping of vegetation from multispectral satellite data. *Remote Sensing of Environment*, 48, 191–204.

Prakash, A. and Gens, R. 2010. Remote sensing of coal fires. In: Stracher, G. B., Prakash, A., Sokol, E. V. (eds.) *Coal and Peat Fires: A Global Perspective, Vol. 1: Coal—Combustion and Geology*. Elsevier, Oxford, U.K.

Prakash, A. and Gupta, R. P. 1999. Surface fires in Jharia Coalfield, India—Their distribution and estimation of area and temperature from TM data. *International Journal of Remote Sensing*, 20, 1935–1946.

Quattrochi, D. A., Prakash, A., Evena, M., Wright, R., Hall, D. K., Anderson, M., Kustas, W. P., Allen, R. G., Pagano, T., and Coolbaugh, M. F. 2009. Thermal remote sensing: Theory, sensors, and applications. In: Jackson, M. (ed.) *Manual of Remote Sensing 1.1: Earth Observing Platforms & Sensors*. ASPRS, Bethesda, MD, 550 pp.

Riaño, D., Chuvieco, E., Salas, J., and Aguado, I. 2003. Assessment of different topographic corrections in Landsat-TM data for mapping vegetation types. *IEEE Transactions on Geoscience and Remote Sensing*, 41, 1056–1061.

Scheidt, S., Ramsey, M., and Lancaster, N. 2008. Radiometric normalization and image mosaic generation of ASTER thermal infrared data: An application to extensive sand sheets and dune fields. *Remote Sensing of Environment*, 112, 920–933.

Schott, J. R., Hook, S. J., Barsi, J. A., Markham, B. L., Miller, J., Padula, F. P., and Raqueno, N. G. 2012. Thermal infrared radiometric calibration of the entire Landsat 4, 5, and 7 archive (1982–2010). *Remote Sensing of Environment*, 122, 41–49.

Schott, J. R., Salvaggio, C., and Vochok, W. J. 1988. Radiometric scene normalization using pseudo-invariant features. *Remote Sensing of Environment*, 26, 1–16.

Shimada, M. and Ohtaki, T. 2010. Generating large-scale high-quality SAR mosaic datasets: Application to PALSAR data for global monitoring. *IEEE Journal of Selected Topics in Applied Earth Observations and Remote Sensing*, 3, 637–656.

Singh, A. 1989. Digital change detection techniques using remotely sensed data. *International Journal of Remote Sensing*, 10, 989–1103.

Small, D. 2011. Flattening Gamma: Radiometric terrain correction for SAR imagery. *IEEE Transactions on Geoscience and Remote Sensing*, 49, 3081–3093.

Sobrino, J. A., Jiménez-Muñoz, J. C., Sòria, G., Romaguera, M., Guanter, L., Moreno, J., Plaza, A., and Martínez, P. 2008. Land surface emissivity retrieval from different VNIR and TIR sensors. *IEEE Transactions on Geoscience and Remote Sensing*, 46, 316–327.

Sobrino, J. A. and Raissouni, N. 2000. Toward remote sensing methods for land cover dynamic monitoring: Application to Morocco. *International Journal of Remote Sensing*, 21, 353–366.

Sobrino, J. A., Raissouni, N., Simarro, J., Nerry, F., and François, P. 1999. Atmospheric water vapour content over land surfaces derived from the AVHRR data. Application to the Iberian Peninsula. *IEEE Transactions and Geoscience and Remote Sensing*, 37, 1425–1434.

Teillet, P. M., Barker, J. L., Markham, B. L., Irish, R. R., Fedosjevs, G., and Storey, J. C. 2001. Radiometric cross-calibration of the Landsat-7 ETM+ and Landsat-5 TM sensors based on tandem data sets. *Remote Sensing of Environment*, 78, 39–54.

Teillet, P. M., Guindon, B., and Goodeonugh, D. G. 1982. On the slope-aspect correction of multispectral scanner data. *Canadian Journal of Remote Sensing*, 8, 84–106.

Themistocleous, K., Hadjimitsis, D. G., Retalis, A., and Chrysoulakis, N. 2012. Development of a new image based atmospheric correction algorithm for aerosol optical thickness retrieval using the darkest pixel method. *Journal of Applied Remote Sensing*, 6, 063538.

Valor, E. and Caselles, V. 2005. Validation of the vegetation cover method for land surface emissivity estimation. In: Caselles, V., Valor, E., and Coll, C. (eds.) *Recent Research Developments in Thermal Remote Sensing*. Research Signpost, Kerala, India, pp. 1–20.

Vanonckelen, S., Lhermitte, S., and Van Rompaey, A. 2013. The effect of atmospheric and topographic correction methods on land cover classification accuracy. *International Journal of Applied Earth Observation and Geoinformation*, 24, 9–21.

Vicente-Serrano, S. M., Pérez-Cabello, F., and Lasanta, T. 2008. Assessment of radiometric correction techniques in analyzing vegetation variability and change using time series of Landsat images. *Remote Sensing of Environment*, 112, 3916–3934.

Vincini, M. and Reeder, D. 2000. Minnaert topographic normalization of Landsat TM imagery in rugged forest areas. International Archives of Photogrammetry and Remote Sensing, Vol. XXXIII, Part B7, Amsterdam, the Netherlands.

Wan, Z. and Dozier, J. 1996. A generalized split-window algorithm for retrieving land-surface temperature from space. *IEEE Transactions on Geoscience and Remote Sensing*, 34, 892–905.

Wang, D. D., Morton, D., Masek, J., Wu, A. S., Nagol, J., Xiong, X. X., Levy, R., Vermote, E., and Wolfe, R. 2012. Impact of sensor degradation on the MODIS NDVI time series. *Remote Sensing of Environment*, 119, 55–61.

Wessman, C. A. 1987. Analysis of Landsat Thematic Mapper imagery over UW arboretum and Blackhawk Island. PhD Dissertation, University of Wisconsin-Madison, Madison, WI.

Xiong, X. X., Wenny, B. N., and Barnes, W. L. 2009. Overview of NASA Earth Observing Systems Terra and Aqua moderate resolution imaging spectroradiometer instrument calibration algorithms and on-orbit performance. *Journal of Applied Remote Sensing*, 3, 032501.

Yuan, D. and Elvidge, C. D. 1996. Comparison of relative radiometric normalization techniques. *ISPRS Journal of Photogrammetry and Remote Sensing*, 51, 117–126.

Zhou, Q., Li, B., and Chen, Y. 2011. Remote sensing change detection and process analysis of long-term land use change and human impacts. *AMBIO*, 40, 807–818.

6

Satellite Data Degradations and Their Impacts on High-Level Products

Dongdong Wang
University of Maryland

Acronyms and Definitions

AOD	Aerosol optical depth
AVHRR	Advanced very-high-resolution radiometer
BRDF	Bidirectional reflectance distribution function
ECT	Equator crossing time
EM	Electromagnetic
ETM+	Enhanced thematic mapper plus
MODIS	Moderate Resolution Imaging Spectroradiometer
MS	Mirror side
NDVI	Normalized difference vegetation index
NIR	Near infrared
NOAA	National Oceanic and Atmospheric Administration
RSB	Reflective solar bands
RVS	Response versus scan angle
SD	Solar diffuser
SDSM	Solar diffuser stability monitor
SPOT	Satellite pour l'Observation de la Terre
SZA	Solar zenith angle
TIROS	Television Infrared Observation Satellite Program
TM	Thematic mapper
TOA	Top of atmosphere

6.1 Introduction

It has been half a century since the launch of the first-generation Earth observation satellites, such as former Soviet Union's Sputnik 1 and the United States' Television Infrared Observation Satellite Program (TIROS)-1 (Tatem et al., 2008). Compared with traditional observation approaches, one important advantage of satellite remote sensing is its global mapping capability. Satellite data have thus become an unprecedented source to study global environmental changes (e.g., Kaufman et al., 2002; Nemani et al., 2003). Thanks to the continuity of satellite missions, the length of remote sensing data record is continuously increasing. Currently, we have 30+ years archive of the National Oceanic and Atmospheric Administration (NOAA) advanced very-high-resolution radiometer (AVHRR) at spatial resolution suitable for continental or global studies (Beck et al., 2011) and Landsat TM/ETM+ data at ecosystem scales (Markham and Helder, 2012). Entering the new century, data records from new-generation remote sensors such as the Satellite Pour l'Observation de la Terre (SPOT) VEGETATION are also lengthening. More than one decade of Moderate Resolution Imaging Spectroradiometer (MODIS) data become available (2000–present) and serve as valuable resource for the atmospheric (King et al., 2003), terrestrial (Justice et al., 2002), and oceanic (Esaias et al., 1998) research community with well-designed spectral configuration and improved radiometric and geometric accuracy (Justice et al., 1998).

With progress in sensor design and manufacturing techniques, reliability of radiometric calibration has improved substantially. Detection of subtle trends in environmental variables is extremely sensitive to calibration drifts of satellite sensors (Datla et al., 2004). Errors in geometric coregistration will also affect the analysis of multitemporal data and thus detection of changes and trends (Townshend et al., 1992). Stable, consistent, and reliable time series data with extended temporal coverage and accurate geolocation information are greatly needed to study interannual variability and long-term trends of global environments.

Very-high-level efforts have been devoted to assure the absolute and relative accuracy of radiative quantities measured by remote sensors. Particularly, various calibration approaches (e.g., prelaunch, onboard, vicarious, cross-platform) are employed to convert sensors' signature (e.g., digital number) to accurate values of energy emitted and scattered by the Earth system (Dinguirard and Slater, 1999). Before launch of satellites, sensors' characteristics of spectral, radiometric, and spatial responses are measured in labs at various levels to produce prelaunch calibration metrics. The actual calibration coefficients hardly remain constant because of the impacts from the launching process and the sensor degradation during the operation in space (Mekler and Kaufman, 1995). To monitor sensitivity changes of sensors in radiometric response, many satellite sensors are equipped with onboard calibration units, such as internal lamp, solar diffuser (SD), and blackbody. For sensors without onboard calibrators, observations over pseudo-invariant targets are typically used to adjust calibration coefficients so that sensor degradation can be taken into account. Such technique of vicarious calibration is also applied to sensors with onboard calibration devices as an additional source to verify performance of satellite detectors and their calibration systems (Wu et al., 2008).

Despite the various efforts, it is still a challenge to maintain the accuracy and precision of radiometric calibration due to various reasons discussed here. The harsh space environment where satellite sensors operate makes it extremely hard to keep the onboard calibration devices function ideally as designed. For example, the performance of internal illuminating sources and diffusers may degrade over time as well (Helder et al., 1998). Vicarious calibration is independent on degradation of onboard instruments, but it has its own uncertainties and limitations. Stable atmospheric and surface conditions are usually assumed at the calibration sites, although both tend to change. In addition, undetected cloud or cloud shadow and effects of bidirectional reflectance distribution function (BRDF) are among other major sources of uncertainty. As a result, satellite data records are prone to contain time-dependent radiometric drifts of some levels even after such sophisticated calibration efforts have been made. Substantial degradations have been found to exist in both sensors without onboard calibrators such as AVHRR (Wu and Zhong, 1994) and those with onboard calibration devices as Landsat/TM (Helder et al., 1998) and Terra/MODIS (Wang et al., 2012).

Degradation in satellite signature (i.e., top-of-atmosphere [TOA] observations) will eventually translate into errors and bias in high-level products through the satellite data processing chain, if the issue of data degradation is not well addressed in the design of retrieval algorithms. Due to the time-dependence nature of data degradations, their impacts will be especially evident in analysis of time series data, such as trend detection and investigation of interannual variability. In such cases, subtle changes of environmental parameters of interest will be obscured by artificial trends embedded in the time series of degraded data. For instance, analysis of four AVHRR Normalized Difference Vegetation Index (NDVI) datasets revealed inconsistent trends over Europe, Africa, and the Sahel (Beck et al., 2011). Difference in handling sensor degradation can be an important factor of such discrepancies. Comparison of NDVI trends derived from Aqua/MODIS and Terra/MODIS showed contradictory results over boreal North America (Wang et al., 2012). Further data analysis suggested this be caused by the degradation issue in Terra/MODIS Collection 5 data. Similarly, inconsistent trends were observed in aerosol (Levy et al., 2010) and ocean color (Djavidnia et al., 2010) when comparing products derived from Aqua/MODIS and Terra/MODIS.

In addition to radiometric responses, other characteristics of satellites and sensors may also change with time. For example, orbits of sun-synchronous satellites may drift and cause shift in solar zenith angle (SZA) (Ignatov et al., 2004). This can lead to artifact drifts in time series of some satellite products, similar to effects of sensor degradation. Thus, we here treat orbit drift as one type of data degradations as well. The issues of sensor degradation exist in not only optical sensors (visible and near infrared [NIR] channels) but also other remote sensing data such as thermal data (Tonooka et al., 2005). In this chapter, we will mainly focus on the degradation of optical sensors, introduce some common degradation issues, discuss their impacts on high-level products, and summarize approaches, methods, and techniques to address these degradation issues.

6.2 Common Issues of Data Degradation

This section briefly summarizes three types (Table 6.1) of common degradation issues of satellite data that consist of progressive changes in radiometric response and drifts in satellite orbital

TABLE 6.1 Major Types of Data Degradation and Their Impacts on Satellite Products

Types	Summary	Impacts	Solution	Example	Reference
Sensor degradation without onboard calibration unit	Response of detectors changes with time due to the harsh environment where satellites operate.	Inaccurate radiance and time-dependent artifacts in time series	Using pseudo-invariant targets on the ground to adjust calibration coefficients (vicarious calibration)	NOAA AVHRR	Molling et al. (2010)
Degradation of onboard calibration unit	Characteristics of onboard calibration device themselves degrade with time.	Inaccurate radiance and time-dependent artifacts in time series	Combining calibration with additional sources, such as lunar observation and pseudo-invariant targets	MODIS/ Terra	Wu et al. (2013)
Orbital drift	Orbits of sun-synchronous satellites drift and cause changes in ECT.	Gradual shift in SZA and variations in surface reflectance due to BRDF	Postcorrection of time series data using statistical or BRDF models	NOAA AVHRR	Privette et al. (1995)

parameters and two types of radiometric degradations (one with onboard calibration and one without onboard calibration).

6.2.1 Radiometric Degradation without Onboard Calibration Systems

Unlike ground-based measuring instruments, spaceborne sensors are usually physically inaccessible after satellites are launched. It is thus impossible to monitor the stability of their responses by periodically calibrating them with traditional approaches of laboratory measurement. To achieve reliable performance, extremely stable materials are typically used to manufacture satellite detectors. Calibration coefficients which convert satellite signature (e.g., digital number) to radiance or other radiative variables are measured in laboratory before launching. The sensors without onboard calibration units do not have mechanisms to update the coefficients on orbit. However, the characteristics of satellite detectors tend to change since the moment when the prelaunch coefficients are measured. The radiometric response of sensors will degrade due to the impacts from the launching process and the harsh space environment (Rao and Chen, 1995).

Visible and NIR bands of AVHRR do not have onboard calibration systems. Staylor (1990) evaluated the calibration stability of AVHRR visible bands onboard NOAA 6, 7, and 9 satellites by monitoring their performance over the Libyan Desert. By assuming a fixed degradation rate, an exponential degradation model was used to fit AVHRR time series (Staylor, 1990). The annual degradation rates of Channel 1 of NOAA 6, 7, and 9 were found to be 0%, 4%, and 6%, respectively (Staylor, 1990). Masonis and Warren (2001) used data of the same bands onboard NOAA 9, 10, and 11 over Greenland to estimate their degradation rates. They found that the sensor *degradation* is the reduced sensitivity to electromagnetic (EM) signals. NOAA 11 had an annual increased rate of 2.3% in radiometric gain (Masonis and Warren, 2001). Similarly, MODIS is also found to have increased gains in Band 1 and 2 (Wu et al., 2013).

Due to the existence of radiometric degradation, measurements from such sensors cannot be directly used in quantitative studies (Gorman and McGregor, 1994; Gutman, 1999) that require accurate values of radiance to quantify variables of interest. For such sensors, in practice, vicarious calibration is usually employed to account for the time-variant drifts of radiometric response (Dinguirard and Slater, 1999). After launching, calibration coefficients are updated by benchmarking observations over various types of stable surfaces, such as desert (Rao and Chen, 1995), ocean and high cloud (Vermote and Kaufman, 1995), and snow and ice (Loeb, 1997). In addition, the observation can also be adjusted by comparing with other well-calibrated satellite images such as MODIS (Heidinger et al., 2002) or airborne data (Smith et al., 1988). Besides the calibration of individual bands, variable of interest (e.g., NDVI) can also be directly adjusted by a similar procedure of vicarious calibration (Los, 1998).

Different vicarious calibration practices choose different calibration sites and use various forms of degradation equations. Lots of factors contribute to uncertainties in such methods.

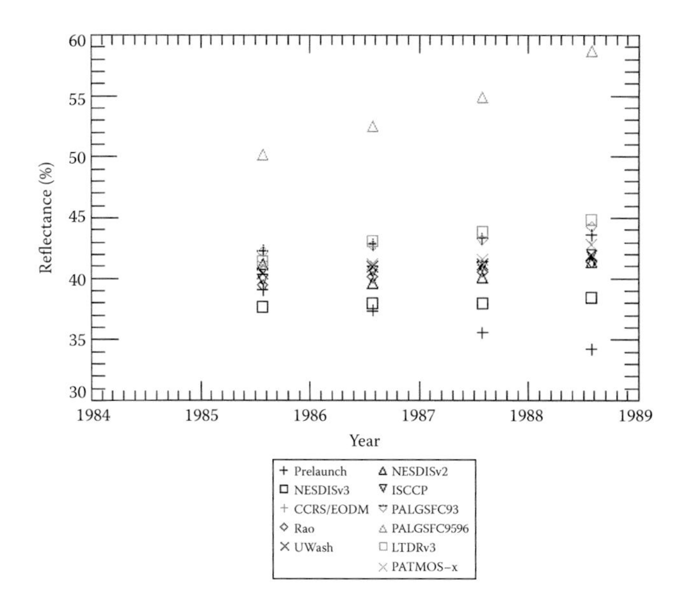

FIGURE 6.1 Mean reflectance of July over the Libyan Desert from NOAA9/AVHRR Channel 1 using various versions of calibration coefficients. The prelaunch coefficients show a clear decreasing trend, caused by issue of sensor degradation. Many efforts have been devoted to account for such degradation through vicarious calibration. However, large discrepancies exist among results from different methods. (Adapted from Figure 1 of Molling, C.C. et al., *Int. J. Remote Sens.*, 31, 6519, 2010.)

Reflectivity of pseudo-invariant targets is hardly a constant. Though the selected target is relatively isotropic in terms of its reflectivity, BRDF effects are a major source of uncertainties. Moreover, changes in atmospheric conditions can cause errors, especially aerosol from volcano eruption and water vapor effects for NIR bands. Issues of subpixel or undetected cloud cannot be negligible either. As a result, the derived calibration coefficients may contain large errors. Molling et al. (2010) compared AVHRR reflectances of a desert region calculated from 10 groups of calibration coefficients and found that the difference among them can be as large as 20% (Figure 6.1).

Ideally, vicarious calibration is expected to account for all the effects of sensor degradation, so that the satellite data after vicarious calibration can be free of problems of sensor degradation. However, as we mentioned, the vicarious calibration has its disadvantages and limitations. An alternative strategy of onboard calibration is in great need.

6.2.2 Radiometric Degradation with Onboard Calibration Device

Given the previously mentioned problems of vicarious calibration, an alternative way is to continuously monitor and correct sensor degradation on orbit. Actually, many sensors have units designed for onboard calibration, for example, SD for MODIS (Xiong and Barnes, 2006), internal calibrator for thematic mapper (TM) (Chander and Markham, 2003), and two additional devices (Full Aperture Solar Calibrator, Partial Aperture Solar Calibrator) for enhanced thematic mapper plus (ETM+) (Markham et al., 2004).

Theoretically, a well-designed onboard calibration system should be able to account for the sensor degradation and correct it by using the updated calibration coefficients.

For instance, MODIS is equipped with a sophisticated onboard calibration system for its visible, NIR, and shortwave infrared bands (also known as reflective solar bands [RSB]) (Xiong et al., 2010). SD and solar diffuser stability monitor (SDSM) are two key devices of this system (Xiong and Barnes, 2006). The radiometric response of the MODIS detector is calibrated by viewing SD, which has known reflecting characteristics. The degradation of SD is monitored by SDSM. It is actually a type of radiometer that determines SD degradation by observing the Sun and SD alternately (Xiong et al., 2007). In this process, spectral, angular, and mirror-side (MS) dependency of reflectance and radiometric response has been well taken into account. Moreover, a periodic view of the moon surface is used as an additional calibration source to independently check the stability of radiometric calibration (Xiong et al., 2010). Normally, such well-designed onboard calibration system is able to update the calibration coefficients as needed to compensate the degradation of the detector, so that reliable calibration results can be generated during the lifetime of the sensor. As a matter of fact, MODIS/Aqua is proved to achieve a very high level of calibration stability. A combination of four lines of evidence indicated that RSB reflectance derived from MODIS/Aqua has changed less than 1% during its decadal operation (Wu et al., 2013).

However, the other sensor of twin MODIS, MODIS/Terra, was found to have a serious issue of calibration drifts in some SRB bands (Franz et al., 2008; Wu et al., 2013). According to the original strategy, the SD door opens only when onboard calibration is in progress to reduce the exposure of SD to the Sun. Due to an SD anomaly of Terra/MODIS, the door remains open after 2003 (Xiong and Barnes, 2006). Terra/MODIS has thus experienced much worse degradation problems than Aqua/MODIS since then. Due to the difference of viewing geometry between onboard calibration and observing the Earth, stability of SD BRDF is essential for SDSM to track its degradation (Wu et al., 2013). As a result, bidirectionality of SD reflectance and changes in response versus scan angle (RVS) are major sources of uncertainties in calibrating MODIS in flight (Xiong et al., 2007). Because of its excessive exposure, the changes in BRDF of SD are believed to be the major cause of the calibration drift of MODIS/Terra (Wu et al., 2013). This degradation is found to be dependent on view zenith angle, spectral band, and MS. The shorter wavelength bands have the most series degradation. The stability of Terra/MODIS blue band reflectance changed as much as 7% for nadir observations (Figure 6.2) (Wang et al., 2012). With the lesson learned, the C6 reprocessing of MODIS/Terra data will use stable observations of the moon and desert to correct the errors in radiometric calibration (Wenny et al., 2010). The calibration drift issue of MODIS/Terra is expected to be reduced substantially through reprocessing (Wu et al., 2013).

Landsat 5/TM is another example of onboard calibration drift. Landsat 5 initially used onboard units to provide scene-by-scene calibration coefficients. However, the performance of the internal tungsten lamps of Landsat 5 failed to remain constant (Helder et al., 1998). The vicarious calibration and onboard

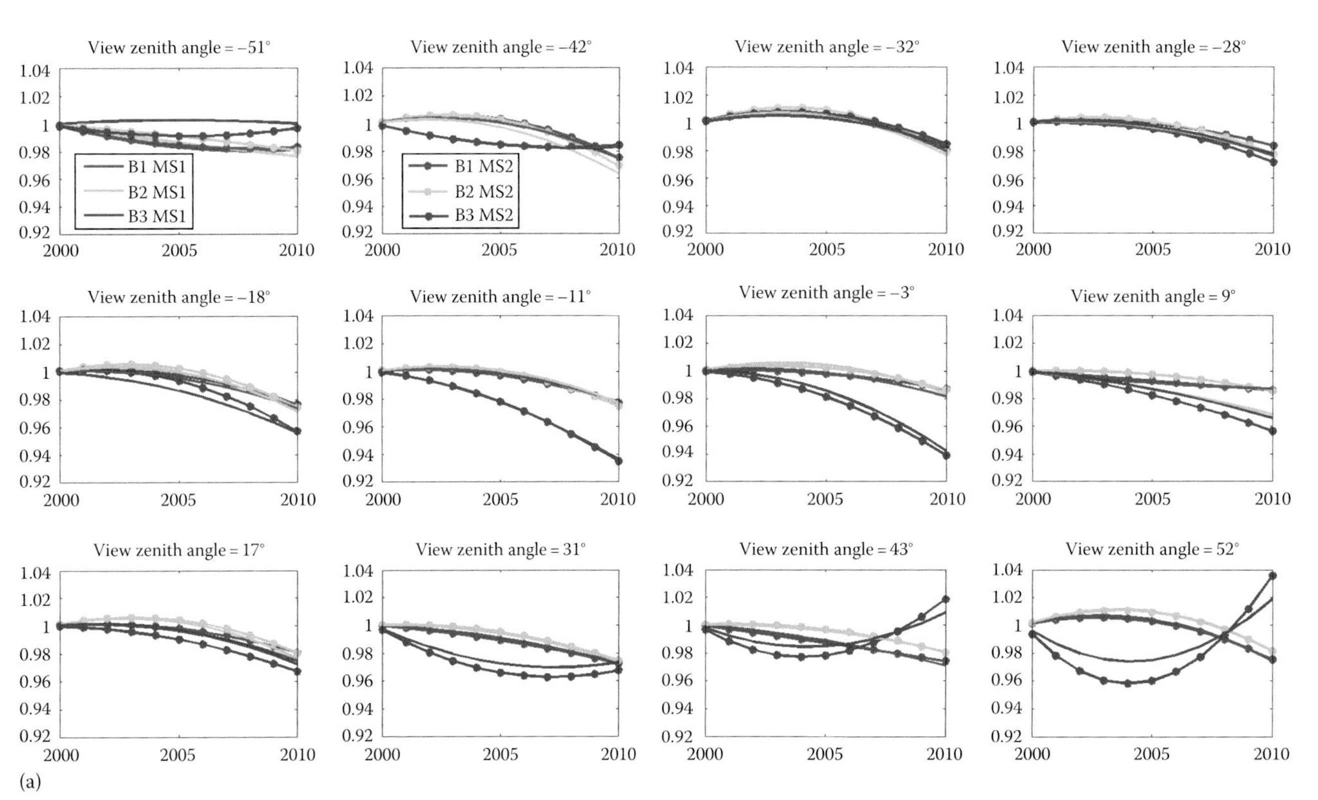

FIGURE 6.2 Degradation of TOA reflectance at three MODIS bands (B1, red; B2, NIR; and B3, blue) for (a) Terra/MODIS. *(Continued)*

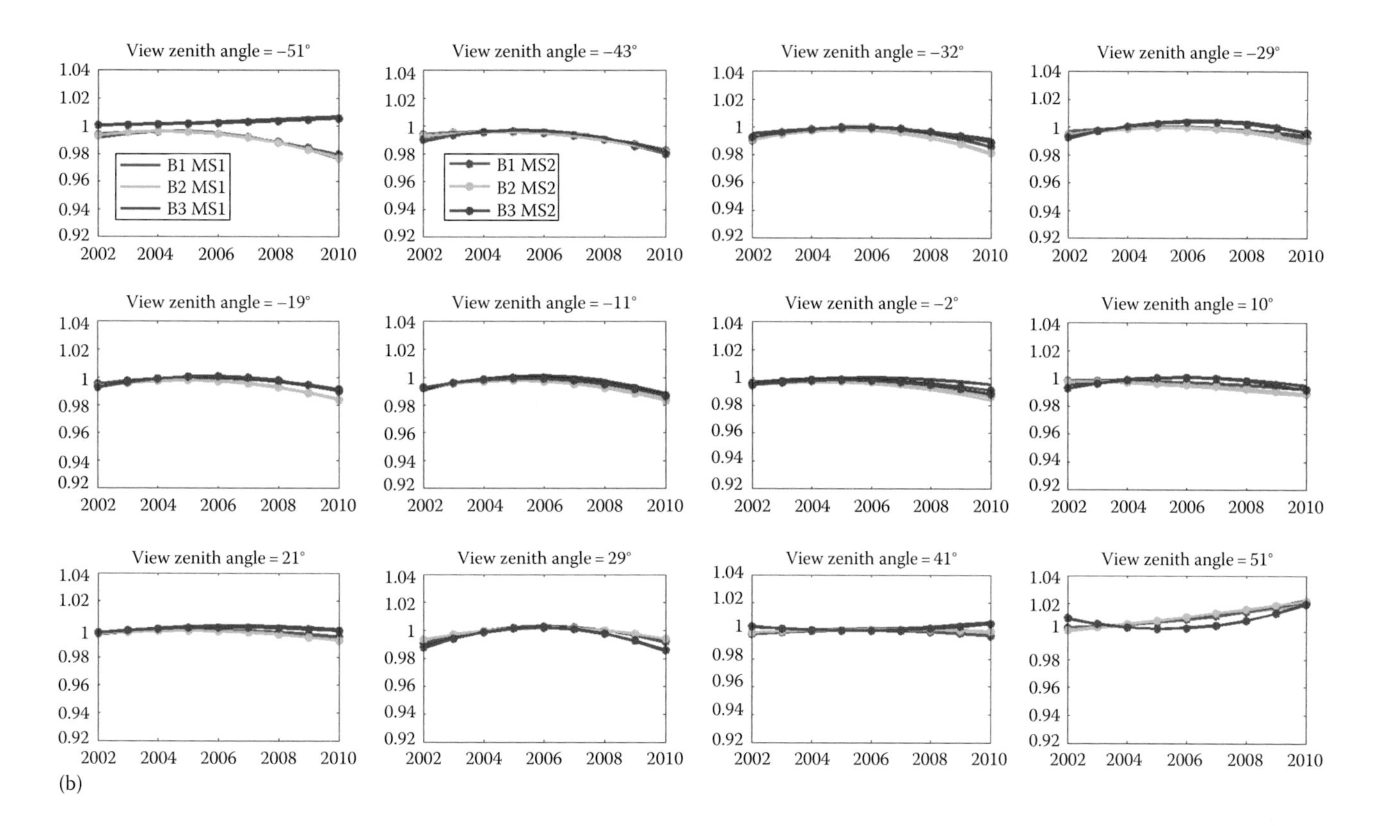

FIGURE 6.2 (*Continued*) Degradation of TOA reflectance at three MODIS bands (B1, red; B2, NIR; and B3, blue) for (b) Aqua/MODIS. The degradation rates are dependent on view zenith angle, spectral band, and MS. (Adapted from Figure S1 of Wang, D.D. et al., *Remote Sens. Environ.*, 119, 55, 2012.)

calibration suggested that there existed trends in sensor response (Thome et al., 1997). Scene-by-scene calibration coefficients derived from the onboard calibrator are not reliable anymore (Chander and Markham, 2003). As a result, a new calibration method based on monitoring desert and cross-calibration with Landsat 7/ETM+ is developed (Chander et al., 2007). Compared to Landsat 5/TM, ETM+ is rather stable. Annual degradation rates of visible and NIR bands of ETM+ are smaller than 0.4% (Markham et al., 2004).

6.2.3 Orbit Drift of Sun-Synchronous Satellites

Sun-synchronous is a feature of many polar-orbiting Earth observation missions. Such satellites will always cross the equator at the same local time to minimize effects of diurnal changes in variables of interest. Besides, this also assures that the observations have relatively similar illumination conditions. Due to BRDF effects, fixed illumination geometry will make reflectance comparable with each other. However, the designated orbits will change during the operation due to various perturbations (Ignatov et al., 2004). One straightforward way to address the issue is to accordingly adjust the satellite orbit from time to time. However, it will be a problem for satellites without the capability of adjusting flight on orbit, for example, NOAA heritage satellites. For NOAA morning satellites, equator crossing time (ECT) becomes gradually earlier. For afternoon satellites, the overpassing time moves toward later hours. ECT can change as long as several hours during the lifetime of NOAA satellites (Figure 6.3). From 1988 to 1994, NOAA 11 is the primary data source of AVHRR NDVI long-term data archive, because of the failure of its replacement satellite (Gutman, 1999). ECT of NOAA11 became 3 h later after its 6 years' use.

The change in crossing time will translate into shift of SZA. The angular changes are dependent on latitudes. Higher

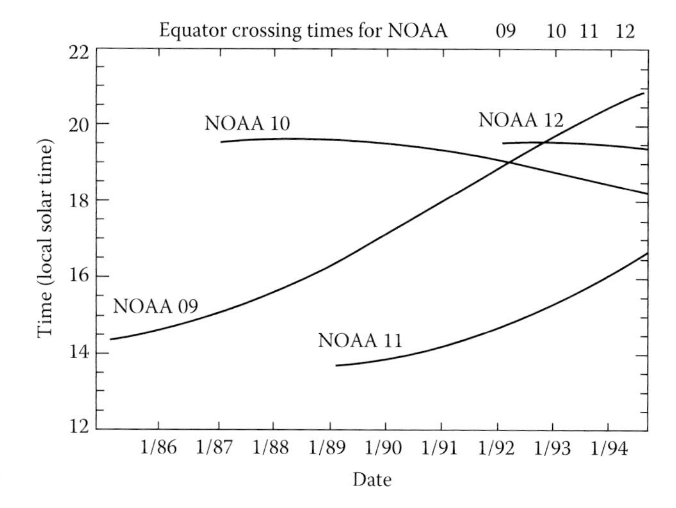

FIGURE 6.3 Changes in ECT for four NOAA satellites. Orbits are predicted from the historical ephemeris data. (Adapted from Figure 1 of Privette, J.L. et al., *Remote Sens. Environ.*, 53, 164, 1995.)

latitudes see greater seasonal variability but relative smaller shift of SZA, whereas equatorial area experiences substantial change of SZA (Privette et al., 1995). The progressive orbit drift will add a layer of noise into the temporal signals for some applications. The impacts mainly are in two aspects. ECT changes will affect features with diurnal variations, for example, cloud (Devasthale et al., 2012). The other issue is the changes in viewing geometry. Correction of atmospheric effects, for example, scattering of molecule and aerosol and absorption of water vapor, is dependent on the length of path and viewing geometry. In addition, surface reflectance is also dependent on viewing geometry because of surface BRDF effects. In Section 6.3.2, we use AVHRR data as example to illustrate how orbit shifts may affect NDVI products.

6.3 Impacts of Data Degradations on High-Level Products

6.3.1 MODIS/Terra Degradation

As we learned in Section 6.2.2, SRB of MODIS/Terra has experienced noticeable issues of calibration drifts, mainly caused by changes of bidirectional reflectance of SD. The actual degradation rates are dependent on spectral bands, view zenith angle, and MS. Generally speaking, bands of shorter wavelength and near nadir observations have the worst problem of degradation. Blue bands are important inputs to algorithms to generate products of ocean color and aerosol loadings. Large discrepancies are observed for the two products derived from Terra/MODIS and Aqua/MODIS. Sensor degradation of Terra/MODIS is likely to be a major of such difference. Besides, some products such as NDVI may also be impacted by the degradation issue of blue bands through the processing chain of MODIS products, although they do not directly use blue bands as inputs.

In a simulation study, Wang et al. (2012) analyzed how degradation of Terra/MODIS blue band affects its NDVI product and how the embedded errors in NDVI impact data analysis such as trend detection. As shown in Figure 6.2, TOA reflectance of Terra/MODIS red and NIR bands does not have serious issue of degradation. However, the MODIS NDVI algorithm needs surface reflectance of red and NIR bands as inputs. MODIS uses the dense dark vegetation algorithm to atmospherically correct reflectance of red and NIR bands (Kaufman et al., 1997), which heavily depends on aerosol optical depth (AOD) information derived from the blue band. Using the actual degradation rates of blue, red, and NIR bands at TOA as inputs, the induced errors in AOD and NDVI are simulated for various combinations of view zenith angle, aerosol loadings, and land surface types (Figure 6.4). The simulated results are consistent with the observed difference between Aqua/MODIS and Terra/MODIS corresponding products.

Although the annual degradation rate in the high-level NDVI products is only with the magnitude of 10^{-3}, it has great implication for the study of global vegetation change with the decadal time series of MODIS NDVI, because this is close to natural changing rates of vegetation without disturbance such

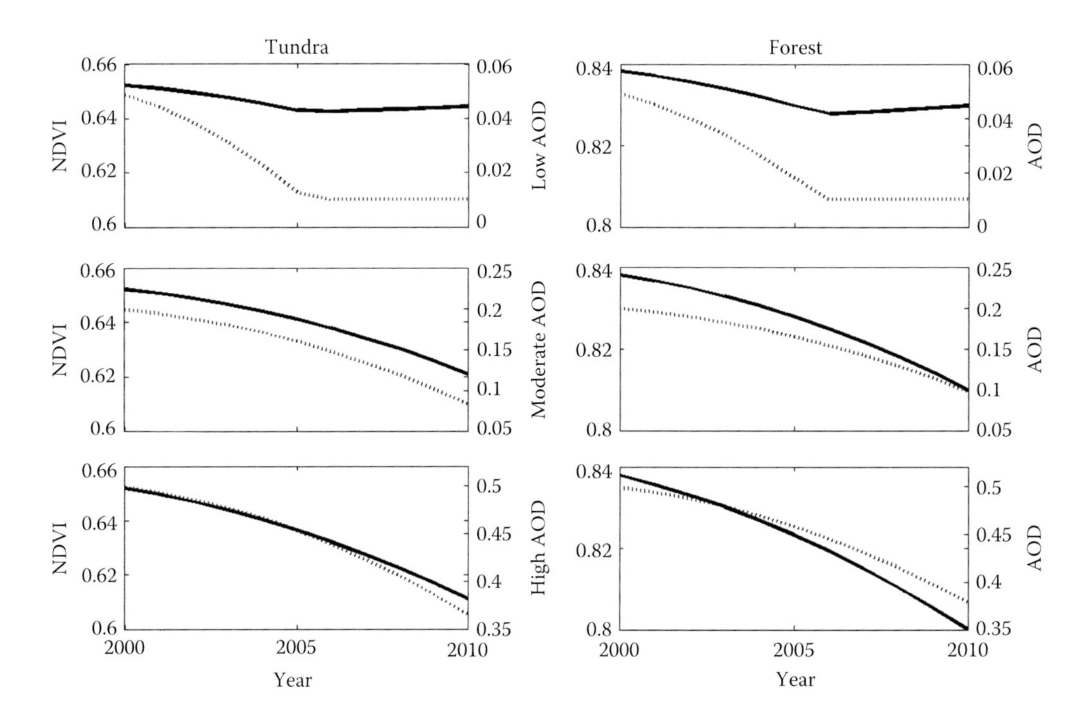

FIGURE 6.4 Temporal changes of NDVI (solid lines) and AOD (dashed lines) from 2000 to 2010, simulated with the degradation of Terra/MODIS as inputs. Two biomes (tundra [left] and forest [right]) and three levels of aerosol loadings (0.05, 0.20, and 0.50) are used in simulation. (Adapted from Wang, D.D. et al., *Remote Sens. Environ.*, 119, 55, 2012.)

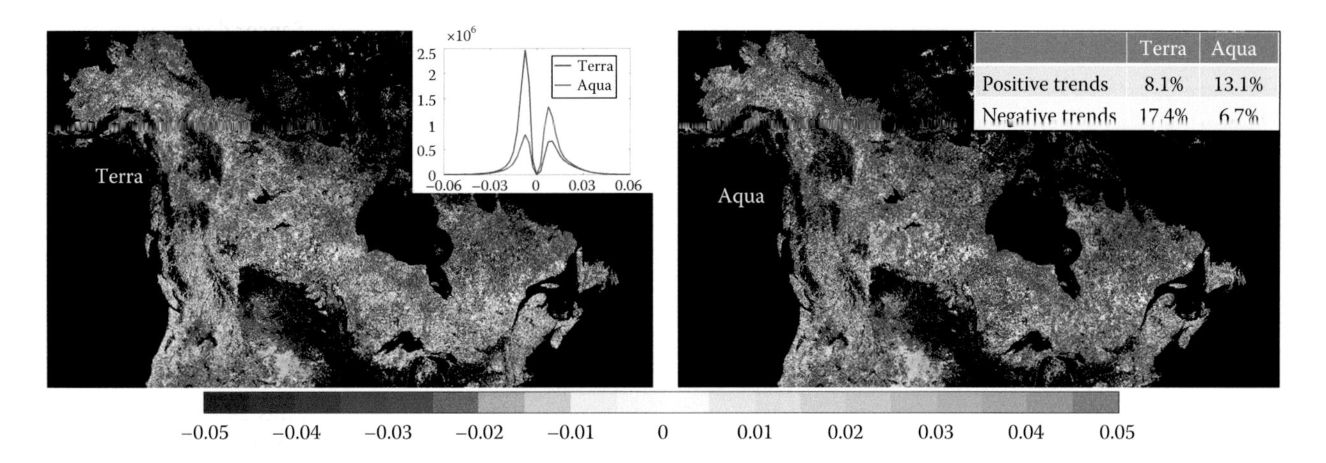

FIGURE 6.5 Boreal North America maps of NDVI trends between 2002 and 2010 calculated from the two actual MODIS products (Terra, left, and Aqua, right). White pixels are those recently disturbed by fire, wood harvest, etc. Black is nontundra or forest area. (Adapted from Wang, D.D. et al., *Remote Sens. Environ.*, 119, 55, 2012.)

as fire, insect burst, and logging. Under such circumstances, the Monte Carlo simulation suggested that it is extremely difficult to reliably detect trends of NDVI from the decadal MODIS data. Figure 6.5 shows the boreal North America maps of NDVI trends between 2002 and 2010 derived from Aqua/MODIS and Terra/MODIS, respectively. Historical disturbance datasets were used to exclude pixels disturbed by various natural and anthropogenic factors. The two maps display similar patterns over some area where the change rates are large. However, there exists a systematic difference between the trends derived from the two datasets. As a result, Aqua data suggest two thirds of significant trends are positive trends where Terra data show opposite statistics (Wang et al., 2012). Due to the potential impact of Terra/MODIS C5 data on trend analysis, it is suggested that the user should use Aqua/MODIS data in analyzing interannual variability before reprocessed MODIS data are available.

6.3.2 AVHRR Orbit Drift

Orbit drift will not always be a problem for data users. If the users' models can explicitly handle data variations in viewing geometry, their results will not be affected by the effects of orbit drift. For example, in calculation of NDVI, if BRDF-adjusted reflectance can be used, orbit drift will not necessarily lead to artificial changes in NDVI time series. However, due to the sparse temporal sampling of AVHRR, it is hard to model BRDF from AVHRR observations. The NDVI products typically directly use directional reflectance at the viewing geometry as inputs. As a ratio of two bands, NDVI is less affected by the variations in viewing geometry. Nevertheless, orbit drift still produces spurious variations on top of vegetation dynamics (Latifovic et al., 2012). The impacts of orbit drift on NDVI are dependent on latitude, season, and land cover type. Privette et al. (1995) simulated the changes of TOA NDVI caused by the orbit drift of the NOAA 11 satellite. The drops in NDVI are most significant in boreal winter and the equator. After 5 years on orbit, the decrease of NDVI caused by orbit drift can be as great as 10% at the equator (Figure 6.6) (Privette et al., 1995).

To remove effects of orbit drift, various kinds of algorithms have been developed. Some are based on statistical analysis to isolate signals caused by changes in SZA (Sobrino et al., 2008). Some utilize prior knowledge of vegetation BRDF to estimate NDVI difference induced by angular variations (Bacour et al., 2006). Los et al. (2005) developed a BRDF-based correction approach without assumption of biome-dependent BRDF shapes. Taking a first-order approximation, the NDVI equation in terms of Ross–Li kernels (K_L and K_R) is reduced to (Los et al., 2005)

$$NDVI(\theta_v,\theta_s,\Delta\phi) = k_i' + k_g' K_L + k_v' K_R$$

Even in this reduced form, it is difficult to estimate the kernel parameters (k_i', k_g', and k_v') because of the limited temporal sampling of AVHRR data. The authors further divided NDVI variability into two parts: those caused by phenology and those by BRDF effects. Using both simulated and actual data, the authors demonstrated that the effects of BRDF on NDVI can be reduced by 50%–85% through their method.

Impacts of NOAA satellite orbit drift are usually coupled with other issues such as sensor degradation of AVHRR (Staylor, 1990) and insufficient atmospheric correction (Nagol et al., 2009). So data quality of AVHRR time series is heavily dependent on approaches of data processing and analysis. Large differences in both absolute value and temporal change exist among analysis using different versions of AVHRR NDVI datasets (Beck et al., 2011). Since the aim of data correction is to isolate vegetation variability from noise and artifacts in NDVI time series, an alternative way is to treat all the artifacts originating from data degradation together and correct them all at once (Jiang et al., 2008; Latifovic et al., 2012).

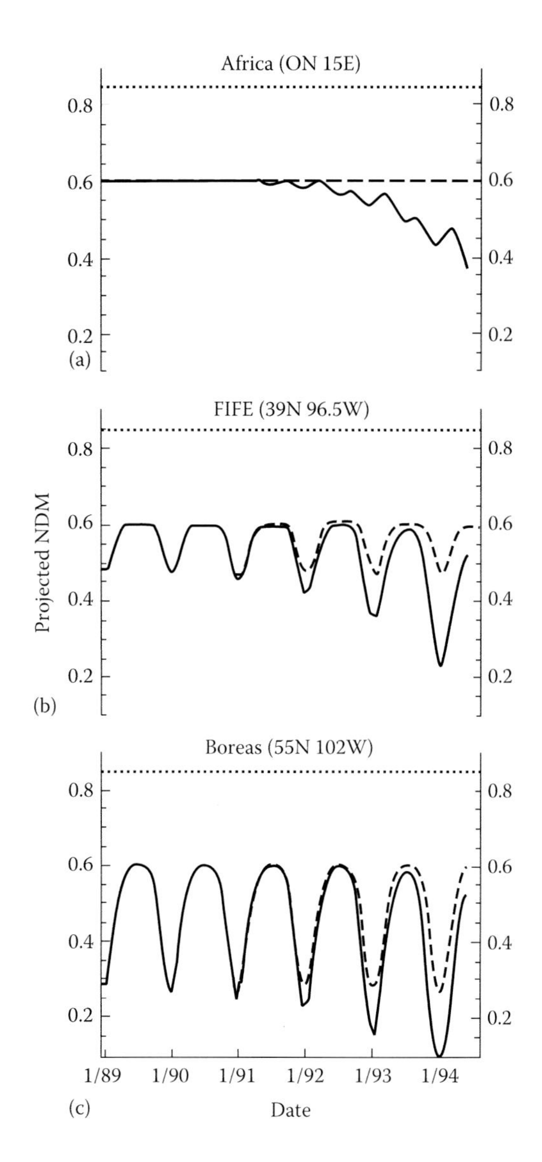

FIGURE 6.6 Impacts of orbit drift on TOA NDVI time series of NOAA11 AVHRR. Dotted lines are top of canopy NDVI with fixed illumination. Dashed lines are TOA NDVI without orbit drift. Solid lines are simulated TOA NDVI with NOAA11 orbit drift as input. (Adapted from Figure 5 of Privette, J.L. et al., *Remote Sens. Environ.*, 53, 164, 1995.)

6.4 Concluding Remarks

Analysis of long-term satellite data record has become an important tool to investigate how the Earth system changes over time and how its components interact with each other. Detection of subtle changes of global environment prefers consistent and stable time series data without artifact drifts. Unfortunately, spaceborne remote sensing data suffer from various issues of degradation during satellite operation in the harsh space environment. This chapter summarizes some common phenomena of data degradations and discusses their impacts on high-level satellite products.

Changes in radiometric response of spaceborne sensors cause time-variant drifts in TOA radiance, which translate into artificial trends in high-level products of environmental variables. For sensors without onboard calibration devices, vicarious calibration is a key step to account for such artifacts in the remote sensing data. However, lots of environmental factors make it a challenge to eliminate the effects of data degradation only through vicarious calibration. So onboard calibration systems are designed to address such problems. However, reliability of onboard calibration is also affected by its own degradation. The onboard calibration system needs collaboration with other calibration methods. A combination of onboard and vicarious calibration and utilization of multiple independent data sources, such as lunar view (Barnes et al., 2001; Cao et al., 2009), is a promising pathway to improve reliability of long-term data archive.

Any measurements have uncertainties, including remote sensing data. Users should always keep uncertainties of satellite data in mind when using them in quantitative studies. Inaccurate absolute calibration leads to bias or errors in satellite retrievals. Degradation of satellite data causes time-variant drift in long-term time series of satellite products. Caution should be exercised when drawing conclusion regarding long-term trends from data with possible artificial drifts. Meanwhile, efforts to evaluate accuracy of absolute calibration as well as long-term stability of calibration are greatly needed for climate change study.

References

Bacour, C., Breon, F.M., and Maignan, F. (2006). Normalization of the directional effects in NOAA-AVHRR reflectance measurements for an improved monitoring of vegetation cycles. *Remote Sensing of Environment, 102*, 402–413.

Barnes, R.A., Eplee, R.E., Schmidt, G.M., Patt, F.S., and McClain, C.R. (2001). Calibration of SeaWiFS. I. Direct techniques. *Applied Optics, 40*, 6682–6700.

Beck, H.E., McVicar, T.R., van Dijk, A.I.J.M., Schellekens, J., de Jeu, R.A.M., and Bruijnzeel, L.A. (2011). Global evaluation of four AVHRR-NDVI data sets: Intercomparison and assessment against Landsat imagery. *Remote Sensing of Environment, 115*, 2547–2563.

Cao, C.Y., Vermote, E., and Xiong, X.X. (2009). Using AVHRR lunar observations for NDVI long-term climate change detection. *Journal of Geophysical Research: Atmospheres, 114*, D20105.

Chander, G., and Markham, B. (2003). Revised Landsat-5 TM radiometric calibration procedures and postcalibration dynamic ranges. *IEEE Transactions on Geoscience and Remote Sensing, 41*, 2674–2677.

Chander, G., Markham, B.L., and Barsi, J.A. (2007). Revised Landsat-5 thematic mapper radiometric calibration. *IEEE Geoscience and Remote Sensing Letters, 4*, 490–494.

Datla, R., Emery, B., Ohring, G., Spencer, R., and Wielicki, B. (2004). Stability and accuracy requirements for satellite remote sensing instrumentation for global climate change monitoring. In S.A. Morain, and A.M. Budge (Eds.), *Post-Launch Calibration of Satellite Sensors*. A.A. Balkema Publishers, Leiden, the Netherlands.

Devasthale, A., Karlsson, K.G., Quaas, J., and Grassl, H. (2012). Correcting orbital drift signal in the time series of AVHRR derived convective cloud fraction using rotated empirical orthogonal function. *Atmospheric Measurement Techniques, 5*, 267–273.

Dinguirard, M. and Slater, P.N. (1999). Calibration of space-multispectral imaging sensors: A review. *Remote Sensing of Environment, 68*, 194–205.

Djavidnia, S., Melin, F., and Hoepffner, N. (2010). Comparison of global ocean colour data records. *Ocean Science, 6*, 61–76.

Esaias, W.E., Abbott, M.R., Barton, I., Brown, O.B., Campbell, J.W., Carder, K.L., Clark, D.K. et al. (1998). An overview of MODIS capabilities for ocean science observations. *IEEE Transactions on Geoscience and Remote Sensing, 36*, 1250–1265.

Franz, B.A., Kwiatkowska, E.J., Meister, G., and McClain, C.R. (2008). Moderate Resolution imaging spectroradiometer on Terra: Limitations for ocean color applications. *Journal of Applied Remote Sensing, 2*, 1–17.

Gorman, A.J. and McGregor, J. (1994). Some considerations for using AVHRR data in climatological studies 2. Instrument performance. *International Journal of Remote Sensing, 15*, 549–565.

Gutman, G.G. (1999). On the use of long-term global data of land reflectances and vegetation indices derived from the advanced very high resolution radiometer. *Journal of Geophysical Research: Atmospheres, 104*, 6241–6255.

Heidinger, A.K., Cao, C.Y., and Sullivan, J.T. (2002). Using moderate resolution imaging spectrometer (MODIS) to calibrate advanced very high resolution radiometer reflectance channels. *Journal of Geophysical Research: Atmospheres, 107*(D23), 4702. doi:10.1029/2001JD002035.

Helder, D., Boucyk, W., and Morfitt, R. (1998). Absolute calibration of the Landsat Thematic Mapper using the internal calibrator. In *1998 IEEE International Geoscience and Remote Sensing Symposium Proceedings*, Seattle, WA, July 6–10, pp. 2716–2718.

Ignatov, A., Laszlo, I., Harrod, E.D., Kidwell, K.B., and Goodrum, G.P. (2004). Equator crossing times for NOAA, ERS and EOS sun-synchronous satellites. *International Journal of Remote Sensing, 25*, 5255–5266.

Jiang, L., Tarpley, J.D., Mitchell, K.E., Zhou, S., Kogan, F.N., and Guo, W. (2008). Adjusting for long-term anomalous trends in NOAA's global vegetation index data sets. *IEEE Transactions on Geoscience and Remote Sensing, 46*, 409–422.

Justice, C.O., Townshend, J.R.G., Vermote, E.F., Masuoka, E., Wolfe, R.E., Saleous, N., Roy, D.P., and Morisette, J.T. (2002). An overview of MODIS land data processing and product status. *Remote Sensing of Environment, 83*, 3–15.

Justice, C.O., Vermote, E., Townshend, J.R.G., Defries, R., Roy, D.P., Hall, D.K., Salomonson, V.V. et al. (1998). The moderate resolution imaging spectroradiometer (MODIS): Land remote sensing for global change research. *IEEE Transactions on Geoscience and Remote Sensing, 36*, 1228–1249.

Kaufman, Y.J., Tanre, D., and Boucher, O. (2002). A satellite view of aerosols in the climate system. *Nature, 419*, 215–223.

Kaufman, Y.J., Tanre, D., Remer, L.A., Vermote, E.F., Chu, A., and Holben, B.N. (1997). Operational remote sensing of tropospheric aerosol over land from EOS moderate resolution imaging spectroradiometer. *Journal of Geophysical Research: Atmospheres, 102*, 17051–17067.

King, M.D., Menzel, W.P., Kaufman, Y.J., Tanre, D., Gao, B.C., Platnick, S., Ackerman, S.A., Remer, L.A., Pincus, R., and Hubanks, P.A. (2003). Cloud and aerosol properties, precipitable water, and profiles of temperature and water vapor from MODIS. *IEEE Transactions on Geoscience and Remote Sensing, 41*, 442–458.

Latifovic, R., Pouliot, D., and Dillabaugh, C. (2012). Identification and correction of systematic error in NOAA AVHRR long-term satellite data record. *Remote Sensing of Environment, 127*, 84–97.

Levy, R.C., Remer, L.A., Kleidman, R.G., Mattoo, S., Ichoku, C., Kahn, R., and Eck, T.F. (2010). Global evaluation of the Collection 5 MODIS dark-target aerosol products over land. *Atmospheric Chemistry and Physics, 10*, 10399–10420.

Loeb, N.G. (1997). In-flight calibration of NOAA AVHRR visible and near-IR bands over Greenland and Antarctica. *International Journal of Remote Sensing, 18*, 477–490.

Los, S.O. (1998). Estimation of the ratio of sensor degradation between NOAA AVHRR channels 1 and 2 from monthly NDVI composites. *IEEE Transactions on Geoscience and Remote Sensing, 36*, 206–213.

Los, S.O., North, P.R.J., Grey, W.M.F., and Barnsley, M.J. (2005). A method to convert AVHRR Normalized Difference Vegetation Index time series to a standard viewing and illumination geometry. *Remote Sensing of Environment, 99*, 400–411.

Markham, B.L. and Helder, D.L. (2012). Forty-year calibrated record of earth-reflected radiance from Landsat: A review. *Remote Sensing of Environment, 122*, 30–40.

Markham, B.L., Thome, K.J., Barsi, J.A., Kaita, E., Helder, D.L., Barker, J.L., and Scaramuzza, P.L. (2004). Landsat-7 ETM+ on-orbit reflective-band radiometric stability and absolute calibration. *IEEE Transactions on Geoscience and Remote Sensing, 42*, 2810–2820.

Masonis, S.J. and Warren, S.G. (2001). Gain of the AVHRR visible channel as tracked using bidirectional reflectance of Antarctic and Greenland snow. *International Journal of Remote Sensing, 22*, 1495–1520.

Mekler, Y., and Kaufman, Y.J. (1995). Possible causes of calibration degradation of the advanced very high-resolution radiometer visible and near-infrared channels. *Applied Optics, 34*, 1059–1062.

Molling, C.C., Heidinger, A.K., Straka, W.C., and Wu, X.Q. (2010). Calibrations for AVHRR channels 1 and 2: Review and path towards consensus. *International Journal of Remote Sensing, 31*, 6519–6540.

Nagol, J.R., Vermote, E.F., and Prince, S.D. (2009). Effects of atmospheric variation on AVHRR NDVI data. *Remote Sensing of Environment, 113*, 392–397.

Nemani, R.R., Keeling, C.D., Hashimoto, H., Jolly, W.M., Piper, S.C., Tucker, C.J., Myneni, R.B., and Running, S.W. (2003). Climate-driven increases in global terrestrial net primary production from 1982 to 1999. *Science, 300*, 1560–1563.

Privette, J.L., Fowler, C., Wick, G.A., Baldwin, D., and Emery, W.J. (1995). Effects of orbital drift on advanced very high-resolution radiometer products—Normalized difference vegetation index and sea surface temperature. *Remote Sensing of Environment, 53*, 164–171.

Rao, C.R.N., and Chen, J. (1995). Intersatellite calibration linkages for the visible and near-infrared channels of the advanced very high-resolution radiometer on the NOAA-7, NOAA-9, and NOAA-11 spacecraft. *International Journal of Remote Sensing, 16*, 1931–1942.

Smith, G.R., Levin, R.H., Abel, P., and Jacobowitz, H. (1988). Calibration of the solar channels of the NOAA-9 AVHRR using high altitude aircraft measurements. *Journal of Atmospheric and Oceanic Technology, 5*, 631–639.

Sobrino, J.A., Julien, Y., Atitar, M., and Nerry, F. (2008). NOAA-AVHRR orbital drift correction from solar zenithal angle data. *IEEE Transactions on Geoscience and Remote Sensing, 46*, 4014–4019.

Staylor, W.F. (1990). Degradation rates of the AVHRR visible channel for the NOAA-6, NOAA-7, and NOAA-9 spacecraft. *Journal of Atmospheric and Oceanic Technology, 7*, 411–423.

Tatem, A.J., Goetz, S.J., and Hay, S.I. (2008). Fifty years of earth-observation satellites—Views from space have led to countless advances on the ground in both scientific knowledge and daily life. *American Scientist, 96*, 390–398.

Thome, K., Markham, B., Barker, J., Slater, P., and Biggar, S. (1997). Radiometric calibration of Landsat. *Photogrammetric Engineering and Remote Sensing, 63*, 853–858.

Tonooka, H., Palluconi, F.D., Hook, S.J., and Matsunaga, T. (2005). Vicarious calibration of ASTER thermal infrared bands. *IEEE Transactions on Geoscience and Remote Sensing, 43*, 2733–2746.

Townshend, J.R.G., Justice, C.O., Gurney, C., and McManus, J. (1992). The impact of misregistration on change detection. *IEEE Transactions on Geoscience and Remote Sensing, 30*, 1054–1060.

Vermote, E., and Kaufman, Y.J. (1995). Absolute calibration of AVHRR visible and near-infrared channels using ocean and cloud views. *International Journal of Remote Sensing, 16*, 2317–2340.

Wang, D.D., Morton, D., Masek, J., Wu, A.S., Nagol, J., Xiong, X.X., Levy, R., Vermote, E., and Wolfe, R. (2012). Impact of sensor degradation on the MODIS NDVI time series. *Remote Sensing of Environment, 119*, 55–61.

Wenny, B.N., Sun, J., Xiong, X., Wu, A., Chen, H., Angal, A., Choi, T. et al. (2010). MODIS calibration algorithm improvements developed for Collection 6 Level-1B. In *Proceedings. SPIE 7807*, Earth Observing Systems XV, 78071F (August 28, 2010).

Wu, A., Angal, A., Xiong, X., and Cao, C. (2008). Monitoring MODIS calibration stability of visible and near-IR bands from observed top-of-atmosphere BRDF-normalized reflectances over Libyan Desert and Antarctic surfaces. *Proceedings of SPIE 7081*, 708113.

Wu, A. and Zhong, Q. (1994). A method for determining the sensor degradation rates of NOAA AVHRR channels 1 and 2. *Journal of Applied Meteorology, 33*, 118–122.

Wu, A.S., Xiong, X.X., Doelling, D.R., Morstad, D., Angal, A., and Bhatt, R. (2013). Characterization of Terra and aqua MODIS VIS, NIR, and SWIR spectral bands' calibration stability. *IEEE Transactions on Geoscience and Remote Sensing, 51*, 4330–4338.

Xiong, X., Sun, J., Barnes, W., Salomonson, V., Esposito, J., Erives, H., and Guenther, B. (2007). Multiyear on-orbit calibration and performance of Terra MODIS reflective solar bands. *IEEE Transactions on Geoscience and Remote Sensing, 45*, 879–889.

Xiong, X.X., and Barnes, W. (2006). An overview of MODIS radiometric calibration and characterization. *Advances in Atmospheric Sciences, 23*, 69–79.

Xiong, X.X., Sun, J.Q., Xie, X.B., Barnes, W.L., and Salomonson, V.V. (2010). On-orbit calibration and performance of aqua MODIS reflective solar bands. *IEEE Transactions on Geoscience and Remote Sensing, 48*, 535–546.

IV

Vegetation Index Standardization and Cross-Calibration of Data from Multiple Sensors

7

Inter- and Intrasensor Spectral Compatibility and Calibration of the Enhanced Vegetation Indices

Tomoaki Miura
University of Hawaii at Manoa

Kenta Obata
National Institute of Advanced Industrial Science and Technology

Javzandulam T. Azuma
University of Hawaii at Manoa

Alfredo Huete
University of Technology, Sydney

Hiroki Yoshioka
Aichi Prefectural University

Acronyms and Definitions

AERONET	Aerosol Robotic Network
AOT	Aerosol optical thickness
ARVI	Atmospherically resistant vegetation index
ASTER	Advanced Spaceborne Thermal Emission and Reflection Radiometer
AVHRR	Advanced Very High Resolution Radiometer
CAI	Cloud and aerosol imager
EO-1	Earth Observing-1
EOS	Earth Observing System
ESA	European Space Agency
ETM+	Enhanced Thematic Mapper Plus
EVI	Enhanced Vegetation Index
EVI2	Two-band Enhanced Vegetation Index (without a blue band)
GLI	Global imager
GMFR	Geometric mean functional regression
GPP	Gross primary productivity
HRVIR	Haute Résolution dans le Visible et l'Infra-Rouge
LAI	Leaf Area Index
MERIS	Medium-Resolution Imaging Spectrometer
MISR	Multi-angle Imaging SpectroRadiometer
MODIS	Moderate Resolution Imaging Spectroradiometer
NASA	National Aeronautics and Space Administration
NDVI	Normalized Difference Vegetation Index
NIR	Near-infrared
NOAA	National Oceanic and Atmospheric Administration
NPP	National Polar-orbiting Partnership
OLI	Operational Land Imager
PAC	Partial atmosphere correction
RMSD	Root mean square difference
SAVI	Soil-Adjusted Vegetation Index
SeaWiFS	Sea-Viewing Wide Field-of-View Sensor
SGLI	Second-generation Global Imager
SPOT	Système Pour l'Observation de la Terre
sRMPD	Systematic square root of mean product difference
TM	Thematic Mapper
TOA	Top of the atmosphere
TOC	Top of canopy
uRMPD	Unsystematic square root of mean product difference
VGT	VEGETATION
VIIRS	Visible Infrared Imaging Radiometer Suite

7.1 Introduction

The Enhanced Vegetation Index (EVI), an index developed for the National Aeronautics and Space Administration (NASA) Earth Observing System (EOS) Moderate Resolution Imaging Spectroradiometer (MODIS) mission (Huete et al., 2002), has been shown useful across a wide range of terrestrial vegetation studies. These include climate–vegetation interactions (Saleska et al., 2007; Ponce Campos et al., 2013), gross primary productivity (GPP) (Sims et al., 2008; Chen et al., 2011b; Guanter et al., 2014), drought impact assessment (Song et al., 2013), land cover classification (Friedl et al., 2010), and phenology (Zhang et al., 2006). The EVI has also been derived from sensors other than MODIS and used in characterizing terrestrial vegetation conditions and dynamics. These include Système Pour l'Observation de la Terre (SPOT) VEGETATION (VGT) for GPP estimation (e.g., Xiao et al., 2003, 2004), Landsat Thematic Mapper (TM) for floristic diversity assessment (Cabacinha and de Castro, 2009), and Landsat Enhanced Thematic Mapper Plus (ETM+) and IKONOS for Leaf Area Index (LAI) estimation (Soudani et al., 2006).

Recently, several new satellite sensors have successfully been launched and placed in orbit as data continuity missions to existing ones, all of which have spectral bands suitable for the EVI. Suomi National Polar-orbiting Partnership (NPP) Visible Infrared Imaging Radiometer Suite (VIIRS) is a moderate-resolution sensor to continue the data streams from the National Oceanic and Atmospheric Administration (NOAA) Advanced Very High Resolution Radiometer (AVHRR) sensor series and from EOS MODIS (Changyong et al., 2014). PROBA-V is a European Space Agency (ESA)-owned satellite mission launched to continue the moderate-resolution data stream of SPOT VGT (Sterckx et al., 2013). Landsat 8 Operational Land Imager (OLI) provides continuation of Landsat type, medium spatial resolution data with a greater number of spectral bands and higher radiometric capabilities (Irons et al., 2012). In particular, the EVI is included as one of the VIIRS standard operational products (Vargas et al., 2013).

It is thus of great importance to develop a good understanding of EVI spectral compatibilities across sensors and also of significant interest to investigate whether the EVI data record begun with EOS MODIS can be temporally and spatially extended with other satellite sensors for improved monitoring capabilities and climate science studies. This chapter focuses on the EVI and discusses intersensor spectral compatibility of the EVI. We also discuss intersensor spectral compatibility of a two-band version of the EVI without a blue band (EVI2) (Jiang et al., 2008) and its intrasensor compatibility with the EVI. Specific objectives of this chapter are to

1. Present a comprehensive review of inter- and intrasensor spectral compatibility of the EVI and EVI2
2. Evaluate the atmospheric impact on spectral compatibility of the EVI and EVI2 across sensors
3. Discuss cross-sensor calibration methodologies for EVI and EVI2 spectral compatibility

7.2 EVIs

The EVI is a three-band index, requiring reflectances of the near-infrared (NIR) (ρ_{NIR}), red (ρ_{red}), and blue (ρ_{blue}) bands. The index was designed to optimize the vegetation signal through a decoupling of the canopy background signal, with a reduction in atmospheric aerosol influences and with improved sensitivity in high biomass region, which complements the conventional, Normalized Difference Vegetation Index (NDVI) (Huete et al., 2002). The EVI (dimensionless) takes the form

$$EVI = G \cdot \frac{\rho_{NIR} - \rho_{red}}{\rho_{NIR} + C_1 \cdot \rho_{red} - C_2 \cdot \rho_{blue} + L} \quad (7.1)$$

where

L (reflectance unit, dimensionless) is the canopy background brightness adjustment factor that came from the Soil-Adjusted Vegetation Index (SAVI) (Huete, 1988)
C_1 (dimensionless) and C_2 (dimensionless) are the coefficients of the aerosol resistance term adapted from the Atmospherically Resistant Vegetation Index (ARVI) (Kaufman and Tanré, 1992)
G (dimensionless) is the gain factor that adjusts the EVI dynamic range to a comparable one to that of the NDVI

The input reflectances in Equation 7.1 need to be corrected for a partial atmosphere (i.e., molecular scattering and gaseous absorption effects) or the total atmospheric effects including aerosols. Miura et al. (2001) demonstrated the effectiveness of the EVI in reducing residual aerosol contaminations in the total-atmosphere corrected reflectances that arise due to the highly spatially and temporally variable nature of aerosol loadings and properties, and uncertainties in the aerosol loading estimations. The coefficients adopted in the MODIS EVI algorithm are $L = 1$, $C_1 = 6$, $C_2 = 7.5$, and $G = 2.5$ (Huete et al., 1997), which can be used for both of the aforementioned input reflectance types.

While many studies found that the EVI was advantageous in vegetation productivity assessments and characterization of vegetation biophysical properties (e.g., an improved surrogate measure of GPP; Rahman et al., 2005), several potential issues have been reported for the EVI. First, by its very design, the EVI is limited to a sensor system with a blue band in addition to a red and NIR bands (Equation 7.1). The EVI is not applicable to, for example, SPOT Haute Résolution dans le Visible et l'Infra-Rouge (HRVIR), EOS Advanced Spaceborne Thermal Emission and Reflection Radiometer (ASTER), or AVHRR. Second, the EVI algorithm can produce faulty values for snow-/ice-contaminated observations that have higher blue and red reflectances than NIR reflectance (Huete et al., 2002; Didan and Huete, 2006; Vargas et al., 2013). Finally, Fensholt et al. (2006) suggested that the consistency of EVI values across different sensors may be more problematic than that of the NDVI due to variable and more difficult atmospheric correction schemes of the blue reflectance as described later (Section 7.4).

In order to address the aforementioned issues, a two-band version of the EVI without a blue band, or EVI2, was developed, which is functionally equivalent to the EVI although slightly more prone to aerosol noise (Jiang et al., 2008). Jiang et al. (2008) noted that the blue band in the EVI is rather aimed at reducing noise and uncertainties associated with highly variable atmospheric aerosols than at providing additional biophysical information on vegetation.

The EVI2 was derived by optimization of the Linear Vegetation Index (LVI) to attain the best similarity with the EVI, particularly when atmospheric effects are insignificant and data quality is good (Jiang et al., 2008). The LVI was a newly proposed index for the EVI2 derivation that incorporated the soil-adjustment factor of SAVI with a linearity-adjustment factor. The derived EVI2 (dimensionless) is of the form

$$EVI2 = G \cdot \frac{\rho_{NIR} - \rho_{red}}{\rho_{NIR} + C \cdot \rho_{red} + L} \tag{7.2}$$

where

$G = 2.5$ (dimensionless)

$C = 2.4$ (dimensionless)

$L = 1.0$ (reflectance unit, dimensionless) for the MODIS spectral bands (Jiang et al., 2008)

Jiang et al. (2008) used MODIS reflectance data extracted from 40 globally distributed sites to obtain these coefficients. The consistency between EVI and EVI2 across various land cover types demonstrated that their similarity was independent of land cover. Comparisons of EVI and EVI2 time series further revealed that their similarity was seasonally independent. The EVI2 (Equation 7.2) has been adopted as the EVI backup algorithm, which is used as a substitute for the EVI for snow-/ice-contaminated pixels, in the Collection 6 MODIS VI Product suite.

7.3 Multisensor Applications of EVI and EVI2

A growing number of studies have used the EVI from MODIS and from other sensors (Table 7.1). Although not as extensive as the EVI, the number of studies that used the EVI2 has been growing (Table 7.2). Tables 7.1 and 7.2 list a sample of these studies for the purpose of showing a range of applications of the EVI and EVI2.

While most of these studies used EVI or EVI2 data from a single sensor, several studies used or evaluated the utilities of multisensor EVI or EVI2. Soudani et al. (2006) examined the feasibility of IKONOS, ETM+, and SPOT HRVIR for LAI estimation and whether retrieved LAI would be comparable across the 3 sensors for 28 coniferous and deciduous temperate forest stands in France. Various vegetation indices including the EVI were used as estimators of LAI. For a lack of a blue band with the HRVIR sensor, the EVI was computed only from ETM+ and IKONOS data, and EVI differences between the two sensors were found to be larger for higher LAI (LAI range ≈0–7).

In Pfeifer et al. (2012), the EVI2 was obtained from atmospherically corrected Landsat ETM+ and SPOT HRVIR data over a large number of sites where in situ LAI measurements were made in the Eastern Arc Mountains and their catchment areas in Kenya and Tanzania. ETM+ EVI2 and HRVIR EVI2 were treated as a single dataset and used together as a predictor variable to derive an empirical LAI estimation equation. Pfeifer et al. (2012) also examined the NDVI and the modified Soil-Adjusted Vegetation Index (MSAVI2) (Qi et al., 1994) along with the EVI2.

Kobayashi et al. (2007) compared multisatellite reflectance, NDVI, and EVI to elucidate the capabilities of satellite-based phenology monitoring of larch forests in eastern Siberia. Although nearly all the analyses were focused on individual band reflectances and NDVI, they compared EVI temporal profiles from MODIS and VGT at a larch forest site and found that MODIS EVI and VGT EVI had very similar seasonality.

Although it is not a multisensor EVI application, Huettich et al. (2009) used a Landsat ETM+ and MODIS multiresolution dataset along with local scale in situ botanical survey data for vegetation mapping in a dry savanna ecosystem in Namibia. The Landsat multispectral data were used first to locate homogeneous patches and then to scale up the in situ botanical information, which were in turn used to obtain training data in MODIS EVI temporal signature-based classification.

Recently, Kim et al. (2014) developed a long-term (>30 years) record of spring frost day (SFD) and spring frost damage day (SFDD) metrics over the conterminous United States by integrating a satellite microwave remote sensing record of daily landscape freeze–thaw (FT) status and an EVI2-based phenology record of start of season (SOS) and day of peak (DOP) canopy cover metrics. This long-term EVI2 dataset used in their study was derived from calibrated and temporally overlapping MODIS, AVHRR, and VGT satellite data.

7.4 Multisensor Compatibility of EVI and EVI2: A Review

As in the case of the NDVI, the EVI and EVI2 from different sensors are likely subject to systematic differences due to differences in sensor/platform characteristics and product generation algorithms. The spectral bandpass is one key sensor characteristic that varies widely among sensors (Figure 7.1). Although focused on the NDVI, Swinnen and Veroustraete (2008) provide a comprehensive list of factors that would potentially impact intersensor compatibility between SPOT VGT and NOAA AVHRR.

Unlike the NDVI, only a limited number of studies have addressed the issue of multisensor EVI and/or EVI2 compatibility (Table 7.3). Most of these studies used MODIS EVI or EVI2 as a reference in examining intersensor EVI and/or EVI2 compatibility, as these indices were originally designed for MODIS. Many of these studies examined spectral compatibility across different sensor bandpasses using hyperspectral data (e.g., Miura and Yoshioka, 2011).

While all these studies primarily investigated compatibility of the EVI across sensors or of the EVI2 among different sensors using atmospherically corrected, surface reflectance data,

TABLE 7.1 List of Sensors and Application Fields of EVI

Sensors	Application Fields	Geographic Regions	References
MODIS	GPP estimation	North America	Sims et al. (2008)
			Xiao et al. (2005a)
			Yang et al. (2007)
			Zhang et al. (2013)
		Central Asia	Li et al. (2007)
		Africa	Sjostrom et al. (2011)
		Global	Guanter et al. (2014)
	Net ecosystem exchange (CO_2 and water vapor)	Europe	Houborg and Soegaard (2004)
	Crop yield prediction	Southeast Asia	Son et al. (2013)
	Evapotranspiration	North America	Nagler et al. (2005)
			Yang et al. (2006)
		Australia	Guerschman et al. (2009)
	Aboveground biomass estimation	South America	Anaya et al. (2009)
	Soil classification	Middle East	Hengl et al. (2007)
	Rooting depth estimation	South America	Ichii et al. (2007)
	Site index	North America	Waring et al. (2006)
	Vegetation type mapping	Southern Africa	Huettich et al. (2009)
		Central Africa	Betbeder et al. (2014)
	Species richness	Eastern Asia	John et al. (2008)
	Canopy conductance	North America, Europe, and Australia	Yebra et al. (2013)
	Ecosystem respiration	North America and Europe	Jaegermeyr et al. (2014)
	Phenology/drought response	North America	Mendez-Barroso et al. (2009)
			Moran et al. (2014)
		South America	Huete et al. (2006)
		North America and Australia	Ponce Campos et al. (2013)
VGT	GPP estimation	North and South America	Xiao et al. (2004)
			Xiao et al. (2005b)
	Land cover classification	East Asia	Boles et al. (2004)
MISR[a]	LAI estimation	North America	Pocewicz et al. (2007)
Landsat TM/ETM+	Vegetation dynamics	Western Asia	Wittenberg et al. (2007)
	LAI, aboveground biomass, yield estimations	North America	Zhao et al. (2007)
	Flux footprint analysis, vegetation type mapping	North America	Chen et al. (2011a)
	Floristic diversity	South America	Cabacinha and de Castro (2009)
IKONOS	LAI estimation	Europe	Soudani et al. (2006)

[a] Multiangle imaging spectroradiometer.

TABLE 7.2 List of Sensors and Application Fields of EVI2

Sensors	Application Field	Geographic Region	References
MODIS	Crop yield estimation	North America	Bolton and Friedl (2013)
	GPP estimation	Northern high latitude	Schubert et al. (2012)
		North America	Wu (2012)
	Net ecosystem exchange (CO_2)	North America	Rocha and Shaver (2011)
	Phenology	North America	Yang et al. (2012)
			Kim et al. (2014)
		South America	Morton et al. (2014)
Landsat TM/ETM+	Vegetation type mapping	Europe	O'Connell et al. (2013)
	LAI estimation	North America	Liu et al. (2012)
		East Africa	Pfeifer et al. (2012)
HRVIR	LAI estimation	East Africa	Pfeifer et al. (2012)
Quickbird	Burn severity mapping	North America	Boelman et al. (2011)
Radiation sensor	Greenness and LAI	North America	Rocha and Shaver (2009)

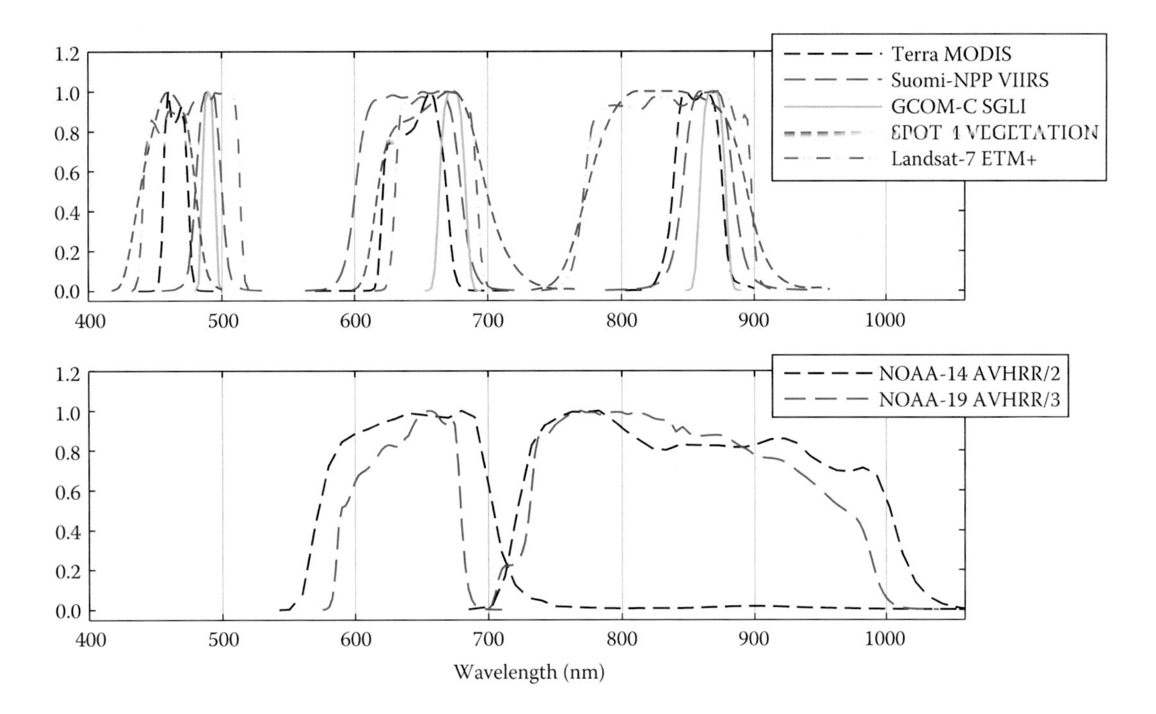

FIGURE 7.1 Normalized spectral response curves of blue, red, and NIR bands of selected satellite sensors. The AVHRR sensors do not have a blue band.

several studies evaluated two additional details. First, Kim et al. (2010) examined intrasensor EVI-to-EVI2 compatibility for VGT and Sea-Viewing Wide Field-of-View Sensor (SeaWiFS) bandpasses and intersensor MODIS EVI-to-AVHRR EVI2 compatibility. Miura et al. (2008) and Yamamoto et al. (2012) also examined EVI-to-EVI2 compatibility, but for ASTER and MODIS. Second, Miura and Yoshioka (2011) assessed the impact of different atmospheric correction schemes on cross-sensor EVI2 compatibility using a scenario where total-atmosphere-corrected EVI2 from MODIS and VGT were compared with partial-atmosphere-corrected EVI2 from AVHRR and VIIRS. In the following, we highlight key findings of these studies.

Ferreira et al. (2003) investigated the utility of spectral vegetation indices in monitoring seasonal dynamics of Brazilian savanna vegetation formations. Airborne hyperspectral data were acquired and convolved to the MODIS and ETM+ bandpasses and converted to the EVI. Intersensor comparisons of seasonal dynamics, based on spectral band-pass properties, showed that the simulated ETM+ EVI had better seasonal discrimination capability than that of MODIS. They attributed this finding to the closer proximity between the ETM+ red and NIR band centers.

Fensholt et al. (2006) evaluated the quality of MODIS EVI, Medium-Resolution Imaging Spectrometer (MERIS) EVI, and VGT EVI using in situ measurements and their intersensor compatibility using actual satellite products at a grassland site in Senegal. A good agreement between the EVI from satellite and from in situ measured MODIS data was found, indicating an accurate atmospheric correction of the MODIS red, NIR, and blue spectral bands. On the other hand, the consistency of the EVI from MERIS and VGT with MODIS EVI was not as good as was found among the NDVI from these sensors. Fensholt et al. (2006) attributed this reduced EVI consistency to different, sensor-specific atmospheric correction schemes used in blue band reflectance retrievals.

Miura et al. (2008) examined the compatibility of ASTER EVI2 and MODIS EVI using actual satellite data extracted from randomly located, globally distributed locations. A robust linear relationship was found between the two satellite indices with a mean difference of 0.012 EVI units (ASTER minus MODIS).

Kim et al. (2010) performed a band decomposition analysis in which band-pass contributions to observed cross-sensor VI difference were identified. They found that disparities in blue bandpasses were the primary cause of EVI differences between MODIS and other coarse-resolution sensors. The highest compatibility was found between VIIRS and MODIS EVI2, while AVHRR EVI2 was the least compatible to MODIS.

Miura and Yoshioka (2011) also empirically examined intersensor EVI and EVI2 relationships using MODIS as a reference, but for a larger number of sensors. They also found that spectral band-pass differences resulted in systematic differences on intersensor EVI2 relationships with MODIS, but the relationships were very linear. Intersensor EVI relationships between MODIS and other sensors were linear for most of sensors, although several sensors had either curve-linear or incompatible relationships with MODIS EVI, all of which have spectrally very different blue band from the MODIS blue band-pass.

TABLE 7.3 List of Multisensor EVI/EVI2 Compatibility Studies

References	Data Analyzed	Sensor/VI Pair
Ferreira et al. (2003)	Airborne hyperspectral data	MODIS EVI vs. ETM + EVI
Fensholt et al. (2006)	Actual sensor data	MODIS EVI vs. MERIS EVI
	Actual sensor data	MODIS EVI vs. VGT EVI
Miura et al. (2008)	Actual sensor data	MODIS EVI vs. ASTER EVI2
Kim et al. (2010)	EO-1[a] Hyperion data	MODIS EVI vs. VIIRS EVI
		MODIS EVI vs. VGT EVI
		MODIS EVI vs. SeaWiFS[b] EVI
	EO-1 Hyperion data	MODIS EVI2 vs. VIIRS EVI2
		MODIS EVI2 vs. VGT EVI2
		MODIS EVI2 vs. SeaWiFS EVI2
		MODIS EVI2 vs. AVHRR/2 EVI2
Miura and Yoshioka (2011)	EO-1 Hyperion data	MODIS EVI vs. VGT EVI
		MODIS EVI vs. GLI[c] EVI
		MODIS EVI vs. VIIRS EVI
		MODIS EVI vs. SeaWiFS EVI
		MODIS EVI vs. SGLI[d] EVI
	EO-1 Hyperion data	MODIS EVI2 vs. AVHRR/2 EVI2
		MODIS EVI2 vs. AVHRR/3 EVI2
		MODIS EVI2 vs. VGT EVI2
		MODIS EVI2 vs. GLI EVI2
		MODIS EVI2 vs. VIIRS EVI2
		MODIS EVI2 vs. SeaWiFS EVI2
		MODIS EVI2 vs. CAI[e] EVI2
		MODIS EVI2 vs. SGLI EVI2
	EO-1 Hyperion data	MODIS EVI2 vs. AVHRR/2 EVI2
		MODIS EVI2 vs. VIIRS EVI2
		VGT EVI2 vs. AVHRR/2 EVI2
		VGT EVI2 vs. VIIRS EVI2
Yamamoto et al. (2012)	Actual sensor data	MODIS EVI vs. ASTER EVI2
		MODIS EVI vs. MODIS/ASTER EVI
Obata et al. (2013)	Actual sensor data	MODIS EVI vs. VIIRS EVI

[a] EO-1: Earth observing-1.
[b] SeaWiFS: Sea-Viewing Wide Field-of-View Sensor.
[c] GLI: global imager.
[d] SGLI: second-generation global imager.
[e] CAI: cloud and aerosol imager.

7.5 Atmospheric Impact on Inter- and Intrasensor Spectral Compatibility of EVI and EVI2

In this section, we present a spectral compatibility analysis conducted to fill in some of the knowledge gaps identified earlier, namely, the impact of operational atmospheric corrections on inter- and intrasensor spectral compatibility of the EVI and EVI2. We specifically evaluated (1) the effects of residual aerosols on intersensor EVI-to-EVI, intersensor EVI2-to-EVI2, and intrasensor EVI2-to-EVI relationships and (2) the effects of different atmospheric correction schemes on intersensor EVI2 compatibility between AVHRR and other sensors. The analysis was conducted by band-pass simulations with a global set of Hyperion hyperspectral data (Miura et al., 2013).

7.5.1 Materials and Methods

7.5.1.1 Hyperion Data and Preprocessing

The dataset used in Miura et al. (2013) and adapted here consisted of 37 Level 1R EO-1 Hyperion hyperspectral scenes obtained over 15 globally distributed Aerosol Robotic Network (AERONET) sites (Holben et al., 2001; Pearlman et al., 2003; Middleton et al., 2010). Hyperion is a push broom sensor with 256 pixels in the cross-track direction (~7.7 km swath at 30 m spatial resolution), acquiring Earth-reflected radiation from 400 to 2500 nm at 10 nm nominal sampling intervals (Pearlman et al., 2003). Level 1R Hyperion scenes provide radiometrically corrected and calibrated radiance values and are corrected for the interspectrometer misregistration (USGS, 2011).

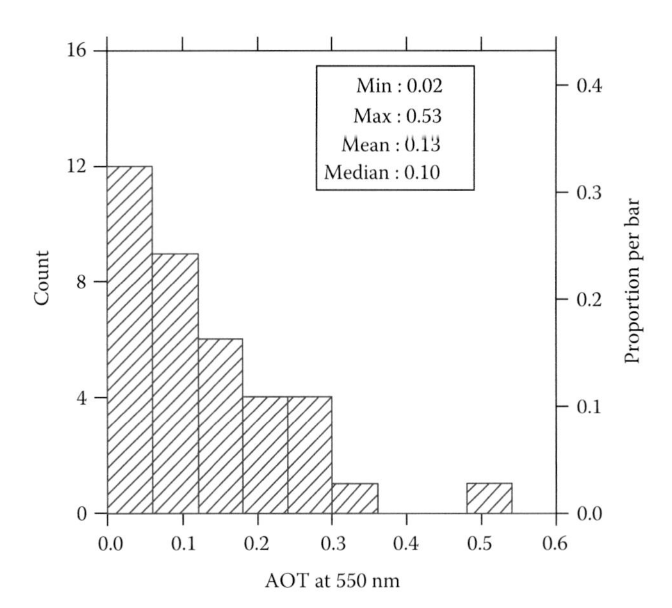

FIGURE 7.2 Histogram and univariate statistics of atmospheric AOT at 550 nm of Hyperion scenes used in this study. (Adapted and reprinted with permission from Miura, T. et al., *IEEE Trans. Geosci. Remote Sens.*, 51, 1349, 2013.)

The sites/scenes represented 13 out of the 17 International Geosphere-Biosphere Programme land cover types, with the 4 cover types not represented being evergreen broadleaf forests, permanent wetlands, snow and ice, and water bodies. Coincident Level 2 AERONET data (±1 h of the Hyperion acquisition time) were available for all of the scenes. Aerosol optical thickness (AOT) at 550 nm ranged from 0.02 (clean) to 0.53 (turbid) for this dataset (Figure 7.2). See Table 1 of Miura et al. (2013) for additional details of this Hyperion dataset.

The Hyperion scenes were spectrally and spatially convolved to simulate top-of-the-atmosphere (TOA) reflectances of the VIIRS, MODIS, second-generation global imager (SGLI), VGT, AVHRR/3, AVHRR/2, and ETM+ bands at 1 km spatial resolution. The spectral response functions of these spectral bands (Figure 7.1) were interpolated to Hyperion band center wavelengths for each Hyperion pixel (Pearlman et al., 2003). The MODIS point spread function, which is, to the first order, triangular in the scan direction and rectangular in the track direction, was assumed for the spatial convolution (Wolfe et al., 2002).

7.5.1.2 Simulation of Atmospheric Correction

The "6S" radiative transfer code was used to atmospherically correct the convolved Hyperion scenes (Vermote et al., 2006; Kotchenova and Vermote, 2007). For each scene, the 6S code was constrained with the corresponding in situ AERONET atmospheric measurements, view and solar zenith, relative azimuth angles at the time of the image acquisition, and the site elevation obtained from GTOPO30.

Three atmospheric corrections were applied. First, all the scenes were corrected for total atmospheric effects including aerosols, that is, TOA reflectances were reduced to surface or top-of-canopy (TOC) reflectances. Second, all the scenes were corrected for total atmospheric effects, but with incorrect AOT values (AERONET-measured AOT values ± 0.05 AOT at 550 nm) to simulate residual aerosol effects in the retrieved surface reflectances. Finally, we applied a partial atmosphere correction (PAC) on the AVHRR band-pass reflectances where all atmospheric effects except for aerosols were accounted for. The EVI and EVI2 were computed from each of these retrieved surface reflectances (i.e., TOC reflectances, TOC reflectances subject to residual aerosol effects, and PAC reflectances).

Approximately, sixty 1 km pixels were randomly selected and extracted from each scene, totaling ~2000 pixels over the 37 scenes. This random sample dataset was used for the following analyses.

7.5.1.3 Statistical Difference Analysis

Three statistical measures of difference were employed to quantitatively evaluate and compare inter- and intrasensor compatibility of the EVI and EVI2: the root mean square difference (RMSD), systematic square root of mean product difference (sRMPD), and unsystematic square root of mean product difference (uRMPD) (Ji and Gallo, 2006):

$$\text{RMSD} = \sqrt{\frac{1}{n}\sum_{i=1}^{n}(X_{1,i} - X_{2,i})^2} \tag{7.3}$$

$$\text{sRMPD} = \sqrt{\frac{1}{n}\sum_{i=1}^{n}\left[(X_{1,i} - X_{2,i})^2 - \left|\hat{X}_{1,i} - X_{1,i}\right|\left|\hat{X}_{2,i} - X_{2,i}\right|\right]} \tag{7.4}$$

$$\text{uRMPD} = \sqrt{\frac{1}{n}\sum_{i=1}^{n}\left(\left|\hat{X}_{1,i} - X_{1,i}\right|\left|\hat{X}_{2,i} - X_{2,i}\right|\right)} \tag{7.5}$$

where

- X_1 and X_2 are the EVI or EVI2 for sensor-1 and sensor-2, respectively
- $\hat{X}_1$ and $\hat{X}_2$ are the corresponding points of X_1 and X_2 on the regression line that is the best estimate of the primary trend of the X_1 vs. X_2 scatter

sRMPD and uRMPD decouple RMSD into its systematic and unsystematic components, respectively; sRMPD measures how far the scatter (trend or regression line) is from the line of $X_1 = X_2$, or the bias errors, and uRMPD measures the magnitude of data scattering (secondary variation) about the regression line (Ji and Gallo, 2006). Following the protocol developed by Ji and Gallo (2006), regression lines were estimated using the geometric mean functional regression (GMFR), which considers that both variables of interest are subject to error. Unlike the ordinary least squares regression, the derived regression line is symmetric about X_1 and X_2, or invertible.

7.5.2 Results

7.5.2.1 Intersensor EVI

EVI differences for all possible combinations among the five sensors of VIIRS, MODIS, SGLI, VGT4, and ETM+ for both atmospheric corrections with and without AOT errors are plotted in Figure 7.3 and summarized in Table 7.4 (for GMFR equations see Table 7.A.1 in Appendix 7.A). All sensor pairs examined here were subject to systematic differences of which magnitudes varied among sensor pairs. The systematic differences basically linearly increased or decreased with EVI values although some curve linearity was seen for those between VIIRS and MODIS, VGT4, or ETM+ (Figure 7.3a, c, and d, respectively). The smallest and largest systematic differences (in terms of sRMPD) were observed for MODIS vs. ETM+ and SGLI vs. VGT4, respectively (Table 7.4). The magnitudes of unsystematic differences also changed among sensor pairs, and uRMPD were the smallest for VGT4 vs. ETM+ and the largest for VIIRS vs. SGLI and ETM+ (Figure 7.3 and Table 7.4).

Overlapped symbols in Figure 7.3 indicate that intersensor EVI differences remained nearly the same when the input surface reflectances were subject to residual aerosol contaminations. For all the sensor pairs examined here, RMSD, sRMPD, and uRMPD of intersensor EVI were the same whether the input surface reflectances were subject to AOT errors or not (Table 7.4). These results indicate that the atmospheric resistance of the EVI functions well for intersensor EVI.

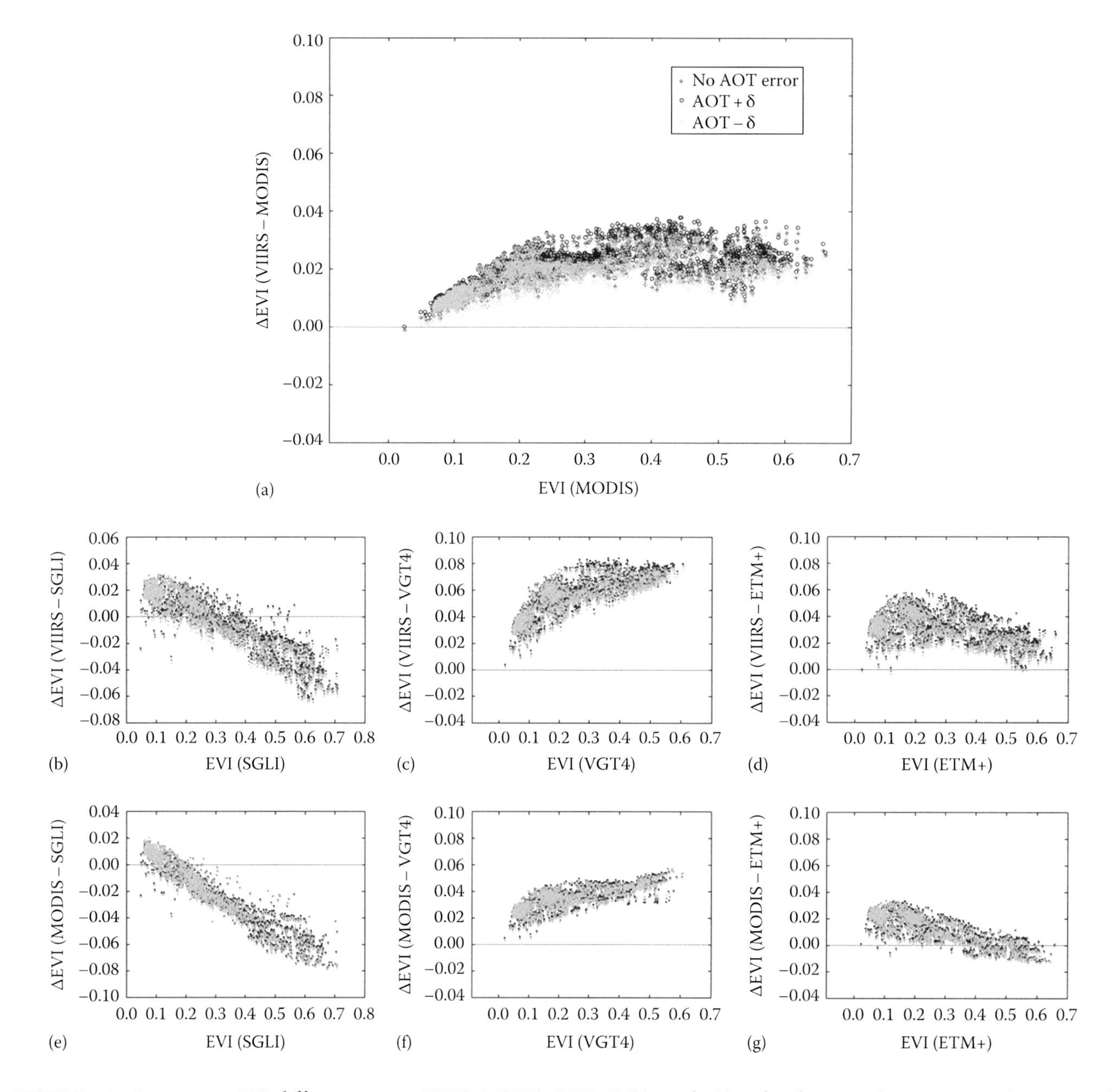

FIGURE 7.3 Inter-sensor EVI difference across VIIRS, MODIS, SGLI, VGT4, and ETM+ bandpasses subject to atmospheric aerosol correction error. *(Continued)*

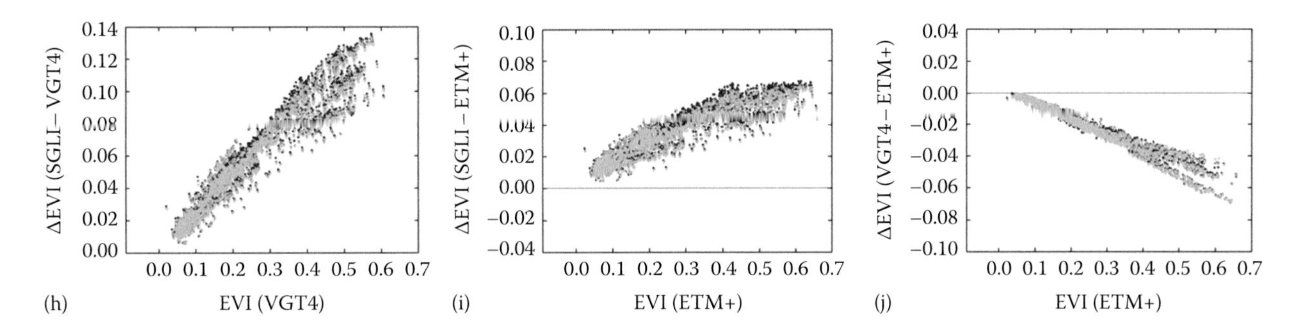

FIGURE 7.3 (*Continued*) Inter-sensor EVI difference across VIIRS, MODIS, SGLI, VGT4, and ETM+ bandpasses subject to atmospheric aerosol correction error.

TABLE 7.4 Statistics for Intersensor Spectral Difference: EVI

	TOC with No AOT Error			TOC with AOT Error of ±0.05			Ratio (TOC Error/TOC No Error)		
	RMSD	sRMPD	uRMPD	RMSD	sRMPD	uRMPD	RMSD	sRMPD	uRMPD
VIIRS vs. MODIS	0.019	0.019	0.005	0.020	0.019	0.005	1.0	1.0	1.0
VIIRS vs. SGLI	0.021	0.019	0.009	0.021	0.019	0.009	1.0	1.0	1.0
VIIRS vs. VGT4	0.054	0.053	0.008	0.054	0.053	0.008	1.0	1.0	1.0
VIIRS vs. ETM+	0.033	0.032	0.009	0.034	0.032	0.009	1.0	1.0	1.0
MODIS vs. SGLI	0.031	0.030	0.008	0.031	0.030	0.008	1.0	1.0	1.0
MODIS vs. VGT4	0.035	0.034	0.005	0.035	0.034	0.005	1.0	1.0	1.0
MODIS vs. ETM+	0.018	0.017	0.006	0.018	0.017	0.006	1.0	1.0	1.0
SGLI vs. VGT4	0.060	0.060	0.008	0.060	0.060	0.008	1.0	1.0	1.0
SGLI vs. ETM+	0.035	0.035	0.006	0.035	0.035	0.006	1.0	1.0	1.0
VGT4 vs. ETM+	0.026	0.025	0.003	0.026	0.026	0.003	1.0	1.0	1.0

7.5.2.2 Intersensor EVI2

Intersensor EVI2 relationships were similar to those of EVI for the sensor pairs examined here (Figure 7.4). They were, however, very different for VIIRS vs. MODIS in that the intersensor EVI2 relationship was very linear and their biases were, on average, negative and smaller than those of EVI (Figure 7.4a).

For the sensor pairs studied here, intersensor EVI2 differences were basically smaller than those of EVI when the input reflectances were not subject to errors. RMSD, sRMPD, and uRMPD for intersensor EVI2 relationships ranged from 0.003 to 0.035, 0.002 to 0.034, and 0.002 to 0.008, respectively (Table 7.5) (for GMFR equations see Table 7.A.2 in Appendix 7.A), whereas the corresponding values for intersensor EVI relationships were 0.02–0.06 (RMSD), 0.019–0.06 (sRMPD), and 0.003–0.009 (uRMPD) (Table 7.4). Table 7.6 lists the ratios of these three different statistics for every sensor pairs. One can observe that, except for MODIS vs. VGT4 and MODIS vs. ETM+, all the ratios were less than 1.0, indicating smaller systematic and unsystematic, and overall differences for intersensor EVI2 relationships.

On the other hand, unlike the EVI, intersensor EVI2 differences varied when the input reflectances were subject to residual aerosol contaminations due to atmospheric correction errors. In Figure 7.4, it can be seen that EVI2 differences formed three separate trends, corresponding to the three correction error scenarios for all the sensor pairs. It can also be seen as a 20% or more increase in uRMPD upon the contaminations (the smallest uRMPD ratio in Table 7.5 being 1.2). The resultant uRMPD of intersensor EVI2 relationships were nearly the same or larger (0.9–1.7 times) than the corresponding uRMPD of intersensor EVI relationships that were found insensitive to residual aerosol contaminations (the most right column in Table 7.6). The residual aerosol contaminations did not impact sRMPD because we used symmetric errors to perturb EVI2. These results indicate that the atmospheric resistance of the EVI with a blue band is advantageous in deriving intersensor EVI relationships.

In order to evaluate and confirm the atmospheric resistance of the EVI against the EVI2, differences between EVI with and without residual contaminations as well as those between EVI2 with and without residual contaminations are plotted in Figure 7.5 and their RMSD summarized in Table 7.7. For the five sensor bandpasses, the EVI successfully reduced residual aerosol errors to less than ±0.005 (Figure 7.5) with RMSD of 0.002 or less (Table 7.7). EVI2 errors exceeded 0.015 with RMSD of 0.005 or more for the AOT estimation errors of ±0.05.

7.5.2.3 Intrasensor EVI2 vs. EVI

EVI2-to-EVI differences with and without residual aerosol errors are plotted as a function of EVI in Figure 7.6 and their statistics summarized in Table 7.8 (for GMFR equations see Table 7.A.3 in Appendix 7.A), in order to analyze intrasensor EVI2-to-EVI compatibilities for the sensors other than MODIS. Intrasensor EVI2 vs. EVI relationships of all sensor bandpasses were subject to systematic differences although their magnitudes changed

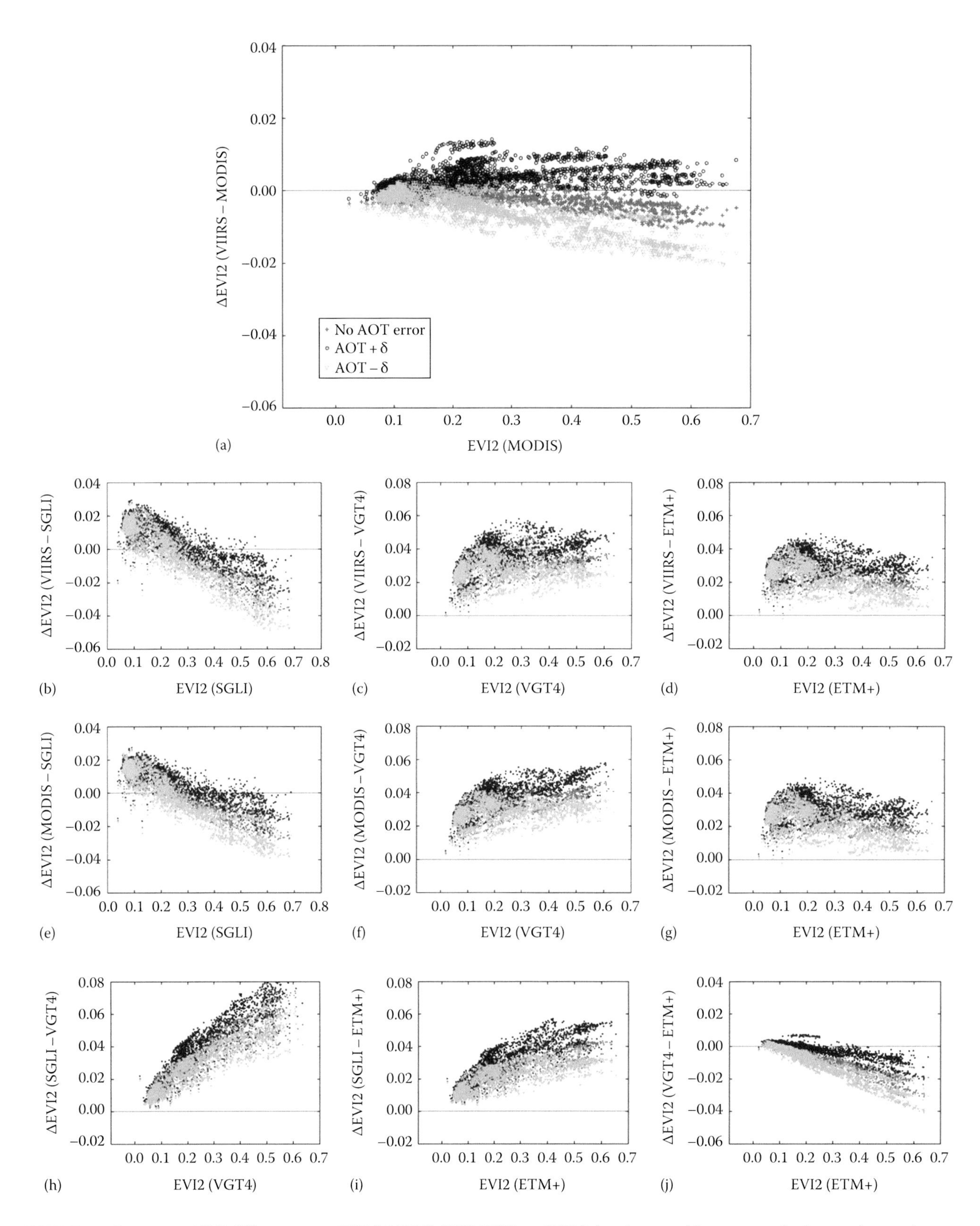

FIGURE 7.4 Inter-sensor EVI2 difference across VIIRS, MODIS, SGLI, VGT4, and ETM+ bandpasses subject to atmospheric aerosol correction error.

TABLE 7.5 Statistics for Intersensor Spectral Difference: EVI2

	TOC with No AOT Error			TOC with AOT Error of ±0.05			Ratio (TOC Error/TOC No Error)		
	RMSD	sRMPD	uRMPD	RMSD	sRMPD	uRMPD	RMSD	sRMPD	uRMPD
VIIRS vs. MODIS	0.003	0.002	0.002	0.006	0.002	0.006	2.1	0.8	3.2
VIIRS vs. SGLI	0.015	0.014	0.007	0.016	0.014	0.009	1.1	1.0	1.3
VIIRS vs. VGT4	0.031	0.031	0.007	0.032	0.031	0.009	1.0	1.0	1.3
VIIRS vs. ETM+	0.026	0.025	0.008	0.027	0.025	0.010	1.0	1.0	1.2
MODIS vs. SGLI	0.013	0.012	0.005	0.014	0.012	0.008	1.1	1.0	1.5
MODIS vs. VGT4	0.033	0.032	0.006	0.034	0.033	0.008	1.0	1.0	1.4
MODIS vs. ETM+	0.027	0.026	0.006	0.028	0.027	0.008	1.0	1.0	1.3
SGLI vs. VGT4	0.035	0.034	0.006	0.036	0.035	0.008	1.0	1.0	1.3
SGLI vs. ETM+	0.026	0.026	0.005	0.027	0.026	0.008	1.0	1.0	1.4
VGT4 vs. ETM+	0.010	0.009	0.002	0.011	0.009	0.006	1.1	1.0	2.6

TABLE 7.6 Statistics for Intersensor Spectral Difference: EVI2 to EVI Ratio

	TOC with No AOT Error			TOC with AOT Error of ±0.05		
	RMSD	sRMPD	uRMPD	RMSD	sRMPD	uRMPD
VIIRS vs. MODIS	0.2	0.1	0.4	0.3	0.1	1.2
VIIRS vs. SGLI	0.7	0.7	0.7	0.8	0.7	0.9
VIIRS vs. VGT4	0.6	0.6	0.8	0.6	0.6	1.0
VIIRS vs. ETM+	0.8	0.8	0.8	0.8	0.8	1.0
MODIS vs. SGLI	0.4	0.4	0.7	0.5	0.4	1.0
MODIS vs. VGT4	1.0	0.9	1.1	1.0	1.0	1.5
MODIS vs. ETM+	1.5	1.6	1.0	1.6	1.6	1.4
SGLI vs. VGT4	0.6	0.6	0.8	0.6	0.6	1.0
SGLI vs. ETM+	0.7	0.7	0.9	0.8	0.8	1.2
VGT4 vs. ETM+	0.4	0.4	0.7	0.4	0.4	1.7

from sensor to sensor. They were the largest for SGLI with sRMPD of 0.015–0.016 and the smallest for ETM+ with sRMPD of 0.006–0.007 (Table 7.8). These sRMPD values were basically all less than those observed for intersensor EVI or EVI2 relationships, except for that of VIIRS EVI2 vs. MODIS EVI2 that had the smallest sRMPD and that of VGT4 EVI2 vs. ETM+ EVI2.

Unsystematic variations of these intrasensor relationships were about the same across all the sensors with uRMPD ranging from 0.007 to 0.008 without residual aerosols and from 0.008 to 0.01 with residual aerosols (Table 7.8). The magnitudes of uRMPD were comparable to those of intersensor EVI2 relationships (Table 7.5).

7.5.2.4 Intersensor Compatibility with AVHRR EVI2

Lastly, we present intersensor compatibility between TOC EVI2 and PAC AVHRR EVI2. No operational aerosol correction of AVHRR data has been implemented, and thus, this provides a realistic scenario on EVI2 compatibility/continuity with AVHRR. For comparisons, AVHRR TOC EVI2 was also analyzed.

Figure 7.7a and b show example plots for VIIRS TOC EVI2 vs. AVHRR/3 TOC EVI2 and for VIIRS TOC EVI2 vs. AVHRR/3 PAC EVI2 differences, respectively. Whether corrected for total or partial atmosphere, AVHRR/3 EVI2 had a linear relationship with VIIRS EVI2. However, AVHRR/3 PAC EVI2 had larger systematic and unsystematic differences than the TOC counterpart against VIIRS EVI2. sRMPD and uRMPD for the former were 0.028 and 0.007, respectively, whereas those for the latter were 0.041 and 0.011, respectively (Table 7.9). The same trends were observed for all the other sensors when paired with AVHRR/3 or AVHRR/2 (Table 7.9). AVHRR EVI2 had the largest differences with SGLI and the smallest differences with VGT4 (Table 7.9).

AVHRR/3 and AVHRR/2 EVI2 showed very high compatibility regardless of the level of atmospheric correction as far as both sensor data had the same levels of atmospheric correction (Figure 7.7a and b). Although they were subject to some systematic differences (sRMPD ~ 0.015), they were subject to very small unsystematic variations (uRMPD ~ 0.002), and their relationships were nearly independent of atmospheres, indicating the compatible sensitivity of their spectral bandpasses to atmospheric (aerosol) effects.

In summary, intersensor EVI relationships were insensitive to residual aerosol contaminations due to EVI's atmospheric resistance. The EVI2 showed higher multisensor compatibility than the EVI, with smaller bias and unsystematic errors, which was however only when the input reflectances were not subject to residual aerosol contaminations. Intrasensor EVI vs. EVI2 relationships were subject to the same magnitudes of unsystematic errors as intersensor EVI2 relationships. Intrasensor EVI2-to-EVI relationships were subject to significant biases for some sensors. This indicated the need of EVI and/or EVI2 coefficient adjustments, which were optimized for the MODIS bandpasses by design. Backward compatibility with AVHRR EVI2 would be a feasible option only when aerosol source of variability are reduced in AVHRR EVI2 either by atmospheric correction or by temporal compositing to select the cleanest observations.

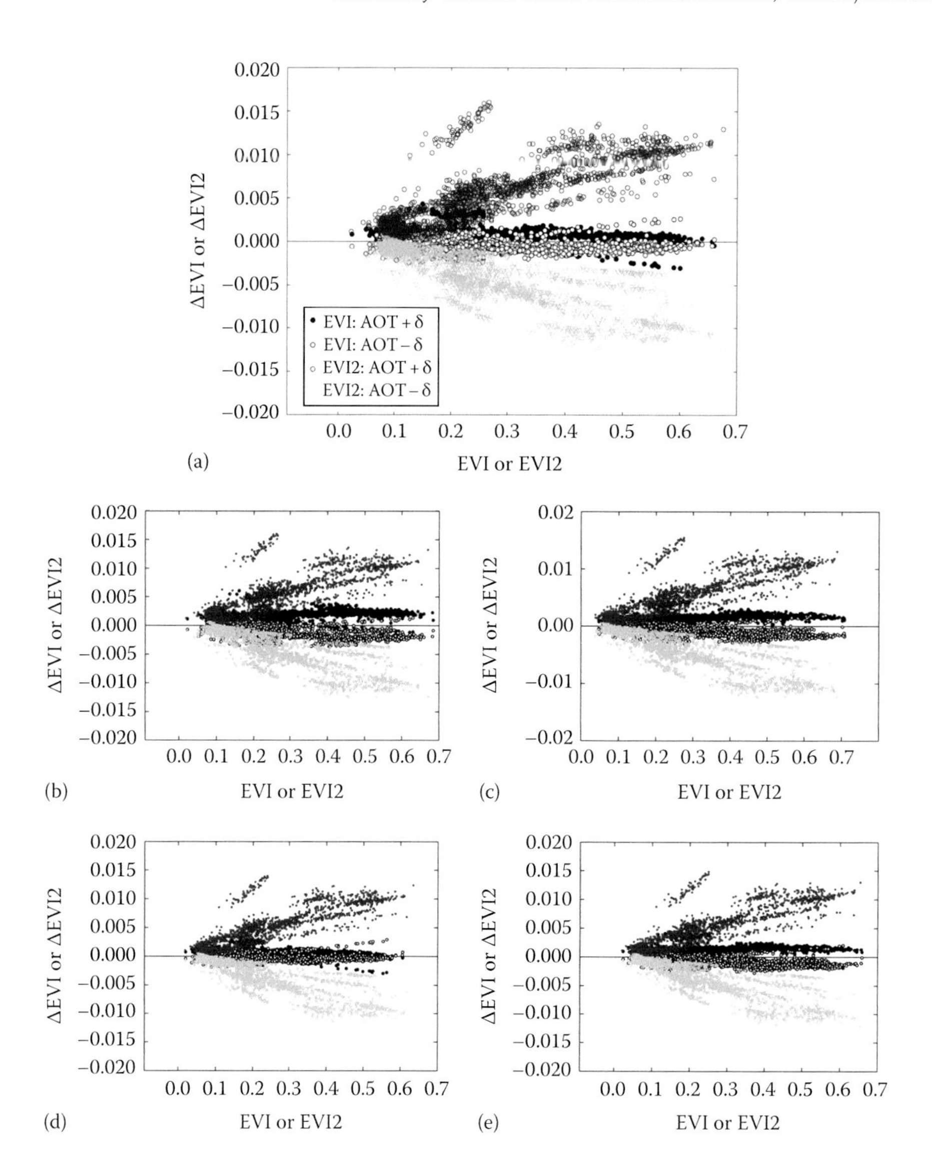

FIGURE 7.5 Comparison of EVI and EVI2 resistance to atmospheric aerosol correction error for five sensor bandpasses. (a) MODIS, (b) VIIRS, (c) SGLI, (d) VGT4, and (e) ETM+.

TABLE 7.7 Comparison between EVI and EVI2 Resistance to Residual Aerosol Effects

RMSD	EVI	EVI2	EVI2-to-EVI Ratio
VIIRS	0.002	0.006	3.5
MODIS	0.001	0.006	6.8
SGLI	0.001	0.006	4.5
VGT4	0.001	0.005	7.6
ETM+	0.001	0.005	4.8

7.6 Intersensor Calibration of EVI and EVI2

A small number of studies have evaluated and/or proposed intersensor calibration for the EVI and EVI2. These cross-calibration methods can be divided into empirical (Kim et al., 2010) and theoretical (Yoshioka et al., 2012; Obata et al., 2013) approaches.

Kim et al. (2010) used a simple linear model and derived EVI-to-EVI and EVI2-to-EVI2 spectral translation equations between MODIS and select medium-resolution sensors:

$$X_{\text{Sensor}} = a \cdot X_{\text{MODIS}} + b + \varepsilon \tag{7.6}$$

where

X is the EVI or EVI2

The subscript, *Sensor*, refers to either VIIRS, VGT4, SeaWiFS, or AVHRR/2

ε is the unexplained residual error

Kim et al. (2010) also developed intrasensor EVI-to-EVI2 translation equations with the simple linear model for the VIIRS, VGT4, and SeaWiFS bandpasses:

$$\text{EVI2}_{\text{Sensor}} = a \cdot \text{EVI}_{\text{Sensor}} + b + \varepsilon \tag{7.7}$$

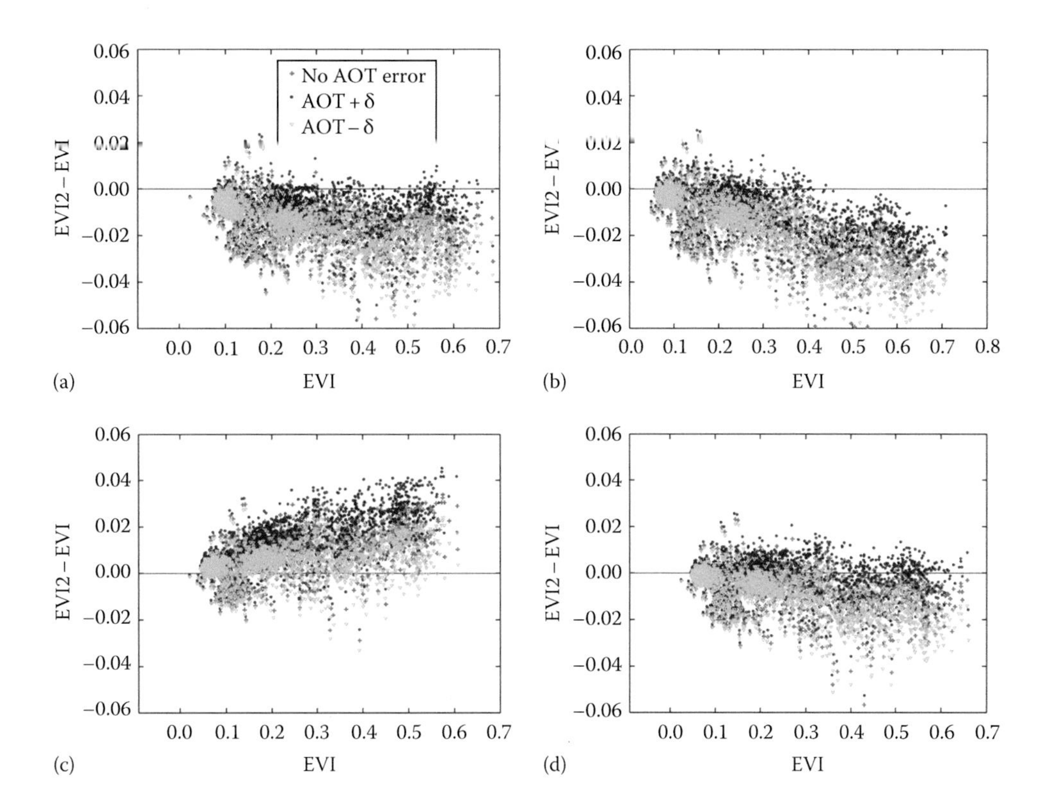

FIGURE 7.6 Intrasensor EVI vs. EVI2 differences subject to atmospheric aerosol correction error for four sensor bandpasses. (a) VIIRS, (b) SGLI, (c) VGT4, and (d) ETM+.

TABLE 7.8 Difference Statistics for Intrasensor EVI2 and EVI Compatibility

	TOC with No AOT Error			TOC with AOT Error of ±0.05			Ratio (TOC Error/TOC No Error)		
	RMSD	sRMPD	uRMPD	RMSD	sRMPD	uRMPD	RMSD	sRMPD	uRMPD
VIIRS	0.016	0.014	0.008	0.016	0.013	0.010	1.02	0.96	1.14
SGLI	0.018	0.016	0.008	0.018	0.015	0.010	1.01	0.97	1.16
VGT4	0.012	0.010	0.007	0.013	0.010	0.008	1.11	1.05	1.22
ETM+	0.010	0.007	0.007	0.011	0.006	0.009	1.08	0.93	1.18

Mean absolute deviations/differences (MAD) were used to evaluate translation results:

$$\mathrm{MAD} = \frac{1}{n}\sum_{i=1}^{n}\left|X_{1,i} - X_{2,i}\right| \tag{7.8}$$

where X's are the EVI, EVI2, translated (predicted) EVI, or translated (predicted) EVI2 using the derived simple linear equations and two of the four variables are selected and inserted into Equation 7.8 depending on the quantity of interest. Hyperion data collected along a tropical forest–savanna ecogradient in Brazil were used in all of these derivations and evaluations of cross-calibration equations.

The derived, simple linear spectral translation equations performed well on the Hyperion-simulated band-pass data (Table 7.10). For both the EVI and EVI2, MAD were reduced to 0.005 or less (in EVI/EVI2 units) upon the translations for most cases, which corresponded to 31%–87% reductions in MAD (Table 7.10). For the EVI-to-EVI2 spectral calibration, the equations did not perform as good as those for the intersensor calibration. Reductions in MAD were 35% or less and the largest MAD of 0.0154 was observed for VGT4 (Table 7.10).

Yoshioka et al. (2012) and Obata et al. (2013) analytically derived EVI and/or EVI2 spectral translation equations. Both studies used the vegetation isoline equations in the derivations, which are based upon the physics of vegetation–photon interactions (Yoshioka, 2004).

Yoshioka et al. (2012) used the vegetation isoline equations first to eliminate NIR and blue reflectances from the EVI equation or NIR reflectance from the EVI2 equation and then to relate two EVI or EVI2 using the red reflectances. Derived is a spectral transformation equation that relates two EVI or two EVI2:

$$v_a = G \cdot \frac{h_1 v_b - h_2}{h_3 v_b - h_4} \tag{7.9}$$

where v_a and v_b are the EVI or EVI2 for sensors a and b, respectively. The four coefficients, h_i ($i = 1, \ldots, 4$), are actually functions,

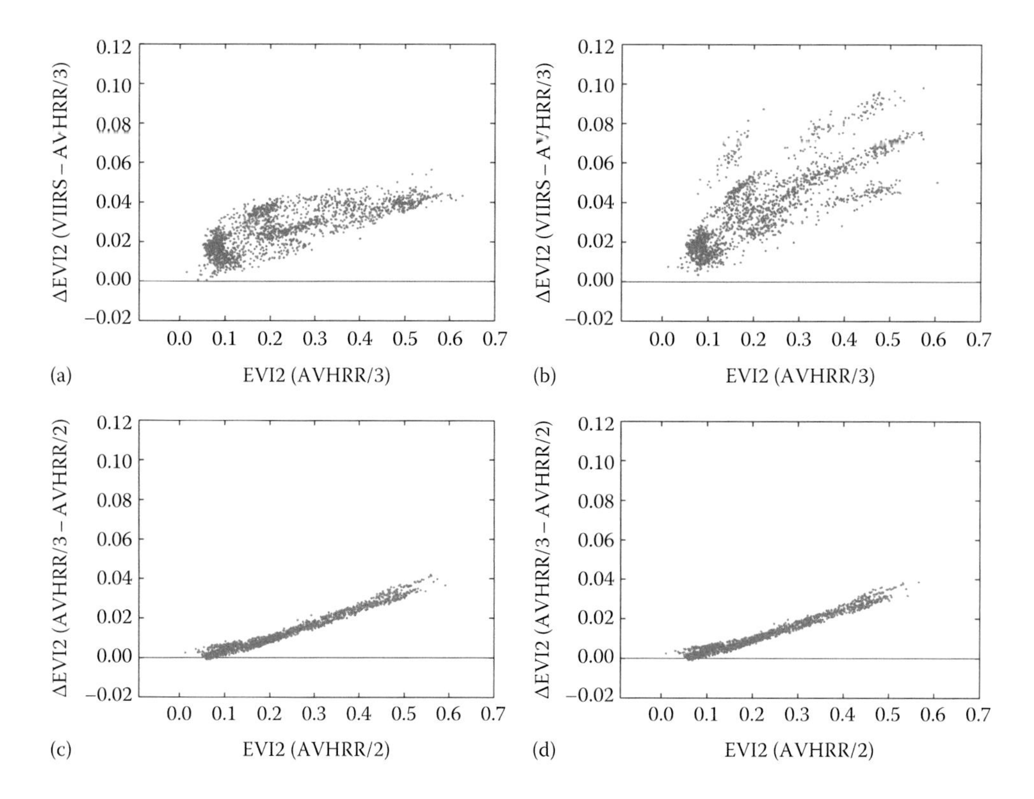

FIGURE 7.7 Intersensor EVI2 difference under two atmospheric correction scenarios: (a) VIIRS vs. AVHRR/3 for TOC, (b) VIIRS vs. AVHRR/3 for PAC, (c) AVHRR/3 vs. AVHRR/2 for TOC, and (d) AVHRR/3 vs. AVHRR/2 for PAC.

TABLE 7.9 Difference Statistics of Intersensor AVHRR EVI2 Compatibility

	TOC vs. TOC			TOC vs. PAC			Ratio (PAC/TOC)		
EVI2 vs. AVHRR EVI2	RMSD	sRMPD	uRMPD	RMSD	sRMPD	uRMPD	RMSD	sRMPD	uRMPD
VIIRS vs. AVHRR/3	0.029	0.028	0.007	0.042	0.041	0.011	1.5	1.5	1.7
MODIS vs. AVHRR/3	0.031	0.030	0.006	0.044	0.043	0.011	1.4	1.4	1.8
SGLI vs. AVHRR/3	0.035	0.034	0.008	0.049	0.047	0.013	1.4	1.4	1.6
VGT4 vs. AVHRR/3	0.008	0.007	0.004	0.018	0.015	0.010	2.2	2.2	2.2
ETM+ vs. AVHRR/3	0.014	0.013	0.005	0.027	0.025	0.010	1.9	1.9	1.9
VIIRS vs. AVHRR/2	0.043	0.042	0.006	0.055	0.054	0.010	1.3	1.3	1.8
MODIS vs. AVHRR/2	0.045	0.045	0.006	0.058	0.057	0.010	1.3	1.3	1.8
SGLI vs. AVHRR/2	0.050	0.050	0.008	0.063	0.062	0.012	1.3	1.2	1.5
VGT4 vs. AVHRR/2	0.018	0.017	0.005	0.030	0.029	0.010	1.7	1.7	1.9
ETM+ vs. AVHRR/2	0.027	0.027	0.006	0.040	0.038	0.010	1.4	1.4	1.7
AVHRR/3 vs. AVHRR/2	0.015	0.015	0.002	0.014[a]	0.014[a]	0.001[a]	0.9	0.9	1.0

See Appendix 7.A.4 for corresponding GMFR equations.
[a] These statistics are for AVHRR/3 PAC EVI2 vs. AVHRR/2 PAC EVI2.

and their values change with vegetation amount, and soil and atmosphere conditions.

In Obata et al. (2013), MODIS red, NIR, and blue reflectances were expressed as a function of VIIRS red, NIR, and blue reflectances, respectively, using the vegetation isoline equations. Substituting MODIS reflectances with the functions resulted in a MODIS-compatible EVI, $\hat{v}_m$:

$$\hat{v}_m = G \cdot \frac{\rho_{NIR} - K_1\rho_{red} + K_2}{\rho_{NIR} + K_1C_1\rho_{red} - K_3C_2\rho_{blue} + K_4} \quad (7.10)$$

where K_i ($i = 1,\ldots, 4$) are the spectral transformation coefficients. As h_i in Equation 7.9, K_i are functions and change with vegetation, soil, and atmosphere conditions.

Obata et al. (2013), however, found via optimization for both model-simulated and actual satellite datasets that a single set of coefficient values can be derived with a reasonable level of accuracy. Their simulation dataset was generated with the PROSPECT+SAIL canopy reflectance model (Feret et al., 2008) and the 6S atmospheric radiative transfer model. The actual MODIS-VIIRS dataset was populated by extracting pairs of

TABLE 7.10 Comparison of Mean Absolute Deviation before and after Spectral Translation Using Simple Linear Cross-Calibration Equation

EVI intersensor spectral compatibility with MODIS			
	MAD for EVISensor vs. EVI_{MODIS}	MAD for EVI_{Sensor} vs. pred. EVI_{Sensor}	Reduction (%)
VIIRS	0.0118	0.0052	56
VGT4	0.0207	0.0042	80
SeaWiFS	0.0108	0.0070	35
EVI2 intersensor spectral compatibility with MODIS			
	MAD for EVI2Sensor vs. $EVI2_{MODIS}$	MAD for $EVI2_{Sensor}$ vs. pred. $EVI2_{Sensor}$	Reduction (%)
VIIRS	0.0027	0.0018	33
VGT4	0.0255	0.0033	87
SeaWiFS	0.0048	0.0033	31
AVHRR/2	0.0411	0.0059	86
EVI-to-EVI2 intrasensor compatibility			
	MAD for $EVI2_{Sensor}$ vs. EVI_{Sensor}	MAD for EVI2Sensor vs. pred. $EVI2_{Sensor}$	Reduction (%)
VIIRS	0.0108	0.0077	29
VGT4	0.0236	0.0154	35
SeaWiFS	0.0103	0.0091	12

Source: Adapted from Kim, Y. et al., *J. Appl. Remote Sens.*, 4, 043520, 2010.

Aqua MODIS and VIIRS daily nadir-view surface reflectance spectra on the same date (in August 2013) at the same coordinates over North America. Refer to Obata et al. (2013) for further details on these datasets.

For the simulated dataset, the mean of δ_1 (MODIS EVI minus VIIRS EVI) and δ_2^* (MODIS EVI minus MODIS-compatible VIIRS EVI) were −0.008 and 0.0001, respectively, and variability of δ_2^* was much smaller than that of δ_1 (RMSD changed from 0.01 to 0.0017, an 83% reduction) (Obata et al., 2013). For the actual dataset, the mean and RMSD of δ_2^* were 0.001 and 0.023, a 95% and 41% reduction, respectively, from those of δ_1 (−0.022 and 0.039) (Figure 7.8).

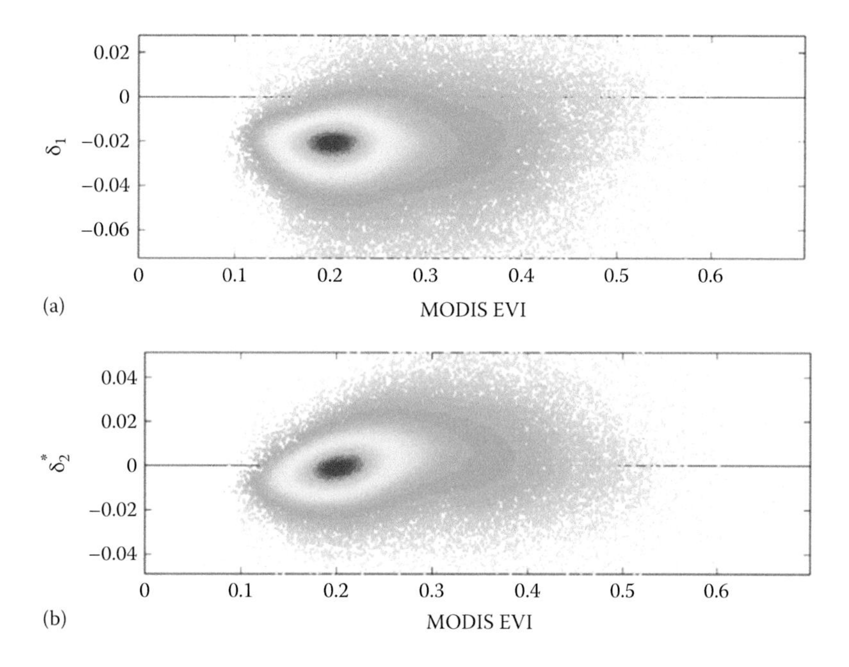

FIGURE 7.8 Density plot of EVI differences against MODIS EVI with actual MODIS and VIIRS data. (a) MODIS EVI minus VIIRS EVI (δ_1) against MODIS EVI (v_m) and (b) MODIS EVI minus MODIS compatible VIIRS EVI using optimized parameters (δ_2^*) against MODIS EVI (v_m). Ranges of y axis for the density plots were 0.1. (Adapted and reprinted with permission from Obata et al., *J. Appl. Remote Sens.* 7, 073467, 2013.)

7.7 Discussions and Future Directions

In this chapter, inter- and intrasensor compatibilities of the EVI and EVI2 were reviewed with a focus on spectral, atmospheric, and calibration issues. Overall, the EVI2 showed higher intersensor spectral compatibilities than the EVI. The highest spectral compatibility was observed between MODIS EVI2 and VIIRS EVI2, and AVHRR/2 EVI2 and AVHRR/3 EVI2 although both intersensor EVI2 relationships were subject to systematic differences. When the residual aerosol effects were considered, however, intersensor spectral compatibilities of the EVI2 and EVI were at the comparable level. The blue band in the EVI is advantageous in reducing residual aerosol effects, but brings an added complication in intersensor compatibilities. Likewise, the higher intersensor EVI2 compatibilities do not necessarily translate into higher absolute accuracies of the EVI2.

Among the limited number of studies and available cross-calibration methods, the approach of Obata et al. (2013) appears to be the most promising. The most significant advantage of the approach would be that the translation technique simultaneously optimizes for band-pass differences and for the EVI coefficients, which cannot be accomplished with the simple linear regression approach. The potential pitfall in the approach is a nonexistence of the unique set of coefficient values, as it requires nonlinear regression to find the optimum coefficient values.

Intersensor spectral compatibility of AVHRR EVI2 with other sensors is challenging, in particular with MODIS, VIIRS, and SGLI of which red and NIR spectral bands are much narrower than those of AVHRR (i.e., Figure 7.1). Backward compatibility of these sensors with AVHRR EVI2 would be a feasible option only when aerosol source of variability are reduced in AVHRR data either by atmospheric correction or by temporal compositing to select cleanest observations. Another approach would be to generate more than one EVI/EVI2 long-term data record in parallel, for example, one solely from the AVHRR sensor series and another by merging MODIS and VIIRS EVI products. Intercomparisons of the two data records over the current overlapping period can be used to develop a methodology for analyzing historical AVHRR EVI2 time series.

A large number of empirical investigations with actual satellite data are needed and the logical next step. For cross-sensor compatibility and calibration of actual satellite data, additional factors need to be taken into consideration, including differences in spatial resolution, overpass time (morning vs. afternoon), and sun-target-viewing geometry (sun and view zenith angle differences due to different sensor platform orbits) (e.g., Sims et al., 2011; Morton et al., 2014).

Appendix 7.A Inter- and Intrasensor Spectral Relationships for EVI and EVI2

The following equations were estimated for the Hyperion dataset (without AOT errors) described in Section 7.5.1 by GMFR (Tables 7.A.1 through 7.A.4).

TABLE 7.A.1 Intersensor EVI vs. EVI Spectral Relationships

Y vs. *X*	Intercept[a]	Slope[a]	R^2
VIIRS vs. MODIS	0.010	1.033	0.999
VIIRS vs. SGLI	0.029	0.893	0.997
VIIRS vs. VGT4	0.033	1.083	0.997
VIIRS vs. ETM+	0.038	0.975	0.997
MODIS vs. SGLI	0.019	0.864	0.998
MODIS vs. VGT4	0.023	1.048	0.999
MODIS vs. ETM+	0.027	0.944	0.999
SGLI vs. VGT4	0.005	1.213	0.998
SGLI vs. ETM+	0.010	1.092	0.999
VGT4 vs. ETM+	0.004	0.900	1.000

[a] Useful for trend analysis; use with caution on actual satellite data.

TABLE 7.A.2 Intersensor EVI2 vs. EVI2 Spectral Relationships

Y vs. *X*	Intercept[a]	Slope[a]	R^2
VIIRS vs. MODIS	0.001	0.989	1.000
VIIRS vs. SGLI	0.021	0.918	0.998
VIIRS vs. VGT4	0.026	1.018	0.998
VIIRS vs. ETM+	0.031	0.971	0.998
MODIS vs. SGLI	0.020	0.929	0.999
MODIS vs. VGT4	0.025	1.030	0.999
MODIS vs. ETM+	0.030	0.982	0.998
SGLI vs. VGT4	0.006	1.108	0.999
SGLI vs. ETM+	0.011	1.057	0.999
VGT4 vs. ETM+	0.005	0.954	1.000

[a] Useful for trend analysis; use with caution on actual satellite data.

TABLE 7.A.3 Intrasensor EVI2 vs. EVI Relationships

EVI2 vs. EVI	Intercept[a]	Slope[a]	R^2
VIIRS	−0.006	0.973	0.997
SGLI	0.002	0.946	0.998
VGT4	0.000	1.036	0.998
ETM+	0.000	0.977	0.998

[a] Useful for trend analysis; use with caution on actual satellite data.

TABLE 7.A.4 Intersensor Relationships for AVHRR EVI2

	TOC vs. TOC			TOC vs. PAC		
Y vs. *X*	Intercept[a]	Slope[a]	R^2	Intercept[b]	Slope[b]	R^2
VIIRS vs. AVHRR/3	0.013	1.058	0.998	0.008	1.148	0.999
MODIS vs. AVHRR/3	0.012	1.070	0.998	−0.013	1.236	0.997
SGLI vs. AVHRR/3	−0.009	1.152	0.997	−0.017	1.115	0.999
VGT4 vs. AVHRR/3	−0.013	1.039	0.999	−0.023	1.169	0.998
ETM+ vs. AVHRR/3	−0.018	1.089	0.999	0.007	1.199	0.995
VIIRS vs. AVHRR/2	0.011	1.119	0.994	0.006	1.212	0.995
MODIS vs. AVHRR/2	0.010	1.132	0.994	−0.016	1.305	0.993
SGLI vs. AVHRR/2	−0.011	1.219	0.993	−0.019	1.178	0.995
VGT4 vs. AVHRR/2	−0.015	1.099	0.995	−0.025	1.234	0.995
ETM+ vs. AVHRR/2	−0.021	1.152	0.995	−0.004	1.073	1.000
AVHRR/3 vs. AVHRR/2	0.009	1.135	0.998	−0.004[c]	1.071[c]	1.000

[a] Useful for trend analysis; use with caution on actual satellite data.
[b] Useful for trend analysis; do not use on actual satellite data.
[c] The intercept and slope values are for AVHRR/3 PAC EVI2 vs. AVHRR/2 PAC EVI2.

References

Anaya, J.A., E. Chuvieco, and A. Palacios-Orueta. 2009. Aboveground biomass assessment in Colombia: A remote sensing approach. *Forest Ecol. Manage.* 257:1237–1246. doi:10.1016/j.foreco.2008.11.016.

Betbeder, J., V. Gond, F. Frappart, N. Baghdadi, G. Briant, and E. Bartholome. 2014. Mapping of central Africa forested wetlands using remote sensing. *IEEE J. Sel. Topics Appl. Earth Observ.* 7:531–542. doi:10.1109/JSTARS.2013.2269733.

Boelman, N.T., A.V. Rocha, and G.R. Shaver. 2011. Understanding burn severity sensing in Arctic tundra: Exploring vegetation indices, suboptimal assessment timing and the impact of increasing pixel size. *Int. J. Remote Sens.* 32:7033–7056. doi:10.1080/01431161.2011.611187.

Boles, S.H., X.M. Xiao, J.Y. Liu, Q.Y. Zhang, S. Munkhtuya, S.Q. Chen, and D. Ojima. 2004. Land cover characterization of Temperate East Asia using multi-temporal VEGETATION sensor data. *Remote Sens. Environ.* 90:477–489. doi:10.1016/j.rse.2004.01.016.

Bolton, D.K. and M.A. Friedl. 2013. Forecasting crop yield using remotely sensed vegetation indices and crop phenology metrics. *Agric. For. Meteorol.* 173:74–84. doi:10.1016/j.agrformet.2013.01.007.

Cabacinha, C.D. and S.S. de Castro. 2009. Relationships between floristic diversity and vegetation indices, forest structure and landscape metrics of fragments in Brazilian Cerrado. *Forest Ecol. Manage.* 257:2157–2165. doi:10.1016/j.foreco.2009.02.030.

Changyong, C., F.J. De Luccia, X. Xiaoxiong, R. Wolfe, and W. Fuzhong. 2014. Early on-orbit performance of the Visible Infrared Imaging Radiometer Suite onboard the Suomi National Polar-Orbiting Partnership (S-NPP) satellite. *IEEE Trans. Geosci. Remote Sens.* 52:1142–1156. doi:10.1109/TGRS.2013.2247768.

Chen, B., N.C. Coops, D. Fu, H.A. Margolis, B.D. Amiro, A.G. Barr, T.A. Black et al. 2011a. Assessing eddy-covariance flux tower location bias across the Fluxnet-Canada Research Network based on remote sensing and footprint modelling. *Agric. For. Meteorol.* 151:87–100. doi:10.1016/j.agrformet.2010.09.005.

Chen, M., Q. Zhuang, D.R. Cook, R. Coulter, M. Pekour, R.L. Scott, J.W. Munger, and K. Bible. 2011b. Quantification of terrestrial ecosystem carbon dynamics in the conterminous United States combining a process-based biogeochemical model and MODIS and AmeriFlux data. *Biogeosciences.* 8:2665–2688. doi:10.5194/bg-8-2665-2011.

Didan, K. and A. Huete. 2006. MODIS Vegetation Index Product Series Collection 5 Change Summary. http://landweb.nascom.nasa.gov/QA_WWW/forPage/MOD13_VI_C5_Changes_Document_06_28_06.pdf, last accessed on July 13, 2014.

Fensholt, R., I. Sandholt, and S. Stisen. 2006. Evaluating MODIS, MERIS, and VEGETATION—Vegetation indices using in situ measurements in a semiarid environment. *IEEE Trans. Geosci. Remote Sens.* 44:1774–1786. doi:10.1109/TGRS.2006.875940.

Feret, J.-B., C. François, G.P. Asner, A.A. Gitelson, R.E. Martin, L.P.R. Bidel, S.L. Ustin, M. le, Guerric, and S. Jacquemoud. 2008. PROSPECT-4 and 5: Advances in the leaf optical properties model separating photosynthetic pigments. *Remote Sens. Environ.* 112:3030–3043. doi:10.1016/j.rse.2008.02.012.

Ferreira, L.G., H. Yoshioka, A. Huete, and E.E. Sano. 2003. Seasonal landscape and spectral vegetation index dynamics in the Brazilian Cerrado: An analysis within the Large-Scale Biosphere-Atmosphere Experiment in AmazÙnia (LBA). *Remote Sens. Environ.* 87:534–550. doi:10.1016/j.rse.2002.09.003.

Friedl, M.A., D. Sulla-Menashe, B. Tan, A. Schneider, N. Ramankutty, A. Sibley, and X. Huang. 2010. MODIS Collection 5 global land cover: Algorithm refinements and characterization of new datasets. *Remote Sens. Environ.* 114:168–182. doi:10.1016/j.rse.2009.08.016.

Guanter, L., Y. Zhang, M. Jung, J. Joiner, M. Voigt, J.A. Berry, C. Frankenberg et al. 2014. Global and time-resolved monitoring of crop photosynthesis with chlorophyll fluorescence. *Proc. Natl. Acad. Sci. U. S. A.* 111:E1327–E1333. doi:10.1073/pnas.1320008111.

Guerschman, J.P., A.I.J.M. Van Dijk, G. Mattersdorf, J. Beringer, L.B. Hutley, R. Leuning, R.C. Pipunic, and B.S. Sherman. 2009. Scaling of potential evapotranspiration with MODIS data reproduces flux observations and catchment water balance observations across Australia. *J. Hydrol.* 369:107–119. doi:10.1016/j.jhydrol.2009.02.013.

Hengl, T., N. Toomanian, H.I. Reuter, and M.J. Malakouti. 2007. Methods to interpolate soil categorical variables from profile observations: Lessons from Iran. *Geoderma.* 140:417–427. doi:10.1016/j.geoderma.2007.04.022.

Holben, B.N., D. Tanré, A. Smirnov, T.F. Eck, I. Slutsker, N. Abuhassan, W.W. Newcomb et al. 2001. An emerging ground-based aerosol climatology: Aerosol Optical Depth from AERONET. *J. Geophys. Res.* 106:12067–12097.

Houborg, R.M. and H. Soegaard. 2004. Regional simulation of ecosystem CO_2 and water vapor exchange for agricultural land using NOAA AVHRR and Terra MODIS satellite data. Application to Zealand, Denmark. *Remote Sens. Environ.* 93:150–167. doi:10.1016/j.rse.2004.07.001.

Huete, A., K. Didan, T. Miura, E.P. Rodriguez, X. Gao, and L.G. Ferreira. 2002. Overview of the radiometric and biophysical performance of the MODIS vegetation indices. *Remote Sens. Environ.* 83:195–213. doi:10.1016/S0034-4257(02)00096-2.

Huete, A.R. 1988. A soil-adjusted vegetation index (SAVI). *Remote Sens. Environ.* 25:295–309.

Huete, A.R., K. Didan, Y.E. Shimabukuro, P. Ratana, S.R. Saleska, L.R. Hutyra, W.Z. Yang, R.R. Nemani, and R. Myneni. 2006. Amazon rainforests green-up with sunlight in dry season. *Geophys. Res. Lett.* 33:L06405.

Huete, A.R., H.Q. Liu, K. Batchily, and W.J. van Leeuwen. 1997. A comparison of vegetation indices over a global set of TM images for EOS-MODIS. *Remote Sens. Environ.* 59:440–451. doi:10.1016/S0034-4257(96)00112-5.

Huettich, C., U. Gessner, M. Herold, B.J. Strohbach, M. Schmidt, M. Keil, and S. Dech. 2009. On the suitability of MODIS time series metrics to map vegetation types in dry savanna ecosystems: A case study in the Kalahari of NE Namibia. *Remote Sens.* 1:620–643. doi:10.3390/rs1040620.

Ichii, K., H. Hashimoto, M.A. White, C. Potters, L.R. Hutyra, A.R. Huete, R.B. Myneni, and R.R. Nemanis. 2007. Constraining rooting depths in tropical rainforests using satellite data and ecosystem modeling for accurate simulation of gross primary production seasonality. *Global Change Biol.* 13:67–77. doi:10.1111/j.1365-2486.2006.01277.x.

Irons, J.R., J.L. Dwyer, and J.A. Barsi. 2012. The next Landsat satellite: The Landsat Data Continuity Mission. *Remote Sens. Environ.* 122:11–21.

Jaegermeyr, J., D. Gerten, W. Lucht, P. Hostert, M. Migliavacca, and R. Nemani. 2014. A high-resolution approach to estimating ecosystem respiration at continental scales using operational satellite data. *Global Change Biol.* 20:1191–1210. doi:10.1111/gcb.12443.

Ji, L. and K. Gallo. 2006. An agreement coefficient for image comparison. *Photogramm. Eng. Remote Sens.* 72:823–833.

Jiang, Z., A.R. Huete, K. Didan, and T. Miura. 2008. Development of a two-band enhanced vegetation index without a blue band. *Remote Sens. Environ.* 112:3833–3845. doi:10.1016/j.rse.2008.06.006.

John, R., J. Chen, N. Lu, K. Guo, C. Liang, Y. Wei, A. Noormets, K. Ma, and X. Han. 2008. Predicting plant diversity based on remote sensing products in the semi-arid region of Inner Mongolia. *Remote Sens. Environ.* 112:2018–2032. doi:10.1016/j.rse.2007.09.013.

Kaufman, Y.J. and D. Tanré. 1992. Atmospherically Resistant Vegetation Index (ARVI) for EOS-MODIS. *IEEE Trans. Geosci. Remote Sens.* 30(2):261–270.

Kim, Y., A.R. Huete, T. Miura, and Z. Jiang. 2010. Spectral compatibility of vegetation indices across sensors: A band decomposition analysis with Hyperion data. *J. Appl. Remote Sens.* 4:043520. doi:10.1117/1.3400635.

Kim, Y., J.S. Kimball, K. Didan, and G.M. Henebry. 2014. Response of vegetation growth and productivity to spring climate indicators in the conterminous United States derived from satellite remote sensing data fusion. *Agric. For. Meteorol.* 194:132–143.

Kobayashi, H., R. Suzuki, and S. Kobayashi. 2007. Reflectance seasonality and its relation to the canopy leaf area index in an eastern Siberian larch forest: Multi-satellite data and radiative transfer analyses. *Remote Sens. Environ.* 106:238–252. doi:10.1016/j.rse.2006.08.011.

Kotchenova, S.Y. and E.F. Vermote. 2007. Validation of a vector version of the 6S radiative transfer code for atmospheric correction of satellite data. Part II. Homogeneous Lambertian and anisotropic surfaces. *Appl. Opt.* 46:4455–4464.

Li, Z., G. Yu, X. Xiao, Y. Li, X. Zhao, C. Ren, L. Zhang, and Y. Fu. 2007. Modeling gross primary production of alpine ecosystems in the Tibetan Plateau using MODIS images and climate data. *Remote Sens. Environ.* 107:510–519. doi:10.1016/j.rse.2006.10.003.

Liu, J., E. Pattey, and G. Jego. 2012. Assessment of vegetation indices for regional crop green LAI estimation from Landsat images over multiple growing seasons. *Remote Sens. Environ.* 123:347–358. doi:10.1016/j.rse.2012.04.002.

Mendez-Barroso, L.A., E.R. Vivoni, C.J. Watts, and J.C. Rodriguez. 2009. Seasonal and interannual relations between precipitation, surface soil moisture and vegetation dynamics in the North American monsoon region. *J. Hydrol.* 377:59–70. doi:10.1016/j.jhydrol.2009.08.009.

Middleton, E.M., P.K.E. Campbell, S.G. Ungar, L. Ong, Q. Zhang, K.F. Huemmrich, D.J. Mandl, and S.W. Frye. 2010. Using EO-1 Hyperion images to prototype environmental products for HyspIRI. In *2010 IEEE International Geoscience and Remote Sensing Symposium (IGARSS)*, June 25–30, Honolulu, HI. pp. 4256–4259. doi:10.1109/IGARSS.2010.5648946.

Miura, T., A.R. Huete, H. Yoshioka, and B.N. Holben. 2001. An error and sensitivity analysis of atmospheric resistant vegetation indices derived from dark target-based atmospheric correction. *Remote Sens. Environ.* 78:284–298. doi:10.1016/S0034-4257(01)00223-1.

Miura, T., J.P. Turner, and A.R. Huete. 2013. Spectral compatibility of the NDVI across VIIRS, MODIS, and AVHRR: An analysis of atmospheric effects using EO-1 Hyperion. *IEEE Trans. Geosci. Remote Sens.* 51:1349–1359. doi:10.1109/TGRS.2012.2224118.

Miura, T. and H. Yoshioka. 2011. Hyperspectral data in long-term, cross-sensor continuity studies. In *Hyperspectral Remote Sensing of Vegetation*. P.S. Thenkabail, J.G. Lyon, and A. Huete, eds. Taylor & Francis, New York. pp. 611–633.

Miura, T., H. Yoshioka, K. Fujiwara, and H. Yamamoto. 2008. Intercomparison of ASTER and MODIS surface reflectance and vegetation index products for synergistic applications to natural resource and environmental monitoring. *Sensors.* 8:2480–2499.

Moran, M.S., G.E. Ponce-Campos, A. Huete, M.P. McClaran, Y. Zhang, E.P. Hamerlynck, D.J. Augustine et al.. 2014. Functional response of U.S. grasslands to the early 21st-century drought. *Ecology.* 95:2121–2133. doi:10.1890/13-1687.1.

Morton, D.C., J. Nagol, C.C. Carabajal, J. Rosette, M. Palace, B.D. Cook, E.F. Vermote, D.J. Harding, and P.R.J. North. 2014. Amazon forests maintain consistent canopy structure and greenness during the dry season. *Nature.* 506:221–224. doi:Letter.

Nagler, P.L., R.L. Scott, C. Westenburg, J.R. Cleverly, E.P. Glenn, and A.R. Huete. 2005. Evapotranspiration on western U.S. rivers estimated using the Enhanced Vegetation Index from MODIS and data from eddy covariance and Bowen ratio flux towers. *Remote Sens. Environ.* 97:337–351. doi:10.1016/j.rse.2005.05.011.

O'Connell, J., J. Connolly, E.F. Vermote, and N.M. Holden. 2013. Radiometric normalization for change detection in peatlands: A modified temporal invariant cluster approach. *Int. J. Remote Sens.* 34:2905–2924. doi:10.1080/01431161.2012.752886.

Obata, K., T. Miura, H. Yoshioka, and A.R. Huete. 2013. Derivation of a MODIS-compatible enhanced vegetation index from visible infrared imaging radiometer suite spectral reflectances using vegetation isoline equations. *J. Appl. Remote Sens.* 7:073467. doi:10.1117/1.JRS.7.073467.

Pearlman, J.S., P.S. Barry, C.C. Segal, J. Shepanski, D. Beiso, and S.L. Carman. 2003. Hyperion, a space-based imaging spectrometer. *IEEE Trans. Geosci. Remote Sens.* 41:1160–1173.

Pfeifer, M., A. Gonsamo, M. Disney, P. Pellikka, and R. Marchant. 2012. Leaf area index for biomes of the Eastern Arc Mountains: Landsat and SPOT observations along precipitation and altitude gradients. *Remote Sens. Environ.* 118:103–115. doi:10.1016/j.rse.2011.11.009.

Pocewicz, A., L.A. Vierling, L.B. Lentile, and R. Smith. 2007. View angle effects on relationships between MISR vegetation indices and leaf area index in a recently burned ponderosa pine forest. *Remote Sens. Environ.* 107:322–333. doi:10.1016/j.rse.2006.06.019.

Ponce Campos, G.E., M.S. Moran, A. Huete, Y. Zhang, C. Bresloff, T.E. Huxman, D. Eamus et al. 2013. Ecosystem resilience despite large-scale altered hydroclimatic conditions. *Nature.* 494:349–352. doi:10.1038/nature11836.

Qi, J., A. Chehbouni, A.R. Huete, Y.H. Kerr, and S. Sorooshian. 1994. A modified soil adjusted vegetation index. *Remote Sens. Environ.* 48:119–126.

Rahman, A.F., D.A. Sims, V.D. Cordova, and B.Z. El-Masri. 2005. Potential of MODIS EVI and surface temperature for directly estimating per-pixel ecosystem C fluxes. *Geophys. Res. Lett.* 32:L19404. doi:10.1029/2005GL024127.

Rocha, A.V. and G.R. Shaver. 2009. Advantages of a two band EVI calculated from solar and photosynthetically active radiation fluxes. *Agric. For. Meteorol.* 149:1560–1563. doi:10.1016/j.agrformet.2009.03.016.

Rocha, A.V. and G.R. Shaver. 2011. Burn severity influences post-fire CO_2 exchange in arctic tundra. *Ecol. Appl.* 21:477–489. doi:10.1890/10-0255.1.

Saleska, S.R., K. Didan, A.R. Huete, and H.R. da Rocha. 2007. Amazon forests green-up during 2005 drought. *Science.* 318:612–612. doi:10.1126/science.1146663.

Schubert, P., F. Lagergren, M. Aurela, T. Christensen, A. Grelle, M. Heliasz, L. Klemedtsson, A. Lindroth, K. Pilegaard, T. Vesala, and L. Eklundh. 2012. Modeling GPP in the Nordic forest landscape with MODIS time series data—Comparison with the MODIS GPP product. *Remote Sens. Environ.* 126:136–147. doi:10.1016/j.rse.2012.08.005.

Sims, D.A., A.F. Rahman, V.D. Cordova, B.Z. El-Masri, D.D. Baldocchi, P.V. Bolstad, L.B. Flanagan et al. 2008. A new model of gross primary productivity for North American ecosystems based solely on the enhanced vegetation index and land surface temperature from MODIS. *Remote Sens. Environ.* 112:1633–1646. doi:10.1016/j.rse.2007.08.004.

Sims, D.A., A.F. Rahman, E.F. Vermote, and Z. Jiang. 2011. Seasonal and inter-annual variation in view angle effects on MODIS vegetation indices at three forest sites. *Remote Sens. Environ.* 115:3112–3120. doi:10.1016/j.rse.2011.06.018.

Sjostrom, M., J. Ardo, A. Arneth, N. Boulain, B. Cappelaere, L. Eklundh, A. de Grandcourt et al. 2011. Exploring the potential of MODIS EVI for modeling gross primary production across African ecosystems. *Remote Sens. Environ.* 115:1081–1089. doi:10.1016/j.rse.2010.12.013.

Son, N.T., C.F. Chen, C.R. Chen, L.Y. Chang, H.N. Duc, and L.D. Nguyen. 2013. Prediction of rice crop yield using MODIS EVI-LAI data in the Mekong Delta, Vietnam. *Int. J. Remote Sens.* 34:7275–7292. doi:10.1080/01431161.2013.818258.

Song, Y., J.B. Njoroge, and Y. Morimoto. 2013. Drought impact assessment from monitoring the seasonality of vegetation condition using long-term time-series satellite images: A case study of Mt. Kenya region. *Environ. Monit. Assess.* 185:4117–4124. doi:10.1007/s10661-012-2854-z.

Soudani, K., C. Francois, G. le Maire, V. Le Dantec, and E. Dufrene. 2006. Comparative analysis of IKONOS, SPOT, and ETM+ data for leaf area index estimation in temperate coniferous and deciduous forest stands. *Remote Sens. Environ.* 102:161–175. doi:10.1016/j.rse.2006.02.004.

Sterckx, S., S. Livens, and S. Adriaensen. 2013. Rayleigh, deep convective clouds, and cross-sensor desert vicarious calibration validation for the PROBA-V mission. *IEEE Trans. Geosci. Remote Sens.* 51:1437–1452. doi:10.1109/TGRS.2012.2236682.

Swinnen, E. and F. Veroustraete. 2008. Extending the SPOT-VEGETATION NDVI time series (1998–2006) back in time with NOAA-AVHRR data (1985–1998) for Southern Africa. *IEEE Trans. Geosci. Remote Sens.* 46:558–572. doi:10.1109/TGRS.2007.909948.

USGS. 2011. EO-1 (Earth Observing-1). http://eros.usgs.gov/#/Find_Data/Products_and_Data_Available/ALI [Jan 31, 2012].

Vargas, M., T. Miura, N. Shabanov, and A. Kato. 2013. An initial assessment of Suomi NPP VIIRS vegetation index EDR. *J. Geophys. Res. Atmos.* 118:1–16. doi:10.1002/2013JD020439.

Vermote, E., D. Tanré, J.L. Deuzé, M. Herman, J.J. Morcrette, and S.Y. Kotchenova. 2006. Second Simulation of a Satellite Signal in the Solar Spectrum—Vector (6SV) User Guide Version 3.

Waring, R.H., K.S. Milner, W.M. Jolly, L. Phillips, and D. McWethy. 2006. Assessment of site index and forest growth capacity across the Pacific and Inland Northwest USA with a MODIS satellite-derived vegetation index. *Forest Ecol. Manage.* 228:285–291. doi:10.1016/j.foreco.2006.03.019.

Wittenberg, L., D. Malkinson, O. Beeri, A. Halutzy, and N. Tesler. 2007. Spatial and temporal patterns of vegetation recovery following sequences of forest fires in a Mediterranean landscape, Mt. Carmel Israel. *CATENA.* 71:76–83. doi:10.1016/j.catena.2006.10.007.

Wolfe, R.E., M. Nishihama, A.J. Fleig, J.A. Kuyper, D.P. Roy, J.C. Storey, and F.S. Patt. 2002. Achieving sub-pixel geolocation accuracy in support of MODIS land science. *Remote Sens. Environ.* 83:31–49. doi:10.1016/S0034-4257(02)00085-8.

Wu, C. 2012. Use of a vegetation index model to estimate gross primary production in open grassland. *J. Appl. Remote Sens.* 6:063532. doi:10.1117/1.JRS.6.063532.

Xiao, X., B. Braswell, Q. Zhang, S. Boles, S. Frolking, and B.I.I.I. Moore. 2003. Sensitivity of vegetation indices to atmospheric aerosols: Continental-scale observations in Northern Asia. *Remote Sens. Environ.* 84:385–392.

Xiao, X.M., D. Hollinger, J. Aber, M. Goltz, E.A. Davidson, Q.Y. Zhang, and B. Moore. 2004. Satellite-based modeling of gross primary production in an evergreen needleleaf forest. *Remote Sens. Environ.* 89:519–534. doi:10.1016/j.rse.2003.11.008.

Xiao, X.M., Q.Y. Zhang, D. Hollinger, J. Aber, and B. Moore. 2005a. Modeling gross primary production of an evergreen needleleaf forest using modis and climate data. *Ecol. Appl.* 15:954–969. doi:10.1890/04 0470.

Xiao, X., Q. Zhang, S. Saleska, L. Hutyra, C. De, Plinio, S. Wofsy, S. Frolking, S. Boles, M. Keller, and B.I.I.I. Moore. 2005b. Satellite-based modeling of gross primary production in a seasonally moist tropical evergreen forest. *Remote Sens. Environ.* 94:105–122.

Yamamoto, H., T. Miura, and S. Tsuchida. 2012. Advanced Spaceborne Thermal Emission and Reflection Radiometer (ASTER) Enhanced Vegetation Index (EVI) products from Global Earth Observation (GEO) Grid: An assessment using Moderate Resolution Imaging Spectroradiometer (MODIS) for synergistic applications. *Remote Sens.* 4:2277–2293. doi:10.3390/rs4082277.

Yang, F., K. Ichii, M.A. White, H. Hashimoto, A.R. Michaelis, P. Votava, A.-X. Zhu, A. Huete, S.W. Running, and R.R. Nemani. 2007. Developing a continental-scale measure of gross primary production by combining MODIS and AmeriFlux data through Support Vector Machine approach. *Remote Sens. Environ.* 110:109–122. doi:10.1016/j.rse.2007.02.016.

Yang, F., M.A. White, A.R. Michaelis, K. Ichii, H. Hashimoto, P. Votava, A.-X. Zhu, and R.R. Nemani. 2006. Prediction of continental-scale evapotranspiration by combining MODIS and AmeriFlux data through support vector machine. *IEEE Trans. Geosci. Remote Sens.* 44:3452–3461. doi:10.1109/TGRS.2006.876297.

Yang, X., J.F. Mustard, J. Tang, and H. Xu. 2012. Regional-scale phenology modeling based on meteorological records and remote sensing observations. *J. Geophys. Res. Biogeo.* 117:G03029. doi:10.1029/2012JG001977.

Yebra, M., A. Van Dijk, R. Leuning, A. Huete, and J.P. Guerschman. 2013. Evaluation of optical remote sensing to estimate actual evapotranspiration and canopy conductance. *Remote Sens. Environ.* 129:250–261. doi:10.1016/j.rse.2012.11.004.

Yoshioka, H. 2004. Vegetation isoline equations for an atmosphere-canopy-soil system. *IEEE Trans. Geosci. Remote Sens.* 42:166–175.

Yoshioka, H., T. Miura, and K. Obata. 2012. Derivation of relationships between spectral vegetation indices from multiple sensors based on vegetation isolines. *Remote Sens.* 4:583–597. doi:10.3390/rs4030583.

Zhang, X.Y., M.A. Friedl, and C.B. Schaaf. 2006. Global vegetation phenology from Moderate Resolution Imaging Spectroradiometer (MODIS): Evaluation of global patterns and comparison with in situ measurements. *J. Geophys. Res.* 111:G04017. doi:10.1029/2006JG000217.

Zhang, Y., M. Susan Moran, M.A. Nearing, G.E. Ponce Campos, A.R. Huete, A.R. Buda, D.D. Bosch et al. 2013. Extreme precipitation patterns and reductions of terrestrial ecosystem production across biomes. *J. Geophys. Res. Biogeosci.* 118:148–157. doi:10.1029/2012JG002136.

Zhao, D., K.R. Reddy, V.G. Kakani, J.J. Read, and S. Koti. 2007. Canopy reflectance in cotton for growth assessment and lint yield prediction. *Eur. J. Agron.* 26:335–344. doi:10.1016/j.eja.2006.12.001.

8

Toward Standardization of Vegetation Indices

Michael D. Steven
University of Nottingham

Timothy J. Malthus
Commonwealth Scientific and Industrial Research Organisation

Frédéric Baret
National Institute of Agronomic Research

Acronyms and Definitions

6S	Second Simulation of a Satellite Signal in the Solar Spectrum—an atmospheric correction model
AATSR	Advanced Along-Track Scanning Radiometer
ASTER	Advanced Spaceborne Thermal Emission and Reflection Radiometer
ATSR2	Along-Track Scanning Radiometer 2
AVHRR	Advanced very-high-resolution radiometer
BRDF	Bidirectional reflectance distribution function
CEOS	Committee on Earth Observation Satellites
CHRIS	Compact High-Resolution Imaging Spectrometer on the PROBA satellite
DMC	Disaster Monitoring Constellation
DN	Digital number
ETM+	Enhanced Thematic Mapper (Landsat)
EVI	Enhanced Vegetation Index
fAPAR	The fraction of photosynthetically active radiation absorbed by a vegetation canopy
Formosat	Name of a commercial high-resolution satellite
IKONOS	Name of a commercial high-resolution satellite
IRS	Indian Remote Sensing Satellite
ISO	International Organization for Standardization
Kompsat	Korean Multipurpose Satellite
LAI	Leaf area index
Landsat	A system of earth observation satellites in operation since 1972
LISS	Linear Imaging Self-Scanning Sensor (IRS)
MERIS	MEdium-Resolution Imaging Spectrometer
MISR	Multi-angle Imaging SpectroRadiometer
MODIS	MOderate-Resolution Imaging Spectrometer
MODIS-TIP	Two-stream inversion package applied to MODIS products by the European Joint Research Centre
MODTRAN	MODerate-resolution atmospheric TRANsmission—an atmospheric correction model
MSAVI	Modified Soil-Adjusted Vegetation Index
MSG/SEVIRI	Meteosat Second Generation/Spinning Enhanced Visible and Infrared Imager
MSS	Multispectral Scanner (Landsat)
NASA	National Aeronautics and Space Administration
NDVI	Normalized difference vegetation index
NOAA	National Oceanic and Atmospheric Administration

OrbView	Name of a commercial high-resolution satellite
OSAVI	Optimized Soil-Adjusted Vegetation Index
PAR	Photosynthetically Active Radiation
POLDER	POLarization and Directionality of the Earth's Reflectances—an optical imaging radiometer
PROSAIL	A combination of the PROSPECT leaf optical properties model with the SAIL canopy bidirectional reflectance model
PROSPECT	A model of leaf optical properties
QuickBird	Name of a commercial high-resolution satellite
SAIL	Scattering by Arbitrarily Inclined Leaves—a canopy bidirectional reflectance model
SAVI	Soil-Adjusted Vegetation index
SeaWIFS	Sea-Viewing Wide Field-of-View Sensor
SPOT	Système Probatoire d'Observation de la Terre
SPOT-VEG	The VEGETATION sensor on the SPOT satellite
TM	Thematic Mapper (Landsat)
TSAVI	Transformed Soil-Adjusted Vegetation Index
VEGETATION	1 km resolution monitoring instrument on the SPOT satellite
Venμs	Vegetation and Environment monitoring on a New MicroSatellite
VI	Vegetation index

8.1 Introduction: Vegetation Indices and Their Uses

The concept of the vegetation index (VI) is one of the lasting success stories of terrestrial remote sensing. The physiological and anatomical characteristics of vegetation give rise to distinctive spectral features that allow its presence to be detected in any environment and with suitable precautions permit the properties of the vegetation canopy to be inferred from the reflected spectrum. Healthy vegetation absorbs visible (especially red) light via chlorophyll and other pigments. In the near-infrared (NIR) where no absorbers are active, light is strongly reflected by foliage because the juxtaposition of cells, essentially containing water, with air spaces between, creates a strongly scattering medium (Gates et al. 1965; Gausman and Allen 1973). The resulting reflectance spectrum (Figure 8.1) is highly characteristic and is recognizable as the spectral signature of vegetation even when distorted by other environmental variables. When used to monitor growing vegetation, the spectral signal of vegetation is mixed with that of soil or other backgrounds, which tend to show a much flatter response across this spectral region. The result is that the NIR tends to increase with vegetation cover, for example as a crop grows, while the red reflectance decreases, ending at 100% cover with a spectrum close (but not identical) to that of a single leaf.

The key factor in the development of VIs is the increasing contrast between the reflectance in the two bands. To encapsulate this contrast, early work identified a variety of combinations

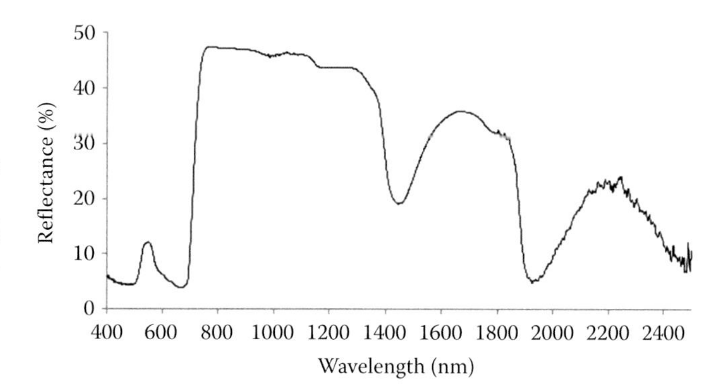

FIGURE 8.1 Laboratory-measured reflectance spectrum of a bean leaf. (From the Author.) The jitters at longer wavelengths are instrumental noise.

of the NIR and visible bands, of which the best known by far (although by far from the best) is the Normalized Difference Vegetation Index (NDVI), defined as

$$\text{NDVI} = \frac{\rho_{\text{NIR}} - \rho_{\text{red}}}{\rho_{\text{NIR}} + \rho_{\text{red}}} \tag{8.1}$$

where ρ is the spectral reflectance (dimensionless) in the NIR or red spectral band. The NDVI is functionally equivalent to the simple ratio of the two bands (Perry and Lautenshlager 1984). However, its formulation ensures that the value of the NDVI ranges strictly from −1 to +1, which is computationally more convenient than the ratio, which has no upper bound. A wide variety of alternative formulations have been used, for example, the Soil-Adjusted Vegetation Index (SAVI) (Huete 1988), defined as

$$\text{SAVI} = (1+L)\frac{(\rho_{\text{NIR}} - \rho_{\text{red}})}{(\rho_{\text{NIR}} + \rho_{\text{red}} + L)} \tag{8.2}$$

The formulation of SAVI differs from NDVI by including a *soil calibration factor L* that adjusts for variability in the index introduced by soil reflectance characteristics. Although Huete (1988) found that the optimal value of L varied with vegetation density, a midrange value of 0.5 was found to provide effective correction for variations due to soil background across the full range of densities. Later variants on this approach varied the value of L, with Rondeaux et al. (1996) proposing Optimized SAVI (OSAVI) with $L = 0.16$ on the basis of an optimization across a range of agricultural soils. Others include the Modified Soil-Adjusted Vegetation Index (MSAVI) (Qi et al. 1994), which employs a self-adjusted value of L based on the spectral reflectance data themselves, and the Transformed Soil-Adjusted Vegetation Index (TSAVI) (Baret and Guyot 1991), where L is based on information about the soil characteristics. Most VIs are reformulations based on the same two spectral bands, but some, such as the Enhanced Vegetation Index (EVI) (Huete et al. 1997), introduce additional spectral information, usually in an attempt to reduce atmospheric sensitivity. Each index has its advocates and many have particular merits, but key features are that they are intrinsically dimensionless and that they all ultimately share the characteristic of using the contrast between NIR and red reflectance as the primary measure of vegetation. For most of this discussion, the term *vegetation index* (VI) will be used generically to apply to all of them.

8.1.1 Vegetation Index Applications

The applications of VIs are based on their ability to measure foliage density in a consistent manner across a wide range of vegetation types. In early field studies VIs were successfully related to leaf area index (LAI) defined as the total (single-sided) area of leaf per unit area of ground (m^2/m^2, i.e., dimensionless), canopy chlorophyll ($g\ m^{-2}$), wet and dry biomass ($g\ m^{-2}$), the fraction of ground covered by leaves (dimensionless), primary productivity ($g\ m^{-2}\ day^{-1}$), the fraction of photosynthetically active radiation absorbed by a vegetation canopy (fAPAR) (dimensionless), and other variables (Tucker 1977; Tucker et al. 1979; Holben et al. 1980; Steven et al. 1983). As all of these factors are measures of foliage density, they tend to be highly intercorrelated in any individual study. However, these variables are hierarchically linked by the process of canopy photosynthesis that converts absorbed PAR to fixed energy in biomass (Figure 8.2). The reflection of light by a plant canopy is largely determined by the total area of leaf and the projections of that area toward both the source of illumination and the detection device. This introduces a dependence on leaf angle distribution (Verhoef 1985) and to a lesser extent other factors such as clumping (Gower et al. 1999). As they involve similar projections, this is closely connected to the way in which leaf area and angle combine to determine leaf cover fraction or PAR absorption (Steven et al. 1986), so that as indicated in Figure 8.2 the response of a VI relates most directly to these variables. An illustration of this link is that in spite of variations with time of day, Pinter (1993) found that the relationship of fAPAR with

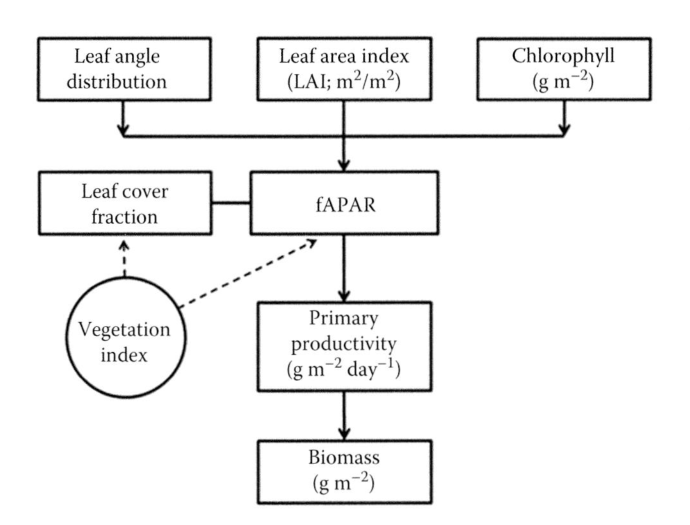

FIGURE 8.2 Hierarchical diagram showing the relationships between various measures of canopy density and their link with vegetation indices.

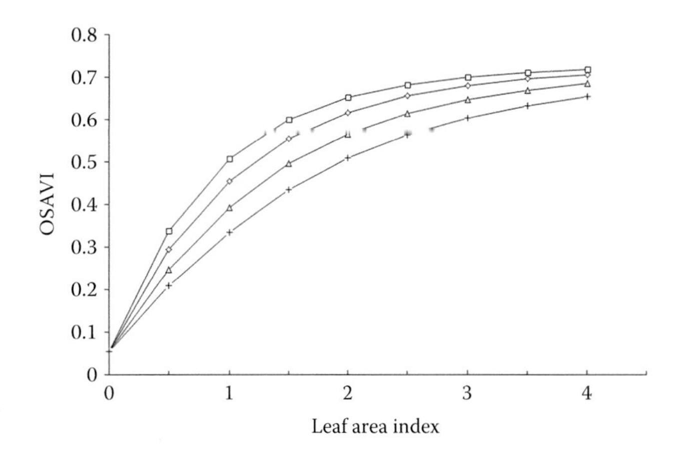

FIGURE 8.3 Sensitivity of OSAVI to leaf angle distribution for a range of leaf area indices, for ellipsoidal distributions with mean leaf angles of 30° (□), 45° (◊), 57.3° (Δ), and (65°) +. All the parameters are dimensionless. (Adapted from Steven, M.D., *Remote Sens. Environ.*, 63, 49, 1998.)

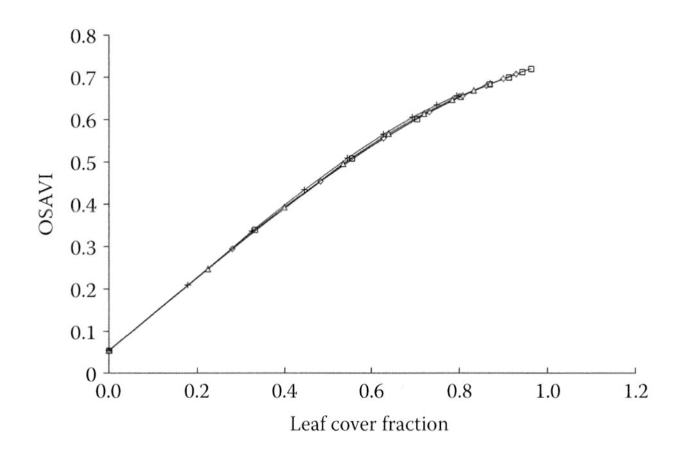

FIGURE 8.4 Sensitivity of OSAVI cover estimates to leaf angle distribution. Parameters as for Figure 8.3. (Adapted from Steven, M.D., *Remote Sens. Environ.*, 63, 49, 1998.)

various VIs was independent of illumination angle. Figure 8.3 shows the modeled relationship between OSAVI (representative here of VIs generally) and LAI for canopies with different leaf angle distributions, while Figure 8.4 shows that the relationships essentially collapse onto a single curve when expressed as functions of leaf cover fraction (Steven 1998). The same behavior can be found with NDVI, where the merger of curves for different leaf angle distributions is slightly less tight, although not enough to introduce serious variations. The implication of this merger is that there is redundancy in the biophysical parameters and that VIs can be used to measure foliage density in a way that is largely independent of canopy structure and, to some extent, species. Moreover as leaf cover fraction is itself closely related, through similar dependence on leaf area and structure, to the interception of solar radiation by plant canopies (Steven et al. 1986), VIs lend themselves to the estimation of light capture by vegetation (or fAPAR). Following Monteith (1977) who established the model outlined in Figure 8.2, showing that light capture was the key determinant of the conversion of solar energy to biomass, remote sensing of this variable leads to direct estimation of pro ductivity in crops (Steven et al. 1983; Wiegand et al. 1991). The same argument allows VIs to be applied in large-scale monitoring of vegetation dynamics (Goward et al. 1993), with a range of applications in agriculture (Maselli et al. 2000) or ecology (Pettorelli et al. 2005). Direct estimation of fAPAR on a global scale is now the focus of several major space programs, using the MOderate-Resolution Imaging Spectrometer (MODIS), MEdium-Resolution Imaging Spectrometer (MERIS), Sea-Viewing Wide Field-of-View Sensor (SeaWIFS), the two-stream inversion package applied to MODIS products by the European Joint Research Centre (MODIS-TIP), the VEGETATION sensor on the Système Probatoire d'Observation de la Terre satellite (SPOT-VEG), and Advanced Very-High-Resolution Radiometer (AVHRR) systems (Picket-Heaps et al. 2014).

8.2 The Need for Standards

All sciences commence with a lengthy period of exploration, with a diversity of idiosyncratic approaches until the difficulties of rationalizing different methodologies lead to a movement for standardization. The benefits of standardization that are particularly pertinent to remote sensing are interoperability of systems and data continuity. The use of standards also helps to generate authority for standardized products and with increasing use, a greater familiarity with their capabilities and limits.

A persistent issue in vegetation monitoring is the acquisition of sufficient data to capture the dynamics of plant growth. Plant growth requires water, usually supplied by rainfall, so the more productive vegetated regions are frequently cloudy, obstructing the view of satellite sensors (Heller 1961). The data acquisition problem for satellite systems has broadly been resolved in two ways: at higher resolutions by the development of systems such as SPOT or IKONOS with pointable cameras that can target particular sites several times within the satellite repeat cycle and at lower resolutions with more frequent data acquisition by selectively compositing over periods of 10 days or more (Gutman 1991; van Leeuwen et al. 1999). Alternatively, vegetation monitoring can be performed by aircraft that can fly beneath the cloud deck (Jaggard and Clark 1990) or by monitoring systems mounted on mobile farm machinery for precision agriculture. All these solutions introduce their own problems. In particular, both pointing and compositing tend to increase the range of viewing angles and to a lesser extent solar angles, used in the VI product. Monitoring under cloud requires additional data to normalize for incident solar irradiance and measures different reflectance characteristics corresponding to the multidirectional diffuse illumination of the target. Corrections can be made for angular effects using a model of the bidirectional reflectance distribution function (BRDF), but require *a priori* knowledge of the vegetation type (Steven 1998; Bacour et al. 2006).

A complementary approach is to combine data from more than one system, sometimes referred to as the use of a virtual constellation (CEOS 2006; Martínez-Beltrán et al. 2009). Long-term environmental analysis may also require meta analysis of data from a range of systems (Boyd and Foody 2011). Key to these approaches is the adoption of a set of operating standards for the systems to be combined. The increasing focus on long-term continuity of vegetation observations, particularly for monitoring environmental change at larger scales, has led to considerable interest in back-calibrating data from earlier systems such as Landsat (operational since 1972) and AVHRR (since 1978), as near as possible to current standards, to establish the long-term baseline (Brown et al. 2006; Samain et al. 2006; Swinnen and Veroustrate 2008). Precise calibration of the instruments is required and attention to variations in BRDF associated with different orbital characteristics (Röder et al. 2005; Teillet et al. 2006; Martínez-Beltrán et al. 2009). Thus, standardization is required first for individual sensors, to account for variability in calibration and other observational parameters over time. Second, standardization is required between sensors to allow the interoperability of different systems. Coupled with these instrumental issues is the need to specify standard observing conditions, with unambiguous correction procedures to account for deviations from the standard. Achievement of these goals would in principle allow consistent monitoring of vegetation across time and across spatial scales. However, although standards for many activities within the field of remote sensing are being promoted by Committee on Earth Observation Satellites (CEOS), International Organization for Standardization (ISO), National Aeronautics and Space Administration (NASA), and other agencies, no concerted attempt has yet been made to standardize VIs.

8.2.1 Vegetation Index Formula

At present, a wide range of VIs are generated by different earth observation systems and data users. Le Maire et al. (2004) evaluated 61 indices for their sensitivity to chlorophyll across different types of leaves, and Agapiou et al. (2012) evaluated 71 established indices for distinguishing archaeological crop marks. Neither list is exhaustive. As well as highlighting the inadequacies of NDVI, the enormous proliferation of indices indicates both the need for a standard and the difficulty of achieving one. Although a number of indices have been developed that are claimed to represent specific properties of vegetation—water content, chlorophyll, carotenoids, and radiation use efficiency (Peñuelas and Filella 1998)—these usually require high spectral resolution and are not the concern here. In general, the different broadband VIs are highly interrelated and are usually used to represent the same vegetation parameters. In most cases, their differences lie in the generality of the data that have been used to test them and the degree to which they can suppress factors extraneous to the estimation of vegetation characteristics, such as soil or atmospheric effects. However, different indices do have distinct responses and may yield differing estimates of derived parameters: Boyd et al. (2011) found that estimates of phenological event dates such as the start and end of the growing season differed by several days when different indices were used, with uncertainties from these differences "as large as those arising from climatic perturbations."

8.2.2 Alternatives to Standard Vegetation Indices

While a standard VI would therefore seem desirable, it is nonetheless worth discussing the alternatives. If standards are developed prematurely, there is a danger of lock-in to a standard that proves to be inadequate. To some extent, this is the case with the NDVI. Many papers have pointed out its excessive sensitivity to soil characteristics (Huete 1988; Rondeaux et al. 1996), but having been adopted as an operational product from the early 1980s (Townshend and Justice 1986), it has become the *de facto* standard for later systems; a recent web search (June 2014) reported over 5000 academic papers on this index. One alternative to VIs is to estimate vegetation parameters directly by inverting a vegetation canopy reflectance model with earth observation data. As the basis for productivity estimation, fAPAR has been recognized as an essential climate variable for global modeling and is already a routine product of the MODIS, MERIS, VEGETATION, and Meteosat Second Generation/Spinning Enhanced Visible and Infrared Imager (MSG/SEVIRI) systems. Gobron et al. (2008) evaluated the effects of radiometric uncertainties on the MERIS product and estimated that errors in fAPAR estimation should be ≤ 0.1, but D'Odorico et al. (2014) found inconsistencies, particular in forest, when comparing three fAPAR algorithms over Europe. Similar issues were found by Picket-Heaps et al. (2014) who tested six alternative fAPAR products over Australia. They suggest that current fAPAR products are not reliable enough to be fed into biogeochemical process models or used in data fusion approaches. A further difficulty with these products is that they rely on different types of data input including, in most cases, multiangular reflectance data. While greater reliability may be achievable with future systems or better models, it would not be realistic to generate the equivalent product from historic data where key inputs are missing. However, a VI, although less directly applicable as an input to process models (it remains just an index), can within limits be standardized.

8.3 Sources of Variation in Vegetation Indices

Although VIs respond primarily to foliage density, however expressed, they are critically affected by a range of other factors. Van Leeuwen (2006) distinguishes between uncertainties related to input parameters, VI formulation, and product generation issues such as compositing rules. Most of the discussion here relates to the first of these categories. Table 8.1 classifies the sources of extraneous variation into environmental, observational, and instrumental categories, as discussed further as follows.

8.3.1 Soil Background

Dependence on soil reflectance is inherent in the formulation of VIs as measures of vegetation–soil contrast. With the NDVI, darker soils will tend to amplify the vegetation component of the signal, while brighter soils will tend to suppress it (Huete 1988; Rondeaux et al. 1996). Field studies that relate NDVI to measures of foliage density are almost invariably conducted on a single soil type and almost inevitably achieve relationships with high correlations; but subsequent attempts to transfer the relationships to other environments are often disappointing due to changes in the background effect of soil type. Individual soils also decrease in brightness with wetting (Bowers and Hanks 1965; Rondeaux et al. 1996). Soil effects in NDVI can be as high as 50% of the dynamic range, but are considerably reduced in the SAVI range of indices (Rondeaux et al. 1996), corresponding to a maximum error of about 0.05 in the estimation of leaf cover fraction.

8.3.2 Atmospheric Effects

Atmospheric effects change the radiances measured so that the top-of-atmosphere reflectances generate a different VI from that observed at the surface. It is essential to point out here that to standardize a VI requires the index to be based on surface reflectances as defined in Equation 8.1 or 8.2. Any other measurement, such as top-of-atmosphere reflectance, represents a combination of signals from both vegetation and atmosphere, but using equivalent surface reflectance data eliminates many of the errors and allows comparison of indices (Guyot and Gu

TABLE 8.1 Sources of Extraneous Variation in Vegetation Indices

Source of Variation	Approach to Overcome	Key References
Environmental factors		
Soil background	Soil-adjusted indices	Huete (1988) and Rondeaux et al. (1996)
Atmosphere	Atmospheric modeling	Vermote et al. (1997) and Berk et al. (1998)
Observation parameters		
Solar angle	Canopy BRDF modeling	Jacquemoud et al. (2009) and Vermote et al. (2009)
Viewing angle		
Instrumental factors		
Pixel size	Aggregation, scale modeling	Martínez-Beltrán et al. (2009) and Obata et al. (2012)
Spectral bands	VI intercalibration	Steven et al. (2003); this chapter

1994). Measurements must also be calibrated according to best practice (Price 1987). Many studies in the past have applied the VI concept rather loosely, some even using uncalibrated digital numbers (DNs) to compute the index. Such formulations can provide strong correlations with vegetation density measures in individual studies, but cannot easily be compared with studies using different instruments or formulations. Zhou et al. (2009) found large sensor-dependent differences between NDVIs for various systems depending on whether they were DN based, radiance based, or reflectance based. Hadjimitsis et al. (2010) found a mean difference of 18% between uncorrected and atmospherically corrected NDVI values and more modest, but still troublesome, differences in a range of other indices. Miura et al. (2013) evaluated atmospheric effects on intersensor compatibility and showed that atmospheric correction to top of canopy led to the greatest consistency between systems. Corrections for atmospheric effects to retrieve the surface reflectance are therefore a strict necessity. A variety of empirical and model-based approaches are available (Mahiny and Turner 2007). The Second Simulation of a Satellite Signal in the Solar Spectrum (6S) (Vermote et al. 1997) and MODerate-resolution atmospheric TRANsmission (MODTRAN) (Berk et al. 1998) atmospheric models are widely used, but difficulties can arise in acquisition of the atmospheric parameters required to make the correction, particularly in remote locations, with uncertainties associated mainly with aerosol content (Nagol et al. 2009). Fortunately, atmospheric effects in VIs are somewhat limited by the ratio construction of most indices coupled with the fact that the visible and NIR bands are to a large extent affected in similar ways by atmospheric aerosol; Steven (1998) found that OSAVI estimates of leaf cover fraction for mid-latitude summer monitoring conditions were relatively insensitive to quite large errors in modeled atmospheric parameters, resulting in no more than 4–5% error in the cover estimate.

8.3.3 Directional Effects

The complex structure of vegetation canopies as assemblages of leaves and other components suspended above the surface gives rise to a more or less complex BRDF that leads to a VI dependence on solar and viewing angle. The BRDF of vegetation is largely controlled by the relative fractions of illuminated soil, foliage, and shadow visible in a given direction (Guyot 1990; Schluessel et al. 1994). At large incidence angles, the scene will appear to be fully vegetated, even when LAI is small, whereas the nadir view will show a greater proportion of soil and maximizes sensitivity to leaf area and canopy structure. In general, variations of reflected radiance with either solar or viewing angle are greatest for erectophile leaf angle distributions and least for planophile distributions. Verrelst et al. (2008) showed that VIs showed large variations with viewing angle (and, by the principle of reciprocity, solar angle) that are dependent on both index and vegetation type. Sims et al. (2011) and Moura et al. (2012) also found that directional effects were index sensitive, being greater for EVI than for NDVI. If uncorrected, solar and view angle effects may vary with orbital drift of long-term satellites (Tucker et al. 2005; Brown et al. 2006). Angular effects also occur as a result of the compositing approaches used to generate cloud free vegetation products from global datasets (van Leeuwen et al. 1999). Where the vegetation characteristics are known, angular effects on VIs can be estimated by canopy models (Shultis 1991; Steven 1998; Verrelst et al. 2008; Jacquemoud et al. 2009), and as the solar and viewing angles are known precisely for any observation, the residual errors are not large. For global-scale vegetation monitoring, where it is impractical to apply individual models to specific vegetation types, Vermote et al. (2009) demonstrated that BRDF corrections could be made on the basis of an assumption that the shape of the BRDF varies more slowly than the magnitude of the reflectance. The shape can then be quantified by two parameters *R* and *V* (related in broad terms to vegetation and roughness) that can be derived by inversion of the time series.

BRDF effects can also occur with cloudiness, topography, and atmospheric state. These factors change the relative contribution of diffuse solar irradiance that comprises a distribution of incident angles (Steven 1977; Steven and Unsworth 1980) interacting with different parts of the BRDF and generating spectral and angular reflectances different to those from the unidirectional direct solar irradiance. By comparing canopy measurements made under full sunlight and simulated cloud (entirely diffuse illumination), Steven (2004) found differences up to 0.15 in NDVI between canopies in sunlight and shade, but demonstrated that the canopy spectra under standard conditions could be reconstructed to a precision of 10%–15% from the shaded measurements. Overall, these studies indicate that although angular responses are problematic and complex, they are susceptible to correction by modeling the BRDF.

8.3.4 Pixel-Size Effects

A number of studies have commented on the effects of pixel aggregation when integrating data from multiple sources (van Leeuwen 2006; Tarnavsky et al. 2008; Martínez-Beltrán et al. 2009; Munyati and Mboweni 2013). A fundamental issue is that VIs are not linear functions of reflectance, so that spatial averages of the VI are not precisely equivalent to VI values calculated on the basis of averaging the radiances or reflectances (Figure 8.5). The difference is not large in the example shown, but would in principle increase with greater heterogeneity of the surface and with the number of pixels to be aggregated. Obata et al. (2012) developed a theoretical basis for dealing with scaling effects based on monotonic behavior of the effect as a function of spatial resolution. Martínez-Beltrán et al. (2009) indicate that in practice this nonlinearity has not been found to be the major issue and that geolocation uncertainty is a more serious source of error in intersensor comparison. However, Munyati and Mboweni (2013) found that aggregation in areas of sparse, patchy vegetation could lead to underestimation of productivity.

Red

0.1	0.15	0.2
0.15	0.2	0.25
0.2	0.25	0.3

NIR

0.6	0.55	0.5
0.6	0.55	0.5
0.6	0.55	0.5

Averages 0.200 0.550

NDVI from average = 0.467

NDVI

0.71	0.57	0.43
0.60	0.47	0.33
0.50	0.38	0.25

Average NDVI = 0.471

FIGURE 8.5 Hypothetical illustration on the effect of NDVI nonlinearity on aggregation of nine pixels into one. The mean of the NDVIs across a 3 × 3 pixel area (lower box) is not identical to the NDVI calculated from the mean reflectance values (upper two boxes).

8.3.5 Spectral Band Effects

In addition to variability of the VI itself, any standardization protocol must also account for differences between measurement systems. Even when instruments are precisely calibrated and all the proper corrections are applied for BRDF and atmospheric effects, indices from the various measurement systems differ systematically due to differences in the position, width, and shape of the wavebands used (Gallo and Daughtry 1987; Guyot and Gu 1994). Band placement does not critically affect the behavior or strength of relationships with vegetation density, but can give rise to large differences in the fitted coefficients (Lee et al. 2004; Soudani et al. 2006). Differences in index are inevitable because different sensors measure different parts of the vegetation–soil reflectance spectrum and consequently respond in differing degrees to the biophysical variables concerned. In addition, Teillet and Ren (2008) found that spectral differences are generated by the spectral dependence of atmospheric gas transmittance. However, it transpires that in most cases these differences are more quantitative than qualitative, with almost perfect correlations between indices recorded by different systems (Steven et al. 2003, 2007). This study proposes the application of the relationships established in these papers as a key step in the standardization of VIs.

8.4 Vegetation Index Intercalibration Approaches

Intercomparison of VIs can be performed by direct comparison of measurements made by different instruments, by simulation from model spectra, or by partial simulation (of band responses) from field reflectance spectra. Direct comparisons are limited to particular instruments and are subject to errors due to nonsimultaneity, inexact image coregistration, and atmospheric effects; they also fail to distinguish spectral band effects from sensor calibration errors. Although model spectra are highly versatile, particularly for sensitivity tests, their direct application requires supreme confidence that the model parameterization captures all the important sources of variation. Simulation from data has the advantage in this context in that the spectral data are realistic, the simulated results are not limited to particular instruments, and the intercomparisons are all made on a common dataset.

8.4.1 Intercalibration of Vegetation Indices after Steven et al. (2003)

The analysis by Steven et al. (2003, 2007) used a database of high-resolution spectra of vegetation canopies to simulate NIR and visible band responses of particular instruments and then compared VIs as measured by different simulated observing systems. The database, described in more detail by Steven et al. (2003), comprised a set of 166 nadir-viewing bidirectional reflectance spectra measured over canopies of contrasting architecture (sugar beet and maize) in the United Kingdom and France in experiments conducted in 1989 and in 1990. The plant canopies had a full range of LAI and soil backgrounds, while contrasting leaf colors were achieved by treatments with disease or diluted herbicide. Spectral band responses were simulated by convolving the top-of-canopy spectral radiance data with the full spectral response function of each of the sensors tested and normalizing with the corresponding convolved data for the reference panel used in the field, adjusted for its true reflectance. The spectral response functions were found in the literature, or from the web, or obtained by personal communication and were digitized every 1 nm to match the spectral data (Figure 8.6). The operators of the OrbView-2 and OrbView-3 systems were unwilling to release the spectral response functions of their instruments, so Steven et al. (2007) tested two alternative models: a box function across the nominal wavelength range and a Gaussian fitted so that the nominal waveband limits were the half-power points. The wavebands for Venμs, which had not been precisely defined, were also modeled with a Gaussian on the basis of the developer's advice. The simulated band reflectances were then applied to compute

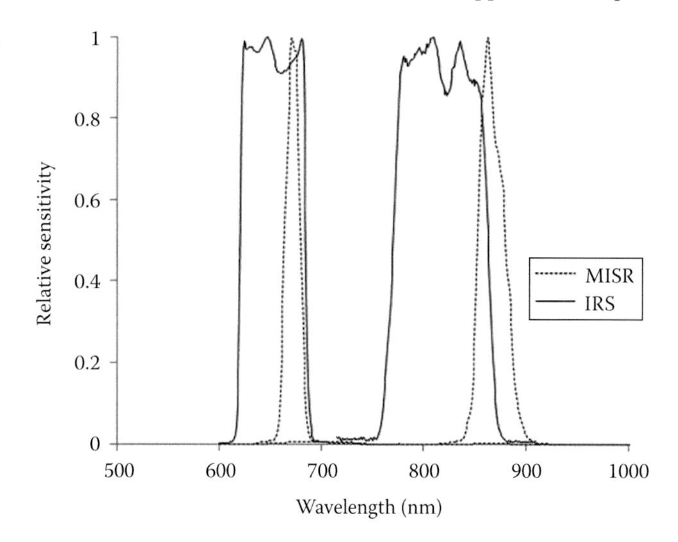

FIGURE 8.6 Example of spectral response functions for two of the systems simulated. (Adapted from Steven, M.D. et al., *Remote Sens. Environ.* 88, 412, 2003.)

the VIs NDVI, SAVI, and OSAVI for the different systems as well as for a hypothetically ideal narrow-band sensor pair based on narrow-bands at 670 and 815 nm, proposed as a standard. With narrow bands, the reflectance values are no longer sensitive to bandwidth or spectral response function. However, to reduce the effects of instrumental noise in the original database, bandwidths of 20 nm centered on the nominal wavelengths were used to determine the reflectances for the standard bands. The VIs from the different simulated systems were then compared.

Figures 8.7 and 8.8 show NDVI values for two systems compared with the corresponding NDVIs for the proposed standard bands at 670 and 815 nm. VIs from all systems were highly linearly correlated. NOAA8 and the Compact High-Resolution Imaging Spectrometer (CHRIS) as shown here represent the extremes of slope, with NDVI for NOAA8 being as much 19% lower than the standard on the same target. The strength and linearity of the correlations (minimum $r^2 = 0.984$) means that NDVIs recorded by different observation systems can be intercalibrated to a degree of precision of about ±1%. Steven et al. (2003) tabulated two-way intersystem conversion coefficients for 15 operational systems as well as the standard narrow band sensor pair. In a later update, Steven et al. (2007) extended the intercalibration of VIs forward to include conversion coefficients for orbiting sensor systems launched since the 2003 paper and backward to include historical variations in the NOAA AVHRR system, a total of 41 systems (Table 8.2). Two-way interconversion tables are no longer practical for the large number of systems involved, so for simplicity the conversions are given relative to an NDVI based on the standard pair of narrowbands at 670 and 815 nm. These bands were originally chosen to be close to

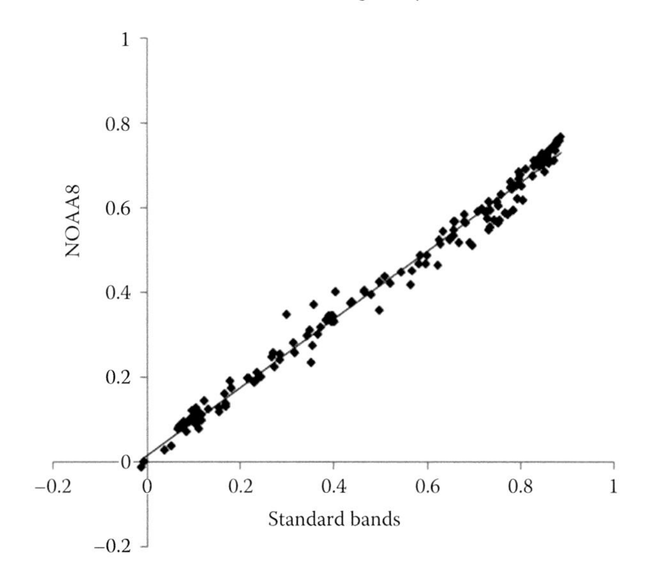

FIGURE 8.7 Regression of NDVI based on NOAA8 bands against the standard bands 670 ± 10 and 815 ± 10 nm. (Adapted from Steven, M.D. et al., Intercalibration of vegetation indices—An update, in M.E. Schaepman, eds., *10th International Symposium on Physical Measurements and Spectral Signatures in Remote Sensing*, International Archives of the Photogrammetry, Remote Sensing and Spatial Information Sciences, Vol. XXXVI, Part 7/C50, ISPRS, Davos (CH), pp. 1682–1777.)

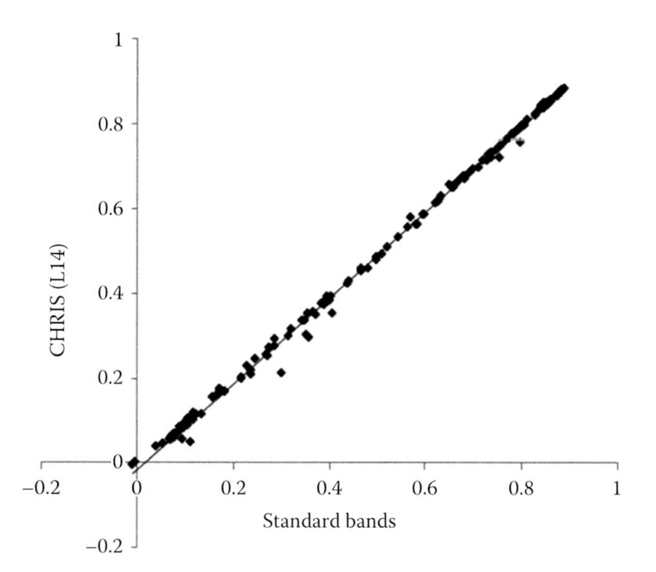

FIGURE 8.8 Regression of NDVI based on the CHRIS near-infrared (L14) and red bands against the standard bands. (Adapted from Steven, M.D. et al., Intercalibration of vegetation indices—An update, in M.E. Schaepman, eds., *10th International Symposium on Physical Measurements and Spectral Signatures in Remote Sensing*, International Archives of the Photogrammetry, Remote Sensing and Spatial Information Sciences, Vol. XXXVI, Part 7/C50, ISPRS, Davos (CH), pp. 1682–1777.)

the optimum, maximizing the NDVI value, although in fact the NDVI for the CHRIS system using the L14 NIR band (Figure 8.8) does have a slightly greater dynamic range.

Linear regressions for SAVI, OSAVI, and NDVI differ in slope and intercept by no more than about 0.0008 and 0.013 respectively, across the whole range of indices. As these differences are considerably less than typical errors of measurement, a single conversion table is adequate for the range of VI formulations considered. It is also possible to convert from one operational system to another using Table 8.2 to convert first to the standard as an intermediate stage and then from the standard to the second system. On examples tested, the error in this two-stage process, as compared with direct conversion between the systems, was up to 0.01 in slope and 0.007 in intercept.

Table 8.2 allows VI data from any of the listed systems to be corrected to the corresponding VI for the standard pair of spectral bands. To convert a VI from an operational system VI_{op} to the standard, VI_{std}, Equation 8.3 is applied, using the slope and intercept values from column B. To convert from the standard to a particular operational system, Equation 8.4 is applied, using the slope and intercept values from column A:

$$VI_{std} = VI_{op} \times [slope]_B + [intercept]_B \tag{8.3}$$

$$VI_{op} = VI_{std} \times [slope]_A + [intercept]_A \tag{8.4}$$

For the OrbView systems where the detailed spectral response functions were unavailable, there are two results in Table 8.2, corresponding to the alternative assumptions applied. For

TABLE 8.2 Conversion Coefficients for VIs (NDVI, SAVI, and OSAVI) from Different Systems (Steven, Malthus, and Baret 2007) Based on Simulations from the Database of the Full NIR and Visible Spectral Responses of Each System Where Alternative Options for a System Exist, Both Are Shown

	A		B	
	Sensor vs. Standard		Standard vs. Sensor	
Satellite Sensor	Intercept	Slope	Intercept	Slope
ALI	−0.005	0.965	0.006	1.034
ASTER, using band 3B	−0.001	0.933	0.003	1.068
ASTER, using band 3N	0.000	0.933	0.002	1.068
ATSR2/AATSR	0.008	0.968	−0.006	1.030
CHRIS, using band L14	−0.015	1.009	0.016	0.989
CHRIS, using band L15	0.005	0.991	−0.004	1.007
DMC	0.006	0.954	−0.005	1.046
Formosat	0.002	0.936	0.000	1.065
IKONOS	−0.010	0.870	0.015	1.144
IRS	0.005	0.950	−0.004	1.050
Kompsat	0.004	0.942	−0.003	1.058
Landsat 5 TM	0.005	0.938	−0.003	1.063
Landsat 7 ETM+	0.003	0.957	−0.002	1.041
Landsat MSS	0.029	0.883	−0.024	1.115
MERIS	0.008	0.983	−0.008	1.016
MISR	0.005	0.985	−0.005	1.014
MODIS	0.017	0.935	−0.015	1.065
MSG/SEVIRI	0.012	0.926	−0.010	1.076
NOAA10	0.003	0.854	0.001	1.160
NOAA11	0.015	0.831	−0.011	1.188
NOAA12	0.015	0.844	−0.012	1.173
NOAA13	0.017	0.835	−0.014	1.184
NOAA14	0.016	0.837	−0.013	1.180
NOAA15	0.016	0.902	−0.014	1.100
NOAA16	0.017	0.897	−0.015	1.107
NOAA17	0.016	0.904	−0.014	1.098
NOAA18	0.017	0.905	−0.014	1.097
NOAA6	0.021	0.850	−0.018	1.163
NOAA7	0.015	0.857	−0.012	1.155
NOAA8	0.015	0.807	−0.012	1.226
NOAA9	0.015	0.839	−0.012	1.179
OrbView-2, using block function	0.005	0.989	−0.004	1.009
OrbView-2, using Gaussian	0.005	0.982	−0.005	1.016
OrbView-3, using block function	0.002	0.937	0.000	1.063
OrbView-3, using Gaussian	0.002	0.857	0.001	1.159
POLDER	0.005	0.985	−0.005	1.014
QuickBird	0.000	0.909	0.002	1.096
SeaWIFS	0.005	0.982	−0.004	1.016
SPOT2 Hrv2	0.012	0.921	−0.011	1.081
SPOT4 Hrv2	0.010	0.917	−0.008	1.085
SPOT5	0.010	0.928	−0.008	1.073
Venµs, using band B10 with Gaussian	−0.012	0.984	0.013	1.015
Venµs, using band B11 with Gaussian	0.007	0.967	−0.006	1.032

OrbView-2, both methods give comparable results so that adjustment to the standard can be made to better than 1% precision. For OrbView-3, however, the difference is about 8%, indicating that in the absence of further information, this system is unsuitable for applications requiring intercalibration with others.

8.4.2 Validation of Cross-Sensor Conversion

Although direct intercomparisons of sensors suffer from greater errors and limitations than the simulation approach applied in Section 8.4.1, they are valuable in validating the findings. Steven et al. (2003) reported image-based comparisons between SPOT High-Resolution Visible (HRV) and Landsat Thematic Mapper (TM) data and between Along-Track Scanning Radiometer 2 (ATSR-2) and AVHRR. In both studies, pixels were aggregated across quasi-uniform targets to minimize coregistration errors and the empirical results were in reasonable agreement (± 0.03) with the simulated cross-sensor calibrations. Studies by Hill and Aifadopoulou (1990) and Guyot and Gu (1994) also generated coefficients that supported the findings, but previous simulations by Gallo and Daughtry (1987) were significantly different, although similarly based on simulation from field data. More recent studies have included both direct comparisons of image data and various forms of simulation. The quality of image comparisons in the literature is however quite variable, with many suffering from substantial differences in image acquisition time, different solar or viewing angles, insufficient aggregation to overcome coregistration errors, or other problems. Such studies may show differences between sensors but have insufficient precision to test conversion factors. Selected results are discussed as follows.

Martínez-Beltrán et al. (2009) compared Enhanced Thematic Mapper (ETM+), TM, Linear Imaging Self-Scanning Sensor (LISS), Advanced Spaceborne Thermal Emission and Reflection Radiometer (ASTER), QuickBird, and AVHRR data on selected sites in southeastern Spain covering a wide range of surface types (Figure 8.9). All images were near-nadir viewing and the image pairs compared were no more than a few hours apart. Although comparisons were of top-of-atmosphere NDVI, without atmospheric correction, they found that with sufficient aggregation to reduce noise, their cross-sensor relationships were linear, reasonably precise, and in good agreement with the results of Steven et al. (2003). However, Figure 8.9 shows the difficulty of establishing reliable cross-sensor relationships by direct comparison. In spite of the use of aggregation and homogenous areas for comparison, substantial differences in coefficients remain.

Ji et al. (2008) evaluated differences between AVHRR and MODIS in two years of data over the conterminous United States. In addition to their own findings that the differences are substantial and 20% systematic, they compare their results with 17 previous cross-sensor studies. Gallo et al. (2005) compared NDVI values for MODIS and AVHRR over the United States for identical 16-day compositing periods and found linear relationships between NDVI values from different sensors. Their regression slopes differ from Steven et al. (2003) by no more than 0.02 indicating good agreement within the limits of the data; although the compositing process can introduce a systematic upward bias in NDVI (Goward et al. 1993), this would probably be similar for both systems. Fensholt et al. (2006) compared MERIS, MODIS, and VEGETATION products on grass savannah in Senegal using wide-angle *in situ* measurements with band radiometers designed to approximate the relevant bands. They report generally good agreement with Steven et al. (2003) but with higher MERIS sensitivity to vegetation than predicted. However, the accuracy of their comparisons depends on the degree to which the *in situ* sensor bands match those of the satellite instruments. In addition, as they noted in their study, wide-angle measurements would exaggerate VIs, particularly in the middle of the range, although, given the extensive overlap of the spectral bands, the angular response effect is likely to be very similar for the different systems compared.

Trishchenko et al. (2002) and D'Odorico et al. (2013) have applied quadratic correction factors for intersensor corrections. There is no doubt that a quadratic function will provide a better statistical fit to any given set of data, even data that appear strongly linear as in Figures 8.7 and 8.8, but fitting a curve to such data makes the coefficients relatively unstable as the additional coefficient tends to counteract the previous ones, and it is unclear whether the extra coefficient is justified by a general gain in accuracy when applied to independent data. Swinnen and Veroustraete (2008) compared the effect of corrections according to Steven et al. (2003) and the polynomial corrections of Trishchenko et al. (2002). In general, the corrections by Steven et al. (2003) gave the greater reduction of error, except for open or sparse grassland with low NDVI, where values were overcorrected, while the method of Trishchenko et al. (2002) provided a more consistent correction across all land cover classes.

Song et al. (2010) found linear conversions between AVHRR Global Inventory Modeling and Mapping Studies (GIMMS) and SPOT-VGT (VEGETATION instrument) NDVI values, but the coefficients varied regionally across China. Figure 8.10 shows a map of average differences between the GIMMS and VGT products. Mountainous regions (C) show both positive and negative differences, while the borderline between regions A and B, where the differences are larger, and regions C and D, where they are lower, corresponds to a major climatic and geological boundary. When rates of change of seasonally integrated NDVI were computed, there were substantial differences between the two systems (Figure 8.11).

Miura et al. (2006) compared NDVI values for a number of systems on Hyperion hyperspectral image data over tropical forest and savannah in Brazil. They applied a similar approach to Steven et al. (2003, 2007), combining the atmospherically corrected data with spectral response functions to simulate surface measured radiance in various bands. Although relationships

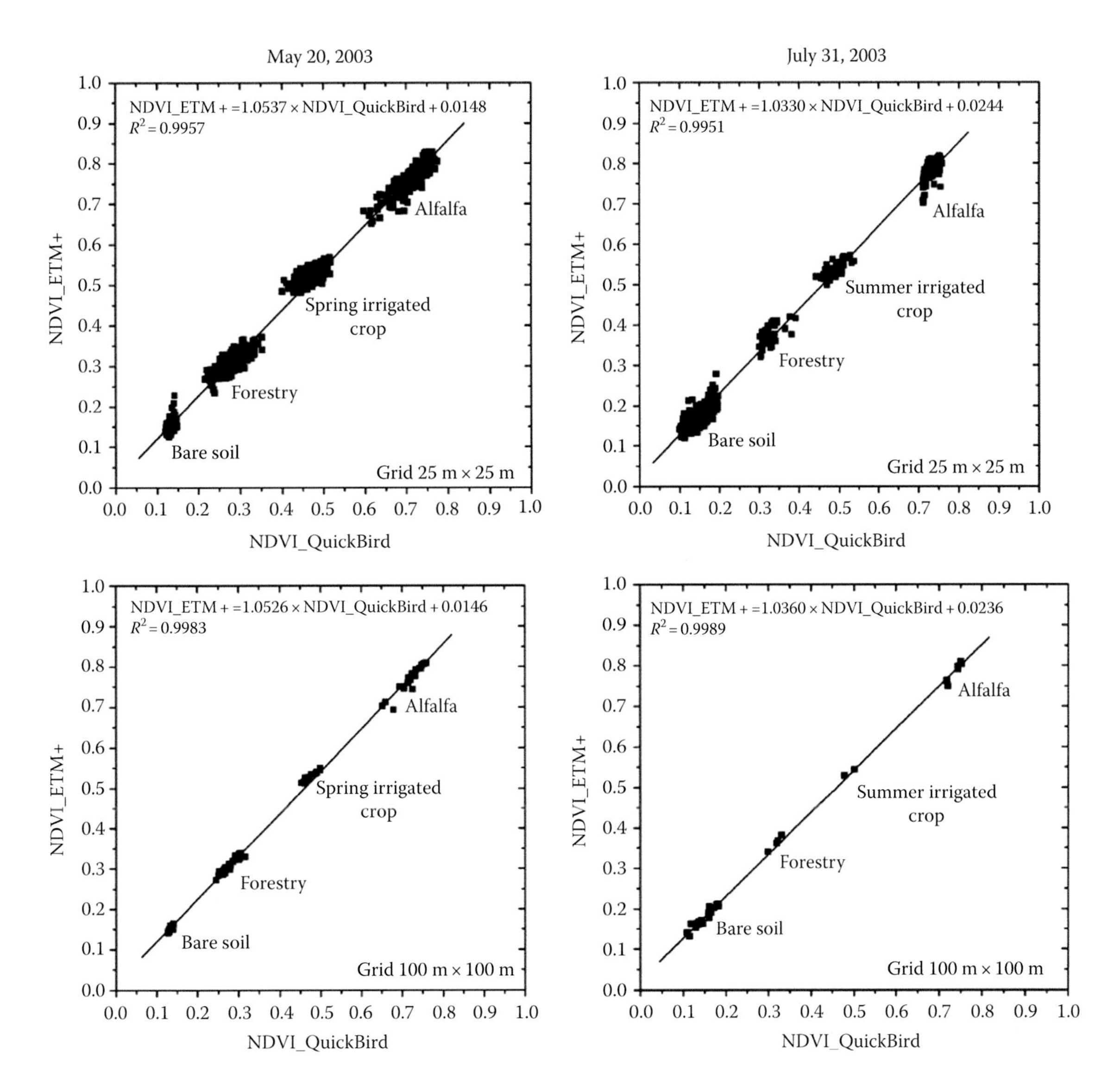

FIGURE 8.9 Intercomparisons of ETM+ and QuickBird NDVI in homogeneous zones at 25 m and 100 m scales. The data for July 31, 2003, include a prior conversion from TM to ETM+ NDVI. (Adapted from Martínez-Beltrán, C. et al., *Int. J. Remote Sens.*, 301355, 2009.)

between simulated radiances in paired bands were found to be land cover dependent, the relationships for NDVI were independent of land cover. However, they were sufficiently nonlinear to require fitting a quadratic function for an adequate conversion between systems.

Intersensor conversions have also been evaluated by modeling approaches. Van Leeuwen et al. (2006) simulated NDVI from AVHRR, MODIS, and Visible Infrared Imaging Radiometer Suite (VIIRS) using the scattering by arbitrarily inclined leaves (SAIL) model with a wide range of LAI. The model was parameterized with data inputs from spectral libraries of vegetation (two spectra), soil (two spectra), and snow (one spectrum). Their result for NOAA16 versus MODIS is within 0.01 of values predicted from Table 8.1, but their prediction for NOAA14 has a slope 0.03 higher. Gonsamo and Chen (2013) used a coupled PROSPECT + SAIL (or PROSAIL) and 6S model to generate synthetic data corresponding to 21 satellite sensors. Their simulations included a range of leaf characteristics, LAI from 0 to 6 with a fixed spherical leaf angle distribution, a range of solar and viewing angles, and eight distinct backgrounds. They found good agreement with both Steven et al. (2003) and van Leeuwen et al. (2006), but less so with other studies that had used quadratic fitting. Soudani et al. (2006) also found good agreement with Steven et al. (2003) for PROSAIL simulations of IKONOS, ETM+, and SPOT NDVIs from forest canopies; the maximum differences in slope and intercept were 0.383 and 0.0183, respectively. D'Odorico et al. (2013) assessed the general effectiveness of synthetic calibration data generated by radiative transfer modeling for cross-sensor calibration. In general, airborne spectral measurements gave better results, but model-generated data were found to be a good substitute for regional or global monitoring.

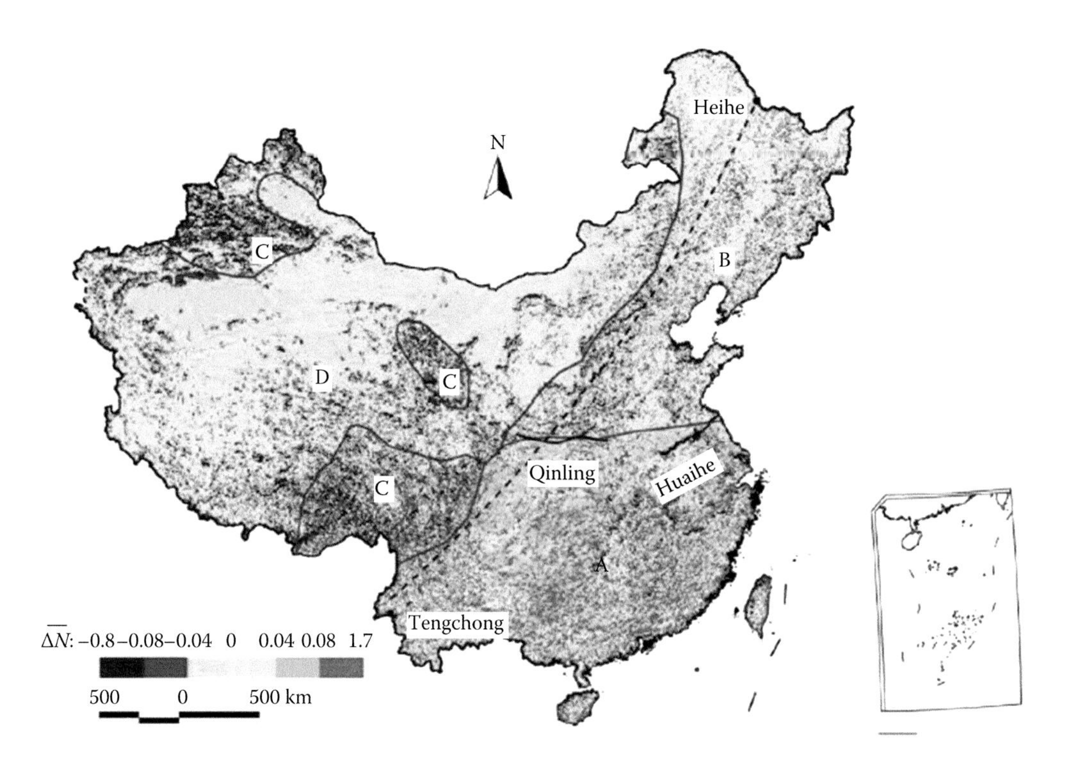

FIGURE 8.10 Map of mean annual differences of VGT and GIMMS NDVI for 1998–2003 in China. (Adapted from Song, Y. et al., *Int. J. Remote Sens.*, 31, 2377, 2010).

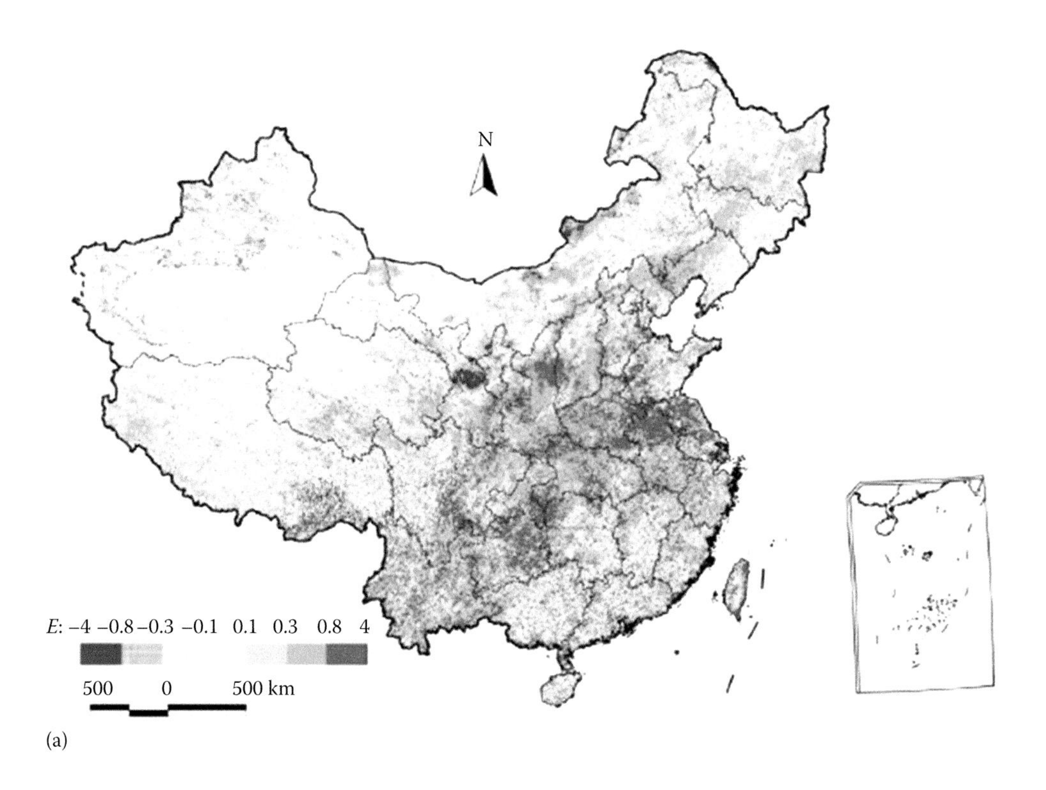

FIGURE 8.11 Rate of change (year^{-1}) of seasonally integrated NDVI in China from 1998 to 2006 for (a) VGT and (b) GIMMS systems. (Adapted from Song, Y. et al., *Int. J. Remote Sens.*, 31, 2377, 2010.)

(*Continued*)

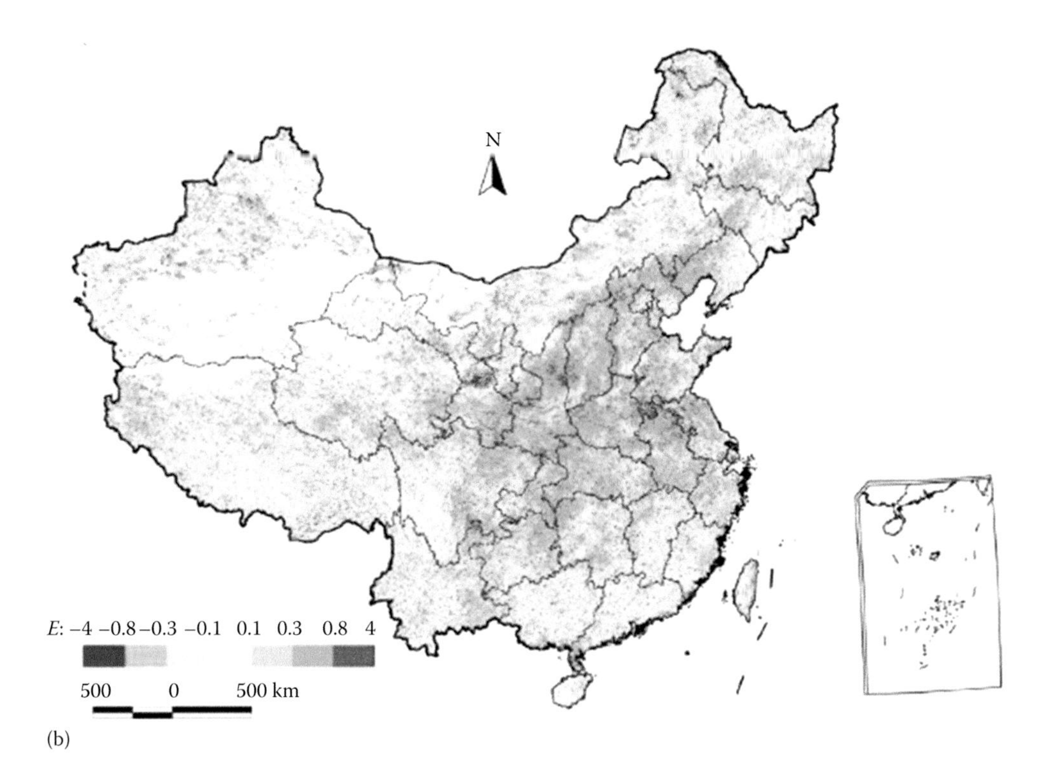

FIGURE 8.11 (*Continued*) Rate of change (year^{-1}) of seasonally integrated NDVI in China from 1998 to 2006 for (a) VGT and (b) GIMMS systems. (Adapted from Song, Y. et al., *Int. J. Remote Sens.*, 31, 2377, 2010.)

8.5 Vegetation Index Standardization

In proposing a standard for VIs, some general principles should be considered.

Principle 1—Standardization should be applicable to vegetation indices from all systems, including the very earliest, to meet the aim of providing a long-term baseline of global vegetation measurements. For this reason, the index chosen should be based on just two bands, the red and the NIR, in spite of improvements in performance reported with incorporation of additional band information.

Principle 2—Standards should be capable of evolving over time. A VI standard must be traceable and, if possible, reversible, as new knowledge or improved modeling may require readjustment of past data. With an evolving standard, complete metadata are essential to allow later adjustments, and adopters of the standard should specify precisely what corrections and procedures have been applied in adjusting measured data.

Principle 3—Standardization procedures should be modular, so that corrections for spectral bands, soil background, BRDF, and other factors are performed separately. Modularity improves the ability to trace (and if necessary reverse) processes and errors in the system. The sequence of correction procedures should also be considered as not all the operations are strictly commutative. Radiometric corrections—calibration and atmospheric correction—should come first, followed by corrections for illumination and viewing geometry as these may vary to some extent with the particular bands of the earth observation system. Finally, the VI should be calculated and adjustment made for the spectral bands as described in Section 8.4.1 (Equation 8.3).

8.5.1 Proposals for Vegetation Index Standards

In practice, atmospheric corrections are usually combined in a single model with calibration. The 6S (Vermote et al. 1997) and MODTRAN (Berk et al. 1998) models are both widely used and appropriate here. Callieco and Dell'Acqua (2011) investigated differences between these models, but found that significant effects were confined to wavelengths shorter than 500 nm, so the choice of model should have minimal impact on VIs. Such models also account for variations in earth–sun distance and the effects of solar and viewing angle on the atmospheric effects. Variations in solar output are not modeled, but are typically of the order of 0.1% and can safely be ignored.

A BRDF model is needed to account for effects of viewing and illumination directions. The PROSAIL model (Jacquemoud et al. 2009) is widely respected and sufficiently versatile for this purpose. To standardize for bidirectional effects, it is also necessary to adopt standard solar and viewing angles. Bacour et al. (2006) proposed standardizing on viewing at nadir with a solar zenith angle of 40°. Conveniently, these angles approximate to the average measurement conditions that apply to the canopy spectral database used by Steven et al. (2003, 2007), so adoption of these angles as standard would ensure a self-consistent system. The solar angle proposed is also a reasonable mid-value for summer viewing conditions at mid-latitudes. The bidirectional effects depend on vegetation type that can be modeled by PROSAIL on the basis of leaf angle distribution. If this information is unavailable, or impractical to implement such as in global-scale applications, corrections based on a spherical leaf angle distribution are recommended here.

The instrument-specific VI can next be calculated from the corrected surface reflectance values. Adjustment should then be made to determine the corresponding index for the standard bands—670 and 815 nm are proposed here—by applying Equation 8.3 with the appropriate conversion coefficients from Table 8.2.

Multiple standards for indices may be necessary, at least in the short term. The NDVI is so well established that its continuance is unavoidable, in spite of its deficiencies. An index adjusted to minimize sensitivity to variable soil should also be included, but there are many claimants. The simplest is the SAVI defined in Equation 8.2 (Huete 1988), with a value of 0.5 for the adjustment factor L. The advantage of the SAVI formulation is that the adjustment factor eliminates the need for specific calibration to different soils (Huete and Liu 1994). Reevaluation by Rondeaux et al. (1996) of a range of L values found that a value of 0.16 gave a slightly better performance, particularly when the soil types in the analysis were restricted to those likely to be found in agriculture and the OSAVI was proposed incorporating this value. In the analysis by Rondeaux et al. (1996), the variance in the index due to soil type was reduced from 7.5% for NDVI to 1.1% and 1.7% for SAVI and OSAVI, respectively. The case for OSAVI was largely based on its performance with a restricted set of agricultural soil types where it reduced the soil signal to 0.06% with the residual error distributed evenly across the range of foliage cover. With the wider dataset that might be more applicable for global application, SAVI and MSAVI were slightly better performers. SAVI (with $L = 0.5$) is recommended here over MSAVI for its greater simplicity and its known compatibility with the conversion coefficients in Table 8.2. Both OSAVI and the original SAVI are based on analyses of relatively narrow datasets, and further studies over a wider range of soil types may lead to better understanding of the errors and an improved general index, but a perfect soil adjustment is not attainable, and residual errors of the order of 5% can be expected in the estimation of fractional vegetation cover (Rondeaux et al. 1996).

A summary of these proposals is shown in Table 8.3.

8.5.2 Limits to Standardization

Finally, it must be recognized that there are limits to the ability to correct and standardize VIs. The potential pitfalls of failing to apply limits to the operational parameters are illustrated in Figure 8.12, which represents the average of MODIS EVI observations over two-month periods around the solstices. On close examination, there appears to be a slight increase in EVI along the Arctic Circle in the middle of the northern winter. This is emphatically not a result of increased vegetation! At this time of year in this location, the sun is grazing the horizon and as a result of its low altitude is severely depleted in shorter wavelengths by Rayleigh scattering in the atmosphere. The effect of this selective depletion is that although the sun would appear red to a ground observer, it is actually relatively weak in the red compared to the NIR. Sunlight reflected from snow therefore has an exaggerated NIR: red ratio that enhances the VI. We recommend this image as a puzzle and a warning for students of remote sensing! The problem is easily avoided: the spectral balance of solar irradiation is conservative once solar elevation exceeds 10° (Monteith and Unsworth 1990). Variations in spectral irradiance can probably be modeled with reasonable accuracy within a few degrees of the horizon, but until this can be reliably demonstrated, observations beyond a solar zenith angle of 80° should be excluded. The same limit applies to observation angles, as selective Rayleigh scattering works in exactly the same way on the upwelling reflected radiance.

8.6 Discussion

There is a general difficulty with both atmospheric correction and BRDF modeling that the models require input data for parameterization. Atmospheric correction parameters are difficult to obtain

TABLE 8.3 Summary of Specific Proposals Made in This Chapter for Vegetation Index Standards

Standard Parameters	Value	Remarks
VI formula	SAVI	Preferred
	NDVI	Tolerated
Spectral bands		
Near infrared	815 nm	Nominal bandwidth 20 nm, centered on these wavelengths
Red	670 nm	
View zenith angle	0°	After Bacour et al. (2006)
Solar zenith angle	40°	
Atmospheric correction	To surface reflectance	6S or MODTRAN routines recommended[a]
BRDF corrections	To standard angles	PROSAIL recommended with spherical leaf angle distribution by default[a]

[a] These are interim recommendations and do not preclude alternative or improved procedures.

for remote areas, but Steven (1998) found that OSAVI estimates were affected to a very minor extent by errors in the model assumptions. In a global monitoring context, it would also be possible to incorporate satellite aerosol observations into a routine correction procedure. For BRDF modeling, the leaf angle distribution is critical. In a simulation of the effect of variable SPOT viewing angles, Steven (1998) found variations in OSAVI up to 4% relative to vertical viewing. This difference corresponded to 30° leftward tilt of the SPOT camera and larger differences will occur at greater angles. However, as the off-nadir effect can be estimated by the model, it should not of itself be a major source of error, while errors associated with the assumed leaf angle distribution can be expected to be second order relative to the overall magnitude of the off-nadir effect.

Since the paper by Steven et al. (2003), a number of studies have applied the coefficients provided to convert for cross-sensor effects (Sun et al. 2007; Fisher et al. 2008; Pouliot et al. 2009; Tang et al. 2009; Zhang et al. 2009; Propastin and Erasmi 2010; Ouyang et al. 2010, 2012). Others have tested the conversion equations against data or modeling approaches as described in Section 8.4.2. The validation studies discussed provide general support for the idea of intercalibrating VIs and, in many cases, support for the linear coefficients provided by Steven et al. (2003) and given in extended form in Table 8.2. The accuracy suggested by these studies is of the order of ±0.03 in the slope. This is probably about the limit of accuracy for a validation study in this context. More pertinent here is the relative precision of the conversions, which is about ±0.01.

Despite this support, the generality of the intersensor relationships should be considered. Song et al. (2010) found that

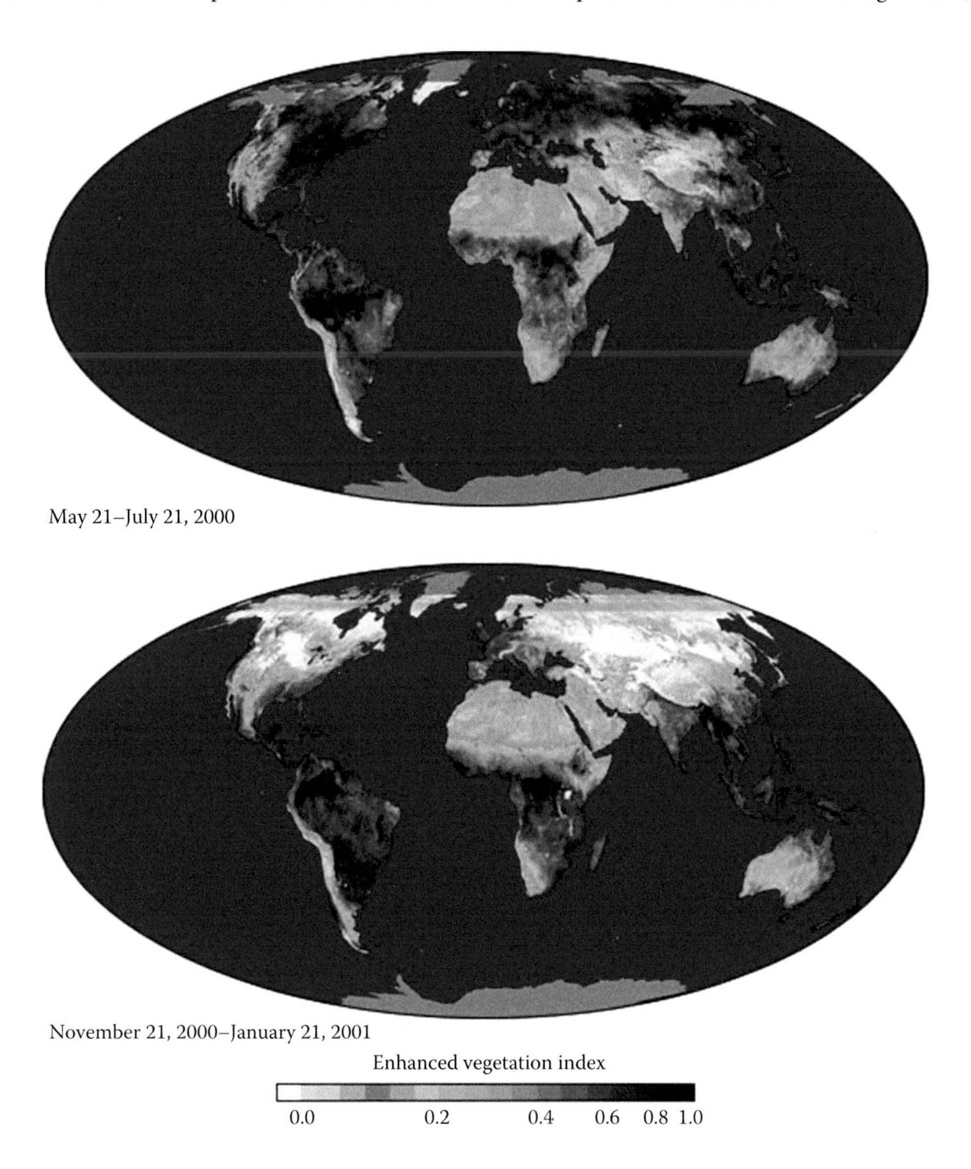

FIGURE 8.12 Composited seasonal MODIS EVI observations illustrating the effects of extreme solar angles on the index. (Adapted from http://earthobservatory.nasa.gov/IOTD/view.php?id=2033.)

differences between AVHRR GIMMS and SPOT VEGETATION NDVI were both land cover and seasonally dependent. However, it is difficult to assess whether such variations are fundamentally linked to the basic observations or an artifact of the compositing and processing procedures applied to generate one or other of the datasets. It is also possible that these dependencies, and the earlier finding by Swinnen and Veroustraete (2008) that the equations of Steven et al. (2003) overcorrected low values, may be related to the high variability of NDVI with respect to soil. D'Odorico et al. (2013) referred to earlier studies that mention land cover dependence, but these studies related to differences in VIs rather than their ratio. Miura et al. (2006), as well as the study by D'Odorico et al. (2013) itself, showed that conversion to NDVI takes out most of the land cover dependence that is found in the individual bands.

A further concern is that the cross-sensor relationships have been established on a relatively narrow base of data. The database used by Steven et al. (2003, 2007) comprised 166 independent measurements of spectral reflectance of strictly agricultural canopies. Other studies have even fewer measurements, and while model simulations may generate large synthetic datasets, their empirical support base is often very narrow. In addition, many of the studies discussed in Section 8.4.2 have applied the same models, so similar findings are to be expected. Forest cover classes, where deep shadow may be a significant component of the target, are not well represented in the datasets, although the study by Soudani et al. (2006) suggests that the coefficients of Table 8.2 still provide a good correction. As noted by Steven et al. (2003), previous studies have also found senescent biomass, or litter to be a significant factor. The results of Miura et al. (2006), which required a quadratic correction between sensors, suggest that the relations established in Table 8.2 are not universal. A broader database might help to resolve these issues, but as Miura et al. (2006) pointed out, observations with different spectral bands are inherently different and may introduce bias into downstream products. The same caveat applies to correction for BRDF effects. Nevertheless, standardization is imperative for a range of issues and does not preclude use of the data in original format for specific purposes. Steven et al. (2003) argued that restriction of the intercalibration database to cultivated vegetation may actually have allowed a tighter statistical relationship to be established than would have occurred with a broader, more representative dataset. Differences between indices are in most cases quite small, and applying conversions based on agricultural datasets may help to maintain relationships with fAPAR and related parameters that have been well validated in such environments.

8.7 Conclusions

The outcome of this study is that corrections are possible, to an acceptable degree of precision, for the main sources of variation in VIs: soil background, atmospheric, BRDF, and spectral band effects. The uncertainty of these corrections is about 5% for soil, when using a SAVI (Rondeaux et al. 1996); 3% for poorly characterized atmospheric aerosol amount (Steven 1998); 4% for off-nadir view angle effects up to 30°, if poorly characterized (Steven 1998); and about 1% for spectral band effects. The corresponding uncertainties in downstream products such as fAPAR will be slightly larger. Greater errors can also be expected at larger angles of view or solar zenith angles, and if errors combine malignly, but the overall effect of atmospheric and BRDF errors can be reduced substantially by application of simple well-established models and appropriate input data. Further studies are needed to extend the database for comparison of sensors and assessment of BRDF effects, to assess the errors of modeling and to validate the procedures and their effect on VIs and downstream products. It is anticipated that the errors will become better characterized as procedures evolve. Nevertheless, there is sufficient evidence to proceed now. Standardization will have immense benefits for vegetation monitoring, both on short-timescale regional studies where the constellation approach is applied to monitor the dynamics of specific land covers and in large-scale studies of environmental history where records from early satellites are merged with modern data.

References

Agapiou, A., D. G. Hadjimitsis, and D. D. Alexakis. 2012. Evaluation of broadband and narrowband vegetation indices for the identification of archaeological crop marks. *Remote Sens* 4:3892–3919.

Bacour, C., F. M. Bréon, and F. Maignan. 2006. Normalization of the directional effects in NOAA–AVHRR reflectance measurements for an improved monitoring of vegetation cycles. *Remote Sens Environ* 102:402–413.

Baret, F. and G. Guyot. 1991. Potentials and limits of vegetation indices for LAI and APAR assessment. *Remote Sens Environ* 35:161–173.

Berk, A., L. S. Bernstein, G. P. Anderson, P. K. Acharya, D. C. Robertson, J. H. Chetwynd, and S. M. Adler-Golden. 1998. MODTRAN cloud and multiple scattering upgrades with application to AVIRIS. *Remote Sens Environ* 65:367–375.

Bowers, S. A. and R. J. Hanks. 1965. Reflection of radiant energy from soil. *Soil Sci* 100:130–138.

Boyd, D. S., S. Almond, J. Dash, P. J. Curran, and R. A. Hill. 2011. Phenology of vegetation in Southern England from Envisat MERIS terrestrial chlorophyll index (MTCI) data. *Int J Remote Sens* 32:8421–8447.

Boyd, D. S. and G. M. Foody. 2011. An overview of recent remote sensing and GIS based research in ecological informatics. *Ecol Informatics* 6:25–36.

Brown, M., J. E. Pinzón, K. Didan, J. T. Morisette, and C. J. Tucker. 2006. Evaluation of the consistency of long-term NDVI time series derived from AVHRR, SPOT-Vegetation, SeaWiFS, MODIS, and Landsat ETM+ sensors. *IEEE Trans Geosci Remote Sens* 44:1787–1793.

Callieco, F. and F. Dell'Acqua. 2011. A comparison between two radiative transfer models for atmospheric correction over a wide range of wavelengths. *Int J Remote Sens* 32:1357–1370.

CEOS. 2006. The CEOS virtual constellation concept V0.4, Committee on Earth Observation Satellites. http://igos-cryosphere.org/docs/CEOS_constellations.doc (accessed July 25, 2014).

D'Odorico, P., A. Gonsamo, A. Damm, and M. E. Schaepman. 2013. Experimental evaluation of Sentinel-2 spectral response functions for NDVI time-series continuity. *IEEE Trans Geosci Remote Sens* 51:1336–1348.

D'Odorico, P., A. Gonsamo, B. Pinty, N. Gobron, N. Coops, E. Mendez, and M. E. Schaepman. 2014. Intercomparison of fraction of absorbed photosynthetically active radiation products derived from satellite data over Europe. *Remote Sens Environ* 142:141–154.

Fensholt, R., I. Sandholt, and S. Stisen. 2006. Evaluating MODIS, MERIS, and VEGETATION vegetation indices using in situ measurements in a semiarid environment. *IEEE Trans Geosci Remote Sens* 44:1774–1786.

Fisher, J. B., K. P. Tu, and D. D. Baldocchi. 2008. Global estimates of the land-atmosphere water flux based on monthly AVHRR and ISLSCP-II data, validated at 16 FLUXNET sites. *Remote Sens Environ* 112:901–919.

Gallo, K., L. Ji, B. Reed, J. Eidenshink, and J. Dwyer. 2005. Multiplatform comparisons of MODIS and AVHRR normalized difference vegetation index data. *Remote Sens Environ* 99:221–231.

Gallo, K. P. and C. S. T. Daughtry. 1987. Differences in vegetation indices for simulated Landsat-5 MSS and TM, NOAA-9 AVHRR and SPOT-1 sensor systems. *Remote Sens Environ* 23:439–452.

Gates, D. M., H. J. Keegan, J. C. Schleter, and V. R. Weidner. 1965. Spectral properties of plants. *Appl Opt* 4:11–20.

Gausman, H. W. and W. A. Allen. 1973. Optical parameters of leaves of 30 plant species. *Plant Physiol* 52:57–62.

Gobron, N., B. Pinty, O. Aussedat, M. Taberner, O. Faber, F. Mélin, T. Lavergne, M. Robustelli, and P. Snoeij. 2008. Uncertainty estimates for the FAPAR operational products derived from MERIS—Impact of top-of-atmosphere radiance uncertainties and validation with field data. *Remote Sens Environ* 112:1871–1883.

Gonsamo, A. and J. M. Chen. 2013. Spectral response function comparability among 21 satellite sensors for vegetation monitoring. *IEEE Trans Geosci Remote Sens* 51:1319–1335.

Goward, S. N., D. G. Dye, S. Turner, and J. Yang. 1993. Objective assessment of the NOAA global vegetation index data product. *Int J Remote Sens* 14:3365–3394.

Gower, S. T., C. J. Kucharik, and J. K. M. Norman. 1999. Direct and indirect estimation of leaf area index, f_{APAR}, and net primary production of terrestrial ecosystems. *Remote Sens Environ* 70:29–51.

Gutman, G. 1991. Vegetation indices from AVHRR: an update and future prospects. *Remote Sens Environ* 35:121–136.

Guyot, G. 1990. Optical properties of vegetation canopies. In M. D. Steven and J. A. Clark (eds), *Applications of Remote Sensing in Agriculture*. Butterworths, London, U.K., pp. 19–43.

Guyot, G. and X. F. Gu. 1994. Effect of radiometric corrections on NDVI—Determined from SPOT HRV and Landsat TM data. *Remote Sens Environ* 49:169–180.

Hadjimitsis, D. G., G. Papdavid, A. Agapiou, K. Themistocleous, M. G. Hadjimitsis, A. Retalis, S. Michaelides, N. Chrysoulakis, L. Toulios, and C. R. I. Clayton. 2010. Atmospheric correction for satellite remotely sensed data intended for agricultural applications: impact on vegetation indices. *Nat Hazards Earth Syst Sci* 10:89–95.

Heller, J. 1961. *Catch 22*. Simon and Schuster, New York.

Hill, J. and D. Aifadopoulou. 1990. Comparative analysis of Landsat-5 TM and SPOT HRV-1 data for use in multiple sensor approaches. *Remote Sens Environ* 34:55–70.

Holben, B. N., C. J. Tucker, and C. J. Fan. 1980. Spectral assessment of soybean leaf area and leaf biomass. *Photogramm Eng Remote Sens* 46:651–656.

Huete, A. R. 1988. A soil adjusted vegetation index (SAVI). *Int J Remote Sens* 9:295–309.

Huete, A. R. and H. Q. Liu 1994. An error and sensitivity analysis of the atmospheric- and soil-correcting variants of the NDVI for the MODIS-EOS. *IEEE Trans Geosci Remote Sens* 32:897–905.

Huete, A. R., H. Q. Liu, K. Batchily, and W. van Leeuwen. 1997. A comparison of vegetation indices over a global set of TM images for EOSMODIS. *Remote Sens Environ* 59:440–451.

Jacquemoud, S., W. Verhoef, F. Baret, C. Bacour, P. J. Zarco-Tejada, G. P. Asner, C. François, and S. L. Ustin. 2009. PROSPECT + SAIL models: a review of use for vegetation characterization. *Remote Sens Environ* 113:S56–S66.

Jaggard, K. W. and C. J. Clark. 1990. Remote sensing to predict the yield of sugar beet in England. In M. D. Steven and J. A. Clark (eds), *Applications of Remote Sensing in Agriculture*. Butterworths, London, U.K., pp. 201–208.

Ji, L., K. Gallo, J. C. Eidenshink, and J. Dwyer. 2008. Agreement evaluation of AVHRR and MODIS 16-day composite NDVI data sets. *Int J Remote Sens* 29:4839–4861.

Lee, K-S., W. B. Cohen, R. E. Kennedy, T. K. Maiersperger, and S. T. Gower. 2004. Hyperspectral versus multispectral data for estimating leaf area index in four different biomes. *Remote Sens Environ* 91:508–520.

Le Maire, G., C. François, and E. Dufrêne. 2004. Towards universal broad leaf chlorophyll indices using PROSPECT simulated database and hyperspectral reflectance measurements. *Remote Sens Environ* 89:1–28.

Mahiny, A. S. and B. Turner. 2007. A comparison of four common atmospheric correction methods. *Photogramm Eng Remote Sens* 73:361–368.

Martínez-Beltrán, C., M. A. O. Jochum, A. Calera, and J. Meliá. 2009. Multisensor comparison of NDVI for a semi-arid environment in Spain. *Int J Remote Sens* 30:1355–1384.

Maselli, F., S. Romanelli, L. Bottai, and G. Maracchi. 2000. Processing of GAC NDVI data for yield forecasting in the Sahelian region. *Int J Remote Sens* 21:3509–3523.

Miura, T., A. Huete, and H. Yoshioka. 2006. An empirical investigation of cross-sensor relationships of NDVI and red/near-infrared reflectance using EO-1 Hyperion data. *Remote Sens Environ* 100:223–236.

Miura, T., J. P. Turner, and A. R. Huete. 2013. Spectral compatibility of the NDVI across VIIRS, MODIS and AVHRR: an analysis of atmospheric effects using EO-1 Hyperion. *IEEE Trans Geosci Remote Sens* 51:1349–1359.

Monteith, J. L. 1977. Climate and the efficiency of crop production in Britain. *Philos Trans R Soc Lond Ser B*, 281:277–294.

Monteith, J. L. and M. H. Unsworth. 1990. *Principles of Environmental Physics*, 2nd edn., Edward Arnold, London, U.K.

Moura, Y. M., L. S. Galvão, J. R. dos Santos, D. A. Roberts, and F. M. Breunig. 2012. Use of MISR/Terra data to study intra- and inter-annual EVI variations in the dry season of tropical forest. *Remote Sens Environ* 127:260–270.

Munyati, C. and G. Mboyeni. 2013. Variation in NDVI values with change in spatial resolution for semi-arid vegetation: a case study in Northwestern South Africa. *Int J Remote Sens* 34:2253–2267.

Nagol, J., E. F. Vermote, and S. D. Prince. 2009. Effects of atmospheric variation on AVHRR NDVI data. *Remote Sens Environ* 113:392–397.

Obata, K., T. Miura, and H. Yoshioka. 2012. Analysis of the scaling effects in the area-averaged fraction of vegetation cover retrieved using an NDVI-isoline-based linear mixture model. *Remote Sens* 4:2156–2180.

Ouyang, W., F. H. Hao, A. K. Skidmore, T. A. Groen, A. G. Toxopeus, and T. Wang. 2012. Integration of multi-sensor data to assess grassland dynamics in a Yellow River sub-watershed. *Ecol Indic* 18:163–170.

Ouyang, W., F. H. Hao, C. Zhao, and C. Lin. 2010. Vegetation response to 30 years hydropower cascade exploitation in upper stream of Yellow River. *Commun Nonlinear Sci Numer Simul* 15:128–1941.

Peñuelas, J. and I. Filella. 1998. Visible and near-infrared techniques for diagnosing plant physiological status. *Trends Plant Sci* 3:151–156.

Perry, C. R. and L. F. Lautenshlager. 1984. Functional equivalence of spectral vegetation indices. *Remote Sens Environ* 14:169–182.

Pettorelli, N., J. O. Vik, A. Mysterud, J-M. Gaillard, C. J. Tucker, and N. C. Stenseth. 2005. Using the satellite-derived NDVI to assess ecological responses to environmental change. *Trends Ecol Evol* 20:503–510.

Picket-Heaps, C. A., J. G. Canadell, P. R. Briggs, N. Gobron, V. Haverd, M. J. Paget, B. Pinty, and M. R. Raupach. 2014. Evaluation of six satellite-derived Fraction of Absorbed Photosynthetic Active Radiation (FAPAR) products across the Australian continent. *Remote Sens Environ* 140:241–256.

Pinter, P. J. 1993. Solar angle independence in the relationship between absorbed PAR and remotely sensed data for alfalfa. *Remote Sens Environ* 46:19–25.

Pouliot, D., R. Latifovic, and I. Olthof. 2009. Trends in vegetation NDVI from 1 km AVHRR data over Canada for the period 1985–2006. *Int J Remote Sens* 30:149–168.

Price J. C. 1987. Calibration of satellite radiometers and the com parison of vegetation indices. *Remote Sens Environ* 21:15–27.

Propastin, P. and S. Erasmi. 2010. A physically based approach to model LAI from MODIS 250 m data in a tropical region. *Int J App Earth Obs Geoinf* 12:47–59.

Qi, J., A. S. Chehbouni, A. R. Huete, Y. H. Kerr, and S. Sorooshian. 1994. A modified soil adjusted vegetation index. *Remote Sens Environ* 48:119–126.

Röder, A., T. Kuemmerle, and J. Hill. 2005. Extension of retrospective datasets using multiple sensors. An approach to radiometric intercalibration of Landsat TM and MSS data. *Remote Sens Environ* 95:195–210.

Rondeaux, G., M. D. Steven, and F. Baret. 1996. Optimization of soil adjusted vegetation indices. *Remote Sens Environ* 55:95–107.

Samain, O., B. Geiger, and J-L. Roujean. 2006. Spectral normalisation and fusion of optical sensors for the retrieval of BRDF and Albedo: Application to VEGETATION, MODIS, and MERIS data sets. *IEEE Trans Geosci Remote Sens* 44:3166–3179.

Schluessel, G., R. E. Dickinson, J. L. Privette, W. J. Emery, and R. Kokaly. 1994. Modeling the bidirectional reflectance distribution function of mixed finite plant canopies and soil. *J Geophys Res-Atmos* 99(D5):10577–10600.

Shultis, J. K. 1991. Calculated sensitivities of several optical radiometric indices for vegetation canopies. *Remote Sens Environ* 38:211–228.

Sims, D. A., A. F. Rahman, E. F. Vermote, and Z. Jiang. 2011. Seasonal and inter-annual variation in view angle effects on MODIS vegetation indices at three forest sites. *Remote Sens Environ* 115:3112–3120.

Song, Y., M. Ma, and F. Veroustraete. 2010. Comparison and conversion of AVHRR GIMMS and SPOT VEGETATION NDVI data in China. *Int J Remote Sens* 31:2377–2392.

Soudani, K., C. François, G. le Maire, V. Le Dantec, and E. Dufrêne. 2006. Comparative analysis of IKONOS, SPOT, and ETM+ data for leaf area index estimation in temperate coniferous and deciduous forest stands. *Remote Sens Environ* 102:161–175.

Steven, M. D. 1977. Standard distributions of clear sky radiance. *Q J Roy Meteorol Soc* 103:457–465.

Steven, M. D. 1998. The sensitivity of the OSAVI vegetation index to observational parameters. *Remote Sens Environ* 63:49–60.

Steven, M. D. 2004. Correcting the effects of field of view and varying illumination in spectral measurements of crops. *Precis Agric* 5:51–68.

Steven, M. D., P. V. Biscoe, and K. W. Jaggard. 1983. Estimation of sugar beet productivity from reflection in the red and infrared spectral bands. *Int J Remote Sens* 4:325–334.

Steven, M. D., P. V. Biscoe, K. W. Jaggard, and J. Paruntu. 1986. Foliage cover and radiation interception. *Field Crops Res* 13:75–87.

Steven, M. D., T. J. Malthus, and F. Baret. 2007. Intercalibration of vegetation indices—An update. In M. E. Schaepman, S. Liang, N. Groot, and M. Kneubühler (eds), *10th International Symposium on Physical Measurements and Spectral Signatures in Remote Sensing*, International Archives of the Photogrammetry, Remote Sensing and Spatial Information Sciences, Vol. XXXVI, Part 7/C50, ISPRS, Davos (CH), pp. 1682–1777.

Steven, M. D., T. J. Malthus, F. Baret, H. Xu, and M. Chopping. 2003. Intercalibration of vegetation indices from different sensor systems. *Remote Sens Environ* 88:412–422.

Steven, M. D. and M. H. Unsworth. 1980. The angular distribution and interception of diffuse solar radiation below overcast skies. *Q J Roy Met Soc* 106:57–61.

Sun, Z., Q. Wang, Z. Ouyang, M. Watanabe, B. Matsushita, and T. Fukushima. 2007. Evaluation of MOD16 algorithm using MODIS and ground observational data in winter wheat field in North China plain. *Hydrol Proc* 21:1196–1206.

Swinnen, E. and F. Veroustraete. 2008. Extending the SPOT-VEGETATION NDVI time series (1998–2006) back in time with NOAA-AVHRR data (1985–1998) for Southern Africa. *IEEE Trans Geosci Remote Sens* 46:558–572.

Tang, Q., S. Peterson, R. H. Cuenca, Y. Hagimoto, and D. P. Lettenmaier. 2009. Satellite-based near-real-time estimation of irrigated crop water consumption. *J Geophys Res* 114:D05114, doi:10.1029/2008JD010854.

Tarnavsky, E., S. Garrigues, and M. Brown. 2008. Multiscale geostatistical analysis of AVHRR, SPOT-VGT, and MODIS global NDVI products. *Remote Sens Environ* 112:535–549.

Teillet, P. M., B. L. Markham, and R. R. Irish. 2006. Landsat cross-calibration based on near simultaneous imaging of common ground targets. *Remote Sens Environ* 102:264–270.

Teillet, P. M. and X. Ren. 2008. Spectral band difference effects on vegetation indices derived from multiple satellite sensor data. *Can J Remote Sens* 34:159–173.

Townshend, J. R. G. and C. O. Justice. 1986. Analysis of the dynamics of African vegetation using the normalised difference vegetation index. *Int J Remote Sens* 7:1435–1445.

Trishchenko, A. P., J. Cihlar, and Z. Li. 2002. Effects of spectral response function on surface reflectance and NDVI measured with moderate resolution satellite sensors. *Remote Sens Environ* 81:1–18.

Tucker, C. J. 1977. Spectral estimation of grass canopy variables. *Remote Sens Environ* 6:11–26.

Tucker, C. J., J. H. Elgin, and J. E. McMurtrey. 1979. Temporal spectral measurements of corn and soybean crops. *Photogramm Eng Remote Sens* 45:643–653.

Tucker, C. J., J. E. Pinzon, M. E. Brown, D. A. Slayback, E. W. Pak, R. Mahoney, E. F. Vermote, and N. El Saleous. 2005. An extended AVHRR 8-km NDVI dataset compatible with MODIS and SPOT vegetation NDVI data. *Int J Remote Sens* 26:4485–4498.

Van Leeuwen, W. J. D. 2006. Spectral vegetation indices and uncertainty: Insights from a user's perspective. *IEEE Trans Geosci Remote Sens* 44:1931–1933.

Van Leeuwen, W. J. D., A. R. Huete, and T. W. Laing. 1999. MODIS vegetation index compositing approach: A prototype with AVHRR data. *Remote Sens Environ* 69:264–280.

Van Leeuwen, W. J. D., B. J. Orr, S. E. Marsh, and S. M. Herrmann. 2006. Multi-sensor data continuity: uncertainties and implications for vegetation monitoring applications. *Remote Sens Environ* 100:67–81.

Verhoef, W. 1985. Light scattering by leaf layers with application to canopy reflectance modeling: the SAIL model. *Remote Sens Environ* 16:125–141.

Vermote, E., C. O. Justice, and F-M. Bréon. 2009. Towards a generalised approach for correction of the BRDF effect in MODIS directional reflectances. *IEEE Trans Geosci Remote Sens* 47:898–908.

Vermote, E. F., D. Tanré, J. L. Deuze, M. Herman, and J. J. Morcrette. 1997. Second simulation of the satellite signal in the solar spectrum: an overview. *IEEE Trans Geosci Remote Sens* 35:675–686.

Verreslst, J., M. E. Schaepman, B. Koetz, and M. Kneubűhler. 2008. Angular sensitivity analysis of vegetation indices derived from CHRIS/PROBA data. *Remote Sens Environ* 112:2341–2353.

Wiegand, C. L., A. J. Richardson, D. E. Escobar, and A.H. Gerbermann. 1991. Vegetation indices in crop assessments. *Remote Sens Environ* 35:105–119.

Zhang, Y., M. Xu, J. Adams, and X. Wang. 2009. Can Landsat imagery detect tree line dynamics. *Int J Remote Sens* 30:1327–1340.

Zhou, X., H. Guan, H. Xie, and J. L. Wilson. 2009. Analysis and optimization of NDVI definitions and areal fraction models in remote sensing of vegetation. *Int J Remote Sens* 30:721–751.

V

Image Processing Methods and Approaches

9

Digital Image Processing: A Review of the Fundamental Methods and Techniques

Sunil Narumalani
University of Nebraska

Paul Merani
University of Nebraska

Acronyms and Definitions

ACORN	Atmospheric CORrection Now
AGB	Aboveground Biomass
AOI	Area of interest
APAR	Absorbed photosynthetic active radiation
ATREM	ATmospheric REMoval
BR	Band ratio
BV	Brightness value
CALMIT	Center for Advanced Land Management Information Technologies
DEM	Digital elevation model
DOQQ	Digital orthophoto quarter quadrangles
EM	Electromagnetic
EPA	Environmental Protection Agency
EVI	Enhanced Vegetation Index
FLAASH	Fast line-of-sight atmospheric analysis of spectral hypercubes
FT	Fourier transform
GCP	Ground control point
GIS	Geographic information systems
GPS	Global positioning system
ISODATA	Iterative self-organizing data analysis
LAI	Leaf Area Index
MSS	Multispectral scanner
NDVI	Normalized Difference Vegetation Index
NIR	Near-infrared
OIF	Optimum Index Factor
OLI	Operational land imager
PC	Principal component
PCA	Principal component analysis
RGB	Red, green, blue
SDSS	Spatial decision support systems
SR	Simple ratio
TbM	Three-band Model
TM	Thematic mapper
USGS	United States Geological Survey
UTM	Universal Transverse Mercator
VBV	Vegetation biophysical variable
VI	Vegetation Index
VNIR	Visible near-infrared
WDRVI	Wide Dynamic Range Vegetation Index

9.1 Introduction

The modern era of commercial digital remote sensing began with the launch of Landsat 1 in 1972 (Schowengerdt, 1997). For the first time ever, a satellite system was able to provide high-spectral-quality (4-bands), medium (80 m)-spatial-resolution data that could be analyzed by public experts for numerous terrestrial applications. Initially, much of the *digital processing* of these data could be done only by large, expensive computer systems that lay in the purview of government agencies, and only large-format hard-copy images were made available for visual or analog analysis. With the technological evolution and computer revolution in the early 1970s, computing power began to slowly trickle down to the mainstream such as educational institutions and research centers across the United States and Europe. As computing capabilities and their usage increased, their application to remotely sensed data also increased, and for almost two decades, from the mid-1970s to the early 1990s, there continued to be sustained development and improvement of algorithms for extracting image data.

Remotely sensed data acquired in or converted to digital format are most often subject to analyses and information extraction using image processing techniques developed over the past 50 years. The process encompasses a myriad of algorithms that an analyst can apply to get meaningful information from an image. This chapter describes some of the basic image processing methods applied to remotely sensed data.

Remote sensing instruments (or sensors) detect and record electromagnetic (EM) energy. There are two types of senor systems, *active sensors*, which emit energy directed toward a target of interest and then record the return, and *passive sensors* that record energy *either* emitted by or reflected from an object. Data may be acquired in single (panchromatic) or multiple (multi-/hyperspectral) bands of the EM spectrum. In either case, the data appear in a matrix of *x-columns* by *y-rows*, with each square of the matrix referenced as a *pixel* and have a gray-scale value that indicates the EM energy recorded for that pixel (commonly referred to as the *brightness value* (BV)). The range of BVs recorded in each band across the image is dependent on the radiometric resolution of the sensor system (e.g., values ranging from 0 to 255 would be recorded for a sensor system with an 8-bit radiometric resolution). Jensen (2005) notes that most remote sensing studies are based on developing deterministic relationships between the EM signals recorded in various bands of the spectrum and the chemical or biophysical properties of the features being investigated.

9.2 Image Quality Assessment: Basic Statistics and Histogram Analysis

Many remote sensing data are of high quality. However, on occasions, errors (or noise) are introduced into the data by numerous factors such as the environment (e.g., atmospheric scattering), random or systematic malfunction of the sensor system (e.g., an uncalibrated detector creates striping), or improper processing of the raw data prior to actual data analysis (e.g., inaccurate analog-to-digital conversion). Therefore, one of the initial tasks of an image analyst should be to assess its quality and statistical characteristics (Jensen, 2005). This is normally accomplished by

- Examining the frequency of occurrence of individual BVs in the image displayed in a histogram
- Sample visual analysis of individual pixel BVs at specific locations or within a geographic area
- Computing univariate descriptive statistics to determine if there are unusual anomalies in the image data
- Computing multivariate statistics to determine the amount of between-band correlation (e.g., to identify redundancy)

Some of the basic statistical information that an analyst may find useful is found in measures of central tendency such as the mean, standard deviation, and variance. These statistics provide information about the range of BVs in each band, the relationship of the BVs in each band, representation of spectral characteristics of features being examined, and an indicator of values that can be used for image enhancement (e.g., a histogram stretch using minimum–maximum values in a given band). However, for an in-depth analysis, other statistical measures such as *variance* and *correlation* may be required to provide an insight into the data quality and redundancy.

Remote sensing–derived spectral measurements for each pixel often change together in some predictable fashion because objects or features exhibit spectral behavioral patterns across the bands. For example, clear deep water would have low, steadily declining BVs across the blue, green, and red portions of the spectrum, until it reaches near zero in the near-infrared (NIR). If there is no relationship between the BVs in one band and that of another for a given pixel, it may imply that the values are mutually independent (e.g., reflective spectra versus temperature as observed in thermal infrared) or there may be an anomalous observation for a given feature (e.g., sedimentation present in deep water). In most cases, spectral measurements of individual pixels may not be independent, and a measure of their mutual interaction is reflected in the *covariance* or the joint variation of two variables about their common mean.

To estimate the degree of interrelation between variables in a manner not influenced by measurement units, the *correlation coefficient* is commonly used. The correlation between two bands of remotely sensed data, $r_{k,l}$, is the ratio of their covariance ($\text{cov}_{k,l}$) to the product of their standard deviations ($s_k s_l$); thus,

$$r_{k,l} = \frac{\text{cov}_{k,l}}{s_k s_l}$$

If we square the correlation coefficient ($r_{k,l}$), we obtain the *sample coefficient of determination* (r^2), which expresses the proportion of the total variation in the values of "band l" that can be accounted for or explained by a linear relationship with the values of the random variable "band *k*." Thus, a correlation coefficient (r_{kl}) of 0.70 results in an R^2 value of 0.49, meaning that 49% of the total variation of the values of "band *l*" in the sample is accounted for by a linear relationship with values of "band *k*." In order to optimize usage of multiple bands, scientists prefer higher variance and lower correlation. Furthermore, the correlation and covariance information can be used for analysis by advanced image processing functions such as principal component analysis (PCA) and image classification.

9.2.1 Histogram

The histogram is the most fundamental and useful graphical representation of information content in an image. It tabulates frequencies of occurrences of each BV, displays them graphically, and provides information on *contrast* within each band. Peaks and valleys often correspond to the dominant land cover types in an image, and the information can be converted into *percent* representation to highlight information content by masking out BVs with high frequency (or vice versa) (Figure 9.1). Basically, histograms may provide the user with information on the quality of the image (e.g., high-contrast, low-contrast, bimodal, and multimodal) and are often used to enhance imagery—for example, brightening up darker areas in an image or conversely darkening up extremely bright areas of an image (Figure 9.2).

9.3 Image Enhancement

To improve the appearance of an image for visual analysis or at times even for subsequent computer analysis, an analyst may prefer to apply select algorithms. A basic suite of algorithms that aid in enhancing an image include image reduction and magnification, spatial and spectral profiles, image contrasting, density slicing, and composite generation (Table 9.1). Higher order image enhancement techniques include band ratioing, spatial filtering, edge enhancement, and spectral image transformation.

Reduction and magnification operations are used to adjust the image scale visually in order to provide either a regional perspective (i.e., display of an entire scene at a small scale) or a zoom-in of an area of interest (AOI) for closer examination (Figure 9.3). They allow an analyst to derive image coordinates (*x*, *y* locations) and per-pixel spectral data of features across, and gain an understanding of the spatial distribution of objects across the landscape.

An analyst may further their understanding of the image landscape by deriving *spatial* and *spectral* profiles along user-specified transects (Jensen, 2005). The pixels that lie along that transect can be measured and displayed to compare the spectral (BVs) or spatial differences (coordinate space). Multiple transects may be used to determine spatial patterns or trends. Transects can also be used to assist in density slicing an image or a portion of it (Figure 9.4).

Density slicing is a pseudocolor enhancement technique normally applied to a single-band monochrome. It is considered an effective way of highlighting different but apparently homogeneous areas within an image by *slicing* the range of grayscale values (e.g., 0–255) and assigning different colors to each of those slices (Figure 9.5). This technique is often used in conjunction with a vegetation index (VI) such as the Normalized Difference Vegetation Index (NDVI) to highlight variations in the density of biomass.

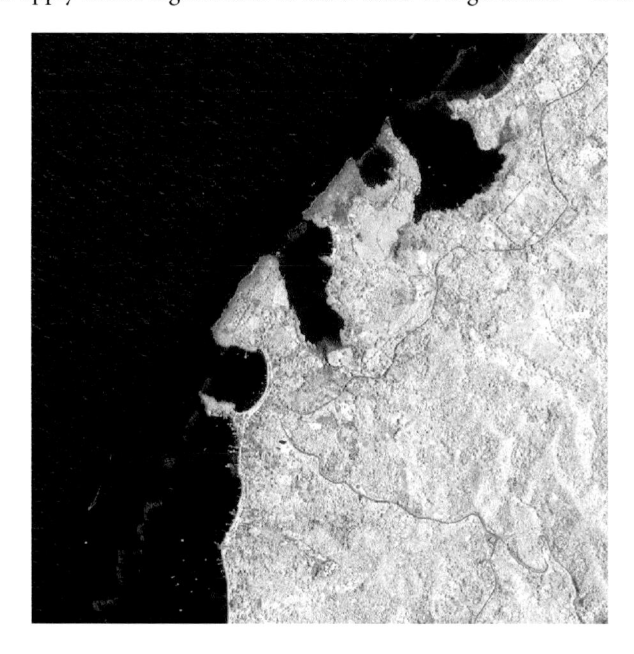

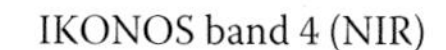

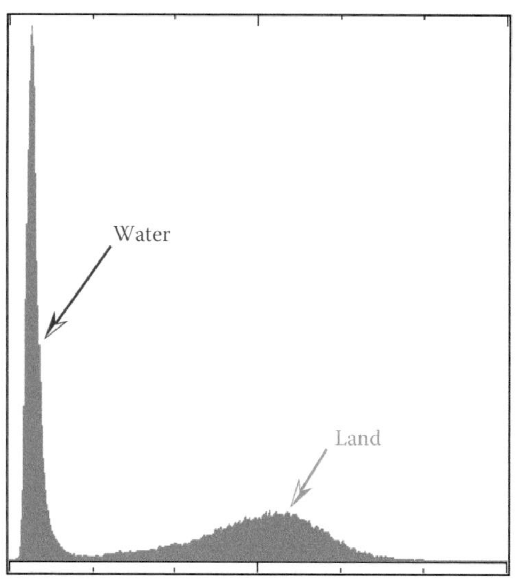

FIGURE 9.1 Histogram of the near-infrared band (band 4) of the IKONOS sensor system. The near-infrared band is useful for land/water delineation, and this is evident in the histogram where clear deep water pixels have very low reflectivity while the land pixels record higher BVs in the band. (Image courtesy of the Center for Advanced Land Management Information Technologies (CALMIT), University of Nebraska, Lincoln, NE.)

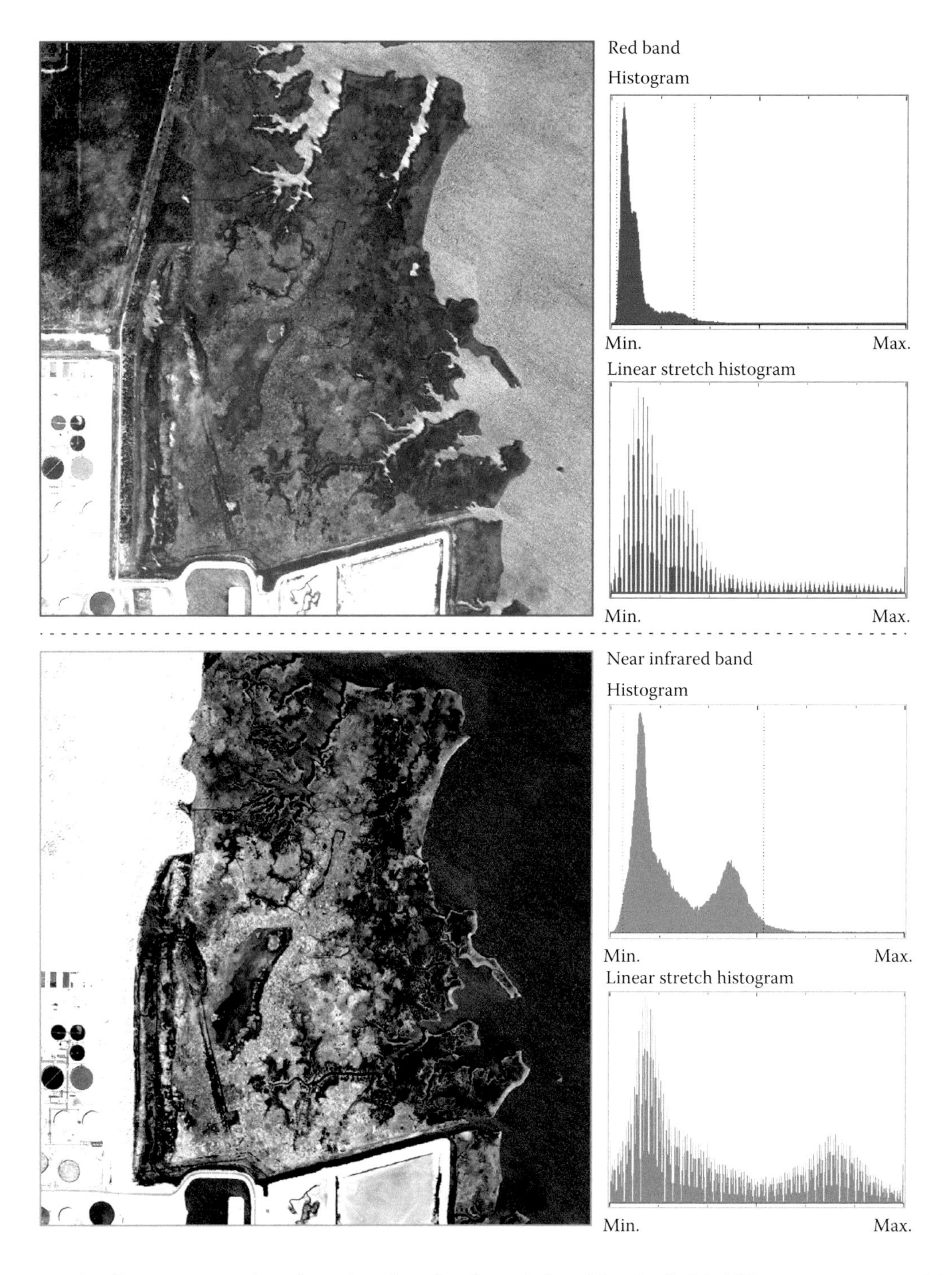

FIGURE 9.2 Example of histogram stretch performed on the red and near-infrared bands of a WorldView 2 image acquired for the Grand Bay, MS area. (Image courtesy of the Center for Advanced Land Management Information Technologies (CALMIT), University of Nebraska, Lincoln, NE.)

Another effective method or visual analysis is composite generation. This method utilizes the three planes of a computer's display device (red (R), green (G), and blue (B)) and allows the analyst to place different bands of a multispectral image into various planes. For example, to generate a true color composite from Landsat Operational Land Imager (OLI) image, one would insert bands 4, 3, and 2 into the R, G, and B planes respectively, thus generating a true color image (Figure 9.6a). Similarly, to display a false color composite of the same image, OLI bands 5, 4, and 3 would be placed in the R, G, and B planes (Figure 9.6b). In Figure 9.6b, the NIR band (OLI band 5) is placed into the red plane of the display and is often used for vegetation analysis because of high spectral reflectance of vegetation in the near- and mid-infrared portions of the spectrum.

TABLE 9.1 Image Enhancement Techniques, Their Effects, and Examples of Application

Image Enhancement Technique	Effect on Image	Example of Application
Image magnification	Zooms into the area of interest or closer observation of feature	
Image reduction	Zooms out either partially or completely from an image to provide a large-area perspective	Enable a geographic or spatial analysis of an entire landscape
Spatial profiles	Draw a transect across the area of interest	Changes in features across the transect (e.g., forest to grasslands, to water)
Spectral profiles	Draw a transect across the area of interest	Changes in spectral signature observed along the transect as variations in features occur
Density slicing	Color coding a band based on BV ranges	Highlight specific features or a rapid visualization of potential land cover observed in the selected band
Image composites	Representation of land cover features in various colors based on the band combinations used	Geologic highlighting of a mineral may use a combination of certain bands based on the spectral properties of that mineral

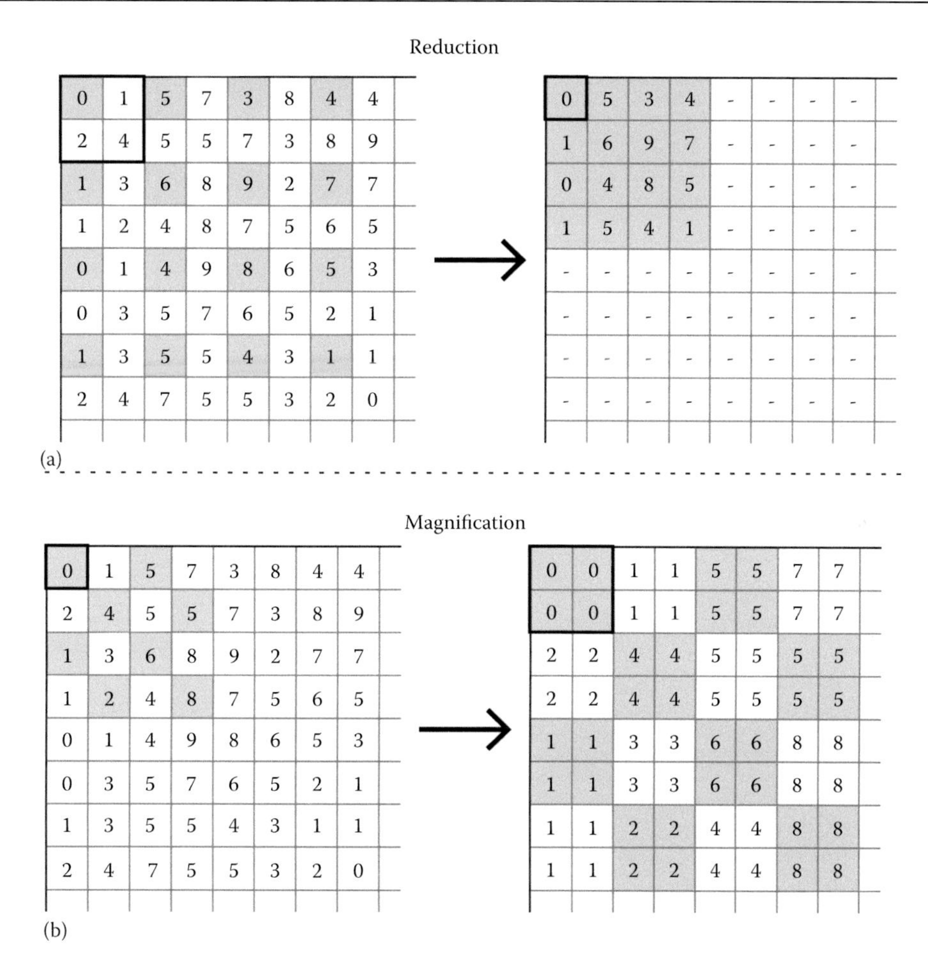

FIGURE 9.3 Concept diagram showing the pixels displayed when (a) 2× image reduction and (b) 2× image magnification are applied.

9.4 Image Preprocessing

Fureder (2010) states that the operational use of remote sensing data is often limited due to sensor variation, atmospheric effects, as well as topographically induced illumination effects. *Image preprocessing* is a preparatory phase that, in principle, improves image quality as the basis for later analyses that will extract information from the image. Often known as image restoration, the process produces a corrected image that is as close as possible, both geometrically and radiometrically, to the radiant energy characteristics of the actual scene. This requires that internal and external errors be determined and corrected for. Internal errors are created by the sensor itself and are generally systematic and stationary (i.e., predictable and constant).

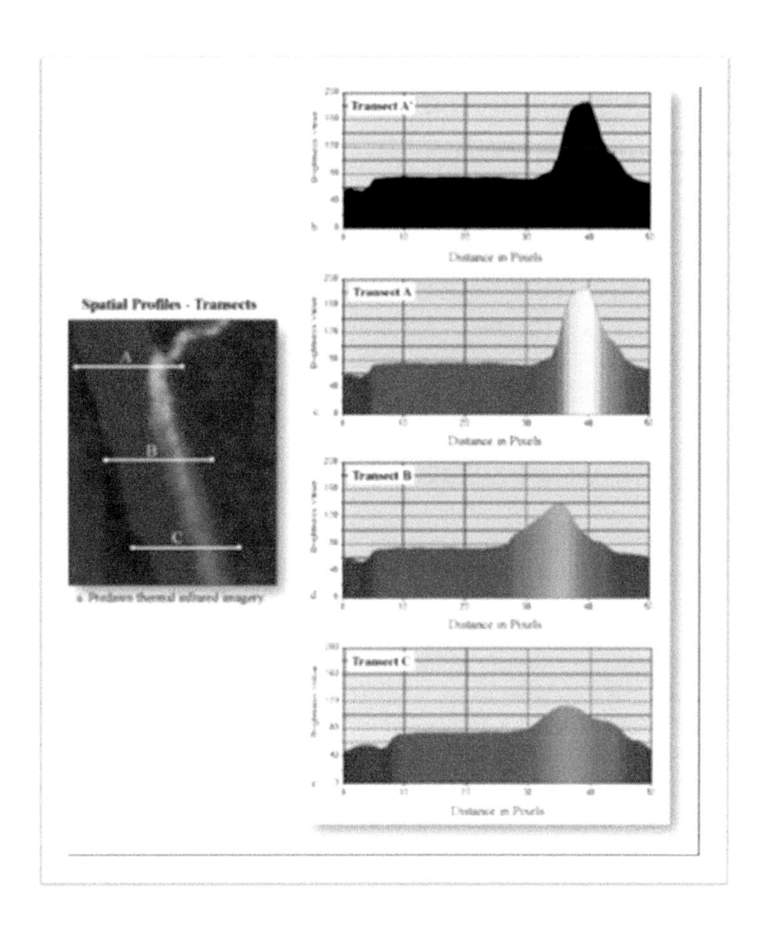

FIGURE 9.4 Example of spatial profiles acquired for the Savannah River over three transects along the channel. (Adapted from Jensen, J.R., *Introductory Digital Image Processing: A Remote Sensing Perspective*, Prentice Hall, Upper Saddle River, NJ, 2005.)

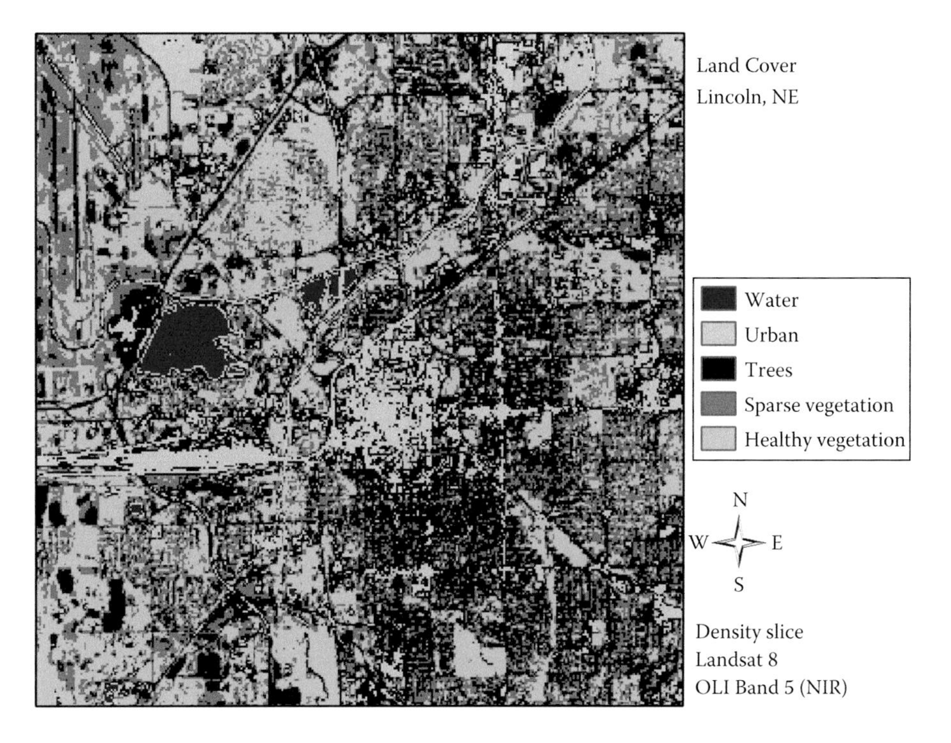

FIGURE 9.5 An example of a density-sliced image acquired by the Landsat OLI system of Lincoln, NE. (Image courtesy of the Center for Advanced Land Management Information Technologies (CALMIT), University of Nebraska, Lincoln, NE.)

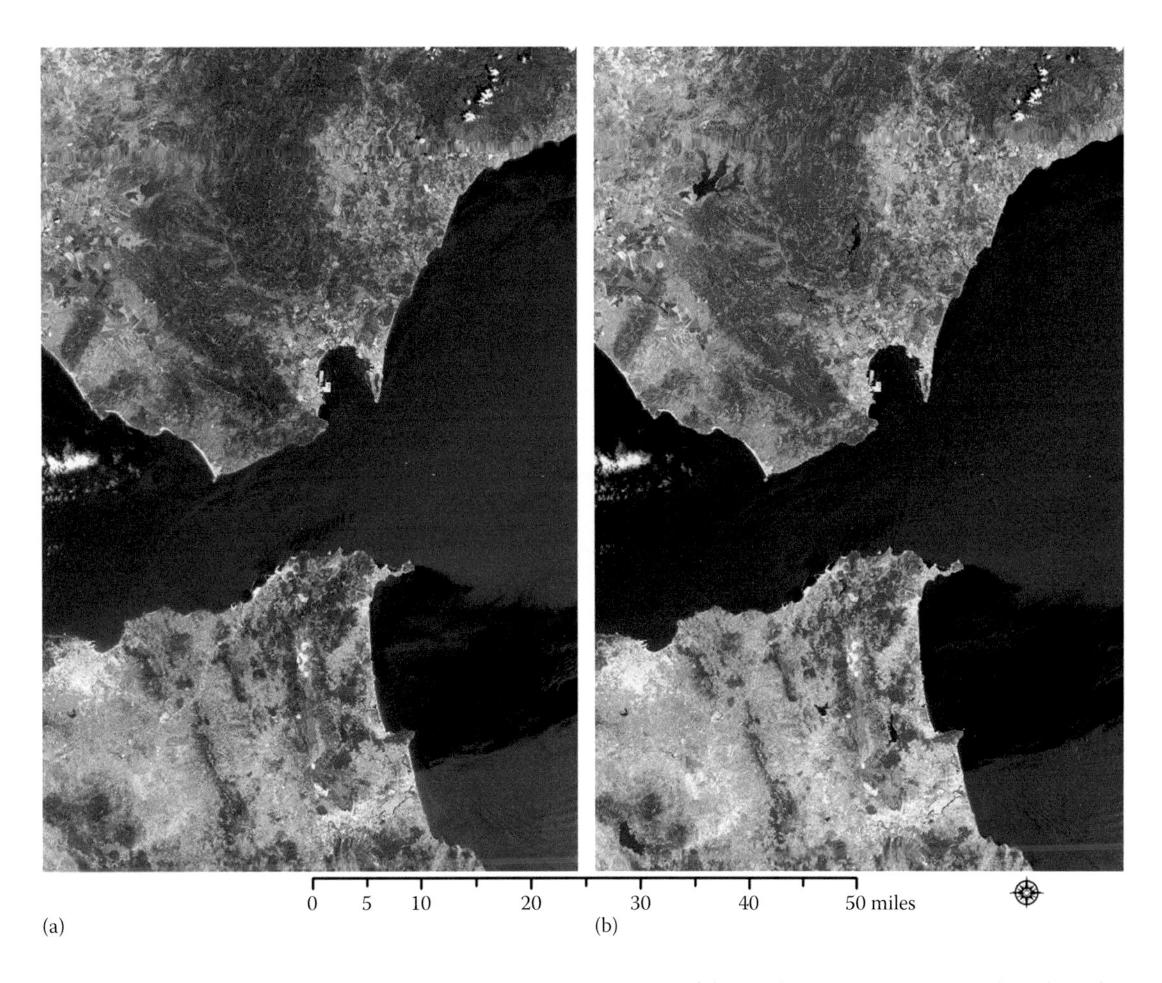

FIGURE 9.6 (a) True color image of the Gibraltar area in the Mediterranean Sea of the Landsat OLI sensor system. (b) Color infrared image of the same region. Note that healthy green vegetation appears bright red, urban areas shades of white, and water (depending on the turbidity) from blue (turbid water) to dark gray (deep, clear water). (Image courtesy of the Center for Advanced Land Management Information Technologies (CALMIT), University of Nebraska, Lincoln, NE.)

External errors are due to perturbations and modulation of scene characteristics—that is, they are variable and are corrected by relating points on ground to sensor measurements.

9.4.1 Radiometric Correction

As radiation passes through the atmosphere, it undergoes several different processes, including absorption, scattering, attenuation, and transmission (Schowengerdt, 1997; Jensen, 2005). Various methods of atmospheric correction can be applied ranging from detailed modeling of the conditions during data acquisition, to simple calculations based solely on the image data. Broadly, atmospheric correction can be divided into two types: (1) absolute and (2) relative. Absolute radiometric corrections turn the BVs into scaled surface reflectance values (Du et al., 2002) and attempt to model the atmosphere, as it would exist at the time of image acquisition. Several radiative transfer models have been developed as part of the absolute radiometric correction efforts, including the Atmospheric CORrection Now (ACORN) by ImSpec (2002), ATmospheric REMoval (ATREM) program by the University of Colorado (Tanre et al., 1986), and Fast Line-of-Sight Atmospheric Analysis of Spectral Hypercubes (FLAASH) by Exelis (formerly Research Systems, 2003). It is important to note that application of these algorithms for a given scene and date requires knowledge of the spectral profile and atmospheric properties for that date and time. This information is extremely difficult to acquire; however, these models can provide a close approximation of the reflectance for the scene sans the atmosphere, versus an atmospherically uncorrected scene.

Relative atmospheric correction is often used if an analyst wants to normalize the BV among the various bands of a single scene or normalize multi-date imagery to a single/standard scene selected among the dataset (Jensen, 2005). An example of the former method is to examine the observed BVs in an area of shadow or for a very dark object (such as a large clear lake or an asphalt surface) and determine the minimum value. The correction is applied by subtracting the minimum observed value, determined for each specific band, from all pixel values in each respective band. Because scattering is wavelength dependent, the minimum values will vary from band to band. This method is based on the assumption that the reflectance from these features, if the atmosphere is clear, should be very small (if not zero). If values are much greater than zero, then they are considered to have resulted from atmospheric scattering.

Multi-date image normalization techniques involve the selection of a base image and then transforming the spectral characteristics of all other images to have the same radiometric scale as the base image (Jensen, 2005). The method involves the selection of pseudo-invariant features (i.e., radiometric ground control points (GCPs)) from the base image, identifying the BVs of the same features across all the multi-date imagery and normalizing them to the base image. The pseudo-invariant features need to meet specific spectral and spatial criteria (Eckhardt et al., 1990; Hall et al., 1991; Jensen et al., 1995).

In addition to atmospheric effects, the landscape elements such as slope and aspect can cause radiometric distortion of the signal received by the sensor system (Jensen, 2005). Teillet et al. (1982) describe four topographic correction methods, including (1) cosine correction, (2) Minnaert correction, (3) statistical–empirical correction, and (4) C-correction. Each correction method is based on illumination and requires a digital elevation model (DEM) to determine how much illumination each pixel receives relative to its topography in the landscape. Much research continues on the removal of topographic effects from the scene. For example, Civco (1989) identifies several considerations including matching the DEM spatial resolution to that of the image, overcorrection of topographic effect because of the Lambertian surface assumption, ignoring that the diffuse component also illuminates the topography, strong anisotropy of apparent reflectance (Leprieur et al., 1988), and consideration to wavelength and deeply shadowed areas (Kawata et al., 1988). Readers should refer to more detailed discussions of radiometric correction in Chapters 3 and 4.

Noise in an image may be due to irregularities or errors that occur in the sensor response and/or data recording and transmission (van der Meer et al., 2009). Common forms of noise include systematic striping or banding and dropped lines. Early Landsat MSS data had substantial striping due to variations and drift in the response over time of the six MSS detectors. The *drift* was different for each of the six detectors, causing the same brightness to be represented differently by each detector (Fureder, 2010). The overall appearance was thus a *striped* effect. The corrective process made a relative correction among the six sensors to bring their apparent values in line with each other. Dropped lines occur when there are system errors that result in missing or defective data along a scan line and is often *corrected* by replacing the line with the pixel values in the line above or below, or with the average of the two.

9.4.2 Geometric Correction

The stability of a remote sensing platform, the curvature of the earth, sensor orientation, topography, and other factors cause geometric distortion in an image. Consequently, geometric correction is applied to remove these distortions so that the image is planimetrically (x, y) correct and the displacement of objects as well as scale variations are minimized or removed entirely (Aronoff, 2005). *Geometrically corrected imagery* can be used to extract accurate distance, area, and direction (bearing) information, and any information derived from such images can be related to other thematic information in a geographic information systems or spatial decision support systems (Jensen, 2005).

Remotely sensed imagery collected from airborne or spaceborne sensors often contains internal and external geometric errors (Jensen, 2005). These can be *systematic* (predictable) or *nonsystematic* (random), and generally, systematic geometric error is easier to identify and correct than random geometric error. Some of these errors can be corrected by using ephemeris of the platform and known internal sensor distortion characteristics. Commercial satellite data (e.g., SPOT Image, Landsat, QuickBird, GeoEye, and others) already have much of the *systematic error* removed. Other errors can be corrected only by matching image coordinates of physical features recorded by the image to the geographic coordinates of the same features collected from a map or global positioning system (GPS).

Internal geometric errors are introduced by the remote sensing system itself or in combination with Earth rotation or curvature characteristics. These distortions are often systematic (predictable) and may be identified and corrected using prelaunch or in-flight platform ephemeris (i.e., information about the geometric characteristics of sensor and the Earth at data acquisition). Geometric distortions in imagery that can sometimes be corrected through analysis of sensor characteristics and ephemeris data include

- Skew caused by Earth rotation effects
- Scanning system–induced variation including ground resolution cell size, relief displacement, and tangential scale distortion

External geometric errors are usually introduced by phenomena that vary in nature through space and time. The most important external variables that can cause geometric error in remote sensor data are random movements of the remote sensing platform at the time of data collection (i.e., altitude and attitude changes). Unless otherwise processed, however, *unsystematic random error* remains in the image, making it non-planimetric. To correct for these errors, two common geometric correction procedures are used to make the digital remote sensor data of value:

- Image-to-map rectification
- Image-to-image registration

Image-to-map rectification is the process by which the geometry of an image is made planimetric. Whenever accurate area, direction, and distance measurements are required, image-to-map geometric rectification should be performed; however, it may not remove all the distortion caused by topographic relief displacement in an image. The *image-to-map rectification* process normally involves selecting well-identified GCPs on an image and associating them with their planimetrically correct map counterparts (i.e., GCPs from a paper or digital map for which geographic coordinates can be derived—e.g., meters northing and easting in a Universal Transverse Mercator (UTM) map projection). Alternatives to obtaining accurate GCP map coordinate information for image-to-map rectification include the following:

- *Hard-copy planimetric maps* (e.g., USGS 7.5 min 1:24,000-scale topographic maps) where GCP coordinates are extracted using simple ruler measurements or a coordinate digitizer.
- *Digital planimetric maps* (e.g., the USGS digital 7.5 min topographic map series) where GCP coordinates are extracted directly from the digital map on the screen.
- *Digital orthophotoquads* that are already geometrically rectified (e.g., USGS digital orthophoto quarter quadrangles.
- *GPS instruments* that may be taken into the field to obtain the coordinates of objects to within ±20 cm if the GPS data are differentially corrected.

Once GCP map coordinates are obtained for several points on an image, spatial interpolation algorithms are applied to transform the image coordinates to the map coordinates, thus making the image planimetrically correct.

- Polynomial equations are used to convert the file coordinates into rectified map coordinates.
- Depending on the distortion of the imagery, complex (higher-order) polynomial equations may be required to express the needed transformation.
- The degree of complexity of the polynomial is expressed as the order of the polynomial (i.e., the highest exponent used in the polynomial).

Intensity interpolation arises from the fact that there is no one-to-one relationship between the input and output pixel location. Therefore, a new BV has to be assigned to the rectified pixel. Three methods for such *resampling* (Figure 9.7) are as follows:

- Nearest neighbor
- Bilinear interpolation
- Cubic convolution

The nearest neighbor algorithm assigns the BV of the closest input x, y to the output x, y. This method maintains the integrity of the data and is not computationally intensive. Unfortunately, the output image may not be aesthetically pleasing as it has a *block appearance*. Bilinear interpolation derives the new BV of the pixel based on the weighted value of the four pixels nearest to those in the original image. This method produces a *smoother* image, but has a slight impact on the integrity of the data. Cubic convolution determines the output BV based on the weighted values of 16 input pixels surrounding the location of the original pixel. This method produces a *smoother* image but is computationally intensive and may have a considerable impact on the integrity of the data (Figure 9.8; Table 9.2).

Image-to-image registration is the translation and rotation alignment process by which two images of like geometry and of the same geographic area are positioned coincident with respect to one another so that corresponding elements of the same ground area appear in the same place on the registered images (Chen and Lee, 1992; Jensen, 2005). This type of geometric correction is used when it is *not* necessary to have each pixel assigned a unique x, y coordinate in a map projection. For example, we might want to make a cursory examination of two images obtained on different dates to see if any change has taken place. In such a case, we need to *register* only the two (or more images) to a *base* image (selected among the available dataset) and perform a rapid visual analyses of the data.

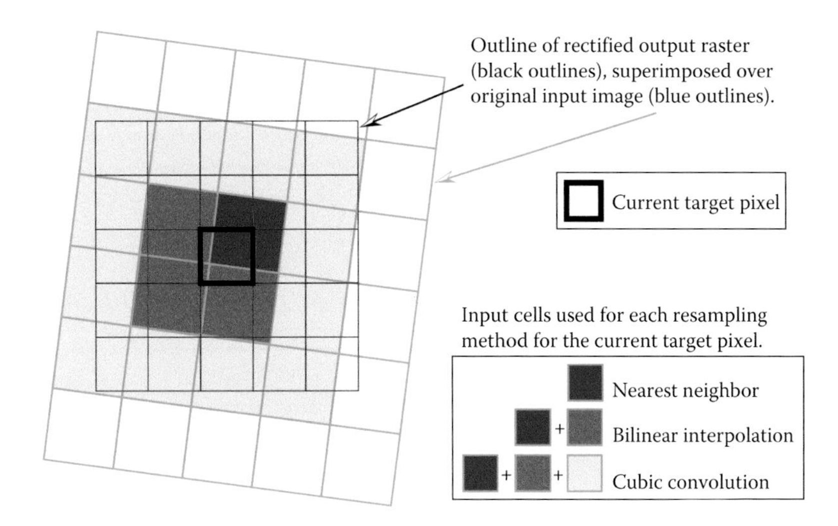

FIGURE 9.7 Schematic diagram showing the comparison between nearest neighbor, bilinear interpolation, and cubic convolution algorithms for intensity interpolation. (Adapted from http://marswiki.jrc.ec.europa.eu/wikicap/index.php/Resampling_techniques_in_image_processing. Accessed June 18, 2014.)

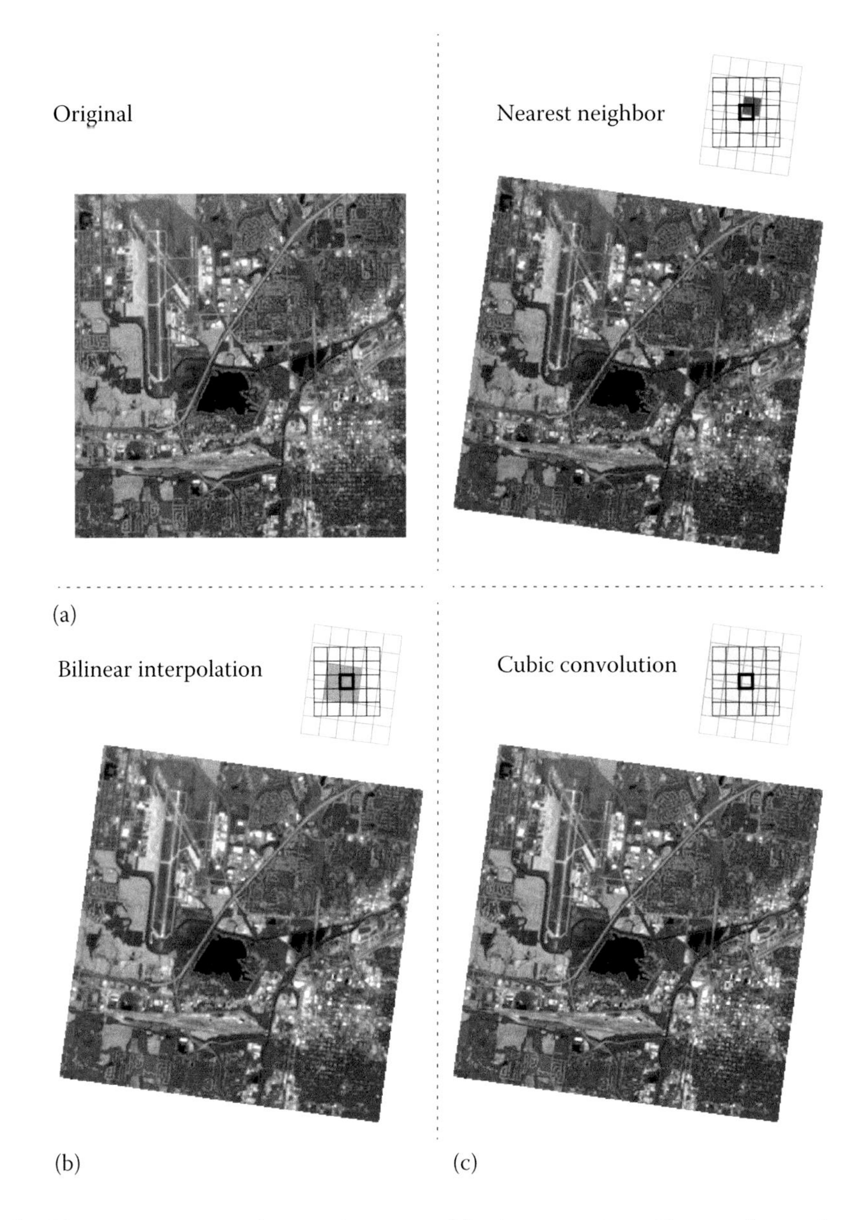

FIGURE 9.8 An example of three-intensity interpolation techniques used for the geometric rectification of remotely sensed data: (a) the nearest-neighbor algorithm that uses the closest pixel location BV to assign to the corrected data; (b) the bilinear interpolation that uses a weighted average of the four spatially closest pixels to assign a BV to the corrected data; and (c) the cubic convolution algorithm that uses the weighted average of the 16 closest pixels to determine the BV of the rectified pixel. (Image courtesy of the Center for Advanced Land Management Information Technologies (CALMIT), University of Nebraska, Lincoln, NE.)

TABLE 9.2 Geometric Correction Methods and Their Comparative Advantages and Disadvantages

Intensity Interpolation Technique	Pixels Used for Interpolation	Advantages	Disadvantages
Nearest neighbor	1	Maintains data integrity	Blocky appearance
		Not computationally intensive	
Bilinear interpolation	4	Minimal disruption to data integrity	Contrast may be reduced
		Smooth appearance	BVs are interpolated
Cubic convolution	16	Improved appearance	BVs are highly manipulated

9.5 Principal Component Analysis

The majority of remote sensing data are acquired in many different bands leading to the generation of vast quantities of data. Because of the spectral proximity of some bands in a multispectral dataset and certainly in the case of hyperspectral imagery, there is often a high degree of correlation between bands, implying that there may be similar information content between them. For example, Landsat Thematic Mapper (TM) bands 2 and 3 (green and red, respectively) typically have similar visual appearances because reflectances for the same cover types are almost equal. Image transformation techniques based on complex processing of the statistical characteristics of multiband datasets can be used to reduce this redundancy and correlation between bands. One such transformation is called PCA whose objective is to reduce the dimensionality (i.e., the number of bands) in the data and compress as much of the information in the original bands into fewer bands. The *new* bands that result from this statistical procedure are called components. The process attempts to statistically maximize the amount of information (or variance) from the original data into the least number of *useful* new components.

PCA transforms the axes of the multispectral space such that it coincides with the directions of greatest correlation. Each of these new axes are orthogonal to one another; that is, they are at right angles and the component images are arranged such that the greatest amount of variance (or information) within the original dataset is contained within the first component and the amount of variance decreases with each component (Jensen, 2005; Figure 9.9). Transformation of original data on X_1 and X_2 axes onto PC_1 and PC_2 axes requires transformation coefficients that can be applied in linear fashion to original pixel values (Figure 9.9). These new axes are called first PC. The second PC is perpendicular (orthogonal) to PC_1 (Gonzalez, 2014). Subsequent components contain decreasing amounts of the variance found in the dataset.

By computing the correlation between each band and each PC, it is possible to determine how each band *loads* or is associated with each PC. A linear combination of original BV and factor scores (eigenvectors) produces the new BV for each pixel of every PC. It is often the case that the majority of the information contained in a multispectral dataset can be represented by the first three or four PCA components. Higher-order components may be associated with noise in the original dataset.

9.6 Spatial Filtering

For any given image, or part thereof, there are changes in BVs throughout the scene. The number of changes in BVs per unit distance for any particular part of the image is called spatial frequency—that is, the *roughness* of the tonal variations occurring in an image (Jensen, 2005). Figure 9.10(a) and (b) demonstrates the differences between low-frequency (*less* roughness) and high-frequency (*more* roughness) images. In a low-frequency area, the changes in BVs are subtle over the given area, while the opposite is true in a high-frequency image.

To extract quantitative information, *local* operations are performed (spatial filtering), and the BV of a given pixel is modified based on the values of neighboring pixels.

Depending on the features to be extracted, filters can be applied to an image. A filter (or a convolution mask/kernel) is a moving window function that defines a small sub-window with a dimension of 3 × 3 or larger and usually with odd-numbered dimensions (e.g., 3 × 3, 5 × 5, and 7 × 7). An example of a 3 × 3 window is shown in Figure 9.11, with pixels numbered from the top left. In this example, pixel $C_{2,2}$ in the window is the center pixel, and odd-numbered window sizes ensure that there is always a center pixel in the sub-window.

Filtering involves computing a weighted average of the pixels in the moving window. The choice of weights determines how the filter affects the image. A window of weight values is called a convolution kernel. Multiplying each pixel in the moving window by

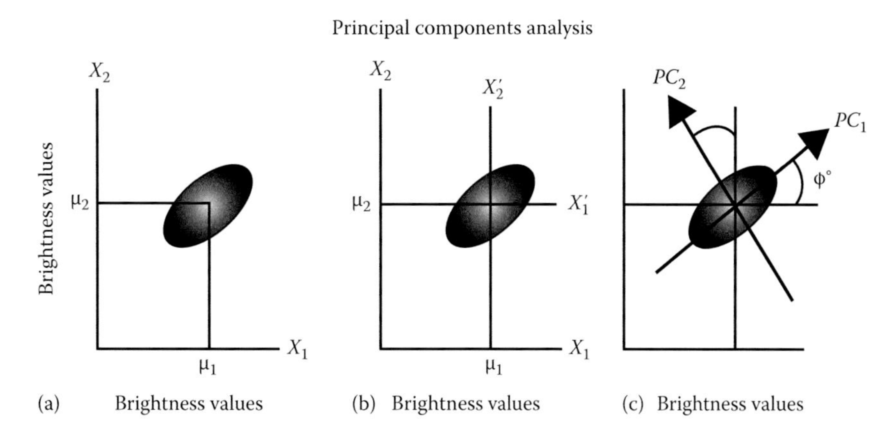

FIGURE 9.9 Concept diagram illustrating the PCA process: (a) the cluster of BVs from two bands of an image, (b) a new coordinate system defined by the X', and (c) the PCA transformation that occurs by rotating to the new axis, which is orthogonal to the original X' axis. The new axes are no longer the bands of the original image, but derivative components from those data. (Adapted from Jensen, J.R., *Introductory Digital Image Processing: A Remote Sensing Perspective*, Prentice Hall, Upper Saddle River, NJ, 2005.)

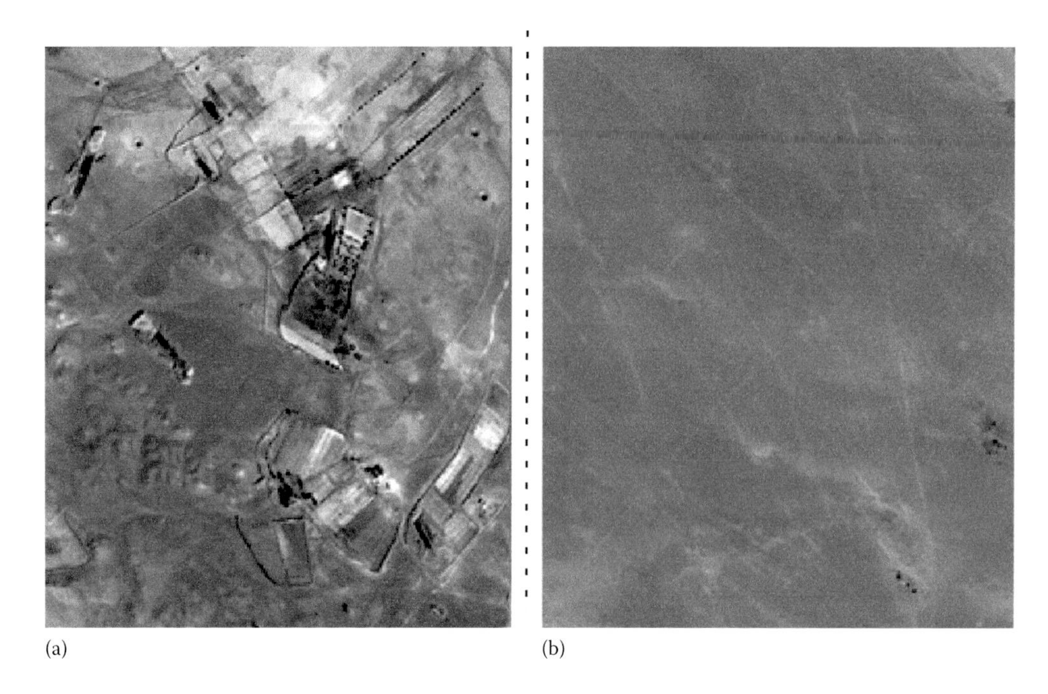

FIGURE 9.10 Examples of (a) high-frequency image and (b) low-frequency image.

$C_{1,1}$	$C_{1,2}$	$C_{1,3}$
$C_{2,1}$	$C_{2,2}$	$C_{2,3}$
$C_{3,1}$	$C_{3,2}$	$C_{3,3}$

FIGURE 9.11 Schematic of 3 × 3 convolution kernel. Values are assigned to each cell depending on the type of information that an analyst may wish to extract.

its weight and summing all the products yield a new value for the center pixel (Figure 9.12). The values used in a convolution kernel define whether the filter is *low pass* or *high pass*.

Low-pass filters are designed to *emphasize low-frequency features and de-emphasize the high-frequency components of an image*. Thus, information changing very fast across a landscape (e.g., in an urban area) will be subdued while low-frequency information (e.g., grassland, water) is preserved. Low-pass filters are excellent for retaining low-frequency information and are useful for removing *noise* (such as speckle) in an image (Jensen, 2005). Low-pass filters make similar cover areas appear uniform and can be useful for boundary detection (Figure 9.13). Conversely, low-pass filters do not preserve edges, and larger window sizes lead to greater smoothing.

High-pass filters emphasize the detailed high-frequency components of an image and deemphasize the more general low-frequency information. They enhance image details (infrequent information) and are useful where lower-frequency information tends to *hide* parts of the scene of interest, for example, roads in an urban scene. When building a high-pass filter, the center pixel of the kernel is given more weight. Consequently, if it is an edge, then the pixel will be greatly enhanced because edges have higher pixel values.

Spatial filtering methods can also be used to remove noise in the data (e.g., striping or speckles). Jensen (2005) used the Fourier transform (FT) on Landsat TM data to remove striping in a coastal area near Al Jubail, Saudi Arabia. The process involved the computation of an FT for the study area, modifying the resulting FT by selectively removing the points associated with the striping and then computing a reverse FT to derive a clean, destriped image. Destriping of an image can also be done by running a low-pass and a high-pass filter on an image and then adding the filter outputs. The resulting image will have its striping removed or minimized (USGS, 2007). Table 9.3 broadly describes the types of filters used on remotely sensed data and their utility.

9.7 Band Ratioing and Vegetation Indices

Another useful image processing technique exploits the relationships among the BVs of different bands, of image features. Mathematical expressions are applied to image bands in order to extract thematic information. These expressions may be a *simple ratio* (SR), or complex equations, and are generally developed to target a specific feature of interest. Many such algorithms have been developed to highlight characteristics of land cover such as vegetation, soil, water, and urban areas, and the information extracted can be applied to a wide

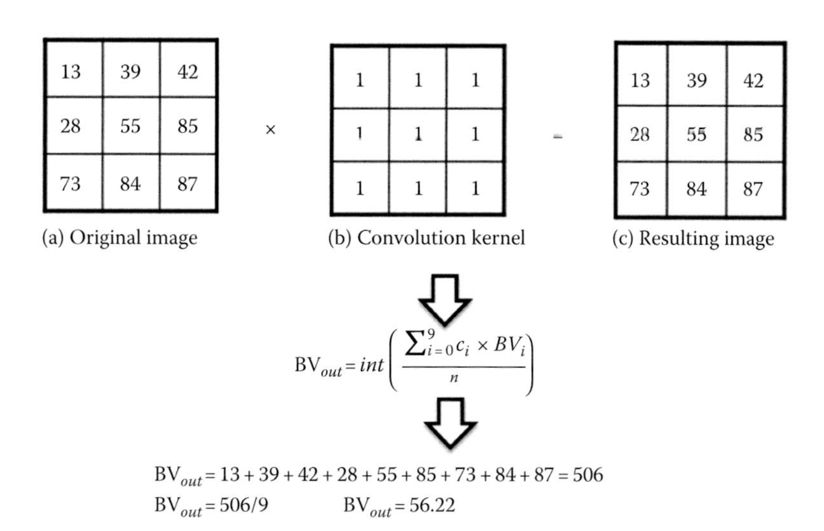

FIGURE 9.12 Schematic diagram showing the process and results of applying a low filter. The original center cell value was (a) 55. With the application of a convolution kernel (b) with equal weights for all 9 cells, an output image (c) is created. Further, the formula described earlier sums the values of the nine cells and produces the average value of 56.22. Rounded off to the nearest whole number, the new value of the center cell will be 56.

range of analyses (see Table 9.4 for examples of ratios and vegetation indices).

9.7.1 Band Ratio

The BVs of specific targets of interest vary from image to image depending on environmental factors, including topography, slope of target surface, aspect ratio, solar angle, seasonal changes, atmospheric conditions, water content, substrate conditions, or shadowing. Such factors may significantly increase or decrease BVs relative to what would be expected in laboratory conditions. This may make complex image analysis functions such as classification, feature discrimination, and change detection, difficult to perform. However, certain ratio transformations applied to two or more spectral bands can minimize such effects. In addition, these ratios may generate unique information, not otherwise attainable, through visual image analysis techniques. The mathematical expression of the band ratio is

$$\mathrm{BR}, P_x = \frac{\mathrm{BV}, P_x B_x}{\mathrm{BV}, P_x B_y}$$

where BR, P_x is the output value for a *pixel* (P_x) using the BVs of two bands: *band x* (B_x) and *band y* (B_y). One obvious problem becomes clear that BR, $P_x = 0$ is a possible outcome. There are several methods to address this, however, including assigning a value of 1 to any BV with a value of 0 or adding a small value to the denominator if it equals zero (such as 0.1).

While band ratios provide a new series of BVs to evaluate, it is not always easy to determine which bands should be used to provide information on specific targets of interest. Sometimes, the decision process may be as simple as the analyst displaying the results of multiple ratios and choosing the resulting dataset that appears most visually appealing or informative. However, there are widely used techniques for determining optimum bands for ratioing, such as the Optimum Index Factor and the Sheffield Index (Chavez et al., 1984; Sheffield, 1985).

9.7.2 Vegetation Index

Vegetation is a critical component of the health and condition of the Earth's natural environment. The U.S. Environmental Protection Agency (EPA) cites vegetation characteristics and biophysical variables (such as biomass and percentage of cover) and key indicators of ecosystem health (EPA/600/s-05, 2010). Because of this, vegetation studies have been a popular subject of remote sensing research since the 1960s. Scientists have modeled biophysical characteristics of vegetation using digital imagery since the data became available and continue to do so today (Hardisky et al., 1986; Gross and Klemas 1988; Gross et al., 1993; Zhang et al., 1997, 2009; Spanglet et al., 1998; Mishra et al., 2012). Many of these studies involve the use of *vegetation indices*, complex mathematical equations applied to image bands to measure the relative *greenness* of image features, from which meaningful information may be extracted of the composition and characteristics of vegetation. Such *vegetation biophysical variables* (VBVs) may include, but are not limited to, above-ground green biomass (AGB), absorbed photosynthetic active radiation, concentration of chlorophyll (or other leaf pigment), leaf area index (LAI), and percent of substrate covered by vegetation (vegetation fraction). Often, these biophysical properties can be key indicators of vegetation health, ecosystem health, and other critical ecological factors.

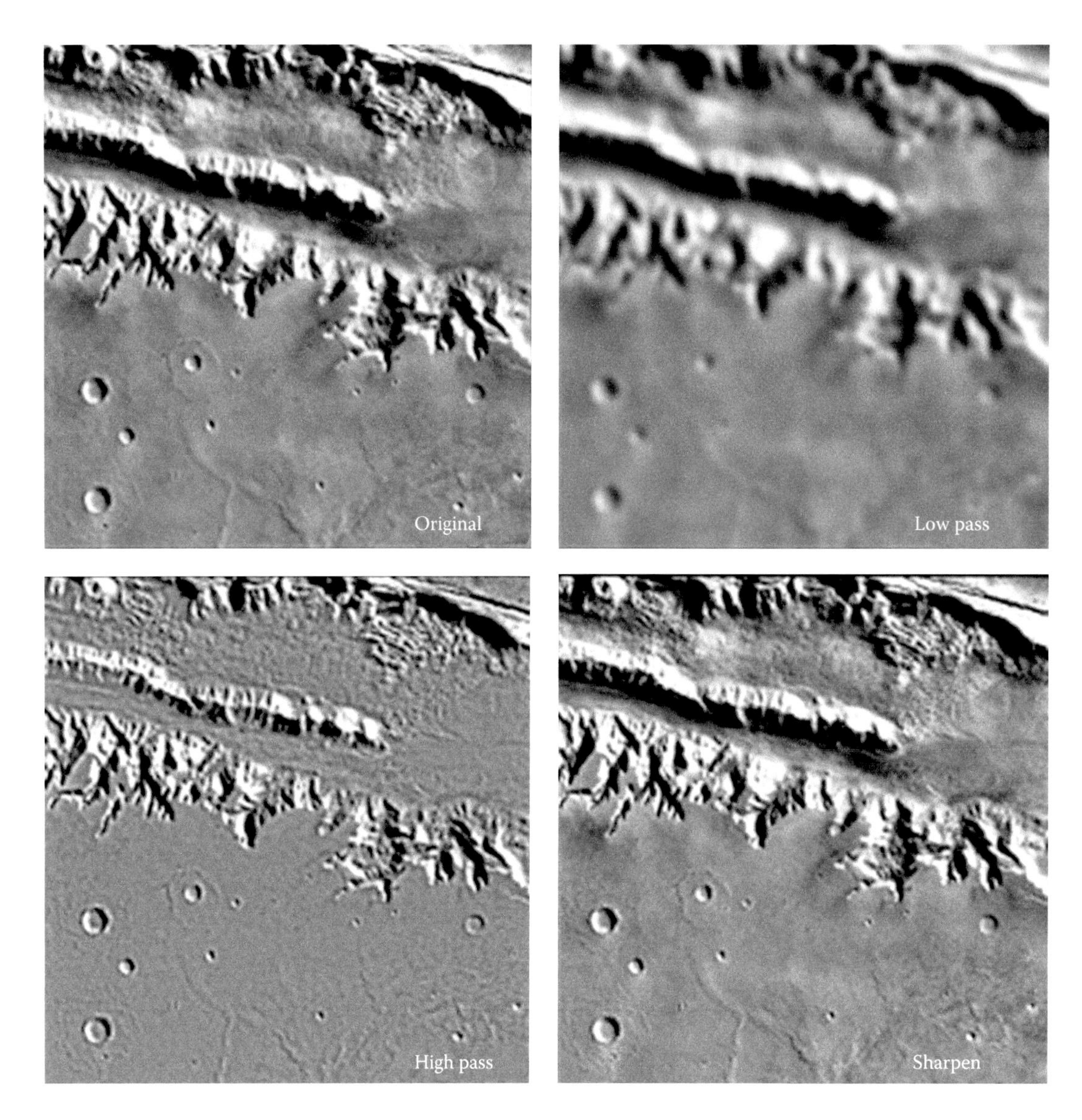

FIGURE 9.13 Comparison of application of a low-pass filter versus a high-pass filter versus a sharpening filter. (Adapted from USGS, The power of spatial filters, http://astrogeology.usgs.gov, accessed June 20, 2014, 2007.)

TABLE 9.3 Filtering Techniques and Their Utility in Analyzing Remotely Sensed Data

Filtering Technique	What It Does	Filter Examples
High frequency	Allows high-frequency information to pass through	Enhancing structural details
	Suppresses low-frequency information	Bring out boundaries and edges
	Edges are sharp and small features stand out	
	Large features look suppressed	
Low frequency	Allows low-frequency information to pass through	Highlight larger features
	Suppresses high-frequency information	Bring out information in larger features
	Edges get subdued	
	Larger features are enhanced	
	Smaller features begin to get smoothed	
Edge enhancement	Detect edges/boundaries between features	Aid in automated feature extraction
		Useful for geologic information, urban areas, boundaries, etc.
Frequency domain	Converts data from spatial to frequency domain	Enhancement, compression
		Noise removal, image restoration
		Textural classification, quality assessment

TABLE 9.4 Examples of Some of the Common Ratio and Vegetation Indices

Index	Formula	Source
Simple ratio	$SR = \frac{\rho_{red}}{\rho_{nir}}$	Birth and McVey (1968)
Normalized Difference Vegetation Index	$NDVI = \frac{(\rho_{nir} - \rho_{red})}{(\rho_{nir} + \rho_{red})}$	Rouse et al. (1974)
Soil-adjusted Vegetation Index	$SAVI = \frac{(\rho_{nir} - \rho_{red})}{(\rho_{nir} + \rho_{red} + a) \times (1 + a)}$	Huete (1988)
Green Normalized Difference Vegetation Index	$GNDVI = \frac{(\rho_{nir} - \rho_{green})}{(\rho_{nir} + \rho_{green})}$	Buschmann and Nagel (1993)
Green Atmospherically Resistant Vegetation Index	$GARI = \frac{\rho_{nir} - [\rho_{green} - (\rho_{blue} - \rho_{red})]}{\rho_{nir} - [\rho_{green} + (\rho_{blue} - \rho_{red})]}$	Gitelson et al. (1996)
Enhanced Vegetation Index	$EVI = 2.5 \times \frac{(\rho_{nir} - \rho_{red})}{(\rho_{nir} + 6(\rho_{red}) - 7.5(\rho_{blue}) + 1)}$	Huete et al. (1996)
Visible Atmospherically Resistant Index	$VARI = \frac{(\rho_{green} - \rho_{red})}{(\rho_{green} + \rho_{red} - \rho_{blue})}$	Gitelson et al. (2002)
Wide Dynamic Range Vegetation Index	$WDRVI = \frac{a \times (\rho_{nir} - \rho_{red})}{a \times (\rho_{nir} + \rho_{red})}$	Gitelson (2004)
Three-band Model	$TbM = [\rho(\lambda_1)^{-1} - \rho(\lambda_2)^{-1}] \times \rho(\lambda_3)$	Gitelson et al. (2006)
Enhanced Vegetation Index 2	$EVI2 = 2.5 \times \frac{(\rho_{nir} - \rho_{red})}{(\rho_{nir} + 2.4(\rho_{red}) + 1)}$	Jiang et al. (2008)

In order to maximize the ability to extract meaningful information, a VI should have four characteristics (Running et al., 1994; Huete and Justice, 1999; Gitelson et al., 2006). *First*, the VI must be sensitive to VBVs of interest. It is helpful if the sensitivity demonstrates a predictable relationship between index and VBV, preferably, a linear relationship that is applicable across a wide range of vegetation conditions, substrates, and species-types. *Second*, the impact of external variables such as atmospheric interaction, solar angle, and viewing angle must be minimized. This is necessary to compare multiple datasets with consistent spatial and temporal conditions. *Third*, the impact of internal variables such as canopy architecture, substrate, phonological changes, and nonphotosynthetic canopy components should be minimized. Such internal variables can contribute substantially to the recorded spectral response of vegetation in digital imagery and coarse spatial resolution may contribute to poor results when comparing multiple image datasets. *Fourth*, accuracy assessment must be tied to a specific measureable VBV such as AGB, LAI, or vegetation fraction.

Since the 1960s, several vegetation indices have been developed, though some of these provide redundant information content (Jensen, 2005). Subtle differences in algorithms are often adopted due to variability in sensor specifications and/or target characteristics. Vegetation, however, is a spectrally unique surface feature due to the chlorophyll absorption and leaf reflectance characteristics in the visible through near-infrared (VNIR) regions of the spectrum. Many of the VIs developed early target the inverse relationship between red and NIR reflectance. However, algorithms have been successfully developed using other characteristics of vegetation reflectance in a variety of environments and/or sensor specifications (Viña et al., 2011).

9.7.3 Simple Ratio

Birth and McVey (1968) described the *SR*, one of the first documented VIs that provides a simple formula for measuring the ratio of red reflectance (ρ_{red} in% or dimensionless) to NIR reflectance (ρ_{nir}):

$$SR = \frac{\rho_{red}}{\rho_{nir}}$$

Green vegetation strongly reflects incident irradiation in the NIR region (40%–60%) while absorbing up to 97% in the red region (Gitelson, 2004). As vegetation *greenness* declines, red reflectance increases and NIR reflectance decreases. By computing the ratio of red to NIR, this relationship can be quantified.

9.7.4 Normalized Difference Vegetation Index

One of the most noticeable problems with the SR is that it is not normalized, making it difficult to compare results among different studies. Rouse et al. (1974) addressed this issue with the NDVI. NDVI is functionally equivalent to SR, and comparison plots reveal no scatter between SR and NDVI.

$$\text{NDVI} = \frac{(\rho_{nir} - \rho_{red})}{(\rho_{nir} + \rho_{red})}$$

NDVI is widely applied to spectral and image data for monitoring, analyzing, and mapping VBVs. There are several characteristics of NDVI that contribute to its utility and continuing popularity among vegetation experts:

- Seasonal and phonological changes in vegetation can be monitored.
- Normalized data make comparisons more reliable.
- Ratioing reduces some cases of multiplicative noise cause by differences in solar angle, shadows, and topographic variations.

Conversely, a major disadvantage to NDVI is the nonlinear nature of the relationship between NDVI values and many VBVs. The index becomes saturated at high levels, and as VBVs increase, NDVI shows little variation.

9.7.5 Enhanced Vegetation Index 1 and 2

Several VIs are tailored to specific sensors and may be *tuned* to maximize the results of analysis at specific resolution characteristics. An example of this is the Enhanced Vegetation Index (EVI), which was developed by Huete et al. (2002) specifically for application to MODIS data. EVI is similar to NDVI; however, it includes several coefficients in the equation to account for atmospheric scattering and to reduce the saturation effects of NDVI at high values.

$$\text{EVI} = 2.5 \times \frac{(\rho_{nir} - \rho_{red})}{(\rho_{nir} + 6(\rho_{red}) - 7.5(\rho_{blue}) + 1)}$$

An updated version of the index, Enhanced Vegetation Index 2 (EVI2), was developed later by Jiang et al. (2008) for use with datasets that did not have sensitivity in the blue region of the spectrum.

$$\text{EVI2} = 2.5 \times \frac{(\rho_{nir} - \rho_{red})}{(\rho_{nir} + 2.4(\rho_{red}) + 1)}$$

9.7.6 Wide Dynamic Range Vegetation Index

Gitelson (2003) proposed a simple adjustment to NDVI to compensate for the high-end saturation. The *Wide Dynamic Range Vegetation Index* (WDRVI) applies a weighted coefficient (*a*) to NDVI with a value of 0.1–0.2 to *linearize* the index relationship to VBVs.

$$\text{WNDVI} = \frac{a \times (\rho_{nir} - \rho_{red})}{a \times (\rho_{nir} + \rho_{red})}$$

9.7.7 Three-Band Model

Most VIs focus on chlorophyll absorption characteristics. Gitelson et al. (2006) proposed an index that may be optimizable for other pigments and potentially other features of interest. This *three-band model* (*TbM*) requires the use of three spectral bands that must be identified as follows:

- Band 1 (λ_1): The band that is most sensitive to changes in VBV.
- Band 2 (λ_2): The band that is the most insensitive to changes in VBV.
- Band 3 (λ_3): The band that accounts for backscattering/noise among samples.

$$\text{TbM} = \left[\rho(\lambda_1)^{-1} - \rho(\lambda_2)^{-1}\right] \times \rho(\lambda_3)$$

TbM generates a linear relationship and, by accounting for backscatter/noise, normalizes the relationship between VBV and index value. Additionally, TbM has the potential to be applied to other features of interest, including non-vegetation surface types (Figure 9.14).

9.8 Image Classification

With the advent of digital image data and computer technology, algorithms were developed to enable the extraction of land use/land cover and biophysical information directly from the images (Jensen et al., 2009.) Over the past five decades, numerous algorithms have been developed to aid analysts in their image interpretation processes. Broadly, these algorithms are based on parametric statistics (assuming normally distributed data), nonparametric statistics (does not require normally distributed data), and nonmetric (capable of operating on real-valued data and nominal scaled data) (Jensen et al., 2009). Jensen (2005) and Lu and Weng (2007) provide an extensive taxonomic overview of several image classification methods, and this section will broadly review some of the more common methods used for thematic information extraction.

In general, image classification can be accomplished using either supervised or unsupervised techniques. Within each of these broad categories exist a host of different methods that can be applied toward specific imagery (e.g., high spatial resolution, hyperspectral) and/or have been developed as technological capabilities have advanced (e.g., artificial neural network (ANN) classification and object-oriented classification). Supervised classification categorizes every image pixel into one of several predefined land-type classes (Jensen et al., 2009). The process requires several steps and includes selecting the land-type categories, preprocessing, defining training data, automated pixel assignment, and accuracy assessment (see Sabins, 1987; Jensen, 2005; Jensen et al., 2009). Various supervised classification algorithms exist including the maximum-likelihood classifier (Strahler, 1980; Foody et al., 1992;

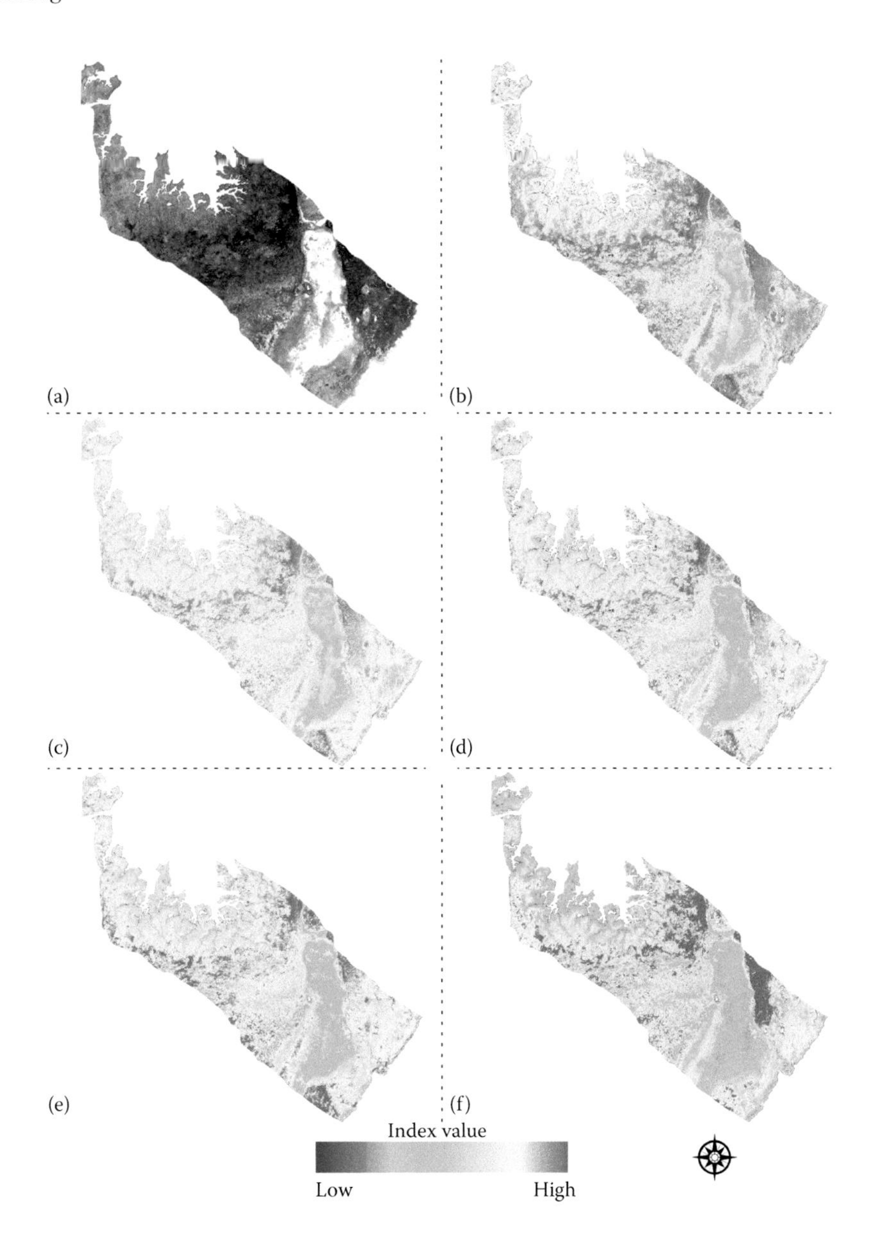

FIGURE 9.14 A suite of selected vegetation indices applied to an image acquired by the Airborne Imaging Spectrometer for Applications (AISA) *Eagle* acquired over Grand Bay, MS. (a) Original aerial image, (b) simple ratio, (c) Normalized Difference Vegetation Index, (d) Wide Dynamic Range Vegetation Index, (e) Enhanced Vegetation Index, and (f) three-band model. (Image courtesy of the Center for Advanced Land Management Information Technologies (CALMIT), University of Nebraska, Lincoln, NE.)

Atkinson and Lewis, 2000), nearest-neighbor classification (James, 1985; Hardin, 1994), decision tree classifiers (DeFries and Chan, 2000; Russell and Norvig, 2003; Jensen 2005), object-oriented classification (AtMost classification processes (Haralick and Shapiro, 1985; Yan et al., 2006; Chen et al., 2009) (Figure 9.15), and ANNs (Gopal and Woodcock, 1996; Hardin, 2000; Jensen et al., 2000)).

Unsupervised classification methods generally partition the spectral data of an image into feature space with a minimal input from the analyst. Operating under various constraints specified by the user (e.g., number of clusters, spectral and spatial search radius, bands used, and iterations defined), an unsupervised classification algorithm will search for natural groupings (or clusters) and produce a map of the number of predefined clusters. These clusters can subsequently be assigned to previously defined information classes (e.g., land cover categories) through an iterative process (Figure 9.16). Because there is a high likelihood that there will be clusters of *mixed* categories, the user may have to employ other methods to parse such clusters and minimize the undefined areas in an image. For example, mixed clusters from the original image may be masked out of the image prior to running an iteration of the unsupervised algorithm using a different combination of inputs (e.g., the number of bands included). The user may

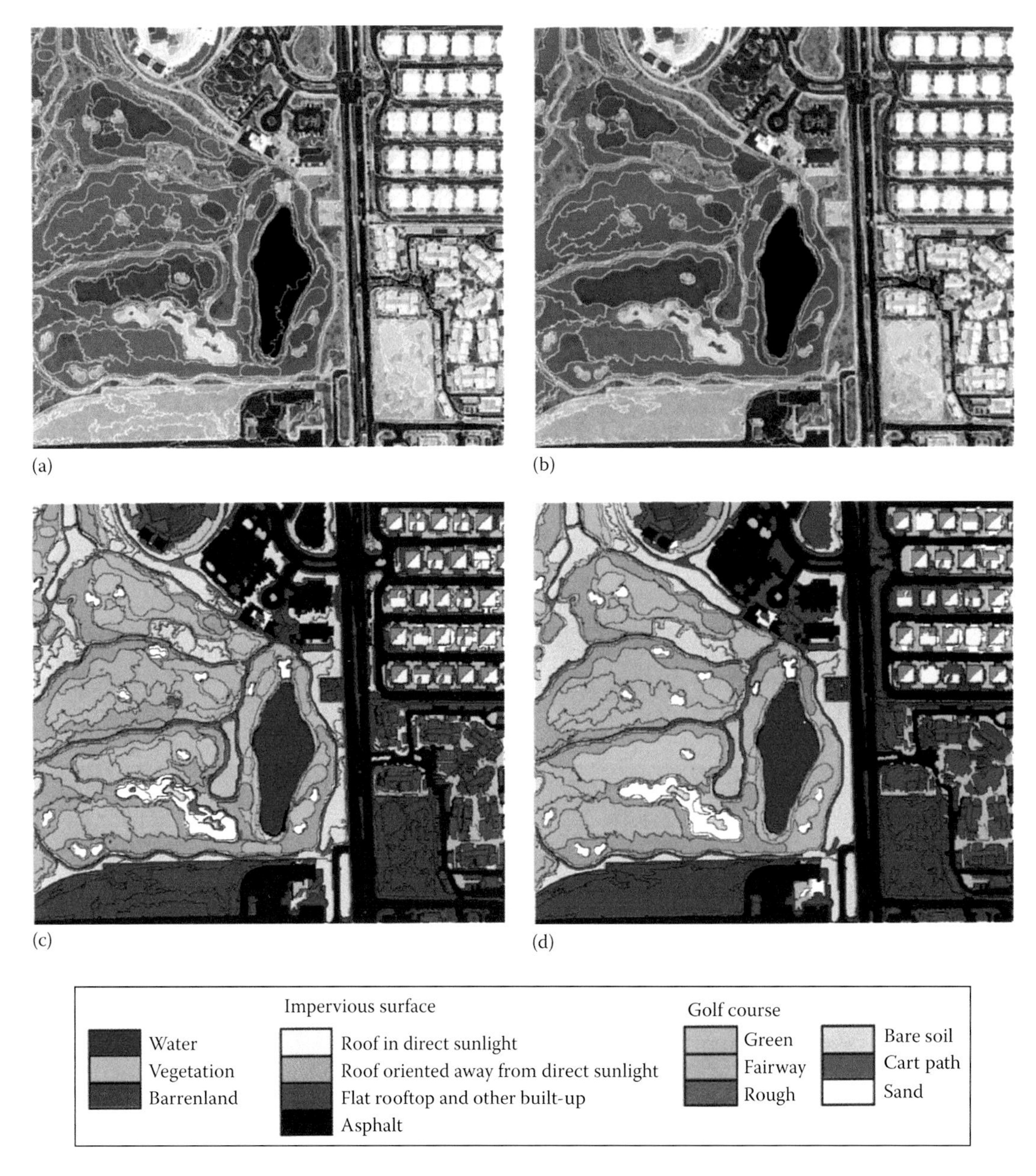

FIGURE 9.15 Example of object-based classification at two different image segmentation scales. The PAN-sharpened QuickBird high-spatial-resolution (61 × 61 cm) multispectral imagery of Las Vegas, NV, was acquired on May 18, 2003. (a) Segmentation scale 100, (b) segmentation scale 150, (c) classification at scale 100, and (d) classification at scale 150. (Adapted from Jensen, J.R. et al., Image classification, in *The SAGE Handbook of Remote Sensing*, Warner, T.A., Nellis, M.D., and Foody, G.M., eds., 2009, pp. 269–281.)

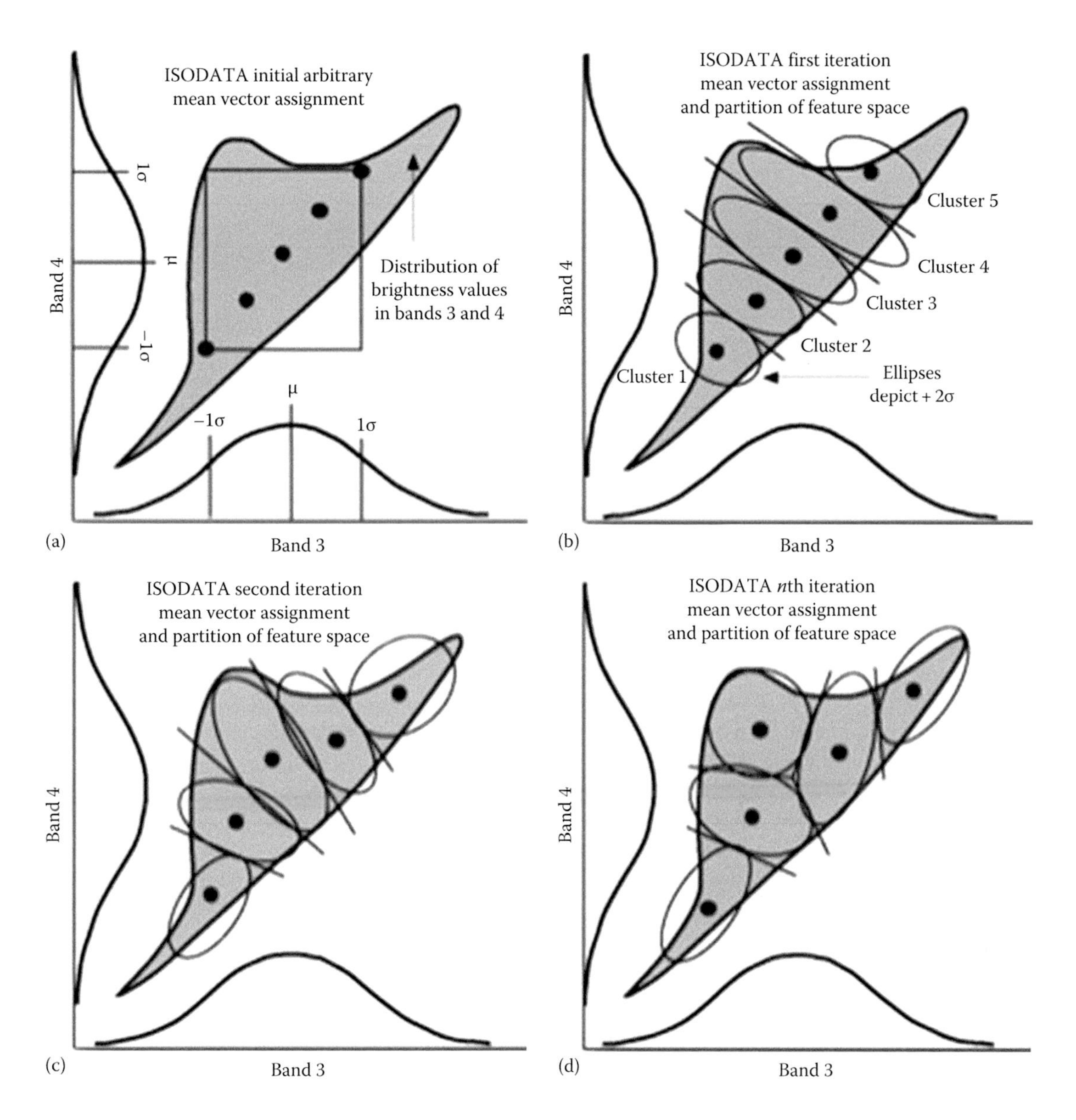

FIGURE 9.16 Conceptual diagram showing how the ISODATA algorithm functions: (a) initial distribution of five hypothetical mean vectors using $\pm 1\sigma$ in both bands as the beginning and ending points, (b) in the first iteration, each candidate pixel is compared to each cluster mean and assigned to the cluster whose mean is closest in Euclidean distance, (c) during the second iteration, a new mean is calculated for each cluster based on the actual spectral locations of the pixels assigned to each cluster, instead of the initial arbitrary calculation. This involves analysis of several parameters to merge and/or split clusters. After the mean vectors are selected, every pixel in the scene is assigned to one of the new clusters, (d) this split/merge–assign process continues until there is little change in class assignment between iterations (the T threshold is reached) or the maximum number of iterations is reached (M). (Adapted from Jensen, J.R., *Introductory Digital Image Processing: A Remote Sensing Perspective*, Prentice Hall, Upper Saddle River, NJ, 2005.)

also apply supervised classification algorithms to glean information from the mixed clusters. Jensen (2005) describes the two common clustering methods including the chain method and the iterative self-organizing data analysis (ISODATA). The building of clusters by either of these algorithms does not require any a priori knowledge. However, once the analyst works interactively with the clusters to assign them into the various land cover categories, considerable familiarity with the study area is needed.

9.9 Future Trends

Sensor systems and image processing software/hardware will continue to evolve and improve. Their utility has already been proven by the vast array of applications that image data and its subsequent information extraction have been used in. With the problems of the twenty-first century being focused on issues such as climate change, atmospheric conditions, environmental degradation, natural hazards, population growth, urbanization, resource scarcity, and many others, it is inevitable that remote sensing data will aid toward a better understanding of these problems. Future sensors will continue the progression toward more comprehensive and accurate Earth science measurements (Hartley, 2003). Increased spatial, spectral, radiometric, and temporal resolutions will provide scientists new levels of detail, and new algorithms will be developed to extract the relevant information with improved accuracies. Furthermore, technological revolutions in the miniaturization of electronics, stabilization of optical systems, efficient power sources, minimizing size and weight of senor systems, etc., will lead to changes in the design of the sensors, uninterrupted data collection, very high data quality, and orbital stability of systems.

The fusion of close range and *in situ* remote sensor data with satellite/airborne imagery and other geospatially derived information has already transformed analytical capabilities by facilitating multiscale studies of phenomena. With massive biogeophysical data volumes being generated, and the computational challenges for their analyses and visualization, image processing is a major player in the Big Data arena where new technologies are being sought to process large quantities of data within tolerable time frames. In addition, there is an urgent need for realistic and computationally simple surface radiation models for inversion of land surface variables from satellite data (Jensen et al., 2007). Liang (2003, 2007) have introduced several physically based inversion algorithms for estimating biogeophysical variables (e.g., LAI, fractional vegetation coverage, broadband albedo, etc.), and further research continues for the derivation of a suite of algorithms that will enable a simple but accurate determination of land surface variables.

References

Aronoff, S., 2005. *Remote Sensing for GIS Managers*. Redlands, CA: ESRI Press.

Birth, G. S. and G. R. McVey, 1968. Measuring color of growing turf with a reflectance spectrophotometer. *Agro J*, 60:640–649.

Chavez, P. C., Guptill, S. C., and J. A. Bowell, 1984. Image processing techniques for Thematic Mapper Data. *Proc Am Soc Photogramm*, 2:728–743.

Chen, L. and L. Lee, 1992. Progressive generation of control frameworks for image registration. *Photogramm Eng Remote Sens*, 58(9):1321–1328.

Civco, D. L., 1989. Topographic normalization of Landsat Thematic Mapper digital imagery. *Photogramm Eng Remote Sens*, 55(9):1303–1309.

DeFries, R. S. and J. C. Chan, 2000. Multiple criteria for evaluating machine learning algorithms for land cover classification. *Remote Sens Environ*, 74:503–515.

Du, Y., Tiellet, P. M., and J. Cihlar, 2002. Radiometric normalization of multitemporal high-resolution satellite images with quality control for land cover change detection. *Remote Sens Environ*, 82:123.

Eckhardt. D. W., Verdin, J. P., and G. R. Lyford, 1990. Automated update of an irrigated lands GIS using SPOT HRV imagery. *Photogramm Eng Remote Sens*, 56(11):1515–1522.

Exelis, 2003. Atmospheric Correction Module: QUAC and FLAASH User's Guide, https://www.exelisvis.com/portals/0/pdfs/envi/Flaash_Module.pdf (accessed June 8, 2014).

Foody, G. M., Campbell, N. A., Trodd, N. M., and T. F. Wood, 1992. Derivation and applications of probabilistic measures of class membership form the maximum-likelihood classification. *Photogramm Eng Remote Sens*, 58(9):1335–1341.

Fureder, R., 2010. Topographic correction of satellite images for improved LULC classification in alpine areas. In *10th International Symposium on High Mountain Remote Sensing Cartography*, Kathmandu, Nepal, pp. 187–194.

Gitelson, A., 2004. Wide Dynamic Range Vegetation Index for remote quantification of biophysical characteristics of vegetation. *J Plant Phys*, 131:165–173.

Gitelson, A. A., Kaufman, Y. J., Stark, R., and D. Rundquist, 2002. Novel algorithms for remote estimation of vegetative fraction. *Remote Sens Environ*, 80:76–87.

Gitelson, A. A., Keydan, G. P., and M. N. Merzlyak, 2006. Three-band model for noninvasive estimation of chlorophyll, carotenoids, and anthocyanin contents in higher plant leaves. *Geophys Res Lett*, 33:L11402.

Gonzalez, L., 2014. Principal component analysis, http://wiki.landscapetoolbox.org/doku.php/remote_sensing_methods:principal_components_analysis (accessed June 23, 2014).

Gopal, S. and C. E. Woodcock, 1996. Remote sensing of forest change using artificial neural networks. *IEEE Trans Geosci Remote Sens*, 34:398–404.

Gross, M. and V. Klemas, 1988. Remote sensing of biomass of salt marsh vegetation in France. *Int J Remote Sens*, 9(3):397–408.

Hall, F. G., Strebel, D. E., Nickeson, J. E., and S. J. Goetz, 1991. Radiometric rectification: Toward a common radiometric response among multidate, multisensor images. *Remote Sens Environ*, 35:11–27.

Haralick, R. M. and L. G. Shapiro, 1985. Image segmentation techniques. *Comp Vis Graph Image Proc*, 29:100–132.

Hardin, P. J., 1994. Parametric and nearest-neighbor methods for hybrid classification. *Photogramm Eng Remote Sens*, 60:1439–1448.

Hardin, P. J., 2000. Neural networks versus nonparametric neighbor-based classifiers for semisupervised classification of Landsat Thematic Mapper imagery. *Opt Eng*, 39:1898–1908.

Hardisky, A., Gross, M., and V. Klemas, 1986. Remote sensing of coastal wetlands. *Bioscience*, 37(7):453–460.

Hartley, J., 2003. Earth remote sensing technologies in the twenty-first century. In *Proceedings*, *International Geoscience and Remote Sensing Symposium*, Toulouse, France.

Huete, A., Didan, K., Miura, T., Rodriquez, P., Gao, X., and L. Ferreira, 2002. Overview of the radiometric and biophysical performance of the MODIS vegetation indices. *Remote Sens Environ*, 83:195–213.

ImSpec, LLC, 2002. ACORN 4.0 user's guide, http://www.aigllc.com/pdf/acorn4_ume.pdf (accessed June 8, 2014).

James, M., 1985. *Classification Algorithms*. New York: Wiley-Interscience.

Jensen, J. R., 2005. *Introductory Digital Image Processing: A Remote Sensing Perspective*. Upper Saddle River, NJ: Prentice Hall.

Jensen, J. R., Fang, Q., and J. Minhe, 2000. Predictive modeling of coniferous forest age using statistical and neural network approaches applied to remote sensor data. *Int J Remote Sens*, 20:2805–2822.

Jensen, J. R., Im, J., Hardin, P., and R. R. Jensen, 2009. Image classification. In *The SAGE Handbook of Remote Sensing*, Eds. Warner, T. A., Nellis, M. D., and Foody, G. M., pp. 269–281. London, U.K.: SAGE.

Jensen, J. R., Rutchey, K., Koch, M., and S. Narumalani, 1995. Inland wetland change detection in the Everglades Water Conservation Area 2A using a time series of normalized remotely sensed data. *Photogramm Eng Remote Sens*, 61(2):199–209.

Jiang, Z., Juete, A., Didan, R., and T. Miura, 2008. Development of a two-band Enhanced Vegetation Index without a blue band. *Remote Sens Environ*, 112(10):3833–3845.

Kawata, Y., Ueno, S., and T. Kusaka, 1988. Radiometric correction for atmospheric and topographic effects on Landsat MSS images. *Int J Remote Sens*, 9(4):729–748.

Liang, S., 2003. *Quantitative Remote Sensing of Land Surfaces*. New York: Wiley.

Liang, S., 2007. Recent developments in estimating land surface biogeophysical variables from optical remote sensing. *Prog Phys Geogr*, 31(5):501–516.

Lu, D. and Q. Weng, 2007. A survey of image classification methods and techniques for improving classification performance. *Int J Remote Sens*, 28(5):823–870.

Mishra, D., Cho, J., Ghosh, S., Fox, A., Downs, C., Merani, P., Kirui, P., Jackson, N., and S. Mishra, 2012. Post-spill state of the marsh: Impact of the Gulf of Mexico oil spill on the health and productivity of Louisiana salt marshes. *Remote Sens Environ*, 118:176–185.

Rouse, J., Haas, R., Schell, J., and D. Deering, 1974. Monitoring vegetation systems in the Great Plains with ERTS. In *Third Earth Research Technical Satellite Symposium, Vol. 1: Technical Presentations*, NASA SP351, NASA, Washington, DC.

Running, S. W., Justice, C. O., Solomonson, V., Hall, D., Barker, J., Kaufmann, Y. J., Strahler, A. H. et al., 1994. Terrestrial remote sensing science and algorithms planned for EOS/MODIS. *Int J Remote Sens*, 15(17):3587–3620.

Russell, S. J. and P. Norwig, 2003. *Artificial Intelligence: A Modern Approach*. Upper Saddle River, NJ: Prentice-Hall, 1080 pp.

Sabins, F. F., Jr., 1987. *Remote Sensing: Principles and Interpretation*. New York: W. H. Freeman.

Schowengerdt, R. A., 1997. *Remote Sensing: Models and Methods for Image Processing*. San Diego, CA: Academic Press.

Sheffield, C., 1985. Selecting band combinations from multispectral data. *Photogramm Eng Remote Sens*, 51(6):681–687.

Spanglet, H., Ustin, S., and E. Rejmankova, 1998. Spectral reflectance characteristics of California subalpine marsh plant communities. *Wetlands*, 18(3):307–319.

Strahler, A., 1980. The use of prior probabilities in maximum likelihood classification of remotely sensed data. *Remote Sens Environ*, 10(2):135–183.

Tanre, D., Deroo, C., Duhaut, P., Herman, M., Morcrette, J., Perbos, J., and P. Y. Deschamps, 1986. *Second Simulation of the Satellite Signal in the Solar Spectrum (6S) User's Guide*. Villeneuve d'Ascq, France: Lab d'Optique Atmospherique, U.S.T. de Lille.

USGS, 2007. The power of spatial filters, http://astrogeology.usgs.gov (accessed June 20, 2014).

van der Meer, F., van der Werff, H. M. A., and S. M. de Jong, 2009. Pre-processing of optical imagery. In *The SAGE Handbook of Remote Sensing*, Eds. Warner, T. A., Nellis, M. D., and Foody, G. M. London, U.K.: SAGE, pp. 229–243.

Viña, A., Gitelson, A., Nguy-Robertson, A., and Y. Peng, 2011. Comparison of different vegetation indices for the remote estimation of green leaf area index of crops. *Remote Sens Environ*, 115:3468–3478.

Yan, G., Mas, J.-F., Maathuis, B. H. P., Xiangmin, Z., and P. M. Van Dijk, 2006. Comparison of pixel-based and object-oriented image classification approaches—A case study in a coal fire area, Wuda, Inner Mongolia, China. *Int J Remote Sens*, 27(18):4039–4055.

Zhang, H., Hu, H., Yao, X., Zheng, K., and Y. Gan, 2009. Estimation of above-ground biomass using HJ-1 hyperspectral images in Hangzhou Bay, China. In *International Conference on Information, Engineering and Computer Science*, Wuhan, China, 2009.

Zhang, M., Ustin, S., Rejmankova, E., and E. Sanderson, 1997. Monitoring Pacific coast salt marshes using remote sensing. *Ecol Appl*, 7(3):1039–1053.

10

Urban Image Classification: Per-Pixel Classifiers, Subpixel Analysis, Object-Based Image Analysis, and Geospatial Methods

Soe W. Myint
Arizona State University

Victor Mesev
Florida State University

Dale A. Quattrochi
NASA Marshall Space Flight Center

Elizabeth A. Wentz
Arizona State University

Acronyms and Definitions

ASTER	Advanced Spaceborne Thermal Emission and Reflection Radiometer
ETM+	Enhanced Thematic Mapper Plus
Geary's C	A spatial autocorrelation index
GEOBIA	Geographic Object-Based Image Analysis
GeoEye	High-resolution satellite operated by GeoEye company
GIS	Geographic information system or science
GLCM	Gray-level co-occurrence matrix
ICAMS	Image characterization and modeling system
IKONOS	High-resolution satellite operated by GeoEye
LiDAR	Light Detection and Ranging
MESMA	Multiple endmember spectral mixture analysis
MIR	Mid-infrared portion of the electromagnetic spectrum
Moran's I	A spatial autocorrelation index
MSS	Multispectral scanner
NIR	Near-infrared portion of the electromagnetic spectrum
NDVI	Normalized difference vegetation index
OBIA	Object-based image analysis
OLI	Operational Land Imager
Pan	Panchromatic
QuickBird	High-resolution satellite operated by DigitalGlobe
TIR	Thermal infrared portion of the electromagnetic spectrum
USGS	United States Geological Survey
V-I-S	Vegetation, impervious, and soil
Vis	Visible portion of the electromagnetic spectrum

10.1 Introduction

Remote-sensing methods used to generate base maps to analyze the urban environment rely predominantly on digital sensor data from spaceborne platforms. This is due in part from new sources of high-spatial-resolution data covering the globe, a variety of multispectral and multitemporal sources, sophisticated statistical and geospatial methods, and compatibility with geographic information system or science (GIS) data sources and methods. The goal of this chapter is to review the four groups of classification methods for digital sensor data from spaceborne platforms; per-pixel, subpixel, object-based (spatial-based), and geospatial methods. Per-pixel methods are widely used methods that classify pixels into distinct categories based solely on the spectral and ancillary information within that pixel. They are used for simple calculations of environmental indices (e.g., normalized difference vegetation

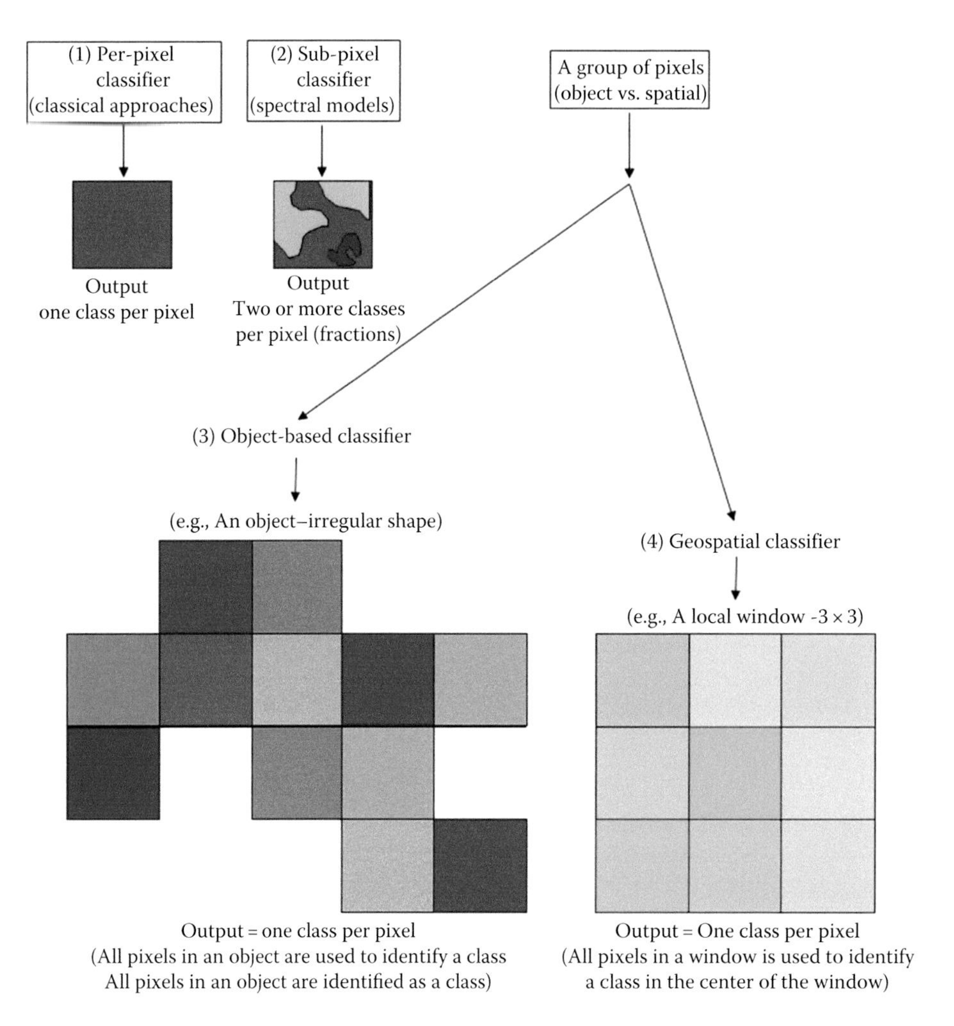

FIGURE 10.1 Overview of the four main classification groups.

index [NDVI]) to sophisticated expert systems to assign urban land covers (Stefanov et al., 2001). Researchers recognize, however, that even with the smallest pixel size, the spectral information within a pixel is really a combination of multiple urban surfaces. Subpixel classification methods therefore aim to statistically quantify the mixture of surfaces to improve overall classification accuracy (Myint, 2006a). While within-pixel variations exist, there is also significant evidence that groups of nearby pixels have similar spectral information and therefore belong to the same classification category. Object-oriented methods have emerged that group pixels prior to classification based on spectral similarity and spatial proximity. Classification accuracy using object-based methods shows significant success and promise for numerous urban applications (Myint et al., 2011). Like the object-oriented methods that recognize the importance of spatial proximity, geospatial methods for urban mapping also utilize neighboring pixels in the classification process. The primary difference though is that geostatistical methods (e.g., spatial autocorrelation methods) are utilized during both the pre- and postclassification steps (Myint and Mesev, 2012).

Within this chapter, each of the four approaches is described in terms of scale and accuracy classifying urban land use and urban land cover and for its range of urban applications. We demonstrate the overview of the four main classification groups in Figure 10.1, while Table 10.1 details the approaches with respect to classification requirements and procedures (e.g., reflectance conversion, steps before training sample selection, training samples, spatial approaches commonly used, classifiers, primary inputs for classification, output structures, number of output layers, and accuracy assessment). The chapter concludes with a brief summary of the methods reviewed and the challenges that remain in developing new classification methods for improving the efficiency and accuracy of mapping urban areas.

10.2 Remote-Sensing Methods for Urban Classification and Interpretation

Urban areas comprised a heterogeneous patchwork of land covers and land uses that are juxtaposed so that classification of specific classes using remote-sensing data can be problematic. Derivation of classification methods for urban landscape features has evolved in tandem with increasing spatial, spectral, and temporal resolutions of remote-sensing instruments (e.g., from 90 m Landsat multispectral scanner [MSS] to 30 m to the Landsat Enhanced Thematic Mapper Plus [ETM+] and Operational Land Imager [OLI] data

TABLE 10.1 Classification Procedures and Characteristics of the Four Main Classification Groups

	Per-Pixel	Subpixel	Object-Based	Geospatial
Reflectance conversion	Not required	Necessary	Not required	Not required
Additional step before training sample selection	No	No	Segment image into objects	No
Training samples	Irregular polygons that cover multiple pixels representing selected land cover classes	Spectra of selected endmembers	Segmented objects that cover multiple pixels representing selected land cover classes	Square windows that cover multiple pixels representing selected land cover classes
Commonly used spatial approaches	GLCM	No	GLCM	Fractal, Geary's C Moran's I, Getis index Fourier transforms, Lacunarity index, wavelet transforms
Widely used classifiers	Maximum likelihood, Mahalanobis distance, minimum distance, regression tree, neural network Bayesian	Linear spectral mixture, multiple regression, regression tree, neural network Bayesian, MESMA	Nearest neighbor	Mahalanobis distance, minimum distance
Primary input for classification	Training samples are used to identify land cover classes	End member spectra are used to quantify fractions of classes	All pixels in each object identified as one of the training sample classes	All pixels in each window are used to identify one class and the winner class is assigned to the center of the local window
No. of output layer	One layer	Multiple layers	One layer	One layer
Output structure	One class per pixel	One fraction per pixel per class	One class per pixel	One class per pixel
Accuracy assessment	Randomly selected pixels for error matrix	Correlation between predicted and reference fractions	Randomly selected pixels for error matrix (or) object-based accuracy assessment	Randomly selected pixels for error matrix

Note: GLCM, Gray-level (or) spatial co-occurrence matrix; MESMA, Multiple endmember spectral mixture analysis.

and progressing to submeter spatial resolution products available from commercial systems such as 0.34 m Geoeye) to achieve more robust digital classification schemes. This evolution of classification techniques, however, does not imply that one method is better than another. As with the type of satellite remote-sensing data that are employed for analyses, the application of a specific algorithm for classification of urban land cover and land use is depend upon what the user's objectives are and what level of detail, frequency, and sensors are required for the anticipated or resulting output products. Table 10.2 shows urban remote-sensing applications with regard to spatial, temporal, and sensor resolutions.

10.2.1 Per-Pixel Methods

Scale is indelible when conducting per-pixel classifications. The spatial resolution of the sensor dictates the classification type, range, and accuracy of urban land use and urban land cover. That is because individual urban features are rarely the same size as pixels, nor are they conveniently rectangular in shape. Add temporal scale representing rapid urban activity and per-pixel classifications become even more removed from reality. Refining the spatial resolution and reducing the area of the pixel do not necessarily lead to improvements in classification accuracy and may even introduce additional spectral noise, especially when pixels are smaller than urban features. In all, the ideal situation that each pixel can be identified to represent conclusively one and only one land cover type has now long been abandoned. So too the perfect relationship between the pixel and the field of view, which assumes reflectance, is recorded entirely and uniformly from within the spatial limits of individual pixels (Figure 10.1).

Regardless, the appeal of per-pixel or hard classifications remains, predominantly because they produce crisp and convenient thematic coverages that can be easily integrated with raster-based GIS models (Table 10.1). Composite models and methodologies containing information from remotely sensed sources are critical for revising databases and for producing comprehensive query-based urban applications. To preserve this relationship with GIS, the quality of per-pixel classifications must be monitored not only using conventional determination of accuracy based on comparisons with more reliable reference data but also in relation to levels of suitability or *scale of appropriateness*. Both were evident in the USGS hierarchical scheme (Anderson et al., 1976) using the much-cited 85% as a general guideline for the accuracy of urban features, which subsequently established a benchmark for researchers to attain and supersede using a variety of statistical and stochastic per-pixel techniques. Some of these focused exclusively on maximizing computational class separability, using the traditional maximum likelihood algorithm (Strahler, 1980) and the more recent support vector machines (Yang, 2011), while others developed methodologies that imported extraneous information when aggregating spectrally similar pixels (Mesev, 1998), by incorporating contextual relationships (Stuckens et al., 2000) or by measuring pixel interconnectivity (Barr and Barnsley, 1997). In both, classification

TABLE 10.2 Urban Remote-Sensing Classifications with Regard to Spatial, Temporal, and Sensor Resolutions

	Urban Features	Urban Process	Spatial Resolution	Temporal Resolution	Sensor Resolution
Micro scale: Individual measurements	Building unit (roofs: flat, pitch) (material: tile, natural/metal, synthetic)	Type and architecture Density	1–5 m	1–5 years	Pan–Vis–NIR
	Vegetation unit (tree, shrub)	Type and health Nature	0.25–5 m	1–5 years	Pan–NIR
	Transport unit (width: road lanes, sidewalk) (material: asphalt, concrete, composite)	Infrastructure Mobility and access	0.25–30 m	1–5 years	Pan–Vis–NIR
Macro scale: Aggregation of imperviousness, greenness, soil and water	Residential neighborhood	Suburbanization Gentrification, poverty, crime, racial segregation, etc.	1–5 km	1–10 years	VIS–NIR–TIR
	Industrial/commercial zone	Land use zoning Storm water flow Heat island effect	1–5 km	1–10 years	VIS–NIR–TIR
	Nonbuilt urban	Environmental concerns Beautification Public space	1–5 km	1–10 years	VIS–NIR–TIR
	Urban area	Centrality and sprawl Flow and congestion Sustainability	5–100 km	1–10 years	VIS–NIR–TIR–MIR–Radar

accuracy typically improves only marginally, simply because there is an inherent numerical limitation to the extent individual pixel values can comprehensively represent the multitude of true urban features within the rigid confines of their regular-sized pixel limits (Fisher, 1997).

However, within these numerical limits, per-pixel classification accuracy can be consistently high if the appropriate spatial resolution (i.e., pixel size) is identified with respect to the suitable level of urban detail (Table 10.2). Such ideas of scale appropriateness can be traced back to Welch (1982) and have since been widely accepted as an important part of the class training process. But the decision is far from trivial and must also consider the appropriate scale of analysis (Mesev, 2012). Consider a continuous scale that can be conceptualized by levels of measurement from remote sensor data, ranging from the representation of atomistic urban features (building, tree, sidewalk, etc.) at the microscale to the representation of aggregate urban features (residential neighborhoods, industrial zones, or even complete urban areas) at the macroscale. Microurban remote sensing by per-pixel classification remains highly tenuous (even using meter and submeter resolutions from the latest sensors), and any reliable interpretation is extracted directly from the spatial orientation of pixels—in a similar vein to conventional interpretation of aerial photography, but with lower clarity and with limited stereoscopic capabilities. However, the spectral heterogeneity problem is less restrictive at the macroscale of analysis where classified pixels, instead of measuring individual urban objects, can be aggregated to represent a generalized view of urban areas, including total imperviousness, approximate lateral growth, and overall greenness. It is at this scale of analysis that many types of urban processes, such as sprawl, congestion, poverty, land use zoning, storm water flow, and heat islands, can be studied simultaneously across an entire urban area as part of the search for theories of livability and sustainability. In sum, per-pixel classifications produce simple and convenient thematic maps of urban land use and land cover that can be incorporated into GIS models. The spatial resolution of the remote sensor, however, limits their accuracy away from mapping individual urban features with any level of pragmatic precision and toward more traditional macroscales of generalized land cover combinations reminiscent of the timeless vegetation, impervious, and soil (V-I-S) model (Ridd, 1995).

10.2.2 Subpixel Methods

If locational and thematic accuracy of urban representation from remote sensing is paramount, per-pixel classifications can be modified statistically to measure spectral mixtures representing multiple land cover classes within individual pixels. These are termed subpixel algorithms or soft classifications because pixels are no longer constrained to representing single classes, but instead represent various proportions of land cover classes, which are conceptually more akin to the spatial and compositional heterogeneity of urban configurations (Ji and Jensen, 1999; Small, 2004). The debate is on which approach, per-pixel or subpixel, can again be tied to the scale of urban analysis. For example, the measurement of impervious surfaces is particularly amenable to subpixel classification because pixels can represent a continuum of imperviousness, from total coverage (downtown areas and industrial estates) to scant dispersion intermingled with biophysical land covers (city parks). Extensive research has been devoted to more precise quantification of impervious surfaces, and other urban land covers at subpixel level, such as linear mixture models (Rashed et al., 2003; Wu and Murray, 2003), background removal spectral mixture analysis (Ji and Jensen, 1999; Myint, 2006a), Bayesian probabilities (Foody et al., 1992; Mesev, 2001; Eastman and Laney, 2002; Hung and Ridd, 2002), artificial neural network (Foody and Aurora, 1996; Zhang and Foody, 2001), normalized spectral mixture analysis (Wu and Yuan, 2007; Yuan and Bauer, 2007), fuzzy c-means methods (Fisher and Pathirana, 1990; Foody, 2000), multivariate statistical analysis (Bauer et al., 2004; Yang and Liu, 2005; Bauer et al., 2007), and regression trees (Yang et al., 2003a,b; Homer et al., 2007).

Among these, linear spectral mixture analysis, regression analysis, and regression trees have had a wider appeal because they are theoretically and computationally simpler, as well as more prevalent in many commercial software packages. However, the success of measuring urban land cover types using linear techniques is dependent on identifying spectrally pure endmembers, preferably using reference samples collected in the field (Adams et al., 1995; Roberts et al., 1998, 2012). Although Weng and Hu (2008) derived moderate accuracy levels from employing linear spectral mixture analysis using ASTER and Landsat ETM+ sensor imagery, they discovered that artificial neural networks were also capable of performing nonlinear mixing of land cover types at the subpixel level (Borel and Gerstl, 1994; Ray and Murray, 1996). Another limitation with linear spectral mixture classifiers is that they do not permit the number of endmembers to be greater than the number of spectral bands (Myint, 2006a). In response, a multiple endmember spectral mixture analysis (MESMA) has been developed to identify many more endmember types to represent the heterogeneous mixture of urban land cover types (Rashed et al., 2003; Powell et al., 2007; Myint and Okin, 2009). Diagrams demonstrating linear spectral mixture analysis and MESMA are provided in Figures 10.2 and 10.3, respectively.

Two challenges dominate the research efforts to improve subpixel analysis methods for urban settings. The first challenge is pixel size. Identifying endmembers for all classes in images with large to medium pixels in urban areas is difficult given the heterogeneous nature of urban areas. In small spatial distances (e.g., <30 m), surfaces rapidly change from impervious, to grass, to building. The smaller pixel size (e.g., 1 m or submeter), however, is not always the optimal solution. While pixels may not reflect a mixture of the desired endmembers (e.g., a combination of asphalt and grass), reflectance from unwanted features begins to appear that needs to be filtered (e.g., oil surfaces and automobiles in asphalt, chimneys and air conditions on rooftops). The second limitation is that it is almost impossible to identify all possible endmembers in a study area and classification accuracy can be degraded by the potential presence of unknown classes or unidentified classes (e.g., the asphalt and rooftop examples mentioned earlier). This is because the classifier is based on the assumption that the sum of the fractional proportions of all possible endmembers in a pixel is equal to one. Although this type of modeling is conceptually more

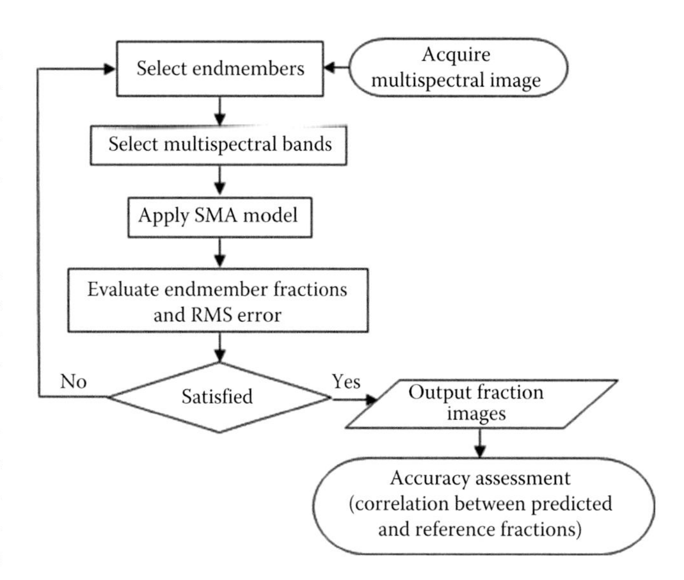

FIGURE 10.2 Spectral mixture analysis.

representative of urban land cover, from a practical standpoint, it nonetheless perpetuates the mixed pixel problem and presents thematic and semantic limitations to urban land classification schemes. In other words, output from subpixel analysis produces fractional classes that are more difficult to integrate with GIS data and may even limit their portability for comparisons across space and through time.

10.2.3 Object-Based Methods

With the representational limitations of purely spectrally based per-pixel and subpixel classifications, it was only a matter of time before the shift to the spatial domain gained momentum. Even from a purely intuitive standpoint, finer resolution (i.e., smaller pixels or large cartographic scale) imagery exhibits higher levels of detailed features that mimic the heterogeneous nature of urban areas. This greater level of spatial detail invariably also leads to many more uncertain spectral classes—known as noise—which can be true but potentially unwanted urban features such as chimneys or manhole covers. Assuming spectral noise is reduced, images with spatial resolutions ranging from about 0.25 to 5 m have the potential to help identify urban structures necessary to perform many urban applications, including estimation of population based on the number of dwellings of different housing types, residential water use, predicting energy consumption, urban heat island, outdoor water use, solar energy use, and storm water pollution modeling (Jensen and Cowen, 1999).

Conceptually, spatial- or object-based approaches are most applicable to high-spatial-resolution remote-sensing data, where objects of interest are larger than the ground resolution element, or pixel. Urban objects may be vegetated features of urban landscapes (e.g., trees, shrubs, and golf course) or anthropogenic features (e.g., buildings, pools, sidewalks, roads, and canals). With regard to mapping categorical data or identifying land

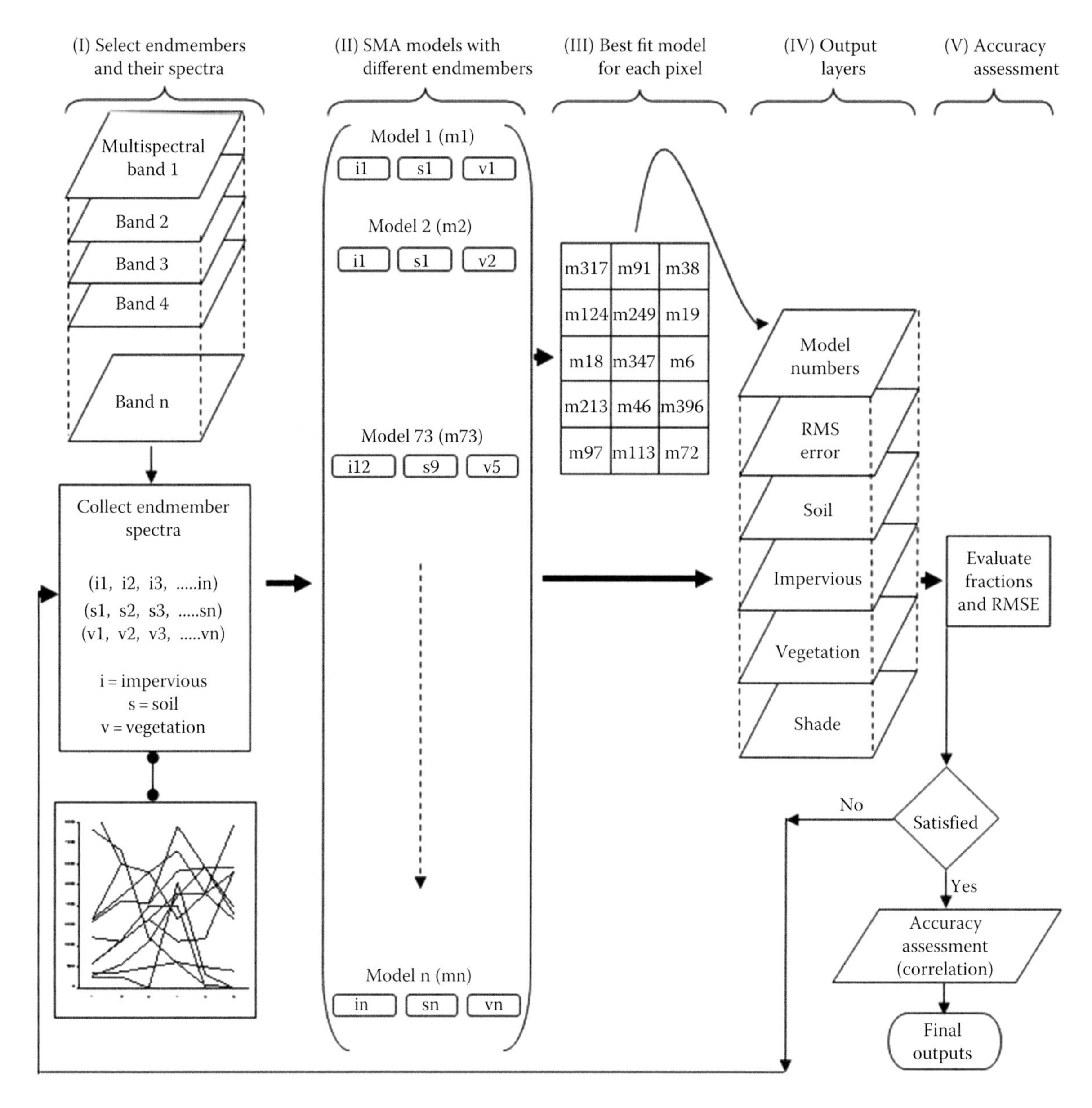

FIGURE 10.3 Multiple endmember spectral mixture analysis.

use land cover classes, remotely sensed image analysis started to shift from pixel-based (per-pixel) to object-based image analysis or geospatial object-based image analysis (GEOBIA) around the year 2000 (Blaschke, 2010). The object-centered classification prototype starts with the generation of segmented objects at multiple scales (Desclee et al., 2006; Navulur, 2007; Im et al., 2008; Myint et al., 2008). To demonstrate, Walker and Briggs (2007) employed an object-oriented classification procedure to effectively delineate woody vegetation in an arid urban ecosystem using high-spatial-resolution true-color aerial photography (without the near-infrared band) and achieved an overall accuracy of 81%. Hermosilla et al. (2012) developed two object-based approaches for automatic building detection and localization using high-spatial-resolution imagery and light detection and ranging (LiDAR) data. Stow et al. (2007) further developed object-based classification by taking advantage of the spatial frequency characteristics of multispectral data and then measuring the proportions of V-I-S subobjects to identify residential land use in Accra, Ghana (they documented an overall accuracy of 75%). In another study by Zhou et al. (2008), postclassification change detection based on the object-based analysis of multitemporal high spatial resolution produced even higher accuracies of 92% and 94%, while Myint and Stow (2011) demonstrated the effectiveness of object-based strategies based on decision rules (i.e., membership functions) and nearest neighbor classifiers on high-spatial-resolution QuickBird multispectral satellite data over the city of Phoenix. These are further supported by Myint et al. (2011) who directly compared the accuracy from object-based classifications (90%) with more traditional spectral-based classifications (68%). The land cover classes that the authors identified for this particular study include buildings, other impervious surfaces (e.g., roads and parking lots), unmanaged soil, trees/shrubs, grass, swimming pools, and lakes/ponds. The study selected 500 sample points that led to approximately 70 points per class (7 total classes) using a stratified random sampling approach for the accuracy assessment of two different subsets of QuickBird over Phoenix. To be consistent and for precise comparison purposes, they applied the same sample points generated for the output generated by the object-based classifier as the output produced by the traditional classification technique (i.e., maximum likelihood).

In general, spectrally similar signatures such as dark/gray soil, dark/gray rooftops, dark/gray roads, swimming pools/blue color rooftops, and red soil/red rooftop remain problematic even with object-based approaches. Furthermore, the most commonly used object-oriented software (Definiens or eCognition) is required to perform a tremendous number of segmentations of objects from all spectral bands using various scale parameters. There is no universally accepted method to determine an optimal level of scale (e.g., object size) to segment objects, and a single scale may not be suitable for all classes. The most feasible approach may be to select the bands for membership functions at the scale that identifies the class with variable options and analyze them heuristically on the display screen. Given that the nearest neighbor classifier and decision rule available in the object-based approach are nonparametric approaches, they are independent of the assumption that data values need to be normally distributed. This is advantageous, because most data are not normally distributed in many real-world situations. Another advantage of the object-based approach is that it allows additional selection or modification of new objects (training samples) at iterative stages, until the satisfactory result is obtained. However, the object-based approach has a significant problem when dealing with a remotely sensed data over a fairly large area since computer memory needs to be used extensively to segment tremendous numbers of objects using multispectral bands. This is true even for fine spatial resolution data with fewer bands (e.g., QuickBird) over a small study area when requiring smaller-scale parameters (smaller objects). Figure 10.4 shows segmented images at scale level 25, 50, and 100 using a subset of a QuickBird image over Phoenix. Figure 10.5 demonstrates how hierarchical image segmentation delineates image objects at various scales.

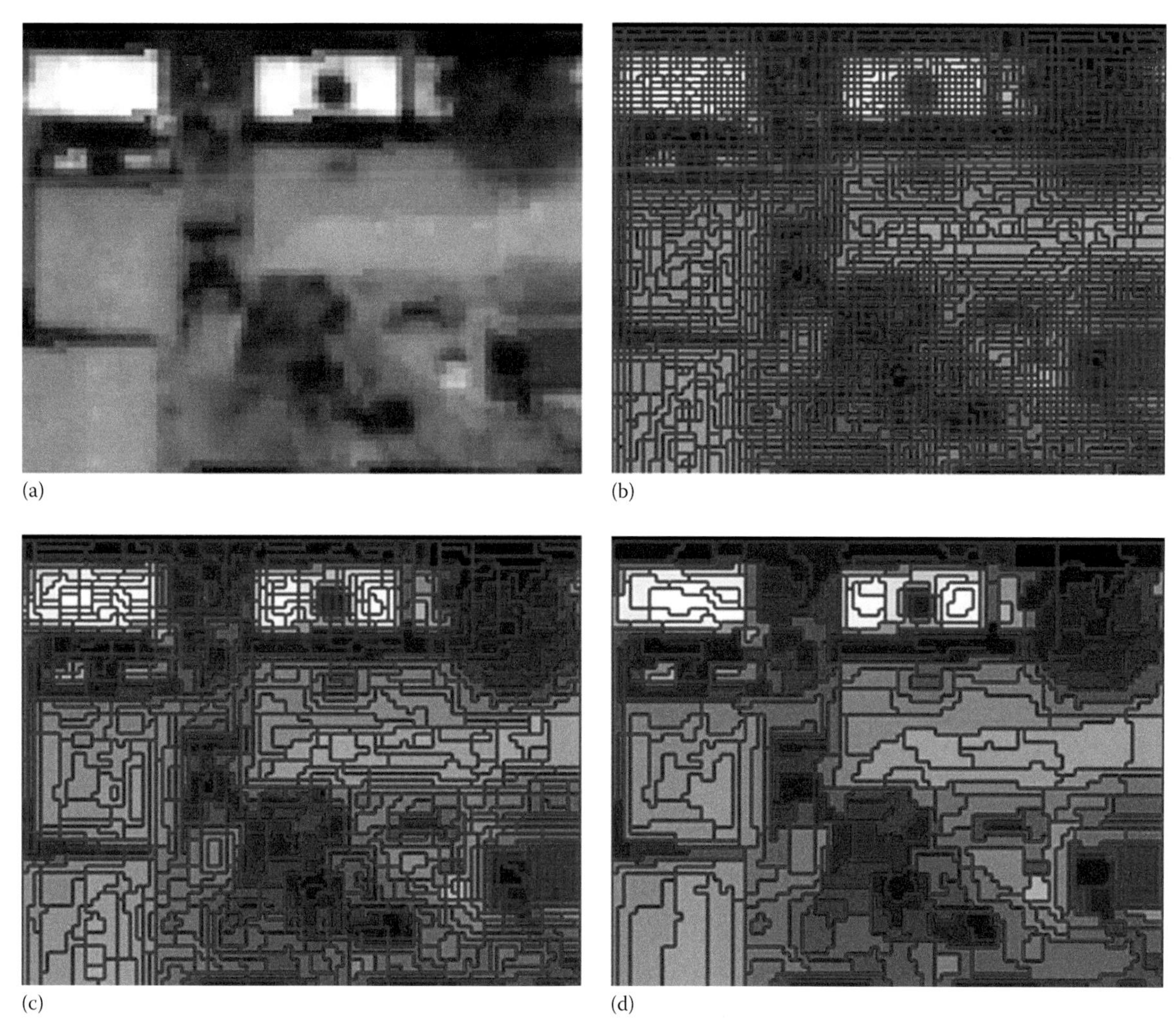

FIGURE 10.4 A subset image and segmented images at different scales. (a) Original subset, (b) level 1 (scale parameter 25), (c) level 2 (scale parameters 50), and (d) level 3 (scale parameter 100).

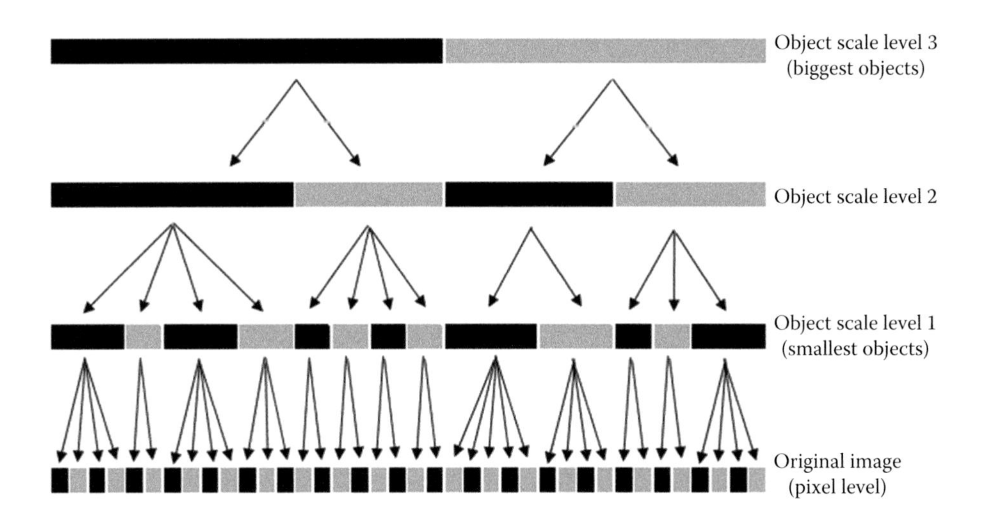

FIGURE 10.5 Image objects at each image scale level. Level 3 = 100, level 2 = 50, and level 1 = 25.

10.2.4 Geospatial Methods

Texture plays an important role in the human visual system for pattern recognition and interpretation. For image interpretation, pattern is defined as the overall spatial form of related features, where the repetition of certain forms is a characteristic pattern found in many cultural objects and some natural features. Local variability in remotely sensed data, which is part of texture or pattern analysis, can be characterized by computing the statistics of a group of pixels, for example, standard deviation, coefficient of variance or autocovariance, or the analysis of fractal similarities or autocorrelation of spatial relationships. There have been some attempts to improve the spectral analysis of remotely sensed data by using texture transforms in which some measure of variability in digital numbers is estimated within local windows, for example, the contrast between neighboring pixels (Edwards et al., 1988), standard deviation (Arai, 1993), or local variance (Woodcock and Harward, 1992). One commonly used statistical procedure for interpreting texture uses an image spatial co-occurrence matrix, which is also known as a gray-level co-occurrence matrix (GLCM) (Franklin et al., 2000). There are a number of texture measures, which could be applied to spatial co-occurrence matrices for texture analysis (Peddle and Franklin, 1991). For instance, Herold et al. (2003) proposed a method based on using landscape metrics to classify IKONOS sensor images, which in turn is compared to a GLCM. Liu et al. (2006) further contrasted spatial metrics, GLCM, and semivariograms in terms of urban land use classification.

Lam et al. (1998) demonstrated how fractal dimensions yield quantitative insight into the spatial complexity and information contained in remotely sensed data. Quattrochi et al. (1997) went further and created a software package known as the image characterization and modeling system to explore how the fractal dimension is related to surface texture. Fractal dimensions were also analyzed by Emerson et al. (1999) who used the isarithm method and Moran's I and Geary's C spatial autocorrelation measures to observe the differing spatial structure of the smooth and rough surfaces in remotely sensed images. In terms of other geospatial techniques, De Jong and Burrough (1995) and Woodcock et al. (1988) implemented variograms to measurements derived from remotely sensed images to quantitatively describe urban spatial patterns. Myint and Lam (2005a,b) and Myint et al. (2006) developed a number of lacunarity approaches to characterize urban spatial features with completely different texture appearances that may share the same fractal dimension values. Both studies report that lacunarity can be considered more effective in comparison to fractal approaches for urban mapping.

The geospatial methods described so far may not provide satisfactory accuracies when they are applied to the classification of urban features from fine spatial resolution remotely sensed images. That is, mainly because most of them focus primarily on coupling features and objects at a single scale and cannot determine the effective representative value of particular texture features according to their directionality, spatial arrangements, variations, edges, contrasts, and the repetitive nature of object and features. There have been a number of reports in spatial frequency analysis of mathematical transforms, which provide solutions using multiresolution analysis. Recent developments in spatial/frequency transforms such as the Fourier transform, Wigner distribution, discrete cosine transform, and wavelet transform have all provided sound multiresolution analytical tools (Bovik et al., 1990; Zhu and Yang, 1998).

Of all transformation approaches, wavelets play the most critical part in texture analysis. Wavelets are part of spatial- and frequency-based classification approaches, and a local window plays an important role in measuring and characterizing spatial arrangements of objects and features. Homogeneity, size of regions, characteristic scale, directionality, and spatial periodicity are important issues that should be considered to identify local windows when performing wavelet analysis (Myint, 2010). From a computational perspective, the ideal window size is the smallest size that also produces the highest accuracy (Hodgson, 1998). The accuracy should increase with a larger local window size since it contains more information than a smaller window size and therefore provides more complete coverage of spatial variation, directionality, and spatial periodicity of a particular texture. However, minimization of local window size is also important in spatial-based urban image classification techniques since a larger window size tends to cover more urban land cover features and consequently creates mixed boundary pixels or mixed land cover problems. However, some spatial and frequency approaches such as wavelet dyadic decomposition approaches require large window sizes to capture spatial information at multiple scales (Myint, 2006b). The potential solution to this problem would be to employ a multiscale overcomplete wavelet analysis using an infinite scale decomposition procedure. This is because a large spatial coverage or a large local window is not needed to describe a spatial pattern. Furthermore, this approach can measure different directional information of anisotropic features at unlimited scales, and it is designed to normalize and select effective features to identify urban classes. Myint and Mesev (2012) employed a wavelet-based classification method to identify urban land use and land cover classes using different decision rule sets and spatial measures and demonstrated the effectiveness of wavelets. However, the current wavelet-based classification system with the dyadic wavelet approach is limited by the fact that higher-level subimages are just a quarter of the preceding image. In general, smaller window size is generally thought to yield higher accuracy in geospatial-based image classification because if the window is too large, much spatial information from two or more land cover classes could create a mixed boundary problem. Further research is required to consider an overcomplete wavelet approach that can generate spatial arrangements of objects and features at any scale level for urban mapping. Such an approach could potentially be applicable to any land use/land cover system at any resolution or scale because it can effectively use any window size. Figure 10.6 shows how wavelet approaches work in comparison to other geospatial approaches in urban mapping.

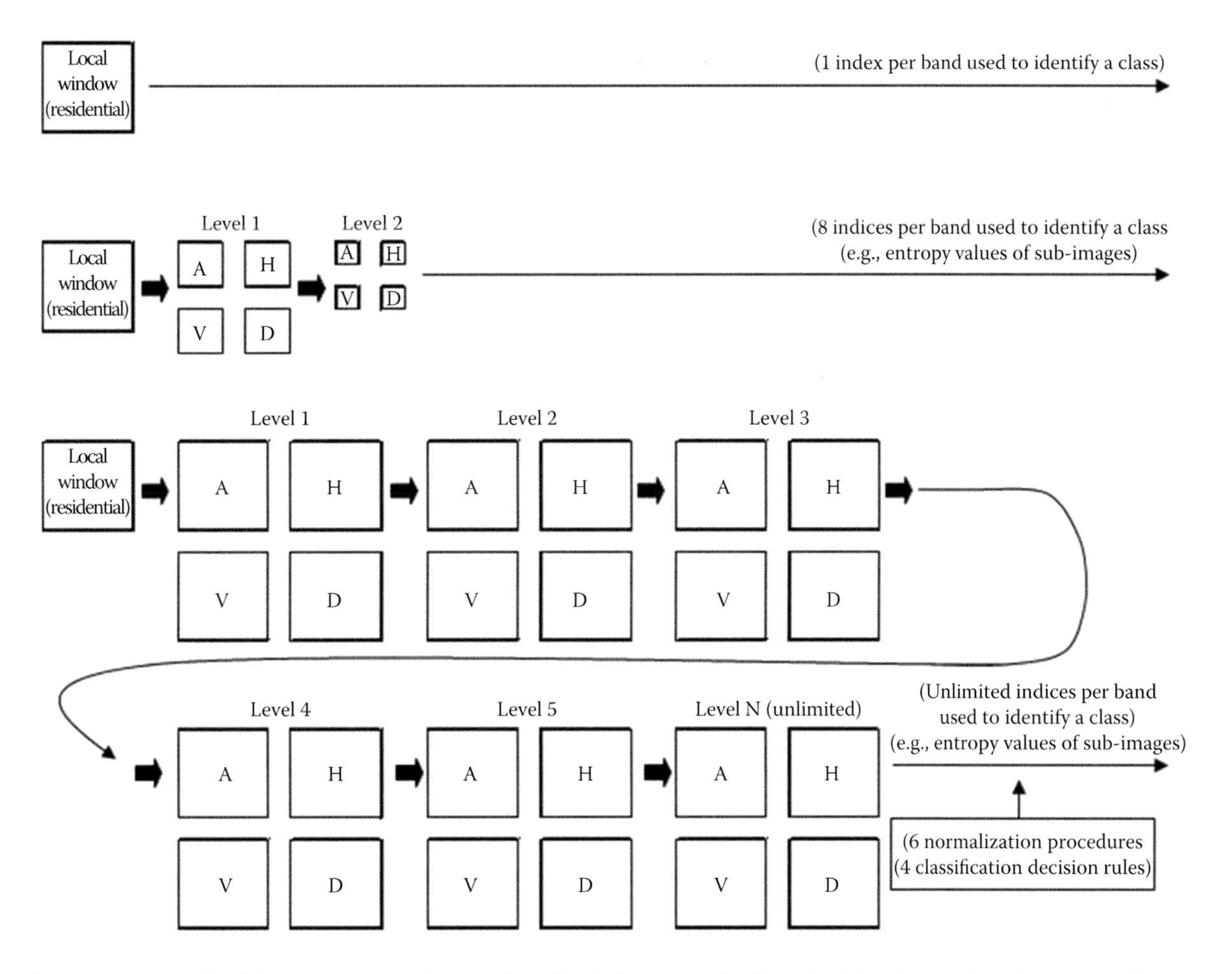

FIGURE 10.6 An example of feature vectors or indices (32 × 32 window or a subset) used to identify an urban class using other geospatial approaches, the dyadic wavelet approach, and the overcomplete wavelet approach. *Note:* Subimages at level two in the dyadic approach reach the suggested minimum dimension (8 × 8 pixels) since any subimages smaller than eight pixels may not contain any useful spatial information. A subimage at a higher level is exactly the same as its original size at the preceding level in the overcomplete approach. It should also be noted that the level of scale with the overcomplete approach is unlimited. A, approximation texture; H, horizontal texture; V, vertical texture; D, diagonal texture. (a) Other geospatial approaches (multiscale or singlescale) (e.g., fractal, spatial autoC, Cooccurrence matrix, G index, lacunarity), (b) dyadic wavelet approach, and (C) overcomplete wavelet approach.

10.3 Concluding Remarks

Interpreting urban land cover from the data captured by remote sensors remains a conceptual and technical challenge. Accuracy levels are typically lower than the interpretation of more naturally occurring surfaces. However, huge strides have been made with the formulation of statistical models that help disentangle the spectral and spatial complexity of urban land covers. Whereas per-pixel classification has stood the test of time (primarily for pragmatic reasons, especially when integrated with GIS-handled datasets), developments in subpixel, object-based, and geospatial techniques have begun, at last, to reproduce the geographical configuration and compositional texture of urban structures. These developments are further tempered by conceptual developments that now consider the *appropriateness of scale* (understanding the level of urban structural measurements) and the *appropriateness of time* (understanding the lag between urban process and urban structure). Both are critical for measuring the rate of urban change, not simply the amount of lateral growth but also the juxtaposition of land use within existing urban limits. Further research will only improve our use of remote sensor data for measuring urban patterns and in turn will complement our understanding of key urban processes.

References

Adams, J. B., Sabol, D. E., Kapos, V., Almeida-Filho, R., Roberts, D. A., Smith, M. O., and Gillespie, A. R. 1995. Classification of multiple images based on fractions of endmembers: Application to landcover change in the Brazilian Amazon, *Remote Sensing of Environment*, 52, 137–154.

Anderson, J. R., Hardy, E. E., Roach, J. T., and Witmer, R. E. 1976. A land use and land cover classification system for use with remote sensor data. U.S. Geological Survey Professional Paper, 964. http://landcover.usgs.gov/pdf/anderson.pdf. Accessed March 27, 2015.

Arai, K. 1993. A classification method with a spatial-spectral variability. *International Journal of Remote Sensing*, 14, 699–709.

Barr, S. and Barnsley, M. A. 1997. A region-based, graph-theoretic data model for the inference of second-order thematic information from remotely-sensed images. *International Journal of Geographical Information Science*, 11, 555–576.

Bauer, M. E., Heinert, N. J., Doyle, J. K., and Yuan, F. 2004. Impervious surface mapping and change monitoring using satellite remote sensing. *Proceedings of the ASPRS 2004 Annual Conference*, May 24–28, Denver, CO.

Bauer, M. E., Loeffelholz, B. C., and Wilson, B. 2007. Estimating and mapping impervious surface area by regression analysis of Landsat imagery. In Q. Wang (Ed.), *Remote Sensing of Impervious Surfaces*, pp. 3–20. Boca Raton, FL: CRC Press.

Blaschke, T. 2010. Object-based image analysis for remote sensing. *ISPRS International Journal of Photogrammetry and Remote Sensing*, 65, 2–16.

Borel, C. C. and Gerstl, S. A. W. 1994. Nonlinear spectral mixing models for vegetative and soil surfaces. *Remote Sensing of Environment*, 47, 403–416.

Bovik, A. C., Clark, M., and Geisler, W. S., 1990. Multichannel texture analysis using localized spatial filters. *IEEE Transactions on Pattern Analysis and Machine Intelligence*, 12, 55–73.

De Jong, S. M. and Burrough, P. A. 1995. A fractal approach to the classification of Mediterranean vegetation types in remotely sensed images. *Photogrammetric Engineering and Remote Sensing*, 61, 1041–1053.

Desclée, B., Bogaert, P., and Defourny, P. 2006. Forest change detection by statistical object-based method. *Remote Sensing of Environment*, 102, 1–11.

Eastman, J. R. and Laney, R. M. 2002. Bayesian soft classification for sub-pixel analysis: A critical evaluation. *Photogrammetric Engineering and Remote Sensing*, 6811, 1149–1154.

Edwards, G., Landry, R., and Thompson, K. P. B. 1988. Texture analysis of forest regeneration sites in high-resolution SAR imagery. *Proceedings of the International Geosciences and Remote Sensing Symposium (IGARSS 88)*, ESA SP-284, pp. 1355–1360. European Space Agency, Paris, France.

Emerson, C. W., Lam, N. S. N., and Quattrochi, D. A. 1999. Multi-scale fractal analysis of image texture and pattern. *Photogrammetric Engineering and Remote Sensing*, 65, 51–61.

Fisher, P. 1997. The pixel: A snare and a delusion. *International Journal of Remote Sensing*, 18, 679–685.

Fisher, P. F. and Pathirana, S., 1990. The evaluation of fuzzy membership of land cover classes in the suburban zone. *Remote Sensing of Environment*, 34, 121–132.

Foody, G. M., 2000. Estimation of sub-pixel land cover composition in the presence of untrained classes. *Computers and Geosciences*, 26, 469–478.

Foody, G. M. and Aurora, M. K. 1996. Incorporating mixed pixels in the training, allocation and testing of supervised classification. *Pattern Recognition Letters*, 17, 1389–1398.

Foody, G. M., Campbell, N. A., Trodd, N. M., and Wood, T. F. 1992. Derivation and applications of probabilistic measures of class membership from the maximum-likelihood classification. *Photogrammetric Engineering and Remote Sensing*, 58, 1335–1341.

Franklin, S. E., Hall, R. J., Moskal, L. M., Maudie, A. J., and Lavigne, M. B. 2000. Incorporating texture into classification of forest species composition from airborne multispectral images. *International Journal of Remote Sensing*, 21, 61–79.

Hermosilla, T., Ruiz, L. A., Recio, J. A., and Cambra-López, M. 2012. Assessing contextual descriptive features for plot-based classification of urban areas. *Landscape and Urban Planning*, 106, 124–137.

Herold, M., Liu, X., and Clarke, K. C. 2003. Spatial metrics and image texture for mapping urban land use. *Photogrammetric Engineering and Remote Sensing*, 69, 991–1001.

Hodgson, M. E. 1998. What size window for image classification? A cognitive perspective. *Photogrammetric Engineering and Remote Sensing*, 64, 797–807.

Homer, C., Dewitz, J., Fry, J., Coan, M., Hossain, N., Larson, C., Herold, N., McKerrow, A., VanDriel, J. N., and Wickham, J. 2007. Completion of the 2001 national land cover database for the conterminous United States. *Photogrammetric Engineering and Remote Sensing*, 73, 337–341.

Hung, M. and Ridd, M. K. 2002. A subpixel classifier for urban land-cover mapping based on a maximum-likelihood approach and expert system rules. *Photogrammetric Engineering and Remote Sensing*, 68, 1173–1180.

Im, J., Jensen, J. R., and Hodgson, M. E. 2008. Object-based land cover classification using high posting density lidar data. *GIScience and Remote Sensing*, 45, 209–228.

Jensen, J. R. and Cowen, D. C. 1999. Remote sensing of urban/suburban infrastructure and socio-economic attributes. *Photogrammetric Engineering and Remote Sensing*, 65, 611–622.

Ji, M. and Jensen, J. R. 1999. Effectiveness of subpixel analysis in detecting and quantifying urban impervious from Landsat Thematic Mapper Imagery. *Geocarto International*, 14, 33–41.

Lam, N. S. N, Quattrochi, D., Qui, H., and Zhao, W. 1998. Environmental assessment and monitoring with image characterization and modeling system using multiscale remote sensing data. *Applied Geographic Studies*, 2, 77–93.

Liu. X., Clarke, K. C., and Herold, M. 2006. Population density and image texture: A comparison study. *Photogrammetric Engineering and Remote Sensing*, 72, 187–196.

Mesev, V. 1998. The use of census data in urban image classification. *Photogrammetric Engineering and Remote Sensing*, 64, 431–438.

Mesev, V. 2001. Modified maximum likelihood classifications of urban land use: Spatial segmentation of prior probabilities. *Geocarto International*, 16, 41–48.

Mesev, V. 2012. Multiscale and multitemporal urban remote sensing. *ISPRS International Archives of the Photogrammetry, Remote Sensing & Spatial Information Sciences*, XXXIX-B2, Melbourne, Australia, pp. 17–21.

Myint, S. W. 2006a. Urban vegetation mapping using sub-pixel analysis and expert system rules: A critical approach. *International Journal of Remote Sensing*, 27, 2645–2665.

Myint, S. W. 2006b. A new framework for effective urban land use land cover classification: A wavelet approach. *GIScience and Remote Sensing*, 43, 155–178.

Myint, S. W. 2010. Multi-resolution decomposition in relation to characteristic scales and local window sizes using an operational wavelet algorithm. *International Journal of Remote Sensing*, 31, 2551–2572.

Myint, S. W., Giri, C. P., Wang, L., Zhu, Z., and Gillette, S. 2008. Identifying mangrove species and their surrounding land use and land cover classes using an object oriented approach with a lacunarity spatial measure. *GIScience and Remote Sensing*, 45, 188–208.

Myint, S. W., Gober, P., Brazel, A., Grossman-Clarke, S., and Weng, Q. 2011. Per-pixel versus object-based classification of urban land cover extraction using high spatial resolution imagery. *Remote Sensing of Environment*, 115, 1145–1161.

Myint, S. W. and Lam, N. S. N. 2005a. A study of lacunarity-based texture analysis approaches to improve urban image classification. *Computers, Environment, and Urban Systems*, 29, 501–523.

Myint, S. W. and Lam, N. S. N. 2005b. Examining lacunarity approaches in comparison with fractal and spatial autocorrelation techniques for urban mapping. *Photogrammetric Engineering and Remote Sensing*, 71, 927–937.

Myint, S. W. and Mesev, V. 2012. A comparative analysis of spatial indices and wavelet-based classification. *Remote Sensing Letters*, 3, 141–150.

Myint, S. W., Mesev, V., and Lam, N. S. N. 2006. Texture analysis and classification through a modified lacunarity analysis based on differential box counting method. *Geographical Analysis*, 38, 371–390.

Myint, S. W. and Okin, G. S. 2009. Modelling land-cover types using multiple endmember spectral mixture analysis in a desert city. *International Journal of Remote Sensing*, 30, 2237–2257.

Myint, S. W. and Stow, D. 2011. An object-oriented pattern recognition approach for urban classification. In X. Yang (Ed.), *Urban Remote Sensing, Monitoring, Synthesis and Modeling in the Urban Environment*, pp. 129–140. Chichester, U.K.: John Wiley & Sons, Ltd. DOI: 10.1002/9780470979563.

Navulur, K. 2007. *Multispectral Image Analysis Using the Object-Oriented Paradigm*. Boca Raton, FL: CRC Press/Taylor and Frances Group.

Peddle, D. R. and Franklin, S. E. 1991. Image texture processing and data integration for surface pattern discrimination. *Photogrammetric Engineering and Remote Sensing*, 57, 413–420.

Powell, R. L., Roberts, D. A., Dennison, P. E., and Hess, L. L. 2007. Sub-pixel mapping of urban land cover using multiple endmember spectral mixture analysis: Manaus, Brazil. *Remote Sensing of Environment*, 106, 253–267.

Quattrochi, D. A., Lam, N. S. N., Qiu, H., and Zhao, W. 1997. Image characterization and modeling system (ICAMS): A geographic information system for the characterization and modeling of multiscale remote sensing data. In D. A. Quattrochi and M. F. Goodchild (Eds.), *Scale in Remote Sensing and GIS*, pp. 295–308. Boca Raton, FL: CRC Press.

Rashed, T., Weeks, J. R., Roberts, D., Rogan J., and Powell, R. 2003. Measuring the physical composition of urban morphology using multiple endmember spectral mixture models. *Photogrammetric Engineering and Remote Sensing*, 69, 1011–1020.

Ray, T. W. and Murray, B. C. 1996. Nonlinear spectral mixing in desert vegetation. *Remote Sensing of Environment*, 55, 59–64.

Ridd, M. K. 1995. Exploring a V-I-S vegetation-impervious surface-soil model for urban ecosystems analysis through remote sensing: Comparative anatomy for cities. *International Journal of Remote Sensing*, 16, 2165–2186.

Roberts, D. A., Gardner, M., Church, R., Ustin, S., Scheer, G., and Green, R. O. 1998. Mapping chaparral in the Santa Monica Mountains using multiple endmember spectral mixture models. *Remote Sensing of Environment*, 65, 267–279.

Roberts, D. A., Quattrochi, D. A., Hulley, G. C., Hook, S. J., and Green, R. O. 2012. Synergies between VSWIR and TIR data for the urban environment: An evaluation of the potential for the hyperspectral Infrared Imager (HyspIRI) decadal survey mission. *Remote Sensing of Environment*, 117, 83–101.

Small, C. 2004. The Landsat ETM+ spectral mixing space. *Remote Sensing of Environment*, 93, 1–17.

Stow, D., Lopez, A., Lippitt, C., Hinton, S., and Weeks, J. 2007. Object-based classification of residential land use within Accra, Ghana based on QuickBird satellite data. *International Journal of Remote Sensing*, 28, 5167–5173.

Stefanov, W. L., Ramsey M. S., and Christensen, P. R. 2001. Monitoring urban land cover change: An expert system approach to land cover classification of semiarid to arid urban centers. *Remote Sensing of Environment*, 77(2), 173–185.

Strahler, A. H. 1980. The use of prior probabilities in maximum likelihood classification of remotely sensed data. *Remote Sensing of Environment*, 10, 135–163.

Stuckens, J., Coppin, P. R., and Bauer, M. 2000. Integrating contextual information with per-pixel classification for improved land cover classification. *Remote Sensing of Environment*, 71, 282–296.

Walker, J. S. and Briggs, J. M. 2007. An object-oriented approach to urban forest mapping with high-resolution, true-color aerial photography. *Photogrammetric Engineering and Remote Sensing*, 73, 577–583.

Welch, R. A. 1982. Spatial resolution requirements for urban studies. *International Journal of Remote Sensing*, 3, 139–146.

Weng, Q. and Hu, X. 2008. Medium spatial resolution satellite imagery for estimating and mapping urban impervious surfaces using LSMA and ANN. *Transactions on Geoscience and Remote Sensing*, 46, 2387–2406.

Woodcock, C. and Harward, V. J. 1992. Nested-hierarchical scene models and image segmentation. *International Journal of Remote Sensing*, 13, 3167–3187.

Woodcock, C. E, Strahler, A. H., and Jupp, D. L. B. 1988. The use of variograms in remote sensing: I. Scene models and simulated images. *Remote Sensing of Environment*, 25, 323–348.

Wu, C. and Murray, A. 2003. Estimating impervious surface distribution by spectral mixture analysis. *Remote Sensing of Environment*, 84, 493–505.

Wu, C. and Yuan, F. 2007. Seasonal sensitivity analysis of impervious surface estimation with satellite imagery. *Photogrammetric Engineering and Remote Sensing*, 73, 1393–1401.

Yang, L., Huang, C., Homer, C. G., Wylie, B. K., and Coan, M. J. 2003a. An approach for mapping large-area impervious surfaces: Synergistic use of Landsat-7 ETM+ and high spatial resolution imagery. *Canadian Journal of Remote Sensing*, 29, 230–240.

Yang, L., Xian, G., Klaver, J. M., and Deal, B. 2003b. Urban land-cover change detection through sub-pixel imperviousness mapping using remotely sensed data. *Photogrammetric Engineering and Remote Sensing*, 69, 1003–1010.

Yang, X. 2011. Parameterizing support vector machines for land cover classification. *Photogrammetric Engineering and Remote Sensing*, 77, 27–37.

Yang, X. and Liu, Z. 2005. Use of satellite-derived landscape imperviousness index to characterize urban spatial growth. *Computers, Environment and Urban Systems*, 29, 524–540.

Yuan, F. and Bauer, M. E. 2007. Comparison of impervious surface area and normalized difference vegetation index as indicators of surface urban heat island effects in Landsat imagery. *Remote Sensing of Environment*, 106, 375–386.

Zhang, J. and Foody, G. M. 2001. Fully-fuzzy supervised classification of sub-urban land cover from remotely sensed imagery: Statistical and neural network approaches. *Photogrammetric Engineering and Remote Sensing*, 22, 615–628.

Zhou, W. Q., Troy, A., and Grove, M. 2008. Object-based land cover classification and change analysis in the Baltimore metropolitan area using multitemporal high resolution remote sensing data. *Sensors*, 8, 1613–1636.

Zhu, C. and Yang, X. 1998. Study of remote sensing image texture analysis and classification using wavelet. *International Journal of Remote Sensing*, 13, 3167–3187.

11

Image Classification Methods in Land Cover and Land Use

Mutlu Özdoğan
University of Wisconsin–Madison

Acronyms and Abbreviations

GIS	Geographic Information System
USGS	United States Geological Survey
LULC	Land Use Land Cover
EMR	Electromagnetic Radiation
LACIE	Large Area Crop Inventory Experiment
SPOT	Satellite Pour l'Observation de la Terre (Earth observation satellite)
NLCD	National Land Cover Database
MODIS	Moderate Resolution Imaging Spectroradiometer
GLCC	Global Land Cover Characteristics
FAO	UN Food and Agricultural Organization
LCCS	Land Cover Classification System
FROM-GLC	Fine Resolution Observation and Modeling Global Land Cover
NIR	Near Infrared
MLC	Maximum Likelihood Classification
MD	Minimum Distance Classification
KMC	K-Means Clustering
AVHRR	Advanced Very High Resolution Radiometer
ANN	Artificial Neural Networks
CT	Classification Trees
SVM	Support Vector Machines
ART	Adaptive Resonance Theory
MLP	Multi-layered Perceptron
ARTMAP	Adaptive Resonance Theory Multi-layered Perceptron
LIBSVM	A Library for SVM classification and Regression
SVC	Support Vector Classification
SVR	Support Vector Regression
OBIA	Object Based Image Analysis
USDA	United States Department of Agriculture
NDVI	Normalized Difference Vegetation Index
TOA	Top of Atmosphere
BT	Brightness Temperature
WELD	Web-Enabled Landsat Data
UMD	University of Maryland

11.1 Introduction

Today, information on land cover and land use is almost exclusively derived from remotely sensed observations at various spatial and temporal scales. The advantages of these observations include, but not limited to, synoptic view, availability of spectral bands that help distinguish land surface properties, archived temporal record, and digital nature. The principle form of deriving land-cover information from remotely sensed images is classification. In the context of remote sensing, classification refers to the process of translating observations into land-cover categories with clearly defined biogeophysical function. For example, a typical land-cover map may contain categories like forest, water, agriculture, and so on. These maps are then used in a growing number of environmental applications, from resource management to global change studies. To this end, the purpose of this chapter is to review existing and emerging image classification methods applied to remote sensing.

This chapter has the following outline. First definitions for land cover and land use as used throughout the chapter are given. Then, advantages and disadvantages of remote sensing with respect to land-cover mapping are discussed. In the main body of the document, existing as well as emerging methods for land-cover mapping are described. The main conclusions and the future of land-cover mapping are provided in the last section.

11.1.1 Definitions

Land cover is defined as describing the physical status of the Earth's land surface. Land use, on the other hand, describes the use of that land cover for a particular purpose. For example, in the case of a forested area, the *forest* land-cover label would be used to describe the fact that the land is occupied by a group of trees that constitute a forest stand without any reference to the use of that forest, while the land-use label could include its usage status, for example, industrial forest for wood production. In this context, image classification methods applied to remotely sensed data allow classification of Earth's surface into *only* land-cover categories, all of which then can be interpreted into land-use classes. For the rest of this chapter, these definitions will be used and primarily, the land-cover classification with remotely sensed data will be emphasized.

11.1.2 Advantages and Limitations of Remote Sensing for Map Making

In the last four decades, both academic-and application-oriented studies have clearly established both the advantages and limitations of remote sensing for deriving land-cover information (Table 11.1). In terms of advantages, remote sensing offers a relatively cheap and rapid method of acquiring information over a large geographical area. For example, a single Landsat scene covers over 30,000 km^2 of land surface, producing data at several spectral bands at fairly high spatial resolutions. When subjected to classification, these data could reveal land-cover information at the scale of a standard topo sheet (~1:25,000 scale). Even with the cost of ground verification, this is very economical. Second, remote sensing allows unobtrusive means to acquire observations, elucidating the Earth surface discretely by means of electromagnetic radiation. Third, remotely sensed observations offer a synoptic view, which can be described as observations that give a broad view of the Earth's surface at a particular time. As a result, regional phenomena, which are invisible from the ground, are clearly visible, such as geological structures or forests instead of the trees. Fourth, remote sensing provides one of few means to obtain data from inaccessible areas. Fifth, presence of archival observations allows rapid assessment of land-cover change, or cheap ways to update/construct base maps without the need for detailed land surveys. Finally, the (mostly) digital nature of data make it possible to use computers for fast processing and combining results with other datasets within a Geographic Information system (GIS).

While there are clear advantages to use remote sensing for extracting land-cover information, remote sensing also has important limitations that need to be specified. Perhaps the biggest limitation of remote sensing is that it is oversold (Jensen, 2006). More specifically, remote sensing is often (incorrectly) thought of having the ability to observe/identify/locate objects with excessive or unwarranted enthusiasm. Secondly, by definition, remotely sensed observations are not direct, and provide only a manifestation of the object of interest, made possible by electromagnetic radiation. This requires that the observations be translated into information of interest and calibrated against reality. However, this calibration is never exact: a classification error of 10% is considered excellent. Third, different land surfaces may yield very similar observations, making them difficult to separate. On the other hand, a single land-cover category may be manifested in multiple different ways in observations, leading to incorrect classifications. Fourth, depending on the observing platform, observations may contain noise associated with atmosphere and sensor characteristics. In certain cases, these unwanted information must be accounted for. Finally, the spatial resolution of satellite imagery may be too coarse for detailed mapping and for distinguishing small contrasting areas. Rule of thumb: a land use must occupy at least 16 pixels (picture elements, cells) to be reliably identified by automatic methods. However, new satellites are being proposed with 1 m resolution, these will have high data volume but will be suitable for land-cover mapping at a detailed scale.

Despite these disadvantages, remote sensing continues to be the primary data source for deriving land-cover and land-use information worldwide, from local to global studies. The land-cover and land-use maps generated from remotely sensed observations also span a wide varieties of themes from crop type identification to forest composition mapping to wetland detection, just to name a few. As our population grows and more pressure is placed on the natural resources, remote sensing will continue to play a significant role to monitor these resources. Both the existing tools and those to be developed in the near future will allow us to achieve this goal more accurately and efficiently.

TABLE 11.1 Advantages and Limitations of Remote Sensing for Land-Cover Studies

Advantages	Disadvantages
Low cost and rapid	Overemphasis on what it can
Up-to-date	Inflated accuracies
Large area coverage	Not a direct sample need to be calibrated
Synoptic view	Large measurement uncertainty
Regular revisit period	Unintended measurements (e.g., clouds)
Digital nature of data for integration in GIS	Lack of observations in certain regions
Data acquisition over inaccessible areas	Spatial/temporal resolution mismatch
Unobtrusive data collection	Difficult to interpret in some cases
Spectral information in the nonvisible	Need specific training to process/interpret
Derive a map as the final result	

11.2 Image Classification in the Context of Land-Use/Land-Cover Mapping

11.2.1 Historical Perspective

Many environmental and natural resource management questions require accurate and timely information on land cover and land use. To this end, many states and agencies today systematically collect, update, and disseminate this most fundamental form of land information. However, it is accurate to suggest that remote sensing has had a much longer history of data collection than we have been making maps using these data. While history of remote sensing is reserved for elsewhere (e.g., Chapter 1 of "Remotely Sensed Data Characterization, Classification, and Accuracies"), we provide a short history of land-cover mapping using remote sensing as it sets the stage for the following discussion.

In many ways, the scientific establishment of modern-era land-cover and land-use mapping with remote sensing can be traced back to early 1970s. In anticipation of civilian space-based remote sensing observations, a number of federal agencies in the United States formed an interagency Steering Committee on Land Use information and Classification early in 1971. The formation and subsequent work of this committee resulted in first standards for mapping land cover (Anderson, 1971). The land-use classification system to be used with remote sensor data, initially proposed by James R. Anderson and published in final form in 1976, was designed to place major reliance on remote sensing and used a system of hierarchically defined categories (Anderson et al., 1976). These developments along with the launch of Landsat I, II, and III are considered to have built the foundation of modern-day land-cover mapping with remote sensing.

Several developments occurred from the 1970s to present (Table 11.2). On the technical side, the U.S. Geological Survey (USGS) released its entire satellite data archive at no cost, making it possible to study state's land-use/land-cover (LULC) on a repeated and low-cost basis. At the same time, new computer based machine learning algorithms have found their way into regional remote sensing investigations that help to provide improved accuracies for map products. On the application side, the environment and the natural resources of our state are under increased pressure from population growth, pollution, and bioenergy prospects.

11.2.2 Methods

Numerous classification algorithms have been developed since the early 1970s when first dedicated land observations became available. These algorithms range from visual interpretation of printed or digital color images to advanced machine learning algorithms that emulate human learning behavior. In this section, we divide the individual methods into two main subcategories: those that rely on manual interpretation and those rely on automation. Note that the traditional division of classification methods into supervised and unsupervised logic is not used here as that topic is extensively treated in many textbooks and tutorials. The decision to use manual versus automated categories stems the very purpose of this chapter: to provide an in-depth review of image classification methods. With this in mind, the supervised versus the unsupervised logic is considered as part of the described classification methods.

11.2.2.1 Approaches That Rely on Manual Interpretation

Overwhelming majority of land-cover classification performed on satellite data today are based on some form of automation. However, manual interpretation of satellite data still plays an important role in image categorization, especially in large state and international organizations such as the United Nations Food

TABLE 11.2 Historical Developments in Land-Cover Land-Use Mapping from Remote Sensing

1920–1930	Development and early applications of aerial photography
1935–1945	WWII—Extensive applications of nonvisible portions of the EMR
1960–1970	First use of the term *remote sensing*
1970–1980	Rapid advances in image processing
1971	Interagency Steering Committee on Land Use Information and Classification in the United States
1974	The Large Area Crop Inventory Experiment (LACIE)
1975	Development of the neural network backpropagation algorithm by Paul Werbos
1976	Publication of James R. Anderson's landmark land use classification system
1970–1980	Launch of early Landsat satellites
1983	Development of the Global vegetation and land-use database by Elaine Matthews
1984	Launch of Landsat 4
1986	Launch of SPOT
1992	Development of the first National Land Cover Dataset (NLCD)
1993	Development of the original SVM algorithm by Vladmir N. Vapnik
1997	Release of the Global Land Cover Characteristics (GLCC) database
1999	Launch of Landsat 7 and MODIS
2000	Development of the FAO Land Cover Classification System (LCCS)
2000	Deep Learning in the neural networks community gets traction
2013	Launch of Landsat 8
2013	Release of the Finer Resolution Observation and Monitoring Global Land Cover (FROM-GLC) database by Chinese researchers

and Agricultural Organization (FAO). The principal advantage of manual (or visual) interpretation of satellite data for the purpose of land-cover mapping is the accuracy. Humans are extremely good at pattern recognition and categorization, even in very complex landscapes with multiple, perhaps overlapping categories. However, the visual interpretation of satellite images can be a complex process, going beyond what is contained in an image. In fact, successful image interpretation is only possible through the iterative process of recognition and interpretation of objects (Figure 11.1), both of which rely heavily on the knowledge and the expertise of the analyst in charge of the analysis (Albertz, 2007).

In a remotely sensed image, the objects are defined in terms of the way they reflect or emit radiation that give rise to their recorded colors and shapes. Thus, visual elements of tone, shape, size, pattern, texture, and association are the available tools that aid recognition and interpretation of the objects in an image.

Perhaps what separates today's manual interpretation of satellite data for mapping land cover is the medium in which the interpretation is made. Early work with satellite imagery relied on visual interpretation of large hard-copy satellite image prints to identify and map land-cover elements. Today's visual image interpretation occurs all in digital format, using the idea of digitizing boundaries of land-cover elements directly on a computer using within a GIS system. Regardless of the analysis environment, several known factors significantly influence the outcomes. First, the red and the near infrared (NIR) portions of the electromagnetic spectrum provide tremendous information for recognizing land-cover elements. Second, the use of multiple images of the same area increases classification accuracy, because the phenology of the land cover is captured. Third, studies involving visual interpretation of satellite data benefit from radiometric enhancements and manipulations in the form of spectral indices. For example, compared to automated techniques of image classification, studies involving visual interpretation of satellite data report superior performance, particularly for area estimation problems.

On the down side, manual interpretation of satellite data is expensive, involving thousands of analyst hours. This is one reason, why large organizations such as various state mapping agencies choose manual interpretation as a method of national land-cover inventories in many developing countries (e.g., Travaglia et al., 2001). Moreover, land-cover maps derived from visual interpretation of satellite data may not contain finest levels of categorical detail due to inherent limitations of analyst's knowledge and experience. For example, in an agricultural landscape, identification of irrigated versus nonirrigated fields may be possible by recognizing shapes associated with center pivot irrigation: identification of crop types grown on that field is less straightforward. On the other hand, manual interpretation of image data appears to be more cost effective if using only sampling units and enumeration as the point of area estimates. However, these applications do not provide wall-to-wall maps that are often required for environmental analyses.

11.2.2.2 Approaches That Rely on Automated Classification

Image classification for the purpose of extracting land-cover information can also be achieved in an automated way with the help of computers. While there are many interpretations of the word *automatic* in the context of classification, what is being referred to here is the idea that only a sample of the landscape need to be known and not the entire population. In other words, unlike manual interpretation in which the knowledge of the whole landscape is needed, automated classification only requires information on a subsample of the landscape. In this context, the idea of automated classification applies to both unsupervised and supervised classification logic. In the unsupervised approach, the relationship between the observed pattern (the feature space) and a predetermined number of statistical clusters (the information classes) is established with the help of an algorithm. It is often applied in situations where prior knowledge of the ground cover is not readily available. Ultimately, the statistical clusters are assigned to land-cover labels and it is here where idea of automation comes into play. That is, the analyst only uses a sample of known associations between the information clusters and actual land-cover labels.

In the case of supervised classification, on the other hand, the user collects a set of learning samples to train a classifier to identify the class label of every pixel in the image. Once again this is what is being referred to automation: only a fraction of the population is needed to estimate the characteristics of every individual in the population: the pixels that constitute the landscape. Another way of looking at this is that there is a large return on investment (training data or known samples): knowledge on less than 5% of the landscape is used to extract information on the entire landscape in question. Note that the accuracy of unsupervised approaches is generally lower than that of supervised methods, especially in complex landscapes where spectral/temporal manifestation of different land-cover categories may be similar.

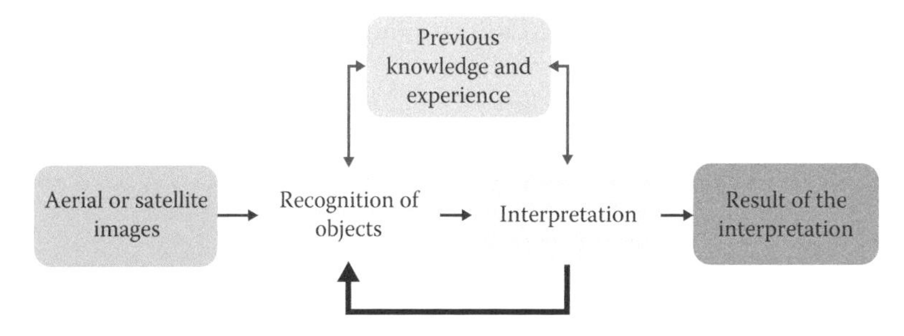

FIGURE 11.1 Schematic presentation of the interpretation process. (Modified from Albertz, D. 2007. Einführung in die Fernerkundung. Grundlagen der Interpretation von Luft- & Satellitenbildern. (Darmstadt) 254pp.)

The last two decades have seen a growing number of automated and advanced image classification methods, both supervised and unsupervised, applied to land-cover mapping across various spatial and temporal scales (Table 11.3). These methods often belong to the nonparametric statistics domain, defined as a set of statistical tools that are not based on parameterized families of probability distributions, and hence make no assumptions about the probability distributions of the variables being evaluated. The principal advantage of the newer methods is their ability to deal with between-class variability that has limited more traditional pattern recognition methods. To this end, the automated approaches are further categorized into those that rely on parametric and nonparametric statistics.

11.2.2.2.1 Parametric Tools

The image classification tools that are parametric in nature assume that the observation matrix comes from a known probability distribution and make inferences about the parameters of the distribution. Many of the traditional classification algorithms, including the maximum likelihood classification (MLC), minimum distance (MD), and, to some degree, K-means clustering (KMC), procedure fall into this category of statistical classifiers. Some of the advantages of parametric algorithms include their simplicity and assumed higher statistical power. From the practical perspective, all off-the-shelf image-processing softwares include these algorithms, thus making them readily available and easy to apply under a wide range of conditions. They also work well, in some cases even better than the new generation of classification techniques as long as the assumptions about the probability distributions of data are valid. However, in many cases, the input satellite data do not conform to these assumptions, rendering parametric classifiers less robust in their predictions. An additional limitation occurs when nontraditional inputs, such as topography or other ancillary information, are part of the feature space as these variables never conform to known probability distributions.

In the MLC procedure, the distribution of each class' learning sample (the training data) is assumed to come from a normal distribution. Using the mean vector and the covariance matrix to characterize a class, the probability that a given pixel belongs to a specific class is calculated (Richards, 2012). Each pixel is then assigned to the class that has the highest probability (hence the maximum likelihood). A prototype land-cover database for the conterminous United States was created by Loveland et al. (1991) by first stratifying vegetated and barren land, then using an unsupervised classification of multitemporal *greenness* data derived from advanced very-high resolution radiometer (AVHRR) imager, and postclassification stratification of classes into homogeneous land-cover regions using ancillary data. Jia and Richards (1994) introduced a simplified MLC technique for handling remotely sensed image data that reduces the processing time and copes with the training of geographically small classes. Hall and Knapp (2000) demonstrated the use of the MLC procedure applied to Landsat data for characterizing the successional and disturbance dynamics of the boreal forest for use in carbon modeling in Canada. Vogelmann et al. (2001) demonstrated the use of statistical tools to map land cover of the continental United States with Landsat data. These earlier studies showed that with smaller number of categories, achieving accuracies better than 80% was possible with the use of parametric classifiers. However, as the number of categories and their complexity increased, these methods did not produce image classification results with better than 75%, particularly for the more refined categories. This notion still holds true today.

In most applications of the MLC, each class is given the same equal likelihood of belonging. However, MLC algorithm allows incorporation of prior probabilities to increase/decrease the likelihood of occurrence for certain classes. For example, Strahler (1980) in an early seminal paper shows how probabilities of occurrence of classes based on separate, independent knowledge such as collateral information datasets (e.g.,

TABLE 11.3 Comparison of Automated Image Classification Algorithms Described Here

	Advantages	Limitations	Typical Accuracies	References
Parametric tools	• Long history • Widespread use • Simple formulation • Fast computation • Small set of parameters	• Normality assumption • Lack of ancillary input use • Limited class separation	• 85%–95% for small class (<5) problems • 755–85% for large class problems	Akaike (1973) Loveland et al. (1991) Jia and Richards (1994) Hall and Knapp (2000) Vogelmann et al. (2001) Mondal et al. (2012) Sun et al. (2013)
Nonparametric tools	• Nondistributional requirement • Improved accuracies • Probability estimates • Ability to ignore unimportant features • Ability to estimate feature importance • Many-to-one feature	• Lack of standard software • Can be a black box • Can be computationally expensive • Some optimized for two class problems • Large parameter space • Specialized knowledge may be needed	• 85%–98% for small class (<5) problems • 75%–90% for large class (>5) problems	Benediktsson et al. (1990) Yoshida and Omatu (1994) Gopal and Woodcock (1996) Friedl and Brodley (1997) Carpenter et al. (1997) Pal and Mather (2003) Rodriguez-Galiano et al. (2012)
Object-based tools	• Minimum mapping unit • Natural class boundaries • Improved accuracy	• Lack of standard software • Large parameter space • Computationally expensive	• 85%–95% for small class (<5) problems • 75%–90% for large class (>5) problems	Woodcock and Harward (1992) Blaschke and Hay (2001) Hay et al. (2001) Benz et al. (2004)

rock type, soil type, topography) can be used in a Bayesian-type classifier (i.e., MLC) to improve classification accuracies. Strahler (1980) also recognized that using prior probabilities with supervised classification algorithms could also bias the posterior probability of a given class especially if there is uncertainty surrounding the information used to prescribe the prior probabilities. Chen et al. (1999) addressed this issue by introducing a subjective confidence parameter (*c*) that conditions the likelihood estimate for the membership of each pixel in a given class. Ranging from 0.0 to 1.0, the condition parameter controls the influence of the ancillary information on the predictions. Although the choice of a value for *c* is subjective, McIver and Friedl (2002) show that a simple heuristic based on objective criteria can be used to prescribe a value for *c* in many situations (Figure 11.2).

The MD procedure is one of the simplest forms of parametric classification algorithms in which the unknown image pixels are categorized by minimizing the distance between the observation matrix of the pixel and the class in multifeature space (Wacker and Landgrebe, 1972). Variants of the MD procedure use different methods to calculate the distance, which is defined as an index of similarity. For example, the Euclidian distance measure is used in cases where the variances of the classes are very different, while the Mahalanobis distance works better in cases where there is correlation between the variables of the feature space (Richards, 2012).

The KMC algorithm, among many, attempts to find a predetermined number of natural groupings or clusters in the data. It is an iterative approach in which the mean of each information class (the cluster) is initialized in some manner (e.g., randomly) and each pixel in the image matrix is assigned to the cluster that is closest to in the feature space. Then new cluster means are calculated based on the current assignment, and an attempt is made to reassign each pixel to its new category based on some similarity index. This procedure is repeated until convergence, which occurs when there is little or no change in assigned cluster means. The KMC method falls into the unsupervised classification domain and is considered to be semiparametric that lies between fully parametric and fully nonparametric statistical domains (Alpaydin, 2004).

11.2.2.2.2 Nonparametric Tools

The nonparametric methods (also called distribution-free methods) do not require the variables in the image matrix to belong to any particular distribution. While there is cornucopia of nonparametric classification techniques, the remote sensing community has adapted the artificial neural networks (ANNs), classification trees (CTs), and support vector machines (SVMs), both from the algorithmic and practical (i.e., availability as an image processing software) perspectives (Hansen et al., 1996; Friedl and Brodley, 1997; Lawrence and Wright, 2001; Pal and Mather, 2003; Rogan et al., 2003; Krishnaswamy et al., 2004). These particular methods, sometimes referred to as machine learning classifiers, have

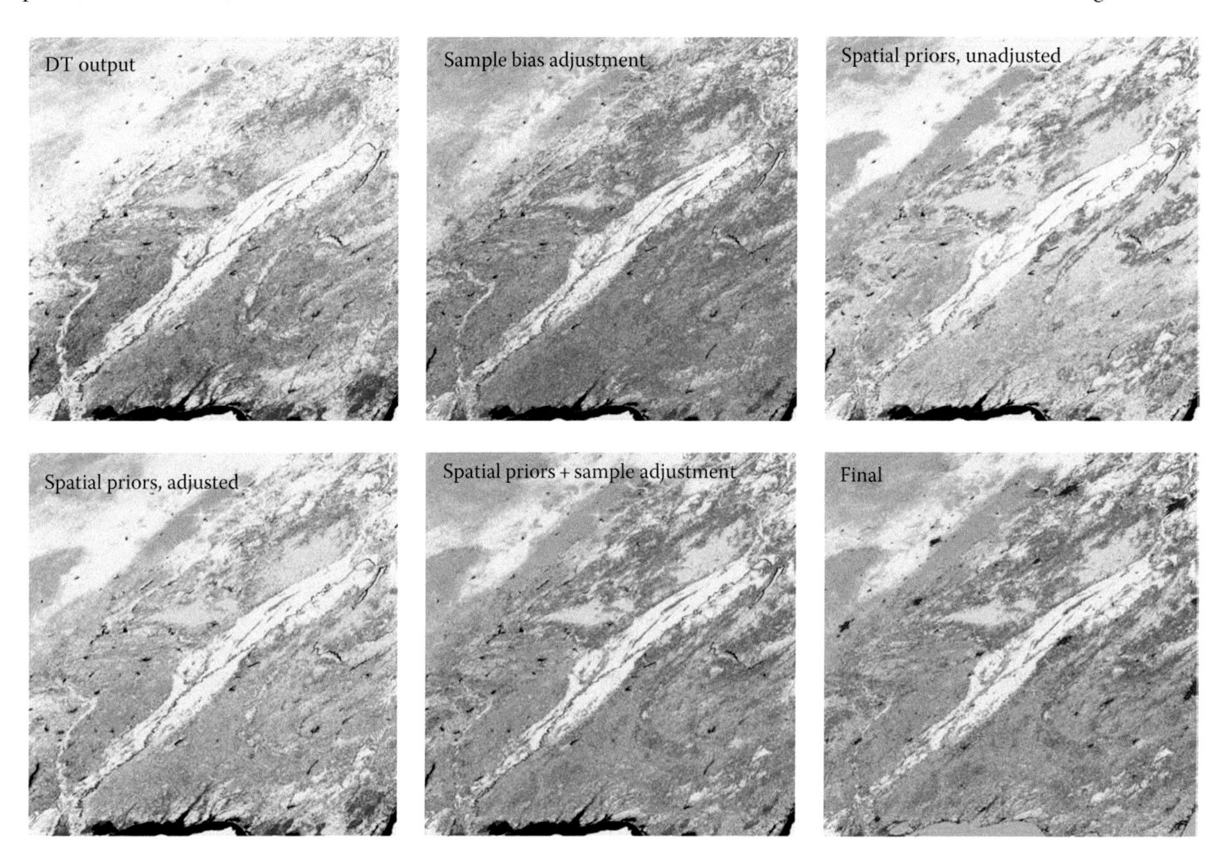

FIGURE 11.2 Classification results for MODIS tile v05h10 (south central United States) outcomes at each stage of processing. (Adapted from Friedl, M.A., et al., *Remote Sensing of Environment*, 114, 168, 2010. With permission.)

been used effectively in a variety of land-cover mapping studies (e.g., Huang and Jensen, 1997; DeFries and Chan, 2000; Friedl et al., 2010). With some exceptions, these classifiers have proven superior to parametric classifiers (e.g., maximum likelihood), with accuracy improving on the order of 10%–20% (Rogan et al., 2002). Their success can be attributed to (1) ability to incorporate class-relevant categorical and continuous observations into the features space; (2) not being constrained by parametric statistical assumptions so that multimodal, noisy, or missing observations are effectively handled; (3) ability to tease apart complex feature spaces; and (4) ability to perform many-to-one classifications where multiple manifestations of the same category are present in the observation matrix.

The ANN technique in remote sensing was introduced over two decades ago by Benediktsson et al. (1990). Since then, a large number of studies have demonstrated their effectiveness in remote sensing image classification. Their potential discriminating power has attracted a great deal of research effort, so many types of neural networks have been developed (Lippman, 1987). Among the more popular implementations of ANN are the backpropagation training algorithm and multilayer perceptron (MLP) approach (e.g., Paola and Schowengerdt, 1995; Atkinson and Tatnall, 1997) and the adaptive resonance theory (ART) approach (Carpenter et al., 1991). MLP networks are architectures in which each node receives inputs from previous layers and information flows in one direction to the output layer (Pratola et al., 2011). The number of nodes in the intermediate layer(s) defines both the complexity and the power of a neural network model to describe underlying relationships and structures inherent in a training dataset (Kavzoglu, 2009). Although ANN classification has been shown to greatly improve accuracy over traditional parametric methods with reduced training sets, the process to implement classification is not straightforward and can be time consuming (Pal and Mather, 2003).

The ART framework describes a number of neural network models, which use supervised and unsupervised learning methods for pattern recognition and prediction (Carpenter et al. 1992). The premise of the ART model is that object identification and recognition generally occur as a result of the interaction of *top-down* observer expectations with *bottom-up* sensory information, in this case information content of satellite and ancillary data. Assuming that the difference between sensation and expectation stays below a set threshold (via the *vigilance parameter*), a pixel will be considered a member of the expected class, thus offering a solution to the problem of plasticity that is, the problem of acquiring new knowledge without disrupting existing one (Carpenter and Grossberg, 2003). Carpenter et al. (1997) applied the Fuzzy ARTMAP, a version of the ART framework, to Landsat data and terrain features for vegetation classification in a challenging environment and reported a fast, reliable, and scalable algorithm overcoming many limitations of backpropagation ANNS, K nearest neighbor algorithms, and MLC. They also report an additional benefit of the ARTMAP method, in which a voting strategy improves prediction and assigns confidence estimates by training the system several times on different orderings of an input set.

A CT classifier takes a different approach to land-cover classification. It breaks an often very complex classification problem into multiple stages of simpler decision-making processes (Breiman et al., 1984; Safavian and Landgrebe, 1991). Depending on the number of variables used at each stage, there are univariate and multivariate decision trees (Friedl and Brodley, 1997). Univariate decision trees have been used to develop land-cover classifications at a global scale (Quinlan, 1993; DeFries et al. 1998; Hansen et al., 2000). Though multivariate decision trees are often more compact and can be more accurate than univariate decision trees (Brodley and Utgoff, 1995), they involve more complex algorithms and, as a result, are affected by a suite of algorithm-related factors (Friedl and Brodley, 1997). An additional benefit of the CT algorithms is their ease of use and computational efficiency (Lawrence and Wright, 2001; Pal and Mather, 2003; Lawrence et al., 2004). Several studies have found CTs to be an acceptable classification method (Hansen et al., 1996; Lawrence and Wright, 2001; Rogan et al., 2003; Krishnaswamy et al., 2004) and have shown improvements in accuracy over traditional parametric classifiers (Friedl and Brodley, 1997; Pal and Mather, 2003).

SVMs are a supervised nonparametric statistical learning technique that is increasingly being used by the remote sensing community (Huang et al., 2002; Mantero et al., 2005; Mountrakis et al., 2011). At the heart of an SVM training algorithm lies the concept of a linear *hyperplane*—an optimal boundary found through an iterative learning procedure that separates the training set into a discrete predefined number of classes while minimizing misclassifications errors (Vapnik, 1979; Zhu and Blumberg, 2002). Several approaches have been developed to improve SVM predictive accuracies using multispectral remote sensing data. These include the soft margin approach (Cortes and Vapnik, 1995) and kernel-based learning (Scholkopf and Smola, 2001) that lead to SVM optimization, although the kernel functions often result in more expensive parameterization (Kavzoglu and Colkesen, 2009).

Prior research has identified at least three benefits of SVMs that make them particularly suitable for remote sensing applications. First, regardless of the size of the learning sample, not all the available examples are used in the specification of the hyperplane. This allows SVMs to successfully handle small training datasets, because only a subset of points—the support vectors—that lie on the margin is used to define the hyperplane (Mantero et al., 2005). Second, unlike many statistical classifiers, SVMs do not make prior assumptions on the probability distribution of the data, which leads to reduction in classification errors when input data do not conform to a required distribution (e.g., Gaussian). Third, SVM-based classification algorithms have been shown to produce generalizable models from a set of input training data, eliminating the notion of overfitting (Montgomery and Peck, 1992).

One of the more popular implementations of SVMs is software called the LIBSVM implementation that provides linear, polynomial (cubic), and radial-basis kernels (Chang and Lin, 2011). This implementation includes C-support vector classification

(C-SVC), ν-support vector classification (ν-SVC), distribution estimation (one-class SVM), ε-support vector regression (ε-SVR), and ν-support vector regression (ν-SVR) formulations. All SVM formulations supported in LIBSVM are quadratic minimization problems. Using the radial-basis kernel classification option, the LIBSVM required only two parameters to be defined: the kernel parameter γ and the cost parameter C (Chang and Lin, 2011). Both of these parameters are data dependent and are identified separately for each footprint/date-pair combination using the grid search option over log-transformed hyperparameters as suggested by Hsu et al. (2001).

Note that SVMs have been shown to perform well given a certain level of noise (i.e., mislabeled training data), but they are not completely impervious to outliers (Vapnik, 1995). While a number of methods have been developed to mitigate the effects of outliers on SVMs (Lin and Wang, 2002; Suykens et al., 2002; Tsujinishi and Abe, 2003; Lin and Wang 2004) they show only incremental improvements over standard SVM methods.

While there is evidence in the literature to show that nonparametric methods perform better than traditional classifiers for mapping land cover, very little attempt has gone into comparing the performance of several nonparametric classifiers. Shao and Lunetta (2012) compared the performance of SVMs, ANNs, and CT for land-cover characterization using MODIS time-series data and investigated the effects of training sample size, sample variability, and landscape homogeneity (purity). Their results indicate a strong relationship between training sample size and classification accuracy but above a set threshold, increasing the number of training data does not lead to an equal increase in performance. They also show that SVMs had superior generalization capability, particularly with respect to small training sample sizes (Figure 11.3). There was also less variability of SVM performance when classification trials were repeated using different training sets.

A growing number of studies continue to show that the nonparametric image classification tools come with undisputed advantages, not least of the improved classification performance. While the more sophisticated nonparametric algorithms still remain in the research domain and require specialized software environment, the rate at which the remote sensing community takes up these methods is high and points to the possibility that they may become the *de facto* choice for image classification.

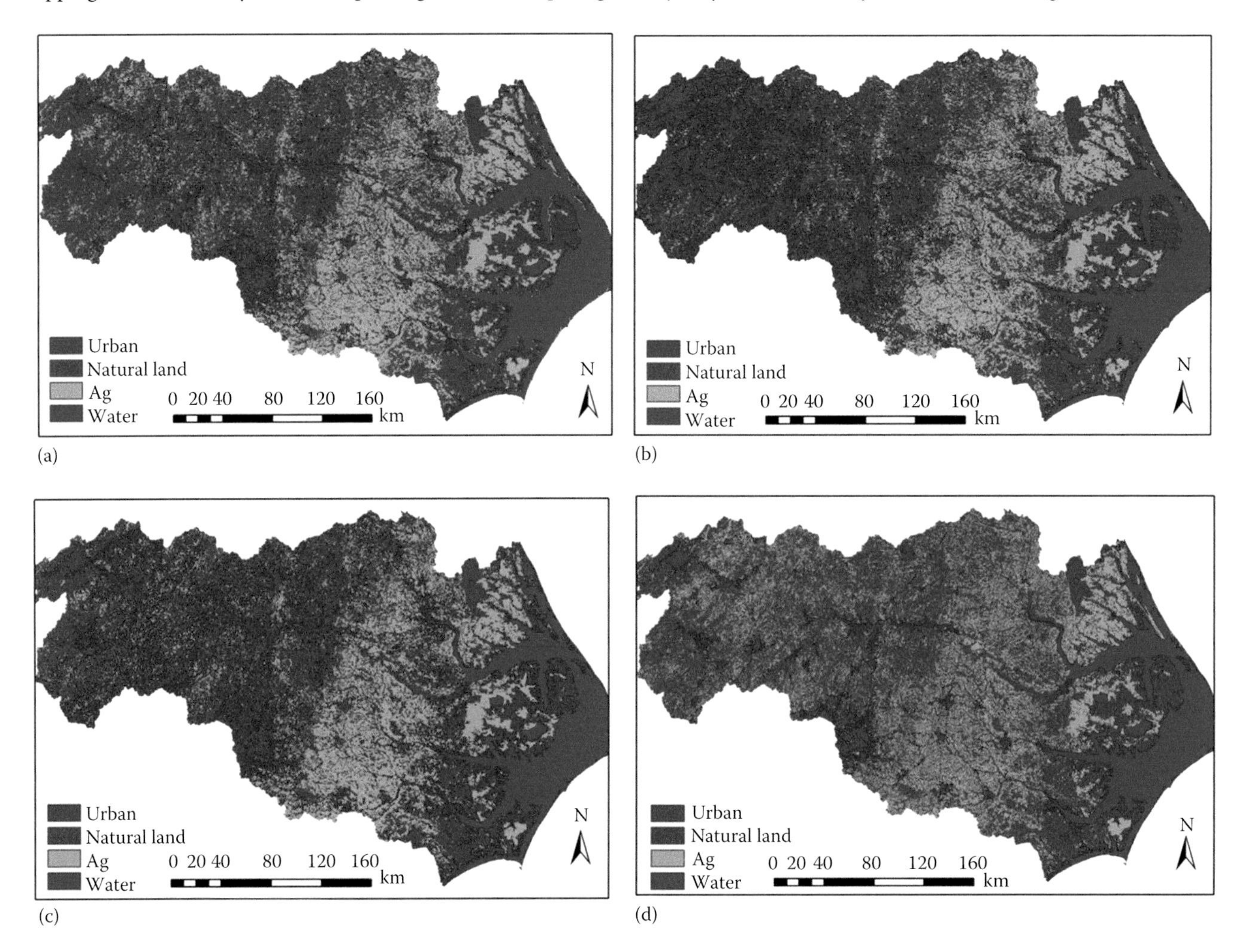

FIGURE 11.3 Comparison of classification results for SVM (a), ANN (b), and CT (c) algorithms. The NLCD 2001 (d) is also included as reference. (Adapted from Shao and Lunetta, 2012. With permission.)

11.2.2.3 Pixel vs. Object Based Classifications

Individual pixels form the smallest unit of analysis when classifying remotely sensed images for the purpose of land-cover mapping. However, it is well known that as arbitrary objects, pixels do not necessarily represent the landscape that is being characterized. Contextual classification methods overcome this issue by incorporating the spatial information into the classification process. In the absence of contextualization, pixel-based classification results tend to contain significant noise as no neighborhood information is being considered (Fisher, 1997). One reason for this is that information content of individual pixels is heavily influenced by the radiance contributed from surrounding pixels, or the Modulation Transfer Function (MTF) effect (Townshend et al., 2000). Moreover, both the landscape structure and the resolving power of the sensor contribute to our ability to identify and map objects of interest in a map. To overcome the errors that result from traditional per-pixel classifiers that ignore contextual properties from surrounding pixels, the remote sensing community has developed a number of image analysis algorithms that go *beyond pixels* and take also into account spatial information (Blaschke, 2010). These new methods include object-based image analyses (OBIA) that have generally yielded higher thematic accuracies than the traditional per-pixel methods (Woodcock and Harward, 1992; Blaschke et al., 2006; Platt and Rapoza, 2008; Lizarazo and Elsner, 2011).

In this chapter, we divide OBIA into three categories, primarily distinguished by the order and the purpose by which the objects are utilized. A common denominator of these methods is image segmentation, purpose of which is to produce a set of nonoverlapping segments (objects, polygons). Although not new in industrial and medical image processing (Haralick and Shapiro, 1985), image segmentation has only two decades of history in geospatial applications (Ryherd and Woodcock, 1996; Blaschke and Hay, 2001; Hay et al., 2005; Esch et al., 2008; Lizaraso and Elsner, 2009) but has been rapidly increasing in the past 10 years.

Image segmentation partitions the image into a set of distinct and uniform (homogenous) regions, in which the criteria of homogeneity are defined by one or more dimensions of the feature space (Blaschke, 2010). It is generally understood that these segments represent meaningful landscape objects (e.g., a forest stand or an agricultural field), but this translation is not always clear at the segmentation stage. What is clear however is that different regions are found at different scales of analysis, placing image segmentation in the realm of multiscale landscape analysis (Hay et al., 2001). The image segments also contain additional spectral (e.g., mean values per band, and also median values, minimum and maximum values, mean ratios, variance, etc.) and spatial information compared to single pixels, and it is often argued that it is this spatial information that provides greater advantage to OBIA using segments (Blaschke and Strobl, 2001; Flanders et al., 2003; Hay and Castilla, 2008).

One question surrounding the OBIA is the scale at which the analysis is made. More specifically, at what scale (or spatial resolution) does the image under consideration lends itself to image segmentation for the purpose of object-based analysis? It turns out the answer is not related only to spatial resolution but also the size of the objects under consideration on the landscape. From the theoretical perspective, the Shannon–Nyquist sampling theorem suggests that to be able reconstruct the original image (i.e., to reconstruct the landscape objects in the image), the spatial sampling rate (or the spatial resolution) of the image has to be higher than twice the highest spatial frequency of the original image (i.e., spatial resolution must be finer than at least half the size of the smallest objects on the landscape) (Blaschke, 2010). From the practical perspective however, a minimum of six pixels per objects has been found to be necessary in order to accurately identify and map objects.

Returning to different kinds of OBIA, in the first category, pixel-based image classification is performed independent of the image segmentation process. More specifically, first a traditional pixel-based classification is performed using the best possible tools and inputs with the highest possible accuracy. Then image segmentation is performed, either on the original image that went into the classification process or on an image with higher spatial resolution of the same area. The key here is to define the minimum mapping unit, defined as the size of the smallest object to be identified on the landscape, based on the spatial resolution of the image used in classification. Finally, the image segments, either in raster or vector format, are merged with the pixel-based classification using the majority rule, in which the most frequently occurring class label is used to label the entire segment (Figure 11.4). While the majority rule works well, there are other ways to collapse individual pixels into objects (polygons).

11.2.2.4 Emerging Methods

11.2.2.4.1 Cloud and Cloud Shadow Masking

The presence of clouds and cloud shadows affects almost all forms and types of analyses conducted with satellite data, and their detection and subsequent removal is often the initial step in most image-processing chains (Simpson and Stitt, 1998; Irish, 2000; Arvidson et al., 2001). For example, they selectively alter the reflectance and transmission of radiation through the Earth's atmosphere and reduce the accuracy of atmospheric correction of images. Land-cover classification involving multiple image composites requires clouds and their shadows to be removed from the scene, before the compositing process. In change detection studies, the presence of clouds and cloud shadows lead to false detection of land-cover change. Calculation of vegetation indices like the normalized difference vegetation index (NDVI) may be biased in locations where clouds and cloud shadows are present. Archiving processes such as scene selection and scene quality assessment too requires cloud information, not only on the total amount but also the location of clouds and cloud shadows in a Landsat scene. In parallel with the growing demand and use of Landsat imagery, there have also been a number of new developments in detecting clouds in Landsat images (e.g., Irish et al., 2006; Oreopoulos et al., 2011; Zhu and Woodcock, 2012; Goodwin et al., 2013).

One of the more robust methods in detecting cloud and cloud shadows was developed by Zhu and Woodcock (2012). This method uses a series of rules based on calculated probabilities

FIGURE 11.4 Object-based analysis example: (a) the original Landsat TM image in true color; (b) the original USDA Cropland Data Layer (c) object-filled USDA Cropland Data Layer in which the objects (agricultural field boundaries) were used to convert pixel-based classification results to a polygon based map.

of temperature, spectral variability, and brightness, using top of atmosphere (TOA) reflectance and brightness temperature (BT) as inputs. The clouds and cloud shadows are treated as 3D objects determined via segmentation of the potential cloud layer and an assumption of a constant temperature lapse rate. The solar illumination and sensor view angles are used to predict possible cloud shadow locations and select the one that has the maximum similarity to cloud shape and size.

11.2.2.4.2 Large Area Mapping

Free and open access to most satellite data (e.g., Landsat, Sentinel) are changing the way remotely sensed images are processed. For example, there is increased interest in automatically processing large volumes of imagery covering large areas of the Earth. These methods rely on availability of a large number of images within and across the years and use both the spectral and the temporal information available in satellite data. While the concept of time series observation for large area land-cover mapping is not new (Hansen et al., 2000; Loveland et al., 2000; Friedl et al., 2002; Bartholomé and Belward, 2005), the new developments mostly apply to medium-resolution (i.e. less than 100 m pixels) images and rely on automated algorithms (Kalensky, 1998; Gong et al., 2013; Sexton et al., 2013; Yu et al., 2013). Also contributing to these developments is the availability of powerful computers and image classification software, mainly based on nonparametric methods that are able to generalize across space and time. For example, development of the Web-Enabled Landsat Data (WELD) (Roy et al., 2010) system has revolutionized the way Landsat data are being processed and temporally aggregated for rapid and efficient large-area applications. However, despite the existence and the availability of maps over large areas, there is still the issue of different land-cover definitions as well as the use of different algorithms. In Figure 11.5, we show forest classification results from three large-area classification projects in Kenya. While there is strong correlation between each map product, notable differences remain. For example, the FROM GLC product (Gong et al., 2013) has a large forest area than the other products. Alternatively, in locations where the UMD tree

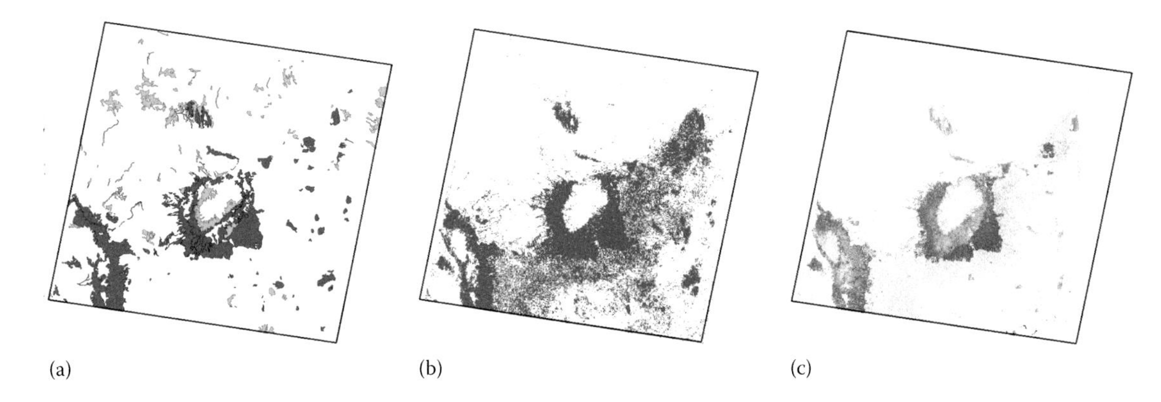

FIGURE 11.5 Comparison of three land cover maps for the forest category in Kenya. (a) is the land cover map from the AfriCover project; (b) forest category of a Landsat based land cover map as part of the FROM GLC project; (c) tree cover map derived from multi-temporal Landsat data. The black polygon outlines the boundaries of a single Landsat WRS2 footprint (path:168 row:60).

cover map (Sexton et al., 2013) shows small tree cover fractions (i.e., less than 50% threshold), the AfricCover product (Kalensky, 1998) maps them as full forest. Of course, this is partly related to class definitions as each of three products uses different descriptions of what a forest is. Nevertheless, data in Figure 11.5 simply highlights the difficulty with which large area land-cover maps are made using medium- to high-resolution satellite data.

11.2.3 Uncertainty Assessment

Accuracy assessment of classification results developed from satellite data (or other sources) is an important but often neglected step in producing land-cover maps. The application of various image classification methods to satellite data produces a map that is hypothesized to represent the land cover in question. Like all hypothesis testing, the next step is to obtain data, preferably from an independent sources, and test the validity of the hypothesis in map form. One reason map validation is either neglected or underestimated stems from the fact that the accuracy assessment —if performed right—is a difficult and costly task. In general, validation is a complex task as maps made from satellite data cover large areas relative to the spatial sampling unit (i.e., pixel or a polygon). Moreover, collecting the reference data to be used in validation is expensive and labor intensive. For this reason, the sampling design to acquire accurate, representative, statistically sound, and independent reference data must be carefully crafted and evaluated against available resources allocated to the validation procedure. Therefore, the key to an effective accuracy assessment is a sound sampling plan that balances accuracy and pragmatism while including both the common and rare categories (Zhu et al., 2000).

It is generally accepted that accuracy assessment using an independent, stratified random sample produces the best evaluation results with respect to statistical rigor and representatives (Congalton, 1991; Stehman and Czaplewski, 1998). In a stratified random design, the map categories often constitute the strata so that each class could be evaluated using an independent random sample. The allocation strategy could be proportionate in which known class areas determine the number of samples per class, or disproportionate in which a set number of samples is assigned in category. One disadvantage of the proportional allocation is that small classes receive a small, and in some cases unrealistic, number of samples. While the question of sample size, both per class and total, has received due attention in the literature (e.g., see Stehman, 2012 for a comprehensive review), the number of samples to be used in the validation effort ultimately depends on the resources available. Note that with the development of free, high-resolution, and accurately located image data, such as those available from Google Earth, some of the cost associated with field work has been shifted to the office in the form of manual interpretation. However, caution must be exercised when using these sources as the independent sample, because complex land-cover categories do not always lend themselves to visual analysis. Moreover, these online sources often represent a single time period corresponding to current or recent past, rendering them less useful for evaluating maps of past time periods. Finally, maps derived from object-based analysis are evaluated at the object level. That is, accuracy assessment is performed using the objects (e.g., polygons) in the independent sample in which the class label of the entire polygon (not the label of an individual pixel inside the polygon) is used for evaluation. In evaluating pixel-based maps, it is recommended that an area corresponding to at least one pixel shift in all directions be included in the analysis to address the geolocation-based errors inherent in all satellite data. For example, a map made from Landsat data with 30 m pixels would require a square-like polygon, at least 90 m on a side, centered on the pixel to be evaluated. Information inside this polygon is then used to label the samples in the independent dataset.

Once the independent sample is acquired, development of the confusion matrix (or error matrix) is the standard practice. The confusion matrix is a specific table that allows assessment of both class-specific and overall accuracies rapidly. The table is also used to quantify the omission and commission errors that allow the specific performance of the classifier in use. In general, when reporting map accuracy, it is recommended that the confusion matrix along with all the other accuracy measures be reported. This allows the map user to draw his/her own conclusions when using the land-cover map.

11.3 The Future

It is the author's opinion that the future of land-cover mapping from satellite data is bright. Contributing factors include: (1) free and unlimited access to satellite observations; (2) development of cheaper and better sensors; (3) availability of dense temporal observations, even at high spatial resolutions; (4) advances in computer hardware and software rooted in artificial intelligence community; (5) ease of access using web-enabled services; and (6) availability of ever-growing number of ancillary datasets on environmental and social variables that in turn help determine the presence or absence of a specific land cover. As our planet is continually pushed to meet the demands of a growing and affluent population, identifying and mapping of its resources will be even more paramount. It is hoped that the new developments in image processing and satellite data availability will allow us to make more accurate and refined land-cover maps to face this challenge.

References

Akaike, H. 1973. Information theory and an extension of the maximum likelihood principle, in *Proceedings of the Second International Symposium on Information Theory*. Petrov, B. and Cazakil, F. eds. Budapest, Hungary: Aakademiai Kidao.

Albertz, D. 2007. Einführung in die Fernerkundung. Grundlagen der Interpretation von Luft- & Satellitenbildern. (Darmstadt) 254pp.

Alpaydin, E. 2004. *Introduction to Machine Learning*. Cambridge, MA: MIT Press.

Anderson, J. R. 1971. Land use classification schemes used in selected recent geographic applications if remote sensing. *Photogrammatic Engineering*, 37(4), 379–387.

Anderson, J. R., Hardy, E. E., Roach, J. T., and Witmer, R. E. 1976. A land use and land cover classification system for use with remote sensor data. USGS Professional Paper 964, 28pp.

Arvidson, T., Gasch, J., and Goward, S. N. 2001. Landsat-7's long-term acquisition plan—An innovative approach to building a global imagery archive. *Remote Sensing of Environment*, 78(1–2), 13–26.

Atkinson, P. M. and Tatnall, A. R. 1997. Introduction: Neural networks in remote sensing. *International Journal of Remote Sensing*, 18, 699–709.

Bartholomé, E. and Belward, A. S. 2005. GLC2000: A new approach to global land cover mapping from Earth observation data. *International Journal of Remote Sensing*, 26(9), 1959–1977.

Benediktsson, J., Swain, P. H., and Ersoy, O. K. 1990. Neural network approaches versus statistical methods in classification of multisource remote sensing data. *IEEE Transactions on Geoscience and Remote Sensing*, 28(4), 540–552.

Benz, U. C., Hofmann, P., Willhauck, G., Lingenfelder, I., and Heynen, M. 2004. Multi-resolution, object-oriented fuzzy analysis of remote sensing data for GIS-ready information. *ISPRS Journal of Photogrammetry and Remote Sensing*, 58, 239–258.

Blaschke, T., Burnett, C., and Pekkarinen, A. 2006. Image segmentation methods for object-based analysis and classification, in S. de Jong and F. van der Meer (eds.), *Remote Sensing Image Analysis: Including the Spatial Domain*. Berlin, Germany: Springer, pp. 211–236.

Blaschke, T. 2010. Object based image analysis for remote sensing. *ISPRS Journal of Photogrammetry and Remote Sensing*, 65, 2–16.

Blaschke, T. and Hay, G. J. 2001. Object-oriented image analysis and scale-space: Theory and methods for modeling and evaluating multi-scale landscape structure. *International Archives of Photogrammetry and Remote Sensing*, 34(Part 4/W5), 22–29.

Blaschke, T. and Strobl, J. 2001. What's wrong with pixels? Some recent developments interfacing remote sensing and GIS. *GIS—Zeitschrift für Geoinformationssysteme*, 14(6), 12–17.

Breiman, L., Friedman, J. H., Olshen, R. A., and Stone, C. J. 1984. *Classification and Regression Trees*. Belmont, CA: Wadsworth International Group.

Brodley, C. E. and Utgoff, P. E. 1995. Multivariate decision trees. *Machine Learning*, 19, 45–77.

Carpenter, G. A., Gjaja, M. N., Gopal, S., and Woodcock, C. E. 1997. ART neural networks for remote sensing: Vegetation classification from Landsat TM and terrain data. *IEEE Transactions on Geoscience and Remote Sensing*, 35(2), 308–325.

Carpenter, G. A., Grossberg, S., and Reynolds, J. H. 1991. ARTMAP: Supervised real-time learning and classification of nonstationary data by a self-organizing neural network. *Neural Networks*, 4, 565–588.

Carpenter, G. A., Grossberg, S., Markuzon, N., Reynolds, J. H., and Rosen, D. B. 1992. Fuzzy ARTMAP: A neural network architecture for incremental supervised learning of analog multidimensional maps. *IEEE Transactions on Neural Networks*, 3, 698–713.

Carpenter, G. A. and Grossberg, S. 2003. Adaptive resonance theory, in M. A. Arbib (ed.), *The Handbook of Brain Theory and Neural Networks*, 2nd edn. Cambridge, MA: MIT Press, pp. 87–90.

Chang, C.-C. and Lin, J.-C. 2011. LIBSVM: A library for support vector machines. *ACM Transactions on Intelligent Systems and Technology*, 2, 1–39.

Chen, M.-H., Ibrahim, J. G., and Yianoutsos, C. 1999. Prior elicitation, variable selection, and Bayesian computation for logistics regression models. *Journal of the Royal Statistical Society B*, 61, 223–242.

Congalton, R. G. 1991. A review of assessing the accuracy of classification of remote sensed data. *Remote Sensing of Environment*, 37, 35–46.

Cortes, C. and Vapnik, V. 1995. Support-vector networks. *Machine Learning*, 20, 273–297.

DeFries, R. S. and Chan, J.C-W. 2000. Multiple criteria for evaluating machine learning algorithms for land cover classification from satellite data. *Remote Sensing of Environment*, 74, 503–515.

DeFries, R. S., Hansen, M., Townshend, J. R. G., and Sohlberg, R. 1998. Global land cover classifications at 8km spatial resolution: the use of training data derived from Landsat imagery in decision tree classifiers. *International Journal of Remote Sensing*, 19, 3141–3168.

Esch, T., Thiel, M., Bock, M., Roth, A., and Dech, S. 2008. Improvement of image segmentation accuracy based on multiscale optimization procedure. *IEEE Geosciences and Remote Sensing Letters*, 5(3), 463–467.

Fisher, P. 1997. The pixel: A snare and a delusion. *International Journal of Remote Sensing*, 18, 679–685.

Flanders, D., Hall-Beyer, M., and Pereverzoff, J. 2003. Preliminary evaluation of eCognition object-based software for cut block delineation and feature extraction. *Canadian Journal of Remote Sensing*, 29(4), 441–452.

Friedl, M. A., McIver, D. K., Hodges, J. C. F., Zhang, X. Y., Muchoney, D., Strahler, A. H., Woodcock, C. E. et al. 2002. Global land cover mapping from MODIS: Algorithms and early results. *Remote Sensing of Environment*, 83, 287–302.

Friedl, M. A., Sulla-Menashe, D., Tan, B., Schneider, A., Ramankutty, N., Sibley, A., and Huang, X. 2010. MODIS Collection 5 global land cover: Algorithm refinements and characterization of new datasets. *Remote Sensing of Environment*, 114, 168–182.

Friedl, M. A. and Brodley, C. E. 1997. Decision tree classification of land cover from remotely sensed data. *Remote Sensing of Environment*, 61(3), 399–409.

Gong, P., Wang, J., Yu, L., Zhao, Y. C., Zhao, Y. Y., Liang, L., Niu, Z. G. et al. 2013. Finer resolution observation and monitoring of global land cover: First mapping results with Landsat TM and ETM+ data. *International Journal of Remote Sensing*, 34(7), 2607–2654.

Goodwin, N. R., Collett, L. J., Denham, R. J., Flood, N., and Tindall, D. 2013. Cloud and cloud shadow screening across Queensland, Australia: An automated method for Landsat TM/ETM+ time series. *Remote Sensing of Environment*, 134, 50–65.

Gopal, S. and Woodcock, C. E. 1996. Remote sensing of forest change using artificial neural networks. *IEEE Transactions on Geoscience and Remote Sensing*, 34(2), 398–404.

Hall, F. G. and Knapp, D. 2000. *BOREAS TE-18 Landsat TM Maximum Likelihood Classification Image of the SSA*, Technical Report Series on the Boreal Ecosystem-Atmosphere Study (BOREAS), Vol. 175. Greenbelt, MD: NASA.

Hansen, M., DeFries, R., Townshend, J. R. G., and Sohlberg, R. 2000. Global land cover classification at 1 km resolution using a decision tree classifier. *International Journal of Remote Sensing*, 21, 1331–1365.

Hansen, M., Dubayah, R., and DeFries, R., 1996, Classification trees: An alternative to traditional land cover classifiers. *International Journal of Remote Sensing*, 17, 1075–1081.

Haralick R. M. and Shapiro L. G. 1985. Survey: Image segmentation techniques. *Computer Vision, Graphics and Image Processing*, 29(1), 100–132.

Hay, G. J., Castilla, G., Wulder, M. A., and Ruiz, J. R. 2005. An automated object-based approach for the multiscale image segmentation of forest scenes. *International Journal of Applied Earth Observation and Geoinformation*, 7(4), 339–359.

Hay, G. J., Marceau, D. J., Dube, P., and Bouchard, A. 2001. A multiscale framework for landscape analysis: Object-specific analysis and upscaling. *Landscape Ecology*, 16(6), 471–490.

Hay, G. J. and Castilla, G. 2008. Geographic Object-Based Image Analysis (GEOBIA): A new name for a new discipline, in T. Blaschke, S. Lang, and G. Hay (eds.), *Object Based Image Analysis*. New York: Springer, pp. 93–112.

Hsu, C.-W., Chang, C.-C., and Lin, C.-J. 2001. A practical guide to support vector classification. Available at http://www.csie.ntu.edu.tw/~cjlin/. Accessed March 20, 2015.

Huang, C., Davis, L. S., and Townshend, J. R. G. 2002. An assessment of support vector machines for land cover classification. *International Journal of Remote Sensing*, 23, 725–749.

Huang, X. and Jensen, J. R. 1997. A machine-learning approach to automated knowledge-base building for remote sensing image analysis with GIS data. *Photogrammetric Engineering and Remote Sensing*, 63(10), 1185–1194.

Irish, R., Barker, J. L., Goward, S. N., and Arvidson, T. 2006. Characterization of the Landsat-7 ETM+ Automated Cloud-Cover Assessment (ACCA) algorithm. *Photogrammetric Engineering and Remote Sensing*, 72(10), 1179–1188.

Irish, R. 2000. Landsat-7 automatic cloud cover assessment algorithms for multispectral, hyperspectral, and multraspectral imagery. *The International Society for Optical Engineering*, 4049, 348–355.

Jensen, J. R. 2006. *Remote Sensing of the Environment: An Earth Resource Perspective*, 2nd edn. Englewood Cliffs, NJ: Prentice Hall, 608pp.

Jia, X. and Richards, J. A. 1994. Efficient maximum likelihood classification for imaging spectrometer data sets. *IEEE Transactions on Geoscience and Remote Sensing*, 32(2), 274–281.

Kalensky, Z. D. 1998. AFRICOVER land cover database and map of Africa. *Canadian Journal of Remote Sensing*, 24(3), 292–297.

Kavzgolu, T. 2009. Increasing the accuracy of neural network classification using refined training data. *Environmental Modelling and Software*, 24(7), 850–858.

Kavzoglu, T. and Colkesen, I. 2009. A kernel functions analysis for support vector machines for land cover classification. *International Journal of Applied Earth Observation and Geoinformation*, 11, 352–359.

Krishnaswamy, J., Kiran, M. C., and Ganeshaiah, K. N. 2004. Tree model based ecoclimatic vegetation classification and fuzzy mapping in diverse tropical deciduous ecosystems using multi-season NDVI. *International Journal of Remote Sensing*, 25(6), 1185–1205.

Lawrence, R., Bunn, A., Powell, S., and M. Zambon. 2004. Classification of remotely sensed imagery using stochastic gradient boosting as a refinement of classification tree analysis. *Remote Sensing of Environment 90*, 331–336.

Lawrence, R. L. and Wright, A. 2001. Rule-based classification systems using classification and regression tree (CART) analysis. *Photogrammetric Engineering and Remote Sensing* 67, 1137–1142.

Lin, C.-F. and Wang, S.-D. 2004. Training algorithms for fuzzy support vector machines with no data. *Pattern Recognition Letters*, 25, 1647–1656.

Lin, C.-F. and Wang, S. D. 2002. Fuzzy support vector machines. *IEEE Transactions on Neural Networks*, 13, 464–471.

Lippman, R. P. 1987. An introduction to computing with neural nets. *IEEE ASSP Magazine*, 4, 2–22.

Lizarazo, I. and Elsner, P. 2009. Fuzzy segmentation for object-based image classification. *International Journal of Remote Sensing*, 30(6), 1643–1649.

Lizarazo, I. and Elsner, P. 2011. Segmentation of Remotely Sensed Imagery: Moving from Sharp Objects to Fuzzy Regions, in Dr. Pei-Gee Ho (ed.), *Image Segmentation*, ISBN: 978-953-307-228-9, InTech, Rijeka, Croatia. Available from: http://www.intechopen.com/books/image-segmentation/segmentation-of-remotely-sensed-imagery-moving-from-sharp-objects-to-fuzzy-regions. Accessed March 20, 2015.

Loveland, T. R., Merchant, J. W., Ohlen, D. O., and Brown, J. F. 1991. Development of a land-cover characteristics database for the conterminous U.S. *Photogrammatic Engineering and Remote Sensing*, 57, 1453–1463.

Loveland, T. R., Reed, B. C., Brown, J. F., Ohlen, D. O., Zhu, Z., Yang, L., and Merchant, J. W. 2000. Development of a global land cover characteristics database and IGBP DISCover from 1 km AVHRR data. *International Journal of Remote Sensing*, 21(6–7), 1303–1330.

Mantero, P., Moser, G., and Serpico, S. B. 2005. Partially supervised classification of remote sensing images through SVM-based probability density estimation. *IEEE Transactions on Geoscience and Remote Sensing*, 43, 559–570.

McIver, D. K. and Friedl, M. A. 2002. Using prior probabilities in decision-tree classification of remotely sensed data. *Remote Sensing of Environment*, 81(2–3), 253–261.

Mondal, A., Kundu, S., Chandniha, S. K., Shukla, R., and Mishr, P. K. 2012. Comparison of support vector machine and maximum likelihood classification technique using satellite imagery. *International Journal of Remote Sensing and GIS*, 1(2), 116–123.

Montgomery, D. C. and Peck, E. A. 1992. *Introduction to Linear Regression Analysis*, 2nd edn. New York: Wiley.

Mountrakis, G., Im, J., and Ogole, C. 2011. Support vector machines in remote sensing: A review. *ISPRS Journal of Photogrammetry and Remote Sensing*, 66, 247–259.

Oreopoulos, L., Wilson, M., and Várnai, T. 2011. Implementation on Landsat data of a simple cloud mask algorithm developed for MODIS land bands. *IEEE Transactions on Geoscience and Remote Sensing*, 8(4), 597–601.

Pal, M. and Mather, P. M. 2003. An assessment of the effectiveness of decision tree methods for land cover classification. *Remote Sensing of Environment*, 86, 554–556.

Paola, J. D. and Schowengerdt, R. A. 1995. A review and analysis of backpropagation neural networks for classification of remotely sensed multi-spectral imagery. *International Journal of Remote Sensing*, 16, 3033–3058.

Platt, R. V. and Rapoza, L. 2008. An evaluation of an object-oriented paradigm for land use/land cover classification. *The Professional Geographer*, 60(1), 87–100.

Pratola, C., Del Frate, F., Schiavon, G., Solimini, D., and Licciardi, G. 2011. Characterizing land cover from X-band COSMO-SkyMed images by neural networks, in *Urban Remote Sensing*, April 11–13. Munich, Germany: IEEE, pp. 49–52.

Quinlan, J. R. 1993. *C4.5 Programs for Machine Learning*. San Mateo, CA: Morgan Kaufmann Publishers.

Richards, J. A. 2012. *Remote Sensing Digital Image Analysis: An Introduction*, 5th edn. New York: Springer Science+Business Media

Rodriguez-Galiano, V., Ghimire, B., Rogan, J., Chica-Olmo, M., and Rigol-Sanchez, J. 2012. An assessment of the effectiveness of a random forest classifier for land-cover classification. *ISPRS Journal of Photogrammetry and Remote Sensing*, 67, 93–104.

Rogan, J., Franklin, J., and Roberts, D. 2002. A comparison of methods for monitoring multitemporal vegetation change using Thematic Mapper imagery. *Remote Sensing of Environment*, 80(1), 143–156.

Rogan, J., Miller, J., Stow, D., Franklin, J., Levien, L., and Fischer, C. 2003. Land cover change mapping in California using classification trees with Landsat TM and ancillary data. *Photogrammetric Engineering and Remote Sensing*, 69(7), 793–804.

Roy, D. P., Ju, J., Kline, K., Scaramuzza, P. L., Kovalskyy, V., Hansen, M. C., Loveland, T. R., Vermote, E. F., and Zhang, C. 2010. Web-enabled Landsat Data (WELD): Landsat ETM+ Composited Mosaics of the Conterminous United States. *Remote Sensing of Environment*, 114, 35–49.

Ryherd, S. and Woodcock, C. E. 1996. Combining spectral and texture data in the segmentation of remotely sensed images. *Photogrammetric Engineering and Remote Sensing*, 62(2), 181–194.

Safavian, S. R. and Landgrebe, D. 1991. A survey of decision tree classifier methodology. *IEEE Transactions on Systems, Man, and Cybernetics*, 21, 660–674.

Scholkopf, B. and Smola, A. J. 2001. *Learning with Kernels*. Cambridge, MA: The MIT Press.

Sexton, J. O., Song, X.-P., Feng, M., Noojipady, P., Anand, A., Huang, C., Kim, D.-H., Collins, K. M., Channan, S., DiMiceli, C., and Townshend, J. R. G. 2013. Global, 30-m resolution continuous fields of tree cover: Landsat-based rescaling of MODIS Vegetation Continuous Fields with lidar-based estimates of error. *International Journal of Digital Earth*, 6(5), 130321031236007. doi:10.1080/17538947.2013.786146.

Shao, Y. and Lunetta, R. 2012. Comparison of support vector machine, neural network, and CART algorithms for the land-cover classification using limited training data points. *ISPRS Journal of Photogrammetry and Remote Sensing*, 70, 78–87.

Simpson, J. J. and Stitt, J. R. 1998. A procedure for the detection and removal of cloud shadow from AVHRR data over land. *Geoscience and Remote Sensing*, 36(3), 880–890.

Stehman, S. V. 2012. Impact of sample size allocation when using stratified random sampling to estimate accuracy and area of land-cover change. *Remote Sensing Letters*, 3(2), 111–120.

Stehman, S. V. and Czaplewski, R. L. 1998. Design and analysis of thematic map accuracy assessment: Fundamental principles. *Remote Sensing of Environment*, 64, 331–344.

Strahler, H. 1980. The use of prior probabilities in Maximum Likelihood Classification of remotely sensed data. *Remote Sensing of Environment*, 10,135–163.

Sun, J., Yang, J., Zhang, C., Yun, W., and Qu, J. 2013. Automatic remotely sensed image classification in a grid environment based on the maximum likelihood method. *Mathematical and Computer Modeling*, 58(3–4), 573–581.

Suykens, J. A. K., Brabanter, J. D., Lukas, L., and Vandewalle, J. 2002. Weighted least squares support vector machines: Robustness and sparse approximation. *Neurocomputing*, 48, 85–105.

Townshend, J. R. G., Huang, C., Kalluri, S. N., DeFries, R. S., Liang, S., and Yang, K. 2000. Beware of per-pixel characterization of land cover. *International Journal of Remote Sensing*, 21(4), 839–843.

Travaglia, C., Milenova, L., Nedkov, R., Vassilev, V., Milenov, P., Radkov, R., and Pironkova, Z. 2001. Preparation of land cover database of Bulgaria through remote sensing and GIS. Environment and Natural Resources Working Paper No. 6, FAO, Rome, Italy, 57pp.

Tsujinishi, D. and Abe, S. 2003. Fuzzy least squares support vector machines for multiclass problems. *Neural Networks*, 16, 785–792.

Vapnik, V. 1979. *Estimation of Dependences Based on Empirical Data*. Moscow, Russia: Nauka, pp. 5165–5184, 27 (in Russian) (English translation: New York: Springer Verlag, 1982).

Vapnik, V. 1995. *The Nature of Statistical Learning Theory*, 2nd edn. New York: Springer.

Vogelmann, J. E., Howard, S. M., Yang, L., Larson, C. R., Wylie, B. K., and Van Driel, J. N. 2001. Completion of the 1990's National Land Cover Data Set for the conterminous United States. *Photogrammetric Engineering and Remote Sensing*, 67, 650–662.

Wacker, A. G. and Landgrebe, D. A. 1972. Minimum distance classification in remote sensing. LARS Technical Reports. Paper 25. http://docs.lib.purdue.edu/larstech/25. Accessed March 20, 2015.

Woodcock, C. E. and Harward, V. J. 1992. Nested-hierarchical scene models and image segmentation. *International Journal of Remote Sensing*, 13, 3167–3187.

Yoshida, T. and Omatu, S. 1994. Neural network approach to land cover mapping. *IEEE Transactions on Geoscience and Remote Sensing*, 32(5), 1103–1109.

Yu, L., Wang, J., and Gong, P. 2013. Improving 30 meter global land cover map FROM-GLC with time series MODIS and auxiliary datasets: A segmentation based approach. *International Journal of Remote Sensing*, 34(16), 5851–5867.

Zhu, G. and Blumberg, D. G. 2002. Classification using ASTER data and SVM algorithms: The case study of Beer Sheva, Israel. *Remote Sensing of Environment*, 80, 233–240.

Zhu, Z., Yang, L., Stehman, S. V., and Czaplewski, R. L. 2000. Accuracy Assessment for the U.S. Geological Survey Regional Land-Cover Mapping Program: New York and New Jersey Region. *Photogrammetric Engineering and Remote Sensing*, 66(12), 1425–1435.

Zhu, Z. and Woodcock, C. E. 2012. Object-based cloud and cloud shadow detection in Landsat imagery. *Remote Sensing of Environment*, 118, 83–94.

12

Hyperspectral Image Processing: Methods and Approaches

Jun Li
Sun Yat-Sen University

Antonio Plaza
University of Extremadura

Acronyms and Definitions

EMPs	Extended morphological profiles
EMPs	Extended morphological profiles
LDA	Linear discriminant analysis
LogDA	Logarithmic discriminant analysis
MLR	Multinomial logistic regression
MLRsubMRF	Subspace-based multinomial logistic regression followed by Markov random fields
MPs	Morphological profiles
MRFs	Markov random fields
PCA	Principal component analysis
QDA	Quadratic discriminant analysis
RHSEG	Recursive hierarchical segmentation
ROSIS	Reflective optics spectrographic imaging system
SVMs	Support vector machines
TSVMs	Transductive support vector machines

12.1 Introduction

Hyperspectral imaging is concerned with the measurement, analysis, and interpretation of spectra acquired from a given scene (or specific object) at a short, medium, or long distance, typically, by an airborne or satellite sensor [1]. The special characteristics of hyperspectral data sets pose different processing problems, which must be necessarily tackled under specific mathematical formalisms [2], such as classification and segmentation [3] or spectral mixture analysis [4]. Several machine learning and image-processing techniques have been applied to extract relevant information from hyperspectral data during the last decade [5,6]. Taxonomies of hyperspectral image-processing algorithms have been presented in the literature [3,7,8]. It should be noted, however, that most recently developed hyperspectral image-processing techniques focus on analyzing the spectral and spatial informations contained in the hyperspectral data in simultaneous fashion [9]. In other words, the importance of analyzing spatial and spectral information simultaneously has been identified as a desired goal by many scientists devoted to hyperspectral image analysis. This type of processing has been approached in the past from various points of view. For instance, several possibilities are discussed by Landgrebe [10] for the refinement of results obtained by spectral-based techniques through a second step based on spatial context. Such contextual classification [11] accounts for the tendency of certain ground cover classes to occur more frequently in some contexts than in others. In certain applications, the integration of high spatial and spectral information is mandatory to achieve sufficiently accurate mapping and/or detection results. For instance, urban area mapping requires sufficient spatial resolution to distinguish small spectral classes, such as trees in a park or cars on a street [12] (Figure 12.1).

However, there are several important challenges when performing hyperspectral image classification. In particular, supervised classification faces challenges related with the unbalance between high dimensionality and the limited number of training samples or the presence of mixed pixels in the data (which may compromise classification results for coarse spatial resolutions). Specifically, due to the small number of training samples and the high dimensionality of the hyperspectral data, reliable estimation of statistical class parameters is a very challenging goal [13]. As a result, with a limited training set, classification

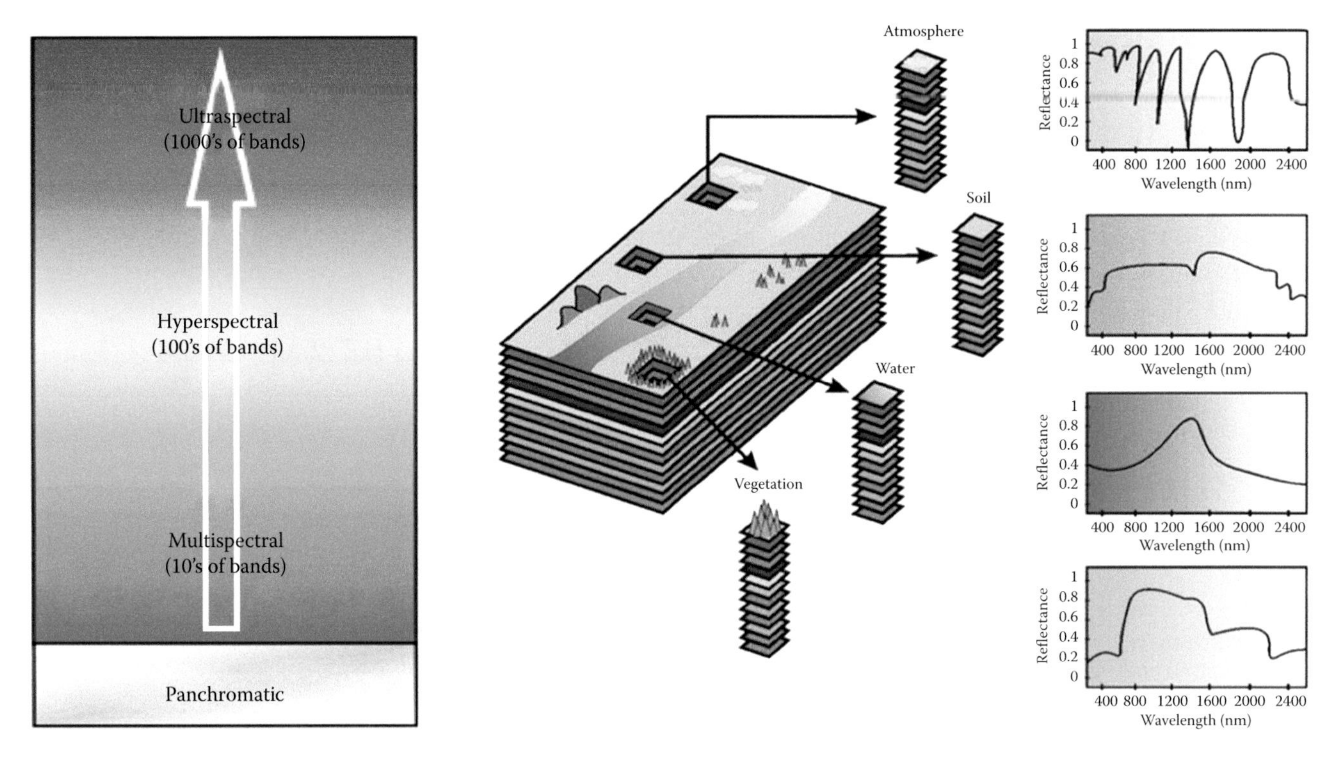

FIGURE 12.1 The challenges of increased dimensionality in remote-sensing data interpretation.

accuracy tends to decrease as the number of features increases. This is known as the Hughes effect [14]. Another relevant challenge is the need to integrate the spatial and spectral information to take advantage of the complementarities that both sources of information can provide. These challenges are quite important for future developments and solutions to some of them have been proposed. Specifically, supervised [15] and semisupervised [16–18] techniques for hyperspectral image classification, strategies for integrating the spatial and the spectral information [19–22], or subspace classifiers [23] that can better exploit the intrinsic nature of hyperspectral data have been quite popular in the recent literature.

Our main goal in this chapter is to provide a seminal view on recent advances in techniques for hyperspectral image analysis that can successfully deal with the dimensionality problem and with the limited availability of training samples *a priori*, while taking into account both the spectral and spatial properties of the data. The remainder of the chapter is structured as follows. Section 12.2 discusses available techniques for hyperspectral image classification, including both supervised and semisupervised approaches, techniques for integrating spatial and spectral information and subspace-based approaches. Section 12.3 provides an experimental comparison of the techniques discussed in Section 12.2, using a hyperspectral data set collected by the ROSIS over the University of Pavia, Italy, which is used here as a common benchmark to outline the properties of the different processing techniques discussed in the chapter. Finally, Section 12.4 concludes the paper with some remarks and hints at the most pressing ongoing research directions in hyperspectral image classification.

12.2 Classification Approaches

In this section, we outline some of the main techniques and challenges in hyperspectral image classification. Hyperspectral image classification has been a very active area of research in recent years [3]. Given a set of observations (i.e., pixel vectors in a hyperspectral image), the goal of classification is to assign a unique label to each pixel vector so that it is well defined by a given class. In Figure 12.2, we provide an overview of a popular strategy to conduct hyperspectral image classification, which is based on the availability of labeled samples. After an optional dimensionality reduction step, a supervised classifier is trained using a set of labeled samples (which are often randomly selected from a larger pool of samples) and then tested with a disjoint set of labeled samples in order to evaluate the classification accuracy of the classifier.

Supervised classification has been widely used in hyperspectral data interpretation [2], but it faces challenges related with the high dimensionality of the data and the limited availability of training samples, which may not be easy to collect in pure form. However, mixed training samples can also offer relevant information about the participating classes [10]. In order to address these issues, subspace-based approaches [23,24] and semisupervised learning techniques [25] have been developed. In subspace approaches, the goal is to reduce the dimensionality of the input space in order to better exploit the (limited) training samples available. In semisupervised learning, the idea is to exploit the information conveyed by additional (unlabeled) samples, which can complement the available labeled samples with a certain degree of confidence. In all cases, there is a clear need to

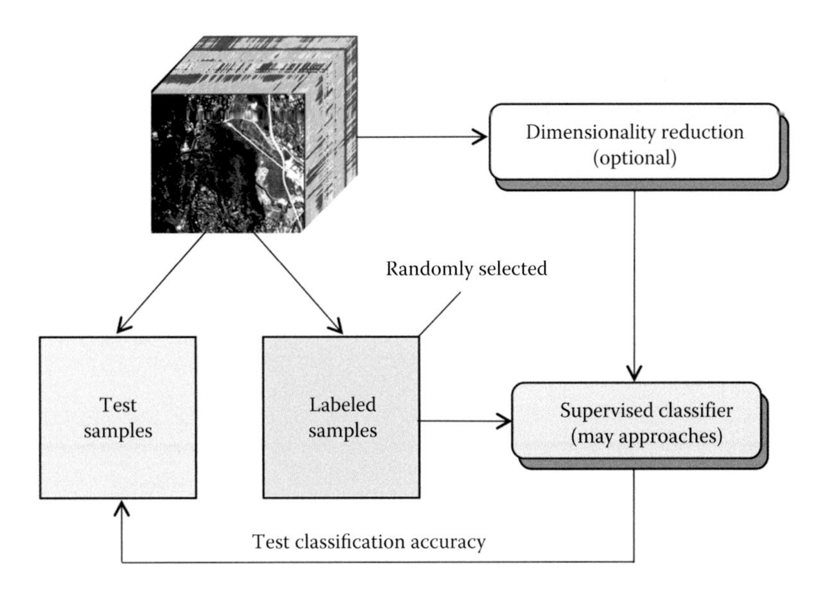

FIGURE 12.2 Standard approach for supervised hyperspectral image classification.

integrate the spatial and spectral information to take advantage of the complementarities that both sources of information can provide [9]. An overview of these different aspects, which are crucial to hyperspectral image classification, is provided in the following subsections.

12.2.1 Supervised Classification

Several techniques have been used to perform supervised classification of hyperspectral data. For instance, in discriminant classifiers, several types of discriminant functions can be applied: nearest neighbor, decision trees, linear functions, or nonlinear functions (see Figure 12.3). In linear discriminant analysis (LDA) [26], a linear function is used in order to maximize the discriminatory power and separate the available classes effectively. However, such a linear function may not be the best choice, and nonlinear strategies such as quadratic discriminant analysis (QDA) or logarithmic discriminant analysis (LogDA) have also been used. The main problem of these classic supervised classifiers, however, is their sensitivity to the Hughes effect.

In this context, kernel methods such as the support vector machine (SVM) have been widely used in order to deal effectively with the Hughes phenomenon [27,28]. The SVM was first investigated as a binary classifier [29]. Given a training set mapped into a Hilbert space by some mapping, the SVM separates the data by an optimal hyperplane that maximizes the margin (see Figure 12.4). However, the most widely used approach in hyperspectral classification is to combine soft margin classification with a kernel trick that allows separation of the classes in a higher-dimensional space by means of a nonlinear transformation (see Figure 12.5). In other words, the SVM used with a kernel function is a nonlinear classifier, where the nonlinear ability is included in the kernel and different kernels lead to different types of SVMs. The extension of SVM to the multiclass cases is usually done by combining several binary classifiers.

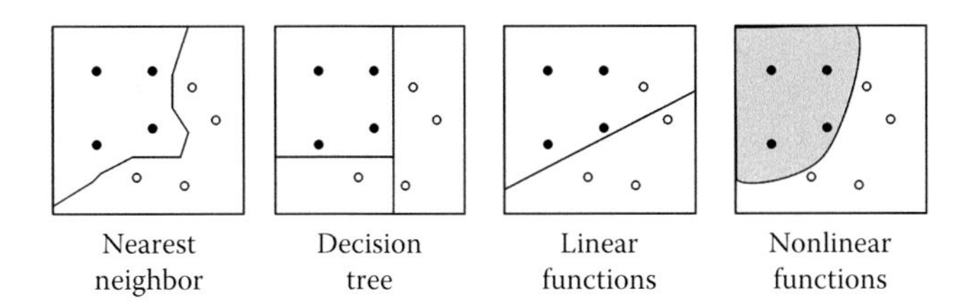

FIGURE 12.3 Typical discriminant functions used in supervised classifiers.

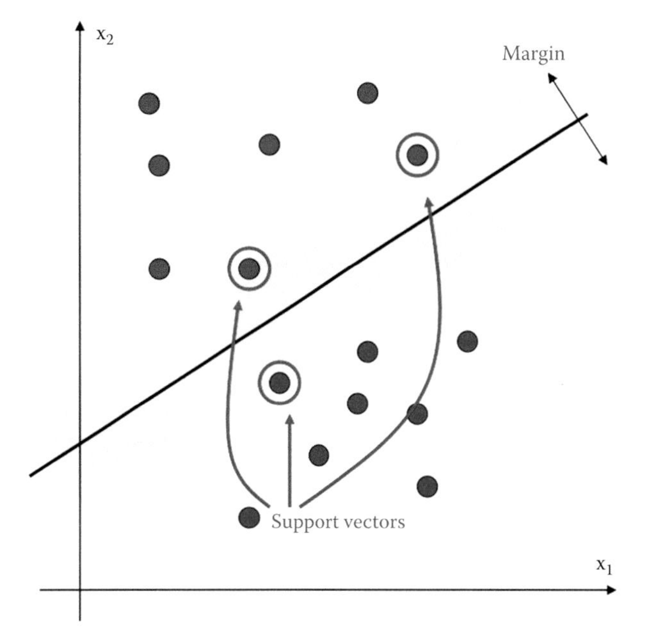

FIGURE 12.4 Soft margin classification with slack variables.

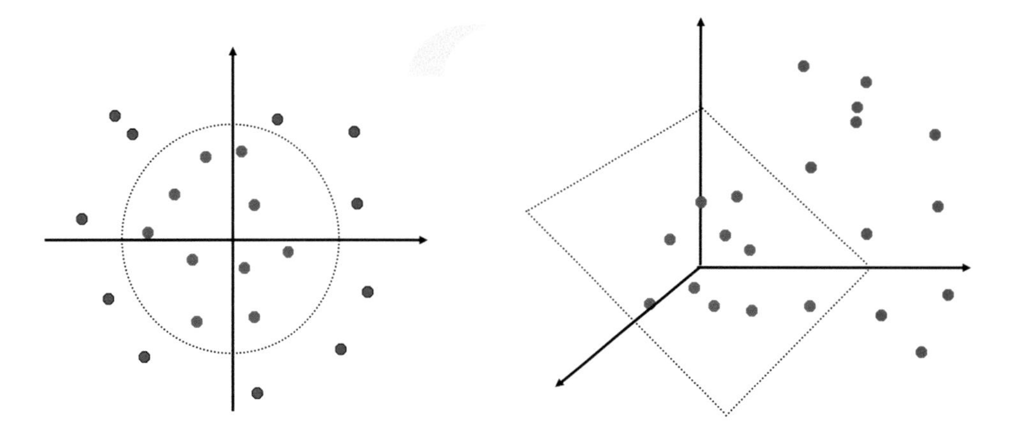

FIGURE 12.5 The kernel trick allows separation of the classes in a higher-dimensional space by means of a linear or nonlinear transformation.

12.2.2 Spectral–Spatial Classification

Several efforts have been performed in the literature in order to integrate spatial–contextual information in spectral-based classifiers for hyperspectral data [3,9]. It is now commonly accepted that using the spatial and the spectral information simultaneously provides significant advantages in terms of improving the performance of classification techniques. An illustration of the importance of integrating spatial and spectral information is given in Figure 12.6. As shown in this figure, spectral–spatial classification (obtained using morphological transformations) provides a better interpretation of classes such as urban features, with a better delineation and characterization of complex urban structures.

Some of the approaches that integrate spatial and spectral information include spatial information prior to the classification, during the feature extraction stage. Mathematical morphology [30] has been particularly successful for this purpose. Morphology is a widely used approach for modeling the spatial characteristics of the objects in remotely sensed images. Advanced morphological techniques such as morphological profiles [31] have been successfully used for feature extraction prior to classification of hyperspectral data by extracting the first few principal components of the data using principal component analysis [13] and then building so-called extended morphological profiles (EMPs) on the first few components to extract relevant features for classification [32].

Another strategy in the literature has been to exploit simultaneously the spatial and the spectral information. For instance, in order to incorporate the spatial context into kernel-based classifiers, a pixel entity can be redefined, simultaneously both in the spectral domain (using its spectral content) and also in the spatial domain, by applying some feature extraction to its surrounding area, which yields spatial (contextual) features, for example, the mean or standard deviation per spectral band. These separated entities lead to two different kernel matrices, which can be easily computed. At this point, one can sum spectral and textural dedicated kernel matrices and introduce the

FIGURE 12.6 The importance of using spatial and spectral information in classification.

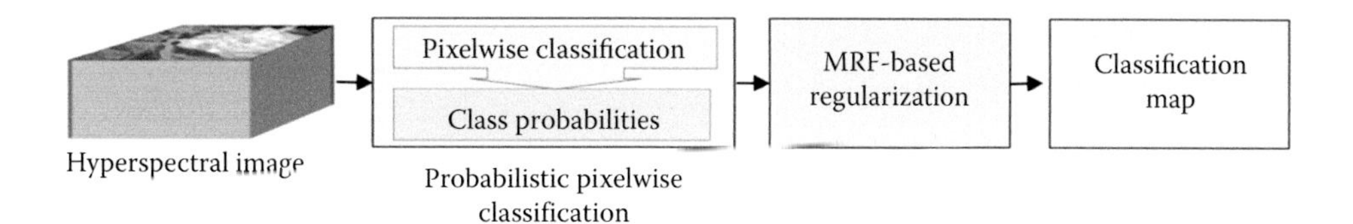

FIGURE 12.7 Standard processing framework using pixel-wise probabilistic classification followed by MRF-based spatial postprocessing.

cross information between textural and spectral features in the formulation. This simple methodology yields a full family of new kernel methods for hyperspectral data classification, defined in [33] and implemented using the SVM classifier, thus providing a composite kernel-based SVM.

Another approach to jointly exploit spatial and spectral information is to use Markov random fields (MRFs) for the characterization of spatial information. MRFs exploit the continuity, in probability sense, of neighboring labels [19,34]. In this regard, several techniques have exploited an MRF-based regularization procedure, which encourages neighboring pixels to have the same label when performing probabilistic classification of hyperspectral data sets. An example of this type of processing is given in Figure 12.7, in which a pixel-wise probabilistic classification is followed by an MRF-based spatial postprocessing that refines the initial probabilistic classification output.

Several other approaches include spatial information as a postprocessing, that is, after a spectral-based classification has been conducted. One of the first classifiers with spatial postprocessing developed in the hyperspectral imaging literature was the well-known extraction and classification of homogeneous objects [10]. Another one is the strategy adopted in [35], which combines the output of a pixel-wise SVM classifier with the morphological watershed transformation [30] in order to provide a more spatially homogeneous classification. A similar strategy is adopted in Ref. [36], in which the output of the SVM classifier is combined with the segmentation result provided by the unsupervised recursive hierarchical segmentation (RHSEG)* algorithm.

12.2.3 Subspace-Based Approaches

Subspace projection methods [23] have been shown to be a powerful class of statistical pattern classification algorithms. These methods can handle the high dimensionality of hyperspectral data by bringing it to the right subspace without losing the original information that allows for the separation of classes. In this context, subspace projection methods can provide competitive advantages by separating classes that are very similar in spectral sense, thus addressing the limitations in the classification process due to the presence of highly mixed pixels. The idea of applying subspace projection methods to improve classification relies on the basic assumption that the samples within each class can approximately lie in a lower-dimensional subspace. Thus, each class may be represented by a subspace spanned by a set of basis vectors, while the classification criterion for a new input sample would be the distance from the class subspace [24].

Recently, several subspace projection methods have been specifically designed for improving hyperspectral data characterization, with successful results. For instance, the subspace-based multinomial logistic regression followed by Markov random fields (MLRsubMRF) method in Ref. [23] first performs a learning step in which the posterior probability distributions are modeled by a multinomial logistic regression (MLR) [37] combined with a subspace projection method. Then, the method infers an image of class labels from a posterior distribution built on the learned subspace classifier and on a multilevel logistic spatial prior on the image of labels. This prior is an MRF that exploits the continuity, in probability sense, of neighboring labels. The basic assumption is that, in a hyperspectral image, it is very likely that two neighboring pixels will have the class same label. The main contribution of the MLRsubMRF method is therefore the integration of a subspace projection method with the MLR, which is further combined with spatial–contextual information in order to provide a good characterization of the content of hyperspectral imagery in both spectral and the spatial domains. As will be shown by our experiments, the accuracies achieved by this approach are competitive with those provided by many other state-of-the-art supervised classifiers for hyperspectral analysis (Figure 12.8).

12.2.4 Semisupervised Classification

A relevant challenge for supervised classification techniques is the limited availability of labeled training samples, since their collection generally involves expensive ground campaigns [38]. While the collection of labeled samples is generally difficult, expensive, and time-consuming, unlabeled samples can be generated in a much easier way. This observation has fostered the idea of adopting semisupervised learning techniques in hyperspectral image classification. The main assumption of such techniques is that new (unlabeled) training samples can be obtained from a (limited) set of available labeled samples without significant effort/cost. This can be simply done by selecting new samples from the spatial neighborhood of available labeled samples, under the principle that it is likely that the new unlabeled samples will have similar class labels as the already available ones.

In contrast to supervised classification, semisupervised algorithms generally assume that a limited number of labeled samples are available *a priori* and then enlarge the training set using unlabeled samples, thus allowing these approaches to address ill-posed problems. However, in order for this strategy to work,

* http://opensource.gsfc.nasa.gov/projects/HSEG/.

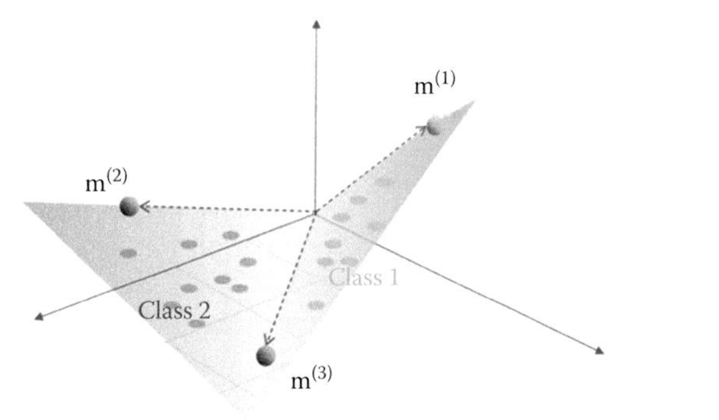

Multinominal logistic regression subspace classifier (MLRsubMRF)

Modelling of spectral information (use a subspace probabilistic MLR classifier tuned to deal with mixed pixels)

Modelling of spatial information (use a multi-level logistic spatial prior to enforce smoothness in the classification)

FIGURE 12.8 Multinomial logistic regression subspace classifier.

several requirements need to be met. First and foremost, the new (unlabeled) samples should be obtained without significant cost/effort. Second, the number of unlabeled samples required in order for the semisupervised classifier to perform properly should not be too high in order to avoid increasing computational complexity in the classification stage. In other words, as the number of unlabeled samples increases, it may be unbearable for the classifier to properly exploit all the available training samples due to computational issues. Further, if the unlabeled samples are not properly selected, these may confuse the classifier, thus introducing significant divergence or even reducing the classification accuracy obtained with the initial set of labeled samples. In order to address these issues, it is very important that the most highly informative unlabeled samples are identified in computationally efficient fashion, so that significant improvements in classification performance can be observed without the need to use a very high number of unlabeled samples.

The area of semisupervised learning for remote-sensing data analysis has experienced a significant evolution in recent years. For instance, looking at machine learning–based approaches, in Ref. [39] transductive support vector machines (TSVMs) are used to gradually search a reliable separating hyperplane (in the kernel space) with a transductive process that incorporates both labeled and unlabeled samples in the training phase. In [40], a semisupervised method is presented that exploits the wealth of unlabeled samples in the image and naturally gives relative importance to the labeled ones through a graph-based methodology. In [41], kernels combining spectral–spatial information are constructed by applying spatial smoothing over the original hyperspectral data and then using composite kernels in graph-based classifiers. In [18], a semisupervised SVM is presented that exploits the wealth of unlabeled samples for regularizing the training kernel representation locally by means of cluster kernels. In [42], a new semisupervised approach is presented that exploits unlabeled training samples (selected by means of an active selection strategy based on the entropy of the samples). Here, unlabeled samples are used to improve the estimation of the class distributions, and the obtained classification is refined by using a spatial multilevel logistic prior. In [16], a novel context-sensitive semisupervised SVM is presented that exploits the contextual information of the pixels belonging to the neighborhood system of each training sample in the learning phase to improve the robustness to possible mislabeled training patterns.

In [43], two semisupervised one-class (SVM-based) approaches are presented in which the information provided by unlabeled samples present in the scene is used to improve classification accuracy and alleviate the problem of free-parameter selection. The first approach models data marginal distribution with the graph Laplacian built with both labeled and unlabeled samples. The second approach is a modification of the SVM cost function that penalizes more the errors made when classifying samples of the target class. In [44], a new method to combine labeled and unlabeled pixels to increase classification reliability and accuracy, thus addressing the sample selection bias problem, is presented and discussed. In [45], an SVM is trained with the linear combination of two kernels: a base kernel working only with labeled examples is deformed by a likelihood kernel encoding similarities between labeled and unlabeled examples and then applied in the context of urban hyperspectral image classification. In [46], similar concepts to those addressed before are adopted using a neural network as the baseline classifier. In [17], a semiautomatic procedure to generate land cover maps from remote-sensing images using active queries is presented and discussed.

Last but not least, we emphasize that the techniques summarized in this section only represent a small sample (and somehow subjective selection) of the vast collection of approaches presented in recent years for hyperspectral image classification. For a more exhaustive summary of available techniques and future challenges in this area, we point interested readers to [47].

12.3 Experimental Comparison

In this section, we illustrate the performance of the techniques described in the previous section by processing a widely used hyperspectral data set collected by ROSIS optical sensor over the urban area of the University of Pavia, Italy. The flight was operated by the Deutschen Zentrum for Luftund Raumfahrt (DLR, the German Aerospace Agency) in the framework of the HySens project, managed and sponsored by the European Union. The image size in pixels is 610 × 340, with very-high-spatial

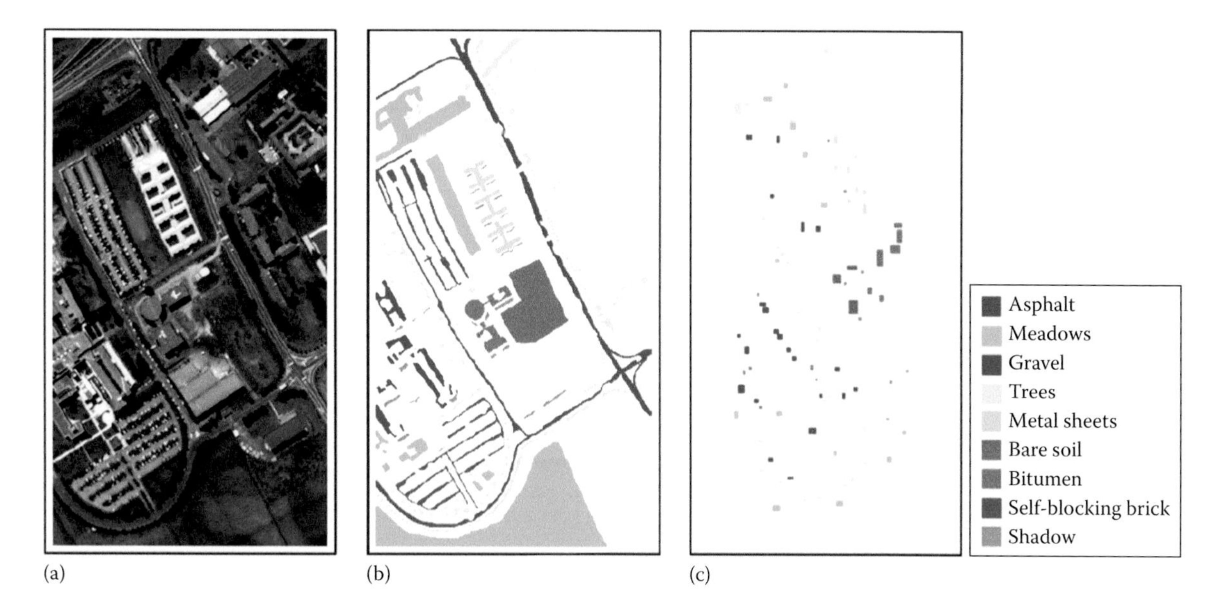

FIGURE 12.9 The ROSIS Pavia University scene used in our experiments. The scene was collected by the ROSIS instrument in the framework of the HySens campaign. It comprises 103 spectral bands between 0.4 and 0.9 μm and was collected over an urban area at the University of Pavia, Italy. (a) ROSIS Pavia hyperspectral image. (b) Ground-truth classes. (c) Fixed training set.

resolution of 1.3 m per pixel. The number of data channels in the acquired image is 103 (with spectral range from 0.43 to 0.86 μm). Figure 12.9a shows a false color composite of the image, while Figure 12.9b shows nine reference classes of interest, which comprise urban features, as well as soil and vegetation features. Finally, Figure 12.9c shows a fixed training set available for the scene, which comprises 3,921 training samples (42,776 samples are available for testing). This scene has been widely used in the hyperspectral imaging community to evaluate the performance of processing algorithms [6]. It represents a case study that integrates a challenging urban classification problem, with a data set comprising high spatial and spectral resolution, and a highly reliable ground truth, with a well-established training set. All these factors have made the scene a standard and an excellent test bed for evaluation of hyperspectral image classification algorithms, particularly those integrating the spatial and the spectral information.

Table 12.1 illustrates the classification results obtained by different supervised classifiers for the ROSIS University of Pavia scene in Figure 12.9a, using the same training data in Figure 12.9c to train the classifiers and a mutually exclusive set of labeled samples in Figure 12.9b to test the classifiers. As shown by Table 12.1, the SVM classifier obtained comparatively superior performance in terms of the overall classification accuracy when compared with discriminant classifiers such as LDA, QDA or LogDA.

TABLE 12.1 Classification Results Obtained for the ROSIS Pavia University Scene by the SVM Classifier as Compared with Several Discriminant Classifiers

Class	Training	Testing	LDA	QDA	LogDA	SVM
Asphalt (6,631)	548	6,083	69.45	67.75	70.89	83.71
Meadows (18,649)	532	18,117	81.92	75.73	76.72	70.25
Gravel (2,099)	265	1,834	39.11	59.79	55.31	70.32
Trees (3,064)	231	2,833	95.07	96.64	96.38	97.81
Metal Sheets (1,345)	375	970	99.41	99.93	100	99.41
Bare Soil (5,029)	540	4,489	46.59	73.49	75.06	92.25
Bitumen (1,330)	392	938	63.31	93.53	83.98	81.58
Self-Blocking Bricks (3,682)	524	3,158	88.29	89.52	87.91	92.59
Shadow (947)	514	433	99.79	99.26	99.79	96.62
Overall accuracy	—	—	77.95	77.95	78.41	80.99

LDA stands for linear discriminant analysis. QDA stands for quadratic discriminant analysis. LogDA stands for logarithmic discriminant analysis. SVM stands for support vector machine. The numbers in the parentheses are the total number of available samples.

TABLE 12.2 Classification Results Obtained for the ROSIS Pavia University Scene by the SVM Classifier as Compared with a Composite SVM Obtained Using the Summation Kernel

Class	Training	Testing	SVM	Composite SVM
Asphalt (6,631)	548	6,083	83.71	79.85
Meadows (18,649)	532	18,117	70.25	84.76
Gravel (2,099)	265	1,834	70.32	81.87
Trees (3,064)	231	2,833	97.81	96.36
Metal Sheets (1,345)	375	970	99.41	99.37
Bare Soil (5,029)	540	4,489	92.25	93.55
Bitumen (1,330)	392	938	81.58	90.21
Self-Blocking Bricks (3,682)	524	3,158	92.59	92.81
Shadow (947)	514	433	96.62	95.35
Overall accuracy	—	—	80.99	87.18

Source: Camps-Valls, L. et al., *IEEE Geosci. Remote Sens. Lett.*, 3, 93, 2006.
The numbers in the parentheses are the total number of available samples.

In a second experiment, we compared the standard SVM classifier with the composite kernel strategy as defined in Ref. [33], which combines spatial and spectral information at the kernel level. After carefully evaluating all possible types of composite kernels, the summation kernel provided the best performance in our experiments as reported in Table 12.2. This table suggests the importance of using spatial and spectral information in the analysis of hyperspectral data.

Although the integration of spatial and spectral information carried out by the composite kernel in Table 12.2 is performed at the classification stage, the spatial information can also be included prior to classification. For illustrative purposes, Table 12.3 compares the classification results obtained by the SVM applied on the original hyperspectral image to those obtained using a combination of the morphological EMP for feature extraction followed by SVM for classification (EMP/SVM).

As shown in Table 12.3, the EMP/SVM provides good classification results for the ROSIS University of Pavia scene, which represent a good improvement over the results obtained using the original hyperspectral image as input to the classifier. These results confirm the importance of using spatial and spectral information for classification purposes, as it was already found in the experimental results reported in Table 12.2.

In order to illustrate other approaches that use spatial information as postprocessing, Tables 12.4 and 12.5, respectively, compare the classification results obtained by the traditional SVM with those found using the strategy adopted in Ref. [35], which combines the output of a pixel-wise SVM classifier with the morphological watershed, and in Ref. [36], in which the output of the SVM classifier is combined with the segmentation result provided by the RHSEG algorithm. As shown in Tables 12.4 and 12.5, these strategies lead to improved classification with regard to the traditional SVM. In fact, Tables 12.2 through 12.5 illustrate different aspects concerning the integration of spatial and spectral information. The results reported in Table 12.2 are obtained by integrating spatial and spectral information at the classification stage. On the other hand, the results reported in Table 12.3 are obtained by using spatial information

TABLE 12.3 Classification Results Obtained for the ROSIS Pavia University Scene by the SVM Classifier as Compared with Those Obtained Using a Combination of the Morphological EMP for Feature Extraction Followed by SVM for Classification (EMP/SVM)

Class	Training	Testing	SVM	EMP/SVM
Asphalt (6,631)	548	6,083	83.71	95.36
Meadows (18,649)	532	18,117	70.25	80.33
Gravel (2,099)	265	1,834	70.32	87.61
Trees (3,064)	231	2,833	97.81	98.37
Metal Sheets (1,345)	375	970	99.41	99.48
Bare Soil (5,029)	540	4,489	92.25	63.72
Bitumen (1,330)	392	938	81.58	98.87
Self-Blocking Bricks (3,682)	524	3,158	92.59	95.41
Shadow (947)	514	433	96.62	97.68
Overall accuracy	—	—	80.99	85.22

The numbers in the parentheses are the total number of available samples.

TABLE 12.4 Classification Results Obtained for the ROSIS Pavia University Scene by the SVM Classifier as Compared with Those Obtained Using the Strategy Adopted in [35], Which Combines the Output of a Pixel-Wise SVM Classifier with the Morphological Watershed

Class	Training	Testing	SVM	SVM + Watershed
Asphalt (6,631)	548	6,083	83.71	94.28
Meadows (18,649)	532	18,117	70.25	76.41
Gravel (2,099)	265	1,834	70.32	69.89
Trees (3,064)	231	2,833	97.81	98.30
Metal Sheets (1,345)	375	970	99.41	99.78
Bare Soil (5,029)	540	4,489	92.25	97.51
Bitumen (1,330)	392	938	81.58	97.14
Self-Blocking Bricks (3,682)	524	3,158	92.59	98.29
Shadow (947)	514	433	96.62	97.57
Overall accuracy	—	—	80.99	86.64

The numbers in the parentheses are the total number of available samples.

TABLE 12.5 Classification Results Obtained for the ROSIS Pavia University Scene by the SVM Classifier as Compared with Those Obtained Using the Strategy Adopted in [36], in Which the Output of the SVM Classifier Is Combined with the Segmentation Result Provided by the Recursive Hierarchical Segmentation (RHSEG) Algorithm Developed by James C. Tilton at NASA's Goddard Space Flight Center

Class	Training	Testing	SVM	SVM + RHSEG
Asphalt (6,631)	548	6,083	83.71	94.77
Meadows (18,649)	532	18,117	70.25	89.32
Gravel (2,099)	265	1,834	70.32	96.14
Trees (3,064)	231	2,833	97.81	98.08
Metal Sheets (1,345)	375	970	99.41	99.82
Bare Soil (5,029)	540	4,489	92.25	99.76
Bitumen (1,330)	392	938	81.58	100
Self-Blocking Bricks (3,682)	524	3,158	92.59	99.29
Shadow (947)	514	433	96.62	96.48
Overall accuracy	—	—	80.99	93.85

The numbers in the parentheses are the total number of available samples.

at a preprocessing step prior to classification. Finally, the results reported in Tables 12.4 and 12.5 correspond to cases in which spatial information is included at a postprocessing step after conducting spectral-based classification. As a result, the comparison reported in this section illustrates different scenarios in which spatial and spectral information are used in complementary fashion but following different strategies, that is, spatial information is included at different stages of the classification process (preprocessing, postprocessing, or kernel level).

After evaluating the importance of including spatial and spectral information, we now discuss the possibility to perform a better modeling of the hyperspectral data by working on a subspace. This is due to the fact that the dimensionality of the hyperspectral data is very high, and often, the data live in a subspace. Hence, if the proper subspace is identified prior to classification, adequate results can be obtained. In order to illustrate this concept, Table 12.6 shows the results obtained after comparing the SVM classifier with a subspace-based classifier such as the MLRsub [23], followed by an MRF-based spatial regularizer.

The idea of applying subspace projection methods relies on the basic assumption that the samples within each class can approximately lie in a lower-dimensional subspace. In the experiments reported in Table 12.6, it can also be seen that spatial information (included as an MRF-based postprocessing) can be greatly beneficial in order to improve classification performance.

Finally, it is worth noting that the results discussed in the previous text are all based on supervised classifiers that assume the sufficient availability of labeled training samples. In case that no sufficient labeled samples are available, semisupervised learning techniques can be used to generate additional unlabeled samples (from the initial set of labeled samples) that can complement the available labeled samples. The unlabeled samples can also be used to enhance subspace-based classifiers in case dimensionality issues are found to be relevant in the considered case

TABLE 12.6 Classification Results Obtained for the ROSIS Pavia University Scene by the SVM Classifier as Compared with a Subspace-Based Classifier Followed by Spatial Postprocessing (MLRsubMRF)

Class	Training	Testing	SVM	MLRsubMRF
Asphalt (6,631)	548	6,083	83.71	93.83
Meadows (18,649)	232	18,117	70.25	94.80
Gravel (2,099)	265	1,834	70.32	71.13
Trees (3,064)	231	2,833	97.81	92.17
Metal Sheets (1,345)	375	970	99.41	100
Bare Soil (5,029)	540	4,489	92.25	98.43
Bitumen (1,330)	392	938	81.58	99.32
Self-Blocking Bricks (3,682)	524	3,158	92.59	95.19
Shadow (947)	514	433	96.62	96.20
Overall accuracy	—	—	80.99	94.10

The numbers in the parentheses are the total number of available samples.

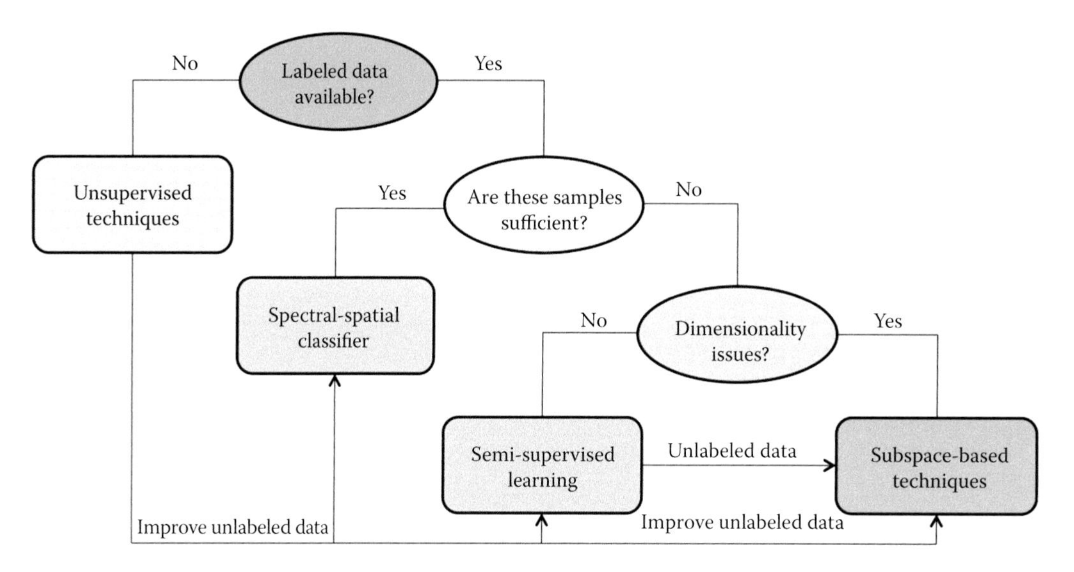

FIGURE 12.10 Summary of contributions in hyperspectral image classification discussed in this chapter.

study. If sufficient labeled samples are available, then the use of a spectral–spatial classifier is generally recommended as spatial information can provide a very important complement to the spectral information. Finally, in case that labeled samples are not available at all, unsupervised techniques need to be used for classification purposes. For instance, a relevant unsupervised method successfully applied to hyperspectral image data is Tilton's RHSEG algorithm.* The different analysis scenarios for classification discussed in this chapter are summarized in Figure 12.10.

12.4 Conclusions and Future Directions

In this chapter, we have provided an overview of recent advances in techniques and methods for hyperspectral image classification. The array of techniques particularly illustrated in this chapter comprises supervised and semisupervised approaches, techniques able to exploit both spatial and spectral information, and techniques able to take advantage of a proper subspace representation of the hyperspectral data before conducting the classification in spatial–spectral terms. These approaches represent a subjective selection of the wide range of techniques currently adopted for hyperspectral image classification [3], which include other techniques and strategies that have not been covered in detail in this chapter for space considerations. Of particular importance is the recent role of sparse classification methods [2,20], which are gaining significant popularity and will likely play an important role in this research area in upcoming years.

Despite the wide arrange of techniques and strategies for hyperspectral data interpretation currently available, some unresolved issues still remain. For instance, the geometry of hyperspectral data is quite complex and dominated by nonlinear structures. This issue has undoubtedly an impact in the outcome of the classification techniques discussed in this section. In order to mitigate this, manifold learning has been proposed [48]. An important property of manifold learning is that it can model

* http://opensource.gsfc.nasa.gov/projects/HSEG/.

and characterize the complex nonlinear structure of the data prior to classification [49]. Another remaining issue is the very high computational complexity of some of the techniques available for classification of hyperspectral data [50]. In other words, there is a clear need to develop efficient classification techniques that can deal with the very large dimensionality and complexity of hyperspectral data.

Acknowledgments

The authors thank Prof. Paolo Gamba, the University of Pavia, and the HySens project for providing the ROSIS data used in the experiments.

References

1. A. F. H. Goetz, G. Vane, J. E. Solomon, and B. N. Rock, Imaging spectrometry for Earth remote sensing, *Science*, 228(4704), 1147–1153, 1985.
2. J. M. Bioucas-Dias, A. Plaza, G. Camps-Valls, P. Scheunders, N. Nasrabadi, and J. Chanussot, Hyperspectral remote sensing data analysis and future challenges, *IEEE Geoscience and Remote Sensing Magazine*, 1(2), 6–36, 2013.
3. M. Fauvel, Y. Tarabalka, J. A. Benediktsson, J. Chanussot, and J. C. Tilton, Advances in spectral-spatial classification of hyperspectral images, *Proceedings of the IEEE*, 101(3), 652–675, 2013.
4. J. M. Bioucas-Dias, A. Plaza, N. Dobigeon, M. Parente, Q. Du, P. Gader, and J. Chanussot, Hyperspectral unmixing overview: Geometrical, statistical and sparse regression-based approaches, *IEEE Journal of Selected Topics in Applied Earth Observations and Remote Sensing*, 5(2), 354–379, 2012.
5. J. Chanussot, M. M. Crawford, and B.-C. Kuo, Foreword to the special issue on hyperspectral image and signal processing, *IEEE Transactions on Geoscience and Remote Sensing*, 48(11), 3871–3876, 2010.
6. A. Plaza, J. M. Bioucas-Dias, A. Simic, and W. J. Blackwell, Foreword to the special issue on hyperspectral image and signal processing, *IEEE Journal of Selected Topics in Applied Earth Observations and Remote Sensing*, 5(2), 347–353, 2012.
7. N. Keshava and J. Mustard, Spectral unmixing, *IEEE Signal Processing Magazine*, 19(1), 44–57, 2002.
8. W.-K. Ma, J. M. Bioucas-Dias, T.-H. Chan, N. Gillis, P. Gader, A. Plaza, A. Ambikapathi, and C.-Y. Chi, A signal processing perspective on hyperspectral unmixing: Insights from remote sensing, *IEEE Signal Processing Magazine*, 31(1), 67–81, 2014.
9. A. Plaza, J. A. Benediktsson, J. Boardman, J. Brazile, L. Bruzzone, G. Camps-Valls, J. Chanussot et al., Recent advances in techniques for hyperspectral image processing, *Remote Sensing of Environment*, 113(Suppl 1), 110–122, 2009.
10. D. A. Landgrebe, *Signal Theory Methods in Multispectral Remote Sensing*. Hoboken, NJ: Wiley, 2003.
11. L. O. Jimenez, J. L. Rivera-Medina, E. Rodriguez-Diaz, E. Arzuaga-Cruz, and M. Ramirez-Velez, Integration of spatial and spectral information by means of unsupervised extraction and classification for homogenous objects applied to multispectral and hyperspectral data, *IEEE Transactions on Geoscience and Remote Sensing*, 43(1), 844–851, 2005.
12. P. Gamba, F. Dell'Acqua, A. Ferrari, J. A. Palmason, J. A. Benediktsson, and J. Arnasson, Exploiting spectral and spatial information in hyperspectral urban data with high resolution, *IEEE Geoscience and Remote Sensing Letters*, 1(3), 322–326, 2004.
13. J. A. Richards and X. Jia, *Remote Sensing Digital Image Analysis: An Introduction*. New York: Springer-Verlag, 2006.
14. G. F. Hughes, On the mean accuracy of statistical pattern recognizers, *IEEE Transactions on Information Theory*, 14(1), 55–63, 1968.
15. G. Camps-Valls and L. Bruzzone, *Kernel Methods for Remote Sensing Data Analysis*. Hoboken, NJ: Wiley, 2009.
16. L. Bruzzone and C. Persello, A novel context-sensitive semisupervised SVM classifier robust to mislabeled training samples, *IEEE Transactions on Geoscience and Remote Sensing*, 47(7), 2142–2154, 2009.
17. J. Muñoz-Marí, D. Tuia, and G. Camps-Valls, Semisupervised classification of remote sensing images with active queries, *IEEE Transactions on Geoscience and Remote Sensing*, 50(10), 3751–3763, 2012.
18. D. Tuia and G. Camps-Valls, Semisupervised remote sensing image classification with cluster kernels, *IEEE Geoscience and Remote Sensing Letters*, 6(2), 224–228, 2009.
19. Y. Tarabalka, M. Fauvel, J. Chanussot, and J. Benediktsson, SVM- and MRF-based method for accurate classification of hyperspectral images, *IEEE Geoscience and Remote Sensing Letters*, 7(4), 736–740, 2010.
20. B. Song, J. Li, M. Dalla Mura, P. Li, A. Plaza, J. M. Bioucas-Dias, J. A. Benediktsson, and J. Chanussot, Remotely sensed image classification using sparse representations of morphological attribute profiles, *IEEE Transactions on Geoscience and Remote Sensing*, 52(8), 5122–5136, 2014.
21. J. Li, P. R. Marpu, A. Plaza, J. M. Bioucas-Dias, and J. A. Benediktsson, Generalized composite kernel framework for hyperspectral image classification, *IEEE Transactions on Geoscience and Remote Sensing*, 51(9), 4816–4829, 2013.
22. J. Li, J. M. Bioucas-Dias, and A. Plaza, Spectral-spatial classification of hyperspectral data using loopy belief propagation and active learning, *IEEE Transactions on Geoscience and Remote Sensing*, 51(2), 844–856, 2013.
23. J. Li, J. Bioucas-Dias, and A. Plaza, Spectral-spatial hyperspectral image segmentation using subspace multinomial logistic regression and Markov random fields, *IEEE Transactions on Geoscience and Remote Sensing*, 50(3), 809–823, 2012.
24. J. Bioucas-Dias and J. Nascimento, Hyperspectral subspace identification, *IEEE Transactions on Geoscience and Remote Sensing*, 46(8), 2435–2445, 2008.

25. I. Dopido, J. Li, P. R. Marpu, A. Plaza, J. M. Bioucas-Dias, and J. A. Benediktsson, Semi-supervised self-learning for hyperspectral image classification, *IEEE Transactions on Geoscience and Remote Sensing*, 51(7), 4032–4044, 2013.
26. T. V. Bandos, L. Bruzzone, and G. Camps-Valls, Classification of hyperspectral images with regularized linear discriminant analysis, *IEEE Transactions on Geoscience and Remote Sensing*, 47(3), 862–873, 2009.
27. F. Melgani and L. Bruzzone, Classification of hyperspectral remote sensing images with support vector machines, *IEEE Transactions on Geoscience and Remote Sensing*, 42(8), 1778–1790, 2004.
28. G. Camps-Valls and L. Bruzzone, Kernel-based methods for hyperspectral image classification, *IEEE Transactions on Geoscience and Remote Sensing*, 43, 1351–1362, 2005.
29. B. Schölkopf and A. Smola, *Learning with Kernels: Support Vector Machines, Regularization, Optimization and Beyond*. Cambridge, MA: MIT Press, 2002.
30. P. Soille, *Morphological Image Analysis, Principles and Applications*, 2nd edn. Berlin, Germany: Springer-Verlag, 2003.
31. M. Pesaresi and J. Benediktsson, A new approach for the morphological segmentation of high-resolution satellite imagery, *IEEE Transactions on Geoscience and Remote Sensing*, 39(2), 309–320, 2001.
32. J. Benediktsson, J. Palmason, and J. Sveinsson, Classification of hyperspectral data from urban areas based on extended morphological profiles, *IEEE Transactions on Geoscience and Remote Sensing*, 43(3), 480–491, 2005.
33. G. Camps-Valls, L. Goméz-Chova, J. Muñoz-Marí, J. Vila-Francés, and J. Calpe-Maravilla, Composite kernels for hyperspectral image classification, *IEEE Geoscience and Remote Sensing Letters*, 3, 93–97, 2006.
34. M. Khodadadzadeh, J. Li, A. Plaza, H. Ghassemian, J. Bioucas-Dias, and X. Li, Spectral–spatial classification of hyperspectral data using local and global probabilities for mixed pixel characterization, *IEEE Transactions on Geoscience and Remote Sensing*, 52(10), 6298–6314, 2014.
35. Y. Tarabalka, J. Chanussot, and J. Benediktsson, Segmentation and classification of hyperspectral images using watershed transformation, *Pattern Recognition*, 43, 2367–2379, 2010.
36. Y. Tarabalka, J. A. Benediktsson, J. Chanussot, and J. C. Tilton, Multiple spectral-spatial classification approach for hyperspectral data, *IEEE Transactions on Geoscience and Remote Sensing*, 48(11), 4122–4132, 2011.
37. D. Böhning, Multinomial logistic regression algorithm, *Annals of the Institute of Statistics and Mathematics*, 44, 197–200, 1992.
38. F. Bovolo, L. Bruzzone, and L. Carlin, A novel technique for subpixel image classification based on support vector machine, *IEEE Transactions on Image Processing*, 19, 2983–2999, 2010.
39. L. Bruzzone, M. Chi, and M. Marconcini, A novel transductive SVM for the semisupervised classification of remote sensing images, *IEEE Transactions on Geoscience and Remote Sensing*, 11, 3363–3373, 2006.
40. G. Camps-Valls, T. Bandos, and D. Zhou, Semi-supervised graph-based hyperspectral image classification, *IEEE Transactions on Geoscience and Remote Sensing*, 45, 3044–3054, 2007.
41. S. Velasco-Forero and V. Manian, Improving hyperspectral image classification using spatial preprocessing, *IEEE Geoscience and Remote Sensing Letters*, 6, 297–301, 2009.
42. J. Li, J. Bioucas-Dias, and A. Plaza, Semi-supervised hyperspectral image segmentation using multinomial logistic regression with active learning, *IEEE Transactions on Geoscience and Remote Sensing*, 48(11), 4085–4098, 2010.
43. J. Muñoz Marí, F. Bovolo, L. Gómez-Chova, L. Bruzzone, and G. Camp-Valls, Semisupervised one-class support vector machines for classification of remote sensing data, *IEEE Transactions on Geoscience and Remote Sensing*, 48(8), 3188–3197, 2010.
44. L. Gómez-Chova, G. Camps-Valls, L. Bruzzone, and J. Calpe-Maravilla, Mean MAP kernel methods for semisupervised cloud classification, *IEEE Transactions on Geoscience and Remote Sensing*, 48(1), 207–220, 2010.
45. D. Tuia and G. Camps-Valls, Urban image classification with semisupervised multiscale cluster kernels, *IEEE Journal of Selected Topics in Applied Earth Observations and Remote Sensing*, 4(1), 65–74, 2011.
46. F. Ratle, G. Camps-Valls, and J. Weston, Semisupervised neural networks for efficient hyperspectral image classification, *IEEE Transactions on Geoscience and Remote Sensing*, 48(5), 2271–2282, 2010.
47. G. Camps-Valls, D. Tuia, L. Gómez-Chova, S. Jiménez, and J. Malo, *Remote Sensing Image Processing*. San Rafael, CA: Morgan and Claypool, 2011.
48. L. Ma, M. Crawford, and J. Tian, Local manifold learning based k-nearest-neighbor for hyperspectral image classification, *IEEE Transactions on Geoscience and Remote Sensing*, 48(11), 4099–4109, 2010.
49. W. Kim and M. Crawford, Adaptive classification for hyperspectral image data using manifold regularization kernel machines, *IEEE Transactions on Geoscience Remote Sensing*, 48(11), 4110–4121, 2010.
50. A. Plaza, J. Plaza, A. Paz, and S. Sanchez, Parallel hyperspectral image and signal processing, *IEEE Signal Processing Magazine*, 28(3), 119–126, 2011.

13

Thermal Infrared Remote Sensing: Principles and Theoretical Background

Claudia Kuenzer
German Remote Sensing Data Center, DFD of DLR

Jianzhong Zhang
Kaufering, Germany

Stefan Dech
German Remote Sensing Data Center, DFD of DLR and University of Wuerzburg

Acronyms and Definitions

°	Degree
μm	Micrometer
AQUA	Platform on which the MODIS sensor is flown
ASTER	Advanced Spaceborne Thermal Emission and Reflection Radiometer
ATI	Apparent thermal inertia
AVHRR	Advanced Very-High-Resolution Radiometer
C	Celsius
DLR	German Aerospace Center
DN	Digital number
ε	Emissivity
ETM+	Enhanced Thematic Mapper Plus
K	Kelvin
kg	Kilogram
LST	Land surface temperature
m	Meter
MODIS	Moderate-Resolution Imaging Spectroradiometer
MSG	Meteosat Second Generation
NASA	National Aeronautics and Space Administration
NIR	Near infrared
nm	Nanometer
NOAA	National Oceanic and Atmospheric Administration
s	Second
SEVIRI	Spinning Enhanced Visible and Infrared Imager
SST	Sea surface temperature
SWIR	Shortwave infrared
T	Temperature
TERRA	Platform on which the MODIS sensor is flown
TIR	Thermal infrared
TM	Thematic Mapper
VIS	Visible
W	Watt
λ	Wavelength

13.1 Introduction

It is temperature that defines the habitat boundaries of plants, animals, and humans. Temperature—among other variables—defines global, regional, and local climate and weather, the rate of sea level rise, the magnitude of forest fire risk, and the reproduction conditions for bacteria and viruses, and it even defines our cultural preferences. Housing style, preferred means of transport, customs of eating, and free-time behavior focusing on the inside or the outside, as well as the attractiveness of tourist destinations, are all defined by temperature. Temperature strongly impacts our well- or malbeing, and extreme cold spills or heat waves have proven to negatively impact local economies and even lead to fatalities.

Remote sensing in the thermal infrared (TIR) wavelength domain is an often neglected subdiscipline within the field of remote sensing. Reasons for this range from a limited availability of spaceborne sensors acquiring data in the TIR domain to the, generally, lower spatial resolution of this data, when compared with other, for example, optical data, to last but not least different data analyses approaches, which are mandatory when dealing with TIR data. Thus, it is the goal of this chapter to provide a comprehensive overview of the principles of remote sensing in the TIR. We address the theoretic background of the TIR domain, including important physical laws; address parameters such as kinetic and radiance temperature, emissivity, and thermal inertia; discuss the preprocessing, the analyses, as well as the validation of thermal data; and present a range of illustrative application examples. Much of this chapter has been reformulated based on Kuenzer and Dech (2013).

In general, electromagnetic radiation is emitted by all objects that have a temperature above 0 K (equals –273°C). Depending on the temperature of the object, it has its peak of electromagnetic emittance in a certain wavelength domain. The peak of emission of our planet, which has an average temperature of about 300 K, is located in the TIR domain at a wavelength of about 9.7 μm (Sabins, 1996; Tipler, 2000). Objects on Earth absorb the sun's incoming radiation and emit corresponding amount at longer wavelengths in the TIR domain.

Several sensors, such as Landsat-5 Thematic Mapper (TM), contain bands (detectors), which are responsive in the thermal domain and record TIR radiation emitted by objects (see Figure 13.1). The thermal imagery then represents the kinetic temperature of objects at a certain spatial resolution. We can differentiate between different categories of temperature. Land surface temperature (LST) describes the temperature at the land surface. It represents the temperature of object's surfaces—be it meadows, house roofs, forest canopies, or inland water bodies. Typical LST-based thermal analyses focus on LST patterns as global or continental scale, at local scale on urban heat island (UHI) studies, assessments of forest fires, grassland fires, gas flares, underground coal fires, or geothermal phenomena, the observation of nuclear accident sites (e.g., Chernobyl, Fukushima), or the monitoring of industry. Contrary to LST, sea surface temperature (SST) is the temperature of the upper water layer of an ocean or large inland sea (Dech et al., 1998). The SST datasets are usually exploited for an improved understanding of global circulations and for analyses in the context of algae blooms or, for example,

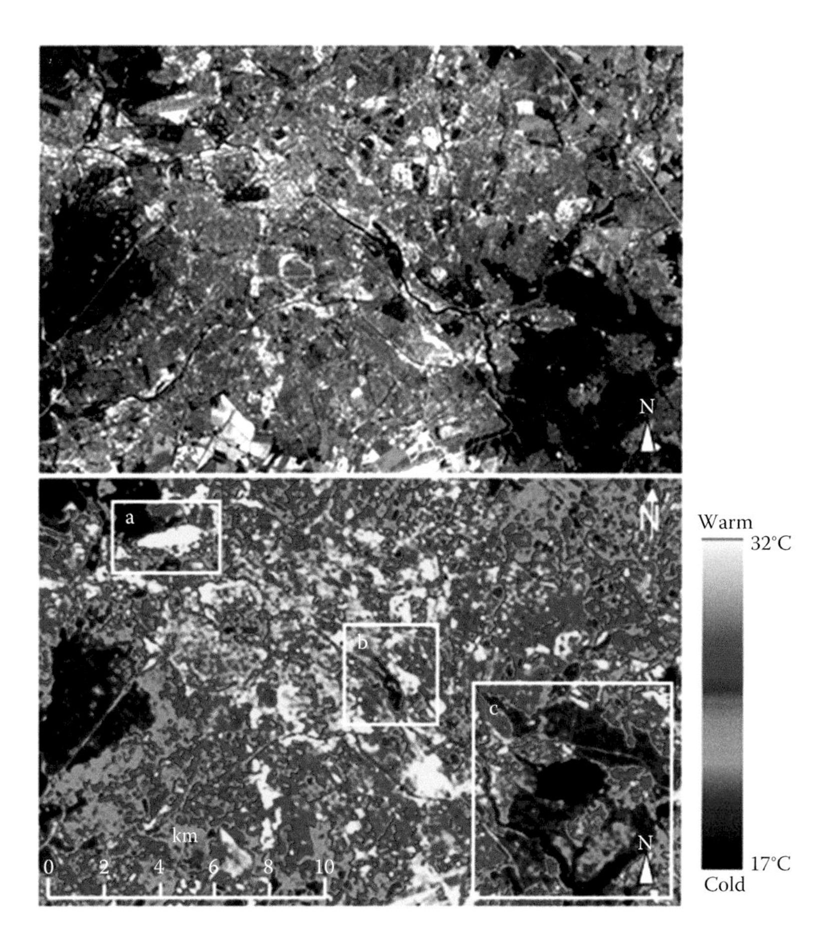

FIGURE 13.1 Thermal daytime land surface temperature image of Berlin, acquired on June 2, 2011, by the Landsat-5 Thematic Mapper sensor. The subset areas depict the Berlin airport "Tegel" (a), the river "Spree" (b), and a large lake named "Müggelsee" and its surrounding forest (c), Water surfaces and densely vegetated areas make up the cooler places within this daytime thermal image, while areas with a high percentage of artificial surfaces appear warmest. Coordinates: UL: 52°32′25N, 13°09′59E; LR: 52°22′20N, 13°44′14E.

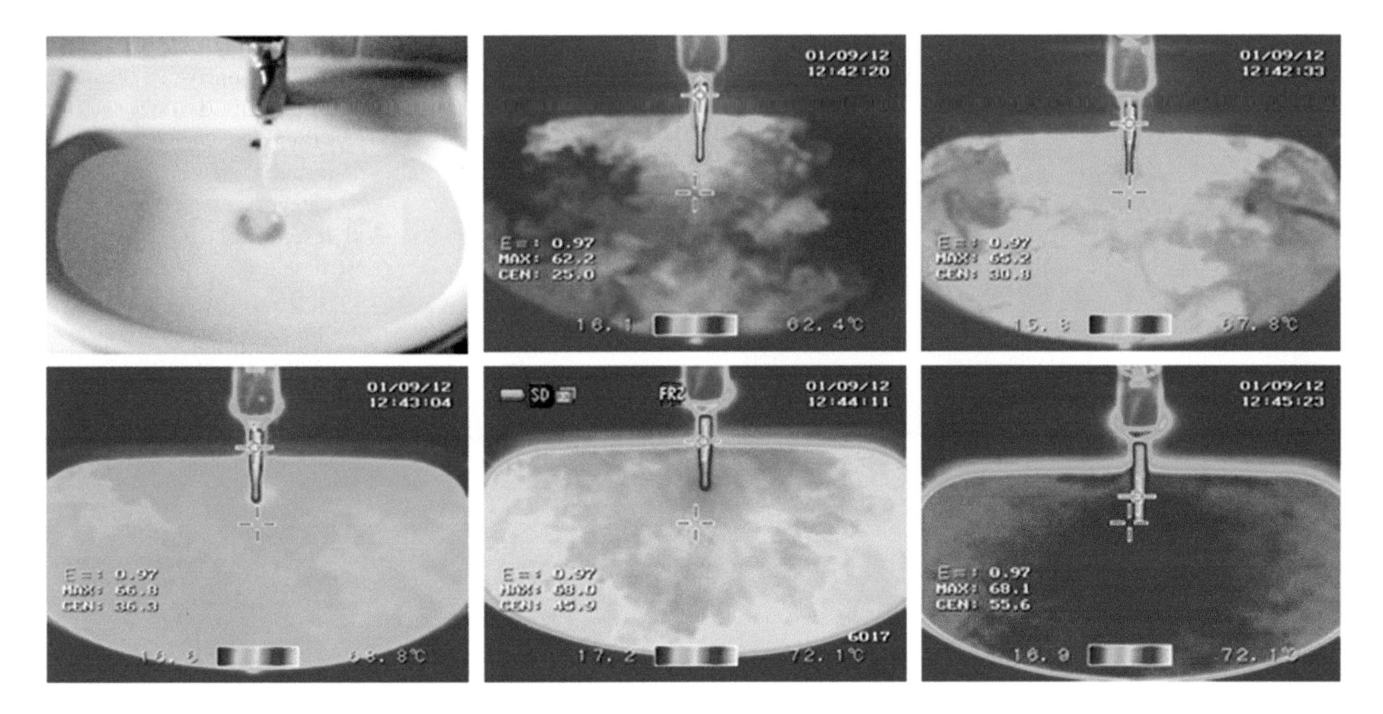

FIGURE 13.2 Landsat Thematic Mapper 5 daytime image subsets depicting an agricultural area in northern Italy, acquired on June 12, 2009 (a); July 14, 2009 (b); and August 31, 2009 (c). The upper row presents true color composites in RGB, whereas the lower row presents the color-coded Thermal infrared band 6.

thermal water pollution (see Figure 13.2 for a handheld camera example).

However, thermal sensor data are also employed outside of the geosciences. In medical imaging, inflamed areas can be detected due to their higher temperature compared with their surroundings. In industry, machine performance is monitored with thermal cameras. Architects use thermal imagery to detect energy leaks in buildings, and also police and the military use thermal cameras for object detection. Several examples for the use of thermal camera imagery are illustrated by Figure 13.3.

A large variety of TIR sensors data is available in data archives, as well as for current tasking (Kuenzer et al., 2013a). Figure 13.4 presents all platforms and instruments that allow for data acquisition in the TIR domain. We can see that especially the fleet of Advanced Very-High-Resolution Radiometer (AVHRR), as well as Landsat sensors, enable a long-term monitoring of our planet in the TIR domain (Frey et al., 2012). Most TIR data are available from American sensors, and good access is especially granted to data of AVHRR, Moderate-Resolution Imaging Spectroradiometer (MODIS), Landsat, and Advanced Spaceborne Thermal Emission and Reflection Radiometer (ASTER). Also numerous Chinese sensors record data in the TIR domain; however, here, data access is not as easy. A detailed overview of TIR-related instruments and their preferred application domain can be found in Table 13.1 as well as in Kuenzer et al. (2013a). Abbreviations can be found in the abbreviations section of this book.

13.2 Principles, Theoretical Background, and Important Laws

13.2.1 Thermal Infrared Domain

Different authors define the TIR domain differently, so an overall valid strict, physical definition does not exist. Sabins (1996) defines the TIR range from about 3 to 13 μm as a wavelength range, in which two important atmospheric windows are located (see Figure 13.5). In the 8–14 μm window, only ozone absorption occurs, which is omitted by most sensors. In the 3–5 μm range, reflected sunlight can still slightly contaminate the emitted thermal signal, so the data have to be interpreted with care. However, according to Lillesand and Kiefer (1994) and Löffler (1994), the TIR domain ranges from 3 to 1000 μm. Common to all authors is that they define TIR remote sensing as the field that deals with emitted radiation, whereas multispectral remote sensing in the visible (VIS) and near-infrared (NIR) domain records reflected radiation.

13.2.2 Important Laws: Planck

Planck's law, or Planck's blackbody radiation law as it is officially termed, describes the electromagnetic radiation emitted by a blackbody at a given wavelength, M, as a function of the blackbody's temperature (Planck, 1900). A blackbody is a theoretical concept and does not exist in reality. It is an ideal radiator, which reemits all energy it absorbs. However, there are surfaces on Earth that show "near blackbody-like"

(a)

(b)

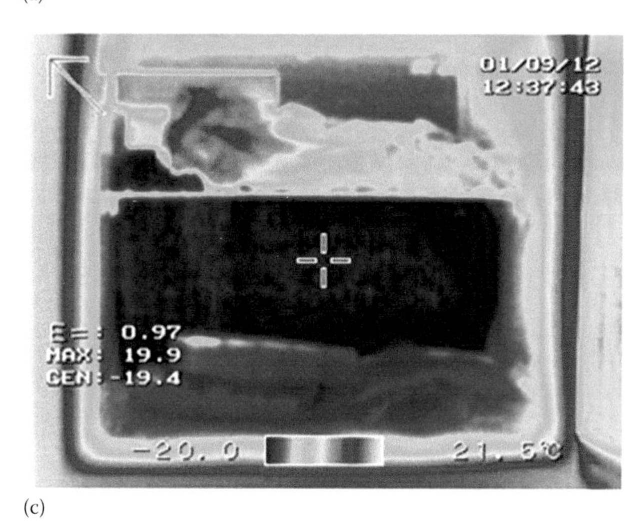

(c)

(d)

FIGURE 13.3 Thermal camera images. (a) Daytime image of a high-rise residential area in Beijing, China, acquired during a cold wave in the winter of 2012. While the outside air and background indicate temperatures of well below 0°C (up to −20°C), it can be seen that heat from the inside of the apartments penetrates through the windows, so that temperatures at the outside of the building reach about −3°C, (b) daytime image of a skyscraper façade acquired in the same residential area: the surface temperature of the façade is well below 0°C (down to −11°C); however, from two windows, heat penetrates to the outside. This heat originates from air-condition systems people install privately above their windows (inside), which are used as cooling devices in summer and heating devices in winter. Temperatures of up to 9.8°C occur outside, (c) picture of an opened freezer. Temperatures in the freezer go down to −20°C. The lower part of the freezer is colder than the upper part (cold air sinks down). On the outside of the freezer, temperatures of up to 21.5°C are reached, (d) tree-lined pathway in the village of Gilching, near Munich. In the evening around sunset, cemented surfaces, as well as vegetation, appear warmest. (Photographs: C. Kuenzer, 2012.)

behavior for certain wavelengths. The Planck formula allows to calculate the emitted radiation, *M*, by inserting a certain wavelength as well as the body's temperature. It also allows deriving a blackbody's temperature, if *M* and the wavelength are known (Equation 13.1):

$$M_\lambda = \frac{2\pi hc^2}{\lambda^5\left(e^{hc/\lambda kT} - 1\right)} \tag{13.1}$$

where

- M_λ is the spectral radiant exitance (W m^{-2} μm^{-1})
- h is Planck's constant (6.626×10^{-34} J s)
- c is the speed of light (2.9979246×10^{8} m s^{-1})
- k is Boltzmann constant (1.3806×10^{-23} J K^{-1})
- T is the absolute temperature (K)
- λ is the wavelength (μm)

The Stefan–Boltzmann law (following) as well as Wien's law (following as well) allows to calculate the total energy a theoretical blackbody radiates, as well as its wavelength of maximum emittance (Tipler, 2000; Walker, 2008).

13.2.3 Important Laws: Stefan–Boltzmann

The Stefan–Boltzmann enables to calculate the total energy a theoretical blackbody radiates, as a function of its temperature. As depicted in Figure 13.6, the emitted radiation is described by

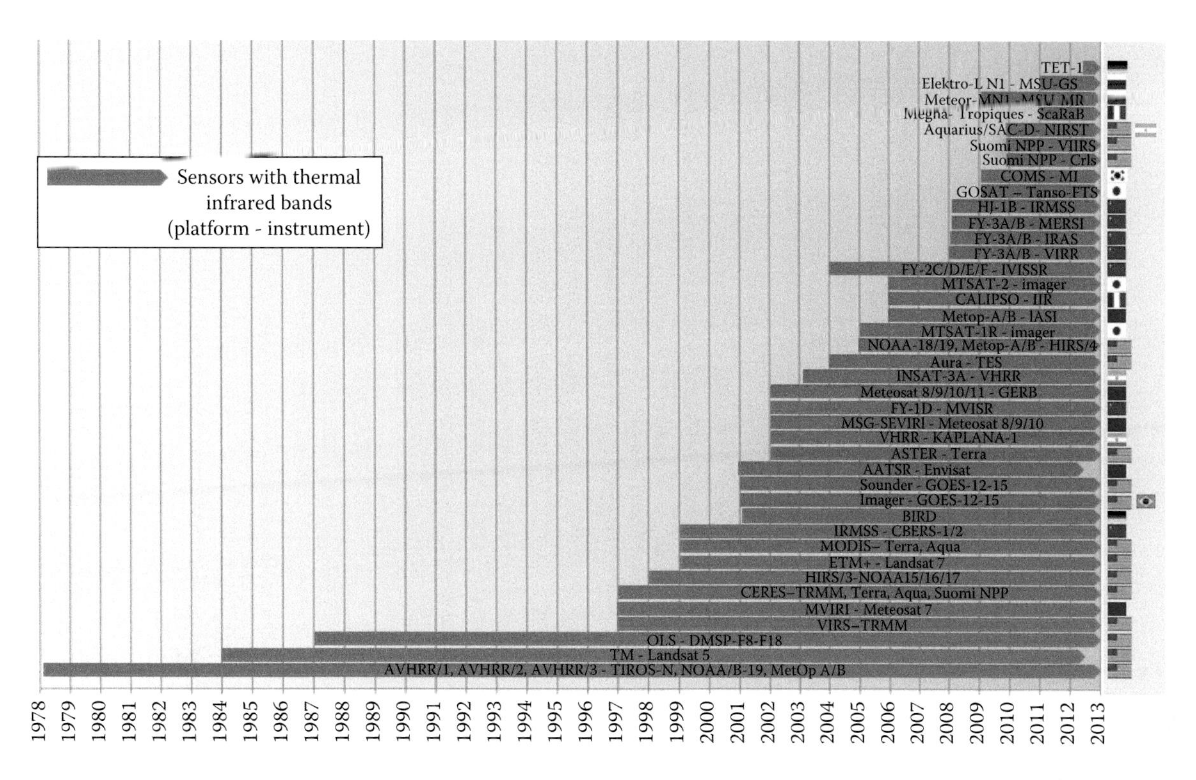

FIGURE 13.4 Satellite sensors with bands sensitive in the thermal infrared until (documentation until the end of 2012). The chart depicts the sensor's start and ending dates, name of the platform and the instrument, and the country of origin.

the area under the radiation curve. The larger the overall energy an object radiates, the higher its temperature. As Formula 13.2 elucidates this, the relationship between temperature and energy is not linear, but is described by a fourth power relationship (see the following equation):

$$T_{\text{RadBB}} = \sigma T_{\text{kin}}^4 \tag{13.2}$$

where

T_{RadBB} is the radiant flux of a blackbody (W m^{-2})
T is the absolute kinetic temperature (K)
σ is Stefan–Boltzmann constant (5.6697×10^{-8} W m^{-2} K^{-4})

13.2.4 Important Laws: Wien

The law of Wien, also sometimes called Wien's displacement law, describes the wavelength of maximum spectral exitance (in μm) as a function of an object's temperature (see Equation 13.3). The hotter an object, the further shifts its maximum exitance toward shorter wavelengths (Tipler, 2000; Heald, 2003; Walker, 2008). This has also already been demonstrated in Figure 13.6. The sun has an average temperature of 5505°C (or 5778 K), and its peak of emission is located in the VIS domain of the electromagnetic spectrum at 0.55 μm. Colder objects, such as Earth, have their peak of emission in the TIR.

$$\lambda_{\max} = \frac{A}{T} \tag{13.3}$$

where

$\lambda_{\max}$ is the wavelength of maximum spectral radiant exitance (μm)
A is Wien's constant (2897.8 μm K)
T is the absolute kinetic temperature (K)

Wien's displacement law can be well demonstrated on multispectral remote-sensing imagery depicting areas of different temperature, including extreme hot spots. Figure 13.7 presents several bands of a Landsat-5 TM image of the Etna volcano in Italy, acquired on July 29, 2001. In Figure 13.7a, we can see the volcano and its main chimney on the top. Clouds of smoke are emanating from the volcano. We can see areas of bare soil on the left flank of the peak, as well as some vegetated patches. Lava flows cannot be observed. In the temperature image displayed in Figure 13.7 on the lower right (band 6 of Landsat-5 TM, representing the TIR domain from 10.4 to 12.5 μm), we can see that the smoke clouds are warm, whereas the flanks of the volcano are much colder. Vegetated areas appear coldest. At the same time, we can observe some linear structures in the upper center of this thermal image, which do not seem to belong to the smoke itself, but cannot further differentiate this. Here, now band 7—the shortwave infrared (SWIR) band 7 (2.09–2.35 μm)—comes into play. As depicted in Figure 13.7c. Here, we can see that lava flows of extreme temperature lead to an elevated signal in this band 7 (Wien's law). The material is so hot that its peak of emission does not occur in the TIR domain from 10.4 to 12.5 μm, but rather in the SWIR. At the same time, the clearly warm clouds (Figure 13.7d) appear dark in Figure 13.7c. The reason is the high water vapor content of the clouds. Furthermore, they are

TABLE 13.1 Typical Sensors and Their Characteristic Used for the Analyses of Land Surface Temperature and Related Applications

Sensor	Spatial Resolution	Revisit Time (Days)	Swath Width	Platform	Agency	Launch Date
ETM+	15–60 m	16	185 km	Landsat 7	USGS/NASA	1999
TM	30–120 m	16	185 km	Landsat 5	USGS/NASA	1984
ASTER	15–90 m	4–16	60 km	Terra	NASA	1999
IRSCAM	40–89 m	26	60 km	CBERS-3	CRESDA/INPE	2012(launch end of 2012)
TIRS	100 m	16	185 km	Landsat 8 (LDCM)	USGS/NASA	2013(launch February 2013)
IRMSS	156 m	26	120 km	CBERS-2	CRESDA/INPE	2003
BIRD	185–370 m	1–3.5	1440 km	BIRD	DLR	2001
MODIS	250–1000 m	1–2	2330 km	Terra, Aqua	NASA	1999, 2002
MERSI	250–1.1 km	1	2800 km	FY-3A, FY-3B	CMA/CNSA	2008, 2010
NIRST	350 m	7	182–1000 km	Aquarius	NASA/CONAE	2011
VIIRS	375–750 m	16	3000 km	Suomi NPP	NASA/NOAA	2011
AATSR	1000 m	35	500 km	Envisat	ESA	2002
AVHRR/1	1.1 km	<1	2600 km	TIROS-N, NOAA6,8,10	NOAA	1978–1986
AVHRR/2	1.1 km	<1	3000 km	NOAA 7,9,11,12,14	NOAA	1981–1994
AVHRR/3	1.1 km	<1	2500 km	NOAA15-19, MetOp A,B	NOAA, EUMETSAT	1998–2012
MSG-SEVIRI	1–3 km	<1	Full earth disk	Meteosat-8/9/19	ESA/EUMETSAT	2002, 2005, 2012
MVISR	1.1 km	3–4	3200 km	FY-1D	CMA/NRSCC	2002
IVISSR	5 km	1	Full earth disk	FY-2D, FY-2E, FY-2F	NRSCC/CAST/ NSMC-CMA	2006,2008,2012
VHRR	8 km	<1	Full earth disk	Kaplana-1, Insat-3A	ISRO	2002, 2003
IASI	30 km	1	2052 km	MetOP-A/B	EUMETSAT	2006, 2012
HIRS/3	20.3 km		2240 km	NOAA 15-17	NOAA	1998, 2000, 2002
HIRS/4	20.3 km		2240 km	NOAA 18/19, MetOp A/B	NOAA, EUMETSAT	2005, 2006, 2009, 2012

Source: Kuenzer, C., Guo, H., Ottinger; M., Dech, S., Spaceborne thermal infrared observation—An overview of most frequently used sensors for applied research, in Kuenzer, C., and Dech, S., eds., *Thermal Infrared Remote Sensing—Sensors, Methods, Applications*, Remote Sensing and Digital Image Processing Series, Vol. 17, 572 pp., pp. 131–148, 2013a.

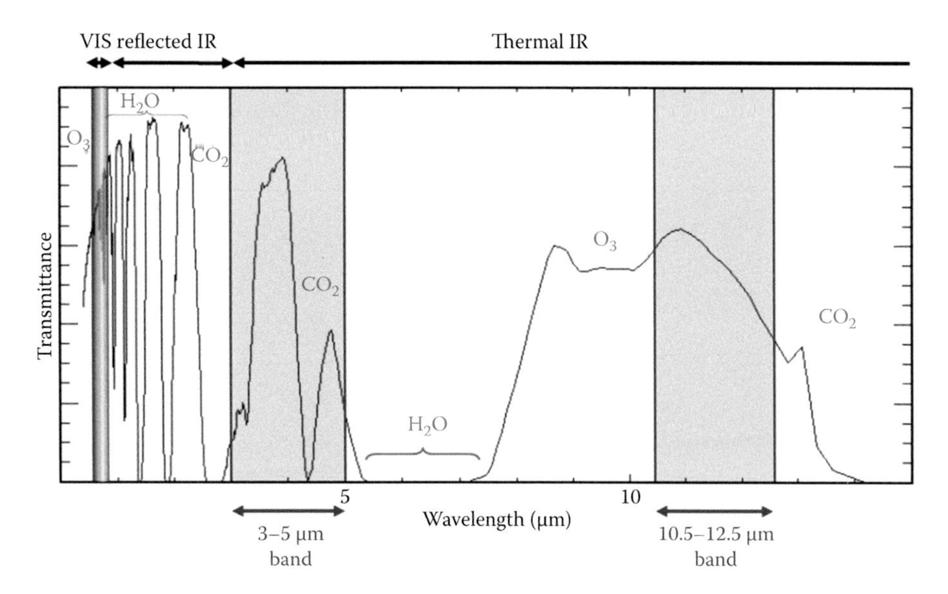

FIGURE 13.5 The diagram depicts the thermal infrared wavelength domain, typical absorption bands induced by gasses and water, and atmospheric transmittance (atmospheric windows). (Modified from Kuenzer, C. and Dech, S., Theoretical background of thermal infrared remote sensing, in Kuenzer, C. and Dech, S., eds., *Thermal Infrared Remote Sensing—Sensors, Methods, Applications*, Remote Sensing and Digital Image Processing Series, Vol. 17, 572 pp., Springer, Dordrecht, pp. 1–26, 2013.)

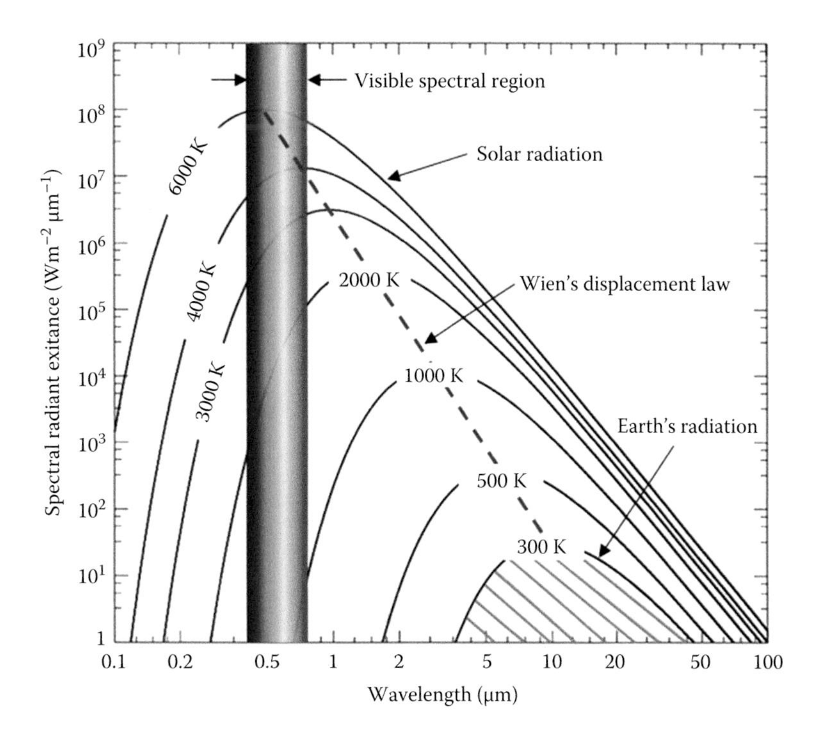

FIGURE 13.6 Blackbody radiation curves for blackbodies with different temperatures, as derived from Planck's equation. Stefan–Boltzmann's equation describes the area under the curves. The rainbow-colored bar marks the VIS region. (Modified from Kuenzer, C. and Dech, S., Theoretical background of thermal infrared remote sensing, in Kuenzer, C. and Dech, S., eds., *Thermal Infrared Remote Sensing—Sensors, Methods, Applications*, Remote Sensing and Digital Image Processing Series, Vol. 17, 572 pp., Springer, Dordrecht, pp. 1–26, 2013.)

not hot enough to elevate the signal in the SWIR; they do have the peak of emittance in the TIR. Creating a color composite of bands 7, 5, and 4 now allows to beautifully depict hottest lava (orange), warm smoke (white), former cooled lava flows (black), and sparse and dense vegetation (blue).

13.2.5 Important Laws: Kirchhoff and the Role of Emissivity

Planck's law describes blackbody radiation, but it has already been stated that blackbodies are fictive objects, as most objects on our planet emit less energy than would be

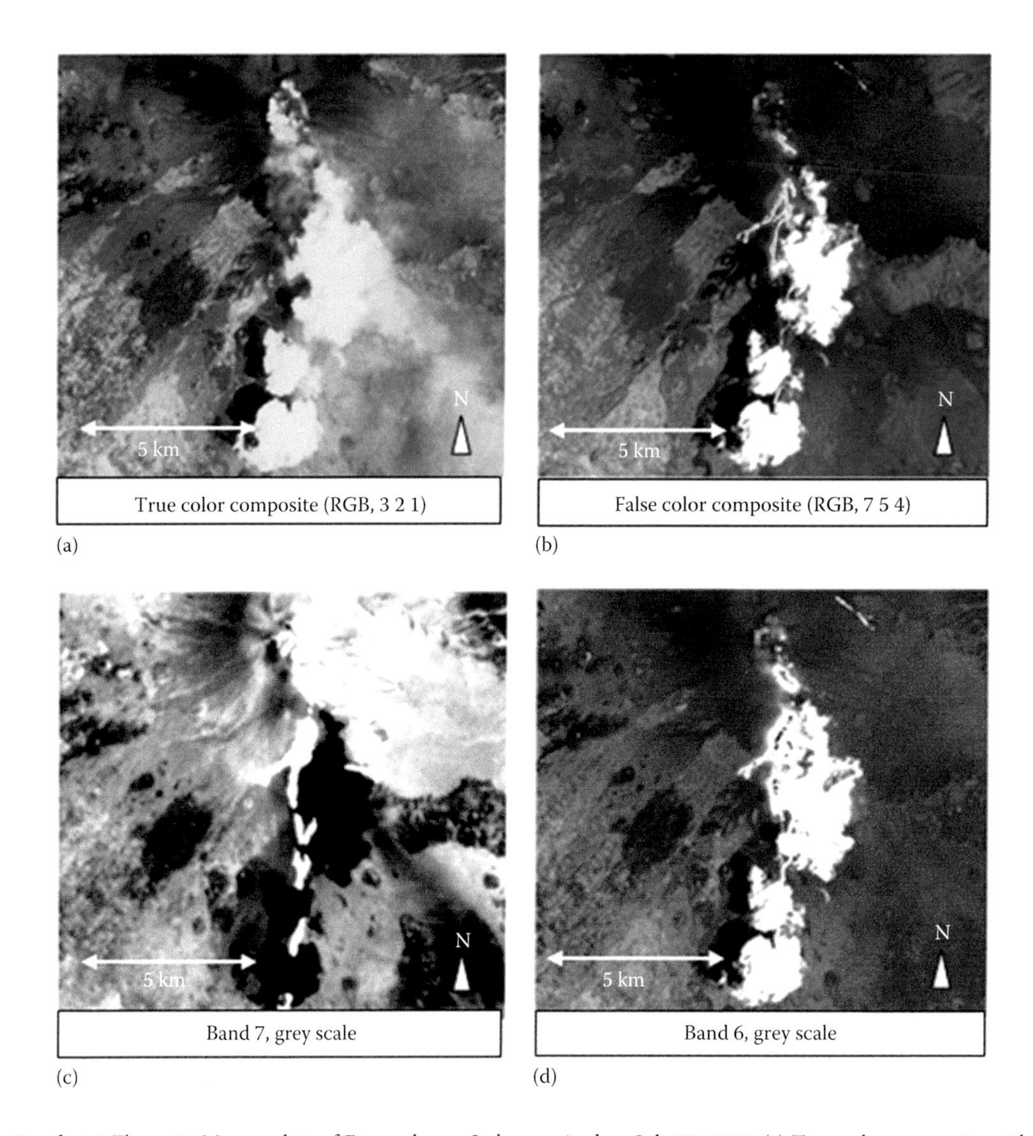

FIGURE 13.7 Landsat-5 Thematic Mapper data of Etna volcano, Italy, acquired on July 29, 2001. (a) True color composite with the red, green, and blue bands displayed in RGB; (b) false color composite with the two SWIR bands 7 and 5 and NIR band 4 displayed in RGB; (c) grayscale image of the shortwave infrared band 7 (2.09–2.35 μm); and (d) grayscale image of the thermal infrared band 6 (10.40–12.50 μm). Coordinates: UL: 37°50′32N, 14°52′11E; LR: 37°39′30N, 15°05′07E.

predicted based on their kinetic temperature. The so-called emissivity coefficient, ($\varepsilon(\lambda)$), is taking this fact into account and is defined at the radiant flux of an object at a given temperature over the radiant flux of a blackbody at the same temperature. A perfect blackbody would emit all radiation it absorbed, and Kirchhoff's law states that—for a blackbody—emittance and absorbance at a given wavelength are equal (Kirchhoff, 1860):

$$\varepsilon(\lambda) = \alpha(\lambda) \tag{13.4}$$

As, according to energy conservation, the sum of absorption (α), reflection (ρ), and transmission (τ) equals 1, and considering Kirchoff's law, we can postulate that

$$\varepsilon(\lambda) + \rho(\lambda) + \tau(\lambda) = 1 \tag{13.5}$$

However, most solid objects (except, e.g., water and leafs) do not transmit radiation. They are opaque, so Equation 13.5 can be reformulated to

$$\varepsilon(\lambda) + \rho(\lambda) = 1 \tag{13.6}$$

This means that an object's (blackbody's) reflectance allows to calculate its emittance and vice versa. Materials with low ε absorb and radiate lower amounts of energy, whereas materials with a high ε absorb large amounts of incident energy and radiate large quantities of energy (Kirchhoff, 1860; Sabins, 1996). Emissivity varies depending on surface type and wavelength, but is not temperature dependent (Becker, 1987). Table 13.2 illustrates the varying emissivities of common surfaces for the wavelength region 8–14 μm (averaged).

TABLE 13.2 Emissivity of Different Surfaces in the 8–14 μm Wavelength Range as Compiled from Different Sources (Own Measurements and Lillesand et al., 2008; Sabins, 1996; Hulley et al., 2009)

Surface	Emissivity at 8–14 μm
Carbon powder	0.98–0.99
Water	0.98
Ice	0.97–0.98
Plant leaves, healthy	0.96–0.99
Plant leaves, dry	0.88–0.94
Asphalt	0.96
Sand	0.93
Basalt	0.92
Wood	0.87
Granite	0.83–0.87
Polished metals, averaged	0.02–0.21
Aluminum foil	0.036

Source: Kuenzer, C. and Dech, S., Theoretical background of thermal infrared remote sensing, in Kuenzer, C. and Dech, S. (eds.) *Thermal Infrared Remote Sensing—Sensors, Methods, Applications*, Remote Sensing and Digital Image Processing Series, Vol. 17, 572 pp., pp. 1–26, 2013.

As can be seen, for all real materials, emissivity is below 1. This means that the radiance temperature, $T_{(rad)}$, measured by a thermal sensor (e.g., handheld thermal camera, airborne thermal scanner, thermal detector on board a satellite), is always lower than the real kinetic (surface) temperature, $T_{(kin)}$ (in which one would measure with a contact thermometer), of the object. This also means that objects with exactly the same kinetic temperature on the ground can exhibit a very different radiant temperature. If we consider a typical land surface with different geologic surfaces, different types of vegetation, and differing moisture conditions and maybe even containing a large variety of construction materials (e.g., urban areas), it is obvious that a remotely acquired thermal image does not represent the true kinetic temperatures of these objects unless the data are corrected for emissivity effects. This is especially crucial for all TIR applications in urban areas as the emissivities of different construction materials vary considerably. Emissivities of metals (e.g., aluminum, tin, copper, in some areas used as roof materials) are extremely low. This will lead to the fact that temperatures appear much lower than the sensed temperatures of the surrounding objects of similar kinetic temperature (see also Figures 13.8 and 13.9). On the contrary, objects with a very high emissivity, such as water and vegetated surfaces, allow for a pretty exact assessment of their kinetic temperature. The conversion of radiance temperature to kinetic temperature can be undertaken as follows:

$$T_{(rad)} = \varepsilon^{(1/4)} * T_{(kin)} \quad (13.7)$$

where

$T_{(rad)}$ is the radiance temperature
$T_{(kin)}$ is the kinetic temperature
ε is the emissivity

Also Figure 13.9 visualized this phenomenon very well, based on an imminent example. A handheld thermal camera photograph is taken. The photograph shows a concrete stairs with a polished metal railing and a male hand with a wedding ring on one finger. The picture was acquired on a warm day in late summer, with an ambient outside air temperature of 22°C. The human hand depicts temperature values of 37°C (as would be expected). Note that even the veins within the hand can be seen (white, hot lines). Also the stairs in the background appears warm. However, the handrail appears at a radiance temperature of −8°C (cross at image center), although it is definitely not that cold. However, as this picture was acquired with a standard

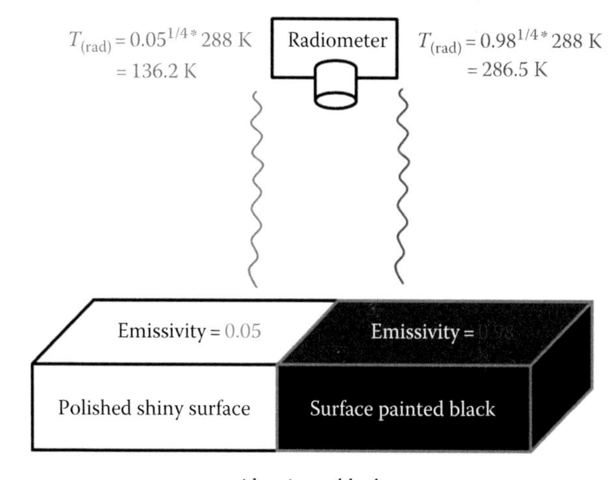

FIGURE 13.8 Emissivity strongly impacts the temperature recorded at a (remote, noncontact) sensor. In this figure modified after Sabins (1996), a block of aluminum with a homogeneous kinetic temperature and a very low emissivity is covered with carbon-rich dark paint (on its right side). This paint has a very different emissivity than the polished aluminum block. Although the object is 15°C warm, it appears as −136.9°C on the uncovered side and as 13.3°C on the painted side (calculation based on Equations 13.2 and 13.7).

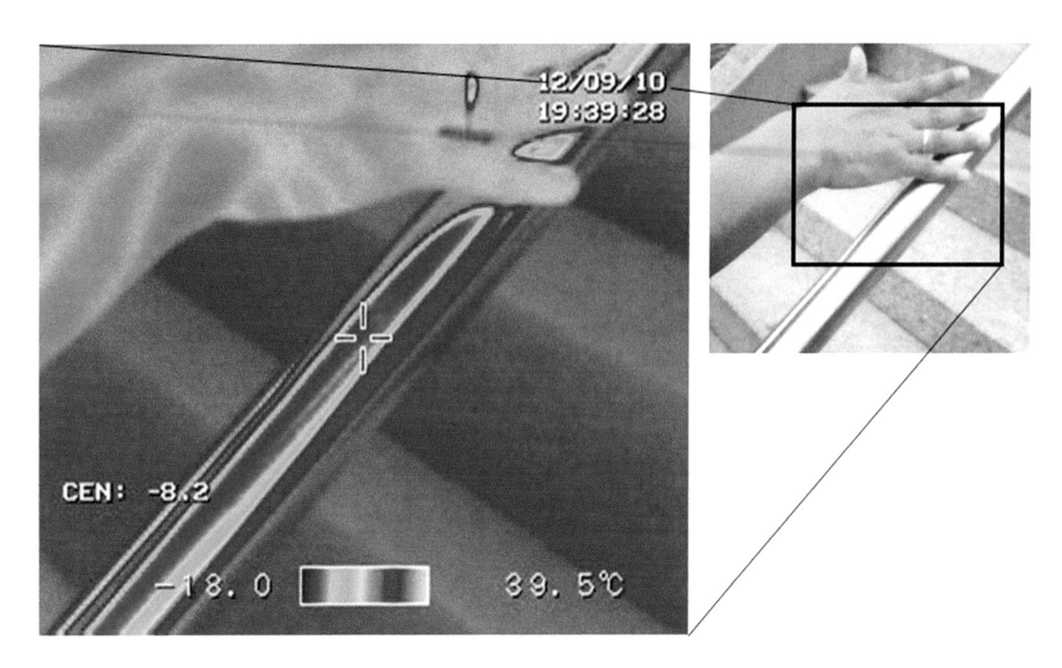

FIGURE 13.9 Impact of emissivity differences on radiance temperature recorded at the sensor of a handheld thermal camera. Picture taken by C. Kuenzer in September, 2012. (Modified from Kuenzer, C. and Dech, S., Theoretical background of thermal infrared remote sensing, in Kuenzer, C. and Dech, S., eds., *Thermal Infrared Remote Sensing—Sensors, Methods, Applications*, Remote Sensing and Digital Image Processing Series, Vol. 17, 572 pp., Springer, Dordrecht, pp. 1–26, 2013.)

preset emissivity value of 1 (so no emissivity correction), the handrail, as well as the gold ring on the person's hand, appears much colder than they actually are.

Now, imagine a remotely sensed thermal image of an area with many metal roofs. The temperatures in the image will all appear much cooler than reality, if not corrected for this emissivity effect. Therefore, thermal imagery in artificial environments, as well as in areas with a large variety of exposed rocks and minerals, has to be handled with care.

13.3 Potential of Diurnal and Time Series of Thermal Infrared Remote-Sensing Data

TIR data have an enormous advantage to other multispectral data; it can be acquired independent of the sun as an illumination source. This means that TIR data can be acquired during the daytime, as well as during the nighttime. Many multispectral sensors either automatically acquire data during the night or can be tasked to acquire nighttime data. MODIS, National Oceanic and Atmospheric Administration (NOAA)-AVHRR, or Meteosat Second Generation–Spinning Enhanced Visible and Infrared Imager (MSG SEVIRI), for example, all acquire TIR data in an automated mode all day long. MSG SEVIRI delivers geostationary data for every 15–30 min at 3 km spatial resolution in the TIR. MODIS (as on board the platforms TERRA and AQUA) for each spot on Earth delivers up to four acquisitions per day at 1 km spatial resolution, which usually cover the morning, the afternoon, an early nighttime images, and a predawn image. Therefore, this sensor holds a large potential for diurnal thermal mapping (Kuenzer et al., 2008). Also, AVHRR delivers a 1 km nighttime LST image for each spot on Earth. Sensors such as Landsat can be tasked to acquire thermal nighttime data on special request. The same procedure could be undertaken with the ASTER sensor. Also, several other TIR sensors depicted in Figure 13.4 can acquire nighttime data upon request with data providers.

Nighttime data have the advantage that influences of solar uneven heating due to topography are minimized. These data are therefore especially suitable to detect thermal anomalies, such as hot spots induced by forest fires, coal fires (Kuenzer et al., 2007, 2008, 2013c; Zhang et al., 2007), peat fires, industry-related hot spots, or geothermal phenomena. Daytime data usually reflect uneven solar heating due to varying sun-sensor object geometry, topography, and thermal inertia and therefore often hinder the extraction of anomalies, or the analyses of time series of thermal data. Especially in predawn, thermal nighttime data are the solar component least accentuated. Figure 13.10 illustrates the differences in thermal daytime and nighttime data based on two subsets of MODIS TIR data acquired at 11:20 a.m. during the daytime as well as at 2:30 a.m. during the night. In the upper image, the snow in the Alps appears coldest. But also lakes, such as the clearly VIS Lake Constance in the upper center of the image or the Swiss and Italian lakes, appear in dark tones (cold). Relief impacts can be clearly observed in the mountainous area of the northwest–southeast trending chain of the Emilia–Romagna. In the nighttime data, it is the lakes, as well as the settled and agricultural areas, that appear warmest, while the Alpine chain and other elevated and forested regions appear colder.

All objects on Earth's surface have their own characteristic diurnal temperature curve (Eastwood et al., 2011). This curve

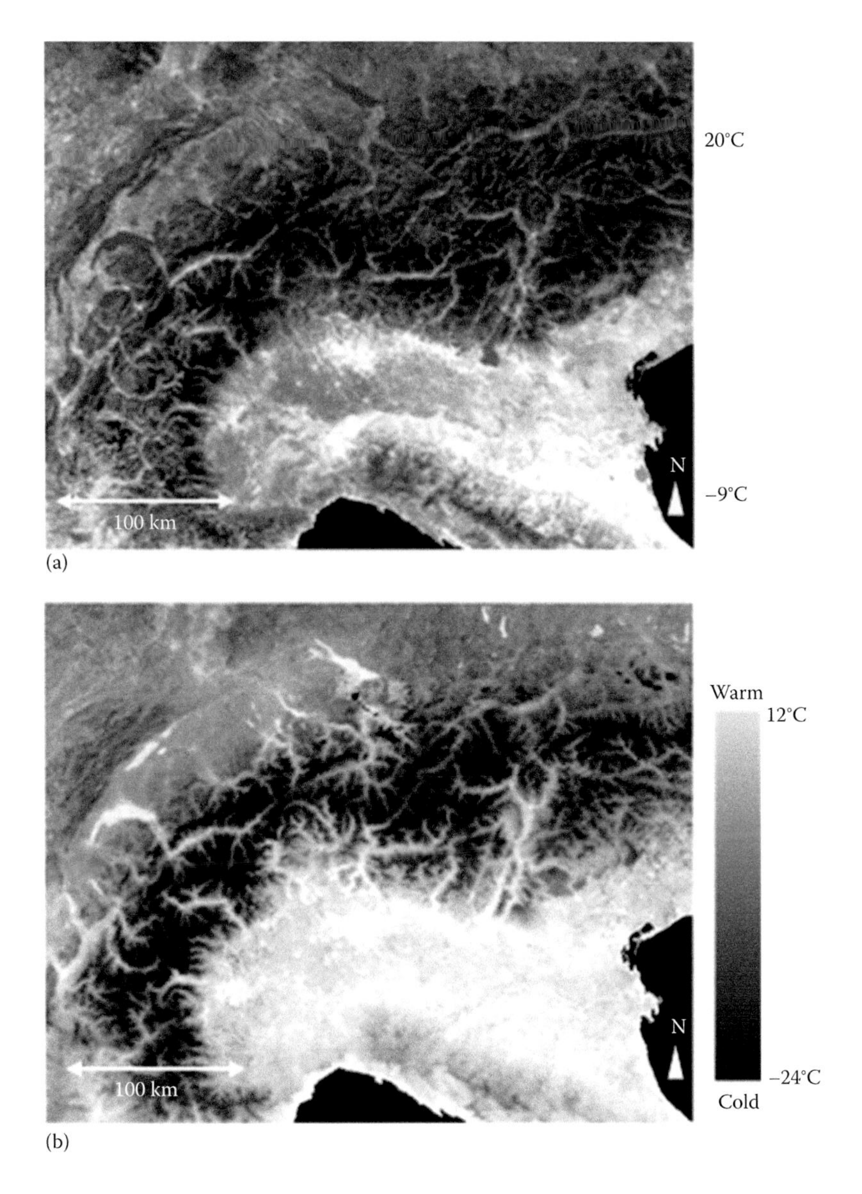

FIGURE 13.10 MYD11_L2 product acquired over northern Italy on February 27 and 28, 2009. (a) Acquired during daytime at 11:20 a.m., (b) acquired at 2:30 a.m. during nighttime. Coordinates: UL: 47°27′12N, 5°45′48E; LR: 44°05′74N, 12°31′51E.

represents the temperature behavior of the object within a 24-h cycle and describes how much and how fast an object heats up and cools down. The diurnal curve of an object depends on the material of the object in the first place—defined by its so-called thermal inertia. But also season (sun-object geometry characterizing strength of illumination), atmospheric disturbances, and—complicating the matter for land surfaces—its exposure (aspect, slope) influence those diurnal curves. Figure 13.11 depicts diurnal temperature curves of dry soil/rock and water. Differences in this example are purely based on material-related properties, described by the thermal inertia. This parameter (measured in $J/m^2/K/s^{0.5}$) is defined as the resistance of a material to heating and is the product of three factors: the energy needed to raise the temperature of a material by 1°C (heat capacity *c*) per mass unit of the substance; the density of a material, *p*; and the thermal conductivity, *k*, of the object:

$$I = \sqrt{c \times p \times k} \tag{13.8}$$

The difference between the highest and lowest temperature in a typical diurnal temperature curve is called ΔT (Kahle et al., 1976). Variations in the thermal inertia (e.g., due to moisture) lead to changes in ΔT. Small ΔTs indicate a high thermal inertia: so a high resistance of the object against temperature changes. Water is a good representative for such a surface. High ΔTs indicate a low thermal inertia—so little resistance of the material against temperature change (Cracknell and Xue, 1996).

Thermal inertia cannot be derived from remote-sensing data directly, but the idea behind it still allows to be employed with remote-sensing data, as minimum and maximum temperature of a surface within the diurnal cycle can be derived from daytime and nighttime thermal images, such as the ones presented

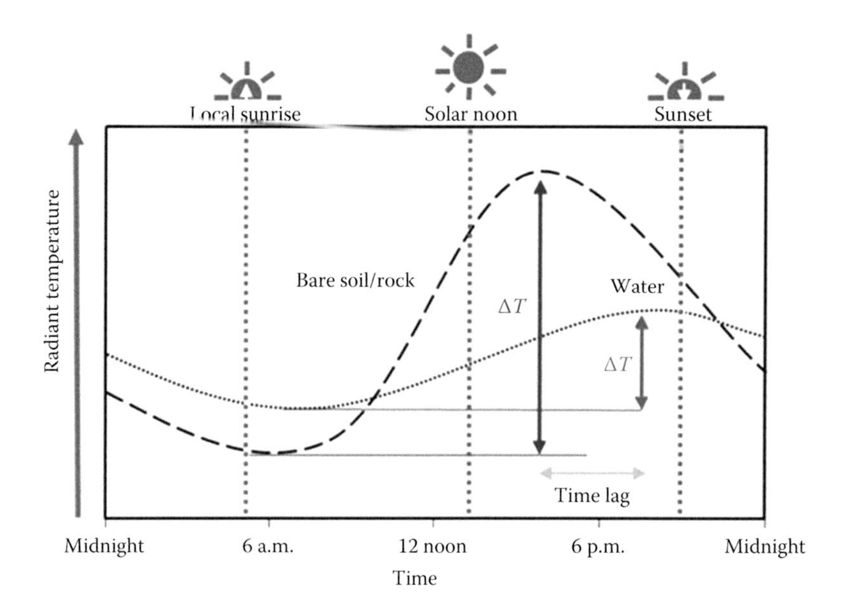

FIGURE 13.11 Diurnal temperature variation of water and dry soil/rock. Each object shows a distinct diurnal temperature cycle determined by the thermal inertia of the object and the history of the incoming solar radiation. (Modified from Lillesand, T.M. et al., *Remote Sensing and Image Interpretation*, 6th edn, John Wiley & Sons, New York, 768 pp., 2008.)

in Figure 13.10. ΔT can then be calculated subtracting the night-time from the daytime temperature. In soils, ΔT, for example, changes under different moisture conditions. Dry soils have a larger ΔT than wet soils. Thus, already in the 1970s did Idso et al. (1975) and Schmugge (1991) employ synthetic ΔT images for surface soil moisture retrieval. The concept of ΔT over time has been extended to the concept of the so-called apparent thermal inertia (ATI). The ATI is defined as

$$\text{ATI} = \frac{(1-A)}{\Delta T} \quad (13.9)$$

Here, A represents the albedo of the pixel in the VIS band. Albedo is included as dark materials absorb more sunlight than light materials—so by including it here, the impact of this effect is compensated for. As one example, Notarnicola et al. (2013) employed ATI data to differentiate varying stages of soil moisture conditions (Kuenzer et al., 2013b). However, working with ΔT and ATI is a difficult task as they can vary depending on an objects bidirectional reflectance distribution function (BRDF) relief-induced variations, shadowing, and impacts of wind. To compensate for relief-induced variations in ΔT, topographic data and information on solar elevation and azimuth need to be integrated into approximated corrections.

Diurnal effects of solar uneven heating were observed in situ on the ground by Zhang and Kuenzer (2007) as depicted in Figure 13.12. Temperatures were measured with a handheld radiometer at a very dense temporal interval of 10 min on a sand dune with slopes to the north, east, south, and west.

As expected, we can see that the east-exposed slope heats up earliest and fastest in the morning (as the sun rises in the east), and also the peak temperature here is reached earliest. Highest temperatures are reached on the southward-facing slope, and peak temperature here occurs more than 1 h later than on the eastern slope. While peak temperature on the eastern and northern slope reaches around 35°C, on the southern slope, 45°C is reached. This demonstrated that one and the same surface material can—at one time step—exhibit completely different temperature depending on aspect and slope. At local Landsat overpass time of 10:30, for example, the temperature of the same object can differ up to 10°C. The sand dune exhibits temperature ranging from 5°C to 45°C over the course of the day.

Solar uneven heating effects, as well as effects of differing sun-sensor-object geometries in LST imagery, thus have to be corrected for—especially when a study focuses on thermal change—detection or even time series analyses (Warner and Chen, 2001).

Prior to these, more complex correction TIR data of course have to be preprocessed like other multispectral data also. Just like optical imagery, data have to be geo-corrected, sensor calibration needs to be undertaken with constantly updated calibration coefficients (establishing a constant relationship between the radiation received at the sensor and the digital number [DN]), and DNs then have to be transferred in object radiance, considering atmospheric effects as well, which is usually undertaken in atmospheric correction software tools (Vidal, 1991).

But, not only diurnal variation is interesting to exploit in multiple TIR acquisitions, but also annual temperature curves of objects can be analyzed (see Figure 13.13). Annual ΔT can be calculated from the temperature difference of an object between the coldest (winter) and the warmest (summer) season.

To define, at which point in time and object is the warmest and the coldest, usually gap-free or gap-filtered annual time series of daily available data are needed. In higher-resolution

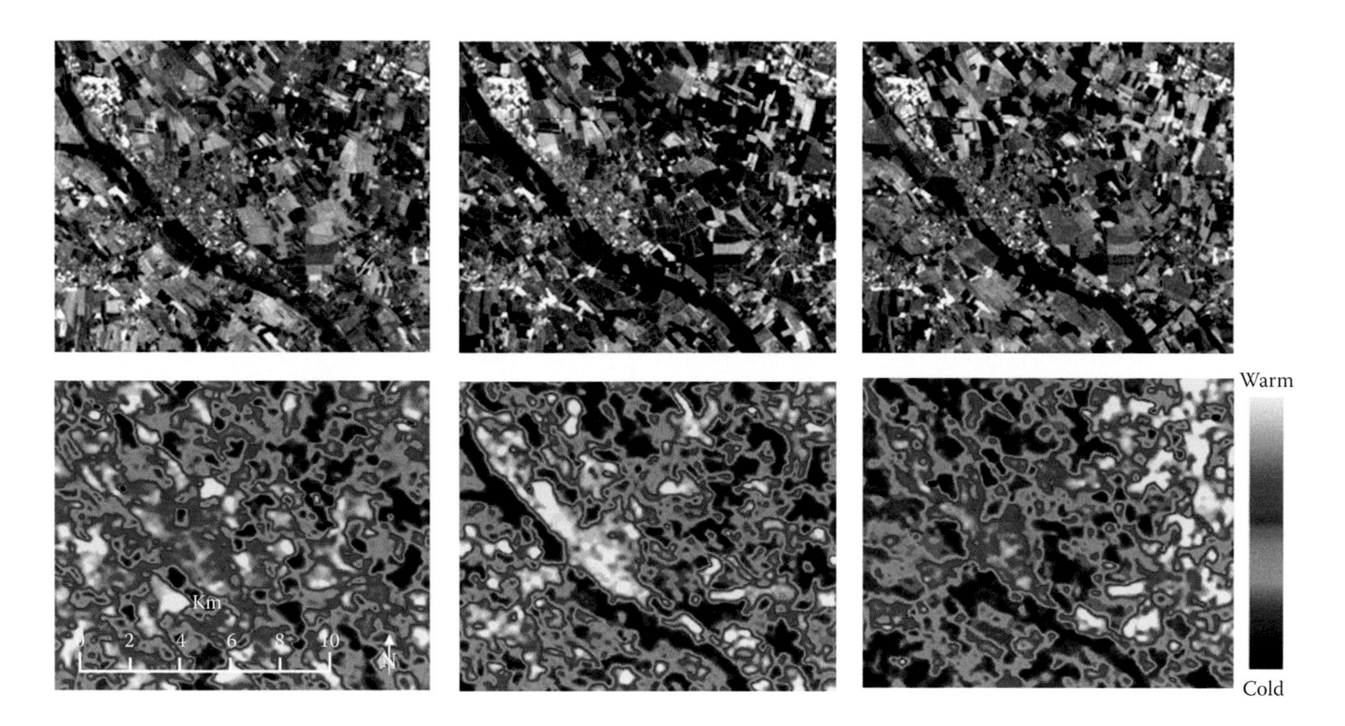

FIGURE 13.12 Diurnal temperature variation as measured on a sand dune, where one and the same material occurs at different aspects (thermal anisotropy). (From Zhang, J. and Kuenzer, C. *J. Appl. Geophys.*, 63, 117, 2007.)

data (e.g., Landsat data, which are only acquired every 16 days in the best case—meaning cloud-free conditions), the search for the highest and lowest annual temperature would be a large challenge. However, even here can images acquired within only a few weeks difference illustrate the strong variability of temperature—with season, but also with season-related land use changes. In Figure 13.13, we can clearly see how agricultural areas turn cooler with the expansion of vegetation covering the underlying soil (Kant et al., 2009). Urban areas usually appear warmest.

The strong variability of thermal data with acquisition data and time is often not considered—not even in studies, where it would be most crucial. Numerous authors study so-called urban heat islands (Streutker, 2003; Tiangco et al., 2008; Schwarz et al., 2011), to assess if a city—due to increased surface sealing—is getting hotter over time due, but do not really consider that actually a large amount of data would be needed, to eliminate the effects of different acquisition dates and times. A comprehensive time series of scenes acquired during the same date and time need to be analyzed to derive a real trend (e.g., possible with MODIS or AVHRR data).

13.4 Application Examples of Thermal Infrared Data Analyses

In the field of TIR analyses, SST analyses are the furthest advanced (see Figure 13.13) (Iwasaki et al., 2008) and one of the few domains in TIR remote sensing, where services are already offered—for example, by the European Space Agency (ESA). Figure 13.13 depicts a mean product of SST derived from Envisat Advanced Along-Track Scanning Radiometer (AATSR) data during August 2004. Gaps exist in areas, which are often cloud covered that they could not be interpolated and filled. Poles are not covered due to the near polar orbit of Envisat. Time series of these products are freely available from ESA and can be analyzed with respect to SST averages, minima, mean, variability, anomalies, and the representation of occurrences of, for example, El Niño in TIR-derived SST data. TIR data acquired over the ocean have the big advantage that scientists do not have to deal with topographic effects of solar uneven heating. Furthermore, the water surface is more or less homogeneous—at least when compared with the patchy mosaics of the land surface. Furthermore, BRDF effects are minimized, and the thermal inertia of water is high, and therefore temperature changes take place relatively slowly.

Figure 13.14 presents a typical application example of TIR data analyses on land. Here, hot spots have been derived over the area of the Niger Delta, Nigeria, Africa, during three time steps and covering the time span from 1986/1987 to 2013. Oil industry in the Niger Delta flares enormous amounts of natural gas. This practice brings with it the harmful release of climate-relevant and toxic gasses and substances, contributing to the severe environmental degradation in the area. Different approaches for thermal anomaly extraction exist, and here an automated, moving window-based approach (Kuenzer et al., 2007) was applied to extract local hot spots of different temperature—an approach clearly superior to simple empirical thresholding. The illustration of hot spot occurrences over time (in this case, gas flares) can depict the dynamics of oil exploiting industry development in the region and can support the designation of especially threatened natural resource or communities.

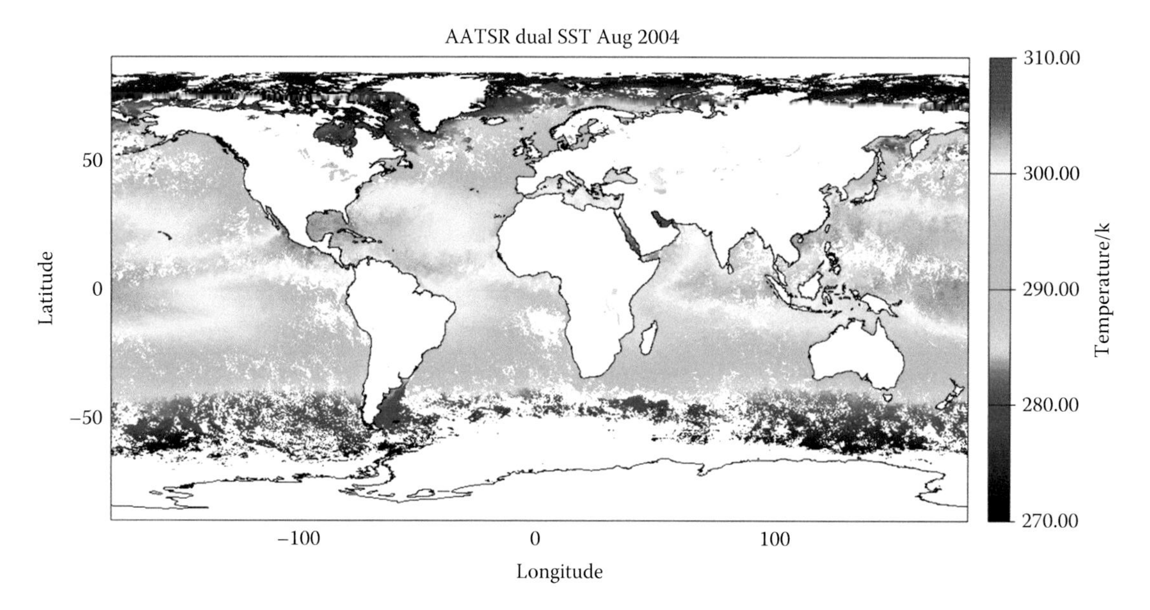

FIGURE 13.13 Monthly average of sea surface temperature as derived from Envisat Advanced Along-Track Scanning Radiometer data at 1 km resolution for August 2004. (Courtesy of the European Space Agency, Noordwijk, the Netherlands.)

The range of TIR-based application studies is very broad, and published studies are numerous. A comprehensive overview of TIR remote sensing is provided in the book *Thermal Infrared Remote Sensing: Sensors, Method, Applications* (Kuenzer and Dech, 2013), to which numerous authors contributed and which contains three parts. Sensor-related chapters focus, among others, on the geometric calibration of thermographic cameras, TIR spectroscopy, challenges and opportunities for unmanned aerial vehicle-borne thermal imaging, or planned new thermal missions, such as National Aeronautics and Space Administration's (NASA) Hyperspectral Thermal Emission Spectrometer, NASA's Hyperspectral Infrared Imager, or thermal remote sensing with small satellites such as Bispectral and Infrared Remote Detection (BIRD), Technologie-Erprobungsträger (TET), and the next-generation Berlin InfraRed Optical System (BIROS). Method-oriented chapters present cross-comparisons of daily LST products from NOAA-AVHRR and MODIS, compare the advantages and shortcomings of the thermal sensors of SEVIRI and MODIS for LST mapping, and discuss methods for improving atmospheric correction of TIR data, or for time series corrections and analyses in TIR data, or novel concepts for the derivation of SST products. Application chapters address approaches to derive urban structure types and address TIR-based mineral mapping, soil moisture derivation, the assessment of vegetation fires, analyses of lava flows, thermal analyses of volcanoes, investigations in underground coal fire regions, and the analyses of geothermal systems.

13.5 Ground Data and Validating Thermal Infrared Data

To validate temperatures derived from TIR data, or to confirm phenomena or patterns observed in this data, ground data collection is a common procedure undertaken (Coll et al., 2005). Validation on past data is of course not possible, as LST varies from hour to hour and from day to day. Therefore, ground data collection activities should be performed during the satellite's overpass. As temperatures change within minutes, there is little time to measure the on-ground surface temperatures of several objects within the scene's footprint. Usually, several people with intercalibrated radiometers, contact thermometers, and thermal cameras are on the ground to measure the kinetic and radiance temperatures of the hottest and coldest objects (see Figure 13.15). Water surfaces are a good target to establish a relationship between ground-measured temperatures and satellite imagery-derived temperatures, as they have a high thermal inertia and do exhibit fast or accentuated temperature changes over time.

Indirect validation via Google Earth can be undertaken for selected applications. For example, the Niger Delta gas flare locations, which were presented in Figure 13.13, were partially validated via checks in high-resolution data Google Earth. The high-resolution data available for most parts of the planet allow to put thermal anomalies detected into the right context. Gas flares, for example, are clearly visible also in high-resolution optical data. If a flame is obscured by smoke, other indirect indicators are burnt vegetation, soot around the flare site, and hot spots can sometimes also be validated by the numerous photographs available in Google Earth. The same applies for forest fire occurrences or grassland fires.

13.6 Discussion and Conclusion

Thermal remote-sensing data are used for a variety of applications such as the assessment of LST, SST, and inland water body temperature; the analyses of land cover–related thermal patterns, such as UHI; the detection of hot spots and thermal anomies resulting from forest fires, coal fires, and industry-related

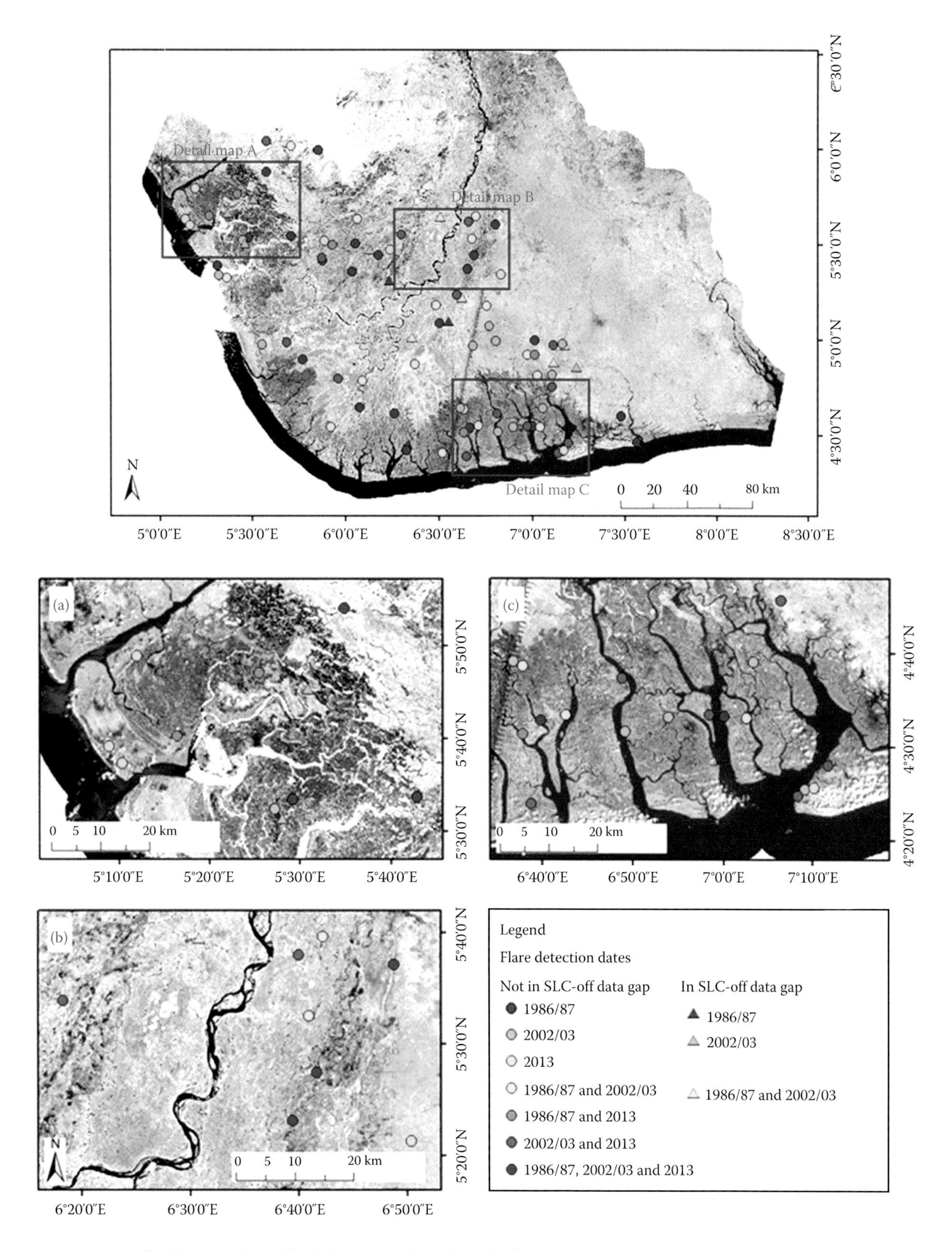

FIGURE 13.14 Gas flare hot spots derived for different years from thermal infrared Landsat data over the course of 27 years (1986/1987 until 2013). Oil industry in the Niger Delta flares enormous amounts of natural gas. This practice brings with it the harmful release of climate-relevant and toxic gasses and substances, contributing to the severe environmental degradation in the area. (Adapted from Kuenzer, C. et al., *Appl. Geogr.*, 53, 354, 2014.)

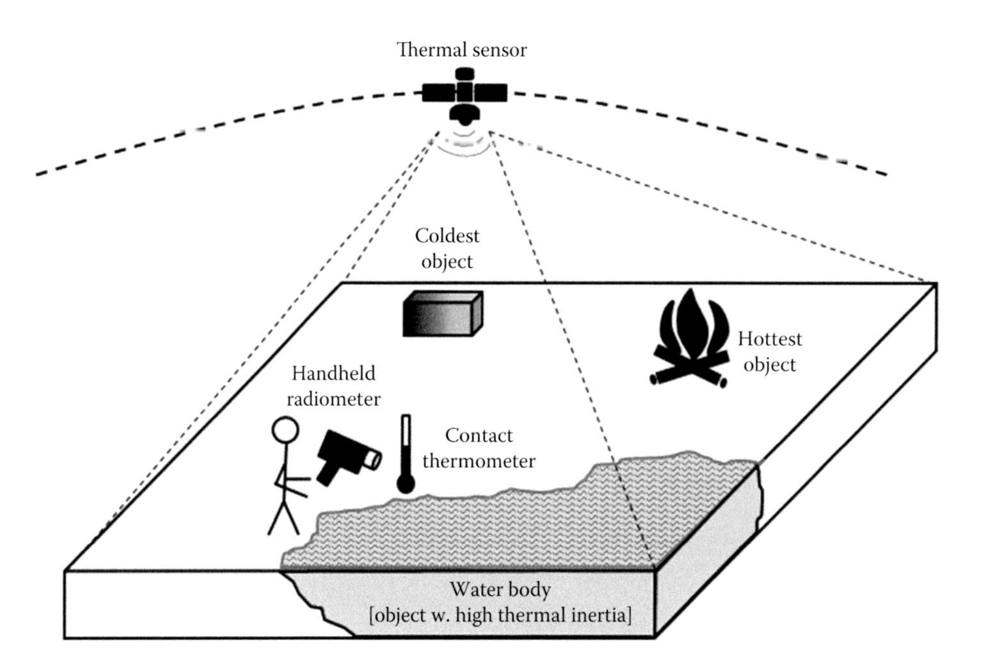

FIGURE 13.15 Ground data collection/validation of thermal infrared data and observed phenomena undertaken during satellite overpass. (Modified from Kuenzer, C. and Dech, S., Theoretical background of thermal infrared remote sensing, in Kuenzer, C. and Dech, S., eds., *Thermal Infrared Remote Sensing—Sensors, Methods, Applications*, Remote Sensing and Digital Image Processing Series, Vol. 17, 572 pp., Springer, Dordrecht, pp. 1–26, 2013.)

activities; and the derivation of moisture conditions via the investigation of diurnal temperature dynamics. While the fleet of satellites with a thermal sensor on board is larger than often assumed, there are only few sensors available, which offer free and easy access to thermal data. However, TIR data offer several advantages. While TIR data—just like multispectral data—are affected by cloud cover, an active illumination source is not needed. Therefore, the data can be acquired during the daytime as well as the nighttime. Thermal data depict a direct physical quantity (in K or °C) and are thus easy to interpret without too much bias. Furthermore, long-term data archives of common medium-resolution sensors with daily acquisition coverage allow for the analyses as decadal, annual, and monthly means, deviations, variability, and trends. Disadvantages of TIR remote sensing are the relatively low spatial resolution of the data. Currently, the highest-resolution spaceborne data have a 100 m pixel size. Thermal sensors—be it airborne or spaceborne—are more costly than, for example, VIS sensors. Thus, the TIR bandwidth is usually the domain that is usually discarded if budget cuts have to be made. At the same time, the TIR community is relatively small, and lobby voices for thermal sensors are thus not so powerful.

This chapter presented an overview of the principles and theoretical background of remote sensing in the TIR domain. We addressed data characteristics and important laws of physics, presented common past and up-to-date TIR instruments, discussed approaches to analyze thermal data, and presented selected application examples. Remote sensing of the TIR is an often neglected discipline of remote sensing. Thermal sensors are very expensive, and due to the longer wavelength compare to optical and NIR data, spatial resolution is always inferior to shorter-wavelength multispectral data. Astonishingly, the new thermal band on board of Landsat-8, launched in early 2013, only offers 100 m spatial resolution—compared to 60 m resolution, which was available with Landsat-7-Enhanced Thematic Mapper Plus (ETM+), launched over a decade earlier in 2001. To receive up to data, suitable for monitoring purposes, the TIR LST community currently has to rely on MODIS and AVHRR data, as well as on Landsat TM and ETM+ data. The soon-to-be-launched Sentinel 3 satellite (foreseen for 2015), including an instrument continuing the AATSR time series, is eagerly expected.

References

Becker F (1987) The impact of emissivity on the measurement of land surface temperature from a satellite. *International Journal of Remote Sensing* 8(10):1509–1522.

Coll C, Caselles V, Galve JM, Valor E, Niclòs R, Sánchez JM, Rivas R (2005) Ground measurements for the validation of land surface temperatures derived from AATSR and MODIS data. *Remote Sensing of Environment* 97:288–300.

Cracknell AP, Xue, Y (1996) Thermal inertia determination from space—A tutorial review. *International Journal of Remote Sensing* 17(3):431–461.

Dech SW, Tungalagsaikhan P, Preusser C, Meisner RE (1998) Operational value-adding to AVHRR data over Europe: Methods, results, and prospects. *Aerospace Science and Technology* 2:335–346.

Eastwood S, Le Borgne P, Péré S, Poulter D (2011) Diurnal variability in sea surface temperature in the Arctic. *Remote Sensing of Environment* 115(10):2594–2602.

Frey C, Kuenzer C, Dech S (2012) Quantitative comparison of the operational NOAA AVHRR LST product of DLR and the MODIS LST product V005. *International Journal of Remote Sensing* 33(22):7165–7183.

Heald, MA (2003) Where is the 'Wien peak?' *American Journal of Physics* 71(12):1322–1323.

Hulley GC, Hook SJ, Baldridge AM (2009) Validation of the North American ASTER Land Surface Emissivity Database (NAALSED) version 2.0 using pseudo-invariant sand dune sites. *Remote Sensing of Environment* 113(10):2224–2233.

Idso SB, Jackson RD, Reginato RJ (1975) Detection of soil moisture by remote surveillance. *American Scientist* 63:549–557.

Iwasaki S, Kubota M, Tomita H (2008) Inter-comparison and evaluation of global sea surface temperature products. *International Journal of Remote Sensing* 29(21):6263–6280.

Kahle AB, Gillespie AR, Goetz AFH (1976) Thermal Inertia Imaging: A new geological mapping tool. *Geophysical Research Letters* 3(1):26–28.

Kant Y, Bharath BD, Mallick J, Atzberger C, Kerle N (2009) Satellite-based analysis of the role of land use/land cover and vegetation density on surface temperature regime of Delhi, India. *Journal of the Indian Society of Remote Sensing* 37(2):201–214.

Kirchhoff G (1860) Ueber das Verhältniss zwischen dem Emissionsvermögen und dem Absorptionsvermögen der Körper für Wärme und Licht. *Annalen derPhysik und Chemie (Leipzig)* 109:275–301.

Kuenzer C, Dech S (2013) Theoretical background of thermal infrared remote sensing. In Kuenzer C and Dech S (eds.) *Thermal Infrared Remote Sensing—Sensors, Methods, Applications*, Remote Sensing and Digital Image Processing Series, Vol. 17, 572 pp, Springer, Dordrecht, pp. 1–26.

Kuenzer C, Gessner U, Wagner W (2013b) Deriving soil moisture from thermal infrared satellite data—Synergies with microwave data. In Kuenzer C, Dech S (eds.) *Thermal Infrared Remote Sensing—Sensors, Methods, Applications*, Remote Sensing and Digital Image Processing Series, Vol. 17, 572 pp, Springer, Dordrecht, pp. 315–330.

Kuenzer C, Guo H, Ottinger M, Dech S (2013a) Spaceborne thermal infrared observation—An overview of most frequently used sensors for applied research. In Kuenzer C, Dech S (eds.) *Thermal Infrared Remote Sensing—Sensors, Methods, Applications*, Remote Sensing and Digital Image Processing Series, Vol. 17, 572 pp, Springer, Dordrecht, pp. 131–148.

Kuenzer C, Hecker C, Zhang J, Wessling S, Wagner W (2008) The potential of multi-diurnal MODIS thermal bands data for coal fire detection. *International Journal of Remote Sensing* 29:923–944.

Kuenzer C, van Beijma S, Gessner U, Dech S (2014) Land surface dynamics and environmental Challenges of the Niger Delta, Africa: Remote sensing based analyses spanning three decades (1986–2013). *Applied Geography* 53:354–368.

Kuenzer C, Zhang J, Li J, Voigt S, Mehl H, Wagner W (2007) Detection of unknown coal fires: Synergy of coal fire risk area delineation and improved thermal anomaly extraction. *International Journal of Remote Sensing* 28:4561–4585.

Kuenzer C, Zhang J, Tetzlaff A, Dech S (2013c) Thermal infrared remote sensing of surface and underground coal fires. In Kuenzer C, Dech S (eds.) *Thermal Infrared Remote Sensing—Sensors, Methods, Applications*, Remote Sensing and Digital Image Processing Series, Vol. 17, 572 pp, Springer, Dordrecht, pp. 429–451.

Lillesand TM, Kiefer RW (1994) *Remote Sensing and Image Interpretation*, 3rd edn., Wiley, New York, 748 pp.

Lillesand TM, Kiefer RW, Chipman JW (2008) *Remote Sensing and Image Interpretation*, 6th edn., John Wiley & Sons, New York, 768 pp.

Löffler E (1994) *Geographie und Fernerkundung*, 3rd edn., Teubner, Stuttgart, Germany, 251 p.

Notarnicola C, Lewinska E, Temimi M, Zebisch, M (2013) Application of the apparent thermal inertia concept for soil moisture estimation in agricultural areas. In Kuenzer C, Dech S (eds.) *Thermal Infrared Remote Sensing—Sensors, Methods, Applications*, Remote Sensing and Digital Image Processing Series, Springer, the Netherlands, Vol. 17, 572 pp.

Planck M (1900) Entropie und Temperatur strahlender Wärme. *Annalen der Physik* 306(4):719–737.

Sabins FF (1996) *Remote Sensing*, 3rd edn., Wiley, New York, 450 pp.

Schmugge TJ, Becker F, Li ZL (1991) Spectral emissivity variations observed in airborne surface temperature measurements. *Remote Sensing of Environment* 35(2):95–104.

Schwarz N, Lautenbach S, Seppelt R (2011) Exploring indicators for quantifying surface urban heat islands of European cities with MODIS land surface temperatures. *Remote Sensing of Environment* 115(12):3175–3186.

Streutker D (2003) Satellite-measured growth of the urban heat island of Houston, Texas. *Remote Sensing of Environment* 85:282–289.

Tiangco M, Lagmay AMF, Argete J (2008) ASTER-based study of the night-time urban heat island effect in Metro Manila. *International Journal of Remote Sensing* 29(10): 2799–2818.

Tipler PA (2000) *Physik*, 3rd edn., Spektrum Akademischer, Verlag, Germany, 1520 pp.

Vidal A (1991) Atmospheric and emissivity correction of land surface temperature measured from satellite using ground measurements or satellite data. *International Journal of Remote Sensing* 12(12):2449–2460.

Walker J (2008) *Fundamentals of Physics*, 8th edn., John Wiley & Sons, New York, 891 pp.

Warner TA, Chen X (2001) Normalization of Landsat thermal imagery for the effect of solar heating and topography. *International Journal of Remote Sensing* 22(5):773–788.

Zhang J, Kuenzer C (2007) Thermal surface characteristics of coal fires 1: Results of in-situ measurements. *Journal of Applied Geophysics* 63:117–134.

Zhang J, Kuenzer C, Tetzlaff A, Oettl D, Zhukov B, Wagner W (2007) Thermal characteristics of coal fires 2: Results of measurements on simulated coal fires. *Journal of Applied Geophysics* 63:135–147.

14

Object-Based Image Analysis: Evolution, History, State of the Art, and Future Vision

Thomas Blaschke
University of Salzburg

Maggi Kelly
University of California, Berkeley

Helena Merschdorf
University of Salzburg

Acronyms and Definitions

API	Application programming interface
ASTER	Advanced Spaceborne Thermal Emission and Reflection Radiometer
ECHO	Extraction and classification of homogeneous objects
ENVI	Environment for visualizing images
GEOBIA	Geographic object-based image analysis
GIS	Geographic information system
GIScience	Geographic information science
GPL	General public license
GPS	Global positioning system
ISI	Institute for Scientific Information
LiDAR	Light detection and ranging
NGO	Nongovernmental organization
OBIA	Object-based image analysis
OGC	Open Geospatial Consortium
PPGIS	Public participation geographic information system
RGB	Red, green, blue color system
RS	Remote sensing
RSGISLib	Remote sensing and GIS software library
SAGA	System for automated geoscientific analyses
UAS	Unmanned aerial systems
VGI	Volunteered geographic information
WoS	Web of science

14.1 Introduction

Remote sensing, what it is and what it can be used for, is laid out in various chapters of this comprehensive book. We may only state here that remote sensing has a short history—when compared to traditional disciplines such as mathematics or physics. Contrarily, we may state that it has a long history when we compare it to recent Internet-based technology like social media or, closer to our field, the tracking of people and moving objects by means of cell phone signals. Remote sensing has been a domain for specialists for many years and to some degree it still is. Similarly, geographic information system (GIS) has for years been a field where professionals worked on designated workstations while not being fully integrated in standard corporate information technology infrastructures. The latter changed more than a decade ago, while for remote sensing only recently, one may still witness remnants of historical developments of Remote Sensing (RS)-specific hardware and software. The dominant concept in remote sensing has been the pixel, while GIS functionality has always been somehow splintered into the raster and vector domains. Blaschke and Strobl (2001) provocatively raised the question "What's wrong with pixels?"

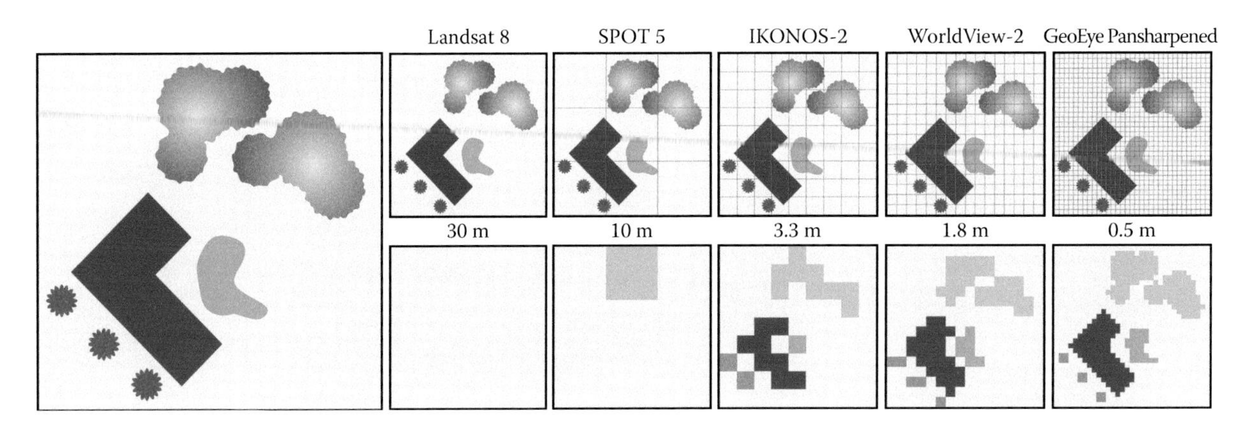

FIGURE 14.1 Objects and resolutions: OBIA methods are associated with the notion of high resolution—whereby *high* has always to be seen in context.

having identified an increasing dissatisfaction with pixel-by-pixel image analysis. Although this critique was not new (Cracknell 1998; see also Blaschke and Strobl 2001; Burnett and Blaschke 2003; Blaschke 2010; Blaschke et al. 2014 for a more thorough discussion), these authors described a need for applications *beyond pixels* and for specific methods and methodologies that support this (Figure 14.1).

Over the last years, the number of applications that conceptually aim for objects—still built on the information of the underlying pixels—rose quickly. Blaschke et al. (2014) identified a high number of relevant publications that use—with some degree of fuzziness in their terminology—the concept of object-based image analysis (OBIA). They even claim that this concept and its instantiation to a particular order of scale—the *geographic* as opposed to applications in medical imaging or cell biology—is a new paradigm in remote sensing. For this level of scale and the *geodomain*, this paradigm is then referred to by some scholars as Geographic Object-Based Image Analysis (GEOBIA), while the generic principles—the multiscale segmentation and object handling—may generically be called OBIA. Other sources use the more generic term of OBIA when referring to the geospatial domain also, and Blaschke et al. refer to Kuhn (1962) stating that an inconsistent use of terminology can be expected for a new paradigm. Nevertheless, it is high time to consolidate this terminology and to support a coherent usage of terms and naming conventions—after having agreed upon the concepts and the conception of the overall approach.

This chapter therefore briefly explains OBIA methods as used in the geospatial domain and elsewhere. We will start from the quest to partitioning geospatial data into meaningful image objects and the needs and possibilities to assessing their characteristics through spatial, spectral, and temporal scale. At its most fundamental level, OBIA requires image segmentation, attribution, classification, and the ability to query and link individual objects (aka segments) in space and time. We will elucidate the evolution of this approach, its relatively short history, and its older origins. Instead of a comprehensive state-of-the-art analysis, we refer to the key literature and try to summarize the core concepts for the reader in an understandable way, with a particular emphasis on a common nomenclature, definitions, and reporting procedures. Ultimately, we will ask where this development will lead to in terms of applications, research questions and needs in education, and training and professional workforce development, and we conclude with the main advances and recommendations for future work.

14.2 History of OBIA

14.2.1 Intellectual Roots

14.2.1.1 Conceptual Foundations

The conceptual foundations of OBIA are rooted in the 1960s with predigital aerial photography. The spatial information found in digital imagery that is harnessed in the object-based approach, for example, image texture, contextual information, pixel proximity, and geometric attributes of features, were discussed in the 1960s as possible components to yet possible automation of photo interpretation. In his seminal work on aerial photography and early remote-sensing applications, Colwell (1965) describes the photo interpretation process as the act of examining photographic images for the purpose of identifying objects and judging their significance. He said that photo interpretation involves the observation of the size, shape, shadow, tone, texture, pattern, and location of the features, as well as the significance of the features, based largely on their *interrelationships* or *association* (Colwell 1965). His assessment of the potential for automation of an object recognition process depended on the capacities of a digital scanner and the ability of an algorithm to assess the differences, in photographic tone, between a *blob* and its surroundings (Colwell 1964, 1965). Colwell was an important advisor on the Landsat 1 mission, and his ideas on extraction of meaningful features transferred to his ambitions for the satellite missions (Colwell 1973).

14.4.1.2 Image Segmentation

Image segmentation is the division of an image into different regions, each having certain properties, and it provides the building blocks of OBIA (Blaschke 2010). The desire expressed by Colwell and others in the 1960s to more automatically delineate meaningful features, objects, or *blobs* in his early terminology launched numerous approaches to image segmentation that rapidly advanced in the 1980s. It is widely agreed that the

TABLE 14.1 Overview of Major Groups of Image Segmentation Techniques

	Main Issues	Strengths	Weaknesses
Thresholding and clustering	Threshold values are applied globally (to the whole image) or locally (applied to subregions)	Most thresholding algorithms are computationally simple. Clustering an image or a raster may be intuitive for a given number of clusters	The results depend on the initial set of clusters and user values or thresholds, respectively
Edge detection	Boundaries of object or regions under consideration and edges are assumed to be closely related, since there are often sharp differences in intensity at the region boundaries	Discontinuities are identified across the array of values studied. Particularly suited for internally relatively homogeneous objects such as buildings, roads, or water bodies	Edges identified by edge detection are often disconnected. To segment an object from an image, however, one needs closed region boundaries. Typically problematic in objects with high internal heterogeneity such as forests
Region growing	Starting from the assumption that the neighboring pixels within one region have similar values, a similarity criterion is defined and applied in to neighboring pixels		Selection of the similarity criterion significantly influences the results

segmentation algorithms implemented in the OBIA software of today owe a debt to theoretical and applied work in the 1970s and 1980s that developed and refined numerous methods for image segmentation (Blaschke et al. 2004; Blaschke 2010). Early key papers for the remote-sensing field include Kettig and Landgrebe (1976) who presented experimental results in segmentation of Landsat 1 (ERTS-1) imagery, and McKeown et al. (1989) who developed a knowledge-based system with image segmentation and classification tools designed for semiautomated photo interpretation of aerial photographs. Key reviews are provided in numerous papers (Fu and Mui 1981; Haralick and Shapiro 1985; Pal and Pal 1993). Building on that work, image segmentation techniques implemented today include those focused on thresholding or clustering, edge detection, region extraction, and growing, and some combination of these has been explored since the 1970s (Fu and Mui 1981; Blaschke 2010) (Table 14.1).

14.2.2 Needs and Driving Forces

With a focus on geospatial data, OBIA has particular needs that were not anticipated by its antecedents. The OBIA methods were driven first by the need to more accurately map multiscaled Earth features with high-spatial-resolution imagery such as the tree, the building, and the field. Following that, the spatial dimension of objects (distances, pattern, neighborhoods, and topologies) was mined for classification accuracy (e.g., Guo et al. 2007). Most recently, the OBIA field has been characterized by discussions of object semantics within fixed or emergent ontologies (Arvor et al. 2013; Yue et al. 2013) and by the need for interoperability between OBIA and GIS and spatial modeling frameworks (Harvey and Raskin 2011; Yue et al. 2013). The OBIA approach has evolved from a method of convenience to what has been called a new paradigm in remote sensing and spatial analysis (Blaschke et al. 2014).

14.2.3 GEOBIA Developments

14.2.3.1 Emergence (1999–2003/2004)

The emergence of OBIA has been written about extensively elsewhere (e.g., Blaschke 2010; Blaschke et al. 2014) and had its largest boost from the availability of satellite imagery of increasing spatial resolution such as IKONOS (1–4 m), QuickBird (resolution), and OrbView (resolution) sensors (launched in 1999, 2001, and 2003, respectively) (Blaschke 2010). This ready availability of high-resolution multiband imagery coincided with increasing awareness in the remote-sensing literature that novel methods to extract meaningful and more accurate results were critically needed. The *business-as-usual* pixel-based algorithms were not reliable with imagery exhibiting high local variability and obvious spatial context (Cracknell 1998; Townshend et al. 2000; Blaschke and Strobl 2001).

Importantly, the software package called eCognition from the company Definiens (subsequently called Definiens Earth Sciences) became commercially available from 2000. This event is marked as a milestone in the emergence of a body of work on OBIA, as it was the first commercially available, object-based, image analysis software (Flanders et al. 2003; Benz et al. 2004; Blaschke 2010), and many peer-reviewed papers during this phase relied on the software. The eCognition software is built on to the approach originally known as Fractal Net Evolution (Baatz and Schäpe 2000; Blaschke 2010) that is not easily nor often described in detail in early papers that relied on the software.

Representative papers demonstrating the utility of the newly released software from this time frame include the following. Flanders et al. (2003) evaluated the object-based approach from eCognition software and classified forest clearings and forest structure elements in British Columbia, Canada, using a Landsat-enhanced thematic mapper plus image. They found that forest clearings as well as forest growth stage, water, and urban features were classified with significantly higher accuracy than using a traditional pixel-based method. With slightly different results, Dorren et al. (2003) also compared pixel- and object-based classification of forest stands using Landsat imagery in Austria. They used eCogntion for the object-based approach and found that while the pixel-based method provided slightly better accuracies, the object-based approach was more realistic and better served the needs of local foresters. Benz et al. (2004) used eCognition to update urban maps (buildings, roofs, etc.) from high-resolution (0.5 m) RGB aerial orthoimages in Austria. Theirs was an early and comprehensive examination of

TABLE 14.2 Overview of Early Application Fields

Application	Images	Comparison to Pixel-Based/Findings	Software
Forest clearings and forest structure (Flanders et al. 2003)	Landsat-enhanced thematic mapper plus image	Significantly higher accuracy of OBIA compared to pixel based	eCognition
Forests (Dorren et al. 2003)	Landsat	Pixel-based method provided slightly better accuracies, but the object-based approach was more realistic and better served the needs of local foresters	eCognition
Update urban maps (Benz et al. 2004)	High-resolution (0.5 m) RGB aerial orthoimages	Comprehensive examination of the use of the software. Discussed, for example, the importance of semantic features and uncertainties in representation	eCognition
Shrub cover and rangeland characteristics (Laliberte et al. 2004)	Historic (1937–1996) scanned aerial photos and a contemporary QuickBird satellite image		eCognition
Correlate field-derived forest inventory parameters and image objects (Chubey et al. 2006)	IKONOS-2 imagery	The strongest relationships were found for discrete land-cover types, species composition, and crown closure	eCognition
Vegetation inventory (Yu et al. 2006)	Digital airborne imaging system imagery	The object-based approach outperformed the pixel-based approach	
Map surface coal fires (Yan et al. 2006)	ASTER image (15 m resolution)	The OBIA approach yielded classifications of marked improvement over the pixel-based approach	

the use of the software, and they discussed numerous aspects of the OBIA approach that are still actively discussed today—for example, the importance of semantic features and uncertainties in representation. Laliberte et al. (2004) used a combination of historic (1937–1996) scanned aerial photos and a contemporary QuickBird satellite image to map shrub cover and rangeland characteristics over time. The eCognition was critical in their workflow. Chubey et al. (2006) used eCognition to segment IKONOS-2 imagery and decision tree analysis to correlate field-derived forest inventory parameters and image objects for forests in Alberta, Canada. They found that the strongest relationships were found for discrete land cover types, species composition, and crown closure. While much work focused on the use of eCognition for high-resolution imagery, not all work in this phase did. Many papers explored the method using Landsat imagery (e.g., Dorren et al. 2003) (Table 14.2).

14.2.3.2 Establishment (2005–2010)

14.2.3.2.1 Accuracy

Many papers during this time frame focused on proving the utility of the new approach and provided comparisons between OBIA and pixel-based classifiers (Yan et al. 2006; Cleve et al. 2008; Maxwell 2010). For example, Yu et al. (2006) used high-spatial-resolution digital airborne imaging system imagery and associated topographic data of the Point Reyes National Seashore in California, United States, for a comprehensive and detailed vegetation inventory at the alliance level. The object-based approach outperformed the pixel-based approach. Yan et al. (2006) compared pixel- and object-based classification of an Advanced Spaceborne Thermal Emission and Reflection Radiometer (ASTER) image (15 m resolution) to map surface coal fires and coal piles. The OBIA approach yielded classifications of marked improvement over the pixel-based approach. Similar results were shown using high-resolution aerial imagery for urban features (Cleve et al. 2008), Landsat imagery, and land cover (Maxwell 2010).

14.2.3.2.2 Applications

From 2005 to 2010, there was a wide net cast around OBIA application areas. Table 14.3 provides an overview of the various application areas, which emerged over these years.

Capturing, attributing, and understanding changing landscapes continues to be a primary research area in remote sensing, and the use of OBIA methods for studying and understanding

TABLE 14.3 Development of OBIA Application Fields

Application Area	Examples
Forests	Flanders et al. (2003)
	Dorren et al. (2003)
Individual trees	Guo et al. (2007)
	De Chant et al. (2009)
Forest stands	Radoux and Defourny (2007)
	Gergel et al. (2007)
Parklands	Rocchini et al. (2006)
	Yu et al. (2006)
Rangelands	Laliberte et al. (2007)
Wetlands and other critical habitat	Bock et al. (2005)
Urban areas	Weeks et al. (2007)
	Cleve et al. (2008)
	Durieux et al. (2008)
Land use and land cover	Maxwell (2010)
Public health	Kelly et al. (2011)
Disease vector habitats	Koch et al. (2007)
	Troyo et al. (2009)
Public health infrastructure (e.g., refugee camps)	Lang and Blaschke (2006)
Hazard vulnerability and disaster aftermath	Al-Khudhairy et al. (2005)
	Gusella et al. (2005)

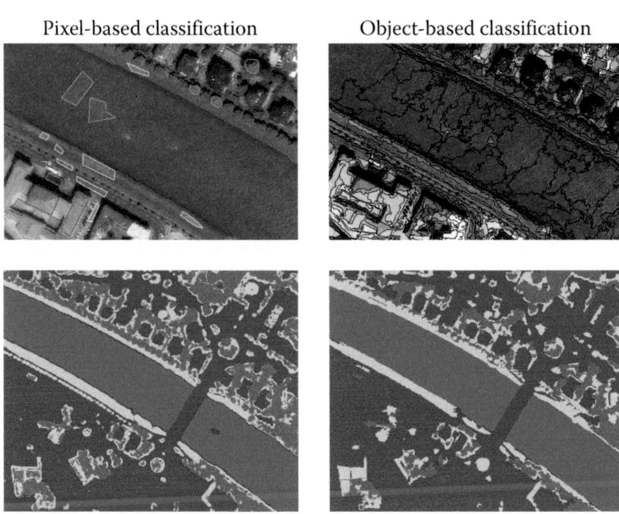

FIGURE 14.2 Object-based versus pixel-based classification.

change were increasingly popular during this period. In a comprehensive review article, Chen et al. (2012) presented a timely overview of the main issues in remote-sensing change detection and suggested reasons for favoring object-based change detection over pixel-based approaches. They suggested that an object-based approach to change detection allows for multiscale analysis to optimize the delineation of individual landscape features, it reduces spurious changes due to high spectral variability in high-spatial-resolution imagery, and the approach also allows for more meaningful ways to evaluate change (Figure 14.2).

14.2.3.2.3 Data Fusion

Data fusion became increasingly common during this phase. The utility of light detection and ranging (LiDAR) data for capturing height, that could be used in both segmentation and classification, was recognized soon after LiDAR became somewhat operational. Pascual et al. (2008) incorporated LiDAR data to help characterize forest stands and structure using OBIA in a complex Pinus sylvestris–dominated forest in central Spain. Zhou and Troy (2008) used LiDAR with high-resolution digital aerial imagery to analyze and characterize the urban landscape structure of Baltimore at the parcel level. Ebert et al. (2009) used optical, LiDAR, and digital elevation models to estimate social vulnerability indicators through the use of physical characteristics and hazard potential. Tullis et al. (2010) found that certain land covers (e.g., forest and herbaceous cover rather than impervious surface) benefited more from a synergy between LiDAR and optical imagery. Image fusion has also involved multiple spatial and spectral resolutions. For example, Walsh et al. (2008) used both QuickBird and Hyperion hyperspectral imagery to map an invasive plant species in the Galapagos Islands. The fusion of multi- and hyperspectral imagery was beneficial.

14.2.3.2.4 Software

During this time frame, papers evolved from naive and sometimes simplistic use of complicated software (e.g., "we used eCognition to segment and classify our imagery") to more nuanced descriptions of methodology. Editorial boards of journals with higher impact factors (e.g., Remote Sensing of Environment) began demanding more explanation in a method section than the use of a software package. The success of the suite of eCognition/Definiens software packages likely prompted rapid development of alternative software for the OBIA workflow. Berkeley Image Seg (Clinton et al. 2010), Visual Learning Systems' Feature Analyst extension for ArcGIS (Visual Learning Systems 2008), System for Automated Geoscientific Analyses (SAGA) (Böhner et al. 2006), Environment for Visualizing Images (ENVI) Feature Extraction (Hölbling and Neubert 2008), ERDAS IMAGINE's objective module (ERDAS 2009), and IDRISI Taiga's segmentation module (Clark Labs 2009) appeared between 2006 and 2010. Use of additional, external software, particularly for the classification step of the OBIA workflow, became increasingly common. For example, many papers discuss the use of decision trees such as classification and regression trees (CART), usually run externally to a software package such as eCognition, in R (http://www.r-project.org/) or See5 (Quinlan 2013), to classify objects. Yu et al. (2006) used this approach to map vegetation alliances in a California reserve; Laliberte et al. (2007) did so with high-resolution data over rangelands, as did Chubey et al. (2006) for forest inventory mapping. Green and Lopez (2007) used CART to label polygons created in eCognition for benthic habitat in Texas, and Stow et al. (2007) used a similar combined approach to map urban areas in Accra, Ghana. Since then, eCognition has implemented a decision tree algorithm for classification.

Next to the commercial software mentioned, several open-source software products have been developed. While earlier attempts may be considered to be more of an academic, not very user-friendly and not well-documented prototypical software such as GeoAida (Bückner et al. 2001), recent open-source developments aim to compete with commercial software such as eCognition, ERDAS, or ENVI in respect to a modern user-friendly GUI and software documentation. InterIMAGE is an open-source and free-access framework for knowledge-based image classification. It is based on algorithms from GeoAida and provides a capacity for customization and extension tools. Costa et al. (2010) describe the InterIMAGE system as a multiplatform framework, implemented for Linux and Windows operational systems (http://www.lvc.ele.puc-rio.br/projects/interimage/).

A more recent development is the Geographic Data Mining Analyst. It bridges GIS and image-processing functionality and includes algorithms for segmentation, feature extraction, feature selection, classification, landscape metrics, and multitemporal methods for change detection and analysis (Körting et al. 2013).

Bunting et al. (2014) developed the open-source platform RSGISLib for data-processing techniques. Users interact with the software through an XML script, where XML tags and attributes are used to parameterize 300 available commands. The developers claim that command options are easily recognizable to the user because of their logical and descriptive names. Through the XML interface, processing chains and batch processing are supported. More recently, a Python binding has been added to Remote Sensing and GIS Software Library (RSGISLib) allowing individual XML commands to be called as Python functions. The software has been released under a GPL3 (General Public License) license and makes use of a number of other open-source software libraries (e.g., Geospatial Data Abstraction Library (GDAL)/OGR); a user guide and the source code are available at http://www.rsgislib.org.

14.2.3.3 Consolidation (Since Around 2010)

Since around 2010, the field has emerged from its earlier stages and is displaying more maturity. Blaschke et al. (2014) raise the discussion that in some ways, this maturity suggests a label of new *paradigm*. From a workshop on OBIA convened at the 2012 GIScience Conference in Columbus, OH, to discuss key theoretical and applied aspects of the approach emerged several important topics for the next decade: integration with GIS, semantics, accuracy, change, standards, and learning from the past. These themes are born out in the literature. There have also been some important developments on the software front. For example, in 2010, Trimble (a company expert in field and mobile technology and one of the leading manufacturers of research and survey grade GPS systems) purchased Definiens Earth Sciences ("Trimble Acquires Definiens' Earth Sciences Business to Expand its GeoSpatial Portfolio": https://www.trimble.com/news/release.aspx?id=061110a), with expectations that the OBIA workflow would be of particular use to mobile mapping, survey, and urban environment reconnaissance. Additionally, there has been increasing use in the remote-sensing world of unmanned aerial systems (UAS) or drones, which provide small footprint, very high-resolution imagery (cm to meter pixel size). Once geometric and radiometric corrections and mosaicking have been applied, these images are routinely being approached with the OBIA workflow. UAS provide the ability for repeated deployment for acquisition of multispectral imagery at high temporal resolution data at very high-spatial resolution. For example, Laliberte et al. (2011) acquired multispectral imagery using UAS and obtained orthorectified, radiometrically calibrated image mosaics for the purpose of rangeland vegetation classification. They relied heavily on an OBIA approach for classification of rangeland classes and achieved relatively high accuracies. Castro et al. (2013) were able to generate weed maps early in the growing season for maize fields by using an unmanned aerial vehicle and OBIA.

The current global explosion of imagery resources at high-temporal and high-spatial resolution is actively changing all aspects of the geospatial enterprise. The ways in which we acquire, store, serve, and generate information from an increasing supply of imagery across domains necessitate the continued development of streamlined OBIA workflows that render imagery useful through geospatial semantics and shared knowledge (Harvey and Raskin 2011; Blaschke et al. 2014). The time-sensitive decision support tasks found in disaster response, for example, which typically make use of rapidly acquired imagery to find targets, are often facilitated currently by human volunteers or *distributed thinking* (Zook et al. 2010). These tasks in the future might be supported by OBIA workflows. And the accelerated pace of geospatial work that accompanies disaster response is increasingly characteristic of science in general than it has ever been in the past. Decisions that routinely waited for annual, seasonal, or monthly data (e.g., forest loss, peak greenness, soil water deficits) can now be made based on data at finer spatial and temporal resolutions (e.g., Hansen et al. 2014). Doubtlessly, future research within OBIA will focus on transferring imagery quickly into comprehensive and web-enabled geographic knowledge bases to be used for decision making (Table 14.4).

14.3 OBIA: A Short Summary of the State of the Art

This section is kept very short and aims to succinctly summarize the main findings from other state-of-the-art reviews, particularly Blaschke (2010) and Blaschke et al. (2014).

14.3.1 Segmentation Is Part of OBIA but Not Married to It

A common denominator of OBIA applications was, and still is, that they are built on image segmentation (see also Burnett and Blaschke 2003; Benz et al. 2004; Liu et al. 2006; Hay and Castilla 2008; Lang 2008). Image segmentation is not at all new (Haralick and Shapiro 1985; Pal and Pal 1993) but has its roots in industrial image processing and was not used extensively in geospatial applications throughout the 1980s and 1990s (Blaschke et al. 2004).

Interestingly, not only independent from most of the OBIA-related developments described in Blaschke (2010) but also triggered by the advent of high resolution satellite imagery, Aplin et al. (1999) and Aplin and Atkinson (2001) developed an approach to segment image pixels using vector field boundaries and to assign subpixel land cover labels to the pixel segments. Subsequently, hard per-field classification, the assignment of land cover classes to fields (land cover parcels) rather than pixels (Aplin et al. 1999), was achieved by grouping and analyzing all land cover labels for all pixels and pixel segments within each individual field. Their approach was somewhat different in a sense that they aimed to classify predefined objects, namely, fields. These developments coincided later with the *OBIA community* when Paul Aplin and Geoff Smith organized

TABLE 14.4 Summary of Historic Effects and OBIA Developments

External Effects/Triggers	OBIA Developments
1972: Landsat 1 and its multispectral sensor set the standard for civilian remote-sensing applications for the next decades	
Late 1970s: image segmentation techniques are developed and are subsequently being used in image processing but not much in geospatial applications	Kettig and Landgrebe (1976) developed the first hybrid classification approach that included neighborhood aspects
Late 1999 and 2000: advent of the first two civilian 1 m resolution satellites mark a new area of high-resolution spaceborne imaging	1999/2000: commercialization of Definiens company and eCognition software
1998/1999: commercial LiDAR systems available June 2003: Orbview-3 high-resolution digital airborne cameras such as the Ultracam (Leberl and Gruber 2003)	July 2001 first scientific workshop on OBIA methods: FE/GIS'2001: Remote sensing: New sensors—innovative methods, Salzburg, Austria (German language) 2002: first book on OBIA in German language based on the 2001 workshop (Blaschke 2002) 2001–2003: first dozen papers in peer-reviewed journals
2004 onward: more high-resolution satellites, decreasing prices of data, higher accessibility	2005: First OBIA-related book for the fast developing Brazilian market (Blaschke and Kux 2005)
2005: Google Earth raised public awareness about remote-sensing imagery and subsequently increased demand for information products	OBIA workshop at the XII Brazilian remote sensing symposium, June 2005, Goiania, Brazil 2006: first OBIA conference in Salzburg, Austria 2007: OBIA workshop at UC Berkeley 2008: GEOBIA international conference in Calgary, Alberta, Canada 2009: Object-based landscape analysis workshop at the University of Nottingham, United Kingdom 2010: GEOBIA international conference in Ghent, Belgium 2012: GEOBIA international conference in Rio de Janeiro, Brazil 2014: GEOBIA international conference in Thessaloniki, Greece

a symposium on "object-based landscape analysis" in 2009 in Nottingham, United Kingdom, and edited a special issue in *International Journal of Geographical Information Science* (Aplin and Smith 2011).

Although most scientists would associate OBIA with segmentation, recent work has shown that some segmentation steps typically involved in OBIA research do not necessarily play a major role, as sometimes postulated in the earlier development of OBIA. See particularly the discussion of Tiede (2014) who in essence decouples OBIA from image processing and Lang et al. (2010, 2014) and their work on concept-related fiat objects, geons, and on *latent phenomena*.

14.3.2 Classification

Blaschke and Strobl (2001) have posed the question "What's wrong with pixels?" and elucidated some shortcomings of a pure per-pixel approach. This was certainly not the first time to highlight the limitations of treating pixels individually based on multivariate statistics. In fact, Kettig and Landgrebe (1976) developed the first algorithm called Extraction and Classification of Homogeneous Objects (ECHO), which at least partially utilizes contextual information. Based on the short history of OBIA in the section before, we may argue that around the turn of the millennium, the quest for objects reached a new dimension. Particularly for high-resolution image, it seems to make much sense to classify segments—rather than pixels. The segments may or may not correspond exactly to the objects of desire. Burnett and Blaschke (2003) called such segments from initial delimitation steps object candidates. They already offer parameters such as size, shape, relative/absolute location, boundary conditions, and topological relationships, which can be used within the classification process in addition to their associated spectral information.

There is increasing awareness that object-based methods make better use of—often neglected—spatial information implicit within remote-sensing images. Such approaches allow for a tightly coupled or even full integration with both vector- and raster-based GIS. In fact, when studying the early OBIA literature for the geospatial domain, it may be concluded that many applications were driven by the demand for classifications, which incorporate structural and functional aspects.

One good example of a comprehensive review is the paper by Salehi et al. (2012). They conducted recent literature and evaluated performances in urban land cover classifications using high-resolution imagery. They analyzed the classification results for both pixel-based and object-based classifications. In general, object-based classification outperformed pixel-based approaches. These authors reason that the cause for the superiority was the use of spatial measures and that utilizing spatial measures significantly improved the classification performance particularly for impervious land cover types.

14.3.3 Complex *Geo-Intelligence* Tasks

Increasingly, OBIA is used beyond simple image analysis tasks such as image classification and feature extraction from one image or a series of images from the same sensor.

Today, terabytes of data are acquired from space- and airborne platforms, resulting in massive archives with incredible information potential. As Hay and Blaschke (2010) argue, it is only recently that we have begun to mine the spatial wealth of these archives. These authors claim that, in essence, we are data rich but geospatial information poor. In most cases, data/image access is constrained by technological, national, and security barriers, and tools for analyzing, visualizing, comparing, and sharing these data and their extracted information are still in their infancy. In the few years since this publication, *big data* have fully arrived in many sciences, and this debate seems not to be OBIA specific from today's point of view.

Furthermore, policy, legal, and remuneration issues related to who owns (and are responsible for) value-added products resulting from the original data sources, or from products that represent the culmination of many different users input (i.e., citizen sensors), are not well understood and still developing. Thus, myriad opportunities exist for improved geospatial information generation and exploitation.

OBIA has been claimed to be a subdiscipline of GIScience devoted to developing automated methods to partition remote-sensing imagery into meaningful image objects and assessing their characteristics through scale (Hay and Castilla 2008). Its primary objective is the generation of geographic information (in GIS-ready format) from which new geo-intelligence can be obtained. Based on this argument, Hay and Blaschke (2010) have defined geo-intelligence as geospatial content in context.

The final theme is intelligence—referring to geo-intelligence—which denotes the *right* (geographically referenced) *information* (i.e., the content) in the *right situation* so as to satisfy a specific query or queries within user-specified constraints (i.e., the context).

Moreno et al. (2010) describe a geographic object-based vector approach for cellular automata modeling to simulate land-use change that incorporates the concept of a dynamic neighborhood. This represents a very different approach for partitioning a scene, compared to the commonly used OBIA segmentation techniques, while producing a form of temporal geospatial information with a unique heritage and attributes.

Lang (2008) provided a more holistic perspective on an image analysis and the extraction of geospatial information or what he called at this time an upcoming paradigm. He started from a review of requirements from international initiatives like Global Monitoring of Environment and Security (now Copernicus), and he discussed in details the concept of *class modeling*. Also, such methods may need further advancement of the required adaptation of standard methods of accuracy assessment and change detection. He introduced the term *conditioned information*. With this term, he addresses processes that entail the creation of new geographies as a flexible, yet statistically robust and (user-) validated unitization of space.

Lang et al. (2014) developed the concept of geons as a strategy to represent and analyze latent spatial phenomena across different geographical scales (local, national, and regional) incorporating domain-specific expert knowledge. The authors exemplified how geons are generated and explored. So-called composite geons represent functional land-use classes, required for regional-planning purposes. They are created via class modeling to translate interpretation schemes from mapping keys. Integrated geons, on the other hand, address abstract, yet policy-relevant phenomena such as societal vulnerability to hazards. They are delineated by regionalizing continuous geospatial data sets representing relevant indicators in a multidimensional variable space. In fact, the geon approach creates spatially exhaustive sets of units, scalable to the level of policy intervention, homogenous in their domain-specific response, and independent from any predefined boundaries. Despite its validity for decision making and its transferability across scales and application fields, the delineation of geons requires further methodological research to assess their statistical and conceptual robustness.

14.4 Ongoing Developments: Influences of OBIA to Other Fields and Vice Versa

14.4.1 GIScience and Remote Sensing

OBIA arguably has its roots firmly in the field of remote sensing. Developments in remote sensing through the decades of the 2000–2010s—including most importantly the widespread availability of high-resolution imagery globally, but also from LiDAR and novel methods of data fusion—have continued this alliance. However, this early grounding of OBIA in theoretical and practical aspects of remote sensing is recently being enhanced through multiple novel interactions with aspects of the GIScience field, and OBIA is poised to develop further from new trends in GIScience.

Since Goodchild (1992) first coined the term GIScience, suggesting it as a manner of dealing with the issues raised by GIS technology by focusing on the unaddressed theoretical shortcomings of conventional GIS, the contents and borders have constantly shifted, especially in light of recent advances in geospatial technologies, including remote sensing (Blaschke and Merschdorf 2014). In order to deal with the special properties of spatial information in an era of Web 2.0 technologies, the field of GIScience has embraced not only classic geographical knowledge and concepts but also increasingly incorporated approaches from other disciplines such as computer science and cognitive sciences (Blaschke and Merschdorf 2014). In turn, other disciplines have recently discovered the potential of GIScience, utilizing its tools and methodologies to serve their own needs and to drastically advance the knowledge base in their own respective fields. Such is not least the case for remote sensing, which has experienced a drastic shift from purely pixel-based methods of image interpretation to the identification of *objects* in remotely sensed imagery by means of OBIA. Hay and Castilla (2006) propose that OBIA is a subdiscipline of GIScience, combining a "unique focus on remote sensing and GI" (Hay and Castilla 2006:1). In this sense, OBIA may be seen as the first in a string of developments leading to the consolidation of GIS and remote sensing, facilitated through the common denominator of GIScience. This implies that current and ongoing developments in the discipline of GIScience may bare a significant impact on the field of remote sensing. Such developments include but are not limited to volunteered geographic information (VGI), ubiquitous sensing, indoor sensing, and the integration of in situ measurements with classic remote-sensing datasets.

Web 2.0 technologies have had a significant impact on GIScience, as they have enabled the bidirectional and participatory use of the Internet (Blaschke and Merschdorf 2014). These technologies go beyond *GIS-centered* assemblages of hardware, software, and functionalities. Wiki-like collective mapping environments, geovisualization APIs, and geotagging may either be based on GIS or they have common denominators in the digital storage, retrieval, and visualization of information based upon its geographic content (Sheppard 2006).

These developments have led to an influx of spatial content, contributed by individual users or groups of users, which nowadays composes a valuable data source in GIS. Such content has been termed as "volunteered geographic information," by Goodchild (2007), and Atzmanstorfer and Blaschke (2013) claim that its full realm of possibilities, in terms of citizens partaking in planning initiatives, yet remains unknown. VGI is not only limited to online applications such as the provision of geotagged photographs on the photo management service Flickr or geolocated messages on the online messaging portal Twitter but also includes the information collected by wireless sensors on common mobile devices. Due to the proliferation of wireless sensors in all sorts of mobile devices, sensory data collection is no longer constrained to few experts equipped with expensive sensors but rather has shifted more into the lay domain. In GIScience, this notion is referred to as ubiquitous sensing and can be used for monitoring activities and locations of users, or groups of users, in near real time. The near real-time capabilities of ubiquitous sensing can assist decision makers in a variety of applications, such as emergency response, public safety, traffic management, environmental monitoring, or public health (Resch 2013). For example, Sagl et al. (2012) utilize the movements of cell phones between pairs of radio cells—termed as handovers—in order to analyze spatiotemporal urban mobility patterns and demonstrate how mobile phone data can be utilized to analyze patterns of real-world events using the example of a soccer match, while Zook et al. (2010) present how a mash-up of various data sources, including both government data and VGI, significantly contributed to disaster relief in Haiti, following the earthquake in 2010.

While VGI is oftentimes a passive by-product, resulting from the use of Web 2.0 technologies and mobile-computing devices, millions of internet users can nowadays choose to actively utilize GIS methodologies and applications by means of public participation geographic information system (PPGIS). Manifestations of such participation can, for instance, be found in the widespread community of users contributing to virtual globes and maps by superimposing new layers, such as street networks or landmarks, or even in disaster relief efforts such as the recent search for the debris of the missing Malaysian Airline flight MH370, which was assisted by tens of thousands of Internet users, who helped in sifting through the vast magnitude of satellite data recorded during the time frame in question.

The contribution of the general public, be it actively by uploading data to virtual globes or maps or passively by utilizing social media platforms such as Twitter or Flickr, has also fuelled the collection of in situ data, such as photos taken at a certain location and values measured there. Such data are particularly valuable in the era of very high-resolution satellite imagery, as well as the subsequent surge of urban remote-sensing applications, such as the mapping of megacities, the monitoring of fast-expanding settlements in developing countries, or the routine monitoring of informal settlements, conducted either by public administration or by commercial companies, as outlined by Blaschke et al. (2011). Based on an extensive literature review, Blaschke et al. (2011) conclude that the increased availability of high-resolution satellite imagery has resulted in a greater demand for timely urban mapping and monitoring. However, remotely sensed imagery, which provides the basis for urban mapping applications, can only provide the bird's-eye view of a given location, neglecting ground information such as the building facades or interiors. With the advent of widely applied Open Geospatial Consortium (OGC) standards, in situ measurement data recorded at ground locations can be integrated with the remote-sensing imagery, providing a more holistic approach to urban-mapping applications (Blaschke et al. 2011). Blaschke et al. (2011) note that although remote sensing and in situ measurements are currently two separate technologies, the strengths of both can be combined by means of sensor webs and OGC standards, potentially producing new and meaningful information (Blaschke et al. 2011). They conclude that "while available information will always be incomplete, decision makers can be better informed through such technology integration, even if loosely coupled" (Blaschke et al. 2011:1768).

Another trend enabled by the recent advances in mobile technology is the concept of indoor sensing, sometimes referred to as indoor geography (Blaschke and Merschdorf 2014). Naturally, remote-sensing imagery can only provide a planar view of the Earth's surface, including natural features, as well as human infrastructure. While LiDAR technology complements the classic 2D imagery with the added dimension of depth, it still doesn't provide any insight as to the contents of buildings. In this sense, indoor sensing may be a future trend in indoor positioning and mapping, whereby sensor fusion will evolve to support indoor locations, paving the way for geoenabled manufacturing (Blaschke and Merschdorf 2014).

14.4.2 Changing Workplace

In the past, remote sensing and GIS were distinctly separated disciplines, whereby remotely sensed imagery was primarily considered as a data source for GIS (Jensen 1996). However, in light of more recent technical and theoretical advancements, these disciplines have begun to consolidate, not least attributed to the quest for tangible objects. The emergence of OBIA as a subdiscipline of GIScience laid a foundation for the use of shared methodologies, and remote sensing was recognized as "one element of an integrated GIS environment, rather than simply an important data source" (Malczewski 1999:20). The bidirectional nature of the relationship between remote sensing and GIS implies that not only advances in remote sensing technology influence the GIS environment but also vice versa. In this sense, we can witness the impact of recent trends in GIScience, described in Section 14.4.1, on the remote-sensing discipline. Especially, the technological advances brought about by the Web 2.0, such as VGI, ubiquitous sensing, or PPGIS, call for new approaches of data integration, with the primary aim of developing more comprehensive and accurate datasets. Such integration can complement the bird's-eye view perspective offered by remotely sensed imagery, with in situ information, which in turn can more efficiently represent dynamic urban environments (Blaschke et al. 2011). To this end, OGC standards can provide the necessary interface for data integration, as is the case for the Global Earth Observing System of Systems, which seamlessly integrates remotely sensed imagery with in situ measurements.

One particular example of an OBIA application as a substitute for GIS overlay is provided by Tiede (2014). GIS-overlay routines usually build on relatively simple data models. Topology is—if at all—calculated on the fly for very specific tasks only. If, for example, a change comparison is conducted between two or more polygon layers, the result leads mostly to a complete and also very complex from–to class intersection. Additional processing steps need to be performed to arrive at aggregated and meaningful results. To overcome this problem, Tiede (2014) presented an automated geospatial overlay method in a topologically enabled (multiscale) framework. The implementation works with polygon and raster layers and uses a multiscale vector/raster data model developed in the OBIA software eCognition. Advantages are the use of the software inherent topological relationships in an object-by-object comparison, addressing some of the basic concepts of object-oriented data modeling such as classification, generalization, and aggregation. Results can easily be aggregated to a change-detection layer; change dependencies and the definition of different change classes are interactively possible through the use of a class hierarchy and its inheritance (parent–child class relationships). The author demonstrates the flexibility and transferability of change comparison for Corine Land Cover data sets. This is only one example where OBIA and GIS are fully integrated, and although this case may be being an exception so far, one field may jeopardize the other field if the fields are seen isolated.

14.4.3 Who Uses OBIA?

In a recent publication, Blaschke et al. (2014) found an increasing number of publications concerned with OBIA in peer-reviewed journals, special issues, books, and book chapters and concluded that OBIA is a new evolving paradigm in remote sensing and to some degree in GIScience also. However, they also noted that the exact terminology used within these publications is distinctly ambiguous, as is characteristic for an emerging multidisciplinary field (Blaschke et al. 2014). Therefore, we herein aim to review the literature databases of the ISI's (Institute for Scientific Information) Web of Science (WoS), as well as Scopus, in an attempt to quantify who uses OBIA, both in terms of countries of origin and contributing field, and to track its presence in literature over the past years.

A search in the WoS database for the phrases *object-based image analysis* or *object-oriented image analysis*, or *OBIA*, or *geographic object-based image analysis, (GEOBIA)*," contained in the title, abstract, or keywords, returns a total of 451 articles (April 17, 2014). When analyzing which countries the publications primarily come from, we determined that the highest number of publications is contributed by the United States, accounting for 24% of all publications; followed by the People's Republic of China with a 14% contribution; Germany contributing 12%; Austria 8%; Canada 7%; Australia, Brazil, and Netherlands 6%, respectively; and Italy and Spain with 4% each, just to name the top 10 contributing countries. This shows that while the United States is the main contributor, accounting for nearly a quarter of all publications returned in the search, many other smaller countries also make a noteworthy contribution. In particularly remarkable is the 8% contribution made by Austria, which has only a fraction of the population (approx. 8.5 million) compared to most other countries represented within the top 10. Compared to the leading country—United States—Austria has merely 2.7% of the population but has 33% as many publications. Such a comparison becomes even more extreme when made with China, the second largest contributor, whereby Austria has only 0.6% as many inhabitants but accounts for 57% as many publications. This shows that there may be certain research clusters in certain countries, which largely contribute to OBIA/GEOBIA research, rather than all countries contributing relatively to their population (Figure 14.3).

A further analysis consisting of the research areas contributing to OBIA/GEOBIA reveals that the largest contribution is made by remote sensing, accounting for 61% of all publications. The second largest contributor, namely, imaging science, accounts for only 31%, followed by geology with a share of 27%. A full chart of the top 10 contributing fields is depicted in Figure 14.4.

When assessing the publication years, it is notable that the number of publications on the topic of OBIA/GEOBIA has drastically increased over the last 5 years, whereby 22% were written in 2013 alone, as compared to >16% prior to 2008. The first OBIA publication indexed in ISI's WoS database dates back to 1985, preceding the second OBIA publication by 10 years, and at least 20 years prior to a steady incline in the number of publications (Figure 14.5).

When the same search is conducted in the Scopus database (same phrases searched for in title, abstract, and keywords), a total of 586 publications are returned (April 17, 2014). The discrepancy in terms of numbers of publications as compared to the WoS database can be attributed to the fact that Scopus contains a broader range of document types, such as notes, short surveys, and in press articles, while the WoS database only contains peer-reviewed journal articles, conference proceedings, reviews, and editorials, all of which are additionally included in Scopus.

Although including a slightly greater number in overall publications, the trends revealed in the Scopus data are largely in line with those depicted in the WoS data. Some discrepancies were found in terms of research areas, which, however, may largely be down to the different naming conventions utilized by both databases (e.g., the top contributing discipline to OBIA/GEOBIA in the Scopus database is "Earth and Planetary Sciences," with a total of 315 publications or 54%, which corresponds to the largest WoS

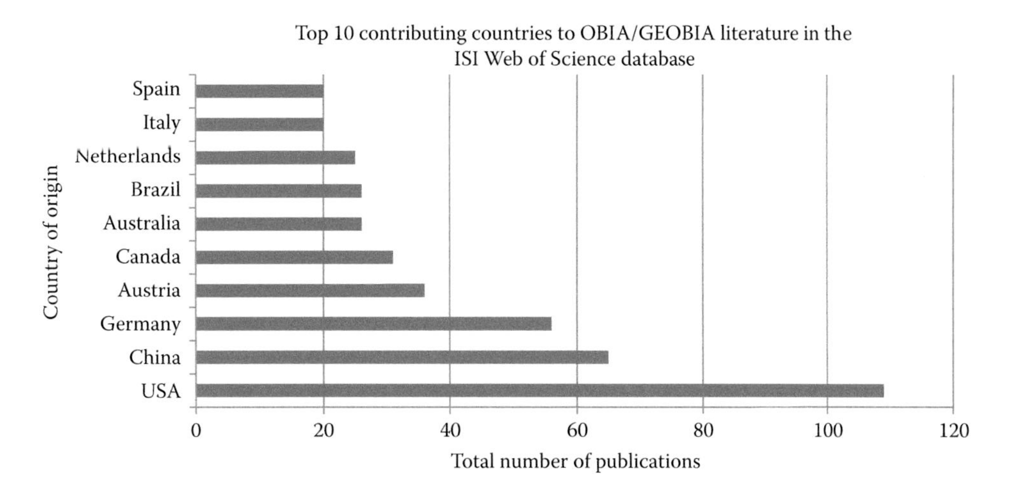

FIGURE 14.3 Top 10 contributing countries to the OBIA/GEOBIA literature in the Web of Science database and their respective contributions.

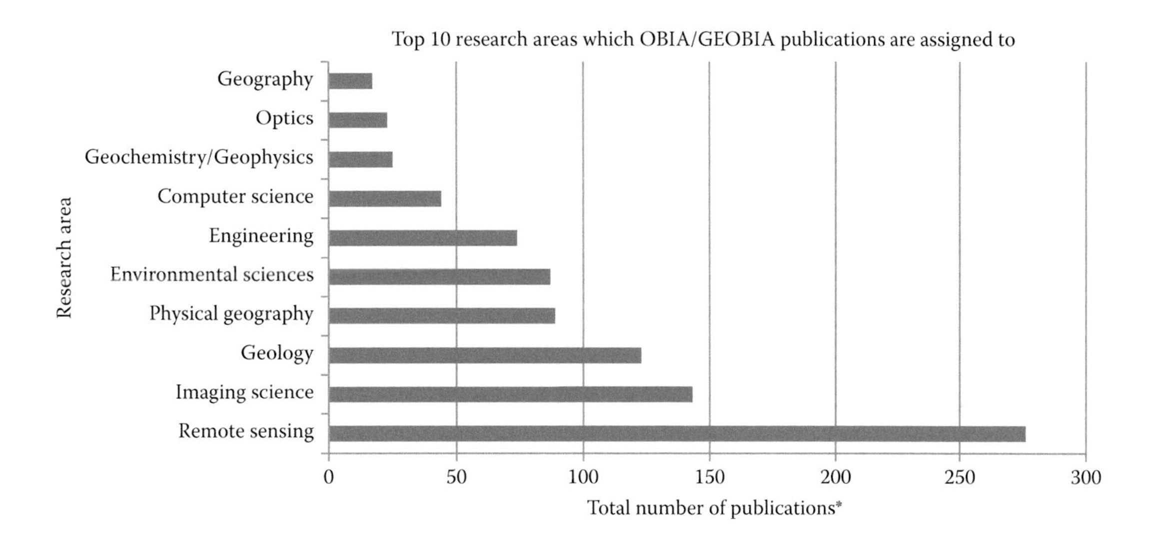

FIGURE 14.4 Main research areas for OBIA/GEOBIA publications in ISI's Web of Science database. *The total numbers add up to more than the total of 451 publications due to the fact that some multidisciplinary publications may have been assigned to more than one research area.

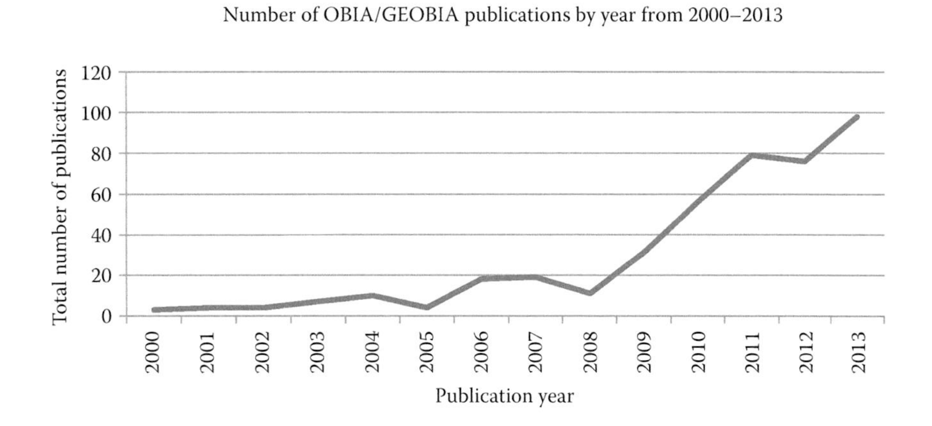

FIGURE 14.5 Number of OBIA/GEOBIA publications by publication year from 2000 to 2014, as indexed in ISI's Web of Knowledge database.

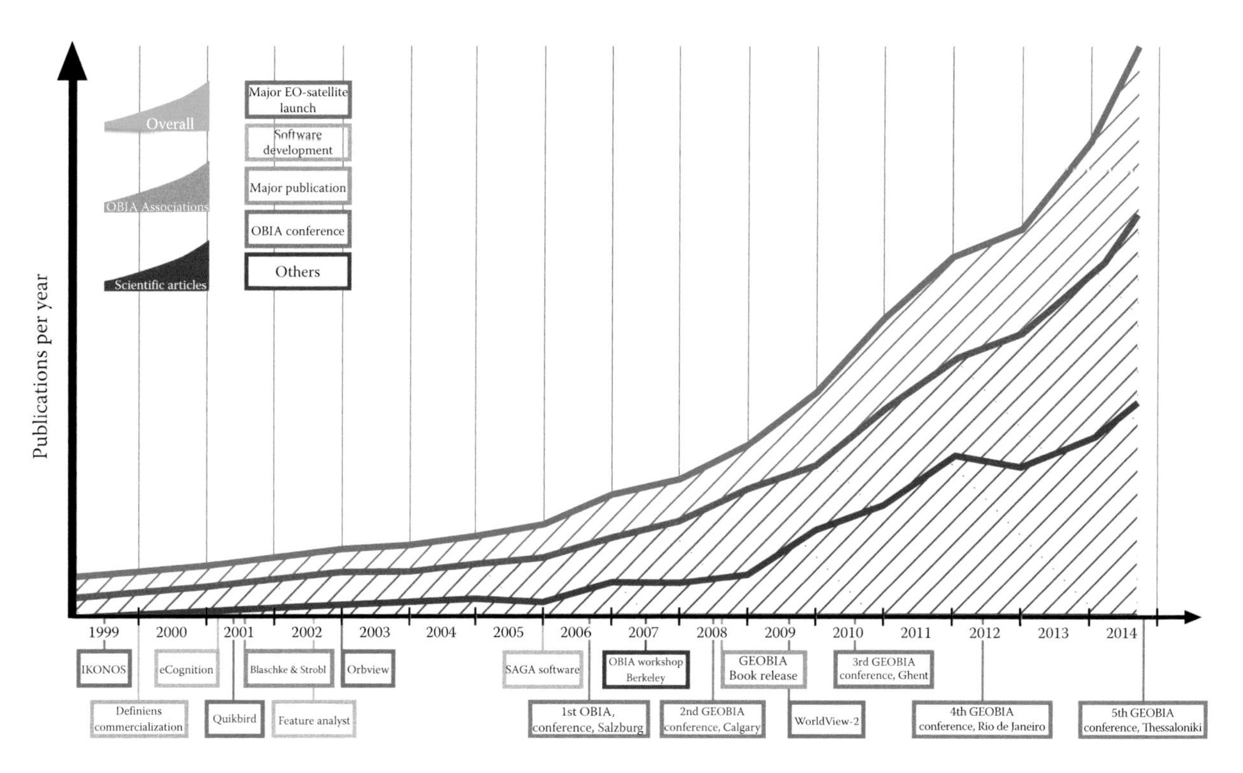

FIGURE 14.6 Milestone timeline of OBIA/GEOBIA development from the late 1990s 'until today.

contributor of "Remote Sensing"). Furthermore, both the publication year timeline and the contributing countries roughly correspond to the results obtained from the analyses of the WoS data.

In conclusion, when analyzing the literature, and some key milestone events and publications, the rise of OBIA/GEOBIA can be clearly traced through the course of the last decade and a half. This is depicted in the timeline shown in Figure 14.6, which exemplifies how both technological and methodological advances gave birth to object-oriented approaches and, according to Blaschke et al. (2014), to a new paradigm in remote sensing although it must be clearly stated that in absolute terms *classic* per-pixel methods are represented way more in publications at the moment.

14.5 Concluding Perspectives

14.5.1 New Paradigm: The Need for a Common Nomenclature and Sound Methodologies

OBIA has certainly arrived at the sciences. While the first years of the development were characterized by a lack of high-quality peer-reviewed scientific publications, the last few years witnessed a sharp increase in such articles. Some of them are remarkably highly cited such as the review paper by Blaschke (2010). Much of the excitement about this new methodology or paradigm has to do with the increasing availability of high-resolution datasets, which can now be used to produce information and, in particular, information on demand or *conditioned information*. Some predict that researchers, policy makers, citizen groups, and private institutions might use information contributed by ordinary people for any number of purposes, including emergency response, mobilizing activist efforts, monitoring environmental change, filling gaps in existing spatial databases, or identifying and addressing needs and problems in urban neighborhoods.

OBIA has developed a rich array of approaches for grappling with the challenges associated with high-resolution data. One remaining task is to standardize terms across methods and methodologies being used. While Blaschke et al. (2014) argue that this is very common for a new paradigm, it is nevertheless troublesome. OBIA needs to urgently harmonize and streamline the terms being used. Otherwise, a widespread recognition from other fields may be hindered.

14.5.2 Toward a Civilian Geo-Intelligence

We do not exactly know how the future will look like. One possible development can best be illustrated by the power and the innovative potential of the *object-by-object change comparison framework* (Tiede 2014) described before. This framework yields flexible, transferable, and highly complex change comparisons that can be visualized or calculated and aggregated to higher level composite objects. Here, the geon concept of Lang et al. (2014) comes into play: as briefly described, it also allows for the creation of more conceptual objects that may represent latent phenomena, which are not directly mappable.

Hay and Blaschke (2010) suggested the term (civilian) geo-intelligence. Since then, the number of technical developments and the number of documented applications, which may support the hypothesis of locational intelligence, have clearly grown. As discussed earlier, Lang (2008) laid some theoretical foundations for the concept of *conditioned information*, and Lang et al. (2014) developed the concept of geons, which may also serve as units to characterize and delimitate latent phenomena.

An area for future research emerges from a wider set of organizational changes within the software industry such as the *software as a service* paradigm. This is a significant development in the organization and deployment of remote-sensing image analysis for the professional and advanced users. It may also open opportunities for nonexpert users in remote sensing in general and for OBIA in particular. Tiede et al. (2012) presented an OBIA geoprocessing service that integrates OBIA methods into a geoprocessing service. This development was—to our best knowledge—the first integration of an eCognition-based OBIA application into an interactive WebGIS geoprocessing environment.

Interestingly, the emergence of OBIA has not been generating a substantial quantity of critical reflection neither about the technology as such nor about the wider scientific and technological implications of this paradigm for various user groups, both geographically and seen along an educational ladder (students–graduates–professionals in private industry and academia).

Another future research area concerns the remote sensing and GIScience practices of nonprofessional actors, not as outsourced operatives for research institutions but as private actors or NGOs following their own agendas. In remote sensing in general, much more than in GIScience, the vast majority of existing literature investigates widely agreed scientific or commercially interesting problems and reflects both the focus of an Anglophone research community looking primarily back in time and a focus primarily on activities in the global North by state actors. Although we did not carry out a severe literature study, we may speculate that OBIA researchers may be a little bit less Anglophone dominated than the general remote-sensing community.

14.5.3 Epistemological and Ontological Challenges

We may claim here for the field of remote sensing that some long-known principles about technological determinism (McLuhan 1964 who basically claimed that humans shape their tools and they in turn shape humans) may become more obvious today because its practical and theoretical implications are now much faster discovered. Nevertheless, the process of the social shaping of technology can be long term, interactive, and sometimes conflict ridden (Rohracher 2003).

Like GIS, which has for some years been decried as *ontologically shallow* and insufficient to the task of comprehending the many epistemological points of difference among users (Schuurman 1999), remote-sensing literature offers very little in regard to its ontological and epistemological foundation. Without doubt, remote-sensing principles have solid foundations in physics. Only through the amalgamation with GIS-principles with OBIA the need for a theoretical, that is, epistemological and ontological, establishment increases. As long as the pixel is more or less the only subject of studies and, more importantly, as long as objects of interest are smaller than or similar in size compared to the pixels, such questions may not be urgent. With the advent of high-resolution imagery, the question "What's wrong with pixels?" (Blaschke and Strobl 2001) is valid to be asked. In fact, concerns about the appropriate use of technology in the application of remote-sensing data suggest that nonexpert users involved in interpretation tasks may gain a relatively sophisticated understanding not just of what the technology can do but of the processes involved in visualizing and disseminating findings via interactive representations and WebGIS.

We refer to Pickles (2004) who contends that the contingent nature of technical outcomes from GIS use is often overlooked, and the exploitation of some groups, particularly those with less access to technology, becomes a real possibility. He also emphasizes how important it is "to study maps in human terms, to unmask their hidden agendas, to describe and account for their social embeddedness and the way they function as microphysics" (Pickles 2004, p. 181).

Lastly, we may call for a relaxation of a potential friction between OBIA and per-pixel approaches. There are dozens, most likely more than a hundred, of scientific papers that compare both methods. Nevertheless, the future may not be dominated by an *either-or* question. Rather, we should be cautious about abandoning too hastily the concepts and terminologies of the *old* paradigm with reference to its dazzling object of cognition in this debate—the pixel. The pixel is a technical construct that may be useful in many cases from a technical, that is, data acquisition, point of view but sometimes also as a cognitional prerogative. In this sense, the aforementioned question "what's wrong with pixels" (Blaschke and Strobl 2001) may appear in a less unfavorable light—for the latter, the pixels.

References

Aplin, P., Atkinson, P. M., and Curran, P. J. (1999). Fine spatial resolution simulated satellite sensor imagery for land cover mapping in the UK. *Remote Sensing of Environment*, 68, 206–216.

Aplin, P. and Atkinson, P. M. (2001). Sub-pixel land cover mapping for per-field classification. *International Journal of Remote Sensing*, 22(14), 2853–2858.

Aplin, P. and Smith, G. M. (2011). Introduction to object-based landscape analysis. *International Journal of Geographical Information Science*, 25(6), 869–875.

Arvor, D., Durieux, L., Andrés, S., and Laporte, M. A. (2013). Advances in geographic object-based image analysis with ontologies: A review of main contributions and limitations from a remote sensing perspective. *ISPRS Journal of Photogrammetry and Remote Sensing*, 82, 125–137.

Atzmanstorfer, K. and Blaschke, T. (2013). Geospatial web: A tool to support the empowerment of citizens through e-participation? In C. Nunes Silva (ed.), *Handbook of Research on E-Planning: ICTs for Urban Development and Monitoring* (pp. 144–171). Hershey, PA: IGI-Global.

Baatz, M., Hoffmann, C., and Willhauck, G. (2008). Progressing from object-based to object-oriented image analysis. In T. Blaschke, Lang, S., and Hay, G. J. (eds.), *Object-Based Image Analysis*. Berlin, Germany: Springer.

Baatz, M. and Schäpe, A. (2000). Multiresolution segmentation: an optimization approach for high quality multi-scale image segmentation. *Angewandte Geographische Informationsverarbeitung XII* (pp. 12–23).

Benz, U. C., Hofmann, P., Willhauck, G., Lingenfelder, I., and Heynen, M. (2004). Multi-resolution, object-oriented fuzzy analysis of remote sensing data for GIS-ready information. *ISPRS Journal of Photogrammetry and Remote Sensing*, 58(3), 239–258.

Benz, U. C., Hofmann, P., Willhauck, G., Lingenfelder, I., and Heynen, M. (2008). Multi-resolution, object-oriented fuzzy analysis of remote sensing data for GIS-ready information. *ISPRS Journal of Photogrammetry and Remote Sensing*, 58(3–4), 239–258.

Blaschke, T. (2010). Object based image analysis for remote sensing. *ISPRS International Journal of Photogrammetry and Remote Sensing*, 65(1), 2–16.

Blaschke, T. (ed.). (2002). *Fernerkundung und GIS: Neue Sensoren—Innovative Methoden*. Karlsruhe, Germany: Wichmann Verlag.

Blaschke, T., Burnett, C., and Pekkarinen, A. (2004). New contextual approaches using image segmentation for object-based classification. In F. De Meer and de Jong, S. (eds.), *Remote Sensing Image Analysis: Including the Spatial Domain* (pp. 211–236). Dordrecht, the Netherlands: Kluwer Academic Publishers.

Blaschke, T., Hay, G. J., Kelly, M., Lang, S., Hofmann, P., Addink, E., Feitosa, R., van der Meer, F., van der Werff, H., Van Coillie, F., and Tiede, D. (2014). Geographic object-based image analysis: A new paradigm in remote sensing and geographic information science. *ISPRS International Journal of Photogrammetry and Remote Sensing*, 87(1), 180–191.

Blaschke, T., Hay, G. J., Weng, Q., and Resch, B. (2011). Collective sensing: Integrating geospatial technologies to understand urban systems—An overview. *Remote Sensing*, 3(8), 1743–1776.

Blaschke, T., Lang, S., and Hay, G. J. (eds.). (2008). *Object-Based Image Analysis, Spatial Concepts for Knowledge-Drives Remote Sensing Applications*. Lecture Notes in Geoinformation and Cartography. Berlin, Germany: Springer-Verlag.

Blaschke, T., Tiede, D., and Lang, S. (2006). An object-based information extraction methodology incorporating a-priori spatial information. Paper presented at the *ESA Conference*, Madrid, Spain.

Blaschke, T. and Kux, H. (2005). *Sensoriamento Remoto e SIG acançados: Novos sistemas sensores métodos inovadores*. Sao Paulo, Brazil: Oficina de Textos.

Blaschke, T. and Lang, S. (2006). Object based image analysis for automated information extraction—A synthesis. Paper presented at the *Measuring the Earth II ASPRS Fall Conference*, San Antonio, TX, November 6–10.

Blaschke, T. and Merschdorf, H. (2014). Geographic Information Science as a multidisciplinary and multi-paradigmatic field. *Cartography and Geographic Information Science*, 41(3), 196–213.

Blaschke, T. and Strobl, J. (2001). What's wrong with pixels? Some recent developments interfacing remote sensing and GIS. *GIS—Zeitschrift für Geoinformationssysteme*, 14(6), 12–17.

Bock, M., Rossner, G., Wissen, M., Remm, K., Langanke, T., Lang, S., T., Klug, H., Blaschke, T., and Vrscay, B. (2005). Spatial indicators for nature conservation from european to local scale. *Environmental Indicators*, 5(4), 322–328.

Bunting, P., Clewley, D., Lucas, R. M., and Gillingham, S. (2014). The Remote Sensing and GIS Software Library (RSGISLib). *Computers & Geosciences*, 62, 216–226.

Burnett, C. and Blaschke, T. (2003). A multi-scale segmentation/object relationship modelling methodology for landscape analysis. *Ecological Modelling*, 168(3), 233–249.

Böhner, J., Blaschke, T., and Montanarella, L. (2008). SAGA—Seconds out. *Hamburger Beiträge zur Physischen Geographie und Landschaftsökologie* (vol. 19). Hamburg, Germany.

Bückner, J., Pahl, M., Stahlhut, O., and Liedtke, C.-E. (2001). GEOAIDA a knowledge based automatic image data analyser for remote sensing data. Paper presented at the *ICSC Congress on Computational Intelligence Methods and Applications—CIMA*, Bangor, U.K.

Camargo, F. F., Almeida, C. M., Costa, G. A. O. P., Feitosa, R. Q., Oliveira, D. A. B., Heipke, C., and Ferreira, R. S. (2012). An open source object-based framework to extract landform classes. *Expert Systems with Applications*, 39(1), 541–554.

Castro, A. I. D., López Granados, F., Gómez-Candón, D., Peña Barragán, J. M., Novella, C., José, J., and Jurado-Expósito, M. (2013). In-season site-specific control of cruciferous weeds at broad-scale using quickbird imagery. 9th European Conference on Precision Agriculture ECPA (July 7–11, 2013).

Chen, G., Hay, G. J., Carvalho, L. M., and Wulder, M. A. (2012). Object-based change detection. *International Journal of Remote Sensing*, 33(14), 4434–4457.

Chubey, M. S., Franklin, S. E., and Wulder, M. A. (2006). Object-based analysis of Ikonos-2 imagery for extraction of forest inventory parameters. *Photogrammetric Engineering and Remote Sensing*, 72(4), 383–394.

Cleve, C., Kelly, M., Kearns, F. R., and Moritz, M. (2008). Classification of the wildland–urban interface: A comparison of pixel-and object-based classifications using high-resolution aerial photography. *Computers, Environment and Urban Systems*, 32(4), 317–326.

Clinton, N., Holt, A., Scarborough, J., Yan, L., and Gong, P. 2010. Accuracy assessment measures for object-based image segmentation goodness. *Photogrammetric Engineering and Remote Sensing*, 76, 289–299.

Colwell, R. N. (1964). Aerial photography—A valuable sensor for the scientist. *American Scientist*, 52(1), 16–49.

Colwell, R. N. (1965). The extraction of data from aerial photographs by human and mechanical means. *Photogrammetria*, 20(6), 211–228.

Colwell, R. N. (1973). Remote sensing as an aid to the management of earth resources. *American Scientist*, 61(2), 175–183.

Costa, G. A. O. P., Feitosa, R. Q., Fonseca, L. M. G., Oliveira, D. A. B., Ferreira, R. S., and Castejon, E. F. (2010). Knowledge-based interpretation of remote sensing data with the InterIMAGE system: Major characteristics and recent developments. Paper presented at the *Third International Conference on Geographic Object-Based Image Analysis (GEOBIA 2010)*, June 29–July 2, Ghent, Belgium.

Cracknell, A. P. (1998). Synergy in remote sensing—What's in a pixel? *International Journal of Remote Sensing*, 19(11), 2025–2047.

De Chant, T. and Kelly, M. (2009). Individual object change detection for monitoring the impact of a forest pathogen on a hardwood forest. *Photogrammetric Engineering & Remote Sensing*, 75(8), 1005–1013.

Dorren, L. K., Maier, B., and Seijmonsbergen, A. C. (2003). Improved Landsat-based forest mapping in steep mountainous terrain using object-based classification. *Forest Ecology and Management*, 183(1–3), 31–46.

Durieux, L., Lagabrielle, E., and Nelson, A. (2008). A method for monitoring building construction in urban sprawl areas using object-based analysis of Spot 5 images and existing GIS data. *ISPRS Journal of Photogrammetry and Remote Sensing*, 63(4), 399–408.

Eastman, J. R. (2009). *IDRISI Taiga Guide to GIS and Image Processing*. Worcester, MA: Clark Labs Clark University.

Ebert, A., Kerle, N., and Stein, A. (2009). Urban social vulnerability assessment with physical proxies and spatial metrics derived from air-and spaceborne imagery and GIS data. *Natural Hazards*, 48(2), 275–294.

ERDAS. (2009). Remote sensing digital image processing software Tutorial.

Flanders, D., Hall-Beyer, M., and Pereverzoff, J. (2003). Preliminary evaluation of eCognition object-based software for cut block delineation and feature extraction. *Canadian Journal of Remote Sensing*, 29(4), 441–452.

Fu, K.-S. and Mui, J. (1981). A survey on image segmentation. *Pattern Recognition*, 13(1), 3–16.

Gergel, S. E., Stange, Y., Coops, N. C., Johansen, K., and Kirby, K. R. (2007). What is the value of a good map? An example using high spatial resolution imagery to aid riparian restoration. *Ecosystems*, 10(5), 688–702.

Goodchild, M. F. (1992). Geographical information science. *International Journal of Geographic Information Systems*, 6, 31–45.

Goodchild, M. F. (2007). Citizens as sensors: The world of volunteered geography. *GeoJournal*, 69(4), 211–221.

Green, K. and Lopez, C. (2007). Using object-oriented classification of ADS40 data to map the benthic habitats of the state of Texas. *Photogrammetric Engineering and Remote Sensing*, 73(8), 861.

Guo, Q., Kelly, M., Gong, P., and Liu, D. (2007). An object-based classification approach in mapping tree mortality using high spatial resolution imagery. *GIScience & Remote Sensing*, 44(1), 24–47.

Hansen, M. C., Egorov, A., Potapov, P. V., Stehman, S. V., Tyukavina, A., Turubanova, S. A., and Bents, T. (2014). Monitoring conterminous United States (CONUS) land cover change with web-enabled landsat data (WELD). *Remote sensing of Environment*, 140, 466–484.

Haralick, R. M. and Shapiro, L. (1985). Survey: Image segmentation techniques. *Computer Vision, Graphics, and Image Processing*, 29, 100–132.

Harvey, F. and Raskin, R. G. (2011). *Spatial Cyberinfrastructure: Building New Pathways for Geospatial Semantics on Existing Infrastructures Geospatial Semantics and the Semantic Web* (pp. 87–96). New York: Springer.

Hay, G. J. and Blaschke, T. (2010). Special issue: Geographic object-based image analysis (GEOBIA). *Photogrammetric Engineering and Remote Sensing*, 76(2), 121–122.

Hay, G. J. and Castilla, G. (2006). Object-based image analysis: Strengths, weaknesses, opportunities and threats (SWOT). *International Archives of Photogrammetry, Remote Sensing and Spatial Information Sciences*, 36, 4–5.

Hay, G. J. and Castilla, G. (2008). Geographic Object-Based Image Analysis (GEOBIA): A new name for a new discipline. In T. Blaschke, Lang, S., and Hay, G. J. (eds.), *Object Based Image Analysis* (pp. 93–112). New York: Springer.

Hölbling, D. and Neubert, M. (2008). ENVI Feature Extraction 4.5. Snapshot. In *GIS Business* (pp. 48–51). Heidelberg, Germany: abcverlag GmbH.

Jensen, J. R. (1996). *Introductory Digital Image Processing: A Remote Sensing Perspective*. Englewood Cliffs, NJ: Prentice-Hall Inc.

Kettig, R. L. and Landgrebe, D. A. (1976). Classification of multispectral image data by extraction and classification of homogeneous objects. *IEEE Transactions on Geoscience and Remote Sensing*, 14(1), 19–26.

Koch, B., Heyder, U., and Weinacker, H. (2006). Detection of individual tree crowns in airborne LiDAR data. *Photogrammetric Engineering and Remote Sensing*, 72(4), 357–363.

Koch, D. E., Mohler, R. L., and Goodin, D. G. (2007). Stratifying land use/land cover for spatial analysis of disease ecology and risk: an example using object-based classification techniques. *Geospatial Health*, 2(1), 15–28.

Kuhn, T. S. (1962). *The Structure of Scientific Revolutions*. Chicago, IL: The Chicago University Press.

Körting, T. S., Garcia Fonseca, L. M., and Câmara, G. (2013). GeoDMA—Geographic data mining analyst. *Computers & Geosciences*, 57, 133–145.

Laliberte, A. S., Fredrickson, E. L., and Rango, A. (2007). Combining decision trees with hierarchical object-oriented analysis for mapping arid rangelands. *Photogrammetric Engineering and Remote Sensing*, 73(2), 197–207.

Laliberte, A. S., Goforth, M. A., Steele, C. M., and Rango, A. (2011). Multispectral remote sensing from unmanned aircraft: Image processing workflows and applications for rangeland environments. *Remote Sensing*, 3(11), 2529–2551.

Laliberte, A. S., Rango, A., Havstad, K. M., Paris, J. F., Beck, R. F., McNeely, R., and Gonzalez, A. L. (2004). Object-oriented image analysis for mapping shrub encroachment from 1937 to 2003 in southern New Mexico. *Remote Sensing of Environment*, 93(1–2), 198–210.

Lang, S. (2008). Object-based image analysis for remote sensing applications: Modeling reality—Dealing with complexity. In T. Blaschke, Lang, S., and Hay, G. J. (eds.), *Object-Based Image Analysis* (pp. 1–25). New York: Springer.

Lang, S., Albrecht, F., Kienberger, S., and Tiede, D. (2010). Object validity for operational tasks in a policy context. *Journal of Spatial Science*, 55(1), 9–22.

Lang, S., Kienberger, S., Tiede, D., Hagenlocher, M., and Pernkopf, L. (2014). Geons–domain-specific regionalization of space. *Cartography and Geographic Information Science*, 41(3), 214–226.

Lang, S. and Blaschke, T. (2006). Bridging remote sensing and GIS—What are the main supporting pillars? *International Archives of Photogrammetry, Remote Sensing and Spatial Information Sciences*, XXXVI-4/C42.

Leberl, F. and Gruber, M. (2003). Flying the new large format digital aerial camera Ultracam. *Photogrammetric Week*, 3, 67–76.

Liu, Y., Li, M., Mao, L., Xu, F., and Huang, S. (2006). Review of remotely sensed imagery classification patterns based on object-oriented image analysis. *Chinese Geographical Science*, 16(3), 282–288.

Malczewski, J. (1999). *GIS and Multicriteria Decision Analysis*. New York: John Wiley & Sons.

Maxwell, S. K. (2010). Generating land cover boundaries from remotely sensed data using object-based image analysis: Overview and epidemiological application. *Spatial and Spatio-Temporal Epidemiology*, 1(4), 231–237.

McKeown Jr., D. M., Harvey, W. A., and Wixson, L. E. (1989). Automating knowledge acquisition for aerial image interpretation. *Computer Vision, Graphics, and Image Processing*, 46(1), 37–81.

McLuhan, M. (1964). *Understanding Media*. New York: McGraw-Hill.

Moreno, N., Wang, F., and Marceau, D. J. (2010). A geographic object-based approach in cellular automata modeling. *Photogrammetric Engineering and Remote Sensing*, 76(2), 183–191.

Opitz, D. and Blundell, S. (2008). Object recognition and image segmentation: the Feature Analyst® approach. In Blaschke, T., Lang, S., Hay, G.J. (eds.), *Object-Based Image Analysis* (pp. 153–167). Berlin/Heidelberg, Germany: Springer.

Pal, R. and Pal, K. (1993). A review on image segmentation techniques. *Pattern Recognition*, 26(9), 1277–1294.

Pascual, C., García-Abril, A., García-Montero, L. G., Martín-Fernández, S., and Cohen, W. (2008). Object-based semi-automatic approach for forest structure characterization using LiDAR data in heterogeneous *Pinus sylvestris* stands. *Forest Ecology and Management*, 255(11), 3677–3685.

Pickles, J. (2004). *A History of Spaces: Cartographic Reason, Mapping, and the Geo-Coded World*. New York: Psychology Press.

Quinlan, R. (2013). Data Mining Tools See5 and C5.0. http://www.rulequest.com/see5-info.html.

Radoux, J. and Defourny, P. (2007). A quantitative assessment of boundaries in automated forest stand delineation using very high resolution imagery. *Remote Sensing of Environment*, 110(4), 468–475.

Resch, B. (2013). *People as Sensors and Collective Sensing-Contextual Observations Complementing Geo-Sensor Network Measurements Progress in Location-Based Services* (pp. 391–406). Berlin, Germany: Springer.

Rocchini, D., Perry, G. L., Salerno, M., Maccherini, S., and Chiarucci, A. (2006). Landscape change and the dynamics of open formations in a natural reserve. *Landscape and Urban Planning*, 77(1), 167–177.

Rohracher, H. (2003). The role of users in the social shaping of environmental technologies. *Innovation*, 16(2), 177–192.

Sagl, G., Loidl, M., and Beinat, E. (2012). A visual analytics approach for extracting spatio-temporal urban mobility information from mobile network traffic. *ISPRS International Journal of Geo-Information*, 1(3), 256–271.

Salehi, B., Ming Zhong, Y., and Dey, V. (2012). A review of the effectiveness of spatial information used in urban land cover classification of VHR imagery. *International Journal of Geoinformatics*, 8(2), 35–51.

Schuurman, N. (1999). Critical GIS: Theorizing an emerging science. *Cartographica*, 36(4), 1–101.

Sheppard, E. (2006). Knowledge production through critical GIS: Genealogy and prospects. *Cartographica*, 40, 5–21.

Silva, T. S. F., Costa, M. P. F., and Melack, J. M. (2010). Spatial and temporal variability of macrophyte cover and productivity in the eastern Amazon floodplain: A remote sensing approach. *Remote Sensing of Environment*, 114(9), 1998–2010. doi: 10.1016/j.rse.2010.04.007.

Stow, D., Lopez, A., Lippitt, C., Hinton, S., and Weeks, J. (2007). Object-based classification of residential land use within Accra, Ghana based on QuickBird satellite data. *International Journal of Remote Sensing*, 28(22), 5167.

Tiede, D. (2014). A new geospatial overlay method for the analysis and visualization of spatial change patterns using object-oriented data modeling concepts. *Cartography and Geographic Information Science*, 41(3), 227–234.

Tiede, D., Huber, J., and Kienberger, S. (2012). Implementation of an interactive WebGIS-based OBIA geoprocessing service. In *Proceedings International Conference on Geographic Object-Based Image Analysis* (Vol. 4, pp. 402–406), Rio de Janeiro, Brazil, May 7–9.

Townshend, J. R. G., Huang, C., Kalluri, S. N. V., Defries, R. S., Liang, S., and Yang, K. (2000). Beware of per-pixel characterization of land cover. *International Journal of Remote Sensing*, 21(4), 839–843.

Troyo, A., Fuller, D. O., Calderon Arguedas, O., Solano, M. E., and Beier, J. C. (2009). Urban structure and dengue incidence in Puntarenas, Costa Rica. *Singapore Journal of Tropical Geography*, 30(2), 265–282.

Tullis, J. A., Jensen, J. R., Raber, G. T., and Filippi, A. M. 2010. Spatial scale management experiments using optical aerial imagery and LiDAR data synergy. *GIScience & Remote Sensing*, 47, 338–359.

Walsh, S. J., McCleary, A. L., Mena, C. F., Shao, Y., Tuttle, J. P., González, A., and Atkinson, R. (2008). QuickBird and Hyperion data analysis of an invasive plant species in the Galapagos Islands of Ecuador: implications for control and land use management. *Remote Sensing of Environment*, 112(5), 1927–1941.

Weeks, J. R., Hill, A., Stow, D., Getis, A., and Fugate, D. (2007). Can we spot a neighborhood from the air? Defining neighborhood structure in Accra, Ghana. *GeoJournal*, 69(1–2), 9–22.

Wulder, M. A., White, J. C., Masek, J. G., Dwyer, J., and Roy, D. P. (2011). Continuity of Landsat observations: Short term considerations. *Remote Sensing of Environment*, 115(2), 747–751.

Yan, G., Mas, J.-F., Maathuis, B. H. P., Xiangmin, Z., and Van Dijk, P. M. (2006). Comparison of pixel-based and object-oriented image classification approaches—A case study in a coal fire area, Wuda, Inner Mongolia, China. *International Journal of Remote Sensing*, 27(18), 4039–4055.

Yu, Q., Gong, P., Clinton, N., Kelly, M., and Schirokauer, D. (2006). Object-based detailed vegetation classification with airborne high spatial resolution remote sensing imagery. *Photogrammetric Engineering and Remote Sensing*, 72(7), 799–811.

Yue, P., Di, L., Wei, Y., and Han, W. (2013). Intelligent services for discovery of complex geospatial features from remote sensing imagery. *ISPRS Journal of Photogrammetry and Remote Sensing*, 83, 151–164.

Zhou, W. and Troy, A. (2008). An object-oriented approach for analysing and characterizing urban landscape at the parcel level. *International Journal of Remote Sensing*, 29(11), 3119–3135.

Zook, M., Graham, M., Shelton, T., and Gorman, S. (2010). Volunteered geographic information and crowdsourcing disaster relief: A case study of the Haitian earthquake. *World Medical and Health Policy*, 2(2), 7–33.

15

Geospatial Data Integration in OBIA: Implications of Accuracy and Validity

Stefan Lang
University of Salzburg

Dirk Tiede
Salzburg University
University of Salzburg

Acronyms and Definitions

3-D	Three-dimensional
AI	Artificial intelligence
APPS	Adaptive per-parcel segmentation
CBD	Convention on Biological Diversity
EC	European Commission
EO	Earth observation
ESA	European Space Agency
ESRI	Environmental Systems Research Institute
EU	European Union
FCA	Formal Concept Analysis
GEO	Group of Earth Observation
GEOBIA	Geographical Object-Based Image Analysis
GI(S)	Geographic Information (System/s)
GMES	Global Monitoring for Environment and Security
LCCS	Land Cover Classification System
LIST	Landscape Interpretation Support Tool
OBAA	Object-based accuracy assessment
OBCD	Object-based change detection
OBIA	Object-based image analysis
OFA	Object fate analysis
OWL	Ontology Web Language
RMS	Root mean square
SDI	Spatial Data Infrastructure
SIA	Spatial image analysis
SIAM	Satellite image automatic mapper
SII	Spatial Information Infrastructure
SPOT	Satellite Pour l'Observation de la Terre
SWEET	Semantic Web for Earth and Environmental Terminology
UN	United Nations
VH(S)R	Very high (Spatial) Resolution
GSD	Ground samle distance
WFD	Water Framework Directive

15.1 Conditioned Information

The advancement of feature recognition, scene reconstruction, and advanced image analysis techniques facilitates the extraction of thematic information, for policy making support and informed decisions. As a strong driver, the ubiquitous availability of image data strives to match the ever-increasing need for updated geo-information. In response, we need methods that exploit image information not only more comprehensively, but also more effectively. Object-based image analysis (OBIA), with its explicit focus on objects and their relationships, enables to address and model new dimensions of target classes, enriching information extracted from image data.

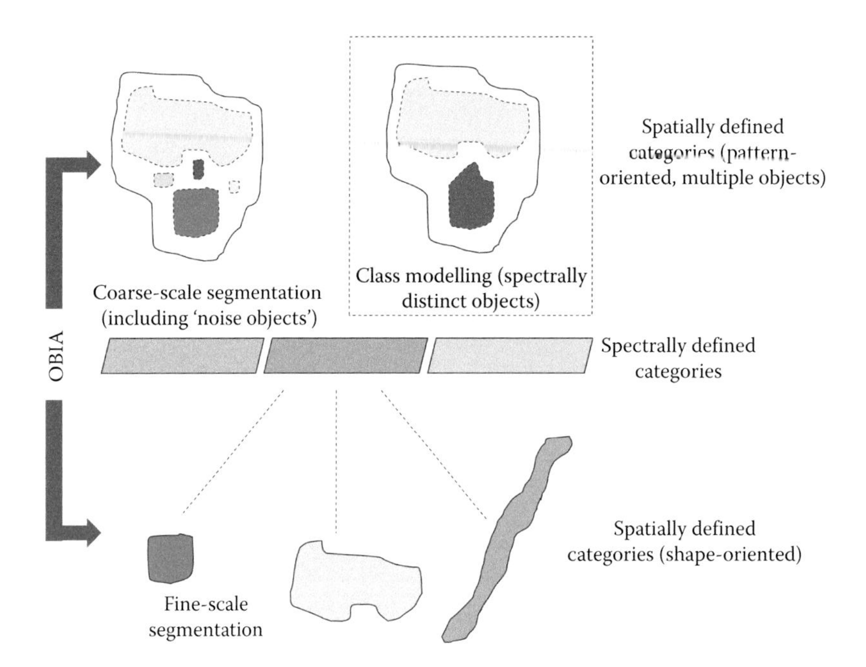

FIGURE 15.1 Extended set of categories (target classes) using spatial in addition to spectral features (see text for further explanations).

15.1.1 OBIA in Support to Geospatial Information Needs

Object-based image analysis (OBIA), as the name indicates, operates on objects representing *real-world* entities as the constituents of our geographic reality. Spatial key features (basically size and shape) are intrinsic parameters to be considered in the information extraction process, making spatial characteristics an additional feature domain in classification. Using regionalization techniques for image segmentation, there is a complementarity (one could say a trade-off) between spectral and spatial similarities, that is, between color and neighborhood. In other words, the spatial constraint balances the spectral behavior, which leads to (scalable) generalization and a reduction of the so-called salt-and-pepper effect (Bischof et al. 1992, Blaschke and Strobl 2001). In OBIA, the category that a group of picture elements is assigned to depends on both spatial and spectral characteristics. This may sound like a restriction narrowing down classification power, but it actually *extends* the set of target classes to be addressed. More precisely, it opens another dimension of potential target classes (Lang 2008), on two additional semantic levels next to the level of spectral classes (see Figure 15.1). This comprises, first, all subcategories or instances of a spectral class that are defined by shape features on individual object level: a class <water> may split into <lake | river> depending on its length/width ratio. A class <built-up> may split into several subcategories of a village typology based on footprint physiognomy. Second, and even more crucial, we find spatial properties in terms of relations among objects. We will come back to this issue in Section 15.2, when discussing the term *class modeling*. In a nutshell, the integration of object relationships (including relative coverage, distance, and in particular topological features) allows for addressing complex, composite target classes on high semantic level.

So, although *object-based* suggests objects to be the fundamental elements to base the analysis on, OBIA goes beyond objects as isolated items distributed over space. In capturing and utilizing the manifold relationships among these objects, we can use additional features in classification that exceed classical quantitative measures. Topological and hierarchical relationships are more qualitative features that apply in a more gradual way than measurable features such as object size, or spectral or spatial distance. The trend set by OBIA—some may call it a paradigm shift, cf. Blaschke et al. (2014)—from pixels (as technically defined units), via objects (as spatially manifested and conceptually ascertained entities), toward relational patterns (as multiscalar qualities of objects), is just taking off, opening a plethora of possible new ways of flexible spatial representations.

The explicit focus on spatial properties and relationships of OBIA* enables to address high-level semantic classes, including spatial composite classes that *emerge* (in terms of relevance) on a certain phenomenon scale. Here we clearly move beyond scientific curiosity, with an enabling tool to meet the ever-increasing demand for geospatial information from the perspective of various space-related policies (Lang et al. 2009b). The term *space-related* is used in a broad sense to comprise all policies with a spatial component that are—directly or indirectly—relying on critical geospatial information updates. Such policies entail more and more societal and environmental domains across geographic scale domains (i.e., from local to continental), for example, a local policy for climate change adaption, a regional policy of sustainable urban development, the European Union (EU)

* In order to highlight the relationships between objects rather than emphasizing objects as such, we may use a more generic term: *spatial image analysis*. Still, since OBIA is an established term in the community, we use it throughout the chapter.

policy on integrated water management (Water Framework Directive, WFD), or a global policy on reducing biodiversity loss (United Nations Convention on Biological Diversity, UN-CBD). The primary source is increasingly supplied by imagery and respective information exploitation strategies, summarized by the term earth observation (EO). EO is, say, a (semi-)political expression (Lang 2008) that highlights the capacity of satellite (and other) sensor technology for the purpose of updated information provision in civil application domains. EO infrastructure (i.e., space and ground infrastructure for satellite imagery, sensor networks) and the analysis capabilities are bundled to support societal benefits and the political ambition of such policies. Recent examples of large-scale international endeavors are Group of Earth Observation (GEO) and the European ESA-EC conjoint initiative Copernicus (formerly Global Monitoring for the Environment and Security (GMES)). In order to reach a consistent approach to monitoring all kinds of environmental and societal conditions, considerable efforts are made to investigate opportunities EO methodologies may hold to facilitate the reporting requirements and the evaluation of intervention options (Lang et al. 2009b). EO technology may contribute significantly toward achieving the objectives of multilateral agreements by (1) increasing knowledge about the underlying processes, (2) supporting the efficient management and monitoring of tasks, and thereby (3) contributing to more effectiveness of conventions or treaties. It goes beyond the scope of this chapter, to list all the information needs in the various societal domains, as well as the phenomenon scale and respective policy scale and link these to existing (and near-future) EO capacity. For an overview, see Zeil et al. (2008). Such scoping is done when designing new satellite missions and the suite of onboard instruments. The Copernicus Sentinel family (starting with the recent launch of Sentinel-1A) is designed in a way to be as flexible and *ubiquitous* in application as possible.

The diverse demand requires not only adequate observation systems but also highly capable analysis tools and methods to effectively transform the enormous amount of data into meaningful and ready-to-use information. OBIA may prove this capability in the following (nonexclusive) areas:

- OBIA, through its multiscale option (Marceau 1999, Hay et al. 2001), enables to adapt the scale inherent in the source data to the phenomenon scale and thereby to the policy scale. In combination with an appropriate resolution level of the used imagery, this gives high flexibility in terms of identifying and working on a commensurate analysis scale.
- The object-oriented data model intrinsically linked to the OBIA approach allows for complex and hierarchical class models, transferable and adjustable between application domains and geographic areas (Tiede 2014). This is not to say that OBIA enables fully automated procedures, but the adjustment of once established class model architectures is a minor effort as compared to manual grouping and reclassifications (Tiede et al. 2010).
- OBIA not only produces objects, but is also able to integrate existing spatial units (i.e., digital geographic features) in the analysis process. In many domains, the update of (existing) geospatial information is more critical than the provision of new, spatially disintegrated units (Tiede et al. 2007). Geodata from SDIs may be used as a reference data set, to a lesser degree as fixed and exclusive set of boundaries, but as a *pool* of potential boundaries to adapt results on, in terms of both spatial reference and generalization level. Remote sensing products are often stand-alone or even *isolated* products, not readily capable to match with existing geodata repositories (see Section 15.3). This applies in terms of both nomenclature (thematic agreement) and generalization (geometrical match). Interestingly, by explicitly addressing the spatial component in the analysis process, we can improve both!

Spatial data policies like the European INSPIRE directive (2007/2/EC) seeks to provide a common baseline for interoperable data usage, to overcome technological constraints via standardization and opening the way to a content-driven debate. Information provision, whether spatial or nonspatial, requires compatibility to existing data pools. SDIs—or more appropriate: spatial information infrastructures (SIIs)—have the particularity to ensure a spatial match as well, by adhering to generalization principles, and telling modifications from different representations. This not only is limited to spatial or thematic disagreement, but also relates to the fitness of information products derived from different sources. Decision makers should be able to interpret the information unambiguously, a requirement usually attached with attributes like interpretability, communicability, spatial explicitness, and clarity about the assumptions. Often, it seems, basic requirements like *usability* are neglected in remote sensing products so that decision makers face a notorious dilemma (Lang et al. 2009b): an overwhelming blessing of satellite data, poured out generously to meet the prescribed monitoring schemes, but hardly fit for integration in stakeholders' daily businesses and established workflows (ibid.).

Let us consider an example (see Figure 15.2) that will be discussed in greater detail in Section 15.3): the representation of different agricultural parcels derived by image segmentation from SPOT-5 data, even if *perfectly* delineated, is a *no-go* for users because the analysis scale does not match with the policy scale. The regional development plan in the Stuttgart Metropolitan area in Southwest Germany requires a baseline geometry representing so-called biotope complexes (Schumacher et al. 2007). These are functional composed spatial units whose outlines should match digital cadastral boundaries. The automated segmentation shown in Figure 15.2a provides neither of it. By means of a combined strategy called "spatially constraint class modeling," Tiede et al. (2010) were able to deliver the appropriate geospatial information (Figure 15.2b).

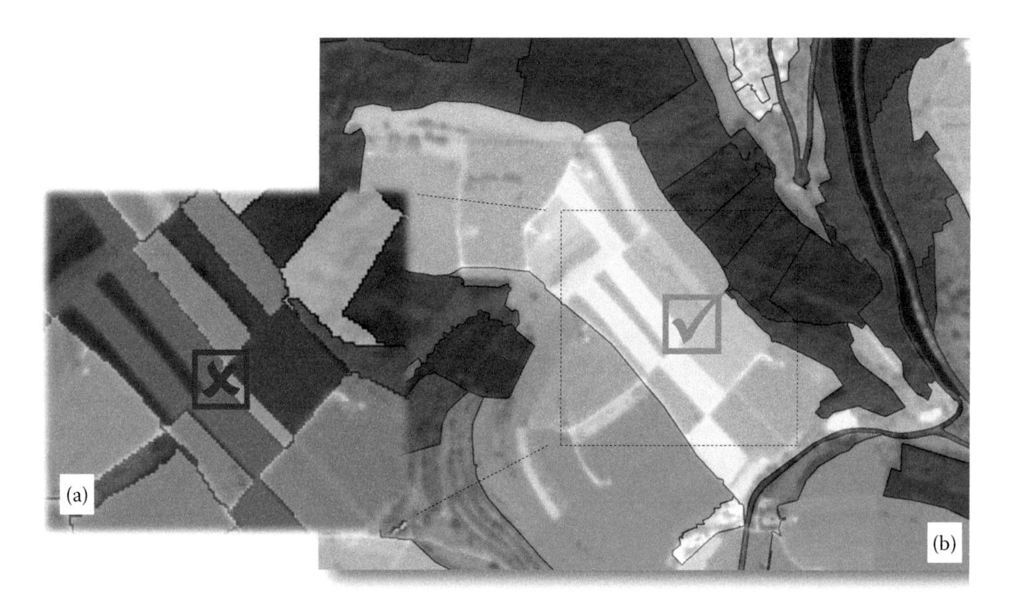

FIGURE 15.2 Understanding others' object understanding. (a) Unconditioned information, simple delineation of land cover objects, (b) conditioned information, aggregated biotope complex.

15.1.2 Enriched Information through OBIA

As pointed out by several authors (Blaschke et al. 2007) at the launching phase of the Copernicus/GMES initiative, EO-based monitoring systems need to be capable of transforming complex scene content into ready-to-use information, in a reliable, transferable, effective, and—desirably—cost-efficient manner. Global commitments, directives, and policies have a pronounced demand for timely, accurate, and conditioned geo-spatial information (Lang et al. 2008), valorizing the load of raw data collected from various monitoring systems. Naturally, as it seems, a technically and spatially more literate user community asks for advanced geo-spatial products and expresses their needs accordingly. With an increased overall consciousness to prevailing societal challenges, the need for targeted information, too, is raising, both in amount and in ambition (toward efficiency, quality, etc.) (Lang et al. 2009a).

The advancement in feature recognition and image analysis techniques facilitates the extraction of rich thematic information, to support informed decision making irrespective of a particular application field. The term *rich* relates to semantic content that relates to requirements from particular application fields or policy contexts, beyond generic land cover classification schemes (Lang et al. 2009a). With increasing capacities, both on the imaging and on the processing side of observation systems, high expectations (Lang 2008) were raised from all sides, including academia, service providers, and users to the recent generation of very-high-resolution (VHR) data (Aschbacher 2002) such as (by today) QuickBird, WorldView, and GeoEye. Even though shortcomings may exist in data provision and handling, data acquired as such through EO missions are abundant, so that users as well as service providers are challenged by integrating and valorizing them. Lang et al. (2009b) use the term "information conditioning" for the act of turning unstructured (and redundant) information into ready-to-use information, whereby ideally a full match between the products provided and the user demand, plus its fitness in existing workflows, is achieved. OBIA, also called GEOBIA (Hay and Castilla 2008) for highlighting the geographic application context, provides the toolbox, capable of treating users' needs individually by utilizing intelligent tools and algorithms for automatically delineating, extracting, and categorizing geographic objects from remotely sensed images (Lang et al. 2010a). Methodologically, OBIA seeks to accommodate various geo-spatial concepts, remote sensing principles (spectral behavior), and the wealth of experience from visual interpretation and manual delineation (Hay and Castilla 2008).

Decision makers and other policy implementation bodies do have a clear mission to follow; in other words: informed decisions are mostly embedded in any kind of operational workflow, that is, the decisions to be taken are hardly independent from any existing institutional setting. In addition, each decision has spatial constraints, often linked to political boundaries. This kind of double constriction (both space- and mandate/task-related) is called the *policy scope* (Lang et al. 2008). "Conditioned information (ibid.) is the result of a process to fulfill the user demand in technological and conceptual sense and has undergone any kind of fitness check for operational use ('user validation')" (p. 3) (Lang et al. 2010a). Conditioned information, in other words, means the full match between geo-spatial information provided and the accommodation of this information in their established workflows.

Another key aspect in information conditioning is the issue of scale. According to hierarchy theorists (Allen and Starr 1982), scale is defined as the "period of time or space over which signals are integrated to give message." Converting imagery to information ideally follows the principles of communication models. The role of image interpretation or classification is a kind of translation support channelizing and conditioning the oft overwhelming (unstructured) information residing in images. Scale is an intrinsic aspect in communication, when signals need to be filtered and integrated to give message. Here we need to match the phenomenon scale in general and the policy scale in particular. The flexibility of current digital representations turns a fixed cartographic scale on maps to a visualization scale confining dynamically the level of detail of displayed information (Montello 2001). An important step, whether in traditional paper map production or dynamic on-screen visualizations, is generalization. Generalization refers to the reduction of redundant detail with coarser scale, a filtering effect that makes representations commensurate to the scale of interest (simplification, enhancement, selection). In digital representations, generalization can be achieved by (1) algorithmic smoothing of lines or polygon outlines by dropping a selected number of vertices; (2) selective display of raster cells or pixels with decreasing zoom level; and (3) manual, context-specific omissions or exaggerations to suppress or emphasize certain cartographic aspects.

OBIA generally supports the idea of multiscale representation (Hay et al. 2001) by enabling nested hierarchies of image objects in several scales. In this process, generalization is a particular challenge (see later) implying a trade-off between spatial embeddedness and appropriate scale: either the object hierarchy is strictly fitting boundary-wise, that is, outlines of higher-scale objects are not generalized (scale-adaptive strategy), or boundaries are generalized according to the scale level and the object hierarchy is not spatial explicit any longer (scale-specific strategy).

The challenges touched upon briefly in the previous paragraphs extend to concept and methods of assessing and validating the results. Among others, this applies for specific object-based approaches to accuracy assessment and change detection (Lang 2008). *Object based* implies that accuracy or changes are assessed not only thematically but also spatially: this refers to the way how objects are delineated, that is, their spatial representation (position and precision of boundary), and to what degree these representations change over time.

15.2 Object Validity

The transformation and reconstruction of an image scene into ready-to-use geospatial information matching the conceptual understanding and the practical needs of users are the primary aim of OBIA. In this advanced process of image understanding, image objects may be correctly classified but not *valid*. This may sound contradictory, but the enabling by OBIA to delineate spatially flexible units challenges the classical concept of binary accuracy assessment (right or wrong). In addition to the assigned label or category, we need to evaluate the way how units are delineated. This implies issues of scale, generalization, as well as representation (including visualization). We subsume the entirety of aspects to be considered in this process, under the term *object validity*.

15.2.1 Color and Form: Elements of Image Understanding

The guiding principle of OBIA is to represent complex scene content in such a way that the imaged reality is best understood and a maximum of the captured information is revealed, extracted, and conveyed to those who need it. The term *image understanding* (Pinz 1994, Zlatof et al. 2004) refers to the reconstruction of the imaged scene as a complete process of transforming reflectance values into symbolic representations and, ultimately, to valid information. Ideally, the target symbology is a fixed set of *symbols* that are to be detected in an image. This applies, for example, to optical character recognition where a scanned document is analyzed for the occurrence of a given set of characters (letters, numbers, and special characters). Their appearance may vary within certain ranges, but in each language or script, there is a fixed and limited set of characters to target on. We can unambiguously assess the success rate of this image reconstruction by counting the correctly identified characters. Once characters are rightly extracted, a word processor is able to interpret their positions relative to each other and eventually identify words, which can be further *understood* by a spell-checker or even be read out by the machine. A visually impaired person, not able to read the content of a letter, may consider such image understanding capabilities as a *full match* between the information provided and his or her need (while any interpretation error can be clearly spotted out as well). Characters to represent letters, words, etc., are standardized representations (just as notes are for musical events) as a means to encode and decode information in a (thereby regulated) communication process. In other words, the process of information exchange uses unequivocal, even standardized symbology (*fonts*), and is thereby well-posed and automatable.

Returning to remotely sensed images, we face the problem that there is no such fixed target symbology through which we could communicate content and exchange information in a standardized way. As close as possible, we get to this in applying discrete color levels, which are intersubjectively comprehensible and used consensually. But even if color categories exist, color remains a continuous phenomenon whose discretization is per convention and not standardized. A pixel (picture element, i.e., the smallest, technically defined unit) appears in a certain color representing an integrated signal of the spectral reflectance of a small portion of the Earth surface. The higher the spectral resolution (number of bands, n) and the number of quantization levels (radiometric resolution, *bit-depth*), the more *colors* a pixel can adopt, as first the dimensionality of the feature space increases and second the number of possible steps.

But what do colors mean? How do we reach from sub-symbolic level (colors) to a symbolic one (label)? The integrated signal stored in a pixel represents physical conditions of the ground (*land cover*), so we can link color with physical models using spectral signatures. Classification schemes such as the land cover classification system (LCCS) (Di Gregorio and Jansen 2005) use spectral signatures to convert reflectance values (via colors) into nominal classes, that is, labels. With additional bands, more and better physical models of geographic features can be incorporated in a classification scheme (e.g., spectral behavior of vegetation in the IR spectrum). The assignment of a pixel to a nominal class allows for classical site—(i.e., pixel-) specific accuracy assessments (Congalton and Green 1998). The remaining ambiguity in the class assignments can be dealt with using, for example, fuzzy rules (Zadeh 1965) in both the classification and evaluation process. Since a pixel is a fixed unit, the assessment of a class assignment focuses on the color-based labeling of this particular pixel. If *true* ground data, that is, independent reference information from the field, is by any reason (inaccessibility, security, and time/costs) impaired or not possible at all, visual inspection of the classification is an alternative (Campbell 2002). Visual accuracy assessment assumes there is an agreement (*full match*) between the color impression on screen and the evaluator's experience of physical models and the spectral behavior of various land cover types in color space. Usually, a visual inspector considers, to various degree, spatial context in taking the decision (see Figure 15.3).

Ideally, physical conditions on the ground are the only influence of a pixel's integrated signal, and different signals would indicate only a change in these conditions. In reality, however, all kinds of disturbances occur such as atmospheric influences and topographic effects (Richter and Schläpfer 2002) viewing angle of a VHR sensor, etc., that make the same land cover captured twice look differently. While color is the most direct (and unambiguous) property of remote sensing images to be used at a per-pixel basis, other features of geographic phenomena like size, form, appearance, and orientation are likewise important. Such features cannot be drawn from single pixels (Blaschke and Strobl 2001), but need to consider spatial context, partly even spatial explicit measures. Quantitative parameters can be utilized (such as area, perimeter, shape, and texture) of pixel aggregates to characterize these features, but a standardized scheme is even more difficult to obtain due to the interrelated character of such features (see later). Spatial features allow for additional differentiation among target classes on (higher) symbolic level: a river is an elongated water body, a lake more compact. Such features can be obtained once pixel aggregates are organized in segments or image objects (Benz et al. 2004). When co-assessing color and form, some subtle variances in color are sacrificed to the overall spatial property of the aggregate. A lake may be considered a lake due to its prevailing polygonal form even if there are some subtle variances in color tone or even color due to different water depths or presence of vegetation, etc. The automated extraction of boundaries is a function of the image resolution that controls average size, number, and complexity

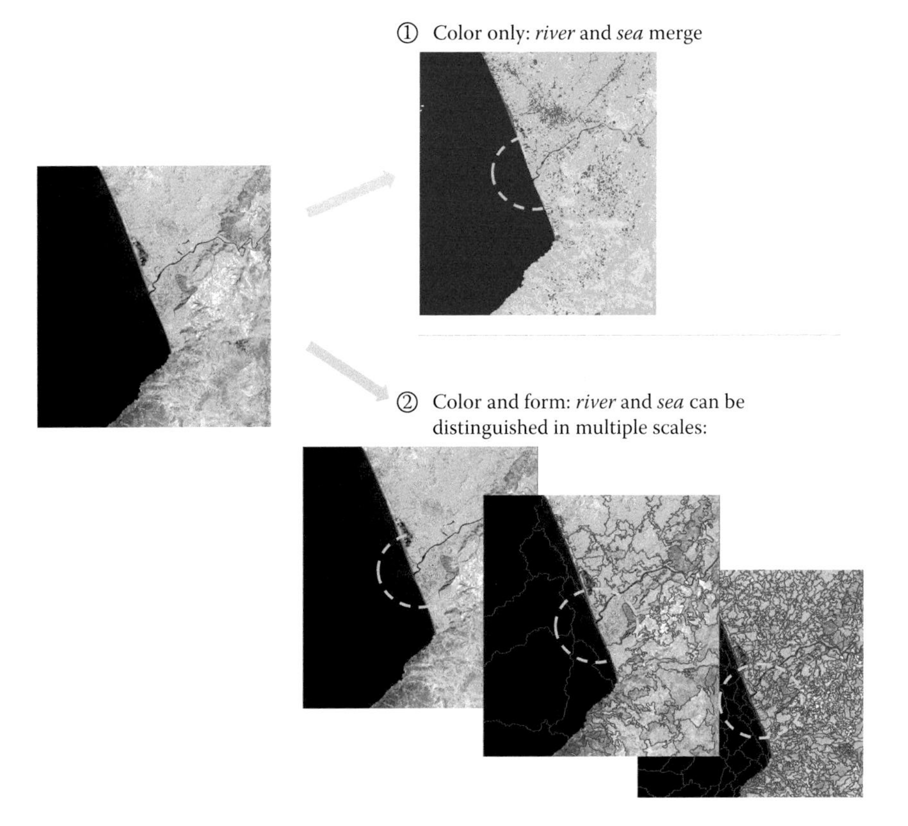

FIGURE 15.3 Shape is a highly distinguishable feature of land cover classes similar in color (here large river/sea). (Image source: Landsat 8 OLI (Nov 2013), US Geological Survey.

of the boundary of the segmented objects. An interpreter inherently applies generalization and adapts the delineation according to the mapping scale and the level of detail prescribed by the classification scheme. Higher-level semantics usually imply a coarser scale (Belgiu et al. 2014). In OBIA, multiresolution segmentation is a strategy to perform image segmentation at several scale levels so to represent the respective multilevel content of a scene scale adaptively (i.e., without generalization) or scale specifically (i.e., with generalization). As we will discuss later, point-based methods fail to assess the spatial dimension of object delineations. Lucieer and Stein (2002) distinguish between existential and extensional uncertainty of delineated objects.

To treat color and form separately suggests independence of both aspects. A (remote sensing) image understanding system (IUS) needs to integrate both aspects, according to models of human vision (see following section). An IUS that follows the multistage paradigm of human vision by Marr (1982) starts with pre-attentive vision as a first stage and then works along these building blocks to compose more complex target classes. There are several strategies to accomplish this first stage; two of them—with more communalities than differences—are shortly discussed here. The first is a symbolic (i.e., meaningful) (pre-)classification reflecting pre-attentive vision (Baraldi and Broschetti 2012) according to color similarity based on physical models of spectral behavior as implemented in the SIAM™ (Satellite Image Automatic Mapper) software (Baraldi 2011). Pixels are preclassified by rule-based clustering in color space. Segments can be built by grouping neighboring pixels assigned to the same symbolic (i.e., meaningful) color levels. The initial preclassification aims to reach on a semantic level of (descriptive pre-)classes, while leaving the spectral discernibility as high as possible: depending on the aggregation level, an 8-bit 6-band Landsat-7 imagery with 256^6 quantization levels can be, for example, transformed in a preclassification layer between 18 and 96 classes. This process lowers the number of quantization levels significantly. While this is desirable from the perspective of deductive classification, it leads to a predefined segmentation result (by grouping neighboring pixels of the same (color) class), that no longer adheres to the principles of regionalization, as described earlier. Here the second strategy comes into play. Utilizing the strength of regionalization to consider both spatial and spectral similarity simultaneously, multiscale segmentation provides image objects by spatial classification (Wise et al. 2001) at first hand. These sub-symbolic segments are then classified according to their spectral and spatial behavior into (higher) semantic classes. The building blocks used in the first strategy are (strictly) spectrally homogeneous pixel aggregates—no matter of what spatial characteristics—while in the second strategy, we have image regions—fairly homogeneous, but spatially contiguous and scaled to feed the IUS (Lang 2005). The first strategy uses pre-existing knowledge on (potentially exhaustive) spectral signatures, and the second requires (empirically determined) parameterization.

15.2.2 Human vs. Machine Vision

Great advances have been made in computer vision, but the entire potential of human vision has not been achieved so far it seems. While biophysical principles like retinal structure and functioning and singular processes such as the cerebral reaction are analytically known, we still lack the bigger *picture* on human perception as a whole (Blaschke et al. 2014). OBIA tries to externalize basic principles of human perception (Lang 2008), so it is important to understand how we deal with imaged information in various scales, how we manage to relate recognized objects to each other with ease, and how we understand complex scene contents readily. "The ultimate benchmark of OBIA is human perception" (Lang 2008) (page 7), but in fact, it is difficult to explicitly describe what happens when we look at an image and suddenly *see* something (Blaschke et al. 2014).

Pattern recognition—not necessarily the interpretation—works without major effort (Eysenck and Keane 1995, Tarr and Cheng 2003). Human perception is a complex matter of filtering relevant signals from noise (Lang 2008), a selective processing of detailed information, and, finally, experience. With respect to visual information processing, Marr (1982) provided a conceptual framework of a three-leveled structure. According to Marr's paradigm, we can distinguish between the following explanatory levels: (1) the computational level, which is related to the purpose, logic, and strategy of perception; (2) the algorithmic level dealing with issues of implementation and the details of a proper transfer between input and output; and (3) the hardware level being concerned with the physical realization of the representation identified in (1) and its processing algorithms derived in (2). The computational level within visual processing is characterized according to Marr (1982) by several stages of visual representation of an image's content. These stages (or *sketches*) are considered to proceed in a subsequent manner providing more and more detailed information about the scene being viewed. The primal sketch involves a 2-D description of the radiometric behavior of pixels and pixel aggregates and their geometrical properties. The pure spectral differentiation of gray shades and color tones makes up the basic level (raw primal sketch), whereas the grouping of spectrally like pixels into geometrical units represents the full primal sketch. The latter is built upon quasi-homogeneous regions (blobs) or bounding areas of high contrast (contours or edges). The raw primal sketch marks the stage where image primitives such as blobs, edge segments, and their low-level descriptions are produced. They are then ordered and organized into higher-level place tokens (Marr 1982) or perceptual chunks (Bruce and Green 1990). This is performed within the process of perceptual organization, which aims at logically grouping a perceived pattern and to transfer it to meaningful symbolic representation. Touching the issue of scale, we can consider relatively small place tokens to cause a certain granularity (Julesz 1975), which is a visual effect of a regular distribution of small enough elements.

When moving from image perception to image interpretation, experience comes into play. Findings in neuropsychology

(Spitzer 2000) explain that data processing by our senses is a kind of vector coding of signals in a high-dimensional feature space. *Experience* can be thought of as a certain cluster or nexus in the feature space. When signals are perceived, they are compared to this experience nexus, which compare to a model to reach from (1) raw data, through (2) patterns or aggregates of color and form that are structured in various levels, to (3) relationships between object concepts in an image. Still, human perception is likely far from linear (Gorte 1998), and this may suggest that more than one model is used to construct meaning from an image (Lang et al. 2004). Image interpretation, when dealing with unfamiliar perspective and scale, requires *multiobject recognition* in a rather abstracted mode, and the interpreter needs to understand the whole scene. This includes the physical properties of objects viewed in an image, but also their affordance which refers to an object's values and meanings towards an interpreter. The skilled visual interpreter may recognize some features instantly and others by matching the visual impression against experience or examples listed in an interpretation key (Blaschke et al. 2014)

The *gestalt* approach (Wertheimer 1925) has established general principles (factors, *laws*) that can be observed when certain patterns or figures are examined. Following the Ehrenfels criterion, a *gestalt* shows emergent properties. Gestalt theorems have a strong predictive (though not an explanatory) potential of how we perceive structures and patterns. Based on the factor of *proximity*, single elements within a scene are grouped together if they are close enough to each other, depending on the scale of observation. The factor of *good gestalt* assumes that simple, well-shaped geometric figures are more readily to be perceived than more complex ones. Both factors suggest that a manual interpreter tends to group similar elements into a larger one and to close lines straight over a gap, something that is also reflected by the factor of *good continuation* (Lang 2005) (Figure 15.4).

15.2.3 Class Modeling

Multiscale segmentation, for example, realized by the *multiresolution segmentation* algorithm (Baatz and Schäpe 2000), generates homogeneous image objects in a nested hierarchy. Spectral information is aggregated while loss of detail is minimized (Drăgut et al. 2014) and higher-level objects' internal heterogeneity is increasing, leading to more functional homogeneity in the sense of Spence (1961). Using this strategy, some composite classes (e.g., forest stands composed by tree species with similar spectral behavior) can be directly delineated by multiresolution segmentation (Strasser and Lang 2015). More often, however, composite classes such as orchard field, mixed arable land, riparian forest, suburban area, and informal settlement, bear a degree of internal heterogeneity that exceeds the capability of state-of-the-art segmentation algorithms (Lang et al. 2010a). The internal homogeneous building blocks consist of different land cover, thus being spectrally diverse (e.g., grassland and crop patches), and there is no *a priori "true"* delineation. Lang et al. (2014) use the term composite geon for such target classes, whose composition is scale depended. Body plans are used to represent the targeted composite objects that are intersubjectively perceived by experts as functional *bona fide* units (Smith 1994). Class modeling (Tiede et al. 2010) can be employed to topologically describe spatial constellations of a set of subunits in a way that image information is structured into hierarchical divisions based on ontology-like rule sets that employ relational features.

Tiede et al. (2010) use a class modeling approach as a supervised regionalization technique based on spectrally homogeneous elementary units (spectrally *bona fide*), which implies iterative segmentation and classification steps. The initial segmentation and preliminary classification of the basic units is only the first step in the modeling process. Based on these

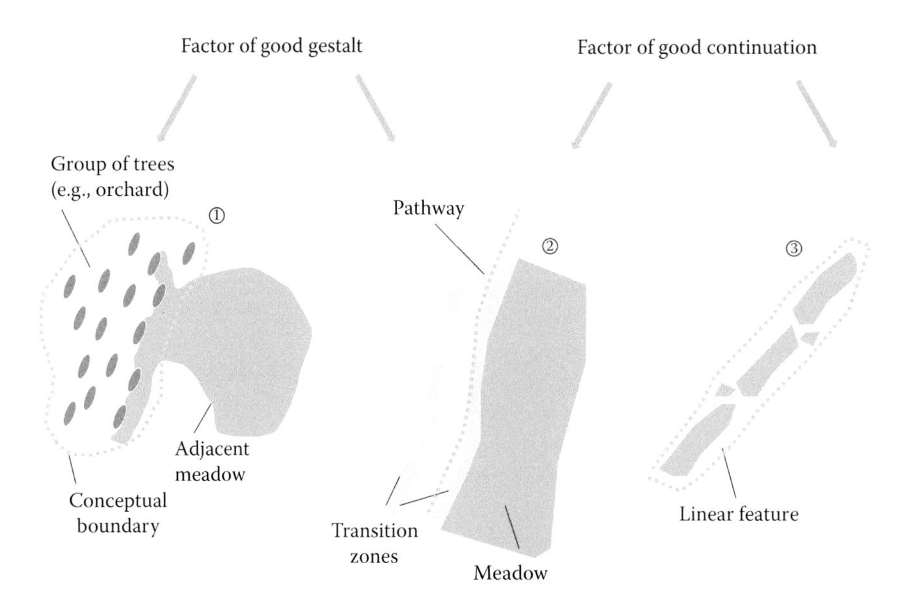

FIGURE 15.4 Gestalt theorems applicable to image analysis problems.

(a)

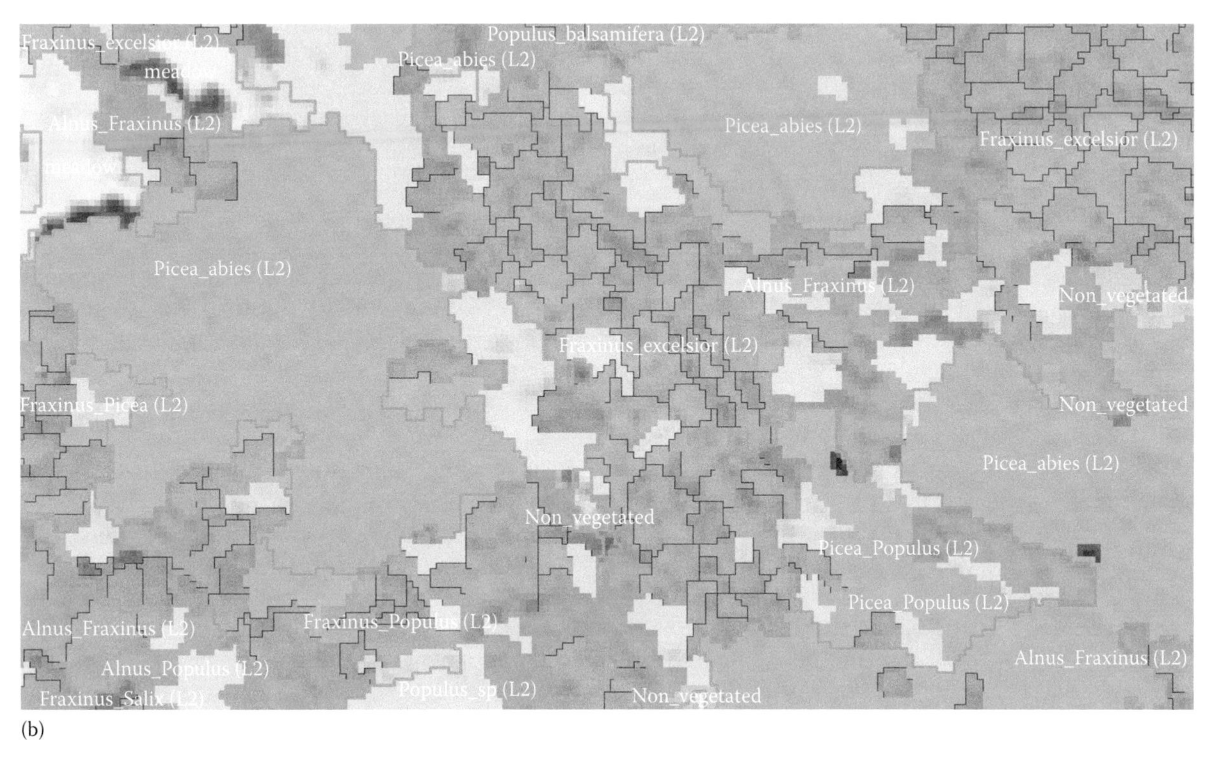

(b)

FIGURE 15.5 (a) SPOT-derived land use units to fully validated composite geons; (b) initial forest patch/generalized forest patch.

initial classes, additional expert knowledge (here based on the mapping key) is formalized in rule sets and/or the specific use of auxiliary data sets is defined. The result is a cyclic process for a generation of composite objects (functionally homogeneous, structurally heterogeneous, see Figure 15.5a), overcoming the limitations of object delineations using region-based segmentation algorithms. In the described building block case of limited, well-distinguishable subunits, class modeling aims at (re-)assembling familiar (i.e., from everyday experience) or expert-known (i.e., trained) target units, according to the gestalt principle of *good gestalt*. Otherwise, our perception works in a more subtle way, when we intrinsically generalize detail. For example, when analyzing a piece of (natural) forest, we would rather perceive quasi-homogeneous forest stands according to their specific composition of prevailing and accompanying tree species. This can be seen as an intermediary stage between a pure texture-based impression (where individual elements are difficult to discern) and a building block composition (where elements are few in number and intersubjectively perceivable). This is what gestalt theorists would call the factor of *granularity*. When fine granularity enables the segmentation algorithm to find higher hierarchical boundaries (Strasser and Lang 2015), the number of elementary units can increase significantly (Figure 15.5b).

Interpretation keys for mapping and survey are designed by experts and for experts (Lang et al. 2004). Yet an automated interpretation of an entire scene in a wall-to-wall manner is still away from being operational. The intrinsic and intuitive knowledge of a skilled interpreter is hard to force into rule systems. The criterion of being *relatively homogeneous* governs any mental organization of reality, manual interpretation, or image analysis, whereas it remains a relative measure and a matter of disposition depending on the target scale dimension.

15.2.4 Validity of Object Delineation and Classification

Object identification implies—next to the cognitive skills of interpreting color and form and linking them to biophysical properties—a grasp of the relevance of the extracted information for a certain purpose. *Object validity* (Lang et al. 2010a) has been defined as the degree of fitness of object delineations for operational tasks in a policy context. Comprising object representations in both the *bona fide* as well as the *fiat* domain, OBIA poses new challenges for evaluating object validity in an operational context (geon concept; also see Chapter 22, Volume I). According to Smith (1994), objects can be differentiated into (1) concrete, tangible objects with a visible physical boundary (*bona fide*) and (2) objects that lack a physical border, thus not visible in the landscape (*fiat*). Examples are administrative boundaries or artificial, concept-related object boundaries constructed by human fiat. In remote sensing, we treat a spectrally homogeneous image segment as a *bona fide* object. This, of course, implies a scale issue. As we leave *crisp bona fide* objects behind, the binary decision between *correct* and *false* labeling gets blurred, while the use of, for example, fuzzy rule sets (Zadeh 1965) only partly solves this. The more complex or concept-related a specific category is, the more likely a binary assessment will fail. Thus, Lang et al. (2010a) propose to use the terms *appropriateness* or *validity* to accommodate the fact that such quality criteria are more a continuum with certain thresholds to be set, instead of a binary decision between *right* or *wrong*.

Object recognition has been described by theoretical concepts ranging from template matching theories over feature analysis to structural descriptions (Bruce and Green 1990), the latter being used to describe the components of a configuration and to make explicit the arrangement of these. The matching of extracted visual information with the stored structural description is referred to as perceptual classification (Eysenck and Keane 1995). Semantic classification and naming are subsequent stages and involve retrieval of functions and object associates (Lang et al. 2009a). One of the striking capabilities of human perception is to tell signal from noise, in other words to distinguish information against a *simplified* environment (Bruce and Green 1990). By experience or training, we continuously feed our implicit knowledge with explicit knowledge derived from formal learning situations (e.g., spectral behavior of stressed vegetation). Artificial intelligence distinguishes knowledge in procedural and structural knowledge. Procedural knowledge is concerned with the specific computational functions and can be represented by a set of rules. Structural knowledge implies the way concepts of a domain are interrelated: in our case, that means how far links between image objects and *real-world* geographic features (Hay and Castilla 2008) are established. It is characterized by high semantic contents and therefore by far more difficult to tackle with. Structural knowledge can be organized in knowledge organizing systems such as semantic networks (Liedtke et al. 1997) as graphical representations, or more mathematically by, for example, formal concept analysis (Ganter and Wille 1996). Within image analysis, semantic nets and frames (Pinz 1994) offer a formal framework for semantic knowledge representation using an inheritance concept (*is part of, is more specific than, is instance of*). In recent years, ontologies have gained a lot of intention to formalize knowledge. Semantic webs allow for a representation of structural knowledge in different semantic levels, that is, hierarchical complexities as a top-level ontology. Ontologies are increasingly used in the GIS and remote sensing community, including geospatial knowledge representation (De Martino and Albertoni 2011), geographic information retrieval (Lutz and Klien 2006), image processing and analysis (Tönjes et al. 1999), semantic sensor networks (Kuhn 2009), and geo-data interoperability (Reitsma et al. 2009). The Semantic Web for Earth and Environmental Terminology ontologies, for example, provides an upper-level ontology for Earth system science comprising thousands of terms from the Earth system science domain using OWL (Ontology Web Language) (Raskin and Pan 2003).

While in visual interpretation, cognitive processes run more or less simultaneously, within OBIA, this process is split and organized along a procedural processing line, and these compartments can be decomposed and *controlled* individually (Lang et al. 2010) (see Figure 15.6).

Machine-based knowledge representation and hierarchy theory (i.e., the principal understanding of decomposability of complex scene content) complement each other by enriching automated image analysis with geospatial scale concepts. The rule-based intelligence of a production system can be enriched by learning algorithms, empowering the classification system to improve itself (with increasing number of classification tasks), but to some degree obscuring the transparency (and potentially the transferability) of the rule base as such. The spatial relationships among (basic) objects are useful in modeling higher-level (i.e., composed) object classes. These have to match with the *truth* in the sense of matching the epistemological significance

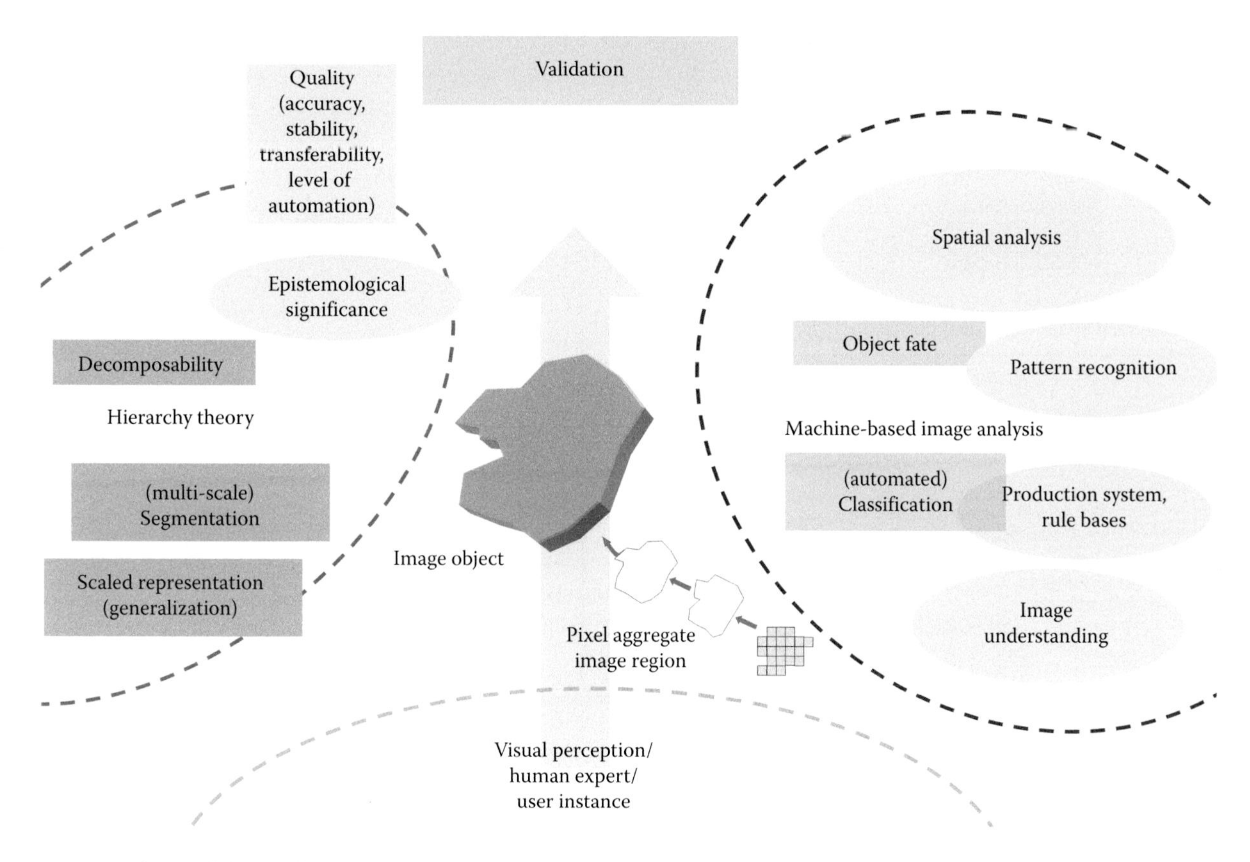

FIGURE 15.6 Object validity in object-based image analysis (Lang et al. 2010). Starting from pixel aggregates (image regions) as initial products of image portioning, a set of routines may turn suitable segments into image objects that (1) are considered appropriate for the respective context of use and (2) are adhering to general principles of image segmentation such as scale domain, spatial coherence, nonoverlap, and the like. (Adapted from Blaschke, T. et al., *Int. J. Photogramm. Remote Sens.*, 87(1), 180, 2014.)

that underlies the classification problem. Simple quality criteria (like any accuracy measure) should be applied under consideration of the process lineage of the object provision and a general understanding on the epistemological nature of the target objects (Lang et al. 2010b).

Table 15.1 summarizes particularities of validation tasks within OBIA for *bona fide* objects and composite. Specific details are discussed by Lang et al. (2010a). Main methodological challenges are the ones attached to optimizing the process of OBIA-specific automation and machine assistance within each category. The issue of reference data varies between different types of object classes. Label verification refers to the external process of confirming the assigned category. The validation of boundary delineation as such is scale dependent, as well as policy related. It includes the issue of degree of freedom in delineation when it comes to the grouping of elementary units. Finally, the automation of the delineation as such is characterized.

15.2.5 Multiple-Stage Validation

Semiautomated classification of complex land use/cover units faces a high number of degrees of freedom (Lang et al. 2010a). Therefore, the validation of modeled composite classes stands

TABLE 15.1 Aspects of Object Validity

	Bona Fide Objects	Modeled Composite Objects
Example(s)	Land cover classes, for example, meadow, field, forest	Biotope complexes, for example, mixed arable land and grassland area
Main methodological challenge	Scale-specific delineation, boundary generalization	Meet requirements in terms of appropriateness and matching key
Real-world reference	Existing and recordable in field	Perceived by expert vision, agreement with mapping key, and existing planning units (here cadastral data)
Label verification	Binary or fuzzified	In stages, appropriateness as a term to be operationalized
Boundary validation	Scale-dependent, low degree of freedom	Expert-based, high degree of freedom
Policy driven	On an elementary level: yes	Yes, high expectations in EO-based techniques
Delineation	Automated	Automated, partly manual regrouping

Source: Modified from Lang, S. et al., *J. Spat. Sci.*, 55(1), 9, 2010a.

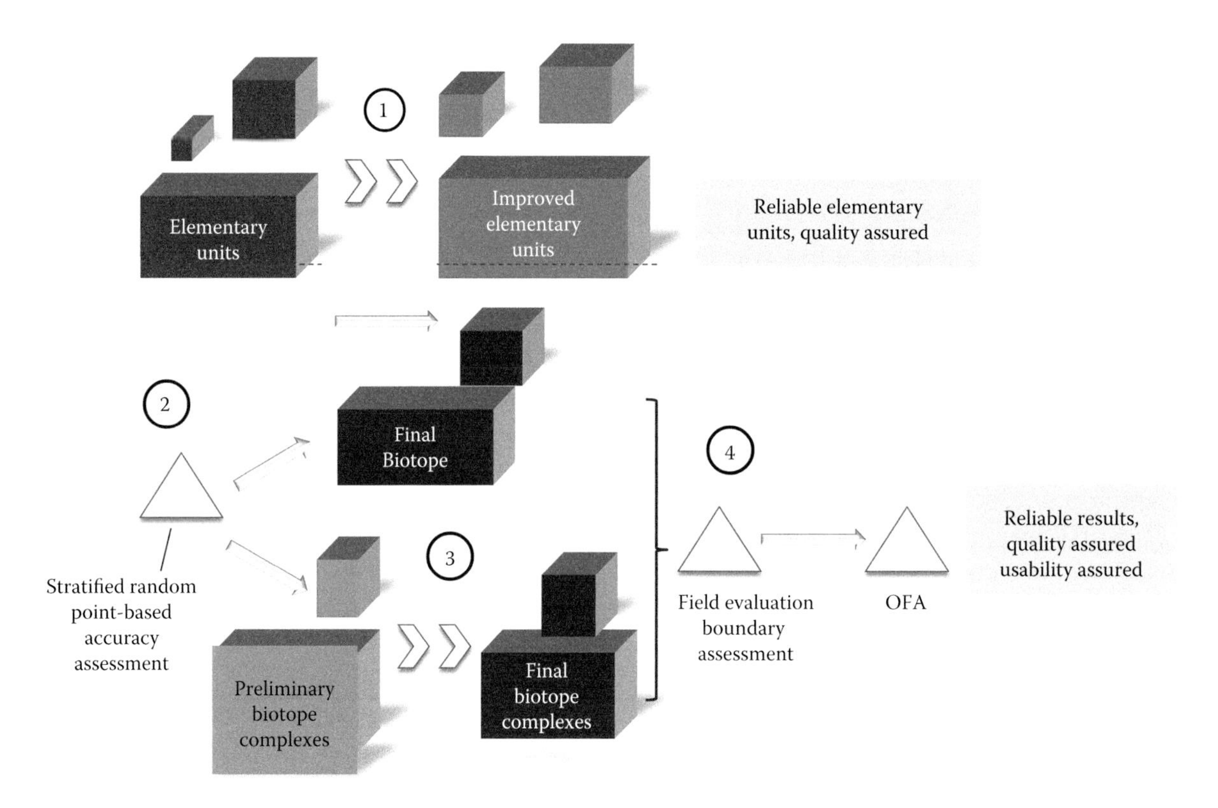

FIGURE 15.7 Multistage validation of composite classes. (Adapted from Tiede, D. et al., *Photogramm. Eng. Remote Sens.*, 193, 2010.)

between traditional methods of quantifying accuracy and more qualitative validation measures (matching scale and relationship patterns). Validating OBIA class modeling faces the problem of which step in the cyclic approach to validate. The validation sequence, as proposed by Tiede et al. (2010), includes verification, improvement, quality assurance, reliability, and even usability to ensure the product fully satisfies the user requirements. That results in a four-stage validation procedure (Figure 15.7):

- Statistical point-based accuracy assessment based on stratified random sampling (Congalton and Green 1998) to validate the elementary (*bona fide*) units, which were based on spectral homogeneous units.
- A first expert-based validation cycle including both the composite classes (in this case biotope complexes) with spectral heterogeneity and their dependent basic units. This was part of a hybrid modeling approach (on-screen expert assessment).
- A second expert-based validation cycle included the quality of the modeling approach not only regarding the given mapping key as compared to field mappings by domain experts, but also considering geometric accuracy of the delineated boundaries.
- Evaluation of inaccuracies introduced by the policy-demanded matching of delineated biotope complex boundaries to cadastral data (scale-gap). Object-based fate analysis (OFA, see later) has been applied on individual object level.

15.3 Object Fate

Image objects as real-world objects are subject to change. Changes may occur more gradually (varying quality) or more abruptly (altering category or label). At the same time, image objects may change their extent, location, or composition. Even topological relationships between neighboring objects may be modified over time. Object-based change detection (OBCD) strives to assess such changes in terms of type, intensity, size, and shape, as well as horizontal and vertical relations of image objects. Thereby, OBIA faces the challenge to tell real changes from generalization effects or any kind of *noise* or *disturbance* occurring in the imagery. A similar problem arises in object-based accuracy assessment (OBAA) when evaluating the quality (accuracy, robustness, etc.) of OBIA classifications, as the same effects may render an OBIA result and a reference data layer difficult to compare. The concept of object fate analysis (OFA) provides a strategy to deal with such challenges.

15.3.1 From Static to Dynamic Change Concepts

Geographic features, when observed over time, are subject to change—and so are image objects that represent them. So while it seems a straightforward task to extend established change detection methods to the object-based domain, there are some additional challenges and specifics to what we may call object-based change detection (OBCD) (Lang 2008, Chen et al. 2012).

Change, in fact, is a more challenging issue in OBIA than in pixel-based analysis, due to the reasons already discussed with *object validity*: an image object can change in label and extent, usually in both: the case that a particular geographic feature (a lake, a forest, a town) changes its category within exactly the same boundary is rather rare. So the static *raster* we apply in a per-pixel analysis is no longer valid. There is not the single pixel converted from meadow to sealed, but the expanding settlement that eats up grassland patches in the surrounding. Blaschke (2005) has proposed a typology of object geometry changes considering different categories (existence-related, size- and shape-related, and location-related changes). In reality, we are facing a combination of all of these basic categories of geometric changes (Lang et al. 2009a). In addition, when an object is shrinking or expanding, the surrounding is inevitably affected. Raza and Kainz (2001) include subdivision and amalgamation of objects in their list of spatiotemporal characteristics of parcels. Other studies specifically investigate topological relations to mathematically evaluate the identity of two object hypotheses when comparing two different data sets (Straub and Heipke 2004). They present eight topological relations being reducible to the principal relationships: disjoint, equal, overlap, and containment. The latter (including touch and cover) utilizes a margin concept.

Practically, when changes are detected in multitemporal imagery, such spatial implications are often neglected so that established methods for spatially explicit object-based change analysis are still rare (Schöpfer et al. 2008). Map-to-map comparisons using raster overlay techniques are site specific, but not object specific; they refer to changes in pixels but not of objects. Vector overlays with physical intersections of the boundaries produce complex geometry with sliver polygons (Chen et al. 2012). Visual comparisons, on the other hand, are powerful but subjective and time consuming (Schöpfer et al. 2008).

Next to the simultaneity of change in label and extent, there are other reasons why there is no straightforward solution for object-based change. Image objects can vary due to different representations, while those geographic features they represent remain constant. A forest patch may be represented in different complexities in terms of the outer limit boundaries (Weinke et al. 2008) and its internal structure, and the same applies to other features like settlements, wetlands, and the like. The question arises whether a lack of spatial congruence of corresponding image objects is caused by data mismatches, different representations, or real change. More specific, there are several reasons for spatially inconsistent features, which are inherent to the representation itself (Lang et al. 2009a) and not related to the geographic features under concern (*pseudo-changes*):

- Image misregistrations due to poor spatial referencing may cause shifts in corresponding objects and pseudo-boundaries when overlaying those.
- Differing image characteristics of multitemporal or multiseasonal data sets in terms of viewing angles, sun illumination, etc.
- Image data of different spatial resolutions are combined and/or different segmentation algorithms were employed.
- Segmentation results are compared to visual interpretations and manual delineations, applying different scales and outline complexities (partly ignoring subtle changes due or outliers due to generalization effects).

This has implications on the assessment of OBIA outcomes, for example, when comparing a segmented and classified result with a visual interpretation map. Here the issue of spatial disagreement or *spatial error* (Radoux and Defourny 2008) arises, as object boundaries may mismatch by simply applying a diverging (or no) generalization strategy. Obviously, the quality of the segmentation, that is, *segmentation goodness* (Clinton et al. 2010, Hernando et al. 2012), is a critical factor in this process. Van Coillie et al. (2008) proposed a Purity Index measuring the common area between segmentation objects and reference objects. Vector-based measures (area-based, location-based, and a combination of both) were analyzed by Clinton et al. (2010), calculating the similarities between segments and training objects. Two goodness measures for optimal segmentation were proposed by Johnson and Zhixiao (2011): global intrasegment homogeneity (the variance of each object) and intersegment heterogeneity (how similar a region is to its neighbors).

In the following section, we shall have a closer look into how to analyze differences in objects by investigating spatial relationships among corresponding objects. We thereby abstract from the issue whether this is a product of object transition (change over time) or an outcome of different object representations or delineations. Since spatial relations are various and appear in reality in different combinations, there is a demand for ready-to-use solutions that are able to structure and categorize these.

15.3.2 Application Scenario #1: Object-Based Information Update

The need for updating existing information through remote sensing–based technology is undoubted and still increases with the availability of especially high spatial resolution data. In operations workflows, the update of information from image data is hampered by the lack of full integration into existing geo-spatial infrastructures. This issue is discussed for decades (Ehlers 1990), and the integration of GIS and remote sensing is progressing. But some problems are still not solved for many cases. For example, the operational workflow for a planning authority to integrate a—even very accurate and based on VHSR data—land cover classification in existing vector data sets (e.g., cadastral data) is still limited, if the boundaries of the land cover classes do not match the spatial characteristics of the geo-data in use. The geometric matching of the different data sets is an issue of scale/resolution, the related issue of boundary complexity, but also a principal problem of conversion between different data models (usually raster to vector) (Merchant and Narumalani 2009). To overcome the latter, expectations have been placed on the ability of segmentation-based approaches to overcome the problem of mismatches that are due to the use of different data models (raster vs. vector) by delivering *GIS-ready information* (Benz et al. 2004) from remote sensing data. However, as Tiede et al. (2010) pointed out, such a full integration of image-derived spatial information is not a trivial task: the problem is residing lesser in the raster–vector conversion itself, but more in the integration of information from different (spatial) resolutions. Coarser image resolution than the existing geodata to be updated results in objects with boundaries often not matching existing ones. Also the contrary, very high resolution imagery (VHRI) imagery as information source to update lower-resolution (i.e., coarser scale, generalized) vector data results in objects often too complex regarding shape and boundaries. Smoothening and generalization routines need to be applied, which can lead—according to their nondeterministic operation—to unsatisfactory match with existing boundaries. Refer to Walter and Fritsch (1999) and Buthenuth et al. (2007) for a discussion about matching of different (vector) data models and the integration of raster and vector data.

To summarize, specific challenges are faced when existing boundaries are considered in a process of information update. We shortly presented a collection of instances on how to perform OBIA for monitoring purposes, aiming at the provision of up-to-date information under the explicit consideration of existing geo-spatial data. In Figure 15.9 the different cases are grouped into three categories. *Boundaries* refer to existing polygon or line data that are used for adaptive parcel-based segmentation. A specific case of using cadastre information for generating polygon data sets has been dubbed "adaptive per parcel segmentation" by Tiede et al. (2007).

In a study for a planning association (*Verband Region Stuttgart*, an association of local authorities in the Stuttgart region, Germany), both problems were addressed (Tiede et al. 2010). The requirements to support regional planning based on EO data encompassed the delineation of biotope complexes as target units. To be fed into the existing data structure of the planning authority (*cadastre-conform*), the biotope complexes should preserve boundaries of the digital cadastre map in cases where a change in biotope complex type is also reflected in the cadastral data, but remove boundaries within the same biotope complex. In addition, biotope boundaries not reflected in the cadastral data need to be integrated.

As a trade-off between costs and quality, image data of 5 m GSD were used for the analysis, which introduced a certain scale-gap, especially for newly introduced boundaries, not yet reflected in the cadastral map. A twofold strategy was applied to tackle the problem, by using (1) the digital cadastral data as spatial constraints in the initial object building step (using an adaptive per-parcel approach) and (2) the adjustment and validation of newly introduced rasterized boundaries by comparison with the target vector geometry. In the first step, we flagged either boundaries of the digital cadastral map to be preserved if they were also representing biotope complex boundaries or boundaries to be removed if they are within

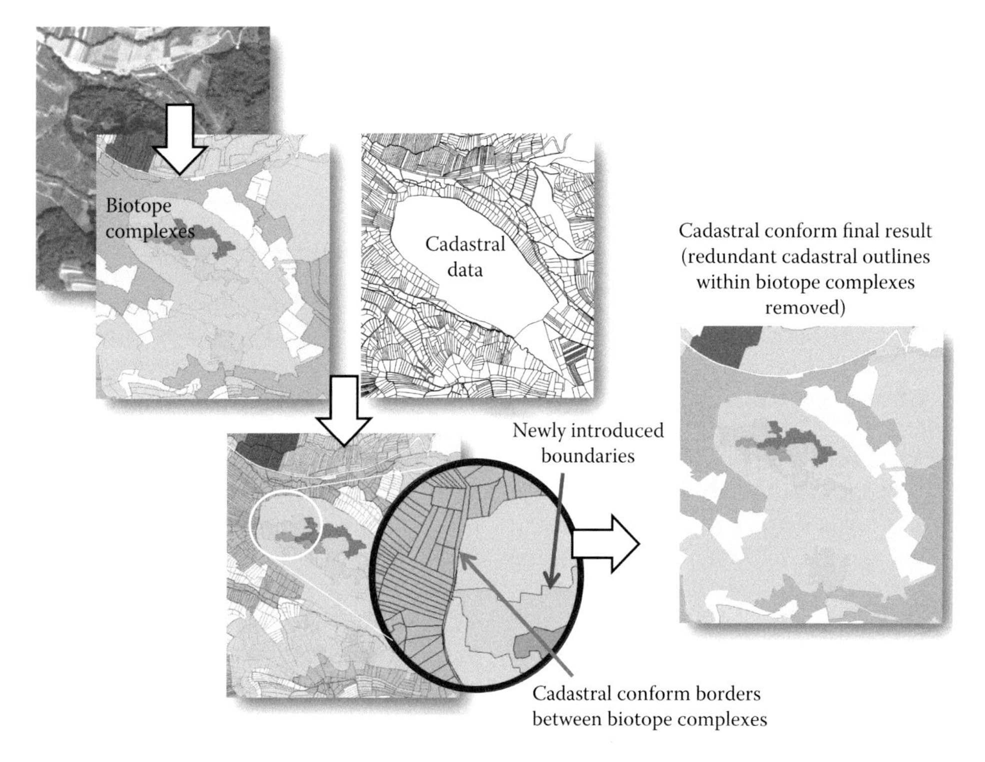

FIGURE 15.8 Cadastre-conform extraction of biotope complex units from EO data: Cadastral data are retained, where biotope complexes are changing; redundant cadastral boundaries within the same biotope complex are eliminated.

the same biotope complex. By this, the scale-gap of the image data compared to the existing data is avoided. In the second step—to integrate newly introduced boundaries from the image-based biotope complex delineation—it was necessary to perform an adjustment in regard to the same generalization/smoothness level of the existing data (instead of a rasterized boundary). The latter was achieved using a combination of established GI-routines and additionally programmed solutions. Figure 15.8 schematically shows the three occurring problems that were addressed to reach the requirements for an accurate and compatible data set for administrative purposes: (1) The replacement of biotope complex boundaries with corresponding cadastral boundaries by considering a spatial displacement tolerance (*scale-gap*) according to the pixel size of the image data. (2) Merging of cadastral polygons within a biotope complex (removing of cadastral boundaries). (3) The introduction of new boundaries (based on an OBIA-based biotope complex delineation), not reflected in the cadastral data set but which represents changes of biotope complexes. The newly introduced boundaries were finally smoothed and generalized using standard GI procedures (Douglas and Peucker 1973, Bodansky et al. 2002).

15.3.3 Object Fate Analysis

OFA is a method proposed by Lang (2005) and discussed by Schöpfer et al. (2008) for investigating spatial relationships between corresponding objects in two different representations. Rephrased and put from an individual image object's perspective, this would reflect the *fate* of this particular object and its representation (see Figure 15.10). Object fate may be caused by real change captured in data from different points of time. Or otherwise, it may root in differences in object generation by using segmentation algorithms or visual analysis, heterogeneous data reference data sets from other sources (Schöpfer et al. 2008), etc. In reality, we often face a combination of change and representation-induced divergences.

OFA has been implemented in a tool called *Landscape Interpretation Support Tool* (LIST) (Weinke et al. 2007, Lang et al. 2009a), as an extension for ESRI's ArcGIS software. Following the concept of parent and child relationships, two vector layers are used to represent the specific *fate* of corresponding objects. Thereby only the geometry is considered, assuming that corresponding objects retain their label (see later for an extension of OFA to the thematic dimension). To overcome spatial uncertainty in image

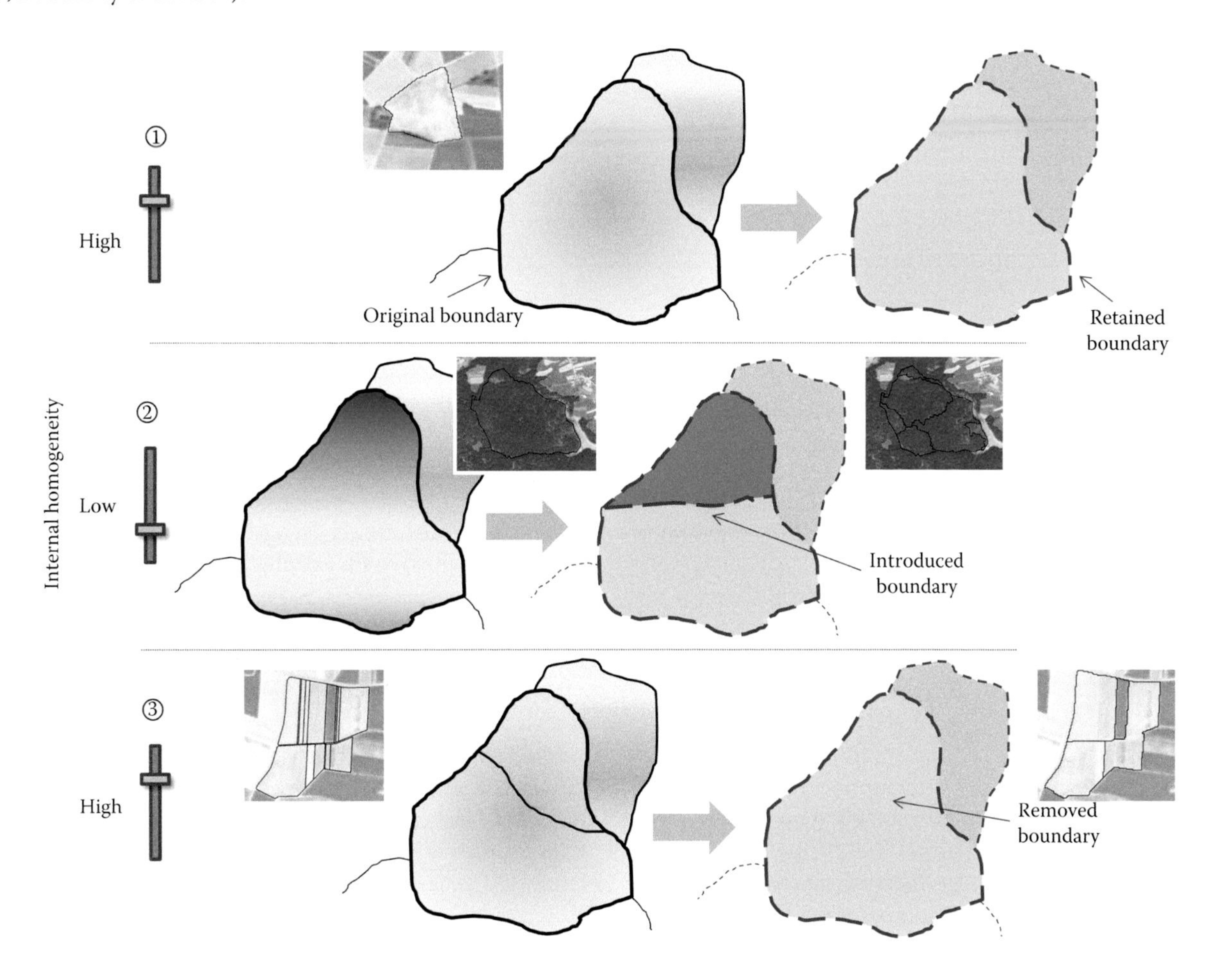

FIGURE 15.9 Illustration of differentiated boundary treatment. Additional cases (not shown): boundaries are generalized, fractalized, shifted, found in the surroundings. Case 1: existing boundaries are retained (limited internal heterogeneity); case 2: new boundaries are introduced—internal variance larger than (given) threshold; and case 3: boundaries are removed—internal variance larger than (given) threshold.

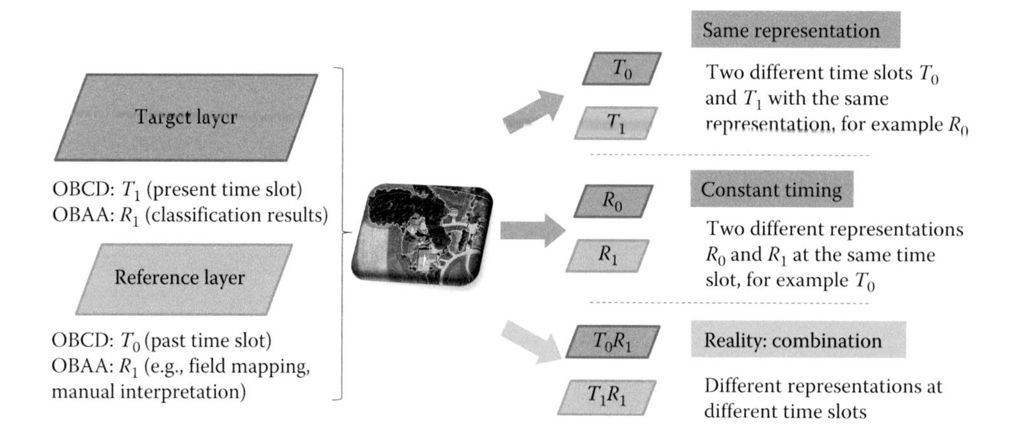

FIGURE 15.10 Object fate considering aspects of different status (change) and representation.

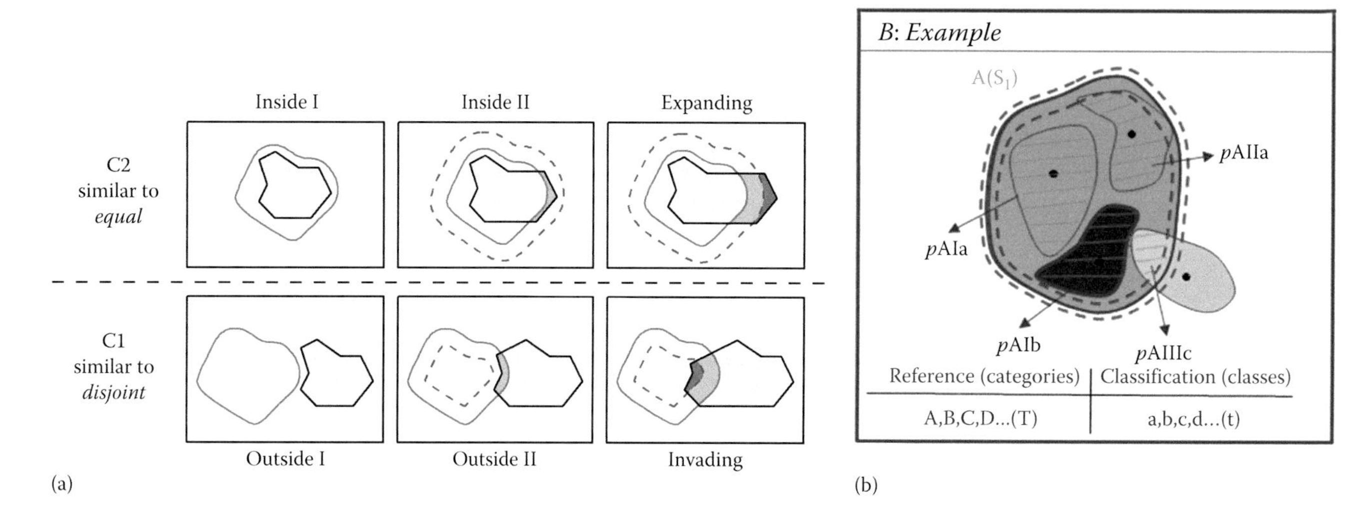

FIGURE 15.11 Spatial and thematic assessment in extended OFA. (a) OFA categories depending on mutual spatial relations of two objects (outlined, shaded). (b) Differentiated OFA categories considering object classification. (Modified from Hernando, A. et al., *Int. J. Appl. Earth Obs. Geoinf.*, 19, 214, 2012.)

objects, an error band (positive and negative buffer around the object boundary) is used. The size of the buffer is specified either manually or as a function relative to object size. A *virtual* overlay characterizes spatial relationships via the relative positions of centroids, retaining the original object boundaries. This distinguishes between weak overlap (*similar to disjoint*) and strong overlap (*similar to identical*). The resulting categories of object relationships are discussed by Schöpfer et al. (2008) and shown in a simplified version in Figure 15.11. Object relationships are divided into either C1 cluster or C2 cluster by evaluating whether the centroid of the classification object fell inside the reference object (upper row: *inside* or *good I, inside II, expanding*) or not (lower row: *outside* or *not interfering I, outside II, invading*). There are six relationships; five of them are relevant for further analysis when two corresponding objects at least partly overlap. The relationship *not interfering I* assumes that objects do not *correspond* any longer. Two stability measures, *object (or offspring) loyalty* and *interference* (Lang et al. 2009a), were derived based on the relationships between objects.

Hernando et al. (2012) extended OFA to the thematic dimension by assessing results from a detailed OBIA classification of forest stands as compared to an existing forest stand mapping derived from visual air-photo interpretation. By combining spatial and thematic aspects (see Figure 15.11b), an OFA matrix was established. First, each classified object (classes *a, b, c, d,... t*) was compared to the corresponding reference object with S_T the overlapping area for each of the reference categories (*A, B, C, D...T*) and S_{Total} the total reference area for all *w* reference categories (*A, B, C...w....T*). The matrix is established by listing all relationships between all category/class shares for each of the five relationship types. Each cell of this matrix is filled by a value r_{TCt} as the sum of the relative areas of the *t* classified objects considering an OFA type.

15.3.4 Object Linking

Traditional automated change detection approaches are limited if, for example, co-registration errors are present or if an object changed over time (changes especially in size/form or

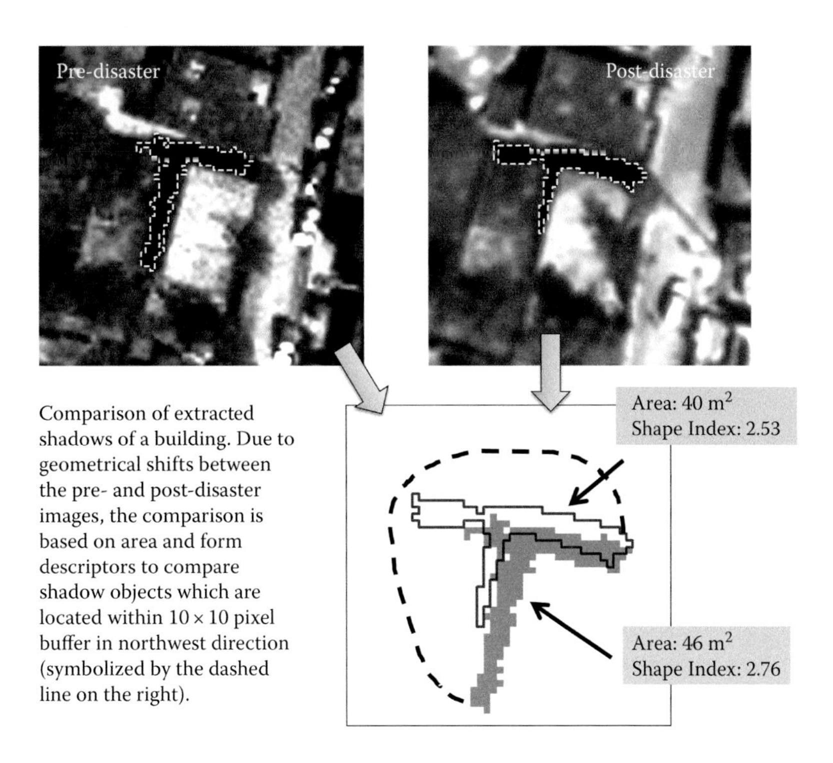

FIGURE 15.12 Conceptual illustration of the object-linking approach. Here, comparing extracted shadow objects from an earthquake pre- and post-disaster images—despite a geometrical shift between the images—to indicate if the shadow is considered to be still in existence (here, no indication of destruction of the building since the area and form descriptors are quite similar). (Adapted from Tiede, D. et al., *Photogramm. Eng. Remote Sens.*, 9, 933, 2011.)

it disappeared). Such change detection approaches, looking at pixel value changes only neglecting potential geometrical shifts, require precise coregistration of multitemporal images. (Lu et al. 2004). Following a study about the robustness of different pixel-based change detection algorithms by Sundersan et al. (2007), the registration error between the images should not exceed 0.2 pixels (RMS) for satisfactory performance of the algorithms.

When comparing objects created from different sensor data or time epochs. we are able to compensate to some degree the requirement for image co-registration. This includes a generalization issue in the object generation process, comparing statistical values per object instead of single pixels, comparing also geometric properties). Nevertheless, a strict vertical object hierarchy, as often used in OBIA frameworks, hampers sophisticated analysis, because of the intersection of boundaries hampers treating multitemporal data similar as kind of a pseudo-hierarchy.

Human interpreters usually compensate positional shifts between data sets when recognizing similar objects. Such a *loose* coupling can be mimicked using an object-linking approach to track objects over time or to analyze changes per object instead of, for example, intersections of objects (Tiede et al. 2011, Hofmann and Blaschke 2012). Object linking relies on GI methods adopted in a topologically enabled OBIA framework It establishes spatial relationships between objects incorporating shifts introduced through positional errors using (multi)directional buffering and can be enriched by the comparison of geometric object properties (like form and size).

Figure 15.12 shows the implementation of an object-linking approach for shadows casted by buildings, before and after an earthquake (Tiede et al. 2011). The class-constrained object-by-object comparison is able to compare objects of different classifications including geometrical buffering in x or y directions, that is, a geometrical shift between the dates can be compensated (or to track moving objects). Since such a linking of objects—including the neighborhood and not only the direct spatial overlap—can lead to 1:n or even n:n matches, not only class constraints (matching of objects of the same classes only) are important, but also the comparison of geometrical object properties (e.g., form and size). By this, the degree of change or the degree of *movement* can be estimated. In the following, an application scenario is described using an object-linking approach in the case of automated rapid information extraction after a disaster (here damage indication after an earthquake)

15.3.5 Application Scenario #2: Rapid Information Extraction

Specific automated image analysis techniques based on VHR images, for example, satellite images (QuickBird, GeoEye, and Worldview), have reached a status of maturity to be utilized in application domains crucially depending on reliability and timeliness (Lang et al. 2010a). Several studies have shown that different approaches can lead to successful applications in the field of damage assessment using change detection methodologies, for example, comparing of pre- and post-event images. Gusella et al.

(2005) demonstrated the use of QuickBird data for quantifying the number of buildings that collapsed after the Bam earthquake in Iran, 2003. Also using QuickBird data, Pesaresi et al. (2007) showed that very high accuracies allow for rapid damage assessment of built-up structures in tsunami-affected areas based on a multicriteria recognition system within a restricted set of general assumptions (e.g., built-up structures cast a shadow, and collapsed built-up structures no longer cast a shadow, and they leave debris on the ground). Vu and Ban (2010) developed a context-based automated approach for earthquake damage mapping relying on debris area identification. In addition, the calculation has been speed-optimized by parallel processing implementation showing good results for the test area, after the Sichuan earthquake in China, 2008. Ehrlich et al. (2008) and Brunner et al. (2010) demonstrate the usefulness of applications with the additional use of synthetic aperture radar data for damaged building assessment also for the Sichuan earthquake area. Three-dimensional information has been taken into account to detect damaged built-up structures after earthquakes by Turker and Cetinkaya (2005) with the help of stereo-images, but for rapid damage assessment shortly after an earthquake, the necessity of capturing a pair of fitting stereo-images is usually not feasible. A broad overview of automated techniques for earthquake damage detection is given by Chini (2009) and Rathje and Adams (2008).

Still, such approaches are rarely applied on an operational level in the relief phase shortly after a disaster. One reason might be that the process critically depends on a good quality of the preprocessed data, that is, mainly a high reliability of the image (co)registration. For most of the automated change detection algorithms, this is a crucial point (Lu et al. 2004, Sundersan et al. 2007). Right after a disaster, there is often a lack of high-quality pre- and post-disaster VHSR imagery (Rathje and Adams 2008). After the Haiti earthquake in January 2010, the availability of VHSR satellite data was outstanding compared to previous disasters. Data have even been provided to the public for free (also outside the International Charter Space and Major Disasters) by companies like Google or DigitalGlobe. One of the effects was that a tremendous amount of *crowd-sourced* information was provided in the aftermath of the earthquake. Nevertheless, these data sets are often not well suited for automated analysis methods, due to reasons such as lacking metadata, non-documented or insufficient preprocessing, and missing spectral NIR bands.

A strategy for automated extraction of damage indication from very high–spatial resolution satellite imagery was presented for the Haitian town of Carrefour after the January 2010 earthquake (Tiede et al. 2011). Damaged buildings are identified by changes in their shadows compared in pre- and post-event data. The approach builds on OBIA concepts to extract the relevant information about damage distribution (collapsed buildings). The lacking quality of the pre- and post-disaster imagery usually available directly after an event (in this case, the shift between the multitemporal images was up to 5 m and varied throughout the images) was bypassed using the described object-linking approach between objects extracted from the different images. This approach allows to overcome sub-pixel co-registration of images (especially if the image source and the conducted preprocessing steps are not known) in rapid information extraction workflows as in the context of disaster events (see Figure 15.13). For the area of Carrefour, the new methodology produced positively validated results in acceptable processing time and was, to our best knowledge, the only automated damage assessment

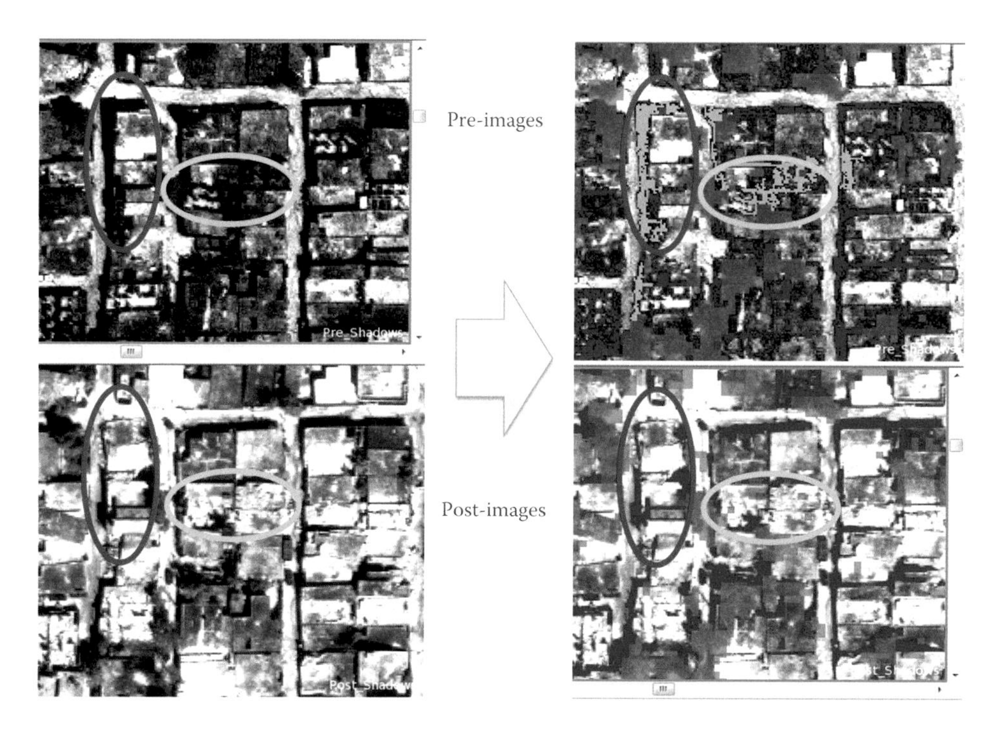

FIGURE 15.13 Object linking for relating collapsed buildings.

method that delivered relevant results to requesting relief organizations in the first days after the Haiti earthquake.

15.4 Conclusions

In this chapter, we have discussed implications of geodata integration in the context of OBIA and geospatial information update from an interdisciplinary, applied perspective. We have shown how the advancement of feature recognition and advanced image analysis techniques facilitates the extraction of thematic information for decision support. OBIA techniques were discussed in light of transforming and reconstructing image scenes into ready-to-use geospatial information matching the conceptual understanding and the practical needs of users. This implies issues of scale, generalization, as well as representation (including visualization); thus, we subsumed the entirety of aspects to be considered in this process, under the term *object validity*. Since image objects mimicking real-world objects are subject to change, OBCD was reviewed in terms of type, intensity, size and shape, as well as horizontal and vertical relations of image objects. OBAA faces similar challenges when evaluating the quality (accuracy, robustness, etc.) of OBIA classifications. The concept of OFA and the object linking approach provide strategies on how to deal with such challenges. We have rounded up the chapter with two representative application scenarios to illustrate the maturity of the discussed conceptual items in a practical context.

References

Allen, T. F. H. and T. B. Starr. 1982. *Hierarchy*. Chicago, IL: University of Chicago Press.

Aschbacher, J. 2002. Monitoring environmental treaties using earth observation. In *VERTIC Verification Yearbook*. London, U.K.: VERTIC, pp. 171–186.

Baatz, M. and A. Schäpe. 2000. *Multiresolution Segmentation: An Optimization Approach for High Quality Multi-Scale Image Segmentation*. Salzburg, Austria: Wichmann Verlag.

Baraldi, A. 2011. Satellite Image Automatic Mapper™ (SIAM™). A turnkey software button for automatic near-real-time multi-sensor multi-resolution spectral rule-based preliminary classification of spaceborne multi-spectral images. *Recent Patents on Space Technology* 2011(1): 81–106.

Baraldi, A. and L. Broschetti. 2012. Operational automatic remote sensing image understanding systems: Beyond Geographic Object-Based and Object-Oriented Image Analysis (GEOBIA/GEOOIA). Part 1: Introduction. *Remote Sensing* 4: 2694–2735.

Belgiu, M., B. Hofer, and P. Hofmann. 2014. Coupling formalized knowledge bases with object-based image analysis. *Remote Sensing Letters* 5(6): 530–538.

Benz, U., P. Hofmann, G. Willhauck, I. Lingenfelder, and M. Heynen. 2004. Multi-resolution, object-oriented fuzzy analysis of remote sensing data for GIS-ready information. *ISPRS Journal of Photogrammetry and Remote Sensing* 58: 239–258.

Bischof, H., W. Schneider, and A. Pinz. 1992. Multispectral classification of Landsat-images using neural networks. *IEEE Transactions on Geoscience and Remote Sensing* 30: 482–490.

Blaschke, T. 2005. Towards a framework for change detection based on image objects. Göttinger Geographische Abhandlungen. In *Remote Sensing and GIS for Environmental Studies*, S. Erasmi, B. Cyffka, and M. Kappas (eds.). University of Göttingen, Göttingen, Germany, p. 113.

Blaschke, T. and J. Strobl. 2001. What's wrong with pixels? Some recent developments interfacing remote sensing and GIS. *Zeitschrift für Geoinformationssysteme* 14(6): 12–17.

Blaschke, T. et al. 2007. GMES: From research projects to operational environmental monitoring services. *ISPRS Workshop on High-Resolution Earth Imaging for Geospatial Information*, Hannover, Germany.

Blaschke, T. et al. 2014. Geographic object-based image analysis: A new paradigm in remote sensing and geographic information science. *International Journal of Photogrammetry and Remote Sensing* 87(1): 180–191.

Bodansky, E., A. Griboy, and M. Pilouck. 2002. Smoothing and compression of lines obtained by raster-to-vector: Conversion. *Graphics Recognition: Algorithms and Applications* 2390: 256–265.

Bruce, V. and P. R. Green. 1990. *Visual Perception*. East Sussex, U.K.: Lawrence Erlbaum Associates.

Brunner, D., G. Lemoine, and L. Bruzzone. 2010. Earthquake damage assessment of buildings using VHR optical and SAR imagery. *IEEE Transactions on Geoscience and Remote Sensing* 48(5): 2403–2420.

Buthenuth, M., G. Gösseln, M. Tiedge, C. Heipke, U. Lipeck, and M. Sester. 2007. Integration of heterogeneous geospatial data in a federated database. *ISPRS Journal of Photogrammetry and Remote Sensing* 62(5): 328–346.

Campbell, J. B. 2002. *Introduction to Remote Sensing*. New York: The Guilford Press.

Chen, G., G. J. Hay, and B. St-Onge. 2012. A GEOBIA framework to estimate forest parameters from lidar transects, Quickbird imagery and machine learning: A case study in Quebec, Canada. *International Journal of Applied Earth Observations and Geoinformation* 15: 28–37.

Chini, M. 2009. Earthquake damage mapping techniques using SAR and optical remote sensing satellite data. In *Advances in Geoscience and Remote Sensing*, G. Jedlovec (ed.). InTech—Open Access Publisher, pp. 269–278.

Clinton, N., A. Holt, J. Scarborough, L. Yan, and P. Gong. 2010. Accuracy assessment measures for object-based image segmentation goodness. *Photogrammetric Engineering and Remote Sensing* 76(3): 289–299.

Congalton, R. G. and K. Green. 1998. *Assessing the Accuracy of Remotely Sensed Data: Principles and Practices*. Boca Raton, FL: Lewis Publishers.

De Martino, M. and R. Albertoni. 2011. A multilingual/multicultural semantic-based approach to improve data sharing in an SDI for nature conservation. *International Journal of Spatial Data Infrastructures Research* 6: 206–233.

Di Gregorio, A. and L. J. M. Jansen. 2005. *Land Cover Classification System (LCCS): Classification Concepts and User Manual*. Rome, Italy: Food and Agriculture Organization of the United Nations.

Douglas, D. H. and T. K. Peucker. 1973. Algorithms for the reduction of the number of points required to represent a digitized line or its caricature. *Cartographica* 10(2): 112–122.

Drăgut, L., O. Csillik, C. Eisank, and D. Tiede. 2014. Automated parameterisation for multi-scale image segmentation on multiple layers. *ISPRS Journal of Photogrammetry and Remote Sensing* 88: 119–127.

Ehlers, M. 1990. Remote sensing and geographic information systems: Towards integrated spatial information processing. *IEEE Transactions on Geoscience and Remote Sensing* 28(4): 763–766.

Ehrlich, D., H. D. Guo, K. Molch, J. W. Ma, and M. Pesaresi. 2008. Identifying damage caused by the 2008 Wenchuan earthquake from VHR remote sensing data. *International Journal of Digital Earth* 4: 309–326.

Eysenck, M. W. and M. T. Keane. 1995. *Cognitive Psychology: A Student's Handbook*. East Sussex, U.K.: Psychology Press.

Ganter, B. and R. Wille. 1996. *Die Formale Begriffsanalyse*. Berlin, Germany: Springer.

Gibson, J. J. 1979. *The Ecological Approach to Visual Perception*. Boston, MA: Houghton Mifflin.

Gorte, B. 1998. *Probabilistic Segmentation of Remotely Sensed Images*. ITC Publication Series, Twente, the Netherlands.

Gusella, L., B. J. Adams, G. Bitelli, C. K. Huyck, and A. Mognol. 2005. Object-oriented image understanding and post-earthquake damage assessment for the 2003 Bam, Iran, earthquake. *Earthquake Spectra* 21(S1): 225–238.

Hay, G. J. and G. Castilla. 2008. Geographic object-based image analysis (GEOBIA): A new name for a new discipline. In *Object-Based Image Analysis: Spatial Concepts for Knowledge-Driven Remote Sensing Applications*, T. Blaschke, S. Lang, and G. J. Hay (eds.). Berlin, Germany: Springer.

Hay, G. J., D. J. Marceau, P. Dubé, and A. Buchard. 2001. A multiscale framework for landscape analysis: Object-specific analysis and upscaling. *Landscape Ecology* 16(6): 471–490.

Hernando, A., D. Tiede, F. Albrecht, and S. Lang. 2012. Spatial and thematic assessment of object-based forest stand delineation using an OFA-matrix. *International Journal of Applied Earth Observations and Geoinformation* 19: 214–225.

Hofmann, P. and T. Blaschke. 2012. Object based change detection using temporal linkages. *Proceedings of the 4th GEOBIA Conference 2012*, Rio de Janeiro, Brazil.

Johnson, B. and X. Zhixiao. 2011. Unsupervised image segmentation evaluation and refinement using a multi-scale approach. *ISPRS Journal of Photogrammetry and Remote Sensing* 66: 473–483.

Julesz, B. 1975. Experiments in the visual perception of texture. *Scientific American* 212: 38–48.

Kuhn, W. 2009. A functional ontology of observation and measurement. In *GeoSpatial Semantics*, K. Janowicz, M. Raubal, and S. Levashkin (eds.). Berlin, Germany: Springer, Vol. 5892, pp. 26–43.

Lang, S. 2005. Image objects and landscape objects: Interpretation, hierarchical representation and significance. Salzburg University, Salzburg, Austria.

Lang, S. 2008. Object-based image analysis for remote sensing applications: Modeling reality—Dealing with complexity. In *Object-Based Image Analysis—Spatial Concepts for Knowledge-Driven Remote Sensing Applications*, T. Blaschke, S. Lang, and G. J. Hay (eds.). Berlin, Germany: Springer, pp. 3–28.

Lang, S., F. Albrecht, S. Kienberger, and D. Tiede. 2010a. Object validity for operational tasks in a policy context. *Journal for Spatial Science* 55(1): 9–22.

Lang, S., C. Burnett, and T. Blaschke. 2004. Multi-scale object-based image analysis: A key to the hierarchical organisation of landscapes. *Ekologia* 23(Suppl.): 1–9.

Lang, S., S. Kienberger, D. Tiede, M. Hagenlocher, and L. Pernkopf. 2014. Geons—Domain-specific regionalization of space. *Cartography and Geographic Information Science* 41(3): 214–226.

Lang, S., E. Schöpfer, and T. Langanke. 2009a. Combined object-based classification and manual interpretation—Synergies for a quantitative assessment of parcels and biotopes. *Geocarto International* 24(2): 99–114.

Lang, S., D. Tiede, F. Albrecht, and P. Füreder. 2010b. Automated techniques in rapid geospatial reporting—Issues of object validity. In *VALgEO 2010*, Ispra, Italy, C. Corbane, D. Carrion, M. Broglia, and M. Pesaresi (eds.), pp. 65–75.

Lang, S., D. Tiede, D. Hölbling, P. Füreder, and P. Zeil. 2009b. Conditioning land-use information across scales and borders. In *Geospatial Crossroads @ GI_Forum' 09*, A. Car, G. Griesebner, and J. Strobl (eds.). Heidelberg, Germany: Wichmann, pp. 100–109.

Lang, S., P. Zeil, S. Kienberger, and D. Tiede. 2008. Geons—Policy-relevant geo-objects for monitoring high-level indicators. *Proceedings of the GI Forum, Geospatial Crossroads @ GI_Forum' 08*, Salzburg, Austria, A. Car and J. Strobl (eds.). pp. 180–186.

Liedtke, C. E., J. Bückner, O. Grau, S. Growe, and R. Tönjes. 1997. AIDA: A system for the knowledge based interpretation of remote sensing data. *Third International Airborne Remote Sensing Conference*, Copenhagen, Denmark.

Lu, D., P. Mausel, and E. Moran. 2004. Change detection techniques. *International Journal of Remote Sensing* 25: 2365–2401.

Lucieer, A. and A. Stein. 2002. Existential uncertainty of spatial objects segmented from satellite sensor imagery. *IEEE Transactions on Geoscience and Remote Sensing* 40(11): 2518–2521.

Lutz, M. and E. Klien. 2006. Ontology based retrieval of geographic information. *International Journal of Geographical Information Science* 20(3): 233–260.

Marceau, D. J. 1999. The scale issue in the social and natural sciences. *Canadian Journal of Remote Sensing* 25: 347–356.

Marr, D. 1982. *Vision*. New York: W.H. Freeman.

Merchant, J. W. and S. Narumalani. 2009. Integrating remote sensing and geographic information system. In *The SAGE Handbook of Remote Sensing*, T. A. Warner, M. D. Nellis, and G. M. Foody (eds.). London, U.K.: SAGE Publications Ltd., pp. 257–268.

Montello, D. R. 2001. *Scale in Geography: International Encyclopedia of the Social & Behavioural Sciences*. Oxford, U.K.: Pergamon Press, pp. 13501–13504.

Pesaresi, M., A. Gerhardinger, and F. Haag. 2007. Rapid damage assessment of built-up structures using VHR satellite data in tsunami-affected areas. *International Journal of Remote Sensing* 28(13): 3013–3036.

Pinz, A. 1994. *Bildverstehen*. Vienna, Austria: Springer.

Radoux, J. and P. Defourny. 2008. A framework for the quality assessment of object-based classification. In *Object-Based Image Analysis—Spatial Concepts for Knowledge-Driven Remote Sensing Applications*, T. Blaschke, S. Lang, and G. J. Hay (eds.). Berlin, Germany: Springer, pp. 257–271.

Raskin, R. and M. Pan. 2003. Semantic web for earth and environmental terminology (sweet). *Proceedings of the Workshop on Semantic Web Technologies for Searching and Retrieving Scientific Data*, Sanibel Island, FL.

Rathje, E. M. and B. J. Adams. 2008. The role of remote sensing in earthquake science and engineering: Opportunities and challenges. *Earthquake Spectra* 24(2): 471–492.

Raza, A. and W. Kainz. 2001. An object-oriented approach for modeling urban land-use changes. *Journal of the Urban and Regional Information Association* 14(1): 37–55.

Reitsma, F. L. J., S. Ballard, W. Kuhn, and A. Abdelmoty. 2009. Semantics, ontologies and eScience for the geosciences. *Computers and Geosciences* 35: 706–709.

Richter, R. and D. Schläpfer. 2002. Geo-atmospheric processing of airborne imaging spectrometry data, part 2: Atmospheric/topographic correction. *International Journal of Remote Sensing* 23: 2631–2649.

Schöpfer, E., S. Lang, and F. Albrecht. 2008. Object-fate analysis—Spatial relationships for the assessment of object transition and correspondence. In *Object-Based Image Analysis—Spatial Concepts for Knowledge-Driven Remote Sensing Applications*, T. Blaschke, S. Lang, and G. J. Hay (eds.). Berlin, Germany: Springer, pp. 785–801.

Schumacher, J., S. Lang, D. Tiede, D. Hölbling, J. Rietzke, and J. Trautner. 2007. Einsatz von GIS und objekt-basierter Analyse von Fernerkundungsdaten in der regionalen Planung Methoden und erste Erfahrungen aus dem Biotopinformations- und Managementsystem (BIMS) Region Stuttgart. In *Angewandte Geoinformatik 2007*, J. Strobl, T. Blaschke, and G. Griesebner (eds.). Heidelberg, Germany: Wichmann, pp. 703–708.

Smith, B. 1994. Fiat objects. *Eleventh European Conference on Artificial Intelligence*, Amsterdam, the Netherlands, N. Guarino, L. Vieu, and S. Pribbenow (eds.).

Spence, N. A. 1961. A multifactor uniform regionalization of British Countries on the basis of employment data for 1961. *Regional Studies* II 2: 87–104.

Spitzer, M. 2000. *Geist im Netz. Modelle für Lernen, Denken und Handeln*. Heidelberg, Germany: Spektrum Akademischer Verlag.

Strasser, T. and S. Lang. In press. Object-based class modelling for multi-scale riparian forest habitat mapping. *International Journal of Applied Earth Observations and Geoinformations* 37: 29–37.

Straub, B. M. and C. Heipke. 2004. Concepts for internal and external evaluation of automatically delineated tree tops. *IntArchPhRS* 26(8): 62–65.

Sundersan, A., P. K. Varshney, and M. K. Arora. 2007. Robustness of change detection algorithms in the presence of registration errors. *Photogrammetric Engineering and Remote Sensing* 73: 165–174.

Tarr, M. J. and Y. D. Cheng. 2003. Learning to see faces and objects. *Trends in Cognitive Science* 7(1): 23–30.

Tiede, D. 2014. A new geospatial overlay method for the analysis and visualization of spatial change patterns using object-oriented data modeling concept. *Cartography and Geographic Information Science* 41: 227–234.

Tiede, D., S. Lang, F. Albrecht, and D. Hölbling. 2010. Object-based class modeling for cadastre constrained delineation of geo-objects. *Photogrammetric Engineering and Remote Sensing* 76(2): 193–202.

Tiede, D., S. Lang, P. Füreder, D. Hölbling, C. Hoffmann, and P. Zeil. 2011. Automated damage indication for rapid geospatial reporting. An operational object-based approach to damage density mapping following the 2010 Haiti earthquake. *Photogrammetric Engineering and Remote Sensing* 9: 933–942.

Tiede, D., M. Möller, S. Lang, and D. Hölbling. 2007. Adapting, splitting and merging cadastral boundaries according to homogenous LULC types derived from Spot 5 dat. *ISPRS Journal of Photogrammetry and Remote Sensing* 36(3): 99–104.

Tönjes, R., S. Growe, J. Bückner, and C. E. Liedtke. 1999. Knowledge-based Interpretation of remote sensing images using semantic nets *Photogrammetric Engineering and Remote Sensing* 65: 811–821.

Turker, M. and B. Cetinkaya. 2005. Automatic detection of earthquake-damaged buildings using DEMs created from pre- and post-earthquake stereo aerial photographs. *International Journal of Remote Sensing* 26(4): 823–832.

Van Coillie, F. M. B., R. P. C. Verbeke, and R. R. De Wulff. 2008. Semi-automated forest stand delineation using wavelet based segmentation of very high resolution optical imagery. In *Object-Based Image Analysis for Remote Sensing Applications: Modeling Reality-Dealing with Complexity*, T. Blaschke, S. Lang, and G. J. Hay (eds.). Berlin, Germany: Springer, pp. 237–256.

Vu, T. T. and Y. Ban. 2010. Context-based mapping of damaged buildings from high-resolution optical satellite images. *International Journal of Remote Sensing* 31(13): 3411–3425.

Walter, V. and D. Fritsch. 1999. Matching spatial data sets: A statistical approach. *International Journal of Geographical Information Science* 13(5): 445–473.

Weinke, E., S. Lang, and M. Preiner. 2008. Strategies for semi-automated habitat delineation and spatial change assessment in an Alpine environment. In *Object-Based Image Analysis—Spatial Concepts for Knowledge-Driven Remote Sensing Applications*, T. Blaschke, S. Lang, and G. J. Hay (eds.). Berlin, Germany: Springer, pp. 711–732.

Weinke, E., S. Lang, and D. Tiede. 2007. *Landscape Interpretation Support Tool* (*LIST*). Shaker Verlag, Graz, Austria.

Wertheimer, M. 1925. *Drei Abhandlungen zur Gestalttheorie* (*in German*). Erlangen, Germany: Palm & Enke.

Wise, S., R. Haining, and J. Ma. 2001. Providing spatial statistical data analysis functionality for the GIS user: The SAGE project. *International Journal of Geographical Information Science* 3(1): 239–254.

Zadeh, L. A. 1965. Fuzzy sets. *Information and Control* 8(3): 338–353.

Zeil, P., H. Klug, and I. Niemeyer. 2008. GIS and remote sensing: Monitoring Environmental conventions and agreements. *BICC Brief*, Bonn, 37: 50–56.

Zlatof, N., B. Tellez, and A. Baskurt. 2004. Image understanding and scene models: A generic framework integrating domain knowledge and Gestalt theory. *International Conference on Image Processing, ICIP '04*, Singapore.

16

Image Segmentation Algorithms for Land Categorization

James C. Tilton
Goddard Space Flight Center

Selim Aksoy
Bilkent University

Yuliya Tarabalka
Sophia-Antipolis Méditerranée

Acronyms and Definitions

AA	Average accuracy
BSMSE	Band sum mean squared error
DMP	Derivative of the morphological profile
HSeg	Hierarchical segmentation
HSWO	Hierarchical step-wise optimization
ICM	Iterated conditional modes
IOER	Institute for Ecological and Regional Development
MP	Morphological profile
MRF	Markov random field
MSE	Mean squared error
MSF	Minimum spanning forest
NDVI	Normalized Difference Vegetation Index
OA	Overall accuracy
PV	Plurality vote
RBF	Radial basis function
RHSeg	Recursive hierarchical segmentation
ROSIS	Reflective optics system imaging spectrometer
SAM	Spectral angle mapper
SAR	Synthetic aperture Radar
SE	Structuring element
SVM	Support vector machines

16.1 Introduction

Image segmentation is the partitioning of an image into related meaningful sections or regions. Segmentation is the key first step for a number of image analysis approaches. The nature and quality of the image segmentation result is a critical factor in determining the level of performance of these image analysis approaches. It is expected that an appropriately designed image segmentation approach will provide a better understanding of a landscape and/or significantly increase the accuracy of a landscape classification. An image can be partitioned in several ways, based on numerous criteria. Whether or not a particular image partitioning is useful depends on the goal of the image analysis application that is fed by the image segmentation result.

The focus of this chapter is on image segmentation algorithms for land categorization. Our image analysis goal will generally be to appropriately partition an image obtained from a remote sensing instrument on board a high flying aircraft or a satellite circling the earth or other planet. An example of an earth remote sensing application might be to produce a labeled map that divides the image into areas covered by distinct earth surface covers such as water, snow, types of natural vegetation, types of rock formations, types of agricultural crops, and types of other man-created development. Alternatively, one can segment the land based on climate (e.g., temperature, precipitation) and elevation zones. However, most image segmentation approaches do not directly provide such meaningful labels to image partitions. Instead, most approaches produce image partitions with generic labels such as region 1, region 2, and so on, which need to be converted into meaningful labels by a post-segmentation analysis.

An early survey on image segmentation grouped image segmentation approaches into three categories (Fu and Mui, 1981): (1) characteristic feature thresholding or clustering, (2) boundary detection, and (3) region extraction. Another early survey (Haralick and Shapiro, 1985) divides region extraction into several region growing and region split-and-merge schemes.

Both of these surveys note that there is no general theory of image segmentation, most image segmentation approaches are *ad hoc* in nature, and there is no general algorithm that will work well for all images. This is still the case even today.

We start our image segmentation discussion with spectrally based approaches, corresponding to Fu and Mui's characteristic feature thresholding or clustering category. We include here a description of support vector machines (SVMs), as a supervised spectral classification approach that has been a popular choice for analyzing multispectral and hyperspectral images (images with several tens or even hundreds of spectral bands). We then go on to describe a number of spatially based image segmentation approaches that could be appropriate for land categorization applications, generally going from simpler approaches to more complicated and more recently developed approaches. Here our emphasis is guided by the prevalence of reported use in land categorization studies. We then take a brief look at various approaches to image segmentation quality evaluation and include a closer look at a particular empirical discrepancy approach with example quality evaluations for a particular remotely sensed hyperspectral data set and selected image segmentation approaches. We wrap up with some concluding comments and discussion.

16.2 Spectrally Based Segmentation Approaches

The focus of this section is on approaches that are mainly based on analyses of individual pixels. These approaches use an initial labeling of pixels using unsupervised or supervised classification methods, and then try to group neighboring pixels with similar labels using some form of postprocessing to produce segmentation results.

16.2.1 Thresholding-Based Algorithms

Thresholding has been one of the oldest and most widely used techniques for image segmentation. Thresholding algorithms used for segmentation assume that the pixels that belong to the objects of interest have a property whose values are substantially different from those of the background, and they aim to find a good set of thresholds that partition the histogram of this property into two or more nonoverlapping regions (Sezgin and Sankur, 2004). While the spectral channels can be directly used for thresholding, other derived properties of the pixels are also commonly used in the literature. For example, Akcay and Aksoy (2011) used thresholding of the red band to identify buildings with red roofs, Bruzzone and Prieto (2005) performed change detection by thresholding the difference image, Rosin and Hervas (2005) used thresholding of the difference image for determining landslide activity, Aksoy et al. (2010) applied thresholding to the normalized difference vegetation index (NDVI) for segmenting vegetation areas, and Unsalan and Boyer (2005) combined thresholding of NDVI and a shadow–water index to identify potential building and street pixels in residential regions.

Selection of the threshold values is often done in an *ad hoc* manner usually when a single property is involved, but optimal values can also be found by employing exhaustive or stochastic search procedures that look for the values that optimize some criteria on the shape or the statistics of the histogram such as minimization of the within-class variance and maximization of the between-class variance (Otsu, 1979). A stochastic search procedure is particularly needed for finding multiple thresholds where an exhaustive search is not computationally feasible due to the combinatorial increase in the number of candidate values. For example, a recent use of multilevel thresholding for the segmentation of Earth observation data is described in Ghamisi et al. (2014), where a particle swarm optimization-based stochastic search algorithm was used to obtain a multilevel thresholding of each spectral channel independently by maximizing the corresponding between-class variance.

Even when the selected thresholds are obtained by optimizing some well-defined criteria on the distributions of the properties of the pixels, they do not necessarily produce operational image segmentation results because they suffer from the lack of the use of spatial information as the decisions are independently made on individual pixels. Thus, thresholding is usually applied as a preprocessing algorithm, and various postprocessing methods such as morphological operations are often applied to the results of pixel-based thresholding algorithms as discussed in the following section.

16.2.2 Clustering-Based Algorithms

The clustering-based approaches to image segmentation aim to make use of the rich literature on data grouping and/or partitioning techniques for pattern recognition (Duda et al., 2001). It is intuitive to pose the image segmentation problem as the clustering of pixels, and thus, pixel-based image analysis techniques in the remote sensing literature have found natural extensions to image segmentation. In the most widely used methodology, first, the spectral feature space is partitioned, and the individual pixels are grouped into clusters without regard to their neighbors, and then, a postprocessing step is applied to form regions by merging neighboring pixels having the same cluster label by using a connected components labeling algorithm.

The initial clustering stage commonly employs well-known techniques such as *k*-means (Aksoy and Akcay, 2005), fuzzy *c*-means (Shankar, 2007), and their probabilistic extension using the Gaussian mixture model estimated via expectation–maximization (Fauvel et al., 2013). Since no spatial information is used during the clustering procedure, pixels with the same cluster label can either form a single connected spatial region or can belong to multiple disjoint regions that are assigned different labels by the connected components labeling algorithm. This reduces the significance of the difficulty of the user's *a priori* selection of the number of clusters in many popular clustering algorithms as there is no strict correspondence between the initial number of clusters and the final number of image regions. However, it still has a high potential of producing an oversegmentation consisting of noisy results with isolated pixels having labels different from those of their neighbors due to the lack of the use of spatial data.

Therefore, a follow-on postprocessing step is used to produce a smoother and spatially consistent segmentation by converting the pixel-based clustering results into contiguous regions. A popular approach is to use an additional segmentation result (often also an oversegmentation) and to use a majority voting procedure for spatial regularization by assigning each region in the oversegmentation a single label that is determined according to the most frequent cluster label among the pixels in that region (Fauvel et al., 2013). An alternative approach is to use an iterative split-and-merge procedure as follows (Aksoy et al., 2005; Aksoy, 2006):

1. Merge pixels with identical labels to find the initial set of regions and mark these regions as foreground.
2. Mark regions with areas smaller than a threshold as background using connected components analysis.
3. Use region growing to iteratively assign background pixels to the foreground regions by placing a window at each background pixel and assigning it to the class that occurs the most in its neighborhood.

This procedure corresponds to a spatial smoothing of the clustering results. The resulting regions can be further processed using mathematical morphology operators to automatically divide large regions into more compact subregions as follows:

1. Find individual regions using connected components analysis for each cluster.
2. For all regions, compute the erosion transform and repeat the following:
   a. Threshold erosion transform at steps of three pixels in every iteration.
   b. Find connected components of the thresholded image.
   c. Select subregions that have an area smaller than the threshold.
   d. Dilate these subregions to restore the effects of erosion.
   e. Mark these subregions in the output image by masking the dilation using the original image.

   Repeat the steps a-e until no more subregions are found.
3. Merge the residues of previous iterations to their smallest neighbors.

Even though we focused on producing segmentations using clustering algorithms in this section, similar postprocessing techniques for converting the pixel-based decisions into contiguous regions can also be used with the outputs of pixel-based thresholding (Section 16.2.1) and classification (Section 16.2.3) procedures (Aksoy et al., 2005).

It is also possible to pose clustering, and the corresponding segmentation, as a density estimation problem. A commonly used algorithm that combines clustering with density estimation and segmentation is the mean shift algorithm (Comaniciu and Meer, 2002). Mean shift is based on nonparametric density estimation where the local maxima (i.e., modes) of the density can be assumed to correspond to clusters. The algorithm does not require *a priori* knowledge of the number of clusters in the data and can identify the locations of the local maxima by a set of iterations. These iterations can be interpreted as the shifting of points toward the modes where convergence is achieved when a point reaches a particular mode. The shifting procedure uses a kernel with a scale parameter that determines the amount of local smoothing performed during density estimation. The application of the mean shift procedure to image segmentation uses a separate kernel for the feature (i.e., spectral) domain and another kernel for the spatial (i.e., pixel) domain. The scale parameter for the spectral domain can be estimated by maximizing the average likelihood of held-out data. The scale parameter for the spatial domain can be selected according to the amount of compactness or oversegmentation desired in the image, or can be determined by using geospatial statistics (e.g., by using semivariogram-based estimates) (Dongping et al., 2012). Furthermore, agglomerative clustering of the mode estimates can be used to obtain a multiscale segmentation.

16.2.3 Support Vector Machines

Output of supervised classification of pixels can also be used as input for segmentation techniques. In recent years, SVMs and the use of kernels to transform data into a new feature space where linear separability can be exploited have been proposed. The SVM method attempts to separate training samples belonging to different classes by tracing maximum-margin hyperplanes in the space where the samples are mapped. SVM have shown to be particularly well suited to classify high-dimensional data (e.g., hyperspectral images) when a limited number of training samples are available (Vapnik, 1998; Camps-Valls, 2005). The success of SVM for pixel-based classification has led to its subsequent use as part of image segmentation methods. Thus, we discuss the SVM approach in detail in the following text.

SVMs are primarily designed to solve binary tasks, where the class labels take only two values: 1 or −1. Let us consider a binary classification problem in a B-dimensional space $\mathbb{R}^B$, with N training samples, $x_i \in \mathbb{R}^B$, and their corresponding class labels $y_i = \pm 1$ available. The SVM technique consists in finding the hyperplane that maximizes the margin, that is, the distance to the closest training data points in both classes (see Figure 16.1). Noting $\mathbf{w} \in \mathbb{R}^B$ as the vector normal to the hyperplane and $b \in \mathbb{R}$ as the bias, the hyperplane H is defined as

$$\mathbf{w} \cdot x + b = 0, \quad \forall x \in H.$$

If $x \notin H$, then

$$f(x) = \frac{|\mathbf{w} \cdot x + b|}{||\mathbf{w}||}$$

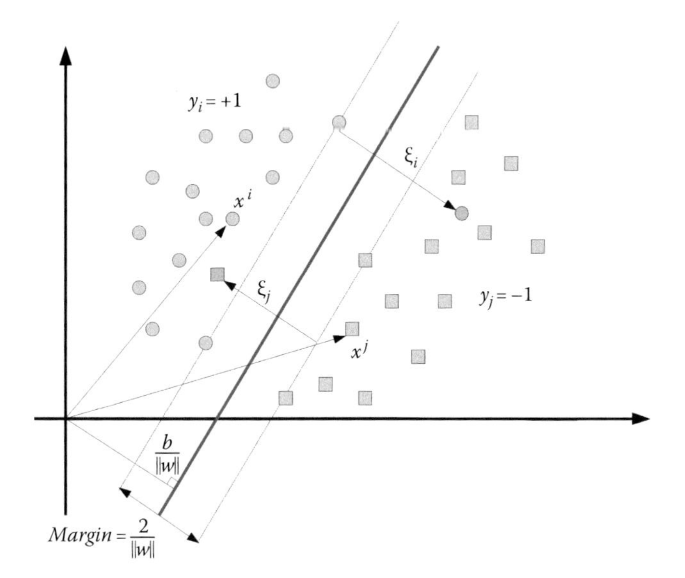

FIGURE 16.1 Schematic illustration of the SVM binary classification method. There is one nonlinearly separable sample in each class.

defines the distance of the sample x to H. In the linearly separable case, such a hyperplane must satisfy

$$y_i(\mathbf{w} \cdot x_i + b) > 1, \quad \forall i \in [1, N]. \tag{16.1}$$

The optimal hyperplane is the one that maximizes the margin $2/||\mathbf{w}||$. This is equivalent to minimizing $||\mathbf{w}||/2$ and leads to the following quadratic optimization problem:

$$\min\left[\frac{||\mathbf{w}||^2}{2}\right], \quad \text{subject to (16.1).} \tag{16.2}$$

To take into account nonlinearly separable data, slack variables ξ are introduced to deal with misclassified samples (see Figure 16.1). Equation 16.1 becomes

$$y_i(\mathbf{w} \cdot x_i + b) > 1 - \xi_i, \quad \xi_i \geq 0, \forall i \in [1, N]. \tag{16.3}$$

The final optimization problem is formulated as

$$\min\left[\frac{||\mathbf{w}||^2}{2} + C\sum_{i=1}^{N}\xi_i\right], \quad \text{subject to (16.3),} \tag{16.4}$$

where the constant C is a regularization parameter that controls the amount of penalty. This optimization problem is typically solved by quadratic programming (Vapnik, 1998). The classification is further performed by computing $y_u = \text{sign}(\mathbf{w} \cdot x_u + b)$, where $(\mathbf{w}, b)$ are the hyperplane parameters found during the training process, and x_u is an unseen sample.

One can notice that the pixel vectors in the optimization and decision rule equations always appear in pairs related through a scalar product. These products can be replaced by nonlinear functions of the pairs of vectors, essentially projecting the pixel vectors in a higher-dimensional space $\mathbb{H}$ and thus improving linear separability of data:

$$\begin{aligned} &\mathbb{R}^B \rightarrow \mathbb{H}, \\ &x \rightarrow \Phi(x), \\ &x_i \cdot x_j \rightarrow \Phi(x_i) \cdot \Phi(x_j) = K(x_i, x_j). \end{aligned} \tag{16.5}$$

Here, $\Phi(\cdot)$ is a nonlinear function to project feature vectors into a new space, $K(\cdot)$ is a *kernel* function, which allows one to avoid the computation of scalar products in the transformed space $[\Phi(x_i) \cdot \Phi(x_j)]$ and thus reduces the computational complexity of the algorithm. The kernel K must satisfy Mercer's condition (Burges, 1998). The Gaussian radial basis function (RBF) kernel is the most widely used for remote sensing image classification:

$$K_{Gaussian}(x_i, x_j) = \exp[-\gamma||x_i - x_j||^2], \tag{16.6}$$

where γ is the spread of the RBF kernel.

To solve the K-class problem, various approaches have been proposed. Two main approaches combining a set of binary classifiers are defined as (Smola, 2002) follows:

- *One versus all*: K binary classifiers are applied on each class against others. Each pixel x_i is assigned to the class with the maximum output $f(x_i)$.
- One versus one: $K(K - 1)/2$ binary classifiers are applied on each pair of classes. Each pixel is assigned to the class winning the maximum number of binary classification procedures.

As a conclusion, SVMs directly exploit the geometrical properties of data, without involving a density estimation procedure. This method has proven to be more effective than other nonparametric classifiers (such as neural networks or the k-nearest neighbor classifier (Duda et al., 2001)) in terms of classification accuracies, computational complexity, and robustness to parameter setting. SVM can efficiently handle high-dimensional data, exhibiting low sensitivity to the Hughes phenomenon (Hughes, 1968). Finally, it exhibits good generalization capability, fully exploiting the discrimination capability of available training samples.

Pixel-based supervised classification results, such as those obtained using an SVM classifier, are often given as input to segmentation procedures that aim to group the pixels to form contiguous regions as discussed in the following sections.

16.3 Spatially Based Segmentation Approaches

We cannot possibly discuss myriad of spatially based image segmentation approaches that have been proposed and developed over the years. Instead, we will focus on approaches that have achieved demonstrated success in remote sensing land

TABLE 16.1 Algorithmic Basis of Image Segmentation Approaches in Remote Sensing Oriented Image Analysis Packages

Name	Website or Reference	Algorithmic Basis
BerkeleyImgseg	Clinton et al. (2010)	Region growing, region merging
Ceasar	Cook et al. (1996)	Simulated annealing
eCognition Developer[a]	Trimble Geospatial Imaging (2014) and Baatz and Schape (2000)	Region growing
ENVI feature extraction	Exelis Visual Information Solutions (2015) and Robinson et al. (2002)	Edge-based (full lambda-schedule algorithm for region merging)
Extended watershed EWS	Li and Xiao (2007)	Multichannel watershed transformation
Image WS for Erdas Imagine	Sramek and Wrbka (1997)	Hierarchical watershed
InfoPACK	Cook et al. (1996)	Simulated annealing
PARBAT	Lucieer (2015) and Lucieer (2004)	Region growing
RHSeg	Tilton (1998) and Tilton et al. (2012)	Region growing and spectral clustering
SCRM	Castilla et al. (2008)	Watershed and region merging
SegSAR	Sousa (2014)	Hybrid (edge/region oriented)
SEGEN	Gofman (2006)	Region growing
GRASS GIS	GRASS Development Team (2015)	Region growing and watershed
IDRISI	Clark Labs (2015)	Watershed
Orfeo Toolbox	CNES (2015)	Region growing, watershed, level sets, mean shift

Note: Most of these image segmentation approaches were evaluated in a series of papers by the Leibniz IOER group.

[a] Was Definiens Developer—but the remote sensing package is now marketed by Trimble, and the Definiens product is now oriented to biomedical image analysis.

categorization applications. A compilation of such approaches can be found in a series of papers published by a research group based at the Leibniz Institute for Ecological and Regional Development (IOER) that present comparative evaluations of image segmentation approaches implemented in various image analysis packages (Meinel and Neubert, 2004; Neubert et al., 2006, 2008; Marpu et al., 2010). Table 16.1 provides a summary listing of most of the remote sensing–oriented image analysis packages whose image segmentation approach was evaluated in these papers, plus image segmentation approaches from three additional notable remote sensing–oriented image analysis packages (GRASS GIS, IDRISI, and the Orfeo Toolbox).

We note from Table 16.1 that region growing is the most frequent image segmentation approach utilized by these remote sensing–oriented image analysis software packages. Further, several packages combine region growing with other techniques (RHSeg with spectral clustering, SCRM and GRASS GIS with watershed, and SegSAR with edge detection). Watershed segmentation (an approach based on region boundary detection) is the next most popular approach. Simulated annealing is often utilized in analysis packages oriented toward analyzing synthetic aperture radar (SAR) imagery data (Ceasar and InfoPACK).

The next several sections describe various spatially based image segmentation approaches, starting with region-growing algorithms and continuing with texture-based algorithms, morphological algorithms, graph-based algorithms, and MRF-based algorithms.

16.3.1 Region-Growing Algorithms

In the region-growing approach to image segmentation, an image is initially partitioned into small region objects. These initial small region objects are often single image pixels, but can also be $n \times n$ blocks of pixels or another partitioning of the image into small spatially connected region objects. Then, pairs of spatially adjacent region objects are compared and merged together if they are found to be similar enough according to some comparison criterion. The underlying assumption is that region objects of interest are several image pixels in size and relatively homogeneous in value. Most region-growing approaches can operate on either grayscale, multispectral, or hyperspectral image data, depending on the criterion used to determine the similarity between neighboring region objects.

A very early example of region growing was described in Muerle and Allen (1968). Muerle and Allen experimented with initializing their region-growing process with region objects consisting of 2×2 up to 8×8 blocks of pixels. After initialization, they started with the region object at the upper left corner of the image and compared this region object with the neighboring region objects. If a neighboring region object was found to be similar enough, the region objects were merged together. This process was continued until no neighboring region objects could be found that were similar enough to be merged into the region object that was being grown. Then, the image was scanned (left-to-right, top-to-bottom) to find an unprocessed region object, that is, a region object that not yet been considered as an initial object for region growing or merged into a neighboring region. If an unprocessed region object was found, they conducted their region-growing process from that region object. This continued until no further unprocessed region objects could be found, upon which point the region-growing segmentation process was considered completed.

Many early schemes for region growing, such as Muerle and Allen's, can be formulated as logical predicate segmentation, defined in Zucker (1976) as follows:

A segmentation of an image X can be defined as a partition of X into R disjoint subsets $X_1, X_2,..., X_R$, such that the following conditions hold:

1. $\cup_{i=1}^{R} X_i = X$.
2. X_i, $i = 1, 2,..., R$ are connected.
3. $P(X_i)$ = TRUE for $i = 1, 2,..., R$.
4. $P(X_i \cup X_j)$ = FALSE for $i \neq j$, where X_i and X_j are adjacent.

$P(X_i)$ is a logical predicate that assigns the value TRUE or FALSE to X_i, depending on the image data values in X_i.

These conditions are summarized in Zucker (1976) as follows: the first condition requires that every picture element (pixel) must be in a region (subset). The second condition requires that each region must be connected, that is, composed of contiguous image pixels. The third condition determines what kind of properties each region must satisfy, that is, what properties the image pixels must satisfy to be considered similar enough to be in the same region. Finally, the fourth condition specifies that any merging of adjacent regions would violate the third condition in the final segmentation result.

Several researchers in this early era of image segmentation research, including Muerle and Allen, noted some problems with logical predicate segmentation. For one, the results were very dependent on the order in which the image data were scanned. Also, the statistics of a region object can change quite dramatically as the region is grown, making it possible that many adjacent cells that were rejected for merging early in the region-growing process would have been accepted in later stages based on the changed statistics of the region object. The reverse was also possible, where adjacent cells that were accepted for merging early in the region-growing process would have been rejected in later stages.

Subsequently, an alternate approach to region growing was developed that avoids these problems, and that approach eventually came to be referred to as *best merge region growing*. An early version of best merge region growing, hierarchical step-wise optimization (HSWO), is an iterative form of region growing, in which the iterations consist of finding the most optimal or best segmentation with one region less than the current segmentation (Beaulieu and Goldberg, 1989). The HSWO approach can be summarized as follows:

1. Initialize the segmentation by assigning each image pixel a region label. If a pre-segmentation is provided, label each image pixel according to the pre-segmentation. Otherwise, label each image pixel as a separate region.
2. Calculate the dissimilarity criterion value, d, between all pairs of spatially adjacent regions, find the smallest dissimilarity criterion value, T_{merge}, and merge all pairs of regions with $d = T_{merge}$.
3. Stop if no more merges are required. Otherwise, return to step 2.

HSWO naturally produces a segmentation hierarchy consisting of the entire sequence of segmentations from initialization

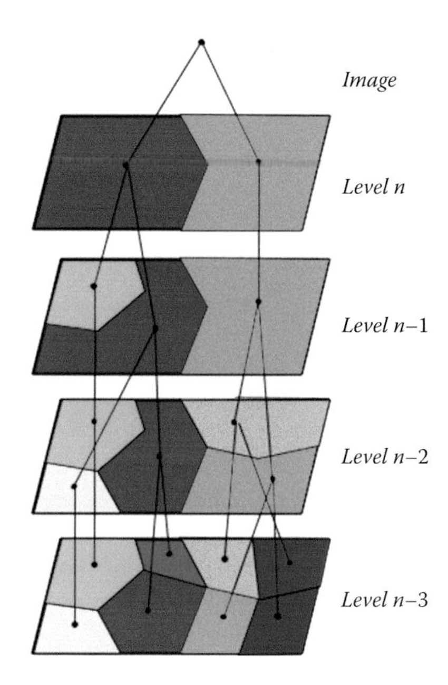

FIGURE 16.2 The last four levels of an n-level segmentation hierarchy produced by a region-growing segmentation process. Note that when depicted in this manner, the region-growing process is a *bottom-up* approach.

down to the final trivial one-region segmentation (if allowed to proceed that far). For practical applications, however, a subset of segmentations needs to be selected out from this exhaustive segmentation hierarchy. At a minimum, such a subset can be defined by storing the results only after a preselected number of regions are reached and then storing selected iterations after that such that no region is involved in more than one merge between stored iteration until a two-region segmentation is reached. A portion of such a segmentation hierarchy is illustrated in Figure 16.2. (The selection of a single segmentation from a segmentation hierarchy is discussed in Section 16.3.1.5.)

A unique feature of the segmentation hierarchy produced by HSWO and related region-growing segmentation approaches is that the segment or region boundaries are maintained at the full image spatial resolution for all levels of the segmentation hierarchy. The region boundaries are coarsened in many other multilevel representations.

Many variations on best merge region growing have been described in the literature. As early as 1994, Kurita (1994) described an implementation of the HSWO form of best merge region growing that utilized a heap data structure (Williams, 1964) for efficient determination of best merges and a dissimilarity criterion based on minimizing the mean squared error between the region mean image and original image. Further, several of the image segmentation approaches studied in the previously referenced series of papers published by the Leibniz Institute are based on best merge region growing (see Table 16.1). As we will discuss in more detail later, the main differences between most of these region-growing approaches are the dissimilarity criterion employed and, perhaps, some control logic designed to remove small regions or otherwise tailor the segmentation output.

16.3.1.1 Seeded Region Growing

Seeded region growing is a variant of best merge region growing in which regions are grown from preselected seed pixels or seed regions. Adams and Bischof (1994) is an early example of this approach. In seeded region growing, the best merges are found by examining the pixels adjacent to each seed pixel of region formed around each seed pixel. As described by Adams and Bischof (1994), the region-growing process continues until each image pixel is associated with one of the preselected seed pixels or regions.

16.3.1.2 Split-and-Merge Region Growing

In the split-and-merge approach, the image is repeatedly subdivided until each resulting region has a sufficiently high homogeneity. Examples of measures of homogeneity are the mean squared error between the region mean and the image data values, or the region standard deviation. After the region-splitting process converges, the regions are grown using one of the previously described region-growing approaches. This approach is more efficient when large homogeneous regions are present. However, some segmentation detail may be lost. See Horowitz and Pavlidis (1974), Cross et al. (1988), and Strasters and Gerbrands (1991) for examples of this approach.

16.3.1.3 Hybrid of Spectral Clustering and Region Growing

Tilton (1998) and Tilton et al. (2012) describe a hybridization of HSWO best merge region growing with spectral clustering, called HSeg (for hierarchical segmentation). We remind the reader that HSWO is performed by finding a threshold value, T_{merge}, equal to the value of a dissimilarity criterion of the most similar pair of spatially adjacent regions, and then merging all pairs of regions that have dissimilarity equal to T_{merge}. HSeg adds to HSWO a step following each step of adjacent region merges in which all pairs of spatially nonadjacent regions are merged that have dissimilarity $\leq S_w T_{merge}$, where $0.0 \leq S_w \leq 1.0$ is a factor that sets the priority between spatially adjacent and nonadjacent region merges. Note that when $S_w = 0.0$, HSeg reduces to HSWO.

Unfortunately, the inclusion of the step in HSeg of merging spatially nonadjacent regions adds significantly to the computational requirements of this image segmentation approach. This is because comparisons must now be made between all pairs of regions instead of just between pairs of spatially adjacent regions. A recursive divide-and-conquer approximation of HSeg (called RHSeg), with a straightforward parallel implementation, was developed to help overcome this problem (Tilton, 2007). The computational requirements of this approach were further reduced by a refinement in which the nonadjacent region merging was limited to between regions of a minimum size, P_{min} (Tilton et al., 2012). In this refinement, the value of P_{min} is dynamically adjusted to keep the number of *large regions* (those of at least P_{min} in size) to a range that substantially reduces the computational requirements without significantly affecting the image segmentation results.

16.3.1.4 Dissimilarity Criterion and Specialized Control Logic

Muerle and Allen (1968) experimented with various criteria for determining whether or not pairs of region objects were similar enough to be merged together. They concluded that an optimal criterion would be a threshold of a function of the mean and standard deviation of the gray levels of the pixels contained in the compared region objects.

In our image segmentation research, we have implemented and studied several region merging criteria in the form of dissimilarity criteria (Tilton, 2013). Included among these dissimilarity criteria are criteria based on vector norms (the 1-, 2-, and ∞-norms), criteria based on minimizing the increase in mean squared error between the region mean image and the original image data, and a criterion based on the Spectral Angle Mapper (SAM) criterion (Kruse et al., 1993). We briefly describe these dissimilarity criteria here.

The dissimilarity criterion based on the 1-norm of the difference between the region mean vectors, u_i and u_j, of regions X_i and X_j, each with B spectral bands, is

$$d_{1\text{-norm}}(X_i, X_j) = \|u_i - u_j\|_1 = \sum_{b=1}^{B} |\mu_{ib} - \mu_{jb}|, \quad (16.7)$$

where μ_{ib} and μ_{jb} are the mean values for regions i and j, respectively, in spectral band b, that is, $u_i = (\mu_{i1}, \mu_{i2}, \ldots, \mu_{iB})^T$ and $u_j = (\mu_{j1}, \mu_{j2}, \ldots, \mu_{jB})^T$.

The dissimilarity criterion based on the 2-norm is

$$d_{2\text{-norm}}(X_i, X_j) = \|u_i - u_j\|_2 = \left[\sum_{b=1}^{B} (\mu_{ib} - \mu_{jb})^2\right]^{1/2}, \quad (16.8)$$

and the dissimilarity criterion based on the ∞-norm is

$$d_{\infty\text{-norm}}(X_i, X_j) = \|u_i - u_j\|_\infty = \max(|\mu_{ib} - \mu_{jb}|, b = 1, 2, \ldots, B). \quad (16.9)$$

As noted here, a criterion based on mean squared error minimizes the increase in mean squared error between the region mean image and the original image data as regions are grown. The sample estimate of the mean squared error for the segmentation of band b of the image X into R disjoint subsets $X_1, X_2, \ldots, X_R$ is given by

$$MSE_b(X) = \frac{1}{N-1} \sum_{i=1}^{R} MSE_b(X_i), \quad (16.10)$$

where N is the total number of pixels in the image data and

$$MSE_b(X_i) = \sum_{x_p \in X_i} (\chi_{pb} - \mu_{ib})^2 \quad (16.11)$$

is the mean squared error contribution for band b from segment X_i. Here, x_p is a pixel vector (in this case, a pixel vector in data subset X_i), and χ_{pb} is the image data value for the bth spectral band of the pixel vector, x_p. The dissimilarity function based on a measure of the increase in mean squared error due to the merge of regions X_i and X_j is given by

$$d_{BSMSE}(X_i, X_j) = \sum_{b=1}^{B} \Delta MSE_b(X_i, X_j), \quad (16.12)$$

where

$$\Delta MSE_b(X_i, X_j) = MSE_b(X_i \cup X_j) - MSE_b(X_i) - MSE_b(X_j). \quad (16.13)$$

BSMSE refers to *band sum MSE*. Using (16.11) and exchanging the order of summation, (16.13) can be manipulated to produce an efficient dissimilarity function based on aggregated region features (for the details, see Tilton (2013)):

$$d_{BSMSE}(X_i, X_j) = \frac{n_i n_j}{(n_i + n_j)} \sum_{b=1}^{B} (\mu_{ib} - \mu_{jb})^2. \quad (16.14)$$

The dimensionality of the d_{BSMSE} dissimilarity criteria is equal to the square of the dimensionality of the image pixel values, while the dimensionality of the vector norm-based dissimilarity criteria is equal to the dimensionality of the image pixel values. To keep the dissimilarity criteria dimensionalities consistent, the square root of d_{BSMSE} is often used.

The SAM criterion is widely used in hyperspectral image analysis (Kruse et al., 1993). This criterion determines the spectral similarity between two spectral vectors by calculating the *angle* between the two spectral vectors. An important property of the SAM criterion is that poorly illuminated and more brightly illuminated pixels of the same color will be mapped to the same spectral angle despite the difference in illumination. The spectral angle θ between the region mean vectors, u_i and u_j, of regions X_i and X_j is given by

$$\theta(u_i, u_j) = \arccos\left(\frac{u_i \cdot u_j}{\|u_i\|_2 \|u_j\|_2}\right) = \arccos\left(\frac{\sum_{b=1}^{B} \mu_{ib}\mu_{jb}}{\left(\sum_{b=1}^{B} \mu_{ib}^2\right)^{1/2} \left(\sum_{b=1}^{B} \mu_{jb}^2\right)^{1/2}}\right), \quad (16.5)$$

where μ_{ib} and μ_{jb} are the mean values for regions i and j, respectively, in spectral band b, that is, $u_i = (\mu_{i1}, \mu_{i2}, \ldots, \mu_{iB})^T$ and $u_j = (\mu_{j1}, \mu_{j2}, \ldots, \mu_{jB})^T$. The dissimilarity function for regions X_i and X_j, based on the SAM distance vector measure, is given by

$$d_{SAM}(X_i, X_j) = \theta(u_i, u_j). \quad (16.6)$$

Note that the value of d_{SAM} ranges from 0.0 for similar vectors up to $\pi/2$ for the most dissimilar vectors.

A problem that can often occur with basic best merge region-growing approaches is that the segmentation results contain many small regions. We have found this to be the case when employing dissimilarity criteria based on vector norms or SAM, but not a problem for dissimilarity criteria based on minimizing the increase in mean squared error. This is because the mean squared error criterion has a factor, $n_i n_j/(n_i - n_j)$, where n_i and n_j are the number of pixels in the two compared regions, that biases toward merging small regions into larger ones. We have found it useful to add on a similar *small region merge acceleration factor* to the vector norm and SAM-based criterion when one of the compared regions is smaller than a certain size. See Tilton et al. (2012) for more details.

Implementations of best merge region growing often add special control logic to reduce the number of small regions or otherwise improve the final classification result. An example of this is SEGEN (Gofman, 2006), which uses the vector 2-norm (otherwise known as Euclidean distance) for the dissimilarity criterion. As noted in Tilton et al. (2012), SEGEN is a relatively pure implementation of best merge region growing, optimized for efficiency in performance, memory utilization, and image segmentation quality. SEGEN adds a number of (optional) procedures to best merge region growing, among them a low-pass filter to be applied on the first stage of the segmentation and outlier dispatching on the last stage. The latter removes outlier pixels and small segments by embedding them in neighborhood segments with the smallest dissimilarity. SEGEN also provides several parameters to control the segmentation process. A set of *good in average* control values is suggested in Gofman (2006).

The best merge region-growing segmentation approach employed in eCognition Developer utilizes a dissimilarity function that balances minimizing the increase in heterogeneity, f, in both color and shape (Baatz and Schape, 2000; Benz et al., 2004):

$$f = w_{color} \cdot \Delta h_{color} + w_{shape} \cdot \Delta h_{shape}, \quad (16.17)$$

where w_{color} and w_{shape} are weights that range in value from 0 to 1 and must mutually sum up to 1.

The color component of heterogeneity, Δh_{color}, is defined as follows:

$$\Delta h_{color} = \sum_{c} w_c (n_{merge} \cdot \sigma_{c,merge} - (n_1 \cdot \sigma_{c,1} + n_2 \cdot \sigma_{c,2})), \quad (16.18)$$

where n_x is the number of pixels, with the subscript x = *merge* referring to the merged object, and subscripts x = 1 or 2 referring to the first and second objects considered for merging. σ_c refers to the standard deviation in channel (spectral band) c, with the same additional subscripting denoting the merged or pair of considered objects. w_c is a channel weighting factor.

The shape component of heterogeneity, Δh_{shape}, is defined as follows:

$$\Delta h_{shape} = w_{compt} \cdot \Delta h_{compt} + w_{smooth} \cdot \Delta h_{smooth}, \quad (16.19)$$

where

$$\Delta h_{compt} = n_{merge} \cdot \frac{l_{merge}}{\sqrt{n_{merge}}} - \left(n_1 \cdot \frac{l_1}{\sqrt{n_1}} + n_2 \cdot \frac{l_2}{\sqrt{n_2}} \right), \quad (16.20)$$

$$\Delta h_{smooth} = n_{merge} \cdot \frac{l_{merge}}{b_{merge}} - \left(n_1 \cdot \frac{l_1}{b_1} + n_2 \cdot \frac{l_2}{b_2} \right), \quad (16.21)$$

where
l is the perimeter of the object
b is the perimeter of the object's bounding box

The weights w_c, w_{color}, w_{shape}, w_{smooth}, and w_{compt} can be selected to best suit a particular application.

The level of detail, or scale, of the segmentation is set by stopping the best merge region-growing process when the increase if heterogeneity, f, for the best merge reaches a predefined threshold value. A multiresolution segmentation can be created by performing this process with a set of increasing thresholds.

In eCognition Developer, other segmentation procedures can be combined with the multiresolution approach. For example, spectral difference segmentation merges neighbor objects that fall within a user-defined maximum spectral difference. This procedure can be used to merge spectrally similar objects from the segmentation produced by the multiresolution approach.

16.3.1.5 Selection of a Single Segmentation from a Segmentation Hierarchy

Some best merge region-growing approaches such as HSWO and HSeg do not produce a single segmentation result, but instead produce a segmentation hierarchy. The best merge region-growing segmentation approach employed in eCognition Developer can also produce a segmentation hierarchy. A segmentation hierarchy is a set of several image segmentations of the same image at different levels of detail in which the segmentations at coarser levels of detail can be produced from simple merges of regions at finer levels of detail. In such a structure, an object of interest may be represented by multiple segments in finer levels of detail and may be merged into a surrounding region at coarser levels of detail. A single segmentation level can be selected out of the segmentation hierarchy by analyzing the spatial and spectral characteristics of the individual regions and by tracking the behavior of these characteristics throughout different levels of detail. A manual approach for doing this using a graphical user interface is described in Tilton (2003, 2013). A preliminary study on automating this approach is described in Plaza and Tilton (2005), where it was proposed to automate this approach using joint spectral/spatial homogeneity scores computed from segmented regions. An alternate approach for making spatially localized selections of segmentation detail based on matching region boundaries with edges produced by an edge detector is described in Le Moigne and Tilton (1995).

Tarabalka et al. (2012) proposed a modification of HSeg through which a single segmentation output is automatically selected for output from the usual segmentation hierarchy. The idea is similar to the previously described seeded region growing. The main idea behind the marker-based HSeg algorithm consists in automatically selecting seeds, or *markers*, for image regions and then performing best merge region growing with an additional condition: two regions with different marker labels cannot be merged together (see Figure 16.3). The authors proposed to choose markers of spatial regions by analyzing results of a probabilistic supervised classification of each pixel and by retaining the most reliably classified pixels as region seeds.

An alternative algorithm that produces a final segmentation by automatically selecting subsets of regions appearing in different levels in the segmentation hierarchy is described in Section 16.3.3.2.

16.3.2 Texture-Based Algorithms

Along with spectral information, textural features have also been heavily used for various image analysis tasks, including their use as features in thresholding-based, clustering-based, and region-growing-based segmentation in the literature. In this section, we focus on the use of texture for *unsupervised* image segmentation. In the following, first, we discuss some particular examples that involve texture modeling for segmenting natural landscapes and man-made structures using local image properties and, then, present recent work on generalized texture algorithms that aim to model complex image structures in terms of the statistics and spatial arrangements of simpler image primitives.

One of the common uses of texture in remote sensing image analysis is the identification of natural landscapes. For example, Epifanio and Soille (2007) used morphological transformations

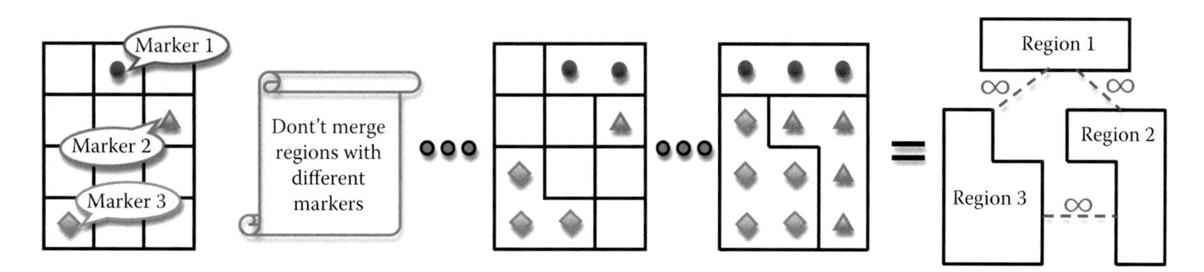

FIGURE 16.3 Scheme illustrating the marker-based HSeg algorithm.

such as white and black top-hat by reconstruction and thresholded image gradients with an unsupervised clustering procedure where the texture prototypes were automatically selected based on the dissimilarities between the feature vectors of neighboring image windows to segment vegetation zones and forest stands, and Wang and Boesch (2007) combined an initial color-based oversegmentation step with a threshold-based region merging procedure that used wavelet feature statistics inside candidate image regions to delineate forest boundaries.

Man-made structures also exhibit particular characteristics that can be modeled using textural features. For example, Pesaresi et al. (2008) proposed a rotation invariant anisotropic texture model that used contrast features computed from gray level co-occurrence matrices and used this model to produce a built-up presence index. The motivation behind the use of the contrast-based features was to exploit the relationships between the buildings and their shadows. Sirmacek and Unsalan (2010) and Gueguen et al. (2012) used a similar idea and modeled urban areas using spatial voting (smoothing) of local feature points extracted from the local maxima of Gabor filtering results and corner detection results, respectively. All of these results can be converted to a segmentation output by thresholding the corresponding urban area estimates.

An open problem in image segmentation is to identify the boundaries of regions that are not necessarily homogeneous with respect to low-level features such as color and texture that are extracted from individual pixels or from small local pixel neighborhoods. Recent literature includes several examples that can be considered as generalized texture algorithms that aim to model heterogeneous image content in terms of the spatial arrangements of relatively homogeneous image primitives. Some of this work has considered the segmentation of particular structures that have specific textural properties. For example, Dogrusoz and Aksoy (2007) aimed the segmentation of regular and irregular urban structures by modeling the image content using a graph where building objects and the Voronoi tessellation of their locations formed the vertices and the edges, respectively, and the graph was clustered by thresholding its minimum spanning tree so that organized (formal) and unorganized (informal) settlement patterns were extracted to model urban development. Some agricultural structures such as permanent crops also exhibit specific textural properties that can be useful for segmentation. For example, Aksoy et al. (2012) proposed a texture model that is based on the idea that textures are made up of primitives (trees) appearing in a near-regular repetitive arrangement (planting patterns) and used this model to compute a regularity score for different scales and orientations by using projection profiles of multiscale isotropic filter responses at multiple orientations. Then, they illustrated the use of this model for segmenting orchards by iteratively merging neighboring pixels that have similar regularity scores at similar scales and orientations.

More generic approaches have also been proposed for segmenting heterogeneous structures. For example, Gaetano et al. (2009) started with an oversegmentation of atomic image regions and then performed hierarchical texture segmentation by assuming that frequent neighboring regions are strongly related. These relations were represented using Markov chain models computed from quantized region labels, and the image regions that exhibit similar transition probabilities were clustered to construct a hierarchical set of segmentations. Zamalieva et al. (2009) used a similar frequency-based approach by finding the significant relations between neighboring regions as the modes of a probability distribution estimated using the continuous features of region co-occurrences. Then, the resulting modes were used to construct the edges of a graph where a graph mining algorithm was used to find subgraphs that may correspond to atomic texture primitives that form the heterogeneous structures. The final segmentation was obtained by using the histograms of these subgraphs inside sliding windows centered at individual pixels and by clustering the pixels according to these histograms. As an alternative to graph-based grouping, Akcay et al. (2010) performed Gaussian mixture-based clustering of the region co-occurrence features to identify frequent region pairs that are merged in each iteration of a hierarchical texture segmentation procedure.

In addition to using co-occurrence properties of neighboring regions to exploit statistical information, structural features can also be extracted to represent the spatial layout for texture modeling. For example, Akcay and Aksoy (2011) described a procedure for finding groups of aligned objects by performing a depth-first search on a graph representation of neighboring primitive objects. After the search procedure identified aligned groups of three or more objects that have centroids lying on a straight line with uniform spacing, an agglomerative hierarchical clustering algorithm was used to find larger groups of primitive objects that have similar spatial layouts. The approach was illustrated in the finding of groups of buildings that have different statistical and spatial characteristics that cannot be modeled using traditional segmentation methods. Another approach for modeling urban patterns using hierarchical segmentations extracted from multiple images of the same scene at various resolutions was described by Kurtz et al. (2012) where binary partition trees were used to model image data and tree cuts were learned from user-defined segmentation examples for interactive partitioning of images into semantic heterogeneous regions.

16.3.3 Morphological Algorithms

Mathematical morphology has been successfully used for various tasks such as image filtering for smoothing or enhancement, texture analysis, feature extraction, and detecting objects with certain shapes in the remote sensing literature (Soille and Pesaresi, 2002; Soille, 2003). Morphological algorithms have also been one of the most widely used techniques for segmenting remotely sensed images. These approaches view the two-dimensional image data that consist of the spectral channels or some other property of the pixels as an imaginary topographic relief where higher-pixel values map to higher imaginary elevation levels (see Figure 16.4). Consequently, differences in the elevations

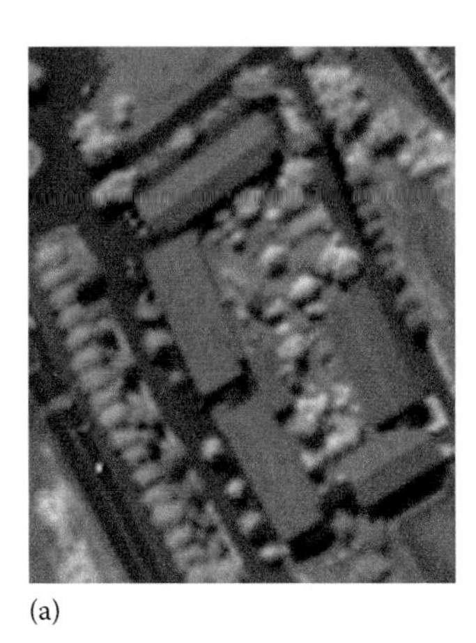
(a)

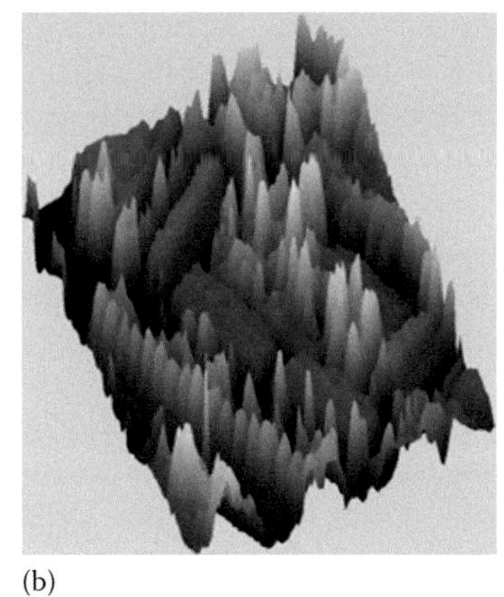
(b)

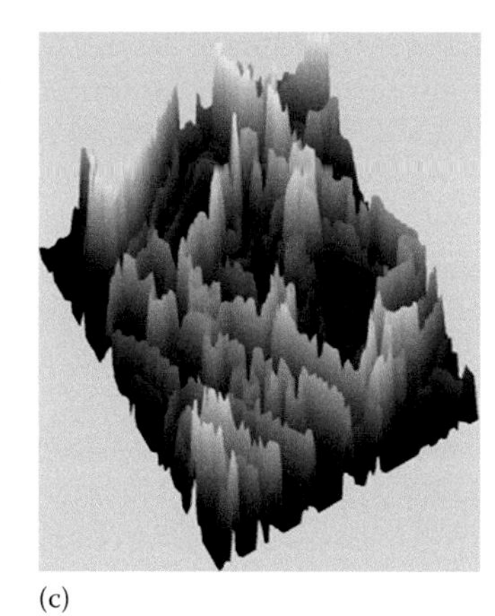
(c)

FIGURE 16.4 Illustration of mapping of the two-dimensional image data that consist of the spectral channels or some other property of the pixels as an imaginary topographic relief so that higher-pixel values map to higher imaginary elevation levels. (a) An example spectral band. (b) The spectral values viewed as a three-dimensional topographic relief. (c) Gradient of the spectral data at each pixel viewed as a three-dimensional topographic relief.

of the pixels in a spatial neighborhood can be exploited to partition those pixels into different regions. Two morphological approaches for segmentation have found common use in the literature: watershed algorithms and morphological profiles (MPs). These approaches are described in the following sections. Other approaches using mathematical morphology for image segmentation, and particularly for producing segmentation hierarchies, can be found in Soille (2008), Soille and Najman (2010), Ouzounis et al. (2012), and Perret et al. (2012).

16.3.3.1 Watershed Algorithms

The watershed algorithm divides the imaginary topographic relief into catchment basins so that each basin is associated with one local minimum in the image (i.e., individual segments), and the watershed lines correspond to the pixel locations that separate the catchment basins (i.e., segment boundaries). Watershed segmentation can be simulated by an immersion process (Vincent and Soille, 1991). If we immerse the topographic surface in water, the water rises through the holes at the regional minima with a uniform rate. When two volumes of water coming from two different minima are about to merge, a dam is built at each point of contact. Following the immersion process, the union of all those dams constitutes the watershed lines. A graph-theoretical interpretation of the watershed algorithm can be found in Meyer (2001). Couprie et al. (2011) describe a common framework that unifies watershed segmentation and some other graph-based segmentation algorithms that are described in Section 16.3.4.

The most commonly used method for constructing the topographic relief from the image data to be segmented is to use the gradient function at each pixel. This approach incorporates edge information in the segmentation process and maps homogeneous image regions with low-gradient values into the catchment basins and the pixels in high-contrast neighborhoods with high-gradient values into the peaks in the elevation function. The gradient function for single-channel images can easily be computed using derivative filters. Multivariate extensions of the gradient function can be used to apply watershed segmentation to multispectral and hyperspectral images (Aptoula and Lefevre, 2007; Li and Xiao, 2007; Noyel et al., 2007; Fauvel et al., 2013).

A potential problem in the application of watershed segmentation to images with high levels of detail is oversegmentation when the watersheds are computed from raw image gradient where an individual segment is produced for each local minimum of the topographic relief. Preprocessing or postprocessing methods can be used to reduce oversegmentation. For example, smoothing filters such as the mean or median filters can be applied to the original image data as a preprocessing step. Alternatively, the oversegmentation produced by the watershed algorithm can be given as input to a region merging procedure for postprocessing (Haris et al., 1998).

Another commonly used alternative to reduce the oversegmentation is to use the concept of dynamics that are related to the regional minima of the image gradient. A regional minimum is composed of a group of neighboring pixels with the same value where the pixels on the external boundary of this group have a greater value. When we consider the image gradient as a topographic surface, the dynamic of a regional minimum can be defined as the minimum height that a point in the minimum has to climb to reach a lower regional minimum (Najman and Schmitt, 1996). The h-minima transform can be used to suppress the regional minima with dynamics less than or equal to a particular value h by performing geodesic reconstruction by erosion of the input image f from $f + h$ (Soille, 2003). When it is difficult to select a single h value, it is common to create a multiscale segmentation by using an increasing sequence of h values. The multiscale watershed segmentation generates a set of nested partitions where the partition at scale s is obtained as the watershed segmentation of the image gradient whose regional minima with dynamics less than or equal to s are eliminated by using the h-minima transform. First, the initial partition is calculated as the classical watershed corresponding to all local minima. Next, the two catchment basins having a dynamic of 1 are merged with their neighbor catchment basins at scale 1. Then, at each scale s, the minima with dynamics less than or equal to s are filtered, whereas the minima with dynamics greater than s remain the same or are extended. This continues until the last scale corresponding to the largest dynamic in the gradient image. Figure 16.5 illustrates the use of the h-minima transform for suppressing regional minima for obtaining a multiscale watershed segmentation.

Yet another popular approach for computing the watershed segmentation without a significant amount of oversegmentation is to use markers (Meyer and Beucher, 1990). Marker-controlled watershed segmentation can be defined as the watershed of an input image transformed to have regional minima only at the marker locations. Possible methods for identifying the markers include manual selection or selection of the pixels with high confidence values at the end of pixel-based supervised classification (Tarabalka et al., 2010a). Given a marker image f_m that consists of pixels whose value is 0 at the marker locations and a very large value in the rest of the image, the minima in the input image f can be rearranged by using minima imposition. First, minima can be created only at the locations of the markers by taking the point-wise minimum between $f + 1$ and f_m. Note that the resulting image is lower than or equal to the marker image. The second step of the minima imposition is the morphological reconstruction by erosion of the resulting image from the marker image f_m. Finally, watershed segmentation is applied to the resulting image. It is also possible to produce a multiscale segmentation by applying marker-controlled watershed segmentation to the input image by using a decreasing set of markers. Marker selection is also discussed in Section 16.3.4.

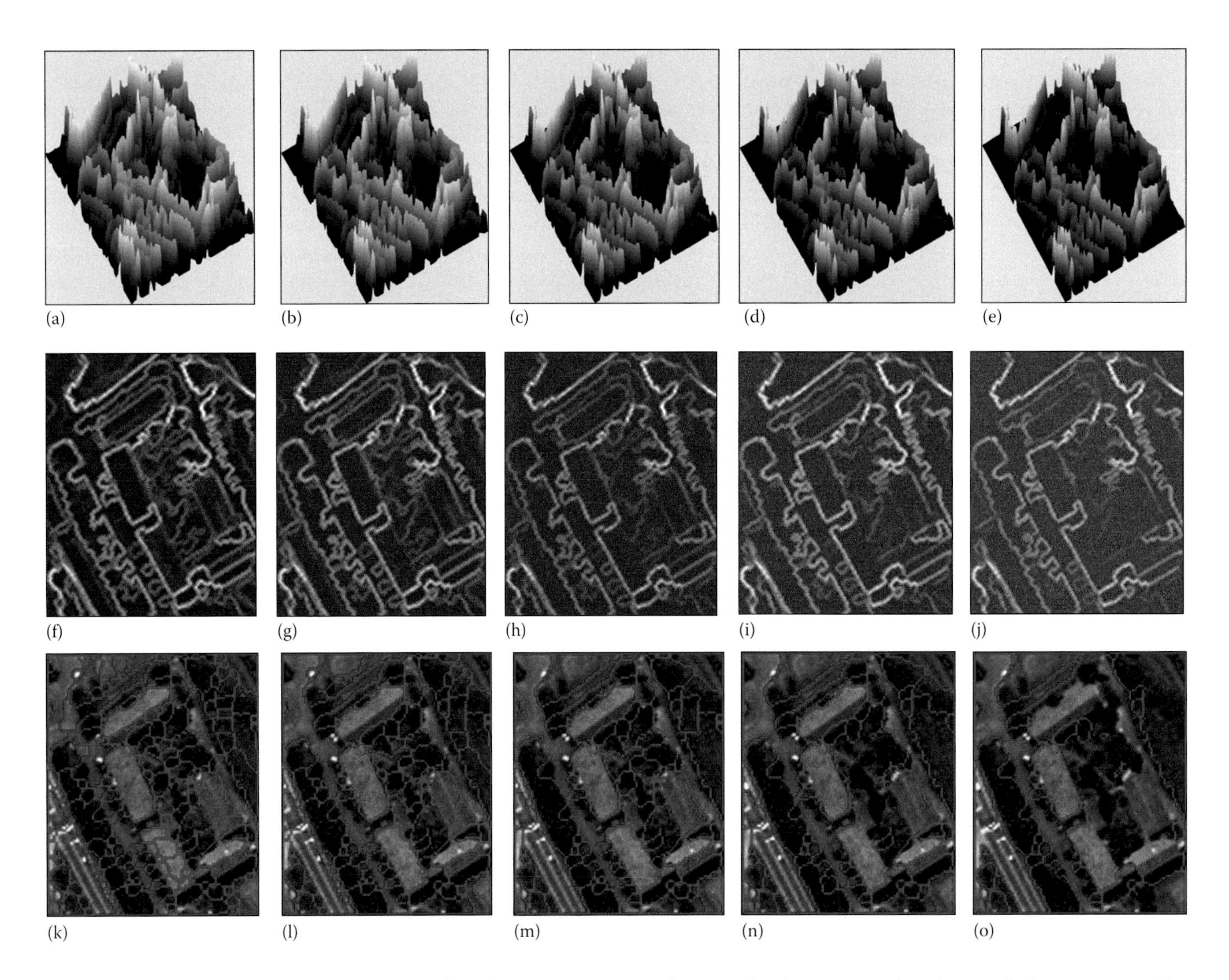

FIGURE 16.5 Illustration of the h-minima transform for suppressing regional minima for obtaining a multiscale watershed segmentation. The columns represent increasing values of h, corresponding to decreasing amount of detail in the gradient data. The first row shows the gradient information at each pixel as a topographic relief. The second row shows the gradient data as an image. Brighter values represent higher gradient. The third row shows the segmentation boundaries obtained by the watershed algorithm in red.

16.3.3.2 Morphological Profiles

The image representation called morphological profiles was popularized in the remote sensing literature by Pesaresi and Benediktsson (2001). The representation uses the morphological residuals between the original image function and the composition of a granulometry constructed at multiple scales. The proposed approach makes use of both classical morphological operators, such as opening and closing, and recent theoretical advances such as leveling and morphological spectrum to build the MP.

The fundamental operators in mathematical morphology are erosion and dilation (Soille, 2003). Both of these operators use the definition of a pixel neighborhood with a particular shape called a structuring element (SE) (e.g., a disk of radius of three pixels). The erosion operator can be used to identify the image locations where the SE fits the objects in the image and is defined as the infimum of the values of the image function in the neighborhood defined by SE. The dilation operator can be used to identify the pixels where the SE hits the objects in the image and is defined as the supremum of the image values in the neighborhood defined by SE. These two operators can be combined to define other operators. For example, the opening operator, which is defined as the result of erosion followed by dilation using the same SE, can be used to cut the peaks of the topographic relief that are smaller than the SE. On the other hand, the closing operator, which is defined as the result of dilation followed by erosion using the same SE, can be used to fill the valleys that are smaller than the SE.

The morphological operations are often used with the non-Euclidean geodesic metric instead of the classical Euclidean metric (Pesaresi and Benediktsson, 2001). The elementary geodesic dilation of f (called the marker) under g (called the mask) based on SE is the infimum of the elementary dilation of f (with SE) and g. Similarly, the elementary geodesic erosion of f under g based on SE is the supremum of the elementary erosion of f (with SE) and g. A geodesic dilation (respectively, erosion) of size k can also be obtained by performing k successive elementary geodesic dilations (respectively, erosions). Next, the reconstruction by dilation (respectively, erosion) of f under g is obtained by the iterative use of an elementary geodesic dilation (respectively, erosion) of f under g until idempotence is achieved. Then, the opening by reconstruction of an image f can be defined as the reconstruction by dilation of the erosion under the original image. Similarly, the closing by reconstruction of the image f can be defined as the dual reconstruction by erosion of the dilation above the original image.

The advantage of the reconstruction filters is that they do not introduce discontinuities and, therefore, preserve the shapes observed in the input images. Hence, the opening and closing by reconstruction operators can be used to identify the sizes and shapes of different objects present in the image such that opening (respectively, closing) by reconstruction preserves the shapes of the structures that are not removed by erosion (respectively, dilation), and the residual between the original image and the result of opening (respectively, closing) by reconstruction, called the top-hat (respectively, inverse top-hat, or bot-hat) transform, can be used to isolate the structures that are brighter (respectively, darker) than their surroundings.

However, to determine the shapes and sizes of all objects present in the image, it is necessary to use a range of different SE sizes. This concept is called granulometry. The MP of size $(2k + 1)$ can be defined as the composition of a granulometry of size k constructed with opening by reconstruction (opening profile), the original image, and an antigranulometry of size k constructed with closing by reconstruction (closing profile) using a sequence of k SEs with increasing sizes. Then, the derivative of the morphological profile (DMP) is defined as a vector where the measure of the slope of the opening–closing profile is stored for every step of an increasing SE series (see Figures 16.6 and 16.7 for the illustration of opening and closing profiles, respectively).

Pesaresi and Benediktsson (2001) used DMP for image segmentation. They defined the size of each pixel as the SE size at which the maximum DMP is achieved. Then, they defined an image segment as a set of connected pixels showing the greatest value of the DMP for the same SE size. That is, the segment label of each pixel is assigned according to the scale corresponding to the largest derivative of its profile. This scheme works well in images where the structures are mostly flat so that all pixels in a structure have only one derivative maximum. A potential drawback of this scheme is that neighborhood information is not used at the final step of assigning segment labels to pixels. This may result in an oversegmentation consisting of small noisy segments in very high spatial resolution images with non-flat structures where the scale with the largest value of the DMP may not correspond to the true structure.

Akcay and Aksoy (2008) proposed to consider the behavior of the neighbors of a pixel while assigning the segment label for that pixel. The method assumes that pixels with a positive DMP value at a particular SE size face a change with respect to their neighborhoods at that scale. As opposed to Pesaresi and Benediktsson (2001) where only the scale corresponding to the greatest DMP is used, the main idea is that a neighboring group of pixels that have a similar change for any particular SE size is a candidate segment for the final segmentation. These groups can be found by applying connected components analysis to the DMP at each scale. For each opening and closing profile, through increasing SE sizes from 1 to m, each morphological operation reveals connected components that are contained within each other in a hierarchical manner where a pixel may be assigned to more than one connected component appearing at different SE sizes. Each component is treated as a candidate meaningful segment (see Figures 16.6 and 16.7). Using these segments, a tree is constructed where each connected component is a node and there is an edge between two nodes corresponding to two consecutive scales if one node is contained within the other. Leaf nodes represent the components that appear for SE size 1. Root nodes represent the components that exist for SE size m.

After forming a tree for each opening and closing profile, the goal is to search for the most meaningful connected components among those appearing at different scales in the segmentation hierarchy. Ideally, a meaningful segment

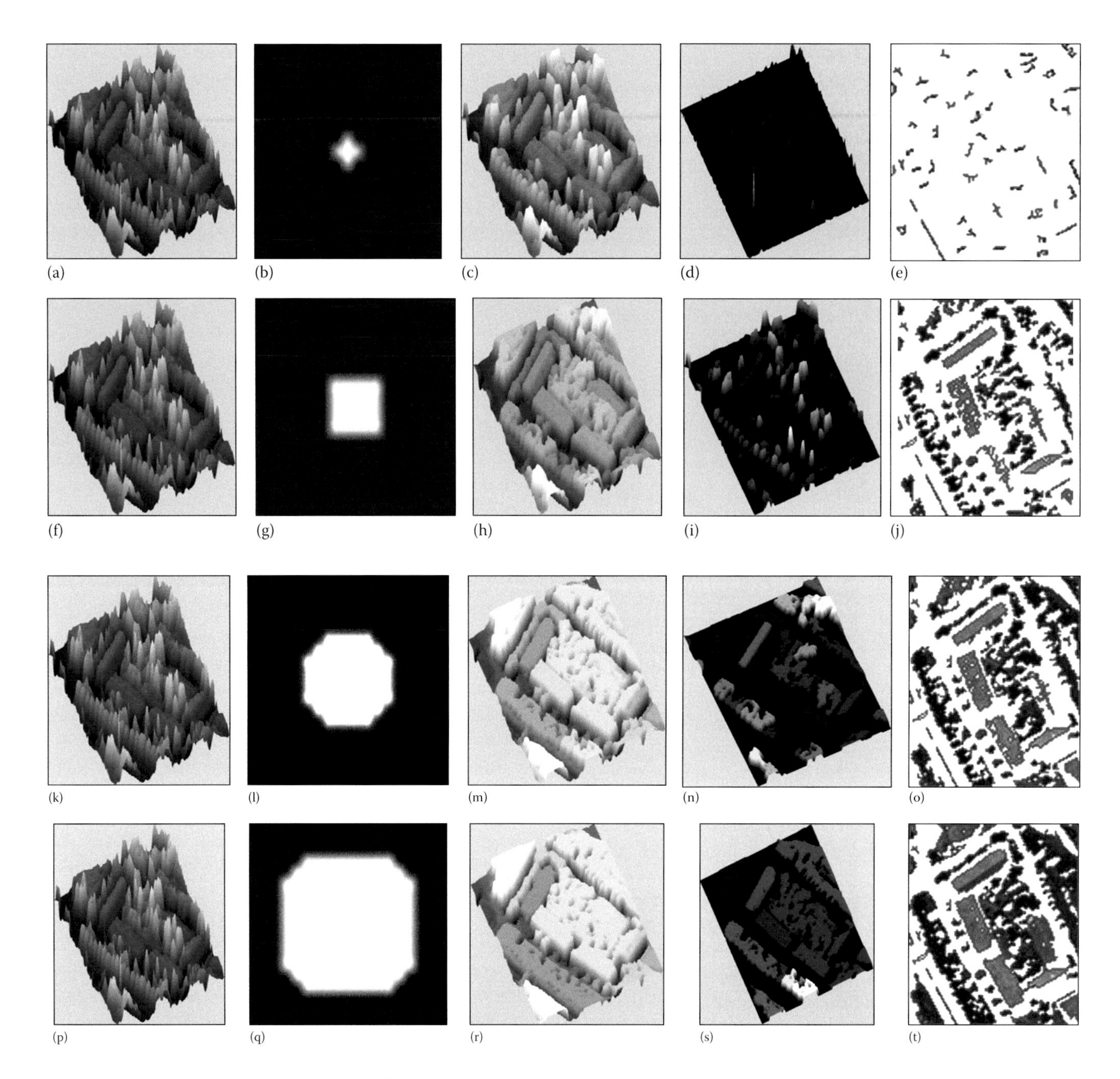

FIGURE 16.6 Illustration of the opening profile obtained using increasing SE sizes. Each row shows the results for an increasing SE series. The first column shows the input spectral data as a topographic relief. The second column shows the SEs used. The third column shows the result of opening by reconstruction of the topographic relief with the corresponding SEs. The fourth column shows the derivative of the opening morphological profile. The fifth column shows the boundaries of the connected components having a nonzero derivative profile for the corresponding SE for a multiscale segmentation

is expected to be spectrally as homogeneous as possible. However, in the extreme case, a single pixel is the most homogeneous. Hence, a segment is also desired to be as large as possible. In general, a segment stays almost the same (both in spectral homogeneity and in size) for some number of SEs and then faces a large change at a particular scale either because it merges with its surroundings to make a new structure or because it is completely lost. Consequently, the size of interest corresponds to the scale right before this change. In other words, if the nodes on a path in the tree stay homogeneous until some node n, and then the homogeneity is lost in the next level, it can be said that n corresponds to a meaningful segment in the hierarchy. With this motivation, to check the meaningfulness of a node, Akcay and Aksoy (2008) defined a measure consisting of two factors: spectral homogeneity, which is calculated in terms of the difference of the

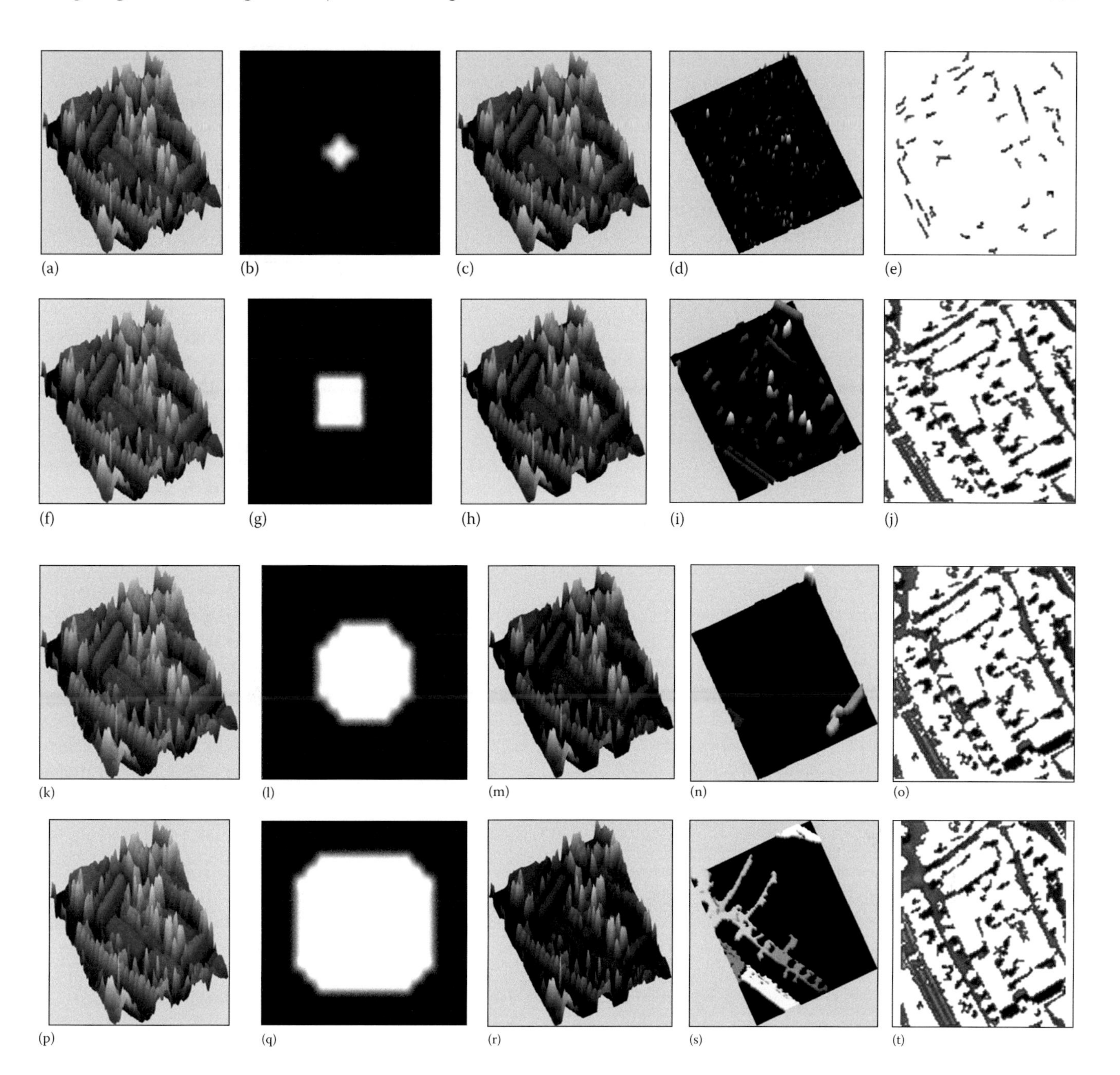

FIGURE 16.7 Illustration of the closing profile obtained using increasing SE sizes. Each row shows the results for an increasing SE series. The first column shows the input spectral data as a topographic relief. The second column shows the SEs used. The third column shows the result of closing by reconstruction of the topographic relief with the corresponding SEs. The fourth column shows the derivative of the closing morphological profile. The fifth column shows the boundaries of the connected components having a nonzero derivative profile for the corresponding SE for a multiscale segmentation.

standard deviation of the spectral features of the node and its parent, and neighborhood connectivity, which is calculated using sizes of connected components. Then, starting from the leaf nodes (level 1) up to the root node (level m), this measure is computed at each node, and a node is selected as a meaningful segment if it is highly homogeneous and large enough on its path in the hierarchy (a path corresponds to the set of nodes from a leaf to the root).

After the tree is finalized, each node is regarded as a candidate segment for the final segmentation. Given the goodness measure of each node in the hierarchy, the segments that optimize this measure are selected by using a two-pass algorithm that satisfies the following conditions. Given N as the set of all nodes and P as the set of all paths in the tree, the algorithm selects $N^* \subseteq N$ as the final segmentation such that any node in N^* must have a measure greater than all of its descendants, any two nodes in N^* cannot be on the same path (i.e., the corresponding segments cannot overlap in the hierarchical segmentation), and every path must include a node that is in $N^\star$ (i.e., the segmentation must cover the whole image). The first pass finds the nodes having a measure greater than all of their descendants in a bottom-up traversal. The second pass selects the most meaningful nodes having the largest measure on their corresponding paths of the tree in a top-down traversal. The details of the algorithm can be found in Akcay and Aksoy (2008). Even though the algorithm was illustrated using a tree constructed from a DMP, it is a generic selection algorithm in the sense that it can be used with other hierarchical image partitions, such as the ones described in Section 16.3.1, and can be applied to specific applications by defining different goodness measures for desired image segments (e.g., see Genctav et al. (2012) for an application of this selection algorithm to an hierarchical segmentation produced by a multiscale watershed procedure).

16.3.4 Graph-Based Algorithms

Graph-based segmentation techniques gained popularity in recent years. In the graph-based framework, the image is modeled by a graph, where nodes typically represent individual pixels or regions, while edges connect spatially adjacent nodes. The weights of the edges reflect the (dis)similarity between the neighboring pixels/regions linked by the edge. The general idea is then to find subgraphs in this graph, which correspond to regions in the image scene. The early graph-theoretic approaches for image segmentation were described in Zahn (1971), where a minimum spanning tree was used to produce connected groups of vertices, and Narendra and Goldberg (1977), where directed graphs have been employed to define regions in edge-detected images. In this section, we will review two algorithms that have been successfully applied for remote sensing applications: optimal spanning forests and normalized cuts.

16.3.4.1 Optimal Spanning Forests

The optimal spanning forest segmentation is based on the minimum spanning tree algorithm introduced by Kruskal (1956) and Prim (1957). It was employed for segmentation of remote sensing images in Tarabalka et al. (2009) and Skurikhin (2010).

We denote an image undirected graph as $G = (V, E, W)$, where each pixel is considered as a vertex $v \in V$, each edge $e_{i,j} \in E$ connects a couple of vertices i and j corresponding to the neighboring pixels. Furthermore, a weight $w_{i,j}$ is assigned to each edge $e_{i,j}$, which indicates the degree of dissimilarity between two pixels connected by this edge. Different dissimilarity measures can be used to compute weights of edges, such as vector norms, SAM, or spectral information divergence (Tarabalka et al., 2010a).

Given a connected graph $G = (V, E)$, a spanning tree $T = (V, E_T)$ of G is a connected graph without cycles such that $E_T \subset E$. A spanning forest $F = (V, E_F)$ of G is a nonconnected graph without cycles such that $E_F \subset E$. Given a graph $G = (V, E, W)$, the *minimum spanning tree* is defined as a spanning tree $T^\star = (V, E_{T^\star})$ of G such that the sum of the edge weights of $T^\star$ is minimal among all the possible spanning trees of G. The *minimum spanning forest* (MSF) rooted on a set of m distinct vertices $\{t_1, \ldots, t_m\}$ consists in finding a spanning forest $F^\star = (V, E_{F^\star})$ of G, such that each distinct tree of $F^\star$ is grown from one root t_i, and the sum of the edge weights of $F^\star$ is minimal among all the spanning forests of G rooted on $\{t_1, \ldots, t_m\}$.

The MSF-based segmentation typically consists of two steps:

1. The objective of this step is to select a *marker*, or region seed, for each spatial object in the image. Such region seeds $\{t_1, \ldots, t_m\}$ can be manually selected from image pixels via interactive image analysis software; however, automatic marker selection is highly desirable. Markers are often defined by automatically searching flat zones (i.e., connected components of pixels of constant intensity value), zones of homogeneous texture, or image extrema (Soille, 2003). Tarabalka et al. (2010a) proposed to perform a supervised probabilistic classification of each pixel (i.e., compute probabilities for each pixel to belong to each of the land categories of interest) and to choose the most reliably classified pixels as markers of spatial regions.
2. Image pixels are grouped into an MSF, where each tree is rooted on a marker. To compute an MSF, an additional root vertex r is added and is connected by the null-weight edges to the marker vertices t_i. The minimum spanning tree of the constructed graph induces an MSF in G, where each tree is grown on a marker vertex t_i; the MSF is obtained after removing the vertex r. The two most commonly used algorithms for computing a minimum spanning tree are Prim's (1957) and Kruskal's (1956) algorithms.

The watershed transform described in the previous section can be efficiently built by computing an MSF rooted on the image minima (Cousty et al., 2009). For this purpose, an ultrametric flooding distance has to be used to compute weights of edges (Meyer, 2005). This distance is defined as the minimal level of flooding for which two pixels belong to the same lake.

Meyer (2005) showed that an MSF can also be efficiently computed from a minimum spanning tree of image pixels, without introducing an additional root vertex r, as depicted in Figure 16.8. This algorithm is useful if the initial markers can be modified (e.g., suppression and addition of markers during interactive segmentation). Given the minimum spanning tree T of a graph $G = (V, E, W)$ and a set of markers $\{t_1, \ldots, t_m\}$, the edges of T are first sorted in the order of their decreasing weights and are considered one after another. Suppose that e is the edge currently under consideration. The edge e belongs to a sub-tree of T. Suppressing e will cut this tree into two smaller sub-trees; if each of them contains at least one marker, then the suppression of e is validated (this is the case in Figure 16.8c through e, where an edge has been suppressed each time); if at least one of the subtrees does not contain a marker, then the edge e is reintroduced. The process stops when each of the created sub-trees contains one and only one marker. This algorithm outputs an MSF with

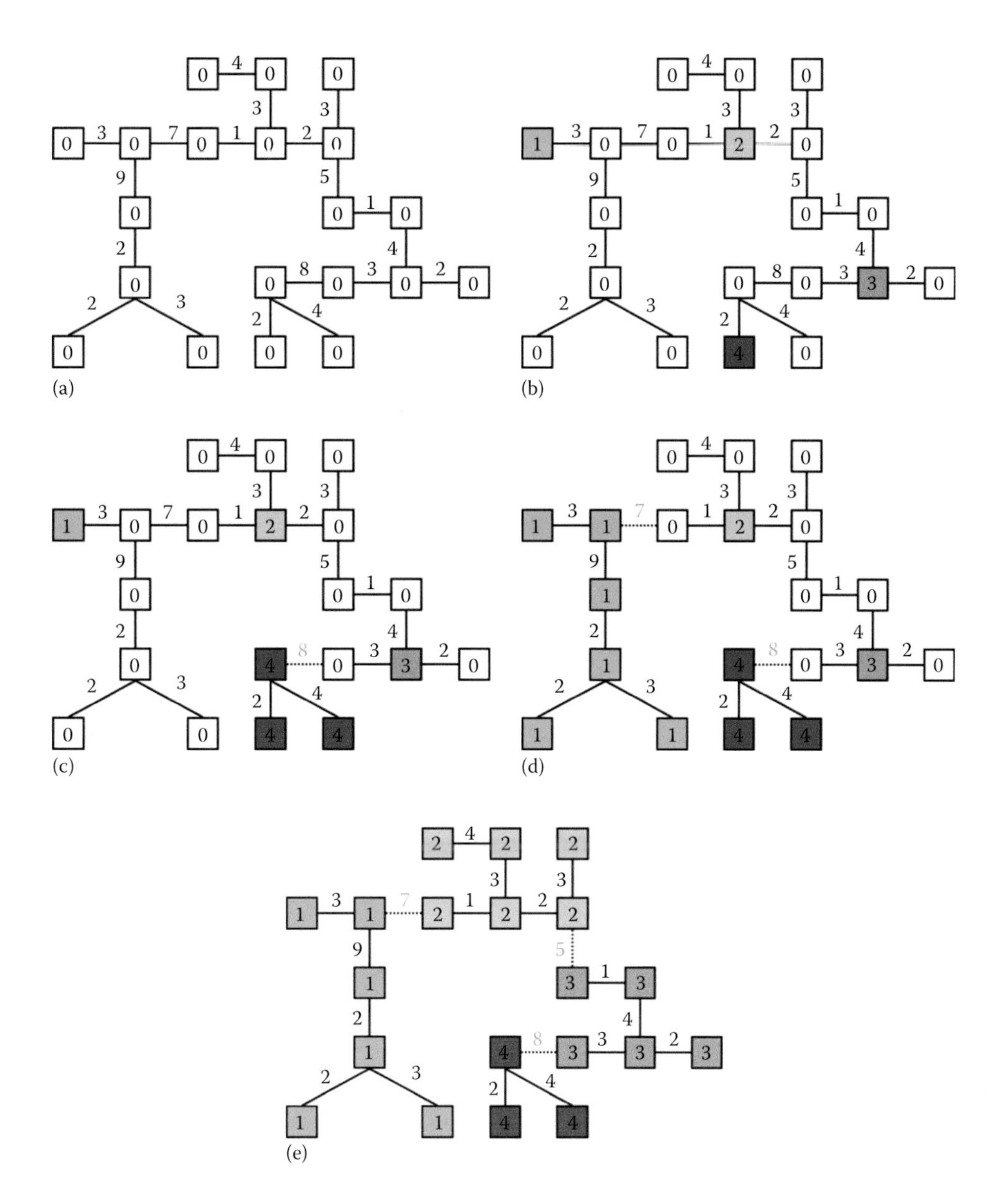

FIGURE 16.8 Example of construction of an MSF rooted on markers from a minimum spanning tree. (a) The initial minimum spanning tree. (b) Four markers defined by the colored nodes. (c)–(d) Illustration of the construction of the MSF from the four markers by highest weight edge suppression. (e) Final MSF, where each tree has the color of its marker.

one tree rooted in each marker. It was applied in Bernard et al. (2012) for segmentation of hyperspectral remote sensing images. If the markers of the image regions cannot be reliably found, a similar algorithm can be iteratively applied, by suppressing the edge of the minimum spanning tree with the highest weight at each iteration until convergence. A threshold for the edge weight can control in this case the convergence.

16.3.4.2 Normalized Cuts

The normalized cuts segmentation method introduced by Shi and Malik (2000) aims at partitioning the image in the way to minimize the similarity between adjacent regions while maximizing the similarity within the regions. 16.9 shows an example graph, where the pixels in group X are strongly connected with high similarities between adjacent pixels, shown as thick red lines, as are the pixels in group Y. The connections between groups X and Y, shown as blue lines, are much weaker. A normalized cut between these two groups separates them into two clusters.

The *cut* between two groups X and Y is computed as the sum of the weights of all edges being cut:

$$cut(X,Y) = \sum_{i \in X, j \in Y} w_{i,j}, \tag{16.22}$$

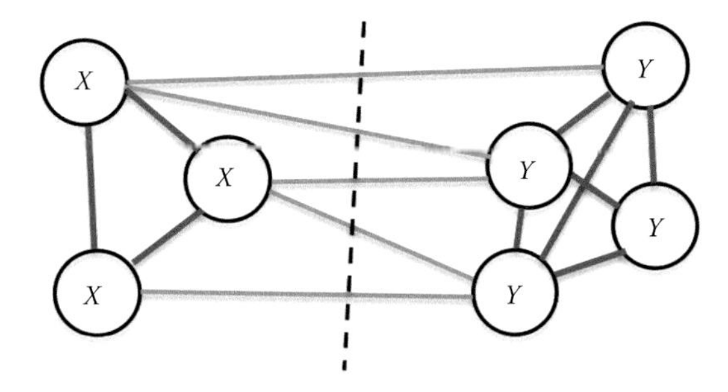

FIGURE 16.9 Example of a weighted graph and its normalized cut (shown as a dashed line).

where the weights of the edges between two vertices i and j measure the similarity between the corresponding pixels (or regions). The optimal bipartitioning of the graph is the one that minimizes this cut value, that is, to find the *minimum cut* of a graph. However, because the value of the cut computed by Equation 16.22 increases with the number of edges separating two partitions, using a minimum cut as a segmentation criterion favors keeping in one group isolated pixels or small sets of isolated nodes in the graph.

To avoid such partitioning of small sets of nodes into separate groups, Shi and Malik (2000) proposed a new measure of disassociation between two groups. The new measure, called a *normalized cut*, computes the cut cost as a fraction of the total edge connections to all the nodes in the graph:

$$Ncut(X,Y) = \frac{cut(X,Y)}{assoc(X,V)} + \frac{cut(X,Y)}{assoc(Y,V)}, \quad (16.23)$$

where $assoc(X,V) = \sum_{i \in X, v \in V} w_{i,v}$ is the sum of all weights from nodes in X to all nodes in the graph, and $assoc(Y, V)$ is similarly defined. The normalized cut defined by Equation 16.23 better reflects the fitness of a particular segmentation, because it seeks to cut a collection of edges that are weak relatively to all of the edges both inside and emanating from each of the regions. Shi and Malik (2000) proposed an efficient algorithm based on a generalized eigenvalue problem to optimize the normalized cut criterion.

Numerous works employed a normalized cut algorithm for segmentation of remote sensing images, such as Grote et al. (2007) and Sung and He (2009). Normalized cuts were also applied in combination with other techniques. For instance, Jing et al. (2010) first used watershed algorithm to find initial segments, and then the normalized cuts technique grouped these segments into the final regions.

16.3.5 MRF-Based Algorithms

Markov random fields (MRFs) are probabilistic graphical models that conceptually generalize the notion of Markov chain (Moser et al., 2013; Wang et al., 2013). They provide a flexible tool to include spatial context into image analysis schemes in terms of minimization of suitable energy functions. While earlier algorithms for optimizing MRF energy, such as iterated conditional modes (ICM) and simulated annealing (Solberg et al., 1996; Tarabalka et al., 2010b), were time consuming, more advanced methods, such as graph cuts (Boykov et al., 2001; Li et al., 2012), provided powerful alternatives from both theoretical and computational viewpoints, resulting in a growing use of the MRF-based segmentation techniques.

For land categorization applications, MRFs are usually applied in the framework of image classification, where the output is a landcover map, with every region assigned to one of the thematic classes. The commonly used MRF energy function in this case is computed as a linear combination of two terms:

$$E(L) = E_{data} + E_{smooth}. \quad (16.24)$$

The first term $E_{data} = \sum_{i=1}^{n} V_i(L_i)$ is related to pixelwise information, and it measures for each pixel the disagreement between a prior probabilistic model and the observed data. Thus, individual potentials $V_i(L_i)$ measure a penalty for a pixel i ($i = 1, 2, \ldots, n$) to have a label L_i. This term is often formulated in terms of the probability density function of feature vectors conditioned to the related class label:

$$V_i(L_i) = -\ln(p(x_i \mid L_i)). \quad (16.25)$$

This probability density can be estimated based on the available training samples for each landcover class. A discussion of the main approaches for this estimation problem can be found in Duda et al. (2001).

The second contribution $E_{smooth} = \sum_{i \sim j} W_{i,j}(L_i, L_j)$ expresses interaction between neighboring pixels, thus exploiting image spatial context. $i \sim j$ denotes a pair of spatially adjacent pixels, and $W_{i,j}(L_i, L_j)$ is an interaction term for these pixels. Most works in remote sensing image classification use a Potts model (Tarabalka et al., 2010b; Li et al., 2012) to compute this spatial term, which favors spatially adjacent pixels to belong to the same class (or spatial region):

$$W_{i,j}(L_i, L_j) = \beta(1 - \delta(L_i, L_j)), \quad (16.26)$$

where $\delta(\cdot)$ is the Kronecker function ($\delta(a, b) = 1$ for $a = b$ and $\delta(a, b) = 0$ otherwise), and β is a positive constant parameter that controls the importance of spatial smoothing. This model tends to deteriorate classification results at the edges between landcover classes and near small-scale details. In order to preserve the border in the output thematic map, *edge* functions have been proposed and integrated in the spatial energy term, such as Tarabalka et al. (2010b):

$$W_{i,j}(L_i, L_j) = \beta(1 - \delta(L_i, L_j)) \frac{t}{t + |\rho_{i|j}|}, \quad (16.27)$$

where

$\rho_{i|j}$ is the gradient value of the pixel i in the direction of j
t is a parameter controlling the fuzzy edge threshold

To optimize the MRF energy, different methods were proposed and applied for remote sensing applications, as described in the following subsections.

16.3.5.1 Simulated Annealing and Iterated Conditional Modes

The main idea of simulated annealing described in Kirkpatrick et al. (1983) is to iteratively propose a change from the current configuration of pixel labels (region or class labels) L to the new randomly generated configuration of labels, and to probabilistically decide if this change is accepted or not. This procedure yields configurations of labels with the lower MRF energy.

Stewart et al. (2000) applied MRF-based method with simulated annealing optimization for segmentation and classification of SAR data. They used Gamma distribution to model the data energy term and a Potts model to compute the spatial term. Tarabalka et al. (2010b) applied simulated annealing technique for the classification of hyperspectral remote sensing images. They performed probabilistic SVM classification to derive the data term and employed the spatial energy contribution with the fuzzy edge function described previously. At each iteration, a new class label L_i^{new} was randomly selected for a randomly selected pixel x_i, and a new local energy $E^{new}(x_i)$ was computed. If the variation of the energy $\Delta E = E^{new}(x_i)-E(x_i)<0$, the new class label is accepted. Otherwise, the new class assignment is accepted with the probability $p = \exp(-\Delta E/T)$, where T is a global parameter called temperature. The optimization begins at a high temperature, which is gradually lowered as the iteration procedure proceeds. This helps to avoid converging to local minima.

The ICM optimization algorithm proposed in Besag (1986) is conceptually simpler and computationally less expensive. It iteratively considers every pixel x_i and assigns a (region or class) label to this pixel, which minimizes the local energy centered at x_i. Solberg et al. (1996) applied the ICM optimization for multisource classification of optical, SAR, and geographic information systems images. Farag et al. (2005) employed the ICM method for hyperspectral image classification. The density functions in this case were estimated by mean field-based SVM regression. The drawback of the ICM algorithm is that it is suboptimal and converges only to a local minimum of the energy function.

16.3.5.2 Graph Cuts

Boykov and Jolly (2001) were the first to apply the graph-cut optimization technique proposed by Greig et al. (1989) to binary image segmentation. This algorithm is computationally efficient, and it gives a globally optimal solution to the binary segmentation problem. In the approach of Boykov and Jolly (2001), all pixels are connected by t-links to the additional two nodes, called source and sink, respectively, as depicted in Figure 16.10. The weights of these edges are computed as the potentials $V_i(L_i)$ (see Equation 16.25), so that pixels that are more compatible with the foreground or background region get stronger connections to the respective source or sink. The e-links between adjacent pixels are assigned weights $W_{i,j}(L_i,L_j)$. The resulting MRF energy (see Equation 16.24) is optimized by solving the minimum-cut/maximum-flow problem, yielding a binary segmentation map.

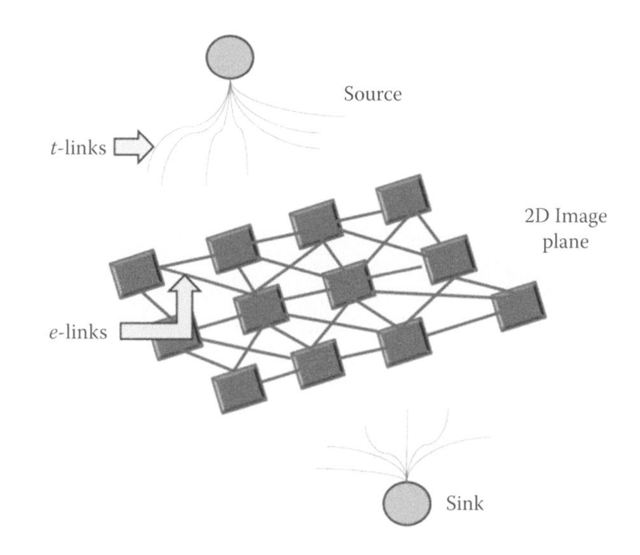

FIGURE 16.10 Mapping of an image to the graph for the graph-cut optimization.

Boykov et al. (2001) proposed two efficient approximation algorithms based on graph-cuts (α-expansion and α–β-swap) to solve multi-label classification problems. In remote sensing, Denis et al. (2009) applied a graph-cut-based algorithm for regularization of SAR images. Li et al. (2012) and Tarabalka and Rana (2014) used an α-expansion optimization of the MRF energy for the segmentation of hyperspectral images.

16.4 Image Segmentation Quality Evaluation

We noted in the introduction to this chapter that there is no general theory of image segmentation, and most image segmentation approaches are *ad hoc* in nature. This is also the case with image segmentation quality evaluation. In practice, image segmentation quality is best evaluated by how well the goals of the analysis utilizing the segmentation are served.

Several general approaches to quantitative segmentation quality evaluation are listed and described in the series of papers published by the research group based at the Leibniz IOER mentioned in the introduction to this chapter, in particular Neubert et al., (2006). Among the evaluation approaches discussed and employed are the following:

1. *Fragmentation and area-fit-index* (Strasters and Gerbrands, 1991; Lucieer, 2004): These measures address over- and under-segmentation by analyzing the number of segmented and reference regions.
2. *Geometric features circularity and geometric features shape index* (Yang et al., 1995; Neubert and Meinel, 2003): These measures address the shape conformity between segmentation and reference regions.

3. *Empirical evaluation function* (Borsotti et al., 1998): This measure addresses how uniform a feature is within segmented regions.
4. *Entropy-based evaluation function and a weighted disorder function* (Zhang and Gerbrands, 1994): This measure addresses the uniformity within segmented regions using entropy as a criterion of disorder.
5. *Fitness function* (Everingham et al., 2002): This measure addresses multiple criteria and parameterizations of algorithms by a probabilistic fitness/cost analysis.

Neubert et al. (2006) also note that the vast majority of quantitative image segmentation approaches are empirical discrepancy methods that analyze the number of misclassified pixels in relation to a reference classification. In the remainder of this section, we describe and demonstrate such an approach to image segmentation quality evaluation. The first step in this evaluation approach is to perform a pixelwise classification of an image data set. In our example, we create our pixelwise classification using the SVM classifier. Then, a region classification is obtained by assigning each spatially connected region from the segmentation result to the most frequently occurring class within the region. The SVM classifier and this *plurality vote* (PV) region-based classification approach are described in more detail in Tilton et al. (2012).

Our test data set is the University of Pavia data set that was recorded by the Reflective Optics System Imaging Spectrometer (ROSIS) over the University of Pavia, Pavia, Italy. The image is 610 × 340 pixels in size, with a spatial resolution of 1.3 m. The ROSIS sensor has 115 spectral channels, with a spectral range of 0.43–0.86 μm. The 12 noisiest channels were removed, and the remaining 103 spectral bands were used in this experiment. See Tarabalka et al. (2010) for more details on this data set. The reference data contain nine ground cover classes: asphalt, meadows, gravel, trees, metal sheets, bare soil, bitumen, bricks, and shadows. A three-band false color image of this data set and the ground reference data are shown in Figure 16.11.

We note that this data set is not rigorously geo-referenced. North is roughly toward the bottom of the image, and the area covered is 793 m × 442 m.

The classification accuracy results are listed in Table 16.2 for the pixelwise SVM classification and the PV region-based classification approach for several region-growing approaches. Figure 16.12 shows the corresponding classification maps. The classification maps for the PV region-based classification approach appear smoother than the pixelwise SVM classification, with the D8- and SEGEN-based classification appearing the smoothest. All of the PV region-based classification accuracies are substantially higher that the accuracy of the SVM pixelwise classification, with the D8, SEGEN, and HSeg PV region-based classifications notably more accurate than the PV region-based classifications based on Muerle Allen or HSWO segmentation. The HSeg-based classification has a marginally higher classification than the other region-growing approaches.

Table 16.2 and Figure 16.6 also show classification results obtained by using the graph-based segmentation approaches. The MSF segmentation with PV was performed as described in

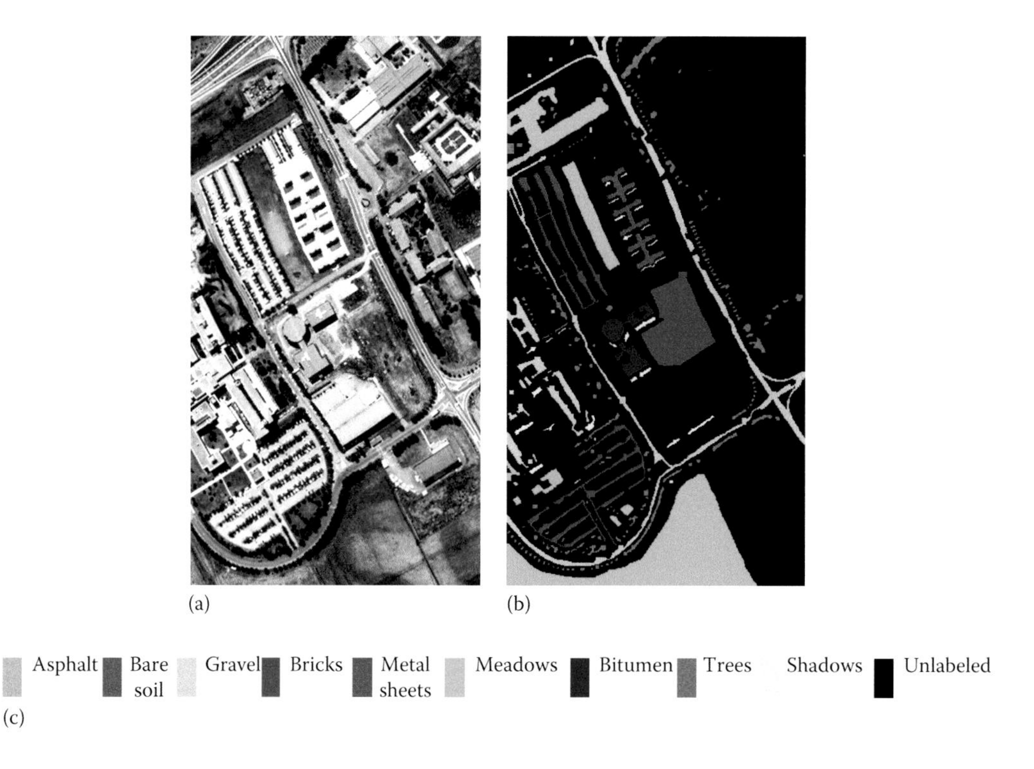

FIGURE 16.11 (a) Three-band false color image of the University of Pavia hyperspectral data set (RGB = bands 56, 33, and 13). (b) Reference data. (c) Color key. The training data for the SVM classifier and segmentation optimization are two separate randomly selected subsets of the reference data.

TABLE 16.2 Comparison of Percentage Classification Accuracies on the University of Pavia Hyperspectral Data Set for Per-Pixel SVM and with the Plurality Vote (PV) Region-Based Classification Method for the Region Muerle Allen, HSWO, Definiens 8.0 (D8), SEGEN, and HSeg

	SVM	Muerle Allen	HSWO + PV	D8 + PV	SEGEN + PV	HSeg + PV ($S_w = 0.3$)	MSF + PV	Graph-Cut + PV
OA	89.03	95.35	95.38	97.54	98.09	98.35	96.99	98.49
AA	89.56	95.26	95.50	97.26	97.95	98.15	97.01	97.73
κ	85.46	93.78	93.83	96.71	97.45	97.79	95.99	98.47

HSeg was performed with small region merge acceleration. Results are also included for two graph-based approaches—the minimum spanning forest (MSF) with plurality vote (PV) method and MRF-based method using the α-expansion graph-cut algorithm. Percentage classification accuracies in terms of OA, AA, and Kappa coefficient (κ).

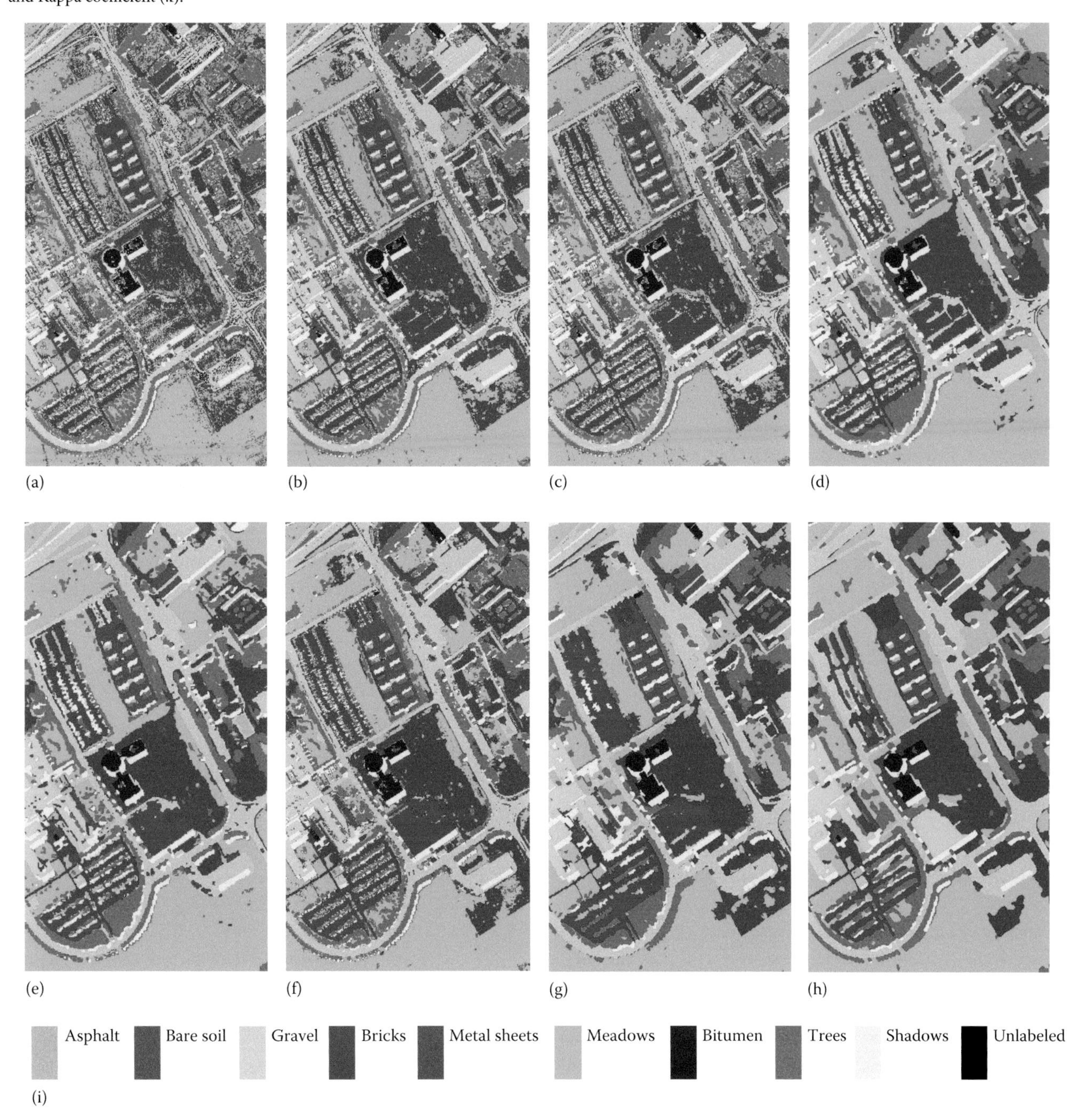

FIGURE 16.12 (a) SVM classification. (b) PV classification with Muerle Allen segmentation. (c) PV classification with HSWO segmentation. (d) PV classification with D8. (e) PV classification with SEGEN segmentation. (f) PV classification with HSeg segmentation ($S_w = 0.3$). (g) Minimum spanning forest (MSF) with plurality vote (PV) classification map. (h) Classification map of the α-expansion graph-cut algorithm. (i) Color key.

Tarabalka et al. (2010a), by using the *L*1 norm between spectral vectors to compute weights of edges between adjacent pixels. For the graph-cut method, probabilistic SVM classification was applied to derive the data energy term, and a Potts model was used to express the spatial term. We fixed the parameter $\beta = 1.5$ as suggested in Tarabalka et al. (2010b). The MRF-based method using the α-expansion optimization yields the highest overall classification accuracy.

The classifications are evaluated in terms of overall accuracy (OA), average accuracy (AA), and the kappa coefficient (κ). OA is the percentage of correctly classified pixels, AA is the mean of the class-specific accuracies, and κ is the percentage of agreement (correctly classified pixels) corrected by the number of agreements that would be expected purely by chance (e.g., see Richards and Jia, 1999).

16.5 Concluding Remarks

Land categorization applications are quite varied. They range from land cover monitoring (e.g., croplands, forests, wetlands, and urbanization), snow and ice mapping, geology and mineral exploration, and even wildfire and agricultural burn monitoring. Each land categorization will have its particular image analysis needs that will be best served by a particular class of image segmentation algorithm. Generally, region-growing approaches may be best suited for finding homogeneous regions, while if locating region borders are more important, watershed approaches may be more effective. Region-growing and morphological approaches that produce segmentation hierarchies would be most appropriate if analysis of the image data at different scales is desired.

We have demonstrated through a simple PV classification approach that utilizing image segmentation can dramatically improve image classification results both qualitatively and quantitatively. Another advantage of image segmentation is that image segmentation defines spatial objects with particular shapes and sizes that can be utilized to perform a deeper, object-based analysis of the image scene than that is possible with pixel-based analysis. In the context of land categorization, this form of analysis has led to the establishment of the new subdiscipline of Geographic Object-Based Image Analysis (GEOBIA) and a popular series of biennial conferences starting with GEOBIA 2006 in Salzburg, Austria—and most recently with GEOBIA 2014 in Thessaloniki, Greece (Gitas et al., 2014).

We have described and discussed a wide range of image segmentation approaches developed over the past century, focusing on those approaches most applicable to the analysis of remotely sensed imagery data for land categorization. The wide range of approaches included in our discussion and the ongoing active research are a clear indication that there still is no general theory of segmentation and that most of the successful image segmentation approaches are rather *ad hoc* in nature. But the wide range of approaches provides a rich menu to choose from for image analysis practitioners in tailoring their image analysis approach to their particular application.

References

Adams, R. and L. Bischof. 1994. Seeded region growing. *IEEE Trans. Pattern Anal. Mach. Intell.*, 16(6):641–646.

Akcay, H. G., S. Aksoy, and P. Soille. 2010. Hierarchical segmentation of complex structures. *Proceedings of ICPR*, Istanbul, Turkey.

Akcay, H. G. and S. Aksoy. 2008. Automatic detection of geospatial objects using multiple hierarchical segmentations. *IEEE Trans. Geosci. Remote Sens.*, 46(7):2097–2111.

Akcay, H. G. and S. Aksoy. 2011. Detection of compound structures using hierarchical clustering of statistical and structural features. *Proceedings of IGARSS*, Vancouver, British Columbia, Canada.

Aksoy, S., H. G. Akcay, and T. Wassenaar. 2010. Automatic mapping of linear woody vegetation features in agricultural landscapes using very-high resolution imagery. *IEEE Trans. Geosci. Remote Sens.*, 48(1):511–522.

Aksoy, S., I. Z. Yalniz, and K. Tasdemir. 2012. Automatic detection and segmentation of orchards using very high-resolution imagery. *IEEE Trans. Geosci. Remote Sens.*, 50(8):3117–3131.

Aksoy, S., K. Koperski, C. Tusk, G. Marchisio, and J. C. Tilton. 2005. Learning Bayesian classifiers for scene classification with a visual grammar. *IEEE Trans. Geosci. Remote Sens.*, 43(3):581–589.

Aksoy, S. 2006. Spatial techniques for image classification. In: C. H. Chen, ed. *Signal and Image Processing for Remote Sensing*. Boca Raton, FL: CRC Press, pp. 491–513.

Aksoy, S. and H. G. Akcay. 2005. Multi-resolution segmentation and shape analysis for remote sensing image classification. *Proceedings of the Second International Conference of Recent Advances in Space Technologies*, Istanbul, Turkey.

Aptoula, E. and S. Lefevre. 2007. A comparative study on multivariate mathematical morphology. *Pattern Recognit.*, 40(11):2914–2929.

Baatz, M. and A. Schape. 2000. Multiresolution segmentation: An optimizing approach for high quality multi-scale segmentation. In: J. Strobl and T. Blaschke, eds. *Angewandte Geographich Informationsverarbeitung*, XII. Heidelberg, Germany: Wichmann, pp. 12–23.

Beaulieu, J.-M. and M. Goldberg. 1989. Hierarchy in picture segmentation: A stepwise optimal approach. *IEEE Trans. Pattern Anal. Mach. Intell.*, 11:150–163.

Benz, U., P. Hofmann, G. Wilhauck, I. Lingenfelder, and M. Heynen. 2004. Multi-resolution, objected-oriented fuzzy analysis of remote sensing data for GIS-ready information. *ISPRS J. Photogramm. Remote Sens.*, 58(3/4):239–258.

Bernard, K., Y. Tarabalka, J. Angulo, J. Chanussot, and J. A. Benediktsson. 2012. Spectral-spatial classification of hyperspectral data based on a stochastic minimum spanning forest approach. *IEEE Trans. Image Proc.*, 21(4):2008–2021.

Besag, J. 1986. On the statistical analysis of dirty pictures. *J. R. Stat. Soc.*, 48(3):259–302.

Borsotti, M., P. Campadelli, and R. Schettini. 1998. Quantitative evaluation of color image segmentation results. *Pattern Recognit. Lett.*, 19(8):741–747.

Boykov, Y., O. Veksler, and R. Zabih. 2001. Fast approximate energy minimization via graph cuts. *IEEE Trans. Pattern Anal. Mach. Intell.*, 23(11):1222–1239.

Boykov, Y. and M.-P. Jolly. 2001. Interactive graph cuts for optimal boundary and region segmentation of objects in N-D images. *Proceedings of ICCV*, Vol. 1, Vancouver, British Columbia, Canada, pp. 105–112.

Bruzzone, L. and D. F. Prieto. 2005. Automatic analysis of the difference image for unsupervised change detection. *IEEE Trans. Geosci. Remote Sens.*, 38(3):1171–1182.

Castilla, G., G. J. Hay, and J. R. Ruiz. 2008. Size-constrained region merging (SCRM): An automated delineation tool for assisted photointerpretation. *PE&RS*, 74(4):409–419.

Clark Labs. 2015. Clark Labs – Image Processing. http://www.clarklabs.org/products/idrisi-image-processing.cfm, last accessed March 31, 2015.

Clinton, N., A. Holt, J. Scarborough, L. Yan, and P. Gong. 2010. Accuracy assessment measures for object-based image segmentation goodness. *Photogrammetric Engineering and Remote Sensing*, 76(3):289-299.

CNES. 2015. Start using Orfeo Toolbox | Orfeo ToolBox. https://www.orfeo-toolbox.org/start/, last accessed March 31, 2015.

Comaniciu, D. and P. Meer. 2002. Mean shift: A robust approach toward feature space analysis. *IEEE Trans. Pattern Anal. Mach. Intell.*, 24(5):603–619.

Cook, R., I. McConnell, D. Stewart, and C. J. Oliver. 1996. Segmentation and simulated annealing. In: G. Franceschetti, F. S. Rubertone, C. J. Oliver, and S. Tajbakhsh, eds. *Microwave Sensing and Synthetic Aperture Radar, Taormina, Italy; Proceedings of SPIE*, Vol. 2958. Bellingham, WA: SPIE, pp. 30–35.

Couprie, C., L. Grady, L. Najman, and H. Talbot. 2011. Power watershed: A unifying graph-based optimization framework. *IEEE Trans. Pattern Anal. Mach. Intell.*, 33(7):1384–1399.

Cousty, J., G. Bertrand, L. Najman, and M. Couprie. 2009. Watershed cuts: Minimum spanning forests and the drop of water principle. *IEEE Trans. Pattern Anal. Mach. Intell.*, 31(8):1362–1374.

Cross, A. M., D. C. Mason, and S. J. Dury. 1988. Segmentation of remotely sensed image by a split-and-merge process. *Int. J. Remote Sens.*, 9:1329–1345.

Denis, L., F. Tupin, J. Darbon, and M. Sigelle. 2009. SAR image regularization with fast approximate discrete minimization. *IEEE Trans. Image Proc.*, 18(7):1588–1600.

Dogrusoz, E. and S. Aksoy. 2007. Modeling urban structures using graph-based spatial patterns. *Proceedings of IGARSS*, Barcelona, Spain, pp. 4826–4829.

Dongping, M., C. Tianyu, C. Hongyue, L. Longxiang, Q. Cheng, and D. Jinyang. 2012. Semivariogram-based spatial bandwidth selection for remote sensing image segmentation with mean-shift algorithm. *IEEE Geosci. Remote Sens. Lett.*, 9(5): 813–817.

Duda, R. O., P. E. Hart, and D. G. Stork. 2001. *Pattern Classification*, 2nd edn. New York: Wiley.

Epifanio, I. and P. Soille. 2007. Morphological texture features for unsupervised and supervised segmentations of natural landscapes. *IEEE Trans. Geosci. Remote Sens.*, 45(4):1074–1083.

Everingham, M., H. Muller, and B. Thomas. 2001. Evaluating image segmentation algorithms using monotonic hulls in fitness/cost space. In: T. Cootes and C. Taylor, eds. *Proceedings of the 12th British Machine Vision Conference*, Manchester, U.K., pp. 363–372.

Exelis Visual Information Solutions. 2015. ENVI – Environment for Visualizing Images (using ENVI). http://www.exelisvis.com/docs/using_envi_Home.html, last accessed March 31, 2015.

Farag, A. A., R. M. Mohamed, and A. El-Baz. 2005. A unified framework for MAP estimation in remote sensing image segmentation. *IEEE Trans. Geosci. Remote Sens.*, 43(7):1617–1634.

Fauvel, M., Y. Tarabalka, J. A. Benediktsson, J. Chanussot, and J. C. Tilton. 2013. Advances in spectral-spatial classification of hyperspectral images. *Proceedings of the IEEE*, 101(3):652–675.

Fu, K. S. and J. K. Mui. 1981. A survey on image segmentation. *Pattern Recognit.*, 13:3–16.

Gaetano, R., G. Scarpa, and G. Poggi. 2009. Hierarchical texture-based segmentation of multiresolution remote-sensing images. *IEEE Trans. Geosci. Remote Sens.*, 47(7):2129–2141.

Genctav, A., S. Aksoy, and S. Onder. 2012. Unsupervised segmentation and classification of cervical cell images. *Pattern Recognit.*, 45(12):4151–4168.

Ghamisi, P., M. S. Couceiro, F. M. L. Martins, and J. A. Benediktsson. 2014. Multilevel image segmentation based on fractional-order Darwinian particle swarm optimization. *IEEE Trans. Geosci. Remote Sens.*, 52(5):2382–2394.

Gitas, I., G. Mallinis, P. Patias, D. Stathakis, and G. Zalidis (Guest eds). 2014. *GEOBIA 2014, Advancements, Trends and Challenges, Fifth Geographic Object-Based Image Analysis Conference*, Thessaloniki, Greece, May 21–24, 2014. Special issue of the *S East. Eur. J. Earth Observ. Geomat.*, 3(2S):1–768.

Gofman, E. 2006. Developing an efficient region growing engine for image segmentation. *Proceedings of ICPR*, Hong Kong, China, pp. 2413–1416.

GRASS Development Team. 2015. GRASS GIS – Home. http://grass.osgeo.org/, last accessed March 31, 2015.

Greig, D., B. Porteous, and A. Seheult. 1989. Exact maximum a posteriori estimation for binary images. *J. R. Stat. Soc. B*, 51(2):271–279.

Grote, A., M. Betenuth, M. Gerke, and C. Heipke. 2007. Segmentation based on normalized cuts for the detection of suburban roads in aerial imagery. *Urban Remote Sensing Joint Event*, Paris, France, pp. 1–5.

Gueguen, L., P. Soille, and M. Pesaresi. 2012. A new built-up presence index based on density of corners. *Proceedings of IGARSS*, Munich, Germany.

Haralick, R. M. and L. G. Shapiro. 1985. Survey: Image segmentation techniques. *Comput. Vis. Graph. Image Proc.*, 29(1):100–132.

Haris, K., S. N. Efstratiadis, N. Maglaveras, and A. Katsaggelos. 1998. Hybrid image segmentation using watersheds and fast region merging. *IEEE Trans. Image Proc.*, 7(12): 1684–1699.

Horowitz, S. L. and T. Pavlidis. 1974. Picture segmentation by a directed split-and-merge procedure. *Proceedings of the Second International Joint Conference on Pattern Recognition*, Copenhagen, Denmark, pp. 424–433.

Jing, W., J. Hua, and W. Yubin. 2010. Normalized cut as basic tool for remote sensing image. *Proceedings of International Conference on ICISS*, Gandhinagar, India, pp. 247–249.

Kirkpatrick, S., C. D. Gelatt, Jr., and M. P. Vecchi. 1983. Optimization by simulated annealing. *Science*, 220(4598):671–680.

Kruse, F. A., A. B. Lefkoff, J. W. Boardman, K. B. Heidebrecht, A. T. Shapiro, P. J. Barloon, and A. F. H. Goetz. 1993. The spectral image processing system (SIPS)—Interactive visualization and analysis of imaging spectrometer data. *Remote Sens. Environ.*, 44(2/3):145–163.

Kruskal, J. 1956. On the shortest spanning tree of a graph and the traveling salesman problem. *Proc. Am. Math. Soc.*, 7:48–50.

Kurita, T. 1994. An efficient agglomerative clustering algorithm for region growing. *Proceedings of MVA, IAPR Workshop on Machine Vision Applications*, Kawasaki, Japan, pp. 210–213.

Kurtz, C., N. Passat, P. Gancarski, and A. Puissant. 2012. Extraction of complex patterns from multiresolution remote sensing images: A hierarchical top-down methodology. *Pattern Recognit.*, 45(2):685–706.

Le Moigne, J. and J. C. Tilton. 1995. Refining image segmentation by integration of edge and region data. *IEEE Trans. Geosci. Remote Sens.*, 33(3):605–615.

Li, J., J. M. Bioucas-Dias, and A. Plaza. 2012. spectral-spatial hyperspectral image segmentation using subspace multinomial logistic regression and Markov random fields. *IEEE Trans. Geosci. Remote Sens.*, 50(3):809–823.

Li, P. and B. Xiao. 2007. Multispectral image segmentation by a multichannel watershed-based approach. *Int. J. Remote Sens.*, 28(19):4429–4452.

Lucieer, A. 2004. Uncertainties in segmentation and their visualisation. PhD thesis, International Institute for Geo-Information Science and Earth Observation (ITC) and University of Utrecht, Utrecht, the Netherlands.

Lucieer, A. 2015. Homepage of Arko Lucieer – Parbat. http://parbat.lucieer.net/, last accessed March 31, 2015.

Marpu, P. R., M. Neubert, H. Herold, and I. Niemeyer. 2010. Enhanced evaluation of image segmentation results. *J. Spat. Sci.*, 55(1):55–68.

Meinel, G. and M. Neubert. 2004. A comparison of segmentation programs for high resolution remote sensing data. *Proceedings of Commission IV, XXth ISPRS Congress*, Instanbul, Turkey, pp. 1097–1102.

Meyer, F. 2001. An overview of morphological segmentation. *Int. J. Pattern Recognit. Artif. Intell.*, 15(7):1089–1118.

Meyer, F. 2005. Grey-weighted, ultrametric and lexicographic distances. In: C. Ronse, L. Najman, and E. Decenciere, eds. *Mathematical Morphology: 40 Years On*. Dordrecht, the Netherlands: Springer, pp. 289–298.

Meyer, F. and S. Beucher. 1990. Morphological segmentation. *J. Vis. Commun. Image Represent.*, 1(1):21–46.

Moser, G., S. B. Serpico, and J. A. Benediktsson. 2013. Landcover mapping by Markov modeling of spatial-contextual information in very-high-resolution remote sensing images. *Proc. IEEE*, 101(3):631–651.

Muerle, J. L. and D. C. Allen. 1968. Experimental evaluation of techniques for automatic segmentation of objects in a complex scene. In: G. C. Cheng, ed. *Pictorial Pattern Recognition*. Washington, DC: Thompson, pp. 3–13.

Najman, L. and M. Schmitt. 1996. Geodesic saliency of watershed contours and hierarchical segmentation. *IEEE Trans. Pattern Anal. Mach. Intell.*, 18(12):1163–1173.

Narendra, P. M. and M. Goldberg. 1977. A graph-theoretic approach to image segmentation. *Proceedings of the IEEE Computer Society Conference on Pattern Recognition and Image Processing*, Troy, NY, pp. 248–256.

Neubert, M., H. Herold, and G. Meinel. 2006. Evaluation of remote sensing image segmentation quality—Further results and concepts. *Proceedings of the First International Conference on OBIA*, Salzburg, Austria.

Neubert, M., H. Herold, and G. Meinel. 2008. Assessing image segmentation quality—Concepts, methods and application. In: T. Blaschke, S. Lang, and G. J. Hay, eds. *Object-Base Image Analysis: Spatial Concepts for Knowledge-Driven Remote Sensing Applications*. Berlin, Germany: Springer-Verlag.

Neubert, M. and G. Meinel. 2003. Evaluation of segmentation programs for high resolution remote sensing applications. *Proceedings of Joint ISPRS/EARSeL Workshop on High Resolution Mapping from Space 2003*, Hannover, Germany.

Noyel, G., J. Angulo, and D. Jeulin. 2007. Morphological segmentation of hyperspectral images. *Image Anal. Stereol.*, 26(3):101–109.

Otsu, N. 1979. A threshold selection method from gray-level histograms. *IEEE Trans. Syst. Man Cybernet.*, SMC-9(1):62–66.

Ouzounis, G. K., M. Pesaresi, and P. Soille 2012. Differential area profiles: Decomposition properties and efficient computation. *IEEE Transactions on Pattern Analysis and Machine Intelligence*, 34(8):1533–1548.

Perret, B., S. Lefevre, C. Collet, and E. Slezak. 2012. Hyperconnections and hierarchical representations for grayscale and multiband image processing. *IEEE Trans. Image Proc.*, 30(7):14–27.

Pesaresi, M., A. Gerhardinger, and F. Kayitakire. 2008. A robust built-up area presence index by anisotropic rotation-invariant textural measure. *IEEE JSTARS*, 1(3):180–192.

Pesaresi, M. and J. A. Benediktsson. 2001. A new approach for the morphological segmentation of high-resolution satellite imagery. *IEEE Trans. Geosci. Remote Sens.*, 39(2):309–320.

Plaza, A. J. and J. C. Tilton. 2005. Automated selection of results in hierarchical segmentations of remotely sensed hyperspectral images. *Proceedings of IGARSS*, Vol. 7, Seoul, Korea, pp. 4946–4949.

Prim, R. 1957. Shortest connection networks and some generalizations. *Bell Syst. Tech. J.*, 36:1389–1401.

Richards, J. A. and X. Jia. 1999. *Remote Sensing Digital Image Analysis: An Introduction*. New York: Springer-Verlag.

Robinson, D. J., N. J. Redding, and D. J. Crisp. 2002. Implementation of a fast algorithm for segmenting SAR imagery. In: Scientific and Technical Report. Defence Science and Technology Organization, Edinburgh, South Australia, Australia.

Rosin, P. L. and J. Hervas. 2005. Remote sensing image thresholding methods for determining landslide activity. *Int. J. Remote Sens.*, 26(6):1075–1092.

Sezgin, M. and B. Sankur. 2004. Survey over image thresholding techniques and quantitative performance evaluation. *J. Electron. Imaging*, 13(1):146–168.

Shankar, B. U. 2007. Novel classification and segmentation techniques with application to remotely sensed images. In: V. W. Marek, E. Orlowska, R. Slowinski, and W. Ziarko, eds. *Transactions on Rough Sets VII*. New York: Springer-Verlag, pp. 295–380.

Shi, J. and J. Malik. 2000. Normalized cuts and image segmentation. *IEEE Trans. Pattern Anal. Mach. Intell.*, 22(8):888–905.

Sirmacek, B. and C. Unsalan. 2010. Urban area detection using local feature points and spatial voting. *IEEE Geosci. Remote Sens. Lett.*, 7(1):146–150.

Skurikhin, A. N. 2010. Patch-based image segmentation of satellite imagery using minimum spanning tree construction. *Proceedings of GEOBIA*, Ghent, Belgium.

Soille, P. 2003. *Morphological Image Analysis*. New York: Springer-Verlag.

Soille, P. 2008. Constrained connectivity for hierarchical image partitioning and simplification. *IEEE Trans. Pattern Anal. Mach. Intell.*, 30(7):1132–1145.

Soille, P. and L. Najman. 2010. On morphological hierarchical representations for image processing and spatial data clustering. *International Workshop on Applications of Discrete Geometry and Mathematical Morphology*, Istanbul, Turkey, pp. 43–67.

Soille, P. and M. Pesaresi. 2002. Advances in mathematical morphology applied to geoscience and remote sensing. *IEEE Trans. Geosci. Remote Sens.*, 40(9):2042–2055.

Solberg, A. H. S., T. Taxt, and A. K. Jain. 1996. A Markov random field model for classification of multisource satellite imagery. *IEEE Trans. Geosci. Remote Sens.*, 34(1):100–113.

Sousa, Jr., M. A. 2014. SegSAR – Image Segmentation. https://sites.google.com/site/segsar2/home, last accessed March 31, 2015.

Sramek, M. and T. Wrbka. 1997. Watershed based image segmentation—An effective tool for detecting landscape structure. *Digital Image Processing and Computer Graphics (DIP'97), Proceedings of SPIE*, Vol. 3346, Vienna, Austria, pp. 227–235.

Stewart, D., B. Blacknell, A. Blake, R. Cook, and C. Oliver. 2000. Optimal approach to SAR segmentation and classification. *IEE Proc. Radar Sonar Navig.*, 147(3):134–142.

Strasters, K. and J. Gerbrands. 1991. Three-dimensional segmentation using a split, merge and group approach. *Pattern Recognit. Lett.*, 12:307–325.

Sung, F. and J. He. 2009. The remote-sensing image segmentation using textons in the Normalized Cuts framework. *Mechatronics and Automation: International Conference on ICMA*, Changchun, China, pp. 1877–1881.

Tarabalka, Y., J. A. Benediktsson, J. Chanussot, and J. C. Tilton. 2010. Multiple spectral-spatial classification approach for hyperspectral data. *IEEE Trans. Geosci. Remote Sens.*, 48(11):4122–4132.

Tarabalka, Y., J. C. Tilton, J. A. Benediktsson, and J. Chanussot. 2012. A marker-based approach for the automated selection of a single segmentation from a hierarchical set of image segmentations. *IEEE JSTARS*, 5(1):262–272.

Tarabalka, Y., J. Chanussot, and J. A. Benediktsson. 2009. Classification of hyperspectral images using automatic marker selection and minimum spanning forest. *Proceedings of IEEE WHISPERS*, Grenoble, France, pp. 1–4.

Tarabalka, Y., J. Chanussot, and J. A. Benediktsson. 2010a. Segmentation and classification of hyperspectral images using minimum spanning forest grown from automatically selected markers. *IEEE Trans. Syst. Man Cybern. B: Cybern.*, 40(5):1267–1279.

Tarabalka, Y., M. Fauvel, J. Chanussot, and J. A. Benediktsson. 2010b. SVM- and MRF-based method for accurate classification of hyperspectral images. *IEEE Geosci. Remote Sens. Lett.*, 7(4):736–740.

Tarabalka, Y. and A. Rana. 2014. Graph-cut-based model for spectral-spatial classification of hyperspectral images. *Proceedings of IGARSS*, Quebec City, Quebec, Canada.

Tilton, J. C., Y. Tarabalka, P. Montesano, and E. Gofman. 2012. Best merge region-growing segmentation with integrated nonadjacent region-object aggregation. *IEEE Trans. Geosci. Remote Sens.*, 50(11):4454–4467.

Tilton, J. C. 1998. Image segmentation by region growing and spectral clustering with a natural convergence criterion. *Proceedings of IGARSS*, Seattle, WA.

Tilton, J. C. 2003. Analysis of hierarchically related image segmentations. *Proceedings of IEEE Workshop on Advances in Techniques for Analysis of Remotely Sensed Data*, Greenbelt, MD, pp. 60–69.

Tilton, J. C. 2007. Parallel implementation of the recursive approximation of an unsupervised hierarchical segmentation algorithm. In: A. J. Plaza and C. Chang, eds. *High Performance Computing in Remote Sensing*. New York: Chapman & Hall, pp. 97–107.

Tilton, J. C. 2013. RHSeg user's manual: Including HSWO, HSeg, HSegExtract, HSegReader, HSegViewer and HSegLearn, Version 1.59. Available via email request to James.C.Tilton@nasa.gov

Trimble Geospatial Imaging. 2014. eCognition Developer. http://www.ecognition.com/suite/ecognition-developer, last accessed March 31, 2015.

Unsalan, C. and K. L. Boyer. 2005. A system to detect houses and residential street networks in multispectral satellite images. *Comput. Vis. Image Underst.*, 98(3):423–461.

Vincent, L. and P. Soille. 1991. Watersheds in digital spaces: An efficient algorithm based on immersion simulations. *IEEE Trans. Pattern Anal. Mach. Intell.*, 13(6):583–598.

Wang, C., N. Komodakis, and N. Paragios. 2013. Markov Random field modeling, inference and learning in computer vision and image understanding: A survey. *Comput. Vis. Image Understand.*, 117:1610–1627.

Wang, Z. and R. Boesch. 2007. Color- and texture-based image segmentation for improved forest delineation. *IEEE Trans. Geosci. Remote Sens.*, 45(10):3055–3062.

Williams, J. W. 1964. Heapsort. *Commun. ACM*, 7(12):347–348.

Yang, L., F. Albregsten, T. Lonnestad, and P. Grottum. 1995. A supervised approach to the evaluation of image segmentation methods. *Proceedings of CAIP, Lecture Notes on Computer Science*, Vol. 970, Prague, Czech Republic, pp. 759–765.

Zahn, C. T. 1971. Graph-theoretic methods detecting and describing gestalt clusters. *IEEE Trans. Comput.*, C-20(1):68–86.

Zamalieva, D., S. Aksoy, and J. C. Tilton. 2009. Finding compound structures in images using image segmentation and graph-based knowledge discovery. *Proceedings of IGARSS*, Cape Town, South Africa.

Zhang, Y. J. and J. J. Gerbrands. 1994. Objective and quantitative segmentation evaluation and comparison. *Signal Proc.*, 39:3–54.

Zucker, S. W. 1976. Region growing: Childhood and adolescence. *Comput. Graph. Image Proc.*, 5:382–399.

17

LiDAR Data Processing and Applications

Shih-Hong Chio
National Chengchi University

Tzu-Yi Chuang
National Taiwan University

Pai-Hui Hsu
National Taiwan University

Jen-Jer Jaw
National Taiwan University

Shih-Yuan Lin
National Chengchi University

Yu-Ching Lin
National Defense University

Tee-Ann Teo
National Chiao Tung University

Fuan Tsai
National Central University

Yi-Hsing Tseng
National Cheng Kung University

Cheng-Kai Wang
National Cheng Kung University

Chi-Kuei Wang
National Cheng Kung University

Miao Wang
National Cheng Kung University

Ming-Der Yang
National Chung Hsing University

Acronyms and Definitions

3D	Three dimensional
ALS	Airborne laser scanning
ASPRS	American Society of Photogrammetry and Remote Sensing
CBH	Crown base height
CF	Constant fraction
CG	Center of gravity
DBH	Diameter at breast height
DBMS	Database management system
DEM	Digital elevation model
DG	Direct georeferencing
DSM	Digital surface model
DTM	Digital terrain model
FDNs	Fixed distance neighbors
GCPs	Ground control points
GIS	Geographical information system
GLAS	Geoscience Laser Altimeter System
GNSS	Global Navigation Satellite System

GPS	Global positioning system
GRCS	Ground reference coordinate system
ICESat	Ice Cloud and Land Elevation Satellite
ICP	Iterative Closest Point
IMU	Inertial measurement unit
INS	Inertial navigation system
ITS	Intelligent transportation system
k-NNs	*k*-nearest neighbors
LiDAR	Light Detection and Ranging
LS3D	Least squares 3D surface matching
MA	Maximum
OGC	Open Geospatial Consortium
PCA	Principal component analysis
POS	Positioning and orientation systems
QA/QC	Quality assessment and quality control
RANSAC	Random sample consensus
RMSD	Root mean square difference
RMSE	Root mean square error
SDMBS	Spatial DBMS
TH	Threshold
TIN	Triangulated irregular networks
TLS	Terrestrial laser scanning
VLR	Variable length record
WB	Wavelet based
VMNS	Voxel-based marked neighborhood searching
Voxel	Volume element
ZC	Zero crossing of the first deviation

17.1 Introduction

Light detection and ranging (LiDAR), also called laser scanning, is an effective remote sensing system for acquiring three-dimensional (3D) information about scanned objects, which has been widely applied in a broad range of disciplines since it was first developed less than two decades ago (Baltsavias, 1999a; Krabill et al., 2000). A LiDAR system integrates several accurate optical, electronic, mechanic, timing, and georeferencing units that make it enable to acquire 3D point measurements rapidly. LiDAR systems are mainly classified into two types based on the motions that occur when they collect data. The first category includes static terrestrial (or ground-based) LiDAR systems (usually mounted at a tripod) with a variety of scanning ranges, which are also called 3D laser scanners. The primary component of a terrestrial LiDAR system is a laser ranging unit to obtain the distance to the target. Integrating with one or two rotating mirror or prism that changes the direction of emitting laser pulses, the LiDAR system can scan and measure the distances to the surrounding objects. By calculating the obtained range and the scanning angle, the 3D coordinates of the scanned target are obtained. Because the mirror rotates in rapid speed with small-angle variation when the LiDAR operates, a huge number of dense and accurate point measurements, often called point cloud, of the surfaces of scanned objects are obtained (Figure 17.1).

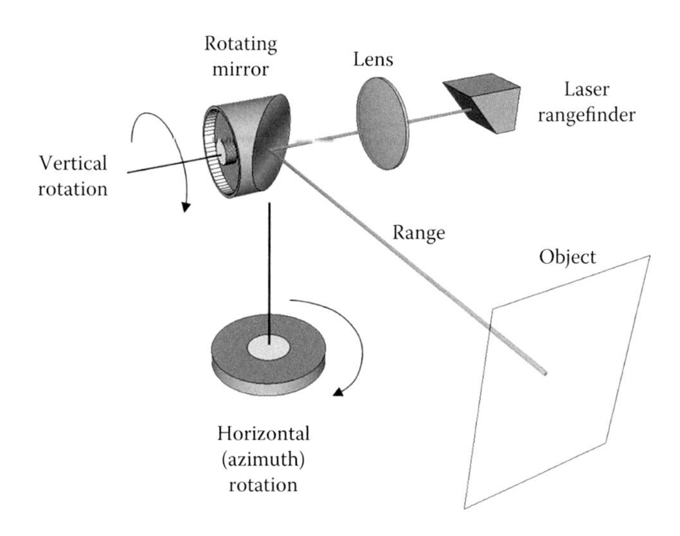

FIGURE 17.1 The measurement mechanism of a terrestrial laser scanner.

The second category includes airborne LiDAR systems (also called airborne laser scanning [ALS]), mobile terrestrial LiDAR systems, or other LiDAR systems boarded on moving platforms (Baltsavias, 1999b; Petrie and Toth, 2008a,b). These systems offer the capability of acquiring 3D information about the scanned objects at the mapping coordinate system by means of direct georeferencing (DG), which is enabled by the combination of a global positioning system (GPS) unit and an inertial measurement unit (IMU) unit (together, these are also known as positioning and orientation systems [POS]) (Baltsavias, 1999a; Krabill et al., 2000), as shown in Figure 17.2. The spaceborne LiDAR systems can be regarded in the second category. However, data processing consideration for such systems is somewhat different. One of the reasons is that the footprint size and the sampling interval of the spaceborne systems are significantly larger, for which they have been called large footprint systems. For example, the Geoscience Laser Altimeter System aboard the Ice Cloud and Land Elevation Satellite has the footprint size of 60 m and sampling interval of 172 m (Zwally et al., 2002). The geometry details obtained from spaceborne LiDAR systems have smoother appearance, which prevents many of the processing logic designed for the small-footprint systems to be directly applied. For brevity and conciseness, these large footprint systems are not included in this chapter.

LiDAR measures the surfaces of objects by accurately recording the time of flight of the laser pulse, which is transmitted from the laser ranging system and then returned back as it reflects off the scanned surface, and the scan angle of the laser pulse, which the laser is directed to, with respect to the laser ranging system. The distance between the laser ranging system and the scanned object can be obtained by halving the multiplication of the speed of light and the round trip time of the laser pulse, and when this is combined with the scan angle of the laser pulse, the 3D coordinates of the point measurement at a local coordinate system can be obtained. For a LiDAR system installed on a

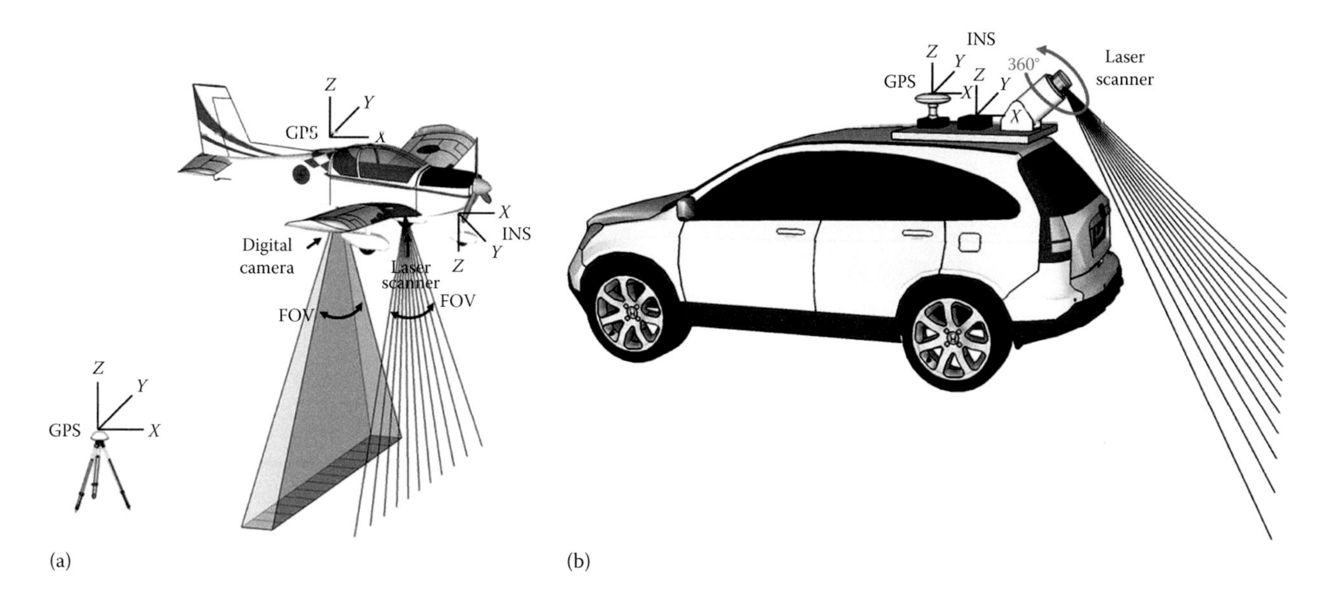

FIGURE 17.2 (a) Airborne and (b) vehicle-borne LiDAR systems.

moving platform, the POS provides highly accurate data on the position and attitude of the platform for DG. The resulting 3D information about the scanned objects is then registered in the same mapping coordinate system as the POS. The choice of the mapping coordinate system depends on the scope of utility for the LiDAR data (Fornaciai et al., 2010). Figure 17.3a and b shows the point clouds of a city area and a forest area, and Table 17.1 lists the characteristics of ALS, terrestrial laser scanning (TLS), and mobile TLS. Table 17.2 lists the main applications, strengths, and limitations of ALS, TLS, and mobile TLS.

Airborne LiDAR systems have been widely employed for large area or nationwide applications, in many cases replacing photogrammetric approaches, as the process flow of LiDAR data is relatively straightforward and less time consuming. There are two major technological benefits of airborne LiDAR that have contributed to its popularity. The first is that LiDAR can acquire 3D information of homogeneous surfaces, which is a difficult task for photogrammetry due to the lack of textures for successful stereo matching. The other is that it is capable of acquiring multiple measurements at different distances with only one pulse. For forested areas, it is very useful to be able to acquire 3D information about not only the top of the canopy but also the ground surfaces beneath it and the structures within it. As a comparison, photogrammetry tends to be used only to measure the top canopy, because the ground surfaces are only partially visible or completely invisible from aerial photographs.

Careful data processing must be carried out in order to fully exploit the point measurements obtained from the scanned objects. This chapter covers the issues of data quality assessment and quality control (QA/QC), data management, point cloud feature extraction, and full-waveform data processing of the LiDAR data. In addition, applications for digital elevation model (DEM) and digital surface model (DSM) generation and 3D city modeling are also described.

In Section 17.2, the concept of error budget is introduced and implemented as the main tool for the QA/QC of the LiDAR point measurements. Internal and external controls are then suggested to ensure that the data derived from the LiDAR system meet the requirement of a specific application.

The number of point measurements resulting from a normal LiDAR scan project, either airborne or terrestrial, can easily exceed several tens of millions. Due to the nature of these 3D point measurements, which lack an efficient spatial index to easily identify neighboring points and quickly access any points of interest, the LiDAR data are difficult to visualize, edit, and process. Section 17.3 thus presents some concepts that can be used to establish a spatial index for point clouds.

Assigning meaningful attribute to the point clouds can significantly increase the usage of the LiDAR data. In an urban setting, extracting spatial features from the point clouds, such as points, lines, and surfaces, can be quite useful as the results can be readily adopted by 3D city modeling. Section 17.4 gives detailed information regarding this kind of feature extraction from LiDAR data. Furthermore, several examples of using LiDAR point clouds for 3D city modeling are presented in Section 17.5.

The complete temporal history of the laser return signal is called the full waveform. In practice, the waveform is processed on the fly, and only a few meaningful signals (i.e., returns) are extracted while the waveform data are then discarded. Some of the early LiDAR systems were only able to extract two returns for each waveform. More *r* systems normally provide four returns, that is, the first, second, third, and last, for each laser pulse. An unlimited number of returns for each laser pulse is possible, provided the full-waveform data and a decent postprocessing algorithm are available. However, handling such a massive amount of data is rather difficult, and thus this approach has not been widely embraced by the community (e.g., manufacturers, service providers, and end users).

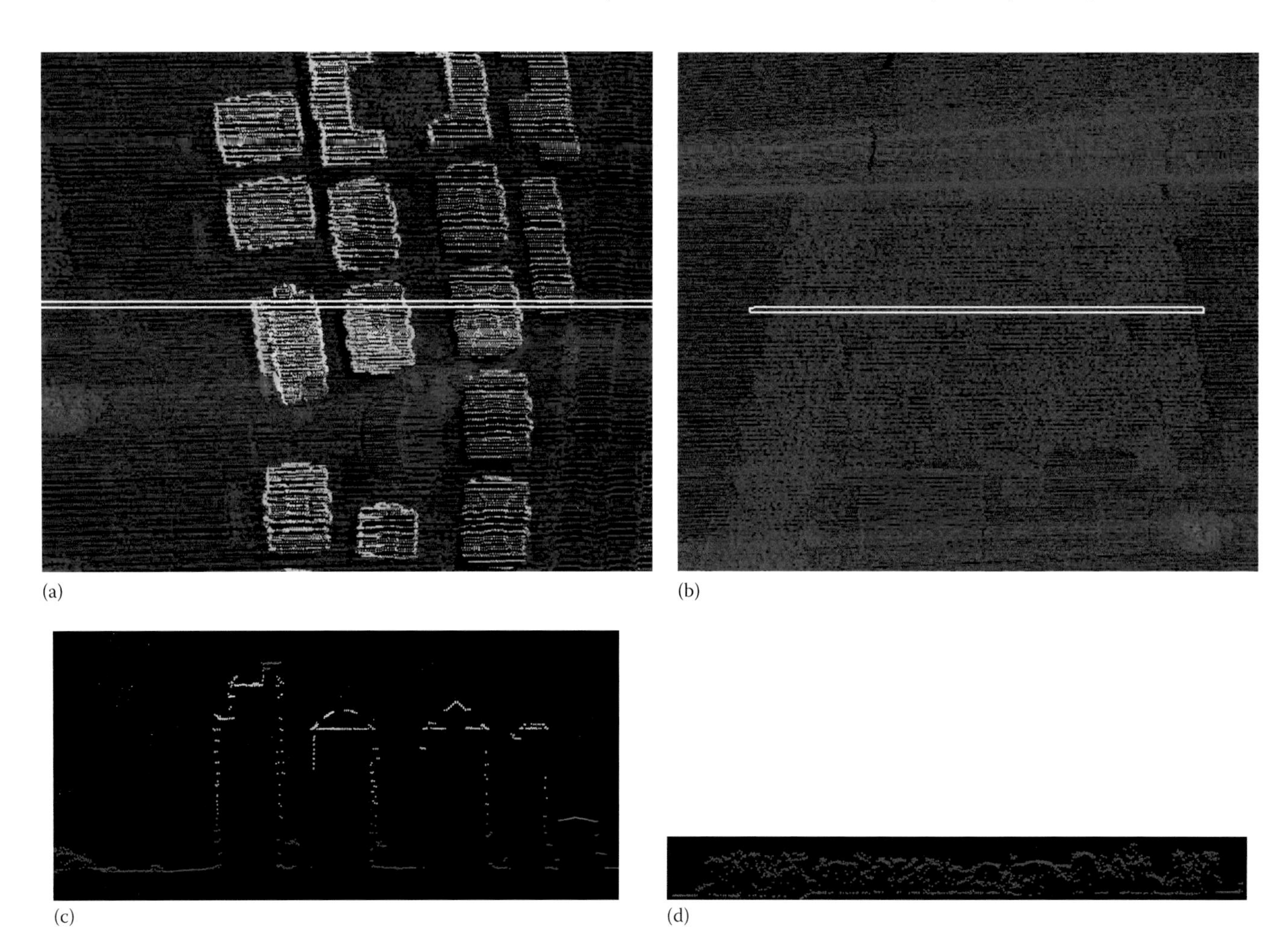

FIGURE 17.3 (Up) Point clouds and (down) profile of (a) a city area and (b) a forest areas. Color represents the height of the points.

TABLE 17.1 Characteristics and Applications of ALS, TLS, and Mobile TLS

System	ALS	TLS	Mobile TLS
Footprint size	Small (~20–40 cm in diameter).	Extreme small (~2 cm in diameter).	Extreme small (~2 cm in diameter).
Waveform data	Available from more recent systems.	Available from limited systems.	Available from limited systems.
Georeferencing	Direct.	Indirect.	Direct.
Measuring distance	200–3000 m.	<1 km.	<1 km.
Data characteristic	The sunlit surfaces are scanned, where most of the point measurements are distributed on the top surfaces of scanned objects.	Point measurements are distributed on the side of the scanned objects. Usually extremely high detail of the surface can be obtained within the scan range of the static station.	Point measurements are distributed on the side of the scanned objects, which are close to the path of the moving platform.
Applications	DEM, DSM, DBM, CHM, land-cover classification, city modeling, road network, biomass.	DBH, building modeling.	DBH, city modeling, road network.
Survey area	Nationwide.	Single or multiple objects with small vicinity.	City-wide.

Nevertheless, processing the full-waveform data is suggested for areas where the penetration of LiDAR is unsatisfactory, and Section 17.6 explains the benefits provided by the full-waveform data.

Generating DEM using LiDAR point clouds is the main purpose of obtaining airborne LiDAR data. To ensure a good-quality DEM, the ground points must be identified from the whole LiDAR point clouds and several algorithms that have been designed to achieve this. The methodology used can be as simple as treating all of the last returns as ground points. While this simple solution may be sufficient for a bare surface or in a city setting, it is likely to fail in forested areas, since many of the last returns may not represent the ground surface (being merely close to it). To obtain good-quality DEM in an efficient manner, the LiDAR points are filtered by a morphology-based algorithm, which considers the connectivity of the nearby points, and this

TABLE 17.2 Recording Fields for Point Format 10 of LAS 1.4

Item	Format	Size
X	Long	4 bytes
Y	Long	4 bytes
Z	Long	4 bytes
Intensity	Unsigned short	2 bytes
Return number	4 bits (bits 0–3)	4 bits
Number of returns (given pulse)	4 bits (bits 4–7)	4 bits
Classification flags	4 bits (bits 0–3)	4 bits
Scanner channel	2 bits (bits 4–5)	2 bits
Scan direction flag	1 bit (bit 6)	1 bit
Edge of flight line	1 bit (bit 7)	1 bit
Classification	Unsigned char	1 byte
User data	Unsigned char	1 byte
Scan angle	Short	2 bytes
Point source ID	Unsigned short	2 bytes
GPS time	Double	8 bytes
Red	Unsigned short	2 bytes
Green	Unsigned short	2 bytes
Blue	Unsigned short	2 bytes
NIR	Unsigned short	2 bytes
Wave packet descriptor index	Unsigned char	1 byte
Byte offset to waveform data	Unsigned long	8 bytes
Waveform packet size in bytes	Unsigned long	4 bytes
Return point waveform location	Float	4 bytes
$X(t)$	Float	4 bytes
$Y(t)$	Float	4 bytes
$Z(t)$	Float	4 bytes

is then followed by manual inspection/editing to ensure all ground points are reasonably identified. The production of DEM is finally realized by interpolating the identified ground points. Section 17.7 provides a detailed explanation of DEM generation from LiDAR data.

The characteristics of the point clouds obtained from the ground-based and mobile LiDAR systems, for example, the pattern of point distribution, the variation of point density within the LiDAR data, and the lack of echo information, are different from those from airborne LiDAR systems. The processing strategy used for such data should take into account these differences and make any necessary adjustments. The processing of the ground-based and mobile LiDAR point clouds is thus described in Section 17.8.

17.2 LiDAR Data Quality Assessment and Control

Due to the nature of LiDAR systems, every single point that is generated usually contains no redundant measurements, and there is no associated measure that can be used to evaluate the quality of point clouds, except for the nominal precision that manufacturers claim. Therefore, a well-defined set of QA/QC procedures is needed before embarking on any aspect of LiDAR data processing. This chapter thus introduces basic components of LiDAR systems and relevant error budgets, and this is followed by a discussion of common approaches to the quality assessment and control of LiDAR data.

17.2.1 System Components

The components of LiDAR systems differ with regard to different LiDAR platforms, which are classified into two categories, namely, mobile platforms, for example, airborne or vehicle-based LiDAR systems, and static platforms, for example, terrestrial LiDAR mounted on a tripod. The system structure of static LiDAR platforms, which integrate state-of-the-art laser ranging and scanning for the rapid, highly dense, and precise acquisition of 3D point clouds (Petrie and Toth, 2008a), is relatively simple as compared to mobile ones. Mobile LiDAR systems are based on the combination of a laser scanning system and a DG system. The laser scanning system can be subdivided into three key units: the optomechanical scanner, the ranging unit, and the control processing unit; the DG system composed of a GPS/Global Navigation Satellite System (GNSS) and an inertial navigation system (INS) measures the sensor's position and orientation directly with respect to a referenced coordinate system. The combination of GPS/INS technologies may also be referred to as a GPS-aided INS or a POS. Each of these technologies alone has limitations, but the integration of GPS and INS is a powerful solution for DG. The DG system and scanning sensor must be in a rigidly fixed position with respect to each other and calibrated within the reference frame of the platform for meaningful results. During a scanning task, the DG system records position and orientation data and also records a corresponding time tag for each laser scan. The DG postprocessing software interpolates the position and orientation of the laser reference point at each time tag. The 3D ground coordinates of every laser return can then be computed using these data and the range measured by the laser.

17.2.2 LiDAR Error Budget

The LiDAR systems introduced earlier face problems with regard to the random and systematic errors that may occur in each component, as well as due to their integration (Filin, 2001). Random errors include those related to position and orientation measurements from the DG system, mirror angles, and ranges, and these are based on the precision of the instrumental measurements within a mobile system. On the other hand, systematic errors are mainly caused by biases in the mounting parameters related to the system components, as well as those in the system measurements resulting from biases related to the range and mirror angle in a laser scanner system. Moreover, systematic errors are also caused by differences in the troposphere and ionosphere, as well as multipath, INS initialization and misalignment errors, and gyro drifts in a DG system. Lichti and Licht (2006) presented a systematic error model that consisted of 19 coefficients grouped into two categories: physical and empirical. The physical group comprises known error sources such as rangefinder offset, cyclic errors, collimation axis error, trunnion axis error, and vertical circle index error. The empirical terms lack ready physical explanation but nonetheless model significant errors. Typically, errors of a static LiDAR platform come principally from the laser scanning system, while those of a mobile LiDAR platform are the combined effects of the laser scanning system and DG system. A detailed description of random and systematic errors of LiDAR systems is given in Huising and Pereira (1998), Baltsavias (1999b), Schenk (2001), Latypov and Zosse (2002), and Habib et al. (2008, 2009). A summary of laser scanning and DG errors is given in the following paragraphs.

Laser scanner errors: There are a number of factors affecting the accuracy of a laser scanner. Reshetyuk (2006) classified scanning errors into three major categories, which highlight the influence of the scanned object on the accuracy of the related point clouds. The first category is instrumental system errors, based on the range and angular measurements, and these vary with different scanning devices. The second category is object-related errors, and these are related to the reflectance properties of the object's surface, due to several factors such as material properties, laser wavelength, polarization, surface color, moisture, roughness, and temperature. The last category is environmental errors, which affect laser beam propagation in the atmosphere, causing both distortion and attenuation of the returned signal. The degree of attenuation depends on the wavelength, temperature, pressure, microscopic particles in the air, and weather conditions. Other factors influencing laser beam propagation are reflection and atmospheric turbulence, caused by the beam wandering from its initial direction and Gaussian wave-front distortion, called beam intensity fluctuation. Based on these characteristics, it can be understood that the longer the range, the greater the expected errors.

Direct georeferencing errors: With regard to positioning performance, most errors in a GPS/GNSS system are dependent on the operating conditions and setup (Morin, 2002). These errors, such as atmospheric errors, multipath effects, poor satellite geometry, baseline length, and loss of lock, have a direct impact on the resulting positioning accuracy of a GPS/GNSS system, and most of the related factors are difficult to predict and describe via a mathematic model. Positioning noise leads to similar amounts of noise in the derived point cloud. As for orientation errors, the overall accuracy of navigation attitude depends on the quality of IMU. Errors in the LiDAR return position due to attitude errors are directly proportional to the range from the scanner to target. As a result, an IMU with higher accuracy is normally required for fixed wing operations compared to helicopter or ground-based data collection, due to the increased target range (Glennie, 2007). When taking the integration of the various sensor components into consideration, the boresighting angles and the physical mounting angles between an IMU and a laser scanning system are usually the main sources of systematic errors, and thus careful calibration of these is required. Moreover, due to the fact that the center of observations from the laser scanner and the origin of the navigation system cannot be colocated, the precise offset or lever arm between the two centers must be known in order to accurately determine the laser scanner measurements. Since the physical measurement origin of the navigation system or laser scanner assembly cannot be directly observed, the lever-arm offset must be obtained indirectly by a calibration procedure. Biases in the lever-arm offsets can then lead to constant shifts in the derived point cloud.

17.2.3 Quality Assessment

Quality assessment encompasses management activities that are carried out prior to data collection to ensure that the derived data are of the quality demanded by the user. These management controls cover the calibration, planning, implementation, and review of data collection activities. In consideration of the potential errors mentioned earlier, calibration of LiDAR systems is prerequisite for reducing the effects of these on the accuracy of acquired point clouds. Ground-based LiDAR systems are calibrated by scanning a field of distributed control targets, while airborne LiDAR systems are calibrated by flying over a set of ground calibration targets that have been precisely surveyed and well mapped. However, there is currently no standard procedure for a calibration task, and thus each LiDAR manufacturer may have its own calibration scheme for its own products, based on specific parameters. Further discussions on LiDAR system calibration can be found in Schenk (2001), Burman (2002), Filin (2003), Kager (2004), and Skaloud and Lichti (2006). One problem is that careful calibrations require some raw measurements, such as navigation data, mirror angles, and ranges, and in general, LiDAR systems do not supply such raw measurements, and thus quality control procedures are needed to assess system performance after the data have been collected.

17.2.4 Quality Control

Quality control procedures can be realized by internal and external checks. The internal measures are used to check the relative consistency of the LiDAR data, while the external measures verify the absolute quality of the LiDAR data by checking its compatibility with an independently collected and more accurate data set.

17.2.4.1 Internal Quality Control

LiDAR data are usually acquired from different scanning viewpoints or different strips; so that the relative consistency of point clouds within the overlapping areas can be assessed through the correspondence of conjugate features. This check is commonly conducted by comparing interpolated range or intensity images derived from the overlapping areas or by comparing the conjugate features extracted from corresponding data sets. The degree of coincidence of the extracted features can be used as a measure of the data quality and to detect the presence of systematic biases. A well-presented demonstration in internal quality control can be found in Habib et al. (2010).

17.2.4.2 External Quality Control

A common approach to external checks involves checkpoint analysis using specially designed LiDAR targets. The targets are then extracted from the range and intensity LiDAR imagery using a segmentation procedure. The coordinates of the extracted targets are then compared with the surveyed coordinates using a root mean square error (RMSE) or root mean square difference (RMSD) indicator. The former usually refers to comparing the estimated results with error-free data, while the latter compares the estimated results with erroneous reference data, and both share the same expression. For more details regarding this approach, one can refer to Csanyi and Toth (2007). On the other hand, in addition to employing special control targets, features within point clouds can also be used as a measure of external quality control. In such case, the derived features can be compared with independently collected control entities, such as point, line, and planes over the same area. To this end, checkpoints situated on smoothing surfaces are usually adopted, and the quality of the vertical components is the main factor to be evaluated.

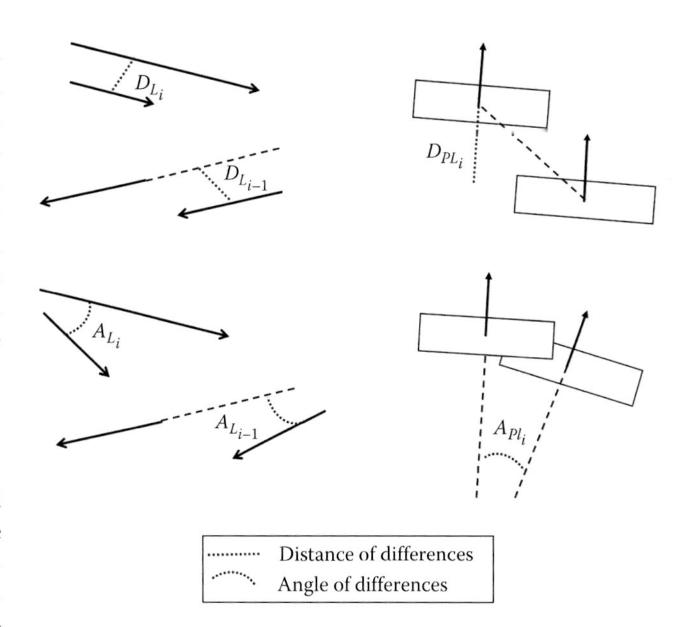

FIGURE 17.4 Illustration of distance and angle differences when using line and plane features.

As shown in Figure 17.4, in addition to the conventional way of assessing positional discrepancies for point features, angle and distance measures can also be utilized to assess the quality, including both internal and external indicators for line and plane features, if applicable (Chuang, 2012).

Registration of terrestrial LiDAR data sets and strip adjustment of airborne LiDAR point clouds are two major ways to provide QA/QC for final geospatial products. In the case of TLS, it is normal to acquire multiple-point clouds from different standpoints for a complete scene of objects. The manipulation of these data should thus be preceded by having them registered relative to the same reference frame. On the other hand, strip adjustment of airborne LiDAR data serves as a means of producing a best-fit surface through an adjustment process that compensates for small misalignments between adjacent data sets (Filin and Vosselman, 2004; Tao and Li, 2007). The discrepancies among strips are typically caused by the varied performance of the georeferencing components and thus show systematic patterns that are more visible in overlap areas rich in objects of simpler geometric shapes, as illustrated in Figure 17.5.

In summary, the quality assessment of LiDAR systems is restricted to the availability of the raw measurements, which

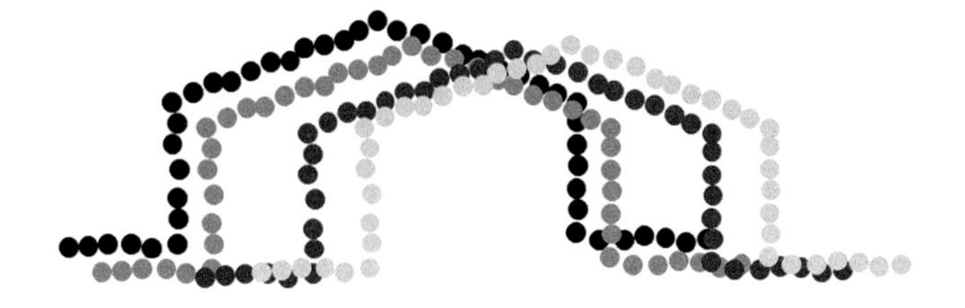

FIGURE 17.5 Illustration of strip discrepancies of airborne LiDAR point clouds.

are not usually revealed to the end user. Consequently, quality control is an essential procedure to ensure that the data derived from a LiDAR system is able to meet the requirements of a specific application.

17.3 LiDAR Data Management

17.3.1 Storage of LiDAR Point Cloud Data

A LiDAR data set mainly contains the 3D coordinates of observed points, which are commonly called point cloud data representing the spatial properties of scanned objects. The reflected laser intensity of each point is also recorded as the radiometric property. Many LiDAR systems, particularly for airborne LiDAR systems, may record additional information for each point, such as return number, number of returns, scan direction, scan angle, and GPS time tag.

Each manufacturer has developed a particular LiDAR data format for storage, and a standard format is thus needed for data exchange. The simplest way to exchange data is to rewrite it in a text (ASCII) file, although this requires much more storage space than a binary file. In 2003, the American Society of Photogrammetry and Remote Sensing (ASPRS) published a format standard, known LAS 1.0, for LiDAR data exchange (ASPRS, 2013). The LAS format is now used as the standard format of LiDAR data for both users and hardware and software manufacturers. Waveform data can also be included in the two latest versions of this format, announced in 2010 and 2011.

The LAS file format provides several optional ways of recording different point data contents. The optional fields include those for the data related to GPS time tag, color (red, green, blue), waveform package, and near infrared. Users can choose appropriate formats depends on their specific aims. The LAS 1.4 specification provides up to 11 point formats (ID: 0–10). Table 17.2 shows the recorded fields of point format 10, which contains all optional fields.

In addition to point data, the LAS format also contains different numbers of variable length records (VLRs) for storing additional information about the point cloud. VLRs contain various types of data, including projection information, metadata, waveform packet information, and user application data.

Although LAS files store data in binary format and consume less storage space than text files, they are not compact. Isenburg (2013) thus developed a lossless compression scheme for LAS files to reduce the storage space down to less than 25% of the original.

Although the LAS format was specially designed for recording airborne LiDAR data, it is also perfectly good for recording the data produced by more recent, mobile LiDAR systems. However, ground-based LiDAR data are often stored in specially designed proprietary file formats, in order to take advantage of the particular features of the instruments and to aid in the subsequent data processing that occurs using on the software that came with the hardware. Such software usually allows advanced users to export the required attributes of the point data into a text file in order to carry out data processing using custom-developed software. In practice, ground-based LiDAR data may also be stored in LAS format, and the instrument-related data may be stored as VLRs in the related data file. However, users must know the format of the special VLRs in advance to correctly read and use the recorded data.

For the convenience of data management, airborne LiDAR data are usually stored in one file per strip, and ground-based LiDAR data are stored in one file per scan. However, the way in which mobile LiDAR data is stored is usually decided by the user.

17.3.2 Organization and Generalization of LiDAR Point Cloud

The data of scanned points in a LiDAR data file are originally saved in a sequence based on the scanning time. Retrieving a point cloud of a local area or a cluster of neighboring points can thus be very inefficient, especially when there is a large amount of data. Organization and generalization of point clouds are thus important practical issues for LiDAR data processing.

A scheme of spatial indexing is usually proposed for the organization of LiDAR data. Point clouds in a specified region of interest (ROI) or a cluster of neighboring points at a location can be retrieved efficiently through the use of spatial indices. In order to visualize a huge LiDAR data set efficiently, the spatial indexing approach should be extended to allow a data set to be organized in a hierarchical, multiresolution fashion. Furthermore, a database management system (DBMS) may be needed to handle the long-term collection of LiDAR data.

17.3.2.1 Spatial Indexing

The purpose of using spatial index of LiDAR point clouds is to establish the relationships among neighboring points, so that it is easier to search for and retrieve of points of interest in a large data set. LiDAR data processing, such as feature extraction and DEM generation, often requires the derivation of meaningful information from the relationships among neighboring points or closely distributed points in a 3D space. However, searching for specific points and their neighbors in a sequentially stored point data set is a time-consuming and inefficient task. Some ground-based LiDAR systems store points in 2D array fashion. While this means that points obtained from adjacent laser pulses can be easily accessed from the file, these points may not be located near to each other in 3D space. Beside time concern, memory space of computer is another important issue. In addition, the computer being used to process these data may not have enough memory to load all the point data in a file to search and process it. However, these problems may be overcome if a spatial index for points is used.

Building a spatial index for a point cloud requires the assistance of spatial data structures, such as triangulated irregular networks (TINs), trees, and grids. The distribution of airborne LiDAR point clouds is similar to a 2.5D data set, so

TIN (Chen et al., 2006; Pu and Vosselman, 2006) and 2D grids (Chen et al., 2007) are often used to handle the neighborhood of points. While ground-based and mobile LiDAR point clouds have a 3D distribution, they are usually handled using 3D grids (or volume elements, known as voxels) (Bucksch et al., 2009; Gorte and Pfeifer, 2004; Wang and Tseng, 2011). Region quadtrees and octrees, which decompose the space of point distribution into regular subspaces, are commonly used to provide regular spatial indexing for data access and to determine the fixed distance neighbors of points. On the other hand, a point *k*-d tree is often used to determine the *k*-nearest neighbors of points (Rabbani, 2006). In practice, because most grids are empty in a 3D grid for a ground-based or mobile LiDAR point cloud, a region octree structure may be used to record the 3D grids to reduce the storage space (Wang and Tseng, 2011).

17.3.2.2 Hierarchical Representation of LiDAR Data

Before starting the data processing task, it is always helpful for users to realize the contents of the LiDAR point clouds by visually inspecting both the outlines and details of the data. Software for viewing LiDAR data thus need the ability to efficiently display a huge amount points on different scales, with a small scale used for the outlines of the point cloud, and a large one for the details. As with an image pyramid that provides hierarchically multiresolution subimages of the original image, a hierarchical representation of LiDAR point clouds can be achieved using a similar approach. At a grid-organized point cloud, a representative point, whose coordinates are the average of the points inside the grid, can be obtained for each grid. These representative points form the first level of generalization of the original point cloud, and the complete generalization hierarchy is then established level by level based on this. Figure 17.6 shows the example of hierarchical representation of a point cloud at different levels.

An additional index file is required to store the resulting generalized hierarchy and spatial index. Based on the viewing scale, representative points of a certain interest level can be loaded from the index file and displayed individually. To view the original point cloud at a large scale, the visible points inside the viewing window can be loaded efficiently through the spatial index stored in the index file. In practice, a 2D hierarchy can be simply saved like an image pyramid. In contrast, the 3D hierarchy organized with an octree structure can be saved as a file in the form of a linear tree (Samet, 1990).

Organizing and generalizing a LiDAR point cloud requires extra time, and the results require additional storage space. However, this work only needs to be done one time, and it can then benefit all subsequent data processing and viewing.

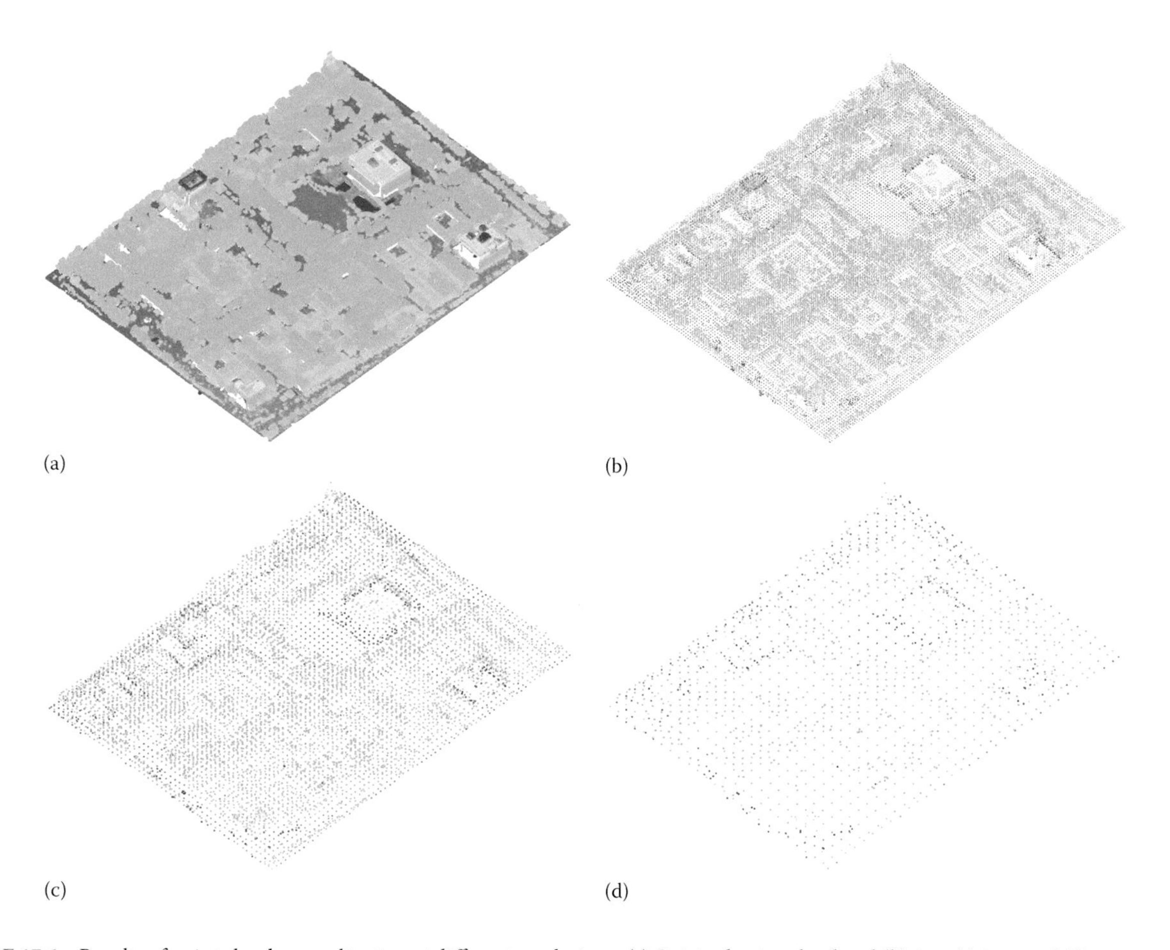

FIGURE 17.6 Results of point cloud generalization at different resolutions. (a) Original point cloud and (b) 4 m, (c) 8 m, and (d) 16 m.

17.3.3 LiDAR Database

As the use of LiDAR systems for collecting spatial data getting easier and popular, more and more LiDAR data are collected and accumulated. Thus, the management of the huge amount of LiDAR point clouds becomes an important and inevitable issue. The most intuitive way to manage such huge data is using the well-developed DBMS, which is equipped with many handy functions such as query, backup, and sorting for manipulating data.

There are two major different ways in using DBMS for point cloud management. The first method partitions point clouds into unified grids and stores the points in each grid separately. The metadata of each grid is also recorded as intermediate for accessing point data. The second method employs the spatial DBMS (SDMBS) to store point clouds. Each point is recorded individually as a spatial element in the SDBMS that makes allowance for optimization of LiDAR processing (Lewis et al., 2012). The generalized hierarchy of point clouds that is used for visual inspection purpose can also be stored at DBMS.

17.4 LiDAR Point Cloud Feature Extraction

Being the points sampling from the object surfaces, LiDAR point clouds contain rich spatial information about the scanned targets. Feature extraction of LiDAR point clouds is the recognition of specific geometric shapes or more general smooth surfaces in the point data (Vosselman et al., 2004) and is the primary procedure to the interpretation of the contents of the point clouds. The purpose of LiDAR point cloud feature extraction is to improve the automation of identifying objects of interest and their characteristics from complicated LiDAR data sets. As LiDAR data are composed of discrete point clouds, and there are no relationships among neighboring points, conventional spectral, spatial (texture-based), and photogrammetric feature extraction methods designed for images may not produce satisfactory results if directly applied to the processing and analysis of LiDAR point cloud data. Several algorithms have thus been developed specifically for the feature extraction of LiDAR point clouds and full-waveform LiDAR data. In addition, the appropriate method for extracting features from LiDAR data may vary significantly according to the targets of interest and the characteristics of the data. For example, the features required to reconstruct tree canopy models may be very different from those required for building model reconstruction. Similarly, full-waveform LiDAR data consist of features that are not available in conventional LiDAR point clouds.

The features embedded in LiDAR data may be derived from the following properties of LiDAR point clouds:

- Location (x, y, z)
- Intensity
- Echo (return) number
- Waveform

The features that are derived or extracted may simply be points, lines, plans, surfaces, shapes, or other geometric characteristics. On the other hand, the features may also directly represent objects of interest (such as a tree canopy, building model, and traffic sign) and physical properties of the objects.

In urban areas, ALS and TLS are effective data sources for city modeling, especially for the reconstruction of 3D building models. An important issue in building reconstruction from LiDAR data is to filter out ground and occlusion (such as trees) points. Several algorithms have been developed specifically for ground filtering of LiDAR data, and there are thoroughly reviewed in Meng et al. (2010). This filtering is often achieved by employing partitioning algorithms to segment the point clouds and then collecting the segmentation attributes to classify the segments (Chen et al., 2008; Rabbani et al., 2006; Sampath and Shan, 2007; Vosselman, 2009; Zhang and Whitman, 2005).

In addition to 3D coordinates, LiDAR data also consist of intensity information. Some applications may treat the intensity data as images (after rasterization and resampling) and employ spectral or spatial image analysis algorithms to extract features. For example, a spectral analysis of LiDAR points was used to identify and map volcano lava flows (Mazzarini et al., 2007). Spectral and spatial analyses of LiDAR intensity data were also successfully applied to land-cover classification (Im et al., 2008), extraction of building footprints (Chen et al., 2008; Zhao and Wang, 2014), and other applications.

17.4.1 Spatial Features in LiDAR Data

Because of the blind operating manner of LiDAR systems, the spatial features of the scanned objects are implicitly contained in the point cloud. Before they can be used in GIS systems, for example, explicit and simple geometrical spatial elements like points, lines, and surfaces must be extracted from the point clouds for object reconstruction and modeling.

The shapes of 3D objects, especially man-made ones, are composed of simple geometric elements, like points, lines, and surfaces. Being blind remote sensing instruments, the point and line features of objects are difficult to directly and accurately measure by LiDAR systems. In contrast, points distributed on the surfaces of such objects can aid in the extraction of surface features. Point and line features can then be obtained indirectly from the intersection of neighboring surface features.

Methods to extract different spatial features from LiDAR data should thus be designed according to the characteristics of the features of interest. Based on the extraction method used, the spatial features in LiDAR point clouds are classified into three categories, including fitting features, intersection features, and boundary features. Fitting features are obtained by fitting a point set to geometric elements, like surfaces, lines, and points. Intersection features are obtained from the intersection of neighboring surfaces or line features. Boundary features appear at the boundary of surface or line features. All three of these are described in more detail, in the following:

- *Fitting features*: Fitting features are formed if points evenly distribute close to a specific geometric model, such as surfaces or lines. The least squares estimation that minimizes

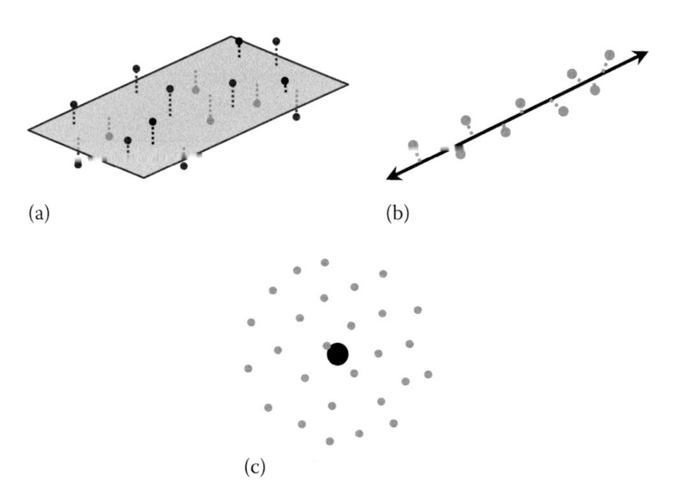

FIGURE 17.7 Basic ideas of (a) fitting plane, (b) fitting line, and (c) gravity center of points.

the squared sum of the normal distance from each point to the selected geometric model is a reasonable method to extract fitting features from a point cloud. Least squares estimation can be used to determine cosurface or colinear features. The centroid of a point set, which is the center position of the points, can also be categorized as a particular fitting feature. However, this can be obtained directly by calculating the average coordinates of points, without the need for least squares estimation. Figure 17.7 shows the basic ideas of fitting surfaces, fitting lines, and gravity centers.

- *Intersection features*: Intersection features are the intersections of existing features, which are usually fitting features. Fitting features should thus be extracted before obtaining intersection features. Intersection features include lines and points. Intersection lines can be generated by intersecting two nonparallel surfaces, while intersection points can be obtained by intersecting three nonparallel surfaces, a surface and a line, or two lines. Figure 17.8 shows the basic ideas of several cases of intersection features.
- *Boundary features*: Boundary features appear at the boundary of a point set. If the point set fits to a surface feature, the boundary lines and points form the boundary feature. If the point set fits to a line feature, the two end points of the line are boundary features. Therefore, fitting features must be extracted first, and then the boundary features are determined based on the information contained in the point set, as shown in Figure 17.9.

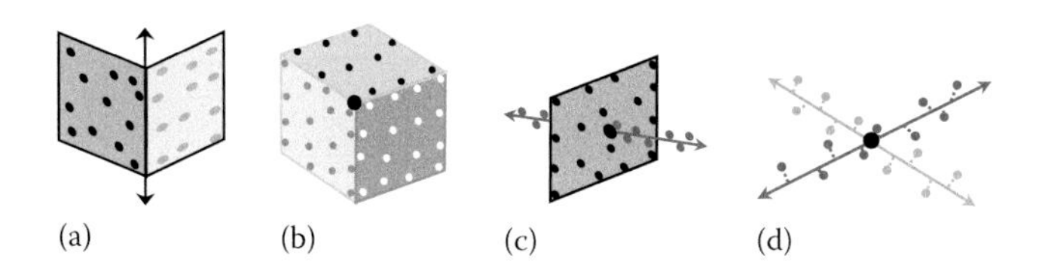

FIGURE 17.8 Basic ideas of intersection of (a) two surfaces, (b) three surfaces, (c) a surface and a line, and (d) two lines.

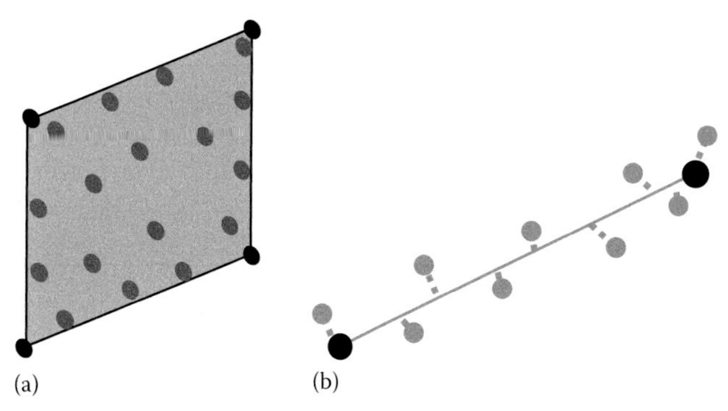

FIGURE 17.9 Basic ideas of boundary features. (a) Boundary lines and points of a fitting plane and (b) end points of a fitting line.

17.4.2 Methods for the Extraction of Spatial Features

LiDAR point clouds represent the outside appearances of the scanned objects. The shapes of some objects are simple, like pipelines in an industrial installation (Rabbani, 2006), and some are complex, like the terrain of the earth and buildings in a city. If the shape of an object, like cylinder or sphere, can be formulated with a simple equation, it can be extracted directly by determining its geometric parameters using the least squares fitting algorithm (Shakarji, 1998). Complex shapes that cannot be easily formulated are often disassembled into several plane features using segmentation algorithms. The objects are then reconstructed and modeled using the extracted features.

In the context of building reconstruction, lines and surfaces are fundamental elements to shape 3D building models. However, extracting line and surface features from LiDAR data is a difficult task, because the point data are usually randomly distributed in the 3D point clouds and lack any topological connections. There are two common approaches to extracting 2D or 3D line features from LiDAR data. The first is to extract 3D line segments directly from the point cloud using scale and rotation invariant point and object features, as demonstrated in Gross and Thoennessen (2006) and Brenner et al. (2013). The other is to intersect identified adjacent 3D planar surfaces to generate line segments (Habib et al., 2005). Since the inherit properties of the distribution of LiDAR points benefit the extraction of surface features, the rest of this section will discuss methods for surface, and especially planar surface, extraction.

17.4.2.1 Determining Feature Parameters Using the Least Squares Fitting Algorithm

The spatial features of simple shapes in LiDAR point clouds can be extracted using the model-based method that simply determines their geometric parameters. Because LiDAR points contain random noises, points are distributed close to the object surfaces with small undulations. The least squares fitting algorithm is often used to determine the geometric parameters. The object is to minimize the square sums of the normal distance between points and the shape. Linearization and iterated calculation are required if the equation of the shape is nonlinear. The linearization of some shapes can be found in Shakarji (1998).

Among the various shapes of geometric elements, planar surfaces are the simplest case of 3D surfaces. The determination of plane parameters for a point set using the least squares fitting algorithm can be transformed to the principal component analysis of points. The solution is then solved directly using eigensystem analysis, without iteration (Shakarji, 1998; Weingarten et al., 2004).

Two different types of strategies can be applied for curved lines and surfaces. The first is utilizing a semiautomatic, model-driven (or a combination of model- and data-driven) approach to compare point segments with known primitives based on the shapes or characteristic parameters. For example, a cylinder can be described by five parameters, while a sphere can be defined by four parameters. However, directly analyzing data in the parametric space may be time consuming and the results may not be reliable. Therefore, processes for reducing the parametric dimensionality or incorporating additional information are often introduced to improve the performance of curved feature extraction. For example, extraction of a cylinder can be separated into two parts: cylinder axis direction and circle plane, while the extraction of a sphere can also utilize the normal vector and other constraints (Vosselman et al., 2004). One thing to note is that this type of feature extraction scheme usually requires a priori knowledge about the objects of interest.

The other strategy is to directly fit the points into parametric surfaces. Figure 17.10 demonstrates an example of constructing curved roof surfaces from TLS and ALS point clouds. TLS data acquired from multiple stations (Figure 17.10a) and airborne LiDAR point clouds (Figure 17.10b) were registered first. The roof boundaries were then generated as cubic spline curves using RANSAC and curve fitting algorithms. Finally, the roof surfaces were approximated as ruled surfaces from the boundary curves (Figure 17.10c). The advantage of this type of approach is that it is flexible and can deal with complicated shapes and objects, although usually with simplified results.

The features within LiDAR point clouds may be implicit and difficult to extract directly. Therefore, incorporating additional information in the feature extraction process by fusing LiDAR point clouds with other data sets has also been proposed, with successful results reported in different applications. Pu and Vosselman (2009b) combined TLS point clouds with images to reconstruct detailed building facade models. Integration

(a)

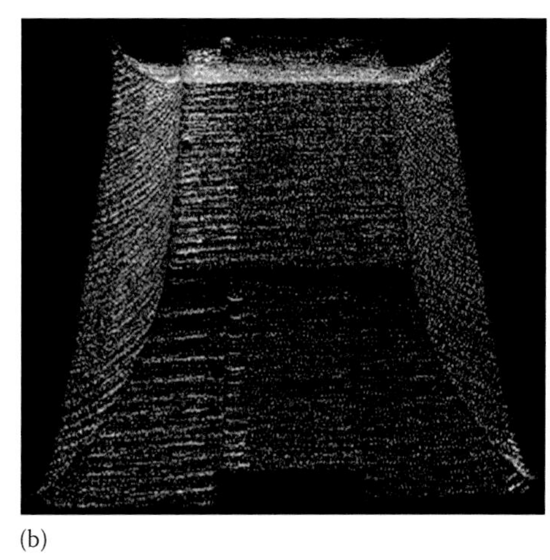

(b)

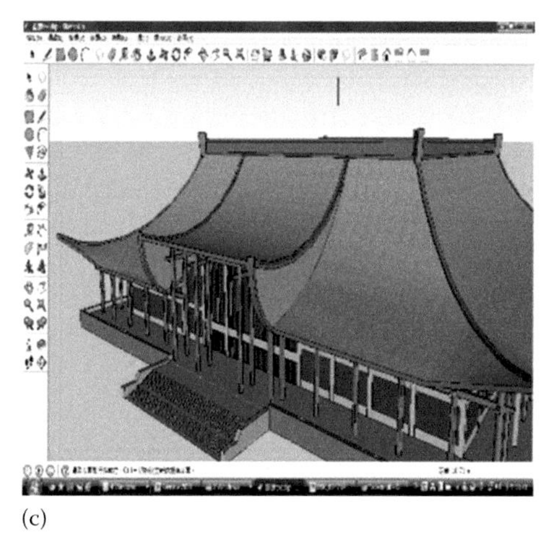

(c)

FIGURE 17.10 Parametric surface reconstruction from point clouds. (a) TLS data, (b) ALS point clouds, and (c) curved roofs constructed as ruled surfaces.

of ground plans or large-scale vector maps and ALS data was developed to reconstruct polyhedral building models (e.g., Chen et al., 2008; Vosselman and Dijkman, 2001). Similar techniques were also applied to generate detailed 3D road models (Chen and Lo, 2009). The idea of the data fusion approach is to take advantage of the explicit information (building layouts, road boundaries, and so on) in images and vector data sets to help extract corresponding features from LiDAR point clouds.

17.4.2.2 Extraction of Plane Features Using Segmentation Algorithm

In most cases, the appearances of objects are too complex to be formulated by a simple equation. Reconstruction of these objects from LiDAR point clouds requires disassembling the surfaces of an object into several simple geometric elements and then modeling the object using the extracted results. Many studies use plane features as the primary geometric elements for reconstructing 3D objects from LiDAR point clouds, for the following reasons:

1. The distribution of LiDAR point clouds benefits the extraction of surface features.
2. A general surface is too complex to model using a mathematical function.
3. The shapes of most artificial objects are composed of planar surfaces.
4. Detecting the lines and vertices of structures is not trivial, due to the scanning mechanism of LiDAR systems.
5. The lines and vertices of structures can be obtained from the intersections of neighboring planes.
6. General surfaces can be obtained from the union of neighboring planar patches.

Although determining a plane feature for a point set is easy, as described previously, automatic extraction of all possible plane features from a large amount of points is difficult, and is not a trivial task. Segmentation, a data-driven method, is the most popular approach for the extraction of plane features from LiDAR point clouds (Filin and Pfeifer, 2006; Rabbani et al., 2006; Sithole, 2005; Wang and Tseng, 2010). The strategy of partitioning or segmenting point clouds and extracting linear and planar features has been proved to be an effective approach for identifying building parts, such as roof facets and facades. For example, after segmenting mobile TLS point clouds into planar faces, vertical wall features can be further extracted according to the inclination angles of the segmented planar faces (Rutzinger et al., 2009). Secondary features, such as windows, doors, and curtains, can also be extracted successfully and accurately based on the sizes, relative positions, orientations, and other characteristics of the segmented planar surfaces (Pu and Vosselman, 2009a).

Segmentation algorithms deal with the coherence and proximity of points (Melzer, 2007). In other word, points that are distributed closely and have similar geometric properties are grouped together by the segmentation algorithm. Coplanarity is used as the coherence property of points for the extraction of plane features. The segment results are groups of coplanar points, each of which represents a plane feature.

Many successful segmentation algorithms have been developed for extracting spatial features form digital images (Gonzalez and Woods, 1992), and the segmentation algorithms for extracting spatial features from LiDAR point clouds are mainly adopted from these methods. Based on the strategy used to deal with the coherence and proximity of the points, LiDAR point cloud segmentation algorithms can be classified into three categories: clustering, region growing, and split and merge.

Clustering algorithms perform at the attribute space to group points of similar properties into clusters (Filin, 2004). The normal vector of the plane is the necessary attribute for plane feature extraction. Some algorithms also employ additional attributes to raise the success rate. Because the grouping of points does not consider the distance between points, the points in each of the clusters have to be separated into neighbor point groups according to the proximity criterion. The tensor voting algorithm (Schuster, 2004) and the 3D Hough transform (Vosselman et al., 2004) are two types of clustering algorithm.

Region growing algorithms (Hoover et al., 1996; Rabbani, 2006; Vosselman et al., 2004) start with the selection of a set of coherent and neighboring points as the seed of a point group. The neighboring points of the seed are added to the point group one by one if they satisfy the coherence criterion, until no more neighboring points can be added to the group. Then, another seed is selected and the growing procedure continues until all points are processed.

Split-and-merge algorithms (Wang and Tseng, 2010) include two parts: the split and merge processes. The split process starts with the examination of the coherence of all points of the point cloud. If the points cannot satisfy the coherence criterion, the space is split into eight equal-size subspaces. The same procedure is then performed on the points contained in each subspace, until all points satisfy the coherence criterion. During the merge step, the points contained in neighboring subspaces are merged if they satisfy the coherence criteria. The merge procedure then continues until no more points can be merged and the segmentation process is complete. In this method, the proximity criterion of points is used during both the split and merge processes.

The profile segmentation algorithm (Sithole, 2005), which is adopted from the scan line segmentation algorithm designed for range image segmentation (Jiang and Bunke, 1994), is a special kind of split-and-merge algorithm. In this method, the space of a point cloud is sliced into connected or cross profiles in different directions. Each profile is treated as a scan line, and the adopted scan line segmentation algorithm is performed on the contained points to obtain collinear point groups. The collinear points at neighboring profiles are then merged to form coplanar point groups.

After the plane features are extracted, line and point features can be obtained by the intersection of neighboring plane features. With some constraints, for example, the included angle, neighboring planes can also be merged to form curved surfaces. However, a general curved surface, like a terrain, is difficult to express using a simple equation. TIN meshes are thus used with most GIS software to represent general curved surfaces.

17.4.2.3 Extraction of Line and Plane Features Using the Hough Transform

The Hough transform is a classical method of surface extraction. Points belonging to a 2D straight line can be represented as a group of (r, θ) values in the Hough space, where r is the distance between the line and the origin and θ is the angle of the vector orthogonal to the line and pointing toward the half upper plane. Similarly, points on a 3D planar surface can be described as a collection of (ρ, θ, ϕ) in a 3D Hough space represented in spherical coordinates (Equation 17.1):

$$(\theta,\phi) \rightarrow \rho = \cos\theta\cos\phi x + \sin\theta\cos\phi y + \sin\phi z \qquad (17.1)$$

17.4.2.4 Extraction of Plane Features Using RANSAC

Another popular approach for plane extraction from LiDAR point clouds is to employ the random sample consensus (RANSAC) algorithm, which was proposed by Fischler and Bolles (1981). RANSAC is a resampling technique to estimate model parameters and is designed to deal with data sets containing a large portion of outliers. RANSAC starts with the smallest set of the data and proceeds to enlarge this data set with consistent data points (Fischler and Bolles, 1981). A typical RANSAC-based algorithm for extracting a planar feature from a LiDAR point cloud is outlined in Algorithm 1.

Algorithm 1 RANSAC for plane extraction

1. Randomly select three points.
2. Construct the plane model (solve the parameters of the plane equation).
3. Calculate the distance of a point to the plane, d_i, for all points.
4. Find inliers (points whose d_i is less than a predefined threshold).
5. If the ratio of the inliers to the total number of points is greater than a predefined threshold, reconstruct the plane based on all the identified inliers and terminate the process.
6. Otherwise, repeat steps 1 through 5 until reaching the maximum number of iterations, N.

The final plane reconstruction (step 5 in Algorithm 1) is usually based on a least squares estimation with all identified inliers. The MA number of iterations, N, should be high enough that there is a probability p, which is usually set as 0.99, that at least one set of randomly selected samples does not include any outlier. Let u be the probability that a selected point is an inlier and $v = 1 - u$ is the probability of an outlier, then

$$1 - p = (1 - u^3)^N \qquad (17.2)$$

and

$$N = \frac{\log(1-p)}{\log(1-(1-v)^3)} \qquad (17.3)$$

The disadvantage of the classical RANSAC algorithm is that it may be very time consuming. However, limiting the selection of samples from presegmented regions of point clouds will significantly improve the efficiency of RANSAC process. For example, when extracting the roof facets or wall facades of a building, the selection of points should be focused on regions with the same normal orientation.

In conclusion, feature extraction is an essential step of LiDAR processing and analysis. The features of interest vary with different targets and applications. From the feature extraction point of view, LiDAR point clouds provide abundant information, and the methods used for extraction are as important as the data itself. The algorithm chosen depends on the characteristics of the data, the objects of interest, and the aim of the application. There is unlikely to be a single method or piece of software that can adequately address the varied needs of all users, and thus it is necessary to explore the possibility of different algorithms to identify the appropriate methodology for the feature extraction of LiDAR point clouds, as well as data fusion with other geospatial data sets.

17.5 Three-Dimensional City Modeling from LiDAR Data

17.5.1 Properties of LiDAR Data in a City Area

The importance of 3D city modeling is increasing due to rapid urbanization and the need for accurate 3D spatial information for urban planning, construction, and management. City modeling is mainly based on images and LiDAR point clouds. The uniqueness of the data characteristics indicates different perspectives for city modeling. Image sensors provide spectral information that can be used to derive the well-defined 3D corner and 3D linear features, which are implicit in stereo images. LiDAR systems provide abundant 3D shape information for reconstruction of city model. Figure 17.11 compares the images of a building taken by an image sensor and produced by a LiDAR system. The ridge line is oversaturated in the image, while the ridge line can easily be distinguished from shaded LiDAR triangles. The benefits of using LiDAR point clouds for city modeling include (1) high vertical accuracy for even low texture surfaces, (2) the ability to directly and accurately obtain the 3D shape information, (3) the ability to establish the nonplanar surfaces of objects in a city, and (4) the ability to provide different viewpoints (e.g., top view and front view) for object interpretation. LiDAR thus has great potential with regard to producing 3D spatial information for a city model.

Airborne LiDAR acquires data from the air to the ground and is used to obtain the 3D points on building rooftop and object surface. On the other hand, terrestrial LiDAR usually acquire the 3D points on building façade and object surface. The scanning distance and beam divergence angle of airborne LiDAR is larger than terrestrial LiDAR and consequently the point density of airborne LiDAR is lower than terrestrial LiDAR. Therefore,

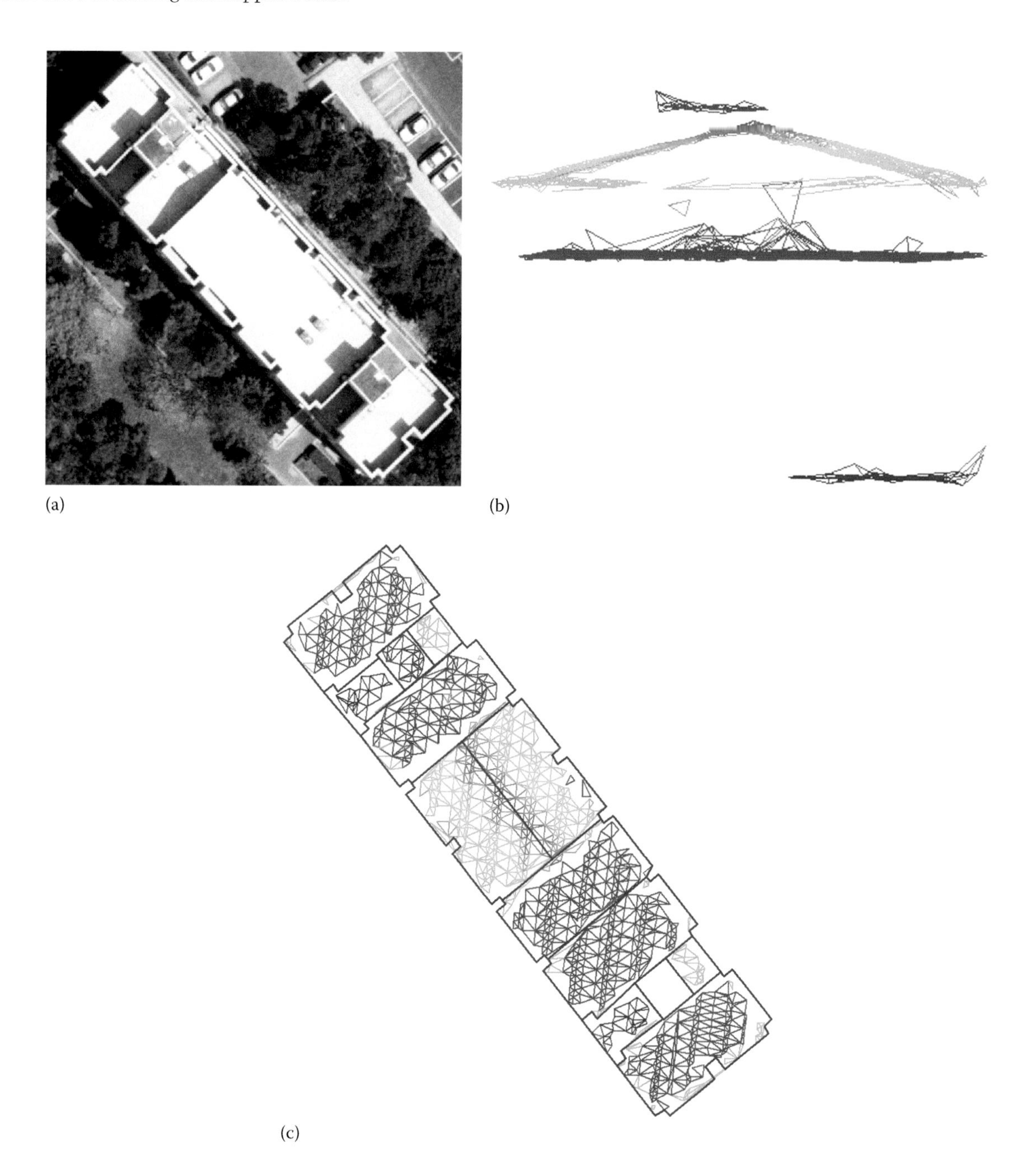

FIGURE 17.11 An example of a building ridge line in an aerial image and airborne LiDAR point clouds: (a) aerial image, (b) LiDAR triangle mesh, and (c) horizontal view of LiDAR triangle mesh.

airborne LiDAR is more suitable for city-scale object reconstruction, while terrestrial LiDAR is suitable for building-scale detailed object reconstruction. For example, the OGC CityGML (OGC, 2012) LOD1 and LOD2 city objects can be reconstructed by airborne LiDAR, while the terrestrial LiDAR is usually used for LOD3 and LOD4 city objects.

17.5.2 Object Reconstruction Strategies

Object reconstruction strategies can be classified into three categories, that is, model-driven, data-driven, and hybrid approaches (Brenner, 2005). The model-driven approach is a top–down strategy that starts with a hypothetical building model, which is verified by the consistency of the model with the existing data. This method needs to define a database of building primitives, and thus a parametric model is generally used, as shown in Figure 17.12a. The data-driven approach is a bottom–up strategy in which the building features, such as point, linear, and planar features, are extracted at the beginning, then grouped into a building model through a hypothesizing process, as shown in Figure 17.12b. Finally, the hybrid approach integrates the ideas of both model-driven and data-driven methods.

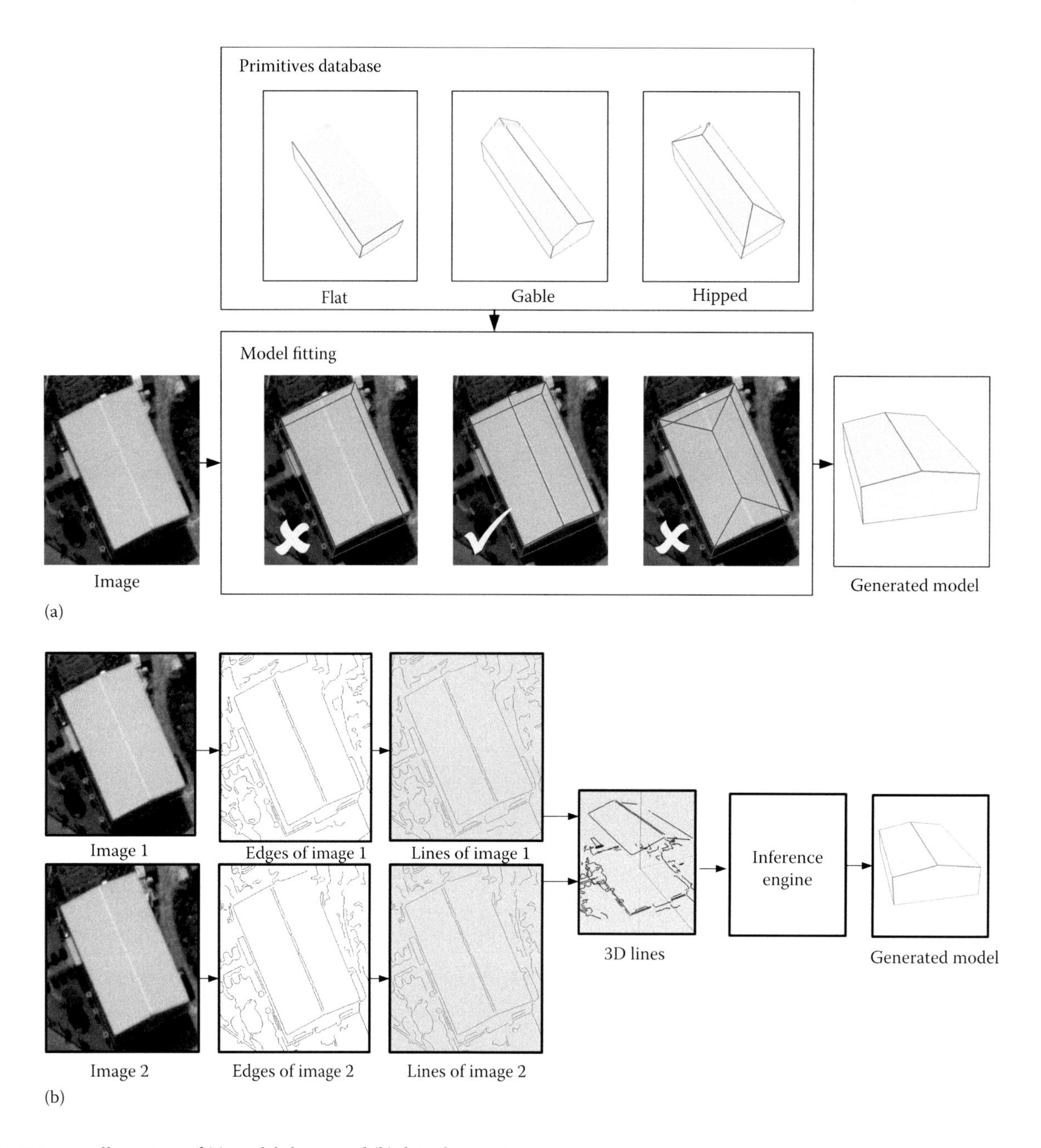

FIGURE 17.12 Illustration of (a) model-driven and (b) data-driven strategies.

17.5.3 Building Extraction

Buildings are the most attractive elements in a 3D city. There are three major steps to establish a building model, and these are detection, reconstruction, and attribution (Gruen, 2005). Building detection is composed of various methods to detect ROIs for subsequent building reconstruction. Building reconstruction is the determination of the 3D geometrical description of buildings located in a given ROI. Finally, building attribution assigns the depiction of building properties, such as type of building, semantic attribution, and textures. All three steps are described in more detail, in the following.

17.5.3.1 Building Detection

The role of building detection is to extract the location of regions where there are buildings. Once a building's location is established, the reconstruction process can be focused in a specific region rather than considering the whole data set. This advantage not only saves computing time but also reduces ambiguity. The idea of building detection is carried out using different characteristics, such as spectral, texture, shape, and roughness, to separate buildings and nonbuildings. A general procedure to obtain a building region from LiDAR includes the following steps: (1) generate aboveground objects by subtracting the digital terrain model (DTM) from the DSM; (2) calculate the different features for each object (an object's features can be shape, texture, roughness, echo ratio, and so on); and (3) separate building and nonbuilding regions based on the object's features using a classifier.

If LiDAR point clouds are combined with a multispectral image, then building regions can be detected by simultaneously considering spectral and shape information. The detection rate may reach 80% when the automatic approach is adopted, although this figure could be worse if the point cloud density is too low or the area of the building is too small. For example, Rottensteiner et al. (2005) used LiDAR data with an average point distance of 1.2 m and multispectral images with 0.5 m spatial resolution to perform building detection. For a building larger than 40 m^2, the detection rate was between 50% and 90%, but for a building smaller than 40 m^2, the detection rate was lower than 50%. Point density and building size are two major factors that will influence the detection rate. Increasing point density may thus improve the accuracy of building detection. Figure 17.13 shows an example of building detection using a LiDAR and multispectral image.

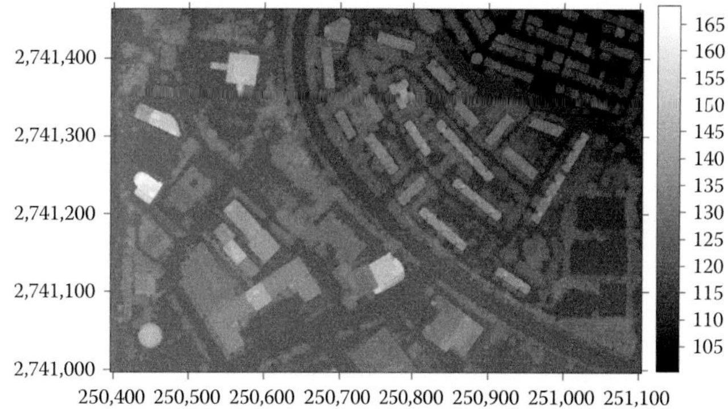

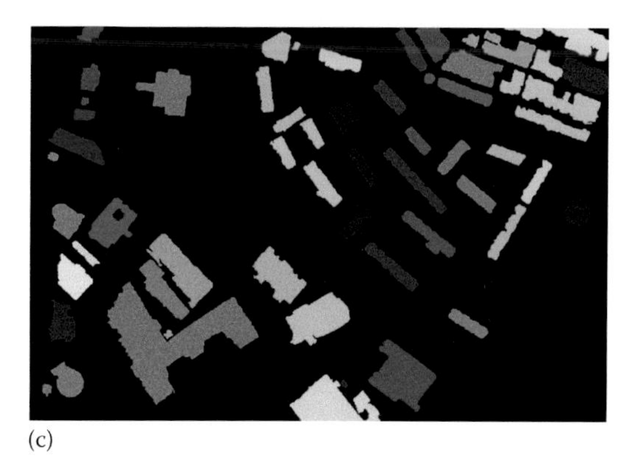

FIGURE 17.13 An example of building detection: (a) LiDAR digital surface model, (b) multispectral image, and (c) detected building regions.

17.5.3.2 Building Reconstruction

The objective of building reconstruction is to build up the geometry of a building. The irregular point clouds need to be structuralized for further processing. There are different data structures for LiDAR data in building reconstruction. The first one uses point data to perform the 3D Hough transform to extract the building models from LiDAR point clouds (Hoffman, 2004). The second one applies TIN data to analyze the planar parameters (Chen et al., 2008). The third one groups and linearizes the building models based on grid data.

The general methods used for building reconstruction based on LiDAR data can be classified into two types. In the first type, planar features are extracted and then the extracted planes are used to derive the line features (Sampath and Shan, 2007). The building model is obtained by integrating the planar and linear features. In the second type, linear features are first extracted and then the extracted lines are used to trace the building polygons. The building model is obtained by shaping the top of the building polygons (Hu, 2003). Each approach has less planimetric accuracy when a regular density (e.g., 1–2 pts./m^2) is employed. Figure 17.14 shows examples of building reconstruction using LiDAR contours and the boundary regularization method.

17.5.3.3 Building Attribution

A building model is an object in a geographical information system (GIS). It consists of spatial information and attributes. Building reconstruction is used to shape the spatial information of a building, while building attribution is used to assign the attributes of a building model. Building attributes enable the 3D analysis to produce more valuable results. There are three kinds of attributes. The first can be generated from the data itself, like floor number, area, and volume. The second are the semantic attributes, which can be obtained from exterior data, like a topographic map. Semantic attributes include the names of buildings and materials of buildings. The third attributes are based on the building texture, which make the building model more photo-realistic.

17.5.4 Road Extraction

A 3D road network is one of the important infrastructures needed for an intelligent transportation system and GIS, and these are applied to transportation management, maintenance, planning, analysis, and navigation. The traditional 2D road centerlines and

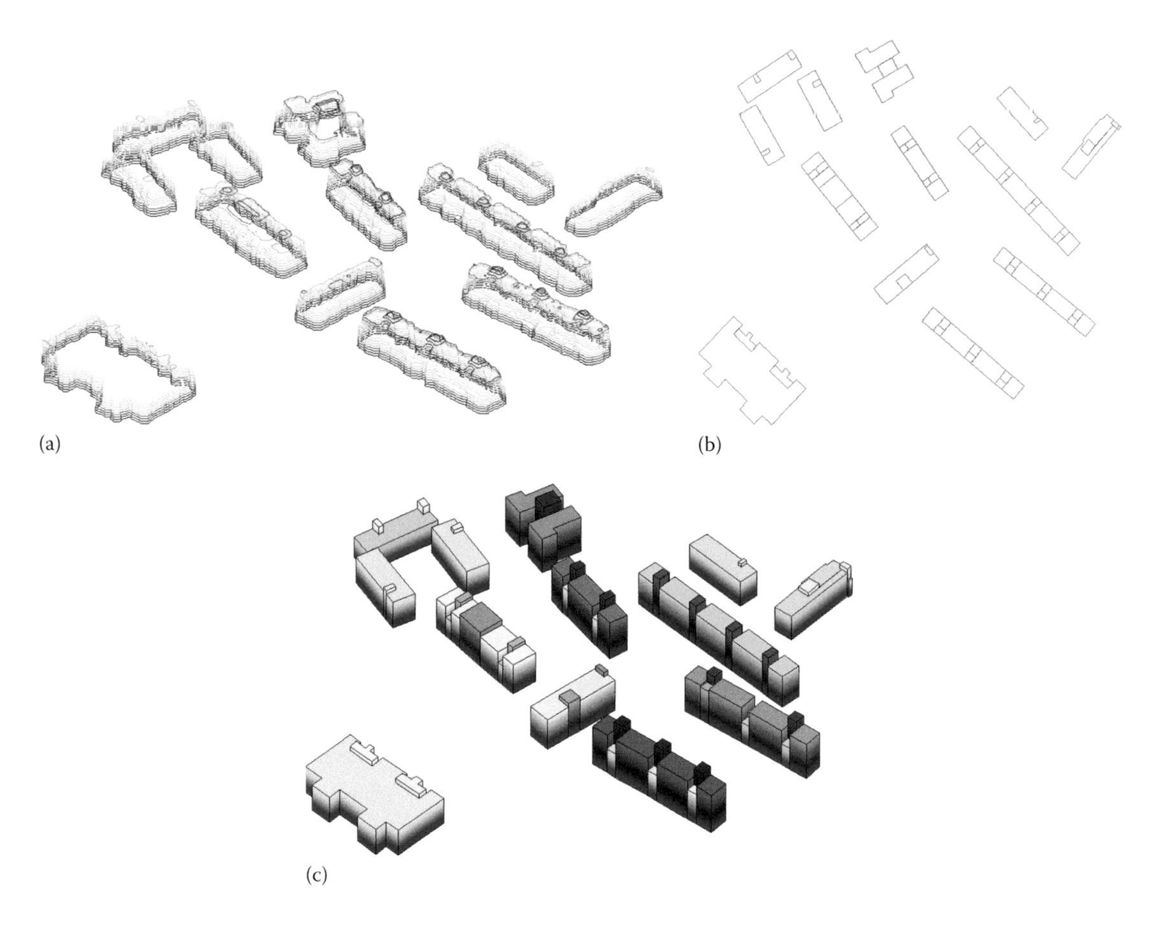

FIGURE 17.14 Examples of building reconstruction: (a) building contours, (b) building boundaries after regularization, and (c) building models.

2D road boundaries are insufficient to represent the 3D reality of actual road systems, especially multilayered ones. Since LiDAR data may provide a large number of 3D points on the road surface, these can be used to generate 3D road models (Oude Elberink and Vosselman, 2009). Road extraction also includes three major steps, that is, detection, reconstruction, and attribution. Road detection is used to detect the 2D road region, while road reconstruction is to shape the 3D road surface. Road attribution is then used to assign the attributes to the road models.

Road detection from LiDAR data usually analyzes the LiDAR intensity, flatness, and continuity. The near infrared of the LiDAR system may have a low return signal in asphalt-covered region, and this is an important attribute that can be used to detect roads. The road design should meet the related regulations, which means the flatness and slope of roads can be estimated to detect road regions. Finally, the continuity of a road network is another constraint that should be considered (Hu, 2003). Note that LiDAR points are irregular points that do not always model the step edge or roadside. The road boundaries extracted from LiDAR points are thus not as accurate as the road boundaries obtained from vector maps. An alternative approach to road detection is adopting the reliable 2D road boundaries from existing topographic maps, with LiDAR points then used for road surface shaping.

In road reconstruction, the major tasks are surface modeling and making connections between road segments (Chiu et al., 2013). Once the road regions are detected, the 3D road surface modeling process will focus on them. A general procedure of 3D road surface modeling includes the following steps: (1) remove the nonroad points based on road profile fitting, (2) carry out 3D road surface fitting by using polynomial functions, and (3) check the consistency of 3D road segments. The main challenge in road reconstruction is the shaping of multilayer roads. Because of the occlusion of multilayer roads, an inference engine is needed to restore the missing parts under the upper roads. Figure 17.15 shows an example of multilayer road extraction using LiDAR points (Chen and Lo, 2009).

17.6 Full-Waveform Airborne LiDAR

17.6.1 Introduction

Recent advances in laser scanning technology have led to the development of new airborne LiDAR commercial systems, called waveform LiDAR systems. A waveform LiDAR system is capable of recording the complete waveform of a return signal (Mallet and Bretar, 2009), which is the digitized intensity of the backscattered signals obtained from the surfaces illuminated by an emitted laser beam. In addition to range measurements, other physical features of the illuminated surfaces can also be derived from the waveform data (Guo et al., 2011). For example, the echo width, amplitude, and backscatter cross section obtained from waveforms have proven to be useful for object classification of urban (Alexander et al., 2010) and vegetation (Heinzel and Koch, 2011; Neuenschwander et al., 2009) areas. Compared with conventional LiDAR systems, a waveform scanner provides additional information about the physical characteristics and geometric structures of the illuminated surfaces. A waveform data set can be used to generate finer point clouds than the data set originally provided by the system, so that it improves the interpretation of physical measurements. It is also possible to retrieve some missing points by detecting surface responses known as echoes in waveforms, and the detection process is especially useful with regard to weak and overlapping echoes. This ability is very significant for some applications of airborne LiDAR and especially for the study of forest areas.

With improvements in data storage capacity and processing speed, waveform airborne LiDAR systems have become

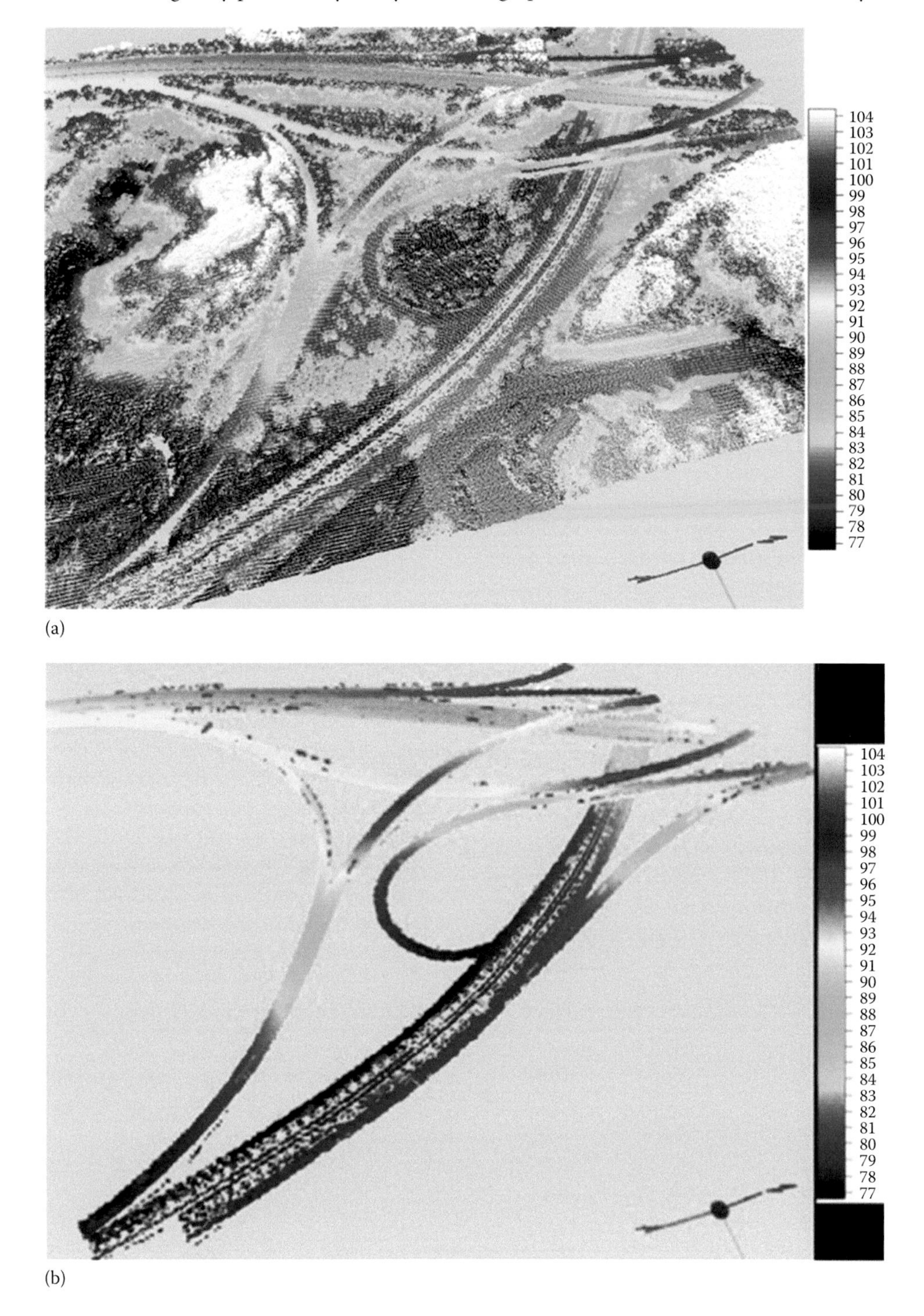

FIGURE 17.15 An example of road extraction: (a) LiDAR points, (b) road regions. *(Continued)*

(c)

FIGURE 17.15 (*Continued*) An example of road extraction: (c) road models.

increasingly available and first appeared on the commercial market in 2004 (Hug et al., 2004). Table 17.3 shows the technical specifications of waveform airborne LiDAR systems manufactured by Leica, Riegl, Optech, and Trimble. Such systems store the entire waveform of each received pulse, and thus the sum of reflections from all intercepted surfaces within the laser footprint is preserved. Compared to discrete-return airborne systems, which only provide a single range distance to a target, waveform systems record the entire time history of laser pulses with a high-resolution sampling interval. Moreover, such waveforms are able to store more details about illuminated targets than discrete systems are. Users can then apply their own pulse detection methods to identify return signals in a more effective manner. The waveform attributes, which represent characteristics of each scanned object, can then be extracted and used as criteria for distinguishing the points into different types.

17.6.2 Waveform Data Analysis

The waveform data record a series of return signals related to each individual laser shot. The data sampling interval is usually 1 ns, which corresponds to a 15 cm interval along the path of the laser beam. The intensity of the backscattered signal is then quantified within a range of specified digital levels (e.g., 8 bits), so that it is finally converted to a digital data stream. In general, it is understood that the higher the sampling rate and the signal-quantified levels are, the more details of the raw analogue signal that are kept.

As the waveform data contain valuable information about the object distribution and reflectance along the laser path, a specially designed process of waveform data analysis is needed to extract the information contained in the waveform data. Echo detection and waveform feature extraction are two common processes of waveform data analysis, which will both be

TABLE 17.3 Technical Specifications of Waveform Laser Scanners Produced by Leica, Riegl, Optech, and Trimble

Manufacturer	Leica ALS70	Leica ALS60	Riegl LMS Q780i	Riegl LMS Q680i	Optech ALTM Pegasus	Optech ALTM 3100	Trimble Harrier 68i	Trimble Harrier 56
Beam deflection	Oscillating mirror	Oscillating mirror	Rotating polygon	Rotating polygon	Oscillating mirror	Oscillating mirror	Rotating polygon	Rotating polygon
Flying height (m)	1600–5000	200–5000	<4700	<1600	300–2500	<2500	30–1600	<1000
Laser wavelength (nm)	1064	1064	1064	1550	1064	1064	1556	1550
Pulse width (ns)	5–6	5	3	3	7	8	3	<4
Pulse rate (kHz)	250–500	<50	100–400	<400	<400	<200	<400	<240
Beam divergence (mrad)	0.22	0.22	0.25	0.5	0.2	0.3 or 0.8	≤0.5	0.3 or 1.0
Field of view (degrees)	0–75	75	0–60	±30	75	±25	45–60	45 or 60
Sampling interval (ns)	1	1	1	1	1	1	1	1

introduced in this section. However, before this the shape of the received waveform is first discussed.

17.6.2.1 Shape of Received Waveform

As the shape of a received waveform varies subject to differences in the target structures (e.g., the distribution of illuminated objects and surface reflectance along the laser path), it is necessary to study the types of waveforms caused by some typical target structures. Based on the size of the focused surface geometry within the laser beam (footprint, *d*; wavelength, λ), Jutzi et al. (2005) categorized illuminated targets into three types, which were macro, meso, and micro structures. Macro structures are considered to have a more extended surface than the footprint *d* (e.g., building roofs). Meso structures are considered to have a less extended surface than the footprint *d*, but much greater than wavelength λ. Such structures result in a mixture of different range values, for example, in an area with small elevated objects, a slanted plane, large roughness, or vegetation. Finally, micro structures are considered to have a surface less extended than the wavelength λ. It is also noted that the shapes of the received waveforms vary depending on different surface structures illuminated, especially for meso structures.

Jutzi and Stilla (2006) further examined the relationship between surface structures and the shape of waveforms (Figure 17.16). A sloped surface or an area with randomly distributed small objects causes the deformation (widening) of the backscattered waveform with one distinct peak (Figure 17.16b and e). With an increase in the height difference between two elevated surfaces, the two peaks of a waveform are increasingly close together (Figure 17.16d).

Moreover, Wagner et al. (2006) pointed out that when the two scattering clusters are at a distance comparable to or smaller than the range resolution, which is the pulse duration of the LiDAR system, a waveform with a scattering cluster is presented. Jutzi and Stilla (2006) claimed that the waveform of overlapping responses at the distance of ≤0.85 times pulse width becomes a single widened peak and is unlikely to be separated as two returns by peak detection. It is also apparent that when considering the effects from various reflectance properties of multiple returns within the travel path of the laser pulse, the deformation of the received waveforms will be more complex than for those affected by surface geometry alone. As a result, since reflected laser pulses are distorted by surface variations within the footprint, the shapes of received waveforms important information in relation to surface roughness, slope, and reflectivity (Gardner, 1992).

17.6.2.2 Echo Detection

As full-waveform systems provide an entire time history of laser pulses, users are able to apply their own pulse detection methods to detect successive returns and determine the elapsed time. It even becomes possible to design a combined pulse detection method becomes possible to extract information of interest for specific applications (Wagner et al., 2004). For this purpose, Harding (2009) pointed out that the laser pulse width, detector sensitivity, and response time, the system's signal-to-noise performance, detection TH, and implementation of the ranging electronics all influence the ability to detect discrete returns. Wagner et al. (2004) reported that the characteristics of the effective scattering cross section, object distance, and noise level can affect the performance of the detectors. For received pulses with a complex form, the number and arrival timing of the returns are critically dependent on the detection methods employed.

Many echo detection algorithms are available for waveform data. Wagner et al. (2004) compared the results of using methods based on the TH, center of gravity (CG), MA, zero crossing of the first deviation (ZC), and constant fraction (CF). It was found that the ZC algorithm has the best discrimination with regard to detecting overlapping echoes among the five algorithms, especially when the noise is at minimum. The CF algorithm achieved worse results than the ZC algorithm, while the TH, CG, and MA algorithms failed at resolving overlapping echoes.

Current state-of-the-art echo detection techniques use small-footprint full-waveform data, for example, the correlation (Levanon and Mozeson, 2004), deconvolution (Jutzi and Stilla, 2006), wavelet-based (WB), and Gaussian decomposition methods. Since the waveforms can be considered as a convolution between system waveform and apparent cross section of

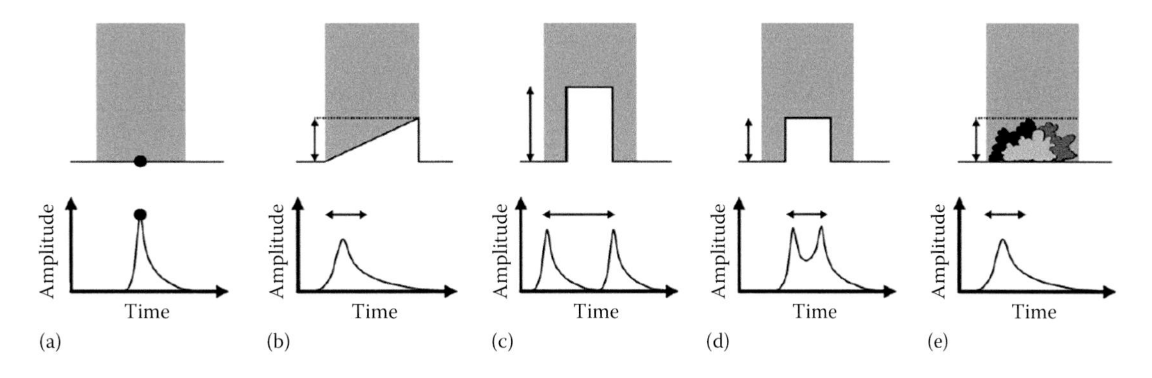

FIGURE 17.16 Surface structure and waveform: (a) plane surface, (b) sloped surface, (c) two significantly different elevated areas, (d) two slightly different elevated areas, and (e) randomly distributed small objects. (Adapted from Jutzi, B. and Stilla, U., *ISPRS J. Photogramm. Remote Sens.*, 61(2), 95, 2006.)

scatters (Wagner et al., 2006), the surface response (cross section of scatters) can be recovered by deconvolution algorithms (Jutzi and Stilla, 2006; Wu et al., 2011). However, prior knowledge such as the original laser pulse shape needs to be known, and sampling of the return signal also needs to be sufficient for such approaches to be applied, and for many of today's LiDAR systems, these prerequisites are not met (Morsdorf et al., 2009). The WB echo detector is based on a wavelet transformation algorithm, which measures the similarity between the signal and the wavelet. A wavelet (i.e., a Gaussian wavelet for detecting echoes of LiDAR waveforms) is thus chosen if its shape is similar to that of the echoes being detected. The wavelet can be scaled by a scale factor and shifted along the signal time domain by a translation factor, and so echoes can be located by detecting significant peaks in the wavelet coefficients (the results of wavelet transform). Gaussian decomposition is widely used to model small-footprint waveforms. With this method, it is assumed that pulses are transmitted with a Gaussian-like distribution (i.e., the impulse response is Gaussian) and that the received signal is a sum of individual Gaussian distributions (Hofton et al., 2000; Wagner et al., 2006). Fitting Gaussian functions to waveform data provides each return with a parametric description, which can be used to store pulse shape information and decrease the effect of noise. In this way, the Gaussian parameters for each return, which are the amplitude, temporal position of Gaussian peak, and Gaussian width, can be extracted.

The Gaussian model is a classical curve fitting algorithm (Wagner et al., 2006) as described in the following equation:

$$f_j(x_i) = A_j \exp\left(-\frac{(x-\mu)^2}{2W_j^2}\right) \tag{17.4}$$

where

A_j and W_j are the pulse amplitude and width of the laser pulse
μ_j is the center of the echo

Roncat et al. (2011) and Cheng and Tsai (2014) both proposed fitting waveforms into higher-order spline curves to better describe the laser waveform patterns. However, high-order spline curve fitting requires more intensive computation, and waveform fitting is usually based on the peak locations of the echo signals. However, since most full-waveform laser scanners provide only a limited number of signal peak locations, certain subtle but useful waveform features may be overlooked. Lu and Tsai (2013) thus adopted derivative analysis (Tsai and Philpot, 1998, 2002) to detect the occurrence of local peak positions more accurately and to generate a more complete waveform of the returned laser pulse.

17.6.2.3 Waveform Feature Extraction

With the development of full-waveform LiDAR, waveform-related characteristics have also become valuable features for LiDAR analysis and applications. Once the waveform has been decomposed, related waveform features can be derived from the results. Typical surface features extracted from Gaussian parameters are range, roughness, and reflectance (Jutzi and Stilla, 2005), and thus the estimated parameters provide a direct physical interpretation of the surface targets. Additionally, Wagner et al. (2006) adapted the radar equation order to convert the received power into a backscatter cross section. As this is a measure of how much energy is scattered backward toward the sensor and is very useful for comparing the physical quantities between different surveys, Wagner et al. (2008) suggested that backscatter cross sections should be also considered as a standard product in laser scanning. The related waveform attributes are as follows:

1. *Pulse width*: The pulse width can be described by the standard deviation W_j in Equation 17.4. It can be a quantitative indicator to evaluate the extent of the pulse broadening. Several studies into small-footprint waveforms have observed that the pulse width of vegetation points is generally larger than that of terrain points (Persson et al., 2005; Wagner et al., 2008). Forest terrain also tends to exhibit larger pulse widths than open terrain (Mücke, 2008), with trees and meadows also generating larger pulse widths than buildings (Stilla and Jutzi, 2008). The distinction of smooth surfaces from plants, bushes, trees, and even short hedges thus seems to be relatively clear (Wagner et al., 2008). This implies that pulse width can be an additional factor used to discriminate between ground returns and vegetation returns, although Stilla and Jutzi (2008) suggested that additional information, such as the 3D geometrical relationships of the returns, is required to classify each return pulse as a specific surface type.
2. *Amplitude*: The amplitude can be represented by A_j in Equation 17.4. The estimated amplitude for each return by postprocessing waveforms provides reflectance information about the illuminated targets. However, without calibration of such data, the information is too noisy to be utilized for classification purposes. It has been observed that there is a large overlap among the amplitude histograms obtained from different land-cover classes (Ducic et al., 2006; Mücke, 2008). At low amplitudes, the reliability for the estimate of pulse width can be problematic (Wagner et al., 2008). Since the emitted power is constant, pulse broadening of received returns also results in a reduction of the amplitude.
3. *Backscatter cross section*: The backscatter cross section is one of the unique features available from full-waveform LiDAR data. The backscatter cross section provides information in relation to the range and scattering properties of the targets (Wagner et al., 2006), and it depends strongly on the number of returns (Wagner et al., 2008). For each individual backscatter cross section for each return, it has been observed that single returns produce stronger estimates than multiple returns. Grass and gravel can be distinguished well, with a small overlay of cross-sectional values estimated. Vegetation returns generally produce

TABLE 17.4 Strengths/Limitations of Waveform and Multireturn LiDAR Point Cloud

Waveform Point Cloud	Multireturn Point Cloud
Number of return is unlimited.	Number of return is limited.
Detection of error or missing echoes become possible.	Echo detection is a black box operation. Detection of error or missing cannot be evaluated.
Dead zone problem can be eased by a user developed detector.	Multireturn system often suffers from LiDAR dead zone problem (overlapping echoes problem).
More features (waveform features) can be attached to the point cloud.	Only one additional information (intensity) can be attached to each point.

lower cross-sectional values than terrain returns, which may be useful to discriminate terrain from vegetation points (Wagner et al., 2008). In addition, the backscatter cross section varies within forest canopies, which may be useful for tree species classification (Wagner et al., 2008). The backscatter cross-sectional feature (σ) of a laser echo can be computed from the reflectance (ρ), range (R), and laser beam divergence angle (β) of the target (Wagner et al., 2006), as described in the following equation:

$$\sigma = \pi \rho R^2 \beta^2 \quad (17.5)$$

Equation 17.5 represents the general form of the apparent cross section of each surface within the laser footprint. For a Gaussian waveform, the equation can be rewritten as

$$\sigma = C_{cal} R^4 A_j W_j \quad (17.6)$$

$$C_{cal} = \frac{\rho \pi \beta^2}{R^2 A_j W_j} \quad (17.7)$$

where C_{cal} is a calibration constant that can be obtained through laboratory experiments (Alexander et al., 2010).

4. *Backscatter coefficient* γ: The backscatter coefficient is the backscatter cross section per unit-illuminated area. Alexander et al. (2010) found that the backscatter coefficient is more useful than the amplitude and backscatter cross section for discriminating road and grass. Wagner (2010) stated that it is helpful to employ the backscattering coefficient γ when comparing data sets acquired by different sensors and/or different flight campaigns.

17.6.3 Applications

Waveform LiDAR data has been used to explore weak and overlapping returns that were missed by the on-the-fly detection process of LiDAR systems. The point clouds extracted from the waveform data can then better describe the sensed landscape. The canopy height (Hancock et al., 2011; Wang et al., 2013), forest biomass (Clark et al., 2011; Yao et al., 2011), and DTM can thus be improved.

The land-cover classification is another important application of waveform LiDAR data. The waveform features introduced in Section 17.6.2.3 have been demonstrated in 3D point cloud classification. Integrations with other information (e.g., geometric and spectral information) provide even more criteria that can aid in achieving advanced applications, such as forestry investigation (Buddenbaum et al., 2013) and building modeling.

Comparing between waveform and multireturn LiDAR point clouds, the main applications are not much difference. For example, DEM generation, building reconstruction, and forest parameter estimation can be done by using both kinds of point clouds. However, a major improvement for waveform point cloud is that the landscape or the target can be described in more detail. Thus, the detection or the reconstruction of the interested targets can be better accomplished. Besides, the waveform feature is another advantage for waveform point cloud applications, especially on the purpose of point cloud classification. Table 17.4 lists the strengths and limitations of waveform LiDAR point cloud and multireturn point cloud.

17.7 DEM and DSM Generation from Airborne LiDAR Data

17.7.1 Introduction

A DEM is a digital model (or the height information) representing the ground surface of a terrain. In general, the ground surface means the earth's surface without any objects like plants, buildings, and other man-made structures. However, permanent earthwork structures, such as dams and road embankments, are usually counted as the ground surface. In contrast, a DSM, in general, represents the visual surface of the earth, including all the objects on it. However, in most cases, we do not treat the surface of a water body as part of DSM. Figure 17.17 shows the general definitions of DEM and DSM.

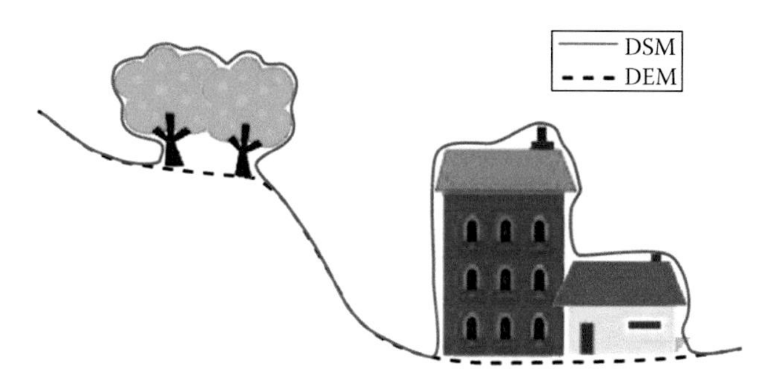

FIGURE 17.17 The general definitions of DEM and DSM.

A DEM or DSM can be represented as a raster (a grid of squares data structure) or as a vector-based TIN.

Airborne LiDAR data have proven effective for the generation of high-resolution and high-accuracy DEMs and DSMs of various terrestrial environments. In general, a DEM is formed with the ground points extracted from a data set of LiDAR point clouds, while a DSM is formed with the surface (first-return) points extracted from LiDAR data. The extracted ground and surface points are treated as measured points to form DEMs and DSMs (Mongus and Žalik, 2012).

Accurate DEMs are often used to derive morphometric parameters (e.g., slope, aspect, and surface ruggedness) for the morphological analysis of an area of interest. Morphological analysis is an important tool for characterizing environmental changes in both temporal and spatial scales. For example, the data derived from DEMs allow the identification of morphologically homogeneous zones, for example, the identification of different volcanic edifices, structural domains, and evolutionary stages (Norini et al., 2004). DEMs have been applied to reveal the variation in elevation of hills and river beds for landslide and inundation investigations focusing on Chenyulan River (Yang et al., 2011). High-resolution DEMs generated by LiDAR have now become more readily available, and advances in computing technology make it now possible to produce spatially explicit, fully distributed hydraulic and hydrological models and hydrogeomorphological assessments (Biron et al., 2013; Vaze et al., 2010).

17.7.2 Data Processing for DEM and DSM Generation

DEM generation usually employs a filter (or a classifier) to separate ground points from nonground points. Popular filters, such as those based on TINs (Axelsson, 2000) and morphology (Pingel et al., 2013), are mainly developed by exploiting the detailed geometrical description of the ground topology in LiDAR point clouds. For DSM generation, the surface points can be obtained by choosing all of the first-returning echoes of the LiDAR data. The process of DSM generation is much easier than that of DEM generation.

The overall data processing for DEM and DSM generation using airborne LiDAR data includes preprocessing, boresight calibration, and point cloud generation, strip adjustment, ground and surface point filtering, and interpolation to form DEM and DSM. A LiDAR data set may be contaminated by systematic errors, if the procedures of boresight calibration and strip adjustment are not well performed. The point density of the source data constrains the resolution of the generated DEMs and DSMs (Florinsky, 1998). Accordingly, the resolution of DEM and DSM is constrained mostly by the density of the input terrain data. McCullagh (1988) suggested that the number of grid cells should be similar to the number of terrain data in a covered area.

Since the extracted ground points are treated as measured points to form DEMs, the accuracy of the generated DEMs is subject to the correctness of ground point extraction. Under this circumstance, ground objects, which prevent laser beams from reaching the ground surface, may significantly affect the quality of DEMs. Given the decreased penetration capability of LiDAR systems in dense building or forest areas, the filtering process may mistakenly register nonground points as ground points, or vice versa. This error depends on the parameter setup of the specific filter employed. To remedy the misclassification of point clouds and obtain a reliable DEM, a manual editing procedure is conducted in the registration of point clouds to the correct class. However, this manual editing procedure requires visual inspection of the thin profiles of point clouds for the whole survey area and thus is extremely time consuming and costly.

Different levels of DEM quality can therefore be related to the LiDAR DEM generation procedures, which are subject to the processes of calibrating possible systematic errors, checking misclassification errors, and maintaining topographical features. In accordance with the typical demands of users, DEM products are usually categorized into three levels, as follows:

1. *Level 1 DEM*: DEMs generated by means of a fast and automatic procedure, without any quality control process, belong to this category. The products are not verified via a careful strategy for the removal of systematic errors and are thus likely to involve a number of mistakes. As the land-cover objects of the observed region become denser and more complicated, so the proportion of the inappropriate filtered points of the DEM could become higher.
2. *Level 2 DEM*: DEMs generated by means of a careful double-check procedure with a quality assessment and control process belong to this category. The procedures of LiDAR system calibration and strip adjustment are usually required to reduce the systematic errors. In addition, a procedure of manual editing is also needed to improve the quality of ground point extraction (filtering) and ensure the reliability of final products.
3. *Level 3 DEM*: If the level 2 DEMs are generated with consideration of water boundaries and geomorphological features, then they belong to this category. To meet this requirement, the filtered point cloud should be manually inspected and edited to improve the precision of interpolation by adding certain points of water boundaries and geomorphology features (e.g., ridges, valleys, steep slopes).

17.7.3 DEM Quality Assessment and Control

The process of DEM quality assessment and control (QA/QC) is to ensure the quality of the generated DEMs through a series of verification procedures. The quality of a DEM may be affected by three major factors, including the quality of the original LiDAR point cloud, the quality of the filtering, and the quality of the DEM interpolation (if applicable, including thinning and break line modeling) (Pfeifer and Mandlburger, 2008). These three factors affect the DEM quality at different stages of the generation procedure. The QA/QC must be performed step by step on each stage of the DEM generation workflow. For many practical projects, aerial photos of the survey area are also acquired during the LiDAR point cloud acquisition. An orthophoto of the survey area can then be produced based on these photos and the LiDAR point clouds, which can provide helpful visual clues for manual point cloud classification. The QA/QC procedure should thus take into account the use of aerial photos. An effective the DEM QA/QC procedure may include the following six items:

1. Verification of airborne LiDAR mission plan: to be performed before the scanning operation, including inspection of flight plan and LiDAR system calibration report. The nominal point density should be checked according to the flight plan.
2. Verification of ground control survey: the quantity, distribution, and quality of ground control points (GCPs) used for LiDAR strip adjustment.
3. Verification of LiDAR points and strip adjustment: to check the LiDAR strip coverage and overlaps, overall point density achieved, and external and internal errors of the strips.
4. Verification of LiDAR point cloud filtering: including visual inspections and field survey checks.
5. Verification of aerial photography acquisition: to check overall image coverage and quality.
6. Verification of orthophoto: to check the quality of the orthoimages, including the overall image continuity and possible distortion of ground objects.

The quality index of a DEM production depends on the requirements of the project. For general purposes, the quality index provided in the USGS LiDAR Base Specification Version 1.0 (Heidemann, 2012) is an appropriate reference.

At forest area, airborne LiDAR laser may penetrate the gap between leaves and detect the ground points under tree canopy, which is difficult to be obtained from aerial photos. Therefore, DEMs produced from LiDAR data may have higher accuracy than that from photogrammetry. At an earlier study, the height accuracy of DEM derived from airborne LiDAR data is about ±10 cm while that derived from photogrammetry is about ±25 cm (Kraus and Pfeifer, 1998).

17.8 Terrestrial/Vehicle-Borne LiDAR Data Processing

17.8.1 Terrestrial LiDAR Data Processing

Terrestrial laser scanners (TLSs) are stationary, ground-based LiDAR systems that scan surrounding objects. The data captured are also point clouds, which are distributed as layers corresponding to the surfaces of scanned objects. A growing number of applications, such as change detection and deformation analysis (Lim et al., 2005; Monserrat and Crosetto, 2008; Santibanez et al., 2012), rapid modeling of industrial infrastructure, architecture, agriculture, construction or maintenance of tunnels and mines, facility management, and urban and regional planning, have been carried out using TLS and demonstrated that it is a key surveying tool for capturing and modeling highly detailed geospatial data.

The main advantages of TLS are as follows: (1) the direct measurement of 3D coordinates, (2) the high degree of automation, (3) the easy-to-use hardware, and (4) a massive sampling capability (Monserrat and Crosetto, 2008).

17.8.1.1 Properties of Terrestrial LiDAR Data

The measurement mechanism of a TSL is similar to that of a total station, a surveying instrument used for angle and distance measurement. The difference is that a TSL allows automated and near-simultaneous measurements of hundreds or thousands of nonspecific points in the area surrounding the position where the instrument is set up. In Figure 17.1, a slant range by the laser rangefinder and two orthogonal angles by angular encoders in the horizontal and vertical directions are measured simultaneously. The point clouds are then captured point by point by rotating the reflecting mirrors whose horizontal/vertical angles are gradually increased, with a measuring rate of 1000 Hz or more. These simultaneous measurements of distance and angle are carried out in a highly automated manner (Petrie and Toth, 2008b).

The TSL can be mounted on a tripod for fixed positions or on a vehicle (or a moving platform) for mobile mapping. The first case also refers to static scanning, in which the exterior orientation of the platform is constant for one scan position, and 2D coverage in the angular domain is performed by rotating components of the device. In the second case, the scanning is performed by a univariate beam deflection unit, and area-wise data acquisition is established by the movement of the scanning platform, that is, dynamic scanning is achieved (Pfeifer and Briese, 2007). Many different kinds of static and dynamic TSLs that are used for topographic mapping and modeling operations are described in Petrie and Toth (2008b).

17.8.1.2 Data Processing Strategy

The 3D point cloud captured with a TLS contains detailed geometric information, but further data processing is still needed for certain applications. In general, the processing of TLS LiDAR data can be divided into three procedures: georeferencing, feature extraction, and visualization. The task of georeferencing is to transform the original point cloud, registered in a local coordinate system, into a ground reference coordinate system (GRCS). The task of feature extraction converts the point cloud data to be meaningful information for further processing, and the types of derivative features are subject to the involved applications, such as object classification and surface and building reconstruction. Visualization and interactive operations implemented on a computer are usually needed for manual interpretation and extraction of meaningful object features. The traditional field work of mapping surveys can now be performed on a computer system with the TLS data.

17.8.1.2.1 Registration and Georeferencing

The point clouds needed for 3D object modeling may be acquired from different scan stations, although the local point clouds must be transformed into a common coordinate system for further processing. If only the focal object itself is of interest, it is sufficient to determine the relative orientation between scans using registration. If the object also has to be placed in a superior coordinate system, absolute orientation is also needed. If the superior coordinate system is earth fixed, this then becomes a georeferencing task (Pfeifer and Briese, 2007).

One of the most popular methods of registration is the ICP (iterative closest point) algorithm developed by Besl and McKay (1992). Some variations and improvements based on ICP methods have been proposed by Chen and Medioni (1992), Zhang (1994), Okatani and Deguchi (2000), and Segal et al. (2009). The ICP is based on the search for pairs of nearest points between two scans, and the rigid body transformation is estimated and applied to the points of one scan. The ICP procedure is iterated until convergence is achieved. Another method for point cloud registration using least squares matching was proposed by Gruen and Akca (2005). The least squares 3D surface matching (LS3D) algorithm estimates the transformation parameters between two or more fully 3D surfaces, using the generalized Gauss–Markov model, minimizing the sum of the squares of the Euclidean distances between the surfaces. This formulation makes it possible to match arbitrarily oriented 3D surfaces simultaneously, without using explicit tie points.

The georeferencing of TLS data is a procedure of coordinate transformation in which the 3D point cloud is transformed to a GRCS, and the method of georeferencing used may be direct or indirect (dos Santos et al., 2013). The DG method is based on integrating additional POS sensors, such as a GNSS and an IMU, so that the platform's position and orientation at each moment of data acquisition can be determined accurately. The DG of the TLS data has been investigated primarily for terrestrial mobile mapping (Haala et al., 2008; Hunter et al., 2006; Talaya et al., 2004). In indirect georeferencing, a set of presurveyed GCPs are required to transform one or multiple scans into a superior coordinate system. In this case, the 3D similarity transformation model is regularly utilized, and the transformation parameters are estimated using least squares adjustment.

17.8.1.2.2 Point Cloud Organization and Segmentation

TSL scans contain detailed geometric information but still require interpretation of the data to make it useable for mapping purposes. Point cloud organization and segmentation are the early steps in LiDAR data processing, which are not directly linked to an application (Pfeifer and Briese, 2007).

Modern static and kinematic TLS have the ability to acquire point cloud data with large amount of points. A well-organized data structure for TLS point clouds will be helpful for the acceleration of the data storage, processing, and visualization. The commonly used data structures include TIN and grid and octree data structures (Elseberg et al., 2013; Wang and Tseng, 2011).

Segmentation and clustering can also be used to organize discrete points into homogeneous groups (Pfeifer and Briese, 2007), and many algorithms have been proposed for extracting planar surfaces from point clouds using segmentation methods for model reconstruction. Usually, one of three distinct methods is employed for segmenting points: region growing (Dold and Brenner, 2004; Hoffman et al., 2002; Pu and Vosselman, 2006), clustering of features (Biosca and Lerma, 2008; Filin, 2002; Filin and Pfeifer, 2006; Hoffman, 2004; Lerma and Biosca, 2005), or model fitting (Bauer et al., 2003; Boulaassal et al., 2007; Bretar and Roux, 2005).

17.8.1.2.3 Visualization

The visualization system is the foundation for several interactive analysis tools for quality control, extraction of survey measurements, and the extraction of isolated point cloud features (Kreylos et al., 2008). Staiger (2003) treated the visualization of point cloud data into six different ways:

1. Point clouds in a 3D projection, with a color- or grayscale-coded representation of intensity, are often used as a first visual check of the acquired data.
2. Point clouds can also be combined with derived geometrical elements.
3. *True* orthophotos are realized by the fusion of digital images (point information) and the registered point clouds (geometry).
4. 3D contour plans.
5. 3D models.
6. *Virtual flights* through the modeled scene.

17.8.2 Object Extraction from Vehicle-Borne LiDAR Data

Vehicle-borne LiDAR technology enables the real-time capture of high-resolution 3D spatial information, which is not possible with static terrestrial LiDAR scanning survey technology. This approach is an important supplement to photogrammetry and remote sensing.

Vehicle-borne and static terrestrial LiDAR data are different in three respects (Boulaassal et al., 2011). First, they are different in terms of the level of accuracy that can be obtained. Because the vehicle-borne LiDAR data require synchronization of the positioning and orientation components, the resulting accuracy is less than that of static LiDAR data. Secondly, because the system operates on a moving platform, the density and resolution of vehicle-borne LiDAR data are significantly lower than those seen with the stationary method, and so fewer details of objects are obtained. Moreover, the number of points acquired by vehicle-borne LiDAR systems is often higher than that acquired by several successive stationary terrestrial LiDAR stations.

Because vehicle-borne LiDAR systems are operated on roads when collecting data, they record numerous points of various objects on and nearby the road. This point data can then be used to reconstruct and model road surfaces, guardrails, pavements, utility facilities or polelike features (e.g., power poles, traffic sign poles, and light poles), building façades, bushes, and trees, and this is especially useful for obtaining details of objects in an urban area. The features on the road surface, for example, road markings, can also be extracted from these data.

Since vehicle-borne LiDAR systems move on a road, all the objects around the road that appear in the field of view of the scanner can reflect the laser beam and generate points, although this also means that the system has limited views in some directions. Consequently, the collected data may often not fully cover the target objects of interest, making feature extraction and identification very difficult. Furthermore, when the vehicle is moving at high speed during data collection, the point and scan line intervals may become rather large and make further data processing very complex. Certain strategies designed to handle vehicle-borne LiDAR data thus need to be adopted to extract, reconstruct, and model the objects along or nearby roads.

First, LiDAR data should be organized in advance using auxiliary data structures, such as scan lines, 2D grids, or 3D grids (voxels). The LiDAR data are then classified into point clusters of various objects, like road surfaces, building faces, and utility poles. It is often useful to classify LiDAR data into road surface points and nonroad surface points based on the knowledge of the actual scene. For example, the point height on road surfaces should usually be lower than that on nonroad surfaces. Additionally, most of the points collected from the road surface will have the same height, and the density of road points is higher than that seen in other places due to the small range from the scanned road surface to the scanner. The height of the road surface varies smoothly, with a very small deviation along the width of the road (Manandhar and Shibasaki, 2001). Based on the road surface points, the road surface can then be modeled by a TIN or road boundaries.

Nonroad surface points may be located on the surfaces of polelike features (e.g., power poles, traffic sign poles, and light poles), building façades, or trees. Some important knowledge about the objects in the scanning scene should thus be generalized in order to extract and model the objects from discrete and incomplete LiDAR data in a complicated scanning environment. In general, geometrical constraints for objects of interest are often employed. For example, the building façades should be vertical to the ground or at least be planar. To avoid the interference from irrelevant points with regard to the extraction of object points for modeling, the algorithms developed for extracting object points for object model reconstruction have to be robust, and knowledge about the scanner can sometimes be used to achieve this. For example, the scanning mechanism of the scanner can be considered. If the scan line is perpendicular to the ground, the points on the scan lines can be segmented into several vertical lines belonging to the walls. These vertical lines can then be grouped into one plane belonging to the wall.

Manandhar and Shibasaki (2001, 2002), Goulette et al. (2006), and Li et al. (2004) discussed the aforementioned concepts with regard to feature extraction, reconstruction, and modeling. More specifically, Manandhar and Shibasaki (2001, 2002) and Goulette et al. (2006) organized LiDAR data by scan lines before processing, while Li et al. (2004) organized LiDAR data into 2D grids before processing it. In addition to organizing LiDAR data with scan lines or 2D grids, additional information can be attributed into 2D grids for advanced classification. Douillard et al. (2009) developed a cell-wise semantic classification approach for ground cells. This involves the use of color imagery to classify cells into one of two classes: asphalt or grass. Classification of ground cells proceeds according to three main steps: (1) generation of ROIs in the image, (2) feature extraction within each ROI, and (3) feature-based classification.

In addition, organizing LiDAR data using voxels before further data processing is another good strategy. Aijazi et al. (2013) presented a method to classify urban scenes based on a supervoxel segmentation of sparse 3D data obtained from LiDAR sensors. The 3D point cloud is first segmented into voxels, which are then characterized by several attributes transforming them into supervoxels. These are joined together by using a link-chain method rather than the usual region growing algorithm to explore objects. These objects are then classified using geometrical models and local descriptors. Schmitt and Vogtle (2009) converted a raw irregular point cloud into regular voxels and then extracted planar features extracted by merging adjacent voxels with collinear normal vectors. Wu et al. (2013) presented a new voxel-based marked neighborhood searching (VMNS) method for efficiently identifying street trees and deriving their morphological parameters from vehicle-borne LiDAR point cloud data. The VMNS method consists of six technical components: voxelization, calculating values of voxels, searching and marking neighborhoods, extracting potential trees, deriving morphological parameters, and eliminating polelike objects other than trees. The method was validated and evaluated through two case studies, with the results showing that the completeness and correctness of the proposed method for street tree detection are both over 98%. The derived morphological parameters, including tree height, crown diameter, diameter at breast height (DBH), and crown base height, were also in a good agreement with the field measurements. This method provides an effective tool for extracting various morphological parameters for individual street trees from vehicle-borne LiDAR point cloud data.

17.9 Conclusions

LiDAR is a promising approach for the fast and robust acquisition of 3D information from scanned surfaces and has been widely used in various applications. For example, airborne LiDAR systems can obtain nationwide DSM and DEM data, while ground-based and mobile LiDAR systems have been used more often in urban areas for 3D city modeling.

Similar to all other remote sensing methods, careful handling and processing of LiDAR data are required to ensure high quality results. The standard procedures provided in this chapter should be suitable for general applications. However, for more specific applications, for example, very-high-density point clouds and DEM data for a very dense forest, it is recommended that consultations with professional service providers are carried out ahead of time, as some technical issues may arise that are not considered here.

References

Aijazi, A.K., Checchin, P., and Trassoudaine, L., 2013. Segmentation based classification of 3D urban point clouds: A super-voxel based approach with evaluation. *Remote Sensing*, 5(4): 1624–1650.

Alexander, C., Tansey, K., Kaduk, J., Holland, D., and Tate, N.J., 2010. Backscatter coefficient as an attribute for the classification of full-waveform airborne laser scanning data in urban areas. *ISPRS Journal of Photogrammetry and Remote Sensing*, 65(5): 423–432.

ASPRS, 2013. LASer (LAS) file format exchange activities, http://www.asprs.org/Committee-General/LASer-LAS-File-Format-Exchange-Activities.html (last date accessed: November 11, 2013).

Axelsson, P., 2000. DEM generation from laser scanner data using adaptive TIN models. *International Archives of Photogrammetry and Remote Sensing*, 33(Part B4): 110–117, Amsterdam, the Netherlands.

Baltsavias, E.P., 1999a. Airborne laser scanning: Basic relations and formulas. *ISPRS Journal of Photogrammetry and Remote Sensing*, 54(2–3): 199–214.

Baltsavias, E.P., 1999b. Airborne laser scanning: Existing systems and firms and other resources. *ISPRS Journal of Photogrammetry and Remote Sensing*, 54(2–3): 164–198.

Bauer, J., Karner, K., Klaus, A., Zach, C., and Schindler, K., 2003. Segmentation of building models from dense 3D point-clouds. *27th Workshop of the Ausman Association for Pattern Recognition,* 253–258.

Besl, P.J. and McKay, N.D., 1992. A method for registration of 3-D shapes. *IEEE Transactions on Pattern Analysis and Machine Intelligence*, 14(2): 239–256.

Biosca, J.M. and Lerma, J.L., 2008. Unsupervised robust planar segmentation of terrestrial laser scanner point clouds based on fuzzy clustering methods. *ISPRS Journal of Photogrammetry and Remote Sensing*, 63(1): 84–98.

Biron, P.M., Choné, G., Buffin-Bélanger, T., Demers, S., and Olsen, T., 2013. Improvement of streams hydro-geomorphological assessment using LiDAR DEMs. *Earth Surface Processes and Landforms*, 38(15): 1808–1821.

Boulaassal, H., Landes, T., and Grussenmeyer, P., 2011. 3D modelling of facade features on large sites acquired by vehicle based laser scanning. *Archives of Photogrammetry, Cartography and Remote Sensing edited by Polish Society for Photogrammetry and Remote Sensing*, 22: 215–226.

Boulaassal, H., Landes, T., Grussenmeyer, P., and Tarsha-Kurdi, F., 2007. Automatic segmentation of building facades using terrestrial laser data. *International Archives of Photogrammetry, Remote Sensing and Spatial Information Sciences*, 36(Part 3) : 65–70.

Brenner, C., 2005. Building reconstruction from images and laser scanning. *International Journal of Applied Earth Observation and Geoinformation*, 6(3–4): 187–198.

Brenner, M., Wichmann, V., and Rutzinger, M., 2013. Eigenvalue and Graph-based object extraction from mobile laser scanning point clouds, *ISPRS Annals of the Photogrammetry, Remote Sensing and Spatial Information Sciences*, II-5(W2): 55–60.

Bretar, F. and Roux, M., 2005. Hybrid image segmentation using LiDAR 3D planar primitives. *International Archives of Photogrammetry, Remote Sensing and Spatial Information Sciences*, 16: 72–78.

Bucksch, A., Lindenbergh, R., and Menenti, M., 2009. *SkelTre-Fast Skeletonisation for Imperfect Point Cloud Data of Botanic Trees, Eurographics Workshop on 3D Object Retrieval.* Eurographics, München, Germany, p. 8.

Buddenbaum, H., Seeling, S., and Hill, J., 2013. Fusion of full-waveform LiDAR and imaging spectroscopy remote sensing data for the characterization of forest stands. *International Journal of Remote Sensing*, 34(13): 4511–4524.

Burman, H., 2002. Laser strip adjustment for data calibration and verification. *International Archives of Photogrammetry and Remote Sensing*, 34(Part 3A/B): 67–72.

Chen, L., Teo, T., Kuo, C., and Rau, J., 2008. Shaping polyhedral buildings by the fusion of vector maps and LiDAR point clouds. *Photogrammetric Engineering and Remote Sensing*, 74(5): 1147–1157.

Chen, L.-C. and Lo, C.-Y., 2009. 3D road modeling via the integration of large-scale topomaps and airborne LIDAR data. *Journal of the Chinese Institute of Engineers*, 32(6): 811–823.

Chen, L.C., Teo, T.A., Hsieh, C.H., and Rau, J.Y., 2006. Reconstruction of building models with curvilinear boundaries from laser scanner and aerial Imagery. *Lecture Notes in Computer Science*, 4319: 24–33.

Chen, Q., Gong, P., Baldocchi, D., and Xie, G., 2007. Filtering Airborne Laser Scanning Data with Morphological Methods. *Photogrammetric Engineering and Remote Sensing*, 73(2): 175–185.

Chen, Y. and Medioni, G., 1992. Object modelling by registration of multiple range images. *Image and Vision Computing*, 10(3): 145–155.

Cheng, Y-H and F. Tsai, 2014. Spline curve fitting of full-waveform Lidar data and feature extraction for land-cover classification, in Proc. 2014 International Symposium on Remote Sensing, Apr. 16–18, 2014, Busan, Korea.

Chiu, C.M., Teo, T.A., and Chen, C.T., 2013. Three-dimensional modelling of multilayer road networks using road centerlines and airborne LiDAR data. *The International Symposium on Mobile Mapping Technology 2013*, May 1–3, Tainan, Taiwan, pp. CD-ROM.

Chuang, T.Y., 2012. *Feature-Based Registration of LiDAR Point Clouds*, National Taiwan University, Taipei, Taiwan.

Clark, M.L., Roberts, D.A., Ewel, J.J., and Clark, D.B., 2011. Estimation of tropical rain forest aboveground biomass with small-footprint LiDAR and hyperspectral sensors. *Remote Sensing of Environment*, 115(11): 2931–2942.

Csanyi, N. and Toth, C.K., 2007. Improvement of LIDAR data accuracy using LIDAR specific ground targets. *Photogrammetric Engineering and Remote Sensing*, 73(4): 385–396.

Dold, C. and Brenner, C., 2004. Automatic matching of terrestrial scan data as a basis for the generation of detailed 3D city models. *International Archives of Photogrammetry, Remote Sensing and Spatial Information Sciences*, 35(B3): 1091–1096.

dos Santos, D.R., Dal Poz, A.P., and Khoshelham, K., 2013. Indirect georeferencing of terrestrial laser scanning data using control lines. *The Photogrammetric Record*, 28(143): 276–292.

Douillard, B., Brooks, A., Ramos, F., and Durrant-Whyte, H., 2009. Combining laser and vision for 3D urban classification. *Proceedings of Neural Information Processing Systems Conference (NIPS),* Vancouver, B.C., Canada.

Ducic, V., Hollaus, M., Ullrich, A., Wagner, W., and Melzer, T., 2006. 3D vegetation mapping and classification using full-waveform laser scanning. *Workshop on 3D Remote Sensing in Forestry*, Vienna, Austria, pp. 211–217.

Elseberg, J., Borrmann, D., and Nüchter, A., 2013. One billion points in the cloud—An octree for efficient processing of 3D laser scans. *ISPRS Journal of Photogrammetry and Remote Sensing*, 76: 76–88.

Filin, S., 2001. Calibration of airborne and spaceborne laser altimeters using natural surfaces. PhD thesis, The Ohio State University, Columbus, OH, 140pp.

Filin, S., 2002. Surface clustering from airborne laser scanning data. *International Archives of Photogrammetry and Remote Sensing*, 34(Part 3A/B): 119–124.

Filin, S., 2003. Recovery of systematic biases in laser altimeters using natural surfaces. *International Archives of Photogrammetry and Remote Sensing*, 35(WG III/3), Graz, Austria, 65–70.

Filin, S., 2004. Surface classification from airborne laser scanning data. *Computers & Geosciences*, 30: 1033–1041.

Filin, S. and Pfeifer, N., 2006. Segmentation of airborne laser scanning data using a slope adaptive neighborhood. *ISPRS Journal of Photogrammetry and Remote Sensing*, 60(2): 71–80.

Filin, S. and Vosselman, G., 2004. Adjustment of airborne laser altimetry strips. *International Archives of Photogrammetry and Remote Sensing*, 35(B3): 285–289.

Fischler, M.A. and Bolles, R.C., 1981. Random sample consensus: A paradigm for model fitting with applications to image analysis and automated cartography. *Communications of the ACM*, 24(6): 381–395.

Florinsky, I.V., 1998. Combined analysis of digital terrain models and remotely sensed data in landscape investigations. *Progress in Physical Geography*, 22(1): 33–60.

Fornaciai, A., Bisson, M., Landi, P., Mazzarini, F., and Pareschi, M.T., 2010. A LiDAR survey of Stromboli volcano (Italy): Digital elevation model-based geomorphology and intensity analysis. *International Journal of Remote Sensing*, 31(12): 3177–3194.

Gardner, C.S., 1992. Ranging performance of satellite laser altimeters. *IEEE Transactions on Geoscience and Remote Sensing*, 30(5): 1061–1072.

Glennie, C., 2007. Rigorous 3D error analysis of kinematic scanning LIDAR systems. *Journal of Applied Geodesy*, 1(3): 147–157.

Gonzalez, R.C. and Woods, R.E., 1992. *Digital Image Processing*, Addison-Wisley, Reading, MA, 716pp.

Gorte, B. and Pfeifer, N., 2004. Structuring laser-scanned trees using 3D mathematical morphology. *International Archives of Photogrammetry and Remote Sensing*, 35(B5): 929–933.

Goulette, F., Nashashibi, F., Abuhadrous, I., Ammoun, S., and Laurgeau, C., 2006. An integrated on-board laser range sensing system for on-the-way city and road modelling. *Proceedings of the ISPRS Commission I Symposium, "From Sensors to Imagery,"* Paris, France, Vol. 3, p. 43.

Gross, H. and Thoennessen, U., 2006. Extraction of lines from laser point clouds. *International Archives of Photogrammetry, Remote Sensing and Spatial Information Sciences,* Vol. 36, pp. 86–91.

Gruen, A., 2005. Towards photogrammetry 2025. *Photogrammetric Week*, Stuttgart, Germany.

Gruen, A. and Akca, D., 2005. Least squares 3D surface and curve matching. *ISPRS Journal of Photogrammetry and Remote Sensing*, 59(3): 151–174.

Guo, L., Chehata, N., Mallet, C., and Boukir, S., 2011. Relevance of airborne LiDAR and multispectral image data for urban scene classification using Random Forests. *ISPRS Journal of Photogrammetry and Remote Sensing*, 66(1): 56–66.

Haala, N., Peter, M., Kremer, J., and Hunter, G., 2008. Mobile LiDAR mapping for 3D point cloud collection in urban areas—A performance test. *The International Archives of the Photogrammetry, Remote Sensing and Spatial Information Sciences*, 37: 1119–1127.

Habib, A., Al-Durgham, M., Kersting, A., and Quackenbush, P., 2008. Error budget of LiDAR systems and quality control of the derived point cloud. *Proceedings of the XXI ISPRS Congress, Commission I*, Vol. 37, Beijing, China, pp. 203–209.

Habib, A., Bang, K., Kersting, A.P., and Lee, D.-C., 2009. Error budget of LiDAR systems and quality control of the derived data. *Photogrammetric Engineering and Remote Sensing*, 75(9): 1093–1108.

Habib, A., Ghanma, M., Morgan, M., and Al-Ruzouq, R., 2005. Photogrammetric and LiDAR data registration using linear features. *Photogrammetric Engineering and Remote Sensing*, 71(6): 699–707.

Habib, A., Kersting, A.P., Bang, K.I., and Lee, D.-C., 2010. Alternative methodologies for the internal quality control of parallel LiDAR strips. *IEEE Transactions on Geoscience and Remote Sensing*, 48(1): 221–236.

Hancock, S., Disney, M., Muller, J.P., Lewis, P., and Foster, M., 2011. A threshold insensitive method for locating the forest canopy top with waveform LiDAR. *Remote Sensing of Environment*, 115(12): 3286–3297.

Harding, D., 2009. Pulsed laser altimeter ranging techniques and implications for terrain mapping. In: J. Shan and C.K. Toth (eds.), *Topographic Laser Ranging and Scanning Principles and Processing*, CRC Press, Boca Raton, FL, pp. 173–194.

Heidemann, H.K., 2012. *Lidar Base Specification Version 1.0.* US Geological Survey Techniques and Methods, U.S. Geological Survey, Reston, VA, 63pp.

Heinzel, J. and Koch, B., 2011. Exploring full-waveform LiDAR parameters for tree species classification. *International Journal of Applied Earth Observation and Geoinformation*, 13(1): 152–160.

Hoffman, A.D., 2004. Analysis of TIN-structure parameter spaces in airborne laser scanner data for 3-D building model generation. *International Archives of Photogrammetry, Remote Sensing and Spatial Information Sciences*, 35(Part B3): 302–307.

Hoffman, A.D., Maas, H.G., and Streilein, A., 2002. Knowledge-based building detection based on laser scanner data and topographic map information. *International Archives of Photogrammetry and Remote Sensing*, 34(Part 3A/B) September 9–13, Graz, Austria, pp. 169–174.

Hofton, M.A., Minster, J.B., and Blair, J.B., 2000. Decomposition of laser altimeter waveforms. *IEEE Transactions on Geoscience and Remote Sensing*, 38(4): 1989–1996.

Hoover, A. et al., 1996. An experimental comparison of range image segmentation algorithms. *IEEE Transactions on Pattern Analysis and Machine Intelligence*, 18(7): 673–689.

Hu, Y., 2003. Automated extraction of digital terrain models, roads and buildings using airborne LiDAR data. PhD thesis, University of Calgary, Calgary, Alberta, Canada, 223pp.

Hug, C., Ullrich, A., and Grimm, A., 2004. Litemapper-5600—A waveform-digitizing LiDAR terrain and vegetation mapping system. *International Archives of Photogrammetry, Remote Sensing and Spatial Information Sciences*, 36(Part 8): W2.

Huising, E.J. and Pereira, L.M.G., 1998. Errors and accuracy estimates of laser data acquired by various laser scanning systems for topographic application. *ISPRS Journal of Photogrammetry and Remote Sensing*, 53(5): 245–261.

Hunter, G., Cox, C., and Kremer, J., 2006. Development of a commercial laser scanning mobile mapping system–StreetMapper. *International Archives of the Photogrammetry, Remote Sensing and Spatial Information Sciences*, 36.

Im, J., Jensen, J.R., and Hodgson, M.E., 2008. Object-based land cover classification using high-posting-density LiDAR data. *GIScience & Remote Sensing*, 45(2): 209–228.

Isenburg, M., 2013. LASzip: Lossless compression of LiDAR data. *Photogrammetric Engineering and Remote Sensing*, 79(2): 209–217.

Jiang, X. and Bunke, H., 1994. Fast segmentation of range images into planar regions by scan line grouping. *Machine Vision and Applications*, 7(2): 115–122.

Jutzi, B., Neulist, J., and Stilla, U., 2005. High-resolution waveform acquisition and analysis of laser pulses. *Measurement Techniques*, 2(3.1): 2.

Jutzi, B. and Stilla, U., 2005. Measuring and processing the waveform of laser pulses. *In: Proc. 7th Optical 3-D Measurement Techniques. FIG/IAG/ISPRS*, Vienna, Austria, *3–5* October *2005, pp. 194–203.*

Jutzi, B. and Stilla, U., 2006. Range determination with waveform recording laser systems using a Wiener Filter. *ISPRS Journal of Photogrammetry and Remote Sensing*, 61(2): 95–107.

Kager, H., 2004. Discrepancies between overlapping laser scanner strips—Simultaneous fitting of aerial laser scanner strips. *International Archives of Photogrammetry, Remote Sensing and Spatial Information Sciences*, 35(B1): 555–560.

Krabill, W. et al., 2000. Airborne laser II.'Assateague National Seashore Beach. *Photogrammetric Engineering & Remote Sensing*, 66(1): 65–71.

Kraus, K. and Pfeifer, N., 1998. Determination of terrain models in wooded areas with airborne laser scanner data. *ISPRS Journal of Photogrammetry and Remote Sensing*, 53(4): 193–203.

Kreylos, O., Bawden, G., and Kellogg, L., 2008. Immersive visualization and analysis of LiDAR data. *Advances in Visual Computing*, 5358: 846–855.

Latypov, D. and Zosse, E., 2002. LIDAR data quality control and system calibration using overlapping flight lines in commercial environment. *ACSM-ASPRS 2002 Annual Conference*, April 22–26, Washington, DC.

Lerma, J. and Biosca, J., 2005. Segmentation and filtering of laser scanner data for cultural heritage. *CIPA 2005 XX International Symposium*, Torino, Italy.

Levanon, N. and Mozeson, E., 2004. *Radar Signals*, John Wiley & Sons, Hoboken, NJ.

Lewis, P., McElhinney, C.P., and McCarthy, T., 2012. LiDAR data management pipeline; from spatial database population to web-application visualization. *Proceedings of the Third International Conference on Computing for Geospatial Research and Applications*, ACM, New York, NY, p. 16.

Li, B., Li, Q., Shi, W., and Wu, F., 2004. Feature extraction and modeling of urban building from vehicle-borne laser scanning data. *International Archives of Photogrammetry, Remote Sensing and Spatial Information Sciences*, 35: 934–939.

Lichti, D.D. and Licht, M.G., 2006. Experiences with terrestrial laser scanner modelling and accuracy assessment. *International Archives of the Photogrammetry, Remote Sensing and Spatial Information Sciences*, 36(5): 155–160.

Lim, M. et al., 2005. Combined digital photogrammetry and time-of-flight laser scanning for monitoring cliff evolution. *The Photogrammetric Record*, 20(110): 109–129.

Lu, Y.-H. and Tsai, F., 2013. Analysis of full-waveform LiDAR data for land-cover classification. 2013 International Symposium on Remote Sensing May 15–17, Chiba, Japan, pp. CDROM.

Mallet, C. and Bretar, F., 2009. Full-waveform topographic LiDAR: State-of-the-art. *ISPRS Journal of Photogrammetry and Remote Sensing*, 64(1): 1–16.

Manandhar, D. and Shibasaki, R., 2001. Feature extraction from range data. *Paper presented at the 22nd Asian Conference on Remote Sensing*, Vol. 5, Singapore, p. 9.

Manandhar, D. and Shibasaki, R., 2002. Auto-extraction of urban features from vehicle-borne laser data. International Archives of Photogrammetry Remote Sensing And Spatial Information Sciences, 34(4): 650–655.

Mazzarini, F. et al., 2007. Lava flow identification and aging by means of LiDAR intensity: Mount Etna case. *Journal of Geophysical Research: Solid Earth (1978–2012)*, 112(B2).

McCullagh, M., 1988. Terrain and surface modelling systems: Theory and practice. *The Photogrammetric Record*, 12(72): 747–779.

Melzer, T., 2007. Non-parametric segmentation of ALS point clouds using mean shift. *Journal of Applied Geodesy*, 1(3): 159–170.

Meng, X., Currit, N., and Zhao, K., 2010. Ground filtering algorithms for airborne LiDAR data: A review of critical issues. *Remote Sensing*, 2(3): 833–860.

Mongus, D. and Žalik, B., 2012. Parameter-free ground filtering of LiDAR data for automatic DTM generation. *ISPRS Journal of Photogrammetry and Remote Sensing*, 67: 1–12.

Monserrat, O. and Crosetto, M., 2008. Deformation measurement using terrestrial laser scanning data and least squares 3D surface matching. *ISPRS Journal of Photogrammetry and Remote Sensing*, 63(1): 142–154.

Morin, K.W., 2002. Calibration of airborne laser scanners. MSc thesis, University of Calgary, Calgary, Alberta, Canada, 134pp.

Morsdorf, F., Nichol, C., Malthus, T., and Woodhouse, I.H., 2009. Assessing forest structural and physiological information content of multi-spectral LiDAR waveforms by radiative transfer modelling. *Remote Sensing of Environment*, 113(10): 2152–2163.

Mücke, W., 2008. Analysis of full-waveform airborne laser scanning data for the improvement of DTM generation. MSc thesis, Institute of Photogrammetry and Remote Sensing, Vienna University of Technology, Vienna, Austria.

Neuenschwander, A.L., Magruder, L.A., and Tyler, M., 2009. Landcover classification of small-footprint, full-waveform LiDAR data. *Journal of Applied Remote Sensing*, 3(1).

Norini, G., Groppelli, G., Capra, L., and De Beni, E., 2004. Morphological analysis of Nevado de Toluca volcano (Mexico): New insights into the structure and evolution of an andesitic to dacitic stratovolcano. *Geomorphology*, 62(1): 47–61.

OGC, 2012. OGC City Geography Markup Language (CityGML) Encoding Standard, Version 2.0.

Okatani, I.S. and Deguchi, K., 2002. A method for fine registration of multiple view range images considering the measurement error properties. *Computer Vision and Image Understanding*, 87(1–3): 66–77.

Oude Elberink, S.J. and Vosselman, G., 2009. 3D information extraction from laser point clouds covering complex road junctions. *The Photogrammetric Record*, 24(125): 23–36.

Persson, Å., Söderman, U., Töpel, J., and Ahlberg, S., 2005. Visualization and analysis of full-waveform airborne laser scanner data. *International Archives of Photogrammetry, Remote Sensing and Spatial Information Sciences*, 36(3/W19): 103–108.

Petrie, G. and Toth, C.K., 2008a. Airborne and spaceborne laser profilers and scanners. In: J. Shan and C.K. Toth (eds.), *Topographic Laser Ranging and Scanning Principles and Processing*, CRC Press, Boca Raton, FL, pp. 29–85.

Petrie, G. and Toth, C.K., 2008b. Terrestrial laser scanners. In: J. Shan and C.K. Toth (eds.), *Topographic Laser Ranging and Scanning Principles and Processing*, CRC Press, Boca Raton, FL, pp. 87–127.

Pfeifer, N. and Briese, C., 2007. Geometrical aspects of airborne laser scanning and terrestrial laser scanning. *International Archives of Photogrammetry, Remote Sensing and Spatial Information Sciences*, 36(3/W52): 311–319.

Pfeifer, N. and Mandlburger, G., 2008. LiDAR data filtering and DTM generation. In: J. Shan and C.K. Toth (eds.), *Topographic Laser Ranging and Scanning Principles and Processing*, CRC Press, Boca Raton, FL, pp. 307–333.

Pingel, T.J., Clarke, K.C., and McBride, W.A., 2013. An improved simple morphological filter for the terrain classification of airborne LIDAR data. *ISPRS Journal of Photogrammetry and Remote Sensing*, 77: 21–30.

Pu, S. and Vosselman, G., 2006. Automatic extraction of building features from terrestrial laser scanning. *International Archives of Photogrammetry, Remote Sensing and Spatial Information Sciences*, 36(Part 5): 5pp. (on CD-ROM).

Pu, S. and Vosselman, G., 2009a. Building facade reconstruction by fusing terrestrial laser points and images. *Sensors*, 9(6): 4525–4542.

Pu, S. and Vosselman, G., 2009b. Knowledge based reconstruction of building models from terrestrial laser scanning data. *ISPRS Journal of Photogrammetry and Remote Sensing*, 64(6): 575–584.

Rabbani, T., 2006. Automatic reconstruction of industrial installations using point clouds and images. PhD thesis, Delft University of Technology, Delft, the Netherlands, 175pp.

Rabbani, T., van den Heuvel, F.A., and Vosselman, G., 2006. Segmentation of point clouds using smoothness constraint. *International Archives of Photogrammetry, Remote Sensing and Spatial Information Sciences*, 36(5): 248–53.

Reshetyuk, Y., 2006. Calibration of terrestrial laser scanners Callidus 1.1, Leica HDS 3000 and Leica HDS 2500. *Survey Review*, 38(302): 703–713.

Roncat, A., Bergauer, G. and Pfeifer, N., 2011. B-spline deconvolution for differential target cross-section determination in full-waveform laser scanning data. ISPRS Journal of Photogrammetry and Remote Sensing, 66(4): 418–428.

Rottensteiner, F., Trinder, J., Clode, S., and Kubik, K., 2005. Using the Dempster–Shafer Method for the fusion of LIDAR data and multi-spectral images for building detection. *Information Fusion*, 6(4): 283–300.

Rutzinger, M., Elberink, S.O., Pu, S., and Vosselman, G., 2009. Automatic extraction of vertical walls from mobile and airborne laser scanning data. *The International Archives of Photogrammetry, Remote Sensing and Spatial Information Sciences*, 38(Part 3): W8.

Samet, H., 1990. *Applications of Spatial Data Structures: Computer Graphics, Image Processing, and GIS*, Addison-Wesley, Reading, MA, 507pp.

Sampath, A. and Shan, J., 2007. Building boundary tracing and regularization from airborne LiDAR point clouds. *Photogrammetry Engineering & Remote Sensing*, 73(7): 805–812.

Santibanez, S.F., dos Santos, D.R., and Faggion, P.L., 2012. Influence of fitting models and point density sample in the detection of deformations of structures using terrestrial laser scanning. *Applied Geomatics*, 4(1): 11–19.

Schenk, T., 2001. Modeling and analyzing systematic errors in airborne laser scanners. Technical Report. Department of Civil and Environmental Engineering and Geodetic Science, The Ohio State University, OH.

Schmitt, A. and Vogtle, T., 2009. An advanced approach for automatic extraction of planar surfaces and their topology from point clouds. *Photogrammetrie-Fernerkundung-Geoinformation*, 2009(1): 43–52.

Schuster, H.-F., 2004. Segmentation of LIDAR data using the tensor voting framework. *International Archives of Photogrammetry, Remote Sensing and Spatial Information Sciences*, 35: 1073–1078.

Segal, A., Haehnel, D., and Thrun, S., 2009. Generalized-ICP. *Robotics: Science and Systems*, 2: 4.

Shakarji, C.M., 1998. Least-squares fitting algorithms of the NIST algorithm testing system. *Journal of Research of National Institute of Standards and Technology*, 103: 633–641.

Sithole, G., 2005. Segmentation and classification of airborne laser scanner data. PhD thesis, Delft University of Technology, Delft, the Netherlands, 203pp.

Skaloud, J. and Lichti, D., 2006. Rigorous approach to bore-sight self-calibration in airborne laser scanning. *ISPRS Journal of Photogrammetry and Remote Sensing*, 61(1): 47–59.

Staiger, R., 2003. Terrestrial laser scanning technology, systems and applications. *Second FIG Regional Conference Marrakech*, Marrakech, Morocco, p. 1.

Stilla, U. and Jutzi, B., 2008. Waveform analysis for small-footprint pulsed laser systems. In: J. Shan and C.K. Toth (eds.), *Topographic Laser Ranging and Scanning Principles and Processing*, CRC Press, Boca Raton, FL, pp. 215–234.

Talaya, J. et al., 2004. Integration of a terrestrial laser scanner with GPS/IMU orientation sensors. *Proceedings of the XXth ISPRS Congress*, Vol. 35, Istanbul, Turkey, pp. 1049–1055.

Tao, C.V. and Li, J., 2007. *Advances in Mobile Mapping Technology*, ISPRS Series, Vol. 4, Taylor & Francis: London, UK.

Tsai, F. and Philpot, W., 1998. Derivative analysis of hyperspectral data. *Remote Sensing of Environment*, 66(1): 41–51.

Tsai, F. and Philpot, W.D., 2002. A derivative-aided hyperspectral image analysis system for land-cover classification. *IEEE Transactions on Geoscience and Remote Sensing*, 40(2): 416–425.

Vaze, J., Teng, J., and Spencer, G., 2010. Impact of DEM accuracy and resolution on topographic indices. *Environmental Modelling & Software*, 25(10): 1086–1098.

Vosselman, G., 2009. Advanced point cloud processing. *Photogrammetric Week*, Enschede, the Netherlands, 9, pp. 137–146.

Vosselman, G., Gorte, B.G.H., Sithole, G., and Rabbani, T., 2004. Recognising structure in laser scanner point clouds. *International Archives of Photogrammetry and Remote Sensing*, 36(Part 8/W2): 33–38.

Vosselman, G. and Dijkman, S., 2001. 3D building model reconstruction from point clouds and ground plans. *International Archives of Photogrammetry and Remote Sensing*, 34(3/W4): 37–43, Annapolis, MD.

Wagner, W., 2010. Radiometric calibration of small-footprint full-waveform airborne laser scanner measurements: Basic physical concepts. *ISPRS Journal of Photogrammetry and Remote Sensing*, 65(6): 505–513.

Wagner, W., Hollaus, M., Briese, C., and Ducic, V., 2008. 3D vegetation mapping using small-footprint full-waveform airborne laser scanners. *International Journal of Remote Sensing*, 29(5): 1433–1452.

Wagner, W., Ullrich, A., Ducic, V., Melzer, T., and Studnicka, N., 2006. Gaussian decomposition and calibration of a novel small-footprint full-waveform digitising airborne laser scanner. *ISPRS Journal of Photogrammetry and Remote Sensing*, 60(2): 100–112.

Wagner, W., Ullrich, A., Melzer, T., Briese, C., and Kraus, K., 2004. From single-pulse to full-waveform airborne laser scanners: Potential and practical challenges. *International Archives of Photogrammetry and Remote Sensing*, 35(B3): 201–206.

Wang, C. et al., 2013. Wavelet analysis for ICESat/GLAS waveform decomposition and its application in average tree height estimation. *IEEE Geoscience and Remote Sensing Letters*, 10(1): 115–119.

Wang, M. and Tseng, Y.-H., 2010. Automatic segmentation of LiDAR data into coplanar point clusters using an octree-based split-and-merge algorithm. *Photogrammetric Engineering and Remote Sensing*, 76(4): 407–420.

Wang, M. and Tseng, Y.-H., 2011. Incremental segmentation of LiDAR point clouds with an octree-structured voxel space. *The Photogrammetric Record*, 26(133): 32–57.

Weingarten, J.W., Gruener, G., and Siegwart, R., 2004. Probabilistic plane fitting in 3D and an application to robotic mapping. *Proceedings ICRA'04. 2004 IEEE International Conference on Robotics and Automation*, Vol. 1, pp. 927–932.

Wu, B. et al., 2013. A voxel-based method for automated identification and morphological parameters estimation of individual street trees from mobile laser scanning data. *Remote Sensing*, 5(2): 584–611.

Wu, J.Y., van Aardt, J.A.N., and Asner, G.P., 2011. A comparison of signal deconvolution algorithms based on small-footprint LiDAR waveform simulation. *IEEE Transactions on Geoscience and Remote Sensing*, 49(6): 2402–2414.

Yang, M.-D. et al., 2011. Landslide-induced levee failure by high concentrated sediment flow—A case of Shan-An levee at Chenyulan River, Taiwan. *Engineering Geology*, 123(1): 91–99.

Yao, T. et al., 2011. Measuring forest structure and biomass in New England forest stands using Echidna ground-based LiDAR. *Remote Sensing of Environment*, 115(11): 2965–2974.

Zhang, K. and Whitman, D., 2005. Comparison of three algorithms filtering airborne LiDAR data. *Photogrammetric Engineering and Remote Sensing*, 71(3): 313–324.

Zhang, Z., 1994. Iterative point matching for registration of free-form curves and surfaces. *International Journal of Computer Vision*, 13(2): 119–152.

Zhao, T. and Wang, J., 2014. Use of LiDAR-derived NDTI and intensity for rule-based object-oriented extraction of building footprints. *International Journal of Remote Sensing*, 35(2): 578–597.

Zwally, H. et al., 2002. ICESat's laser measurements of polar ice, atmosphere, ocean, and land. *Journal of Geodynamics*, 34(3): 405–445.

VI

Change Detection

18

Change Detection Techniques Using Multisensor Data

Daniela Anjos
National Institute for Space Research

Dengsheng Lu
Michigan State University

Luciano Dutra
National Institute for Space Research

Sidnei Sant'Anna
National Institute for Space Research

Acronyms and Definitions

AG	Agriculture
AG–SS1	Agriculture to initial secondary succession
BS	Bare soil
BS–AG	Bare soil to agriculture
CP	Clean pasture
CP–DP	Clean pasture to dirty pasture
CP–SS1	Clean pasture to initial secondary succession
DP	Dirty pasture
DP–CP	Dirty pasture to clean pasture
DP–SS1	Dirty pasture to initial secondary succession
EO-1 ALI	Earth Observing-1 Advanced Land Imager instrument
ETM+	Enhanced Thematic Mapper Plus
FQ13	Fine-resolution quad-polarization beam mode 13
HH	Single polarization, transmit and receive horizontal signal
HV	Cross polarization, transmit horizontal signal and receive vertical signal
IS	Impervious surface
MODIS	Moderate Resolution Imaging Spectroradiometer
MSS	Multispectral scanner
PF	Primary forest
SPOT HRG	Satellite Pour l'Observation de la Terre—high resolution geometrical
SPOT XS	Satellite Pour l'Observation de la Terre—multispectral
SS1	Initial secondary succession
SS1–BS	Initial secondary succession to bare soil
SS1–SS2	Initial secondary succession to intermediate secondary succession
SS2	Intermediate secondary succession
SVM	Support vector machine
TM	Thematic Mapper
UTM	Universal Transverse Mercator
VH	Cross polarization, transmit vertical signal and receive horizontal signal
VV	Single polarization, transmit and receive vertical signal
WT	Water

18.1 Introduction

Land-cover change has long been an active research topic because of its impacts on climate and environmental changes, as well as its close interactions with human activities (Lu et al., 2014; Volpi et al., 2013). Early publications related to change detection occurred in the 1960s. Lillestrand (1972) and Singh (1989) overviewed the major change detection techniques developed between the 1960s and 1980s. Since then, change detection techniques have been applied to different fields such as forest change (Collins and Woodcock, 1996; Desclée et al., 2006; Huang et al., 2010; Reiche et al., 2013), urban expansion (Borghys et al., 2007; Du et al., 2012; Lo and Shipman, 1990; Yeh and Li, 2001), wetland change (Baker et al., 2007; Houhoulis and Michener, 2000), and even biomass change (Coppin et al., 2001). A large number of change detection techniques have been developed and are summarized in the literature (e.g., Bhagat, 2012; Chen et al., 2012; Coppin et al., 2004; Hussain et al., 2013; Lu et al., 2004, 2014).

Change detection is defined as the process of identifying changes in the state of an object or phenomenon at different times (Lu et al., 2004; Singh, 1989). This process is critical to understand the relationships and interactions between human and natural phenomena to promote better decision making in various situations. Identification and analysis of changes occurring in a particular region during a time interval require use of multitemporal datasets. Because of the unique characteristics of remote sensing data in data collection and presentation, it has become the main data source for detecting land-cover change (Chen et al., 2012). The basic premise of using remote sensing data is that changes in land covers are expressed in changes of the responses obtained by passive or active sensors (Mas, 1999). However, these changes must be separated from changes in spectral responses caused by other factors such as weather, view angle, and moisture content at the time of image acquisition (Singh, 1989).

In general, change detection applications are directly related to the concepts of land-use/land-cover change. In this context, it is noteworthy that there is an important distinction between the terms *land use* and *land cover*. The term *land cover* refers only to the physical footprints in a region, and *land use* refers to human activities performed in a place, consisting of a single coverage or a mosaic for a particular purpose (Barnsley et al., 2001). Treitz and Rogan (2004) claim that land use is an abstract concept related to social, cultural, economic, and political factors, and it is weakly associated with the reflectance properties in remote sensing. Therefore, change detection handles the land-cover changes that present results caused by conversions and modifications. Land-cover conversion is characterized by a full replacement of the land cover by another, such as deforestation or urbanization. Land modification is characterized by subtle changes without the change of land-cover types, such as those caused by selective logging, drought, and forest insects. A large number of previous studies have investigated land-cover conversion (e.g., Guild et al., 2004; Lambin and Strahler, 1994; Li and Yeh, 1998; Mubea and Menz, 2012), but in the past decade, detection of forest disturbances has received increasing attention due to its importance in reducing carbon estimation uncertainty and its impacts on management of environmental conditions and biodiversity (Huang et al., 2013; Kennedy et al., 2007; Thomas et al., 2011; Wilson and Sader, 2002).

Although much research work related to land-cover change detection has been published, the majority of it has used the same sensor data. In reality, collection of the same sensor data at a multitemporal scale is a challenge, especially in moist tropical regions due to the problem of cloud cover (Asner, 2001). Since different sensor data from optical to radar data with different spatial resolutions are available, use of multisensor data for land-cover change detection provides a new but also challenging opportunity (Lu et al., 2014). This chapter aims to provide brief descriptions of key steps used in a change detection procedure, to discuss the challenges of using multisensor data for land-cover change detection, and to provide a case study to explore the use of multisensor data for land-use/land-cover change detection in the Brazilian Amazon.

18.2 Brief Description of Key Steps Used in a Change Detection Procedure

The process for conducting land-cover change detection consists of a chain of the following steps: (1) description of research problems and objectives, (2) determination of the study area and datasets to be used, (3) preprocessing of multitemporal remote sensing data, (4) extraction of suitable variables and selection of proper algorithms for implementing change detection, and (5) evaluation of the change detection results (see Figure 18.1). Lu et al. (2014) provide a brief description of each step used in a change detection procedure. The following subsections discuss four important aspects in change detection—selection of suitable remote sensing variables and algorithms, impacts of scale issues on change detection methods, and challenges of using multisensor data.

18.2.1 Selection of Suitable Remote Sensing Variables

Before selecting potential remote sensing variables for change detection, it is critical to make sure that proper image preprocessing is conducted, including atmospheric and topographic correction, image-to-image registration, and speckle reduction if radar data are used. Detailed descriptions of image preprocessing issues can be found in earlier literature (Chander et al., 2009; Lu et al., 2008b; Shi and Hao, 2013; Stow and Chen, 2002; Vicente-Serrano et al., 2008).

Remote sensing systems and environmental characteristics have important impacts on the design of a change detection procedure (Biging et al., 1999; Jensen, 2005; Lu et al., 2004). Ideally, the same sensor data with the same radiometric and spatial resolution at the same time of the year are needed to eliminate the effects of external sources such as sun angle and phenological differences. However, in reality, selection of the same sensor data in a specific study is difficult, especially in moist tropical regions due to cloud cover (Asner, 2001) and low revisit periods of satellites. If a change detection requires a long period, satellite images may not be available; thus, multisource data consisting of different satellite images, aerial photographs, and existing thematic maps may be used (Groen et al., 2012; Ichii et al., 2003; Li, 2010; Petit and Lambin, 2001; Tian et al., 2013; Walter, 2004). Table 18.1 provides potential variables that may be used in change detection analysis. Selection of suitable variables is a prerequisite for successfully conducting change detection for a specific study (Lu et al., 2014).

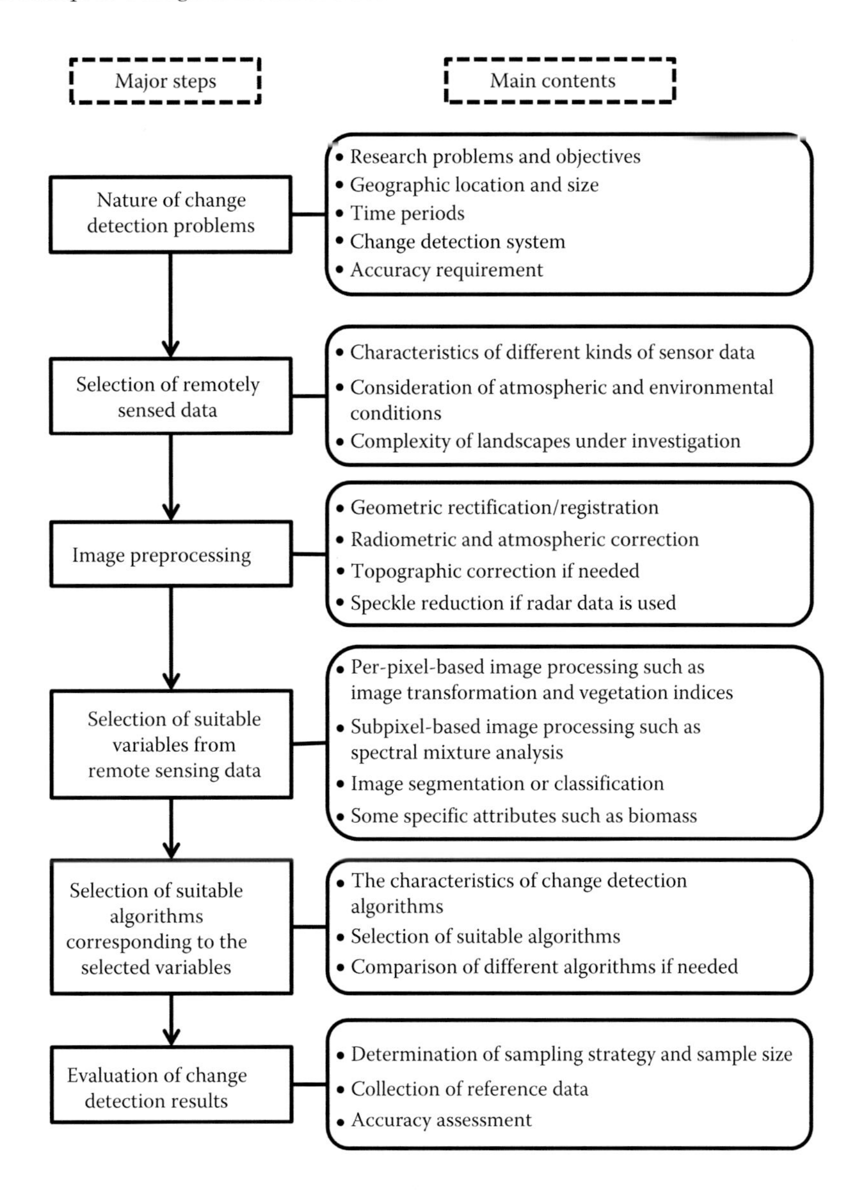

FIGURE 18.1 Major steps used in a general change detection procedure.

Spectral responses (e.g., spectral bands, vegetation indices, and transformed images) may be the most common variables used for change detection, in particular, when medium-spatial-resolution images such as Landsat and Satellite Pour l'Observation de la Terre (SPOT) are used. As more sensor data with different spatial resolutions are available, spectral signatures may not be the optimal variables; other remote sensing features such as spatial and subpixel may better represent the change features. For example, textures and segments developed from very-high-spatial-resolution images could provide better change detection results than spectral responses, whereas subpixel features developed from coarse-spatial-resolution images may be better suitable for land-cover change detection in a large area due to mixed-pixel problems. On the other hand, because of the impacts of external factors such as atmosphere, soil moisture, and vegetation phenology on the spectral signatures, use of spectral responses may produce large spurious change detection results. In this situation, if sufficient training samples are available for conducting land-cover classification, the change detection approaches based on classified images may provide better results than spectral responses.

In the past decade, multisource data are easily obtained; thus, effective use of different source data may improve change detection performance. Much previous research is based on detection of land-cover conversion, but vegetation disturbance is now regarded as an important source of carbon emission to the atmosphere. Accurate detection of vegetation disturbance has gained increasing attention. Due to special characteristics of vegetation types, the change of some forest stand attributes such as leaf area index and biomass may better represent the change of vegetation disturbance due to natural disasters or human-induced activities such as selective logging. Therefore, many potential variables can be used for change detection studies. The key is to identify the variables that can best represent the change detection purpose. Integration of different

TABLE 18.1 Summary of Potential Variables Used in Land-Cover Change Detection Analysis

Variables	Major Characteristics	Advantages	Disadvantages
Spectral features	Spectral responses are the most common variables for change detection. Vegetation indices and transformed images may further improve change detection performance. The majority of change detection techniques are based on the use of spectral features.	Remote sensors capture spectral features of land-cover surfaces. The change in spectral features represents the change of land covers and thus may provide the best change detection results.	Some land-cover types have similar spectral features and thus cannot be effectively used to detect this kind of land-cover change. External factors such as the atmosphere, vegetation phenology, and soil moisture affect the spectral responses.
Spatial information	Texture and segmentation are common methods to use the spatial information inherent in remote sensing data. The spatial features are especially valuable for high-spatial-resolution images and radar data.	Use of spatial features between different land-cover surfaces and reduction of spatial heterogeneity in the same land covers are valuable in improving land-cover classification and change detection performance.	Pure spatial features have relatively poor capability in separating land-cover types, thus lessening effective detection of land-cover changes. The spatial resolution of remote sensing data and complexity of land-cover surfaces in a study area affect the spatial features.
Subpixel information	Mixed pixels are a major problem affecting the change detection because changed areas are small and scattered in different locations. Use of subpixel features such as the fraction images through unmixing the multispectral images has the potential to provide better change detection results.	Subpixel features can provide more accurate results for detecting land-cover modification, especially vegetation disturbance caused by natural disasters or human-induced activities.	Application of subpixel features for change detection is very limited due to the difficulty in developing algorithms to conduct change detection based on the fractional images. The complexity of a study area may affect the development of fraction data.
Thematic information	Separate land-cover classification for each date of imagery is required before conducting change detection. The classified images are then used to examine land-cover change trajectories using the postclassification comparison approach.	The change detection based on classified images can provide detailed change trajectories. The external factors do not affect the change detection quality.	Accurate classification result for each image is required but may be difficult for historical remote sensing data due to lack of reference data.
Biophysical attributes	Some biophysical attributes such as impervious surface area in urban landscapes and leaf area index in forest ecosystems can be derived from remotely sensed data. These variables may better suit specific change detection purposes.	The land-cover change, especially vegetation change, can be more quantitatively evaluated and detected using certain biophysical parameters.	To identify a suitable parameter that better reflects the change of a specific land-cover type is critical; meanwhile, this parameter can be accurately extracted from remotely sensed data.
Multisource data	Different variables from remote sensing or ancillary data (e.g., previously developed thematic maps) may be used in a change detection procedure.	Availability of multisource data provides an alternative method to conduct land-cover change detection, especially when remote sensing data are not available.	Different sources of data have various formats, quality, and spatial resolutions. Improper use of multisource data may produce a large error due to the data quality problems.

variables in a change detection procedure may provide some new insights for more accurately developing the change detection results.

18.2.2 Selection of Suitable Algorithms for Change Detection

In the past four decades, many change detection techniques have been developed, and the majority of them are summarized in the literature (e.g., Bhagat, 2012; Hussain et al., 2013; Lu et al., 2004; Singh, 1989). For example, Lu et al. (2004) summarized over 30 techniques and grouped them into 7 categories: algebra, transformation, classification, advanced models, approaches to geographic information science, visual analysis, and others. Bhagat (2012) summarized 29 techniques grouped into 8 categories: spectral classification, multidate radiometric change, support vector analysis, the hybrid approach, artificial neural network, fusion, object comparison, and the triangle model. Hussain et al. (2013) summarized over 20 approaches, which were grouped into 10 subclasses, and provided a brief description of 15 per-pixel-based change detection techniques and 3 object-based techniques. Recently, the combination or fusion of different change detection methods has received increasing attention for improving change detection performance (Du et al., 2012, 2013). Table 18.2 groups the change detection algorithms into six categories: per-pixel thresholding, per-pixel classification, subpixel, object-oriented, hybrid, and indirect methods, according to the use of variables and corresponding change detection techniques.

Image-processing techniques can be based on per-pixel, subpixel, and segment (or object-oriented) scales and combinations of them; thus, change detection also can be based on these scales. The majority of change detection techniques are based on pixel-based methods (Bhagat, 2012; Hussain et al., 2013; Lu et al., 2004) because of the unique characteristics of remote sensing data. Depending on the purpose of the change detection analysis, many techniques such as image differencing, regression analysis, and principal component analysis are used to detect binary change and nonchange categories; thus, the key step is to identify the optimal thresholds (Lu et al., 2005). The per-pixel thresholding-based approaches can provide only change and

TABLE 18.2 Summary of Six Categories of Change Detection Algorithms

Algorithms	Major Characteristics	Advantages	Disadvantages
Per-pixel thresholding-based methods	Many change detection techniques, such as image differencing, vegetation index differencing, principal component analysis, and regression analysis, belong to this category. The critical steps are to identify suitable variables that can effectively reflect the change of types of interest and to determine suitable thresholds in both tails of the histogram representing the changed areas.	Simple and easy to implement and can provide valuable information for designing a proper procedure for further examining the detailed land-cover change trajectories.	Selection of thresholds is subjective and scene dependent, depending on analyst's skills and familiarity with the study area. Different external factors such as phenology and soil moistures may affect change detection results. This category of methods can only provide change and nonchange detection.
Per-pixel classification-based methods	The postclassification comparison may be the most common approach to detect land-cover trajectories based on the per-pixel-based classification for each date of remote sensing data. Another approach is to conduct direct classification using per-pixel-based algorithms based on the combined multitemporal remote sensing data. Detailed training sample data for each land cover or each change trajectory are required.	The impacts of external factors on change detection results can be avoided. This category of methods can provide detailed land-cover change trajectories.	These methods are criticized for their incapability of solving mixed-pixel problem in medium- or coarse-spatial-resolution images and the high spectral variation within the same land covers in high-spatial-resolution images. The quality of *from–to* change detection results depends on the classification accuracy for each date being analyzed.
Object-based methods	These detection methods can be conducted in three ways: (1) direct comparison of segmentation images at different dates, (2) comparison of classified objects, and (3) image segmentation and classification based on stacked multitemporal images.	This category of methods can reduce the spectral variation within the same land covers; thus, they are preferable for high-spatial-resolution images.	High-quality segmentation image is required. It is important to identify optimal parameters (e.g., minimum value distance, variance factor, minimum size of pixels in a segment) used in a segmentation procedure.
Subpixel-based methods	Different subpixel methods, such as fuzzy classification and spectral mixture analysis, may be used.	This method is valuable for some specific change detection studies such as forest disturbance and urban expansion.	They lack suitable subpixel-based algorithms to detect detailed *from–to* land-cover change trajectories.
Hybrid methods	Hybrid methods may be conducted in two ways: (1) combination of different methods into one change detection procedure and (2) combination of change detection results from different methods into a new result using developed rules or fusion methods at feature or decision level.	The merits of different methods may be integrated into a change detection procedure to produce more accurate results than individual methods.	It is critical to select suitable variables and corresponding algorithms to conduct the change detection procedure, but these methods lack general rules to guide this process.
Indirect methods	Indirect methods identify some biophysical attributes that can effectively reflect the land-cover change. These attributes can be derived from remote sensing data through modeling. Common attributes are, for example, impervious surface in urban landscapes and leaf area index in forest ecosystems.	This method could be valuable for global land-cover change detection using coarse-spatial-resolution images, especially for deforestation and urbanization. Another important application may be the detection of forest disturbance caused by natural or anthropogenic factors by analyzing the change in vegetation structures such as leaf area index or biomass.	The key is to develop accurate biophysical attributes from remote sensing data where the attribute is suitable for use as a variable for change detection analysis. It is also critical to make sure that the uncertainty from the attribute estimation is much less than the true change amount of the corresponding attribute.

nonchange categories, but in reality, detailed *from–to* change trajectories are needed for many studies (Lu et al., 2012, 2013). In this case, the per-pixel-based classification approaches such as the postclassification comparison can meet this requirement. Although per-pixel-based approaches are most commonly used in practice, its accuracy for calculation of change detection areas may be poor, especially when coarse-spatial-resolution images are used due to the complex land-cover composition such as in urban landscapes. Therefore, subpixel-based and object-based approaches are proposed. The spatial resolution of selected remote sensing data, complexity of the study area, and the user's needs may significantly affect the determination of the change detection method. For high-spatial-resolution image, object-based methods may be preferable because of the high spectral variation within the same land cover. In contrast, the subpixel-based methods may be preferable for coarse-spatial-resolution images due to the mixed-pixel problem (Lu et al., 2014). Since different change detection methods have their own merits and characteristics, a hybrid approach consisting of different algorithms may provide better change detection results, but have not been fully examined yet.

18.2.3 Impacts of Scale Issues on the Design of a Change Detection Procedure

Scale is an important concern in designing a proper change detection procedure. Different extents of a study area will directly influence the selection of remote sensing data and associated change detection methods. Table 18.3 provides a summary of potential variables and algorithms suitable for local, regional, national, and global land-cover change detection and discusses major problems that occur at different scales.

At a local scale, accurate change detection results are required because these results may be used as reference data for evaluation of other data from medium- or coarse-spatial-resolution images. Thus, high-spatial-resolution images are needed for this purpose. In practice, use of multitemporal high-spatial-resolution images for change detection is a challenge. As shown in Figure 18.2, the major problems include (1) different shadow sizes between two QuickBird images caused by different sun elevation angles at image acquisition dates; (2) different colors for the same impervious surfaces, as the locations labeled in A, B, C, and D in Figure 18.2a1 and a2; and (3) displacement caused by tall buildings, labeled in A and B in Figure 18.2b1 and b2. Because of these problems, direct use of high-spatial-resolution images for change detection may produce large errors based on traditional per-pixel-based change detection techniques. It is necessary to develop suitable change detection techniques by taking these problems into account.

At a regional scale, medium-spatial-resolution images such as Landsat and SPOT are commonly used for land-cover change detection. Most previous research was based on regional scale, and different change detection techniques as summarized in Table 18.2 can be used. It is difficult to select an optimal change detection technique for a specific study because of the following problems: Which variables as summarized in Table 18.1 should be selected? Which change detection technique should be used considering the complexity of the study area such as urban-dominated or forest-dominated ecosystems? Although much research work has been done to explore remote sensing variables and algorithms for different ecosystems such as forest, urban, wetland, and agriculture, as reviewed in previous literature (e.g., Bhagat, 2012; Hussain et al., 2013; Lu et al., 2004), it is still unclear which one should be used for a specific study. In practice, a comparative analysis of different variables and algorithms is used to identify the best change detection procedure for a specific purpose (Lu et al., 2014). Recently, hybrid approaches that consist of different algorithms or remote sensing data have been used to combine their merits and generate the best change detection result.

Compared with the change detection research at regional scale, studies at national and global scales face greater challenges due to the complexity of land-cover types across large areas and the mixed-pixel problem in the coarse-spatial-resolution images, as illustrated in Figure 18.3. The majority of change detection techniques suitable for medium-spatial-resolution images at a regional scale are not feasible for the coarse-spatial-resolution images at national or global scale; thus, researchers need to develop new change detection techniques that are suitable for coarse-spatial-resolution images. Integration of multiscale and multisource data may provide new insights for developing detection techniques suitable for large-area changes but has not received sufficient attention (Lu et al., 2014).

TABLE 18.3 Impacts of Scale Issues on Change Detection Analysis

Scales	Potential Data	Potential Algorithms	Major Problems
At local scale	Submeter spatial resolution satellite images such as QuickBird and WorldView are preferable to provide accurate and detailed land-cover change detection results.	Application of textures and segmentation-based methods will be valuable.	The following factors will significantly affect the change detection results: shadows; confusion of shadow, dark impervious surfaces, water, and wetland; high spectral variation of the same urban land cover; and spectral confusion among impervious surfaces and bare soils.
At regional scale	Medium-spatial-resolution images such as Landsat and SPOT data; integration of medium-spatial-resolution image and limited high-spatial-resolution images; multisource data.	Almost all change detection techniques are developed at regional scale and can be used to detect land-cover change. Data fusion techniques at different levels such as pixel, feature, and decision can be used to integrate different remote sensing data or combine change detection results from different algorithms to improve change detection performance.	Depending on research purpose and datasets, selection of suitable variables and algorithms is important to provide accurate change detection results.
At national and global scales	Coarse-spatial-resolution images such as time-series MODIS or SPOT datasets.	Integration of MODIS and Landsat images has been used to detect land-cover change.	The complexity of land-cover composition in coarse-spatial-resolution image makes change detection a challenge. The critical task is to solve the mixed-pixel problem and the complexity of land-cover distribution in a large area.

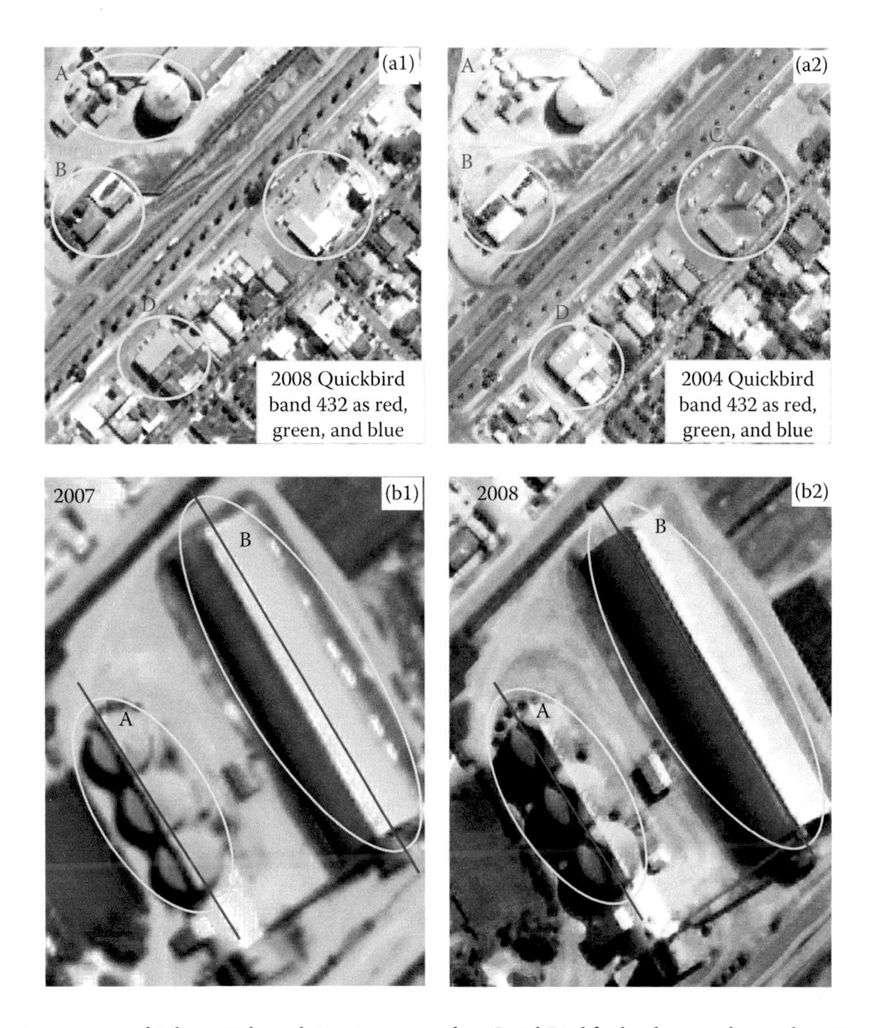

FIGURE 18.2 Problems in using very-high-spatial-resolution images such as QuickBird for land-cover change detection analysis (a1 and a2 show the impacts of different sun elevation angles on the size of shadows cast by tall buildings; b1 and b2 show the impacts of different sun elevation angles on the displacement caused by tall buildings too.)

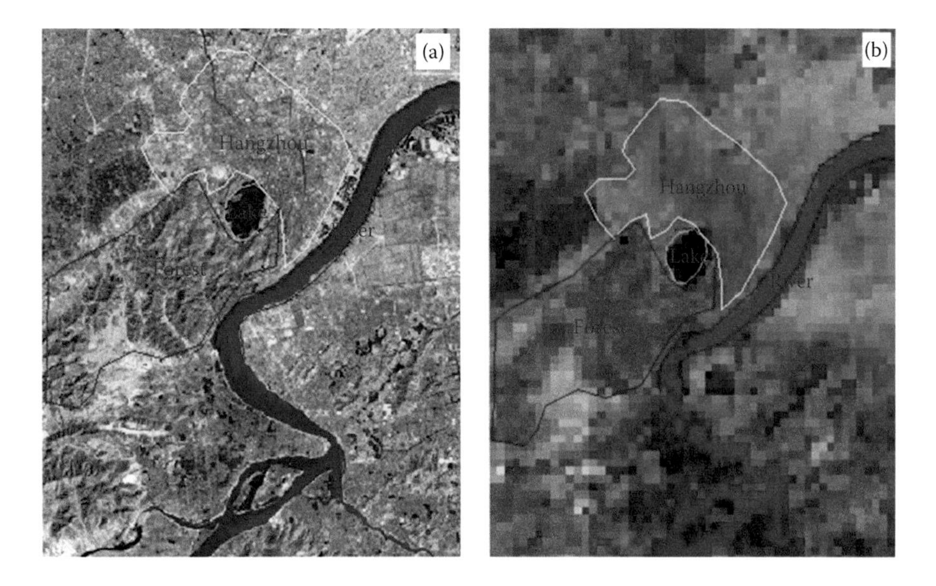

FIGURE 18.3 Problems in using (a) medium- and (b) coarse-spatial-resolution images (Landsat TM vs. MODIS surface reflectance images) for land-cover change detection analysis.

18.2.4 Challenge of Using Multisensor Data for Change Detection

Change detection is usually conducted using the same sensor data with anniversary or near-anniversary dates in order to reduce the impacts of remote sensing data per se and environmental conditions on the change detection results (Lu et al., 2014). However, sometimes, the same sensor data for a specific study area are not available, especially in situations where the revisit period of sensors does not fit the date of interest well or when images suffer from cloud cover (Lu et al., 2008a, 2014). Microwave data represent an alternative to overcome this limitation in tropical regions because they enable the imaging in adverse weather conditions, in addition to other advantages such as the production of images with different types of polarized energy (HH, HV, VV, VH). However, we should take into account that unlike optical images containing information about the reflectance and the physicochemical properties of the targets, microwave data carry information about the structure and dielectric properties of the targets (Jensen, 2007) and emphasize complementary information between optical sensors and microwave data. One possibility for dealing with the data availability problem, particularly in tropical regions, is the use of multisensor data, which greatly increases the acquisition of cloud-free data and may also add information due to the different characteristics of the sensors involved in the process.

Integration of multisensory data is extensively used for improving land-cover classification (Lu et al., 2011; Pereira et al., 2013). The use of different sensor data at multitemporal scales in a change detection procedure presents an alternative to solve the data acquisition problem but presents a great challenge in the development of techniques (Akiwowo and Eftekhari, 2013; Zeng et al., 2010) because different data can have various qualities and spatial resolutions. The different sensor data could be from optical sensors, radar, or a combination of them. Because of the differences in spatial, spectral, and radiometric resolutions, as well as view angles and polarization options for radar data, traditional change detection techniques such as image differencing and principal component analysis are not suitable for different sensors. Previous research has mainly examined the use of multiple optical sensor data for land-cover change detection, such as Landsat Multispectral Scanner (MSS) and Thematic Mapper (TM) (Serra et al., 2003), Landsat Enhanced Thematic Mapper Plus (ETM+), and Satellite Pour l'Observation de la Terre multispectral (SPOT XS) (Deng et al., 2008), Landsat TM, and Satellite Pour l'Observation de la Terre high resolution geometrical (SPOT HRG) (Lu et al., 2008a). However, rarely has research examined the integration of optical and radar data for land-cover change detection (Alberga, 2009; Lu et al., 2014; Reiche et al., 2013) due to their significantly different features. In practice, use of optical and radar data may be needed, especially in tropical moist regions. The challenge is that current change detection techniques cannot effectively handle this problem; thus, new techniques to implement the change detection are required.

18.3 Case Study of Land-Cover Change Detection Using Optical and Radar Data

In order to explore the use of multitemporal multisensor data for land-cover change detection, Radarsat-2 C-band in 2009 and EO-1 ALI (Earth Observing-1 Advanced Land Imager instrument) data from 2013 were used in this research. Figure 18.4 illustrates the flowchart for conducting the change detection. The major steps include image preprocessing of both datasets and change detection using the direct classification approach. As a comparison of the change detection results, a postclassification approach is also conducted.

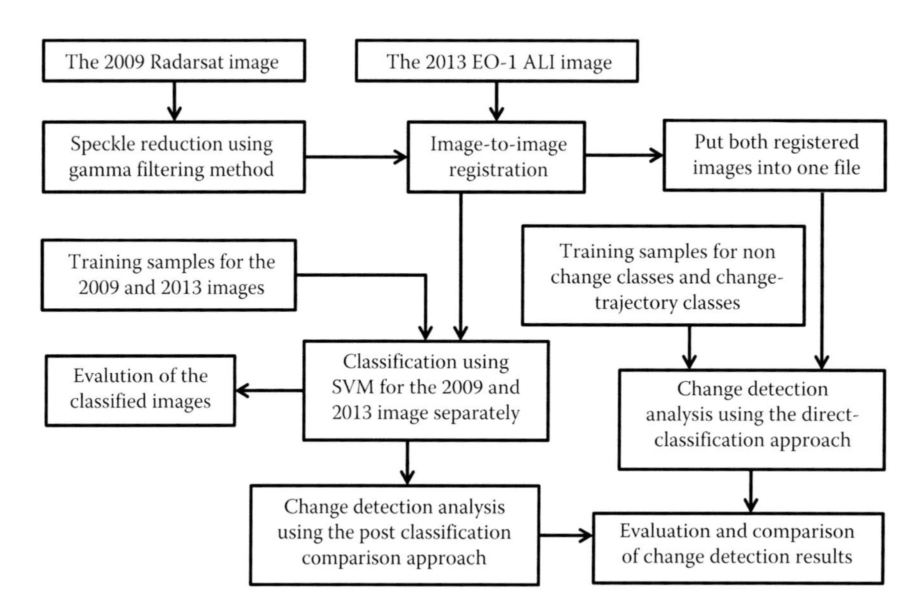

FIGURE 18.4 Framework of land-cover change detection analysis used in this case study.

18.3.1 Study Area

The study area with an extent of approximately 600,000 ha is located near the Tapajos National Forest along the Cuiabá-Santarém highway BR-163 in Belterra Municipality, Pará State, Brazil (Figure 18.5). In the 1970s, the Brazilian government initiated a policy of encouraging the occupation of Brazilian Amazonia and financed the construction of BR-163. Because of this policy and highway construction, deforestation and activities associated with this process such as wood extraction and expansion of agricultural fields and pasture areas started along this highway and adjoining areas. Currently, agriculture and cattle ranching are the main economic activities in this region, generating a variety of land-cover classes including pastures, soybeans, and other grains such as rice, maize, and sorghum. This study area is relatively flat with elevations ranging from 50 to 200 m and has a warm and humid climate. The maximum annual average temperature varies between 31°C and 33°C,

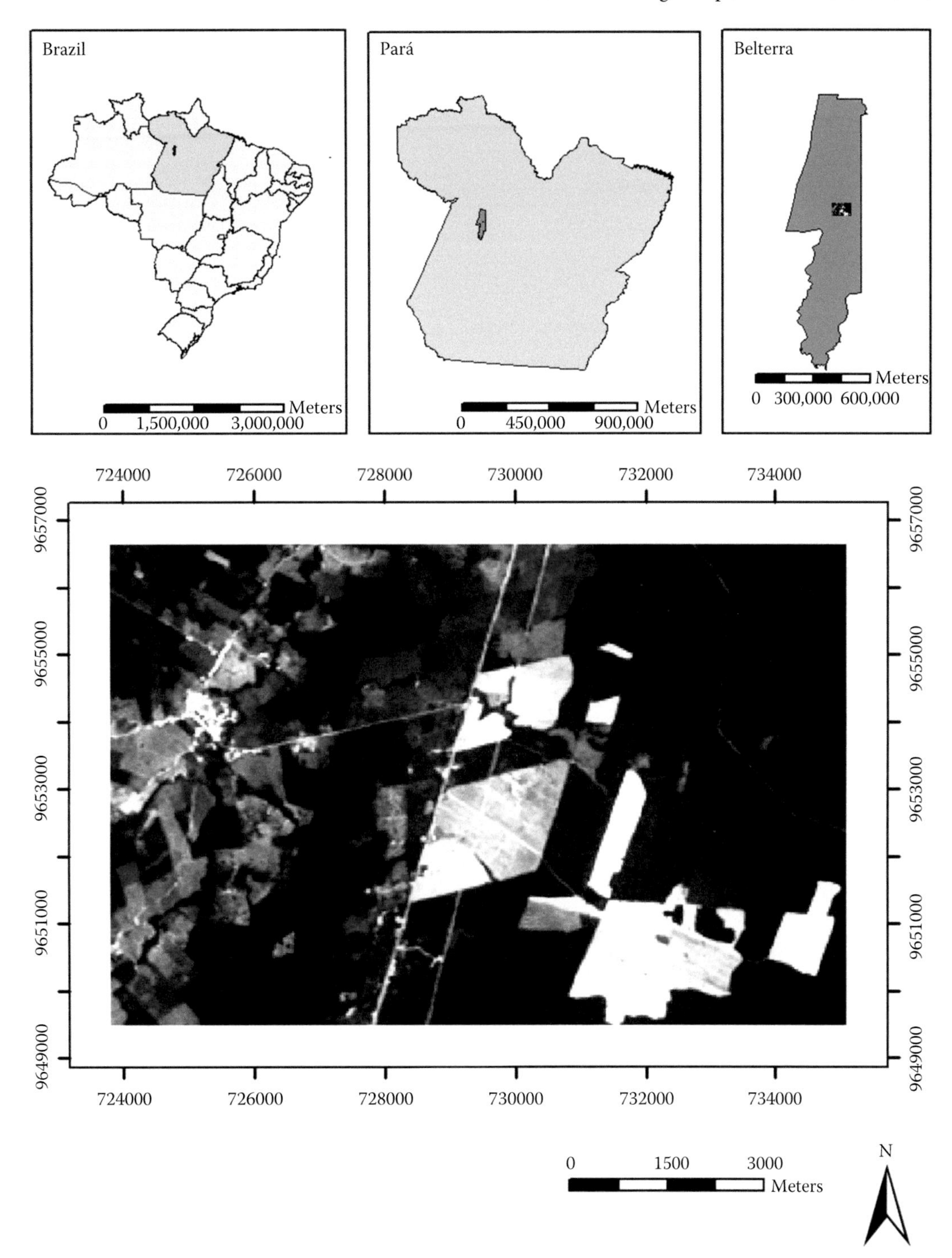

FIGURE 18.5 Location of the study area in Belterra Municipality, Pará State, Brazil.

and the minimum average ranges from 24°C to 25°C. Annual rainfall varies between 1800 and 2800 mm with a clear division in rainfall distribution. The rains are plentiful in the first half of the year and low in intensity during the rest of the year.

18.3.2 Data Collection and Preprocessing

Radarsat-2 C-band, EO-1 ALI, and field survey data were collected and summarized in Table 18.4. More detailed samples collected during the field survey are summarized in Tables 18.5 and 18.6. The Radarsat-2 and EO-1 ALI data were coregistered into the Universal Transverse Mercator (UTM) coordinate system and resampled to a cell size of 5 m. The registration error was 0.49 pixels. The Radarsat-2 data in amplitude with three bands—HH, VV, and (HV + VH)/2—were filtered using a standard gamma filter with a window size of 3 × 3 pixels.

18.3.3 Methodology

18.3.3.1 Land-Use/Land-Cover Classification with a Support Vector Machine

Research on classification algorithms has received great attention in the past four decades. A large number of algorithms, from traditional statistical-based methods such as minimum distance and maximum likelihood to complex nonstatistical algorithms such as neural network, decision tree, and support vector machine (SVM), have been developed for land-cover classification (see Lu and Weng, 2007; Tso and Mather, 2009). In the last decade, nonstatistical-based classification algorithms have been regarded as having more advantages for providing better land-cover classification than traditional statistical-based algorithms, especially when multisensor/source data are used (Li et al., 2012b; Lu et al., 2011). Of the many nonstatistical algorithms, SVM is a fairly recent pattern recognition method that has been drawing attention due to its data distribution independence, good generalization ability, and robustness against the Hughes phenomenon (Camps-Valls and Bruzzone, 2009; Camps-Valls et al., 2008; Mountrakis et al., 2011; Perumal and Bhaskaran, 2009; Vapnik et al., 1996). Therefore, SVM is used in this research for land-cover classification in the moist tropical region.

This method, in its original conception, consists of finding a separation hyperplane between training samples with larger margins. The separating hyperplane is a geometric place where the following linear function is zero:

$$f(\mathbf{x}) = \langle \mathbf{w}, \mathbf{x} \rangle + b, \tag{18.1}$$

where

w represents the orthogonal vector to the hyperplane $f(\mathbf{x}) = 0$

$b/||\mathbf{w}||$ is the distance from the hyperplane to the origin

$\langle \cdot,\cdot \rangle$ denotes the inner product

TABLE 18.4 Datasets Used in Research

Datasets	Major Characteristics	Data Acquisition Dates
Radarsat-2	A full, polarized Radarsat-2 scene in fine-resolution quad-polarization beam mode 13 (FQ13) mode with a pixel size of 4.733 × 4.748 m.	September 9, 2009
EO-1 ALI	Nine spectral bands with 30 m spatial resolution.	October 5, 2013
Field survey	A large number of samples for training and test purposes were collected (see Tables 18.5 and 18.6).	September 2009 August 2013

TABLE 18.5 Sample Pixels Collected in 2009 and 2013 for Use in the Classification of Individual Sensor Data

	Samples from 2009		Samples from 2013	
Class	Training	Test	Training	Test
AG	385	269	2095	1978
BS	742	1786	1747	2211
CP	2134	2115	2360	1900
DP	1004	1952	1720	1702
IS	1063	1063	698	1101
PF	1391	3042	1812	2314
SS1	1272	1281	1059	858
SS2	828	796	844	929
WT	877	1445	1448	1105

Notes: AG, agriculture; BS, bare soil; CP, clean pasture; DP, dirty pasture; IS, impervious surface; PF, primary forest; SS1, initial secondary succession; SS2, intermediate secondary succession; WT, water.

TABLE 18.6 Sample Pixels for Use in the Support Vector Machine-Based Direct Classification

Class	Training	Test
No change		
AG–AG	1246	1106
BS–BS	1850	1278
CP–CP	1070	1299
DP–DP	966	1103
IS–IS	881	1145
PF–PF	1610	4415
SS2–SS2	1111	1264
WT–WT	2422	1713
Change		
AG–SS1	522	221
BS–AG	1568	1665
CP–DP	2078	1933
CP–SS1	892	1082
DP–CP	814	555
DP–SS1	900	1195
SS1–BS	1039	1157
SS1–SS2	416	460

Notes: AG, agriculture; BS, bare soil; CP, clean pasture; DP, dirty pasture; IS, impervious surface; PF, primary forest; SS1, initial secondary succession; SS2, intermediate secondary succession; WT, water.

The parameters of Equation 18.1 are obtained from the following quadratic optimization problem (Theodoridis and Koutrombas, 2008):

$$\sum_{i=1}^{m}\lambda_i - \frac{1}{2}\sum_{i=1}^{m}\sum_{j=1}^{m}\lambda_i\lambda_j y_i y_j \langle \varphi(\mathbf{x}_i), \varphi(\mathbf{x}_j)\rangle,$$

$$\text{subject to} \begin{cases} 0 \le \lambda_i \le C; \quad i = 1,\ldots,m, \\ \sum_{i=1}^{m}\lambda_i y_i = 0, \end{cases} \tag{18.2}$$

where

λ_i represents Lagrange multipliers

$y_i = \{-1, +1\}$ defines the class of $\mathbf{x}_i$, since SVM is a binary classifier

C (cost) acts as an upper bound of λ values

$\varphi(\mathbf{x})$ is a function adopted to remap the input vectors into a higher dimensionality space

The inner product $\varphi(\mathbf{x}_i),\varphi(\mathbf{x}_j)$ is known as the kernel function. The radial basis function (RBF) is the kernel function used in this study (Webb, 2002):

$$\varphi(\mathbf{x}_i), \varphi(\mathbf{x}_j) = e^{(\mathbf{x}_i - \mathbf{x}_j^2/2\sigma^2)}, \quad \sigma \in \mathbb{R}. \tag{18.3}$$

The optimization problem in Equation 18.2 is solved considering a training set $\mathcal{D} = \{(\mathbf{x}_i, y_i) : i = 1,\ldots,l\}$, where $\mathbf{x}_i \in \mathbb{R}^d$. Let $SV = \{\mathbf{x}_i : \lambda_i \neq 0; i = 1,\ldots,l\}$, known as the support vector set. The parameters **w** and b are computed by

$$\mathbf{w} = \sum_{\mathbf{x}_i \in SV}\lambda_i y_i \varphi(\mathbf{x}_i), \tag{18.4}$$

$$b = \frac{1}{\#SV}\sum_{\mathbf{x}_i \in SV} y_i + \sum_{i=1}^{l}\sum_{j=1}^{l}\lambda_i\lambda_j y_i y_j \langle \varphi(\mathbf{x}_i), \varphi(\mathbf{x}_j)\rangle. \tag{18.5}$$

To apply this method to a multiclass problem (problems with more than two classes), it is necessary to adopt multiclass strategies, such as *one against all* or *one against one* (Theodoridis and Koutrombas, 2008).

Before implementing land-cover classification using the SVM approach, it is necessary to determine a suitable classification system. Based on our fieldwork in 2009 and 2013, a classification system consisting of nine land-use/land-cover classes—agriculture (AG), bare soil (BS), clean pasture (CP), dirty pasture (DP), impervious surface (IS), primary forest (PF), initial secondary succession (SS1, up to 7 years of regrowth), intermediate secondary succession (SS2, from 7 to 15 years of regrowth), and water (WT)—was designed for this study area. Training samples as summarized in Table 18.5 were used separately for the land-cover classification based on the 2009 Radarsat C-band and the 2013 EO-1 ALI data. Based on our previous exploration using SVM for land-cover classification in the moist tropical region (Negri et al., 2011, 2012, 2013, 2014) and this experiment, the parameters used in this research were finally selected as follows: penalty parameter of 100 for all classification cases and gamma value in the RBF kernel function of 0.33 for Radarsat classification, 0.11 for EO-1 ALI classification, and 0.083 for the 12-band stacked data using the SVM-based direct classification approach.

18.3.3.2 Change Detection with Postclassification Comparison

Although many change detection techniques have been developed, as summarized in previous literature (e.g., Bhagat, 2012; Chen et al., 2012; Coppin et al., 2004; Hussain et al., 2013; Lu et al., 2004, 2014), the postclassification comparison approach is commonly used to examine land-cover change trajectories and thus used in this research. In order to conduct the land-cover change detection, one critical step is to design the change detection system for this study. Based on our classification results with 9 land-cover types for 2009 and 2013, the postclassification comparison approach can produce 9 nonchange classes and 72 potential change classes. However, for a specific study area, some of the change trajectories are not possible within a 4-year interval or do not exist, as illustrated in Table 18.7.

TABLE 18.7 Potential Land-Cover Change Trajectories at the Study Site in Belterra Municipality, Pará State, Brazil, 2009–2013

	AG	BS	CP	DP	FP	SS1	SS2	WT	IS
AG	N	P	P	P	I	P	I	P	P
BS	P	N	P	P	I	P	I	P	P
CP	P	P	N	P	I	P	I	P	P
DP	P	P	P	N	I	P	I	P	P
FP	P	P	P	P	N	P	I	U	P
SS1	P	P	P	P	I	N	P	U	P
SS2	P	P	P	P	I	P	N	U	P
WT	P	P	P	P	I	P	P	N	P
IS	U	U	U	U	I	U	I	P	N

Notes: AG, agriculture; BS, bare soil; CP, clean pasture; DP, dirty pasture; IS, impervious surface; PF, primary forest; SS1, initial secondary succession; SS2, intermediate secondary succession; WT, water; P, possible; I, impossible; N, nonchange; U, improbable.

In this table, all transitions in the postclassification comparison labeled I or U are left as unclassified.

Based on our field survey, only eight change classes were observed: AG–SS1, BS–AG, CP–DP, CP–SS1, DP–CP, DP–SS1, SS1–BS, and SS1–SS2. In 4 years, the SS1–SS1 nonchange class was not found. This is natural because it is very unlikely that SS1 will remain stable for more than 3 years. Also, no deforestation was found in this area between 2009 and 2013.

18.3.3.3 Change Detection with SVM-Based Direct Classification

Although the postclassification comparison approach can provide detailed land-cover change trajectories, the change detection result depends on the accuracy of individual classification results. This means that the classification results for both dates should have high accuracy in order to produce reliable change detection results (Lu et al., 2014). However, in reality, land-cover classification for historical remote sensing data is often difficult if training sample data are not available or the required land-cover types cannot be effectively separated from the remote sensing data (Lu et al., 2014). In this study, Radarsat-2 C-band data were used, but previous research has indicated that Radarsat-2 C-band cannot provide reliable land-cover classification results and has very poor classification accuracy compared to optical sensor data (Li et al., 2012a; Lu et al., 2011). Therefore, an SVM-based direct classification approach was used in this study to detect land-cover change based on detailed field survey data.

After image preprocessing for both the 2009 Radarsat and the 2013 EO-1 ALI data, these images were stacked into one file. Training samples for the eight change classes and eight nonchange classes, as summarized in Table 18.6, were used to train the SVM classifier. The nonchange classes were used as counterexamples in the classification stage to help avoid the errors of inclusion. This approach has the advantage of avoiding the sum of errors produced when comparing two land-use/land-cover maps. Also, transitions that are not feasible, or impossible, are avoided once only valid transitions are used as reference. The drawback of such a procedure is finding a sufficient and large enough set of all representative transitions and considering all types of nonchange classes at the same time.

18.3.3.4 Evaluation and Comparison of Change Detection Results

The error matrix approach provides detailed assessment of the agreement between the classified result and reference data and provides information about how the misclassification happened (Congalton and Green, 2008). Therefore, this approach is usually used for evaluating land-cover classification based on reference data (Congalton and Green, 2008; Foody, 2002; Van Oort, 2007). From the error matrix, producer's accuracy, user's accuracy, overall classification accuracy, and kappa coefficient are computed (Congalton, 1991; Congalton and Green, 2008; Foody, 2002). The accuracy assessment for historical land-cover classification is difficult due to lack of test samples, and evaluation of the change detection results is especially challenging because of the difficulty in collecting reference data at multitemporal periods (Foody, 2010; Morisette and Khorram, 2000; Olofsson et al., 2013).

In this research, test samples of nine land-cover types for 2009 and 2013 were collected during the field survey and are summarized in Table 18.5. The test samples for eight change classes and eight nonchange classes are summarized in Table 18.6. During implementation of the accuracy assessment of the change detection results using postclassification comparison, many pixels fell into an unclassified/nonobserved class. Therefore, it was necessary to take the unclassified/nonobserved class into account in the accuracy assessment. A modified confusion matrix in which the unclassified/nonobserved pixels were disregarded was used. This kappa value, denominated here as a partial kappa coefficient, could not be directly compared with standard kappa values.

18.3.4 Results

18.3.4.1 Analysis of Classification Results Using a Support Vector Machine

A comparison of the land-cover classification results from 2009 and 2013, as illustrated in Figure 18.6, indicates that obviously, the Radarsat data cannot effectively separate the land-cover types and has much poorer classification accuracy than the EO-1 ALI data. Based on the error matrices, the kappa coefficient for the 2009 Radarsat-based result is only 0.19 and is 0.76 for the 2013 EO-1 ALI–based classification result. The error matrices drawn in Figure 18.7 indicate that only CP can be separated from other land-cover types with an accuracy of 77%, but all other land covers have very low accuracy based on the Radarsat data; in contrast, the EO-1 ALI data can provide good classification for AG, BS, DP, PF, and AT with accuracies of more than 83%, but SS1 and SS2 are highly confused with each other. This conclusion is similar to our previous research conclusions that optical sensor data can provide much better classification results than radar data in the moist tropical regions of the Brazilian Amazon (Lu et al., 2011).

18.3.4.2 Analysis of Change Detection Results Using Postclassification Comparison

The change detection result using postclassification comparison shows very poor accuracy, as illustrated in Figure 18.8, and many pixels are not detected or wrongly detected. The major problem is the poor classification result from the 2009 Radarsat image, thus resulting in a large number of unclassified pixels in which several pairs of from–to classes are considered invalid (i.e., impossible or very unlikely, as indicated in Table 18.7), as observed in black in Figure 18.8. Another problem is the pixels in yellow that correspond to a supposed valid transition but was not observed during fieldwork. These problems imply the difficulty of using postclassification comparisons for land-cover change detection based on optical and radar data.

Figure 18.9 represents a special confusion matrix where two lines are added at the bottom to show the percentage of known test pixels that are supposed to be a certain type of change. Based on this

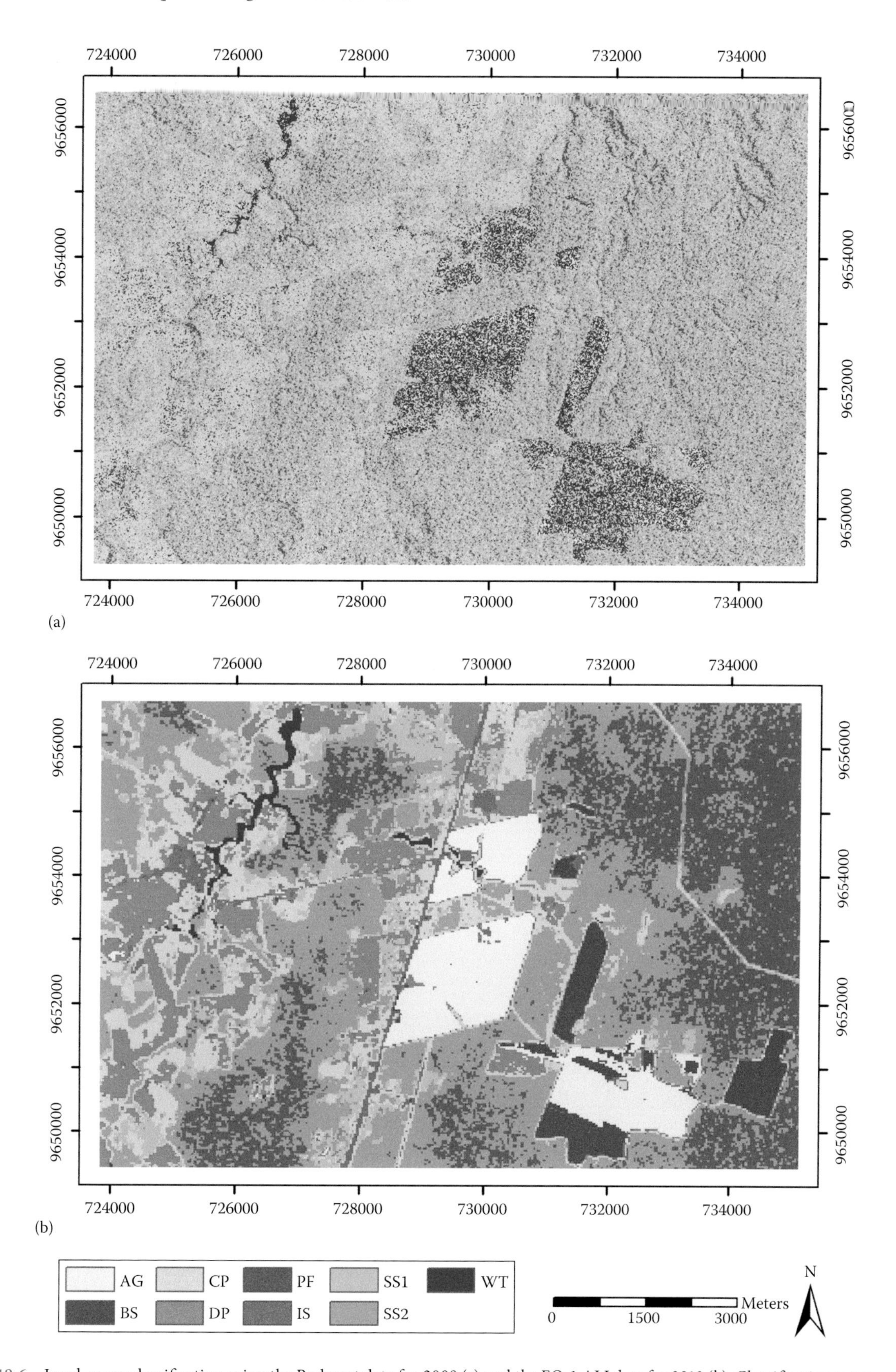

FIGURE 18.6 Land-cover classification using the Radarsat data for 2009 (a) and the EO-1 ALI data for 2013 (b). Classification types are agriculture (AG), bare soil (BS), clean pasture (CP), dirty pasture (DP), impervious surfaces (IS), primary forest (PF), initial secondary succession (SS1), intermediate secondary succession (SS2), and water (WT).

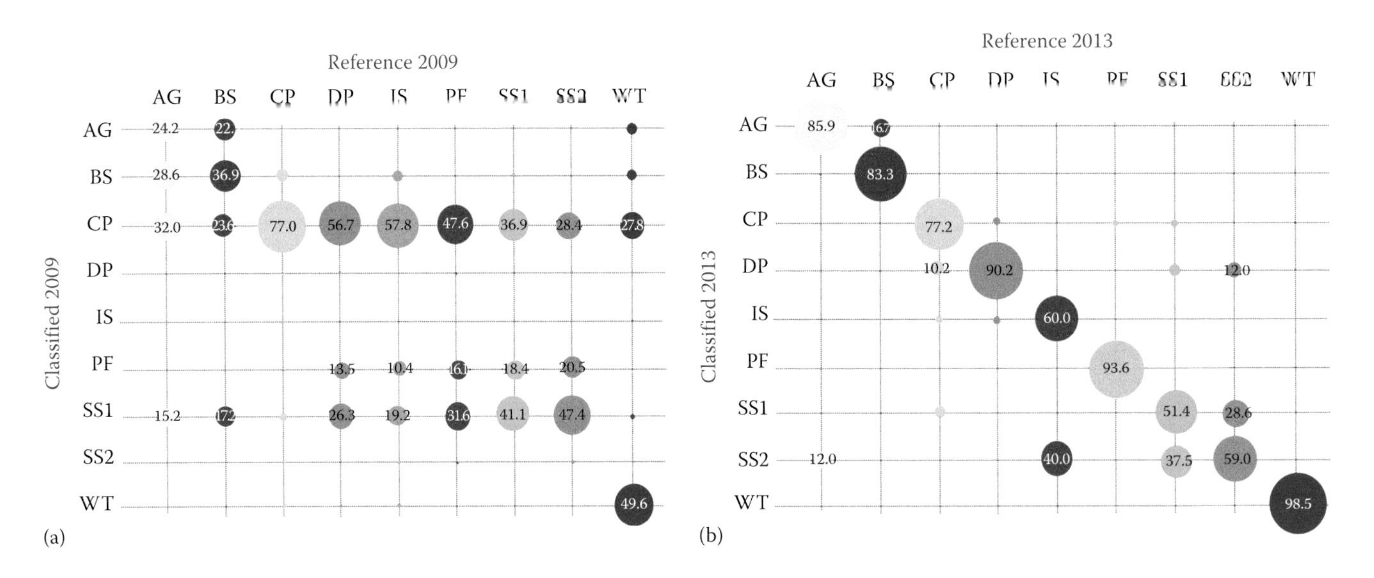

FIGURE 18.7 Confusion matrices for land-cover maps using support vector machine classification based on the Radarsat data in 2009 (a) and the EO-1 ALI data in 2013 (b).

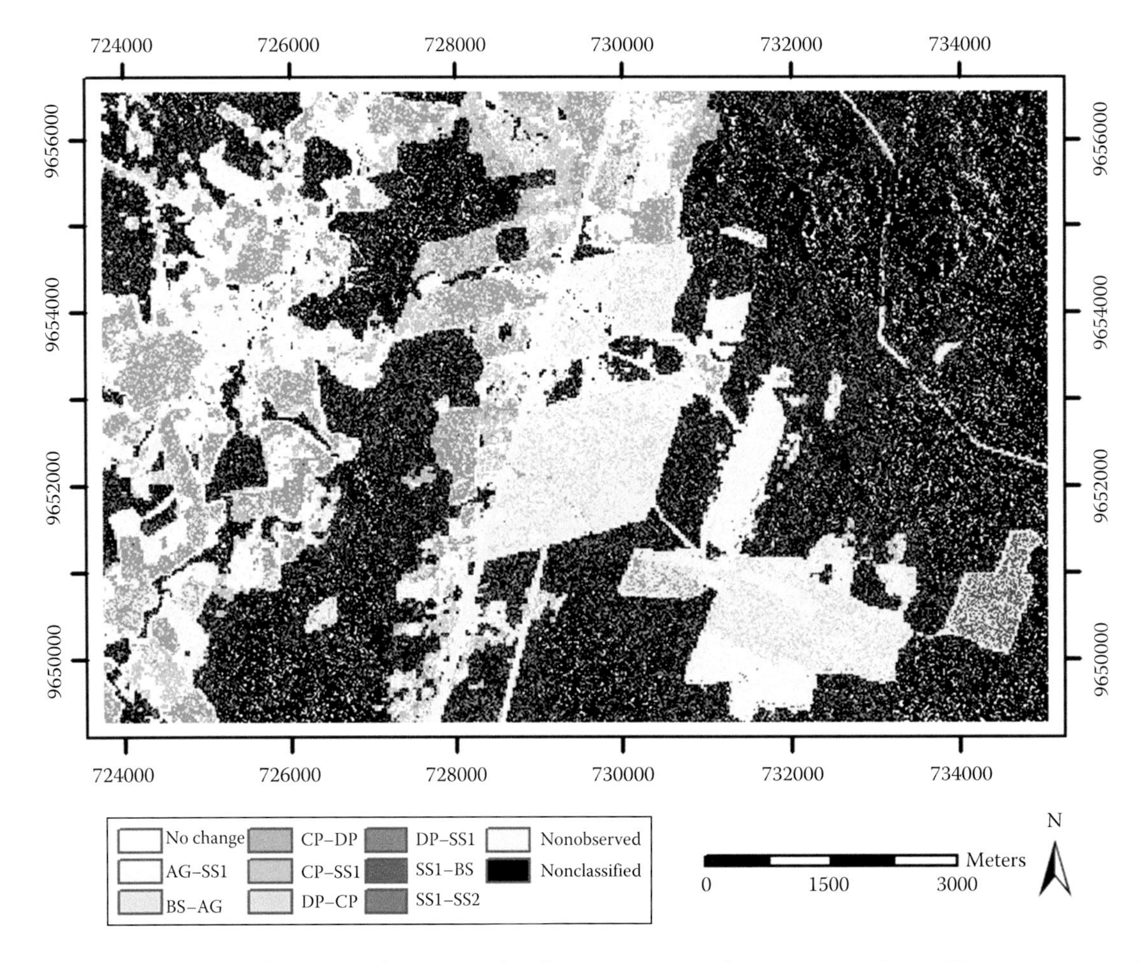

FIGURE 18.8 Land-cover change detection result using postclassification comparison between 2009 and 2013. Change types are agriculture to initial secondary succession (AG–SS1), bare soil to agriculture (BS–AG), clean pasture to dirty pasture (CP–DP), clean pasture to initial secondary succession (CP–SS1), dirty pasture to clean pasture (DP–CP), dirty pasture to initial secondary succession (DP–SS1), initial secondary succession to bare soil (SS1–BS), and initial secondary succession to intermediate secondary succession (SS1–SS2).

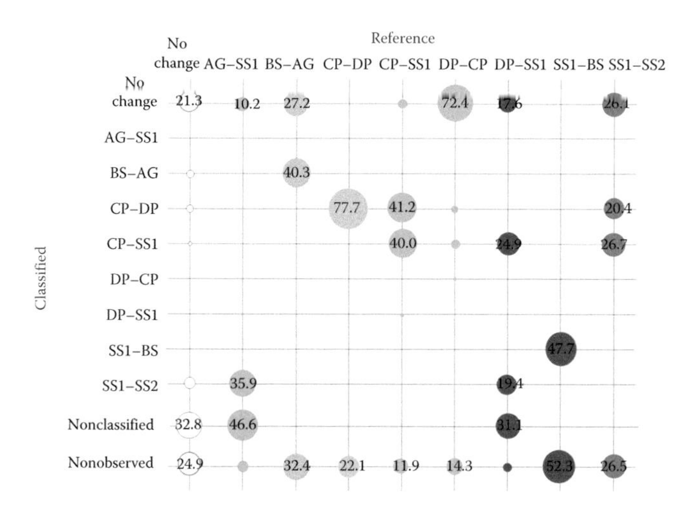

FIGURE 18.9 Confusion matrix for land-cover change map from postclassification comparison between 2009 and 2013.

error matrix, the calculated partial kappa coefficient is only 0.28. One can notice a great number of unclassified pixels, which corresponds to a calculated 22.3% (mostly impossible type), in the area where the majority are nonchange classes. This is due to the fact that Radarsat C-band data saturate very quickly with increasing biomass; thus, most PF and SS2 are misclassified as clean pasture in Radarsat data (Figure 18.7), leading to impossible transitions in the change detection mapping. Overall, the postclassification comparison can only provide the accuracy of CP–DP with 77.7% and BS–AG, CP–SS1, and SS1–BS with accuracies of 40%–47.7%. Other land-cover change trajectories were not detected.

18.3.4.3 Analysis of Change Detection Results Using Direct Classification

The change detection result using SVM-based direct classification is illustrated in Figure 18.10, which highlights the eight change classes. Compared with the change detection result using postclassification comparison (see Figure 18.8), the SVM-based direct classification provided much better spatial distribution patterns of the change detection results. The accuracy assessment results illustrated in Figure 18.11 indicate the reliability of direct classification. Based on the error matrix, the overall accuracy of 87.8% and kappa coefficient of 0.80 were obtained. The change trajectories of AG–SS1, CP–DP, SS1–BS, and SS1–SS2 have high accuracy with over 91%, but the change trajectories of BS–AG and CP–SS1 have relatively poor accuracy with only 63% and 56%, respectively, because of the confusion with nonchange classes.

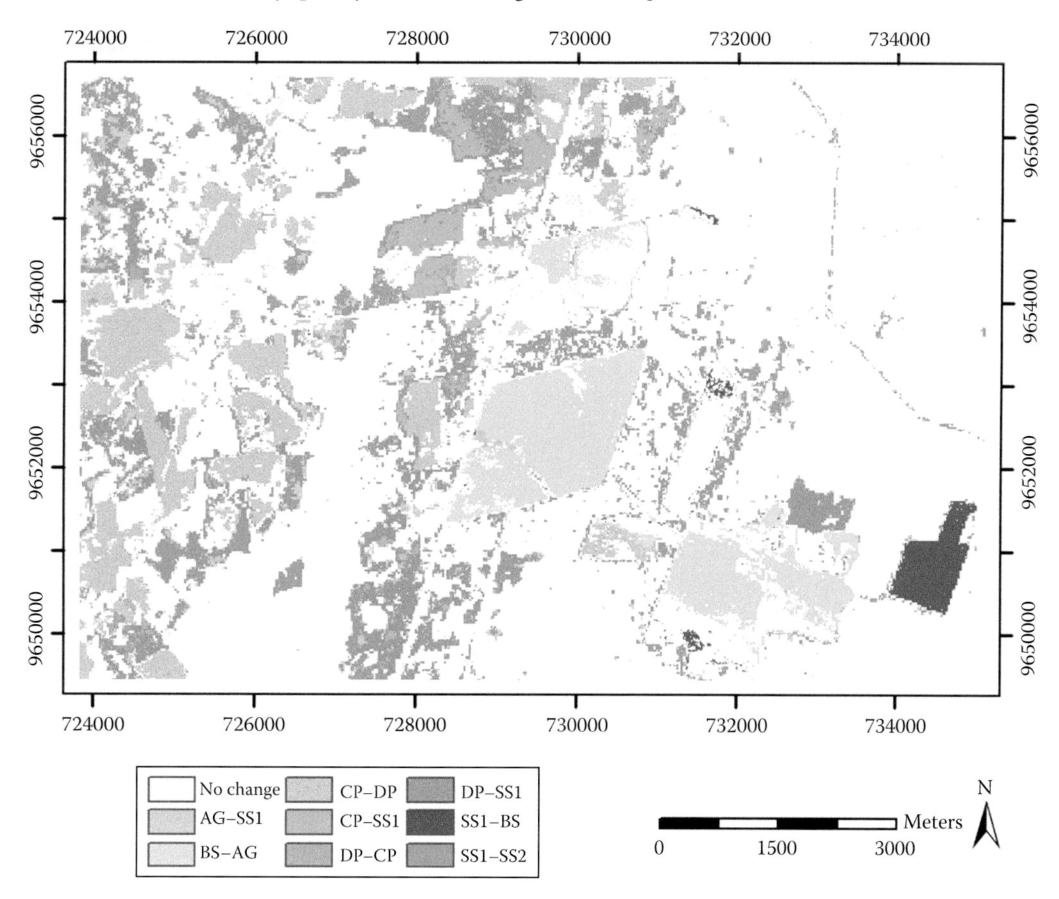

FIGURE 18.10 Land-use/land-cover change distribution using support vector machine-based direct classification between 2009 and 2013. Change types are agriculture to initial secondary succession (AG–SS1), bare soil to agriculture (BS–AG), clean pasture to dirty pasture (CP–DP), clean pasture to initial secondary succession (CP–SS1), dirty pasture to clean pasture (DP–CP), dirty pasture to initial secondary succession (DP–SS1), initial secondary succession to bare soil (SS1–BS), and initial secondary succession to intermediate secondary succession (SS1–SS2).

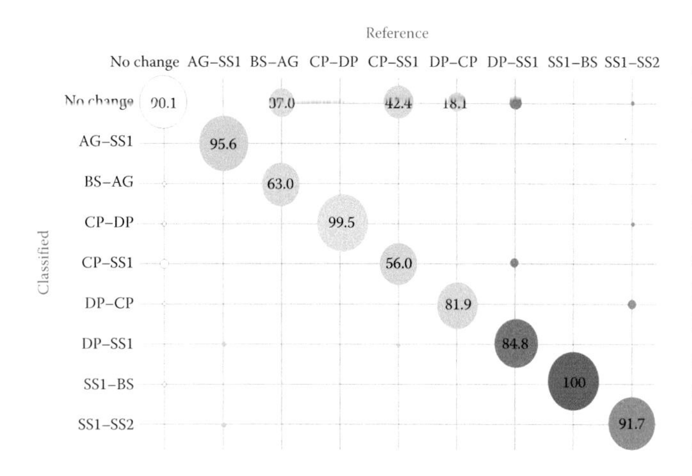

FIGURE 18.11 Confusion matrix for land-cover change result using direct classification between 2009 and 2013.

The good change detection result using SVM-based direct classification can be used to infer the roles of human activities in influencing land-use/land-cover change. For example, the conversion from BS to AG or from SS1 to BS (preparation soils) shows the interference of human activities in the region. However, the conversions from CP to DP or SS1, from DP to SS1, or from AG to SS1 imply poor management of the land, resulting in land degradation.

Although direct classification provides good change detection accuracy, we needed to prove that the change and nonchange classes obtained in the change detection process were indeed the existing classes in the region. Thus, to show the representativeness of this methodology, the 2009 and 2013 land-use/land-cover maps were retrieved from the change map. Each pixel of the land-use/land-cover map for 2009 is related to the initial condition in the change process, and each pixel of the land-use/land-cover map for 2013 contains information related to the late stage. Their accuracies are evaluated using independent test samples (see Table 18.5), and the results for these classifications are shown in Figure 18.12. Comparing this information with the Radarsat classification result as shown in Figure 18.6, the result in Figure 18.12 overcomes the problems in Radarsat data and provides a much improved classification result.

The confusion matrixes for the 2009 and 2013 classification results with nine land-use/land-cover classes are illustrated in Figure 18.13. The kappa coefficients of 0.79 for the 2009 map and 0.78 for the 2013 map are calculated, implying that the 16 classes of change/nonchange used in the change detection process described earlier were able to generate consistent classifications for 2009 and 2013. It is also interesting to observe that the change map has a higher kappa coefficient than the land-use/land-cover map for each year, showing the importance of object correlation in time, which improves the overall accuracy.

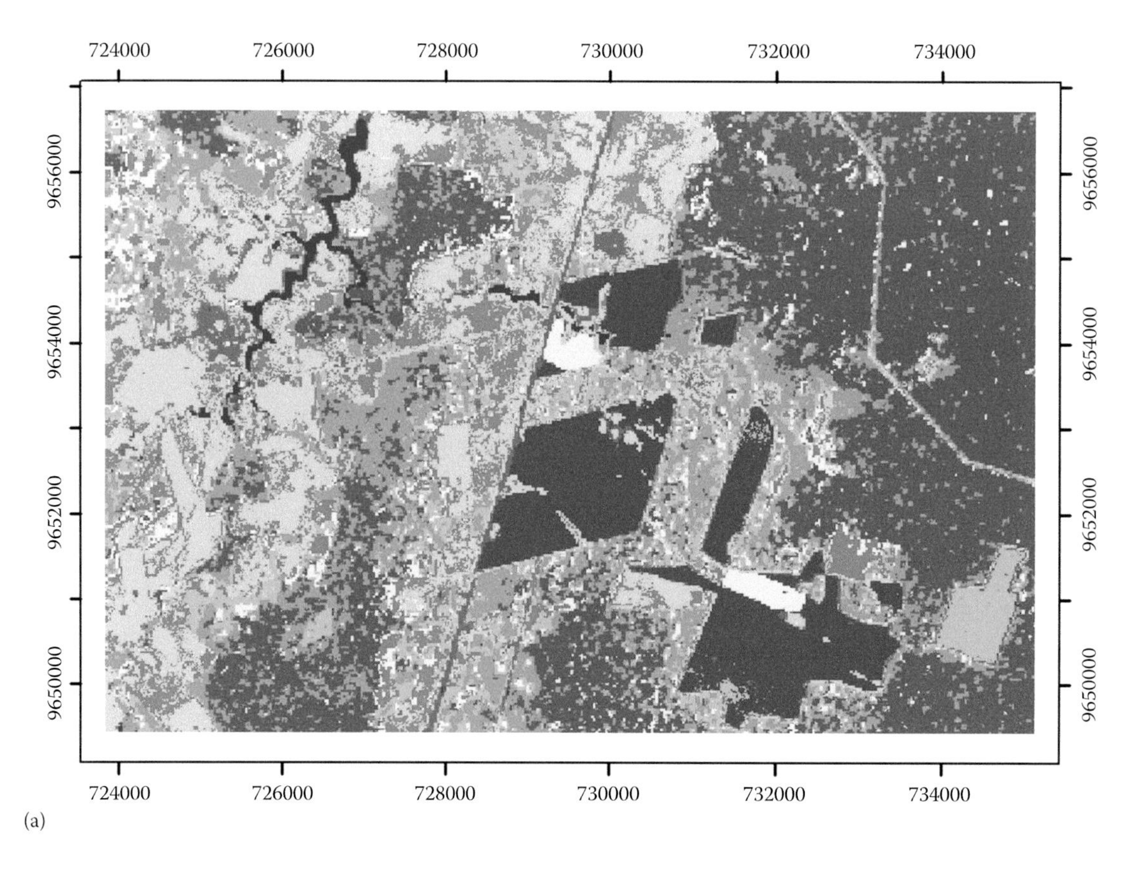

FIGURE 18.12 Land-use/land-cover map in 2009 (a), which are inferred from the change detection results using support vector machine-based direct classification.

(Continued)

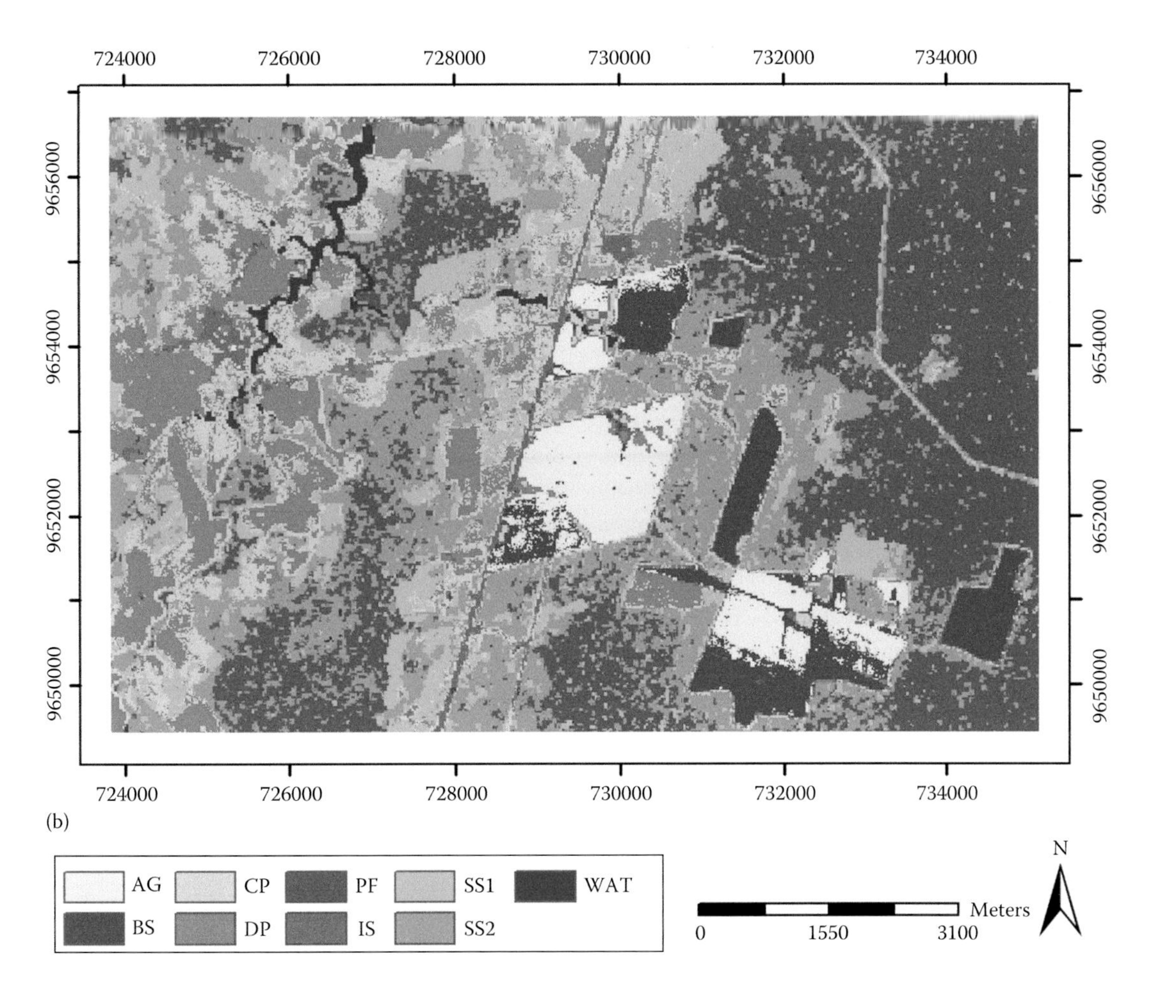

FIGURE 18.12 (*Continued*) Land-use/land-cover map in 2013 (b), which are inferred from the change detection results using support vector machine–based direct classification.

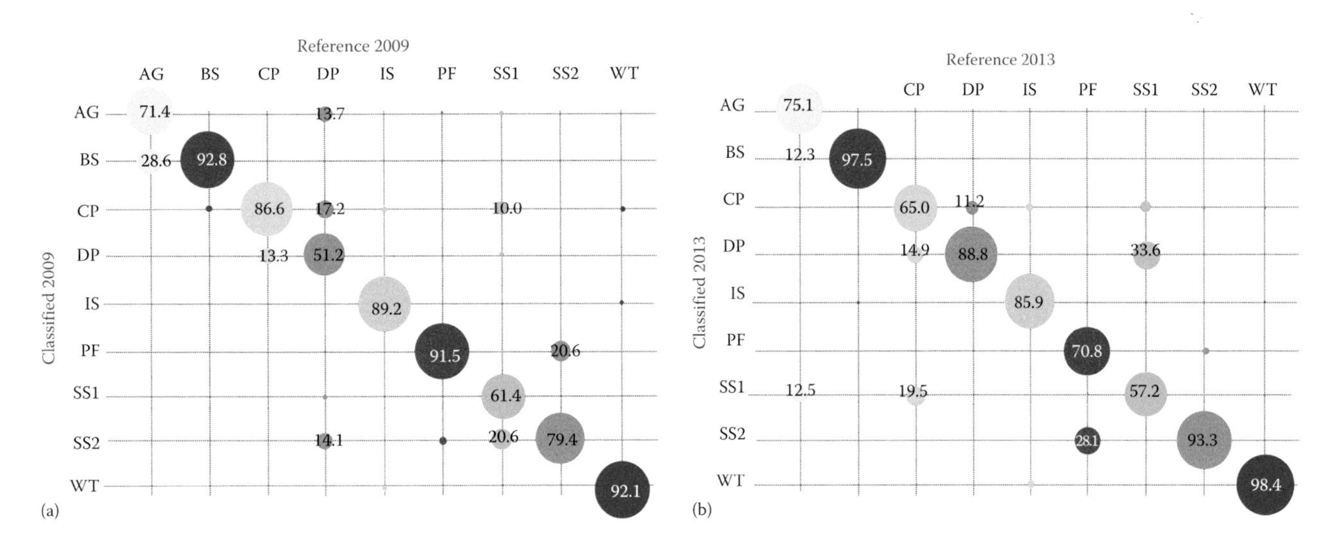

FIGURE 18.13 Confusion matrices for land-cover maps in 2009 (a) and 2013 (b) developed from the results from support vector machine–based direct classification.

18.3.5 Summary of the Case Study

In the moist tropical regions, collection of the same optical sensor data at different dates is difficult due to cloud cover. Use of different sensor data, especially radar data, is needed but is challenging. Traditional techniques such as image differencing and principal component analysis cannot be used for the change detection based on optical and radar data. Although the postclassification comparison approach can be used, the poor land-cover classification using pure radar data makes the change detection results very poor. The SVM-based direct classification used in this research provides a much better change detection result with overall accuracy of 87.8%. One problem of the postclassification comparison technique is the feasibility matrix dependency on the time span considered, which will make the whole process more complex. The direct classification approach can lead to better results because it is supervised not only with change samples but also with nonchange samples. The main drawback of this approach is the difficulty in obtaining a sufficient number of reference samples for change classes from the same region in different years to generate change regions of interest. Another problem is that some important land-cover transitions may be missed without a sound field survey. This research provides a preliminary result of using optical sensor and radar data for change detection without exhaustive examination of different image-processing methods for extraction of optimal variables and different algorithms for identifying the best change detection result. More research is needed to examine how to integrate multisensor data to implement land-cover change detection.

18.4 Conclusions

Change detection is a comprehensive procedure that requires careful consideration for each step, as illustrated in Figure 18.1. When the study area and user's needs are defined, selection of suitable variables and corresponding algorithms becomes the most important aspect. Therefore, much research has concentrated on these topics for a long time. Overall, per-pixel-based change detection techniques have been the most common approaches. As high-spatial-resolution images are now easily available, object-based methods are gaining attention. On the other hand, the importance of rapidly updating the land-cover dynamic change at national and global scales leads to increasing application of coarse-spatial-resolution satellite imagery such as MODIS. Subpixel-based methods are needed to solve the mixed-pixel problem, but new techniques are required to extract the fractional land-cover change in a large area. In order to make full use of the merits of different variables and algorithms, hybrid methods that can effectively integrate these features will become a new research trend in the near future. Another important research topic is the integration of multisensor data for change detection. This is especially valuable in moist tropical regions due to cloud cover. This chapter provides a new exploration for using optical sensor EO-1 ALI and Radarsat data for land-cover change detection in the Brazilian Amazon using direct classification with the assistance of sound field survey data. More research is needed to explore the techniques to effectively integrate multisensor data and multitemporal scales to provide accurate change detection results.

Acknowledgments

The authors acknowledge financial support from the National Council for Scientific and Technological Development (CNPq-Brazil, grants 307666/2011-5, 401528/2012-0, and 211268/2013-5) and facility support from the Center for Global Change and Earth Observations at Michigan State University, United States, and the National Institute for Space Research (INPE–Brazil). LVD also thanks the Canadian Space Agency for providing Radarsat-2 data.

References

Akiwowo, A. and Eftekhari, M. 2013. Feature-based detection using Bayesian data fusion. *International Journal of Image and Data Fusion*, 4(4), 308–323.

Alberga, V. 2009. Similarity measures of remotely sensed multi-sensor images for change detection applications. *Remote Sensing*, 1, 122–143.

Asner, G.P. 2001. Cloud cover in Landsat observations of the Brazilian Amazon. *International Journal of Remote Sensing*, 22, 3855–3862.

Baker, C., Lawrence, R., Montagne, C., and Patten, D. 2007. Mapping wetlands and riparian areas using Landsat ETM+ imagery and decision-tree-based models. *Wetlands*, 26, 465–474.

Barnsley, M.J., Moller-Jensen, L., and Barr, S.L. 2001. Inferring urban land use by spatial and structural pattern recognition. In: Donnay, J.P., Barnsley, M.J., and Longley, P.A., eds. *Remote Sensing and Urban Analysis*. Taylor & Francis Group, London, U.K., pp. 115–144.

Bhagat, V.S. 2012. Use of remote sensing techniques for robust digital change detection of land: A review. *Recent Patents on Space Technology*, 2, 123–144.

Biging, G.S., Chrisman, N.R., Colby, D.R., Congalton, R.G., Dobson, J.E., Ferguson, R.L., Goodchild, M.F., Jensen, J.R., and Mach, T.H. 1999. Accuracy assessment of remote sensing-detected change detection. In: Khorram, S., ed. *Monograph Series*. American Society for Photogrammetry and Remote Sensing (ASPRS), Bethesda, MD, p. 65.

Borghys, D., Shimoni, M., and Perneel, C. 2007. Change detection in urban scenes by fusion of SAR and hyperspectral data. *Proceedings of SPIE, the International Society for Optical Engineering*, 6749, 27–38.

Camps-Valls, G. and Bruzzone, L. 2009. *Kernel Methods in Remote Sensing Image Processing*. John Wiley & Sons, Hoboken, NJ.

Camps-Valls, G., Gómez-Chova, L., Muñoz-Marí, J., RojoÁlvarez, J.L., and Martínez-Ramón, M. 2008. Kernel-based framework for multi-temporal and multi-source remote sensing data classification and change detection. *IEEE Transactions on Geoscience and Remote Sensing*, 46(6), 1822–1835.

Chander, G., Markham, B.L., and Helder, D.L. 2009. Summary of current radiometric calibration coefficients for Landsat MSS, TM, ETM+, and EO-1 ALI sensors. *Remote Sensing of Environment*, 113, 893–903.

Chen, G., Hay, G.J., Carvalho, L.M.T., and Wulder, M.A. 2012. Object-based change detection. *International Journal of Remote Sensing*, 33, 4434–4457.

Collins, J.B. and Woodcock, C.E. 1996. An assessment of several linear change detection techniques for mapping forest mortality using multi-temporal Landsat TM data. *Remote Sensing of Environment*, 56, 66–77.

Congalton, R.G. 1991. A review of assessing the accuracy of classification of remotely sensed data. *Remote Sensing of Environment*, 37(1), 35–46.

Congalton, R.G. and Green, K. 2008. *Assessing the Accuracy of Remotely Sensed Data: Principles and Practice*, 2nd edn. CRC Press, Taylor & Francis Group, Boca Raton, FL, p. 183.

Coppin, P., Jonckheere, I., Nackaerts, K., Muys, B., and Lambin, E. 2004. Digital change detection methods in ecosystem monitoring: A review. *International Journal of Remote Sensing*, 25, 1565–1596.

Coppin, P., Nackaerts, K., Queen, L., and Brewer, K. 2001. Operational monitoring of green biomass change for forest management. *Photogrammetric Engineering & Remote Sensing*, 67, 603–611.

Deng, J.S., Wang, K., Deng, Y.H., and Qi, G.J. 2008. PCA-based land-use change detection and analysis using multi-temporal and multi-sensor satellite data. *International Journal of Remote Sensing*, 29, 4823–4838.

Desclée, B., Bogaert, P., and Defourny, P. 2006. Forest change detection by statistical object-based method. *Remote Sensing of Environment*, 102, 1–11.

Du, P., Liu, S., Gamba, P., Tan, K., and Xia, J. 2012. Fusion of difference images for change detection over urban areas. *IEEE Journal of Selected Topics in Applied Earth Observation and Remote Sensing*, 5, 1076–1086.

Du, P., Liu, S., Xia, J., and Zhao, Y. 2013. Information fusion techniques for change detection from multi-temporal remote sensing images. *Information Fusion*, 14, 19–27.

Foody, G.M. 2002. Status of land cover classification accuracy assessment. *Remote Sensing of Environment*, 80(1), 185–201.

Foody, G.M. 2010. Assessing the accuracy of land cover change with imperfect ground reference data. *Remote Sensing of Environment*, 114, 2271–2285.

Groen, T.A., Fanta, H.G., Hinkov, G., Velichkov, I., Van Duren, I., and Zlatanov, T. 2012. Tree line change detection using historical hexagon mapping camera imagery and Google earth data. *GIScience & Remote Sensing*, 49(6), 933–943.

Guild, L.S., Cohen, W.B., and Kauffman, J.B. 2004. Detection of deforestation and land conversion in Rondonia, Brazil using change detection techniques. *International Journal of Remote Sensing*, 25(4), 731–750.

Houhoulis, P.F. and Michener, W.K. 2000. Detecting wetland change: A rule-based approach using NWI and SPOT-XS data. *Photogrammetric Engineering & Remote Sensing*, 66, 205–211.

Huang, C., Goward, S.N., Masek, J.G., Thomas, N., Zhu, Z., and Vogelmann, J.E. 2010. An automated approach for reconstructing recent forest disturbance history using dense Landsat time series stacks. *Remote Sensing of Environment*, 114, 183–198.

Huang, W., Sun, G., Dubayah, R., Cook, B., Montesano, P., Ni, W., and Zhang, Z. 2013. Mapping biomass change after forest disturbance: Applying LiDAR footprint-derived models at key map scales. *Remote Sensing of Environment*, 134, 319–332.

Hussain, M., Chen, D., Cheng, A., Wei, H., and Stanley, D. 2013. Change detection from remotely sensed images: From pixel-based to object-based approaches. *ISPRS Journal of Photogrammetry and Remote Sensing*, 80, 91–106.

Ichii, K., Maruyama, M., and Yamaguchi, Y. 2003. Multi-temporal analysis of deforestation in Rondônia state in Brazil using Landsat MSS, TM, ETM+ and NOAA AVHRR imagery and its relationship to changes in the local hydrological environment. *International Journal of Remote Sensing*, 24(22), 4467–4479.

Jensen, J.R. 2005. *Introductory Digital Image Processing: A Remote Sensing Perspective*, 3rd edn. Prentice Hall, Upper Saddle River, NJ, p. 526.

Jensen, J.R. 2007. *Remote Sensing of Environment: An Earth Resources Perspective*. Prentice Hall, Upper Saddle River, NJ.

Kennedy, R.E., Cohen, W.B., and Schroeder, T.A. 2007. Trajectory-based change detection for automated characterization of forest disturbance dynamics. *Remote Sensing of Environment*, 110, 370–386.

Lambin, E.F. and Strahler, A.H. 1994. Change-vector analysis: A tool to detect and categorize land-cover change processes using high temporal-resolution satellite data. *Remote Sensing of Environment*, 48, 231–244.

Li, D. 2010. Remotely sensed images and GIS data fusion for automatic change detection. *International Journal of Image and Data Fusion*, 1(1), 99–108.

Li, G., Lu, D., Moran, E., Dutra, L., and Batistella, M. 2012a. A comparative analysis of ALOS PALSAR L-band and RADARSAT-2 C-band data for land-cover classification in a tropical moist region. *ISPRS Journal of Photogrammetry and Remote Sensing*, 70, 26–38.

Li, G., Lu, D., Moran, E., and Sant'Anna, S.J.S. December 14, 2012b. A comparative analysis of classification algorithms and multiple sensor data for land use/land cover classification in the Brazilian Amazon. *Journal of Applied Remote Sensing*, 6(1), 061706.

Li, X. and Yeh, A.G.O. 1998. Principal component analysis of stacked multi-temporal images for monitoring of rapid urban expansion in the Pearl River Delta. *International Journal of Remote Sensing*, 19(8), 1501–1518.

Lillestrand, R.L. 1972. Techniques for change detection. *IEEE Transactions Computers*, C-21, 654–659.

Lo, C.P. and Shipman, R.L. 1990. A GIS approach to land-use change dynamics detection. *Photogrammetric Engineering & Remote Sensing*, 56, 1483–1491.

Lu, D., Batistella, M., and Moran, E. 2008a. Integration of Landsat TM and SPOT HRG images for vegetation change detection in the Brazilian Amazon. *Photogrammetric Engineering & Remote Sensing*, 74, 421–430.

Lu, D., Ge, H., He, S., Xu, A., Zhou, G., and Du, H. 2008b. Pixel-based Minnaert correction method for reducing topographic effects on the Landsat 7 ETM+ image. *Photogrammetric Engineering & Remote Sensing*, 74(11), 1343–1350.

Lu, D., Hetrick, S., Moran, E., and Li, G. 2012. Application of time series Landsat images to examining land use/cover dynamic change. *Photogrammetric Engineering & Remote Sensing*, 78(7), 747–755.

Lu, D., Li, G., and Moran, E. 2014. Current situation and needs of change detection techniques. *International Journal of Image and Data Fusion*, 5(1), 13–38.

Lu, D., Li, G., Moran, E., Dutra, L., and Batistella, M. 2011. A comparison of multisensor integration methods for land-cover classification in the Brazilian Amazon. *GIScience & Remote Sensing*, 48(3), 345–370.

Lu, D., Li, G., Moran, E., and Hetrick, S. 2013. Spatiotemporal analysis of land-use and land-cover change in the Brazilian Amazon. *International Journal of Remote Sensing*, 34(16), 5953–5978.

Lu, D., Mausel, P., Batistella, M., and Moran, E. 2005. Land-cover binary change detection methods for use in the moist tropical region of the Amazon: A comparative study. *International Journal of Remote Sensing*, 26(1), 101–114.

Lu, D., Mausel, P., Brondizio, E., and Moran, E. 2004. Change detection techniques. *International Journal of Remote Sensing*, 25, 2365–2401.

Lu, D. and Weng, Q. 2007. A survey of image classification methods and techniques for improving classification performance. *International Journal of Remote Sensing*, 28(5), 823–870.

Mas, J.F. 1999. Monitoring land-cover change: A comparison of change detection techniques. *International Journal of Remote Sensing*, 20, 139–152.

Morisette, J.T. and Khorram, S. 2000. Accuracy assessment curves for satellite-based change detection. *Photogrammetric Engineering & Remote Sensing*, 66(7), 875–880.

Mountrakis, G., Im, J., and Ogole, C. 2011. Support vector machines in remote sensing: A review. *ISPRS Journal of Photogrammetry and Remote Sensing*, 66(3), 247–259.

Mubea, K. and Menz, G. 2012. Monitoring land-use change in Nakuru (Kenya) using multi-sensor satellite data. *Advances in Remote Sensing*, 1, 74–84.

Negri, R.G., Dutra, L.V., and Sant'Anna, S.J.S. 2012. Support vector machine and Bhattacharyya kernel function for region based classification. *IEEE International Geoscience and Remote Sensing Symposium (IGARSS)*, Munich, Germany, pp. 5422–5425.

Negri, R.G., Dutra, L.V., and Sant'Anna, S.J.S. 2014. An innovative support vector machine based method for contextual image classification. *ISPRS Journal of Photogrammetry and Remote Sensing*, 87, 241–248.

Negri, R.G., Sant'Anna, S.J.S., and Dutra, L.V. 2011. Semi-supervised remote sensing image classification methods assessment. *IEEE International Geoscience and Remote Sensing Symposium (IGARSS)*, Vancouver, Canada, pp. 2939–2942.

Negri, R.G., Sant'Anna, S.J.S., and Dutra, L.V. 2013. A new contextual version of Support Vector Machine based on hyperplane translation. *IEEE International Geoscience and Remote Sensing Symposium (IGARSS)*, Melbourne, Australia, pp. 3116–3119.

Olofsson, P., Foody, G.M., Stehman, S.V., and Woodcock, C.E. 2013. Making better use of accuracy data in land change studies: Estimating accuracy and area and quantifying uncertainty using stratified estimation. *Remote Sensing of Environment*, 129, 122–131.

Pereira, L.O., Freitas, C.C., Sant'Anna, S.J.S., Lu, D., and Moran, E.F. 2013. Optical and radar data integration for land use and land cover mapping in the Brazilian Amazon. *GIScience & Remote Sensing*, 50(3), 301–321.

Perumal, K. and Bhaskaran, R. 2009. SVM-based effective land use classification system for multispectral remote sensing images. *International Journal of Computer Science and Information Security*, 6(2), 97–105.

Petit, C.C. and Lambin, E.F. 2001. Integration of multi-source remote sensing data for land cover change detection. *International Journal of Geographical Information Science*, 15(8), 785–803.

Reiche, J., Souza, C.M., Hoekman, D.H., Verbesselt, J., Persaud, H., and Herold, M. 2013. Feature level fusion of multi-temporal ALOS PALSAR and Landsat data for mapping and monitoring of tropical deforestation and forest degradation. *IEEE Journal of Selected Topics in Applied Earth Observations and Remote Sensing*, 6, 2159–2173.

Serra, P., Pons, X., and Sauri, D. 2003. Post-classification change detection with data from different sensors: Some accuracy considerations. *International Journal of Remote Sensing*, 24, 3311–3340.

Shi, W. and Hao, M. 2013. Analysis of spatial distribution pattern of change-detection error caused by misregistration. *International Journal of Remote Sensing*, 34(19), 6883–6897.

Singh, A. 1989. Digital change detection techniques using remotely-sensed data. *International Journal of Remote Sensing*, 10, 989–1003.

Stow, D.A. and Chen, D.M. 2002. Sensitivity of multitemporal NOAA AVHRR data of an urbanizing region to land-use/land-cover change and misregistration. *Remote Sensing of Environment*, 80, 297–307.

Theodoridis, S. and Koutrombas, K. 2008. *Pattern Recognition*. Academic Press, Burlington, MA, p. 984.

Thomas, N.E., Huang, C., Goward, S.N., Powell, S., Rishmawi, K., Schleeweis, K., and Hinds, A. 2011. Validation of North American forest disturbance dynamics derived from Landsat time series stacks. *Remote Sensing of Environment*, 115, 119–132.

Tian, J., Reinartz, P., dÁngelo, P., and Ehlers, M. 2013. Region-based automatic building and forest change detection on Cartosat-1 stereo imagery. *ISPRS Journal of Photogrammetry and Remote Sensing*, 79, 226–239.

Treitz, P. and Rogan, J. 2004. Remote sensing for mapping and monitoring land-cover and land-use change: An introduction. *Progress in Planning*, 61, 269–279.

Tso, B. and Mather, P.M. 2009. *Classification Methods for Remotely Sensed Data*. Taylor & Francis Group, London, U.K., p. 356.

Van Oort, P.A.J. 2007. Interpreting the change detection error matrix. *Remote Sensing of Environment*, 108(1), 1–8.

Vapnik, V., Golowich, S.E., and Smola, A. 1996. Support vector method for function approximation, regression estimation, and signal processing. In: M.C. Mozer, M.I. Jordan, and T. Petsche, eds., *Advances in Neural Information Processing Systems 9*. [S.l.] MIT Press, Cambridge, MA, pp. 281–287.

Vicente-Serrano, S.M., Pérez-Cabello, F., and Lasanta, T. 2008. Assessment of radiometric correction techniques in analyzing vegetation variability and change using time series of Landsat images. *Remote Sensing of Environment*, 112, 3916–3934.

Volpi, M., Tuia, D., Bovolo, F., Kanevski, M., and Bruzzone, L. 2013. Supervised change detection in VHR images using contextual information and support vector machines. *International Journal of Applied Earth Observation and Geoinformation*, 20, 77–85.

Walter, V. 2004. Object-based classification of remote sensing data for change detection. *ISPRS Journal of Photogrammetry and Remote Sensing*, 58(3–4), 225–238.

Webb, A.R. 2002. *Statistical Pattern Recognition*. John Wiley & Sons, New York, p. 514.

Wilson, E.H. and Sader, S.A. 2002. Detection of forest type using multiple dates of Landsat TM imagery. *Remote Sensing of Environment*, 80, 385–396.

Yeh, A.G. and Li, X. 2001. Measurement and monitoring of urban sprawl in a rapidly growing region using entropy. *Photogrammetric Engineering & Remote Sensing*, 67, 83–90.

Zeng, Y., Zhang, J., van Genderen, J.L., and Zhang, Y. 2010. Image fusion for land cover change detection. *International Journal of Image and Data Fusion*, 1(2), 193–215.

VII

Integrating Geographic Information Systems (GIS) and Remote Sensing in Spatial Modeling Framework for Decision Support

19

Geoprocessing, Workflows, and Provenance

Jason A. Tullis
University of Arkansas

Jackson D. Cothren
University of Arkansas

David P. Lanter
CDM Smith

Xuan Shi
University of Arkansas

W. Fredrick Limp
University of Arkansas

Rachel F. Linck
University of Arkansas

Sean G. Young
University of Iowa

Tareefa S. Alsumaiti
United Arab Emirates University

Acronyms and Definitions

ACSM	American Congress of Surveying and Mapping
API	Application programming interface
CDAT	Climate data analysis tool
CI	Cyberinfrastructure
DBMS	Database management system
FGDC	Federal Geographic Data Committee
HMM	Hidden Markov model
ISO	International Standards Organization
LIDAR	Light detection and ranging
LULC	Land use/land cover
MODAPS	MODIS adaptive data processing system
NCDCDS	National Committee for Digital Cartographic Data Standards
NSDI	National Spatial Data Infrastructure
OGC	Open Geospatial Consortium
OpenMI	Open modeling interface
PROV-DM	PROV data model
REST	Representational state transfer
SOAP	Simple object access protocol
SOC	Service-oriented computing
SOI	Service-oriented integration
SQL	Structured query language
W3C	World Wide Web Consortium
WPS	Web processing service
WSDL	Web services description language
XML	Extensible markup language
XSEDE	eXtreme Science and Engineering Development Environment

19.1 Introduction

Integrated remote sensing and GIS-assisted problem solving now supports a remarkable array of domains (e.g., food and agricultural security, climate change, forest management, heritage preservation, and urban and regional planning) and is being configured in a great variety of technical means. Given the sheer quantity of innovations reported in journals and books (including the Remote Sensing Handbook), any one expert may be keenly aware of only a fraction of the detailed remote sensing and related geospatial methods available to address a given problem statement. Regardless of the remote sensing application under study or review, some reliance (whether implied or reported) is always made upon the *geoprocesses* and *workflows* associated with any geospatial artifacts produced. In the context of a specific geospatial decision support artifact (e.g., a map of predicted crop yield in kg/ha), a record of the specific geoprocesses may be termed geospatial *provenance* (or *lineage*; see Section 19.1.1). This chapter explores how remote sensing–assisted geoprocessing and related GIS workflows have been or may be combined with digital provenance information in order to augment scientific reproducibility, comparison, trust, or to otherwise improve remote sensing–assisted decision support.

Increasingly of interest in computer systems, digital provenance has relatively early geospatial origins that date back to at least the 1980s (e.g., Chrisman 1983), with a definite resurgence around 2009 (e.g., Yue et al. 2010a). The early and expanded geospatial interest and connection to provenance are driven in large part by the question of methodological innovation. For remote sensing and GIS integration to best improve the quality of decision making tools across a range of applications and domains, it seems reasonable that, if possible, such innovation must first be machine recognizable. Unfortunately, many geospatial decision support tools lack suitable means to even replicate their findings, and innovation reported is naturally bracketed by complex questions of accuracy, fitness for use, and a variety of other qualitative and quantitative metrics related to reliability and trust. So, while there is broad conceptual agreement that machine-interpretable source and process history records are vital and may even be scientifically transformative in the modern era, questions remain unanswered on how provenance information may simultaneously benefit multiple domains (including the geospatial domain), and what mechanisms for its digital capture and exchange will most successfully convey those benefits.

There are at least two good reasons to believe that even partial success toward machine-interpretable geospatial process history records will be rewarded. First, correct expert interpretation of the full scope of relevant methods, procedures, algorithms, and expert knowledge is subject to entropy and constitutes an increasingly complex, even daunting companion to the twenty-first-century *big [geospatial] data* (Hey et al. 2009). Second, as remote sensing and other geospatial techniques are communicated in the scientific literature, there is a well-known continuing expectation and scholarly requirement that previously published studies are carefully acknowledged for their relevant achievements and/or limitations. Failure to increasingly harness machine power on these two fronts (but to continue interpretations by experts alone) is probably not a viable long-term option. In a related example from computer systems, Buneman (2013) notes that the underappreciated machine-managed provenance in software version control systems has helped prevent a total disaster in software engineering.

It is clear that absent the kinds of methodological analyses enabled in part through exchange of provenance information, an increasingly data-intensive *geo-cyberinfrastructure* (Di et al. 2013a) renders comprehensive remote sensing–assisted geospatial workflow interpretations, comparisons, and knowledge transfers ever more difficult by experts alone. Furthermore, depending on the geospatial laboratory setting and the capabilities of a given research team, the actual digital methods linked to published materials may overlap significantly with previously reported work, may offer similar results using a more or less computationally efficient means of problem solving, and/or may be idiosyncratic to individual skills and experience. In an integrated geoprocessing, workflow, and provenance cycle, expert refinement of remote sensing–assisted decision support knowledge may be augmented by software agents capable of automated exchange and recognition of innovation (Figure 19.1).

Over the past 25 years, various prototype forms of geospatial provenance have been implemented in shared workflow environments, including those specialized for high-performance capabilities. In spite of the potential of these prototypes, single user/workstation geoprocessing and workflow design continue to be a dominant tradition with many active options (e.g., from Hexagon Geospatial, Exelis Visual Information Solutions, and ESRI). There is therefore a discrepancy between futuristic collaborative goals and the actual state of the art of remote sensing–assisted software. There are also variations in how provenance itself is defined, whether specifically in a remote sensing or geospatial-related forum, or more broadly in computer systems. It therefore seems reasonable to report progress in terms of what the actual computational environments entail and which definitions are implied.

19.1.1 Working Definitions

Though commonly understood in a broad remote sensing and geospatial computation parlance, Wade and Sommer (2006) define *geoprocessing* in the context of the many tools available in one software platform (ESRI's ArcGIS) with an emphasis on input GIS datasets, operations performed, and associated outputs. More generically, its root, *process*, implies an instance of a computer program execution, and this is naturally compatible with a geospatial/remote sensor data processing software context. Of course, identical geospatial computer programs operating on identical input datasets may produce different results as a function of additional configuration parameters. For example, raster-based geoprocessing tools in ESRI's ArcGIS 10 platform allow for an *environment setting* called *Snap Raster*. This setting allows the user to specify the spatial grid on which computations are made. In practice, use of this parameter allows pixels in an output raster layer to be exactly aligned with another raster having the same cell size. To a novice, the resulting subpixel geometric shift may seem inconsequential at the overview scale. However, remote sensing experts know that when geoprocessing tools are chained together into a *workflow* (in the present context, a repeatable

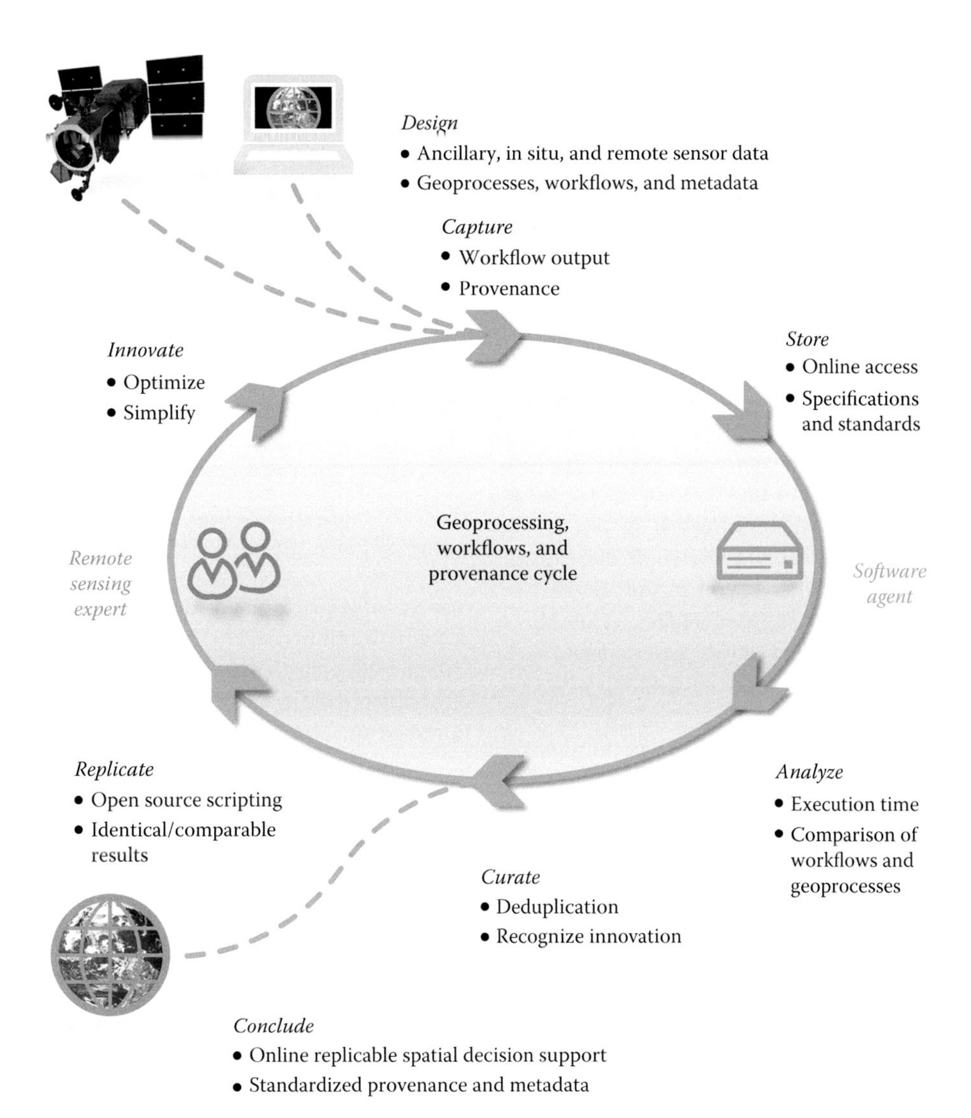

FIGURE 19.1 Integrated geoprocessing, workflows, and provenance may be conceptualized as a positive developmental cycle used to refine remote sensing knowledge before decision support is communicated. Highlighted aspects of this cycle suggest a capacity of remote sensing experts, in conjunction with software agents, to cooperatively capture, store, analyze, curate, replicate, and innovate remote sensing–assisted decision support methods. (Artist image of WorldView-3, Courtesy of DigitalGlobe, 2014.)

sequence of geoprocesses of interest to a person or group), environment settings like *Snap Raster* can affect the logic of a decision support conclusion.

Provenance traces back to 1294 in Old French as a derivative of the Latin *provenire*, and while Merriam-Webster (2014) emphasizes provenance as a *concept* (e.g., ownership history of a painting), Oxford University Press (2014) highlights the *record* of such provenance (Moreau 2010). In the art domain where the term is very well established, provenance entails an artifact's complete ownership history, but ideally will also include artistic, social, and political influences upon the work from its creation to the present day. There is an established research process for obtaining an artifact's trusted provenance, and the information is highly valued, particularly to authenticate real versus fraudulent works (IFAR 2013; Yeide et al. 2001). As a related term, provenance is now increasingly used in a broad range of fields (e.g., archaeology, computer science, forestry, and geology) with usually overlapping definitions.

Computational definitions of provenance are more numerous than in other domains, largely because of (1) the difference between *concepts* of digital records and *actual* digital records, and (2) the variation in software environment such as a database management system (DBMS) versus file-based processing (Moreau 2010). Understanding provenance *within* DBMS queries requires more computationally detailed observations than understanding provenance at a more generalized workflow level (where one step in the workflow may entail multiple database queries). Various traditions further influence how provenance is viewed, for example, whether it is conflated with *metadata* or *trust*, two closely related but distinct concepts (Gil et al. 2010). Given the infrastructural importance of the web in remote sensing–assisted decision support, the following W3C

Provenance Incubator Group's working definition of provenance (in a web resource context) carries significant weight:

> Provenance of a resource is a record that describes entities and processes involved in producing and delivering or otherwise influencing that resource. Provenance provides a critical foundation for assessing authenticity, enabling trust, and allowing reproducibility. Provenance assertions are a form of contextual metadata and can themselves become important records with their own provenance.
>
> **Gil et al. (2010)**

It should be noted that while *provenance* and *lineage* are here used interchangeably, one can argue that there are subtle differences in their meanings. Process history seems to fit more easily with the many definitions attributed to provenance, and lineage implies a kind of genealogy or data pedigree record relative to a remote sensing–assisted decision support artifact. While these semantic differences are not a point of the present focus, each word will appear in its historical context (beginning with lineage). Also, a number of surveys have been conducted on provenance including some with a geoprocessing and workflow flavor. For example, Yue and He (2009) provide a review covering various aspects of geospatial provenance. For a broader perspective, Bose and Frew (2005) provide a review covering provenance in geospatial as well as other domains. More recently, Di et al. (2013b) provide an overview of geoscience data provenance.

19.2 Historical Context

The earliest work in geospatial lineage was spurred in the United States through the formation of the National Committee for Digital Cartographic Data Standards (NCDCDS) by the American Congress of Surveying and Mapping in 1982 (Bossler et al. 2010). In 1988, chaired by Dr. Harold Moellering from Ohio State University, the NCDCDS proposed five fundamental components of a geospatial data quality report, including (1) lineage, (2) positional accuracy, (3) attribute accuracy, (4) logical consistency, and (5) completeness. The NCDCDS described lineage in detail, which they presented as the *first* quality component. Less than a third of their description for lineage follows (Moellering et al. 1988, p. 132):

> The lineage section of a quality report shall include a description of the source material from which the data were derived, and the methods of derivation, including all transformations involved in producing the final digital files. The description shall include the dates of the source material...

As geospatial workflows began to transition from analog to digital environments, it became clear that lineage-implied geoprocesses would need to be tracked from their origins, through revisions to the data, and finally to the output (Moore 1983). Chrisman (1983) noted that unfortunately over its lifetime, lineage information in quality records would be subject to entropy or fragmentation as a result of continuous GIS maintenance. He described *reliability diagrams* (for intelligence and other reliability-sensitive applications) embedded with lineage-related geometry and attributes (e.g., polygons identifying specific aerial photographic sources) and recommended them to be incorporated in typical GIS design. While not typically portrayed as lineage or provenance today, this type of lineage-related geodata, such as DigitalGlobe image collection footprints accessible in Google Earth, is extremely useful for visualization purposes and may resist digital entropy due to established geodata interoperability.

Beyond the challenges presented by digital records of lineages for multiple geodata versions, Langran and Chrisman's (1988) emphasis on multitemporal GIS highlighted additional record complexity that would be required. Nyerges's (1987) discussion on geodata exchange implied that quality metadata (including lineage information) could eventually facilitate geoprocessing design (workflows) with the two being mutually dependent. Others including Grady (1988) reasoned that lineage need not only support records of data quality but could in turn be used to record societal mandates (e.g., legislative drivers of geodata development) in the lineage information. While the existence of these additional complexities and potential requirements for geospatial lineage/provenance did not thwart attempts to forge ahead with possible software solutions, they pointed to significant challenges.

19.2.1 Digital Provenance in Remote Sensing and Geospatial Workflows

Over the last few decades and especially in the last 5 years, there has been significant attention given to understanding lineage/provenance in computer systems, and a variety of formalisms have been developed to understand their role in scientific workflows (e.g., Bose and Frew 2005; Buneman and Davidson 2010; Hey et al. 2009; Simmhan et al. 2005). In the following text, we highlight pioneering digital advances with geospatial lineage (circa 1990s) and more recent geo-cyberinfrastructure advances in provenance (circa 2000s to present).

19.2.1.1 Pioneering Work in Geospatial Lineage

As Chrisman (1986) suggested, "evaluation and judgment of fitness of use must be the responsibility of the user, not the producer. To carry out this responsibility, the user must be presented with much more information to permit an informed decision" (p. 352). Moellering et al. (1988) later emphasized producers' obligation to first document and update the lineage of their data in order to trace all the work (whether analog or digital) from original source materials through the intermediate processes to final digital output. It became obvious that both GIS software and international standards would be needed to facilitate the development and the maintenance of such records.

An early version of ESRI's ARC/INFO Geographic Information System featured a LIBRARIAN module capable of capturing and querying some aspects of geospatial lineage. Using the module's CATALOG command, a database administrator could retrieve information on map production status as well as review time stamps and coordinates of recent map updates (Aronson and Morehouse 1983). In the mid-1980s, the U.S. Geological Survey (USGS) began development of a GIS-linked automated cartographic workflow system called Mark II with partial lineage capabilities. An important part of Mark II's design was its capacity to track the location (e.g., network address) of datasets and their progress from curated archive toward final map products (Anderson and Callahan 1990; Guptill 1987). While it was envisioned this system would play a key role in fulfilling the National Mapping Program's mission through 2000, the agency focus transitioned by the mid-1990s toward GIS data development including the National Map. The first reported development of a system to specifically and directly address geospatial lineage was David Lanter's *Geolineus* project commenced in the late 1980s as part of his doctoral research at the University of South Carolina's Department of Geography (Lanter 1989). As the prototype pioneering work in geospatial lineage/provenance, this is reviewed in detail with added explanation.

Lanter invented a method and means to capture, structure, and process geospatial lineage to determine and communicate the meaning and integrity of the contents of a GIS database (Lanter 1993a). His metadata and processing algorithms track and document remotely sensed and other geodata sources and analytic transformations applied to them to derive new datasets. In addition to differentiating between source and derived datasets, Lanter further distinguished intermediate and product-derived datasets. More concisely, let

$$\text{Datasets} = \{\text{Dataset}_i : i = \text{source}, \text{derived}\},$$

$$\text{Dataset}_{\text{derived}} = \{\text{Dataset}_{\text{derived}.k} : k = \text{intermediate}, \text{product}\}.$$

Source datasets can be the results of in situ sampling and data collection, remote sensing, or ancillary data (e.g., digitization of maps, or thematic data resulting from digital processing of remotely sensed data). Initially, only source datasets are available for geoprocessing and transformation into a derived dataset (Figure 19.2; $n \geq 1$, $m = 0$). Later, new datasets can be generated exclusively from derived datasets ($n = 0$, $m \geq 1$) using spatial analysis transformations such as reclassification, distance measurement (buffering), connectivity, neighborhood characterization, and summary calculations. Alternatively, new datasets can be derived from inputs that include sources, derived, or both ($n + m > 1$) using multi-input transformations such as arithmetic, statistical, and logical overlays, as well as drainage network and viewshed determinations.

Lanter classified datasets into source, intermediate, and product types (Figure 19.2), and related them to one another as inputs and outputs of each data processing step of an analytical application. He gave input datasets *parent* links pointing to output datasets they were used to create (Who am I the parent of?) and provided output datasets *child* links connecting them back to their input datasets (Who am I the child of?). Each parent-and-child relationship was defined as an ordered pair of input and output datasets. Lanter's parent relationship identified the derived output given a source or derived input dataset, while his child relationship would identify a derived or source dataset when given an output dataset.

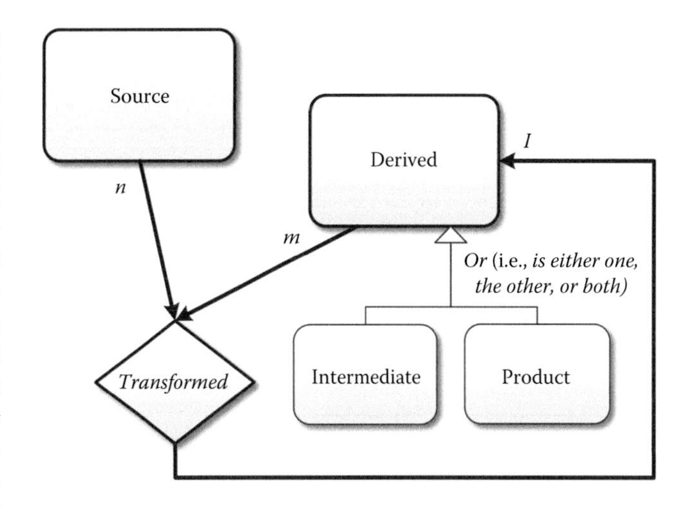

FIGURE 19.2 Relationship among source and derived datasets, where each instance of the latter may be either an intermediate or product dataset, or both.

Child links connecting output datasets to their inputs enabled automatic deduction of which datasets within an analytic database are sources and which are derived (Lanter 1993b). Derived datasets are connected to their inputs by child links, while sources lack such links. Lanter defined his *child* operator to take a derived dataset, access its child links, and identify inputs used to create it. His *Ancestors* algorithm applied the *child* operator and by a recursive function traced the child links to identify datasets used to create a derived dataset, including any sources in the geoprocessing application. Lanter defined the *parent* operator to take a source or derived dataset as input, and access and traverse its parent links to identify all the outputs derived from it. His *Descendants* function recursively traced parent links and identified all datasets derived from a source or other derived input dataset used within a geoprocessing application.

Classification of datasets into source, intermediate, and product paved the way to structuring additional lineage metadata attributes. Lanter used the artificial intelligence *frame* data structure to organize knowledge about the metadata properties of source, intermediate, and product dataset types. Each source dataset was provided a frame for storing source properties such as its name, feature type(s), date(s), responsible agency, scale, projection, and accuracy attributes. He provided each derived dataset with a frame for storing detailed metadata elements about where it is physically stored, the command applied to derive it, the command's parameters, who derived it, and other aspects of its derivation. Lanter saw products as derived datasets that were provided an additional frame for metadata detailing the analysis goal the dataset was intended to meet, intended audience/users of the dataset, when it was released, etc.

More formally, each Dataset_i (i.e., source, intermediate, or product) was provided an ordered list of metadata properties, A_j, such that $A_j = \{A_{j1}, A_{j2},\ldots,A_{jk=f(i)}\}$. Specifically,

$$\text{Dataset}_{\text{source}}A_{\text{source}} = \{\text{Name}, \text{Features}, \text{Data}, \text{Scale}, \text{Projection}, \text{Agency}, \text{Accuracy},\ldots\},$$

$$\text{Dataset}_{\text{intermediate}}A_{\text{intermediate}} = \{\text{Name}, \text{Command}, \text{Parameters}, \text{User}, \text{Date},\ldots\},$$

and

$$\text{Dataset}_{\text{product}}A_{\text{product}} = \{\text{Goal}, \text{Audience}, \text{Release Date}, \text{Intended Use},\ldots\}.$$

Given $w \in \{source, intermediate, product\}$, m a metadata property of w, and a_{wm} a value of A_{wm} then a $\text{dataset}_w = (a_{w1},a_{w2},\ldots, a_{wl})$.

Lanter's lineage metadata structure represented datasets as nodes coupled with source, command, and product properties, and connected them with parent and child links (Figure 19.3). Lanter adapted the *Ancestors* function to respond to lineage queries and to report on data sources and the sequence of processing (i.e., data lineage) applied to sources and intermediates to derive a target dataset (Lanter 1991). He integrated the *Ancestors* function with a rule-based processor that checked the inputs of each user-entered GIS command, determined their related sources, and evaluated their metadata to detect and warn users when they were entering commands that would otherwise combine datasets of incompatible properties such as projections, scales, and dates (Lanter 1989). Lanter subsequently modified the *Descendants* function to automatically generate and run GIS scripts and propagate new source data to update dependent intermediates and products (Lanter 1992a).

19.2.1.1.1 Geolineus

Lanter and Essinger designed the *lineage diagram*, an icon-based flowchart graphical user interface (GUI), to enable users direct interaction with lineage metadata to understand, modify, and maintain their analytical applications and ESRI's ARC/INFO's spatial data contents (Essinger and Lanter 1992; Lanter and Essinger 1991), and implemented it in *Geolineus* (Lanter 1992b)—the first lineage-enabled geospatial workflow system.

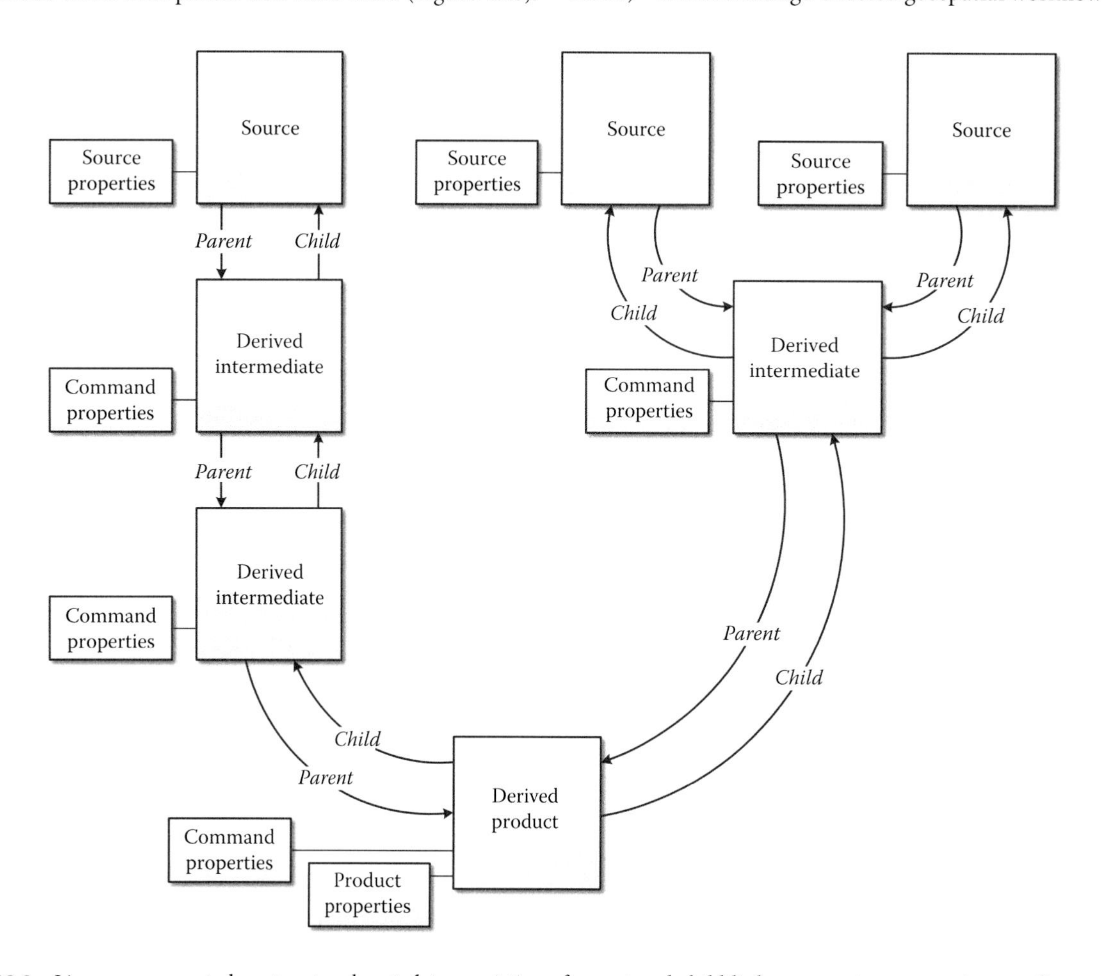

FIGURE 19.3 Lineage represented as structured metadata consisting of parent and child links connecting source, intermediate, and product datasets. While each source possesses a frame containing metadata properties, frames for derived datasets detail the GIS command used in its creation. In addition, derived product datasets possess a frame describing analytic goals, release date, and users.

Geolineus enabled users of ESRI's ARC/INFO and GRID (for image processing) to capture, create, save, exchange, analyze, and reuse lineage metadata to maintain their GIS databases. Geolineus's user interface included a lineage data flow diagram within one panel, coupled with another panel containing its own command line processor in place of the command line processor of ARC/INFO and GRID. As users added source datasets, they were presented with a form to document them, after which they were displayed along the top of the data flow diagram, each with a square icon with a bar at its top. Symbols within the icons would identify if the dataset contained points, lines, polygons, raster grids, and/or value attribute tables. Icons further down the flowchart represent datasets derived with geospatial analysis operations such as CLASSIFY, BUFFER, and INTERSECT. Geolineus would create icons and arrows connecting them to the flowchart automatically as these commands were used. Icons at the bottom of the flowchart signifying products, that is, derived datasets that represent the final step in the geospatial application, each included a bar along its bottom edge (Figure 19.4).

Written in Common LISP, Geolineus used multiprocessing capabilities of UNIX to run the geospatial processing software as a background job while providing its own command line window to the user. As the user would enter a command transforming one or more spatial datasets to derive a new one (e.g., classify, union, and intersect), Geolineus would parse, extract the identities of the input and output datasets and the command and its parameters, and pass the command off to the geospatial processing software running in the background. Geolineus monitored the processing and feedback messages returned from the geospatial processing software and presented them to the user within its own command line window in real time to provide the user with the illusion that they were interacting directly with the geospatial software. The system detected whether the processing successfully completed and, if so, the input/output relationships and command information would be stored within its metadatabase and the data flow diagram dynamically updated with a new icon for the output dataset connected by dotted arrows (labeled with the command) emanating from its data sources. When the final data product was reached, the user could click on its icon and fill in the displayed product form to document the analytical goal it represented (e.g., wells at risk from nearby leaking pipes) and who should be contacted if it was updated or changed.

Geolineus also monitored each dataset in the diagram to determine if it was edited or replaced. If a source or derived dataset was found to be modified, its icon would turn yellow in the diagram. If the dataset needed its topology rebuilt in response to an edit, the polygon or line feature symbol within the icon would turn red. If a derived dataset was potentially out of date because one of the sources it was derived from was edited, its icon would turn orange. Users could click on a source icon to

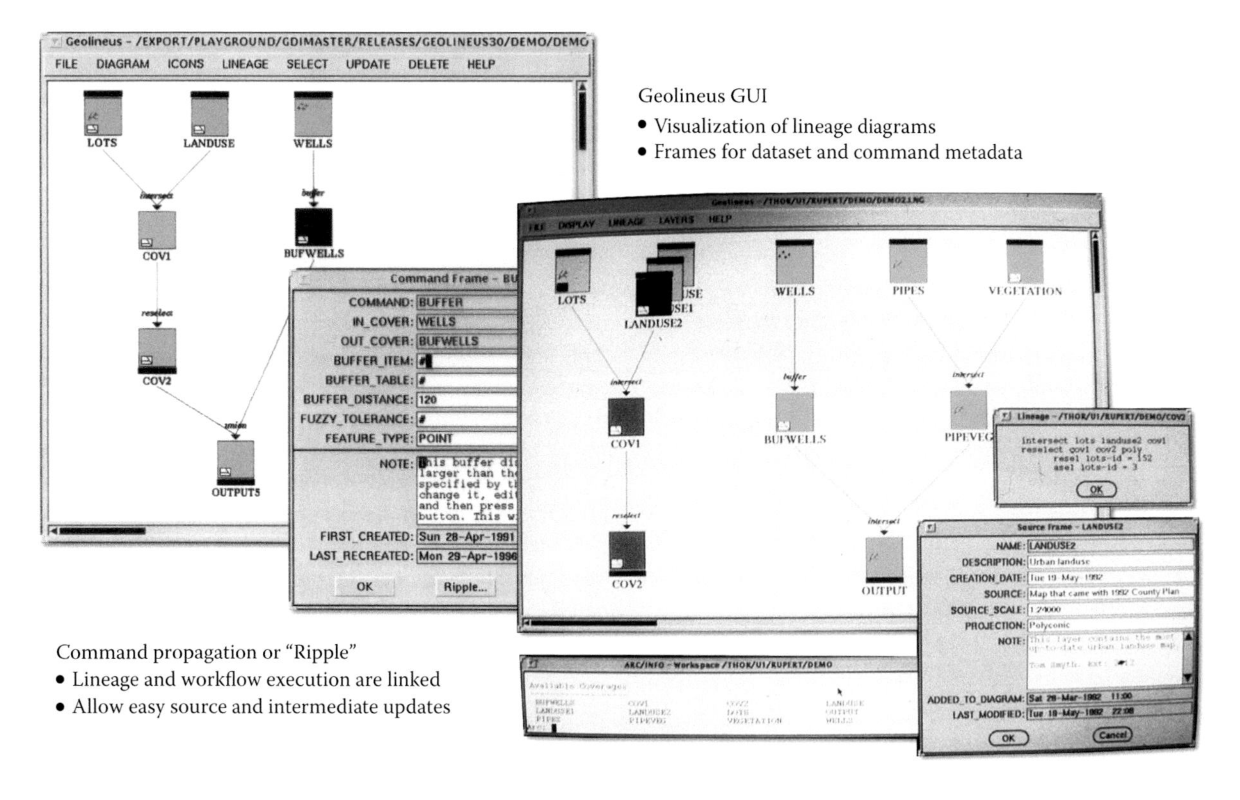

FIGURE 19.4 Examples of Geolineus' interactive lineage diagram GUI. The left screen shot illustrates linkage of a command frame to the BUFWELLS dataset highlighted in black; clicking on the "Ripple" button at the bottom of the command frame propagated changed buffer command parameters throughout the workflow. The right screen shot illustrates the LANDUSE2 dataset's source frame, and commands applied to that source to derive the COV1 and COV2 datasets.

view metadata about what it represented, when it was created, where it came from, and cause a propagation (*ripple*) of its data through sequences of commands updating intermediate and product datasets originally derived and created from it. Users could also click on a derived dataset to rerun the commands necessary to pull new, updated, and modified source data through the flowchart's processing logic and update the derived geodata (Lanter 1994b). Geolineus enabled users to save, exchange, and import lineage metadata in ASCII file format to meet the Federal Geographic Data Committee's (FGDC's) Content Standard for Digital Geospatial Metadata, document exchanged datasets, accompany source datasets, provide logic for use within other instances of the software to reconstitute a derived geospatial database, and serve as reusable analytic application logic templates to snap to replacement source datasets associated with different study areas.

Lanter and Veregin (1991, 1992) modified the lineage metadata to store error measures and demonstrated new algorithms for mathematically modeling how error measures of data sources are transformed and combined through a sequence of spatial analysis functions to determine the quality of a derived spatial analytic product dataset. They added properties to the source frame for storing user-entered measures of data source error, and properties to the command frame for storing derived error measures for each derived dataset. Geolineus's *Ancestors* and *Descendants* functions were modified, enabling them to access error properties of input datasets, select and apply an appropriate error propagation function to derive, store and present the error measure of the derived geospatial dataset as the user typed in their spatial analysis commands. Lanter (1993b) followed this by modifying the lineage metadata and *Ancestors* and *Descendants* functions to use commercial costs of data storage and central processing time to calculate and compare the relative costs of storing versus using lineage metadata to re-derive intermediate and product datasets when needed. The results enabled Geolineus to determine an optimal spatial database configuration and choose which datasets to delete and re-create when needed. Veregin and Lanter (1995) modified the metadata frames and *Ancestors* and *Descendants* functions to demonstrated lineage metadata-based error propagation techniques for identifying the best data source to improve based on cost value per product quality improvement achieved. Geolineus was programmed to systematically vary the error value of each source, iteratively applying mathematical error propagation functions and determining its effect on product quality. Comparing slopes of lines graphing source error versus resulting product enables determination of relative impact each data source has on data product quality.

To help analysts and auditors understand undocumented preexisting analytically derived GIS datasets, Lanter provided Geolineus with capabilities to extract lineage metadata and create a lineage diagram from ARC/INFO log files. Similar to the history list the UNIX operating system recorded user commands into, the ARC/INFO GIS copied user-entered GIS commands into log files, which it stored and maintained within the operating system file system directories or workspaces. Geolineus's "Create from log" option automatically extracted lineage metadata and created a lineage diagram reflecting the commands contained in the log file of a targeted workspace. While the log files contained the name of the dataset and the file system path indicating where the dataset was stored, they did not include other source meta data (i.e., thematic feature type, date, agency, scale, projection accuracy, etc.) necessary for achieving a clear understanding of contents and qualities of each source. To resolve this, analysts and auditors working with Geolineus clicked on the source icons within the lineage diagram, brought up source frames, and filled in missing source metadata if available.

Lanter (1994a) formulated metadata comparison functions that enabled him to automatically determine if two spatial analytic datasets were equivalent and if two geospatial datasets were similar. These were implemented within Geolineus to identify common and unique geospatial data processing conducted in and among multiple GIS workspaces (Lanter 1994b). His search for datasets common to different lineage metadata representations began with a determination of source equivalence. Source datasets were considered equivalent when their source metadata properties were found to have equivalent values, assuming these properties are sufficient to uniquely identify their contents and qualities. This enabled the detection of equivalent and possibly redundant source datasets that are stored in different file system locations but contain equivalent content. Source data equivalence was implemented in Geolineus's "Merge" function, which enabled users to analyze log files of data processing applications run in different workspaces and produce a single unified lineage diagram illustrating their common and unique data sources (Figure 19.5).

In turn, Lanter considered derived datasets equivalent when (1) their input datasets were equivalent and (2) when transformations applied to compute them from their inputs were found to be equivalent. Derived data equivalence was implemented in Geolineus's "Condense" function. Condense enabled Geolineus's users to detect the lineage representations of redundant processing and resulting copies of derived data stored under different names or in different file system locations, remove the redundant data, and consolidate the transformational logic applied in their derivation within the unified metadata and lineage diagram.

Lanter and Surbey (1994) put Geolineus's capabilities to work in the first enterprise GIS database and geoprocessing quality audit. They systematically evaluated the geospatial data sources, products, and geoprocessing applied to derive 40 GIS data products, developed within 14 projects, for eight departments of a large southwestern electric utility. Lanter and Surbey identified 54 data sources among the 806 raster (GRID) and vector (ARC/INFO) GIS datasets produced for the electric utility's decision makers. They interviewed the department's GIS specialists, filled in as much missing source metadata that could be recalled and confirmed, and noted findings about what was unknown about the source data. In addition to assessing adequacy of source data documentation, Lanter and Surbey analyzed the resulting lineage diagrams they created and measured the complexity of spatial analysis logic employed within the 14 GIS application projects.

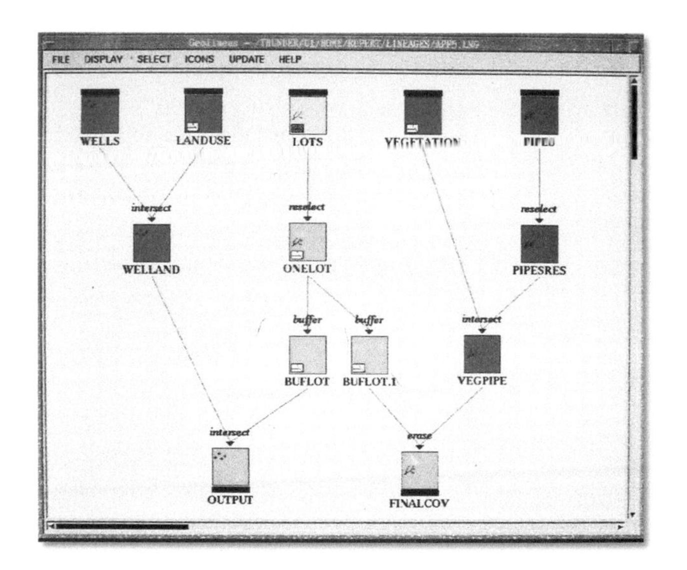

FIGURE 19.5 Geolineus GUI illustrating the results of the "Merge" function unifying two lineage diagrams at their common source LOTS dataset, and "Condense" function which removed redundant processing and derived datasets unifying them at their common intermediate ONELOT dataset. The red mark on the yellow LOTS dataset indicated an edit and need for polygon topology repair, and the orange color in derived datasets reflected the need for changes to be propagated using the "Ripple" function to update the OUTPUT and FINALCOV products.

Lanter (1994a) extended his dataset equivalence tests and formulated a set of source and derived data similarity tests in order to detect patterns of data usage and derivations within workflows. He coupled these with a geospatial data taxonomy (e.g., Anderson et al. 1976) and a GIS command language taxonomy (e.g., Giordano et al. 1994) to generalize analytic logic employed within prior applications of GIS and find common data analysis patterns. Lanter presented a suite of lineage-based metadata analysis methods for detecting and communicating commonalities and differences among particularly useful spatial analysis applications, with the intent of improving geographers' basic understandings of spatial analytic reasoning and to provide a method and means to answer fundamental geographic questions including the following:

- Are there a finite number of spatial relationships studied within and among different GIS applications areas? If so, what are they?
- Within particular applications areas, are certain spatial relationships stressed more than others? If so, what are they?
- Are common patterns of analytic logic used to build up certain complex spatial relationships? If so, what are they?
- Are certain spatial relationships consistently sought at different spatial, thematic, and temporal scales?

19.2.1.1.2 Geo-Opera

Also incorporating geospatial lineage into its design in the 1990s, Geo-Opera was developed as a prototype geoprocessing support or geospatial workflow management system that would enable interoperability, data recovery, process history records, and data version monitoring in commercial GIS (Alonso and Hagen 1997). Geo-Opera was based on a modular architecture composed of interface, process, and database modules. It used its own process scripting language and was based on the OPERA distributed operating system that allowed for data distribution and process scheduling within a local area network. In Geo-Opera, geodata first had to be registered before being utilized, thus mitigating the common problem (that persists today) of lack of source metadata.

19.2.1.2 Expansion of Limited Provenance in Commercial and Public Geoprocessing

As commercial and public (i.e., free and open-source) GIS applications rapidly matured and grew in analytical power, it became necessary to provide a way for users to build and track workflows involving interactions among many complex and varied geoprocessing operations. At least two approaches to create and manage workflows have emerged—graphical block programming and integrated database style querying. The first is essentially a visual interface to programming, while the second approach appeals to users trained in database management. While both enable at least some form of provenance, enterprise database systems can provide record-level transaction management, which, at least in detail, is beyond the scope of this chapter.

By far the most common approach, due in large part no doubt to its ease of use and graphic nature, is graphic block programming approach. Commercial GIS and remote sensing applications such as ESRI's ArcGIS, Hexagon Geospatial's ERDAS IMAGINE, and PCI's Geomatica expose their complex processing tools in this way (e.g., Figure 19.6). The free and open-source GRASS GIS also provides a visual programming environment for both vector and raster operations. Boundless (formerly OpenGeo) is developing a visual programming environment for QGIS, an open-source GIS. All of these environments capture and store some degree of provenance including in some cases important environmental settings that can significantly affect geoprocessing results. It is important to note that visual programming interfaces can normally be bypassed by skilled users familiar with application programming interfaces (APIs) or scripting languages integrated with GIS.

A less common approach is incorporated almost exclusively in enterprise databases that have integrated spatial operators and native spatial data objects. With this level of integration, spatial operators become just another type of operation exposed through (often extended) structured query language (SQL) interfaces. At a minimum, the SQL commands used to manipulate spatial data objects are recorded and may be inspected in a variety of graphical environments. As of mid-2014, most spatially enabled databases have extensive vector operators but limited raster or image operators of particular interest in remote sensing workflows. However, technologies such as the Oracle Spatial and Graph option for Oracle Database 12c now enable image algebra in addition to other remote sensing–oriented capabilities such as LIDAR data processing.

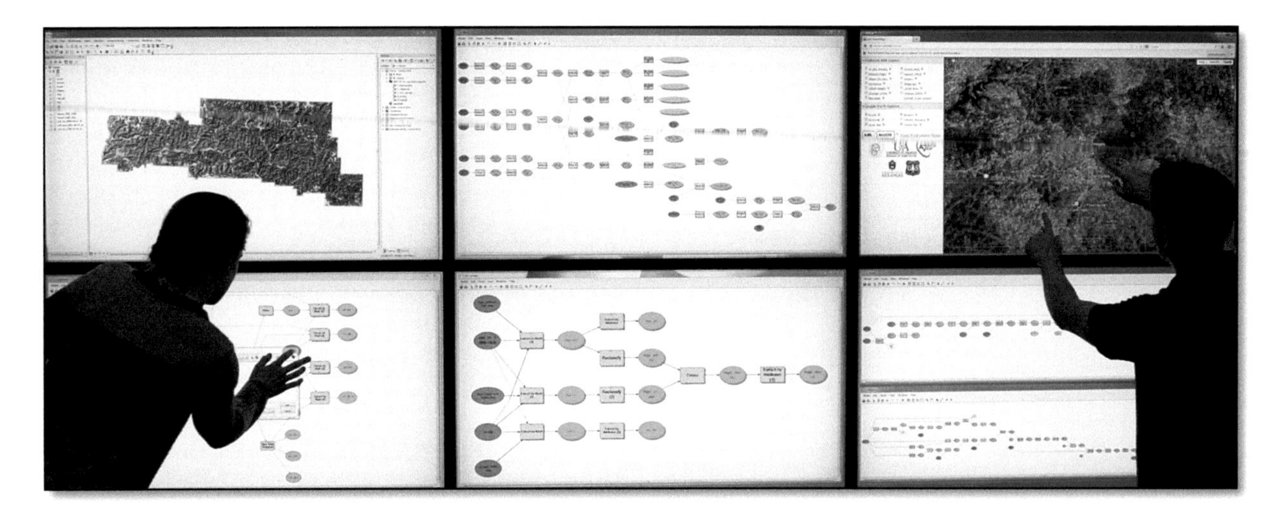

FIGURE 19.6 Geospatial scientists interact with the ASA Hazard Map (Tullis et al. 2012), a remote sensing–assisted silviculture assessment spatial decision support system, and its five downloadable ArcGIS 10 ModelBuilder workflows using a collaborative multitouch display. Each yellow rectangle represents an ArcGIS tool (e.g., for estimating incoming solar radiation using a LIDAR-derived DEM) and, together with inputs, outputs, and other parameters (colored ovals), constitutes a geoprocess. After execution, geoprocesses are marked with shadows that may be cleared only by resetting or changing geoprocess parameters including geoprocessing environment settings. User interaction with shaded geoprocesses effectively provides access to workflow-level provenance information for the most recent execution and facilitates dependent geoprocess updates after any modifications are made.

19.2.1.3 Interest in Provenance as a Component in Geo-Cyberinfrastructure

Cyberinfrastructure (CI) is a concept that has been extensively used since Atkins et al. (2003) *Revolutionizing Science and Engineering through Cyberinfrastructure: Report of the National Science Foundation Blue-Ribbon Advisory Panel on Cyberinfrastructure*. As a common infrastructure for scientific data and computing, a variety of components and topics are involved in CI construction, including hardware, software, network, data, and most importantly people. The development of CI can be traced back to the construction of the TeraGrid infrastructure in the 1990s that was replaced by eXtreme Science and Engineering Development Environment (XSEDE) in 2012. By linking supercomputers through high-speed networks, TeraGrid and XSEDE have provided a powerful computing environment and capability to support petascale to exascale scientific computation.

In a broader and general domain, the internet can be regarded as the CI since all computers can be linked together through the network. When varieties of data and databases can be hosted and connected on the internet, data processing and analytics can be conducted through service-oriented computing (SOC). In early 2000, web service technology was proposed to be the solution for software interoperability. In this vision of interoperable software engineering and integration, a service is an API defined in Web Services Description Language, while communication between the service provider and the service requester is based on the Simple Object Access Protocol (SOAP). Meanwhile, Representational State Transfer (REST) services are based on HTTP protocol using its GET/POST methods for mashup online resources (Fielding 2000). Both SOAP- and REST-based services can be deployed for remote procedure calls. Furthermore, with the advancement of telecommunication infrastructure and technology, wireless networking has been providing another approach for data sharing and network computing, while varieties of sensor networks can be connected through wireless networks.

Today, different computing networks can be linked together. Supercomputers on the XSEDE can be accessed through a web portal, while wireless sensor networks can be accessed on the internet. Such a huge but heterogeneous CI increases the difficulty and complexity for geoprocessing, workflows, and provenance research (Wang et al. 2008). In 2007, the National Science Foundation (NSF) released the DataNet program that would support comprehensive data curation research over the CI, and NSF's Data Infrastructure Building Blocks program "will support development and implementation of technologies addressing a subset of elements of the data preservation and access lifecycle, including acquisition; documentation; security and integrity; storage; access, analysis and dissemination; migration; and deaccession," as well as "cybersecurity challenges and solutions in data acquisition, access, analysis, and sharing, such as data privacy, confidentiality, and protection from loss or corruption" (NSF 2014), which are all topics relevant to the themes in provenance.

19.2.2 Specifications and International Standards for Implementation of Shared Provenance-Aware Remote Sensing Workflows

Since the Moellering et al. (1988) proposal identifying geospatial lineage as the first component in a data quality report, a variety of provenance-related standards have been developed including those at the international level. The most current standard in use is the International Standards Organization's ISO 19115-2, which has been endorsed by the Federal Geographic Data Committee (FGDC; ISO 2009).

19.2.2.1 Metadata Interchange Standards

In the United States, the FGDC has been coordinating the development of the National Spatial Data Infrastructure by developing policies and standards for sharing geographic data. The Content Standard for Digital Geospatial Metadata defines common geospatial metadata about identification information, spatial reference, status information, metadata reference information, source information, processing history information, distribution information, entity/attribute information, and contact information of the geodata creator.

Partially based on the FGDC's 1994 metadata standards, the ISO Technical Committee (TC) 211 published ISO 19115 Metadata Standard, covering a conceptual framework and implementation approach for geospatial metadata generation. ISO/TC 211 suggests that metadata structure and encoding are implemented based on the Standard Generalized Markup Language that has the same format as the Extensible Markup Language (XML). The XML-based ISO metadata standard has exemplified the advantage in implementation covering a variety of elements in standard definition. ISO 19115 Metadata Standards contain a data provenance component in defining the data quality within the metadata. Unfortunately, while Gil et al. (2010) defined provenance in part as "a form of contextual metadata," their emphasis on the clear distinction between provenance and traditional metadata is not reflected in metadata interchange standards for provenance. For instance, geodata cardinality between a land use land cover (LULC) map and its metadata is one to one; in contrast, geodata cardinality between an LULC map and its provenance is potentially one to many, thus leading to much duplicate information in a "provenance as metadata" paradigm.

19.2.2.2 Provenance-Specific (Non-Metadata) Interchange Standards

Provenance-specific (non-metadata) standards have been developed at different levels and in a variety of domains. ISO 8000 has a series of standards that address data quality. ISO 8000-110 specifies requirements that can be checked by computer for the exchange, between organizations and systems, of master data that consists of characteristic data. It provides requirements for data quality, independent of syntax. ISO 8000-120 specifies requirements for capture and exchange of data provenance information and supplements the requirements of ISO 8000-110. ISO 8000-120 includes a conceptual data model for data provenance where a given "*provenance_event* records the provenance for exactly one *property_value_assignment*," and every "*property_value_assignment* has its provenance recorded by one or many *provenance_event* objects."

In order to trace the changing information and the provenance of data (and by implication geodata) over the web, W3C has recently published a series of documents and recommendations (starting with the term PROV) to guide the provenance interchange on the web. Specifically, the current PROV data model for provenance (PROV-DM; Moreau and Missier 2013) "defines a core data model for provenance for building representations of the entities, people and processes involved in producing a piece of data or thing in the world" (Gil and Miles 2013).

To illustrate PROV-DM in a remote sensing and geoprocessing context, the provenance of a 2001–2006 canopy change layer incorporated in the ASA Hazard Map (Jones et al. 2014; Tullis et al. 2012) can be represented using PROV-DM structures. This may be encoded (Figure 19.7; Table 19.1) as agents (e.g., a specific version of PCI Geomatica as a software agent), entities (e.g., a Landsat image clipped to a forest boundary), activities (e.g., ATCOR2 atmospheric correction based on specific calibration and other parameters), and relationships (e.g., wasInfluencedBy to represent the influence of Wang et al. (2007) on the change detection methodology). It is important to note that PROV-DM is extensible such that subtypes of agents, entities, activities, and relationships can be identified as needed for domain-specific applications (Moreau and Missier 2013).

In the geospatial domain, the efforts of the Open Geospatial Consortium (OGC) initially (late 1990s and early 2000s) focused on the development of specifications that encouraged geospatial data interoperability such as the OGC Simple Features Specification. While not directly related to provenance, this effort has led to common ontologies and semantic structures that are foundational to the integration of geoprocessing, workflows, and provenance. In the 2000s, the OGC's attention shifted to web processing and interoperability of various web services. The OpenGIS Web Processing Service (WPS) specification (Schut 2007) has a *lineage* element in defining the request message to *execute* the spatial operation. In case lineage is defined as *true*, the response message from WPS will contain a copy of input parameter values specified in the service request definition. The OGC also developed the Sensor Web Enablement standard, in which the OpenGIS Sensor Model Language (Botts and Robin 2007) has one specific element that documents the observation lineage to describe how an observation is obtained. Elements of a number of earth observation process specifications, such as the Catalogue Services Standard 2.0 Extension Package for ebRIM Application Profile: Earth Observation Products (Houbie and Bigagli 2010), the Sensor Observation Service Interface Standard (Bröring et al. 2012), and others, increasingly have provenance-related components as key elements. The more recent developments in WaterML and the Open Modeling Interface have increasingly emphasized provenance components.

> The purpose of the Open Modeling Interface (OpenMI) is to enable the runtime exchange of data between process simulation models and also between models and other modeling tools such as databases and analytical and visualization applications. Its creation has been driven by the need to understand how processes interact and to

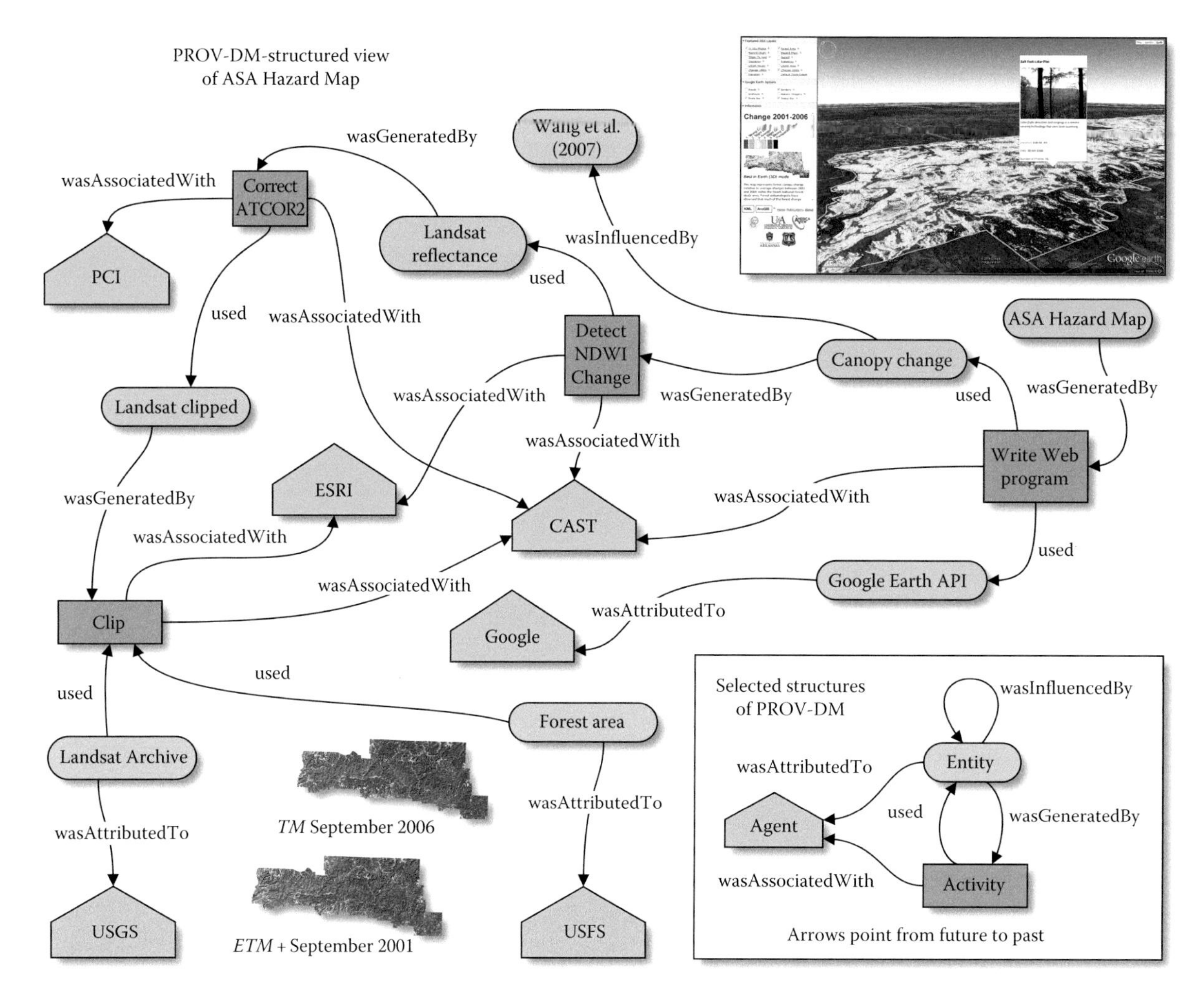

FIGURE 19.7 Selected provenance of the ASA Hazard Map (Jones et al. 2014; Tullis et al. 2012) structured according to W3C's PROV Data Model (PROV-DM; Gil and Miles 2013; Moreau and Missier 2013; Table 19.1). Arrows (relationships) point from future to past, first from the online ASA Hazard Map to its 2001–2006 canopy change layer, then to various agents, entities, and activities involved in the canopy change layer's creation. Some entities (e.g., "Landsat Archive") represent PROV-DM collections of entities (e.g., individual Landsat images available from USGS), and many potential PROV-DM details are not shown.

> predict the likely outcomes of those interactions under given conditions. A key design aim has been to bring about interoperability between independently developed modeling components, where those components may originate from any discipline or supplier. The ultimate aim is to transform integrated modeling into an operational tool accessible to all and so open up the potential opportunities created by integrated modeling for *innovation* and wealth creation.
>
> **Vanecek and Moore (2014, p. ix, emphasis added)**

It is likely that future OGC efforts will increasingly focus on provenance. The OGC is a major participant in EarthCube (2014). In 2011, NSF's Cyberinfrastructure and Geosciences Divisions established the EarthCube community to promote geosciences data discovery and interoperability. The OGC plays a major role in this community, which, as of 2014, has several NSF-funded research and implementation grants pertaining to provenance records in geoprocessing.

19.3 Why Provenance in Remote Sensing Workflows

As Buneman (2013) argues, a "change of attitude" is in order regarding the role for provenance across a range of computer system-supported domains and activities, including (by implication), remote sensing workflows. He makes the comparison between scientific activities where it is considered obvious that such information should be recorded and other domains where there is little or no awareness of process history or its value. He concludes that "we should worry less about what provenance is and concentrate more on what we can do with it once we have it" (p. 11).

TABLE 19.1 Characteristics of PROV-DM Structures Including Core Types and Selected Relationships (Moreau and Missier 2013), Each with an Example Provided from the Provenance of the ASA Hazard Map (Jones et al. 2014; Tullis et al. 2012; Figure 19.7)

PROV-DM Structure	Interpretive Highlights	Example from ASA Hazard Map Provenance
Core types		
Agent	Need not be a person but could also represent an organization or even a specific software process	Center for Advanced Spatial Technologies (CAST) *agent* (organization) at University of Arkansas
Entity	May be physical, digital, or conceptual	Landsat 5 TM *entity* (satellite image) collected on September 15, 2006, over Ozark National Forest
Activity	Involves entities and requires some time to complete	ESRI ArcGIS 10 for Desktop "Extract by Mask" *activity* (software tool) used to clip the Landsat 5 TM imagery to the bounds of the study area), together with environment settings (e.g., "Snap Raster")
Selected relationships		
wasGeneratedBy	Can represent creation of only new entities (that did not already exist)	Clipped Landsat 5 TM image that has been corrected for atmospheric attenuation *was generated by* running the ATCOR2 algorithm
Used	Only implies that usage has begun (but not that it is completed)	A GIS model for detecting oak-hickory forest decline or growth *used* a clipped and atmospherically corrected Landsat ETM+ image collected September 25, 2001
wasAttributedTo	Links an entity to an agent without any understanding of activities involved	The Google Earth API (used to write a web program to generate the ASA Hazard Map) *was attributed to* Google
wasAssociatedWith	Links an activity to an agent	The ATCOR2 algorithm used to correct Landsat TM and ETM+ imagery for atmospheric attenuation *was associated with* PCI Geomatics through their Geomatica 10 platform
wasInfluencedBy	At a minimum, suggests some form of influence between entities, activities, and/or agents; however, highly specific influence may be captured	The 2001–2006 oak-hickory forest canopy change data produced for the ASA Hazard Map *was influenced by* Wang et al. (2007), who used statistical thresholds of change in Landsat-derived normalized difference water index (NDWI) over time to detect oak canopy changes in the Mark Twain National Forest

19.3.1 Remote Sensing Questions That Only Provenance Can Answer

For volumes that contain primarily raw or unprocessed geodata (e.g., imagery telemetered directly from a satellite sensor), provenance (as used in the present context) may not offer much over traditional metadata. However, when looking at geodata products resulting from complex geoprocessing workflows, there is much valuable information that metadata is ill-equipped to capture and store.

There is sometimes confusion concerning what provenance offers in terms of valuable information to an end user over the far more common and better supported (in terms of software integration) metadata. One way to structure such a discussion is to look at some of the questions data users might ask that can be only reasonably answered using (at least in part) detailed provenance information. For instance, one might ask the following regarding a remote sensing–derived product:

1. What was the processing time necessary to create this product, and what system configuration was implemented (including disk, processor, and RAM information)?
2. In what exact order were processing steps taken, and what precise parameters were used during each intermediate step? Was the process completely automated, or were manual steps (such as onscreen digitization) included in the workflow?
3. What datasets, both source and derived, were used to create this product, and how did each contribute to the product?
4. How were errors expressed and propagated during the product's creation? Is the result statistically significant?

In addition to these, several questions could be asked trying to identify the source of errors or anomalies in the data. For example, one might wonder at what point in the geoprocessing did a specific region get assigned null values and why? Using provenance data, it should be possible to analyze two similar data products and compare their processing history to see how and why they differ (Bose and Frew 2005; Lanter 1994a). The opportunity to better understand and manage the complexity of spatial scale in remote sensing–assisted workflows is a further justification for provenance-enabled geoprocessing (Tullis and Defibaugh y Chávez 2009). Finally, provenance-aware systems could be used to enable and support temporal GIS analyses, which require detailed history of a dataset's change over time to properly function (Langran 1988).

The value of provenance tracking and visualization was demonstrated in a study conducted at the Regional Geospatial Service Center at the University of Texas, El Paso (Del Rio and da Silva 2007). In this study, conducted as part of NSF's GEON Cyberinfrastructure project, web services were built to perform geoprocessing tasks (filtering, gridding, and contouring) required to create a contoured gravity map from a raw gravity dataset. Del Rio and da Silva generated multiple contour maps with incorrect parameters (e.g., a grid size parameter larger than important anomalies in the gravity field), and participants in the study were asked to evaluate each contour map with and without provenance information. Without provenance information, subject matter experts were able to detect errors in only 50% of the cases and to explain cause in only 25% of the cases. Nonsubject matter experts fared much worse (11% and 11%). However, when

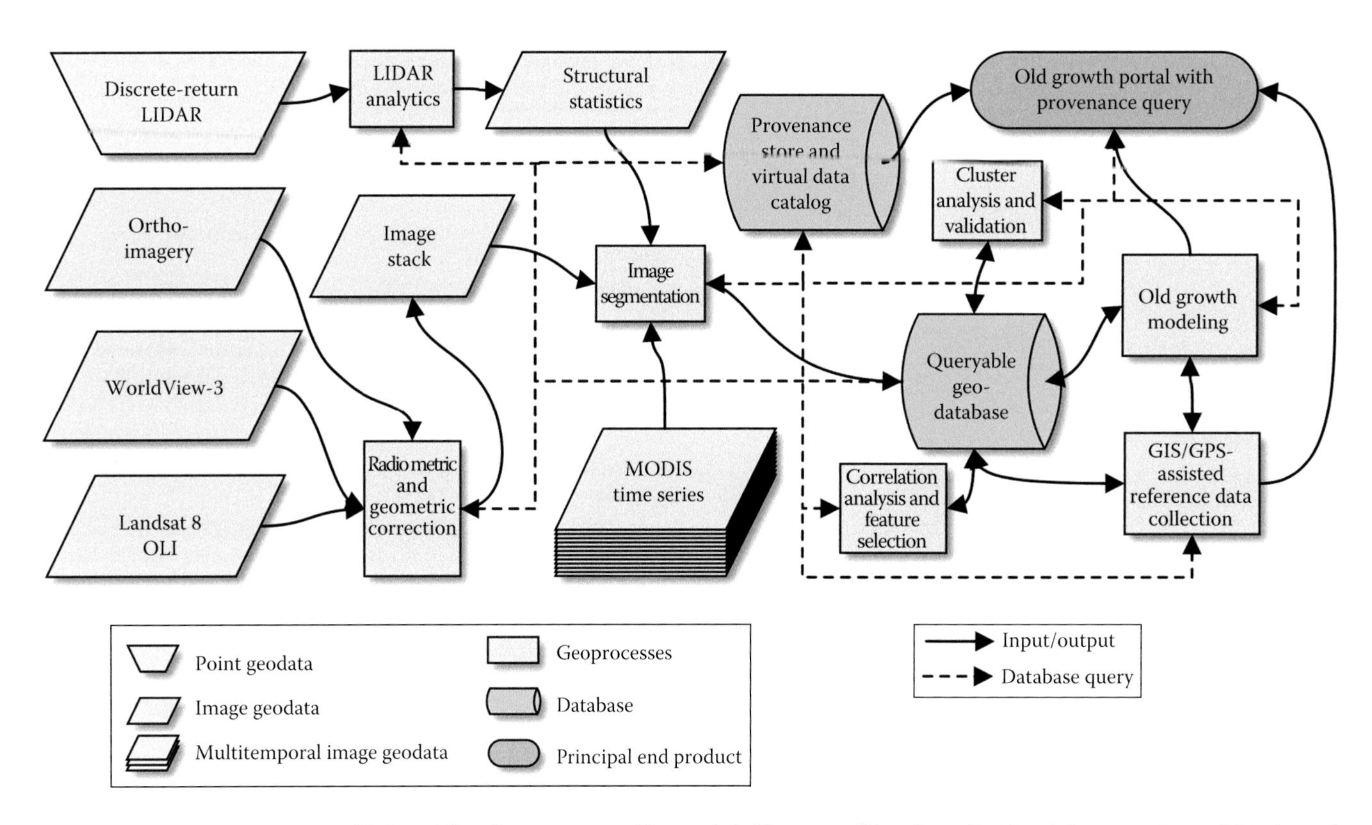

FIGURE 19.8 A provenance-enabled workflow for extracting old-growth bald cypress (*Taxodium distichum*) forest quality and biophysical parameters from airborne LIDAR and orbital multispectral imagery. Such a framework could enable (a) detailed methodological transparency related to old-growth inventory maps published in the web portal, (b) detailed transparency of the accuracy assessment, and (c) autogenerated scripts for the replication of the detailed workflow despite a complicated blend of commercial and open-source workstation-level software, including cyber-enabled high-performance geoprocessing tools.

provenance information was provided and visualized using Del Rio and da Silva's (2007) ProbeIt! (a provenance visualization tool), the subject matter experts were able to detect and explain all the errors. A more impressive result, though, is that provenance improved the ability of the nonsubject matter experts to detect and explain errors by a factor of 7 (78%).

It can be argued that extending a traditional remote sensing workflow to include provenance information offers a number of immediate advantages. For example, a web portal displaying a detailed map of old growth forest in the southeastern United States could include provenance query that enables (1) detailed methodological transparency, (2) detailed transparency of the accuracy assessment, and (3) autogenerated scripts for replication of the detailed workflow even though it includes a complicated blend of commercial and open-source software, including cyber-enabled high-performance geoprocessing tools (Figure 19.8).

19.3.2 Provenance and Trust in the Remote Sensing Process

When provenance of sources, intermediates, and products is captured and maintained throughout the digital remote sensing process life cycle, data quality improves, an audit trail is available for reviewers and users, replication is straightforward, attribution is streamlined, and the interpretability of geodata products is enhanced (Simmhan et al. 2005). However, any one of these advantages by itself is not as critical as establishing trust in the remote sensing process, which is clearly not immune to being oversold (Jensen 2007). Generally speaking, statistical validity is a key to establishing trust in scientific processes and has been extensively developed in remote sensing workflows (Congalton 2010). Simply maintaining an audit trail that demonstrates sound choices in statistical processing parameters and methods is one way to help ensure that trust is not erroneously called into question.

In addition to maintaining an audit trail, statistical techniques can be applied to a given workflow to determine trust, and recent work has focused on assigning a measure of trust by using a workflow's provenance. While not demonstrated in a remote sensing context, a nonstationary hidden Markov model can be used to provide a measure of this trust (Naseri and Ludwig 2013). Provenance can also be useful in helping to determine workflow trust for data on the web. When handling linked data provenance, data authentication may serve as an estimation of data trustworthiness. Uniform resource identifiers and digital signatures can be used as a measure of authentication (Hart and Dolbar 2013).

Traditional indicators of trust may also be used in conjunction with provenance. Recommendation, authority, believability,

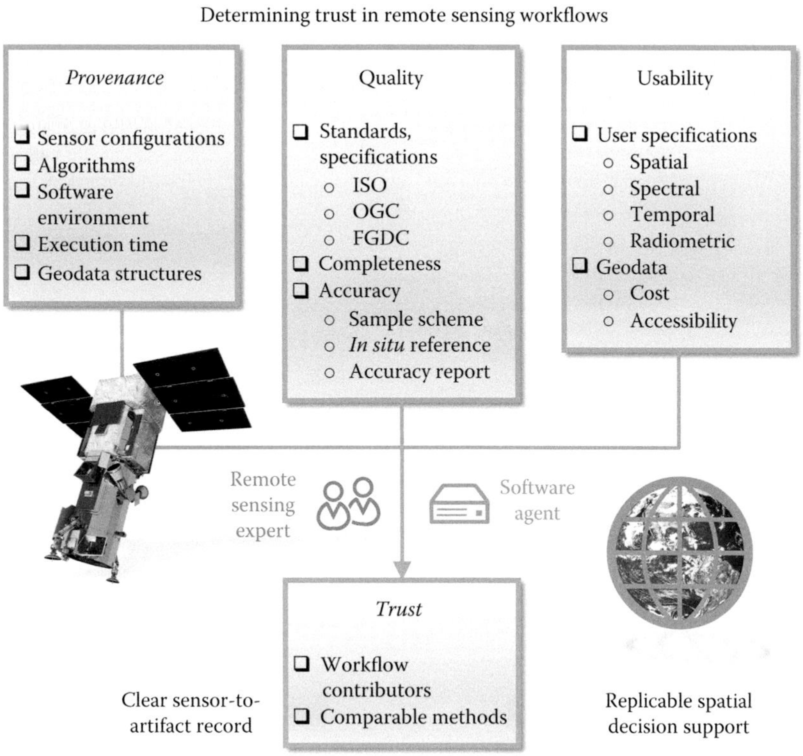

FIGURE 19.9 Provenance, quality, and usability can be used by remote sensing experts to make a subjective decision on workflow trust (Gamble and Goble 2011; Jensen 2005; Malaverri et al. 2012); a sample of their characteristics is shown. As geoprocessing, workflows, and provenance are integrated, software agents can objectively influence replicable spatial decision support. (Artist image of WorldView-3, Courtesy of DigitalGlobe, 2014.)

reputation, and objectivity can all serve as indicators of trust (Gamble and Goble 2011). In addition to provenance, a workflow's quality and usability should also be evaluated when determining trust. As geoprocessing, workflows, and provenance are integrated, software agents can objectively influence replicable spatial decision support (Figure 19.9). Until more quantitative techniques are developed for measuring trust of geographic workflows using provenance, measures of quality and usability used in conjunction with subjective trust indicators should be examined before making a decision to trust a workflow and its lineage.

19.4 Selected Recent and Proposed Provenance-Aware Systems

Many provenance-aware systems have largely been concerned with provenance capture, and this capability is critical for synergistic geoprocessing, workflows, and provenance of interest in remote sensing applications (Figure 19.1). The characteristics of the captured provenance information can greatly influence how it may benefit the remote sensing process. Of particular significance to geospatial applications is the level of provenance detail or granularity. Fine-grain provenance is obtained at a data level and can even deal with individual pixels (Woodruff and Stonebraker 1997), whereas coarse-grain provenance represents the workflow level (Tan 2007) and can thus be used to facilitate scientific reproducibility. As Henzen et al. (2013) point out, the quality of provenance communication is also very important even when presented in a text format. A number of recent and proposed provenance-aware approaches and systems related to remote sensing (Table 19.2) have addressed these and other issues.

19.4.1 General Approaches

19.4.1.1 Inversion

Inversion was developed for fine-grain data lineage and provenance in database transaction and transformation (Buneman et al. 2001; Cui et al. 2000; Woodruff and Stonebraker 1997). Database queries or processing functions that generate a view, table, or new data product can be registered in a database system (or provenance store). Registered database transformations can be inverted so as to trace the lineage between the data product and the sources that derive the product. For example, when a view is created or updated, the inversion method can help identify the source tables from which the view is generated. Although inversion can be applied in data provenance for geospatial data, not all functions are invertible. However, a weak or general inversion could be substituted to approximate the provenance by returning a fraction (or a projection) of the desired provenance. Examples of inversion can be found in some

TABLE 19.2 Characteristics of Selected Provenance-Aware Systems Reported in Remote Sensing and Other Geodata Applications

Geodata Application	Successes	Limitations or Future Work	References
Earth System Science Workbench and ES3			
Track processing of a laboratory's raw satellite imagery (e.g., AVHRR) into higher level products	a. Automates geodata provenance capture from running processes b. Stores provenance in both XML documents and in a searchable online store	Predates recent geodata interoperability standards and specifications	Frew (2004), Frew and Bose (2001), Frew and Slaughter (2008)
MODAPS and OMIDAPS			
Manage MODIS and other NASA satellite imagery and its provenance	a. Automates version tracking of geodata processing algorithms b. Reduces geodata storage via on-demand processing based on a virtual archive	Identifies science community's lack of appreciation for provenance information	Tilmes and Fleig (2008)
Karma			
Capture provenance for Japan's AMSR-E passive microwave radiometer on Aqua	a. Modularizes architecture to facilitate web service interoperability b. Is compatible with open provenance model (OPM) and ISO 19115-2 metadata standards	Requires additional geodata interoperability standards to facilitate geodata (scientific) reproducibility	Conover et al. (2013), Plale et al. (2011), Simmhan et al. (2008)
Data Quality Provenance System			
Assess quality of agricultural mapping based on SPOT satellite imagery	a. Assigns geodata quality index based on provenance information b. Is compatible with OPM and FGDC metadata standards	Needs to address geodata quality dependencies on provenance granularity	Malaverri et al. (2012)
VisTrails			
Model habitat suitability using WorldView-3 and LIDAR-derived forest structure	a. Provides Python-based open-source provenance and workflow management b. Allows key focus on provenance in rapidly changing workflows (e.g., during remote sensing process development)	Designed to be domain generic, VisTrails may have a steep learning curve	Freire et al. (2012), Talbert (2012)
UV-CDAT			
Analyze large-scale remote sensing–derived climate data	a. Built on top of VisTrails with an extensible modularized architecture that supports high-performance workflows b. First end-to-end provenance-enabled tool for large-scale climate research	Future work could adapt UV-CDAT successes in climate change for other geodata application areas	Santos et al. (2012)
GeoPWProv			
Visualize and navigate city planning geodata (e.g., LIDAR-derived elevation data) provenance via a map	a. Allows geodata provenance to be visualized and explored in a map environment b. Provides for several geospatial levels of provenance query (e.g., via a single polygon versus a larger dataset)	Future work could support geoprocessing replication	Sun et al. (2013)

spatial databases including Oracle Database, Microsoft SQL Server, IBM DB2, and Boundless PostGIS. One early approach was implemented for vector operations in Intergraph's (now Hexagon's) Geomedia product.

19.4.1.2 Service Chaining

In a vision of SOC and service-oriented integration (SOI), different web services can be found, composed and invoked to accomplish certain tasks. The sequence of service discovery, composition, and execution looks like a chain, while alternatively, the composition processes can be constructed through different approaches, such as service orchestration or choreography that could be applied in enabling business processes and transactions on the web. Service composition and chaining could represent a workflow in which scientific computation can be implemented through the SOC/SOI approach. Capturing the provenance information within service-oriented workflows has been explored in geospatial applications (Del Rio and da Silva 2007; Yue et al. 2010a,b, 2011), though feasible and convincing approaches for provenance in SOC/SOI need further exploration and investigation.

19.4.1.3 Virtual Data Catalog Service

A virtual data catalog (VDC) is a provenance approach to trace the derivation route of data product and virtual data generated in the workflow in order to enable scientists to reproduce a data product and validate the quality of the workflow and related simulations. The intermediate data may be generated within a workflow but may not exist physically in a database or computer system (e.g., due to storage limitations). For this reason, such data are called virtual data because it is "the representation and manipulation of data that does not exist, being defined only by computational procedures" (Foster et al. 2002, 2003).

Within virtual data systems, such as Chimera, which is a virtual data grid managing the derivation and analysis of data objects, Virtual Data Language (VDL) is developed to define the workflow, while VDC is a service in the virtual data systems. The latter is defined and implemented based on the virtual data schema (VDS). The VDS defines the relevant data objects and relationships, and VDC can be queried by VDL to construct data derivation procedures from which derived data and output can be recomputed (Foster et al. 2002, 2003; Glavic and Dittrich 2007; Simmhan et al. 2005).

19.4.2 Earth System Science Workbench and ES3

The Earth Systems Science Workbench (ESSW) was an early attempt at automated provenance management and storage. It was a nonintrusive system that made use of Perl scripting techniques and Java to store data provenance as XML documents (Frew and Bose 2001). It contained a registry for provenance and a server for making the information searchable on the web. ESSW was followed up by the Earth System Science Server (ES3), which allowed for more flexibility in client-side implementation, but used essentially the same structure as the ESSW (Frew 2004). ES3, unlike many other systems, automatically captures provenance from running processes. It can also create provenance graphs in XML that can then be visualized using third-party tools like yEd (Frew and Slaughter 2008).

19.4.3 MODAPS and OMIDAPS

The MODIS Adaptive Data Processing System (MODAPS) and OMI Data Processing System (OMIDAPS) were designed for use by NASA to manage satellite imagery and provenance from MODIS sensors on the Terra and Aqua satellites, and the OMI sensor on Aura respectively (Tilmes and Fleig 2008). Both systems are operational and use a scripting process to track changes in versions of geodata processing algorithms. Using this technique, there is no need to store workflow iterations because enough information is retained that previous versions of the data can be re-created. Further, these systems periodically are tasked with reprocessing past data using the most up-to-date algorithms to maintain a consistent and improved series of data products. MODAPS in particular makes use of these features to maintain a *virtual archive* with provenance information that persists after a geodata product is deleted, allowing the system to re-create data products on demand rather than keeping extensive archives. Data re-creation as implemented in these systems is unique and is something that could be useful in other geospatial provenance systems.

19.4.4 Karma

Plale et al. (2011) make use of the Karma system designed by Simmhan et al. (2008) to collect provenance data for the Advanced Microwave Scanning Radiometer–Earth Observing System (AMSR–E) flown on the Aqua satellite. One of the biggest benefits of Karma is its modular architecture, which simplifies interoperability with Java and other web services. Karma's architecture for this application consists of an application layer, web service layer, core service layer, and a database layer (Plale et al. 2011). The inclusion of open provenance model (OPM) specifications and XML makes its interoperability extend further (Moreau et al. 2011). Conover et al. (2013) also made use of Karma to retrofit a legacy system for provenance capture. They chose the NASA Science Investigator–led Processing System (SIPS) for the AMSR-E sensor on the Aqua satellite. Their system uses a two-tiered approach that captures provenance for individual data files as well as collections or series, both automatically and via manual entry using the ISO 19115-2 lineage metadata standard. Query and display are handled with a database-driven web interface called the Provenance Browser.

19.4.5 Data Quality Provenance System

Taking into account a source's trustworthiness and the data's age, Malaverri et al. (2012) created a provenance system that allows a quality index to be assigned. This approach is based on a combination of the OPM and FGDC geospatial metadata standards. Criteria considered in the quality index include granularity, accuracy of attribute descriptions, completeness of the data, a logical measure of the data, and spatial positional accuracy. Although measures of trust can be very subjective in nature, in this case requiring a domain experts' input, this approach is somewhat unique in that it attempts to quantify data quality (Malaverri et al. 2012).

19.4.6 VisTrails

VisTrails is a free and open-source scientific workflow and provenance management system (Freire et al. 2012). Written in Python/Qt and designed to be integrated with existing workflow systems, VisTrails has been used in a number of research applications ranging from climate (including the UV-CDAT described later) to ecology and biomedical research. Talbert (2012) created software based on VisTrails to capture the details of habitat suitability and species distribution modeling. One of the major advantages of VisTrails is that as an open-source project built in Python, it is interoperable, easily customizable, and benefits from a large community of developers contributing code. A key focus of VisTrails is rapidly changing workflows. The information contained in how workflows are developed (and change over time) may provide highly valuable insight into the creative and development aspects of the remote sensing process.

19.4.7 UV-CDAT

Climate Data Analysis Tools (CDATs) are cutting-edge domain-specific tools for the climate research community, but they are ill-equipped to handle very large geodata and provenance information. The UV-CDAT is a relatively new provenance system for handling large amounts of climate-based data (Santos et al. 2012). The UV-CDAT uses a highly extensible modular design and makes use of a Visualization Control System and Visualization Toolkit (VTK)/ParaView infrastructure, which allows for high-performance parallel-streaming data analysis and visualization. Its loosely coupled modular design allows for integration with third-party tools such as R and MATLAB® for both analysis and visualization. The UV-CDAT is unique in that it is the first end-to-end application for provenance-enabled analysis and visualization for large-scale climate research. It has already been distributed and is widely used by scientists throughout the climate change field.

19.4.8 GeoPWProv

GeoPWProv is a provenance system specializing in displaying geospatial provenance as an easily accessible interactive map layer. GeoPWProv has the capability to capture provenance at the feature, dataset, service, or knowledge level (Sun et al. 2013). Comparisons can be made between entities in each level or between various levels. In addition to displaying provenance as a map layer, GeoPWProv supports displaying provenance in a workflow or in the more traditional text-based format. Implementation on the client side through use of a browser and *Open Layers* allows for ease of use. GeoPWProv's display of provenance in different formats and at different levels allows for a customizable user experience when evaluating a workflow.

19.5 Conclusions and Research Implications

Integrated geoprocessing, workflows, and provenance may be conceptualized as a positive developmental cycle that enables experts and software agents to capture, store, analyze, curate, replicate, and innovate remote sensing methods. Such integration is increasingly understood as a key to high-quality, replicable remote sensing–assisted spatial decision support. In early discussions in the 1980s, it soon became clear that provenance (or lineage) in particular is a fundamental element in understanding earth observation-related and other geodata quality (Moellering et al. 1988). As commercial GIS accelerated during the early 1990s, the Geolineus project (Lanter 1992b) demonstrated how software dedicated to lineage/provenance capture, management, and visualization can enable such gains as replicable geospatial workflows, automated workflow comparison, data quality modeling, data update management, and increased sharing of expert knowledge of geodata creation. Now with increasingly heightened awareness of provenance in computer systems (Bose and Frew 2005; Ikeda and Widom 2009; Simmhan et al. 2005; Yue et al. 2010a), there has been a maturing appreciation of the need to computationally address provenance capture, management, and exchange in an increasingly big data scenario.

While definitions of geodata provenance have varied, it is quite arguably distinct from and offers unique benefits over traditional metadata in large part because it encompasses process *history*. Regardless of definitions, the application of provenance benefits in remote sensing–assisted decision support workflows cannot be realized without development and demonstration of collaborative software architectures including those in a geo-cyberinfrastructure. Provenance has and will be of increasing interest to and a focus of organizations that create and encourage international specifications and standards (e.g., ISO, W3C, and OGC). As these organizations formulate procedures for the specification of provenance, we will see software developers add this capability to their products in a far more complete implementation than is currently the case. Even before the emerging international standards begin to mature, research is critically needed to demonstrate and fully understand *practical* benefits that user-friendly and integrated geoprocessing, workflows, and provenance can offer. With additional research and development, geospatial provenance has a high potential to benefit quality, trust, and innovation related to remote sensing–assisted spatial decision support.

References

Alonso, G. and C. Hagen. 1997. Geo-opera: Workflow concepts for spatial processes. In *Advances in Spatial Databases*, Springer, Berlin, Germany, pp. 238–258.

Anderson, J.R., E.E. Hardy, J.T. Roach, and R.E. Witmer. 1976. *A Land Use and Land Cover Classification System for Use with Remote Sensor Data*. Geological Survey Professional Paper 964. Washington, DC: U.S. Government Printing Office.

Anderson, K.E. and G.M. Callahan. 1990. The modernization program of the U.S. Geological Survey's National Mapping Division. *Cartography and Geographic Information Systems* 17(3): 243–248.

Aronson, P. and S. Morehouse. 1983. The ARC/INFO map library: A design for a digital geographic database. In *Auto-Carto Six; Proceedings of the Sixth International Symposium on Automated Cartography*, Vol. 1, Ottawa/Hull, Canada, pp. 372–382.

Atkins, D.E., K.K. Droegemeier, S.I. Feldman, H. Garcia-Molina, M.L. Klein, D.G. Messerschmitt, P. Messina, J.P. Ostriker, and M.H. Wright. 2003. Revolutionizing science and engineering through cyberinfrastructure: Report of the National Science Foundation Blue-Ribbon Advisory Panel on Cyberinfrastructure. National Science Foundation: Washington, DC. https://arizona.openrepository.com/arizona/handle/10150/106224. Accessed June 29, 2014.

Bose, R. and J. Frew. 2005. Lineage retrieval for scientific data processing: A survey. *ACM Computing Surveys* 37(1): 1–28.

Bossler, J.D., J.B. Campbell, R.B. McMaster, and C. Rizos, eds. 2010. *Manual of Geospatial Science and Technology*, 2nd edn. Boca Raton, FL: CRC Press.

Botts, M. and A. Robin, eds. 2007. *OpenGIS Sensor Model Language (SensorML) Implementation Specification*. Open Geospatial Consortium. http://portal.opengeospatial.org/files/?artifact_id=21273. Accessed June 24, 2014.

Bröring, A., C. Stasch, and J. Echterhoff, eds. 2012. *OGC Sensor Observation Service Interface Standard.* Open Geospatial Consortium.

Buneman, P., S. Khanna, and W.-C. Tan. 2001. Why and where. A characterization of data provenance. In *International Conference on Database Theory* (*ICDT*), pp. 316–330.

Buneman, P. and S.B. Davidson. 2010. Data provenance—The foundation of data quality. Carnegie Mellon University Software Engineering Institute, Pittsburgh, PA. http://www.sei.cmu.edu/measurement/research/upload/Davidson.pdf. Accessed June 30, 2014.

Buneman, P. 2013. "The Providence of Provenance." In Big Data, edited by G. Gottlob, G. Grasso, D. Olteanu, and C. Schallhart, 7968:7–12. Berlin, Germany: Springer. http://link.springer.com/10.1007/978-3-642-39467-6_3.

Chrisman, N.R. 1983. The role of quality information in the long-term functioning of a geographic information system. *Automated Cartography*, 6: 302–312.

Chrisman, N.R. 1986. Obtaining information on quality of digital data. In *AutoCarto Proceedings of the International Symposium on Computer-Assisted Cartography*, Vol. 1. London, U.K.: Cartography and Geographic Information Society, pp. 350–358.

Congalton, R. 2010. Remote sensing: An overview. *GIScience & Remote Sensing* 47(4): 443–459.

Conover, H., R. Ramachandran, B. Beaumont, A. Kulkarni, M. McEniry, K. Regner, and S. Graves. 2013. Introducing provenance capture into a legacy data system. *IEEE Transactions on Geoscience and Remote Sensing* 51(11): 5098–5014.

Cui, Y., J. Widom, and J.L. Wiener. 2000. Tracing the lineage of view data in a warehousing environment. *ACM Transactions on Database Systems* 25(2): 179–227.

Del Rio, N. and P.P. da Silva. 2007. Probe-It! Visualization support for provenance. In *Advances in Visual Computing*. Springer, Berlin, Germany, pp. 732–741. http://link.springer.com/chapter/10.1007/978-3-540-76856-2_72. Accessed June 24, 2014.

Di, L., P. Yue, H.K. Ramapriyan, and R.L. King. 2013b. Geoscience data provenance: An overview. *IEEE Transactions on Geoscience and Remote Sensing* 51(11): 5065–5072.

Di, L., Y. Shao, and L. Kang. 2013a. Implementation of geospatial data provenance in a web service workflow environment with ISO 19115 and ISO 1911-2 lineage model. *IEEE Transactions on Geoscience and Remote Sensing* 51(11): 5082–5089.

EarthCube. 2014. EarthCube: Transforming geosciences research. http://earthcube.org/. Accessed June 30, 2014.

Essinger, R. and D.P. Lanter. 1992. User-centered software design in GIS: Designing an icon-based flowchart that reveals the structure of ARC/INFO data graphically. In *Proceedings of the 12th Annual ESRI User Conference*, Palm Springs, CA.

Fielding, R.T. 2000. *Architectural Styles and the Design of Network-Based Software Architectures.* Irvine, CA: University of California. http://www.ics.uci.edu/~fielding/pubs/dissertation/top.htm. Accessed June 29, 2014.

Foster, I., J. Vöckler, M. Wilde, and Y. Zhao. 2002. Chimera: A virtual data system for representing, querying, and automating data derivation. In *Proceedings of the 14th International Conference on Scientific and Statistical Database Management*, Los Alamitos, CA IEEE, pp. 37–46. http://ieeexplore.ieee.org/xpls/abs_all.jsp?arnumber=1029704.

Foster, I., J. Vöckler, M. Wilde, and Y. Zhao. 2003. The virtual data grid: A new model and architecture for data-intensive collaboration. In *Proceedings of the First Biennial Conference on Innovative Data Systems Research* (*CIDR*), Vol. 3. Asilomar, CA: Citeseer, p. 12.

Freire, J., D. Koop, E. Santos, C. Scheidegger, C. Silva, and H.T. Vo. 2012. VisTrails. In Brown, A. and Wilson, G., eds. *The Architecture of Open Source Applications: Elegance, Evolution, and a Few Fearless Hacks*, Vol. I. aosabook.org. http://aosabook.org/en/vistrails.html.

Frew, J. 2004. Earth system science server (ES3): Local infrastructure for earth science product management. In *Proceedings of the Fourth Earth Science Technology Conference*, Palo Alto, CA. http://esto.gsfc.nasa.gov/conferences/estc2004/papers/a4p3.pdf. Accessed March 1, 2014.

Frew, J. and P. Slaughter. 2008. ES3: A demonstration of transparent provenance for scientific computation. In J. Freire, D. Koop, and L. Moreau, eds., *Provenance and Annotation of Data and Processes*. Lecture Notes in Computer Science, Vol. 5272. Berlin, Germany: Springer, pp. 200–207. http://link.springer.com/chapter/10.1007/978-3-540-89965-5_21. Accessed June 26, 2014.

Frew, J. and R. Bose. 2001. Earth system science workbench: A data management infrastructure for earth science products. In *Proceedings of the International Conference on Scientific and Statistical Database Management*. Los Alamitos, CA: IEEE Computer Society, pp. 180–189.

Gamble, M. and C. Goble. 2011. Quality, trust, and utility of scientific data on the web: Towards a joint model. In *Proceedings of the Third International Web Science Conference*, Vol. 15. ACM. New York, NY. http://dl.acm.org/citation.cfm?id=2527048. Accessed June 24, 2014.

Gil, Y., J. Cheney, P. Groth, O. Hartig, S. Miles, L. Moreau, and P.P. da Silva, eds. 2010. The foundations for provenance on the web. *Foundations and Trends in Web Science* 2(2–3): 99–241.

Gil, Y. and S. Miles, eds. 2013. PROV Model Primer. W3C. http://www.w3.org/TR/prov-primer/.

Giordano, A., H. Veregin, E. Borak, and D.P. Lanter. 1994. A conceptual model of GIS-based spatial analysis. *Cartographica: The International Journal for Geographic Information and Geovisualization* 31(4): 44–57.

Glavic, B. and K.R. Dittrich. 2007. Data provenance: A categorization of existing approaches. In *Proceedings of the 12th GI Conference on Database Systems in Business, Technology, and Web* (*BTW*), Vol. 7, Aachen, Germany, pp. 227–241.

Grady, R.K. 1988. The lineage of data in land and geographic information systems. In *Proceedings of GIS/LIS'88 American Congress on Surveying and Mapping: Data Lineage in Land and Geographic Information Systems*, Vol. 2, San Antonio, TX, pp. 722–730.

Guptill, S.C. 1987. Techniques for managing digital cartographic data. In *Proceedings of the 13th International Cartographic Conference*, Morelia, Mexico, Vol. 4(16), pp. 221–226.

Hart, G. and C. Dolbar. 2013. *Linked Data: A Geographic Perspective*, 1st edn. CRC Press , London, U.K.

Henzen, C., S. Mas, and L. Bernard. 2013. Provenance information in geodata infrastructures. In *Geographic Information Science at the Heart of Europe, III*. Lecture Notes in Geoinformation and Cartography. Springer International Publishing, New York, NY, pp. 133–151.

Hey, T., S. Tansley, and K. Tolle, eds. 2009. *The Fourth Paradigm: Data-Intensive Scientific Discovery*. Redmond, WA: Microsoft Research.

Houbie, F. and L. Bigagli. 2010. OGC catalogue services standard 2.0 extension package for ebRIM application profile: Earth observation products. http://portal.opengeospatial.org/files/?artifact_id=35528.

IFAR. 2013. *Provenance Guide*. International Foundation for Art Research. New York, NY, http://www.ifar.org/provenance_guide.php. Accessed September 9, 2014.

Ikeda, R. and J. Widom. 2009. Data lineage: A survey. Technical Report. Stanford University InfoLab, Stanford, CA, http://ilpubs.stanford.edu:8090/918/. Accessed September 13, 2013.

ISO 19115-2:2009(E). 2009. Geographic information—Metadata—Part 2: Extensions for imagery and gridded data. International Organization for Standardization, Geneva, Switzerland.

Jensen, J.R. 2005. In K.C. Clarke, ed., *Introductory Digital Image Processing: A Remote Sensing Perspective*, 3rd edn. Prentice Hall Series in Geographic Information Science. Upper Saddle River, NJ: Prentice Hall.

Jensen, J.R. 2007. In K.C. Clarke, eds. *Remote Sensing of the Environment: An Earth Resource Perspective*, 2nd edn. Prentice Hall Series in Geographic Information Science. Upper Saddle River, NJ: Prentice Hall.

Jones, J.S., J.A. Tullis, L.J. Haavik, J.M. Guldin, and F.M. Stephen. 2014. Monitoring oak-hickory forest change during an unprecedented red oak borer outbreak in the Ozark Mountains: 1990 to 2006. *Journal of Applied Remote Sensing* 8(1): 1–13.

Langran, G. and N.R. Chrisman. 1988. A framework for temporal geographic information. *Cartographica* 25(3): 1–14.

Langran, Gail. 1988. "Temporal GIS Design Tradeoffs." In Proceedings of GIS/LIS '88, 890–99. San Antonio, TX: American Congress on Surveying and Mapping.

Lanter, D.P. 1989. *Techniques and Method of Spatial Database Lineage Tracing*. Columbia, SC: University of South Carolina.

Lanter, D.P. 1991. Design of a lineage-based meta-data base for GIS. *Cartography and Geographic Information Systems* 18(4): 255–261.

Lanter, D.P. 1992a. Propagating updates by identifying data dependencies in spatial analytic applications. In *Proceedings of the 12th Annual ESRI User Conference*, Palm Springs, CA.

Lanter, D.P. 1992b. *GEOLINEUS: Data Management and Flowcharting for ARC/INFO, 92-2*. Santa Barbara, CA: National Center for Geographic Information & Analysis. http://www.ncgia.ucsb.edu/Publications/tech-reports/91/91-6.pdf. Accessed June 24, 2014.

Lanter, D.P. 1993a. Method and means for lineage tracing of a spatial information processing and database system. Patent No. 5,193,185. United States Department of Commerce Patent and Trademark Office.

Lanter, D.P. 1993b. A lineage meta-database approach toward spatial analytic database optimization. *Cartography and Geographic Information Systems* 20(2): 112–121.

Lanter, D.P. 1994a. Comparison of spatial analytic applications of GIS. In W. K. Michener, J. W. Brunt, and S. G. Stafford, eds., *Environmental Information Management and Analysis: Ecosystem to Global Scales*. CRC Press , London, U.K.

Lanter, D.P. 1994b. A lineage metadata approach to removing redundancy and propagating updates in a GIS database. *Cartography and Geographic Information Systems* 21(2): 91–98.

Lanter, D.P. and C. Surbey. 1994. Metadata analysis of GIS data processing: A case study. In T.C. Waugh and R.G. Healey, eds., *Advances in GIS Research: Proceedings of the Sixth International Symposium on Spatial Data Handling*. London, U.K.: Taylor & Francis Ltd., pp. 314–324.

Lanter, D.P. and H. Veregin. 1991. A lineage information program for exploring error propagation in GIS applications. In *Proceedings of the 15th Conference of the International Cartographic Association*, Bournemouth, U.K., pp. 468–472.

Lanter, D.P. and H. Veregin. 1992. A research paradigm for propagating error in layer-based GIS. *Photogrammetic Engineering and Remote Sensing* 58(6): 825–833.

Lanter, D.P. and R. Essinger. 1991. *User-Centered Graphical User Interface Design for GIS*. Santa Barbara, CA: National Center for Geographic Information & Analysis, pp. 91–96. http://www.ncgia.ucsb.edu/Publications/tech-reports/91/91-6.pdf. Accessed June 24, 2014.

Malaverri, J.E.G., C.B. Medeiros, and R.C. Lamparelli. 2012. A provenance approach to assess quality of geospatial data. In *27th Symposium on Applied Computing*. Riva del Garda (Trento), Italy: ACM.

Merriam-Webster. 2014. Merriam-Webster Online. http://www.merriam-webster.com/.

Moellering, H., L. Fritz, D. Franklin, R.W. Marx, J.E. Dobson, D. Edson, J. Dangermond et al. 1988. The proposed standard for digital cartographic data. *The American Cartographer* 15(1): 9–140.

Moore, H. 1983. The impact of computer technology in the mapping environment. In *Proceedings of the Sixth International Symposium on Automated Cartography*, Vol. 1. Ottawa/Hull, Ontario/Quebec, Canada: Cartography and Geographic Information Society, pp. 60–68.

Moreau, L., B. Clifford, J. Freire, J. Futrelle, Y. Gil, P. Groth, N. Kwasnikowska et al. 2011. The open provenance model core specification (v1.1). *Future Generation Computer Systems* 27(6): 743–756.

Moreau, L. 2010. The foundations for provenance on the web. *Foundations and Trends in Web Science* 2(2–3): 99–241.

Moreau, L. and P. Missier, eds. 2013. PROV-DM: The PROV data model. http://www.w3.org/TR/2013/REC-prov-dm-20130430/#section-example-two.

Naseri, M. and S.A. Ludwig. 2013. Evaluating workflow trust using hidden Markov modeling and provenance data. In Q. Liu, Q. Bai, S. Giugni, D. Williamson, and J. Taylor, eds., *Data Provenance and Data Management in eScience*. Studies in Computational Intelligence, Vol. 426. Berlin, Germany: Springer-Verlag, pp. 35–58.

NSF. 2014. Data Infrastructure Building Blocks (DIBBs). http://www.nsf.gov/pubs/2014/nsf14530/nsf14530.htm. Accessed June 30, 2014.

Nyerges, T. November 1987. GIS research needs identified during a cartographic standards process: Spatial data exchange. *International Geographic Information Systems Symposium: The Research Agenda* 1: 319–330.

Oxford University Press. 2014. Oxford English Dictionary. http://www.oed.com/.

Plale, B., B. Cao, C. Herath, and Y. Sun. 2011. Data provenance for preservation of digital geoscience data. In A.K. Sinha, D. Arctur, I. Jackson, and L.C. Gundersen, eds., *Societal Challenges and Geoinformatics*. Geological Society of America Special Paper 482. Boulder, CO: Geological Society of America, pp. 125–137.

Santos, E., D. Koop, T. Maxwell, C. Doutriaux, T. Ellqvist, G. Potter, J. Freire, D. Williams, and C.T. Silva. 2012. Designing a provenance-based climate data analysis application. In P. Groth and J. Frew, eds., *Provenance and Annotation of Data and Processes*. Lecture Notes in Computer Science, Vol. 7525. Santa Barbara, CA, pp. 214–219.

Schut, P., ed. 2007. *OpenGIS Web Processing Service*. Open Geospatial Consortium.

Simmhan, Y.L., B. Plale, and D. Gannon. 2005. A survey of data provenance in e-science. *SIGMOD Record* 34(3): 31–36.

Simmhan, Y.L., B. Plale, and D. Gannon. 2008. Karma2: Provenance management for data-driven workflows. *International Journal of Web Services Research* 5(2): 1–22.

Sun, Z., P. Yue, L. Hu, J. Gong, L. Zhang, and X. Lu. 2013. GeoPWProv: Interleaving map and faceted metadata for provenance visualization and navigation. *IEEE Transactions on Geoscience and Remote Sensing* 51(11): 5131–5136.

Talbert, C. 2012. *Software for Assisted Habitat Modeling Package for VisTrails (SAHM: VisTrails) v. 1*. Fort Collins, CO: USGS Fort Collins Science Center. https://www.fort.usgs.gov/products/23403. Accessed August 31, 2014.

Tan, W.-C. 2007. Provenance in databases: Past, current, and future. *Bulletin of the IEEE Computer Society Technical Committee on Data Engineering* 30(4): 3–12.

Tilmes, C. and A.J. Fleig. 2008. Provenance tracking in an earth science data processing system. In J. Freire and D. Koop, eds., *Provenance and Annotation of Data and Processes*. Lecture Notes in Computer Science, Vol. 5272. Berlin, Germany: Springer-Verlag, pp. 221–228. http://ebiquity.umbc.edu/_file_directory_/papers/445.pdf. Accessed September 19, 2014.

Tullis, J.A., F.M. Stephen, J.M. Guldin, J.S. Jones, J. Wilson, P.D. Smith, T. Sexton et al. 2012. Applied silvicultural assessment (ASA) Hazard Map. University of Arkansas Forest Entomology's Applied Silvicultural Assessment, Fayetteville, AR, http://asa.cast.uark.edu/hazmap/. Accessed August 30, 2014.

Tullis, J.A. and J.M. Defibaugh y Chávez. 2009. Scale management and remote sensor synergy in forest monitoring. *Geography Compass* 3(1): 154–170.

Vanecek, S. and R. Moore. 2014. OGC open modelling interface standard, Version 2.0. https://portal.opengeospatial.org/files/?artifact_id=59022. Accessed June 26, 2014.

Veregin, H. and D.P. Lanter. 1995. Data-quality enhancement techniques in layer-based geographic information systems. *Computers Environment and Urban Systems* 19(1): 23–36.

Wade, Tasha, and Shelly Sommer. 2006. A to Z GIS: An Illustrated Dictionary of Geographic Information Systems. Redlands, CA: Esri Press.

Wang, C., Z. Lu, and T.L. Haithcoat. 2007. Using Landsat images to detect oak decline in the Mark Twain National Forest, Ozark Highlands. *Forest Ecology and Management* 240: 70–78.

Wang, S., A. Padmanabhan, J.D. Myers, W. Tang, and Y. Liu. 2008. *Towards Provenance-Aware Geographic Information Systems*. Irvine, CA: ACM. http://acmgis08.cs.umn.edu/papers.html#posterpapers.

Woodruff, A. and M. Stonebraker. 1997. Supporting fine-grained lineage in a database visualization environment. In W.A. Gray and P.-Å. Larson, eds., *Proceedings of the 13th International Conference on Data Engineering*, Birmingham, U.K., pp. 91–102.

Yeide, N.H., K. Akinsha, and A.L. Walsh. 2001. *The AAM Guide to Provenance Research*. Washington, DC: American Association of Museums.

Yue, P., J. Gong, and L. Di. 2010a. Augmenting geospatial data provenance through metadata tracking in geospatial service chaining. *Computers & Geosciences* 36: 270–281.

Yue, P., J. Gong, L. Di, L. He, and Y. Wei. 2010b. Semantic provenance registration and discovery using geospatial catalogue service. *Proceedings of the Second International Workshop on the role of Semantic Web in Provenance Management (SWPM 2010), Shanghai, China.*

Yue, P., Y. Wei, L. Di, L. He, J. Gong, and L. Zhang. 2011. Sharing geospatial provenance in a service-oriented environment. *Computers, Environment and Urban Systems* 35(4): 333–343.

Yue, P. and L. He. 2009. Geospatial data provenance in cyberinfrastructure. In *Proceedings of the 17th International Conference on Geoinformatics*, Fairfax, VA.

20

Toward Democratization of Geographic Information: GIS, Remote Sensing, and GNSS Applications in Everyday Life

Gaurav Sinha
Ohio University

Barry J. Kronenfeld
Eastern Illinois University

Jeffrey C. Brunskill
Bloomsburg University of Pennsylvania

Acronyms and Definitions

FGDC	Federal Geographic Data Committee
Geo-ICT	Geospatial information and communication technology
GIS	Geographic information system
GIScience	Geographic Information Science
GNSS	Global Navigation Satellite System
GPS	Global Positioning System
HOT	Humanitarian OpenStreetMap Team
LBS	Location-based services
LKCCAP	Local Knowledge and Climate Change Adaptation Project
NCGIA	National Center for Geographic Information and Analysis
OSM	OpenStreetMap
PGIS	Participatory GIS
PPGIS	Public participation GIS
SDI	Spatial data infrastructure
UAV	Unmanned aerial vehicle
VGI	Volunteered geographic information

20.1 Rethinking Geographic Information and Technologies in the Twenty-First Century

There are very few academic domains that remain unaffected by the reach of geo-information technologies. Geography and other closely affiliated environmental science disciplines rely heavily on such technologies for their empirical work, but environmental engineering, social science, humanities, health science, and the business communities have also found geo-technologies and the geographic perspective crucial to many of their disciplinary pursuits. There was a time, until very recently, when a statement such as this would suffice to describe the geo-information revolution. Yet, today, the growing influence of geographic/geospatial theory, information, and technology goes far beyond the academe and the workplace—it has become the story of our everyday lives. There is no better example of this than the life story of a poor Indian boy who lost his family in 1986, ending up on a wrong train that took him all the way across India to the bustling city of Calcutta, where he had to beg for a living, but from where ultimately he was transported far away to Australia through adoption by a loving couple, the Brierleys. Saroo Brierley ended up in much more comfortable settings than he would have probably been entitled in the impoverished settings of his earlier life, but he never gave up the urge to find his lost family. He had some memories of his first neighborhood, and that it was within a 14 h train ride to Calcutta, but there was no way to use that information gainfully—until Google Earth was released and satellite imagery and maps made it possible to visually search places anywhere remotely using a computer. Saroo drew a circle to locate all places within 14 h train ride of Calcutta (~1200 km), and while it seemed like finding the proverbial needle in a haystack, his persistence paid off when he recognizes landmarks in Khandwa, Madhya Pradesh, several hundred miles from Kolkata (formerly Calcutta, when Saroo got lost). Saroo eventually went to India and tracked down his biological family, 25 years after he was separated—with mostly Google Earth to thank. The story of his life is now a bestseller book *A Long Way Home* (Brierley, 2014) and a great ode to the power of personal spatial memories and geospatial technologies—especially high-resolution satellite imagery–based mapping services that anybody can use.

Google Earth's success and popularity is symbolic of many different transitions that signify today geographic information creation and consumption is no longer under the tight control of institutions, experts, and the powerful few. Their grips have been loosened incrementally and sometimes disruptively by certain developments that finally have given individuals, novices, and the traditionally disenfranchised a strong say in how places, events, and perceptions are mapped and communicated. We now live in a world where the average person is increasingly aware of the everyday benefits and drawbacks for geo-information and associated technologies. This includes powerful tools that support navigation and exploration, communication, social interaction, public participation, security, and the like. Our increasing reliance on mobile devices (e.g., phones, tablets, and notebooks) is an important catalyst for many of these developments. Such developments also give rise to increased concerns over the quality and quantity of the spatially oriented data we collect and maintaining our privacy as personal communication devices, web services, electronic cards, and surveillance cameras (in developed countries) create *digital shadows* (Klinkenberg, 2007) of our everyday lives. In a short time, we have gone from representing places and features to populating our geospatial databases with our doppelgangers, whose controls, ironically, lie in hands other than our own and who will continue to *roam* within large databases, long after we cease to exist ourselves.

The start of this geo-information revolution goes back to the 1960s, when three distinct clusters of major geo-information technologies started to form the expansive geo-information landscape of today: remote sensing, geographic information system (GIS), and Global Navigation Satellite System (GNSS) technologies. At their core, both remote sensing and GNSSs are geospatial data collection systems, while GISs have multiple functional aspects. The following define each briefly in the context of this chapter:

- Remote sensing is the science of collecting, processing, and interpreting information of the Earth from aircraft or spacecraft equipped with instruments for sensing signals emitted or reflected by the surface (or atmosphere) of the Earth.
- A GIS is an epistemology institutionalized through *software* and *social practices* for processing, managing, analyzing, modeling, visualizing, and communicating about geospatial datasets of various types.
- A GNSS is a suite of satellites in orbit around the Earth transmitting location and timing data that may be used by receivers worldwide to determine location on or near the Earth.

These technologies have been constantly evolving, and lately through synergistic combinations, they have revolutionized every aspect of geographic information creation, analysis, and visualization in all the major sectors of the modern economy: scientific, commercial, educational, and governmental. Their collective contribution to the world economy has grown significantly over time to several tens of billions of dollars (Dasgupta, 2013). Jerome Dobson, a pioneer who helped establish the field of Geographic Information Science (GIScience), envisions remote sensing, GIS, GNSS, and related technologies, collectively as a *macroscope* for viewing large-scale phenomena in finer detail (similar to the microscope that magnifies truly small-scale phenomena or the telescope that magnifies distant, apparently small phenomena) (Dobson, 2011). Over the last decade, the development of newer forms of geospatial information and communication technologies (Geo-ICTs) and Web 2.0 services has played a key role in widening the *scope* of the geospatial macroscope by making the collection, processing, and sharing of geographically referenced information a part of our everyday lives. We afford ourselves the functionalities of these technologies not just in professional contexts but also for managing our personal information. In doing so, we uncover another important context—the fascinating relationship between technology and society. As the German philosopher Martin Heidegger observed, "the essence of technology is by no means anything technical" (Heidegger, 1982: 4, as quoted in Crampton, 2010: 6); technology, science, and society do not evolve independent of one another. As such, questions of society have become increasingly important as geotechnologies have become more personalized.

This chapter complements other chapters of this book with a sociotechnical discussion of how remote sensing, GIS, and GNSS are codependent and will continue to strengthen the foundations of our newly geo-enabled societies. We avoid the deep technical trenches to keep the reader focused on the *macroscopic* trends, since technical discussions of these technologies abound elsewhere. The material presented here is intended to help readers recognize these technologies at work in ordinary life situations—not just in scientific and professional settings (which is already covered extensively in other chapters of this book). With this in mind, the remainder of this chapter is organized as follows. In Section 20.2, we review the historical development of remote sensing, GIS, and GNSS technologies as a context for discussing the democratization of the geospatial domain in the twenty first century. In Section 20.3, we explore several application domains to showcase the diverse ways in which these technologies are being integrated into our everyday lives. In Section 20.4, we provide a critical analysis of the technical challenges in trying to make geo-information and related technologies easily accessible to the public, as well as the political, organizational, social, and ethical issues that must be considered in an increasingly geospatially enabled society. We conclude with a brief summary of this chapter in Section 20.5.

20.2 Toward Democratization of Geo-Information Technologies

For centuries, paper maps were the state-of-the-art technology for representing, analyzing, and communicating geographic information. While they still play a significant role in many spheres of our personal and professional lives today, paper maps are being replaced by services on mobile phones, virtual globes, Google Glass, and other Geo-ICTs. The history and development of spatial representations from prehistoric maps to modern spatially enabled mobile devices are characterized by Goodchild et al. (2007) according to three stages of development, namely, *historic*, *enlightened*, and *contemporary*. During the historic phase, the longest of the three phases, map production was uncoordinated, incomplete, and undertaken by both public and private entities for various human needs. The enlightened phase began toward the beginning of the twentieth century with the advent of state-sponsored mapping initiatives and evolved into the contemporary phase starting in the last decade of the twentieth century. This last phase is ongoing and characterized by increasing emphasis on cooperation and data sharing, as well as movement toward greater access to the modes of constructing and using geographic information. In this chapter, we extend Goodchild et al.'s (2007) original model with a *transition* phase to clearly distinguish the intervening period between the enlightened and contemporary phases. This period started with intellectual debates about how and if GIS can matter to societal development, recognition of the importance of geographic information and technologies by governments, development of Internet technologies, and emergence of the open-source movement. This short period was quite critical in setting the stage for the defining developments of the ongoing contemporary phase, especially those related to the democratization of geographic information.

In this section, we present a representative, but not exhaustive, chronology of developments during the enlightened, transition, and contemporary phases to provide a basic background for reasoning about the events that have fostered the public's awareness and understanding of geo-information. We do not discuss the historical phase hereafter, since our concern is mostly with later geo-information technologies and events that have primarily shaped the twenty-first century democratization of geographic information.

20.2.1 Enlightened Phase (Twentieth Century): State-Sponsored Geo-Information Technologies

The enlightened phase started in earnest with the massive increase in surveying and mapping needs during the two World Wars and with the rise of the scientific method in the early to middle twentieth century. Large-scale investments in geo-information technologies by governments (as detailed in Table 20.1) led to the development of national mapping agencies (both military and civil) with large budgets. The agencies were typically charged with the task of producing high-accuracy maps of the physical and social state of nations at several scales. This fostered the development of several new technologies including aerial photogrammetry, remote sensing with satellites (e.g., Television Infrared Observation Satellite [TIROS], Landsat), automated cartography, GIS, and the first GNSS—the U.S. Navstar GPS. The technological developments and emerging benefits to society and science were unquestionable. They were, however, not equally distributed among nations as most of the developments

TABLE 20.1 Important Geo-Information Technology Developments during the Enlightened Phase

Year	Enlightened Phase Events (1960–1989)
1960	First nonmilitary satellite, TIROS-1, launched by the United States for space-based meteorological observations.
1964	Canada Geographic Information System project launched.
1967	U.S. Census Bureau develops Dual Independent Map Encoding vector topological data model.
1969	ESRI, the leading GIS software and services company today, is founded.
1972	Landsat 1, the first civilian Earth-monitoring satellite, is launched.
1978	Map Overlay and Statistical System, the first full-fledged interactive vector and raster GIS, deployed at many U.S. Federal agencies.
1982	ARC/INFO released by ESRI, to become most popular commercial GIS.
1983	United States announces GPS will be available for civilian sector when operational.
1985	GPS becomes operational.
1985	Graphical Resources Analysis Support System software, a raster focused, open-source GIS, is released by U.S. Army Corps.
1986	The Australian Land Information Council is established.
1987	*Handling Geographic Information* report (Chorley Report) released by the United Kingdom.
1987	National Center for Geographic Information and Analysis (NCGIA) established.
1989	U.S. Census Bureau releases Topologically Integrated Geographic Encoding and Referencing products into the public domain.

occurred in the United States, followed by the United Kingdom and other European countries, and Australia. The United States led the world in topographic mapping, remote sensing, GIS, and GNSS. In addition to government-sponsored initiatives, significant contributions were made by academics and by the commercial sector with the creation of popular cartographic, GIS, and remote-sensing software. The development of commercial software, in particular, facilitated collaborations between government and private entities and led to the creation of a special class of geospatial professionals. At the time, the average person, however, was still dependent on various types of maps produced by the professionals to understand their own spaces. It is for this reason that this era, particularly from the late 1980s onward, has been criticized for creating a hegemonic system that perpetuated the grip of powerful agencies and trained professionals on geographic information. This critique marks the beginning of the transition phase, during which the importance of reconceiving mapping technologies as products of and for society became evident.

20.2.2 Transition Phase (Twentieth Century): Governance and Scholarship for Society

During the transition phase (Table 20.2), some social theorists began to argue that the seductive power of technologies had blinded many users from the more complex sociopolitical realities of technological development (Chrisman, 1987; Taylor, 1990). The epistemological interpretation of GISs (and related technological developments) was considered quite intellectually impoverished since the positivist, and absolute space-focused view of conventional GIS, was in denial of the social roots and impacts of GIS and the social constructivist nature of geographic information (Warf and Sui, 2010). This led to the so-called GIS wars, instigated by social theorists' criticism that GIS was imperialistic and used to subjugate local and indigenous perspectives of space, place, and culture in favor of the majority's perspectives. GIS and its practitioners were accused of helping strengthen hegemonic narratives serving the powerful elite, who spared no thought for marginalization of those on the less fortunate side of the economic, social, and digital divide (Pickles, 1994; Sheppard, 1995; Harley, 2002). Others criticized specifically the military origins and "historical complicity of remote sensing, GIS and GNSS in military, colonial, racist and discriminatory practices" (Crampton, 2010: 7) and sought to expose the role of such technologies as weapons of intelligence and war (Smith, 1992; Cloud, 2002; Monmonier, 2002). These social critiques were ultimately channelized, and a set of critical research issues were identified at the Friday Harbor National Center for Geographic Information and Analysis meeting in 1996 (Harris and Weiner, 1996). The results from this and some other similar engagements provided the foundation for a broad GIS and society research agenda that has thrived since the late 1990s.

A particularly long-lasting, tangible outcome of the GIS and society debates was the rise of grassroots mapping projects led by communities and facilitated by GIS and mapping experts. These projects, which may take on many forms, have been called participatory or community mapping projects, community integrated GISs, public participation GISs (PPGIS), or participatory GISs (PGIS), depending on the goals, participation, mapping process, and technologies used (Craig et al., 2002; Sieber, 2006; Elwood, 2011; Brown, 2012). These methods of geographic information creation and/or use are context and issue driven rather than technology driven (Dunn, 2007). Over the last 20 years, hundreds of such mapping efforts have been undertaken worldwide. The projects, ranging from basic sketch mapping to sophisticated online map-based surveys, have been undertaken to help communities broaden their spatial, environmental, social, and political perspectives, address

TABLE 20.2 Important Geo-Information Technology-Related Developments during the Transition Phase That Paved the Way for Contemporary Processes of Democratization of Geo-Information

Year	Transition Phase Events (1989–1998)
1989	Association for Geographic Information lobbying/advisory group established in United Kingdom.
1990	U.S. Federal Geographic Data Committee is established.
1990	"GIS wars" erupt between social theorists and GIS theorists/practitioners.
1991	Miniature GPS receiver technology becomes widely available.
1992	The term "Geographic Information Science" is coined (Goodchild, 1992).
1992	U.S. Remote Sensing Act of 1992 opens skies to commercial satellites.
1993	The NCGIA "Geographic Information and Society" meeting is held at Friday Harbor, Maine.
1993	European Umbrella Organisation for Geographic Information is established.
1994	U.S. President Bill Clinton signs Executive Order 12906 "Coordinating Geographic Data Acquisition and Access: The National Spatial Data Infrastructure."
1994	OpenGIS Consortium (now Open Geospatial Consortium) is formed.
1995	The U.S. Navstar GPS attains full operational capability.
1995	The Predator unmanned aerial vehicle becomes operational.
1996	MapQuest and MultiMap, the earliest interactive web-mapping services based on government-collected public domain data, are released in the United States and United Kingdom.
1996	NCGIA Initiative 19: GIS and society workshop is organized.
1996	Public participation GIS is defined at an NCGIA workshop at the University of Maine.
1996	U.S. Federal Communications Commission requires wireless carriers to determine and transmit the location of callers who dial 9-1-1.

land ownership disputes, conserve and manage natural resources, document and protect cultural heritage, revitalize neighborhoods, and challenge existing narratives of their community. The projects are fertile grounds for the convergence of cartography, GISs, GNSSs, and remote sensing. Topographic maps, aerial photos, satellite imagery, GNSS collected waypoints and tracks, and GIS-based demographic and environmental data may all contribute to a group's mapping and decision making. This is evidenced by the PPgis.net electronic forum and the Integrated Approaches to Participatory Development (IAPAD) website,* which are worldwide resources for those seeking to develop PGIS/PPGIS/participatory mapping projects, especially in developing countries.

The development of the GIS and society research agenda coincided with the development of a new multidisciplinary field called GIScience, a field that explores the conceptual foundations, design, and the application contexts of geospatial technologies (Goodchild, 1992). While the definition and scope of the field has continued to evolve (Goodchild, 2012; Blaschke and Merschdorf, 2014), the one thing that remains constant is the idea that GIScience research proceeds in at least three dimensions: computer, individual, and society (Longley et al., 2010). The issues of interest to GIScience specialists are diverse, pertaining not just to the computational contexts of geographic data use but also to the cognitive, behavioral, social, legal, and ethical factors that govern the creation, dissemination, and consumption of geographic information. Arguably, society still remains the weakest dimension of GIScience, but compared to the time of origination of the GIS and society debates, there is much better understanding today of the two-way relationship between geo-information technologies and society (Warf and Sui, 2010). This is supported by the success of the PPGIS/PGIS movement and the more recent establishment of critical GIS as an important area of research within GIScience (Schuurman, 1999; O'Sullivan, 2006; Wilson and Poore, 2009).

During the transition phase, several key technological developments occurred to lay the groundwork for broader access and so-called democratization of Geo-ICTs in the current contemporary phase. These included the replacement of older, more cumbersome GPS devices with miniature receivers, operational completion of the GPS satellite system, development of unmanned aerial vehicles (UAVs), and, of course, the popularization of the World Wide Web. The development of open-source GIS, which started with Geographic Resources Analysis Support System (GRASS) in 1985, progressed enough to justify establishment of a consortium of open-source GIS software systems. Meanwhile, the commercial sector began to see the potential market in interactive web-mapping services (WMSs) such as MapQuest and MultiMap.

As a final point of interest, the transition phase should also be recognized for the establishment of government-run agencies, national spatial data infrastructures (SDIs), and related organizations (e.g., U.S. Federal Geographic Data Committee [FGDC]), which were designed to coordinate the collection and sharing of geographic data between government agencies and with the public. Today, SDIs have been implemented in more than 100 countries (Masser, 2011). SDIs are still largely one-way vehicles for sharing government data and still have to evolve to become two-way vehicles, allowing the public to also contribute to the SDI (Budhathoki et al., 2008). Despite this limitation, the existence of these first-generation SDIs made many contemporary phase developments possible. During the transition phase, several private/public groups such as Association for Geographic Information and European Umbrella Organisation for Geographic Information (EUROGI) were also established to further the interests of geographic information communities.

20.2.3 Contemporary Phase (Twenty-First Century): The Vision of Democratization

The contemporary phase began during the first decade of the twenty-first century (see Table 20.3 for a list of important events). Unlike the enlightened phase, in which geo-information technologies were managed by large institutions in a fundamentally top–down hierarchy, the contemporary phase is defined by a much more democratized approach to geo-information. Contemporary uses of GISs and related technologies have become more context sensitive and issue driven, with some (but not nearly enough) recognition to issues of power, commodification, and surveillance (Warf and Sui, 2010). This democratized framework is only partly the result of the government SDIs, academic critiques and research agendas, and grassroots participatory mapping initiatives—a major push came from the rise of several private sector companies that invested heavily in online and mobile mapping technologies. By the start of the new century, government agencies across the world were searching for new ways to provide geospatial services in a time of reduced budgets, Geo-ICTs that dropped the entry barrier for geographic information collection and sharing, and an intellectual shift from mapping static places to representing dynamic activities of people across places over time (Goodchild et al., 2007). In the contemporary phase, governments still play a major role but increasingly as facilitators (e.g., maintaining national SDIs). They are also more inclusive, involving citizens in topographic and resource mapping initiatives, environmental monitoring, disaster preparation, and health and emergency services. The development of inclusive, bottom–up (individual and collaborative) processes for creating and sharing geographic information is made possible by integrating the functionality of traditional geospatial technologies with newer Geo-ICTs (Goodchild, 2007; Sui et al., 2013). The development of technologies like OpenStreetMap (OSM), Microsoft Virtual Earth, Google Earth, and Google Glass are prime examples.

At the end of the twentieth century, a seminal event occurred in 1998, when U.S. Vice President Al Gore presented an inspiring vision called Digital Earth as "a multiresolution, three-dimensional representation of the planet, into which we can embed vast quantities of geo-referenced data" (Gore, 1998). Gore proposed the Digital Earth as a comprehensive information system composed of many distributed components, together providing access to all historical and current information about the entire planet (including the activities of its inhabitants), and supported by modeling tools

* www.iapad.org.

TABLE 20.3 Important Geo-Information Technology-Related Developments That Have Proven to Be Central to the Ongoing Democratization of Geographic Information in the Contemporary Phase

Year	Contemporary Phase Events (1998–Present)
1998	U.S. Vice President Gore presents the Digital Earth vision.
1998	Microsoft launches TerraServer in partnership with U.S. Geological Survey (USGS).
1998	NASA starts the Digital Earth Initiative and creates the web-mapping system data-sharing standard.
1999	The first U.S. commercial satellite, IKONOS, is successfully launched with a very high panchromatic spatial resolution of about 0.82 m.
2000	Selective availability, which degraded Navstar GPS signals for civilians, is switched off.
2001	Keyhole's EarthViewer and GeoFusion's GeoPlayer virtual globes are released.
2001	The infrastructure for spatial information in Europe is launched by the European Union.
2001	Wikipedia, a crowdsourced encyclopedia, is launched.
2002	USGS launches *The National Map* web service.
2002	Friendster, a social networking website, goes online.
2003	U.S. e-government data access initiative "Geospatial One-Stop" goes online.
2003	Urban and Regional Information Systems Association approves Code of Ethics for geospatial industry professionals.
2004	The National Imagery and Mapping Agency is reoriented as National Geospatial-Intelligence Agency.
2004	OpenStreetMap launched as first crowdsourced, public domain street mapping database.
2004	Facebook and Flickr are launched.
2005	The Google Earth, Google Maps, and Microsoft Virtual Earth geobrowsers are released.
2005	Global Earth Observation System of Systems' 10-year implementation plan adopted by intergovernmental Group on Earth Observations.
2006	Twitter is launched.
2006	The term "neogeography" is proposed to describe creation and use of geospatial information by nonexperts.
2007	Concepts of citizens as censors and "volunteered geographic information" (VGI) are promoted and gain a foothold (Goodchild, 2007).
2007	The National Science Foundation promotes a cyberinfrastructure to foster collaborative research and data sharing.
2007	Apple releases the first iPhone and Google releases the Android mobile operating system.
2008	Ushahidi VGI platform launched to track violence during Kenya post-election crisis.
2010	Google announces its autonomous car project.
2013	Google Glass is launched for testing purposes.
2014	The U.S. government relaxes restrictions on satellite imagery to allow image resolutions below 50 cm for commercial purposes.

for predicting future conditions. Gore envisioned the resource as a publicly available one-stop virtual environment in which anybody—child or adult—could explore information about the Earth effortlessly (e.g., on a *magic carpet* that could fly through space and also back in time). In many ways, Gore's presentation of his vision can be seen as the start of the contemporary phase. It encouraged existing initiatives and also stimulated new initiatives in the government and the private sector where. For example, the U.S. National Aeronautics and Space Administration (NASA) initiated the "Digital Earth Initiative" in 1998 and, among other things, created the current WMS standard, crucial for seamless sharing of geographic data and services on the web (Grossner et al., 2008). NASA's "World Wind" was an early virtual globe inspired by the Digital Earth vision. Today, Google's Google Earth, Microsoft's Bing, Environmental Systems Research Institute (ESRI's) ArcGIS Explorer, and SkyGlobe's TerraExplorer are the popular virtual globe platforms.* While NASA does not maintain special units to support this vision, the vision is now promoted by the International Society for Digital Earth, spearheaded by China.† The Digital Earth vision today has evolved significantly since Gore's speech, and today, it is understood more pragmatically as a globally distributed set of technological services and practices serving as a collective geographic knowledge organization and information retrieval system (Craglia et al., 2008; Grossner et al., 2008). In this regard, the virtual globes of today can be imagined as small private digital Earths that offer access to the global Digital Earth system.

The popularity of WMSs and virtual globes established *geobrowsing*, that is, the search for geographic information via a map interface (Lemmens, 2011) as a near ubiquitous phenomenon, with new users, services, and regions still being added regularly. Geobrowsing is an intrinsically geospatial endeavor that is so intuitive that even young school kids can engage in it—because of intelligent obfuscation of certain peculiarities of geographic data (e.g., scale and projection). The value of geobrowsing lies in how it brings mapping down to the level of the everyday user and encourages people to engage with geographic information. For example, geobrowsers can be used to explore places, assist with everyday navigation and way-finding tasks, visualize data, and create new maps and datasets. They highlight the manner in which our everyday activities are tied to space. With the growing popularity of geobrowsing applications on mobile devices (Google Maps was the most frequently downloaded smartphone application in 2014‡), it seems reasonable to suggest that there is a strong desire in people

* See http://en.wikipedia.org/wiki/Virtual_globe for a complete list of virtual globes.

† http://www.digitalearth-isde.org.

‡ www.businessinsider.comgoogle-smartphone-app-popularity-2013–9#infographic (Accessed July 17, 2014).

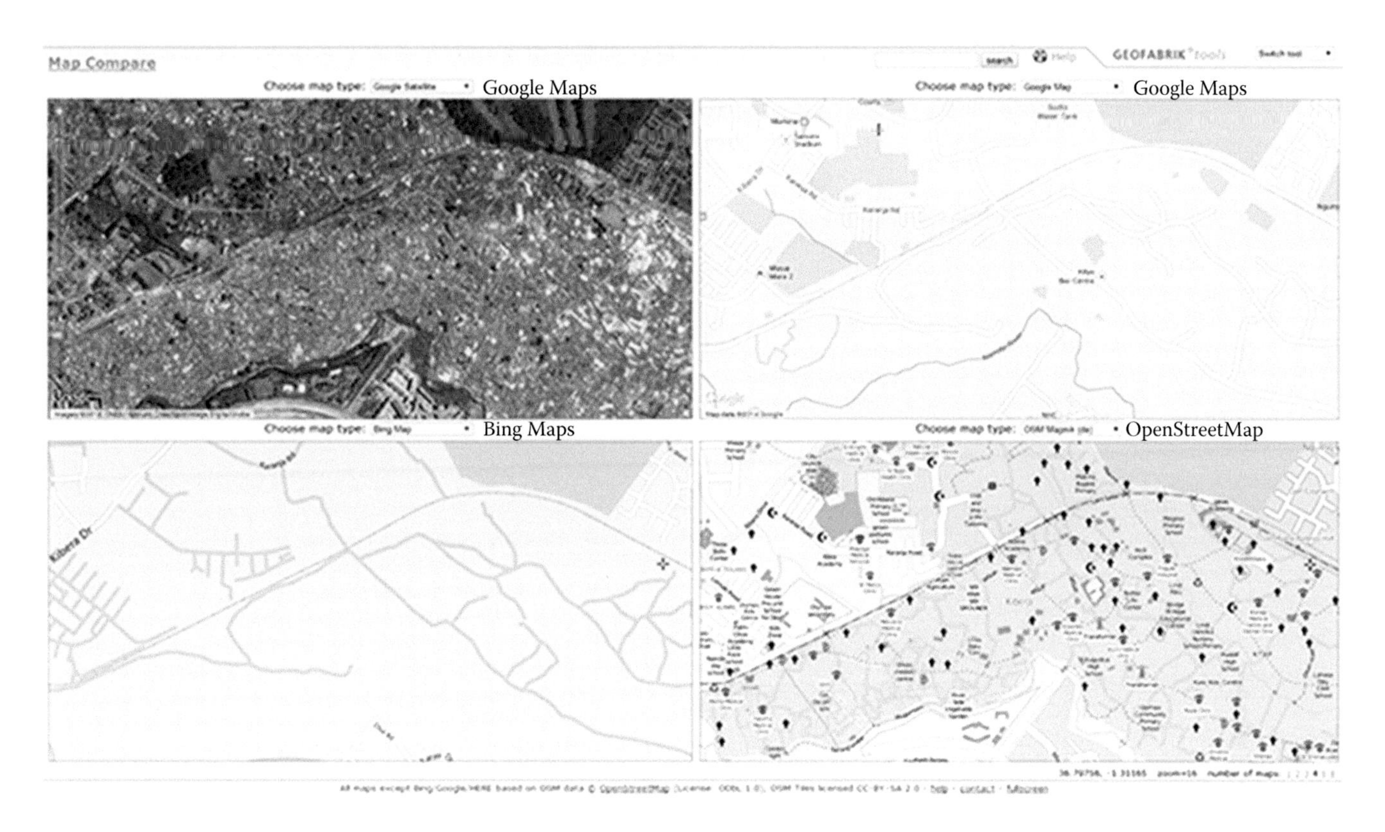

FIGURE 20.1 As the high-resolution satellite image indicates, the slum in Kibera is a heavily populated part of Nairobi. While OSM map captures substantial detail of the structures and roads collected from community-mapping efforts, Google Maps and Bing Maps do not map most structures in Kibera since they are deemed illegal by local authorities. (Screenshot from www.tools.geofabrik.de/mc.)

from all walks of life and professions to directly engage and manipulate geographic information rather than passively absorb it from static maps. Clearly, in the twenty-first century, we have moved on from merely *looking* at maps to *participating and interacting* with geobrowsers.

Fueled partly by the success of the virtual globes, WMSs (i.e., geobrowsers), and easy-to-use GNSS units, the contemporary phase has witnessed an explosion of user-generated geographic information—a phenomenon called *neogeography*. Neogeography refers to the practice of using mapping techniques and tools for and by nonexpert individuals or communities (Turner, 2006). Its applications are generally not analytical or formal but mostly descriptive and visual. The rise of neogeography lies in a synergy that has developed between various technologies (e.g., Geo-ICTs, GNSS-enabled mobile devices, and Web 2.0) and social data collection practices (e.g., crowdsourcing and volunteered geographic information) that allow everyday users to create mash-ups of map services and individualized content (Goodchild, 2007, 2009; Sui et al., 2013; Wilson and Graham, 2013). Evidence of this synergy can be seen in the popularity of crowdsourcing repositories like OSM* and Wikimapia.† OSM is a free-to-use global map database of geographic features built by individual volunteers relying on a mix of resources including satellite imagery, GNSS tracks, and knowledge of place-names. Wikimapia is a comparable database that actually contains many more user-generated entries for place-names than any official list of place-names. These databases are important repositories of socially generated geographic information and, in some countries, may be the only digital maps available to the public for economic or political reasons (see Figure 20.1 for an example). Further, even if official or commercial maps are available, crowdsourced maps offer diversity by recording users' perspectives of space. As Monmonier (1996) states, multiple maps can be made for the same place, with the same data, for the same situation. The diverse nature and versatility of these datasets have been invaluable in efforts to support humanitarian relief following natural disasters and to empower communities through grassroots mapping (as will be discussed in the next section).

Over the last 15 years, many of the barriers to democratization of geographic information have disappeared or at least become less relevant. This is evidenced, to a certain degree, by the Digital Earth initiative and by neogeography efforts to design collaborative, bottom–up processes for creating and sharing geographic information. Yet, despite this progress, efforts to democratize geographic information still have a long way to go—neogeography has limited connection to the academic domain of geography or GIScience (Goodchild, 2009), and its claims of democratization and making geographic information available to anyone, anywhere, and anytime (Turner, 2006) have been shown to be premature and shallow (Haklay, 2013). Still,

* www.openstreetmap.org.
† www.wikimapia.org.

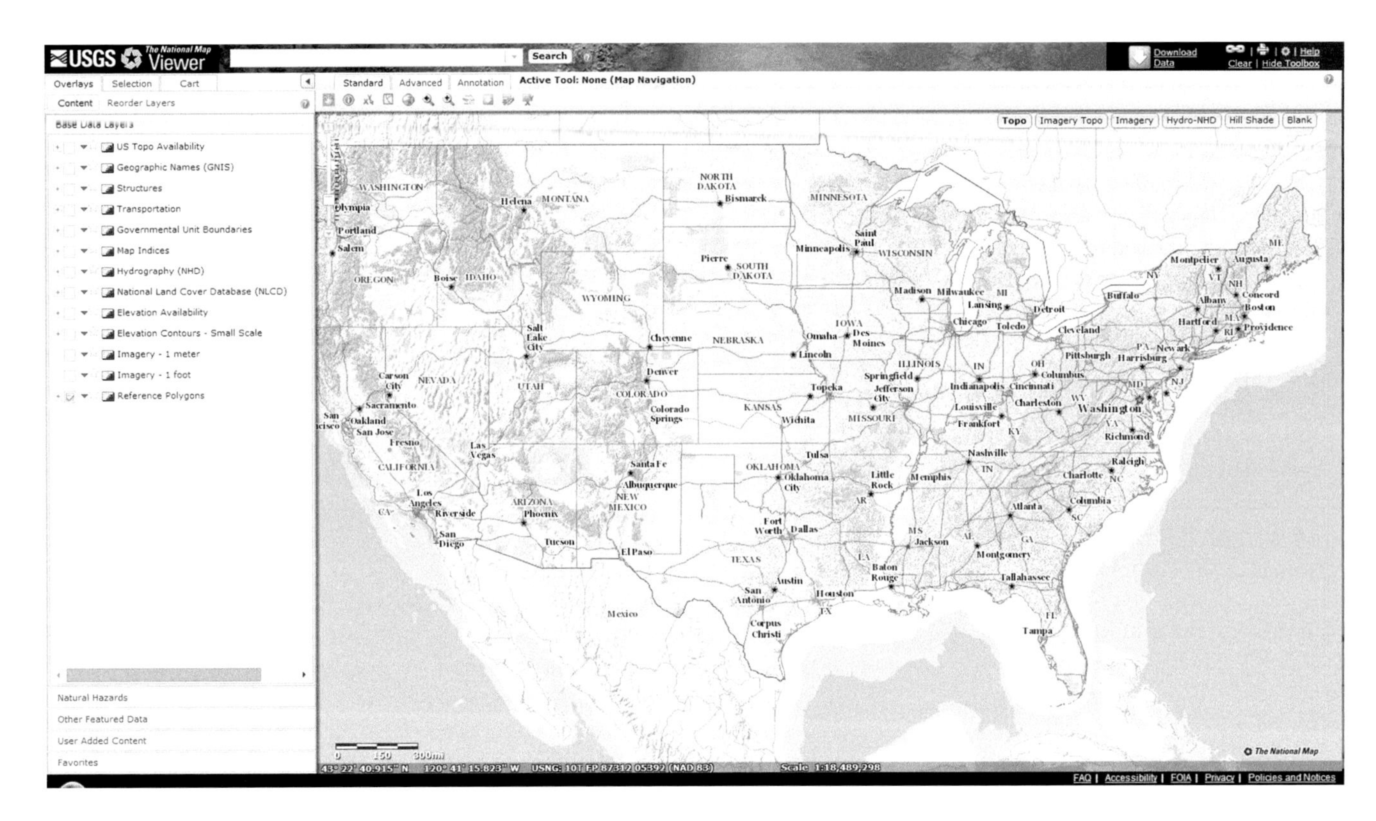

FIGURE 20.2 The USGS National Map provides free topographic mapping and data-downloading services for the entire United States. (Screenshot from www.viewer.nationalmap.gov.)

the developments that define the contemporary phase suggest that the traditional technologies of remote sensing, GISs, and GNSSs *can* be used to empower people and help them view the world from a perspective other than a traditional top–down, institutional frameworks of the recent past.

20.3 Discovering Remote Sensing, GIS, and GNSS Applications in Our Daily Lives

In this twenty-first century contemporary age, the functionalities offered by remote sensing, GISs, and GNSSs are finally becoming increasingly integrated into our daily lives. This integration is largely the result of new Geo-ICTs, as well as an increasing public awareness of the value of *location* as a search parameter when seeking information about the world. Over three decades worth of research exists on the conceptual and technical aspects of efforts to integrate these technologies (Gao, 2002; Mesev, 2007; Merchant and Narumalani, 2009). GNSSs are used to georeference aerial photos and satellite images and are a core component of high-resolution light detection and ranging (LIDAR) remote sensing of the Earth. Both remotely sensed data and GNSS data can be imported and overlaid with other data layers in a GIS environment. In a GIS, these layers can be used to create and analyze a variety of built (e.g., buildings, roads) and environmental (e.g., elevation, hydrography, land use/land cover) datasets. While it is more common to incorporate remotely sensed and GNSS-derived data in a GIS, GIS-derived datasets are also used extensively in the rectification, classification, mapping, and dissemination of remotely sensed imagery. The integration of these technologies is also evident; in these ways, data are often seamlessly distributed to professional users. For example, the U.S. Federal Government's Geospatial One-Stop project provides a single portal to geographic information. One of its services, The National Map* (see Figure 20.2) web service, supports both interactive online visualization and free downloading of a wide array of historical and current datasets including satellite imagery, elevation, land cover, transportation, hydrography, boundaries, and geographic names for the United States.

In addition to the more standard methods for integrating remote sensing, GISs, and GNSSs, outlined earlier, a variety of new increasingly democratized methods have evolved during the contemporary phase of geospatial development. For example, PPGIS projects have taken advantage of low-cost, high-accuracy GNSSs to map local communities and develop representations of features that are not evident on existing maps or from satellite imagery alone. In communities where current and/or high-resolution imagery is not available, amateur aerial photographs have been developed with makeshift aerial cameras consisting of GNSS-enabled digital cameras attached to helium-filled balloons. The aerial cameras have been used to map small neighborhoods at a cost of less than U.S. 35¢ per hectare (Seang et al., 2008). Perhaps, the most obvious example for owners of mobile Geo-ICTs arises in the context of geobrowsing in which high-resolution satellite images and

* www.NationalMap.gov.

aerial photos, made available by a service provider (e.g., Google or Microsoft), are used as base maps for displaying thematic data layers. These integrations are often referred to as *map mash-ups*, or simply *mash-ups*. Similar combinations of imagery and data layers presented in 3D virtual globe software (e.g., Google Earth, NASA World Wind) are also used to create virtual fly-through animations. In addition to government-run geospatial data clearinghouses, public/private clearinghouses like OSM have also been developed, which allow everyday users to contribute and use geospatial data in a community-driven setting. The ability to both edit and download the entire OSM database locally is a tremendous benefit that no other map service provides.

The integration of geo-information data and functionalities provide the backbone for many of the services and tools that are developed by government agencies, nonprofit agencies, and businesses to help people make everyday decisions (Yang et al., 2010; Lemmens, 2011). Such services include but are not limited to emergency response, disaster response and planning, natural resources management, environmental and public health services, weather forecasting, floodplain mapping, precision farming, agricultural services, community planning, property and utility mapping, crime monitoring and analysis, operations research, traffic monitoring, real estate services, and even K-12 education. To explore this topic in greater detail, we present four broad contexts in which Geo-ICTs are interwoven in people's everyday lives: location-based services (LBSs), disaster relief and emergency management, participatory mapping, and participatory sensing of our everyday physical and social environments. In presenting these examples, particular attention is paid to the manner in which these applications facilitate democratization of geographic information.

20.3.1 Location-Based Services

The term location-based service, or LBS, generally refers to software applications that use location as a basis for providing information or performing a service for a user (Zipf and Jöst, 2012). The start of LBSs can be traced to a 1996 U.S. Federal Communications Commission mandate that set a minimum accuracy for determining the location of an emergency E911 call by a wireless device. Although this policy was oriented toward emergency response, resulting improvements in location accuracy have fostered innovation in the marketplace by catalyzing the development of mobile mapping programs and LBSs. As a particularly interesting and futuristic example, consider the recent initiative by Google and others to develop automated cars (see Figure 20.3). In 2010, Google announced that it was working on a project to develop self-driving cars (Thrun, 2010). Not surprisingly, the project leader was also the coinventor of Google's Street View mapping technology. Other automotive companies and universities are also developing prototypes and testing them on roads. Google is ahead of most though, since it has already conducted more than half a million miles of pilot testing. Driverless cars will be legal on the road from January 2015 in the United Kingdom and already so in the U.S. states of California, Florida, and Nevada, with legislation pending in many others (Weiner and Smith, 2014). Autonomous cars are stellar examples of Geo-ICT technology and services integration. Such cars now depend on LIDAR and/or 360° cameras and computer vision technology, and currently data from driving patterns of cars driven by human beings, to create real-time high-resolution 3D models of a car's surroundings. It is estimated that they may need to sense about 1 GB of data *every second* to work optimally today! The GNSS-derived location and LIDAR model also need to be combined with information from a GIS database of static infrastructure (e.g., telephone poles, crosswalks, and traffic lights) to identify all types of static and moving objects (e.g., pedestrians, cyclists) to plot a safe path through space (Fisher, 2013). Similar to cars, lightweight UAVs or drones may be found at the disposal of any citizen in the future for collecting remote-sensing data about any area of interest for business, environmental, recreational, and other purposes. Apart from such futuristic vehicles, smartphones and the recently launched Google Glass are also great platforms for LBS integration.

(a)

(b)

(c)

FIGURE 20.3 (a) A car retrofitted by Google for collecting 360° panoramic street view photos. (Photograph distributed by Wikipedia author, Kowloonese, under a CC-BY 2.0 license. http://en.wikipedia.org/wiki/Google_Street_View#/media/File:GoogleStreetViewCar_Subaru_Impreza_at_Google_Campus.JPG.) (b) A car retrofitted by Google to be tested as part of its self-driving car fleet. (Photograph by Jurvetson, S., modified by Mariordo, distributed under a CC-BY 2.0 license. http://commons.wikimedia.org/wiki/File:Google's_Lexus_RX_450h_Self-Driving_Car.jpg.) (c) Google's prototype of a fully automated self-driving car with numerous sensors but no manual traditional controls (steering wheel and accelerator and brake pedals). (Modified image available from Google.com at http://googleblog.blogspot.com/2014/05/just-press-go-designing-self-driving.html.)

LBSs today represent such an important category of business that it is being fiercely competed over by information technology, telecommunication, mobile phone, and increasingly other types of companies. Today, LBSs are designed to help people make decisions in the contexts of an incredibly wide variety of personal and professional matters (Raper et al., 2007) including but not limited to the following:

- *Navigation*: Turn-by-turn navigation; public transit; traffic and road condition updates; roadside emergency assistance; user-centered route selection; fuel consumption; autonomous car; vehicle to vehicle communication
- *Retail/Business services*: Retail advertising; store locators; shopping aisle information services; retail store/mall maps; real estate services; credit/bank card fraud prevention
- *Recreation*: Online and real-world games; sky gazing; geosocial networking; mobile place guides; location-based augmented reality; travel/tourism services
- *Societal services*: Location-based warnings and alerts; seeking/providing emergency help; disabled people mobility; toll collection; weather services; environmental services; agricultural services; participatory community planning
- *Mapping:* Crowdsourcing points of interest; monitoring environmental conditions; infrastructure maintenance; participatory/citizen sensing; mapping and monitoring personal health; documenting geocoded events; mapping disease outbreaks; finding or tracking people/animals/objects; mash-up services

With such a diverse list of activities, it should be obvious that the commercial sector is agog with LBS innovations and improvements. LBSs almost always combine data and functionalities from GIS, mobile cartography, and GNSS. For example, Google Maps provides several types of map services ranging from locating and describing landmarks, routing directions for different modes of transport (driving, walking, biking, and public transport), traffic patterns, and traffic alerts. In Figure 20.4, a photomontage derived from several screenshots depicts examples of these LBSs as they would look on a single-map view. Bing Maps™, MapQuest, and other mapping sites also provide similar services. Apart from such map or imagery-based LBSs, more specialized map-based LBSs include weather advisory services, field surveying by professionals or by untrained volunteers, and disaster relief work after earthquakes, fires, floods, and storms. LBSs that post real-time alerts about environmental conditions, manage drone delivery services, and support augmented reality Geo-ICTs such as Google Glass should be quite popular soon. The reach of LBSs into both mundane and critical decision-making processes will also continue to get deeper, with technological developments such as 3D mobile cartography, ubiquitous positioning (Mannings, 2008), and ubiquitous computing/Internet of Things (Weiser, 1991) making mobility patterns of people, not just their locations, the focus of LBSs. The implications of such deep LBS integration are obviously both exciting and unsettling at the same time. There are many issues and challenges that will come to the forefront regarding service reliability; information overload; energy use, human–machine interaction; personal

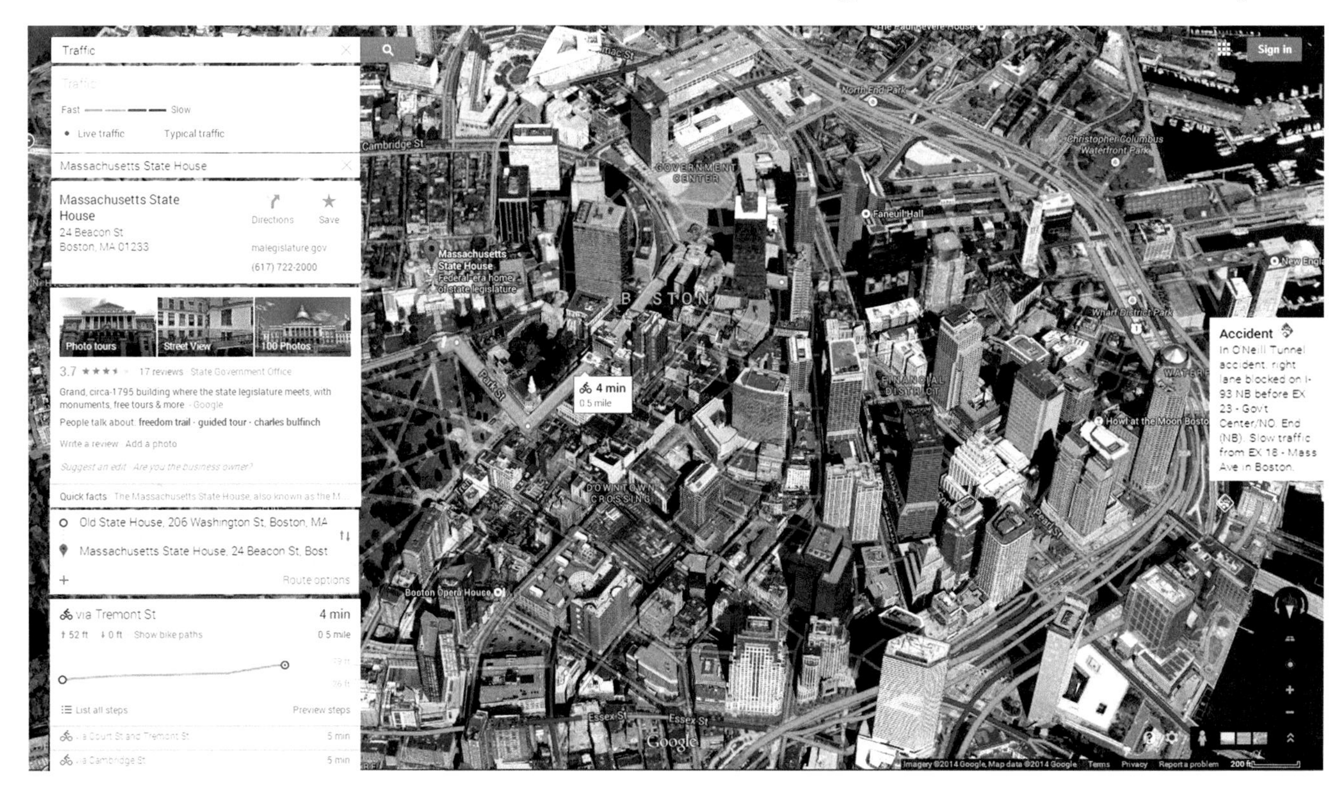

FIGURE 20.4 Examples of various map-based services available through Google Maps (www.maps.google.com). Note that this image is a composite of several computer screenshots, since these services cannot be simultaneously viewed on the same map view (as of August 2014).

privacy; provision of social, environmental, and health services; perception and experience of places; social networking; and cognitive and behavioral modifications.

20.3.2 Disaster Relief and Emergency Management

On November 8, 2013, Super Typhoon Haiyan swept through the Eastern Visayas region of the Philippines, killing 6000 people, displacing 1.9 million and impacting over 9 million (REACH et al., 2014). Even before the typhoon hit, a Humanitarian OpenStreetMap Team (HOT) was created in anticipation to map infrastructure and damage in the affected region. Harnessing the power of global volunteers to quickly evaluate postdisaster satellite imagery, the HOT performed damage assessment and facilitated prioritization of scarce resources. Maps created from the effort were visible to the public within days of the disaster (Buchanan et al., 2013).

The Haiyan HOT was just the latest in a long list of geospatial crowdsourcing efforts for disaster relief. This growth can be traced back at least to Hurricane Katrina in 2005, when a computer programmer in Austin, Texas, created a website that allowed users to post descriptions of local conditions onto a map (Singel, 2005). In 2007, a local radio station contributed to coverage of the San Diego wildfires in California by maintaining a dynamic web map of user-reported conditions (Zook et al., 2010). In 2009, OSM was first used for humanitarian purposes in the Philippines in response to Tropical Storm Ondoy. While crisis mapping* as a concept was beginning to take hold in 2009 (Soden and Palen, 2014), it came to the fore with the formation of HOT following the January 12, 2010, earthquake in Haiti.

In the days that followed the Haiti earthquake, relief workers posted information on locations of trapped survivors, damaged buildings, and triage centers; a global network of volunteers translated information from Haitian Creole, geotagged it, and placed it into online mapping services (created based on remote-sensing imagery) for other relief workers to use. The effort was made possible by the maturation of communities such as OSM, Ushahidi,† Geocommons,‡ and CrisisCamp Haiti (Pool, 2010), the cooperation of private companies (e.g., Google, Microsoft, and GeoEye), government agencies, nongovernmental organizations (NGOs), and Global Earth Observation Catastrophe Assessment Network (GEO-CAN)—a unique voluntary network of more than 600 experts from 23 countries. These entities relied heavily on information produced from the convergence of GIS software and data, GNSS field data, several sources of aerial and satellite imagery, and expert-created information products (Duda and Jones, 2011; Van Aardt et al., 2011). Without any centralized command, these disparate actors and resources converged within hours, and the collective, organic systems that were created in the short time easily surpassed what could ever be achieved by traditional top–down, centralized government efforts.

In a different context, Goodchild and Glennon (2010) discuss the case of volunteered geographic information (VGI) being useful in making quick real-time decisions during four wildfires near Santa Barbara between 2007 and 2009. The volunteers used GNSS-enabled cameras and phones to report information as text, photos, and videos over the Internet. For one wildfire, some citizens were able to access and interpret comparatively fine temporal and spatial resolution imagery from the Moderate Resolution Imaging Spectroradiometer satellite sensor, and, using services such as Google Maps, created several maps that were often more up to date than official maps. For another fire, volunteer map sites were set up to readily synthesize the volunteer and official information as it became available. This real-time lifesaving participatory sensing of an endangered environment bears testimony to the power of democratization of geo-information and placing remote sensing, GISs, GNSSs, and other Geo-ICTs in people's homes.

Until only a few years ago, it was the norm that government and transgovernmental agencies (e.g., United Nations) would lead disaster relief, with NGOs playing a supporting role. Emrich et al. (2010) provide an excellent review of the history of use of remote sensing and GISs in preparing or responding to emergencies. Remote sensing and GISs have been used in all phases of emergency management for multiple decades now. While the early focus was on understanding the physical processes, at the beginning of the twenty-first century, geo-information technologies were being used in all phases of emergency management—though still primarily to support the top–down hazard research, analysis, and disaster response command structure (Emrich et al., 2010; Zook et al., 2010). It is only since 2010, due to technological advances, supporting services, and changing mindsets, that virtually every major disaster has stimulated VGI/crowdsourcing mapping. These have included the Queensland and Australian floods in 2010–2011; the earthquake in Christchurch, New Zealand, in 2011; the earthquake and tsunami in northern Japan in 2011; the 2013 Super Typhoon Haiyan in the Philippines; the 2013 tornadoes in Illinois, United States; and even the search for the missing Malaysian Airlines aircraft MH-370 in 2014, which enlisted millions of volunteers scanning imagery for evidence of the missing aircraft. The trend is likely to become more popular as adoption of mobile devices continues in all parts of the world.

While the *crowd* and volunteer cannot be a replacement for domain experts, these successful relief efforts still clearly demonstrate two things. The first is that crowdsourcing, aided by appropriate online and field Geo-ICTs, can often provide critical geospatial information much faster than traditional methods. The second is that crowdsourcing is highly suited to the large and sudden demand created by natural and man-made disasters. Local volunteer groups and government agencies can also use crowdsourcing to gather crucial information that may be needed by both officials and communities to plan ahead of time for disasters in areas especially prone to storms, floods, landslides, fires, and disease outbreaks. On the other hand, this transition from traditional command-and-control structures to distributed crowdsourcing efforts is not without its own problems and raises numerous technological, organizational, and ethical

* www.crisismappers.net.
† www.ushahidi.com.
‡ www.geocommons.com.

issues that will need to be addressed in the coming years. Some of these issues are touched upon in Section 20.4.

20.3.3 Community Building

The story of crowdsourcing in Haiti after the earthquake could easily have dissipated as the disaster started slipping away from the media spotlight. Fortunately, the HOT that formed initially to respond to the earthquake dug its heels and worked for about a year and half after the earthquake to help people claim ownership of their broken and impoverished communities (Soden and Palen, 2014). The Haitian HOT is recognized today as remarkable for several reasons: it established HOT as a serious techno-humanitarian group, changed the perception of OSM and its contributors, became a quintessential example for champions of neogeography, and blurred the boundary between crowdsourcing, VGI, and PGIS through this long-term collaborative mapping initiative. Above all, HOT inspired Haitians to reclaim legal, cultural, and emotional ownership of the land and codify their local knowledge through geospatial data and maps, thus becoming more resilient against future threats. Undoubtedly, the project will be long cited as the quintessential example of how combining geo-information technologies can produce extremely valuable geo-information for the masses—by the masses.

The mapping success of the Haitian HOT can be traced back to another iconic OSM-led PGIS/collaborative mapping effort—the Map Kibera project.* Kibera, an informal settlement of about 170,000 people in Nairobi, Kenya, has captured worldwide attention in recent years as supposedly the largest slum in Africa. Despite such population, and a tremendous presence of NGOs, it is still invisible on most maps of the Nairobi, appearing only as a large blank space or an uninhabited forest. Similar to most informal settlements, Kibera developed on occupied public lands, and its illegality prevents it from being officially mapped or becoming eligible to receive government services. Its residents had little awareness of its geography or their collective sociopolitical capital until November 2009 when OSM contributors Mikel Maron and Erica Hagen started the Map Kibera project, which helped Kiberans create a community information system to empower themselves (Hagen, 2011). The two project leaders slowly incentivized and built a mapping team of youth by teaching them GNSS mapping, videography, and journalism skills—rather than paying them as NGOs had become accustomed to doing. The achievement of OSM's Map Kibera project can be easily shown by simply comparing satellite imagery for the area with map views from Google Maps, Bing Maps, and OSM. As can be seen in Figure 20.1, Google and Bing map the area with only a few main streets. The emptiness in these officially sanctioned maps stands in stark contrast to the dense settlement visible from the satellite imagery and OSM's map, which is being used by local residents to conceptualize their neighborhoods, demand services from the government, and fight many social problems. Based on the success of the Map Kibera project, mapping efforts have now spread to other slums in Nairobi and elsewhere in Africa and Asia.

* www.mapkibera.org.

It should be kept in mind that OSM facility and infrastructure mapping projects are only one kind of approach to PGIS. Another approach is to pair trained geospatial professionals with local communities to help them record their spatial histories and explore and communicate about community problems through maps. Currently, one of the authors of this chapter (Sinha) is part of an international collaborative research project called the Local Knowledge and Climate Change Adaptation Project (LKKCAP),† under the auspices of which several community mapping projects were organized in Mwanga District, Tanzania. LKCCAP was funded by the U.S. National Science Foundation to explore the relationship between local knowledge, institutions, and climate change adaptation practices embedded in rural livelihoods of Tanzania. The community-mapping projects were designed to let village residents explore social, political, and environmental dimensions of livelihood adaptation to climate change. The project involved sketch mapping (Figure 20.5) and GPS field mapping, with collected data being digitized into GIS databases and represented on poster-sized maps to help the community hold discussions with district officials and aid agencies.

We have focused on OSM and rural mapping projects to highlight that geo-information technologies are currently being used by nonprofessionals, even in the poorest of communities. PGIS/PPGIS and collaborative mapping are, obviously, equally relevant and popular in more prospering communities as well. Community concerns about neighborhood planning, natural resource conservation, climate change, flood mapping, environmental pollution, water scarcity, economic revitalization, infrastructure development, social discrimination, substance abuse, indigenous culture preservation, and many similar issues become the starting point for the PPGIS/PGIS projects. Special online collaborative spatial decision support systems have been developed for many studies (Jankowski, 2011). Various social networking platforms and online community planning forums, such as PlaceSpeak,‡ MindMixer,§ and MetroQuest,¶ may be used in combination with OSM, Google Earth, Google Map Maker™, and other Geo-ICTs to engage communities, planners, and governments in more developed countries (and also some urban communities in developing countries). As Geo-ICTs, such as augmented reality, indoor positioning and mapping, and mobile ground remote-sensing devices, become more common in developed societies, communities will be able to generate 3D virtual and augmented reality models of their neighborhoods—resulting in seamless boundaries between virtual and real-world experiences of places.

20.3.4 Participatory Sensing

Participatory sensing, a mapping process in which citizens and community groups engage in sensing and documenting their daily activities, has become more common in recent years. The scope of participatory sensing ranges from individuals making personal

† www.tzclimadapt.ohio.edu.

‡ www.placespeak.com.

§ www.mindmixer.com.

¶ www.metroquest.com.

FIGURE 20.5 A participatory sketch mapping session in Kirya village located in the Mwanga District of Tanzania for helping villagers map water infrastructure and ecological resources.

observations to the combination of data from hundreds, or even thousands, of individuals that reveals patterns across an entire city (Goldman et al., 2009). Participatory sensing leads to incredibly rich models of environments and spaces that people define through their activities (Sagl et al., 2014). Participatory sensing is a good example of citizen science, since the projects are generally designed for a specific purpose, often by researchers interested in learning about a phenomenon. It is this characteristic that distinguishes it from agenda-free VGI or politically motivated PPGIS/PGIS projects. The most common type of collaborative sensing and mapping involves the use of citizens' mobile devices, which act as nodes of a sensor network to collectively measure any number of interesting geocoded environmental parameters such as traffic characteristics, air quality, noise levels, luminosity, temperature, humidity, and radiation (Burke et al., 2006). Participatory sensing projects such as these have been fostered by advances in mobile phone technology, increased access to remote-sensing imagery and relevant spatial datasets, LBSs, and the increasing social appeal of crowdsourcing.

When participatory-sensed data are integrated via mash-ups, or displayed on virtual globes, they can be used to assess environmental variables at multiple spatial scales or in relation to other geographic datasets (e.g., air quality data). For example, as part of the NoiseTube* project, citizens in Paris used mobile phones to measure their personal exposure to noise in their everyday environment and created collective noise maps by sharing their geolocalized and annotated measurements with the community (Maisonneuve et al., 2010). Figure 20.6 maps the results from one such NoiseTube project using Google Earth satellite imagery as a base map, with a pie chart also providing the breakdown for the variously tagged sources of noise pollution by the participants. A different implementation was seen for the Common Scents project of the SENSEable City Laboratory at the Massachusetts Institute of Technology, United States, which equipped Copenhagen citizens' bikes with location aware environmental sensors to develop a mobile sensor network. The project's goal was to gather real-time, fine-grained, air-quality data to allow citizens and local officials to assess environmental conditions (Resch et al., 2009). Another project, called the Personal Environmental Impact Report, used location records from mobile phones to calculate estimates of an individual's impact and exposure to different environments (e.g., carbon monoxide emissions from cars) (Mun et al., 2009). A similar approach based on mobile phone mobility analysis also helped identify and characterize variations in human activity in urban environments (Sagl et al., 2014). In all such projects, citizens are given the ability to pool their data to reveal impacts and exposures in numbers that could be useful to both the individual and city planners.

Participatory sensing also includes projects for monitoring of environments, mapping live traffic, and collecting data and responding to environmental hazards real time, particularly those that need to be monitored real time (e.g., storms, forest fires, floods, volcanic eruptions) and might require evacuation or help from emergency responders within a short period of time after a critical event. The informal network of storm chasers/spotters, operational meteorologists, and television media

* www.noisetube.net.

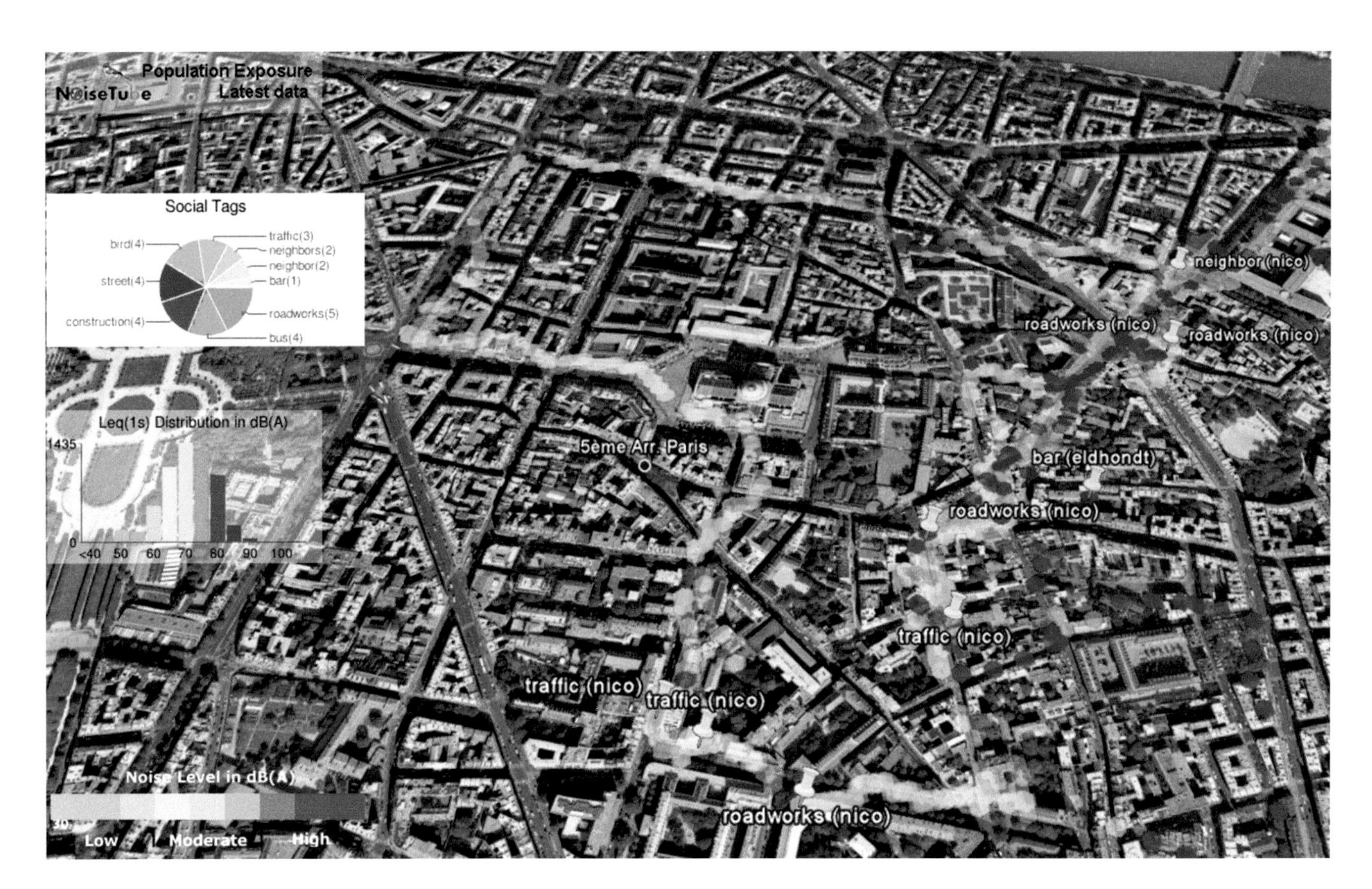

FIGURE 20.6 Map of noise pollution made from participatory sending data collection by the NoiseTube project in Paris, France. Image reproduced with permission from the NoiseTube project. (www.noisetube.net.)

outlets is a classic example of volunteers and experts collaborating monitor weather threats and to help people make decisions in their daily life (Palmer and Kraushaar, 2013). Another project, undertaken by more than 100 volunteers trained by the New Jersey statewide environmental agency, involved the collection of ground truth verification data for a database of 13,000 ephemeral vernal pools (breeding grounds for many endangered or threatened species) that had been derived from satellite imagery (Tulloch 2008). A web-based mapping system was set up to collect the verification data, which must be verified by the state biologists to meet legal requirements. Although the project is far from complete due to lack of budget and inaccessibility of many of the pools, the projects of this nature highlight the potential of using citizens to collect important cultural and scientific data. Thus, environmental collaborative monitoring networks have been proposed (Gouveia and Fonseca, 2008) as a framework for combining traditional environmental monitoring networks with the bottom–up, participatory, and open-source movement.

Personal mobile health monitoring has also become an extremely popular activity/sport for many people in urban centers. This includes the use of personal fitness devices for tracking physiological parameters such as heart rate, perspiration, steps or distance traveled, and velocity. Many devices are also geo-enabled, thereby allowing users to map their physiological parameters in relation to the places they visit during the day. The benefits of geo-enabled personal health monitoring range from casual monitoring of personal fitness to developing early warning systems for high-risk segments of the population. The current trend clearly seems to suggest that such information, contextualized with other forms of static and real-time geo-information about the local environment, will provide an important basis for personalized, governmental/institutional, and environmental health decision making in the future.

20.4 Democratization of Geo-Information: Circumscribing Issues and Challenges

From the start of the geo-information revolution in the 1960s remote sensing, GIS and GNSS technologies have evolved to play an increasingly significant role in our everyday lives. The resoundingly clear message in the previous discussions of LBSs, disaster relief, community mapping, and participatory sensing is that people now have incredibly diverse opportunities to collect and use geographic information to improve their lives. These changes are remarkable and have undoubtedly saved countless lives and empowered large new segments of the global population. At the same time, it is important to recognize that technological revolutions have a tendency to outpace critical theories and institutional structures designed to assure that their use furthers the social and cultural goals of society at large (Chrisman, 1987).

The democratization of Geo-ICTs is also affecting the manner in which data are analyzed and translated into actionable information. In this section, we examine some impediments to democratization and also remind readers that uncritical submission to the easy appeal of trendy Geo-ICTs is regressive and problematic.

20.4.1 Quality of Information and Services

As the task of geospatial data collection has been transferred from geospatial specialists to the general public, many concerns related to quality (accuracy, completeness, geographic diversity, lack of metadata) of the data and metadata have naturally arisen. The lack of obligatory commitment of "free" volunteers or the crowd is a further problem since interest in projects can easily dampen or die, leaving many data collection and mapping efforts incomplete. To objectively assess quality of such volunteered and crowdsourced data, a couple of studies explicitly compared OSM data to authoritative Ordnance Survey data in England (Haklay, 2010) and German Amtliche Topographisch-Kartographische Informationssystem (ATKIS) data for Munich (Fan et al., 2014). The errors were not large (less than 20 m offset for roads in England and 4 m offset and simplification of building footprints in Munich). One way to judge the quality of VGI is the number of peers who have reviewed and edited the content, a principle known as Linus's law (Elwood et al., 2011). Large numbers of peers can yield even higher quality data than obtained by the traditional authoritative sources. This is often evident during emergencies, when fast decisions must be made real time, so that the risks associated with volunteered information are often outweighed by the benefits, as was found in the case of the wildfires near Santa Barbara between 2007 and 2009 (Goodchild and Glennon, 2010). Similarly, McDougall (2012) examined three disasters in Australia, New Zealand, and Japan, respectively, and concluded that crowdsourced mapping provided a unique bottom–up perspective unattainable through the conventional emergency service "command-and-control" structure or implementation. Moreover, postdisaster mapping not just is useful for guiding immediate development efforts but also creates much needed information about places with important long-term use value. For example, as shown in Figure 20.7, Typhoon Haiyan resulted in a dramatic increase in volunteer mapping activity in affected areas, which now have a much higher density of mapped buildings than other places with comparable population densities. Otherwise, only Manila, the political and cultural capital, has a high density of mapped buildings.

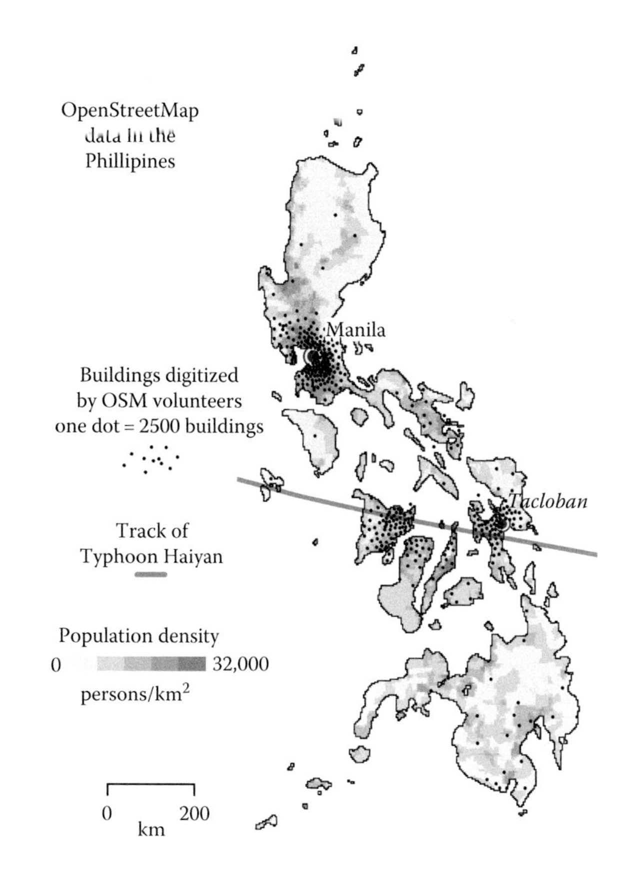

FIGURE 20.7 Map of buildings digitized by OpenStreetMap volunteers in the Philippines until end of July, 2014. Typhoon Haiyan resulted in a dramatic increase in volunteer activity in the affected areas, whereas buildings in other areas, except around Manila, remain sparsely mapped.

Yet, volunteered and crowdsourced datasets do have a lot of quality issues, which, if ignored, lead to faulty decisions. For example, ground truthing of building damage assessment done by volunteers looking at satellite imagery after Typhoon Haiyan in the Philippines found overall classification accuracy to be merely 36%, primarily due to overestimation of building damage (REACH et al., 2014). For the 2011 Christchurch earthquake, a similar analysis found large errors of omission (56% in downtown, 86% in residential areas) but virtually no errors of commission (3% in downtown, 0% in residential areas) (Foulser-Piggot et al., 2013). Reasons for such errors are related to low spatial resolution, lack of predisaster reference imagery, lack of expertise and visual/spatial interpretation skills, imagery georectification errors, and overly simplistic instructions for volunteers. Figure 20.8 shows examples of building digitization errors by volunteers for Typhoon Haiyan HOT.

It would be naïve to expect untrained volunteers with no background in geospatial technology and science to realize or rectify such errors. Champions of neogeography must understand that a lot of expertise is needed to engage in deep geographic thinking and geospatial analysis, quite contrary to the scope of neogeography (Goodchild, 2009). Moreover, the need to process data fast and reliably along with lack of incentives for deep theoretical analysis means that many technologies are being deployed in nonprofessional settings in ways that lead to lowering or relaxation of quality and consistency. This is occurring not only at the level of individual projects but also within the context of institutional and professional norms. A prime example of this is the de facto establishment of Web Mercator as the standard coordinate system for web mapping in all geobrowsers. Established for reasons related to server efficiency, Web Mercator is inappropriate for many (perhaps most) cartographic and quantitative applications (Battersby et al., 2014). Another example is the normalized practice of relying on users to uncover database errors, enabling a faster product release cycle. The botched launch of Apple's iOS6 Maps in 2012, for which Apple CEO Tim Cook had to offer the first open apology

FIGURE 20.8 Screenshot from OSM showing substantial offsets for some buildings digitized by volunteers for Typhoon Haiyan HOT. While the buildings on the right align closely with the satellite image, the three buildings on the left are considerably offset, because volunteers likely had access to different images at different times. (Screenshot from http://www.openstreetmap.org, created July 12, 2014.)

in the history of the company,* demonstrates that this practice can also disenfranchise users who expect high-quality spatial data.

In many cases, the appeal of spatial data is leading to misappropriation of technologies that are not designed for geospatial accuracy. For example, location inference methods used by social networking services, for which location is just a secondary parameter, often rely on coarse methods of positioning, and using such inaccurate locations for applications where precise positions are needed creates unanticipated problems. Worse, many LBSs are not designed properly for efficient decision making, especially for real-time decisions while on the move (Raubal, 2011). One has only to think of the many limitations of automated navigation systems that people routinely experience and have to address based on their own contextual geographic knowledge. Such problems may be ignored in many contexts, but in emergency response situations, inaccurate locations can put lives in danger. Resolutions of such errors are as much about technology improvement as exploring how human cognitive systems function and differ across use contexts and for individuals based on gender, culture, native language, age, and many other specific factors.

20.4.2 Privacy and Confidentiality

The constant demand for geospatial data and increasing use of linked social–spatial data raises concerns about the ability to protect confidentiality and privacy (NRC, 2007). Geosurveillance as a concept has been around from the early 1990s when GIS was first attacked for being an extension of the state for monitoring and surveillance (Pickles, 1991; Smith, 1992). The concept has now evolved to include a much wider spectrum of Geo-ICTs, and the surveillance is not just being done by governments but also powerful corporates, and that too with our implicit acquiescence. The exaflood of big data (Sui et al., 2013) that results from massive data collection efforts is simultaneously a source of exciting new possibilities and a threat to our privacy, and thus, our freedoms. Data mining makes it possible to derive "deep information" on individuals' behaviors in virtual and physical space (Sui and Goodchild, 2011). High-resolution imagery today allows us to explore our world in fascinating detail and simultaneously empowers us to counter secretive government strategies (Perkins and Dodge, 2009).

Yet, much of this big data are owned by either corporations or government intelligence agencies (Sui and Goodchild, 2011)—with a nexus between the two also possible. Online search is currently only possible through commercial search engine platforms, with no practical open-source or nonprofit alternative. Companies benefit from free user-generated content and by monitoring and mining user behavior through advertising and sale of such data to third parties. Should we not be quite concerned that the online information we receive is filtered through commercial search engines and that the digital shadows (Klinkenberg, 2007) we create are not in our control? As the benefits derived from joining social media networks and other Geo-ICTs have increased, so has the willingness of individuals to share geospatial data about themselves and their family and friends, without realizing the implications of their choices. When such information is combined across space, time, and the multiple thematic dimensions of our lives, the analysis

* www.apple.com/letter-from-tim-cook-on-maps.

reveals hidden and surprising patterns about our individual and collective behaviors. Why is that the data miners are not obligated to share such information they find with the very people that unsuspectingly helped generate such information?

It is also important to distinguish the concept of data protection from that of individual privacy, the latter being much more nebulous and culturally dependent. In this new age of the geoweb and Geo-ICTs, the very notion of privacy needs is being reconceptualized to reflect the experienced impacts of the use of such technologies (Elwood and Leszczynski, 2011). A significant problem in this era of instant multimedia communication and proliferation of social networks is that individuals do not realize they may be revealing unwelcome, unauthorized, or even incriminating information about others inadvertently. Moreover, impacts might be several years away in the future, when information from multiple sources is combined to reveal new patterns. Crowdsourcing is another development that presents similar problems of lack of awareness of participants. Despite the bottom-up and open nature of crowdsourced mapping efforts, it is not clear that participants realize that they may be collecting unauthorized information or helping to build databases that may be used by organizations for profit motives, rather than public good. Similarly, in the context of PGIS, community trust needs to be respected and nurtured and it should be transparent what kind of data are being collected and shared, about whom, and if required permissions have been acquired legitimately. This has not always been the case, due to breaches of confidentiality or insensitive protocols for data collection and sharing.

In response to the deluge of technologies and information, an interesting new approach to privacy is that of sousveillance (inverse surveillance), first proposed in Mann et al. (2003) as the philosophy and practice of wearing computing devices visibly to "surveil the surveillers reflectively"—as a form of protest against state and company surveillance measurements. Pervasive sousveillance would mean we are all being constantly monitored by everybody around us, which in itself raises many moral, legal, and ethical problems. Using phones today or Google Glass in the future, and many participatory sensing activities, can be interpreted as forms of sousveillance. In contrast, proponents of the "postprivacy" movement believe that since technology makes it impossible to control ownership of information, society should practically abandon the notion of privacy in favor of a completely transparent society, where everybody knows what's going on most of the time (Brin, 1999; Jarvis, 2011). Social networking giant Facebook's founder and CEO Mark Zuckerberg has even made the claim that people have accepted that privacy is no longer a "social norm" (The Economist, 2013). Perhaps, it is under such assumptions that Facebook (and researchers employed at Cornell University) had intentionally manipulated Facebook news feeds to investigate their influence on users' emotional states (Meyer, 2014). The study drew considerable indignation for Facebook, but the company is hardly unique in its designs to profit from manipulating people's behaviors—any marketing or lobbying campaign with biased information can be accused of such infringements.

If the manipulation of users for experimental purposes strikes many as unethical, the very premise of LBSs may need to be reexamined, since it involves manipulation of content based on user location. For example, while Google's automated car initiative has the potential to offer great benefits to society by preventing accidents, reducing fuel usage and carbon emissions, minimizing traffic congestion, and saving commuting times, the company is not motivated by altruistic reasons, but inspired by the company's long-term goals to collect and eventually monetize information about its customers. Clearly, Google needs and wants to be at the forefront regarding the design and sale of automated driving technology and information systems, the collection and sale of data on driving habits and navigation patterns, and the ability to provide LBSs to drivers. How Google will monetize such information remains to be seen, but one thing is certain—it will create many controversies. Google's Street View technology raised tremendous privacy concerns in several contexts, and the company has been perennially engaged in legal battles and paid fines for alleged invasion of privacy in several countries.* Its latest technology, Google Glass, is already generating similar concerns and polarizing early adopters from the skeptics.

No matter what the technology, the fundamental ethical question for all LBSs is quite simple: how can ethical guidelines be constructed to allow for "legitimate" manipulation of content while excluding more subversive purposes? There is rarely a context in which an LBS technology cannot be subverted to collect information against the will of the user, and the more accurate it gets, the more harm that can be inflicted. Several years ago, Dobson and Fisher (2003) coined the term "geoslavery" to alert us to this kind of misuse of geo-information technologies to spy on people—by government, corporations, and individuals aided by powerful technologies. Several questions arise, therefore, in the quest of democratization and freedom from geoslavery. How do we prevent ourselves from becoming geoslaves and fight the commoditization of our lives? What kind of in-built controls and preventative measures can be provided to users? Can we through our own choices send strong messages about what we can accept and what must change? What legislations need to be passed to rein in companies?

Originally, the concept of geosurveillance was associated only with governments, whose data collection powers far outreach that of corporations. While government may offer recourse against corporations, there may be none against the government itself, especially in nondemocratic countries. Even in the United States, one of the better-functioning democracies, there is currently great concern about military and intelligence uses of the technology, against citizens and foreign nationals. One only has to follow the Edward Snowden leaks of the National Security Agency's documents to realize how much power and technical ability is at the disposal of the intelligence community.(The Guardian, 2013). Research of the U.S. intelligence economy is enough to justify all concerns. The U.S. government is the largest consumer of geospatial data and Geo-ICTs in the world, with most of that budget allocated for intelligence gathering—and, not surprisingly, satellite imagery is the largest component of intelligence budget (Crampton et al., 2014). There also exists a

* www.en.wikipedia.org/wiki/Google_Street_View_privacy_concerns.

massive "contracting nexus" of more than 50,000 contractor companies, universities, and nonprofits, who have received a staggering "3.75 trillion" dollars between 2000 and 2012 from the Department of Defense and intelligence agencies (Crampton et al., 2014). These figures make one thing clear—geo-information technologies and data are being developed and used in the name of national security in diverse ways by diverse entities, while the public is largely unaware of what the data may or may not be used for and who may have access to it.

20.4.3 Empowerment and Equity

Aided by digital mapping technologies and social networking practices of Web 2.0, the neogeography movement has been critical in advancing the democratization of geographic information and broadened the ontological scope of the formal systems for representing and analyzing geographic information. Neogeographical knowledge can be interpreted as arising from contextual personal and communal interpretive interactions of space and place and thus intrinsically about the culture of everyday life (Warf and Sui, 2010). Despite all its successes, neogeography is no panacea to the digital divide between the technological elite and the much larger group of untrained and uncritical laborers who are not empowered by the use of the technology (Dodge and Kitchin, 2013; Haklay, 2013). The "free" labor from neogeography is opportunistic capital for many entrepreneurial companies who benefit directly from such user-generated content through advertising and sale of user-generated information (including users' online behavior), with the users receiving no share of such profits. However, without these laborers, we still would be limited to the sparse and barren maps of Kibera, when the reality, as shown in Figure 20.1, is quite different. This reveals the true meaning of democratization, which is the empowerment of people who would otherwise be underrepresented. Through individual and coordinated organic acts of mapping, neogeographers have put on the map the stories of people who live far on the unprivileged side of the digital divide—those who cannot even read or have access to the basic amenities, those displaced due to war, or those persecuted by their governments. Motivated by curiosity, purpose, and humanitarian instinct, neogeographers have challenged academic researchers to engage in a "place-centric" GIS grounded in qualitative human discourse rather than stay limited to the conventional spatial perspective (Elwood et al., 2013).

Despite neogeographers having been a prime force behind many recent developments, their efforts may fail to serve the original purpose or contribute toward purposes they did not sanction explicitly. Corporations such as Google and Microsoft have retained copyright on volunteer-collected spatial data using their platforms. This has impeded data sharing during humanitarian operations such as the 2011 Haiti earthquake response (Zook et al., 2010). Disasters and other nonurgent crowdsourcing efforts may even be seen as opportunities for data collection for private gain. DigitalGlobe manages the Tomnod* online platform to crowdsource identification of interesting objects and places (often in the context of a disaster) from high-resolution satellite imagery. Volunteers are often recruited for such campaigns through "feel-good" messages, without providing concrete information regarding how the data will be used, who owns the data, or whether it will be made available to the volunteers themselves. At the very least, more transparency is needed to clarify how their data are being used. Google similarly owns all maps and data created by users. Similar to the concerns associated with the dangers to individuals arising from data mining of their personal information from variety of sources, VGI and crowdsourced information can be easily used without permission for enhancing profit margins or, worse, for unethical or illegal purposes—by corporations, governments, and individuals.

Linus' law also implies, unfortunately, that volunteered and crowdsourced information about accessible, popular, or populated places and readily observable phenomena will be more accurate than information about remote places and not so readily observable phenomena. A recent study (Graham et al., 2014) analyzed the geography of Wikipedia articles, only to discover that there is disproportionate amount of information about North America, much of Europe, and heavily populated parts of Asia, and not much about the rest of the world. A large part of the geographic variation was statistically explained by three variables: population, availability of broadband Internet, and number of edits originating from a country. Despite Wikipedia being one of the ten most popular websites in the world, its content reflects existing and possibly also creates new geographies of the digital divide. Other sources of biases in people's perception of places and how they act on such perceptions include lack of cultural diversity, lack of geographic education, limited exposure to places, and disproportionate media attention. For example, Typhoon Haiyan–related mapping efforts exhibited a clear "media effect" since OSM volunteers disproportionately mapped damage in Tacloban City, which was the focus of media coverage (REACH et al., 2014), but ignored other affected places. Since OSM maps were used to guide relief efforts, this may have had consequences as to how much aid was received by different locales. This geographic unevenness in OSM efforts can be observed more generally at the national scale as well. As shown in Figure 20.7, buildings in Manila, being the well-known capital of Philippines, and in areas (especially Tacloban) affected by Typhoon Haiyan have been digitized at a disproportionately higher rate compared to other areas with comparable population density but lacking special appeal.

Yet another form of mapping bias creeps in due to the profit motive and involvement of corporates, which engage in public service often for self-serving public relations purposes. Are recent campaigns sponsored by DigitalGlobe through their Tomnod platform to locate the missing airplane MH370, map damage after the 2013 Illinois tornado outbreak, and identify invasive species in Hawaii examples of corporate good deeds or media attention grabbing? Though both perspectives contain some truth, the bigger picture is that as corporations begin to see public relations benefits of sponsoring VGI efforts for "good causes," they might coopt (perhaps unwittingly) the power to decide which causes are worth sponsoring. There is little doubt that the charitable

* http://www.digitalearth-isde.org.

activities of large corporations such as Google and DigitalGlobe, as well as nonprofit organizations such as OSM, have contributed immensely to recent disaster relief efforts and other causes. While these initiatives should continue to be supported, guidelines should also be developed from within the geospatial community to ensure that volunteers derive tangible benefits from participating in VGI projects and that societal values such as equity, political freedom, information sharing, etc., be supported in such efforts.

20.5 Conclusion

The goal of this chapter has been to diligently cover the brief history of democratization of geographic information and through some selective examples highlight the immense impact it has on our ability to not just function better but completely change the course of our lives. There is no better example of this than Saroo Brierley, whose discovery of his family using Google Earth serves as a shining example of Digital Earth technologies as life-altering solutions and, less radically, as technologies for creating vicarious place experiences and expanding our geographic understanding. Saroo found his family because of these developments, but think of a future where there exist advanced geo-information retrieval programs through which Saroo could express and search automatically places resembling mental map of his neighborhood. Saroo would not have to spend months browsing imagery to find his family, in such a future. Indeed, the democratization of geo-information not just is about people's involvement, and accessibility to information and tools, but must also depends on innovations in computational reasoning, human–computer interfaces (Google Earth's success was primarily due to its ease of use factor), and growth of public information infrastructures. While universities and government settings are better suited for more scientific and socially responsible research, commercial entities have played an important role through technological innovations and expanding the reach of Geo-ICTs to the common citizen.

To come back to Heidegger's quote from earlier in the chapter, technological problems are not technical—and technical strategies cannot alone help us resolve the conflicting demands for data access, data quality, and confidentiality (NRC, 2007). Clearly, technology does not operate in a vacuum and cannot be viewed as devoid of social, political, and economic contexts. However, if there is to be true democratization of geographic information and many of its supporting technologies, citizens must step up to nurture this fledgling democracy. Neogeography, VGI, crowdsourcing, citizen science, Web 2.0, and many other related neologisms capture many such citizen-led empowering forms of technological progress. Unfortunately, these new practices have also created new problems of data quality, devaluation of established knowledge or capital production systems, exploitation of volunteers, and biasing discourses in favor of popular and easily accessible information.

As much as we need technologies to promote our development as individuals and societies, we must also never lose sight of how they are being used by those who control them—be it individuals, communities, corporations, institutions, or governments. As mentioned earlier, information retrieval is practically hostage to commercial search engines, and our online behavior is being harvested for profit purposes. We do benefit from such services through more targeted advertising and more relevant information retrieval, but the downsides are often hidden from us. Even the relationships of large corporations with colleges and universities need to be monitored to limit unethical influences, even if subtly, on teaching, research and administrative decisions of grants, scholarships, endowments, hiring, and free/discounted software made available by companies. Governmental and private geosurveillance may be unavoidable in practice, simply because of geopolitical complexities and ubiquity of monitoring devices, but that is no excuse for different branches of the government not exercising valid checks on each other, and other violators. Furthermore, if knowledge-privileged professionals and intellectuals become accessible to the populace outside their exalted circles, many insidious impacts of Geo-ICTs can be avoided. Ultimately, it falls on all individuals and communities to be vigilant about developments that violate or can help protect our basic freedoms.

At the same time, critical analyses that merely pander to insecurities and fan the politics of hope and fear need to be countered intelligently and collectively. As Monmonier (2002) and Klinkenberg (2007) have reasoned, the reasonable approach is to stay critical, but only to negate the "evils" of technologies—and to ensure that all strata of society are involved in the path to socially positive technologies. There has undoubtedly been considerable progress in geospatial technology, its use, and its discourse in just a few decades, which became the inspiration and provided the content for this chapter on democratization of geographic information. What remains to be seen now is whether the democratization process flourishes in practice and transforms our lives or if it remains a mere cause célèbre in the academic community.

References

Battersby, S. E., Finn, M. P., Usery, E. L., and Yamamoto, K. H. (2014). Implications of Web Mercator and Its Use in Online Mapping. *Cartographica: The International Journal for Geographic Information and Geovisualization*, *49*(2), 85–101.

Blaschke, T. and Merschdorf, H. (2014). Geographic information science as a multidisciplinary and multiparadigmatic field. *Cartography and Geographic Information Science*, *41*(3), 196–213.

Brierley, S. (2014). *A Long Way Home: A Memoir*. New York: Putnam Adult.

Brin, D. (1999). *The Transparent Society: Will Technology Force us to Choose between Privacy and Freedom?* Reading, MA: Basic Books.

Brown, G. (2012). Public participation GIS (PPGIS) for regional and environmental planning: Reflections on a decade of empirical research. *URISA Journal*, *24*(2), 7–18.

Buchanan, L., Fairfield, H., Parlapiano, A., Peçanha, S., Wallace, T., Watkins, D., and Yourish, K. (2013). Mapping the destruction of Typhoon Haiyan. Retrieved from http://www.nytimes.com/interactive/2013/11/11/world/asia/typhoon-haiyan-map.html. Accessed July 14, 2014.

Budhathoki, N., Bruce, B., and Nedović-Budić, Z. (2008). Reconceptualizing the role of the user of spatial data infrastructure. *GeoJournal, 72*(3), 149–160.

Burke, J. A., Estrin, D., Hansen, M., Parker, A., Ramanathan, N., Reddy, S., and Srivastava, M. B. (2006). Participatory sensing. Boulder, CO: Center for Embedded Network Sensing. Retrieved from http://escholarship.org/uc/item/19h777qd. Accessed July 14, 2014.

Chrisman, N. (1987). Design of geographic information systems based on social and cultural goals. *Photogrammetric Engineering and Remote Sensing, 53*(10), 1367–1370.

Cloud , J. (2002). American Cartographic Transformations during the Cold War. *Cartography and Geographic Information Science*, 29(3), 261–282.

Craglia, M., Goodchild, M. F., Annoni, A., Camara, G., Gould, M., Kuhn, W. et al. (2008). Next-generation digital earth: A position paper from the Vespucci initiative for the advancement of geographic information science. *International Journal of Spatial Data Infrastructures Research, 3*, 146–167.

Craig, W., Harris, T., and Weiner, D. (2002): Introduction. In W. Craig, T. Harris, and D. Weiner (eds.), *Community Participation and Geographic Information Systems*, London: Taylor & Francis, 1–16.

Crampton, J. W. (2010). *Mapping: A Critical Introduction to Cartography and GIS: A Critical Introduction to GIS and Cartography*. Malden, MA: Wiley-Blackwell.

Crampton, J. W., Roberts, S. M., and Poorthuis, A. (2014). The new political economy of geographical intelligence. *Annals of the Association of American Geographers, 104*(1), 196–214.

Dasgupta, A. (May, 2013). Economic Value of Geospatial Data: the great enabler. *Geospatial World*. Retrieved from http://geospatialworld.net/paper/business/ArticleView.aspx?aid=30534. Accessed July 14, 2014.

Dobson, J. (2011). Through the macroscope: Geography's view of the world. *ArcNews*, Winter 2011. Retrieved from http://www.esri.com/news/arcnews/winter1112articles/through-the-macroscope-geographys-view-of-the-world.html. Accessed July 14, 2014.

Dobson, J. and Fisher, P. (2003). Geoslavery. *IEEE Technology and Society Magazine*, Spring 2003, 47–52.

Dodge, M. and Kitchin, R. (2013). Crowdsourced cartography: Mapping experience and knowledge. *Environment and Planning A, 45*(1), 19–36.

Duda, K. A. and Jones, B. K. (2011). USGS remote sensing coordination for the 2010 Haiti Earthquake. *Photogrammetric Engineering and Remote Sensing, 77*(9), 899–907.

Dunn, C. (2007). Participatory GIS - a people's GIS? *Progress in Human Geography*, 31(5), 616–637.

Emrich et al. (2011). GIS and Emergency Management. In *The SAGE Handbook of GIS and Society* (pp. 321–343). Thousand Oaks, CA: SAGE Publications Ltd.

Elwood, S. (2011). Participatory approaches in GIS and society research: Foundations, practices, and future directions. In T. Nyerges, H. Couclelis, and R. B. McMaster (eds.), *The SAGE Handbook of GIS and Society* (pp. 381–399). Thousand Oaks, CA: SAGE Publications Ltd.

Elwood, S., Goodchild, M. F., and Sui, D. (2013). Prospects for VGI research and the emerging fourth paradigm. In D. Sui, S. Elwood, and M. Goodchild (eds.), *Crowdsourcing Geographic Knowledge* (pp. 361–375). Dordrecht, the Netherlands: Springer.

Elwood, S., Goodchild, M. F., and Sui, D. Z. (2011). Researching volunteered geographic information: Spatial data, geographic research, and new social practice. *Annals of the Association of American Geographers, 102*(3), 571–590.

Elwood, S. and Leszczynski, A. (2011). Privacy, reconsidered: New representations, data practices, and the geoweb. *Geoforum, 42*(1), 6–15.

Fan, H., Zipf, A., Fu, Q., and Neis, P. (2014). Quality assessment for building footprints data on OpenStreetMap. *International Journal of Geographical Information Science, 28*(4), 700–719.

Fisher, A. (2013, September 18). Inside Google's quest to popularize self-driving cars. *Popular Science*. Retrieved from http://www.popsci.com/cars/article/2013-09/google-self-driving-car. Accessed July 14, 2014.

Foulser-Piggot, R., Spence, R., and Brown, D. (2013). The use of remote sensing for building damage assessment following 22nd February 2011 Christchurch Earthquake: The GEOCAN study and its validation. Cambridge, U.K.: Cambridge Architectural Research, Ltd. Retrieved from http://www.willisresearchnetwork.com/assets/templates/wrn/files/GEOCAN%20Christchurch%20Report.pdf. Accessed July 14, 2014.

Gao, J. (2002). Integration of GPS with remote sensing and GIS: Reality and prospect. *Photogrammetric Engineering & Remote Sensing, 68*(5), 447–453.

Goldman, J., Shilton, K., Burke, J., Estrin, D., Hansen, M., Ramanathan, N. et al. (2009). Participatory Sensing: A citizen-powered approach to illuminating the patterns that shape our world. Washington, DC: Woodrow Wilson International Center for Scholars. Retrieved from http://wilsoncenter.org/topics/docs/participatory_sensing.pdf. Accessed July 14, 2014.

Goodchild, M. F. (1992). Geographical information science. *International Journal of Geographic Information Systems*, 6(1), 31–45.

Goodchild, M. F. (2007). Citizens as sensors: The world of volunteered geography. *GeoJournal, 69*(4), 211–221.

Goodchild, M. F. (2009). NeoGeography and the nature of geographic expertise. *Journal of Location Based Services, 3*(2), 82–96.

Goodchild, M. F. (2012). GIScience in the 21st century. In W. Shi, M. F. Goodchild, B. Lees, and Y. Leung (eds.), *Advances in Geo-Spatial Information Science* (pp. 3–10). Leiden, the Netherlands: CRC Press.

Goodchild, M. F., Fu, P., and Rich, P. (2007). Sharing geographic information: An assessment of the geospatial one-stop. *Annals of the Association of American Geographers, 97*(2), 250–266.

Goodchild, M. F. and Glennon, J. A. (2010). Crowdsourcing geographic information for disaster response: A research frontier. *International Journal of Digital Earth, 3*(3), 231–241.

Gore, A. (1998). The digital earth: Understanding our planet in the 21st century. Los Angeles, CA: California Science Center. Retrieved from http://www.isde5.org/al_gore_speech.htm. Accessed July 14, 2014.

Gouveia, C. and Fonseca, A. (2008). New approaches to environmental monitoring: The use of ICT to explore volunteered geographic information. *GeoJournal, 72*(3–4), 185–197.

Graham, M., Benie, H., Straumann, R. K., and Medhat, A. (2014). Uneven geographies of user-generated information: Patterns of increasing informational poverty. *Annals of the Association of American* Geographers, *104*(4), 746–764.

Grossner, K. E., Goodchild, M. F., and Clarke, K. C. (2008). Defining a digital earth system. *Transactions in GIS, 12*(1), 145–160.

Hagen, E. (2011). Mapping change: Community information empowerment in Kibera (Innovations Case Narrative: Map Kibera). *Innovations: Technology, Governance, Globalization,* 6(1), 69–94.

Haklay, M. (2010). How good is volunteered geographical information? A comparative study of OpenStreetMap and Ordnance Survey datasets. *Environment and Planning B: Planning and Design, 37*(4), 682–703.

Haklay, M. (2013). Neogeography and the delusion of democratisation. *Environment and Planning A, 45*(1), 55–69.

Harley, J. B. (2002). *The New Nature of Maps: Essays in the History of Cartography.* Baltimore, MD: Johns Hopkins University Press.

Harris, T. M. and Weiner, D. (1996). *GIS and Society: The Social Implications of How People, Space, and Environment are Represented in GIS. Scientific Report for the Initiative 19 Specialist Meeting.* (Scientific Report for the Initiative 19 Specialist Meeting No. #96-7). Koinonia Retreat Center, South Haven, MN: National Center for Geographic Information and Analysis (NCGIA).

Heidegger, M. (1982). *The Question Concerning Technology, and Other Essays.* (W. Lovitt, Trans.). New York: Harper Torchbooks.

Jankowski, P. (2011). Designing public participation geographic information systems. In T. Nyerges, H. Couclelis, and R. B. McMaster (eds.), *The SAGE Handbook of GIS and Society.* Thousand Oaks, CA: SAGE Publications Ltd, pp. 417–421.

Jarvis, J. (2011). *Public Parts: How Sharing in the Digital Age Improves the Way We Work and Live.* New York: Simon & Schuster.

Klinkenberg, B. (2007). Geospatial technologies and the geographies of hope and fear. *Annals of the Association of American Geographers, 97*(2), 350–360.

Lemmens, M. (2011). *Geo-information: Technologies, Applications and the Environment.* New York: Springer.

Longley, P. A., Goodchild, M., Maguire, D. J., and Rhind, D. W. (2010). *Geographic Information Systems and Science* (3rd edn.). Hoboken, NJ: Wiley.

Maisonneuve, N., Stevens, M., and Ochab, B. (2010). Participatory noise pollution monitoring using mobile phones. *Information Polity, 15*(1), 51–71.

Mann, S., Nolan, J., and Wellman, B. (2003). Sousveillance: Inventing and using wearable computing devices for data collection in surveillance environments. *Surveillance & Society, 1*(3), 331–355.

Mannings, R. (2008). *Ubiquitous Positioning.* New York: Artech House.

Masser, I. (2011). Emerging frameworks in the information age: The spatial data infrastructure (SDI) phenomenon. In T. Nyerges, H. Couclelis, and R. B. McMaster (eds.), *The SAGE Handbook of GIS and Society* (pp. 271–286). Thousand Oaks, CA: SAGE Publications Ltd.

McDougall (2012). An Assessment of the Contribution of Volunteered Geographic Information During Recent Natural Disasters. In *Spatially enabling government, industry and citizens: research and development perspectives* (pp. 201–214), GSDI Association Press, Needham, MA.

Merchant, J. W. and Narumalani, S. (2009). Integrating remote sensing and geographic information systems. In T. A. Warner, M. D. Nellis, and G. M. Foody (eds.), *The SAGE Handbook of Remote Sensing.* London, U.K.: SAGE Publications Ltd.

Mesev, V. (2007). *Integration of GIS and Remote Sensing.* Hoboken, NJ: Wiley.

Meyer, R. (2014, June 28). Everything we know about Facebook's secret mood manipulation experiment. *The Atlantic.* Retrieved from http://www.theatlantic.com/technology/archive/2014/06/everything-we-know-about-facebooks-secret-mood-manipulation-experiment/373648/. Accessed July 14, 2014.

Monmonier, M. (1996). *How to Lie with Maps* (2nd edn.). Chicago, IL: University of Chicago Press.

Monmonier, M. (2002). *Spying with Maps: Surveillance Technologies and the Future of Privacy.* Chicago, IL: University of Chicago Press.

Mun, M., Reddy, S., Shilton, K., Yau, N., Burke, J., Estrin, D. et al. (2009). PEIR, the personal environmental impact report, as a platform for participatory sensing systems research. In *Proceedings of the Seventh International Conference on Mobile Systems, Applications, and Services* (pp. 55–68). New York: ACM.

National Research Council (NRC) (2007). *Putting People on the Map: Protecting Confidentiality with Linked Social-Spatial Data.* Washington, DC: The National Academies Press.

Nouwt, S. (2008). Reasonable expectations of geo-privacy? *SCRIPTed, 5*(2), 375–403.

O'Sullivan, D. (2006). Geographical information science: Critical GIS. *Progress in Human Geography, 30*(6), 783–791.

Palmer, M. H. and Kraushaar, S. (2013). Volunteered geographic information, actor-network theory, and severe-storm reports. In D. Sui, S. Elwood, and M. Goodchild (eds.), *Crowdsourcing Geographic Knowledge* (pp. 287–306). Dordrecht, the Netherlands: Springer.

Perkins, C. and Dodge, M. (2009). Satellite imagery and the spectacle of secret spaces. *Geoforum*, *40*(4), 546–560.

Pickles, J. (1991). Geography, GIS, and the surveillant society. *Papers and Proceedings of Applied Geography Conferences*, *4*, 80–91.

Pickles, J. (1994). *Ground Truth: The Social Implications of Geographic Information Systems*. New York: The Guilford Press.

Pool, B. (2010). Crisis Camp Haiti: Techno-types volunteer their computer skills to aid quake victims. *Los Angeles Times*. Retrieved from http://articles.latimes.com/2010/jan/16/world/la-fg-haiti-crisiscamp17-2010jan17. Accessed July 14, 2014.

Raper, J., Gartner, G., Karimi, H., and Rizos, C. (2007). Applications of location-based services: A selected review. *Journal of Location Based Services*, *1*(2), 89–111.

Raubal, M. (2011). Cogito Ergo Mobilis sum. In T. Nyerges, H. Couclelis, and R. B. McMaster (eds.), *The SAGE Handbook of GIS and Society* (pp. 159–173). Thousand Oaks, CA: SAGE Publications Ltd.

REACH, American Red Cross, and USAID. (2014). Groundtruthing OpenStreetMap building damage assessment, Haiyan Typhoon, The Philippines, final assessment report. Retrieved from http://www.reach-initiative.org. Accessed July 14, 2014.

Resch, B., Mittlboeck, M., Lipson, S., Welsh, M., Bers, J., Britter, R., and Ratti, C. (2009). Urban sensing revisited—Common scents: Towards standardised geo-sensor networks for public health monitoring in the city. In *Proceedings of the 11th International Conference on Computers in Urban Planning and Urban Management—CUPUM2009*, Hong Kong, China, June 16–18, 2009.

Sagl, G., Delmelle, E., and Delmelle, E. (2014). Mapping collective human activity in an urban environment based on mobile phone data. *Cartography and Geographic Information Science*, *41*(3), 272–285.

Schuurman, N. (1999). Critical GIS: Theorizing an emerging discipline. *Cartographica: The International Journal for Geographic Information and Geovisualization*, *36*(4), 1–108.

Seang, T. P., Mund, J.-P., and Symann, R. (2008). Low cost amateur aerial pictures with balloon and digital camera. MethodFinder. Retrieved from http://methodfinder.net. Accessed July 14, 2014.

Sheppard, E. (1995). GIS and society: Toward a research agenda. *Cartography and Geographic Information Systems*, *22*(1), 5–16.

Sieber, R. (2006). Public Participation Geographic Information Systems: A Literature Review and Framework. *Annals of the Association of American Geographers*, 96(3), 491–507.

Singel, R. (2005). A disaster map "Wiki" is born. *WIRED Magazine*. Retrieved from http://archive.wired.com/software/coolapps/news/2005/09/68743. Accessed July 14, 2014.

Smith, N. (1992). History and philosophy of geography: Real wars, theory wars. *Progress in Human Geography*, *16*(2), 257–271.

Soden, R. and Palen L. (2014). From Crowdsourced Mapping to Community Mapping: The Post-Earthquake Work of OpenStreetMap Haiti. In R. Chiara, L. Ciolfi, D. Martin, and B. Coneien (eds.), COOP 2014: *Proceedings of the 11th International Conference on the Design of Cooperative Systems* (pp. 311–326), 27–30 May 2014, Nice, France.

Sui, D. and Goodchild, M. (2011). The convergence of GIS and social media: Challenges for GIScience. *International Journal of Geographical Information Science*, *25*(11), 1737–1748.

Sui, D., Goodchild, M., and Elwood, S. (2013). Volunteered Geographic information, the exaflood, and the growing digital divide. In D. Sui, S. Elwood, and M. Goodchild (eds.), *Crowdsourcing Geographic Knowledge* (pp. 1–12). Dordrecht, the Netherlands: Springer.

Taylor, P. (1990). Editorial comment: GKS. *Political Geography Quarterly*, 9, 211–212.

The Economist. (2013, November 16). The people's panopticon. *The Economist*. Retrieved from http://www.economist.com/news/briefing/21589863-it-getting-ever-easier-record-anything-or-everything-you-see-opens. Accessed July 14, 2014.

The Guardian. (2013, June 8). The NSA files. Retrieved from http://www.theguardian.com/world/the-nsa-files. Accessed July 14, 2014.

Thrun, S. (2010). What we're driving at. *Google Official Blog*. Commercial. Retrieved from http://googleblog.blogspot.com/2010/10/what-were-driving-at.html. Accessed July 14, 2014.

Tulloch, D. L. (2008). Is VGI participation? From vernal pools to video games. *Geojournal 72(3)*, 161–171.

Turner, A. (2006). *Introduction to Neogeography*. Sebastopol, CA: O'Reilly Media.

Van Aardt, J. A. N., McKeown, D., Faulring, J., Raqueño, N., Casterline, M., Renschler, C. et al. (2011). Geospatial disaster response during the Haiti earthquake: A case study spanning airborne deployment, data collection, transfer, processing, and dissemination. *Photogrammetric Engineering & Remote Sensing*, *77*(9), 943–952.

Warf, B. and Sui, D. (2010). From GIS to neogeography: Ontological implications and theories of truth. *Annals of GIS*, *16*(4), 197–209.

Weiner, G. and Smith, B. W. (2014). Automated driving: Legislative and regulatory action. The Center for Internet and Society. Retrieved from http://cyberlaw.stanford.edu. Accessed July 14, 2014.

Weiser, M. (1991). The computer for the 21st century. *The Scientific American*, *256*(3), 94–104.

Wilson, M. W. and Poore, B. S. (2009). Theory, practice, and history in critical GIS: Reports on an AAG panel session. *Cartographica: The International Journal for Geographic Information and Geovisualization*, *44*(1), 5–16.

Wilson, M.W., and Graham M. (2013). Guest Editorial. *Environment and Planning A*, 45(1), 3–9.

Yang, C., Wong, D., Miao, Q., and Yang, R. (eds.). (2010). *Advanced Geoinformation Science*. Boca Raton, FL: CRC Press.

Zipf, A. and Jöst, M. M. (2012). Location-based services. In W. Kresse and D. M. Danko (eds.), *Springer Handbook of Geographic Information*. New York: Springer, pp. 417–421.

Zook, M., Graham, M., Shelton, T., and Gorman, S. (2010). Volunteered geographic information and crowdsourcing disaster relief: A case study of the Haitian earthquake. *World Medical & Health Policy*, *2*(2), 7–33.

21

Frontiers of GIScience: Evolution, State of the Art, and Future Pathways

May Yuan
University of Texas at Dallas

Acronyms and Definitions

ABM	Agent-based modeling
ACM	Association for Computing Machinery
CA	Cellular automata
COSIT	Conference of Spatial Information Theory
ESRI	Environmental Systems and Research Institute
FGDC	Federal Geographic Data Committee
GIS	Geographic information systems
GIScience	Geographic information science
GWR	Geographically weighted regression
IaaS	Infrastructure as a Service
IEWAT	Information everywhere, any time
LISA	Local indicator of spatial autocorrelation
NCGIA	National Center for Geographic Information and Analysis
NSDI	National spatial data infrastructure
NSF	National Science Foundation (US)
PaaS	Platform as a Service
SaaS	Software as a Service
SIGSPATIAL	Special Interest Group on Spatial Information
UCGIS	University Consortium for Geographic Information Science
VGI	Volunteered geographic information

21.1 Introduction

Geographic information science (GIScience) is the science that underlies geographic information systems (GIS) technology. Roger Tomlinson introduced GIS in his report on computer mapping and analysis to the National Land Inventory in the Canada Department of Agriculture (Tomlinson, 1962). Yet, when GIS is broadly defined as a system that deals with geographic information, it can be traced far back into the time when humans started recording and sharing knowledge about the environment. Before computer-based GIS technology, oral traditions, and maps were primary means to communicate geographic information. Nowadays, GIS technologies are diverse and thriving in mapping, spatial analysis and modeling, location-based services, cyber geographical applications, and spatial crowdsourcing. GIS technologies are now important research tools for research and operations in environmental sciences, biological and agricultural sciences, public health, urban planning, and economic, political, and social studies. GIScience serves the conceptual, theoretical, and computational foundations for these technologies.

Dating to ~6600 BC, the mural found at the Neolithic site of Catalhöyük is considered the world's oldest map (Schmitt et al., 2014). The early adoption of maps is of no surprise. Maps are intuitive and effective ways to give directions, express spatial arrangements of features, and plan spatial activities. Humans made maps long before they invented writing. Likewise, Tomlinson's GIS was motivated by the use of computers to automate map analysis and production. To date, map libraries continue the important role of curating and providing access to atlases, aerial photographs, and spatial data in digital forms, while massive and diverse geographic information is also widely available from government agencies, businesses, organizations, and communities.

While mapping is essential, a GIS also consists of tools to process geospatial data, manage geospatial databases, integrate data through georeferencing and compatible attribute definitions,

TABLE 21.1 GIScience Synonyms

Term	Primary Communities	Special Emphases
Spatial science (or geospatial science to stress the space of interest at a geographic scale)	Geography, statistics, and other domains applying mapping and spatial methods in their fields, such as spatial social science	Mapping, spatial statistics, spatial analysis, spatial modeling
Spatial information science	Computer science and management information science	Spatial computing, spatial database management, and visualization and communication of spatial information
Geoinformatics	Geoscience, computer science, geostatistics	Information computing and management for geoscience data, most commonly with earth science data
Geomatics	Geodetics, survey engineering, geophysics, earth science	Geoid, datums, land surveys, mapping

Note: Generally, these terms are synonyms of GIScience, but they are popular in different communities with some variations in areas of emphases.

analyze embedded spatial patterns, model geographic phenomena and processes, and render data and findings in multiple ways. The technology was initially developed out of application needs, and its conceptual and computational frameworks were fragmented across solutions. GIScience research contributes to developing fundamental frameworks for GIS technologies and takes the technological challenges to improve our understanding of geographic information, processes for geographic knowledge building and communication, and spatial decision support.

This chapter aims to highlight the past, present, and future of GIScience research. As a field of interdisciplinary and multidisciplinary research, GIScience enjoys outstanding advances in both breadth and depth as evidenced by the multitude of names associated with the discipline, such as geospatial science, spatial science, spatial information science, geoinformatics, and geomatics (Table 21.1). Consequently, it is challenging to capture the full scope of research development in the field. What follows reflects the author's perspectives on the evolution, state of the art, and future pathways of GIScience. Since the chapter is focused on GIScience, the discussions here emphasize the key intellectual development of spatial concepts, theories, and computational approaches. GIS applications are not GIScience research and, therefore, are beyond the scope of this chapter. The next section elaborates on the evolution of GIS technologies to GIScience: from the early emphases on the transitions from technological advances in mapping, spatial database building, and inventory and planning applications to scientific inquiries into the nature of geographic information, spatial computing, and geographical understanding. Section 21.3 highlights the active GIScience research directions in cognition, representation, integration, and computation. The chapter concludes with promising pathways for future GIScience development.

21.2 Evolution

Computer-based GIS technology revolutionized the processes of recording and disseminating geographic information and invoked new possibilities to represent, analyze, and compute geography. Since its conception, the term GIS was often referenced exclusively to computer-based GIS. Coppock and Rhind (1991) characterized the early development of computerized GIS into four general phases from 1960 to 1990:

1. *A phase of pioneers* from the early 1960s to 1975. Key leaders included Howard Fisher of the Harvard Laboratory for Computer Graphics, Roger Tomlinson of the Canadian Geographic Information System, and David Bickmore at the Experimental Cartographical Unit in the United Kingdom.
2. *A phase of national drivers* from 1973 to early 1980s. Key agencies included Canada's Department of Agriculture, the United States Bureau of the Census, and the Ordnance Survey in Great Britain. In the United States, GIS technology attracted great interest from many federal agencies such as the Department of Defense, Central Intelligence Agency, US Forest Service, Fish and Wildlife Service, and Department of Housing and Urban Development, as well as state and local governments including California, Maryland, Minnesota, New York, and others.
3. *A phase of commercial dominance* from early to late 1980s, most noticeably the Environmental Systems and Research Institute (ESRI, now Esri) and Integraph. The companies not only developed GIS software packages but also designed and implemented GIS projects for government agencies. These GIS packages were adopted in college courses, and to date, they remain the primary tools for learning GIS and doing GIS projects. In 1988, the United States National Science Foundation (NSF) awarded a grant to establish the National Center for Geographic Information and Analysis (NCGIA) with the University of California at Santa Barbara, State University of New York at Buffalo, and University of Maine. The NSF grant provided $10M dollars for 8 years of NCGIA leadership that transformed GIS to GIScience and resulted in lasting impacts in education and research in the United States and around the world.
4. *A phase of user dominance* since early 1990s with the rise of desktop GIS that emphasized ease of use and promoted wide adoption of GIS technology beyond research universities, large government organizations, and big companies. In 1994, US Executive Order 12906 established the Federal Geographic Data Committee (FGDC) as the executive branch leadership to develop the National Spatial Data Infrastructure (NSDI) marked the first multiagency nation-wide efforts to coordinate GIS data management

and access. The expanded availability of free GIS data stimulated many geospatial research and business opportunities and popularized GIS technology in a wide range of domain applications.

In a short period of 30 years (1960s–1990s), GIS started with a few visionaries who sought ways to use computers for mapping and analyzing geographic data and then grew to a generation of researchers and professionals that brought GIS into mainstream college curricula, government functions, and business operations. With this growth, research efforts went beyond mapping and spatial data handling. Researchers ventured into the unique complexity of geographic information and ensuing challenges in acquiring and using spatial data to understand geographic processes and make spatial predictions. *The International Journal of Geographical Information Systems* (*IJGIS*) was launched in 1987 and was recognized as the primary academic journal in the field (Caron et al., 2008). Goodchild published a landmark paper in *IJGIS*, entitled "Geographic information science" (Goodchild, 1992). The paper highlighted scientific problems unique to geographical data and established the topical content for GIScience.

Since then, many organizations and journals adopted the term "GIScience" over GIS. Efforts of the academic community, with most participants from Geography, established the University Consortium for Geographic Information Science (UCGIS) in 1994 and, through community efforts, defined GIScience as *the development and use of theories, methods, technology, and data for understanding geographic processes, relationships, and patterns. The transformation of geographic data into useful information is central to geographic information science* (UCGIS, 2002, Mark, 2003). It is important to note that GIScience research is not about using GIS technologies to solve scientific problems. This is similar to statistics and mathematics; applications of statistical or mathematical methods to solve a biological problem contribute to biological science, not the sciences of statistics or mathematics.

The early development of GIScience can be attributed to the NCGIA's leadership in a series of initiatives as well as the UCGIS community efforts to identify and articulate research challenges. In 1997, the *International Journal of Geographical Information Systems* was renamed *International Journal of Geographical Information Science*, marking its second decade of publication (Fisher, 2006). However, the tendency to use GIScience as a synonym for GIS was quite common in early 2000 (Mark, 2003) and remains rather persistent today. Many programs offer GIScience courses with the same instructional materials for GIS, and many do not differentiate GIScience research from research using GIS. Nevertheless, leading journals (such as *IJGIS* and *Geoinformatics*) and conferences in GIScience (such as GIScience and ACM-SIGSPATIAL) emphasize papers with contributions to conceptual, theoretical, and computational innovations.

Foundational work in cartography, spatial statistics, and spatial modeling has significantly contributed to the development of GIS, and these continue to be important subjects in GIScience research today. Computer cartography made notable progress in line generalization (Douglas and Peucker, 1973), map generalization (Buttenfield and McMaster, 1991), cartographic label placement (Marks and Shieber, 1991), and interactive digital atlases production (MacEachren, 1998). Landmark spatial studies led to new methods that account for local variations and local processes, such as map algebra (Tomlin, 1994), local indicator of spatial autocorrelation (LISA) (Anselin, 1995), geographically weighted regression (GWR) (Brunsdon et al., 1998), and geoalgebra (Takeyama and Couclelis, 1997). Furthermore, spatial modeling advanced new approaches to simulate hydrological processes (Olivera and Maidment, 1999) and urban systems (Couclelis, 1997, Batty, 2007) by leveraging dynamic methods from other fields, such as distributed modeling, cellular automata (CA), and agent-based modeling (ABM).

Moreover, arguments were made that foundations of GIScience should tie closely to information science (Mark, 2003). Information science studies the means and processes of information transmission among humans and/or computers. Syntactic form, semantic content, and contextual relevance are key elements in determining the value and optimal means of information flows from transmitters to receivers (Worboys, 2003). However, any judgment about value and optimality of the key elements must rely on a common understanding of the domain between transmitters and receivers. Geographic ontologies became an important subject in GIScience research (Agarwal, 2005), and research on spatial ontologies and representation along with other issues related to the nature of geographic information was prominent in NCGIA research initiatives and UCGIS research challenges. Fundamental GIScience research has been promoted through the Conference on Spatial Information Theory (COSIT) starting in 1993 and International Conference on GIScience (GIScience), which began in 2000. Since then, the two conferences have been held in alternate years and locations between Europe and North America. In addition, the Auto-Carto International Symposium on Automatic Cartography and International Symposium on Spatial Data Handling both have a long history as primary academic venues in GIScience. Computer scientists interested in spatial database and information started the Annual Association for Computing Machinery (ACM) Workshop on Advances in Geographic Information Systems in 1993. They successfully expanded the annual workshop to Annual ACM-GIS International Symposium in 1998 and furthermore established the ACM Special Interest Group on Spatial Information (SIGSPATIAL) as the catalyst for research on spatially related information among computer scientists (Samet et al., 2008).

These pioneer efforts established a strong foundation for GIScience. Research has migrated from GIS-enabling computerization of geographic data processing and mapping to GIScience inquiries into the essence of geographic information and epistemology. Goodchild (2014) highlighted research and institutional accomplishments in the 20 years of progress since the introduction of GIScience in 1994. On measurements, research foci shifted from spatial errors in the 1980s to spatial uncertainty in the 1990s. On representation, research advanced from vector/raster in the 1980s and objects/fields in the 1990s to complex object-fields and field-objects in

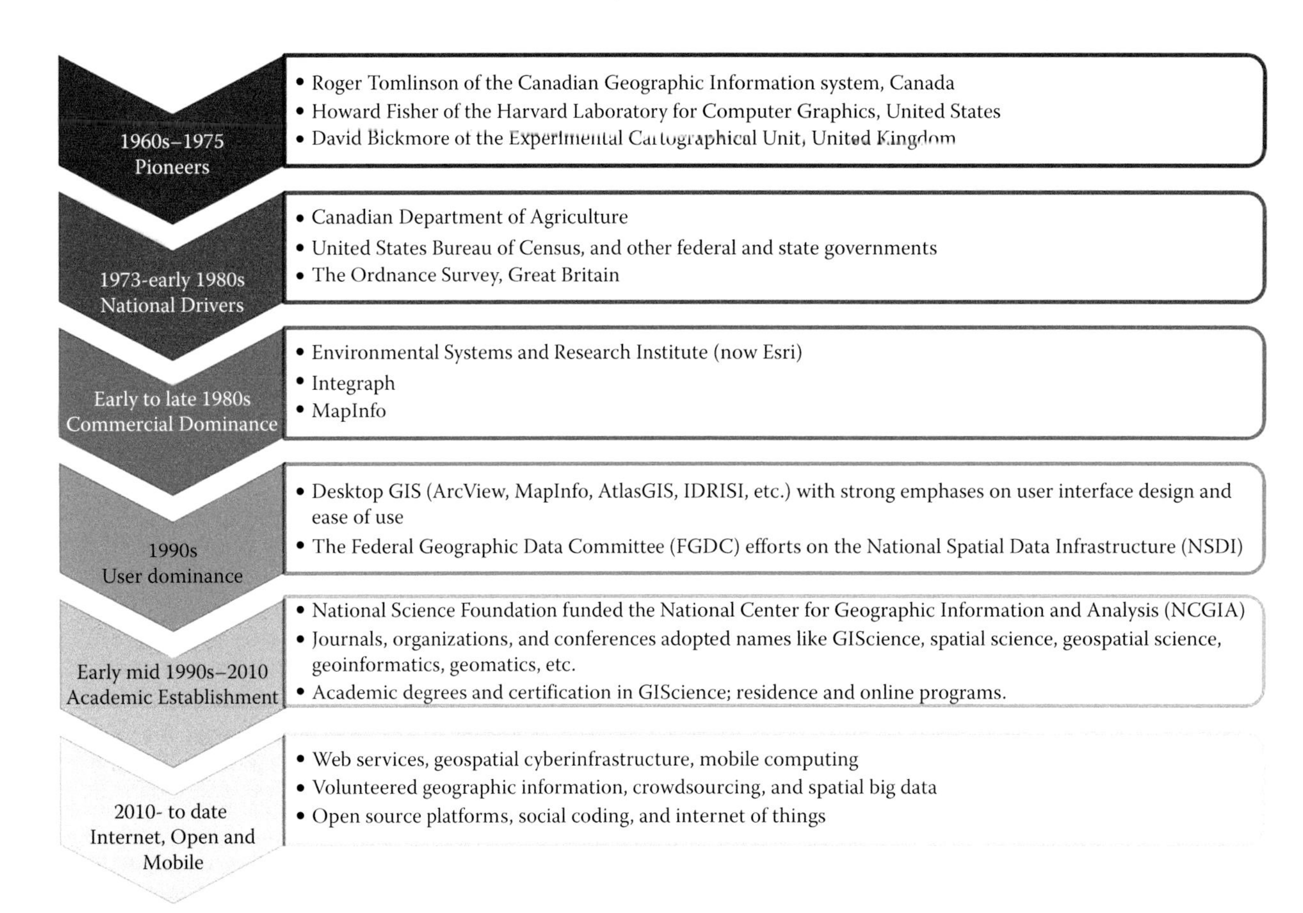

FIGURE 21.1 A summary of the evolution of GI Systems and GIScience.

2000s. GIScience research on analysis progressed from spatial autocorrelation in the 1970s to spatial heterogeneity in the 1990s. These efforts also built a strong foundation that support GIScience research and development into the mainstream of information technology. The GIScience evolution is summarized in Figure 21.1.

Over the years, GIScience research frontiers were articulated through 21 NCGIA research initiatives UCGIS research priorities and Computing Community Consortium Spatial Computing Visioning. Goodchild (2014) listed some of the topics that resulted from discussions in GIScience communities in several venues and summarized in a conceptual framework for GIScience that connects the dimensions of human, society, and computer. Expanded upon his conceptual framework, Figure 21.2 incorporates major developments in cyberinfrastructure and computing that have transformed the interplays among the human, society, and computers as well as how we perceive and understand human, physical, biological, and many other dimensions of reality. There is no shortage of research challenges in GIScience. This becomes evident with a quick search on Google Scholar, which results in more than 5000 publications on the subject. Some fundamental topics remain outstanding and are likely to persist at the core of GIScience research, such as geographic ontologies, space-time representation, spatial algorithms, spatial cognition, geovisualization, and spatial decision support.

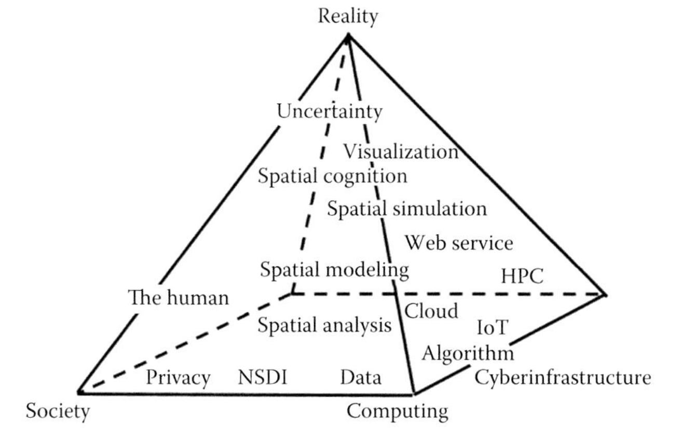

FIGURE 21.2 A conceptual framework for research themes in GIScience, whereas HPC stands for high performance computing, and IoT is the Internet of Things (Modified from Goodchild, M. F. *Journal of Spatial Information Science*, (1), 3, 2014.)

21.3 State of the Art

GIScience continues to evolve with an increasing attention to what is local rather than global, individual rather than aggregated, collaborative rather than authoritative, culturally aware rather universal, open rather than exclusive, and mobile rather than desktop. Moreover, models and methods are being developed to represent and visualize multidimensional and multimedia data. Leveraged by the internet, new GIS platforms are being realized on the World Wide Web, with cyberinfrastructure, and in the cloud. All these developments have profound influences on what is summarized here as 3 A's: *abstraction*, *algorithms*, and *assimilation* throughout GIScience epistemology.

21.3.1 Abstraction

Abstraction takes place at multiple levels in GIScience research. It is concerned with how we conceptualize geographic worlds and spatial problems and subsequently, how we represent, compute, and communicate all the relevant concepts and findings. As spatial data are from different sources, integration can be challenging at each level of abstraction. Perhaps, the most popular abstraction used in GIScience is the so-called data layers (Figure 21.3). While the data-layer abstraction is intuitive, GIScience research examines issues in cognition, ontology, and statistics (e.g., sampling) for better abstraction of reality.

Across all the levels of abstraction, cognitive research helps understand how people learn and organize geographic knowledge. Such cognitive understanding can improve GIS usability and communication. Montello (2009) summarized five main areas of cognitive research in GIScience since 1992: human factors of GIS, geovisualization (including spatialization), navigation systems, cognitive geo-ontologies, spatial thinking and memory, and cognitive aspects of geographic education. Much of the cognitive research confirms the complexity of geographic information and knowledge due to indeterminacy, vagueness, and the interdependency of individuals and geographic context.

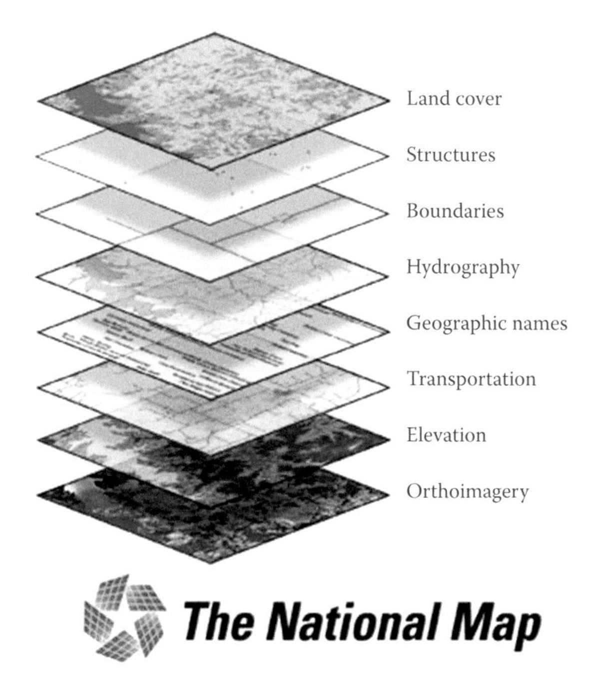

FIGURE 21.3 An example of geographic abstractions: the National Map from the United States Geological Survey.

As a result, geographic categorization and reasoning may vary from person to person or place to place. For example, cognitive geo-ontologies recognize that people may see things differently, and their conceptualizations may vary due to environmental, cultural, or linguistic differences (Mark et al., 2011, Wellen and Sieber, 2013, Turk and Stea, 2014). Such differences have profound implications for information sharing and integration, spatial data infrastructure building, spatial decision support, and many other issues that deal with the usefulness of GIS technologies and intrinsic technological biases.

Information sharing and integration was the initial motivator of ontological research through the rise of the Semantic Web that extends the World Wide Web for people to share and reuse data beyond application boundaries. Ontological approaches are now commonly used to define specifications of geographic abstractions in a problem domain (Jung et al., 2013, Ujang and Rahman, 2013), achieve semantic consistency for data integration and complex query support (Wiegand, 2012), and assure interoperability across systems and over web services (Shi and Nellis, 2013). Different frameworks have been proposed for geo-ontologies. Frank (2001) argued for a tiered ontology to assure consistency constraints based on how different kinds of things are conceptualized and from where they are abstracted. Tier 0 ontologies are for *human-independent reality* where natural laws prevail regardless of human observers. Tier 1 ontologies are for *observations of physical world* with measurements and statistics. Tier 2 ontologies are for *objects with properties* that can be used to identify individuals and determine categorical memberships with necessary and sufficient properties. Tier 3 ontologies are for *social reality* that is subject to social, cultural, and linguistic contexts. Finally, Tier 4 ontologies are for *subjective knowledge*, which may be incomplete or partial, used by individuals or institutions for reasoning or decision making. Couclelis (2010) articulated the need for geographic information constructs as the core of ontologies in GIScience. Her framework centers on an ontological hierarchy to connect intentionality and relevant information. There are seven levels of semantic resolution in the hierarchy. In the order of low to high levels, the semantic levels of resolution include *existence*, *observables*, *similarities*, *simple objects*, *composite objects*, *function*, and *purpose*. She introduced the idea of *semantic contraction* to generalize semantic richness from higher, more complex levels to a lower level of simpler semantics, and *object of discourse* to represent entities as composites of geographic information constructs at the higher levels of the hierarchy. This ontological research expanded our understanding of semantic granularity (Fonseca et al., 2002) and spatial tasks (Wiegand and García, 2007) and laid the foundation for building theories of geographic information.

In addition to ontologies, abstraction also accounts for means by which geographic information can be effectively acquired, analyzed, and communicated. Traditionally, geography is abstracted in forms of data from field surveys, maps, imagery, tables, graphs, and text. Advances have opened new means to acquire geographic information with new kinds of geographic abstraction. For example, data from dynamic geosensor networks (Llaves and Kuhn, 2014), tweets (Tsou et al., 2013), geotagged photos (Samet et al., 2013), and information from various social media (Croitoru et al., 2013; Jiang and Miao, 2014) offer real-time or near real-time environmental and social abstractions that enable detection of events and activities as they unfold. As the geographic world captured by these data is transitory and ephemeral, so is the ensuing geographic abstraction. Volunteered geographic information (VGI), crowdsourced geographic information (Goodchild, 2007; Goodchild and Glennon, 2010), and ambient geospatial information (Stefanidis et al., 2013) commonly condense information entries to point locations. Consequently, geographic abstraction is generally reduced to individuals and collections of points. Spatial synthesis would be more appropriate than analysis to decipher these data.

Besides geosensor and social media data, multimedia data incorporate video, audio, virtual reality, and augmented reality to represent geography (Camara and Raper, 1999). Videos may be interviews, documentary films, or animation of temporal information. Audios may be oral stories, narration by a native speaker, testimonies, songs, or animal sounds. Both video and audio enrich abstractions of geographic reality by enriching the context of spatial abstraction. Virtual reality and augmented reality, usually with 3D visualization, supplement spatial abstraction with videos, audios, photographs, digital documents, and labels in a dynamic context-aware immersive environment. Granularity of geographic abstraction becomes finer or coarser, depending on the user's location and view. Virtual geographic environment (Lin et al., 2013) leverages virtual reality and multidimensional GIS to provide a digital platform for geographic experiments through collaborative visualization and simulation. Collaboration requires shared geographic abstraction of both declarative knowledge and procedural knowledge as the basis for communication and integration, which in turn rests on cognitive and ontological compatibility.

21.3.2 Algorithms

Algorithms are step-by-step procedures for calculations. Here, algorithms are broadly defined as approaches to data processing, analysis, modeling, and simulation. As geographic abstraction shifts emphases to semantics, the development of spatial algorithms also attempts to reveal local meanings and individual behaviors in space and time.

The rise of critical GIS (O'Sullivan, 2006; Schuurman, 2006) reflects the needs to engage social critiques in GIS-based geographic knowledge production in terms of basic concepts, representation, participation, and social implications. VGI and web map services partially address the needs by empowering ordinary citizens to create geographic data and participate in geographic knowledge production. Many critical GIS researchers also echo the criticisms of positivist biases in GIS and advocate for qualitative GIS (Cope and Elwood, 2009) to address the needs to incorporate contextual details and interpretations of the described situation and processes. Broadened GIS methodology and the programming environmental allow qualitative methods that are commonly used by sociologists and humanities scholars, like coding, triangulating source materials, and content analysis in recursive and iterative forms to produce knowledge, such as Geo-Narrative (Kwan and Ding, 2008).

VGI is only one data source available from the Web. There are many crowdsourced systems (Yuen et al., 2011). For geospatial data, crowdsourced systems usually provide web map services or web feature services that support map mash-ups by which geospatial data from remote servers can be visually overlapped in a browser on a client site. As discussed in the abstraction section, the Semantic Web transforms web content to data as Web 2.0. Various social media facilitate crowdsourcing and provide ambient geographic information that can be exploited to recognize social pulses (Croitoru et al., 2013) or validate environmental conditions (See et al., 2013). Crowdsourced data are either directly requested by a project web service such as the "Did you feel it?" web portal by the United States Geological Surveys (USGS) Earthquake Hazards Program* or harvested from social media feeds via application programming interfaces (APIs), such as OpenStreetMap API.† Heipke (2010) provided a good introduction on crowdsourcing geospatial data with highlights of successful projects, the basic technologies and comments on data quality.

Since VGI and crowdsourced data lack statistical sampling schemes and are collected from various sites, researchers need to develop customized algorithms for data preprocessing, mapping, and analysis. Of great challenge is the fact that these data violate most, if not all, sampling assumptions based on which conventional statistical methods are founded. Location information associated with VGI, crowdsourced data, and data from web crawling may be explicitly tagged through GPS readings as latitude, longitude, or other x, y coordinate pairs. Alternatively, location may be implicitly noted in forms of place names or addresses. Addresses can be matched through geocoding against street network databases. For place names, toponym resolution and gazetteer matching will be necessary for georeferencing (Adelfio and Samet, 2013). More generally, conceptual and computational frameworks are being developed to transform text to a rich geospatial data source (Vasardani et al., 2013; Yuan et al., 2014). While several studies showed that VGI and crowdsourced data are timely and at times more representative of geographic reality than authoritative data (Goodchild and Glennon, 2010), most VGI and geospatial crowdsourced projects remain primitive and do not go beyond visualization, animation, and frequency graphing (Batty et al., 2010). Because crowdsourced data collection does not follow any statistical sampling methods, they cannot be applied to established statistical models. Sentiment analysis of postings and messages is often based on keywords without reference to the content. A detailed view of crowdsourcing can be found in Chapter 26 of this book. Chapter 26 also provides many examples of crowdsourcing in spatial sciences.

* http://earthquake.usgs.gov/earthquakes/dyfi.
† http://wiki.openstreetmap.org/wiki/API_v0.6.

Besides VGI and crowdsourced data, GIScience researchers are active in cyberinfrastructure research and cloud computing. CyberGIS integrates GIS and spatial analysis and modeling into cyberinfrastructure that provides high-performance (terra grid) computing and large-scale data repositories (Wang et al., 2013). It transforms GIS from an isolated platform to a cybernetwork of supercomputers, virtual organizations, and massive shared data resources. The fundamental differences in computing platforms require new algorithms for data processing, management, analysis, and modeling, and much has been implemented as middleware. While also taking the advantage of internet information technologies, cloud computing leverages four types of services: Infrastructure as a Service (IaaS), Platform as a Service (PaaS), Software as a Service (SaaS), and Data as a Service (DaaS), with open source resources Hadoop and MapReduce to offer elastic, distributed, and on-demand computing facilities.

The ideas of spatial cloud computing encompass not only utilization of existing cloud services for intensive spatial computing but also the development of data and tool services for geospatial applications that are made accessible over the web (Yang et al., 2011). Cloud computing provides elastic, advanced resources for experimenting with ideas, application development, and app distribution. Location-aware or spatially enabled apps are widely available to map property values, routes, crime incidents, restaurants, and gas stations, for example. Cloud computing opens GIS workflows to tightly connect to resources on the web and transforms GIS into a service that frees users from desktop computers to anywhere and any platform with internet access. CloudGIS emerges as a promising platform for large-scale geospatial computing and opens many opportunities for mobile GIS, geosensors, and spatial big data (Bhat et al., 2011; Shekhar et al., 2012; Fan et al., 2013).

Even before the introduction of big data, many spatial data are big and grow exponentially, especially imagery data and sensor data. The popularity of location-aware devices and geosensors has motivated algorithm development for trajectory analysis, geometrically (Li et al., 2011) or semantically (Yan et al., 2013), among other methods for movement modeling (Long and Nelson, 2012). Intensive observation updates of location-aware and sensor data further challenge hardware and software capacity. CudaGIS is an example of GIS design with GPUs to provide parallel data processing capabilities (Zhang and You, 2012). GPU algorithms are being developed to enable rapid urban simulation (Ma et al., 2008), lidar data processing (Sugumaran et al., 2011), viewshed analysis, and other data-intensive computation (Steinbach and Hemmerling, 2012). Some of the parallel and GPU algorithms have been implemented in open source GRASS GIS for fast spatial computing and rendering (Osterman, 2012). With on-demand IaaS, GPU-based cloud computing has shown to be effective for intelligent transportation management (Wang and Shen, 2011). A detailed view of cloud computing can be found in Chapter 27 of this book. Chapter 27 also gives several examples of where and how cloud computing is used in spatial science.

21.3.3 Assimilation

Assimilation is defined in the Oxford English Dictionary as the action of becoming conformed to or conformity with. In this chapter, assimilation is broadened to processes that bring individual's contributions to the commons for a greater good or to reveal a bigger picture. With the definition, assimilation efforts in GIScience have flourished through Open Source GIS, Social Coding, Open GIS, and Spatial Turns.

Open source GIS, like GRASS, Quantum GIS, and PostGIS, were developed by individuals through community efforts (Neteler and Mitasova, 2008) and have gained significant momentum since 2005. To date, there are more than 350 free and open source GIS software packages available.* Along with the free software are free data and documents to serve as a foundation for building a learning society in which source code, algorithms, and models can be tested and continuously improved upon. Steiniger and Bocher (2009) reviewed 10 free and open source GIS software packages and argued for the use of open source practices and software in research for transparency, testability, and adaptability to other projects. Many diverse open source GIS communities thrived in 2012 (Steiniger and Hunter, 2013). Assimilation of individual's contributions for tool development and code improvement in an open source environment collectively results in richer and better GIS resources for all. Social coding follows a similar idea of collaboration, but instead of working toward a package, social coding can be any project or program codes initiated by individuals rather than a community. Perhaps, the most popular social coding site is Github† where people can freely copy and modify codes to assimilate into other projects. There are many geospatial projects on Github, such as CartoDB, GeoNode, Spatial4J, OSGeo, and geopython. It is noteworthy that Esri is also active at Github with a suite of open source projects.

Another large-scale collaborative assimilation is the R project, an open source environment for statistical computing and graphics built upon the R language developed by Ihaka and Gentleman (1996). R users can study the source code to understand the underlying statistical procedures and assimilate their new modules with existing R methods, which facilitates advances in methodological research and opportunities to submit proposed models from publications for testing and reuse. Over the last 15 years, R has gained strong volunteer support in building various extensions, including packages for spatial statistics, for example, SpatStat (Baddeley and Turner, 2005), GeoXp (Laurent et al., 2009), and Spacetime (Pebesma, 2012). The call for integration of GIS and spatial data analysis (Goodchild et al., 1992) was originally intended to add more spatial analysis capabilities to a GIS. Instead, much success has been realized by assimilating spatial data and methods into R statistics.‡ Currently, R consists

* http://freegis.org.

† http://github.com.

‡ One may use R as a GIS (http://pakillo.github.io/R-GIS-tutorial/. Accessed March 22, 2014).

of a large suite of spatial modules covering raster analysis, interpolation and geostatistics, spatiotemporal simulation models, spatial autocorrelation, spatial econometrics, spatial structure models, spatial Bayesian models, spatiotemporal cluster analysis, and various mapping and graphing tools (Bivand et al., 2013). In addition, efforts are being made to apply R directly to GRASS GIS database files (Bivand, 2000) or port R scripts to Quantum GIS (Solymosi et al., 2010). Similarly, spatial analysis functions are being assimilated into the Python programming environment, most notably PySAL module (Rey and Anselin, 2010), and many spatial functions has been refactored to support parallelization (Rey et al., 2013). Free spatial data analysis packages such as GeoDa, although not open source nor extendable, offer a graphic user interface and tight coupling of GIS and exploratory spatial analysis tools (Anselin et al., 2006).

Assimilation of GIScience into other disciplines led to exciting new approaches, such as spatial ecology, spatial epidemiology, spatial history, spatial humanities, and spatial social sciences. In addition to spatial analysis and modeling, a suite of geospatial online data processing, information services, and computational methods popularizes web mapping and applications. Figure 21.4 illustrates an example of web applications for spatial ecology research that assimilates species, ecological, and environmental data in the Gulf of Mexico (Simons et al., 2013). Location-awareness is now common in research and development in computing and information science (Hazas et al., 2004). Programming libraries are being developed to improve the integration of GIS and remote sensing (Karssenberg et al., 2007; Bunting et al., 2014). Besides mapping and visualization, these spatial turns not only provide new analytical innovations and leveraged space as a problem framing and reasoning framework but also invoked new perspectives to improve understanding in natural sciences (Rosenberg and Anderson, 2011), social sciences (Raubal et al., 2014), and humanities (Bodenhamer, 2013). It is important to make clear that these assimilating efforts are developing new spatially integrated thinking and methodologies, not just applying existing GIS technologies in domain sciences.

21.4 Future Pathways and Concluding Remarks

As spatial abstraction, algorithms, and assimilation continue evolving, GIScience thrives for multiperspective, distributed, and collaborative research across people, platforms, and domain sciences. CyberGIS and Cloud GIS foster high-performance and ubiquitous spatial computing. Both technologies not only accelerate spatial data processing but transform the ways of doing GIScience and developing GIS applications. Wright (2012) sketched *a post-GISystems world where GIS is subsumed into a*

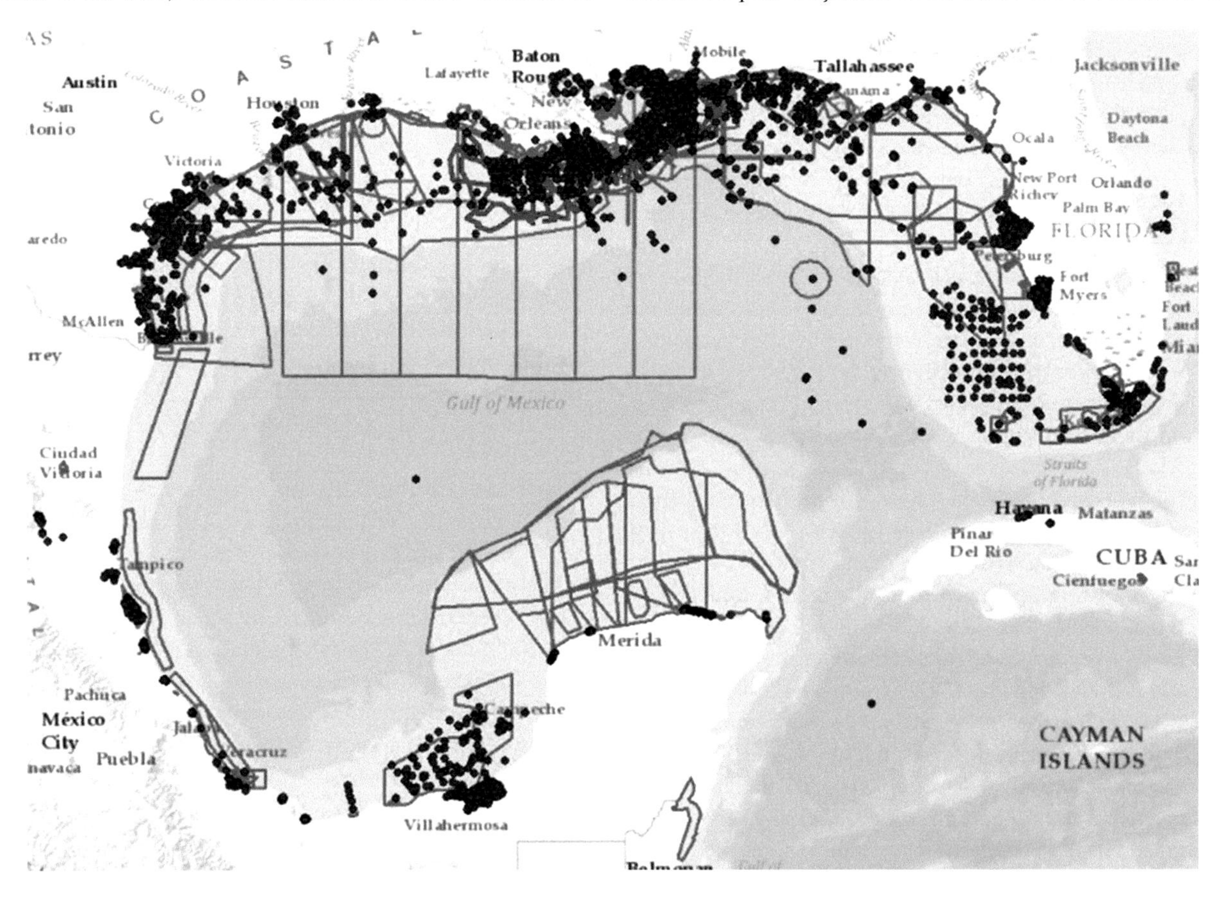

FIGURE 21.4 An example of spatial ecology research in the Gulf of Mexico. Species, ecology, and environmental data for 4092 sites were extracted from 449 references to support meta-analysis of fish habitat and trophic dynamics.

broader framework known simply as 'the web,' divorced from the desktop in a new paradigm (p. 2197). The future of GIScience will manifest itself in the grand scheme of computational, environmental, and social sciences. While time and themes are common axes along which disciplines build knowledge, GIScience distinguishes itself with the emphasis of using space as the first-order principle to acquire, organize, and compute information as well as discover and share knowledge. The distinction was already apparent in early GIS development and initial discussions on GIScience (Goodchild, 1992; Mark, 2003). It will be even more prominent in the big data era when data from location-aware devices continue to grow exponentially in volume and complexity, and spatial contextualization and integration are becoming more effective to sensing making and prediction.

The emphasis of space (e.g., spatiality, location, and situation) will continue to be the focus in pathways for future GIScience development in a world where we have access to needed information everywhere, any time, that is, *an IEWAT world*, enabled through online–offline integration, the Internet of Things, cloud-mobile computing, collaborative information seeking and knowledge building, and integrative cyber–physical–social systems. Clearly, these are also hot topics in the broader computational, environmental, and social sciences. In other words, the future pathways of GIScience are intimately intertwined with those of computational, environmental, and social sciences, and furthermore, GIScience should contribute substantially through understanding of space and use of space to achieve the vision of an IEWAT world.

Recent developments in GIScience have built strong foundations in all the three areas of spatial abstraction, algorithms, and assimilation, as discussed in Section 21.3. The pathways forward for an IEWAT world would extend the three areas into a multiverse of a truly diverse, distributed, and collaborative nature. Every location, every person, and everything is becoming a data producer. Data are from everywhere and anytime with different ontological notions. Algorithms are being developed, coded, modified, and forked by many over the web. Information is being analyzed and synthesized dynamically and continuously to reflect real-time and near real-time situations in the environment and our society. Online and offline computational platforms are being transitioned seamlessly to maximize the efficiency of mobile computing anywhere and anytime. Fully integrative cyber–physical–social systems inform us of the past, present, and future of what things/people are, where they reside, how they work, how they may evolve, where we should go, and what we should do.

To date, a GIS is no longer confined in a computer system or as a software package. GIS is immersed into the greater web computing environment and heading to an IEWAT world of truly ubiquitous spatial computing. Ontological and cognitive understandings of geospatial categorization and reasoning are essential to properly conceptualize geospatial problems and realize geospatial abstractions to connect reality and GIS databases. Spatial programming, web programming, and statistical programming are essential skills to analyze data and develop geospatial solutions. Spatial analysis, spatial data mining, mapping, geovisualization, and visual analytics remain critical to geospatial data exploration, information understanding, and knowledge discovery. Moreover, the pathway that will revolutionize GIScience is heading to the direction in which the common mode of GIScience practices is not confined to conventional research groups but involves scientists, practitioners, and citizens in a collaborative social cloud environment. It will be an IEWAT world of the people, by the people, and for the people.

References

Adelfio, M. D. and H. Samet (2013). GeoWhiz: Toponym resolution using common categories. *Proceedings of the 21st ACM SIGSPATIAL International Conference on Advances in Geographic Information Systems*. ACM.

Agarwal, P. (2005). Ontological considerations in GIScience. *International Journal of Geographical Information Science* **19**(5): 501–536.

Anselin, L. (1995). Local indicators of spatial association—LISA. *Geographical Analysis* **27**(2): 93–115.

Anselin, L., I. Syabri, and Y. Kho (2006). GeoDa: An introduction to spatial data analysis. *Geographical Analysis* **38**(1): 5–22.

Baddeley, A. and R. Turner (2005). Spatstat: An R package for analyzing spatial point patterns. *Journal of Statistical Software* **12**(6): 1–42.

Batty, M. (2007). *Cities and Complexity: Understanding Cities with Cellular Automata, Agent-Based Models, and Fractals*. Cambridge, MA: The MIT Press.

Batty, M., A. Hudson-Smith, R. Milton, and A. Crooks (2010). Map mashups, Web 2.0 and the GIS revolution. *Annals of GIS* **16**(1): 1–13.

Bhat, M. A., R. M. Shah, and B. Ahmad (2011). Cloud computing: A solution to Geographical Information Systems (GIS). *International Journal on Computer Science & Engineering* **3**(2): 594–600.

Bivand, R. S. (2000). Using the R statistical data analysis language on GRASS 5.0 GIS database files. *Computers & Geosciences* **26**(9): 1043–1052.

Bivand, R. S., E. Pebesma, and V. Gómez-Rubio (2013). *Applied Spatial Data Analysis with R*. New York: Springer.

Bodenhamer, D. J. (2013). Beyond GIS: Geospatial technologies and the future of history. In *History and GIS*, von Lunen, A. and Travis, C. (eds.). New York: Springer, pp. 1–13.

Brunsdon, C., S. Fotheringham, and M. Charlton (1998). Geographically weighted regression. *Journal of the Royal Statistical Society: Series D (The Statistician)* **47**(3): 431–443.

Bunting, P., D. Clewley, R. M. Lucas, and S. Gillingham (2014). The Remote Sensing and GIS Software Library (RSGISLib). *Computers & Geosciences* **62**: 216–226.

Buttenfield, B. P. and R. B. McMaster (1991). *Map Generalization: Making Rules for Knowledge Representation*. New York: Longman Scientific & Technical.

Camara, A. S. and J. Raper (1999). *Spatial Multimedia and Virtual Reality*. CRC Press.

Caron, C., S. Roche, D. Goyer, and A. Jaton (2008). GIScience journals ranking and evaluation: An international Delphi study. *Transactions in GIS* **12**(3): 293–321.

Cope, M. and S. Elwood (2009). *Qualitative GIS: A Mixed Methods Approach*. Sage.

Coppock, J. T. and D. W. Rhind (1991). The history of GIS. *Geographical Information Systems: Principles and Applications* **1**(1): 21–43.

Couclelis, H. (1997). From cellular automata to urban models: New principles for model development and implementation. *Environment and Planning B* **24**: 165–174.

Couclelis, H. (2010). Ontologies of geographic information. *International Journal of Geographical Information Science* **24**(12): 1785–1809.

Croitoru, A., A. Crooks, J. Radzikowski, and A. Stefanidis (2013). Geosocial gauge: A system prototype for knowledge discovery from social media. *International Journal of Geographical Information Science* **27**(12): 2483–2508.

Douglas, D. H. and T. K. Peucker (1973). Algorithms for the reduction of the number of points required to represent a digitized line or its caricature. *Cartographica: The International Journal for Geographic Information and Geovisualization* **10**(2): 112–122.

Fan, X., S. Wu, Y. Ren, and F. Deng (2013). An approach to providing cloud GIS services based on scalable cluster. *21st International Conference on Geoinformatics (GEOINFORMATICS)*. IEEE.

Fisher, P. F. (2006). *Classics from IJGIS: Introduction Twenty Years of IJGIS*. CRC Press, pp. 1–6.

Fonseca, F., M. Egenhofer, C. Davis, and G. Câmara (2002). Semantic granularity in ontology-driven geographic information systems. *Annals of Mathematics and Artificial Intelligence* **36**(1–2): 121–151.

Frank, A. U. (2001). Tiers of ontology and consistency constraints in geographical information systems. *International Journal of Geographical Information Science* **15**(7): 667–678.

Goodchild, M. F. (1992). Geographical information science. *International Journal of Geographical Information Systems* **6**(1): 31–45.

Goodchild, M. F. (2007). Citizens as sensors: The world of volunteered geography. *GeoJournal* **69**(4): 211–221.

Goodchild, M. F. (2014). Twenty years of progress: GIScience in 2010. *Journal of Spatial Information Science* (1): 3–20.

Goodchild, M. F. and J. A. Glennon (2010). Crowdsourcing geographic information for disaster response: A research frontier. *International Journal of Digital Earth* **3**(3): 231–241.

Goodchild, M. F., R. Haining, and S. Wise (1992). Integrating GIS and spatial data analysis: Problems and possibilities. *International Journal of Geographical Information Systems* **6**(5): 407–423.

Hazas, M., J. Scott, and J. Krumm (2004). Location-aware computing comes of age. *Computer* **37**(2): 95–97.

Heipke, C. (2010). Crowdsourcing geospatial data. *ISPRS Journal of Photogrammetry and Remote Sensing* **65**(6): 550–557.

Ihaka, R. and R. Gentleman (1996). R: A language for data analysis and graphics. *Journal of Computational and Graphical Statistics* **5**(3): 299–314.

Jiang, B. and Y. Miao (2014). The evolution of natural cities from the perspective of location-based social media. arXiv preprint arXiv:1401.6756.

Jung, C.-T., C.-H. Sun, and M. Yuan (2013). An ontology-enabled framework for a geospatial problem-solving environment. *Computers, Environment and Urban Systems* **38**: 45–57.

Karssenberg, D., K. de Jong, and J. Van Der Kwast (2007). Modelling landscape dynamics with Python. *International Journal of Geographical Information Science* **21**(5): 483–495.

Kwan, M.-P. and G. Ding (2008). Geo-narrative: Extending geographic information systems for narrative analysis in qualitative and mixed-method research. *The Professional Geographer* **60**(4): 443–465.

Laurent, T., A. Ruiz-Gazen, and C. Thomas-Agnan (2009). GeoXp: An R package for exploratory spatial data analysis. TSE Working Paper Series 99-099. Toulouse, France: Toulouse School of Economics.

Li, Z., J. Han, M. Ji, L.-A. Tang, Y. Yu, B. Ding, J.-G. Lee, and R. Kays (2011). MoveMine: Mining moving object data for discovery of animal movement patterns. *ACM Transactions on Intelligent System and Technology* **2**(4): 1–32.

Lin, H., M. Chen, and G. Lu (2013). Virtual geographic environment: A workspace for computer-aided geographic experiments. *Annals of the Association of American Geographers* **103**(3): 465–482.

Llaves, A. and W. Kuhn (2014). An event abstraction layer for the integration of geosensor data. *International Journal of Geographical Information Science* (ahead-of-print): 1–22.

Long, J. A. and T. A. Nelson (2012). A review of quantitative methods for movement data. *International Journal of Geographical Information Science* 1–27.

Ma, C., Y. Qi, Y. Chen, Y. Han, and G. Chen (2008). VR-GIS: An integrated platform of VR navigation and GIS analysis for city/region simulation. *Proceedings of the Seventh ACM SIGGRAPH International Conference on Virtual-Reality Continuum and Its Applications in Industry*. ACM.

MacEachren, A. M. (1998). Cartography, GIS and the world wide web. *Progress in Human Geography* **22**: 575–585.

Mark, D. M. (2003). Geographic information science: Defining the field. *Foundations of Geographic Information Science* 3–18.

Mark, D. M., A. G. Turk, N. Burenhult, and D. Stea (2011). *Landscape in Language: Transdisciplinary Perspectives*. John Benjamins Publishing.

Marks, J. and S. M. Shieber (1991). *The Computational Complexity of Cartographic Label Placement*. Citeseer.

Montello, D. R. (2009). Cognitive research in GIScience: Recent achievements and future prospects. *Geography Compass* **3**(5): 1824–1840.

Neteler, M. and H. Mitasova (2008). *Open Source Software and GIS*. Springer.

Olivera, F. and D. Maidment (1999). Geographic Information Systems (GIS)-based spatially distributed model for runoff routing. *Water Resources Research* **35**(4): 1155–1164.

Osterman, A. (2012). Implementation of the r. cuda. los module in the open source grass gis by using parallel computation on the nvidia cuda graphic cards. *Elektrotehniski Vestnik* **79**(1–2): 19–24.

O'Sullivan, D. (2006). Geographical information science: Critical GIS. *Progress in Human Geography* **30**(6): 783.

Pebesma, E. (2012). Spacetime: Spatio-temporal data in r. *Journal of Statistical Software* **51**(7): 1–30.

Raubal, M., G. Jacquez, J. Wilson, and W. Kuhn (2014). Synthesizing population, health, and place. *Journal of Spatial Information Science* (7): 103–108.

Rey, S. J. and L. Anselin (2010). PySAL: A Python library of spatial analytical methods. In *Handbook of Applied Spatial Analysis*, Springer, pp. 175–193.

Rey, S. J., L. Anselin, R. Pahle, X. Kang, and P. Stephens (2013). Parallel optimal choropleth map classification in PySAL. *International Journal of Geographical Information Science* **27**(5): 1023–1039.

Rosenberg, M. S. and C. D. Anderson (2011). PASSaGE: Pattern analysis, spatial statistics and geographic exegesis. Version 2. *Methods in Ecology and Evolution* **2**(3): 229–232.

Samet, H., M. D. Adelfio, B. C. Fruin, M. D. Lieberman, and J. Sankaranarayanan (2013). PhotoStand: A map query interface for a database of news photos. *Proceedings of the VLDB Endowment* **6**(12): 1350–1353.

Samet, H., W. G. Aref, C.-T. Lu, and M. Schneider (2008). Proposal to ACM for the establishment of SIGSPATIAL, ACM-SIGSPATIAL.

Schmitt, A. K., M. Danišík, E. Aydar, E. Şen, İ. Ulusoy, and O. M. Lovera (2014). Identifying the volcanic eruption depicted in a neolithic painting at Çatalhöyük, Central Anatolia, Turkey. *PloS One* **9**(1): e84711.

Schuurman, N. (2006). Formalization matters: Critical GIS and ontology research. *Annals of the Association of American Geographers* **96**(4): 726–739.

See, L., A. Comber, C. Salk, S. Fritz, M. van der Velde, C. Perger, C. Schill, I. McCallum, F. Kraxner, and M. Obersteiner (2013). Comparing the quality of crowdsourced data contributed by expert and non-experts. *PloS One* **8**(7): e69958.

Shekhar, S., V. Gunturi, M. R. Evans, and K. Yang (2012). Spatial big-data challenges intersecting mobility and cloud computing. *Proceedings of the Eleventh ACM International Workshop on Data Engineering for Wireless and Mobile Access*. ACM.

Shi, X. and M. D. Nellis (2013). Semantic web and service computation in GIScience applications: A perspective and prospective. *Geocarto International* (ahead-of-print): 1–18.

Simons, J. D., M. Yuan, C. Carollo, M. Vega-Cendejas, T. Shirley, M. L. Palomares, P. Roopnarine, L. Gerardo Abarca Arenas, A. Ibanez, and J. Holmes (2013). Building a fisheries trophic interaction database for management and modeling research in the Gulf of Mexico large marine ecosystem. *Bulletin of Marine Science* **89**(1): 135–160.

Solymosi, N., S. E. Wagner, Á. Maróti-Agóts, and A. Allepuz (2010). maps2WinBUGS: A QGIS plugin to facilitate data processing for Bayesian spatial modeling. *Ecography* **33**(6): 1093–1096.

Stefanidis, A., A. Crooks, and J. Radzikowski (2013). Harvesting ambient geospatial information from social media feeds. *GeoJournal* **78**(2): 319–338.

Steinbach, M. and R. Hemmerling (2012). Accelerating batch processing of spatial raster analysis using GPU. *Computers & Geosciences* **45**: 212–220.

Steiniger, S. and E. Bocher (2009). An overview on current free and open source desktop GIS developments. *International Journal of Geographical Information Science* **23**(10): 1345–1370.

Steiniger, S. and A. J. Hunter (2013). The 2012 free and open source GIS software map—A guide to facilitate research, development, and adoption. *Computers, Environment and Urban Systems* **39**: 136–150.

Sugumaran, R., D. Oryspayev, and P. Gray (2011). GPU-based cloud performance for LiDAR data processing. *Proceedings of the Second International Conference on Computing for Geospatial Research and Applications*. ACM.

Takeyama, M. and H. Couclelis (1997). Map dynamics: Integrating cellular automata and GIS through Geo-Algebra. *International Journal of Geographical Information Science* **11**(1): 73–91.

Tomlin, C. D. (1994). Map algebra: One perspective. *Landscape and Urban Planning* **30**(1): 3–12.

Tomlinson, R. (1962). An introduction to the use of electronic computers in the storage, compilation and assessment of natural and economic data for the evaluation of marginal lands. *The National Land Capability Inventory Seminar*. Ottawa, Ontario, Canada: The Agricultural Rehabilitation and Development Admininstration of the Canada Department of Agriculture, p. 11.

Tsou, M.-H., J.-A. Yang, D. Lusher, S. Han, B. Spitzberg, J. M. Gawron, D. Gupta, and L. An (2013). Mapping social activities and concepts with social media (Twitter) and web search engines (Yahoo and Bing): A case study in 2012 US Presidential Election. *Cartography and Geographic Information Science* **40**(4): 337–348.

Turk, A. and D. Stea (2014). David Mark's contribution to ethnophysiography research. *International Journal of Geographical Information Science* (ahead-of-print): 1–18.

UCGIS. (2002). UCGIS bylaws. Retrieved March 6, 2014, from http://ucgis.org/basic-page/laws.

Ujang, U. and A. A. Rahman (2013). Temporal three-dimensional ontology for geographical information science (GIS)—A review. *Journal of Geographic Information System* **5**(3).

Vasardani, M., S. Timpf, S. Winter, and M. Tomko (2013). From descriptions to depictions: A conceptual framework. In *Spatial Information Theory*. Springer, pp. 299–319.

Wang, K. and Z. Shen (2011). Artificial societies and GPU-based cloud computing for intelligent transportation management. *IEEE Intelligent Systems* **26**(4): 22–28.

Wang, S., L. Anselin, B. Bhaduri, C. Crosby, M. F. Goodchild, Y. Liu, and T. L. Nyerges (2013). CyberGIS software: A synthetic review and integration roadmap. *International Journal of Geographical Information Science* **27**(11): 2122–2145.

Wellen, C. C. and R. Sieber (2013). Toward an inclusive semantic interoperability: The case of Cree hydrographic features. *International Journal of Geographical Information Science* **27**(1): 168–191.

Wiegand, N. (2012). *Ontology for the Engineering of Geospatial Systems: Geographic Information Science*. Springer, pp. 270–283.

Wiegand, N. and C. García (2007). A task-based ontology approach to automate geospatial data retrieval. *Transactions in GIS* **11**(3): 355–376.

Worboys, M. F. (2003). Communicating geographic information in context. *Foundations of Geographic Information Science* 33–45.

Wright, D. J. (2012). Theory and application in a post-GISystems world. *International Journal of Geographical Information Science* **26**(12): 2197–2209.

Yan, Z., D. Chakraborty, C. Parent, S. Spaccapietra, and K. Aberer (2013). Semantic trajectories: Mobility data computation and annotation. *ACM Transactions on Intelligent Systems and Technology* (*TIST*) **4**(3): 49.

Yang, C., M. Goodchild, Q. Huang, D. Nebert, R. Raskin, Y. Xu, M. Bambacus, and D. Fay (2011). Spatial cloud computing: How can the geospatial sciences use and help shape cloud computing? *International Journal of Digital Earth* **4**(4): 305–329.

Yuan, M., J. McIntosh, and G. De Lozier (2014). GIS as a narrative generation platform. In *Spatial Narrative and Deep Mapping*, Bodenhamer, J. C. D. and Harris, T (eds.). Indianopolis, IN: Indiana University Purdue University Indianapolis Press.

Yuen, M.-C., I. King, and K.-S. Leung (2011). A survey of crowdsourcing systems. Privacy, security, risk and trust (passat). *2011 IEEE Third International Conference on Social Computing* (*socialcom*). IEEE.

Zhang, J. and S. You (2012). CudaGIS: Report on the design and realization of a massive data parallel GIS on GPUs. *Proceedings of the Third ACM SIGSPATIAL International Workshop on GeoStreaming*. ACM.

22

Object-Based Regionalization for Policy-Oriented Partitioning of Space

Stefan Lang
University of Salzburg

Stefan Kienberger
University of Salzburg

Michael Hagenlocher
University of Salzburg

Lena Pernkopf
University of Salzburg

Acronyms and Definitions

ASEAN	Association of Southeast Asian Nations
AVHRR	Advanced very-high-resolution radiometer
CCCI	Cumulative climate change impact
CRU	Climate Research Unit
DEM	Digital elevation model
DFO	Dartmouth Flood Observatory
EAC	East African Community
EC	European Commission
ECOWAS	Economic Community of West African States
ESP	Estimate scale parameter
EU	European Union
GEOBIA	Geographical object-based Image analysis
GIS	Geographic Information System(s)
IPCC	Intergovernmental Panel on Climate Change
LV	Local variance
MAUP	Modifiable areal unit problem
MODIS	Moderate-resolution imaging spectroradiometer
MOVE	Methods for the improvement of vulnerability assessment in Europe
NIR	Near infrared
NOAA	National Oceanic and Atmospheric Administration
OBA	Object-based analysis
OBIA	Object-based image analysis
OFA	Object fate analysis
PCA	Principal component analysis
R	Red
RBC	Recognition-by-component
RBD	River basin district(s)
SD	Spectral distance
SHP	Shape Index
TIN	Triangulated irregular network
UNEP	United Nations Environment Programme
VBD	Vector-borne disease
VGI	Volunteered geographic information
VHI	Vegetation Health Index
WHO	World Health Organization

22.1 Mapping Latent Phenomena

Latent, that is, not directly observable, spatial phenomena can be operationalized and thus mapped by integrating multiple sets of indicators. Spatial representation of such complex phenomena, in order to be useful to decision makers in different domains, needs to be likewise efficient and robust. Drawing from the experiences in object-based image analysis (OBIA) and the spatial integration of multidimensional image data, regionalization techniques can be applied to map, analyze, and represent these phenomena in a spatial explicit way.

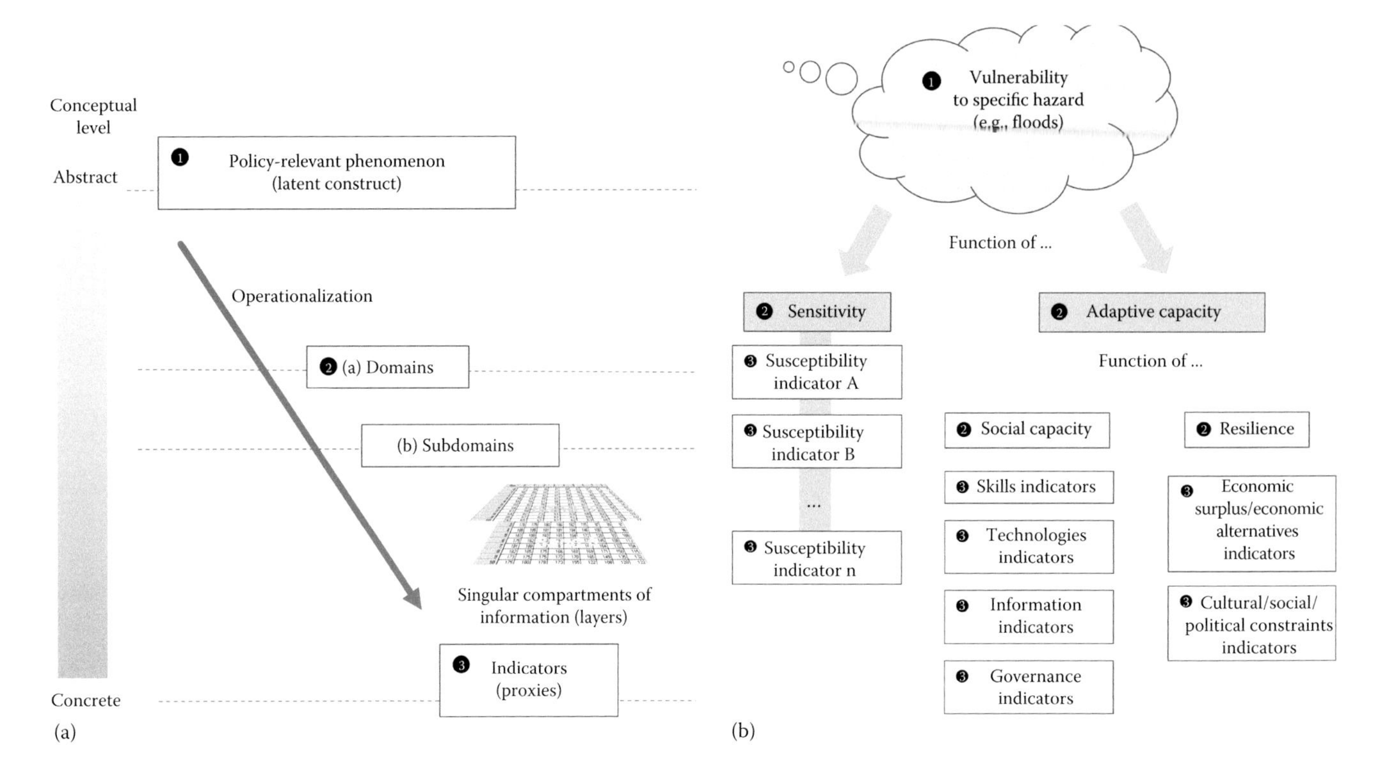

FIGURE 22.1 Operationalization—breaking down latent phenomena to measurable indicators. (a) From abstract, conceptual level to measurable indicators. (b) Example of operationalizing *vulnerability to a specific hazard* (e.g., a flood, an earthquake) and its conceptualization via domains, subdomains, and sets of indicators.

22.1.1 Ambitious Policy Targets Require a Systemic View

The complexity of global challenges faced today entails a holistic, systemic viewpoint, in order to better understand their cause–effect relationships and mutual dependencies. Assessing and monitoring the status of sustainable development policies in a comprehensive and holistic sense relies on integrative, synthesizing techniques (Fiksel 2006, Stahl et al. 2011). The global warming debate, for example, entails an integrated view on the interrelationships among the environmental sphere and the social sphere (including cultural, institutional, and economic aspects) in integrated assessment models (Parson 1994, Weyant et al. 1996). These spheres are complex phenomena themselves with several subspheres to be considered: for example, in the ecological sphere, one may investigate the behavior of ecological sensitivity to changing climatic conditions; in the socio-cultural sphere, one may study social vulnerability to natural hazards. The conceptual operationalization aims at breaking down the complexity of these systems into graspable, and ultimately measurable, compartments. For example, the concept of vulnerability as a latent phenomenon has been operationalized by Kienberger et al. (2009)* in several stages of domains and subdomains, until the level of measurable indicators is reached (see Figure 22.1).

When reaching *down to the bottom* so to speak, we approach the problem by separate aspects, that is, analytically, and likely not able to see the full picture. Measuring and monitoring every contributing aspect separately is fundamental, but more important, though also more challenging, is to capture their cumulative effect (Hagenlocher et al. 2014b). Target-oriented strategic commitments increasingly adopt an integrated assessment approach (Risbey et al. 1996, Jakeman and Letcher 2003). On implementation level, policies and directives on national, regional (EU, ECOWAS, ASEAN, etc.), and global level try to accommodate holistic concepts and integrated assessments, influenced by systems thinking (Capra and Luisi 2014). At the same time, they strive for pragmatic ways to communicate complexity in a *simplified* manner. Comprehensive but integrative, these concepts assume that through systemic behavior, new qualities emerge, which to capture is both a challenge and a chance. Vulnerability, resilience, mitigation, etc. can be considered as systemic properties that require integrative, synthesizing mapping techniques (Capra and Luisi 2014). In other words, spatial analysis techniques that are capable to integrate compartments of information, and overcome the classical, analytical concept of single geospatial data layers are sought (Lang et al. 2014b). A suitable integrative geographic technique for this is regionalization, also known as spatial classification (Wise et al. 2001), that is also the methodological core element of the geon approach, is presented in this chapter.

* Recently the concept has been refined, as discussed in Section 22.3.

22.1.2 Approaching Complex Phenomena with a Spatial Focus

Abstract or complex phenomena that cannot be observed directly are termed *latent* variables, or factors (Byrne 1998). Byrne concludes if "latent variables are not observed directly, [...] they cannot be measured directly" (p. 4). A way out of this dilemma is to "operationally define the latent variable of interest in terms of behavior believed to represent it" (Byrne 1998) (p. 4). The underlying construct is constituted by the direct measurement of the observed (or manifest) variable. Manifest variables are presumed to represent the underlying construct and serve as indicators of it (Byrne 1998). When using such indicators, a conceptual bridge is built between the measurable and the latent part of the phenomenon, a fact that requires careful investigation.

Complex phenomena are, for example, *vulnerability to hazards*, *adaptability to climate change*, *landscape sensitivity to human impact*, and *quality of life*, among others. All these cannot be directly measured with a specific measuring device and hardly reduced to a single indicator. A strategy is to approach such complex properties by aggregating manifest variables or indicators into an index or composite indicator, a procedure that works on multidimensional attribute spaces (Nardo et al. 2005). However, such procedures neglect the true spatial distribution of such properties when using *a priori* geographies (such as neighborhood or district boundaries) that obscure or bias the actual spatial distribution.

Implementing policies on regional, national, or local level that address such complex phenomena ideally adopts a synthetic view on multidomain and multisource geospatial datasets. The fine granularity of recent geographical datasets enables a multipurpose, effective usage for the public sector in fulfilling public tasks (Wise et al. 2001), which are more and more interrelated and integrated. The unitization of space may ideally follow the topic-related functional characteristics of the phenomenon addressed, both in terms of scale and thematic discontinuities. In other words, boundaries should be set where they reflect a significant change in behavior of this phenomenon. An example is the European Water Framework Directive (2000/60/EC), where so-called river basin districts (RBDs) transgress national boundaries and represent catchments rather than existing administrative units. Generalizing from this example, we argue that the spatial variability of the actual phenomenon being addressed requires a more flexible spatial unitization within the geographical policy scope.

22.1.3 Geons—Terminology and Conceptual Background

For providing such policy-related spatial units, a methodological framework has been developed (Lang et al. 2014b). This has been tested in different policy domains such as functional biotope complexes for regional development planning (Tiede et al. 2010), disaster risk reduction, for example, vulnerability units (Kienberger et al. 2009), sensitivity units for strategic environmental assessment (Pernkopf and Lang 2011), and hotspot analysis for climate change adaptation (Hagenlocher et al. 2014b). The framework builds on a workflow to regionalize geodata (i.e., imagery or other continuous geospatial data layers), resulting in a set of *geons*. These are units that synthetically aggregate domain-specific information with a uniform response regarding the complex phenomenon under concern. A geon set represents the spatial explicit distribution of this aggregated information. Table 22.1 and Section 22.3 contain examples to illustrate this. The term *geon* was initially introduced in cognitive psychology by Biederman (1987) in his theory of recognition-by-components (RBC). It is based on the concept of volumetric primitives, defined as geons (geometric ions), and the hypothesis that cognitive objects can be decomposed into basic shapes or components. Geons or gaeons (Peuquet 2002) in Biederman's view are basic volumetric bodies such as cubes, spheres, cylinders, and wedges. Conformity with the original geon concept is discussed by Lang et al. (2008) with regard to (1) the role of generalization for the definition and strength of a geon (though: scale dependent); (2) the significance of the spatial organization of the elements, which leads to emergent properties and specific qualities; (3) the possibility of recovering objects in the presence of occlusions (i.e., data errors, measure failures, lack of data, mismatch of data due to bad referencing). So the concept presented here is related, but not identical to Biederman's idea. The term *geon* has been proposed by Lang (2008) to widen the scope of the original concept and adapt it to the domain of GIScience. Lang et al. (2014b) proposed the following redefinition:

> A *geon* (derived from Greek *gē* (Γῆ) = land, earth and the suffix *-on* = something being) is a type of region, semi-automatically delineated with expert knowledge incorporated, scaled and of uniform response to a phenomenon under space-related policy concern. The aim of generating geons is to map policy-relevant spatial phenomena in an adaptive and expert-validated manner, commensurate to the respective scale of intervention.

The geon approach aims at to delineate regions, but does not relate to administrative regions with predefined, normative boundaries. Instead of the commonly used term "analytical regions" (Duque and Suriñach 2006), we prefer to use the term "synthetic [*sensu* Sui (2011)] regions" to emphasize the integrative character of geons. Thus, geon stands for a systems-theory-driven, scale-dependent *earth-object*. Policy relevant units are based on advanced geodata integration, expert knowledge, and user validation. Lang et al. (2014b) distinguish between two types of geons, composite and integrated geons. We focus on the latter in the present chapter (see Section 22.2). *Geonalytics* are built upon a comprehensive pool of techniques, tools, and methods for (1) geon generation (i.e., transformation of continuous geoinformation into discrete objects by algorithms for

TABLE 22.1 Operationalization of Latent Phenomena in a Certain Policy Scope and Geographical Scale Using a Set of Spatialized Indicators (Dimensionality of Indicator Space = Number of Indicators, Modeling Scale = Minimum Resolution)

Latent Phenomenon (Policy Scope)	Geographic Scale (Modeling Scale)	Domains	Dimensionality of Indicator Space	# of Geons (Geon Size)[a]
Vulnerability units/climate change context	Salzach river catchment (Austria) (1 km^2)	Sensitivity, adaptive capacity (socioeconomic dimension)	52	1462 [36/4.0/3.8]
Vulnerability units/disaster risk reduction context	Salzach river catchment (1 km^2)	Exposure, susceptibility, lack of resilience (social dimension)	16	181 [147/32.6/28.6]
		Economic dimension	6	300 [118/19.7/18.5]
		Environmental dimension	21	314 [376/18.8/38.5]
		Physical dimension	22	248 [147/23.8/22.1]
Vulnerability units/disaster risk reduction context	District of Búzi, Mozambqiue (1 km^2)	Susceptibility, adaptive capacity (social dimension)	11	307 [108/23.4/18.7]
		Economic dimension	6	225 [127/32.0/23.9]
		Environmental dimension	5	391 [120/18.4/33.9]
		Physical dimension	4	213 [801/33.8/105.9]
Sensitivity units/strategic impact assessment context	Marchfeld region, Austria (250 m)	Ability of the ecosystem to resist to external disturbances	9	151 [46.3/5.8/6.5]
		Ecological significance of the affected areas	6	121 [55.3/7.2/9.5]
		Societal significance	6	139 [40.8/6.3/6.8]
Climate change hot spots/ climate change adaptation context	Sahel and western Africa (subcontinental) (~16 km^2)	n/a	4	2,283 [645/63.5/42.1]

The delineated geons are characterized by number (#) as well as size in km^2 (max/mean/std-dev).

interpolation, segmentation, and regionalization); (2) analyzing spatial arrangements and characterize form and spatial organization, investigate spatial emergent properties, and issues of scale; (3) monitoring of modifications and changes and evaluation of the status of geons.

The key issue is spatial regionalization of a complex spatial reality. This reality is represented exhaustively and in a range of interdependent, inherent scales. The strategy of the approach is per se geographically motivated. Geographic location is a key to an integrative assessment of complex and multidimensional phenomena that have a spatial component. With the ever-increasing maturity of Geographic Information Systems (GIS) in its broadest sense, the integrative power of space has been boosted over recent years, also across disciplines, on conceptual, technological, and methodological levels (Lang et al. 2014a). Integrated spatial analysis methods support the shift from a more mechanistic, analytical view, to a more systemic one. This includes a change in perspective from a focus on objects to relationships, from (quantitative) measuring to (qualitative) mapping (Capra and Luisi 2014). Undoubtedly, the ubiquity in spatial data repositories (both public and private, cf. *big data*) and a cross-cultural technology proliferation (e.g., through smart phones) has recently prepared the ground for *spatializing* all kinds of societal or physical phenomena. But there is a gap between the factual capacity of spatial analysis and decision support tools and their actual usage. Usually, socioeconomic indicators are integrated on predefined administrative or regular grid cells (Nardo et al. 2005); the full potential of geographical synthesis is underexploited (Lang et al. 2014a). While this is often due to conventional or pragmatic reasons, we can represent complex phenomena in a more appropriate way.

So just like image segmentation is used to represent tangible real-world objects in multidimensional feature spaces, latent phenomena can be represented by regionalizing multiple sets of spatialized indicators. *Spatialized* in this context means: indicators are represented spatially explicit as continuous fields. What is considered *continuous* is scale depending though. Usually this implies a sampling of the indicator using any kind of tessellation into small units, be it regular (grid cells, hexagons, etc.) or irregular (TINs, enumeration units, etc.). These again may source from data collected on finer level (irregular sensor measurement grids, socioeconomic data collected on household level, etc.) and then interpolated over the area. Often simply the term *spatial* is used for simplification, but strictly speaking *spatial indicators* measure spatial properties (size, distribution, proximity, etc.) of spatial units (Bock et al. 2005).

Since we are looking at latent phenomena that involve spatial (and temporal) variability, we need spatialized indicators presumed to represent the components of such phenomena in a spatially explicit way. The number of such indicators varies depending on the phenomenon studied (Table 22.1). Ecosystem integrity may be simply approximated by a vegetation index relating the red (R) and near-infrared (NIR) bands of a remotely sensed image for indicating tree species diversity and vitality within a deciduous forest. But even if standardization and image calibration is applied and the results are technically robust, most probably, many aspects are missing in this strategy, as many other aspects may contribute to the integrity of the forest ecosystem, like groundwater regime, timber usage and other anthropogenic impacts, climatic conditions, as so on. As pointed out earlier using the example of societal vulnerability to flood hazards, multidimensional problems require conceptual models to operationalize them, a process that is often expert based (Kienberger et al. 2013a). For pragmatic and performance reasons, the number of indicators should not be too high either, so we may aim at a commensurate number of indicators to best represent the phenomenon in a methodologically sound way (Moldan and Dahl 2007).

We try to approximate as much as possible the phenomenon under concern with an integrated set of proxies, to statistically and heuristically turn its latency into aggregates exhibiting gradients and other spatial behavior that we are able to map. In adding a spatial component to the integration of indicators, we arrive at the level of *spatial composite indicators* (Hagenlocher et al. 2014b) or *metaindicators* (Lang et al. 2008). Lang (2008) uses the term *conditioned information* to underline that this process entails the creation of new geographies as a flexible, yet statistically robust and (user) validated unitization of space. The aim is to map complex phenomena to better understand their spatial variability and dynamics over time.

22.2 Domain-Specific Regionalization

Integrated geons are delineated in a semiautomated way incorporating expert knowledge by adhering to statistical robustness and scale optimization. Thus, we are able to map and monitor units of uniform response to the phenomenon under concern, commensurate to the scale of policy intervention measures and stable in their aggregation.

22.2.1 Principles of Regionalization

Within spatial science, the term *regionalization* implies both a top–down (i.e., disaggregating) and a bottom–up (i.e., aggregating) notion. Disaggregating a larger whole into smaller regions is often associated with political will, for which the European Statistical Office uses the term *normative regions* (Eurostat 2006). Scientific regionalization usually follows a more bottom–up strategy applying routines to group neighboring subunits, that is, small geographical units, pixels, raster cells, etc., into larger ones (Figure 22.2). While some regionalization techniques do imply a top–down component (e.g., the *split-and-merge* algorithm), most regionalization methods are implemented in a bottom–up fashion, performing any kind of spatially constrained aggregation method (Duque and Suriñach 2006). Such aggregation is done in a way that the resulting analytical regions are "conveniently related to the phenomena under examination" (Duque and Suriñach 2006, p. 2).

The main objectives of regionalization (Berry 1967) are summarized by Lang et al. (2014b):

1. to organize, visualize, and synthesize the information contained in multivariate spatial data (Long et al. 2010)
2. to reduce data dimensionality (Ng and Han 2002) while minimizing information loss (Nardo et al. 2005)
3. to minimize the effects of outliers or inaccuracies in the data and to facilitate the visualization and interpretation of information in maps
4. to limit the sensitivity due to data fidelity by aggregating the original units (e.g., pixels) into larger zones (Blaschke and Strobl 2001, Wise et al. 2001)

Regionalization is based on the principle of spatial autocorrelation, assuming that neighboring areas tend to have similar properties or uniform behavior (Tobler 1970). Region-building assumes that such uniform behavior exists as long as transitions occur, leading to a different behavior along a certain gradient or boundary. While, due to the principle of spatial autocorrelation, the internal structure of continuous spatial data enables an empirical construction of regions, there is no *a priori* fixed set of regions to be built. As with classification in general, the aggregation of data is to some degree arbitrary (Johnston 1968), and the areal units to be built by spatial aggregation can be done at different scales and in different (though equally plausible) ways (Wise et al. 2001). As with statistical data in general, the problem of aggregation effects in particular applies to spatial studies (modifiable areal unit problem [MAUP]). As Openshaw (1984) points out there are no standards or international conventions for spatial aggregation and it is subject to the "whims and fancies [...] of whoever is doing the aggregation" (p. 3). We will return to this problem and see how we (partly) cope with it.

22.2.2 Integrated Geons

Experiences in analyzing high-fidelity, multispectral imagery using geographic object-based image analysis, (GE-) OBIA (Hay and Castilla 2008, Blaschke 2010), can be transferred to address complex spatial phenomena. The transfer of OBIA techniques for the analysis of nonimage data has been discussed for univariate phenomena such as a digital elevation model (DEM) by Drăgut and Eisank (2011). Kienberger et al. (2009) used object-based analysis (OBA) to regionalize an *n*-dimensional indicator space for assessing socioeconomic vulnerability to flood hazards. Nonimage data in this context are gridded geospatial data provided by interpolation of point samples over space, or based on spatially disaggregated indicators. Principles and conceptual findings from social sciences with technical achievements from OBIA are bridged, to generate *integrated geons* (Lang et al. 2014b). Next to the challenge of incorporating expert knowledge and adhering to statistical robustness and scale optimization, an issue remains in assigning nominal categories or even labels to the generated units and characterize their evolution over time.

Just like image segmentation is used to represent tangible real-world objects, latent phenomena can be represented by regionalizing multiple indicators (Lang et al. 2014a). The indicators may be mapped as singular layers in a GIS and evaluated separately. A common strategy is to approach such complex properties by aggregating variables/indicators toward an index or composite

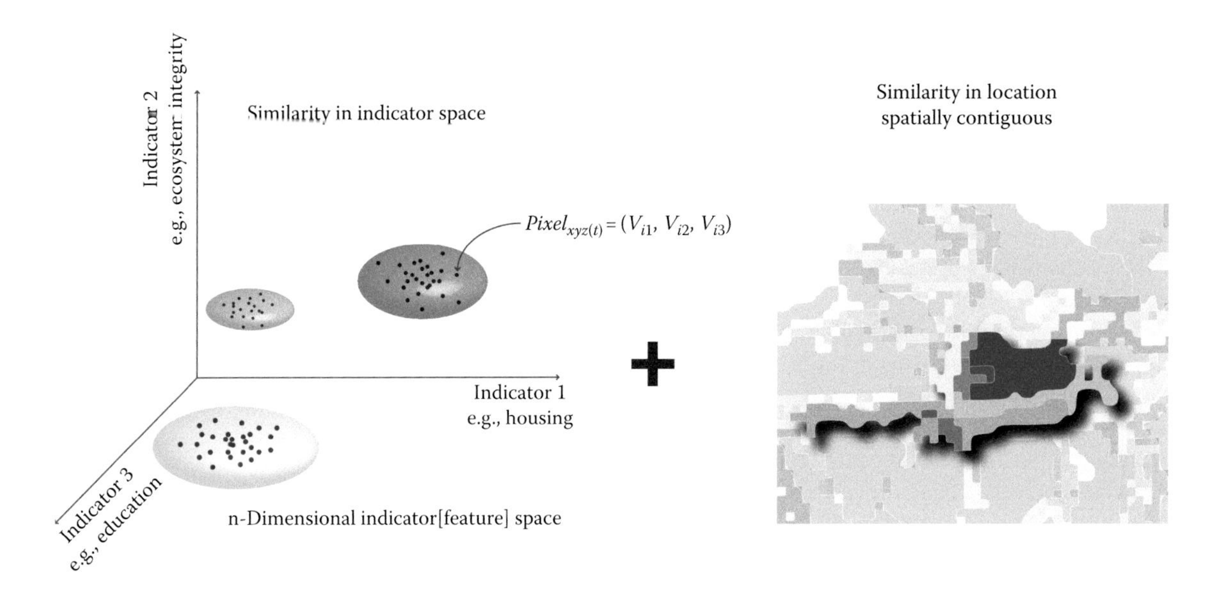

FIGURE 22.2 Regionalization—similarity in attribute and real space. Next to similarity in feature space, the classification of spatial data is controlled by location. Spatially constraint classification is called regionalization.

indicator, a procedure that works on multidimensional attribute spaces (Nardo et al. 2005). However, such procedures neglect the true spatial distribution of such properties when using *a priori* geographies (such as neighborhood or district boundaries) that obscure or bias the actual spatial distribution.

Our strategy is to extend multivariate clustering from attributive to real space, thereby reaching a flexible portioning of real space, (re-)composing new geographies. Just like multispectral imagery can be clustered into homogenous objects by applying segmentation techniques, we represent complex spatial phenomena by regionalizing multidimensional attribute/indicator spaces. This is different to overlaying or intersecting of geospatial layers in treating the spatial dimension as a constraint, but flexible in terms of scale. Regionalization techniques that utilize rule-bases with flexible strategies for geodata integration under consideration of and expert knowledge follow the methodological framework of OBIA.

By applying segmentation routines, we go one step further toward spatial explicitness (see Figure 22.3). *Classical* aggregation that uses a raster overlay with a local operator performed on regular grid cells or given reference units does not change the level of spatial explicitness, while regionalization does. Here, spatial explicitness is not meant in terms of more spatial detail, but spatial synthesis. Regionalization does not increase the spatial detail, but introduces additional spatial characteristics in the aggregation process, and thereby *spatializes* it.

An initial workflow to delineate geons was developed by Kienberger et al. (2009) and applied to assess place-based vulnerability in a climate change context. Since then, the approach has been refined and transferred to various application domains (see Section 22.3), including, for example, hot spot analysis of cumulative climate change impacts (Hagenlocher et al. 2014b), landscape sensitivity in the context of environmental impact assessment for infrastructure projects (Pernkopf and Lang 2011),

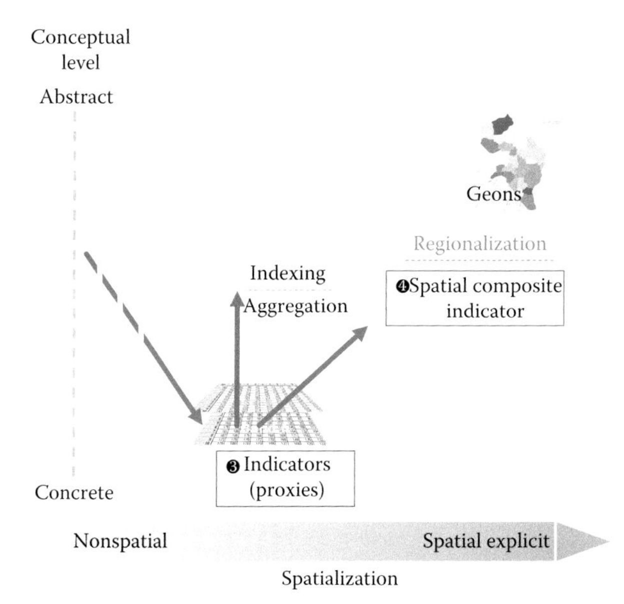

FIGURE 22.3 Regionalization vs. indexing. Note that this is a continuation of Figure 22.1a. Indexing, for example, aggregating on a per-cell basis does not generate new spatial information. Instead, regionalization provides additional spatial-explicit information by its generation of (new) spatial units.

or social vulnerability to malaria (Kienberger and Hagenlocher 2014). Currently available methods to model integrated geons build on four major stages (see Table 22.2). These stages serve as a framework and can be adapted accordingly.

In a *first stage*, the identification and definition of associated concepts of the phenomena under investigation is carried out. Subsequently this is followed in a *second stage* by the identification of indicators and data collection and related preprocessing. In the *third* and *core stage*, the multidimensional indicator framework is integrated through regionalization, which can

TABLE 22.2 Stages of the Geon Workflow and Relevant Aspects to Be Considered

#	Stage	#.n	Key Steps Involved	Issues to Be Considered
1	Definition of conceptual framework	1.1	Conceptual operationalization	• Soundness • Practicability • Expert opinion • Scientific communities/schools, etc.
		1.2	Domains and dimensions	
		1.3	Identification of indicators	• Literature/expert knowledge • Selection criteria: salience, credibility, legitimacy • Data availability
2	Data preprocessing	2.1	Data acquisition	• Scale: global, national, etc. • Data availability: public domain, authority mandate, commercial • Resolution: admin units, grid
		2.2	Preprocessing	• Resampling to continuous grids • Data transformation • Data imputation and outlier treatment • Normalization
		2.3	Multivariate analysis	• Multicollinearity analysis • PCA
		2.4	Sensitivity analysis	• Evaluating the influence of indicator choice, normalization method, aggregating method, and weighing
3	Regionalization	3.1	Indicator weighting	• Equal weights • Expert weighting • Statistical weights
		3.2	Unit delineation	• (Multi)scale assessment and segmentation • Composite index
4	Visualization/geonalytics	4.1	Mapping	• Legend and intervals
		4.2	Explorative analysis	• Shape and size variation, diversity
		4.3	Monitoring	• E.g., using WebGIS solutions

Source: Kienberger, S. et al., *Nat. Hazards Earth Syst. Sci.*, 9, 767, 2009; Kienberger, S. and Hagenlocher, M., *International Journal of Health Geographics*, 13, 29, 2014; Lang et al., *Cartography and Geographic Information Science* 41(3), 214, 2014b.

be built on different weightings of the indicators. This includes sensitivity analysis. In a final and *fourth stage*, the results are appropriately communicated and visualized to the intended user group.

In the following, these four steps are described in detail: latent, multidimensional phenomena are concept driven and need to be defined accordingly. The first stage has a strong link to different theories originating from various disciplines such as sociology, ecology, or geography—to name a few. Such concepts and frameworks may also change over time and within different schools of thought. Therefore, it is important to define an *appropriate* framework depending on the context of the study. Associated to the selection of appropriate theories and useful frameworks, a first set of possible indicators has to be identified. Irrespective of the availability of data, one should consider how the different dimensions of the chosen theoretical framework could be characterized and measured. It is likewise important to have an in-depth understanding of the underlying framework that comprises causal relationships and logical dependencies between the single domains and indicators relevant for operationalizing the phenomena.

Once a preliminary indicator set is identified, it is envisaged in the second stage to establish a quantitative, valid, and representative indicator framework. This includes the collection and quality assurance of different datasets to populate the indicators, as well as different statistical preprocessing routines. Such preprocessing routines are well known in the composite indicator community (see, e.g., Nardo et al. 2005), and can be applied to derive a statistically sound indicator framework. This includes the transformation of the indicators to continuous grids, the transformation from absolute into relative measures, the identification and treatment of outliers, missing data and multicollinearities in the data, as well as the application of normalization techniques to render the different indicator values comparable (e.g., min–max normalization, *z*-score standardization, etc.). Alternatively, value function approaches (Beinat 1997) can be applied, where values are normalized on expert or empirically defined relationships.

Once a final indicator framework is set up, the grid-based indicator data are integrated through regionalization to derive homogenous regions of the investigated phenomena. Geons are delineated using the multiresolution segmentation algorithm of Baatz and

Schäpe (2000) as implemented in the eCognition software environment (Trimble Geospatial). This is a region-based, local mutual best fitting approach that merges image segments according to the gradient of degree of fitting (Baatz and Schäpe 2000). It allows for controlling two complementary criteria of similarity of neighboring segments: likeness in *color* and *form*. The latter refers to the influence of *space* in the regionalization process. Spatial objects can be generated that are rather compact or have rather smooth outlines. A scale-factor enables user-driven control of appropriate scaled representations. Here we shortly reflect the algorithm as discussed earlier by Kienberger et al. (2009): the difference between adjacent objects (Kienberger et al. 2009) is expressed by the spectral distance (SD) of two pixels or objects p_1, p_2 in a feature space:

$$\mathrm{SD} = \sqrt{\sum_{d=1}^{n} (p_1 - p_2)^2}, \tag{22.1}$$

or noted as vector difference for a three-dimensional feature space as

$$\mathrm{SD} = (\vec{v}_1 - \vec{v}_2);$$

$$\text{with } \vec{v}_1 = \begin{pmatrix} d_{11} \\ d_{12} \\ d_{13} \end{pmatrix} \text{ and } \vec{v}_2 = \begin{pmatrix} d_{21} \\ d_{22} \\ d_{23} \end{pmatrix}. \tag{22.2}$$

To optimize the degree of homogeneity between two neighboring pixels or objects, the specific heterogeneity is minimized at every merge. The current degree of fitting (h_{diff}) is characterized by the change in heterogeneity (Equation 22.3):

$$h_{\text{dff}} = h_{\min} - \frac{\mathrm{SD}_1 + \mathrm{SD}_2}{2} \tag{22.3}$$

By weighting heterogeneity with object size, the requirement of producing objects of similar size can be accomplished. Form homogeneity is considered by relating actual boundary length (perimeter) to the perimeter of the most compact form of the same size. This ideal form is a circle, the deviation of which can be expressed by the shape index:

$$\mathrm{SHP} = \frac{p}{2\sqrt{\pi * s}}, \tag{22.4}$$

where

p equals the perimeter
s equals the size of an object

Identifying an appropriate scale factor has been subject to much debate, the common approach is to use *trial and error*. To overcome this subjectivity in object generation, a routine suggests initial parameters for the multiresolution segmentation (Drăgut et al. 2014). The so-called estimation of scale parameter (ESP)

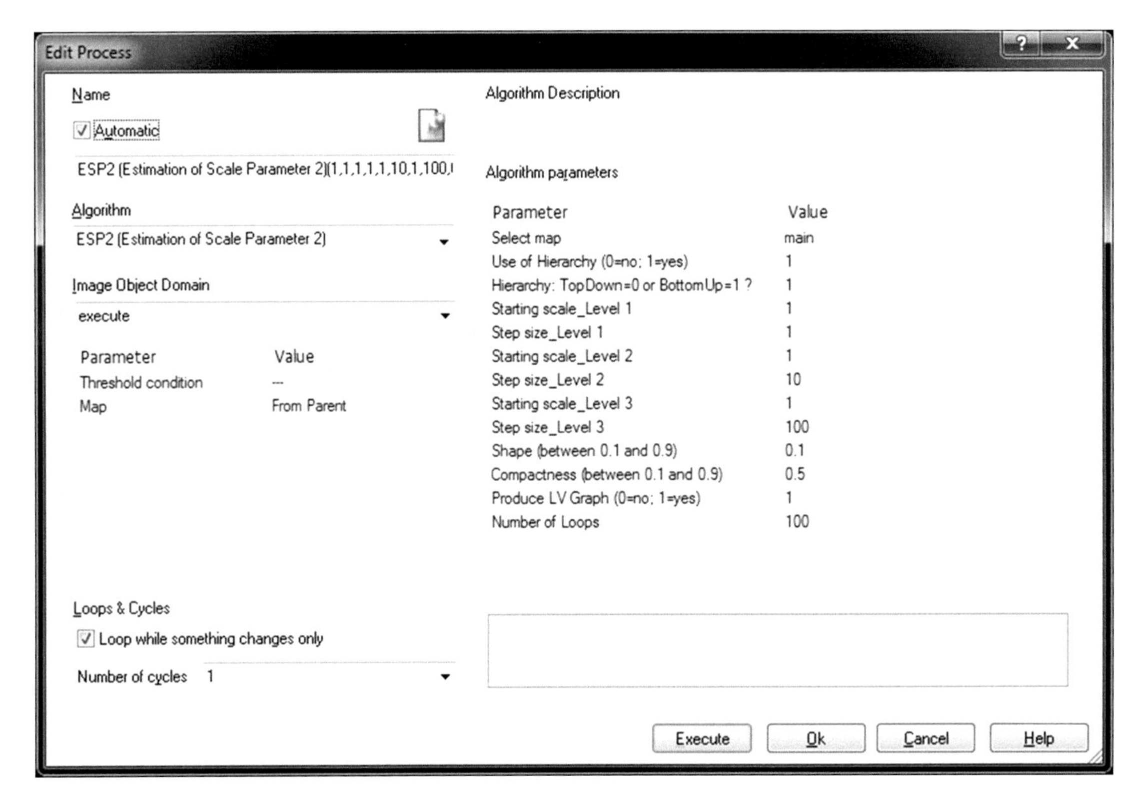

FIGURE 22.4 Graphical user interface of the ESP II tool.

tool is implemented as a generic tool for eCognition software (Figure 22.4).

The tool utilizes local variance (LV) graphs (Woodcock and Strahler 1987) to detect scale domains in geospatial data (Drăguţ et al. 2010). It segments continuous datasets iteratively applying a multiresolution segmentation algorithm, from smaller to coarser scale in a small, constant increment. The mean LV value is computed for each image level that was created as the ratio between the sum of the LVs for each layer ($LV_1 - LV_n$) and the number of layers (n) used in the image segmentation. When a scale level records an LV value that is equal to or lower than the previous value, the iteration ends, and the objects segmented in the previous level are retained (Drăgut et al. 2014).

According to different indicator weights, the algorithm also considers their relative contribution to the resulting composite value. To assign a quantitative value to each of the identified units or regions, a (weighted) vector magnitude is calculated, which measures the vector distance of each unit within its n-dimensional indicator (feature) space (see Equation 22.1). Alternative aggregation approaches, such as weighted sum or geometric mean, are also available, while the vector magnitude emphasizes variability and underlines changes context. Finally, the values of the resulting spatial composite indicator are normalized to a defined classification scheme (e.g., from zero to one) to ease its interpretation. Additionally for each unit, the contribution of the underlying indicators to each integrated geon through different metrics (mean values, standard deviation, etc.) can be assessed.

So far, we discussed how geons are generated by integrating a set of indicators and dimensions. All of the modeling stages discussed introduce uncertainty in the results, since there is a range of plausible alternatives at each stage. If another conceptual framework with a different set of indicators was used, or the experts had weighted the relative importance differently, other units, and thus policy messages, would have been generated. The impact of the normalization method and the regionalization parameters should not be neglected either. For illustrative purposes, we discuss the impact of the final choice of indicators and indicator weights on the modeling outputs in Section 22.3.

We recommend a combination of data-driven (statistical) and normative (expert-based) approaches to generate concept-related *fiat* objects of this kind and to deliver a consolidated output of domain-relevance. This especially applies to the choice of indicators and the setting of weights (Decancq and Lugo 2012). In all cases, the modeling of integrated geons should be complemented with robustness tests to determine to what degree the results are driven by specific indicators or weights. Different approaches to uncertainty and sensitivity analysis have been proposed by the composite indicators community (Nardo et al. 2005, Saisana et al. 2005, Saltelli et al. 2008). Whereas uncertainty analysis quantifies the overall uncertainty in the output as a result of the uncertainties in the model input, sensitivity analysis can be used to evaluate how changes in each individual input parameter affect the final output and how the variation in the output can be apportioned, qualitatively or quantitatively, to different sources of variation in the assumptions. A combination of both practices, uncertainty and sensitivity analysis, provides insights into the robustness of the modeling results. Testing the robustness of geons confronts us with the challenge of changing geometries in combination with changing index values.

The generated units, though considered plausible, do not directly correspond with observable real-world objects. The validity of the spatial extent and the precategorical nature of such units can be indirectly assessed by collecting evidence data on the prevalence of the phenomenon's impact. Examples are damage scenarios or loss estimation scenarios in the case of vulnerability to natural hazards or, in the case of disease vulnerability, to assess the spatial distribution of the disease combined with patient data on socioeconomic background and demographic setting.

In a final stage, integrated geons are visualized. To avoid the categorization and classification of different class ranges, a continuous color scheme is applied. Each unit is visualized based on its composite indicator value, and can be visually compared to the other regions. To allow an interactive exploration of the different regions, and its underlying indicators, integrated geons are best visualized through interactive, web-based mapping tools, for example, Kienberger et al. (2013b).

22.2.3 Monitoring Geons

Once geons are generated in agreement to the policy realm, they can be used to monitor the spatiotemporal dynamics of the considered (latent) phenomenon over time.

Understanding the status quo of a complex spatial phenomenon is one thing, observing its behavior over time, via the dynamics of geons as space-time entities (Peuquet 2002), is another. While the decision for particular intervention may be based on analyzing the situation *as is*, assessing the comparison to one or several previous ones helps well understand its relative meaning. Also, the impact of such intervention requires a repetitive analysis of altering conditions. A multitemporal geon set may provide support in the reassessment and correction of intervention measures. The delineation of geons enables comparable updates as a basic requirement for regular and consistent monitoring, here: against a specific policy background (Lang et al. 2008). Working in a spatial-explicit object-based environment allows for characterizing the shapes of single geons on individual level, as well as their spatial arrangement and distribution on collective level. The spatial variability of the phenomenon under concern may be a key to better understand the overall systemic behavior of it. In addition, assessing the spatiotemporal dynamics and the evolution of a geon set over time allows an evaluation of the underlying trend. Here, as in any change analysis, we need to separate noise or slight modifications from real changes. Dealing with aggregated units rather than single cells helps in overcoming the problem of data fidelity. Originating from OBIA, the concept called object fate analysis (OFA) (Schöpfer et al. 2008, Hernando et al. 2012) can be used to characterize the behavior of geons over time, and their possible transitions to other stages. When

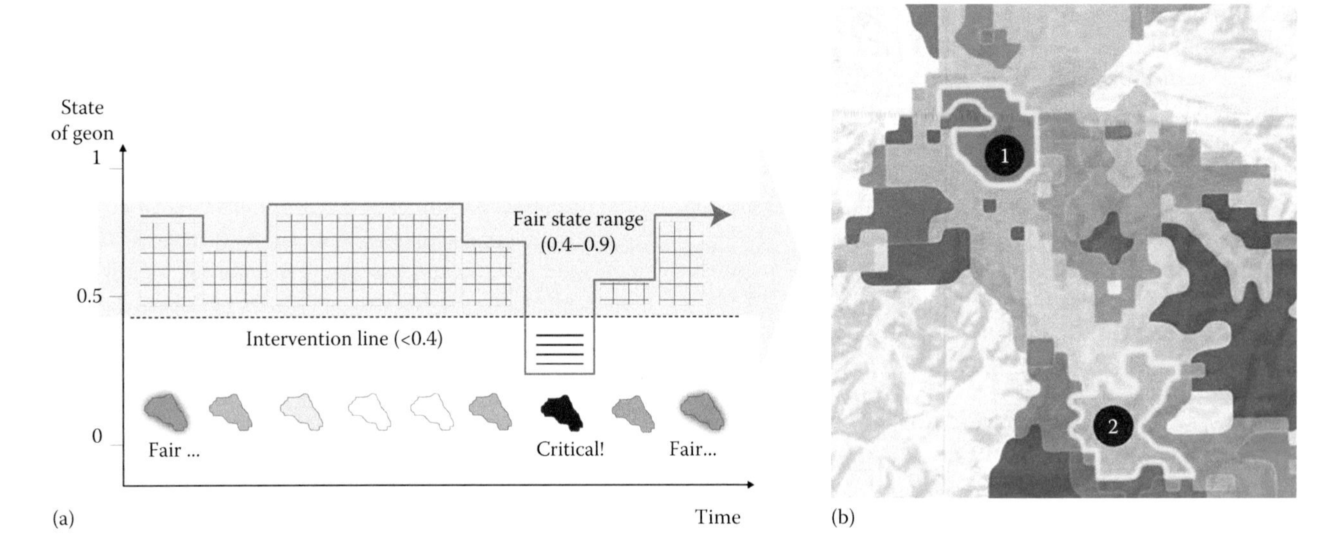

FIGURE 22.5 (a) Monitoring integrated geons supports a scenario of *threshold-based intervention planning*. Fair state range: *natural* dynamic of a geon state through time. Green: geon in fair/good state, no action required, red: below fair-state line, action to be considered, dark red/dashed: intervention required. (b) Two geons (representing *vulnerability to floods*) in different states below fair state. The dark red tone (1) would indicate required measures. See Figure 22.7 for larger extent and additional context. (From Lang, S. et al., *South-Eastern Eur. J. Earth Observ. Geomatics*, 3(2s), 625, 2014.)

monitoring the state of geons over time, we may encounter a certain threshold beyond which we would consider it *favorable*. In other words, no policy intervention is required. But as soon as its state deteriorates and drops beneath that threshold, certain countermeasures have to be taken. We may call this allowable range the fair-state range (see Figure 22.5).

22.2.2.4 MAUP and Scale Dependency

Until today, many assessments of latent spatial phenomena are carried out using administrative boundaries or continuous grids as the final reporting unit. As policies and interventions are often spatially targeted at the administrative level, results reported on administrative units match the scale of current policy interventions. Policy and decision makers on the ground have a long tradition in using them for analysis, benchmarking, reporting, and planning purposes. However, as pointed out by Openshaw (1984), the results and their interpretation have to be treated with caution as they are *biased* by the fact that these units are modifiable, artificial areal units. This holds also true for continuous grids, or pixels, which—like administrative units—do not have an intrinsic geographical meaning (Hagenlocher et al. 2014a).

The MAUP concept has been developed for addressing the drawbacks of mapping any measurement, model outcome, or statistical value on existing geographies (units, e.g., census or other administrative units). In relation to that, ecological fallacy problematizes the fact that all individuals (or subunits) in a given spatial unit are treated collectively, how close (spatially or property wise) they may be to the reported phenomenon (Openshaw 1984). MAUP can also be linked to the observation that smaller units (e.g., pixels in an image, cells in a raster representation, etc.) can be aggregated to larger units quite arbitrarily. While there are an infinite number of possible combinations or groupings, regionalization techniques aim to algorithmically or heuristically group neighboring units according to any kind of similarity criterion. Still, merging neighboring pixels assigned to the same category (class) may depend on the classification routine and/or a given size constraint applied. But there is an inherent reason why these units are generated and they bear any kind of homogenous internal behavior, in particular when compared to neighboring units (basic principle of delineation or demarcation). In the case of integrated geons, the mapping of a spatial composite indicator is embedded in a conceptual framework; its methodological realization, provided the basic principles of multivariate data analysis are obeyed, does provide intersubjectively relevant units. Boundaries are generated where the latent phenomenon under concern changes its behavior. In contrast to administrative units and grids, geons provide a representation of the real spatial distribution of the respective phenomenon (Hagenlocher et al. 2014a). Although geons do not necessarily match the scale of current policy interventions, thus calling for a paradigm shift in intervention planning, they provide a powerful means for planning as their internal heterogeneity is minimized (Lang et al. 2014b).

Integrated geons can be delineated at various scale levels, ranging from local to national, continental or global. It is challenging to find a commensurate scale for finding the respective gradient to represent the phenomenon best (Marceau 1999). Next to the nature of the phenomenon, this is also influenced by the availability of data that are appropriate to be integrated in a regionalization approach, the level of spatial detail that is required, and the scope of the analysis (Hagenlocher et al. 2014b). The three different kinds of scale mentioned by Wu and Li (2006) are likewise important when modeling integrated level geons: the intrinsic scale, the modeling/observational scale, and the policy scale. As we postulate that integrated level geons are policy relevant, we have to consider the relevant scale level at which policies occur. For instance, if a district is responsible for disaster risk reduction activities, decision makers require relevant information on (sub-)district level in a meaningful and disaggregated manner. However, this has to be in line with the intrinsic scale level of the certain phenomena investigated (e.g., where flood occurs; vulnerability at a local level is characterized different than one at the global one). The intrinsic scale level may be difficult to identify in an objective manner as it can also rely on our perception; however, certain assumptions need to be made. This scale level needs to bridge toward the observational and modeling scale. It is essential that the observational scale (e.g., resolution of raw data) and the final modeling scale match, and are valid to establish the bridge between the intrinsic and final policy scale level. These considerations need to be taken into account, when identifying indicators (do they reflect the intrinsic sale at the district level appropriately?), the associated input data and modeling domain (e.g., resolution of raster-based data), and finally, the policy scale level, for which the results will provide decision makers with relevant information. Therefore, such a consistency among the different kinds of scales needs to be maintained in order to provide valid results, while developers of spatial composite indicators should be aware of the implications of spatial scales for analyzing and monitoring latent phenomena (Hagenlocher et al. 2014a).

22.2.4 How To Validate Geons?

User validation is meant in the sense of object validity (Lang et al. 2010) that comprises both policy relevance and expert-proven functional relevance. From a data model point of view, the generated geons are areal (i.e., polygonal) objects. The vector data model suggests crispness of the generated boundaries and soundness of the assigned label. If units are not *a priori*, can they be objective, that is, intersubjectively acceptable? Johnston stated in 1968 that "all approaches to classification are actually subjective," cited in Hancock (1993), so the question arises: do such units have some validity beyond a narrow thematic domain? When delineating integrated geons, we aggregate values of the underlying indicators and their weighted contribution in a multidimensional variable space. At the same time, we deal with spatial constraints that check the similarity of neighboring cells. After that we visualize the resulting composite values: depending on the number of intervals (decile, quantile, etc.), the geon-scape will appear differently, while the delineated units as such remain. Particularly, when aiming at deriving hot spots of these complex phenomena, thresholds need to be defined, which indicate whether or not an object will be marked as a hot spot. Defining these thresholds adds a certain level of subjectivity to the final results, as *a priori* thresholds do not exist and must be defined by either making use of expert knowledge or a needs-driven approach (e.g., specific number of hot spots in an area). A challenge is to assign nominal categories (*labels*) to the generated units, which are at this stage only characterized by value ranges of the computed metaindicator As conceptual links are still missing, currently, we cannot move from ordinal scaled ranks (low, medium, high) to nominal categories, just as we transform continuous elevation data into landform classes (Drăguţ and Eisank 2011).

We argue that the farther we move from *bona fide* objects, the trickier—and more domain subjective—validation gets. Lang et al. (2014b) have characterized integrated geons as concept-related *fiat* entities, which cannot be assessed by established accuracy assessment methods. The term *object validity* (Lang et al. 2010) reflects the limited power of binary assessments in judging thematic accuracy of a given object label. Object validity should ensure a purpose-oriented judgment whether the product meets the users demand.

22.2.5 In Depth: Systemic Areal Units

Systems thinking has been widely referred to in spatial science literature as dealing with scaled representations of continuous data, for example, in multiscale image analysis (Hay et al. 2001) and in addressing areal phenomena over several scales, such as the *scaling ladder* concept (Wu 1999) in landscape ecology. Originally conceptualized in a nonspatial or not explicitly spatial context, systems theory (Bertalanffy 1969) deals with the hierarchical organizations of concrete systems, which on each level exhibit systemic, that is, emergent properties (in fact one may consider it vice versa, a *level* is constituted where emergent properties become obvious). Koestler (1967) coined the term holon to underline the nested behavior of systemic units. The geon concept as a holistic regionalization approach intends to abstract from a specific application domain by explicitly addressing the level of the generated units while not limiting our view on a specific thematic topic or application. The common denominator is to generate units that are not *a priori*, but *demanded* and thus—once provided—of immanent importance for policy-related action. Geons show uniform response regarding the spatial phenomenon under concern, and are (ideally) expert validated regarding practical usability and relevance. They are carriers of integrated spatial information and, we hypothesize, exhibit emergent properties of systemic areal units.

Terminologically, *geo-on* suggests synonymy with *geo-atom* (Goodchild et al. 2007) or even more radical, the "Elementary_geoParticle" (Voudouris 2010). But geons do not claim *atomicity* (Masolo and Vieu 1999) as something undividable. In fact, a geon is not the smallest possible unit in space and/or time as topons or chronons (Couclelis 2010), but the smallest valid one in a certain context ("as small as necessary, as big as possible") (Lang et al. 2014b). In generating an aggregate that has its own meaning and stability, the composition of a geon will start with, but go beyond, a purpose-driven exercise. In the composite indicator community, it is widely acknowledged that integrating single indicators leads to systemic effects (emergent properties) (Nardo et al. 2005). In other words, the variance of a systemic whole is much lower than the sum of the variances of its components. However, raster cells (with composite values) that can represent an Elementary_geoParticle would not qualify as a geon, due to its technically fixed spatial definition. A geon is a spatial object with systemic stability features such as minimized inner variance and gradients toward the outside through vector encoding. Regionalization implies some loss of information (Hancock 1993), redundant detail that acts as noise in the context under concern.

In congruence with systems thinking (Laszlo 1972), we argue that spatial systemic units also bear a dual character in terms of self-assertive tendencies (whole-ness) and self-integrative ones (part-ness) (Lang et al. 2004). This nested behavior can be applied to systemic areal units as well, as a geographical correspondence to holons (Wu 1999). Geons that are composed from spatial elementary units show such systemic properties in bearing certain functional qualities, which are crucial to be treated as a conceptual whole. Geons show a uniform response to a certain phenomenon to be approximated by applying spatial classification of an integrated set of spatialized indicators (Kienberger et al. 2009).

22.3 Case Studies

Here we present several examples how the geon approach was utilized in different application domains including disaster risk reduction, public health, climate change adaptation, and environmental impact assessment.

The case studies section will use practical examples to illustrate the following summary statements:

- Conceptual frameworks build the basis for operationalizing the latent phenomenon addressed; they guide the assessment and foster the reproducibility of results as well as the transferability of the approach to different scales or regions.
- Several plausible methods exist for constructing spatial composite indicators (e.g., normalization, weighting, aggregation); it is important to make the approach transparent and to conduct a sensitivity analysis; for geons, this is still an area under development.
- In some stages, subjective judgment has to be made, which affects the result (selection of indicators, weighting of indicators, regionalization parameter, etc.).
- Transferability across geographic/policy-scales (local to regional) as well as across domains (disaster risk reduction, climate change adaptation, public health, strategic planning), see Figure 22.6.
- The geon approach works with multidomain, multisource datasets and independent from the number of indicators; however, indicators and indicator weights must be chosen carefully to best reflect the idiosyncrasies of the phenomenon that is addressed.
- Validation of latent phenomena is tricky (how to validate the *immeasurable*?), there are no real observations to calibrate the model.

22.3.1 Socioeconomic Vulnerability to Hazards

22.3.1.1 Latent Phenomenon Addressed

Currently much emphasis is put on conceptualizing and mapping vulnerabilities in the context of *disaster risk reduction* and *climate change adaptation*. Here we illustrate an example of how to perform an integrated, holistic assessment of social, economic, and environmental vulnerability to flood hazards. The study has been performed in the Salzach river basin, Austria, and discussed in greater detail by Kienberger et al. (2014) in the context of flood hazards at the catchment scale level for the social, economic, and environmental domains. The Salzach catchment is characterized through its alpine upstream area and highly dynamic river valleys with the city of Salzburg in the downstream area. Recent severe floods have occurred in 2002, 2005, and 2013, affecting large parts of the river catchment. The

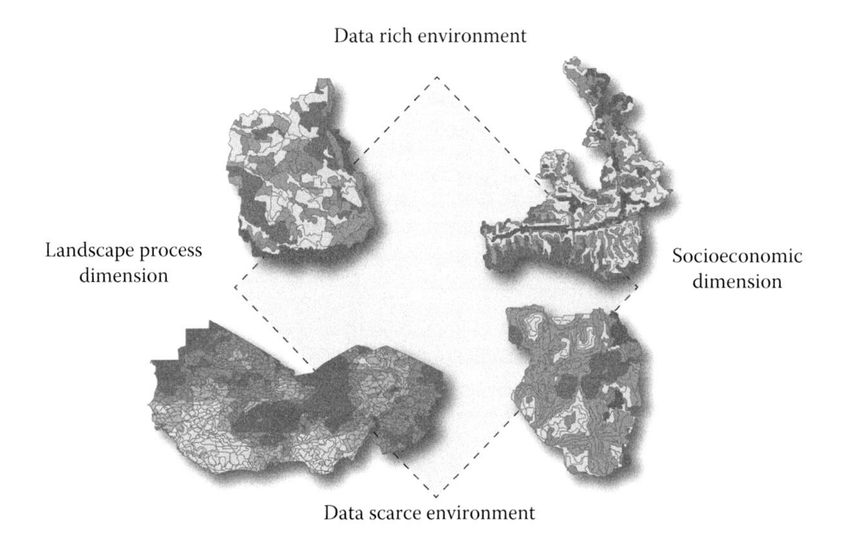

FIGURE 22.6 Geon-scapes—exemplary illustration considering different policy domains and scales, dimensions and data conditions.

majority of the population lives in the city of Salzburg (approx. 150,000 inhabitants) and its urban fringe. The area is highly dynamic with regard to its economic and settlement developments, putting high pressure on adequate spatial planning policies in an area with a limited amount of suitable land.

To achieve the objective to protect people against floods, Austria is currently implementing flood hazard mappings, improvement of river regulations, and technical flood protection measures along with sophisticated early warning systems. Despite these measures, floods are not prevented per se. They allow for enough time to prepare, warn, and evacuate people and minimize economic damage and loss of human lives based on specific flood protection activities.

22.3.1.2 Indicators and Datasets

The study aimed to model homogenous vulnerability regions for the social economic and environmental dimensions. As such, homogenous regions in terms of their degree as well as their inherent characteristic of vulnerability should facilitate the improved identification of place-specific intervention measures for disaster risk reduction.

An expert-based approach was selected by considering the fact that vulnerability is not directly measurable, due to its complex dimension and social construction. In order to model the spatial distribution of a complex phenomenon, established methodologies such as multicriteria decision analysis, Delphi exercises, and regionalization approaches to derive homogenous vulnerability regions were integrated. At the initial stage of the workflow, the concept of vulnerability is addressed, which in this case builds on the adapted MOVE risk and vulnerability framework. An essential step is the identification of indicators and available datasets, the weighting of the different indicators, and its aggregation to subindices. As the final list of appropriate indicators identified from expert choice, literature, and community perspective, five key experts were asked to weigh the indicators through budget allocation methods. Data availability has been an advantage in the Salzach catchment, as important key datasets are available in the spatial data infrastructure established at the Government of Salzburg, as well as the availability of detailed raster-based census data for Austria, which specifically helped to apply the regionalization approach. The data used range from infrastructure data to different socioeconomic parameters such as the size of companies, means of subsistence, age, and workforce in economy sectors, origin, and education level of the population. They originated from the census survey in 2001 and are not only provided on the basis of different administrative units but also for a standardized grid with different available spatial resolutions. To allow comparability, data were normalized through linear min–max normalization within an 8 bit scale range (values with 0–255). The modeling scale level was based on the standardized grid with a regular cell size of 1000 m. In the next step, the modeling of homogenous vulnerability regions for each dimension (as integrated geons) was implemented through the application of a multiresolution segmentation/regionalization approach in combination with a smoothing algorithm (polynomial approximation). Therefore, regions of vulnerability have been delineated that share a commonality with regard to their underlying indicator values as well as a spatial constraint. A vulnerability index was calculated through a weighted vector magnitude (the length of the vector for each region) in the multidimensional indicator space. Final index values were then normalized within a scale range of 0–1, and visualized by applying a centile classification with exponential kernel algorithm.

22.3.1.3 Results

The results allow the identification of spatially explicit regions with different levels of vulnerability (*hot spots*) independent from administrative boundaries. Furthermore, the results provide decision makers with place-specific options for targeting disaster risk reduction interventions that aim to reduce vulnerability and ultimately the risk of impacts from floods. The most vulnerable regions in the three dimensions are located along the Salzach river and its tributaries. However, the most vulnerable region in all the dimensions is the city of Salzburg and its surroundings. These results are due to the density of the built-up area, a big concentration of historic buildings, and widespread infrastructure. Therefore, the highest vulnerability degree with regard to the social dimension is concentrated in the city of Salzburg as one of the largest settlements located along the Salzach river. The economic dimension has its hot spot also in this urban region due to the presence of employment sources, and the city is an important node in the transport network in the country. The environmental dimension shows the highest degree of vulnerability in the stretch of the Salzach river from the city to the north, associated to river fragmentation (Figure 22.7).

22.3.2 Social Vulnerability To Malaria

22.3.2.1 Latent Phenomenon Addressed

Despite the global recession in malaria cases over the past decades, malaria remains the most prevalent vector-borne disease (VBD). Caused by the bite of an infected Anopheles mosquito, malaria resulted in approximately 207 million infections and has causes approximately 627,000 malaria-attributable deaths in 2012 (WHO 2013), primarily among African children. Thus, in addition to modeling transmission potentials, it is of utmost importance for the planning of (preventive) interventions to assess prevailing levels of malaria vulnerability in a spatially explicit manner (Hagenlocher et al. 2014a). Based on an adapted version of the MOVE risk and vulnerability framework, relative levels of social vulnerability to malaria were modeled for the East African Community (EAC) region using the geon approach (Kienberger and Hagenlocher 2014).

22.3.2.2 Indicators and Datasets

Based on the outcomes of a systematic review of literature, the consultation of domain experts, and data availability, (1) a preliminary set of 15 biological and disease-related (e.g., immunity, age, pregnancy, etc.), (2) socioeconomic (e.g., socioeconomic status, poverty, nutritional status, education, etc.), as well as (3) accessibility-related indicators (e.g., access to health services, etc.) were identified. Data

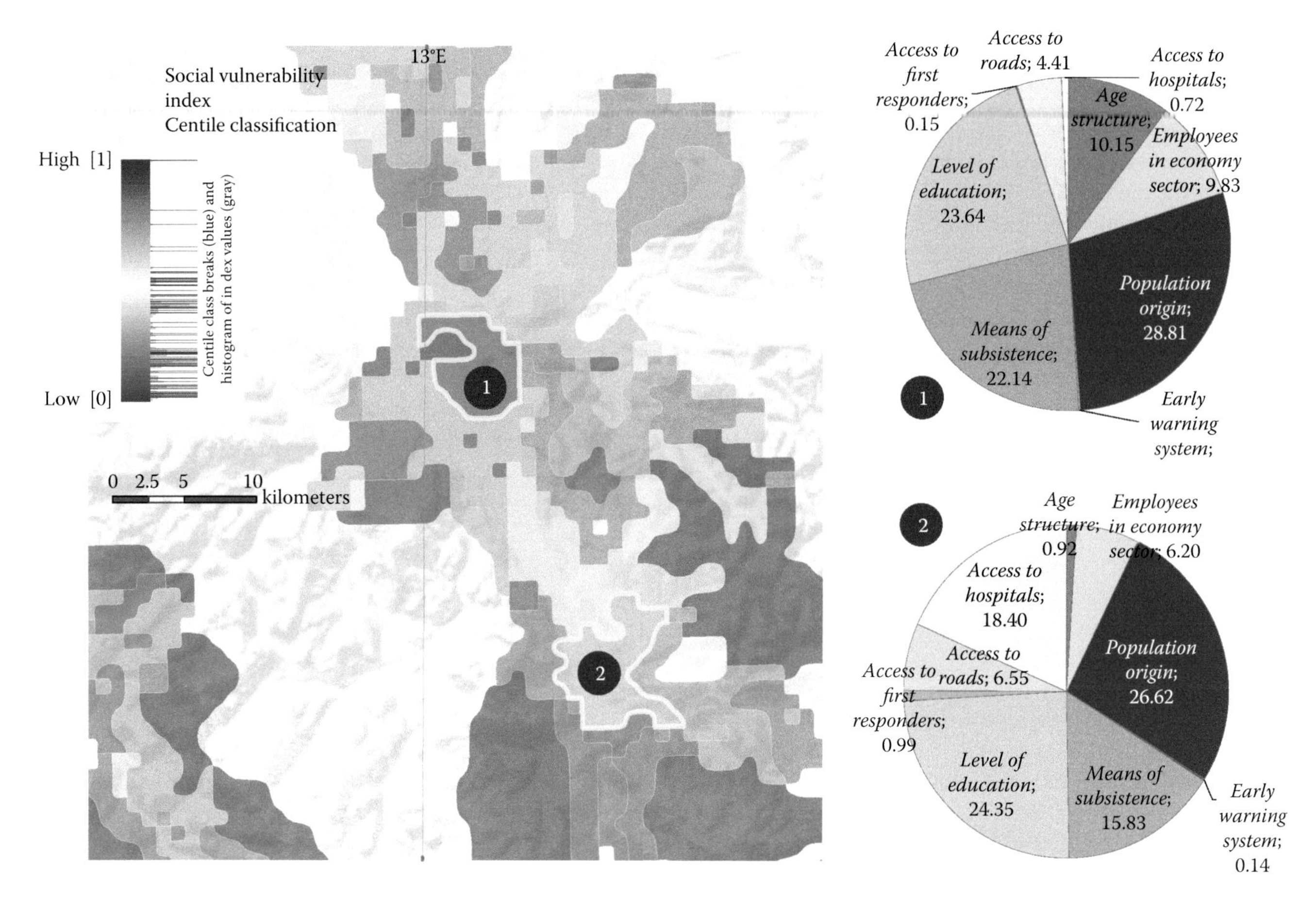

FIGURE 22.7 Geon-scape showing social vulnerability to flood hazards in the Salzach river catchment. Geons score from low to high in range between 0 and 1. See legend. Single geons can be explored in terms of the contributing indicators and their respective contribution to the overall score (see pie charts 1 and 2).

for these indicators was acquired from multiple sources, including remote sensing data (e.g., land use/land cover information, etc.), survey data (e.g., Demographic and Health Survey data), and volunteered geographic information (VGI) as provided by OpenStreetMap. After statistical preprocessing, as described in Section 22.2, one of the 15 indicators was omitted to reduce existing multicollinearities in the data.

22.3.2.3 Results

Figure 22.8 shows the spatial distribution of social vulnerability to malaria for the EAC region. In the map, areas of high vulnerability are displayed in red, while areas of low vulnerability are displayed in blue, indicating high levels of malaria vulnerability in the highland areas where immunity of the population is currently low.

A sensitivity analysis was carried out to assess the impact of (1) the individual indicators on the vulnerability index and (2) the choice of the weighting scheme (expert weights vs. equal weights) on the size and shape of the geons as well as on the final vulnerability index for each geon. The influence of the single indicators on the vulnerability index was evaluated by discarding one indicator at a time, while keeping all other settings, including the geometry of the geons, constant. Figure 22.9 shows the influence of the vulnerability indicators on the vulnerability index. The vulnerability index revealed a high overall robustness with regard to the final choice of vulnerability indicators, with the exception of the immunity indicator that has a marked impact on the composite vulnerability index (Kienberger and Hagenlocher 2014).

Figure 22.10 compares the modeling outputs based on expert weights (Figure 22.10, panel 1) and equal weights (Figure 22.10, panel 2), revealing several interesting findings regarding the geometry of the delineated integrated geons and the vulnerability index. First, the relevance of the immunity indicator, which was ranked one of the most important indicators by the expert, is clearly visible when comparing both maps. The expert-based map clearly shows that vulnerability is high in highland areas where immunity to malaria is generally low. As the impact of the immunity indicator is lowered when assigning equal weights to all indicators, the marked impact of immunity on the vulnerability index, and thus the clear distinction between high levels of vulnerability in the highlands and levels of vulnerability in the lowlands, becomes less distinct when using equal weights. Second, it becomes obvious that the geometry of geons (i.e., their size and shape) is clearly influenced by the weights assigned to the individual indicators. At the same time, it is interesting to see that both approaches clearly demarcate urban centers that differ from

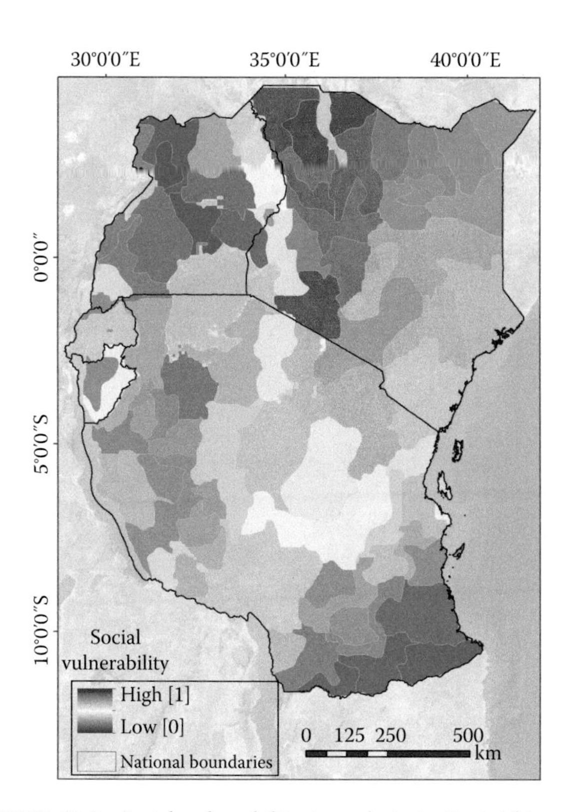

FIGURE 22.8 Social vulnerability to malaria in East Africa. Pie-charts show the contribution of the single vulnerability indicators for each geon. (From Kienberger, S. and Hagenlocher, M., *International Journal of Health Geographics*, 13, 29, 2014.)

their vicinity in terms of both socioeconomic and demographic characteristics, such as Kampala, Uganda, or Nairobi, Kenya.

22.3.3 Landscape Sensitivity

22.3.3.1 Latent Phenomenon Addressed

Major infrastructure plans such as the construction of a new road require an integrative assessment of their likely impacts on the environment. In addition to environmental factors, also social and economic factors should be considered to minimize potentially negative effects. The degree of sensitivity of a particular landscape determines how it is able to accommodate developments. Landscape sensitivity varies both spatially and temporally, and can only be gauged through use of indicators. Pernkopf and Lang (2011) have presented a spatial composite indicator that combines the different aspects of sensitivity in an integrative approach instead of taking them separately, which is a standard practice in environmental impact assessment. The approach was tested in the Marchfeld region, an area of 1000 km^2 in the East of Austria, Europe, where a highway connection was planned between two major cities (Figure 22.11).

22.3.3.2 Indicators and Datasets

Based on a review of environmental reports for different development plans, a set of 21 indicators was selected to model the sensitivity of the landscape. Natural as well as socioeconomic factors were considered in three domains of sensitivity: (1) ability

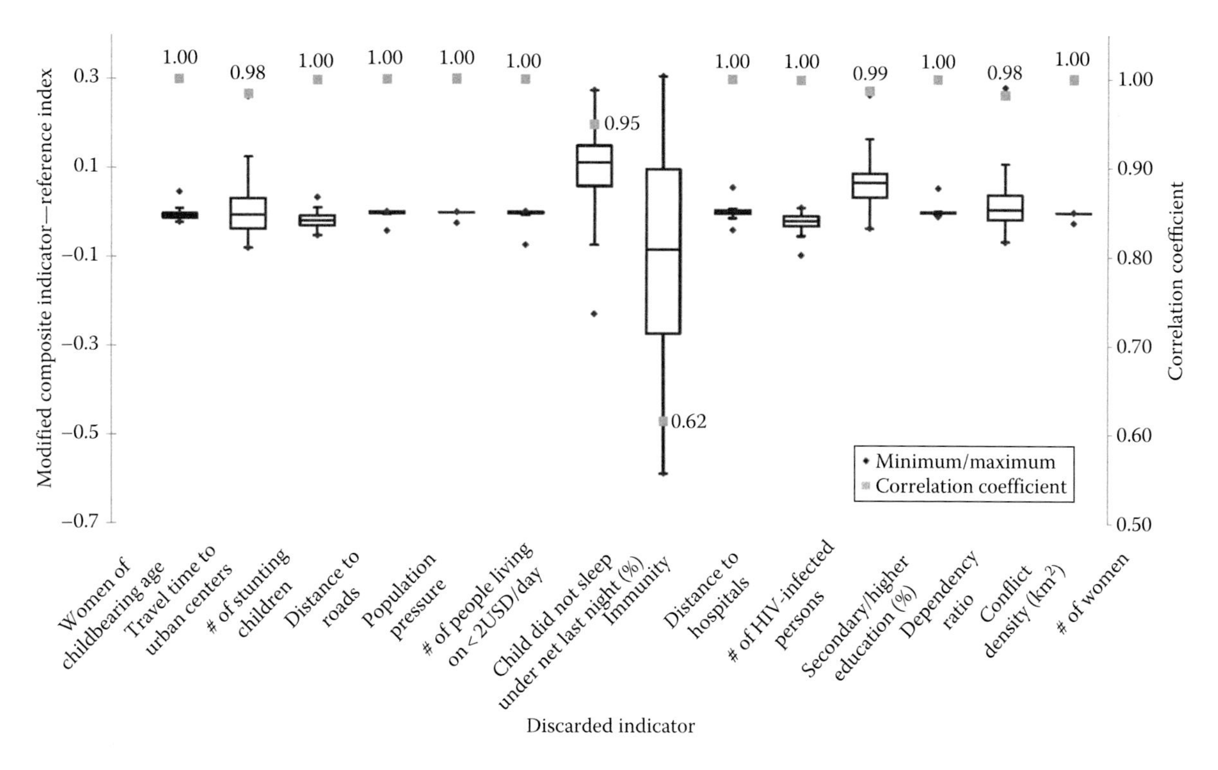

FIGURE 22.9 Influence of the single vulnerability indicators on the vulnerability index. The higher the interquartile range (IQR), the higher the impact of the respective indicator on the final index. (From Kienberger, S. and Hagenlocher, M., *International Journal of Health Geographics*, 13, 29, 2014.)

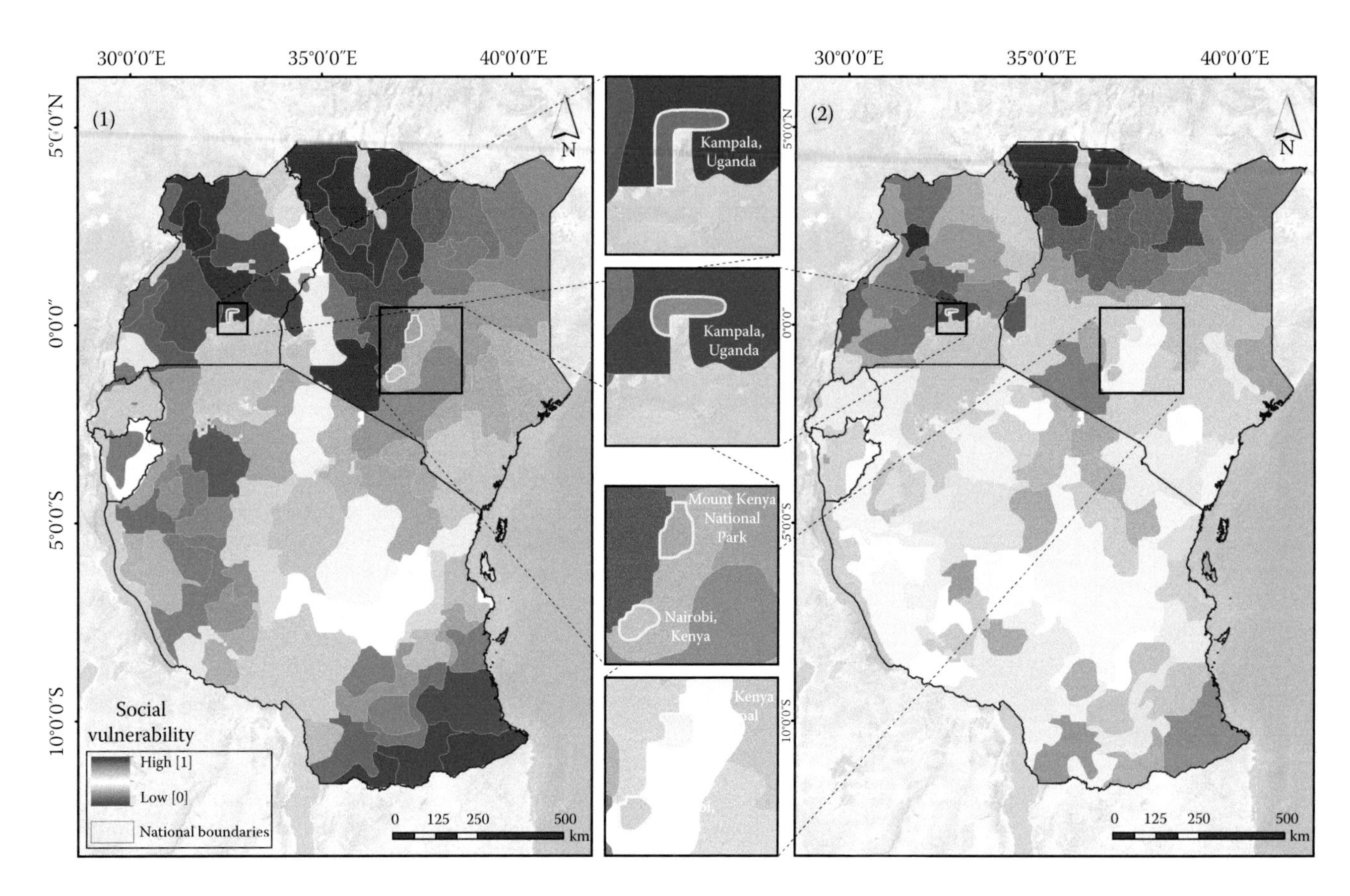

FIGURE 22.10 Evaluating the impact of the choice of the weighting scheme (panel 1: expert weights; panel 2: equal weights) on the modeling outputs.

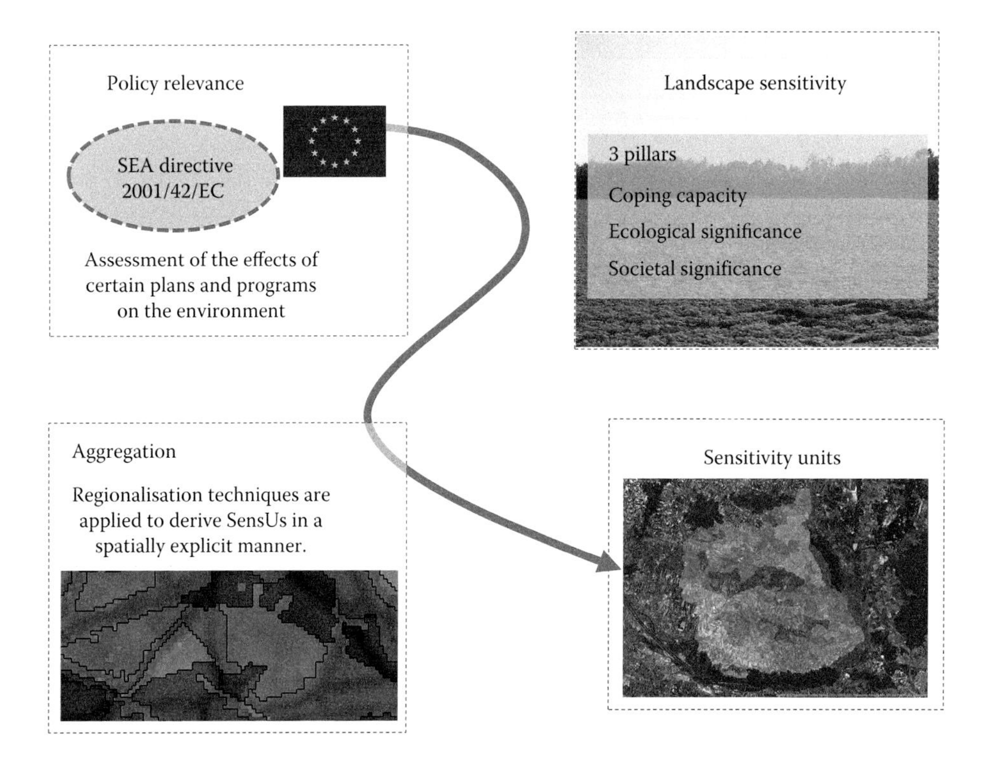

FIGURE 22.11 Using regionalization techniques to model landscape sensitivity in an integrated way.

to resist to external disturbances, (2) ecological significance and (3) societal significance. Each of them consists of 5–10 indicators (e.g., land use, degree of landscape fragmentation, conservation status, and population density), which are weighted by experts according to their relative importance.

22.3.3.3 Results

Figure 22.12 (panel 1) shows the spatial variability of landscape sensitivity to the road infrastructure project in the Marchfeld region. The resulting sensitivity units are discrete spatial units with similar characteristics in terms of their sensitivity to the impact. The results provide aggregated information to decision makers while preserving the detail provided by single indicators. Some areas, even within protected landscapes, are more sensitive than other areas, which is indicated by geon boundaries that are independent from predefined administrative units. It shall, therefore, be possible to indicate particular landscape areas, which would be suitable or less sensitive for the construction of a road and to avoid an undesirable development. The results certainly cannot substitute an in-depth analysis of the impact effects, but support a more integrated and transparent decision-making in environmental impact assessment.

The influence of single indicators on the sensitivity index is shown in Figure 22.12 (panel 2) using the example of agricultural land use. The intensity of land use is considered as one indicator of the societal significance of the landscape. Discarding this indicator has an effect on the resulting geon geometry as well as on the final sensitivity index values. In most areas, modeling outputs are relatively robust to changes in the indicator set, because agricultural land use is only one of 21 equally weighted indicators. Only areas with an intensive land use (wine-growing regions and irrigated areas) show changes in the size and shape of the sensitivity units as well as different index values, which are typically lower when the socioeconomic value of agriculture is not considered in the sensitivity of the landscape to a road infrastructure project. In some areas, with an extensive agricultural land use but other important sensitivity aspects, the index also increases when agricultural land use is not included in the indicator set.

22.3.4 Climate Change Susceptibility (Cumulative CC Impact)

22.3.4.1 Latent Phenomenon Addressed

The recently published IPCC WGII contribution to the Fifth Assessment Report (AR5) highlights that the risk of

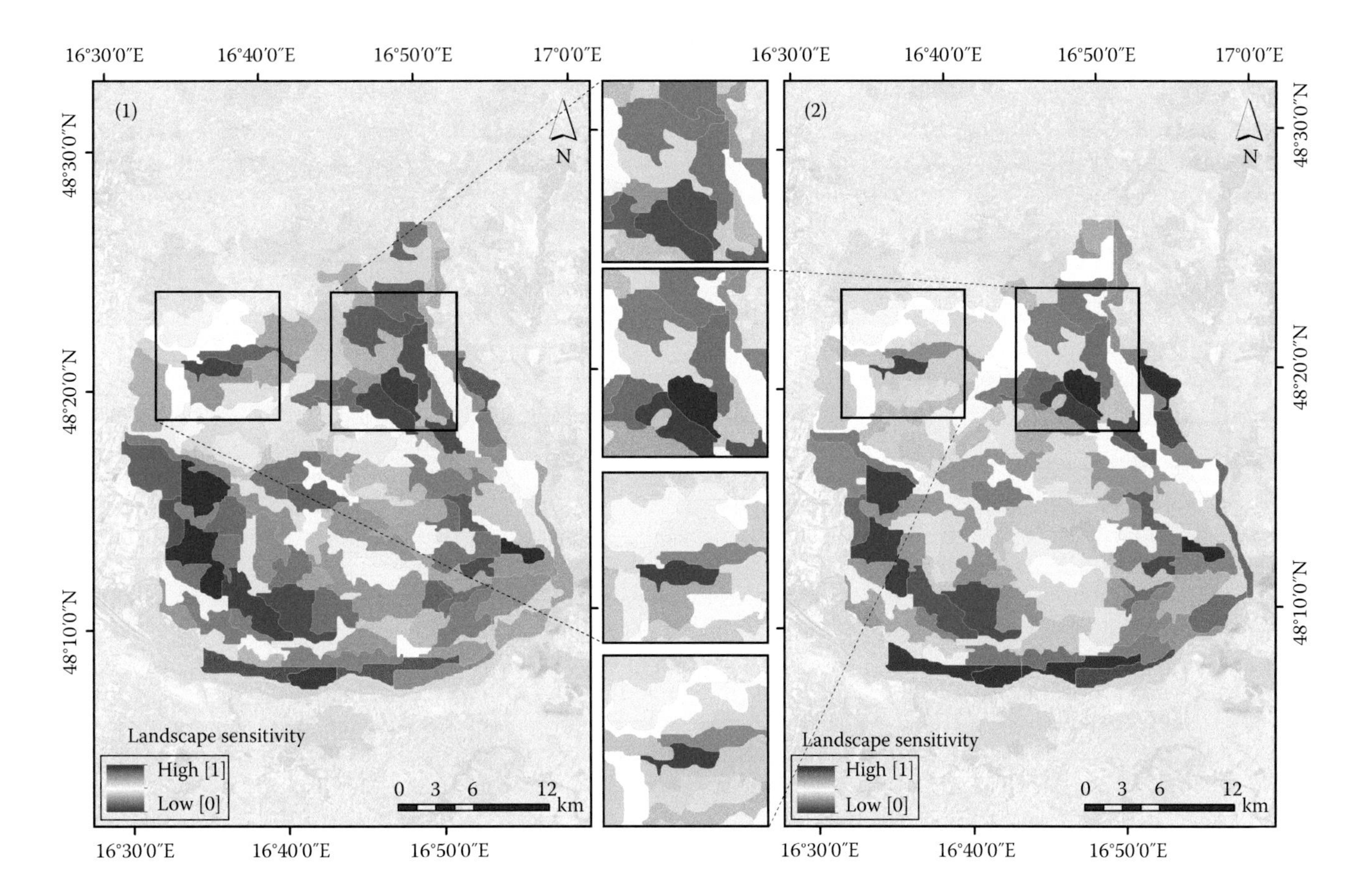

FIGURE 22.12 Evaluating the impact of individual indicators on the modeling outputs (panel 1: use of full indicators set; panel 2: agricultural land use indicator was discarded) on the modeling outputs.

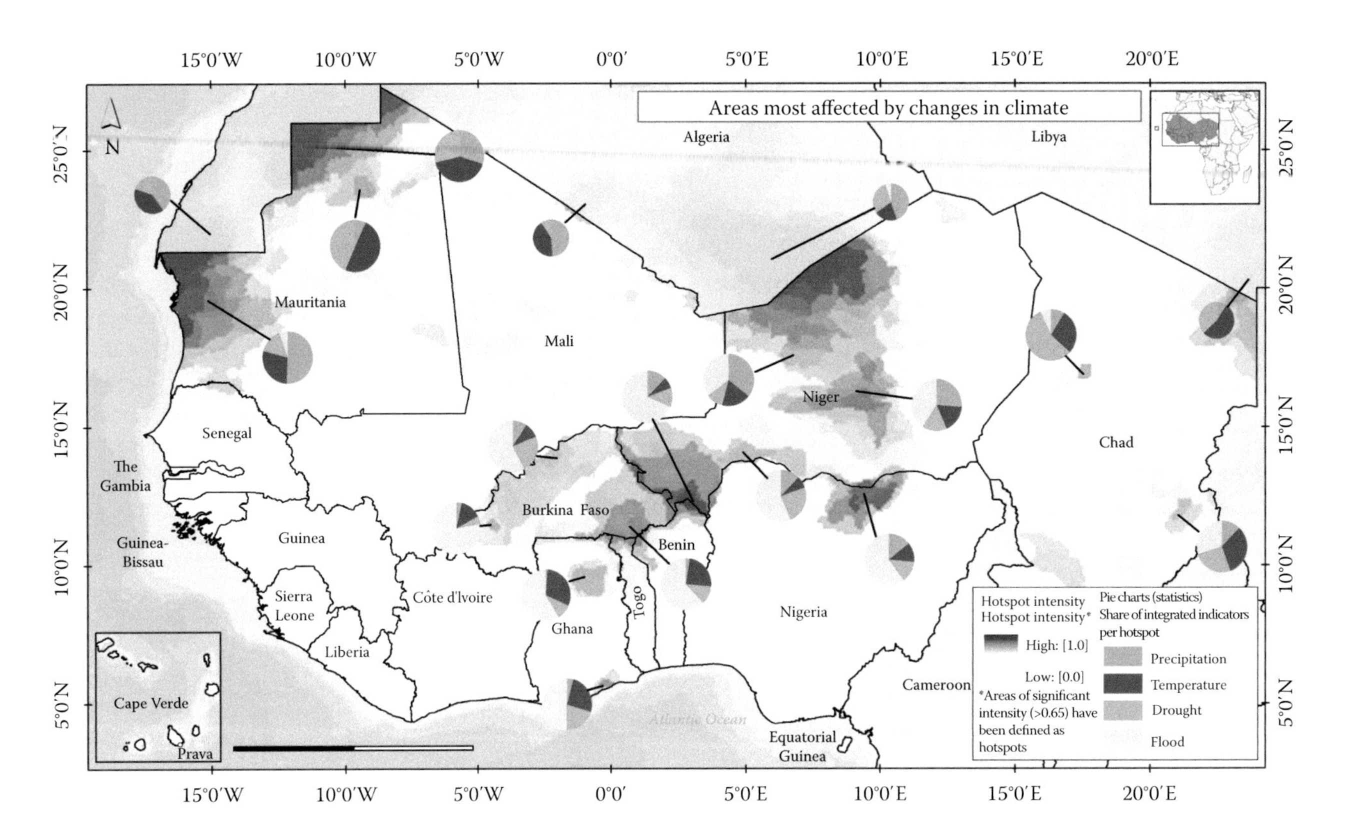

FIGURE 22.13 CCCI in the Sahel and western Africa based on the aggregation of a set of four climate-/hazard-related indicators (temperature, precipitation, drought, and flooding). Hotspots are displayed in red color. (From Hagenlocher et al., 2014b.)

climate-related impacts results from the interaction of climate-related hazards with the vulnerability of exposed human and natural systems (Field et al. 2014). Based on this concept, an integrative, spatial explicit assessment of cumulative climate change impact (CCCI) was carried out using the Sahel and Western Africa as study region. The focus was placed on identifying, mapping, and evaluating *hotspots* of climate change impact to provide conditioned information for targeted climate change adaptation measures (Hagenlocher 2013, Hagenlocher et al. 2014b).

22.3.4.2 Indicators and Datasets

To provide information on the hazards component of the IPCC risk framework, a set of four climate-related datasets was identified in collaboration with domain experts from the United Nations Environment Programme (UNEP), including long-term average seasonal (1) precipitation and (2) temperature trends, as well as frequency of extreme events, such as (3) drought occurrences, and (4) major flood events over the past decades. The analysis was based on time-series of freely available continuous datasets, including remote sensing data. Data on the two essential climate variables, monthly mean temperature and precipitation, was acquired from the Climatic Research Unit (CRU) using their time series (TS 3.0) datasets. These comprise 1224 monthly grids of observed climate for the period from 1901 to 2006, as reported by more than 4000 weather stations around the globe. Data on drought frequency was derived from the NOAA NESDIS-STAR vegetation health index (VHI) dataset, which is based on measurements of the advanced very-high-resolution radiometer (AVHHR) onboard the NOAA satellite, while data on flood frequency was acquired from the Dartmouth Flood Observatory (DFO). They provide an active archive of large flood events that is updated based on remote sensing information (e.g., MODIS, Landsat, etc.) as well as various other sources (news, governmental reports, etc.). In line with the IPCC definition of climate, which was defined as a period of 30 years, the observation period was set to the past 24–36 years (depending on data availability).

22.3.4.3 Results

Following a statistical data preprocessing routine as described in Section 22.2, the four singular climate-related indicators were mapped and visualized. In addition to mapping singular trends in the four indicators, 1233 units showing homogeneous regions of CCCI were delineated using the geon concept. Based on the statistical analysis of the resulting geon-scape, a set of 19 climate change hotspots was identified. These represent areas most affected by changes in climate conditions (precipitation, temperature) and related extreme events (drought and flooding) over the past decades (Hagenlocher et al. 2014b). Next to the location and size of the hotspots, the proportional contribution of each of the four climate-related indicators was analyzed and mapped by means of a pie-chart for each of the hotspots by decomposing each hotspot into its underlying indicators (Figure 22.13).

22.4 Conclusion

In this chapter, we have presented the geon concept as a conceptual approach and methodological toolset to operationalize spatial latent phenomena. We demonstrated that geons are an appropriate means to present geographical information more effectively and thus support efficient policy implementation or intervention planning We showed how geons can be built through integrating dedicated geospatial datasets applying regionalization techniques. We discussed aspects that influence the robustness of geon delineation, also in relation to MAUP. This generic concept reduces complexity by applying a systemic perspective while maintaining the level of integration at the specific level of intervention. By this we treat complex phenomena, such as vulnerability, sensitivity, or climate change impact as a spatial phenomenon emerging from the specific arrangement of units generated by an expert-based partitioning of space.

References

Baatz, M. and A. Schäpe. 2000. *Multiresolution Segmentation: An Optimization Approach for High Quality Multi-Scale Image Segmentation*. Salzburg, Austria, Wichmann Verlag.

Beinat, E. 1997. *Value Functions for Environmental Management*. Dordrecht, the Netherlands, Kluwer Academic Publishers.

Berry, B. J. L. 1967. Grouping and regionalizing: An approach to the problem using multivariate analysis. In *Quantitative Geography*. W. L. Garrison and D. F. Marble (eds.), Northwestern Studies in Geography Evanston, pp. 219–251.

Bertalanffy, L. V. 1969. *General System Theory—Foundations, Developments, Applications*. New York, George Braziller.

Biederman, I. 1987. Recognition-by-components: A theory of human image understanding. *Psychological Review* 94(2): 115–147.

Blaschke, T. 2010. Object based image analysis for remote sensing. *ISPRS International Journal of Photogrammetry and Remote Sensing* 65(1): 2–16.

Blaschke, T. and J. Strobl. 2001. What's wrong with pixels? Some recent developments interfacing remote sensing and GIS. *Zeitschrift für Geoinformationssysteme* 14(6): 12–17.

Bock, M. et al. 2005. Spatial indicators for nature conservation from European to local scale. *Ecological Indicators* 5: 322–338.

Byrne, B. M. 1998. *Structural Equation Modeling with LISREL, PRELIS, and SIMPLIS*. Mahwah, NJ, Lawrence Erlbaum Associates, Inc.

Capra, F. and P. L. Luisi. 2014. *The Systems View of Life: A Unifying Vision*. New York, Cambridge University Press.

Couclelis, H. 2010. Ontologies of geographic information. *International Journal of Geographical Information Science* 24(12): 1785–1809.

Decancq, K. and M. A. Lugo. 2012. Weights in multidimensional indices of wellbeing: An overview. *Econometric Reviews* 32(1): 7–34.

Drăgut, L., O. Csillik, C. Eisank, and D. Tiede. 2014. Automated parameterisation for multi-scale image segmentation on multiple layers. *ISPRS Journal of Photogrammetry* 88: 119–127.

Drăgut, L. and C. Eisank. 2011. Object representations at multiple scales from digital elevation models. *Geomorphology* 129: 183–189.

Dragut, L., D. Tiede, and S. R. Levick. 2010. ESP—A tool to estimate scale parameter for multiresolution image segmentation of remotely sensed data. *International Journal of Geographical Information Science* 24(6): 859–871.

Duque, J. C. and R. R. Suriñach. 2006. Supervised regionalization methods—A survey. Research Institute of Applied Economics Series Title: Working papers, San Diego. 8: 31.

Eurostat. 2006. *Regions in the European Union—Nomenclature of Territorial Units for Statistics NUTS 2010/EU-27*. Luxembourg, Publications Office of the European Union.

Field, C. B. et al. 2014. *Part A: Global and Sectoral Aspects. Contribution of Working Group II to the Fifth Assessment Report of the Intergovernmental Panel on Climate Change*. Cambridge, U.K., Cambridge University Press.

Fiksel, J. 2006. Sustainability and resilience: Toward a systems approach. *Sustainability: Science, Practice, & Policy* 2(2): 14–21.

Goodchild, M. J., M. Yuan, and T. J. Cova. 2007. Towards a general theory of geographic representation in GIS. *International Journal of Geographical Information Science* 21(3): 239–260.

Hagenlocher, M. 2013. Identifying and evaluating hotspots of climate change in the Sahel and Western Africa. In *From Social Vulnerability to Resilience: Measuring Progress toward Disaster Risk Reduction*. S. L. Cutter and C. Corendea (eds.). Bonn, Germany, Publication Series of UNU-EHS, pp. 93–107.

Hagenlocher, M., S. Kienberger, S. Lang, and T. Blaschke. 2014a. Implications of spatial scales and reporting units for the spatial modelling of vulnerability to vector-borne diseases. In *GI_Forum 2014: Geospatial Innovation for Society*. R. Vogler, A. Car, J. Strobl, and G. Griesebner (eds.). Berlin, Germany, Wichmann Verlag, pp. 197–206.

Hagenlocher, M., S. Lang, D. Hölbling, D. Tiede, and S. Kienberger. 2014b. Modeling hotspots of climate change in the Sahel using object-based regionalization of multi-dimensional gridded datasets. *Journal of Selected Topics in Applied Earth Observations and Remote Sensing* 7 (1): 229–234.

Hancock, J. R. 1993. Multivariate regionalization: An approach using interactive statistical visualization. *AutoCarto XI*. Minneapolis, MN, pp. 218–227.

Hay, G. J. and G. Castilla. 2008. Geographic object-based image analysis (GEOBIA): A new name for a new discipline. In *Object-Based Image Analysis: Spatial Concepts for Knowledge-driven Remote Sensing Applications*. T. Blaschke, S. Lang, and G. J. Hay (eds.). Berlin, Germany, Springer.

Hay, G. J., D. J. Marceau, P. Dubé, and A. Buchard. 2001. A multiscale framework for landscape analysis: Object-specific analysis and upscaling. *Landscape Ecology* 16(6): 471–490.

Hernando, A., D. Tiede, F. Albrecht, and S. Lang. 2012. Spatial and thematic assessment of object-based forest stand delineation using an OFA-matrix. *International Journal of Applied Earth Observation and Geoinformation* 19: 214–225.

Jakeman, A. J. and R. A. Letcher. 2003. Integrated assessment and modelling: Features, principles and examples for catchment management. *Environmental Modelling & Software* 18(6): 491–501.

Johnston, R. J. 1968. Choice in classification—The subjectivity of objective methods. *Annals of the Association of American Geographers* 58: 575–589.

Kienberger, S., T. Blaschke, and R. Z. Zaidi. 2013a. A framework for spatio-temporal scales and concepts from different disciplines: The 'vulnerability cube'. *Natural Hazards.*

Kienberger, S., D. Contreras, and P. Zeil. 2014. Spatial and holistic assessment of social, economic, and environmental vulnerability to floods—Lessons from the Salzach River Basin, Austria. In *Assessment of Vulnerability to Natural Hazards: A European Perspective*. J. Birkmann, D. Alexander, and S. Kienberger (eds.). San Diego, CA, Elsevier, pp. 53–74.

Kienberger, S., M. Hagenlocher, E. Delmelle, and I. Casas. 2013b. A WebGIS tool for visualizing and exploring socioeconomic vulnerability to dengue fever in Cali, Colombia. *Geospatial Health* 8(1): 313–316.

Kienberger, S. and M. Hagenlocher. 2014. Spatial-explicit modeling of social vulnerability to malaria in East Africa. *International Journal of Health Geographics*, 13: 29.

Kienberger, S., S. Lang, and P. Zeil. 2009. Spatial vulnerability units—Expert-based spatial modeling of socio-economic vulnerability in the Salzach catchment, Austria. *Natural Hazards and Earth System Sciences* 9: 767–778.

Koestler, A. 1967. *The Ghost in the Machine*. London, U.K., Hutchinson.

Lang, S. 2008. Object-based image analysis for remote sensing applications: Modeling reality—Dealing with complexity. In *Object-Based Image Analysis—Spatial Concepts for Knowledge-Driven Remote Sensing Applications*. T. Blaschke, S. Lang, and G. J. Hay (eds.). Berlin, Germany, Springer, pp. 3–28.

Lang, S., F. Albrecht, S. Kienberger, and D. Tiede. 2010. Object validity for operational tasks in a policy context. *Journal for Spatial Science* 55(1): 9–22.

Lang, S., C. Burnett, and T. Blaschke. 2004. Multi-scale object-based image analysis: A key to the hierarchical organisation of landscapes. *Ekologia* Supplement 23: 1–9.

Lang, S., S. Kienberger, L. Pernkopf, and M. Hagenlocher. 2014a. Object-based multi-indicator representation of complex spatial phenomena. *South-Eastern European Journal of Earth Observation and Geomatics* 3(2s): 625–628.

Lang, S., S. Kienberger, D. Tiede, M. Hagenlocher, and L. Pernkopf. 2014b. Geons—domain-specific regionalization of space. *Cartography and Geographic Information Science* 41(3): 214–226.

Lang, S., P. Zeil, S. Kienberger, and D. Tiede. 2008. Geons—Policy-relevant geo-objects for monitoring high-level indicators. A. Car and J. Strobl (eds.). *GI Forum Salzburg: Geospatial Crossroads @ GI_Forum'08*, Heidelberg, Germany, pp. 180–186.

Laszlo, E. 1972. *The Systems View of the World*. New York, George Braziller.

Long, J., T. Nelson, and M. Wulder. 2010. Regionalization of landscape pattern indices using multivariate cluster analysis. *Environmental Management* 46: 134–142.

Marceau, D. J. 1999. The scale issue in the social and natural sciences. *Canadian Journal of Remote Sensing* 25: 347–356.

Masolo, C. and L. Vieu. 1999. *Atomicity vs. Infinite Divisibility of Space*. London, U.K., Springer, pp. 235–250.

Moldan, B. and A. L. Dahl. 2007. Challenges to sustainability indicators. In *Sustainability Indicators—A Scientific Assessment*. T. Hák, B. Moldan, and A. L. Dahl (eds.). Washington, DC, Island Press, pp. 1–26.

Nardo, M., M. Saisana, A. Saltelli, S. Tarantola, A. Hoffmann, and E. Giovannini. 2005. *Handbook on Constructing Composite Indicators—Methodology and User Guide*. OECD Working Papers. Organisation for Economic Co-Operation and Development, Paris, France.

Ng, R. T. and J. Han. 2002. CLARANS: A method for clustering objects for spatial data mining. *IEEE Transactions on Knowledge and Data Engineering* 14: 1003–1016.

Openshaw, S. 1984. *The Modifiable Areal Unit Problem*. Norwich, U.K.

Parson, E. A. 1994. Searching for integrated assessment: A preliminary investigation of methods and projects in the integrated assessment of global climatic change. *Third Meeting of the CIESIN Harvard Commission on Global Environmental Change Information Policy*, Washington, DC.

Pernkopf, L. and S. Lang. 2011. Spatial meta-indicators: Assessing landscape sensitivity in the context of SEA. *RegioResources Dresden. A Cross-Disciplinary Dialogue on Sustainable Development of Regional Resources*. Dresden, Germany.

Peuquet, D. J. 2002. *Representations of Space and Time*. New York, The Guilford Press.

Risbey, J., M. Kandlikar, and A. Patwardhan. 1996. Assessing integrated assessments. *Climatic Change* 34, 369–395.

Saisana, M., A. Saltelli, and S. Tarantola. 2005. Uncertainty and sensitivity analysis techniques as tools for the quality assessment of composite indicators. *Journal of the Royal Statistical Society* 168(2): 307–323.

Saltelli, A. et al. 2008. *Global Sensitivity Analysis: The Primer*. John Wiley & Sons, West Sussex.

Schöpfer, E., S. Lang, and F. Albrecht. 2008. Object-fate analysis—Spatial relationships for the assessment of object transition and correspondence. In *Object-Based Image Analysis—Spatial Concepts for Knowledge-Driven Remote Sensing Applications*. T. Blaschke, S. Lang, and G. J. Hay (eds.). Berlin, Germany, Springer, pp. 785–801.

Stahl, C., A. Cimorelli, C. Mazzarella and B. Jenkins. 2011. Toward sustainability: A case study demonstrating transdisciplinary learning through the selection and use of indicators in a decision-making process. *Integrated Environmental Assessment and Management* 7(3): 483–498.

Tiede, D., S. Lang, F. Albrecht, and D. Hölbling. 2010. Object-based class modeling for cadastre constrained delineation of geo-objects. *Photogrammetric Engineering & Remote Sensing 76* (2): 193–202.

Tobler, W. 1970. A computer movie simulating urban growth in the Detroit region. *Economic Geography* 46(2): 234–240.

Voudouris, V. 2010. Towards a unifying formalisation of geographic representation: The object-field model with uncertainty and semantics. *International Journal of Geographical Information Science* 24(12): 1811–1828.

Weyant, J. P. et al. 1996. Integrated assessment of climate change: An overview and comparison of approaches and results. In *Climate Change 1995: Economic and Social Dimensions. Contribution of Working Group III to the Second Assessment Report of the Intergovernmental Panel on Climate Change.* P. Bruce, H. Lee, and E. F. Haites (eds.). Cambridge, U.K., Cambridge University Press.

WHO. 2013. World Malaria Report 2013.

Wise, S., R. Haining, and J. Ma. 2001. Providing spatial statistical data analysis functionality for the GIS user: The SAGE project. *International Journal of Geographical Information Science* 3(1): 239–254.

Woodcock, C. E. and A. H. Strahler. 1987. The factor of scale in remote sensing. *Remote Sensing of Environment* 21(3): 311–332.

Wu, J. 1999. Hierarchy and scaling: Extrapolating information along a scaling ladder. *Canadian Journal of Remote Sensing* 25(4): 367–380.

Wu, J. and H. Li. 2006. Concepts of scale and scaling. In *Scaling and Uncertainty Analysis in Ecology: Methods and Applications.* J. Wu, K. B. Jones, H. Li, and O. L. Loucks (eds.). Berlin, Germany, Springer.

VIII

Global Navigation Satellite Systems (GNSS) Remote Sensing

23

Global Navigation Satellite Systems Theory and Practice: Evolution, State of the Art, and Future Pathways

Mohinder S. Grewal
California State University

Acronyms and Definitions

ARNS	Aeronautical Radio Navigation Service
BOC	Binary offset carrier
BPSK	Binary phase-shift keying
BW	Bandwidth
C&P	Coarse align and precision
C/NAV	Commercial navigation
CBOC	Combined BOC
CDMA	Code division multiple access
CNAV	Civil navigation
CS	Commercial service
DGNSS	Differential Global Navigation Satellite System
DOD	Department of Defense
ESA	European Space Agency
EU	European Union
F/NAV	Free navigation
FAA	Federal Aviation Administration
FDMA	Frequency division multiple access
FOC	Final operational capability
GEO	Geostationary Earth orbit
GLONASS	Global Orbiting Navigation Satellite System
GPS	Global positioning system
HOW	Hand over word
HPNS	High-precision navigation signal
I/NAV	Integrity navigation
ICD	Interface control document
IGSO	Inclined geosynchronous satellite orbit
IOV	In-orbit validation
L2C	Civil signal
M code	Military code
MEOs	Medium earth orbits
MMT	Multipath mitigation technology
OS	Open service
P code	Precise code
PPS	Precise positioning service
PRN	Pseudorandom noise
PY code	Precision encrypted code
RHCP	Right hand circular polarization
RNSS	Radio navigation satellite services band
SA	Selective availability
SAR	Search and rescue
SBAS	Space-Based Augmentation System
SNR	Signal-to-noise ratio
SoL	Safety of life
SPNS	Standard precision navigation signal
SPS	Standard positioning service
U.S.	United States

23.1 Introduction

Global navigation satellite system (GNSS) has become almost a household term. GNSS applications track trains, guide planes, and make the world a smaller, more functional planet, as shown in Figure 23.1.

There are currently four Global Navigation Satellite Systems (GNSSs) operating or being developed. The first one is the global positioning system (GPS) developed by the United States (U.S.) Department of Defense (DOD) under its Navstar program. The United States launched its first GPS satellite in 1978, and the system was declared with a final operational capability (FOC) in April 1995. The second GNSS configuration developed is the Global Orbiting Navigation Satellite System (GLONASS), placed in orbit by the Soviet Union and now maintained by the Russian Republic. The first satellite was launched in 1982, and GLONASS was declared an operational system on September 1993. The European Union (EU) and European Space Agency (ESA) are developing the Galileo system, the third GNSS. The first two operational satellites were launched in October 2011, preceded by two Galileo in-orbit validation elements (GIOVE A and B) launched in 2005 and 2008. China's Compass, or BeiDou (Big Dipper), is the fourth GNSS, initially developed in 2000–2003. In December 2012, the BeiDou Navigation Satellite System provided fully operational regional service.

A summary of the signal characteristics of the four GNSSs are given in Figure 23.2. Detailed GPS and GLONASS descriptions are given in Sections 23.2 and 23.3, respectively. Sections 23.4 and 23.5 describe Galileo and BeiDou-2, respectively. Section 23.6 gives conclusions, followed by references.

To improve the accuracy and integrity of the GNSSs, various countries are developing space-based augmentation system (SBAS). SBAS uses geostationary Earth orbit (GEO) satellites to relay corrections and integrity information to users. SBAS is a Differential Global Navigation Satellite System (DGNSS). DGNSSs reduce the errors in GNSS derived positions by using additional data from a reference GNSS receiver at a known location. The most common form of DGNSS involves determining the combined effects of navigation message ephemeris and satellite clock errors (including the effects of propagation) at a reference station and transmitting corrections and integrity in real time to a user's receiver (Grewal 2013) (Table 23.1), (Logsdon 1992).

23.2 GPS

23.2.1 GPS Orbits

The fully operational GPS includes 31 or more active satellites approximately uniformly dispersed around 6 circular orbits with 4 or more satellites each. The orbits are inclined at an angle of 55° relative to the equator and are separated from each other by multiples of 60° right ascension. The orbits are nongeostationary and approximately circular, with radii of 26,560 km and orbital periods of one-half sidereal day (≈11.967 h). Theoretically, three or more GPS satellites will always be visible from most points on the Earth's surface, and four or more GPS satellites can be used to determine an observer's position anywhere on the Earth's surface 24 h/day.

23.2.2 GPS Signals

Each GPS satellite carries a cesium and/or rubidium atomic clock to provide timing information for the signals transmitted by the satellites. Internal clock correction is provided for each satellite clock. Each GPS satellite transmits two spread spectrum, L-band carrier signals—an L1 signal with carrier frequency f_1 = 1575.42 MHz and an L2 signal with carrier frequency f_2 = 1227.6 MHz. These two frequencies are integral multiples, $f_1 = 1540 f_0$ and $f_2 = 1200 f_0$, of a base frequency f_0 = 1.023 MHz.

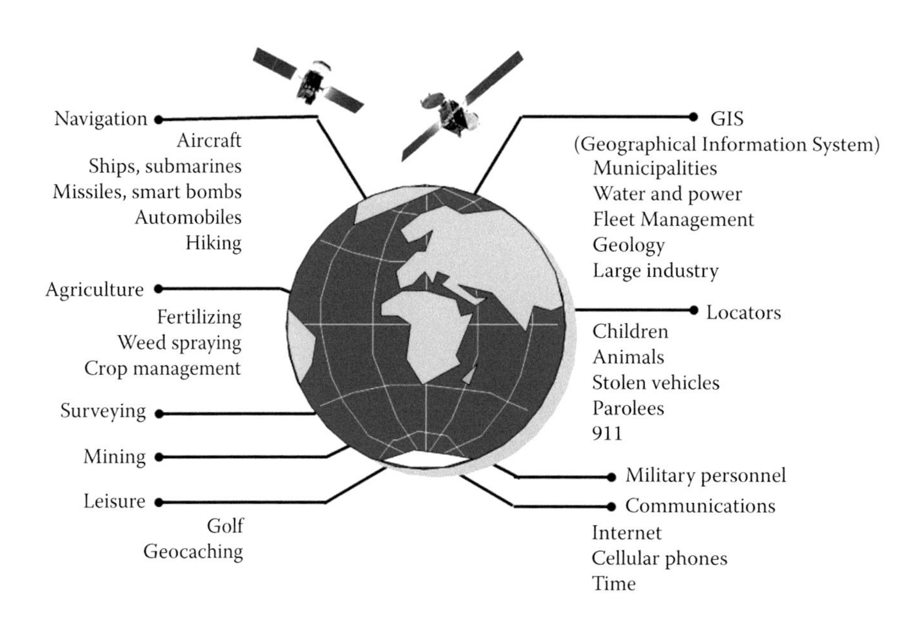

FIGURE 23.1 GNSS applications make the world a smaller more functional planet.

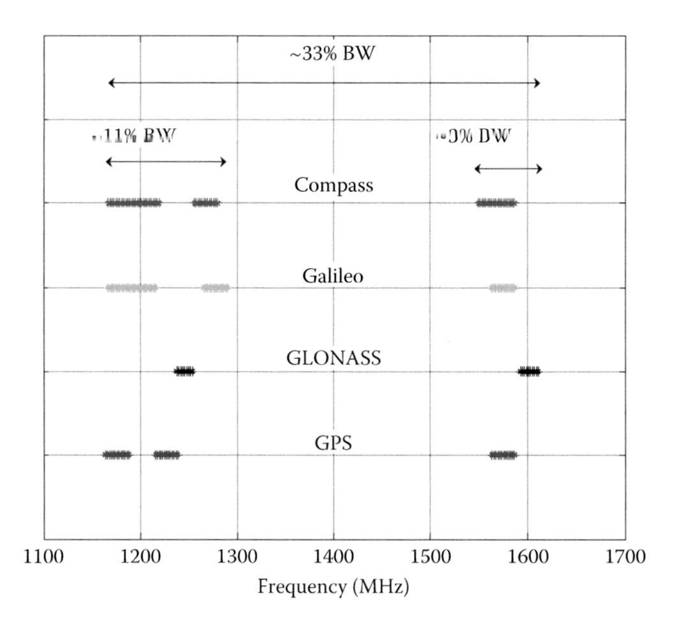

FIGURE 23.2 Comparison of frequency and bandwidth (BW) for various GNSSs.

The L1 signal from each satellite uses *binary phase-shift keying* (BPSK), modulated by two *pseudorandom noise* (PRN) codes in phase quadrature, designated as the coarse acquisition or C/A code and precise or P code. The L2 signal from each satellite is BPSK modulated by only the P code. A brief description of the nature of these PRN codes follows (Rockwell 1991).

23.2.3 Selective Availability

Prior to May 1, 2000, the U.S. Department of Defense deliberately derated the accuracy of GPS for *nonauthorized* (i.e., non-U.S. military) users, using a combination of methods. This selective availability (SA) included pseudorandom time dithering and truncation of the transmitted ephemerides. SA degraded the navigation solution by 100 m horizontally and 156 m vertically. The initial satellite configuration used SA with pseudorandom dithering of the onboard time reference (Janky 1997) only, but this was discontinued on May 1, 2000.

Precise positioning service: Precise positioning service (PPS) is the full-accuracy (within 5 m), single-receiver GPS positioning service provided to the United States and its allied military organizations and other selected agencies. This formal, proprietary service includes access to the unencrypted P code and the removal of any SA effects.

Standard positioning service without SA: Standard positioning service (SPS) provides GPS single-receiver (standalone) positioning service to any user on a continuous, worldwide basis. SPS is intended to provide access only to the C/A code and the L1 carrier.

Standard positioning service with SA: The horizontal-position accuracy, as degraded by SA, advertised as 100 m when in use, the vertical-position accuracy as 156 m, and time accuracy as 334 ns—all at the 95% probability level. SPS also guarantees the user-specified levels of coverage, availability, and reliability.

23.2.4 Modernization of GPS

Since GPS was declared with an FOC in April 1995, applications and the use of GPS have evolved rapidly, especially in the civil sector. See Figure 23.1 for a sampling of applications. As a result, radically improved levels of performance have been reached in positioning, navigation, and time transfer. However, the availability of GPS has also spawned new and demanding applications that reveal certain shortcomings of the present system. Therefore, since the mid-1990s, numerous governmental and civilian committees have investigated these shortcomings and requirements for a more modernized GPS.

The modernization of GPS is a difficult and complex task that requires trade-offs in many areas. Major issues include spectrum needs and availability, military and civil performance, signal integrity and availability, financing and cost containment, and potential competing interests by other GNSSs and developing countries. Major decisions have been made for the incorporation of new civil frequencies, and new civil and military signals that will enhance the overall performance of the GPS.

TABLE 23.1 Comparison of Various GNSSs

	GPS	GLONASS	Galileo	Compass
Number of satellites	31	24	30	35
Orbit radius (km)	26,600	25,510	29,600	19,100
Number of orbit planes when fully developed (global)	6	3	3	Now regional
Orbit plane inclination (°)	55.5	64.8	54	55
Architecture	CDMA	FDMA	CDMA	CDMA
SA	No (SA discontinued in 2000)	No	No	No
Service	Free	Free	Encrypted	Fee
Codes	Coarse align and precision (C&P) same for all PRN	C, P (different codes for each PRN)	C&Psame for all PRN	C&P same for all PRN
Carrier (MHz)	1575.42	1602	1575.42	1561.098
PRN (C/A) code length (chips)	1023	511	4096	2046
Data bit rate (ms)	20	20	4	20

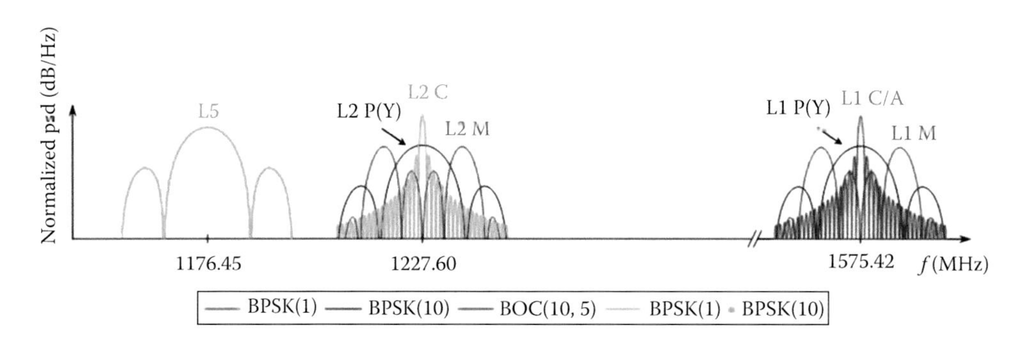

FIGURE 23.3 A modernized GPS signal spectrum shown along with the legacy GPS signals.

GPS IIF and GPS III are being designed under various contracts (Raytheon, Lockheed Martin). These will have a new L2 civil signal and new L5 signal modulated by a new code structure. These frequencies will improve the ambiguity resolution, ionospheric calculation, and C/A code positioning accuracy.

23.2.4.1 Areas to Benefit from Modernization

The areas that could benefit from a modernized GPS are the following:

1. *Robust dual-frequency ionosphere correction capability for civil users*: Since only the encrypted P(Y) code appears on the L2 frequency, civil users have lacked a robust dual-frequency ionosphere. Civil users had to rely on semicodeless tracking of the GPS L2 signal, which is not as robust as access to a full strength unencrypted signal. While civil users could employ a differential technique, this adds complexity to an ionosphere free user solution.
2. *A better civil code*: While the GPS C/A code is a good, simple spreading code, a better civil code would provide better correlation performance. Rather than just turning on the C/A code on the L2 frequency, a more advanced spreading code would provide robust ranging and ionosphere error predictions.
3. *Ability to resolve ambiguities in phase measurements needs improvement*: High-accuracy differential positioning at the centimeter level by civil users requires rapid and reliable resolution of ambiguities in phase measurements. Ambiguity resolution with single-frequency (L1) receivers generally requires a sufficient length of time for the satellite geometry to change significantly. Performance is improved with robust dual-frequency receivers. However, the effective signal-to-noise ratio (SNR) of the legacy P(Y) encrypted signal is dramatically reduced, because the encrypted P code cannot be despread by the civil user.
4. *Dual-frequency navigation signals in the ARNS band*: The Aeronautical Radio Navigation Service (ARNS) band of frequencies are federally protected and can be used for safety-of-life applications. The GPS L1 band is an ARNS band but the GPS L2 band is not. (The L2 band is in the Radio Navigation Satellite Services Band [RNSS] that has a substantial amount of uncontrolled signals in it.) In applications involving public safety, the integrity of the current system can be improved with a robust dual-frequency capability where both GPS signals are within the ARNS bands. This is particularly true in aviation landing systems that demand the presence of an adequate number of high-integrity satellite signals and functional cross-checks during precision approaches.
5. *Improvement is needed in multipath mitigation capability*: Multipath remains a dominant source of GPS positioning error and cannot be removed by differential techniques. Although certain mitigation techniques, such as multipath mitigation technology (MMT), approach theoretical performance limits for in-receiver processing, the required processing adds to receiver costs. In contrast, effective multipath rejection could be made available to all receivers by using new GPS signal designs.
6. *Military requirements in a jamming environment*: The feature of SA was suspended at 8 PM EDT on May 1, 2000. SA was the degradation in the autonomous positioning performance of GPS, which was a concern in many civil applications requiring the full accuracy of which GPS is capable. If the GPS C/A code was ever interfered with, the accuracies that it could be afforded would not be present. Because the P(Y) code has an extremely long period (7 days), it is difficult to acquire unless some knowledge of the code timing is known. P(Y) timing information is supplied by the GPS Hand Over Word (HOW) at the beginning of every subframe. However, to read the HOW, the C/A code must first be acquired to gain access to the navigation message. Unfortunately, the C/A code is relatively susceptible to jamming, which would seriously impair the ability of a military receiver to acquire the P(Y) code. Direct P(Y) acquisition techniques are possible, but these techniques still require information about the satellite position, user position, and clock errors to be successful. Furthermore, an interference on the C/A coded signal, would also effect the P(Y) coded signal in the same frequency band. It would be far better if direct acquisition of a high-performance code were possible, without the need to first acquire the C/A code.

7. Additional power received from the satellite would also help military users operate more effectively. While same advantages can be gained in the user equipment, having an increased power from the satellite could provided added value.
8. *Compatibility and operability with other GNSSs*: With the advances in other GNSSs by other nations, the requirement exists for international cooperation in the development of new GNSS signals to ensure they do not interfere with each other and potentially provide an interoperable combined GNSS service.

23.2.5 Elements of the Modernized GPS

Figure 23.3 illustrates the modernized GPS signal spectrum. The legacy GPS signals (L1 C/A and P(Y), and L2 P(Y)) and the additional modernized signals (L2C, L5, and GPS L1 and L2 M code) are illustrated. The L1C signal is not shown.

A modernized GPS signal spectrum is shown along with the legacy GPS signals.

The bandwidths potentially available for the modernized GPS signals are up to 24 MHz, but the compatibility and power levels relative to the other codes need to be considered. Furthermore, assuming equal received power and filtered bandwidth, the ranging performance (with or without multipath) on a GPS signal, the performance is highly dependent upon the signal's spectral shape (or equivalently, the shape of the autocorrelation function). In this sense, the L1 C/A coded and L2 civil signals are somewhat equivalent in scope, as are the P(Y) and L5 civil signals (albeit with very different characteristics). As we will see, the military M-coded signal and other GNSS codes are different, because they use different subcarrier frequencies and chipping rates. These different subcarriers, in essence, add an aspect known as frequency division multiplexing to the GPS spectrum.

The major elements of these modernized signals are discussed in the following sections.

23.2.5.1 L2 Civil Signal

This new civil signal has a new code structure that has some performance advantages over the legacy C/A code. The L2 civil signal (L2C) signal offers civilian users the following improvements:

1. *Robust dual-frequency ionosphere error correction*: The dispersive delay characteristic of the ionosphere proportional to $1/f^2$ can be estimated much more accurately with this new, full strength signal on the L2 frequency. Thus, civil users can choose to use a semicodeless P(Y) L2 and C/A L1, or a new L2C and C/A L1 technique to estimate the ionosphere.
2. *Carrier phase ambiguity resolution will be significantly improved*: The accessibility of the full strength L1 and L2 signals provides *wide-lane* measurement combinations having ambiguities that are much easier to resolve.
3. *The additional L2C signal will improve robustness in acquisition and tracking*: The new spreading code identified for civil (C) users will provide more robust acquisition and tracking performance.

Originally, modernization efforts considered turning on the C/A code at the L2 carrier frequency (1227.60 MHz) to provide the civilian community with a robust ionosphere correction capability as well as additional flexibility and robustness. However, later in the planning process, it was realized that additional advantages could be obtained by replacing the planned L2 C/A signal with a new L2 civil signal (L2C). The decision was made to use this new signal, and its structure was made public early in 2001. Both the L2C and the new military M code signal (to be described) appear on the L2 carrier orthogonal to the current P(Y).

Like the C/A code, the C code is a PRN code that runs at a 1.023×10^6 cps (chips per second) rate. However, it is generated by 2:1 time-division multiplexing of two independent subcodes, each having half the chipping rate, namely, 511.5×10^3 cps. Each of these subcodes is made available to the receiver by demultiplexing. These two subcodes have different periods before they repeat. The first subcode, the code-moderate (CM), has a moderate length of 10,230 chips, a 20 ms period. The moderate length of this code permits relatively easy acquisition of the signal although the 2:1 multiplexing results in a 3 dB acquisition and data demodulation loss. The second subcode, the code-long (CL), has a length of 707,250 chips, a 1.5 s period, and is data-free. The CM and CL codes are combined to provide the C code at the 1.023 Mcps rate. Navigation data can be modulated on the C code. Provisions call for no data, legacy navigation data at a 50 bps rate, or new civil navigation (CNAV) data at a 25 bps; the CNAV data at a 25 bps rate would be encoded using a rate 1/2 convolutional encoding technique, to produce a 50 sps (symbols per second) data bit stream that could then be modulated onto the L2C signal. With no data, the coherent processing time can be increased substantially, thereby permitting better code and carrier tracking performance, especially at low SNR. The relatively long CL code length also generates smaller correlation sidelobes as compared to the C/A code. Details on the L2 civil signal are given by Fontana et al. (2001).

The existing C/A code at the L1 frequency will be retained for legacy purposes.

23.2.5.2 L5 Signal

Although the use of the L1 and L2C signals can satisfy most civil users, there are concerns that the L2 frequency band may be subject to unacceptable levels of interference for applications involving public safety, such as aviation. The potential for interference arises, because the International Telecommunications Union (ITU) has authorized the L2 band on a coprimary basis with radiolocation services, such as high-power radars. As a result of Federal Aviation Administration (FAA) requests, the Department of Transportation and Department of Defense have called for a new civil GPS frequency, called L5 at 1176.45 MHz in the ARNS of 960–1215 MHz. To gain maximum performance, the L5 spread spectrum codes were selected to have a higher chipping rate and longer period than the C/A codes to allow for better accuracy measurements. Additionally, the L5 signal has two signal components in phase quadrature, one of which will not carry data modulation. The L5 signal will provide the following system improvements:

1. *Ranging accuracy will improve*: Pseudorange errors due to random noise will be reduced below levels obtainable with the C/A codes, due to the larger bandwidth of the proposed codes. As a consequence, both code-based positioning accuracy and phase ambiguity resolution performance will improve.
2. *Errors due to multipath will be reduced*: The larger bandwidth of the new codes will sharpen the peak of the code autocorrelation function, thereby reducing the shift in the peak due to multipath signal components. The eventual multipath mitigation will depend upon the final receiver design and delay of the multipath.
3. *Carrier phase tracking will improve*: Weak-signal phase tracking performance of GPS receivers is severely limited by the necessity of using a Costas (or equivalent-type) PLL to remove carrier phase reversals of the data modulation. Such loops rapidly degrade below a certain threshold (about 25–30 dB Hz), because truly coherent integration of the carrier phase is limited to the 20 ms data bit length. In contrast, the *data-free* quadrature component of the L5 signal will permit coherent integration of the carrier for arbitrarily long periods, which will permit better phase tracking accuracy and lower tracking thresholds.
4. *Weak-signal code acquisition and tracking will be enhanced*: The data-free component of the L5 signal will also permit new levels of positioning capability with very weak signals. Acquisition will be improved, because fully coherent integration times longer than 20 ms will be possible. Code tracking will also improve by virtue of better carrier phase tracking for the purpose of code rate aiding.
5. *The L5 signal will further support rapid and reliable carrier phase ambiguity resolution*: The L5 signal is a full strength, high chipping rate code that will provide high-quality code and carrier phase measurements. These can be used to support various code and carrier combinations for high accuracy carrier phase ambiguity resolution techniques.
6. *The codes will be better isolated from each other*: The longer length of the L5 codes will reduce the size of crosscorrelation between codes from different satellites, thus minimizing the probability of locking onto the wrong code during acquisition, even at the increased power levels of the modernized signals.
7. *Advances Navigation Messaging*: The L5 signal structure has a new civil navigation (CNAV) messaging structure that will allow for increase data integrity.

GPS modernization for the L5 signal calls for a completely new civil signal format (i.e., L5 code) at a carrier frequency of 1176.45 MHz (i.e., L5 carrier). The L5 signal is defined in a quadrature scene where the total signal power is divided equally between in-phase (*I*) and quadrature (*Q*) components. Each component is modulated with a different but synchronized 10,230 chip direct sequence L5 code transmitted at 10.23×10^6 cps (chips per second), the same rate as the P(Y) code, but with a 1 millisecond (ms) period, the same as the C/A code period. The *I* channel is modulated with a 100 sps data stream, which is obtained by applying rate 1/2, constraint length 7, forward error correction (FEC) convolutional coding to a 50 bps navigation data message that contains a 24-bit cyclic redundancy check (CRC). The *Q* channel is unmodulated by navigation data. However, both channels are further modulated by Neuman–Hoffman (NH) synchronization codes, which provide additional spectral spreading of narrowband interference, improve bit and symbol synchronization, and also improve crosscorrelation properties between signals from different GPS satellites. The L5 signal is shown in Figure 23.4 illustrating the modernized GPS (and legacy GPS) signal spectrum.

Compared to the C/A code, the 10-times larger chip count of the *I* and *Q* channel civil L5 codes provides lower autocorrelation sidelobes, and the 10 times higher chipping rate substantially improves ranging accuracy, provides better interference protection, and substantially reduces multipath errors at longer path separations (i.e., long delay multipath). Additionally, these codes were selected to reduce, as much as possible, the crosscorrelation between satellite signals. The absence of data modulation on the *Q* channel permits longer coherent processing intervals in code and carrier tracking loops, with full-cycle carrier tracking in the latter. As a result, the tracking capability and phase ambiguity resolution become more robust.

Further details on the civil L5 signal can be found in references given within parenthesis (Hegarty and Van Dierendonck 1999, Van Dierendonck and Spilker 1999, Spilker and Van Dierendonck 2001, GPS Directorate 2012).

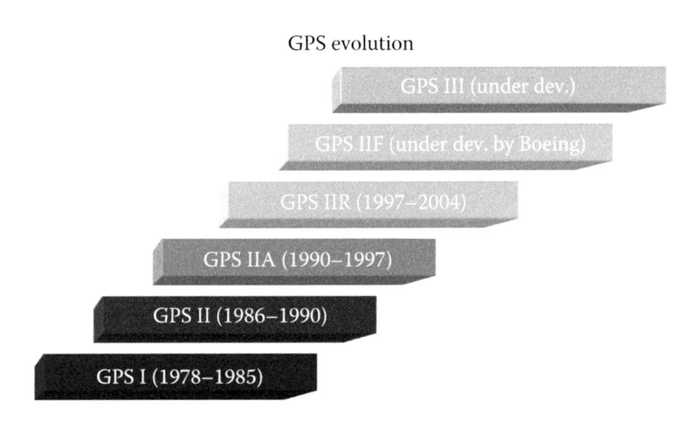

FIGURE 23.4 GPS evolution.

23.2.5.3 M Code

The new military (M) codes will also be transmitted on both the L1 and L2 carrier frequencies. These M codes are based on a new family of split-spectrum GNSS codes for military and new GPS civil signals (Spilker et al. 1998). The M codes will provide the following advantages to military users:

1. *Direct acquisition of the M codes will be possible*: The design of these codes will eliminate the need to first acquire the L1 C/A code with its relatively high vulnerability to jamming.
2. *Better ranging accuracy will result*: As can be seen in Figure 23.3, the M codes have significantly more energy near the edges of the bands, with a relatively small amount of energy near the band center. Since most of the C/A code power is near the band center, potential interference between the codes is mitigated. The effective bandwidth of the M codes is much larger than that of the P(Y) codes, which concentrate most of their power near the L1 or L2 carrier. Because of the modulated subcarrier, the autocorrelation function of the M codes has not just one peak, but several peaks spaced one subcarrier period apart, with the largest at the center. The modulated subcarrier will cause the central peak to be significantly sharpened, significantly reducing pseudorange measurement error.
3. *Error due to multipath will be reduced*: The sharp central peak of the M code autocorrelation function is less susceptible to shifting in the presence of multipath correlation function components.

The M coded signals will be transmitted on the L1 and L2 carriers, with the capability of using different codes on the two frequencies. The M codes are known as binary offset carrier (BOC) encoded signals where the notation of BOC(f_{sx}, f_{cx}) is used where f_{sx} represents the subcarrier multiplier, and f_{cx} represents the code rate multiplier, with respect to a nominal code rate of 1.023 MHz. The M code is a BOC(10, 5) code in which a 5.115 Mcps chipping sequence modulates a 10.23 MHz square wave subcarrier. Each spreading chip subtends exactly two cycles of the subcarrier, with the rising edge of the first subcarrier cycle coincident with initiation of the spreading chip. The spectrum of the BOC(10, 5) code has considerably more relative power near the edges of the signal bandwidth than any of the C/A, P(Y), L2C, and L5 coded signals. As a consequence, the M coded signal has minimal spectral overlap with the other GPS transmitted signals, which permits transmission at higher power levels without mutual interference. The resulting spectrum has two lobes, one on each side of the band center, thereby producing the split-spectrum code. The M code signals are illustrated in Figure 23.10. The M code signal is transmitted in the same quadrature channel as the C/A code (i.e., with the same carrier phase), that is, in phase quadrature with the P(Y) code. The M codes are encrypted and unavailable to unauthorized users. The nominal received power level is −158 dBW at Earth. Additional details on the BOC(10, 5) code can be found in a paper by Barker et al. (2000).

23.2.5.4 L1C Signal

A proposed new L1 civil (L1C) signal is planned for the next generation of GPS SVs. Although the current C/A code is planned to remain on the L1 frequency (1575.42 MHz), the additional L1C signal will add a higher-performance civil signal at the L1 frequency with potential interoperability with other GNSSs. Like the L5 civil signal, the planned L1C signal will have a data-free quadrature component.

The original L1C signal structure considered a pure BOC(1, 1) signal but evolved into a more complex signal that multiplexed two BOC signals. Additional complexities involved the desire to have the L1C signal interoperable with other GNSS signals, as well as, potential intellectual property issues. The L1C signal contains a dataless (i.e., pilot) and data signal component transmitted in quadrature. The L1C signal for GPS that emerged from development is based upon a time-multiplexed BOC (TMBOC) modulation technique that synchronously time-multiplexed the BOC(1, 1) and BOC(6, 1) spreading codes for the pilot component (designated as $L1C_P$), and a BOC(1, 1) modulated signal (designated as $L1C_D$). For both, the BOC codes are generated synchronously at rate of 1.023 MHz and based the Legendre sequence called Weil code. These codes have a period of 10 ms, so 10,230 chips are within one period. Additionally, there is an overlay code that is encoded onto the $L1C_P$ pilot channel. One bit of the overlay code has a duration, and is synchronized to the 10 ms period of the BOC code generators. The overlay code rate is 100 bps and has 1800 bits in an 18 s period.

To generate the TMBOC signal for the $L1C_P$ channel, the BOC(1, 1) and BOC(6, 1) spreading sequences are time-multiplexed. With 33 symbols of a BOC(1, 1) sequence, 4 symbols are replaced with BOC(6, 1) chips. These occur at symbols 0, 4, 6, and 29. Thus, with a 75% of the power planned for distribution in the pilot signal, there will be 1/11th of the power in the BOC(6, 1) component and 10/11th of the power in the BOC(1, 1) component of the carrier.

The L1C signal also has a new navigation message structure, designated as CNAV-2, with three different subframe formats defined. Subframe 1 contains GPS time information (i.e., time of interval [TOI]). Subframe 2 contains ephemeris and clock correction data. Subframe 3 is commutated over various pages and provides less time-sensitive data such almanac, Universal Time, Coordinated (UTC), ionosphere, and can be expended in the future.

The split-spectrum nature of the L1C BOC(1, 1) encoded signal will provide some frequency isolation from the L1 C/A encoded signal. Each spreading chip subtends exactly one cycle of the subcarrier, with the rising edge of the first subcarrier half-cycle coincident with initiation of the spreading chip. The MBOC codes provide a larger RMS bandwidth compared to pure BOC(1, 1).

For many years now, cooperation at the international level has been ongoing to enable the L1C signal to be interoperable with other GNSSs. A combined interoperable signal would allow a user to ubiquitously use navigation signals from different GNSS with known and specified performances attributes.

Additional details on the L1C signal can be found in Spilker et al. 1998, Barker et al. 2000, Spilker and Van Dierendonck 2001, Issler et al. 2004, Betz et al. 2006, and GPS Directorate 2011.

23.2.6 GPS Satellite Blocks

The families of satellites launched prior to recent modernization efforts are referred to as Block I (1978–1985), Block II (1989–1990), and Block IIA (1990–1997); all of these satellites transmit the legacy GPS signals (i.e., L1 C/A and P(Y) and L2 P(Y)). (The United States Naval Observatory has an up-to-date listing of all of the GPS satellites in use today (USNO 2012).)

In 1997 a new family, the Block IIR satellites, began to replace the older Block II/IIA family. The Block IIR satellites have several improvements, including reprogrammable processors enabling problem fixes and upgrades in flight. Eight Block IIR satellites were modernized (designated as Block IIR-M) to include the new military M code signals on both the L1 and L2 frequencies, as well as the new L2C signal on L2. The first modernized Block IIR was launched in September 2005.

To help secure the L5 frequency utilization, one of the Block IIR-M satellites (GPS IIR-20(M)), SV49, was outfitted with a special L5 payload and launched on March 24, 2009. This particular satellite had hardware configuration issues relating to the L5 payload installation and is transmitting a degraded signal. Since that time, the navigation signals have been set unhealthy in the broadcast navigation message.

The Block IIF (i.e., follow-on) family was the next generation of GPS satellites, retaining all the capabilities of the previous blocks, but with many improvements, including an extended design life of 12 years, faster processors with more memory, and the inclusion of the new L5 signal on a third L5 frequency (1176.45 MHz). The first Block IIF satellite was launched in May 2010.

23.2.7 GPS III

The next block of GPS satellites planned is designated as the Block III family, which is still under development. The GPS III block of satellites and associated GPS ground control segment components will represent a major advancement in capabilities for military and civil users. GPS III is planned to include all legacy, and modernized GPS signal components, including the new L1C and L5 signals and add specified signal integrity. The added signal integrity planned for in GPS III may be able to satisfy some of the aviation requirements (FAA 2008). Improvements for military include two high-power spot beams for the L1 and L2 military M code signals, providing 20 dB higher received power over the earlier M code signals. However, in the fully modernized Block III satellites, the M coded signal components are planned to be radiated as physically distinct signals from a separate antenna on the same satellite. This is done in order to enable optional transmission of a spot beam for greater antijam resistance within a selected local region on the Earth. Figure 23.4 shows the evolution of GPS.

23.3 Global Orbiting Navigation Satellite System

A second configuration for global positioning is the GLONASS, placed in orbit by the former Soviet Union, and now maintained by the Russian Republic. GLONASS is the Russian GNSS. The GLONASS has similar operational requirements to GPS with some key differences in its configuration and signal structure. Like GPS, GLONASS is an all-weather, 24 h satellite-based satellite navigation system that has a space, control, and user Segment. The first GLONASS satellite was launched in 1982, and the GLONASS declared an operational system on September 24, 1993.

23.3.1 GLONASS Orbits

The GLONASS satellite constellation is designed to operate with 24 satellites in 3 orbital planes at 19,100 km altitude (whereas GPS uses 6 planes at 20,180 km altitude). GLONASS calls for 8 SVs equally spaced in each plane. The GLONASS orbital period is 11 h 15 min, which is slightly shorter than the 11 h 56 min for a GPS satellite. Because some areas of Russia are located at high latitudes, the orbital inclination of 64.8° is used as opposed to the inclination of 55° used for GPS.

GLONASS has 24 satellites, distributed approximately uniformly in 3 orbital planes (as opposed to 6 for GPS) of 8 satellites each. Each orbital plane has a nominal inclination of 64.8° relative to the equator, and the three orbital planes are separated from each other by multiples of 120° right ascension. GLONASS orbits have smaller radii than GPS orbits, about 25,510 km, and a satellite period of revolution of approximately 8/17 of a sidereal day. A GLONASS satellite and a GPS satellite will complete 17 and 16 revolutions, respectively, around the Earth every 8 days.

Each GLONASS satellite transmits its own ephemeris and system almanac data. Via the GLONASS ground control segment, each GLONASS satellite transmits its position, velocity, and lunar/solar acceleration effects in an ECEF coordinate frame (as opposed to GPS, that encodes SV positions using Keplerian orbital parameters). GLONASS ECEF coordinates are referenced to the PZ-90.02 datum and time reference linked to their National Reference of Coordinated Universal Time UTC(SU) (Soviet Union, now Russia).

23.3.2 GLONASS Signals

The GLONASS system uses frequency-division multiplexing of independent satellite signals. Its two carrier signals corresponding to L1 and L2 have frequencies $f_1 = (1.602 + 9k/16)$ GHz and $f_2 = (1.246 + 7k/16)$ GHz, where $k = 0, 1, 2, \ldots, 23$ is the satellite number. These frequencies lie in two bands at 1.597–1.617 GHz (L1) and 1240–1260 GHz (L2). The L1 code is modulated by a C/A code (chip rate = 0.511 MHz) and by a P code (chip rate = 5.11 MHz). The L2 code is presently modulated only by the P code. The GLONASS satellites also transmit navigational data at a rate of 50 baud. Because the satellite frequencies are distinguishable from each other, the P code and the C/A code are the same for each satellite. The methods for receiving and analyzing GLONASS signals are similar to the methods used for GPS signals (Janky 1997). GLONASS does not use any form of SA.

23.3.2.1 Frequency Division Multiple Access Signals

GLONASS uses multiple frequencies in the L-band and has used frequencies separated by a substantial distance for ionosphere mitigation (i.e., L1 and L2), but these are slightly different than the GPS L1 and L2 frequencies. One significant difference between GLONASS and GPS is that GLONASS has historically used a frequency division multiple access (FDMA) architecture as opposed to the CDMA approach used by GPS.

23.3.2.2 Carrier Components

The GLONASS uses two L-band frequencies, L1 and L2, as defined in the following. The channel numbers for GLONASS signal operation are for

$$f_{K1} = f_{01} + K\Delta f_1$$

$$f_{K2} = f_{02} + K\Delta f_2$$

where

K = channel number ($-7 \leq K \leq +6$)
$f_{01} = 1602M$; $\Delta f_1 = 562.5$ kHz
$f_{02} = 1246M$; $\Delta f_2 = 437.5$ kHz

23.3.2.3 Spreading Codes and Modulation

With the GLONASS signals isolated in frequency, an optimum maximal-length (m-sequence) can be used as the spreading code. GLONASS utilizes two such codes, one standard precision navigation signal (SPNS) at a 0.511 Mcps rate that repeats every 1 ms and a second high-precision navigation signal (HPNS) at a 5.11 Mcps rate, that repeats every 1 s. Similar to GPS, the GLONASS signals utilized BPSK modulation and are transmitted out of a right hand circularly polarized antenna.

23.3.2.4 Navigation Data Format

The format of the GLONASS navigation data is similar to the GPS navigation data format, with different names and content. The GLONASS navigation data format is organized as a Superframe that is made up of frames, where frames are made up of strings. A Superframe has a duration of 150 s and is made up of five frames, so each frame lasts 30 s. Each frame is made up of 15 strings, where a string has a duration of 2 s. GLONASS encodes satellite ephemeris data as immediate data and almanac data as nonimmediate data. There is a time mark in the GLONASS navigation data (last 0.3 s of a string) that is an encode PRN sequence.

23.3.2.5 Satellite Families

While the first series of GLONASS satellites were launched from 1982 to 2003, the GLONASS-M satellites were launched beginning in 2003. These GLONASS-M satellites had improved frequency plans and the accessible signals that are known today. The new generation of GLONASS satellites is designated as GLONASS-K satellites and is considered a major modernization effort by the Russian government. These GLONASS-K satellites plan to transmit the legacy GLONASS FDMA signals as well as a new CDMA format.

Additional details of the GLONASS signal structure and GLONASS-M satellite capabilities can be found in GLONASS (2008).

23.3.3 Next-Generation GLONASS

The satellite for the next generation of GLONASS-K was launched on February 26, 2011, and continues to undergo flight tests. Twenty-four satellites consisting of four GLONASS-M satellites are all transmitting healthy signals. Recently, Russian scientists have proposed a new code-division multiple-access signal format to be broadcast on a new GLONASS L3 signal. Once implemented across the modernizing GLONASS constellation, this will facilitate interoperability with, and eventually interchangeability among, other GNSS signals. The flexible message format permits relatively easy upgrades in the navigation message (GPS World 2013).

23.3.3.1 Code Division Multiple Access Modernization

One of the issues with an FDMA GNSS structure is the interchannel (i.e., interfrequency) biases that can arise within the FDMA GNSS receiver. If not properly addressed in the receiver design, these interchannel biases can be a significant error source in the user solutions. These error sources arise, because the various navigation signals pass through the components within the receiver at slightly different frequencies. The group delay thought these components are noncommon, at the different frequencies, and produced different delays on the various navigation signals, coming from different satellites. This interfrequency bias is substantially reduced (on a comparative basis) with CDMA-based navigation systems, because all of the signals are transmitted at the same frequency. (The relatively small amount of Doppler received from the various CDMA navigation signals is minor when considering the group delay.)

An additional consideration with an FDMA GNSS signal structure is the amount of frequency bandwidth that is required to support the FDMA architecture. CDMA architecture typically has all the signals transmitted at the same carrier frequency, for more efficient utilization of a given bandwidth.

GLONASS has established several separate versions of its GLONASS-K satellites. The first GLONASS-K1 satellite was launched on February 26, 2011, which carried the first GLONASS CDMA signal structure and has been successfully tracked on Earth (Septentrio 2011). The GLONASS-K1 satellite transmits a CDMA signal at a designated L3 frequency of 1202.025 MHz (test signal), as well as the legacy GLONASS FDMA signals at L1 and L2. The CDMA signal from the GLONASS-K1 satellite is considered a test signal. The follow-on generation of satellites is designated as the GLONASS-K2 satellites (Revnivykh 2011). A full constellation of legacy and new CDMA signals are planned for these GLONASS-K2 satellites including plans to transmit its CDMA signal on or near the GPS L1 and L2 frequency bands. The GLONASS KM satellites are in the research phase and plans call for the transmission of the legacy GLONASS FDMA signals, the CDMA signals introduced in the GLONASS-K2, and a new CDMA signal on the GPS L5 frequency.

23.4 Galileo

Galileo is a GNSS being developed by the European Union and the ESA. Like GPS and GLONASS, it is an all-weather, 24 h satellite-based navigation system being designed to provide various services. The program has had three development phases: (1) definition (completed), (2) development of on-orbit validation, and (3) launch of operational satellite, including additional development (ESA 2012). The first Galileo in-orbit validation element (GIOVE) satellite, designated as GIOVE-A, was launched in December 28, 2005, followed by the GIOVE-B on April 27, 2008. The next two Galileo operational satellites were launched on October 21, 2011, to provide additional validation of the Galileo.

23.4.1 Constellation and Levels of Services

The full constellation of Galileo is planned to have 30 satellites in medium earth orbits (MEOs) with an orbital radius of 23,222 km (similar to the GPS orbital radius of 20,180 km). The inclination angle of the orbital plane is 56° (GPS is 55°), with three orbital planes (GPS has 6). This constellation will thus have 10 satellites in each orbital plane.

Various services are planned for Galileo, including an open service (OS), commercial service (CS), public regulated, a safety of life (SoL), and a search and rescue (SAR) service. These services will be supported with different signal structures and encoding formats tailored to support the particular service.

23.4.2 Navigation Data and Signals

Table 23.2 lists some of the key parameters for the Galileo that will be discussed in this section. The table lists the Galileo signals, frequencies, identifiable signal component, its navigation data format, and what service the signal is intended to support. The European Union has published the Galileo Open Service (OS) Interface Control Document that contains significant detail on the Galileo signals in space (EU et al. 2010). All of the Galileo signals transmit two orthogonal signal components, where the in-phase component transmits the navigation data and the quadrature component is dataless (i.e., a pilot). These two components have a power sharing so that the dataless channel can be used to aid the receiver in acquisition and tracking of the signal. All of the individual Galileo GNSS signal components utilize phase-shift keying modulation and right hand circular polarization (RHCP) for the navigation signals.

TABLE 23.2 Key Galileo Signals and Parameters

Signal	Frequency (MHz)	Component	Date	Service
E1	1575.420	E1-B	I/NAV	OS/CS/SOL
		E1-C	Pilot	
E6	1278.750	E6-B	C/NAV	CS
		E6-C	Pilot	
E5	1191.795			
E5a	1176.450	E5a-I	F/NAV	OS
		E5a-Q	Pilot	
E5b	1207.140	E5b-I	I/NAV	OS/CS/SOL
		E5b-Q	Pilot	

To support the various services for Galileo, three different navigation formats are planned for implementation: (1) A free navigation (F/NAV) format to support OS in for the E1 and E5a signals, (2) the integrity navigation (I/NAV) format is planned to support SoL services for the L1 and E5a signals, and (3) a commercial navigation (C/NAV) format is to support CS.

Galileo E1 frequency (same as GPS L1) at 1574.42 MHz will have a split-spectrum type signal around the center frequency. This signal is planned to interoperate with the GPS L1C signal; however, discussions continue on the technical, political, and intellectual property aspects with its implementation. Despite these challenges, the Galileo E1 signal is planned to be a combined BOC (CBOC) signal that is based upon two BOC signals (a BOC(1, 1) and BOC(6, 1) basis component) in phase quadrature. The E1 in-phase component has I/NAV data encoded on it, and the quadrature phase has no data (i.e., pilot).

The Galileo E6 signal is planned to support CS at a center frequency of 1278.750 MHz, with no offset carrier, and a spreading code at a rate of 5.115 MHz (5 × 1.023 MHz). The E6 signal has two signal components in quadrature, where the C/NAV message format is encoded on the in-phase component, and no data is on the quadrature component.

The E5 signal is a unique GNSS signal that has an overall center frequency of 1191.795 MHz, with two areas of maximum power at 1176.450 and 1207.140 MHz. The wideband E5 signal is generated by a modulation technique called AltBOC. The generation of the E5 signal is such that it is actually composed of two Galileo signals that can be received and processed separately, or combined by the user. The first of these two signals within the composite E5 signal is the E5a, centered at 1176.450 MHz (same as the GPS L5). The E5a signal has again two signal components, transmitted in quadrature, where one has data (in-phase) and other does not (quadrature). The F/NAV data is encoded onto the in-phase E5a channel at a 50 sps rate. The second of these two signals within the composite E5 signal is the E5b, centered at 1207.140 MHz. The E5b signal has two signal components, transmitted in quadrature, where one has data (in-phase) and other does not (quadrature). The in-phase channel has the I/NAV message format to support OS, CS, and SOL applications. The data encoded within the I/NAV format will contain important integrity information necessary to support SOL applications.

23.4.3 Updates

The first two Galileo in-orbit validation (IOV) satellites were launched in October 2011 and began broadcasting in December 2011. All Galileo signals were activated on December 17, 2011, simultaneously for the first time across the European GNSS system's three spectral bands known as E_1 (1559–1592 MHz), E_5 (1164–1215 MHz), and E_6 (1260–1300 MHz). The remaining two IOVs were launched in 2012. Current updates are available at ESA website (European Space Agency 2014).

23.5 Compass (BeiDou-2)

The BeiDou Navigation Satellite System is being developed, starting with regional services, and later expanding to global services. Phase I was established in 2000, and Phase II provided for areas in China and its surrounding areas by 2012. Also known as Compass and BeiDou-2, the Chinese BDS started operations in December 2012 and has 14 active satellites in service over the Asia-Pacific region available to general users. When fully deployed by 2020, BDS (Phase III) is expected to comprise a total of 35 satellites offering complete coverage around the globe (GPS World Staff 2013).

23.5.1 Compass Satellites

Compass is the Chinese developed GNSS, where Compass is an English translation for BeiDou. The Chinese Government performed initial development on the Compass in the 2000–2003 timeframe with reference to BeiDou (BD)-1. BeiDou consists of 14 satellites, including 5 GEO satellites and 9 non-geostationary earth orbit (Non-GEO) satellites. The Non-GEO satellites include four medium earth orbit (MEO) and five inclined geosynchronous satellite orbit (IGSO) satellites.

The Compass space segment is called a Dippler Constellation that plans for an eventual 27 MEO, 5 GEOs, and 3 Inclined GEOs (IGSO) satellites. The GEOs are planned for longitude locations at: 58.75°E, 80°E, 110.5°E, 140°E, and 160°E. The system uses its own datum, China Geodetic Coordinate System 2000 (CGCS 2000) and time reference BeiDou Time (BDT) System, relatable to UTC. Of the BD-2 SVs, the first MEO Compass M-1 satellite was launched on April 14, 2007. The GEOs have been launched with an orbital radius of 42,164.17 km. The first IGSO SV was launched on July 31, 2010, with an inclination angle of 55° (same as GPS MEOs). In December 2012, the BeiDou Navigation Satellite System provided fully operational regional service (BeiDou 2014).

The current Interface Control Document (ICD) specifies a B1 GNSS signal at a center frequency of 1561.098 MHz. The architecture is based on a CDMA approach. The spreading codes used are Gold codes, based on 11 stages of delay, running at 2.046 Mcps, so the code will repeat after 1 ms. Navigation data on the MEO and IGSO SVs is at a rate of 50 bps, with a secondary code rate of 1 kbps (China 2011).

23.5.2 Frequency

The nominal carrier frequency is 1561.098 MHz with B1 signal is currently a quadrature phase-shift key (QPSK) in modulation. Compass now has 10 BeiDou-2 satellites operating in its constellation (BeiDou 2011).

23.6 Conclusions

Future developments will provide enhanced safety and services for users. Most future users are anticipated to rely on multiconstellation receivers, that is, that will receive and process GPS, GLONASS, Galileo, and Compass signals, complementing and enhancing each other. A GNSS data integrity channel (GIC) will be provided in the next generation of satellites; for example, in GPS, the GPS IIF and GPS III will provide an integrity channel. In addition, next generations will include airborne monitoring by using redundant measurements to provide signal integrity. New satellites to be built and launched will have more frequencies, higher power, and larger bandwidths. These improvements will reduce the errors such as multipath and atmospheric (ionospheric and tropospheric). The support for various services, including open service (OS), commercial service (CS), public regulated service (PRS), SoL service, and SAR, will be supported with different signal structures and encoding formats, tailored to each service.

Global navigation satellite systems will continue to make the world a smaller, more fully functional planet. Each of the existing systems will individually improve and seek engagement with its neighboring systems to provide a seamless overlay of safe and reliable service around the globe.

References

Barker, B. C., J. W. Betz, J. E. Clark, J. T. Correia, J. T. Gillis, S. Lazar, K. A. Rehborn, and J. R. Straton. Overview of the GPS M Code Signal. January 2000. *Proceedings of the 2000 National Technical Meeting of The Institute of Navigation*, Anaheim, CA, pp. 542–549.

BeiDou Satellite Navigation System News Center. 2014. Article on first anniversary of BeiDou navigation satellite system providing full operational regional service. Available at http://en.beidou.gov.cn/ (accessed February 2014).

Betz, J. W., M. A. Blanco, C. R. Cahn, P. A. Dafesh, C. J. Hegarty, K. W. Hudnut, V. Kasemsri et al. September 2006. Description of the L1C signal. *Proceedings of the 19th International Technical Meeting of the Satellite Division of The Institute of Navigation (ION GPS 2006)*, Fort Worth, TX, pp. 2080–2091.

China Satellite Navigation Office. December 2011. BeiDou Navigation Satellite System, Signal In Space, Interface Control Document (Test Version), December 2011, available at http://www.beidou.gov.cn/attach/2011/12/27/2011 12273f3be6124f7d4c7bac428a36cc1d1363.pdf, visited July 15, 2012.

European Space Agency. 2014. About the European GNSS Evolution Programme. http://www.esa.int/Our_Activities/Navigation/GNSS_Evolution/About_the_European_GNSS_Evolution_Programme (accessed February 10, 2014).

European Union, European Commission, Satellite Navigation, Galileo Open Service. September 2010. Signal-in-space interface control document, OS SIS ICD, Issue 1.1. Available at http://ec.europa.eu/enterprise/policies/satnav/galileo/files/galileo-os-sis-icd-issue1-revision1_en.pdf (accessed February 10, 2014).

ESA, Galileo, 2012, available at http://www.esa/NA/galileo.html, visited July 15, 2012.

FAA, GNSS Evolutionary Architecture Study, Phase I–Panel Report, February 2008, available at http://www.faa.gov/about/office_org/headquarters_office/ato/service_units/techops/navservices/gnss/library/documents/media/GEAS_PhaseI_report_FINAL_15Feb08.pdf, visited July 14, 2012.

Fontana, R. D., W. Cheung, P. M. Novak, and T. A. Stansell. 2001. The new L2 civil signal. *Proceedings of the 14th International Technical Meeting of the Satellite Division of The Institute of Navigation* (*ION GPS 2001*), September, Salt Lake City, UT, pp. 617–631.

Global Navigation Satellite system (GLONASS). 2008. *Interface Control Document L1, L2, Version 5.1*. Russian Institute of Space Device Engineering, Moscow, Russia. Available at www.glonass-ianc.rsa.ru/en/ (accessed February 10, 2014).

GPS Directorate, Systems Engineering & Integration Interface Specification. 2011. Navstar GPS space segment/user segment L1C interface, IS-GPS-800B.pdf. Available at http://www.navcen.uscg.gov/pdf/gps/IS-GPS-800B.pdf (accessed February 10, 2014).

GPS Directorate, Systems Engineering & Integration Interface Specification, IS-GPS-705. 2012. Navstar GPS space segment/user segment L5 interface, IS-GPS-705B. Available at http://www.navcen.uscg.gov/pdf/gps/IS-GPS-705B.pdf (accessed February 10, 2014).

GPS World. 2013. CSR location platforms go live with China's Beidou 2 tracking. Available at http://gpsworld.com/csr-location-platforms-go-live-with-chinas-beidou-2-tracking/ (accessed February 10, 2014).

GPS World Staff. November 2013. New structure for GLONASS nav message. Available at http://gpsworld.com/new-structure-for-glonass-nav-message/ (accessed February 10, 2014). Grewal, M. S. 2012. Space-based augmentation for global navigation satellite systems. *IEEE Transactions on Ultrasonics, Ferroelectrics, and Frequency Control*, 59(3), 497–504.

Grewal, M. S., A. P. Andrews, and C. G. Bartone. 2013. *Global Navigation Satellite Systems, Inertial Navigation, and Integration*. John Wiley & Sons, New York.

Hegarty, C. and A. J. Van Dierendonck. 1999. Civil GPS/WAAS signal design and interference environment at 1176.45 MHz: Results of RTCA SC159 WG1 activities. *Proceedings of the 12th International Technical Meeting of the Satellite Division of The Institute of Navigation* (*ION GPS 1999*), September, Nashville, TN, pp. 1727–1736.

Issler, J. L., L. Ries, J. M. Bourgeade, L. Lestarquit, and C. Macabiau. 2004. Probabilistic Approach of Frequency Diversity as Interference Mitigation Means. *Proceedings of the 17th International Technical Meeting of the Satellite Division of The Institute of Navigation* (*ION GPS 2004*), September, Long Beach, CA, pp. 2136–2145.

Janky, J. M. May 3, 1997. Clandestine location reporting by a missing vehicle. U.S. Patent 5629693.

Logsdon, T. 1992. *The NAVSTAR Global Positioning System*. Van Nostrand Reinhold, New York, pp. 1–90.

Revnivykh, S. 2011. GLONASS Status and Modernization. *Proceedings of the 24th International Technical Meeting of the Satellite Division of The Institute of Navigation* (*ION GPS 2011*), September, Portland, OR, pp. 839–854.

Rockwell International Corporation. July 3, 1991. GPS interface control Document ICD-GPS-200. Satellite Systems Division, Revision B.

Septentrio. April 12, 2011. Septentrio's AsteRx3 receiver tracks first GLONASS CDMA signal on L3. Available at http://www.insidegnss.com/node/2563 (accessed February 10, 2014).

Spilker, J. J, E. H. Martin, and B. W. Parkinson. 1998. A family of split spectrum GPS civil signals. *Proceedings of the 11th International Technical Meeting of the Satellite Division of The Institute of Navigation* (*ION GPS 1998*), September, Nashville, TN, pp. 1905–1914.

Spilker, J. J. and A. J. Van Dierendonck. 2001. Proposed new L5 civil GPS codes. *Navigation, Journal of the Institute of Navigation*, 48(3), 135–144.

United States Naval Observatory (USNO), GPS Operational Satellites (Block II/IIA/IIR/IIR-M/II-F), 2012, ftp://tycho.usno.navy.mil/pub/gps/gpsb2.txt, visited July 14, 2012.

Van Dierendonck, A. J. and J. J. Spilker, Jr. 1999. Proposed civil GPS signal at 1176.45 MHz: In-phase/quadrature codes at 10.23 MHz chip rate. *Proceedings of the 55th Annual Meeting of The Institute of Navigation*, June, Cambridge, MA, pp. 761–770.

24

Global Navigation Satellite System Reflectometry for Ocean and Land Applications

Kegen Yu
Wuhan University

Chris Rizos
The University of New South Wales

Andrew Dempster
The University of New South Wales

Acronyms and Definitions

BRCS	Bistatic radar cross section
CHAMP	Challenging minisatellite payload
COSMIC	Constellation observing system for meteorology, ionosphere, and climate
CYGNSS	Cyclone Global Navigation Satellite System
DMC	Disaster monitoring constellation
ESA	European Space Agency
GEROS-ISS	GNSS reflectometry, radio occultation and scatterometry onboard the International Space Station
GGOS	Global geodetic observing system
GNSS	Global Navigation Satellite System
GNSS-R	GNSS reflectometry
GNSS-RO	GNSS radio occultation
GPS	Global positioning system
GRACE-A	Gravity recovery and climate experiment
IAG	International Association of Geodesy
IERS	International Earth Rotation and Reference Systems Service
IGS	International GNSS Service
ILRS	International Laser Ranging Service
IMU	Inertial measurement unit
ITRF	International Terrestrial Reference Frame
IVS	International VLBI Service for Geodesy and Astrometry
LEO	Low earth orbit
LHCP	Left-hand circularly polarized
LNA	Low-noise amplifier
NASA	National Aeronautics and Space Administration
PARIS	*Passive* Reflectometry and Interferometry System
PDF	Probability density function
PNT	Positioning, navigation, and timing
PRN	Pseudo random noise
RHCP	Right-hand circularly polarized
SAR	Synthetic aperture radar
SSP	Specular scattering point
SSH	Sea surface height
STD	Standard deviation
SWH	Significant wave height
SWP	Significant wave period
TPL	Total path length
UNSW	University of New South Wales

24.1 Introduction

Investigating the use of reflected Global Navigation Satellite System (GNSS) signals for remotely sensing the earth's surface was initiated about two decades ago (Martín-Neira 1993). Remote sensing based on processing and analyzing reflected GNSS signals is commonly termed GNSS reflectometry (GNSS-R). When using the GNSS-R technique to build a remote sensing system, only the receiver needs to be designed and manufactured. The receiver platform (static or mobile; land-based, aircraft, or satellite) needs to be selected based on the specific application. In the case of an aircraft or satellite platform, the direct signal is received via a zenith-looking right-hand circularly polarized (RHCP) antenna, while the reflected signal is received through a nadir-looking left-hand circularly polarized (LHCP) antenna. The reason for such an antenna selection is that the GNSS signals are designed as RHCP; however, when reflected over a ground surface, they are changed to be LHCP. In the case of a land-based platform, either two antennas are used to receive the direct and reflected signals separately or a single antenna is used to capture both the two signals.

The GNSS signals are always available, globally, and the signal structures are typically well known, except for those dedicated to military use. The L-band GNSS signals are sensitive to ground surface parameters so that they can be utilized for remote sensing purposes. Recently, there has been an increase in such investigations by academia and research institutions, partly because this innovative use of the GNSS signals has many potential applications. In particular, space agencies such as NASA and ESA have already funded, or are going to fund, a number of projects/missions that focus on the applications of the GNSS-R. The Cyclone Global Navigation Satellite System (CYGNSS) project is just one example (http://aoss-research.engin.umich.edu/missions/cygnss/), which aims to develop a system using a constellation of eight microsatellites to improve hurricane forecasting especially with regard to the storm intensity. Another example is the ESA's passive reflectometry and interferometry system (PARIS) project (http://www.esa.int/Our_Activities/Technology/PARIS). PARIS can be used as a passive radar altimeter. Different from current radar altimeters, PARIS would measure multiple samples from different tracks and rapidly form images of mid-sized (mesoscale) phenomena such as ocean currents or tsunamis. Geophysical parameters that can be measured using a GNSS-based reflectometry system include those (but not limited to) listed in Table 24.1 (Garrison et al. 2002, Gleason et al. 2005, Gleason 2006, Font et al. 2010, Yu et al. 2012a,b).

This chapter focuses on the GNSS-R for ocean applications including sea surface wind speed estimation and sea surface altimetry and for land applications including soil moisture retrieval and forest change detection. In addition, past and planned satellite missions associated with the GNSS-R as well as the GNSS radio occultation (GNSS-RO) are summarized. Finally, some challenging issues related to the GNSS remote sensing products and services are addressed.

TABLE 24.1 Examples of Ocean and Land Applications of GNSS-R

Ocean Applications	Land Applications
Ocean wind speed and direction estimation	Soil moisture retrieval
Sea surface altimetry	Biomass density estimation
Tropical cyclone intensity estimation	Forest change detection
Sea surface salinity estimation	Land surface classification
Sea wave height characteristic estimation	Snow depth estimation
Oil slick detection	Vegetation height estimation

24.2 Satellite Missions Related to GNSS Remote Sensing

Useful reviews of radio occultation missions can be found in Anthes (2011) and Jin et al. (2011). The concept was first proved by the GPS/MET (GPS/Meteorology) satellite mission in 1995–1997. That satellite took few measurements but was followed by the more productive CHAllenging Minisatellite Payload (CHAMP) and Satellite de Aplicaciones Cientificas-C (SAC-C) satellites. CHAMP provided 8 years of radio occultation data consisting of around 440,000 measurements from February 2001 to October 2008 (Heise et al. 2014). Those missions led to the launch in 2006 of a six-satellite constellation FORMOSAT-3 (Formosa Satellite Mission #3)/COSMIC (Constellation Observing System for Meteorology, Ionosphere, and Climate), which provided 1500–2000 soundings per day, resulting in GPS-RO becoming an operational data source for weather prediction and ionospheric monitoring. Other missions that provided significant quantities of RO data are GRACE-A (Gravity Recovery and Climate Experiment), METOP-A (METeorological Operations), C/NOFS (Communications/Navigation Outage Forecasting System), TerraSAR-X, and TanDEM-X. Table 24.2 (taken from the COSMIC website: http://www.cosmic.ucar.edu/) shows the contributions made by these and other missions to both atmospheric and ionospheric sounding. With the COSMIC constellation degrading, as it has reached its design life, a new FORMOSAT-7/COSMIC-2 constellation is being constructed that has multifrequency, multi-GNSS receivers on board 12 satellites to be launched—6 into low inclination orbits in 2015 and 6 into high inclination orbits in 2018 (http://www.cosmic.ucar.edu/cosmic2/).

Far fewer satellites have been able to perform the GNSS-R. The first was UK-DMC from 2003 to 2011 (Unwin et al. 2003). Others that have been proposed are Techdemosat-1 due for launch Q1 2014 (http://www.sstl.co.uk/Missions/TechDemoSat-1) with a custom reflectometry payload, PARIS, a

TABLE 24.2 Contributions to Atmospheric and Ionospheric Sounding by Mission, from the COSMIC Website

Mission	Total Atm Occs	Total Ion Occs
CHAMP	399,968	303,291
CNOFS	120,588	0
COSMIC	4,039,311	3,707,966
GPSMET	5,002	0
GPSMETSA	4,666	0
GRACE	273,013	132,817
METOPA	993,084	0
SACC	353,808	0
TSX	276,549	0
Total	6,465,989	4,144,074

Source: Anthes, R.A., *Atmos. Measure. Techn.*, 4, 1077, 2011, last updated: December 7, 23:25:02 MST 2013.

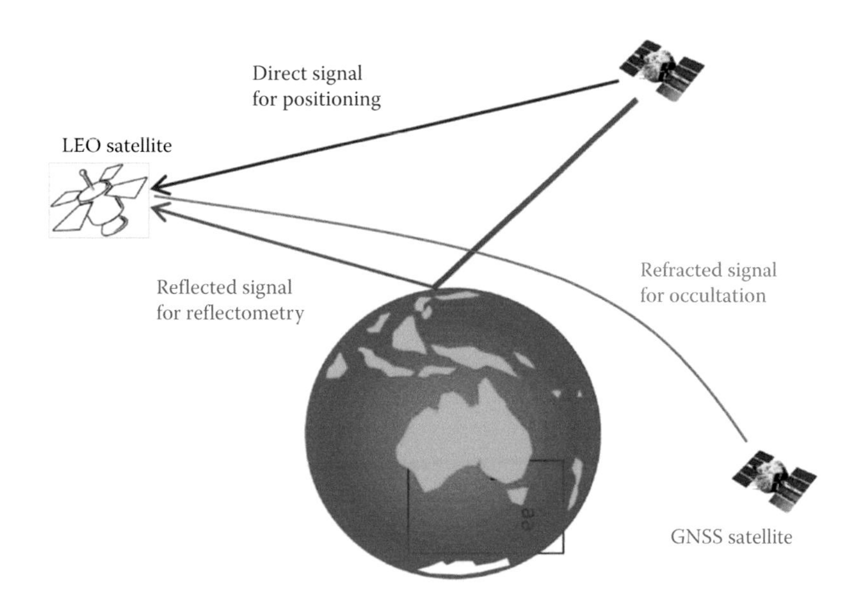

FIGURE 24.1 Illustration of the geometry of the direct, reflected, and refracted GNSS signals and their use for different applications.

dedicated mission for GNSS-R (Martin-Neira et al. 2009, 2011), CYGNSS, a constellation of eight microsatellites, GEROS-ISS (GNSS REflectometry, Radio Occultation and Scatterometry onboard International Space Station) (http://www.ice.csic.es/en/view_project.php?PID=155), and Microsat-1 (Camps et al. 2011). Several of the authors are also from the University of New South Wales (UNSW), which is preparing a CubeSat to carry both GNSS occultation and reflectometry experiments as part of the QB50 mission (http://www.acser.unsw.edu.au/QB50/index.html). Figure 24.1 illustrates that a launched LEO satellite can receive the direct, reflected, and refracted GNSS signals to enable positioning, reflectometry, and occultation applications.

24.3 Ocean Observation

This section discusses two specific GNSS-R-based ocean applications: sea surface wind speed estimation and sea surface height (SSH) estimation.

24.3.1 Sea Surface Wind Speed Estimation

Wind speed retrieval using the GNSS-R is mainly based on the theoretical model of the received signal power reflected over sea surface. Since the wind speed is estimated by observing the wind-driven sea surface wave slope characteristics, the estimation is reliable only when the wind and wave interaction has reached the steady state after the wind has continuously blown the surface in one direction for at least say half an hour. The wind speed estimation accuracy can be better than 2 m/s based on processing the data observed from several GNSS-R airborne experimental campaigns. Some details of the technique are described as follows.

24.3.1.1 Sea Wave Spectrum

Sea surface undulation is a complex process, and sea wave heights change randomly in time and space. Sea surface roughness can be described by a number of parameters including significant wave height (SWH) and significant wave period (SWP). The SWH is defined as the average height of the one-third highest waves, and the SWP is defined as average period of the waves used to calculate the SWH. Alternatively, wave height spectrum and wave direction spectrum can be used to describe the surface roughness. Among the wave height spectral models, the Pierson–Moskowitz model, the JONSWAP model, and the Elfouhaily model are widely studied (Pierson and Moscowitz 1964, Elfouhaily et al. 1997).

The Elfouhaily model that describes the wind-driven wave height spectrum is defined as

$$W(\kappa,\phi)=\frac{\kappa^{-4}}{2\pi}(B_\ell(\kappa)+B_h(\kappa))(1+\Delta(\kappa)\cos(2(\phi-\phi_0)) \quad (24.1)$$

where

κ is the wave number
ϕ is the azimuth angle
ϕ_0 is the wind direction

The long-wave curvature spectrum $B_\ell(\kappa)$ is defined as

$$B_\ell(\kappa)=\frac{3\times10^{-3}U_{10}}{\tilde{c}\sqrt{\Omega}}\exp\left(-\frac{\Omega}{\sqrt{10}}\left(\sqrt{\frac{\kappa}{\kappa_p}}\right)-1\right)\exp\left(-\frac{5}{4}\left(\frac{\kappa_p}{\kappa}\right)^2\right)\gamma^\Gamma \quad (24.2)$$

where U_{10} is the wind speed at a height of 10 m above the sea. Note that the wind speed at a height of 19.5 m above the sea is related to U_{10} by $U_{19.5}=1.026U_{10}$, showing little difference between wind speeds within the vicinity of these heights. Ω is the inverse wave age, which is equal to 0.84 for a well-developed sea (driven by wind). κ_p is the wave number of the dominant waves defined as

$$\kappa_p = \frac{g\Omega^2}{U_{10}^2} \tag{24.3}$$

where g is gravity. The other three parameters in (24.2) are defined as

$$\tilde{c} = \sqrt{\frac{g(1+(\kappa/\kappa_m)^2)}{\kappa}}, \quad \kappa_m = 370$$

$$\Gamma = \begin{cases} 1.7, & 0.83 < \Omega < 1 \\ 1.7 + 6\log(\Omega), & 1 < \Omega < 5 \end{cases} \tag{24.4}$$

$$\gamma = \exp\left(-\frac{1}{2\delta^2}\left(\sqrt{\frac{\kappa}{\kappa_p}} - 1\right)^2\right), \quad \delta = 0.08(1 + 4\Omega^{-3})$$

The short-wave curvature spectrum $B_h(\kappa)$ in the Elfouhaily model is defined as

$$B_h(\kappa) = \frac{c_m \alpha_m}{2\tilde{c}} \exp\left(-\frac{1}{4}\left(\frac{\kappa}{\kappa_m} - 1\right)^2\right) \tag{24.5}$$

where $c_m = 0.23$ and the parameter α_m is determined by

$$\alpha_m = \begin{cases} 10^{-2}\left(1 + \ln\left(\frac{u_f}{c_m}\right)\right), & u_f < c_m \\ 10^{-2}\left(1 + 3\ln\left(\frac{u_f}{c_m}\right)\right), & u_f \geq c_m \end{cases} \tag{24.6}$$

where u_f is the friction velocity that can be iteratively computed by

$$u_f = 0.4U_{10}\left|\left(\ln\left(\frac{10}{b(u_f)}\right)\right)^{-1}\right|, \quad b(u_f) = 0.11 \times 14 \times 10^{-6} u_f^{-1} + \frac{0.48u_f^3\Omega}{gU_{10}} \tag{24.7}$$

The initial value may be chosen as $\sqrt{10^{-3}(0.81 + 0.065U_{10})U_{10}}$.

The mean square slopes of the surface in the upwind direction and in the cross-wind direction are then respectively calculated by

$$mss_x = \int_0^{\kappa_*}\int_{-\pi}^{\pi} \kappa^2 \cos^2\varphi W(\kappa,\phi)\kappa d\phi d\kappa,$$

$$mss_y = \int_0^{\kappa_*}\int_{-\pi}^{\pi} \kappa^2 \sin^2\varphi W(\kappa,\phi)\kappa d\phi d\kappa \tag{24.8}$$

where the wave number cutoff κ_* can be calculated according to

$$\kappa_* = \frac{2\pi}{3\lambda} \tag{24.9}$$

where λ is the wavelength (0.1904 m for the GPS L1 signal). In Garrison et al. (2002), the wave number cutoff is modified as

$$\kappa_* = \frac{2\pi\sin(\theta)}{3\lambda} \tag{24.10}$$

where θ is the incidence angle (complementary to the elevation angle of the satellite). Some details about how to simulate this model and the associated simulation codes can be found in Gleason and Gebre-Egziabher (2009).

24.3.1.2 Sea Surface Scattering

As the GNSS signals arrive at the sea surface, some of the signal energy is absorbed by the seawater, while the other energy is reflected. The reflected signals that travel toward the receiver will be captured by the nadir-looking antenna. Let the positions of the transmitter (on the GNSS satellite) and the receiver (on an aircraft, LEO satellite, or land-based) be (x_t, y_t, z_t) and (x_r, y_r, z_r) respectively. Also define the position of the scattering point on the sea surface as (x_s, y_s, z_s). Then, the distance from the transmitter through the scattering point to the receiver is given by

$$d_{tsr}(x_s, y_s, z_s) = \sqrt{(x_t - x_s)^2 + (y_t - y_s)^2 + (z_t - z_s)^2} + \sqrt{(x_r - x_s)^2 + (y_r - y_s)^2 + (z_r - z_s)^2} \tag{24.11}$$

The specular scattering point (SSP) is the scattering point $(x_{SSP}, y_{SSP}, z_{SSP})$ on the surface where the distance d_{tsr} is minimal. With respect to the signal reflected at the SSP, the signals reflected at other scattering points arrive at the receiver with a delay given by

$$\delta\tau = \frac{d_{tsr}(x_s, y_s, z_s)}{c} - \tau_c \tag{24.12}$$

where

c is the speed of light

$\tau_c = d_{tsr}(x_{SSP}, y_{SSP}, z_{SSP})/c$

Given the transmitter and receiver positions and the delay $\delta\tau$, the scattering points define an ellipse on the surface. That is, at each specific delay, the signals reflected on the ellipse will arrive at the receiver at the same time, supposing that they travel toward the nadir-looking antenna. Figure 24.2 shows an example of the iso-delay when the satellite elevation angle is 63°. The velocity vector of the aircraft is (21.166946, −52.149224, −2.527502 m/s) and the

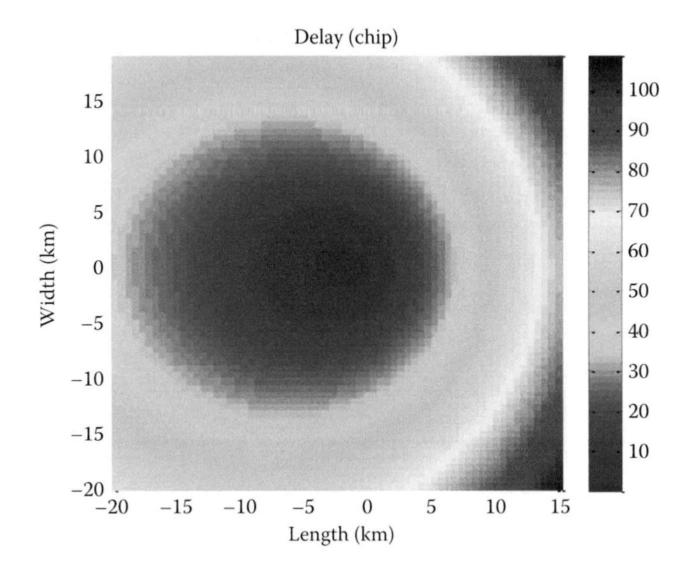

FIGURE 24.2 Example of an iso-delay map for the case where the receiver is carried by an aircraft at a speed of 202.8 km/h and an altitude of 508 m.

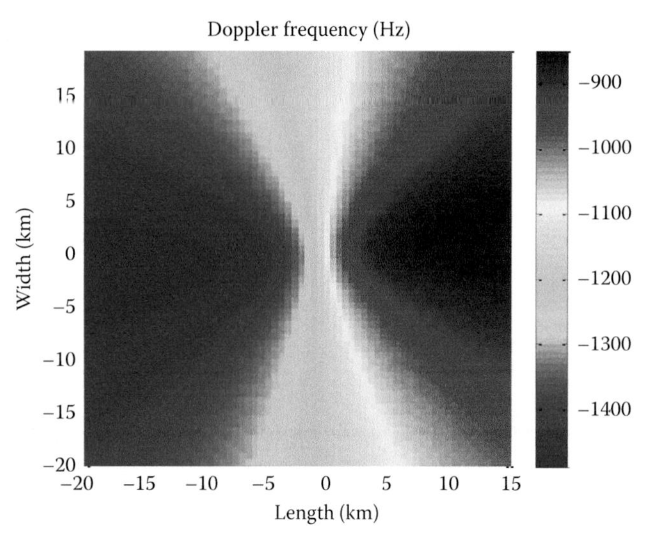

FIGURE 24.3 Example of an iso-Doppler map for the case where the receiver is carried by an aircraft at a speed of 202.8 km/h and an altitude of 508 m.

receiver position is (−33.693117°, 151.2745950°, 508.0884 m) in the WGS84 system. The satellite position is (−52.3194°, 162.2910°, 1.9903e+007 m) and its velocity vector is (−131.75275, −2727.07856, −573.77554 m/s).

Due to the relative movement between the transmitter and the receiver, Doppler frequencies are produced, resulting in the increase or decrease in the signal carrier frequency. Let the velocity vectors of the transmitter and the receiver be $\vec{V}_t$ and $\vec{V}_r$ respectively. The Doppler frequency is determined by

$$f_D = \frac{(\vec{V}_t \cdot \vec{m} - \vec{V}_r \cdot \vec{n})}{\lambda} \tag{24.13}$$

where

- $\vec{m}$ and $\vec{n}$ are the unit vectors of the incident wave and the reflected wave respectively
- "·" denotes the vector dot product

For a given Doppler frequency, Equation 24.13 represents a hyperbola. That is, the signals reflected on such a hyperbola will have the same Doppler frequency. Figure 24.3 shows an example of the iso-Doppler map under the same circumstance as the iso-Doppler map depicted in Figure 24.2. The intersections of the iso-delay lines and the iso-Doppler lines form a network of grids that will be used to determine the power of the reflected signals arriving at the receiver for each pair of relative propagation delay and Doppler frequency.

24.3.1.3 Reflected Signal Power

The reflected signals received via the nadir-looking antenna are first down-converted to IF signals. The code phase offset or delay, and the Doppler frequency associated with the satellite of interest can be estimated based on processing the direct signal received via the zenith-looking antenna through code acquisition and tracking. The carrier frequency of the IF signals is then compensated for, and the resulting baseband signal is correlated with a replica of the PRN code related to a specific satellite. At the central Doppler frequency, the cross-correlation with a sequence of code phases produces a delay waveform (correlation power versus code phase). A delay-Doppler waveform is produced when both a sequence of Doppler frequencies and a sequence of code phases are considered.

Theoretically, the signal power with respect to a code phase and a Doppler frequency can be computed by Zavorotny and Voronovich (2000):

$$Y(\tau, f_D) = \frac{T_i^2 \lambda^2 P_t \eta}{(4\pi)^2} \iint_A \frac{G_t G_r \sigma_0}{R_{ts}^2 R_{sr}^2} \Lambda^2(\tau - \tau_c)\, \mathrm{sinc}^2((f_D - f_c)T_i)\, dA \tag{24.14}$$

where

- T_i is the coherent integration time
- P_t is the transmit power
- η is the atmospheric attenuation
- G_t and G_r are the antenna gains of the transmitting and receiving antennas, respectively
- R_{ts} and R_{sr} are the distance from the transmitter to the scattering point and the distance from the scattering point to the receiver, respectively
- $\Lambda(\tau - \tau c)$ is the triangle correlation function of the PRN code with τ the delay of the replica code
- τ_c is the delay of the received code
- sinc(·) is the sinc function representing the attenuation caused by Doppler misalignment with f_c the Doppler frequency of the received signal
- f_D is the Doppler frequency of the replica signal

A is the effective scattering surface area, and the bistatic radar cross section σ_0 can be calculated according to

$$\sigma_0 = \frac{\pi |\rho|^2}{(\vec{q}_z \vec{q})^4} p\left(-\frac{\vec{q}_\perp}{\vec{q}_z}\right) \quad (24.15)$$

where

- ρ is the polarization-dependent Fresnel reflection coefficient
- $\vec{q}$ is the scattering unit vector that bisects the incident vector and the reflection vector
- $\vec{q}_\perp$ and $\vec{q}_z$ are the horizontal and vertical components of $\vec{q}$ respectively
- $p(\cdot)$ is the probability density function of the surface slope, which may be simply assumed as omnidirectional Gaussian distribution

When performing the double integration in (24.14), the size of the effective scattering area should be appropriately selected. As the flight height increases, the scattering area increases accordingly. Nevertheless, it is not necessary to make the area dimensions too large so as to reduce computational complexity. The contribution of the reflected signals beyond the effective scattering area will be negligible due to the limited antenna beam width and the fact that the power is inversely proportional to the squared distance between the GNSS satellite and the scattering point and between the scattering point and the receiver.

Figure 24.4 shows the decibel delay waveforms based on the theoretical models studied earlier. Six curves correspond to the six different wind speeds (4, 7, 10, 13, 16, and 19 m/s). Four different flight heights were tested: 0.5, 2, 5, and 10 km. In the case of 0.5 km altitude, the six curves are nearly identical, and the spread versus time is rather small. Since the waveforms are insensitive to the wind variation when the flight altitude is less than 0.5 km, it would be inappropriate to use the trailing edge to perform any sea state or wind speed retrieval. As the flight altitude increases, the signal spread increases, and the waveforms distinguish from each other better. Thus, more accurate parameter estimates would be expected.

24.3.1.4 Example

Figure 24.5 shows the five theoretical delay waveforms corresponding to five different wind speeds (3, 4, 5, 6, and 7 m/s) and the measured delay waveform associated with a specific satellite. The data were collected during an airborne experiment at an altitude of about 3 km. The measured waveform is produced through coherent integration of 1 ms IF signals and then noncoherent integration of 1000 such 1 ms waveforms. It can be seen that the measured waveform has a good match with the theoretical waveform of wind speed 4 m/s, which is a good estimate of the real wind speed, which is about 4.3 m/s.

This is just an illustrative example to show how the wind speed is estimated. In practice, a mathematical approach will be employed to automatically produce estimation solutions. Regarding this model-matching approach, a certain number of theoretical waveforms are produced, and a cost function is defined. The theoretical waveform with the minimal cost function is then selected, and the corresponding wind speed is the estimate of the real wind speed. The cost function can be defined as the sum of the squared difference between the theoretical and

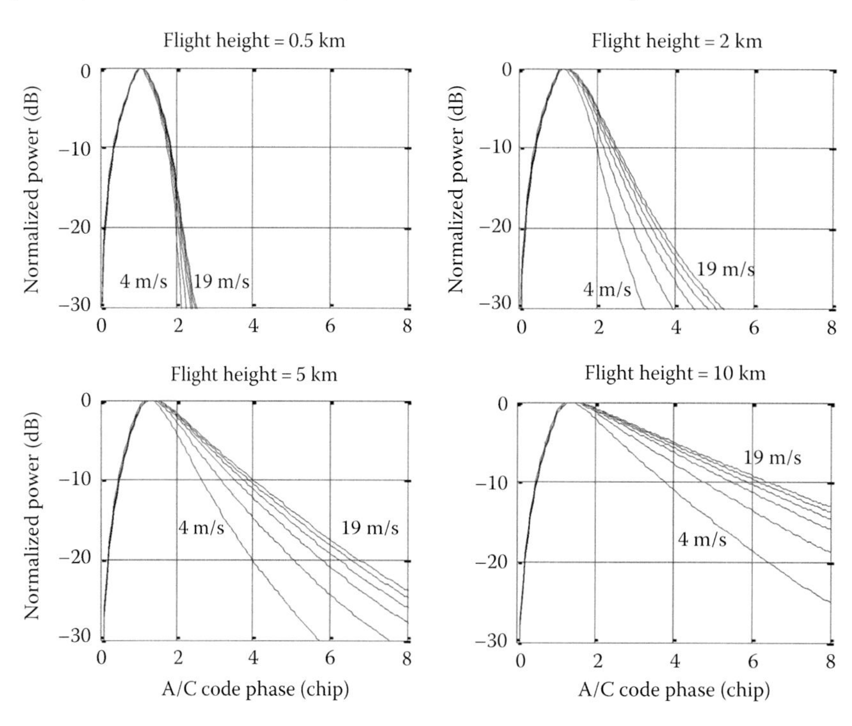

FIGURE 24.4 Normalized correlation powers of the simulated reflected signals using the Elfouhaily wave elevation model with four different flight heights and five different wind speeds (4, 7, 10, 13, 16, and 19 m/s).

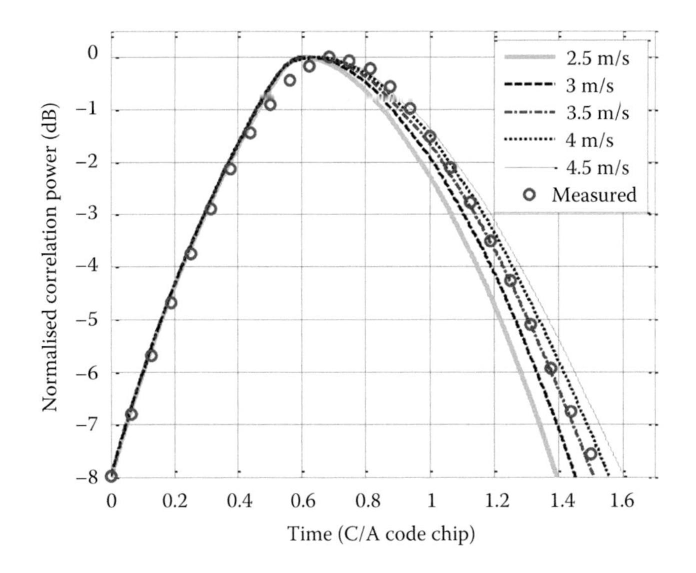

FIGURE 24.5 Wind speed estimation through matching the measured delay waveform with a number of theoretical waveforms.

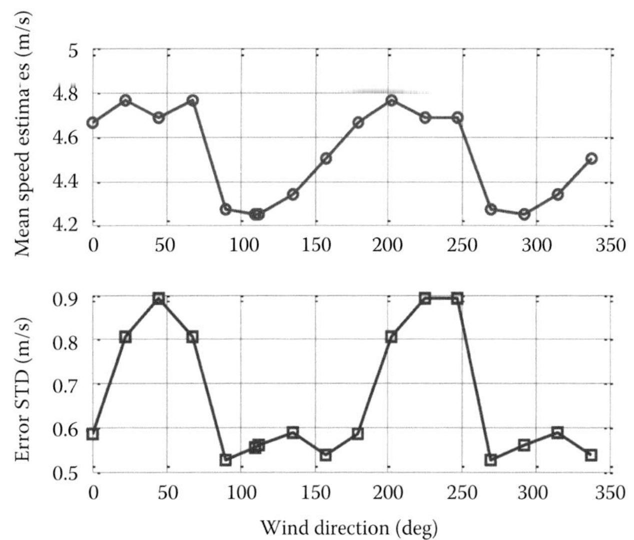

FIGURE 24.6 Mean wind speed estimate versus assumed wind directions using the delay waveform matching method.

measured waveforms, that is, least-squares fitting. However, this method requires the alignment of the two waveforms so that the difference between the two waveforms is minimized. Alternatively, a slope-based method proposed in Zavorotny and Voronovich (2000) can be employed. However, as observed in Figure 24.5, it is rather difficult to distinguish the slopes of neighboring curves related to different wind speeds. The area-based waveform fitting method may be a more suitable option to estimate the wind speed with the following multistep procedure (Yu et al. 2012c):

1. Interpolating both the measured and theoretical waveforms without changing the original data; when using MATLAB®, the library function "INTERP" can be directly used to perform the interpolation.
2. Selecting the cutoff correlation power to retain the waveform above certain power lever so that the slope of the trailing edge does not change abruptly.
3. Calculating the areas of the interpolated waveforms above the cutoff power and calculating the area difference between the measured waveform and each of the theoretical waveforms.
4. Selecting the theoretical waveform that produces the minimum area difference and taking the corresponding wind speed as the estimate.

Figure 24.6 shows the average of wind speed estimates associated with eight satellites when 16 different wind directions are assumed. One observation is that although the mean wind speed estimate varies with the assumed directions, the variation is only about 0.53 m/s. That is, in this case, the variation is not significant. Also shown is the standard deviation (STD) of the wind speed estimation versus the assumed wind directions. The STD ranges from 0.52 to 0.9 m/s. The true wind direction is about 110°, so this STD plot provides some useful information about the wind direction. That is, the wind direction with the smallest STDs may be treated as the actual wind direction. However, the estimation accuracy is rather low, and there exists an ambiguity of 180°. More accurate wind direction estimation may be realized by using the properties of the delay Doppler map as reported in Valencia et al. (2014).

24.3.2 Sea Surface Altimetry

24.3.2.1 Sea Surface Height Calculation

As shown in Figure 24.7, SSH is calculated relative to the surface of the theoretical earth ellipsoid in the WGS84 system, which has zero altitude. The SSH at a specific sea surface point is the distance from the point to the WGS84 earth ellipsoid surface,

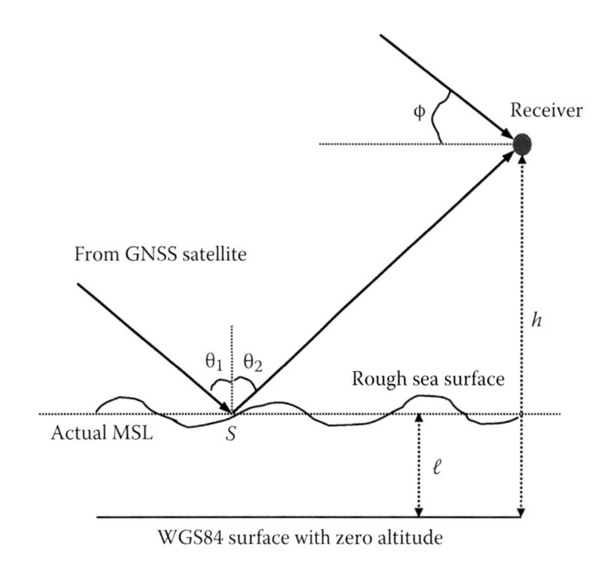

FIGURE 24.7 Geometry of the receiver, WGS84 earth ellipsoid surface, rough sea surface, actual MS, and direct and reflected signal paths.

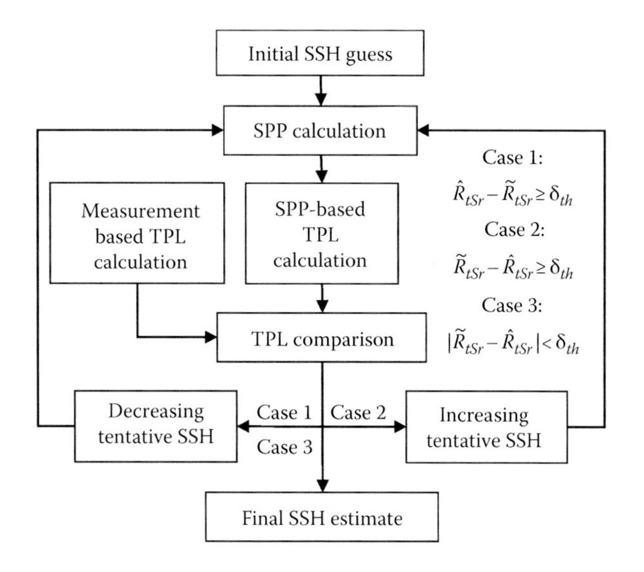

FIGURE 24.8 Flowchart for two-loop iterative SSH calculation. δ_{th} is a small positive number.

and the mean SSH is the average of a large number of these distances over an area of interest.

Figure 24.8 shows the flowchart of how to calculate the SSH using a two-loop iterative method. Specifically, for a given tentative SSH, since both the GNSS satellite position (x_t, y_t, z_t) and the receiver position estimates (x_r, y_r, z_r) are known, the SSP $(\tilde{x}_{SSP}, \tilde{y}_{SSP}, \tilde{z}_{SSP})$ on the tentative sea surface can be readily determined. The SSP estimation is realized by minimizing the total path length (TPL) from the satellite through the SSP and to the receiver, which is given by

$$\tilde{R}_{tSr} = \tilde{R}_{tS} + \tilde{R}_{Sr} \tag{24.16}$$

where

$$\begin{aligned} \tilde{R}_{tS} &= \sqrt{(x_t - \tilde{x}_{SSP})^2 + (y_t - \tilde{y}_{SSP})^2 + (z_t - \tilde{z}_{SSP})^2} \\ \tilde{R}_{Sr} &= \sqrt{(\tilde{x}_{SSP} - x_r)^2 + (\tilde{y}_{SSP} - y_r)^2 + (\tilde{z}_{SSP} - z_r)^2} \end{aligned} \tag{24.17}$$

The TPL can also be estimated using the propagation time (R_{tr}) of the direct signal and the relative delay (τ_{rd}) of the reflected signal, that is,

$$\hat{R}_{tSr} = \hat{R}_{tr} + c\hat{\tau}_{rd} \tag{24.18}$$

where
c is the speed of light
$\hat{\tau}_{rd}$ *is* the estimated relative delay
$\hat{R}_{tr}$ is the estimate of R_{tr} calculated by

$$\hat{R}_{tr} = \sqrt{(x_t - \hat{x}_r)^2 + (y_t - \hat{y}_r)^2 + (z_t - \hat{z}_r)^2} \tag{24.19}$$

where the satellite position is assumed error free, while the receiver position is an estimate. How to estimate the relative delay will be discussed later. Then, as indicated in Figure 24.7, the calculated TPL by (24.16) is compared with the measured TPL by (24.18) to determine whether the tentative SSH should be increased or decreased. The process is terminated once the difference between the two TPLs is smaller than the predefined threshold.

To reduce the computational complexity, a simpler technique may be used. For instance, if $\tilde{R}_{tSr} > \hat{R}_{tSr}$, the tentative surface height is increased by a relatively larger increment such as 40 m. At the next iteration of the outer loop, if $\tilde{R}_{tSr} < \hat{R}_{tSr}$, the increment is decreased by half of the previous increment. In this way, the process will quickly converge to the steady state. Note that the specular reflection must satisfy Snell's law, that is, the two angles (θ_1 and θ_2 in Figure 24.7) between the incoming wave and the reflected wave, separated by the surface normal, must be equal or the difference is extremely small. Thus, the results should be tested to see if this law is satisfied.

From (24.16), the partial derivatives with respect to the coordinates of the specular point can be determined as

$$\frac{\partial R_{tSr}}{\partial u_S} = \frac{u_S - x_r}{R_{rS}} + \frac{u_S - x_t}{R_{tS}}, \quad u_S \in \{x_{SSP}, y_{SSP}, z_{SSP}\} \tag{24.20}$$

which can be rewritten in a vector form as

$$d\vec{S} = \frac{\vec{R} - \vec{S}}{R_{rS}} + \frac{\vec{T} - \vec{S}}{R_{tS}} \tag{24.21}$$

where $\vec{S}$, $\vec{R}$, and $\vec{T}$ are the position vectors of the SSP, the receiver, and the transmitter, respectively. Equation 24.21 is used to generate an iterative solution to the minimum path length. That is, at time instant $n + 1$, the SSP position is updated according to

$$\vec{S}_{n+1} = \vec{S}_n + \kappa d\vec{S} \tag{24.22}$$

where κ is a constant that typically should be set as a larger value as the flight altitude increases. The initial guess of the SSP can be simply the projection of the receiver position onto the surface. At each iteration, a constraint must be applied to restrain the SSP on the surface that is $\tilde{\ell}$ meters above or below the WGS84 ellipsoid surface, which has a zero altitude if $\tilde{\ell}$ is a positive or negative number. That is, the SSP position is scaled according to

$$\vec{S}'_{n+1} = (r_S + \ell)\frac{\vec{S}_{n+1}}{|\vec{S}_{n+1}|} \tag{24.23}$$

where the radius of the earth at the specular point is calculated by

$$r_S = a_{WGS84}\sqrt{\frac{1 - e_{WGS84}^2}{1 - e_{WGS84}^2(\cos\lambda_S)^2}}, \quad \lambda_S = \arcsin\left(\frac{z_S}{|\vec{S}|}\right) \tag{24.24}$$

where
$e_{WGS84} = 0.08181919084262$
$a_{WGS84} = 6{,}378{,}137$ m

Since the altitude of the WGS84 earth's surface is zero, the WGS84 altitude of the SSP is equal to ℓ. Clearly, the altitude of a single SSP cannot be treated as the estimate of the mean SSH. However, a reasonable estimate of the mean surface height will be produced through the generation and subsequent processing of the altitude estimates of many SSPs over a period of time.

24.3.2.2 Calibration

Since the zenith-looking and nadir-looking antennas, the receiver, and the reference point are not in the same position, it is necessary to calibrate the relative delay measurements to remove the effect of these position differences. Note that the reference point may be set at the center of the inertial measurement unit (IMU) provided that such a device is used. Figure 24.9 illustrates the configuration of the devices. The two antennas are connected to the receiver via two cables whose lengths are L_{CR} and L_{CD}. The actual measurement of the relative delay of the reflected signal is given by

$$L_{measured} = AS + SD + L_{DR} - (AU + L_{UR}) + \varepsilon \quad (24.25)$$

where ε is the measurement error. On the other hand, using the reference point position, the relative delay is calculated as

$$L_{calculated} = AS + SC - AC \quad (24.26)$$

As described earlier, the SSH is estimated by comparing the measured and calculated relative delays. However, typically, there is an offset between the calculated and measured delays even in the absence of measurement error. That is,

$$L_{offset} = L_{measured} - L_{calculated} = (SC - SD) + (AU - AC) + (L_{UR} - L_{DR}) \quad (24.27)$$

where the measurement error is ignored.

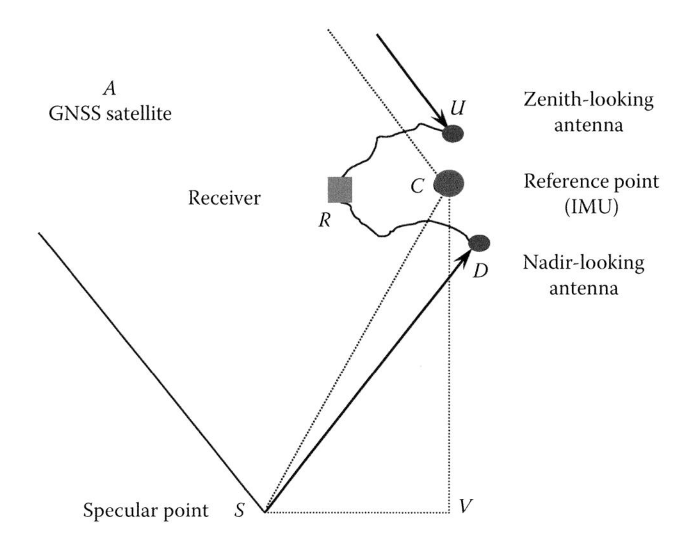

FIGURE 24.9 Geometric relationship between the devices (receiver, antennas, and inertial measurement unit [IMU]) mounted on a LEO satellite or an aircraft.

Therefore, the measured relative delay should be calibrated by subtracting the offset from itself. The lengths of the two cables can be readily measured in advance. In case where an LNA is used to amplify the signal such as captured by the nadir-looking antenna, the path length from D to R will be the sum of the two cable lengths plus the distance between the two connection points of the LNA. The distance from U to C and that from D to C can be manually determined in advance. The positions of points U and C are estimated by the receiver and frame transformation; thus, distances AC and AU can be readily calculated. Calculation of the distances SC and SD requires a knowledge of the SSH, which is unknown in advance. However, initial information about the SSH or previous SSH estimation results can be exploited. The uncertainty in the SSH estimate will affect the calculation of both SC and SD in a very similar way, so that a small SSH error will have a negligible impact. Thus, distance CV can be estimated, and distance SC can be calculated using elevation angle. Calculation of distance SD requires its orientation to determine the angle ∠SDC. In the case where the nadir-looking antenna is fixed directly beneath the reference point, the distance can be simply calculated by

$$SD = \sqrt{SC^2 + CD^2 - 2 \times SC \times CD \times \cos(90 - \phi)} \quad (24.28)$$

Once the distances and the cable lengths in (24.27) are known, the relative delay offset can be readily determined. For instance, suppose that points U and D are directly above and below point C respectively; AC = 20,000 km; satellite elevation angle is 50°; UC = 0.8 m; CD = 0.4 m; CV = 300 m; L_{UR} = 1 m; and L_{DR} = 0.6 m. Then, the unknown distances are obtained from some simple calculations: AU = 19,999.999387 km, SD = 391.32 m, SC = 391.62 m. As a result, the relative relay offset is calculated to be 0.094 m. The SSH estimation error caused by this relative delay offset can be approximated as

$$\delta\ell \approx \frac{L_{offset}}{2\sin\phi} = 6.1\ \text{cm} \quad (24.29)$$

Thus, when the configuration of the devices is arranged properly and the cable lengths are selected based on similar analysis, the SSH error caused by the device configuration will not be large. However, to achieve accurate altimetry, such an error must be compensated for.

24.3.2.3 Relative Delay Estimation Methods

The SSH estimation accuracy is largely dependent on the performance of estimation of the TPL of the reflected signal, from the transmitter through the SSP on the sea surface and to the receiver. The TPL is equal to the sum of the path length of the direct signal and the relative delay of the reflected signal. Since the satellite position is known, the measurement error associated with the path length of the direct signal comes from the receiver position estimation error. When the receiver is given, such an error is typically not reducible, although smoothing may improve the accuracy marginally. Thus, it is important to use a GNSS receiver that can achieve satisfactory position estimation accuracy. When an accurate receiver is used, the relative delay estimation error would be dominant. Therefore, it is vital to reduce this error. The relative delay is estimated by determining the code phase of the direct signal and that of the reflected signal based on the measured correlation waveform.

There are two methods associated with the delay waveform–based method. The first one makes use of the clean code (C/A code), while the second one utilizes the interferometry technique. The clean code method deals with the direct signal and the reflected signal separately and uses only the C/A code to generate the delay waveform. In the interferometry technique, on the other hand, both signals are processed together. That is, the direct signal observed from a zenith-looking antenna is cross-correlated against the reflected one captured by a nadir-looking antenna. High-gain antennas are needed to achieve separability among different satellites. The interferometry technique is intended to exploit either the P(Y) code or military M-code to achieve an accuracy gain at the cost of high gain and directional antennas. In addition, a high-bandwidth front end/receiver is required.

In the case where the zenith-looking antenna is high above the ground, especially when the receiver is mounted on a satellite or on an aircraft, the code phase of the direct GNSS signal can be readily estimated by determining the location of the peak power of the correlation waveform. On the other hand, it may not be easy to obtain an accurate estimate of the desired code phase of the reflected signal forwarded from a rough sea surface. The main reason is that the location of the peak power of the reflected signal would not be the desired code phase of the reflected signal since the peak power location is shifted due to rich multipath propagation. Clearly, using the peak power location to calculate the relative delay would produce a large bias error. The time shift or offset would depend on a number of factors including the surface roughness and the receiver altitude. The peak power location of the delay waveform derivative can be used as the desired code phase of the reflected signal, but the estimate would be biased (Rius et al. 2010). It is observed that the desired code phase of the reflected signal is somewhere between the peak power location of the delay waveform and that of the waveform derivative. However, the exact location of the code phase is typically unknown.

For clarity, it is desirable to explain why the peak power location could shift when a flat sea surface is replaced with a rough sea surface. In the presence of a perfect smooth sea surface, the signal will be reflected only at the SSP and then travels to the receiver. In the presence of a rough sea surface, besides the first path signal, there will be multipath signals arriving at the receiver. Consider the idealized case where there is no noise or error so that the correlation diagram of each path is an isosceles triangle. Suppose that there are J multipath signals whose delays relative to the first path are less than the GNSS code chip width (i.e., half of the correlogram triangle width). Then, the following result exists.

The peak correlation power location of the combined multipath signals will shift from the peak correlation power location of the first path signal provided that

$$\sum_{j=2}^{J} P_j > P_1 \tag{24.30}$$

where P_j is the peak correlation power of the jth path signal.

Since both the first path signal and signals of other paths are reflected signals, the signal power of the second path and a number of following paths can be significant with respect to the first path. Thus, intuitively, (24.30) would always be valid with a rough sea surface. The relationship in (24.30) can be proved mathematically as in Yu et al. (2014).

In the case where the P(Y) code or the military M-code is employed, the peak power location shift can be much smaller so that the shift may be ignored. However, when using the C/A code, the shift needs to be taken into account in the presence of a rough surface. In the absence of information of accurate surface wave statistics, the power ratio–based approach to be discussed next may be employed.

24.3.2.4 Power Ratio–Based Sea Surface Height Estimation: An Example

The power ratio is defined as the ratio of the correlation power at the desired code phase over the peak correlation power of the reflected signal. The desired code phase corresponds to the peak power location when the surface is perfectly smooth. Figure 24.10 is an illustration of the delay waveform of the reflected signal in the presence of a rough surface. $C(\tau_m)$ is the peak power of the

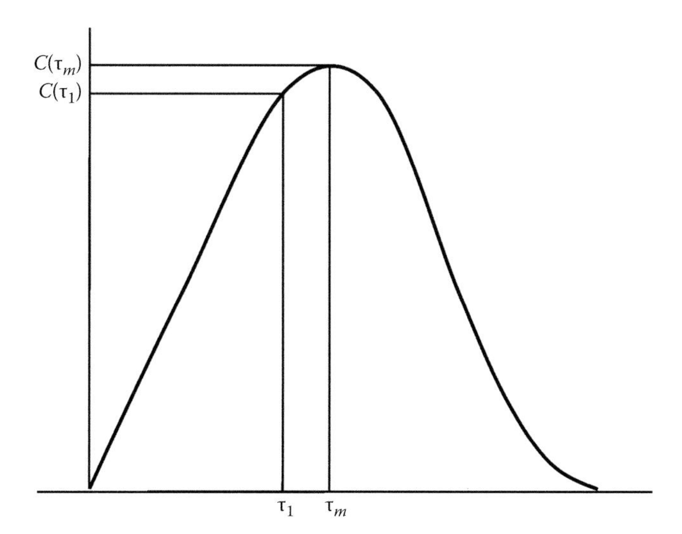

FIGURE 24.10 Illustration of code phases/delays and related correlation power (delay waveform) of reflected signal in the presence of a rough surface.

reflected signal received via the down-looking antenna where τ_m is the time point at which the peak power occurs, while $C(\tau_1)$ is the correlation power at the time point τ_1 where the reflected signal peak power occurs when the surface is perfectly smooth. Since $C(\tau_m)$ can be measured, the time parameter τ_m can be estimated. On the other hand, neither $C(\tau_1)$ nor τ_1 can be simply measured. The power ratio is simply defined as

$$\eta = \frac{C(\tau_1)}{C(\tau_m)} \tag{24.31}$$

Clearly, in the case of a perfectly smooth sea surface, the power ratio is equal to unity. Otherwise, it is less than 1. Given a power ratio and the measured peak power, the power at the desired code phase (τ_1) can be calculated using (24.31), and then τ_1 can be determined using the measured delay waveform. As a consequence, the desired delay of the reflected signal relative to the direct signal can be obtained. The relative delay is then used to calculate the SSH using the method described earlier. A sequence of SSH estimates is produced based on a time series of measured and smoothed delay waveforms. In the case of low-altitude airborne altimetry, the mean SSH could be approximately the same over the surface specular reflection tracks related to several GNSS satellites with the largest elevation angles. Then, sequences of SSH estimates associated with these GNSS satellites can be jointly processed to estimate the desired power ratio and the mean SSH. One method is to define a cost function of the power ratio as

$$\psi(\eta) = \sigma(\eta) \tag{24.32}$$

where σ is the STD of all the sequences of SSH estimates. N power ratio values correspond to N cost function values. The power ratio and the sequence of the SSH estimates or the mean SSH are selected if the corresponding cost function value is the minimum. This criterion selection comes from the consideration that using the desired power ratio would yield estimates that have minimum variations. Figure 24.11 shows an example of the cost function versus the power ratio, which is basically a convex function. In this example, the mean SSH estimate is 23.36 m in the WGS84 system, while the mean SSH measured by a Lidar device is 23.44 m. Measurements related to four satellites as listed in Table 24.3 were used to produce the mean SSH estimate. The average of the mean SSH errors and the root-mean-square mean SSH errors are both 8 cm. This sea surface altimetry method is suited only for the case where the receiver is aircraft-borne. In the case of the satellite-borne receiver, the theoretical model of the received reflected signal power is required to obtain the offset of the peak power location of the reflected signal in the presence of a rough sea surface. However, when the sea surface is calm or the GPS P(Y) code is used, the offset can be negligible, so that measurements associated with a single GNSS satellite can be used to retrieve the mean SSH.

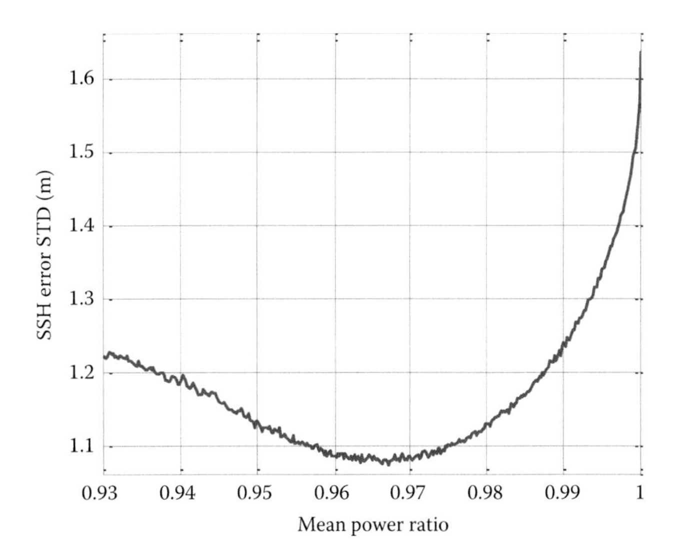

FIGURE 24.11 Cost function (SSH error STD) versus mean power ratio for mean SSH estimation. Data related to four satellites are used.

TABLE 24.3 PRN Numbers, Elevation, and Azimuth Angles of Four Satellites

Satellite (PRN#)	22	18	6	21
Elevation (°)	62.67–62.99	58.14–57.41	50.89–50.92	48.57–47.95
Azimuth (°)	238.19–240.29	148.82–149.94	266.29–268.22	78.87–79.93

24.4 Land Applications

As mentioned earlier in Table 24.2, GNSS-R technique can be used to retrieve a range of land parameters. This section briefly describes how this technique can be employed to estimate soil moisture and detect forest change.

24.4.1 Soil Moisture

Knowledge of soil moisture content is critical for drought and irrigation management, so as to increase crop yields and to gain a better understanding of natural processes linked to the water, energy, and carbon cycles. Soil moisture can be estimated by analyzing surface scattered signals transmitted and received by active radar sensors or natural surface emission detected by microwave radiometers. Alternatively, GNSS-R signals can be employed to estimate soil moisture. Since the soil dielectric constant and reflectivity depend on soil moisture, the variation in the reflected GNSS signal power or signal-to-noise ratio indicates changes in soil moisture (Masters et al. 2004). Using an interferometric method, the power associated with the coherent sum of the direct and reflected GNSS signals can be used to estimate soil moisture. After continuously measuring the total power for a few hours, the location of the power notch over satellite elevation can be observed, and the soil moisture may be estimated (Rodriguez-Alvarez et al. 2009). The total power received can be described as

$$\eta F(\theta)\,|1 + R \cdot \exp(j\phi)|^2 \quad (24.33)$$

where
η is a scaling factor
$F(\theta)$ is the antenna pattern/gain
θ is the incidence angle
R is the reflectivity coefficient
ϕ is the phase difference between the incident and reflected signals

The reflectivity coefficient may be described by a three-layer reflectivity model (air and two soil layers) as

$$\zeta = \exp\left(-\left(\frac{4\pi\sigma_s}{\lambda}\right)^2\right)\frac{r_{i,i+1} + r_{i+1,i+2}\exp(\delta + j2\psi)}{1 + r_{i+1,i+2}^2\exp(\delta + j2\psi)} \quad (24.34)$$

where
σ_s is the soil surface roughness
λ is the signal wavelength (for GPS L1 signal, λ = 19 cm)
$r_{i,i+1}$ and $r_{i+1,i+2}$ are the Fresnel coefficients of the first and second soil layers
δ is the surface roughness correction factor
ψ is a phase term

η may be simply set to be unity only when using the power notch position for soil moisture estimation, while it can be adjusted when using the pattern or shape of the total power. It is an advantage to use existing GPS receivers, installed primarily for geophysical and geodetic applications, to estimate soil moisture (Zavorotny et al. 2010). These receivers, if exploited effectively, could provide a global network for soil moisture monitoring. In addition to the interferometric method, the reflectometry technique also can be employed. For instance, the empirical dielectric model–based method in Wang and Schmugge (1980) may be applied to GNSS-based soil moisture retrieval. In this method, the transition moisture parameter (θ_t) is defined as

$$\theta_t = 0.165 + 0.49\theta_{wt} \quad (24.35)$$

where θ_{wt} is the wilting point moisture that is a function of the percentage of the clay and sand contents. Depending on whether the soil moisture is greater or less than the transition moisture, a specific formula can be used to calculate the soil moisture. This model employs the mixing of either the dielectric constants or the refraction indices of ice, water, rock, and air, and treats the transition moisture value as an adjustable parameter. The dielectric constant of the soil can be estimated by measuring the surface reflectivity. The model is quite general since it was developed by considering many different types of soils. GNSS-based soil moisture estimation is complicated by a number of issues including surface roughness, vegetation canopy, and variation in the percentage of individual soil components. To achieve reliable soil moisture estimation, these issues must be taken into account through processes such as modeling and compensation. Currently, the accuracy of GNSS-R-based soil moisture estimation is generally not as good as that of active sensors and passive microwave radiometers. Most of the GNSS-R experiments conducted for soil moisture retrieval are ground based, and only a few are aircraft-borne. In the case of a satellite-borne receiver, it is still a challenge to infer useful information from GNSS-R measurements.

24.4.2 Forest Change Detection

Forest change detection is another possible application of GNSS remote sensing. Forest change can provide useful information about global climate change impacts and carbon storage, and knowledge of forest change is vital for effective forest management. Received signal strength of the reflected GNSS signals can be employed to distinguish forest conditions from each other, since different surface cover has different reflectivity. A number of signal strength ranges may be defined to be associated with a group of surfaces such as lake/river water, typical dense forest, and cleared area due to logging (Yu et al. 2013). Figure 24.12 shows four ground specular reflection tracks of the reflected signals from four GNSS satellites received by a receiver on an aircraft. The tracks are colorized by the reflected signal power. The low signal power corresponds to the dense forest areas, while the high signal power occurs over three areas marked by A1 (cleared area), A2 (partially cleared area), and A3 (lake water). Accordingly, a surface can be classified by determining within which range its measured strength falls. In addition to signal strength, other signal characteristics such as those related to the observed correlation waveform may be used to enhance surface change detection or surface classification. The major issue related to surface change detection is the dependence of reflectivity on several factors such as soil moisture and surface roughness. Forest change may be quantified by evaluating forest biomass, which depends on the volume of both living and dead trees (leaves, branches, and trunks). The received signal power can be used to determine the scattering coefficient, which, based on simulation (Ferrazzoli et al. 2011), should be a function of the biomass. However, it is a challenging problem to derive a formula that accurately describes the relationship between the scattering coefficient and the biomass.

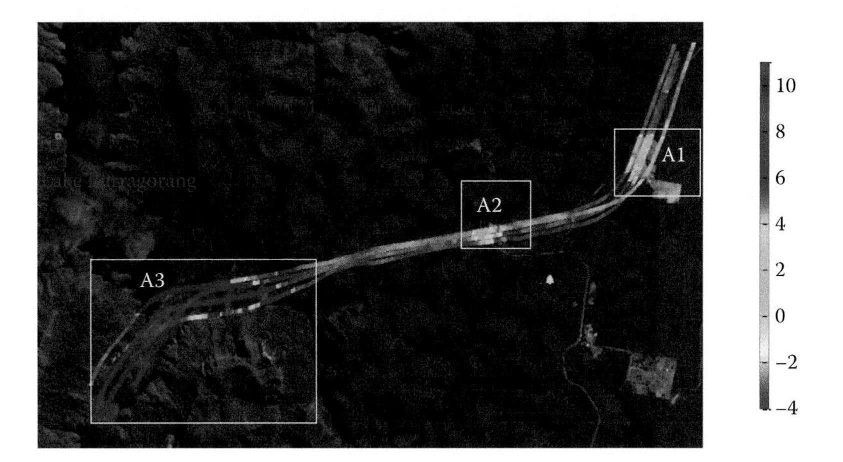

FIGURE 24.12 Ground specular reflection tracks associated with four GPS satellites, colorized by reflected signal power. The picture was generated by Google Earth and GPS Visualizer.

24.5 Challenging Issues and Future Directions

GNSS is a global space-based positioning, navigation, and timing (PNT) capability. To come close to matching the extraordinary success of GNSS-PNT, there are many technological challenges for GNSS remote sensing, especially in the case of the GNSS-R technique. Satellite-borne GNSS remote sensing must demonstrate its value as a reliable, high-quality sensing technology. With the launch of multiple LEO satellites with GNSS-RO- and GNSS-R-capable instrumentation, it seems we are at last close to a renaissance. But are more LEO missions and better algorithms for geophysical parameter extraction sufficient? In the case of GNSS-RO, the answer is yes, because the meteorological/climate community now assimilates GNSS-RO products into their operational and research systems, and is eager for more. However, GNSS-R is still very much a novelty technology as far as the geosciences community is concerned.

The International Association of Geodesy (IAG) has been very successful in launching technique-specific services, see http://www.iag-aig.org. These services aid other geoscientists in their research as well as support important applications in the wider community. Examples of the latter are the contribution of the International GNSS Service (IGS) to precise positioning (http://igs.org), and the International Earth Rotation and Reference Systems Service to the International Terrestrial Reference Frame (http://itrf.ensg.ign.fr). The IGS does generate troposphere and ionosphere parameter products (http://www.igs.org/components/prods.html); however, they are based on observations made from the global ground-based GNSS tracking network (Figure 24.13),

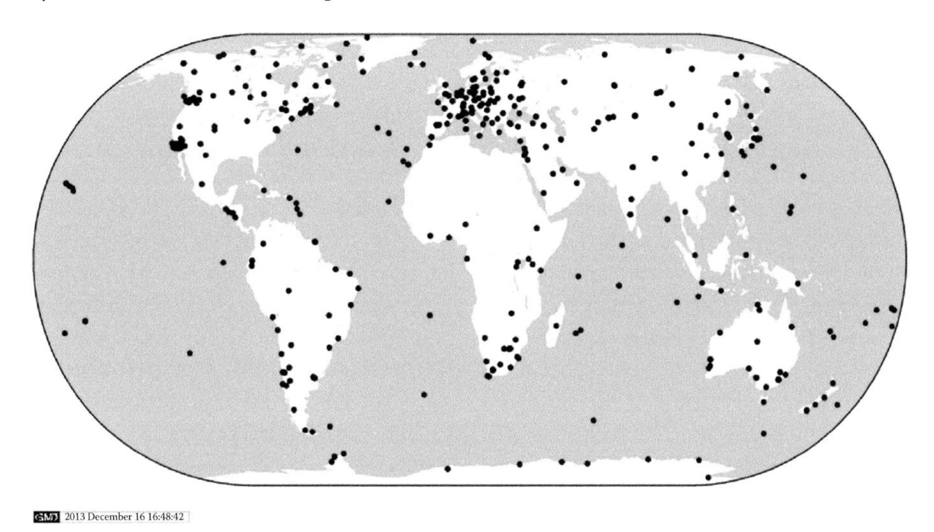

FIGURE 24.13 One of the reasons for the success of IAG services are the permanent geodetic observing networks such as the IGS's global GNSS tracking network, providing accurate, reliable data products for science and society.

not from satellite platforms. There are no geodetic services producing GNSS remote sensing products on a continuous synoptic basis. From the IAG's perspective, the challenging issues are related to *operationalizing* GNSS remote sensing, in all its forms, so that the remotely sensed geophysical parameters (atmospheric, oceanic, wind, soil moisture, biomass, etc.) are included in the suite of geodetic outputs of the Global Geodetic Observing System (http://www.ggos.org).

The nontechnical challenges therefore include the evolution of current GNSS remote sensing science missions to operational services by *increasing* the number of LEO satellites equipped with GNSS receivers to satisfy coverage requirements in space and time; *standardizing* data formats, instrumentation, and calibration, preprocessing, and analysis procedures; and establishing a *coordinating* agency or authority so as to ensure continuous, high-quality product generation and dissemination. The IAG has considerable experience with geometry technique–based services (such as the IGS, ILRS, IVS and IDS) (http://www.iers.org/nn_10880/IERS/EN/Organization/TechniqueCentres/TC.html?__nnn=true); however, it has not yet established any *geodetic imaging* services based on technologies such as synthetic aperture radar, satellite radar altimetry, lidar, or GNSS remote sensing (GNSS-RO, GNSS-R). One of the future challenges is to address this shortcoming, so that GNSS remote sensing can be recognized as a geodetic technique that is making critical contributions to science and society.

24.6 Conclusions

This chapter studied GNSS-based remote sensing with a focus on GNSS-R for both land and ocean applications. In particular, details about GNSS-R-based sea surface wind speed estimation and sea surface altimetry were presented. The wind speed estimation accuracy can be around 1 m/s, and the mean SSH estimation can be of the order of decimeters. In addition to the advantage of low cost, the GNSS-R technique can measure the geophysical parameters over a wide area. The major drawback of the technique may be that the accuracy can be affected considerably by modeling errors. As mentioned in the chapter, a number of dedicated satellite missions associated with GNSS remote sensing have been accomplished, several missions are scheduled, and more missions are expected to be funded. It is envisaged that as a cost-effective technique, GNSS-based remote sensing will play a significant role in a wide range of remote sensing applications.

References

Anthes, R.A. 2011. Exploring Earth's atmosphere with radio occultation: Contributions to weather, climate and space weather. *Atmospheric Measurement Techniques* 4:1077–1103.

Camps, A., J.F. Marchan, E. Valencia, I. Ramos, X. Bosch-Lluis, N. Rodriguez, H. Park et al. July 2011. PAU instrument aboard INTA Microsat-1: A GNSS-R demonstration mission for sea state correction in L-band radiometry. *Proceedings of IGARSS*, Vancouver, Canada, pp. 4126–4129.

Elfouhaily, T., B. Chapron, K. Katsaros, D. Vandemark. 1997. A unified directional spectrum for long and short wind-driven waves. *Journal of Geophysical Research* 102(C7):15781–15796.

Ferrazzoli, P., L. Guerriero, N. Pierdicca, R. Rahmoune. 2011. Forest biomass monitoring with GNSS-R: Theoretical simulations. *Advances in Space Research* 47:1823–1832.

Font, J., A. Camps, A. Borges, M. Martin-Neira, J. Boutin, N. Reul, Y.H. Kerr, A. Hahne, S. Mecklenburg. 2010. SMOS: The challenging sea surface salinity measurement from space. *Proceedings of the IEEE* 98(5):649–655.

Garrison, J.L., A. Komjathy, V.U. Zavorotny, S.J. Katzberg. 2002. Wind speed measurements using forward scattered GPS signals. *IEEE Transactions on Geoscience and Remote Sensing* 40(1):50–65.

GEROS-GNSS REflectometry, Radio Occultation and Scatterometry Onboard International Space Station (GEROS-ISS), http://www.ice.csic.es/en/view_project.php?PID=155, viewed December 9, 2013.

Gleason, S.T. September 2006. Land and ice sensing from low Earth orbit using GNSS bistatic radar. *Proceedings of the International Technical Meeting of the Satellite Division of the Institute of Navigation* (*ION GNSS*), Fort Worth, TX, pp. 2523–2530.

Gleason, S.T., D. Gebre-Egziabher. 2009. *GNSS Applications and Methods*. Artech House, Boston, MA.

Gleason, S.T., S. Hodgart, Y. Sun, C. Gommenginger, S. Mackin, M. Adjrad, M. Unwin. 2005. Detection and processing of bistatically reflected GPS signals from low Earth orbit for the purpose of ocean remote sensing. *IEEE Transactions on Geoscience and Remote Sensing* 43(6):1229–1241.

Global Geodetic Observing System (GGOS), http://www.ggos.org, viewed December 11, 2013.

Heise, S., J. Wickert, C. Arras, G. Beyerle, A. Faber, G. Michalak, T. Schmidt, F. Zus. 2014. Reprocessing and application of GPS radio occultation data from CHAMP and GRACE. In: F. Flechtner, N. Sneeuw, W.-D. Schuh, eds. *Observation of the System Earth from Space—CHAMP, GRACE, GOCE and Future Missions*, GEOTECHNOLOGIEN Science Report No. 20, Springer, Heidelberg, pp. 63–71.

Jin, S., G.P. Feng, S. Gleason. 2011. Remote sensing using GNSS signals: Current status and future directions. *Advances in Space Research* 47(10):1645–1653.

Martín-Neira, M. 1993. A passive reflectometry and interferometry system (PARIS): Application to ocean altimetry. *ESA Journal* 17(14):331–355.

Martin-Neira, M., S. D'Addio, C. Buck, N. Floury, R. Prieto-Cerdeira. 2009. The PARIS in-orbit demonstrator. *Proceedings of IEEE IGARSS*, Vol. 2, Cape Town, South Africa, pp. II-322–II-325.

Martin-Neira, M., S. D'Addio, C. Buck, N. Floury, R. Prieto-Cerdeira. 2011. The PARIS ocean altimeter in-orbit demonstrator. *IEEE Transactions on Geoscience and Remote Sensing* 49(6):2209–2237.

Masters, D., P. Axelrad, S.J. Katzberg. 2004. Initial results of land-reflected GPS bistatic radar measurements in SMEX02. *Remote Sensing of Environment* 92:507–520.

Pierson, W.J., L. Moscowitz. 1964. A proposed spectral form for fully developed wind seas based on the similarity theory of S. A. Kitaigorodskii. Journal of Geophysical Research 69(24):5181–5190.

Rius, A., E. Cardellach, M. Martın-Neira. 2010. Altimetric analysis of the sea-surface GPS-reflected signals. *IEEE Transactions on Geoscience and Remote Sensing* 48(4):2119–2127.

Rodriguez-Alvarez, N., X. Bosch-Lluis, A. Camps, M. Vall-Llossera, E. Valencia, J.F. Marchan-Hernandez, I. Ramos-Perez. 2009. Soil moisture retrieval using GNSS-R techniques: Experimental results over a bare soil field. *IEEE Transactions on Geoscience and Remote Sensing* 47(11):3616–3625.

Unwin, M.J., S. Gleason, M. Brennan. September 2003. The space GPS reflectometry experiment on the UK disaster monitoring constellation satellite. *Proceedings of ION-GPS/GNSS*, Portland, OR.

Valencia, E., V.U. Zavorotny, D.M. Akos, A. Camps. 2014. Using DDM asymmetry metrics for wind direction retrieval from GPS ocean-scattered signals in airborne experiments. *IEEE Transactions on Geoscience and Remote Sensing*, 52(7):3924–3936.

Wang, J.R., T.J. Schmugge. 1980. An empirical model for the complex dielectric permittivity of soils as a function of water content. *IEEE Transactions on Geoscience and Remote Sensing* 18(4):288–295.

Yu, K., C. Rizos, A. Dempster, September 2012a. Error analysis of sea surface wind speed estimation based on GNSS airborne experiment. *Proceedings of ION GNSS*, Nashville, TN.

Yu, K., C. Rizos, A. Dempster. October 2012b. Performance of GNSS-based altimetry using airborne experimental data. *Proceedings of Workshop on Reflectometry Using GNSS and Other Signals of Opportunity*, West Lafayette, IN.

Yu, K., C. Rizos, A.G. Dempster. July 2012c. Sea surface wind speed estimation based on GNSS signal measurements. *Proceedings of International Geoscience and Remote Sensing Symposium (IGARSS)*, Munich, Germany, pp. 2587–2590.

Yu, K., C. Rizos, A. Dempster. 2014. GNSS-based model-free sea surface height estimation in unknown sea state scenarios. *IEEE Journal of Selected Topics in Applied Earth Observations and Remote Sensing*, 7(5):1424–1435.

Yu, K., C. Rizos, A.G. Dempster. July 2013. Forest change detection using GNSS signal strength measurements. *Proceedings of International Geoscience and Remote Sensing Symposium (IGARSS)*, Melbourne, Victoria, Australia, pp. 1003–1006.

Zavorotny, V.U., K.M. Larson, J.J. Braun, E.E. Small, E.D. Gutmann, A.L. Bilich. 2010. A physical model for GPS multipath caused by land reflection: Toward bare soil moisture retrievals. *IEEE Journal of Selected Topics in Applied Earth Observations and Remote Sensing* 3(1):100–110.

Zavorotny, V.U., A.G. Voronovich. 2000. Scattering of GPS signals from the ocean with wind remote sensing application. *IEEE Transactions on Geoscience and Remote Sensing* 38(2):951–964.

25

Global Navigation Satellite Systems for Wide Array of Terrestrial Applications

D. Myszor
Silesian University of Technology and LG Nexera Business Solutions AG

O. Antemijczuk
Silesian University of Technology

M. Grygierek
LG Nexera Business Solutions AG

M. Wierzchanowski
Silesian University of Technology and LG Nexera Business Solutions AG

K.A. Cyran
Silesian University of Technology

25.1 Introduction

Global navigation satellite systems (GNSS) provide positioning services across the globe (Hofmann-Wellenhof et al. 2008). Utilization of these systems gained huge popularity because of many advantages that they provide such as constant availability (although rare cases of downtimes are known), free of charge service, and good positioning accuracy (Gleason and Gebre-Egziabher 2009, 2010, Jeffrey 2010). In general, GNSS might be internally split into three segments: space, control, and users (Prasad and Ruggieri 2005, Groves 2008). Space segment is comprised of satellites, which are constantly orbiting around the globe. Orbits of individual satellites, within particular systems (constellation), are arranged in a way, which allows for observation of at least four satellites, from almost any point of Earth surface, at the same time (Petrovski and Tsujii 2012). Satellites send ceaselessly information with time stamp, precise orbit coordinates (ephemeris), description of satellites constellation (almanac), error corrections as well as health state indicating their operational state (Ward et al. 2005a). GNSS receiver obtains navigation signals and utilizes its internal clock in order to determine delay between time of navigation message transmission and time of message acquisition by the receiver antenna. As a result, pseudorange between transmitting satellite and receiver might be determined (calculated value is called pseudorange, because various errors can influence obtained results) (Kaplan et al. 2005, Grewal et al. 2001, Xu 2007). If signals from four satellites are available, then based on obtained pseudoranges, and ephemeris of satellites, receiver is able to calculate coordinates in three-dimensional space (Gleason and Gebre-Egziabher 2009). It is worthy to mention that contemporary receivers are able to utilize many visible navigation satellites at the same time, and combined data obtained from various GNSS in order to increase the precision of obtained coordinates. System requires high time precision and synchronization of time between satellites, even small errors in time synchronization lead to significant errors at the level of the Earth's coordinates calculations. In order to achieve these tasks, satellites are equipped with several clocks (i.e., a battery of cesium and rubidium atomic clocks) that are regularly synchronized by control segment.

Control segment is composed of installations located at the surface of the Earth. These installations form network of stations scattered around the globe. Following facilities are usually possessed by particular GNSS: master control station as well as backup control station, set of command sites with control antennas, and a battery of monitoring stations. The master control station receives data from monitoring stations; thus, it is able to control proper functioning of navigation system. In general, control segment is responsible for constant monitoring of the navigation satellites' states, supervision of orbital plane drift, sustaining of constellation geometry, taking satellites out of service for maintenance tasks such as orbit modification or software update, and resolution of various anomalies and malfunctions in satellite's function. It also regularly communicates with satellites in order to synchronize their atomic onboard clocks and adjust the ephemeris of satellite orbital model. Every satellite, belonging to GNSS constellation, has a designated lifetime; therefore, control segment is responsible for disposal of retired satellites, launching of new ones on early orbits as well as introduction of these satellites into navigation constellation.

User segment consists of individual, institutional (government, commercial as well as scientific), and military users, which are in the possession of navigation signal receivers. Receivers might be characterized by various levels of accuracy, qualities of internal clocks, multichannel abilities (calculation of coordinates based on many satellites), and possibility of utilization of different GNSS.

25.1.1 Contemporary Global Navigation Satellite Systems

Global positioning system—navigation satellite time and ranging (GPS—NAVSTAR). The system is operated and controlled by the U.S. Department of Defense. Originally, it was created solely for military purposes; however, later, it was opened for civilian applications. With time, improvements were introduced, which allow for achievement of positioning accuracy close to those obtained with military signals (Harte and Levitan 2007). Currently open signals are transmitted on two frequencies; therefore, there is a possibility of autonomous mitigation of ionosphere errors when proper receiver is utilized (Xu 2007). Baseline constellation is composed of 24 slots (Figure 25.1). There are six equally spaced orbital planes, each orbit possess four slots. Initially each slot could possess one satellite; however,

TABLE 25.1 GPS Constellation

Total Number of Satellites in GPS Constellation (June 3, 2014)	32
Operational	30
In commissioning phase	1
In maintenance	1

in order to improve signal availability, especially on areas that are characterized by obstruction reach environment, three slots were expanded that can host two satellites each (fore and aft location within expanded slot are defined). Therefore, 27 navigation satellites could be accommodated in baseline constellation. Constellation can possess additional spare satellites; however, they do not occupy predefined slots (see Table 25.1). Devices are located in the Middle Earth Orbit at an altitude of 20,200 km and orbit the Earth approximately every 12 h. Arrangement of satellites ensures that from almost every point of Earth, at least four satellites are visible at the same time and their elevation angles are greater than 15° relative to the ground (Kim et al. 1998). All satellites utilize the same frequency and code division multiple access (CDMA) technique. Importantly equipment and constellation configuration are continuously improved in order to increase accuracy (see Table 25.4 later in this chapter.).

Russian Global Navigation Satellite System (GLONASS). The system is operated by the Coordination Scientific Information Centre of the Ministry of Defence of the Russian Federation. Similar to GPS, GLONASS was originally created for military purposes and later, it was opened for civilian application. Currently, satellites transmit two types of signals: open one (for civilian purposes) and the encrypted one (for military purposes). Importantly signals for civilian purposes are broadcasted in two bands; thus, ionosphere effects can be mitigated if two band receivers are applied. Baseline constellation is composed of 24 satellites, organized in 3 orbital planes; each orbit possess 8 satellites (Figure 25.2). Fully operational state requires

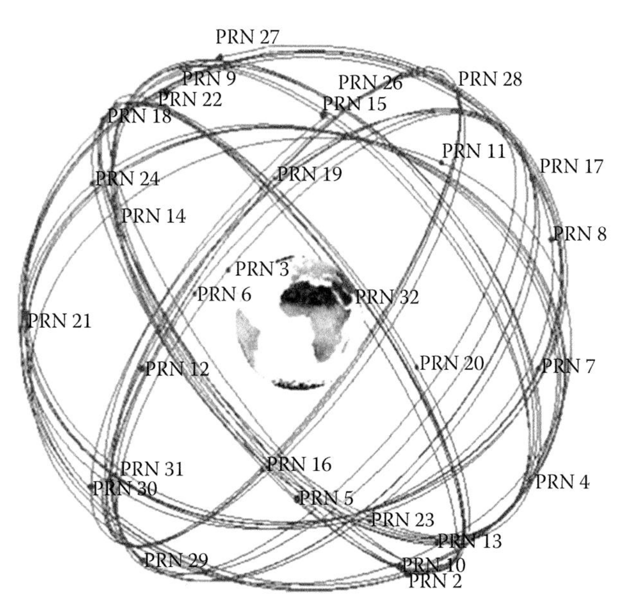

FIGURE 25.1 GPS constellation.

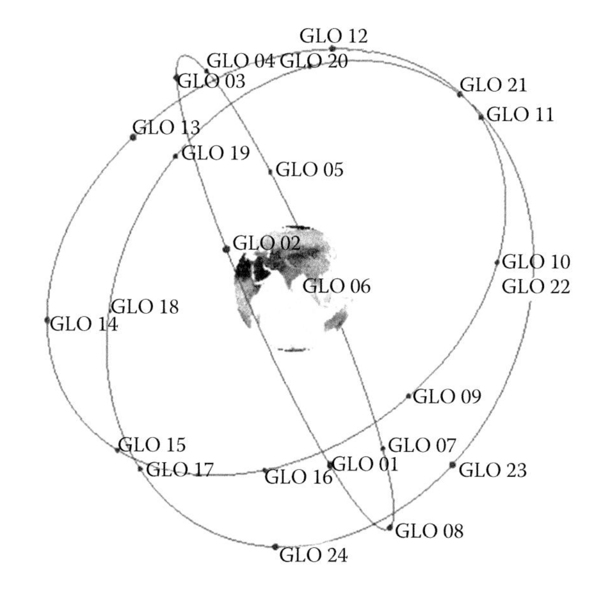

FIGURE 25.2 GLONASS constellation.

TABLE 25.2 GLONASS Constellation

Total Number of Satellites in GLONASS Constellation (June 3, 2014)	29
Operational	24
In commissioning phase	0
In maintenance	0
Spares	4
In flight tests phase	1

21 satellites, additional 3 satellites (called active spares) are maintained for the purpose of replacement of malfunctioning devices (see Table 25.2). Orbital planes are designed in order to obtain higher accuracy than GPS in high altitudes. Devices are located in the Middle Earth Orbit at an altitude of 19,100 km and orbit the Earth approximately every 11 h. Satellites utilize the 15-channel frequency division multiple access technique (Harper 2009).

GALILEO is the European Union's response to military-controlled GNSS systems for which civil accuracy can be degenerated or even disabled in case of conflicts. Such an action could introduce huge disturbances in many fields of industry and human lives; therefore, the decision was made to create civil controlled system that will allow obtaining of precise positioning across the globe. Planned baseline constellation is composed of 30 satellites organized in 3 orbital planes, each orbit possesses 10 satellites (Figure 25.3). System, in order to be fully operational,

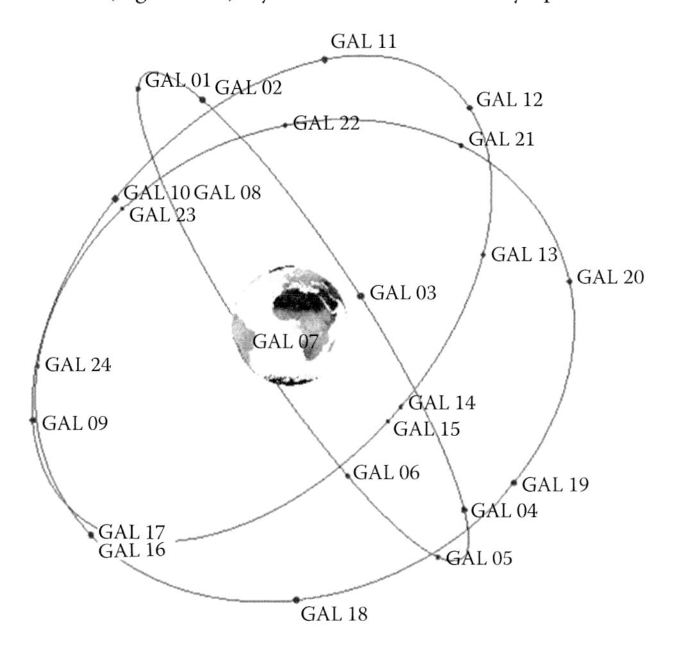

FIGURE 25.3 GALILEO constellation.

requires 27 satellites, additional 3 satellites (called active spare) will be maintained (1 on each orbit) for the purpose of replacement of malfunctioning devices. Such number of satellites will allow to increase the accuracy level in city canyons; in addition, orbital planes in the constellation are designated to obtain better accuracy, at high latitudes, than GPS-based solution; thus, the accuracy of Northern Europe coverage will improve. Current plans assume that fully operational state should be achieved by 2019. Satellites are located in the Middle Earth Orbit at altitude of 23,222 km and orbit the Earth approximately every 15 h. The system is operated by the civilian organization European Space Agency (ESA). GALILEO will provide open signal for all users (as in GPS, two frequencies are available); however, there will be possibility to purchase access to encrypted signal, which will increase the resistance to spoofing and jamming. Interestingly, additional encrypted service will be provided for government agencies. There will be also a possibility of picking up emergency signals transmitted by beacons installed aboard ships, aircrafts, or carried by individuals. Furthermore, system will be able to inform users about signal degradation, such ability is especially important in applications where guaranteed precision is important, for example, plane navigation (Tables 25.3 and 25.4).

25.2 SBAS: Satellite-Based Augmentation System

Precision of calculated coordinates based solely on GNSS is insufficient in many applications; therefore, additional SBASs were created that can improve the accuracy of obtained results. Among them the most popular are Wide Area Augmentation System (WAAS), European Geostationary Navigation Overlay Service (EGNOS), and Multifunctional Satellite Augmentation System (MSAS). WAAS was created by United States; it is designated for GPS only and it is available on the area of North America. EGNOS, on the other hand, was the first complex project in the field of satellite navigation systems, created in cooperation between EU members; it was established in 2005 and the full operational state was achieved in 2009. The purpose of the system was to create augmentation for GPS and GLONASS over the entire Europe (ESA SP-1303 2006). Finally, MSAS was created by Japanese in order to deliver corrections to GPS signals over its area.

In order to ensure the highest positioning accuracy, these systems constantly monitor satellites constellations with the network of ground monitoring stations, calculate errors, and

TABLE 25.3 GNSS Summary

GNSS	No. of Satellites (Base Constellation, without Spares)	Operating Country	Civilian Applications	Military Applications	Accuracy (Perfect Conditions, without Supporting Systems)
GPS	27	USA	True	True	Up to 7 m (Allan 1997)
GLONASS	21	Russian Federation	True	True	Up to 15 m (Miller 2000)
GALILEO	27	European Union	True	True	Constellation is not fully operational

TABLE 25.4 Historical Accuracy of GPS System

Year	1993	1995	2000	2014
Horizontal accuracy (m)	~100 to ~300	~100	~20	~7.8

provide correction data to end users in real time. As a result, significant reduction of positioning errors is obtained, for example, in case of GPS, positioning accuracy can be increased from approximately 10 m when it is calculated without correction services (one frequency receiver) (DoD 2008) to approximately 2 m when EGNOS service is utilized. Interestingly aforementioned augmentation systems are compatible, they were created to increase positioning precision mainly in aviation applications; thus, correction signals are broadcasted through satellites (Figure 25.4). It implies that corrections might be hard to obtain in areas surrounded by high obstacles. In such a case the Internet might be utilized for correction delivery (SISNeT technology in EGNOS, Zinkiewicz et al. 2010, and ongoing work over EGNOS Data Access System [EDAS]) (Table 25.5).

25.3 GNSS Errors

Utilization of satellite navigation systems for measuring of the object position is exposed to a variety of errors that affect the accuracy of designated coordinates (Grewal et al. 2001). Some of the factors that can lead to incorrect results are satellite position errors. They could be caused by various factors such as solar radiation pressure and solar flares, electromagnetic forces (Silva et al. 2002), and relativistic effects. In addition, gravity force of Earth, Sun, Moon, and other celestial bodies as well as oceanic and terrestrial tidal waves influence the satellites positions, thus degrading obtained coordinates. Noteworthy is the fact that geometry of observed satellites in relation to the receiver antenna also has an impact on accuracy of obtained results; dilution of precision (DOP) is a geometric quantity, which describes this relation (Hewlett-Packard 1996). Another issue appears when change in satellites configuration occurs: jump in assessed location can be obtained and accuracy might be compromised (Sukkarieh et al. 1999).

Earth tides, which are small cyclical ground movements caused by gravity forces of other celestial bodies (mainly Sun and Moon), introduce issues when a very precise conversion of coordinates to the location on the map is required. Similar issues are introduced by tectonic movements, which cause constant displacement of reference sites on Earth's surface in relation to reference points set at the level of reference frame, for example, World Geodetic System 1984 (WGS84) (Harper 2009) or its high accuracy version International Terrestrial Reference System (ITRS). When precise measurements are required, magnitude of this error cannot be underestimated. In some areas, this phenomenon can introduce displacement at rate up to 120 mm/year, for example, Great Britain is in constant move with reference to WGS84 and displacement is at the level of 25 mm/year. In order to mitigate this effect, frames of reference that are tied to the ground and fixed in time are introduced. In Europe, the most popular one is European Terrestrial Reference System 1989 (ETRS89), which can be easily converted to ITRS by simple transformation of coordinates that are published by International Earth Rotation Service (IERS). Also Earth's rotator parameters are not constant; therefore, additional source of errors is introduced. As in the previous case, influence of this type of errors can be mitigated by application of corrections, which could be based on data obtained from IERS. It is worthy to mention that all types of conducted calculations, for example, conversions between different reference systems are also prone to precision errors. It happens because of floating point number representation in devices' memory, which usually results in necessity of numerical rounding when extensive calculations are done.

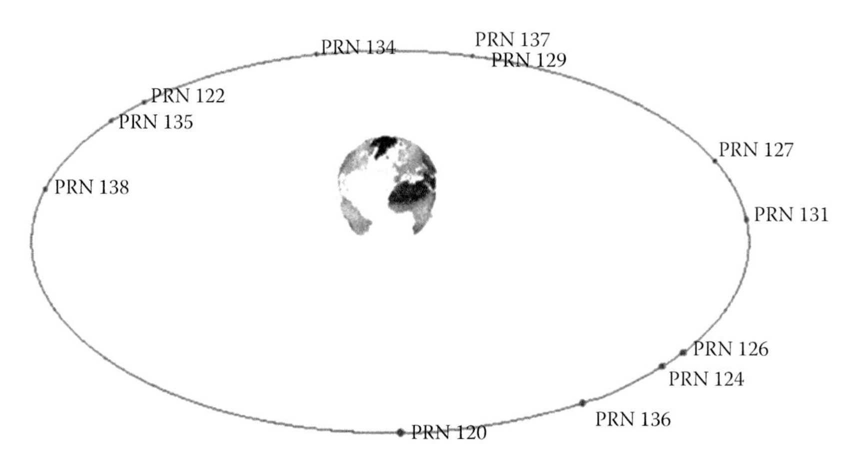

FIGURE 25.4 EGNOSS constellation.

TABLE 25.5 GPS Accuracy When Various Augmentation Systems Are Utilized

GPS	WASS	NDGPS	GDGPS	AGPS	EGNOS	MSAS
L1 = ~7.8 m	~3 m	~10–15 cm	~10 cm	~5 m	~3 m	~0.8–5.3 m

TABLE 25.6 Typical Errors, Their Magnitudes, and Methods of Mitigation

Type of Error	Typical Magnitude of an Error (m)	Methods of Error Mitigation
Time base error	3	Utilization of signals from many satellites
Inaccuracy in positioning of satellite on the orbit	2	Utilization of correction services providing exact location of satellites on the orbit
Satellite orbit geometry errors	2.5	Utilization of correction services that provide precise orbit information
Ionosphere influence	5	Utilization of differential corrections/dual-frequency receivers/models of influence of ionosphere
Troposphere influence	0.5	Utilization of differential corrections/dual-frequency receivers/models of influence of troposphere
Propagation of electromagnetic waves	1	Utilization of proper algorithms

Distortions of propagation of signals transmitted by satellites in Earth's atmosphere, which are mainly caused by ionosphere and troposphere refraction, are another cause of errors (MacGougan 2006). Signal can also be disturbed by high voltage lines and other transmitters (e.g., various GNSS signal jammers). In general, all objects that are blocking direct visual contact between receiver and transmitter (such as trees' leaves, buildings, cliff faces) are also potential source of signal degradation or even its unavailability. When signal is weakened, special software algorithms might be applied, which could improve the ability to obtain the position (Ziedan 2006). In addition, receiver's surrounding environment might introduce the phenomenon of multipath signal propagation (Ward et al. 2005b): it happens when the antenna receives the same signal reflected from various obstacles (i.e., building walls), then multiple copies of the same signal reach receiver antenna at various times, it can introduce the effect of jumping of the coordinates determined by GNSS receiver. Although software incorporated in receivers often tries to detect such a situation and to limit influence of this phenomenon, however, it still might have an effect on the accuracy of obtained results. The requirement of visual contact between receiver and transmitter antennas rules out the possibility of GNSS employment for navigation within buildings.

In another area, errors of transmitting and receiving devices are located. They could be caused by various factors such as instability of internal clocks frequency (most GNSS receivers utilize quartz-based generators, which are easily influenced by temperature fluctuations), internal noises within transmitting and receiving devices, and variability of antennas phase centers. Likewise all electronic devices, navigation satellites are prone to hardware as well as software errors and malfunctions, which could influence acquired positioning level accuracy or even completely disabled services in given areas. In some rare cases, particular GNSS might broadcast wrong information (e.g., 11 h of GLONASS failure in April of 2014), thus rendering the provided services unusable.

For military controlled system such as GPS and GLONASS, there is a possibility of temporal degradation of provided service in the selected areas. Also malicious third parties might jam the navigation signal on selected areas although such a behavior is restricted to an extent in some countries, for example, in the United States. In addition, there is a possibility of spoofing of navigation signal. The purpose of this action is to mislead the receiving device so that it determines its coordinates based on fake signal. Techniques applied in order to fulfill this task might include utilization of simulator of particular GNSS that is connected to the physical antenna, or controlled retransmission of signals transmitted by original navigation satellites. It is worthy to mention that especially civil receivers are prone to this issue. For selected applications (military based), GPS and GLONASS provide additional encrypted services, which reduce vulnerability for spoofing and jamming. GALILEO also would provide encrypted navigation services for selected companies and government agencies such as police, search and rescue operators, humanitarian aid, fire brigades, health services, defense, coastguard, border controls, customs and civil protection units as well as critical infrastructure and networks.

In order to mitigate atmospheric errors, corrections provided by SBAS might be utilized (approximate accuracy 2 m); moreover, there is a possibility of utilization of professional two frequency (in case of GPS or GALILEO) or two band (for GLONASS) receivers, which can autonomously rule out this kind of errors (approximate accuracy 1 m) (Table 25.6).

25.3.1 Troposphere and Ionosphere Errors Mitigation Techniques

During the course of Research on the EGNOS/Galileo in Aviation and Terrestrial Multi-sensors Mobility Applications for Emergency Prevention and Handling (EGALITE) project, our team created application GPS 3D Viewer, which designates coordinates based on RAW GPS data. Various algorithms calculating the influence of troposphere and ionosphere on acquired positions were included. Implemented model of troposphere signal propagation delay is described by the following equation:

$$TC_i = -(d_{hyd} + d_{wet}) \times m(El_i) \tag{25.1}$$

where values of $[d_{hyd}, d_{wet}]$ are calculated based on the receiver antenna height above sea level and are estimated by five meteorological parameters: pressure (P [mbar]), temperature (T [K]), the vapor pressure of water (e [mbar]), change of temperature (β [K/m]), and change in the rate of evaporation (λ [dimensionless]). Value of every meteorological parameter is calculated for

TABLE 25.7 Seasonal Values for GPS Receiver's Troposphere Corrections

Latitude (°)	ΔP (mbar)	ΔT (K)	Δe (mbar)	$\Delta\beta$ (K/m)	$\Delta\lambda$
15° or less	0.00	0.00	0.00	0.00e−3	0.00
30	−3.75	7.00	8.85	0.25e−3	0.33
45	−2.75	11.00	7.24	0.32e−3	0.46
60	−1.75	15.00	5.36	0.81e−3	0.74
75° or more	−0.50	14.50	3.39	0.62e−3	0.30

receiver latitude (φ) and day in astronomical year (D) (which starts on January 1); in addition, average seasonal changes (presented in Table 25.7) are taken into account. Values of particular meteorological parameters, denoted for simplification as (ξ), are calculated with the following formula:

$$\xi(\varphi, D) = \xi_0(\varphi) - \Delta\xi(\varphi) \times \cos\left(\frac{2\pi(D - D_{\min})}{365.25}\right) \quad (25.2)$$

where

$D_{\min}$= 28 for northern latitudes

$D_{\min}$ = 211 for southern latitudes

[ξ_0, $\Delta\xi$] represent mean seasonal fluctuation of values of parameters for latitude of receiver antenna

When latitude is lower than $|\varphi| \leq 15°$ or higher than $|\varphi| \geq 75°$, values of [ξ_0] as well as [$\Delta\xi$] are taken directly from Table 25.8. For latitudes in the range $15° < |\varphi| < 75°$ values of [ξ_0, $\Delta\xi$] are calculated using the following equations:

$$\xi_0(\varphi) = \xi_0(\varphi_i) + [\xi_0(\varphi_{i+1}) - \xi_0(\varphi_i)] \times \frac{\varphi - \varphi_i}{\varphi_{i+1} - \varphi_i} \quad (25.3)$$

$$\Delta\xi(\varphi) = \Delta\xi(\varphi_i) + [\Delta\xi(\varphi_{i+1}) - \Delta\xi(\varphi_i)] \times \frac{\varphi - \varphi_i}{\varphi_{i+1} - \varphi_i} \quad (25.4)$$

Seasonal values:

Mean values:

Delays [Z_{hyd},Z_{wet}], for the level of the sea are calculated with following formulas:

$$Z_{hyd} = \frac{10^{-6} k_1 R_d P}{g_m} \quad (25.5)$$

TABLE 25.8 Mean Values for GPS Receiver's Troposphere Corrections

Latitude (°)	P_0 (mbar)	T_0 (K)	e_0 (mbar)	β_0 (K/m)	λ_0
15° or less	1013.25	299.65	26.31	6.30e−3	2.77
30	1017.25	294.15	21.79	6.05e−3	3.15
45	1015.75	283.15	11.66	5.58e−3	2.57
60	1011.75	272.15	6.78	5.39e−3	1.81
75° or more	1013.00	263.65	4.11	4.53e−3	1.55

$$Z_{wet} = \frac{10^{-6} k_2 R_d}{g_m(\lambda + 1) - \beta R_d} \times \frac{e}{T} \quad (25.6)$$

where

k_1 = 77.604 K/mbar

k_2 = 382,000 K²/mbar

R_d = 287.054 J/kg/K

g_m = 9.784 m/s²

[d_{hyd},d_{wet}] are calculated with

$$d_{hyd} = \left(1 - \frac{\beta H}{T}\right)^{g/R_d\beta} - Z_{hyd} \quad (25.7)$$

$$d_{wet} = \left(1 - \frac{\beta H}{T}\right)^{((g_m(\lambda+1))/R_d\beta)-1} - Z_{wet} \quad (25.8)$$

Parameter (g) is equal to 9.80665 m/s², height of receiver (H) is measured in [m] above the sea level.

Troposphere correction function for the satellite elevation (m (E_i)) is calculated using following equation:

$$m(E_i) = \frac{1.001}{\sqrt{0.002001 + \sin^2(EL_i)}} \quad (25.9)$$

The important factor is: function is incorrect for elevations less than 5°. Finally ith satellite troposphere delay error is equal to

$$\sigma^2_{j\,tropo} = (0.12 \times m(E))^2 \quad (25.10)$$

and vertical troposphere error is σ_{TVE} = 0.12 m.

The influence of Earth's environment on positioning accuracy was presented on 3D graphs. Application responsible for visualization, GPS3D Viewer, gets RAW data from Septentrio PolaRx-3 receiver and calculates antenna position with scheduled frequency. Data presented in this work were sampled with frequency equal to 1 Hz in an experiment lasting 24 h (Cyran et al. 2011). Gathered samples are presented on 3D graph with violet points (see Figures 25.5 through 25.8). Obtained results show clearly that in horizontal plane, GNSS-based coordinates have deviation at the level of 1–2 m; however, accuracy in the vertical plane is worst and, depending on the conditions, could be a few meters. Consecutive figures show influence of taking into account the effect of ionosphere and troposphere on positioning accuracy. Figure 25.6 presents antenna position when troposphere and ionosphere corrections are turned off, while Figures 25.7 through 25.8, present the influence of these corrections on achieved accuracy.

FIGURE 25.5 Influence of Earth's environment on positioning accuracy for troposphere corrections OFF, Ionosphere corrections OFF.

FIGURE 25.6 Influence of Earth's environment on positioning accuracy for troposphere corrections OFF, ionosphere corrections ON.

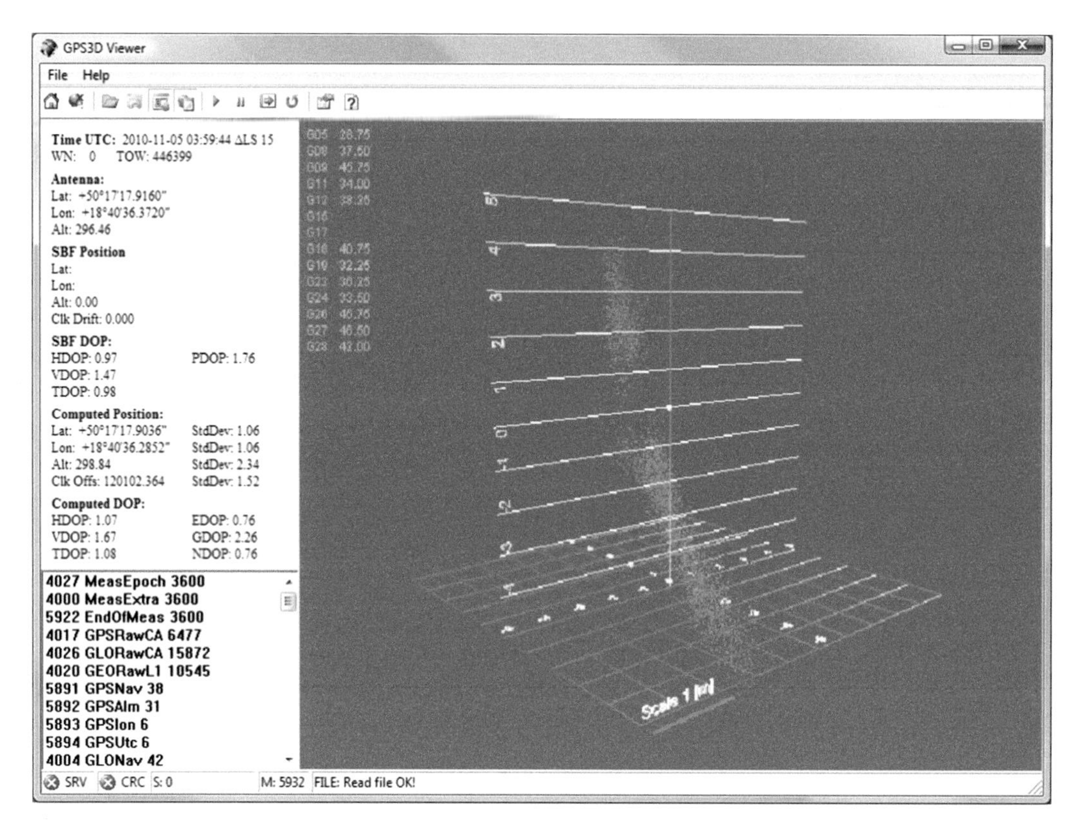

FIGURE 25.7 Influence of Earth's environment on positioning accuracy for troposphere corrections ON, ionosphere corrections OFF.

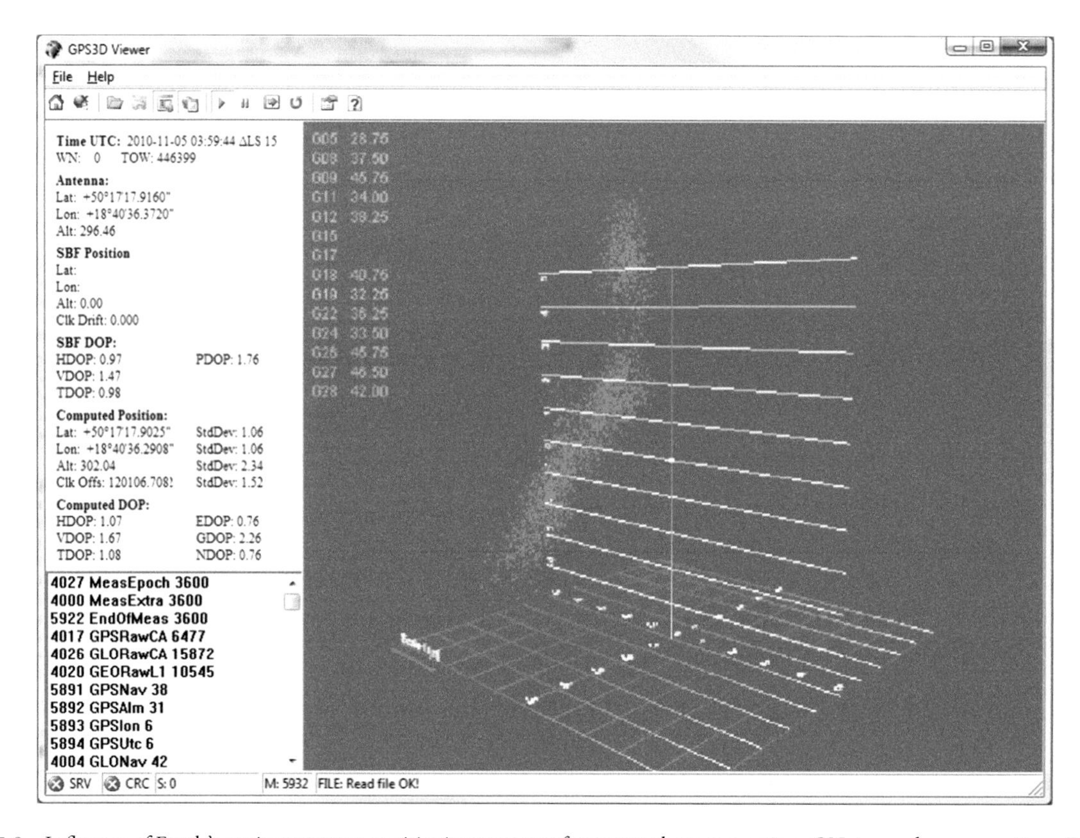

FIGURE 25.8 Influence of Earth's environment on positioning accuracy for troposphere corrections ON, ionosphere corrections ON.

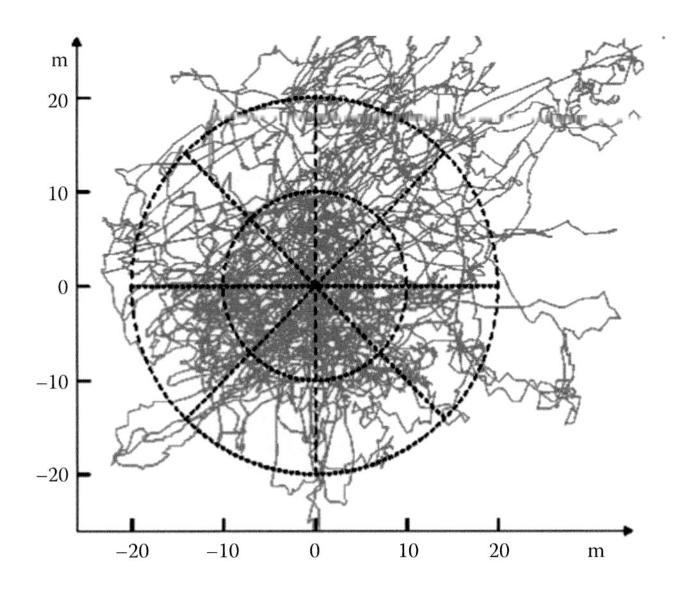

FIGURE 25.9 Graph of coordinates obtained from Samsung Galaxy III internal GNSS receiver, when GPS was utilized (24 h period).

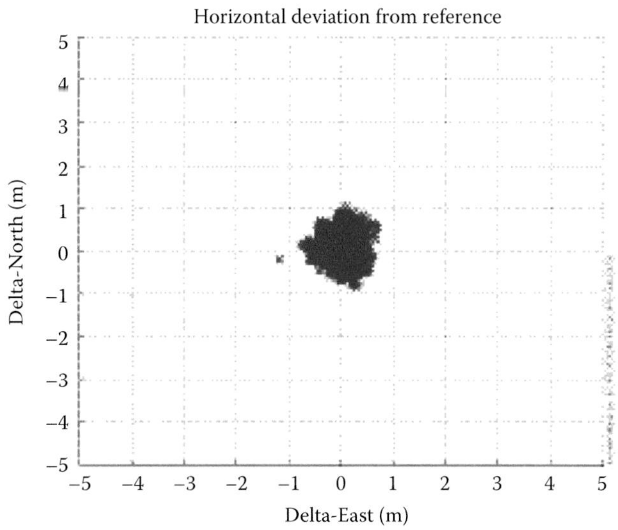

FIGURE 25.10 Graph of coordinates obtained from PolaRx-3 receiver when GPS + SBAS + two frequencies were utilized (24 h period).

25.4 Consumer Grade GNSS Positioning Precision

Typical mobile devices (mobile phones and tablets) available in the market are equipped with multichannel GNSS receivers, which allow for acquisition of the signal from many navigation systems at the same time (GPS, GLONASS, GALILEO); moreover, usually, acquisition of augmentation signal is also enabled. List of typical receivers is presented in Table 25.9.

Samsung Galaxy SIII, as a representative of typical mobile device, was utilized in order to present GNSS precision issues. Special application was prepared and installed in the device. Task of this application was to log co-ordinates of the device to internal database. Outcomes of the experiment were compared with results obtained from Septentrio PolaRx-3 receiver, which is widely utilized in aeronautics. During the experiment, receivers were lying stationary in the same location, in order to provide the same conditions for both devices. Positions of receivers were logged for 24 h, with a frequency of 1 Hz. Based on positioning graphs, estimation of errors was performed; according to results obtained (see Figure 25.9), Samsung Galaxy SIII is characterized by mean error at the level of 15 m while PolaRx-3 is characterized by mean error at the level of 2 m (Figure 25.10). There are many reasons for such a huge difference in precision. The receiver built in Samsung Galaxy SIII obtains positioning data based on one frequency only. In addition, it has small antenna and lower sensitivity, whereas PolaRx-3 is a dual-frequency receiver (GPS L1 and L2 frequency) and it is equipped with lightweight high precision geodetic dual-frequency antenna PolaNt.

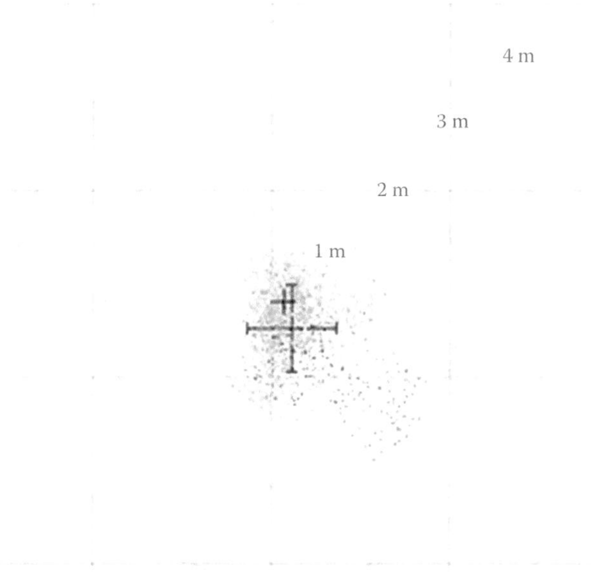

TABLE 25.9 Typical Receiver Chips Utilized in Mobile Devices

Manufacturer	Chipset	Channel Number
SiRF (CSR)	SIRFstarIII	20
SiRF (CSR)	SIRFstarIV	40
MediaTek	MTK3318	51
MediaTek	MTK3339	66
MaxLinear	Mxl800sm	12

FIGURE 25.11 Graph of coordinates obtained from Samsung XPERIA S internal GPS + GLONASS receivers, when GPS and GLONASS were active (24 h period).

There is a possibility of limitation of errors when mobile device is able to utilize more than one GNSS such as Sony XPERIA S, which is equipped with dual system receiver (GPS and GLONASS). Under the same conditions, this device was able to achieve precision at the level of 3 m when both GPS and GLONASS data were taken into account (Figure 25.11). Interestingly when only GLONASS was utilized, the positioning precision was at the level of 18 m (Figure 25.12).

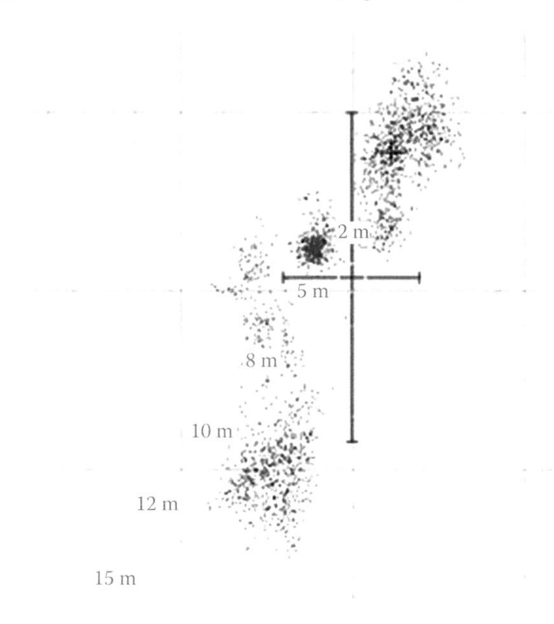

FIGURE 25.12 Graph of coordinates obtained from Samsung XPERIA S internal GLONASS receiver when only GLONASS was active (24 h period).

25.5 Low-Cost Receivers

Nowadays, cheap GNSS receivers, with parameters similar to professional chips utilized by aeronautics, are available in the market. Authors of this chapter conducted studies, in which GNSS receivers, produced by U-Blox company, were employed. To test applicability of these chips, custom device has been developed, schematic is presented in Figure 25.13. GNSS receiver was based on the low-cost U-Blox LEA-6S chip.

During precision tests, receivers were stationary and data were gathered for 24 h. Results presented in Figure 25.14 show that LEA-6S exhibits the precision of 2.5 m when GPS + SBAS reference signal was utilized, it is close to the precision obtained by PolaRx-3 Septentrio receiver. In order to present precision that can be obtained at the level of different GNSS, NEO-7 receiver, which can determine position with GPS, GLONASS, GALILEO, and QZSS, was utilized. Result of survey, of two navigation systems (GPS and GLONASS), for experiments that took 24 h and frequency refresh rate, was equal to 1 Hz; this is presented in Figures 25.15 and 25.16 (note that scale of axis is different).

Importantly, the NEO-7 chip is able to work in the PPP mode (precise point positioning), a method that offers enhanced positioning precision (see Figure 25.17) by utilization of the carrier phase measurements to smooth calculated pseudoranges between receiver and satellites antennas. The algorithm needs continuous carrier phase measurements to be able to smooth the pseudo-range measurements effectively. Additionally, ionosphere corrections like those received from SBAS or from GPS are required. Unfortunately, technology works correctly when un-obscured sky view is available.

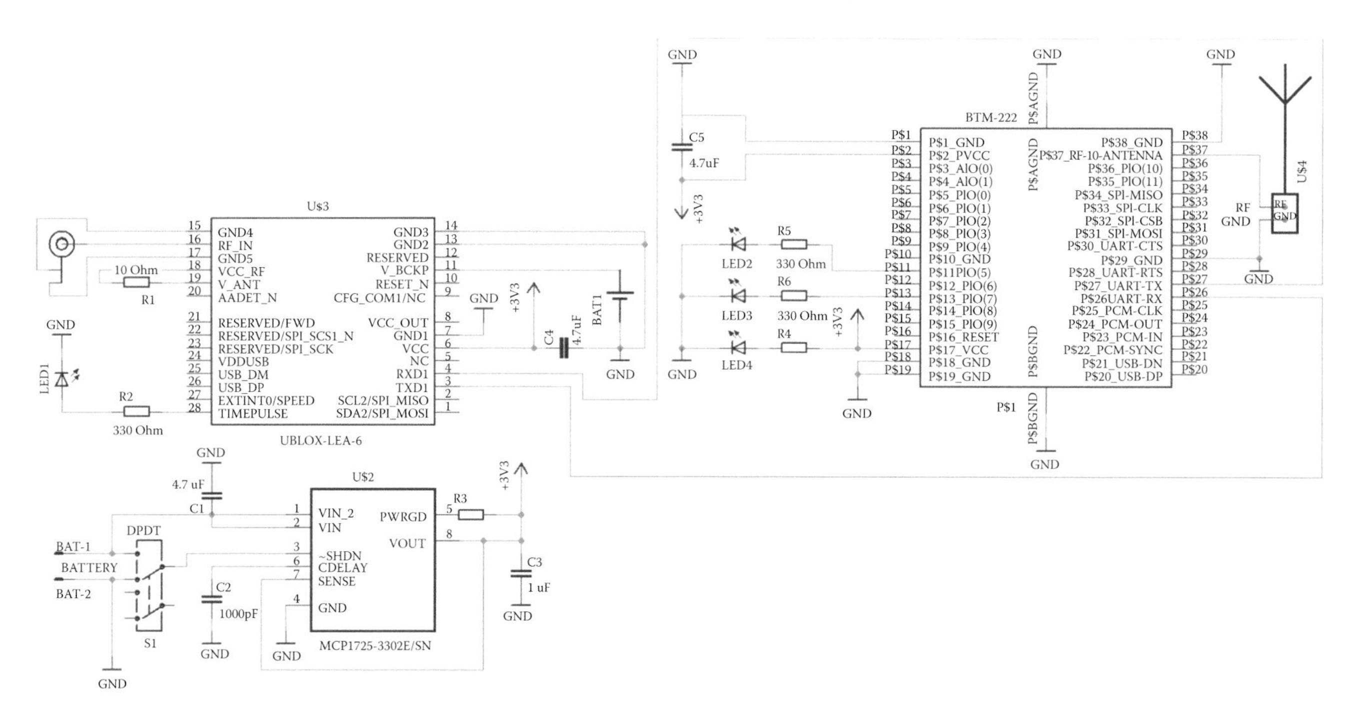

FIGURE 25.13 LEA-6S GPS receiver schematic.

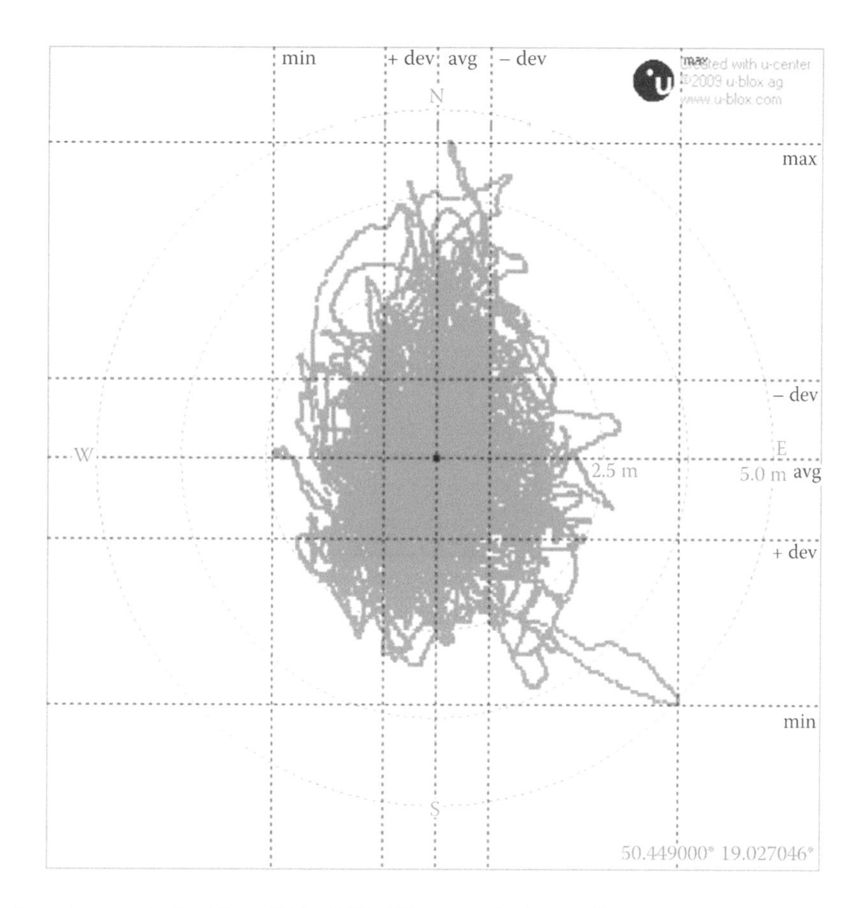

FIGURE 25.14 Graph of coordinates obtained from LEA-6S (GNSS) receiver (24 h period).

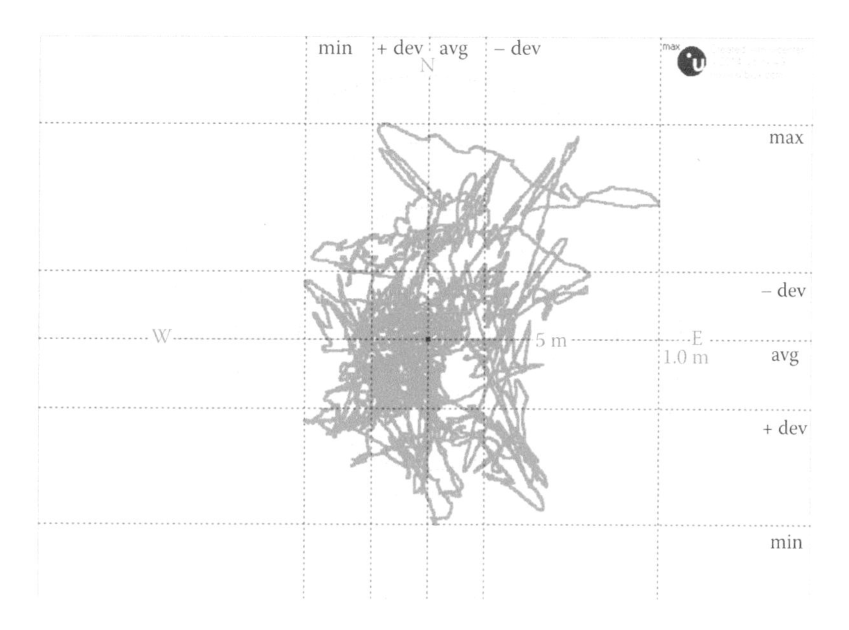

FIGURE 25.15 Graph of coordinates obtained from NEO-7 (GNSS) receiver when GLONASS was active. Sampling frequency 1 Hz (24 h period).

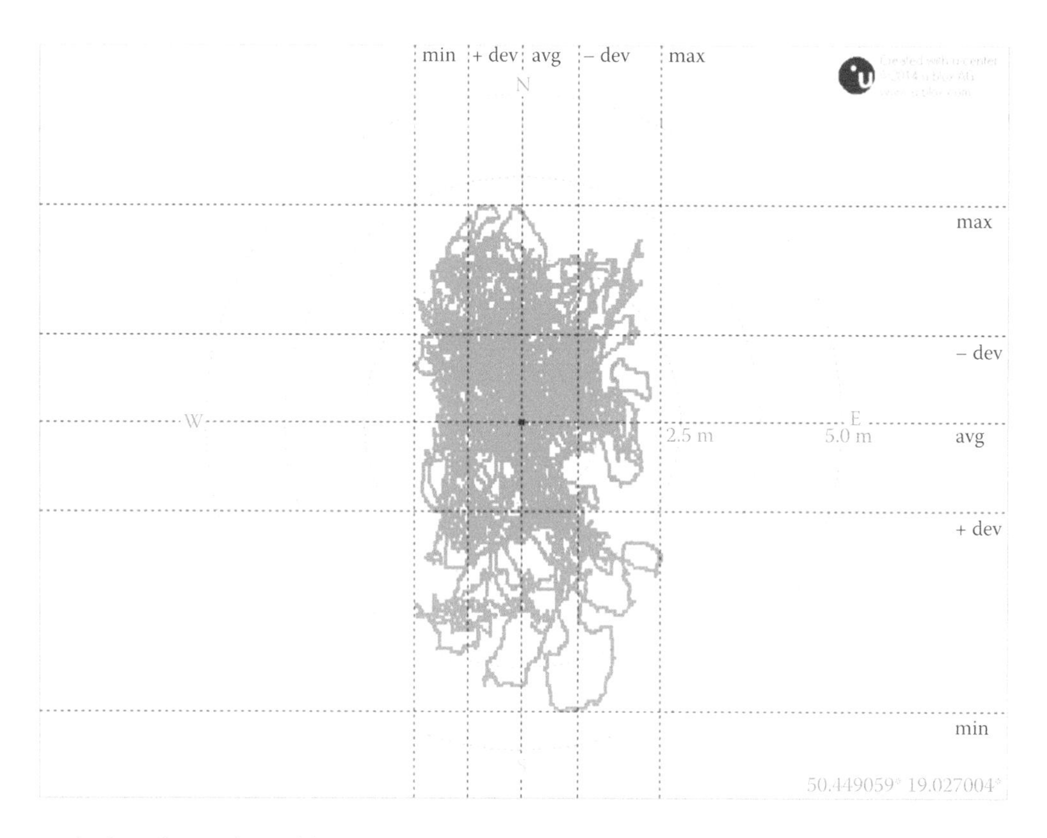

FIGURE 25.16 Graph of coordinates obtained from NEO-7 (GNSS) receiver when GPS was active. Sampling frequency 1 Hz (24 h period).

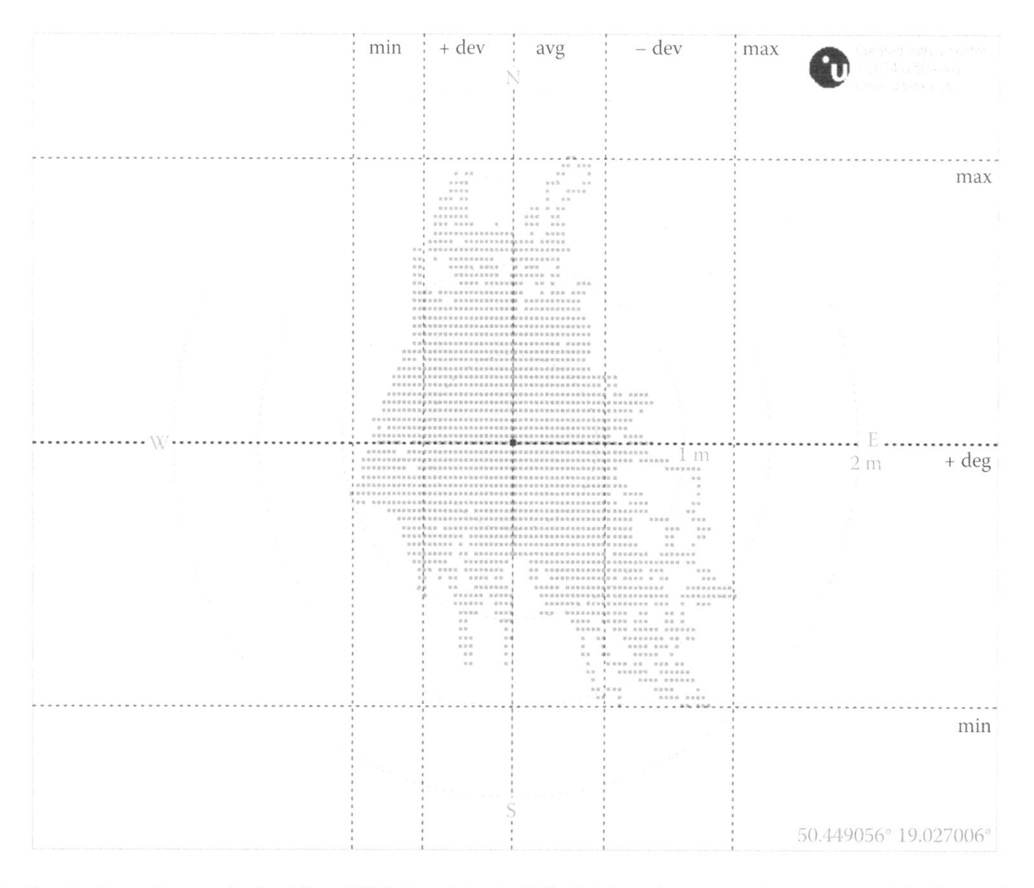

FIGURE 25.17 Graph of coordinates obtained from NEO-7 working in GPS+SBAS mode when PPP was activated (24 h period).

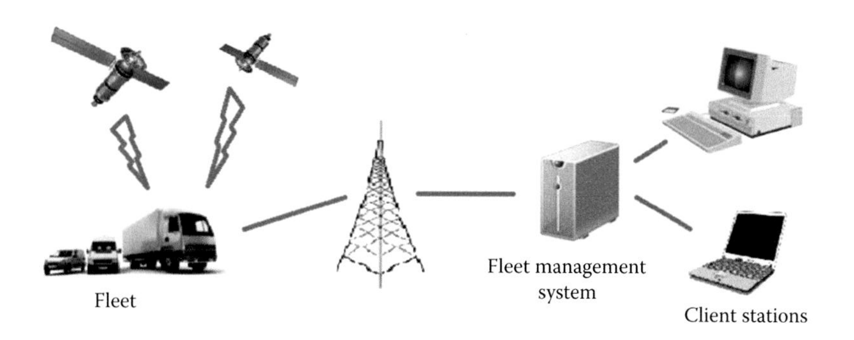

FIGURE 25.18 Fleet management system.

25.6 GNSS Applications

GNSS play important role in various fields (see Table 25.10). Basically, two main types of applications of these services can be distinguished: civilian and military. However ,within these fields, many common categories can be distinguished.

25.6.1 Consumer Applications

GNSS-based devices and applications play important role in everyday life of millions of peoples. The ability of precise determination of location allows for the creation of various applications such as interactive maps with the ability to localize the closest services in the area (such as restaurants, hotels, etc.), automatic location aware reminders (Hariharan et al. 2005), activities trackers that allow for the visualization of people movement and creation of various statistics, automatic photo-taggers that add coordinates to photos, objects trackers (Hasan et al. 2009) (e.g., cars, phones) for determination of location of stolen or lost goods, friend's trackers for searching of acquaintances in the area, and many others. However, the most popular seems to be various navigation applications, which can automatically calculate the best route to the destination location, and present it on the screen as 2D/3D map or in live view mode (Huang et al. 2012). What is more: if the driver misinterprets directions provided by the system, these applications can instantly work out new route. It hugely reduces time of trip preparation. Furthermore, by utilization of proper algorithms that are able to determine the best routes, limitation of the distance that have to be traveled could be achieved. Another useful feature of navigation systems is their ability of utilization of database of speed limits (Bargeton et al. 2010). It could allow for fuel saving, reduction of air pollution, and increase safety on roads.

TABLE 25.10 Application of GNSS

Categories	Applications
Consumer	Precise determination of user location
	Various navigation applications
	Localization of the closest services
	Automatic, location aware, reminders
	Activity trackers
	Automatic photo taggers
Industry	Determination of precise location of equipment and employees
	Location of potential clients (transportation)
	Monitoring of tourist flow (e.g., ski resorts)
	Planning of actions and processes
	Optimization of various processes
Archeology	Surveying of archaeological sites
	Site map creation
Geodesy and cartography	Preparation and update of maps
	Onsite measurements
	Analysis of geological and tectonic movements
Military	Co-ordination of military actions
	Soldiers localization
	Homing missiles
	Detailed map creation
	Rescue operations
Aviation	Primary navigation system or support for ground-based navigation systems

In combination with other services such as wireless internet access, navigation application can check current traffic statistics (e.g., existence of traffic jams on the route, road works) and modify the route on the fly in order to circumvent problematic areas (Bar-Gera 2007). If such data are collected from many users simultaneously, then they can be utilized for traffic pattern recognition and planning of traffic issues solutions. However, in such a case, privacy standards should be maintained, so that users that provide the data can keep anonymity. It might be hard in some cases, because even if the name of the user could not be included in the package sent to the data collecting center, the user can be identified by many different factors, for example, the place of living or place of work. Therefore, various methods of deeper anonymization of data were prepared: for example, it can obtained by simulation of system of induction loops embedded in roads; special location are designated and data are collected only when vehicles drive over such a virtual marker.

Application designated for wearable devices can also benefit from the ability of current location obtainment. These applications can learn human behaviors and predict future user locations. It will give ability of automatic accommodation to user environment, for example, in order to increase productivity (Ashbrook and Starner 2003).

25.6.2 Industry Applications

GNSS-based systems are widely utilized in various workplaces and industrial applications. Employers can hugely benefit from utilization of GNSS technology, especially when it is combined with wireless means of communication. Ability of precise location of equipment as well as employees allows for simplification of control processes and reallocation of resources (e.g., for soil excavation process, in order to control movement of hauling units, loaders, and excavators; Ahn et al. 2011); it can also reduce significantly costs.

Many fields can profit from utilization of this kind of systems, especially those in which employees are working outside of the company headquarters, such as transportation municipal services (street sweeping, snowploughs, garbage collection, road mending), sales departments, emergency services, agriculture, etc. GNSS-based systems are successfully utilized in the area of construction works for the purpose of localization of land and marine construction sites, for example, in offshore oil exploration.

Application of these systems combined with logging of obtained data allows for creation of reports and introduction of various optimizations. Employers gain control, because there is a possibility of validation of realization of task by employees. Routes can be checked and all deviations from the planned path, or omission of planned locations, determined. Current and historical positions of chosen unit can be obtained, detection of entering or leaving of selected area might be performed (so-called geofencing), and various geoalerts might be set. On the other hand, application of GNSS technology can support workers, that is, when salary that was paid is insufficient for the time of work, then readouts obtained from GNSS-based system can confirm or reject legitimacy of worker's complaint. In many cases, utilization of these systems allowed the detection of law violations such as stealing of company's possession, or improper habits such as utilization of company resources for personal purposes. It is worthy to mention that abidance of driving laws can easily be determined, it can reduce the amount of accidents and improve safety. Self-employed persons, such as taxi drivers, can also benefit from utilization of GNSS services. Special GNSS-based dispatching systems, which allow for easy location of passengers and determination of temporal hot spots, allows for optimization of routes, reduction of passenger await time, and increase of driver's incomes (Hou and Chia 2011). In addition, data possessed from systems located in taxis are often utilized by scientists in order to conduct various analyses and researches, for example, automatic land-use classification (Pan et al. 2013), traffic flow patterns, assessment of road quality, etc.

Tourist resorts can automatically acquire information about locations that are visited by guests at various times across the day, tourist moving patterns as well as typical activities in which tourists are involved (Skov-Petersen et al. 2012). It allows for introduction of modifications in various areas such as offered services, resort layouts, travel recommendations presented to the tourists, etc. (Zheng et al. 2009). Ski resorts can monitor skier's flow and accommodate tracks accordingly; also safety might be improved. Tracking can be done through utilization of custom devices or with tourist mobile devices equipped in special application. In order to facilitate sightseeing and surveying of interesting localizations, automatic guide system based on GNSS services can be prepared; it can be combined with aforementioned statistic gathering system so that data can be collected. It is worthy to mention that interesting researches are also done in the field of geotagged photos utilization for the purpose of extraction of travel patterns (Zheng et al. 2012).

Although GNSS proved to be a useful tool in company environment, there are some legal issues, depending on the country or region in which such methods are applied, that should be taken into account at the level of planning and implementation of GNSS-based solutions. In particular, the issues of workers privacy should be taken into account. Also workers' co-ordinates should not be logged in their private time (after work or during brakes) and special zones that are excluded from logging should be established (Inks and Loe 2005, Towns and Cobb 2012).

25.6.3 Transportation Applications

Public transportation benefits hugely from a combination of GNSS with wireless communication abilities. Means of transportation equipped in such devices can constantly communicate their location to the control centre and then information can be posted at bus/tram stops as well as at web page. It can also be distributed to dedicated mobile applications utilized by passengers. Importantly, such system can be integrated with in-vehicle information systems; thus, information about next stop can be presented to the passengers, without driver's action. Further integration, with city traffic control systems, is often implemented, so traffic lights are controlled in a way to give priority for public transportation vehicles (Bain 2011). When actual vehicle position is combined with historical data and information about current state of traffic flow, predicted time of arrival of bus or tram at the line stops can be calculated. As a result, not only public transport controllers can rapidly react at emerging situations and, for example, dispatch additional resources or redirect routes of particular lines, but also passengers can dynamically modify their travel plans and utilize different lines in order to reach their destinations faster. Application of such systems could reduce frustration of the end users in case of delays; in addition, it supports the idea of utilization of public transportation, thus counteracting formation of traffic jams and decrease pollution level. When additional sensors are incorporated, for example, automatic passengers counters, there is a possibility of creation of detailed passenger load statistics for all line's segments. Implementation of such system, which ceaselessly collects data from various means of public transportation, can improve process of schedules modification (in order to obtain better time coverage of some areas of the city), determination of the number of required vehicles, optimization of routes on particular lines, and modification of timetables, so they accommodate real arrival times.

It is worthy to mention that such systems installed in school buses, through automatic notifications about delays, can hugely facilitate lives of parents and increase their trust in the transportation system.

25.6.4 Surveying Applications

GNSS positioning systems play important role in the field of geodesy, archaeology, landscape surveys, and map creation (Ainsworth and Thomason 2003, Bargeton et al. 2010, Bussios et al.). It is a great tool for mapping large areas or difficult-to-access locations (Bargeton et al. 2010). GNSS receivers can be carried by surveyor or mounted in the car and automatically gather coordinate data. The main advantage of the utilization of GNSS technology is the analysis of obtained data can be done on site or raw data, obtained from the receiver, can be recorded and further processed in offline mode. GNSS-based measurements can be done relatively quickly when compared to the utilization of theodolite, tape, or electromagnetic distance measurement methods (Figure 25.19).

Utilization of GNSS technology in the field of archaeology has a long history; it was successfully applied to small-scale projects such as mapping of single archaeological objects (stones/walls etc.) as well as large ones. Application of this technology significantly improves the accuracy of obtained result, it limits the utilization of other geodetic techniques such as tapes measurements, and speeds up the process of coordinate generation. As a result, the reduction in the number of personnel experienced in geodesy measurements involved in the realization of these kinds of projects is possible. The utilization of GNSS technology perfectly fulfills the requirements of the surveying processes: this requires determination of the location of the site in relation to the map, followed by the determination of the location of all points of interest on the site level and determination of the relation between given point and all other points that are localized on analyzed site. In some cases, accuracy provided by pure GNSS solution is insufficient; therefore, in order to increase precision, additional techniques, that can increase accuracy of measurement to centimeters, are often employed. The most popular one is utilization of dedicated GNSS-based reference stations (if such are available on selected area) and application of differential corrections (Cosentino et al. 2005) or Real Time Kinematics (Wei et al. 2010). Deployment of stations is usually done at the national level. For every reference station, very precise coordinates are determined. There are two types of stations: active ones are equipped in stationary GPS receivers, operate all the time, and usually share data through dedicated web sites. They are usually scattered around the area: for example, in Great Britain there were 32 active stations based on GPS (data for December 2002), any point of Great Britain should be within 100 km from the closest active stations; in addition, urban areas are usually covered by more than one active station. If higher level of accuracy is required, passive stations might be employed. This kind of station is usually realized as a ground maker, which does not possess its own GNSS receiver; therefore, user, in order to utilize this station, has to provide its own equipment that is located at passive station site for the time of survey conduction. In Great Britain, there were around 900 such markers; any point of land should be within 20–30 km from the closest marker. Data for differential correction can also be obtained from SBAS.

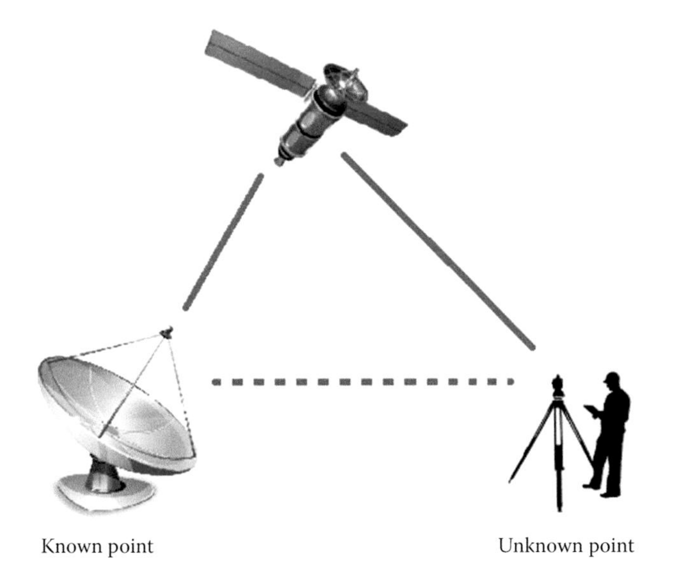

FIGURE 25.19 Utilization of dedicated GNSS-based reference station.

As mentioned earlier, Earth's continents are in constant move: GNSS in combination with fixed reference stations, on the surface of the ground, are great tool for the measurement of such relocations; level of oceans can also be monitored (Yang and Lo 2000). In general, GNSS are valuable tool for measurement of object deformation (e.g., damns), and in engineering survey applications (Frei et al. 1993). Furthermore, scientists investigate ways to utilize sensitivity of GNSS-based devices for measurement of various disturbances in atmosphere. Such systems can be applied in meteorology for determination of the amount of vaporized water in the atmosphere, and current researches point out that they can also be applied to volcanic ash detection; therefore, improvement of the early volcanic ash alerting system in affected areas might be improved (Aranzulla et al. 2013). As a result, flight safety and comfort of population inhabiting these areas might be improved.

25.6.5 Accident and Disaster Recovery Applications

Studies conducted for Germany point out that in 2002, there were more than 45 private cars per 100 inhabitants and on average, every vehicle was characterized by annual travel distance at the level of 12,600 km (Kalinowska and Hartmut 2006). These results show the significance of mobility in contemporary society. Unfortunately, mobility requirement entails exposure of road users to various harmful events. Statistical data gathered for UE reveal that every year, over 1.5 million of people met with accidents; to make matter worse, as a consequence of occurrence of such events, more than 30,000 people annually lost their lives. It is a important social and economic issue. Researches for UE, done in 2000, estimated such costs at the level of 160 billion of euro per annum (Petersen 2000).

Outcomes of researches performed point out that in many cases, quicker delivery of medical help could reduce the number of casualties and decrease the time of postaccident trauma recovery. There are many issues when it comes to emergency services notification: often accident victims or witnesses cannot give precise location of event occurrence, it is usually especially hard in rural areas: the stated location might be incorrect. Another issue is that, in many cases, victims of accident lost their consciousness; however, there might be no witnesses to call for help, such a situation is common in uninhabited areas especially for accidents that occur during night hours. Up to 52% of fatalities could be avoided if victims were located earlier by medical personnel (5%), were transported to the hospital (12%) or advanced trauma centre (32%) soon (Henriksson 2001). The EU estimates that the introduction of efficient system of rapid emergency notification, which will decrease the amount of rescue time and speed up the process of medicine delivery, reduces severity of injures: up to 2500 lives can be saved and 26 billion of euro can be saved per annum.

The aforementioned analysis leads to the proposals of systems, which are responsible for automated detection of accidents and notification of emergency center. One such system was developed during the course of research on EGNOS/Galileo in Aviation and Terrestrial Multisensors Mobility Applications for Emergency Prevention and Handling (EGALITE) project. One of the aims of EGALITE project was the creation of efficient system of emergency handling and prevention base on mobile devices. Contemporary mobile devices (e.g., smart-phones, tablets) have feature-rich items equipped in many sensors (i.e., accelerometers, gyros, magnetometer) and receivers (i.e., GNSS receiver) they are easily accessible and widely available. Moreover, rules introduced by U.S. Federal Communication Commission (FCC) state that in United States, every mobile carrier must be able to determine 911 mobile caller longitude and latitude with a certain accuracy (Yang and Lo 2000). However, precise localization based solely on transmitting tower was, in the time of introduction of this rule, insufficient and it required costly investments in development of infrastructure. It was one of driving forces to equip mobile phones with GPS receivers, which were able to introduce a better positioning precision. Vast popularity of smart-phones allows for the utilization of implemented system by a sufficient fraction of the population without incurring any additional costs. In proposed systems, readouts gathered from accelerometer built in mobile device are analyzed by artificial intelligence based on feed-forward artificial neural networks. Artificial intelligence was trained with real data emerging from test car crashes and nonaccidents car runs. Test cases encompass various types of cars, speeds, and types of roads. As a result, system that is able to achieve high accuracy of accident detection (at the level of test set composed of 18,266,234 samples from accident and nonaccidents runs): sensitivity equal to 0.97 and specificity equal to 1 was created. In the case of accident occurrence, position based on GNSS signal is determined, then coordinates, time of event as well as user ID are sent to the emergency center through GSM network. The application of GNSS technology allows for quick and precise location of accident site, thus contributing to a significant decrease in the emergency service waiting time. In addition, time that is sent to the emergency centre is based on GNSS readouts; it provides precise time synchronization between various devices: as a result, there is a possibility of determination of groups of related events and accurate reconstruction of roads accidents in which many vehicles are involved is facilitated.

In order to automatically determine car environment (for activating or deactivating the application), the NFC module is utilized. An assumption is made that car should be equipped in NFC tag. When device detects the presence of proper NFC card, it automatically switches to active mode; in other cases, inactive mode is chosen. If user's device does not possess NFC capability, there is a possibility of manually choosing the proper mode. Interestingly, GNSS receivers also can be utilized in this area of application. Some creators of similar systems assumed that mobile devices must move with speed higher than some empirically determined threshold value, and for the purpose of speed determination, GNSS readouts were employed. However, such a solution consumes a lot of power, because vehicle position has to be continuously refreshed; in addition, this system will not detect accident when the car is stationary on the crossroad or in the traffic jam (Thompson et al. 2010, White et al. 2011).

In more platform-dependent solution, there is a possibility of integration of mobile accident–detecting application with air bags deployment controller; creation of additional hardware module, which contains high precision accelerometers, is permanently attached to the car's body and communicates with mobile device through Bluetooth technology (Matthews and Adetiba 2011). Such an approach can further reduce amount of power required and increase detection precision.

As mentioned earlier, important issue of utilization of GNSS receiver, built into mobile devices, is limited battery power. Although battery power consumption of specialized GNSS receivers in perfect conditions can be as low as 10 mW (e.g., Sony CXD5600GF), typical power utilization in modern device is higher at the level above 140 mW, while operating GNSS receivers are responsible for significant fraction of battery consumption. Traffic accident–detecting application can be combined with other services such as navigation or traffic jam avoidance system; in that case, constant utilization of GNSS receiver is justified. However, end users can consider continuous employment of GNSS receiver only for the purpose of accident-detecting application as a waste of energy/resources. Researchers try to determine approaches that can decrease utilization of power through application of various time patterns of coordinates obtainment. Theoretically GNSS receiver can be turned off all the time and utilized only in case of accident detection; however, the issue here is the time required for the first fix obtainment. After activation of the receiver, time of determination of location can span from single second up to a few minutes. This process is shorter (single seconds) when receiver possesses current information about almanac, ephemerides, and has synchronized clock with satellite time; however, it extends when such information is not available, and has to be downloaded directly from satellites. For example, for GPS-based system, collection of almanac data takes the longest time (at least 12.5 min in foster conditions, longer if GNSS signal is degraded by obstacles); however, data are valid for several months. If ephemerides have to be updated (which are valid up to 4–6 h; Dorsey et al. 2005), the process can take more than 30 s. When only time synchronization is required, the process of fixing takes a few seconds. Obtaining the first fix can be sped up if receiver is able to utilize wireless services that provide almanac, ephemerides, accurate time, and satellite status, for example, assisted GNSS (A-GNSS) (Harper 2009, Van Diggelen 2009). Importantly, then the position can be calculated within seconds even under poor satellite signal conditions. In addition, utilization of such services can improve positioning accuracy, although GNSS-based readouts without additional corrections should be sufficient for the purpose of emergency site location.

Noteworthy is the fact that GNSS receivers in combination with other sensors (such as pulse oximetry sensor) can support other types of emergency situations such as heart failure (Hashmi et al. 2005).

GNSS play important role in disaster rescue and recovery actions support. Emergency Centre personnel need fresh data about locations of rescue resources in order to make proper decisions in a timely manner (Ameri et al. 2009). Therefore, every rescue unit should be equipped with GNSS receiver that has the ability to automatically send correct location to the Emergency Center. In addition, precise interactive maps of the location, combined with automatic localization, allow for easy navigation in the areas in which characteristic landmarks were hindered by disasters, for example, flood (Van Westen 2000); it can also facilitate introduction of additional support units that are not familiar with terrain (e.g., rescue units from other regions/countries). Pilot programs conducted in Italy showed that if medical personnel are supported by management application, that is, utilizing GNSS readouts, then the response time can be reduced from 16 to 9 min (on average) (Bellini et al. 2013). Furthermore, unmanned autonomic carriers can be sent to the disaster site in order to gather visual data from chosen areas. Navigation of carriers and activation of the camera might be based on the location obtained from GNSS receiver. GNSS-based system can also support early warning of population about disasters such as earthquakes and volcanic eruptions.

25.6.6 Health Researches and Monitoring Applications

Combination of GNSS with mobile devices, which are carried ceaselessly by users, opens up new possibilities for various location services. Interesting application of such services is the trial of assessment of contextual effects, such as environmental conditions, on human's health state, and combination of health hazards with locations (Thierry et al. 2013) (Figure 25.20).

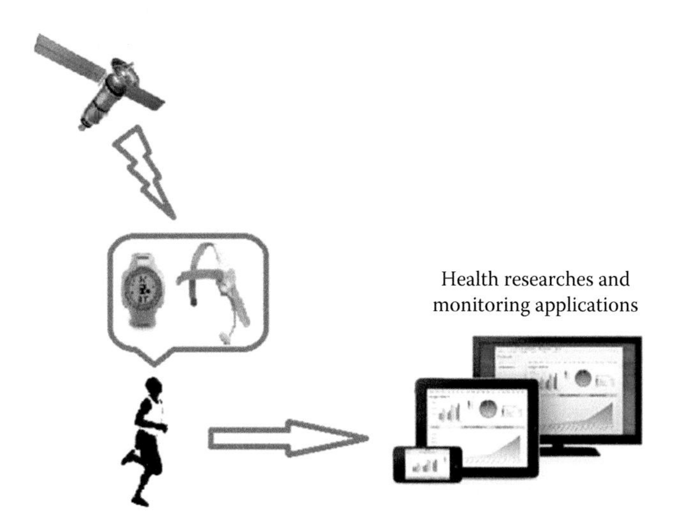

FIGURE 25.20 Application of GNSS for health researches and monitoring applications.

Many studies concerning contextual effects influence on health is focused on human residential area (Thierry et al. 2013); however, people spend a vast amount of time outside their homes. Therefore, the ability of easy collection of co-ordinate time series and combination of these data with information about users' activities opens new possibility to conduct research. However, some big issues with raw GNSS data are a vast amount of information that should be processed and lack of broader context. Therefore, subject of location activity is introduced and specialized algorithms are employed in order to detect activity place location, based on GNSS readouts (e.g., time distant cluster detection, kernel based, etc.). Then data might be analyzed from a wider perspective and influence of series of consecutive activities locations on each other might be analyzed. Such an approach might be useful for various researches, for example, in the determination of causality in epidemiological studies (Thierry et al. 2013). In addition, the determination of life-space can be the base for automatic estimation of older adults' activity level and their health status (Wan and Lin 2013).

25.6.7 Tracking Applications

Other area of GNSS-based tracking is utilized by researchers to track behavior of animals (Zagami et al. 1998). Various animals can be equipped with devices that transmit their locations. Such researches have many practical aspects, for example, leading to better understanding of causes of road collisions with animals or allow for the determination of animals' grazing areas (Rutter et al. 1997). GNSS collars are often coupled with various sensors (e.g., neck movement sensor), in order to improve abilities of automatic determination of animal activity. Issue in this area of application is battery lifetime. Tracking devices should be small and light, they should be virtually unnoticeable for the tracked animal; on the other hand, battery life time depends

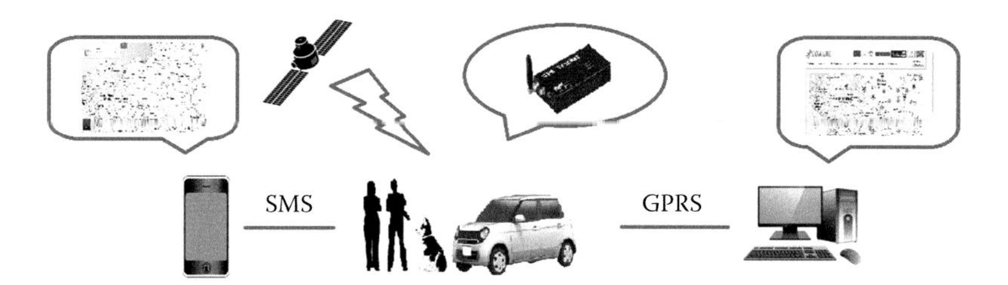

FIGURE 25.21 GNSS are widely utilized in tracking applications.

on its size and weight. When it comes to farm animals, replacement of power source is relatively easy; however, often replacement of batteries in devices designed for wild animals is not possible, because it introduces stress and might cause temporal modification of animal behavior. Therefore, devices utilized for this purpose are characterized by low frequency of coordinates obtained and limited communication with central server (Gottardi et al. 2010) (Figure 25.21).

When it comes to person localization, GNSS-based devices are perfect for systems that are responsible for children tracking. Parents can utilize such devices in order to check whether their descendants are in the proper location, for example, school/home. In combination with correct algorithms, deviations from daily routines can be determined and caretakers can be automatically informed. Similar systems can be utilized in case of persons with dementia or similar illness, that influence their ability of returning to the place of living (Landau et al. 2011). Also prison system could reduce costs, because commitment of minor crimes might be punished with home detention and tracking of convicts can be done with GNSS collars.

Monitored person should be aware about carried GNSS receiver; however, there is ongoing discussion whether caretakers can utilize such devices, without the knowledge or agreement of tracked person.

25.6.8 Unmanned Vehicles Applications

GNSS are utilized in various areas of autonomous and semiautonomous systems. Very promising field is application of this technology in autonomous land vehicle applications, for example, for open-cast mining, agriculture applications (Heraud and Lange 2009, Prakash et al. 2012), cargo, and human transportation (Manabu et al. 2006, Bevly and Cobb 2010, Ozguner et al. 2011) (especially in extreme conditions; Nagashima et al. 2013). Also domain of autonomous robots is developing quickly. Precise control process needs nonstop monitoring of the state of an object; however, refresh rate of typical GNSS receiver varies between 1 and 10 MHz; in addition, precision of particular GNSS is also limited, usually to a few meters. Furthermore, GNSS receivers are subject of errors, especially common are multipath errors and signal loss caused by occlusion of navigation signal by various obstacles, for example, high buildings in city centers. There are many problems to solve, because consequences of errors in autonomous system can be tragic. In order to overcome precision issues, often signals from many satellite navigation systems, supported by SBAS, are processed at the same time. It helps to improve precision; however, still there is a possibility of obtained coordinates getting affected by high errors or event complete loss of GNSS signal in tunnels. Therefore, advanced statistic techniques are applied in order to determine whether coordinates obtained with GNSS should be accepted and utilized. In order to be able to determine the state of an object between successive accepted GNSS readouts, obtain better insight into state of controlled process as well as increase system state sampling frequency and to be able to direct the process despite the fact that some GNSS readouts might be rejected, Internal Navigation Systems (INS) can be utilized. They are composed of additional sensors that give information about state of controlled object; as a result, when proper calculations are done, location of an object can be deduced. This group includes accelerometers, gyros, magnetometers, etc. Therefore, there is a possibility of determination of speed, and angle, of controlled object movement; thus, the calculation of its coordinates based on previously determined position, and results obtained from employed sensors, is possible.

Integrated battery of such supporting devices is referred to as internal measurement unit (IMU). Main advantage of these sensors is high refresh rate (typical 100 Hz; however, it can be more in case of utilization of industrial level devices); in addition, most of them is jamming and deception resistant (Sukkarieh et al. 1998, Zhong et al. 2008). However, a major problem of IMU application is fast accumulation of readout errors that, with time, leads to significant degradation of calculated parameters; therefore, periodical calibration is required. In consequence system controlling process cannot be solely based on IMU readouts. Combination of IMU with GNSS is a great way of obtaining high IMU frequency refresh rate with periodical calibration provided by GNSS. Usually at the heart of such systems Kalman filters are applied (Bullock et al. 2005, Groves 2008), which works in two stages: prediction—responsible for constant determination of coordinates, velocity as well as attitude based on IMU readouts, and estimation—responsible for determination of errors of aforementioned values and correction of their values when valid GNSS readout is available. In terrestrial applications, in order to minimize errors introduced by IMU, there is a possibility of live determination of values of its internal sensors biases (that are influenced by various factors, for example, temperature) which is done every time when the controlled object is stationary. Then obtained results are taken into account in prediction stage. In addition detection of stationary state triggers reset of internally tracked velocity.

Another way of improving of GNSS positioning accuracy is utilization of magnetic markers incorporated in road lanes (Hernandez and Kuo 2003) or application of radio-frequency identification (RFID)-assisted localization in combination with sensor network (Lee et al. 2009). Such systems are especially useful in emergency prevention and unmanned vehicles applications, that is, collision avoidance for vehicular ad hoc network. In such a system, precise location and ability of determination of ranges between cars are very important. Unfortunately, systems solely based on GNSS, even if supported by SBAS corrections, suffer because of precision issue that could be caused by local conditions. It happens because SBAS systems provide corrections, which are valid for large areas; therefore, they cannot perfectly fit to all locations, because they are characterized by various environmental factors (e.g., trees, city canyons, etc.). RFID-assisted concept assumes that roadsides of tracks, on which system is utilized, should be equipped with RFID tags that allow the calculation of GNSS corrections. Then vehicles could improve position precision obtained from GPS with RFID readouts and broadcast corrections through wireless radio to neighborhood cars. As a result, units that are not equipped in RFID readers, or are located on inner lane, outside of range of RFID tags, can also benefit from the system. Such an approach allows obtaining corrections fitted to local conditions, and enables precise determination of relation between moving objects; however, it is worthy to point out that it does not solve the issue of localization when GNSS signal is not available in city canyons or tunnels. Although additional equipment is required, RFID tags are already utilized on many roads, for example, for the purpose of automatic tool collection; therefore. such systems might be based on already existing infrastructure.

25.6.9 Time Synchronization Applications

GNSS services require precise time synchronization between satellites and receivers clocks. Therefore, other applications of such systems emerged. Precise time and frequency synchronization, among different sites on the globe, are important in many areas (Hewlett Packard Company 1996). High-frequency trading utilizes precise time in order to time-stamp transactions (Korreng 2010); mobile telecommunication requires precise time synchronization between base stations so that they can efficiently share common radio spectrum (Sterzbach 1997, Chunping et al. 2003). Precise time is important in all kinds of sensor networks, which collect data from the environment, for example, seismographic researches need time synchronization between sensors scattered around the globe, power companies utilize precise time for efficient energy distribution and pining down issues of disturbances in the network (Fan et al. 2005), physics can precisely measure various values (e.g., one-way light speed) (Gift 2010), observational astronomy needs time synchronization between data-collecting sites. Structural health monitoring (Kim et al. 2012) such as in pipeline transportation can precisely localize malfunction site through utilization of synchronized sensors and measurement of error signal time detection by individual sensors (e.g., acoustic ones) (Tian et al. 2010). Message encryption could also require precise time synchronization (Bahder and Golding 2004).

Utilization of GNSS services allows for achieving of time accuracy at the level of nanoseconds (HP 1996). A huge advantage of solutions based on GNSS is the low cost; as mentioned earlier, access to GNSS signal is free of charge; therefore, only maintenance of receiver infrastructure is required.

25.6.10 Military Applications

GNSS-based system plays important roles in military applications. It might be confirmed by the fact that GPS and GLONASS systems are controlled by army. Among many applications, these systems can be utilized by precise targeting units (Brown et al. 1999), support troops reallocation, and be incorporated into vehicles and soldiers navigation devices. It can hugely improve cooperation between troops and facilitates activities in unfamiliar terrains. GNSS is utilized by the army not only during military operations but also in civilian supporting activities, for example, during disaster recovery. Interestingly UE develops system that will be controlled by civilians and focused mainly on civilian applications; however, services for various government agencies also will be provided (Figure 25.22).

25.6.11 Aeronautic Applications

Contemporary GNSS became popular tool in the area of avionic navigation (Clarke 1998). Utilization of these systems is regulated by regional legislations (Andrade 2001). In general, when pure GPS or GLONASS signals are utilized, GNSS-based navigation equipment that is incorporated in plane is treated as supplementary system. It happens because there is a lack of monitoring of positioning precision within signal broadcasted by navigation satellites; it implies that information about not complying with predetermined accuracy levels, determined solely on readouts from these systems, is missing. In addition, precision of obtained coordinates in vertical plane is worse than those obtained in horizontal planes. Avionics applications require high precision and reliability of GNSS-based positioning in all spatial planes. As mentioned before, in order to overcome these issues, and allow aircraft navigation to be based primarily on GNSS, SBASs were deployed: for example, in Europe, ESA together with air traffic control organizations created EGNOS. These systems provide

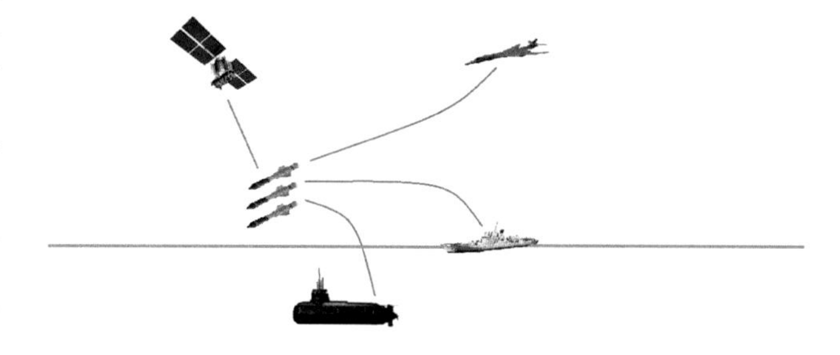

FIGURE 25.22 Various kinds of military equipment utilize GNSS.

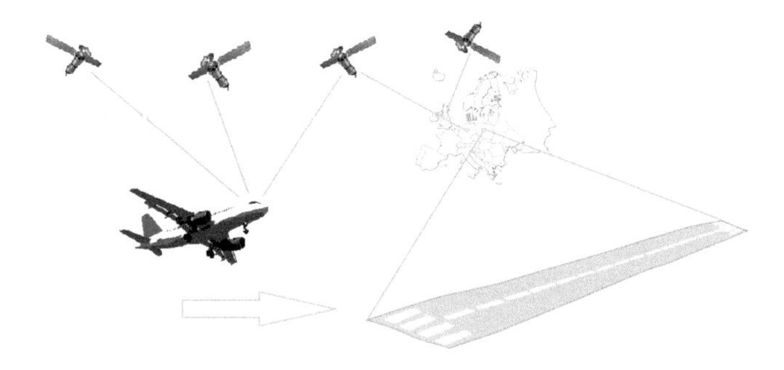

FIGURE 25.23 GNSS allow for determination of precise coordinates of planes.

information about GNSS failures and at the same time increase positioning accuracy. Various organizations (e.g., Institute of Computer Science in Silesian University of Technology) are involved in constant monitoring of performance correctness of GPS, EGNOS, and GLONASS systems. Positioning accuracy, accessibility, and reliability of data provided by GNSS satellites are collected and sent to EGNOS Data Collection Network (EDCN) that is controlled by EUROCONTROL, the organization that is responsible for ceaseless monitoring of accuracy of functionality of GNSS for the purpose of air navigation in European airspace (Figure 25.23).

The huge advantages of these systems are reliability and cost reduction of infrastructure. Therefore, development works are conducted that should result in increase of involvement of GNSS in this field. Application of GNSS improves work of air traffic control, ability of precise location of airplanes leads to avoidance of collision and optimization of routes for reduction of travel time and fuel consumption. Promising area of GNSS-based system incorporation involves support (and in the future automation) of procedure of aircraft approach to landing so-called localizer performance with vertical guidance (LPV) procedures. It facilitates the optimization of landing approaches paths and reduces the amount of financial resources that have to be devoted for airfield infrastructure (e.g., it can rule out necessity of investment in instrument landing system [ILS]), at the same time airplanes can approach runways even in conditions of low visibility. Implementation of LPV procedures has many advantages; some of them are:

- Can be utilized on thousands of airfield across the world (more than 1800)
- Approach path eliminates intermediate phases (dive and drive)
- Approach path is independent of airfield and airplane barometrical equipment
- Reduction of influence of low temperatures and influence of errors in flight instruments settings

Therefore, LPV procedures increase safety of air traffic. In order to perform LPV approach, equipment that meets the specific requirements determined by current legislations, such as dual-frequency GPS receiver compatible with SBAS, as well as special GPS receiver antenna, is required.

25.7 Summary

Utilization of GNSS in various terrestrial applications is characterized by increasing popularity. It is visible in a number of aspects of human lives, various scientific researches are based on data gathered with GNSS; many companies rely on the proper functionality of these systems. The ability of rapid determination of precise location across the globe is widely utilized in military and disaster recovery applications, even artists are trying to use GNSS data as a form of artistic expression (Lauriault and Wood 2009).

The most important factors which allowed GNSS for such a popularity are availability of navigation signals (free of costs, everywhere on Earth), good positioning precision (especially when supported by various augmentation systems), and high reliability of service (i.e., rare cases of downtimes of GPS system functionality). In addition, after achievement of fully operational state by GALILEO system, which will be fully controlled by civilian institutions, new applications of these systems are expected.

Acknowledgments

The research leading to these results has received funding from the PEOPLE Programme (Marie Curie Actions) of the European Union's Seventh Framework Programme FP7/2007–2013/ under REA grant agreement no 285462. We also thank Marcin Paszkuta for support in the figures creation.

References

Ahn, S.M., C. Park, J. Kang, 2011, Application of GPS fleet tracking and stochastic simulation to a lean soil excavation practice, *Proceedings of the 28th ISARC*, Seoul, Korea, pp. 335–336.

Ainsworth, S., B. Thomason, 2003, Where on Earth are we? The Global Positioning System (GPS) in archaeological field survey, Technical Paper, English Heritage Publishing, Swindon, England.

Allan, D.W., 1997, *The Science of Timekeeping*, Hewlett Packard Application Note 1289.

Ameri, B., D. Meger, K. Powert, 2009, UAS applications: Disaster & emergency management, *ASPRS Annual Conference*, Baltimore, MD.

Andrade, A.A.L., 2001, *The Global Navigation Satellite System: Navigating into the New Millennium*, Ashgate, Surrey, U.K., ISBN 9780754618256.

Aranzulla, M., F. Cannavò, S. Scollo, G. Puglisi, G. Immè, 2013, Volcanic ash detection by GPS signal, *GPS Solutions*, 17(4), 485–497.

Ashbrook, D., T. Starner, 2003, Using GPS to learn significant locations and predict movement across, *Personal and Ubiquitous Computing*, 7(5), 275–286.

Bahder, T.B., W.M. Golding, 2004, Clock synchronization based on second-order quantum coherence of entangled photons, *AIP Conference Proceedings*, Glasgow, U.K., Vol. 734, July 25–29, p. 395.

Bain, R.S., 2001, The use of GPS-based automatic vehicle location technologies for bus transit: State of the practice in the USA and lessons for elsewhere, European Transport Conference, September 10–12, Homerton College, Cambridge.

Bar-Gera, H., 2007, Evaluation of a cellular phone-based system for measurements of traffic speeds and travel times: A case study from Israel, *Transportation Research Part C*, 15, 380–391.

Bargeton, A., F. Moutarde, F. Nashashibi, A.S. Puthon, 2010, Joint interpretation of on-board vision and static GPS cartography for determination of correct speed limit, CoRR, abs/1010.3867.

Bellini, P., S. Boncinelli, F. Grossi, M. Mangini, P. Nesi, L. Sequi, 2013, Mobile Emergency, and Emergency Support System for Hospitals in Mobile Devices: Pilot Study, *Journal of medical internet research protocols*, 2(1):e19

Bevly, D.M., S. Cobb, 2010, *GNSS for Vehicle Control*, Artech House, London, U.K., ISBN 9781596933026.

Brown, A.K., G. Zhang, D. Reynolds, 1999, Precision targeting using GPS/internal-aided sensors, *ION 55th Annual Meeting*, June 27–30, Cambridge, MA, pp. 689–695.

Bullock, J.B., M. Foss, G.J. Geier, M. King, 2005, Integration of GPS with other sensors and network assistance, *Understanding GPS: Principles and Applications*, Kaplan, E.D., Hegarty, C.J., eds., Artech House, London, U.K., ISBN 9781580538954.

Bussios, N., Y. Tsolakichy, M. Tsakiri-Strati, O. Goergoula, 2004, Integrated high resolution satellite image, GPS and cartographic data in urban studies, Municipality of Thessaloniki, International Society for Photogrammetry and Remote Sensing 2004, Istanbul, Turkey. July 12–23, pp. 215–221.

Chunping, L., R. Shuangchen, G. Xiangdong, 2003, The programmable logic implementation of GPS/GLONASS clock synchronization, *ASIC'03, Fifth International Conference*, Vol. 2, Beijing, China, October 21–24, pp. 732–735.

Clarke, B., 1998, *Aviator's Guide to GPS*, McGraw-Hill, New York, ISBN 9780070094932.

Cosentino, R.J., D.W. Diggle, M.U. de Haag, C.J. Hegarty, D. Milbert, J. Nagle, 2005, *Understanding GPS: Principles and Applications*, Kaplan, E.D., Hegarty, C.J., eds., Artech House, London, U.K., ISBN 9781580538954.

Cyran, K.A., D. Sokołowska, A. Zazula, B. Szady, O. Antemijczuk, 2011, Data gathering and 3D-visualization at OLEG multiconstellation station in EDCN system, *Proceedings of the 21st International Conference on Systems Engineering*, August 16–18, Las Vegas, NV, pp. 221–226.

DoD, 2008, *Global Positioning System Standard Positioning Service Performance Standard*, 4th edn., U.S. Department of Defense, Washington, DC.

Dorsey, A.J., W.A. Marquis, P.M. Fyfe, E.D. Kaplan, L.F. Wiederholt, 2005, GPS system segments, *Understanding GPS: Principles and Applications*, Kaplan, E.D., Hegarty, C.J., eds., Artech House, London, U.K., ISBN 9781580538954.

ESA SP-1303, The European EGNOS Project, 2006, *EGNOS The European Geostationary Navigation Overlay System—A Cornerstone of Galileo*, ESA Publications Division, Noordwijk, the Netherlands, ISBN: 92-9092-453-5.

Fan, R., I. Chakraborty, N. Lynch, 2005, Clock synchronization for wireless networks principles of distributed systems, Lecture Notes in Computer Science, Higashino, T Ed., Principles of Distributed Systems Lecture Notes in Computer Science Volume 3544, Springer, New York, Vol. 3544, pp. 400–414.

Frei, E., A. Ryf, R. Scherrer, 1993, Use of the global positioning system in dam deformation and engineering surveys, SPN 2.

Gift, S.J.G., 2010, One-way light speed measurement using the synchronized clocks of the global positioning system (GPS), *Physics Essays*, 23(2), 271.

Gleason, S., D. Gebre-Egziabher, 2009, *GNSS Applications and Methods*, Artech House, London, U.K., ISBN 9781596933309.

Gleason, S., D. Gebre-Egziabher, 2010, *Global Navigation Satellite Systems*, McGraw-Hill Education (India) Pvt Limited, Noida, India, ISBN 9780070700291.

Gottardi, E., F. Tua, B. Cargnelutti, M.L. Maublanc, J.M. Angibault, S. Said, H. Verheyden, 2010, Use of GPS activity sensors to measure active and inactive behaviours of European roe deer, *Mammalia*, 74(4), 355–362.

Grewal, M.S., L.R. Weill, A.P. Andrews, 2001, *Global Positioning Systems, Inertial Navigation, and Integration*, John Wiley & Sons, New York, ISBN 978-0-47135-032-3.

Groves, P.D., 2008, *Principles of GNSS, Inertial, and Multi-Sensor Integrated Navigation Systems*, Artech House, London, U.K., ISBN 9781580532556.

Hariharan, R., J. Krumm, E. Horvitz, 2005, Web-enhanced GPS, *First International Conference on Location and Context Awareness LoCA'05*, May 12–13, Bavaria, Germany, pp. 95–104.

Harper, N., 2009, *Server-Side GPS and Assisted-GPS in Java*, Artech House, Incorporated, London, U.K., ISBN 9781607839866.

Harte, L., B. Levitan, 2007, *GPS Quick Course: Technology, Systems and Operation*, Althos, Fuquay Varina, NC, ISBN 9781932813708.

Hasan, K.S., M. Rahman, A.L. Haque, M.A. Rahman, T. Rahman, M.M. Rasheed, 2009, Cost effective GPS-GPRS based object tracking system, *Proceedings of the International MultiConference of Engineers and Computer Scientists*, March 18–20, Hong Kong, China, Vol. 1, pp. 18–20.

Hashmi, N., D. Myung, M. Gaynor, S. Moulton, 2005, A sensor-based web service-enabled emergency medical response system, Workshop on End-to-End, Sense-and-Respond Systems, Applications, and Services, *EESR '05*, Seattle, WA, pp. 25–29.

Henriksson, E., M. Öström, A. Eriksson, 2001, Preventability of vehicle-related fatalities. *Accident analysis and prevention*, 33(4), 467–475.

Heraud, J.A., A.F. Lange, 2009, Agricultural automatic vehicle guidance from horses to GPS, *Agricultural Equipment Technology Conference*, February 9–12, Louisville, KY, pp. 1–67.

Hernandez, J.I., C.Y. Kuo, 2003, Steering control of automated vehicles using absolute positioning GPS and magnetic markers, *IEEE Transactions on Vehicular Technology*, 52(1), 150–161

Hewlett Packard Company, 1996, GPS and precision timing applications, HP Application Note, No 1272.

Hofman-Wellenhof, B., H. Lichtenegger, E. Wasle, 2008, *GNSS—Global Navigation Satellite Systems: GPS, GLONASS, Galileo, and More*, Springer, New York, ISBN 978-3-211-73012-6.

Hou, S.T., F. Chia, 2011, Making senses of technology: A triple contextual perspective of GPS use in the taxi industry, service sciences (IJCSS), *International Joint Conference on Service Sciences*, Taipei, Taiwan, May 25–27.

Huang, J.Y., C.H. Tsai, S.T. Huang, 2012, The next generation of GPs navigation systems, *Communications of the ACM*, 55(3), 84–93.

Inks, S.A., T.W. Loe, 2005, The ethical perceptions of salespeople and sales managers concerning the use of GPS tracking systems to monitor salesperson activity, *Marketing Management Journal*, 15(1), 108.

Jeffrey, C., 2010, *An Introduction to GNSS: GPS, GLONASS, Galileo and Other Global Navigation Satellite Systems*, 1st edn., NovAtel Inc., Calgary, Alberta, Canada, ISBN 978-0-9813754-0-3.

Kalinowska, D., K. Hartmut, 2006, Motor vehicle use and travel behaviour in Germany: Determinants of car mileage, DIW-Diskussionspapiere, No. 602.

Kaplan, E.D., J.L. Leva, D. Milbert, M.S. Pavloff, 2005, Fundamentals of satellite navigation, *Understanding GPS: Principles and Applications*, Kaplan, E.D., Hegarty, C.J., eds., Artech House, London, U.K., ISBN 9781580538954.

Kim, R., T. Nagayama, H. Jo, B.F. Spencer, 2012, Preliminary study of low-cost GPS receivers for time synchronization of wireless sensors, Sensors and Smart Structures Technologies for Civil—Mechanical and Aerospace Systems, 83451A.

Kim I., C. Park, G. Jee, J. G. Lee, 1998, GPS positioning using virtual pseudorange, *Control Engineering Practice*, 6(1), 25–35.

Korreng, M.D., 2010, UTC time transfer for high frequency trading using IS-95 CDMA base station transmissions and IEEE-1588 precision time protocol, *42nd Annual Precise Time and Time Interval (PTTI) Meeting*, November 15–18, Reston, VA, pp. 359–368.

Landau, R., G.K. Auslander, S. Werner, N. Shoval, J. Heinik, 2011, Who should make the decision on the use of GPS for people with dementia? *Aging Mental Health*, 15(1), 78–84.

Lauriault, T.P., J. Wood, 2009, GPS tracings—Personal cartographies, art & cartography, special issue, *The Cartographic Journal*, 46(4), 360–365.

Lee, E.K., S. Yang, S.Y. Oh, M. Gerla, 2009, RF-GPS: RFID assisted localization in VANETs, mobile ad hoc and sensor systems, *MASS '09, IEEE Sixth International Conference*, October 12–15, Macao, China, pp. 621–626.

MacGougan, G.D., 2006, A Short Summary of Tropospheric Math Models By Glenn D. MacGougan (October 11, 2006).

Manabu, O., H. Naohisa, F. Takehiko, S. Hiroshi, 2006, The application of RTK-GPS and steer-by-wire technology to the automatic driving of vehicles and an evaluation of driver behavior, *IATSS RESEARCH*, 30(2), 29–38.

Matthews, V.M., E. Adetiba, 2011, Vehicle accident alert and locator (VAAL), *International Journal of Electrical & Computer Sciences IJECS-IJENS*, 11(02), 35–38.

Miller, K.M., 2000, A review of GLONASS, *The Hydrographic Journal*, 98, 15–21.

Nagashima, K., K. Yamada, A. Tadano, 2013, Driverless Antarctic tractor system, *Hitachi Review*, 62(3), 229–230.

Ozguner, U., T. Acarman, K.A. Redmill, 2011, *Autonomous Ground Vehicles*, Artech House, London, U.K., ISBN 9781608071937.

Pan, G., G. Qi, Z. Wu, D. Zhang, 2013, Land-use classification using taxi GPS traces intelligent transportation systems, *IEEE Transactions*, 14(1), 113–123.

Petersen E.H., 2000, Draft report on the Commission communication to the Council, the European Parliament, the Economic and Social Committee and the Committee of the Regions on the Priorities in EU road safety–Progress report and ranking of actions ((COM(2000) 125–C5 0248/2000–2000/2136(COS)), Committee on Regional Policy, Transport and Tourism.

Petrovski, I.G., T. Tsujii, 2012, *Digital Satellite Navigation and Geophysics: A Practical Guide with GNSS Signal Simulator and Receiver Laboratory*, Cambridge University Press, Cambridge, U.K., ISBN 9780521760546.

Prakash, N.R., D. Kumar, K. Nandan, 2012, An autonomous vehicle for farming using GPS, *International Journal of Electronics and Computer Science Engineering*, 1(3), 1695–1700.

Prasad, R., M. Ruggieri, 2005, *Applied Satellite Navigation Using GPS, GALILEO, and Augmentation Systems*, Artech House, London, U.K., ISBN 9781580538145.

Rutter, S.M., N.A. Beresford, G. Roberts, 1997, Use of GPS to identify the grazing areas of hill sheep, *Computers and Electronics in Agriculture*, 17, 177–188.

Silva, J.M., R.G. Olsen, 2002, Use of global positioning system (GPS) receivers under power-line conductors, *Power Delivery IEEE Transactions*, 17(4), 938–944.

Skov-Petersen, H., R. Rupf, D. Köchli, B. Snizek, 2012, Revealing recreational behaviour and preferences from GPS recordings, MMV6—Stockholm 2012, Session 4E—Recent advances in visitor monitoring: GPS tracking and GIS technology.

Sterzbach, B., 1997, GPS-based clock synchronization in a mobile, *Real-Time Systems*, 12(1), 63–75.

Sukkarieh, S., E.M. Nebot, H.F. Durrant-Whyte, 1998, Achieving integrity in an INS/GPS navigation loop for autonomous land vehicle applications, *Robotics and Automation, IEEE International Conference*, May 20, Leuven, Belgium, Vol. 4, May 16–20, pp. 3437–3442.

Sukkarieh, S., E.M. Nebot, H.F. Durrant-Whyte, 1999, A high integrity IMU/GPS navigation loop for autonomous land vehicle applications, *IEEE Transactions on Robotics and Automation*, 15(3), 572–578.

Thierry, B., B. Chaix, Y. Kestens, 2013, Detecting activity locations from raw GPS data: A novel kernel-based algorithm, *International Journal of Health Geographics*, 12, 14.

Thompson, C., J. White, B. Dougherty, A. Albright, C.D. Schmidt, 2010, Using smartphones to detect car accidents and provide situational awareness to emergency responders, Lecture Notes of the Institute for Computer Sciences, *Social Informatics and Telecommunications Engineering*, 48, 29–42.

Tian, J., J. China, H. Zhao, 2010, Pipeline damage locating based on GPS fiducial clock intelligent control and automation (WCICA), *Eighth World Congress*, July 7–9, Jinan, China, pp. 4161–4164.

Towns, D.M., L.M. Cobb, 2012, Notes on: GPS technology; employee monitoring enters a new era, *Labor Law Journal*, 63(3), 203.

Van Diggelen, F., 2009, *A-GPS: Assisted GPS, GNSS, and SBAS*, Artech House, London, U.K., ISBN-13 978-1-59693-374-3.

Van Westen, C., 2000, Remote sensing for natural disaster management, *International Archives of Photogrammetry and Remote Sensing*, Vol. XXXIII, Part B7, Amsterdam, the Netherlands.

Wan, N., G. Lin, 2013, Life-space characterization from cellular telephone collected GPS data Computers, *Environment and Urban Systems*, 39, 63–70.

Ward, P.W., J.W. Betz, C.J. Hegarty, 2005a, GPS satellite signal characteristics, *Understanding GPS: Principles and Applications*, Kaplan, E.D., Hegarty, C.J., eds., Artech House, London, U.K., ISBN 9781580538954.

Ward, P.W., J.W. Betz, C.J. Hegarty, 2005b, Interference, multipath and scintillation, *Understanding GPS: Principles and Applications*, Kaplan, E.D., Hegarty, C.J., eds., Artech House, London, U.K., ISBN 9781580538954.

Wei, W., S. Baosheng, Z. Kefa, Z. Xia, 2010, The use of GPS for the measurement of details in different areas, *Electronic Commerce and Security (ISECS), Third International Symposium*, July 29–31, Guangzhou, China, pp. 59–62.

White, J., C. Thompson, H. Turner, B. Dougherty, D.C. Schmidt, 2011, Automatic traffic accident detection and notification with smartphones, *Mobile Networks and Applications*, 16(3), 285–303.

Xu, G., 2007, *GPS: Theory, Algorithms and Applications*, 2nd edn., Springer, New York, ISBN 978-3-540-72714-9.

Yang, M., C.F. Lo, 2000, Real-time kinematic GPS positioning for centimeter level ocean surface monitoring, *Proceedings of National Science Council ROC(A)*, 24(1), 79–85.

Zagami J.M., S.A. Pari, J.J. Bussgang, K.D. Melillo, 1998, Providing universal location services using a wireless E911 location network, *IEEE Communications Magazine*, 36(4), 66–71.

Zheng, Y., L. Zhang, X. Xie, W.Y. Ma, 2009, Mining interesting locations and travel sequences from GPS trajectories, *18th International Conference on World Wide Web*, Madrid, Spain, pp. 791–800.

Zheng, Y.T., Z.J. Zha, T.S. Chua, 2012, Mining travel patterns from geotagged photos, *ACM Transactions on Intelligent Systems and Technology TIST*, 3(3), 56:1–56:18.

Zhong, P. et al., 2008, Adaptive wavelet transform based on cross-validation method and its application to GPS multipath mitigation, *GPS Solution*, 12, 109–117.

Ziedan, N.I., 2006, *GNSS Receivers for Weak Signals*, Artech House, Incorporated, London, U.K., ISBN 9781596930520.

Zinkiewicz, D., B. Buszke, M. Houdek, F. Toran-Marti, 2010, SISNeT as a source of EGNOS information: Overview of functionalities and applications, *Fifth ESA Workshop on Satellite Navigation Technologies and European Workshop on GNSS Signals and Signal Processing (NAVITEC)*, December 8–10, Noordwijk, the Netherlands, pp. 1–7.

Figures which were included from external sources, or created with external applications (agreement of external authors were obtained)

SBAS simulator—http://www.iguassu.cz/sbas-sim/by:
IGUAU Software Systems, Evropská 120, Dejvice, 160 00 Praha 6, Czech Republic
Figure 25.1, Figure 25.2, Figure 25.3, Figure 25.4

U-Center software by UBLOX http://www.u-blox.com by:
u-blox AG, Zürcherstrasse 68, 8800 Thalwil, Switzerland
Figure 25.14

PolaRx Control Software http://www.septentrio.com/by:
Septentrio nv, Greenhill Campus, Interleuvenlaan 15G, 3001 Leuven, Belgium
Figure 25.10

Crowdsourcing and Remote Sensing Data

26

Crowdsourcing and Remote Sensing: Combining Two Views of Planet Earth

Fabio Dell'Acqua
Università degli Studi di Pavia

Acronyms and Definitions

CA	California
CIIM	Community internet intensity map
COST	European Cooperation in Science and Technology
DO	Direct Observation
DYFI	"Did You Feel It" service
ENV	Environment theme of the Seventh Framework Programme of the EU
EU	European Union
FP 7	Seventh Framework Programme
GED	Global Exposure Database of GEM
GEM	Global Earthquake Model
GEO	Group on Earth Observations
GEOSS	Global Earth Observation System of Systems
GIS	Geographic Information System
IDCT	GEM inventory data capture tools
LANDSAT	Land Remote Sensing Satellite Program
MMI	Modified Mercalli Intensity scale for earthquakes
OECD	Organisation for Economic Co-operation and Development
OSM	OpenStreetMap
SDI	Spatial Data Infrastructures
SMS	Short Message Service
TCP/IP	Transmission Control Protocol/Internet Protocol
UGC	User-Generated Content
U.S./USA	United States of America
USGS	United States Geological Survey
VGI	Volunteered Geographic Information
WebGIS	Web-based Geographic Information System
ZIP	Zone Improvement Plan (mailing codes)

26.1 Introduction

Although the term *crowdsourcing*, according to the official records, was used for the first time in 2006 in a "Wired" journal paper by Howe (2006), the concept of *on-the-fly* recruiting of a generally large set of contributors to solving a specific problem dates back by centuries. For instance, it is well known that in 1714, the British government offered a money prize (the Longitude Prize) to anyone who could devise a reliable method to compute the longitude a ship was located at (Sobel, 1998); this was an example of entrusting a nonorganized—yet having unexplored intelligent capabilities—pool of people with the task of solving a specific, information-retrieval problem. Probably less widely known is the fact that Toyota, the Japanese car manufacturer, started a competition in 1936 to get its current logo, chosen among 27,000 candidate entries (DesignCrowd, 2010).

Several examples of such calls can be found across history, but it was not until the advent of the World Wide Web, and especially the Web 2.0, that crowdsourcing found the optimal environment for being widely applied in practice. The Internet, indeed, allows recruitment of contributors from virtually any place on the Earth, easy sharing of resources, and immediate transmission of results. This is why the Internet will have an important role in this book chapter. The Internet itself is based on the shared contribution of the infrastructures that physically compose it, yet the term crowdsourcing refers to applications requiring intelligence or otherwise capacity of a human operator to solve problems that machines are still unable to deal with. An example is the so-called "wiki" philosophy, marked in 2001 by the birth of the free encyclopedia named *Wikipedia* (2001) that is maintained by millions of volunteers, and later extended to tens of similarly conceived projects. Although Wikipedia writers work for free, this is not necessarily a rule for all crowdsourced work: in 2005, for example, the Amazon Mechanical Turk was launched, featuring paid, online crowdsourcing services. The name of the service was chosen in reference with the famous mechanical Turkish chess player built in 1770 by Wolfgang Von Kempelen (1734–1804) to shock the Austrian empress Maria Theresa: faking a mechanical chess player, it actually contained a disguised human operator—somehow like the modern realization of crowdsourcing, *hiding* human intelligence behind an electronic interface and infrastructure.

This chapter will describe and analyze how this concept is exploited in the field of remote sensing and Earth observation, where crowdsourcing and the *Citizen Sensor* may represent an effective way to fill in the information gaps left behind by *specialized* remote sensing operation using conventional infrastructure like space-borne and air-borne sensors. The chapter is organized as follows: Section 26.2 introduces the concept of crowdsourcing, Section 26.3 discusses the concepts of user-generated content (UGC), Web 2.0 and social media. Section 26.4 gets closer to the aim of the chapter by presenting the Citizen Sensor and volunteered geographic information (VGI). Section 26.5 presents some examples of implementations of the crowdsourcing concept in a geospatial and remote sensing environment. Section 26.6 closes the chapter while drawing some conclusions.

26.2 What Is Crowdsourcing

As already mentioned, the term *crowdsourcing* was coined by Howe (2006) as a neologism generated on the model of *outsourcing* while implying the meaning of *sourcing* to the *crowd*, that is, entrusting an external provider for the provision of a service, in which the external provider happens to be a group of people not determined a priori rather than a specific, single agent.

After Howe's invention of the term crowdsourcing, several derived terms have been born. Some of the most notable ones are:

1. *Crowd-voting*: Collecting opinions (possibly forced to fit into a predefined set of options) on a given topic or product
2. *Crowd-funding*: Collective funding of a project by many small investors/donors
3. *Creative crowdsourcing*: Coworking on a collective artwork
4. *Wisdom of the crowd*: Merging contributions by individuals, based on the idea that *collective ideas* are better that *individual ideas*, that is, asking many people increases chances of finding the right solution to a problem
5. *Microwork*: Splitting a big task into microtasks entrusted to several human agents (see Amazon Mechanical Turk, 2005)

As we will see in the following, crowdsourcing as a support to Earth observation mainly falls into the categories of crowd-voting and wisdom of the crowd, although new specific concepts like the Citizen Sensor (Sheth, 2009) need to be considered for a complete picture.

Crowdsourcing can thus be conceived as a production model, or problem-solving model, exploiting a sort of *distributed intelligence* or, as we will see for our topic of interest, *distributed data collection*. In its most classical conception, opinions regarding the solving of a given problem or the development of a project is sought from a group of individuals—a *crowd*—not necessarily defined *a priori*, which contribute, individually and/or through mutual interactions, to finding or defining a solution to the proposed problem, possibly subject to the proposer's approval. The solution typically ends up belonging to the company or organization that initially proposed the problem, while the individuals that contributed to the solution are compensated for their work through small money payments, prizes, or even simply moral or intellectual satisfaction.

The advantages of such a model with respect to traditional commercial models are obvious. First of all, the solution may be found at a reasonable cost—or even for zero, in terms of raw labor cost, if the remuneration consists of personal satisfaction as in the case of Wikipedia. Then, the pool of considered, possible solutions—or, more relevant for our *remote sensing* case, of collectable information—is much wider than any single solver or information collector can reach.

On the other hand, the process of crowdsourcing often raises concerns about the quality of the information produced (Allahbakhsh et al., 2013), as contributors might have different levels of skills and expertise, possibly insufficient for performing certain tasks—even simple data collection in many cases requires a minimum degree of knowledge about the observed targets; contributors might also have various and even biased interests and incentives; even malicious activities can be set up to bias the output of a crowdsourcing operation. The usual countermeasures act on the sides of assessing workers' profiles and reputation (Jøsang et al., 2007; Alfaro et al., 2011) and of smartly designing both the task (Dow et al., 2012) and the compensation schemes (Scekic et al., 2012).

26.3 From Web 1.0 to Web 2.0 and Social Media

The Internet as we came to know it when it started becoming popular in the second half of the 1990s was later termed Web 1.0. This term, which was not there at the time, was coined *a posteriori* to point out the difference with the new features that were then emerging within the net, or, better, with the new kind of Internet that was developing.

The term Web 2.0 was indeed coined not too far away in time (O'Reilly, 2007), highlighting a trend that emerged since the beginning of twentieth century but only recently consolidated. Web 2.0 does not mean a new revision of the World Wide Web: it is still based on TCP/IP protocols, and all the elements that characterized the Web since its beginning, like hypertext and linking among contents. Web 2.0 is rather the arriving point of a seamless evolution in terms of approach to online information.

The Web 1.0 is indeed based on static websites, where the user mainly accesses the information (or data) generated by a limited set of other agents. Here, a pretty clear distinction is found between (comparatively few) information contributors on the one side, and (many more) information consumers on the other side. Search engines try and facilitate the access to *the right information* among a flood of less relevant—or irrelevant at all-information. Figure 26.1 depicts the concept of Web 1.0.

The Web 2.0 is instead based on a high level of interaction between the website and the user, which frequently is, at the same time, also a generator of the information contained in the websites, as highlighted in the scheme in Figure 26.2. Web 2.0 allows users to *publish*, that is, make available to anyone requesting to access the website, some self-generated pieces of information. This is the so-called UGC (OECD, 2007), which entails creative effort and nonprofessional, nonstandard labor.

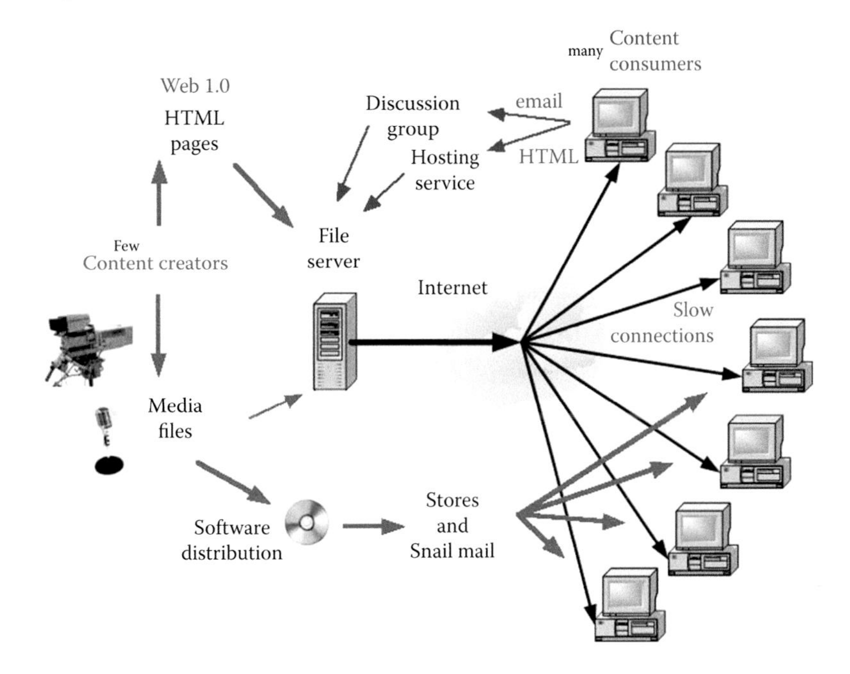

FIGURE 26.1 Schematic concept of Web 1.0. (From http://en.wikiversity.org/wiki/File:Web_1.0_elements.png.)

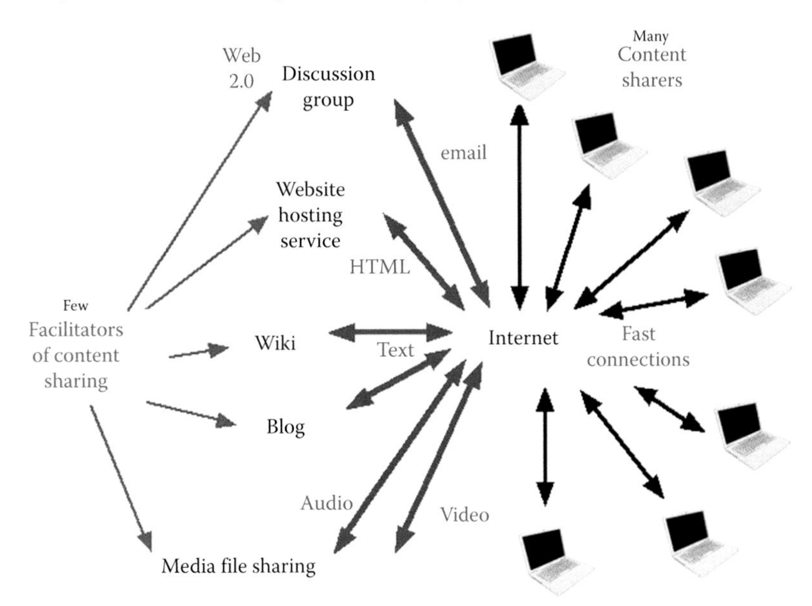

FIGURE 26.2 Schematic concept of Web 2.0. (From http://en.wikiversity.org/wiki/File:Web_2.0_elements.png.)

The means through which this can be done are various and include the so-called social networks, photo and video-sharing websites, blogs, forums, wikis, and other web services. The information can be made completely public, or restrictions can be placed on who can access what, through, for example, registration and/or *linking* among specific individuals or groups.

By uploading material on such sites, commenting material posted by others, and interacting with other individuals, the users generate a large amount of information, which is highly dynamic in that it is continuously amended, topped, and made to evolve including through closed-loop feedback. The phenomenon was spontaneous, still fostered and sped up by the availability of user interfaces, software, and online storage tools, perfectly integrated in the Web, which is increasingly becoming a platform on which it is possible to utilize different applications including remote GIS (WebGIS) ones.

Not only the huge amount of shared information, especially its evolution and the possibility of live interaction, build up a sort of *collective intelligence* (Levy, 1999) or *global brain* (Heylighen, 1997). This also casts a spotlight on Web 2.0 as a possible tool (*the* tool?) to practically implement crowdsourcing. These details are discussed later in this chapter.

The *social media*, as defined by Kaplan and Haenlein (2010), are the expression of massive UGC generation in the framework of the Web 2.0 as an enabling technology. Kaplan and Haenlein (2010) define different types of social media, but the most relevant for our purposes are the following:

1. *Collaborative projects*: Wikis, social bookmarking.
2. *Content communities*: Sharing of content among users.
3. *Social networking sites*: Much less interesting as the contents are not public by default.

In the following, we will see how the social media helps in building the so-called Citizen Sensor Networks.

26.4 Citizen Sensor

Today, the Earth supports about 7 billion people; every year those people, according to statistics (IDC, 2013), purchase millions of *smartphones* with positioning, picture-taking, video- and sound-recording capabilities; the figure for 2012 was 712 million units (IDC, 2013), reporting a 44.1% increase with respect to 2011. It is not difficult to imagine that in a few years practically every inhabited area of the Earth will see the presence of at least one device with capabilities of acquiring environmental data—in a wide sense—and of transmitting it to the *cyberspace*. By the way, with a person using it, which thus happens to fit the definition of Citizen Sensor (Sheth, 2009); whether in an organized or informal way, a distributed set of Citizen Sensors forms a *Citizen Sensor Network*, that is, a distributed network of potential, mobile, connected sensing instruments. The concept of Citizen Sensor, however, finds its natural place, and gains even more sense, in the framework of the VGI concept (Goodchild, 2007), that is, a special case of the more general Web phenomenon of UGC. VGI is defined as *the widespread engagement of large numbers of private citizens...in the creation of geographic information* and it can be seen as the vehicle through which sparsely collected information is conveyed to a single, virtual place where it becomes usable, including by those who created it.

Among the most notable examples of VGI collectors, we find Wikimapia (2007) and OpenStreetMap (2004) websites. The former encourages participants to post comments about georeferenced locations; the latter is an international effort to create a free source of map data through volunteers' efforts.

In Figure 26.3, a comparison between the Google Maps© version and the OpenStreetMap (OSM) version of a map on the same area highlights the amount of information that the Citizen Sensor is willing to contribute to a public VGI repository. Note that this version of the OSM map does not display the 3D building information here, which is however frequently contributed (Over et al., 2010; Goetz and Zipf, 2012).

It is however to be noted that the comparison represented in Figure 26.3 refers to a small-sized town in a developed country, where a large fraction of the population has access to electronic navigation devices and fast Internet connections. One may wonder what is the real coverage to expect when the sight is expanded to a global level, including areas where such favorable conditions are not found at all.

The level of coverage offered by a crowdsourced mapping service like OSM is an interesting issue, but it is difficult to assess for various reasons:

1. No reference data ensuring full coverage of the entire globe is available, against which to measure OSM's coverage—this could actually be solved by subsampling, but there are other issues.
2. No standard definition of items to be included in a map is provided, so a definite response on what is actually mapped versus what *should have been mapped* cannot be given. For instance, an unpaved road could be considered unimportant if the focus is on residential area accessibility, while if the map is to be used in an agricultural context, the same road becomes important; still on the unpaved road, in a developing country, it may be as important as a paved one in a residential neighborhood.
3. Assessing a global coverage of an ever-changing set in diverse environments across the globe like that of the man-made structures that cover the solid Earth surface is in itself a tricky task. Buildings appear so well delineated and separated from each other in Figure 26.3; other sort of *buildings* may not be so easy to tell from each other in a context such as that of, for example, informal settlements in urban areas with social problems.

The statements above naturally do not mean that nothing can be concluded on the coverage and we just have to hope that the area we are interested in is covered with all the information that we need. Local studies provide encouraging results; (Neis et al., 2012; De Leeuw et al. (2012)), yet on limited areas; at a global level, qualitative assessments can be made. A recent, interesting example is the OSM node density map published by Raifer (2014). As can be seen in Figure 26.4, the color-coded density correlates well with the population distribution depicted in Figure 26.5. Low coverage rates

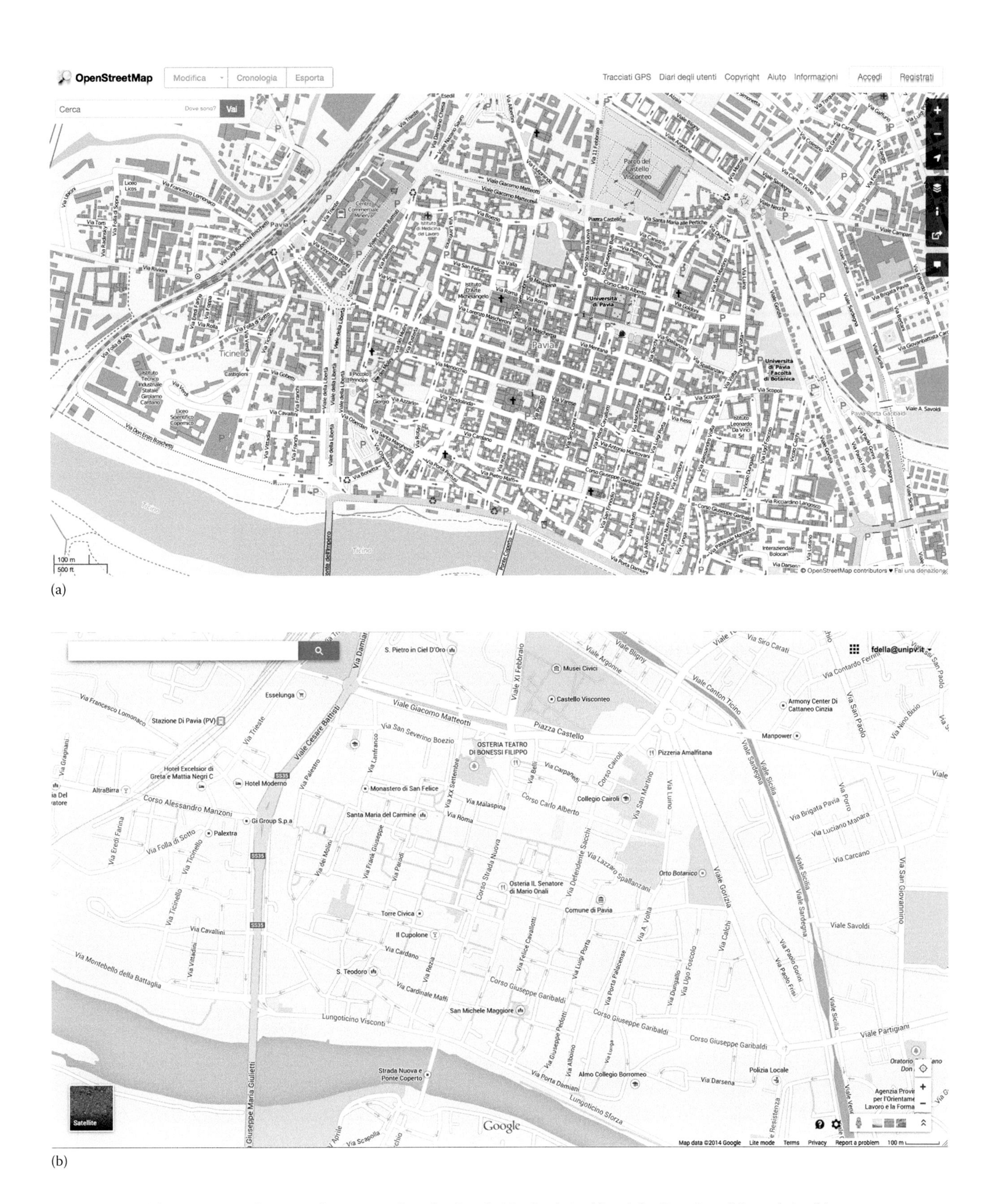

FIGURE 26.3 The city centre of Pavia, Italy, as mapped on the Google Map© website (a) and the OpenStreetMap website (b).

FIGURE 26.4 OSM node density 2014. (From Martin Raifer, CC-BY—source data, OpenStreetMap contributors, ODbL.)

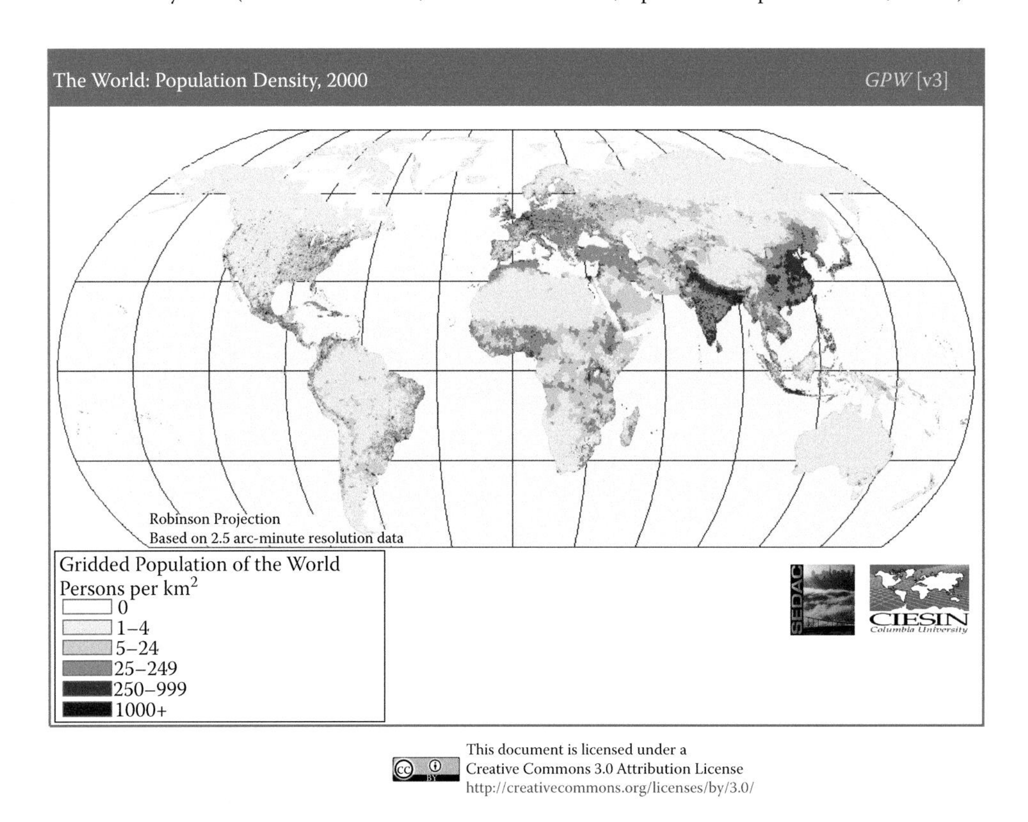

FIGURE 26.5 Global population density according to CIESIN as available from NASA SEDAC. (From Center for International Earth Science Information Network (CIESIN), Columbia University; Centro Internacional de Agricultura Tropical (CIAT), Gridded Population of the World (GPW), Version 3, CIESIN, Columbia University, Palisades, NY, Copyright 2005, available at: http://sedac.ciesin.columbia.edu/gpw.)

are reported in noninhabited (or nearly so) areas such as Saharan Africa, in-land Greenland, or the Australian Outback. All in all, the maps suggest that where one expects to find settlements and thus roads, they are actually found on OSM.

26.5 Current Implementations

In this section, we will see how the concepts illustrated so far find their application in the real world. Various ways of using VGI mechanisms are illustrated here in different fields of applications, including disaster management and land cover mapping.

26.5.1 Seismic Risk: The "Did You Feel It" Service

Although crowdsourcing is potentially very useful to assess damage to buildings and structures in the postdisaster phase (Kerle, 2011), the earliest examples in the field of seismic risk focus on mapping the phenomenon itself rather than its consequences.

The "Did You Feel It" (DYFI; Wald et al., 2011) program was developed by the United States Geological Survey (USGS) in the early 1990s, long before the concept of crowdsourcing came to be formalized by Howe (2006), and can be considered the first notable example of crowdsourcing applied to natural hazard management. DYFI consists of a website (DYFI, 1991) onto which individuals (Citizen Sensors) can report their time-stamped, georeferenced observations regarding a seismic phenomenon that they have experienced. The idea underpinning DYFI is to best exploit the public engagement (an informal Citizen Sensor Network) to generate a map of the event which be useful, in case of a relevant event, to get a prompt input to the emergency procedures, but also to build a historical record of minor quakes to be used as an input to earthquake hazard models. The map is called community internet intensity map (CIIM), and it is effectively a consolidated output of a crowdsourcing mechanism.

A short online questionnaire is proposed to the candidate contributor, including three different sections on context, experience, and effects. The questionnaire was designed to generate reports fitting the modified Mercalli intensity (MMI) through suitable mapping of responses into MMI grades (Dengler and Moley, 1994; Dengler and Dewey, 1998).

Every filled questionnaire will thus result in an estimated MMI grade, linked to the microregion to which the geolocation of the contributor corresponds. Microregions are defined conventionally to correspond with U.S. zip codes as a compromise between statistical significance of the aggregates and spatial resolution. An average value of the linked, estimated MMI grades is computed on every microregion, and attached to it. Visually, the average is turned into a color through a conventional color palette ranging from white (nonperceived) to dark red (extreme). Within a few minutes from the earthquake occurrence, a CIIM starts building on the website, often collecting thousands of contributions for clearly felt quakes (from grade IV upward). Together with the CIIM map, a graph depicting how the average intensity decays with distance is also constructed; examples are shown in Figure 26.6.

The DYFI program has recorded an increasing success. Currently, the total input questionnaire count largely exceeds 1 million units, and the service has been scaled up to world wide coverage, allowing global contributors to feed their observation through the dedicated "outside US" section of the website. In Figure 26.7, the spatial distribution of questionnaires filed between 1991 and 2012 is translated into a map, providing a compelling view of the seismic activity across the North American continent, most interesting for purposes of seismicity on a continental scale.

A few words are in order on the accuracy of DYFI data, because they make an interesting case of how crowdsourcing can be useful not only for qualitative but also for quantitative purposes. As Atkinson and Wald (2007) stated, DYFI data "make up in quantity what they may lack in quality." The authors focused on the 2004 Parkfield, CA, earthquake, comparing MMI data collected through the DYFI system across the affected ZIP codes with the measured earthquake ground motions from local instruments, as cataloged by the USGS's ShakeMap (2004) Web-based database, and estimated using consolidated models. The two data series match very well, not only in terms of general MMI vs. distance trend, but also on more subtle features such as a flattening of attenuation in the distance range from 70 to 150 km. The uncertainty bars of DYFI data appeared a bit wider than those associated with the model-based estimation, but this is a reasonable price to be paid in exchange for an incredibly dense network of *virtual sensors*.

26.5.2 Seismic (and Multi)Risk: Exposure Mapping

The global earthquake model (GEM, 2006) is an international forum where organizations and people come together (virtually, but also physically) to develop, use, and share tools and resources for transparent assessment of earthquake risk. The GEM's collaborative efforts are expected to build a heightened public understanding and awareness of seismic risk, leading to increased earthquake resilience worldwide. One of the most important products of GEM's efforts is OpenQuake (Silva et al., 2013), a suite of open-source software allowing the GEM community to use the data, best practices, and applications collaboratively being developed. The principle scheme of OpenQuake is illustrated in Figure 26.8.

The platform relies on a geographical database containing the exposure data, that is, information on how human beings and goods, at stake in case of disasters, are distributed globally, at different levels of spatial detail. This is called the global exposure database (GED), and its structure is designed to contain information on buildings and people from the country level all the way down to individual buildings. The first version of GED contains aggregate information on population and the number/built area/reconstruction cost of residential and nonresidential buildings at a 1 km resolution. Detailed datasets on single buildings will be available for a selected number of areas and are increasing over time, thanks to crowdsourcing mechanisms. New data are indeed produced through a set of tools termed "the GEM inventory data

USGS Community Internet Intensity Map
Central California
December 22, 2003 11:15:56 PST 35.7058N 121.1017W M6.5 Depth: 7 km ID: nc40148755

Santa Rosa
Laguna
Vallejo
San Francisco
Fremont
San Jose
Merced
Santa Cruz
Salinas
Fresno
Visalia
Delano
Bakersfield
Santa Maria
Palmdale
Santa Clarita
Los Angeles
Long Beach
Murrieta
Oceanside
Encinitas
50 km
38°N
37°N
36°N
35°N
34°N
33°N
122°W
120°W
118°W
21,970 responses in 1,066 ZIP codes (Max CDI = VII)

Intensity	I	II–III	IV	V	VI	VII	VIII	IX	X+
Shaking	Not felt	Weak	Light	Moderate	Strong	Very strong	Severe	Violent	Extreme
Damage	None	None	None	Very light	Light	Moderate	Moderate/Heavy	Heavy	V. Heavy

(a) Processed: Friday September 7 19:06:57 2012

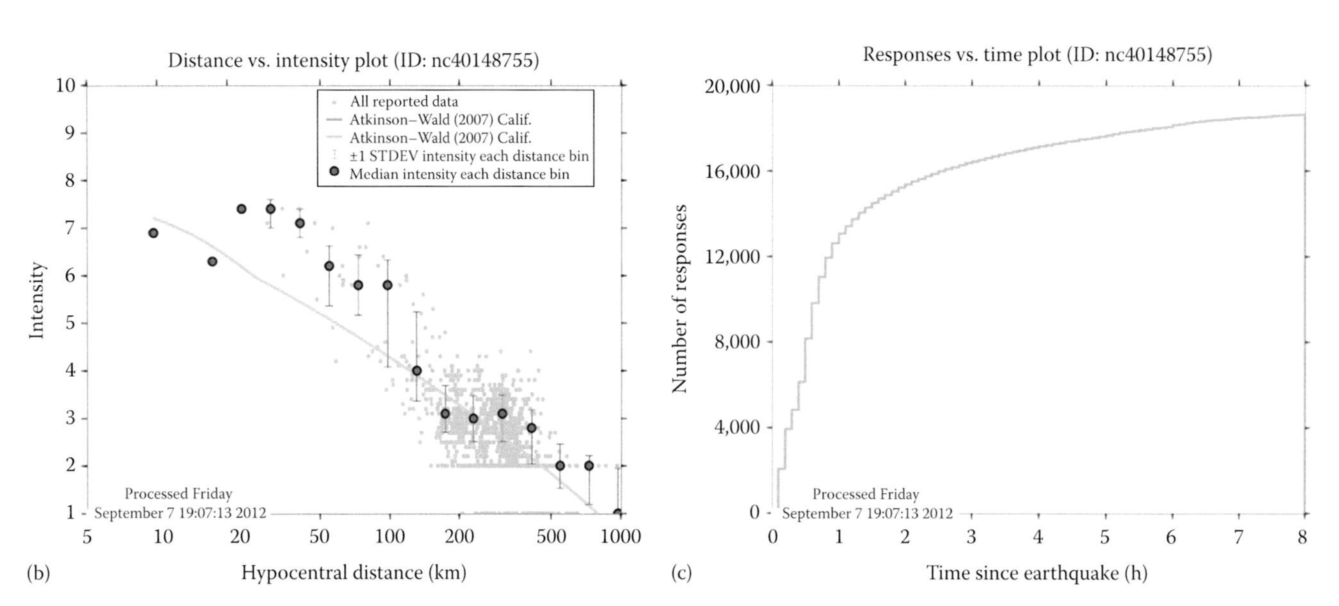

FIGURE 26.6 DYFI products for the for the Central California event on December 22nd, 2003. (a) Color-coded community Internet intensity map. (b) Estimated intensity vs. distance plotting of aggregated responses. (c) Cumulated responses vs. time elapsed. All images from the DYFI website. (From USGS, Reston, VA.)

FIGURE 26.7 Distribution of questionnaires filed between 1991 and 2012 on the DYFI system (from the USGS DYFI web site).

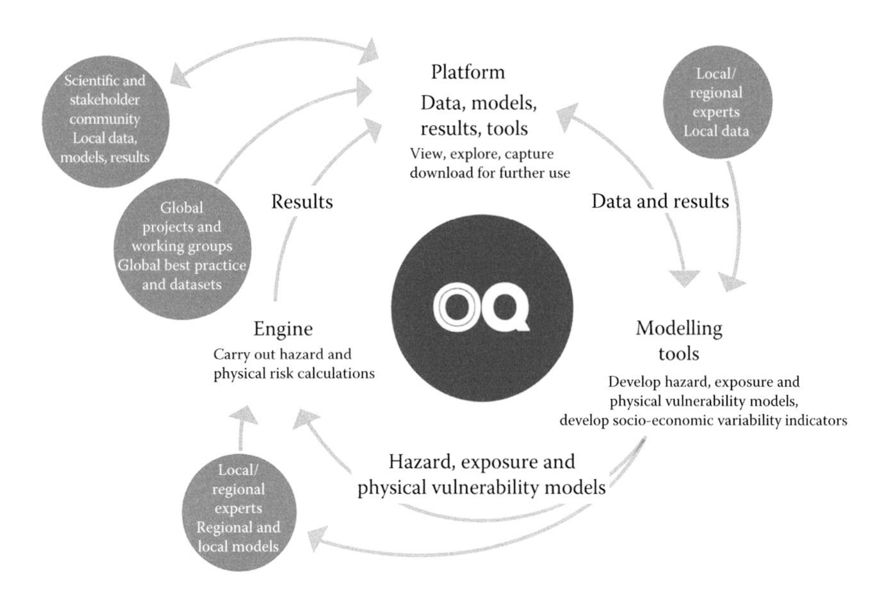

FIGURE 26.8 The OpenQuake system. (From Global Earthquake Model [GEM] Foundation, Pavia, Italy.)

capture tools (IDCT)." The IDCT and the accompanying user protocols are meant to enable users to collect and modify building exposure information, which can be input into the GED. The exposure information can be generated from remote sensing and field observations through the following tools:

- Building data capture application for Android phone or tablet
- Windows tool for field data collection and management
- Tool to develop homogeneous exposure datasets
- QuantumGIS plug-in to extract building footprints from satellite imagery
- Paper forms for building inventory capture

Figure 26.9 depicts the operation principle for GEM—IDCT.

Figure 26.10 represents instead one output of a collaborative digitalization work of building footprints overlaid on a nadir satellite image of L'Aquila, Italy.

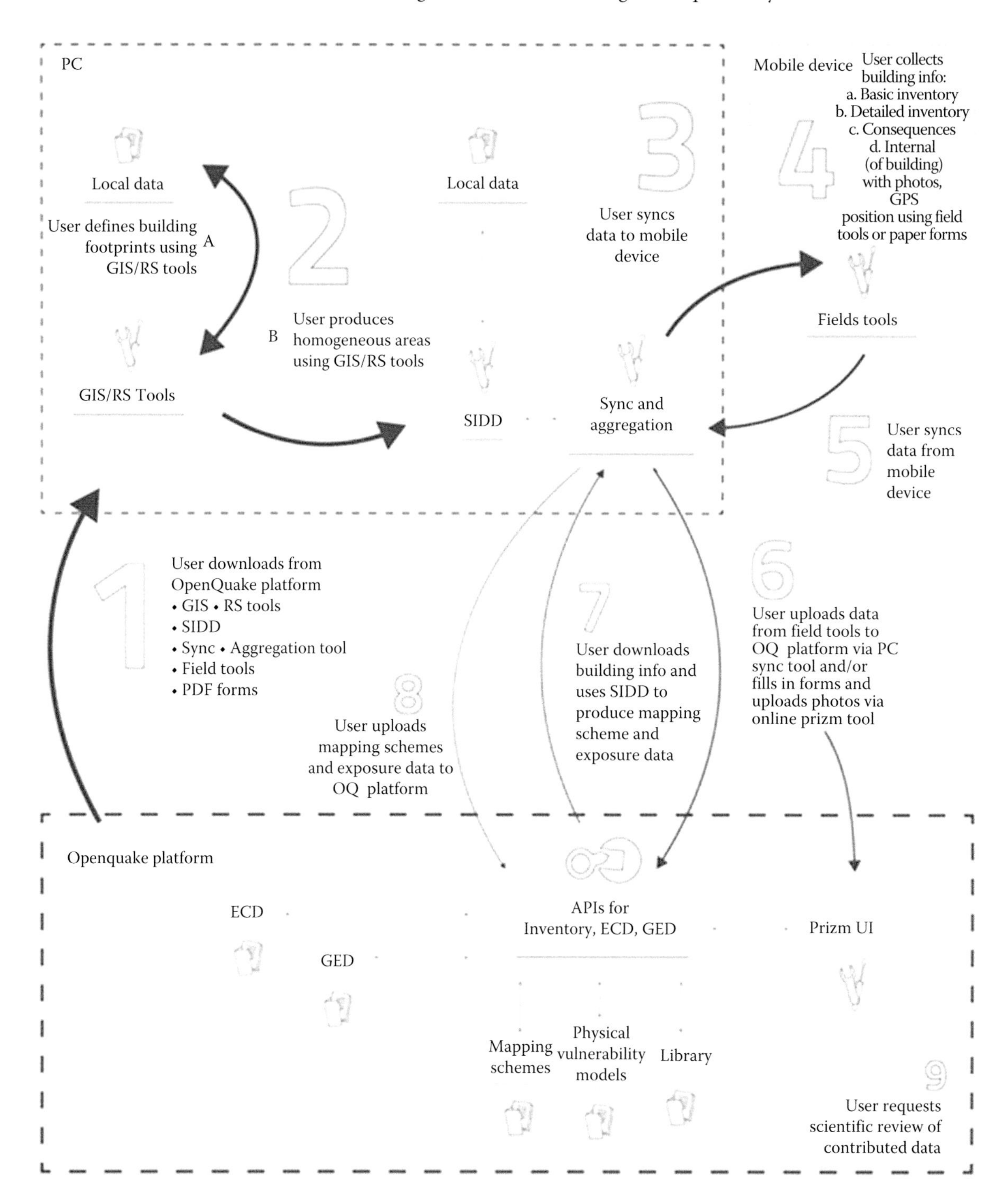

FIGURE 26.9 The operation model for GEM-IDCT. (From Global Earthquake Model [GEM] Foundation, Pavia, Italy.)

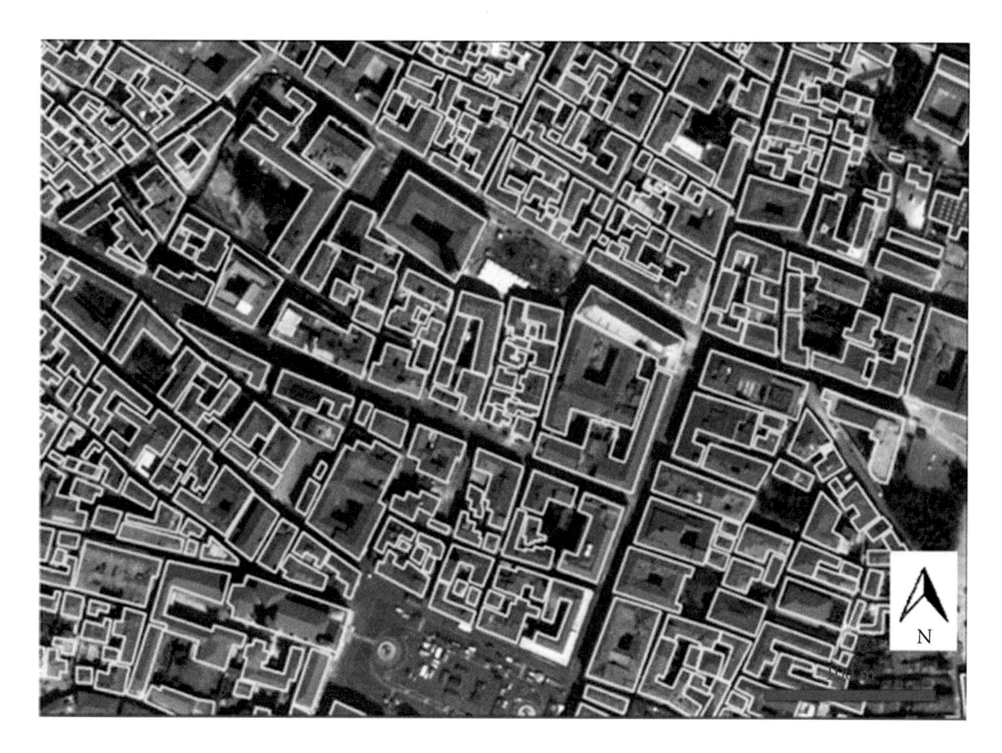

FIGURE 26.10 Results of a collaborative building footprint digitization on L'Aquila, Italy. Yellow lines define boundaries of single buildings. North up. (From Global Earthquake Model [GEM] Foundation, author's reworking, Pavia, Italy.)

Such data can be integrated with field observation through the direct observation (DO) tools running on mobile devices; such tools have been made available even on the Google Play (2014) repository, fully implementing the concept of crowdsourcing. Again, accuracy assessment is made difficult by the scarce availability of actionable ground truth data. Still-to-be-published reports (Iannelli, 2014) on remote sensing IDCT tools convey figures of around 70% accuracy in footprint delineation in the cases of minimal human intervention. (Figure 26.10 refers to manual digitization, which is a different story with accuracy reaching above 90%.)

The case of GEM is particularly interesting, because it realizes the crowdsourcing concept at two different stages of its life cycle: during the definition of risk assessment practices and during collection of hard data that will be used to feed the defined practices.

The first implementation is through GEM Nexus, that is, the GEM collaboration platform where experts and professionals from around the globe find an opportunity of working collectively on state-of-the-art global earthquake risk assessment. The second implementation is through the IDCT set of tools enabling a pool of contributors to generate and feed data into the system from in situ observation or processing of satellite images.

26.5.3 Land Cover Updating

Land cover mapping from multispectral spaceborne data dates back to the beginnings of the LANDSAT satellite series in 1972 (Anderson et al., 1976). It became soon evident that land cover classification from remotely sensed data is far from perfect, even in ideal conditions (Congalton, 1991). Also in this case, crowdsourcing can offer some help, with the Geo-Wiki Project (Fritz et al., 2009). This project aggregates a global network of volunteers willing to help improve the quality of global land cover maps. Each single contributor is requested to review hotspot maps of global land cover and determine, based on what they actually see in Google Earth and their local knowledge, if the land cover maps are correct or incorrect at the spot under review. Georeferenced ground pictures can be uploaded and support the proposed land cover class. Each input is recorded in a database, along with (possible) uploaded photos, to be used in the future for the creation of a new and improved global land cover map. Figure 26.11 shows the interface of Geo-Wiki, where the familiar Google Earth Hi-Res mosaic is overlaid with a much spatially coarser land cover map from the GlobCover (2005) dataset.

A note on accuracy should also be made here. Accuracy assessment of crowdsourced land cover data encounters the same issues that have been pointed out in Chapter 4 for completeness assessment of crowdsourced maps. The issues are made even worse by the higher variability of land cover with respect to, for example, settlement location or road geometry. However, assessments have been attempted, but obviously on datasets whose size is very small compared to the global world coverage. Foody et al. (2013) report a kappa coefficient of around 0.7 for a 10-class Geo-Wiki classification experiment based on labeling of about 300 samples carried out by 65 volunteers, among which the 10 most productive were selected for the final assessment. Similar figures in terms of accuracy (60%–70%) are reported in another Geo-Wiki experiment reported in See et al. (2013) where

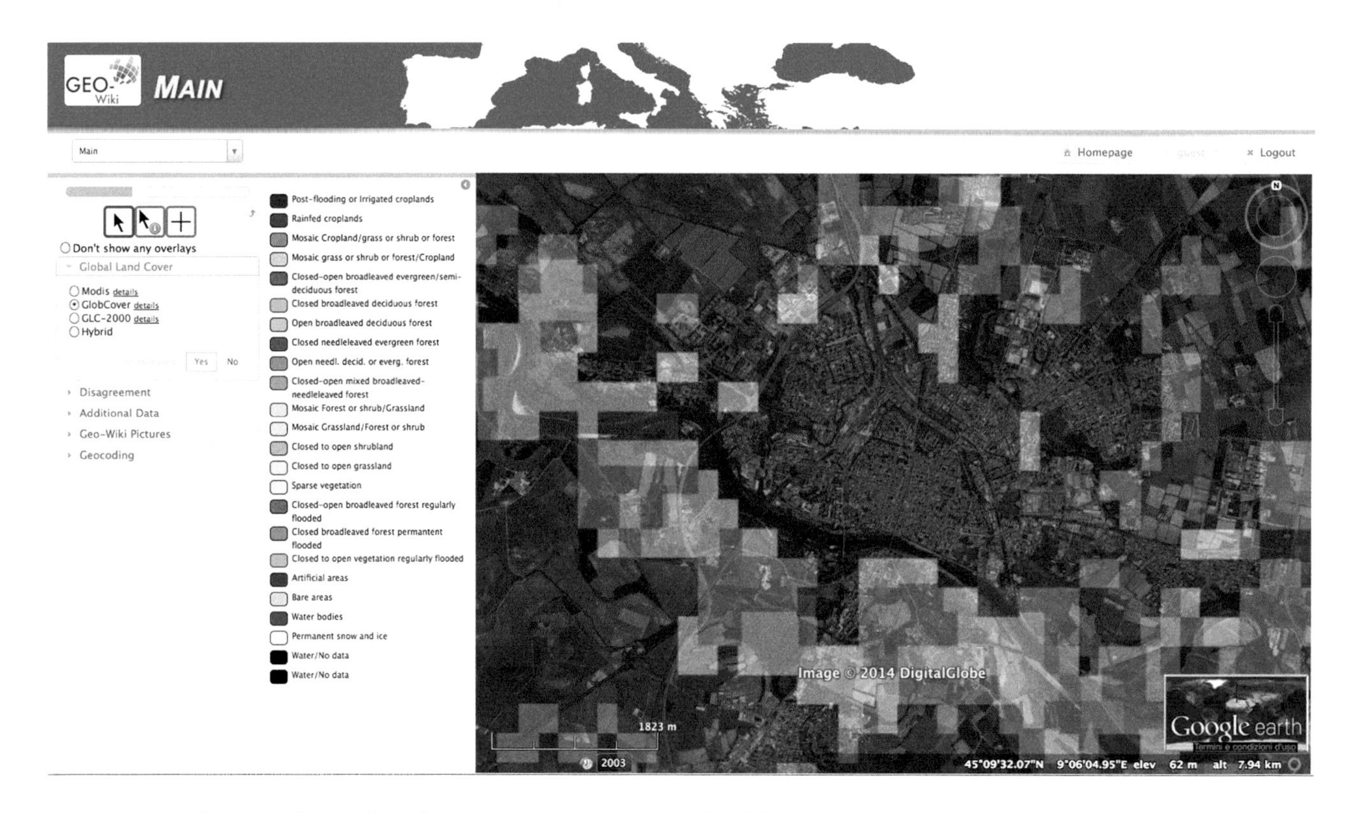

FIGURE 26.11 The Geo-Wiki Interface after connecting as a guest user. The GlobCover land cover map is shown overlaid on the Google Earth© image of an area (Pavia, Italy, in the example) and the user can report on disagreement among the recorded and the actual land cover class by clicking on the appropriate radio button on the left bar. (From http://www.geo-wiki.org/.)

it is also pointed out that the difference between expert and non-expert contributors is narrower than one may expect. Accuracy figures are thus not so high, still, even the combination of disagreeing inputs can bring valuable information (Foody et al., 2013). The situation is probably set to improve as more and more remotely sensed data becomes publicly available as it happened for the LANDSAT repository in 2008.

26.5.4 Research (and Operability) in Progress

As one can imagine, the last word is not said yet on VGI, let alone Web 2.0. While still trying to understand whether the upcoming Web 3.0 (Hendler, 2009) is finally the Semantic Web envisaged by Berners-Lee et al. (2001), on the side of geospatial collaborative work, not only cooperation actions have been established under the COST hat, but also projects are running to improve the effectiveness and productivity of VGI systems.

Examples of cooperation actions are "Mapping and the Citizen Sensor" (COST, 2012a) and ENERGIC (COST, 2012b).

One example of project is the Citizen Observatory Web (COBWEB, 2012) project funded under theme "Environment" of the European Union's Seventh Framework Programme (FP 7). Its main aim is to create a testbed environment that will enable citizens living within biosphere reserves to collect environmental data using mobile devices. The goal is twofold: on the one side, to build a repository of environmental data useful for driving policies and helping decision makers; on the other side, to gain expertise in crowdsourcing/citizen science techniques combined with spatial data infrastructures (SDI), which is still a comparatively new field of development in spatial data science.

The CITI-SENSE project (CITISENSE, 2012) is another EU FP7 ENV project, started in October 2012, aiming at the development of a system for community-based environmental monitoring and information in urban areas using Earth observation and crowdsourcing.

The Citizens' Observatory for Coast and Ocean Optical Monitoring (CITCLOPS, 2012) project is also funded by the EU FP7 under the ENV theme. It uses volunteer inputs from mobile phones on individuals on the spot to help monitoring the color and transparency of natural water, that are important indicators of, for example, the ecosystem health status. An Android app is distributed from the project website, and the job requested is really participatory, in that it is not just about taking a picture of the water surface from one's mobile phone and sending it to the repository; it entails interaction including, for example, selecting the *best matching* color from a range of proposed standard water colors, feeding additional environmental parameters, and *paying off* the contributor by giving specific (though obviously automatic) feedback on the acquired sample. Another FP7 ENV project, WeSenseIt (2012), is also about *Citizen Sensing* of water, but for purposes of flood risk assessment and related capacity building.

Apart from research projects, the state of the art includes an open source platform like Ushahidi (2007), which enables the easy deployment of crowdsourced interactive mapping applications with Web forms, e-mail, short message service (SMS), and Twitter (2006) support. Ushahidi can be freely downloaded and deployed on one's own server or used as an online service hosted by the platform providers itself, that is, Crowdmap. Ushahidi can be accessed on smartphones and tablets through suitable mobile apps. Commercial exploitation of crowdsourcing of satellite and geospatial data is also there, with companies offering services based on crowdprocessing of satellite multispectral images (Tomnod 2010).

The attention level of institutions to the crowdsourcing and Citizen Sensor theme is generally high. Recently, a COST action named "Mapping and the Citizen Sensor" TD1202 (2012) has been approved by the European Science Foundation, and is expected to increase knowledge and practicability of crowdsourcing and VGI provision. The intergovernmental *Group* on *Earth Observations* (GEO, 2005), coordinating efforts to build a Global Earth Observation System of Systems, or GEOSS, in its 2012–2015 WorkPlan, makes several references to fostering crowdsourcing and citizen science observation as additional sources of data and information.

26.6 Conclusions

It is a painful historical fact that, after the enthusiasm in the 1980s and 1990s for the apparently endless reach of information gathering through airborne or space-based remote sensing, a grim disappointment followed in the scientists and user community at large after realizing that remotely sensed data fell short of sustaining the wide range of applications that had been envisaged at the beginning. The foundations of detecting any type of material on the Earth surface by observing it from above had been laid (Clark and Roush, 1984), but translating these principles into something really operational turned out to be tougher than expected. A part of the problem was the inherent difficulty in translating noisy measures of faraway physical quantities into estimates that could be relied upon. Yet, another part was instead connected with the limitations in coverage (information on point x at time y cannot be acquired, because no sensor is there at that time) and visibility of observed features (information on feature z is not accessible, simply, because it is not visible from above).

Both these latter issues can in principle be addressed through integration of crowdsourcing into the production chain of geospatial information. Crowdsourcing, seen as an immersive, distributed network of countless in situ sensors, carries a potential to

- Substantially increase the *a priori* likelihood of *being there* when *it matters*
- Gain accessibility to features that are invisible to the average, faraway nadir or quasinadir sensor, for example, small details, or reflectance of a vertical surface

In addition to the aforementioned points, the introduction of human beings in the loop scales up the observation paradigm from purely quantitative to semantic—even after including and weighing all the implied risks of data pollution, misunderstanding, misinterpretation—is still a big step up.

This is not however the only power of crowdsourcing that comes from the active participation of intelligent humans in a task assigned to them. People are social beings and as such are also oriented toward sharing information and helping one another; the crowdsourcing capitalizes on this orientation, allowing to gather more information (both in terms of quantity and depth), from many more *sensors* and much more quickly.

Acknowledgments

The author especially thanks two persons for their useful inputs: Silvio Dell'Acqua and Pietro Demattei. Silvio Dell'Acqua is the editor of Laputa (2012), a website devoted to the publication of original research on historical and geographical weird facts.

References

Alfaro, L.D. et al. 2011. Reputation systems for open collaboration. *Commun. ACM*, 54(8), 81–87.

Allahbakhsh, M., Benatallah, B., Ignjatovic, A., Motahari-Nezhad, H.R., Bertino, E., Dustdar, S. March–April 2013. Quality control in crowdsourcing systems: Issues and directions. *IEEE Internet Comput.*, 17(2), 76–81. doi: 10.1109/MIC.2013.20.

Amazon Mechanical Turk. 2005. "Amazon Mechanical Turk: Artificial Artificial Intelligence". Online. Available at: https://www.mturk.com/.

Anderson, J.R., Hardy, E.E., Roach, J.T., Witmer, R.E. 1976. *A Land Use and Land Cover Classification System for Use with Remote Sensor Data*. Government Printing Office, Washington, DC, U.S. Geological Survey, Professional Paper 964.

Atkinson, G.M., Wald, D.J. May/June 2007. "Did You Feel It?" intensity data: A surprisingly good measure of earthquake ground motion. *Seismol. Res. Lett.*, 78, 362–368. doi: 10.1785/gssrl.78.3.362.

Berners-Lee, T., Hendler, J., Lassila, O. 2001. The semantic web. *Scient. Am.*, 284(5), 35–43.

CITCLOPS. 2012. The Citizens' observatory for coast and ocean optical monitoring project web site. Online. Available at: http://www.citclops.eu/.

CITISENSE. 2012. The Citizen's observatory community for improving quality of live in cities (CITI-SENSE) project web site. Online. Available at: http://www.citi-sense.eu/.

Clark, R.N., Roush, T.L. July 10, 1984. Reflectance spectroscopy: Quantitative analysis techniques for remote sensing applications. *J. Geophys. Res.: Solid Earth* (1978–2012), 89(B7), 6329–6340.

COBWEB. 2012. The Citizen Observatory Web (COBWEB) project web site. Online. Available at: http://cobwebproject.eu/.

Congalton, R.G. July 1991. A review of assessing the accuracy of classifications of remotely sensed data. *Remote Sens. Environ.*, 37(1), 35–46.

COST. 2012a. European Cooperation in Science and Technology. ICT action: "Mapping and the Citizen Sensor". Action web site: http://www.cost.eu/domains_actions/ict/Actions/TD1202.

COST. 2012b. European Cooperation in Science and Technology. Action: "European Network Exploring Research into Geospatial Information Crowdsourcing: Software and methodologies for harnessing geographic information from the crowd (ENERGIC)". Action web site: http://www.cost.eu/COST_Actions/ict/Actions/IC1203. Accessed March 24, 2015.

De Leeuw, J., Said, M., Ortegah, L., Nagda, S., Georgiadou, Y., DeBlois, M. 2011. An assessment of the accuracy of volunteered road map production in Western Kenya. *Remote Sens.*, 3, 247–256.

Dengler, L.A., Dewey, J.W. 1998. An intensity survey of households affected by the Northridge, California, earthquake of January 17, 1994. *Bull. Seismol. Soc. Am.*, 88, 441–462.

Dengler, L.A., Moley, K. 1994. Toward a quantitative, rapid response estimation of intensities. *Seismol. Res. Lett.*, 65, 48.

DesignCrowd. 2010. 5 Famous Logo Contests—Toyota, Google, Wikipedia & More!. Online. Available at: http://blog.designcrowd.com/article/218/5-famous-logo-contests—toyota-google-wikipedia—more.

Dow, S.P., Kulkarni, A., Klemmer, S.R., Hartmann, B. 2012. Shepherding the crowd yields better work. *Proceedings of the 2012 ACM Conference on Computer Supported Cooperative Work (CSCW'12)*, ACM, 11–15 Feb 2012, Seattle, WA. pp. 1013–1022.

DYFI. 1991. "Did You Feel It" earthquake hazards program. Online. Available at: http://earthquake.usgs.gov/earthquakes/dyfi/ Accessed March 24, 2015.

Foody, G.M., See, L., Fritz, S., Van der Velde, M., Perger, C., Schill, C., Boyd, D.S. 2013. Assessing the accuracy of volunteered geographic information arising from multiple contributors to an internet based collaborative project. *Trans. GIS*, 17(6), 847–860.

Fritz, S., McCallum, I., Schill, C., Perger, C., Grillmayer, R., Achard, F., Kraxner, F., Obersteiner, M. 2009. Geo-Wiki. Org: The use of crowdsourcing to improve global land cover. *Remote Sens.*, 1(3), 345–354. doi: 10.3390/rs1030345. Open access.

GEM. 2006. The global earthquake model. Available at: http://www.globalquakemodel.org/.

GEO. 2005. The group on earth observation web site. Online. Available at: http://www.earthobservations.org/.

GlobCover. 2005. The European Space Agency GlobCover portal. Online. Available at: http://due.esrin.esa.int/globcover/.

Goetz, M., Zipf, A. 2012. Towards defining a framework for the automatic derivation of 3D CityGML models from volunteered geographic information. *Int. J. 3-D Inform. Model.*, 1, 496–507.

Goodchild, M.F. 2007. Citizens as sensors: The world of volunteered geography. *GeoJournal*, 69, 211–221. doi: 10.1007/s10708-007-9111-y.

Google Play. 2014. "IDCT Direct Observation Survey" app. Available on the Internet at: https://play.google.com/store/apps/details?id=org.globalquakemodel.org.idctdo. Accessed March 24, 2015.

Hendler, J. January 2009. Web 3.0 emerging. *Computer*, 42(1), 111–113. doi: 10.1109/MC.2009.30.

Heylighen, F. 1997. Towards a global brain. In: Der Sinn der Sinne, U. Brandes, and C. Neumann, eds. *Integrating Individuals into the World-Wide Electronic Network*. Steidl Verlag, Göttingen, Germany, 1997.

Howe, J. 2006. The rise of crowdsourcing. *Wired Magazine*—Issue 14.06, June 2006. Available at: http://archive.wired.com/wired/archive/14.06/crowds.html. Accessed March 24, 2015

Iannelli, G.C. 2014. "Processing and fusion of multiresolution spaceborne Earth Observation data for assessing exposure and vulnerability to natural disasters risk". PhD Thesis of Gianni Cristian Iannelli. To be published on the Faculty repository at http://www-3.unipv.it/dottIEIE/index.php?pag=italiano/tesi_dottorato.html. Accessed March 24, 2015.

IDC. 2013. International Data Corporation (IDC) worldwide quarterly mobile phone tracker. Online. Available at: http://www.idc.com/tracker/showproductinfo.jsp?prod_id=37.

Jøsang, A., Ismail, R., Boyd, C. 2007. A survey of trust and reputation systems for online service provision. *Decision Supp. Syst.*, 43(2), 618–644.

Kaplan, A.M., Haenlein, M. 2010. Users of the world, unite! The challenges and opportunities of social media. *Business Horiz.*, 53, 59–68.

Kerle, N. 2011. Remote sensing based post-disaster damage mapping—Ready for a collaborative approach? *IEEE Earthzine*. Posted on March 23, 2011 in Articles, Disaster Management Theme, Earth Observation.

Levy, P. 1999. *Collective Intelligence: Mankind's Emerging World in Cyberspace*. Basic Books, New York. ISBN: 9780738202617.

Laputa, 2012. "Nekutima Geografio". Owner and Editor: Silvio Dell'Acqua. Online at: http://www.laputa.it/. English version coming soon. Accessed March 24, 2015.

Neis, P., Zielstra, D., Zipf, A. 2012. The street network evolution of crowdsourced maps: OpenStreetMap in Germany 2007–2011. *Future Internet*, 4, 1–21.

OECD. 2007. *Participative Web and User-Created Content: Web 2.0, Wikis, and Social Networking*. Organisation for Economic Co-operation and Development, Paris, France.

Openstreetmap. 2004. Collaborative mapping web site. Online. Available at: http://www.openstreetmap.org/.

O'Reilly, T. 2007. What is Web 2.0: Design patterns and business models for the next generation of software. *J. Commun. Strat.*, 65, 17.

Over, M., Schilling, A., Neubauer, S., Zipf, A. 2010. Generating web-based 3D city models from OpenStreetMap: The current situation in Germany. *Comput. Environ. Urban Syst.*, 34, 496–507.

Raifer, M. 2014. OSM node density 2014. Map available at: http://tyrasd.github.io/osm-node-density/; explanation available at: http://www.openstreetmap.org/user/tyr_asd/diary/22363. Accessed on August 14, 2014; Re-accessed March 24, 2015.

Scekic, O., Truong, H., Dustdar, S. 2012. Modeling rewards and incentive mechanisms for social BPM. In: A. Barros et al., eds. *Short paper, 10th International Conference on Business Process Management (BPM2012),* September 3–6, 2012, Springer, Tallinn, Estonia, pp. 150–155.

See, L., Comber, A., Salk, C., Fritz, S., van der Velde, M., Perger, C., Schill, C., McCallum, I., Kraxner, F., Obersteiner, M. 2013. Comparing the quality of crowdsourced data contributed by expert and non-experts. *PLoS ONE*, 8(7), e69958. doi: 10.1371/journal.pone.0069958.

ShakeMap. 2004. ShakeMap Archive, M 6.0 event on September 28th, 2004, 17:15:24 UTC. Addressed maps available at: http://earthquake.usgs.gov/eqcenter/shakemap/nc/shake/51147892/; service description available at: http://earthquake.usgs.gov/earthquakes/shakemap/.

Sheth, A. July–August 2009. Citizen sensing, social signals, and enriching human experience. *IEEE Internet Comput.*, 13(4), 87–92, doi: 10.1109/MIC.2009.77

Silva, V., Crowley, H., Pagani, M., Monelli, D., Pinho, R. 2013. Development of the OpenQuake engine, the Global Earthquake Model's open-source software for seismic risk assessment. *Nat. Hazards*, 72(3), 1409–1427. doi: 10.1007/s11069-013-0618-x.

Sobel, D. 1998. *Longitude: The True Story of a Lone Genius Who Solved the Greatest Scientific Problem of His Time*. Fourth Estate Ltd., London, U.K., p. 6. ISBN: 1 85702 571-7.

TD1202. 2012. Information and Communication Technologies (ICT) Action TD1202 "Mapping and the citizen sensor". Available at: http://www.cost.eu/domains_actions/ict/Actions/TD1202.

Tomnod. 2010. Originally a start-up, now owned by DigitalGlobe. Online at: http://www.tomnod.com. Accessed March 24, 2015.

Twitter 2006. The twitter web site. Online. Available at: http://twitter.com/.

Ushahidi. 2007. The Ushaihidi Project web site. Online. Available at: http://ushahidi.com/.

Wald, D.J., Quitoriano, V., Worden, B., Hopper, M., Dewey, J.W. 2011. USGS "Did You Feel It?" Internet-based macroseismic intensity maps. *Ann. Geophys.*, 54(6), 688–707. doi: 10.4401/ag-5354.

WeSenseIt. 2012. "WeSenseIT: Citizen Observatory of Water" project web site. Online. Available at: http://www.wesenseit.eu/.

Wikimapia. 2007. Collaborative mapping web site—Not a part of non-profit Wikimedia foundation. Online. Available at: http://wikimapia.org/.

Wikipedia. 2001. The free on-line encyclopedia. Online. Available at: http://www.wikipedia.org/.

Cloud Computing and Remote Sensing

27

Processing Remote-Sensing Data in Cloud Computing Environments

Ramanathan Sugumaran
The University of Iowa and John Deere

James W. Hegeman
The University of Iowa

Vivek B. Sardeshmukh
The University of Iowa

Marc P. Armstrong
The University of Iowa

Acronyms and Definitions

AWS	Amazon Web Services
CLiPS	Cloud-based LiDAR processing system
CPU	Central processing unit
EC2	Elastic Compute Cloud
EOSDIS	Earth Observing System Data and Information System
GIS	Geographic information system
GPGPU	General-purpose computing on graphics processing units
GPU	Graphics processing unit
HPC	High-performance computing
HTC	High-throughput computing
IaaS	Infrastructure as a service
MODIS	Moderate-Resolution Imaging Spectroradiometer
OCC	Open Cloud Consortium
OGC	Open Geospatial Consortium
PaaS	Platform as a service
Saas	Software as a service
TIN	Triangulated irregular network
UAV	Unpiloted aerial vehicles
USGS	U.S. Geological Survey
VPN	Virtual private network
WMS	Web Map Service

27.1 Introduction

27.1.1 Remote Sensing and Big Data

During the past four decades, scientific communities around the world have regularly accumulated massive collections of remotely sensed data from ground, aerial, and satellite platforms. In the United States, these collections include the U.S. Geological Survey's (USGS) 37-year record of Landsat satellite images (comprising petabytes of data) (USGS, 2011); the NASA Earth Observing System Data and Information System, having multiple data centers and more than 7.5 petabytes of archived imagery (Hyspeed Computing, 2013); and the current NASA systems that record approximately 5 TB of remote-sensing-related data per day (Vatsavai et al., 2012). In addition, new data-capture technologies such as LiDAR are used routinely to produce multiple petabytes of 3D remotely sensed data representing topographic information (Sugumaran et al., 2011). These technologies have galvanized changes in the way remotely sensed data are collected, managed, and analyzed. On the sensor side, great progress has been made in optical, microwave, and hyperspectral remote sensing with (1) spatial resolutions extending from kilometers to submeters, (2) temporal resolutions ranging from weeks to 30 min, (3) spectral resolutions ranging from single bands to

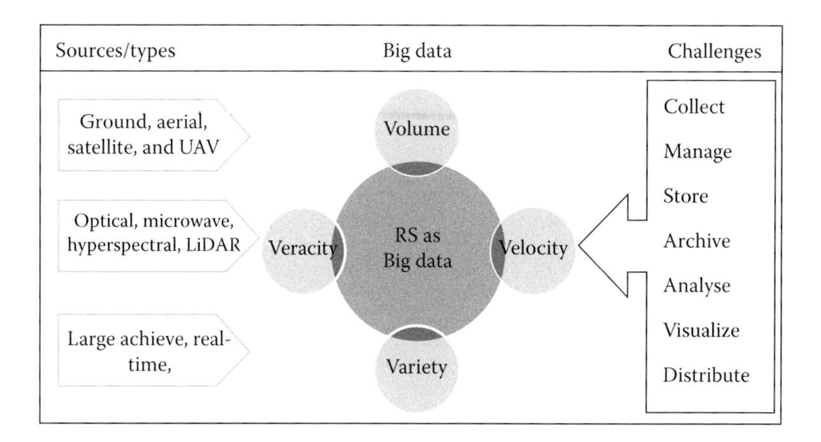

FIGURE 27.1 Remote sensing: big-data sources and challenges.

hundreds of bands, and (4) radiometric resolutions ranging from 8 to 16 bits. The platform side has also seen rapid development during the past three decades. Satellite and aerial platforms have continued to mature and are producing large quantities of remote-sensing data. Moreover, sensors deployed on unpiloted aerial vehicles (UAVs) have recently begun to produce massive quantities of very-high-resolution data.

The technological nexus of continuously increasing spatial, temporal, spectral, and radiometric resolutions of inexpensive sensors, on a range of platforms, along with internet data accessibility is creating a flood of remote-sensing data that can easily be included in what is commonly referred to as "big data." This term refers to datasets that have grown sufficiently large that they have become difficult to store, manage, share, and analyze using conventional software tools (White, 2012). "Big data" are often thought to span four dimensions: volume (data quantity), velocity (real-time processing), variety (source multiplicity), and veracity (data accuracy) (IBM, 2012). Operating hand in glove with Moore's law, the growth of big data is largely a consequence of advances in acquisition technology and increases in storage capacity. Figure 27.1 summarizes the overall sources and challenges presented by big remote-sensing data.

27.1.2 Big-Data Processing Challenges

As the pace of imaging technology has continued to advance, the provision of affordable technology for dealing with issues such as storing, processing, managing, archiving, disseminating, and analyzing large volumes of remote-sensing information has lagged. One major challenge is related to the computational power required to process these massive data sources. Traditionally, desktop computers with single or multiple cores have been used to process remote-sensing data for small areas. In contrast, large- or macroscale remote-sensing applications may require high-performance computing (HPC) technologies; general-purpose computing on graphics processing units (GPGPU); and parallel, cluster, and distributed-computing approaches are gaining broad acceptance (Plaza et al., 2006; González et al., 2009; Simmhan and Ramakrishnan, 2010; Shekhar et al., 2012). Given these architectural advances, the analysis of big data presents new challenges to both cluster-infrastructure software and parallel-application design, and it requires the development of new computational methods. These methods and several articles about the importance of HPC in remote sensing are featured in special journal issues, books, and conferences devoted to this topic (Plaza and Chang, 2007, 2008; Lee et al., 2011; Prasad, 2013).

Graphics processing units (GPUs) have been widely used (in GPGPU applications) to address remote-sensing problems (Chang et al., 2011; Christophe et al., 2011; Song et al., 2011; Yang et al., 2011). Oryspayev et al. (2012) developed an approach to LiDAR processing that used data-mining algorithms coupled with parallel computing technology. A specific comparison was made between the use of multiple central processing units (CPUs) (Intel Xeon Nehalem chipsets) and GPUs (Intel i7 Core CPUs using the NVIDIA Tesla s1070 GPU cards). The experimental results demonstrated that the GPU option was up to 35 times faster than the CPU option. In a similar vein, distributed parallel approaches have also been developed. Haifang (2003) implemented various algorithms using a heterogeneous grid computing environment, and Liu (2010) analyzed the efficiency improved by grid computing for the maximum likelihood classification method. Yue et al. (2010) used cluster computing to solve remote-sensing-image fusion, filtering, and segmentation. Commodity-cluster-based parallel processing of various multispectral and hyperspectral imagery has also been used by various authors (e.g., Plaza et al., 2006).

While HPC environments such as clusters, grid computing, and supercomputers (Simmhan and Ramakrishnan, 2010) can be used, these platforms require significant investments in equipment and maintenance (Ostermann et al., 2010) and individual researchers and many government agencies do not have routine access to these resources. In addition to data volume, the variety and update rate of datasets often exceed the capacity of commonly used computing and database technologies (Wang et al., 2009; Yang et al., 2011; Shekhar et al., 2012). As a result of these limitations, users have begun to search for less-expensive solutions for the development of large-scale data-intensive remote-sensing applications. Cloud computing provides a potential solution to this challenge due to its scalability advantages in data storage and processing and its relatively low cost as compared to user-owned, high-power compute clusters (Kumar et al., 2013). The goal of this chapter is to provide a short review of remote-sensing applications using cloud-computing environments, as well as a case study that provides greater implementation detail. An introduction to cloud computing is provided in Section 27.2, and then in Section 27.3, various applications, including a detailed case study, are described to illustrate the advantages of cloud-computing environments.

27.2 Introduction to Cloud Computing

27.2.1 Definitions

Cloud computing is a vague term, as nebulous as its eponym. *The NIST Definition of Cloud Computing* (Mell and Grance, 2011) defines cloud computing as "a model for enabling ubiquitous, convenient, on-demand network access to a shared pool of configurable computing resources." In research, "cloud computing" is a popular idiom for distributed computing, encompassing the same fundamental concepts—multiplicity, parallelism, and fault tolerance. Distributed computing has come to the forefront as an area of research over the past two decades as dataset sizes have outstripped traditional sequential processing power, even of modern high-performance processors, and as the bottlenecks of large-scale computation have moved outside the CPU (e.g., to storage I/O). To employ distributed or cloud computing means to leverage the additional hardware and computing throughput available from large networks of machines. Because of the challenges inherent in optimizing resource scale and utilization for a problem, two central focuses of cloud computing in practice have been (1) the elasticity of resource provisioning and (2) abstraction layers capable of simplifying these challenges for users. Indeed, as exemplified in an eScience Institute document (2012), it is these two qualities of cloud computing—elasticity and abstraction—that are often most important from the end user's perspective. Thus, while research with cloud computing generally focuses on the distributed scalability of the cloud, the characteristic feature of cloud computing, in practice, is flexibility. There are several related terms associated with cloud computing as explained below.

Cluster computing is a similar, but more limited, model of parallel/distributed computing, in which the machines comprised by the cluster are usually assumed to be tightly connected by a low-latency, private network. This generally implies physical locality of the system itself. The concept of cluster computing was a precursor to today's notion of *HPC* (Lee et al., 2011). Cluster computing differs from cloud computing in its emphasis—a compute cluster is often a localized system dedicated to one particular problem (or class of problems) at a time.

Grid computing can also be thought of as a subset of cloud computing having a slightly different emphasis (Buyya et al., 2009). A compute grid is a distributed system, often encompassing machines physically separated by a large distance, together with a *job scheduler* that allows a user to easily and simultaneously run a certain small set of programs on many different data inputs—the action of a user's program on one such input constitutes a *task*, or *job*. Grid computing is thus the use of a parallel/distributed system for solving a large problem or accomplishing a large collection of tasks that would take too long to complete on a single machine. In grid computing, the problem to be solved can generally be broken down into many nearly identical tasks that can run concurrently and independently on the nodes of the system (see, e.g., Wang and Armstrong, 2003). Once complete, the solutions to, or results of, these tasks are aggregated.

The type of computational problem-solving approach exemplified by grid computing is also known as *high-throughput computing* (HTC), in which similar computations are done independently by a (large) number of processors, and the network interconnect is used primarily for data distribution and results aggregation. Another, distinct, distributed computational paradigm is HPC, which emphasizes slightly different aspects of the system. Whereas HTC emphasizes the problem division, HPC refers to the use of many high-power servers connected by a fast (usually >10 Gbps) network, under any algorithmic paradigm. An HPC algorithm may also qualify as HTC, or it may require much more intermediate communication between processes in order to accomplish the goal. In other words, in HPC, individual processes/processors may need to communicate very rarely, or very often, during the intermediate stages of a computation. The concepts of HTC and HPC thus focus on slightly different qualities of a whole system and are neither identical nor mutually exclusive. In Figure 27.2, an example HTC workflow is depicted. The problem input is divided into many segments, each of which is sent to a node of the system; depending on processing

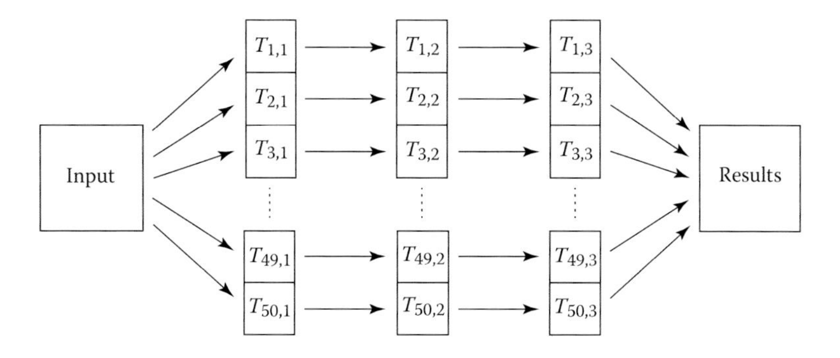

FIGURE 27.2 A typical high-throughput computing workflow model.

capacities, a single node may receive multiple segments. Several tasks (in Figure 27.2, three) may be required to complete the processing of each segment. Thus, in Figure 27.2, any two tasks $T_{i,j}$ with the same first (i) index are distinct but operate (in series) on the same segment of the input, whereas any two tasks with the same second (j) index are identical but operate on different portions of the input. After processing of each segment, the individual results are merged in some manner to form the output. Unlike HTC, there is no canonical diagram for HPC, since HPC is defined more by the power of the computational resources and admits many algorithmic paradigms.

Presently, there are several large-scale commercial options for cloud computing, and many larger institutions have their own distributed systems that users may use as a "cloud." Amazon's Elastic Compute Cloud (EC2) was the first public cloud to be available to large numbers of users over the Internet. Subsequently, Google and Microsoft, as well as several other vendors, have also begun to offer large-scale public cloud services. Since the distributed nature of a cloud is opposite of the traditional mainframe server, most cloud-computing systems in practice run some version of the GNU/Linux operating system. However, Microsoft Windows Server is also an option for cloud systems, and indeed the availability of geographic information system (GIS)-specific software packages makes Windows Server relevant in this domain.

27.2.2 Cloud Paradigms

While a variety of common terms have arisen—*public cloud*, *private cloud*, and *hybrid cloud*—these models of computing are fundamentally the same, differing only in ancillary issues such as security and usability.

A *public cloud* is a cloud system available to the public (or some subset of the public) over the World Wide Web. A public cloud may even rely on Internet infrastructure for "internal" network connectivity. In practice, compute access to most public clouds is available for rent to the general public. Amazon's EC2 is the quintessential public cloud—Amazon Web Services (AWS) was an early leader in providing computing power as a commercial service. Others in the commercial sector have followed suit, such as Google with their Google Cloud Platform. The emergence of public clouds is a good example of an economy of scale—powerful servers and high-performance networks can be expensive and require specialized expertise to administer and support. Many institutions possess problems for which cloud computing is [a part of] an ideal solution, but only the largest have a sufficient quantity of such problems that it makes financial sense to own and manage their own servers. Because of the expense, cloud-computing systems of scale are often only financially practical under high load—smaller organizations without a sufficient volume of computational problems may find themselves only utilizing a private system, say, 50% of the time. In this scenario, the costly overhead looms even larger compared to the cost–benefit analysis of problems solved versus power consumption.

A *private cloud* is just that—a private network of computers owned by the using entity, usually segregated from the public internet. Examples of private clouds include any compute clusters to which access is limited, such as the research clusters commonly operated by and within large universities. Private cloud computing makes sense under many scenarios: (1) For research and development, it may be necessary to have more control over the system than is afforded by the typical public cloud. Oftentimes, complete control of all aspects of a system can only be achieved when the system is private. (2) When it is financially feasible to own a system whose response time would be sufficient for the users, and if the users have a sufficient quantity of computational tasks that the system would be highly utilized, it could make sense to use a private cloud—thus cutting out the middle man from the services.

Finally, the term *hybrid cloud* is sometimes used to refer to an aggregate system of which some portion is owned, or fully controlled, by the user and some portion is available over a public network. A hybrid cloud can make sense when different parts of the cloud are being used for distinctly different tasks and a different cloud-computing model would be ideal for each. For example, an institution with critical data it needs to process may opt to store the data permanently on a smaller, private cloud, where members have full control over data security, but send some portion of the data to a public cloud, on demand, for processing. While a private cloud offers more control to its owners/users, a public cloud can be easier to use and can usually provide greater computational power, with little or no overhead cost. As such, a hybrid cloud model may make sense in practice for many businesses and smaller, short-to-medium-term research endeavors.

27.2.3 Cloud Service Models

Infrastructure as a service: The fundamental concept of outsourcing and commercialization of compute time is referred to as *infrastructure as a service* (IaaS). In this approach, a cloud service provider (for instance, Amazon) owns and operates a collection of networked servers available for rent. At a base level, the cloud service provider provisions rental machines (often virtual machines) with a client's desired operating system, as well as related facilities and tools. For instance, a client might rent compute time from a cloud service provider and request that the rental machines run Fedora Linux. The physical machines, together with the network, Linux kernel, and Fedora distribution constitute the infrastructure, and this system in its entirety is the product provided to the client.

Platform as a service: The concept of a *platform as a service* (PaaS) lies on top of IaaS. For many clients, a compute infrastructure alone is not sufficient for their goals—there is a large gulf between the presence of computing infrastructure and the desired end result. For end users who may have a higher-level abstraction of their computational task(s), an intermediate platform—a computational service that provides more than just an operating system and network and is managed at that higher level—may be appropriate. In PaaS, a software platform is provisioned by the cloud service provider and can be composed of one or more different software layers. One common example is Hadoop, by Apache. Hadoop is an open-source implementation of Google's MapReduce framework that provides the MapReduce computational model along with an underlying distributed file system. In addition, there are a variety of additional layers compatible with Hadoop that can exist on top of the Hadoop–MapReduce platform—a common such layer is the data warehouse Hive, also developed by Apache. For more information on the MapReduce computing framework, see Dean and Ghemawat (2004).

Software as a service: At a higher level than PaaS is *software as a service* (SaaS), in which the cloud service provider furnishes a complete software package designed for a particular domain. The end user need only perform configuration-level tasks on the system before it is ready for use. This model of computational services is well suited for organizations that must perform ubiquitous tasks, such as vehicle location tracking or legal document preparation, and that have no desire for, or commitment to, computational research and development. SaaS is becoming increasingly popular as an operational business model.

27.2.4 Advantages and Limitations of Cloud Computing

On a practical level, the cloud-computing paradigm has many advantages: access to HPC systems; "pay-as-you-go" payment schemes, with few overhead costs; on-demand provisioning of resources; highly scalable/elastic compute and storage resources; and automated data reliability (McEvoy and Schulze, 2008; Watson et al., 2008; Cui et al., 2010; Huang et al., 2010; Rezgui et al., 2013; Yang and Huang, 2013). Note that this last improvement, high data reliability, is different from data security—one of the major challenges in cloud computing. By its very nature of high accessibility/availability, the most scalable type of cloud computing—the use of a public cloud—is inherently less secure than a computing model based on private control of the entire system. The security challenges of using a public cloud are manyfold: depending on privacy requirements, it may not be acceptable that data reside on the cloud service provider's machines; users are reliant on the internal security controls of the service provider to prevent unauthorized data access, and the public network infrastructure that must be traversed for data and compute access may be compromised. More elaborate (and from a performance perspective, costly) security and encryption measures must commonly be taken when using a public cloud, such as the use of a *virtual private network*.

On a fundamental level, the advantages of cloud computing lie in the distributed paradigm of the hardware systems and the prospect of lowered barriers to access through the commoditization of computing power itself. The cloud-computing paradigm provides flexibility, scalability, robustness, and, when managed by a dedicated provider, ease of use. In order to make best use of these new computational opportunities, however, software design must take the distributed/parallel nature of cloud computing into account—the power of cloud computing lies not in any single machine, but in the capacities of the system as a whole. Problems of an HTC nature are natural candidates for solving in the cloud because of the intrinsically scalable aspects of HTC solutions. More generally, for all problem paradigms, the communication requirements of the problem play a role in selecting the appropriate service model when seeking a cloud-computing solution—if substantial communication is required between compute nodes in a particular algorithm, a lower-level service model (IaaS) may be necessary. For example, it may be difficult to find an appropriate, off-the-shelf SaaS-level software implementation for an algorithmic approach that requires compute nodes to communicate heavily. In contrast, in HTC, the focus of the software developer can be narrowed to the proper provisioning of resources; network bandwidth is often a bottleneck only during the dissemination of problem input and gathering of output. Thus, an analysis akin to Amdahl's law plays an important role in determining the practicality of a particular scale of cloud computing for running a specific algorithm or solving a particular problem.

Finally, on issues of cloud security and liability: it should be emphasized that, in part because cloud computing is not an established commodity, many contractual aspects (e.g., various liabilities) are left to the user and provider to agree upon.

27.3 Cloud-Computing-Based Remote-Sensing-Related Applications

This section provides a short summary of remote-sensing applications that use cloud-computing environments as well as a more detailed case study. HPC frameworks (e.g., supercomputers at various research organizations) can potentially be widely adapted for remote-sensing applications (Parulekar et al., 1994; Simmahan and Ramakrishnan, 2010; Lee et al., 2011; Plaza, 2011a, b). The main limitations associated with the use of HPC are as follows: (1) HPC resources are not readily available for large user communities and (2) HPC resources are expensive to acquire and maintain (Ostermann et al., 2010).

Krishnan et al. (2010) evaluated a MapReduce approach for LiDAR gridding on a small private cluster consisting of around 8–10 commodity computers. They investigated the effects of several parameters, including grid resolution and dataset size, on performance. For their software implementation using Hadoop, the authors experimented with Hadoop-specific factors, such as the number of reducers allocated for a problem and the inherent concurrency therein. In their particular study, using quad-core machines with 8 GB of main memory and connected by gigabit Ethernet, they found that doubling the size of their Hadoop cluster from four to eight nodes had little effect on their experimental runtimes. They thus showed that their solution to the task was not strictly compute bound. On the other hand, the authors did note a substantial degradation in performance for their single-node, nondistributed algorithm control (implemented in

C++) when the problem size grew larger than could fit in main memory. Krishnan et al. (2010) concluded that Hadoop could be a useful framework for algorithms that process large-scale spatial data (roughly 150 million points), but that certain serial elements, such as output generation, could still be rate limiting, especially on commodity hardware. Their work also (1) motivates the study of systems with larger memories (this stems from their experience with HPC resources) and (2) demonstrated that the task of designing optimal systems for the processing of massive spatial data is complex.

Project *Matsu* is an open-source project for processing satellite imaginary using a community cloud (Bennett and Grossman, 2012). This project, a collaboration between NASA and the Open Cloud Consortium (OCC), has been developed to process data from NASA's EO-1 satellite and to develop open-source technology for public cloud-based processing of satellite imagery. Most computations were completed using a Hadoop framework running on 9 compute nodes with 54 compute cores and 352GB of RAM. The stated goal of this project (Project Matsu) is (1) to use an open-source cloud-based infrastructure to make high-quality satellite image data accessible through an Open Geospatial Consortium-compliant (OGC-compliant) Web Map Service (WMS), (2) to develop an open-source cloud-based analytic framework for analyzing individual images and collections of images, and (3) to generalize this framework to manage and analyze other types of spatial–temporal data. This project also features an on-demand cloud-based disaster assessment capability through satellite image comparisons. The image comparisons are done via a MapReduce job using a Hadoop-streaming interface. As an example, this project hosts a website that provides real-time information about flood prediction and assessment in Namibia. The final data are served to end users using a standard OGC WMS and Web Coverage Processing Service tools.

Oryspayev et al. (2012) studied LiDAR data reduction algorithms that were implemented using the GPGPU and multicore CPU architectures available on the AWS EC2. This paper tests the veracity of a vertex-decimation algorithm for reducing LiDAR data size/density and analyzes the performance of this approach on multicore CPU and GPU technologies, to better understand processing time and efficiency. The paper documents the performance of various GPGPU and multicore CPU machines including Tesla family GPUs and the Intel's multicore i-CPU series for the data reduction problem using large-scale LiDAR data. The study raises several questions about implementation of spatial-data processing algorithms on GPGPU machines, such as how to reduce overhead during the initialization of devices and how to optimize algorithms to minimize data transfer between CPUs/GPUs.

Eldawy and Mokbel (2013) developed an open-source framework, *SpatialHadoop*, that extends Hadoop by providing native support for spatial data. As an extension of Hadoop, the framework operates similarly—programs are written in terms of map and reduce functions, though the system is optimized to exploit underlying properties and characteristics of spatial data. As case studies, SpatialHadoop has three spatial operations, range queries, *k*-nearest-neighbor queries, and spatial join.

Cary et al. (2009) studied the performance of the MapReduce framework for bulk construction of *R*-trees and aerial image quality computation on both vector and raster data. They deployed their MapReduce implementations using the Hadoop framework on the Google and IBM clouds. The authors presented results that demonstrate the scalability of MapReduce and the effect of parallelism on the quality of the results. This paper also studied various metrics to compare the performance of their implemented algorithms including execution time, correctness, and tile quality. Their results indicate that the appropriate application of MapReduce could dramatically improve task completion times and also provide close to linear scalability. This study motivates further investigation of the MapReduce framework for other spatial-data-handling problems.

Li et al. (2010) studied the integration of data from ground-based sensors with the Moderate-Resolution Imaging Spectroradiometer satellite data using the Windows Azure cloud platform. Specifically, the authors provide a novel approach to reproject input data into timeframe- and resolution-aligned geographically formatted data and also develop a novel reduction technique to derive important new environmental data through the integration of satellite and ground-based data. Slightly modified Windows Azure abstractions and APIs were used to accomplish the reprojection and reduction steps. They suggest that cloud computing has a great potential for efficiently processing satellite data. It should be noted that their current framework doesn't fit into the MapReduce framework since it uses Azure's general queue-based task model.

Berriman et al. (2010) compared various toolkits to create image mosaics and to manage their provenance using both the Amazon EC2 cloud and the Abe high-performance cluster at NCSA, UIUC. They conducted a series of experiments to study performance and costs associated with different types of tasks (I/O bound, CPU bound, and memory bound) in these two environments. Their experiments show that for I/O-bound applications, the most expensive resources are not necessarily the most cost-effective, that data transfer costs can exceed the processing costs for I/O-bound applications on Amazon EC2, and that the resources offered by Amazon EC2 are generally less powerful than those available in the Abe high-performance cluster and consequently do not offer the same levels of performance. They concluded from their results that cloud computing offers a powerful and cost-effective new resource for compute and memory intensive remote-sensing applications.

27.3.1 A Case Study: Cloud-Based LiDAR Processing System

To more completely illustrate an approach to cloud computing of remote-sensing information, in this section we describe a LiDAR application. In this cloud-based LiDAR processing system (CLiPS) project, we use a statewide (Iowa) LiDAR data

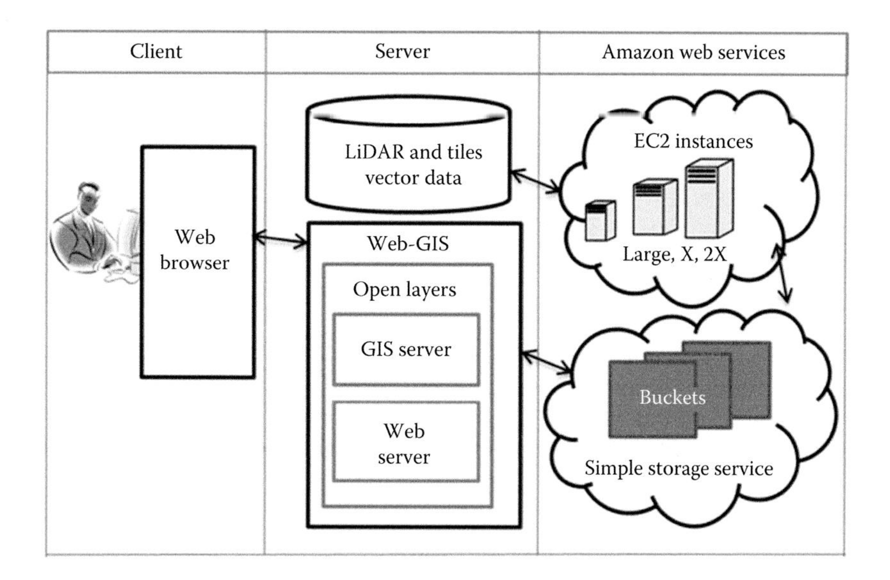

FIGURE 27.3 Overall architecture developed for cloud-based LiDAR processing system.

repository (Iowa DNR, 2009) in which data are distributed to the public as a collection of 34,000 tiles, each covering 4 km^2; this comprises roughly 7 TB of data. In order to process this massive data, CLiPS was designed (Figure 27.3) as a web portal implemented using Adobe's Flex framework along with ESRI's ArcGIS API for Flex (ESRI, 2012; Sugumaran et al., 2014), OpenLayers, and the Amazon EC2 cloud environment. CLiPS use a three-tier client–server model (Figure 27.3). The top tier supports user interaction with the system, the second tier provides process management services, such as monitoring and analysis, and the third tier is dedicated to data and file services. The client-side interface was developed using Flex and the server-side uses custom-built tools, constructed from open-source products.

Figure 27.4 shows the user interface developed for this study. The interactive user interface requires a user to, first, select a region of interest on a map and then select AWS credentials and computing resources as well as a location to which results will be sent for downloading (this is typically an e-mail address). The

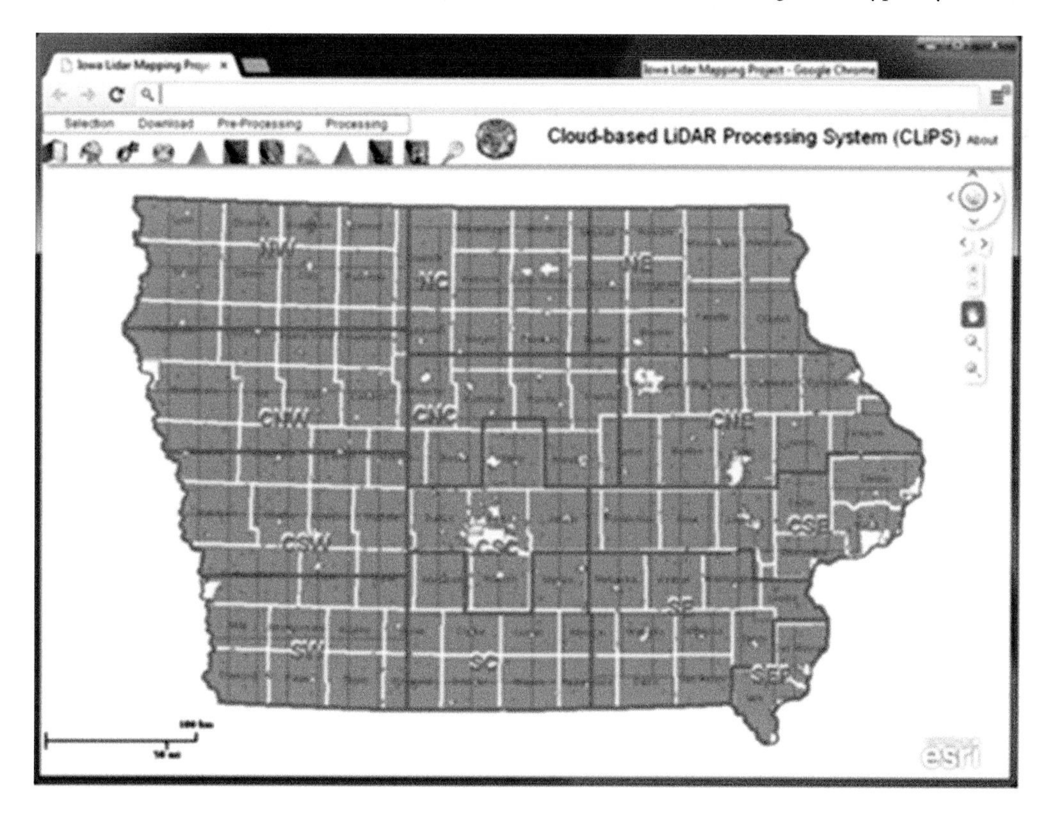

FIGURE 27.4 Cloud-based LiDAR processing system user interface for the state of Iowa, United States.

system was tested by creating a triangulated irregular network (TIN) from a LIDAR point cloud using 18 dataset and processor configurations: three terrain data (flat, undulating, urban), two tile sizes (9 and 25), and three Amazon EC2 processing configurations (large, Xlarge, and double Xlarge). The undulating terrain dataset took more time than the other terrain types for 5 × 5 tile groups, while the urban terrain was the most computationally intensive for the 3 × 3 tile groups used in this study (Sugumaran et al., 2014). The results clearly show that as computer power increases, processing times decrease for all three types of LiDAR terrain data. The various combinations in our evaluations showed that even with up to 25 tiles with varying processing configuration types, each request required less than an hour and cost less than a dollar for data processing (e.g., TIN creation). Moreover, the cost of uploading data from our server and data storage on the cloud was less than 50 dollars. Thus, the overall cost for our test using the Amazon cloud was less than $100, an amount that is affordable by most users.

27.4 Conclusions

It is abundantly clear that as a consequence of technological and sensor licensing improvements, the spatial, spectral, temporal, and radiometric resolution of remote-sensing imagery will continue to increase. This translates into massive quantities of data that must be processed to glean meaningful information that can be used in a variety of decision support and visualization applications. Such data quantities quickly overwhelm the capabilities of even the most powerful desktop systems available now and in the foreseeable future. As a consequence, researchers continue to explore cost-effective, yet powerful, computing environments that can be harnessed for remote-sensing applications. Dedicated HPC systems are expensive to acquire and maintain and have relatively short half-lives. This significantly diminishes this alternative as a practical solution. Instead, the availability of high-speed communication technologies now makes the use of distributed, pay-as-you-go resources an attractive choice for researchers and government agencies, as well as users in the private sector. One term that is used to describe these distributed resources is cloud computing. The cloud provides new patterns for deploying remote-sensing-data processing and provides easy, inexpensive access to servers, elastic scalability, managed infrastructure, and low complexity of deployment.

Despite these significant advantages, the use of cloud computing does have some limitations. First, since data are distributed across networks to distributed, but unknown, locations, security can become problematic. Thus, public cloud resources cannot be used exclusively for applications that require the processing of many types of individual-level information. This limitation, however, is not normally significant for remote-sensing applications. Secondly, different types of spatial algorithms may require considerable amounts of interprocessor communication, particularly in big-data applications, and certain elements of cloud infrastructure may therefore induce processing latencies that may be unacceptable to users.

Acknowledgments

This research was conducted with support from an Amazon Research Grant and USGS-AmericaView projects.

References

Bennett, C. and Grossman, R. 2012. OCC Project Matsu: An open-source project for cloud-based processing of satellite imagery to support the earth sciences. http://matsu.opensciencedatacloud.org (accessed on March 4, 2014).

Berriman, G. B., Deelman, E., Groth, P., and Juve, G. 2010. The application of cloud computing to the creation of image mosaics and management of their provenance. In *SPIE Astronomical Telescopes+ Instrumentation*, International Society for Optics and Photonics, Bellingham WA, Vol. 7740F.

Buyya, R., Yeo, C. S., Venugopal, S., Broberg, J., and Brandic, I. 2009. Cloud computing and emerging IT platforms: Vision, hype, and reality for delivering computing as the 5th utility. *Future Generation Computer Systems*, 25, 599–616.

Cary, A., Sun, Z., Hristidis, V., and Rishe, N. 2009. Experiences on processing spatial data with MapReduce. In *Scientific and Statistical Database Management*, Springer, Berlin, Germany, pp. 302–319.

Chang, C. C., Chang, Y. L., Huang, M. Y., and Huang, B. 2011. Accelerating regular LDPC code decoders on GPUs. *IEEE Journal of Selected Topics in Applied Earth Observations and Remote Sensing* (JSTARS), 4(3), 653–659.

Christophe, E., Michel, J., and Inglada, J. 2011. Remote sensing processing: From multicore to GPU. *IEEE Journal of Selected Topics in Applied Earth Observations and Remote Sensing* (JSTARS), 4(3), 643–652.

Cui, D., Wu, Y., and Zhang, Q. 2010. Massive spatial data processing model based on cloud computing model. In *Computational Science and Optimization CSO, 2010 Third International Joint Conference on Computational Science and Optimization*, May 28–31, Anhui, China, Vol. 2, pp. 347–350.

Dean, J., and Ghemawat, S. 2004. MapReduce: Simplified data processing on large clusters. In *Proceedings of the Sixth Conference on Symposium on Operating Systems Design & Implementation*, December 6–8, San Francisco, CA, Vol. 6, 13pp.

Eldawy, A., and Mokbel, M. F. 2013. A demonstration of SpatialHadoop: An efficient MapReduce framework for spatial data. *Proceedings of the VLDB Endowment*, 6(12), 1230–1233.

eScience Institute, University of Washington. 2012. Understanding cloud computing for research and teaching. http://escience.washington.edu/get-help-now/understanding-cloud-computing-research-and-teaching (accessed on March 4, 2014).

ESRI. 2012. ArcGIS server. http://www.esri.com/software/arcgis/arcgisserver (accessed on March 23, 2012).

González, J. F., Rodríguez, M. C., and Nistal, M. L. 2009. Enhancing reusability in learning management systems through the integration of third-party tools. In *39th IEEE Frontiers in Education Conference, FIE'09*, October 18–21, San Antonio, TX, pp. 1–6.

Haifang, Z. 2003. Study and implementation of parallel algorithms for remote sensing image processing. PhD thesis, National University of Defence Technology, Chasha, China.

Huang, Q., Yang, C., Nebert, D., Liu, K., and Wu, H. 2010. Cloud computing for geosciences: Deployment of GEOSS clearinghouse on Amazon's EC2. In *HPDGIS '10: Proceedings of the ACM SIGSPATIAL International Workshop on High Performance and Distributed Geographic Information Systems*, November 3–5, San Jose, CA, pp. 35–38.

Hyspeed Computing. 2013. Big data and remote sensing—Where does all this imagery fit into the picture? http://hyspeedblog.wordpress.com/2013/03/22/big-data-and-remote-sensing-where-does-all-this-imagery-fit-into-the-picture (accessed on March 2014).

IBM. 2012. Bringing big data to the enterprise. http://www-01.ibm.com/software/data/bigdata (accessed on April 4, 2013).

Iowa DNR. 2009. State of Iowa. http://www.iowadnr.gov/mapping/lidar/index.html, retrieved April 29, 2009 (accessed on April 4, 2013).

Krishnan, S., Bary, C., and Crosby, C. 2010. Evaluation of MapReduce for gridding LIDAR data. In *2010 IEEE Second International Conference on Cloud Computing Technology and Science, CloudCom*, November 30–December 3, Indianapolis, IN, pp. 33–40.

Kumar, N., Lester, D., Marchetti, A., Hammann, G., and Longmont, A. 2013. Demystifying cloud computing for remote sensing application. http://eijournal.com/newsite/wp-content/uploads/2013/06/cloudcomputing.pdf. Accessed on March 4, 2014.

Lee, C. A., Gasster, S. D., Plaza, A., Chang, C. I., and Huang, B. (2011). Recent developments in high performance computing for remote sensing—A review. *IEEE Journal of Selected Topics in Applied Earth Observations and Remote Sensing*, 4.3, 508–527.

Li, J., Humphrey, M., Agarwal, D., Jackson, K., Ingen, C., and Ryu, Y. 2010. "eScience in the cloud: A MODIS satellite data reprojection and reduction pipeline in the windows azure platform." In *IEEE International Symposium on Parallel & Distributed Processing (IPDPS)*, April 19–23, Atlanta, GA, pp. 1–10.

Liu, T. et al. 2010. Remote sensing image classification techniques based on the maximum likelihood method. *FuJian Computer*, (001), 7–8.

McEvoy, G. V., and Schulze, B. 2008. Using clouds to address grid limitations. In *Proceedings of the 6th International Workshop on Middleware for Grid Computing*, December 1–5, Leuven, Belgium.

Mell, P. and Grance, T. 2011. The NIST definition of cloud computing. National Institute of Standards and Technology, Special Publication 800-145.

Oryspayev, D., Sugumaran, R., DeGroote, J., and Gray, P. 2012. LiDAR data reduction using vertex decimation and processing with GPGPU and multicore CPU technology. *Computers & Geosciences*, 43, 118–125.

Ostermann, S., Iosup, A., Yigitbasi, N., Prodan, R., Fahringer, T., and Epema, D. 2010. A performance analysis of EC2 cloud computing services for scientific computing. In *Cloud Computing*, Springer, Berlin, Germany, pp. 115–131.

Parulekar, R. et al. 1994. High performance computing for land cover dynamics. In *Proceedings of the 12th IAPR International Conference on Pattern Recognition, 1994, Vol. 3—Conference C: Signal Processing*, October 9–13, Jerusalem, Israel, IEEE, New York, 1994.

Plaza, A., and Chang, C.-I. 2007. *High Performance Computing in Remote Sensing*, CRC Press, Boca Raton, FL.

Plaza, A., and Chang, C.-I. 2008. Special issue on high performance computing for hyperspectral imaging. *International Journal of High Performance Computing Applications*, 22(4), 363–365.

Plaza, A., Du, Q., Chang, Y.-L., and King, R. L. 2011a. High performance computing for hyperspectral remote sensing. *IEEE Journal of Selected Topics in Applied Earth Observations and Remote Sensing (JSTARS)*, 4(3), 528–544.

Plaza, A., Plaza, J., Paz, A., and Sanchez, S. 2011b. Parallel hyperspectral image and signal processing. *IEEE Signal Processing Magazine*, 28(3), 119–126.

Plaza, A., Valencia, D., Plaza, J., and Martinez, P. 2006. Commodity cluster-based parallel processing of hyperspectral imagery. *Journal of Parallel and Distributed Computing*, 66(3), 345–358.

Prasad. 2013. Special issue on High performance computing in remote sensing. *Remote Sensing*.

Project Matsu, http://matsu.opensciencedatacloud.org/. Accessed on March 4, 2014.

Rezgui, A., Malik, Z., and Yang, C. 2013. High-resolution spatial interpolation on cloud platforms. In *Proceedings of the 28th Annual ACM Symposium on Applied Computing*, March 18–22, Coimbra, Portugal, pp. 377–382.

Shekhar, S., Gunturi, V., Evans, M. R., and Yang, K. 2012. Spatial big-data challenges intersecting mobility and cloud computing. In *Proceedings of the 11th ACM International Workshop on Data Engineering for Wireless and Mobile Access*, May 20–24, Scottsdale, AZ, pp. 1–6.

Simmhan, Y., and Ramakrishnan, L. 2010. Comparison of resource platform selection approaches for scientific workflows. In *Proceedings of the 19th ACM International Symposium on High Performance Distributed Computing*, June 21–25, Chicago, IL, pp. 445–450.

Song, C., Li, Y., and Huang, B. 2011. A GPU-accelerated wavelet decompression system with SPIHT and Reed-Solomon decoding for satellite images. *IEEE Journal of Selected Topics in Applied Earth Observations and Remote Sensing (JSTARS)*, 4(3), 683–690.

Sugumaran, R., Burnett, J., and Armstrong, M. P. 2014. Using a cloud computing environment to process large 3D spatial datasets. In H. Karimi, ed., *Big Data: Techniques and Technologies in Geoinformatics*, CRC Press, Boca Raton, FL, pp. 53–65.

Sugumaran, R., Oryspayev, D., and Gray, P. 2011. GPU-based cloud performance for LiDAR data processing. In *COM.Geo 2011: Second International Conference and Exhibition on Computing for Geospatial Research and Applications*, May 23–25, Washington, DC.

USGS. 2011. Landsat archive. http://landsat.usgs.gov (accessed on April 7, 2013).

Vatsavai, R. R., Ganguly, A., Chandola, V., Stefanidis, A., Klasky, S., and Shekhar, S. 2012. Spatiotemporal data mining in the era of big spatial data: Algorithms and applications. *Proceedings of the First ACM SIGSPATIAL International Workshop on Analytics for Big Geospatial Data*, November 7–9, Redondo Beach, CA, pp. 1–10.

Wang, S., and Armstrong, M. P. 2003. A quadtree approach to domain decomposition for spatial interpolation in grid computing environments. *Parallel Computing*, 29(10): 1481–1504.

Wang, Y., Wang, S., and Zhou, D. 2009. Retrieving and indexing spatial data in the cloud computing environment. In *Proceedings of the First International Conference on Cloud Computing*, December 1–4, Beijing, China, Lecture Notes in Computer Sciences, Vol. 5931, pp. 322–331.

Watson, P., Lord, P., Gibson, F., Periorellis, P., and Pitsilis, G. 2008. Cloud computing for e-Science with CARMEN. In *Second Iberian Grid Infrastructure Conference Proceedings*, May 12–14, Porto, Portugal, pp. 3–14.

White, T. 2012. *Hadoop: The Definitive Guide*, O'Reilly Media, Inc., Sebastopol, CA.

Yang, C. and Huang, Q. 2013. *Spatial Cloud Computing: A Practical Approach*, CRC Press, Boca Raton, FL.

Yang, H., Du, Q., and Chen, G. 2011. Unsupervised hyperspectral band selection using graphics processing units. *IEEE Journal of Selected Topics in Applied Earth Observations and Remote Sensing (JSTARS)*, 4(3), 660–668.

Yue, P., Gong, J., Di, L., Yuan, J., Sun, L., Sun, Z., and Wang, Q. 2010. GeoPW: Laying blocks for the geospatial processing web. *Transactions in GIS*, 14(6), 755–772.

Google Earth for Remote Sensing

28

Google Earth for Remote Sensing

John E. Bailey
University of Alaska Fairbanks

Acronyms and Definitions

API	Application programming interface
ASTER	Advanced Spaceborne Thermal Emission and Reflection Radiometer
AVHRR	Advanced Very High Resolution Radiometer
CPU	Central processing units
DEM	Digital elevation model
EE	(Google) Earth Engine
GE	Google Earth
GEC	Google Earth Community
GUI	Graphical user interface
KML	Keyhole Markup Language
LiDAR	Light Detection and Ranging
MODIS	Moderate Resolution Imaging Spectroradiometer
NAIL	Neighbors Against Irresponsible Logging
NGA	National Geospatial-Intelligence Agency
NGO	Non governmental organization
NOAA	National Oceanic and Atmospheric Administration
OGC	Open Geospatial Consortium
SPOT	Système Pour l'Observation de la Terre
SRTM	Shuttle Radar Topographic Mapping Mission
WYSIWYG	"What you see is what you get"
XML	Extensible Markup Language

28.1 Introduction

A decade has now passed since the release of Google Earth (GE) (Google Inc. 2005). It is hard to fully measure the impact it has had on all areas of society, as it opened possibilities in the world of remote sensing to everyone from the youngest student looking at satellite images for the first time to an experienced professor who was now able to access imagery at a scale and speed never before possible. In 2004 the technology that was to later become GE was called "EarthViewer" (Figure 28.1) and was the proprietary property of Keyhole Inc. It was one of several *virtual globes* (Bailey 2010), computer applications creating a 3D model of Earth or other planets that evolved in the early 2000s as technology caught up with ideas (Bailey and Chen 2011). For many they were seen as realization of a vision that had been laid out in different forms by many from futurists (Fuller 1962) to academics (Bailey et al. 2012), to artists (Eames and Eames 1977) to science fiction writers (Stephenson 1992). However, many credit Al Gore's 1998 address to the California Science Center (Gore 1998) as succinctly verbalizing the concept of a Digital Earth, of which GE ultimately became the most popular actualization.

GE is currently available for a range of operating systems and platforms (Google Inc. 2014a). These can primarily be differentiated into three categories: (1) desktop, (2) mobile, and (3) web (Figure 28.2). The desktop version is a downloadable binary that is currently supported on PC, Mac, and Linux systems. The mobile version is available as Android and iOS Apps but has reduced

FIGURE 28.1 Screenshot of Keyhole's EarthViewer 3D showing the Port of Tokyo near Hamazakibashi Junction.

functionality relative to the desktop version. GE is found on webpages in two different forms: as a WebGL rendering option in Google Maps or as a web plug-in with a JavaScript application programming interface (API). Google announced the release of the desktop version of GE on June 28, 2005, as a free 3D mapping and local search technology (Google Inc. 2005). The capability of importing GIS data was mentioned but the emphasis was on GE as a searchable 3D map rather than as a visualization tool. Content creation was referenced as "easy creation and sharing of annotations among users" (Google Inc. 2005). However, as scientists and remote sensing professionals began to explore with GE, they realized that its greatest potential involved the use of Keyhole Markup Language (KML), the code supported by the application to visualize placemakers, lines, polygons, overlays, and other types of content generated by the user.

This chapter explores how GE has become the go-to application for free and accessible high-resolution satellite imagery (Section 28.2) and how that imagery can be augmented with illustrations (Section 28.3) and used for geospatial storytelling (Section 28.4), by both the professional remote sensing community and the public. It will conclude by exploring the steps in the evolution of Google Geo as an analytical remote sensing toolset (Section 28.5) and will consider what Google's acquisition of their own satellite company might mean for remote sensing using GE.

28.2 Google Earth: Free and Accessible High-Resolution Imagery

GE has integrated high and very high spatial resolution imagery seamlessly for the entire world. The concept has many science and practical applications.

28.2.1 Google Earth Virtual Globe

The core of the GE technology is the rendering of a 3D globe comprised of a combination of terrain data overlain by satellite imagery and aerial photography. The base terrain data are a digital elevation model (DEM) collected by the shuttle radar topographic mapping mission (SRTM) supplemented by other datasets for high latitudes and mountainous regions that requiring higher-resolution data

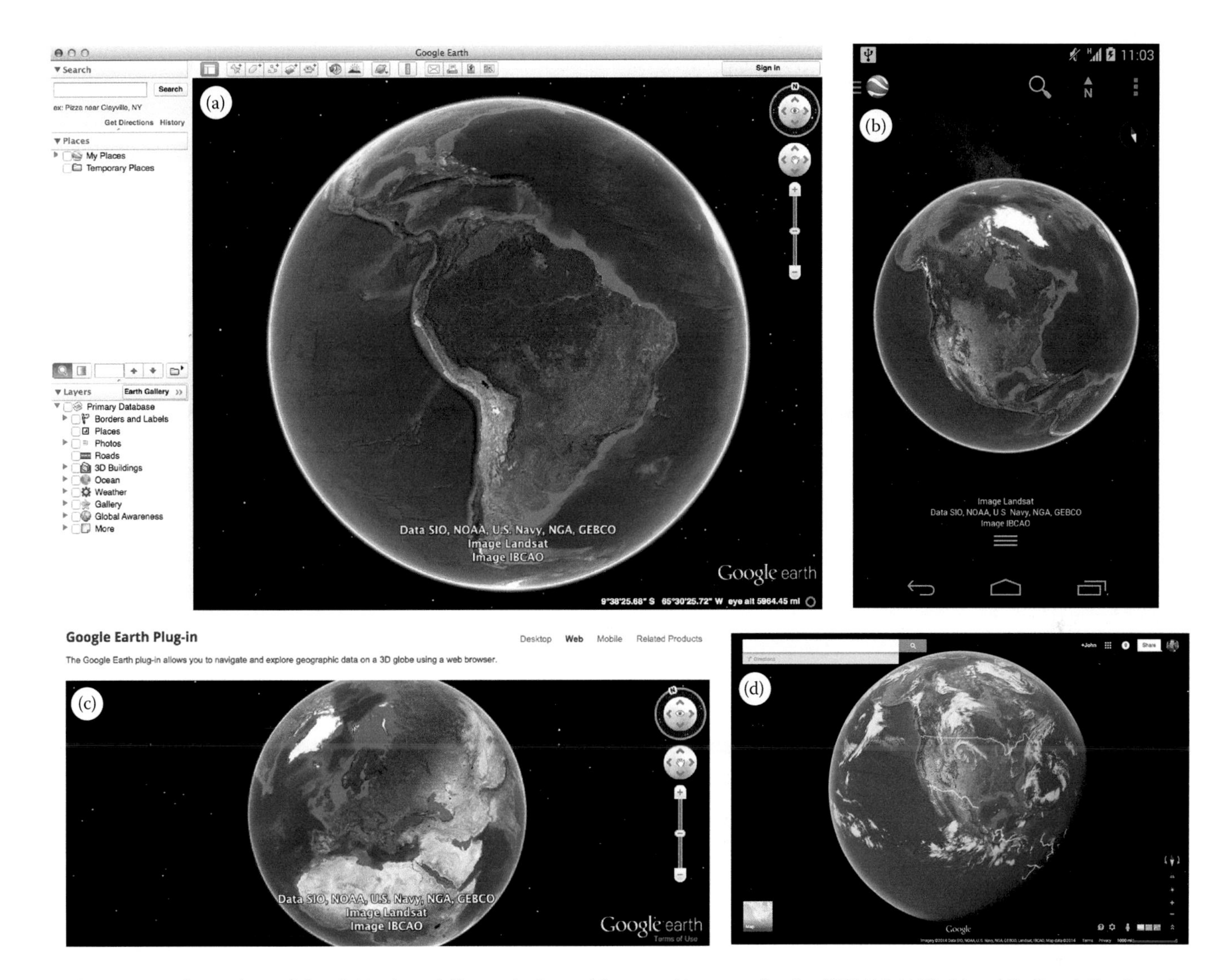

FIGURE 28.2 Screenshots of Google Earth on different platforms: (a) current binary application (GE 7.1.2.2041), (b) mobile Google Earth on the Nexus 7 Android tablet, (c) earth globe in Google Maps, and (d) Google Earth plug-in.

to show the true topography. There are many areas, for example, Western United States, The Swiss Alps, and Canary Islands, where high-resolution light detection and ranging (LiDAR) datasets have been acquired and supplant the SRTM imagery. The original base imagery for GE was Landsat 30 m multispectral data, pan sharpened using Landsat's 15 m panchromatic imagery. In some rural areas, this remains the only available dataset. However, for many regions and most urban centers, very high spatial resolution imagery (VHRI; submeter to 10 m resolution) has been acquired from DigitalGlobe, Système Pour l'Observation de la Terre (SPOT) Image, and other commercial satellite providers.

The modern era of commercially available high-resolution satellite data began with the launches of GeoEye's Ikonos (1999) and DigitalGlobe's Quickbird (2001) satellites. For the first time almost global coverage of Earth at centimeters to meters resolution was available to all who could afford it. At the same time the development of Keyhole's EarthViewer 3D and other virtual globes, technology now possible thanks to the gaming industries needs for high-performance graphics cards and the global expansion of broadband Internet connectivity, created use cases for large volumes of high-resolution data. Many have joked that the first thing new users do with GE is look at their house, but this in itself was an evolutionary step in the world of remote sensing. For the first time hundreds of millions of people were presented with a view of what their home or town looks like from space. The concept that satellites can be used to observe from Earth moved from the realm of science fiction and something scientists do, to any everyday viewpoint for anyone with a computer and an Internet connection.

28.2.2 Google Earth's First Major Application: Hurricane Katrina

An opportunity for GE to display its potential occurred just weeks after the Google announcement of the reworked Keyhole technology (Google Inc. 2005). On August 29, 2005, Hurricane Katrina made landfall in Louisiana, causing devastation across the region (Knabb et al. 2005). Nowhere was the effect felt more than in the city of New Orleans. Much of the city lies below sea level and the high rainfall produced by the hurricane led to protective riverbank levees breaking, and the city was flooded. Thousands of people were trapped in their houses and as the floodwaters rose, people escaped to the roofs of their houses but often without any form of communication.

In order to find those stranded, plan logistics, and work out access routes, the National Oceanic and Atmospheric Administration (NOAA) captured thousands of high-resolution aerial images. This information was invaluable for disaster response efforts, with more than 5 million photos downloaded from the NOAA website each day during the first week (Nourbakhsh et al. 2006). However, processing and navigating this information in a useful manner was a colossal effort and one for which GE was to provide the backbone. The GE engineers worked around the clock to serve the images through the GE client application as KML overlays. These geolocated, stitched together *layers* of data were searchable and easily accessible by federal agencies, disaster response groups, and the general public. For example, this imagery was used to identify intact churches in flood-free suburbs of New Orleans that could used as outlet centers for aid donations (Nourbakhsh et al. 2006). The National Geospatial-Intelligence Agency (NGA) later recognized Google's contribution with a Hurricane Katrina Recognition Award, but more importantly it demonstrated the facility of the application to be more than just a tool for looking at your house.

28.2.3 Google Earth's Numerous Application Possibilities

Despite the obvious benefits for efforts such as crisis response, the ability of every user to become armchair remote sensing analysts also raised concerns as conspiracy theorists now had *evidence* to back up their claims of discovering Atlantis or a hidden alien base (Geens 2009; Google Inc. 2009). But for many communities, the positives far outweighed any negative connotations as these tools opened up exploration of the whole planet for professionals and curious public alike. In particular, it has developed as an observation tool for structural geologists, geomorphologists, volcanologists, and other scientists who require easy access to detailed imagery. In these cases GE has enabled studies at a scale that,

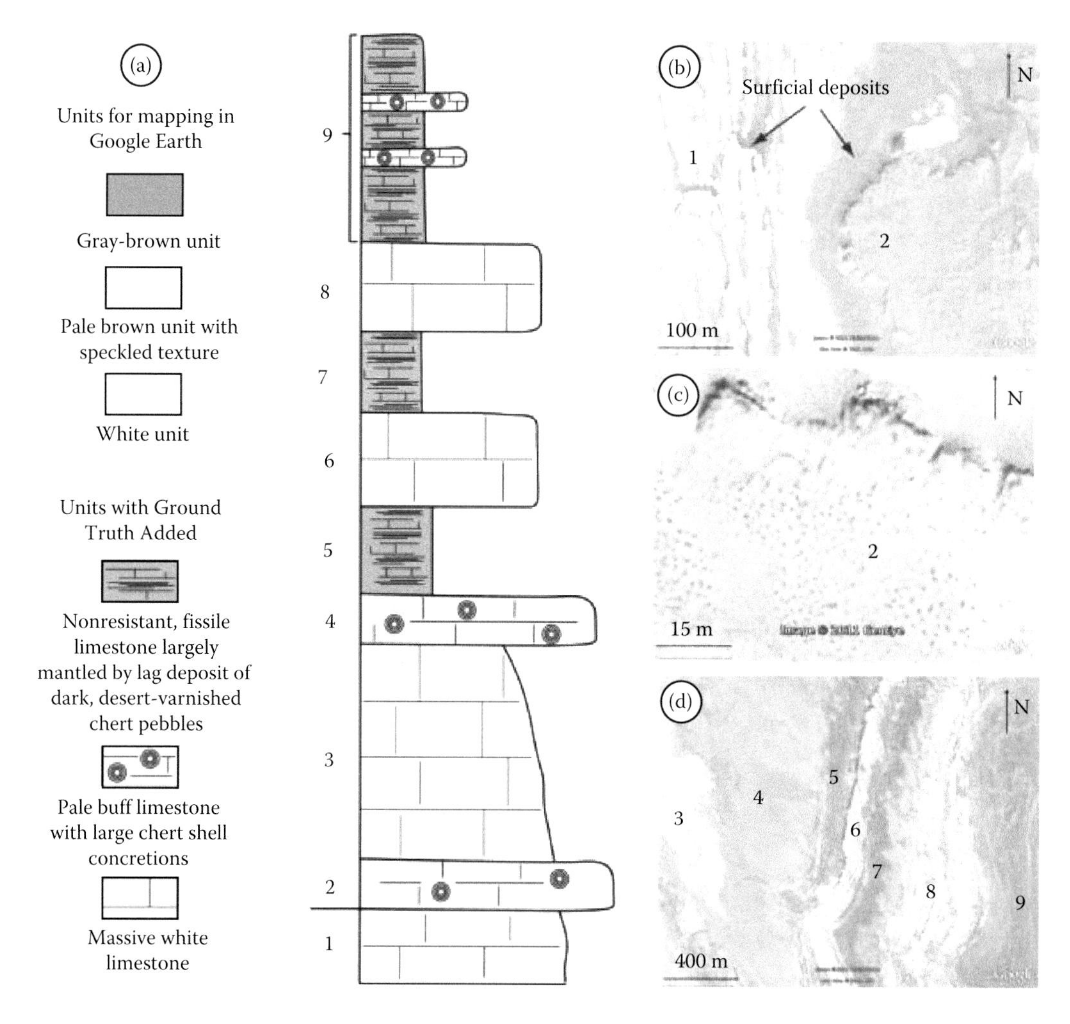

FIGURE 28.3 Tewksbury et al. (2012) used Google Earth to study structural geology in remote areas in Egypt. Using Google Earth imagery, they established a stratigraphy consisting of one subunit in the El-Ruf uf Formation (Unit 1) and eight subunits in the Drunka Formation (Units 2-9).

while technically possible, was not practically obtainable without extensive time-consuming and well-funded efforts (Figure 28.3).

28.3 Evolution of Google Earth in Everyday Life and in Science

28.3.1 Keyhole Markup Language for Easy to Create Geospatial Visualizations

While many different fields of Earth science have used GE's imagery as a tool for landscape studies, it is another component of Keyhole's technology that has made GE revolutionary as a visualization and scientific tool. KML is a computer code that can be generated by users to create and share visualizations in GE (Wernecke 2009). Originally developed by Keyhole Inc., in 2008 it became an international standard of the Open Geospatial Consortium (OGC). KML is an Extensible Markup Language (XML), a computer code that defines a set of rules for encoding documents in a format that is human readable but is also annotated in a way that is syntactically distinguishable from regular text (Figure 28.4). Through KML, GE becomes the canvas on which users can display geospatial information (Bailey 2010; Chen and Bailey 2011; Whitmeyer et al. 2012).

28.3.2 Early Adopters: Bloggers and Journalists

Among early adoption users were bloggers and journalists who grasped the potential for bringing their stories to life (Lubick 2005; Butler 2006a). However, where this broke new ground was rather than GE just being another way to great colorful illustrations, it gave the journalists themselves the power to great dynamic representation of the data. This is what Declan Bulter, a *Nature* reporter and early champion for GE abilities, did with data points showing the spread of avian flu (Butler 2006b,c; Figure 28.5). Although wide acceptance was slower at first, the scientific community also embraced the possibilities offered by GE for both research and education (Grambling 2007; Simonite 2007; Bailey and Chen 2011; Yu and Gong 2012).

```
<?xml version="1.0" encoding="UTF-8"?>
<kml xmlns="http://www.opengis.net/kml/2.2"
xmlns:gx="http://www.google.com/kml/ext/2.2"
xmlns:kml="http://www.opengis.net/kml/2.2"
xmlns:atom="http://www.w3.org/2005/Atom">
<Document>
    <name>City of London</name>
    <Placemark>
        <name>Tower Bridge</name>
        <LookAt>
            <longitude>-0.07590021026462335</longitude>
            <latitude>51.50551181412418</latitude>
            <altitude>0</altitude>
            <heading>0.03906926736428579</heading>
            <tilt>7.368525557807322</tilt>
            <range>625.9559805685249</range>
        </LookAt>
        <Point>
            <coordinates>
                -0.07497789972809921,51.50522132197029,0
            </coordinates>
        </Point>
    </Placemark>
</Document>
</kml>
```

FIGURE 28.4 A simple example of a keyhole markup language (KML) file. Different components of the text have been colored (in this example) for the purposes of identification. The grey text is namespace information, which references the rules of KML file types. The black text creates a container for KML features. The blue text creates a placemark at the given coordinates. The red text defines the default "fly-to" viewpoint for this placemark.

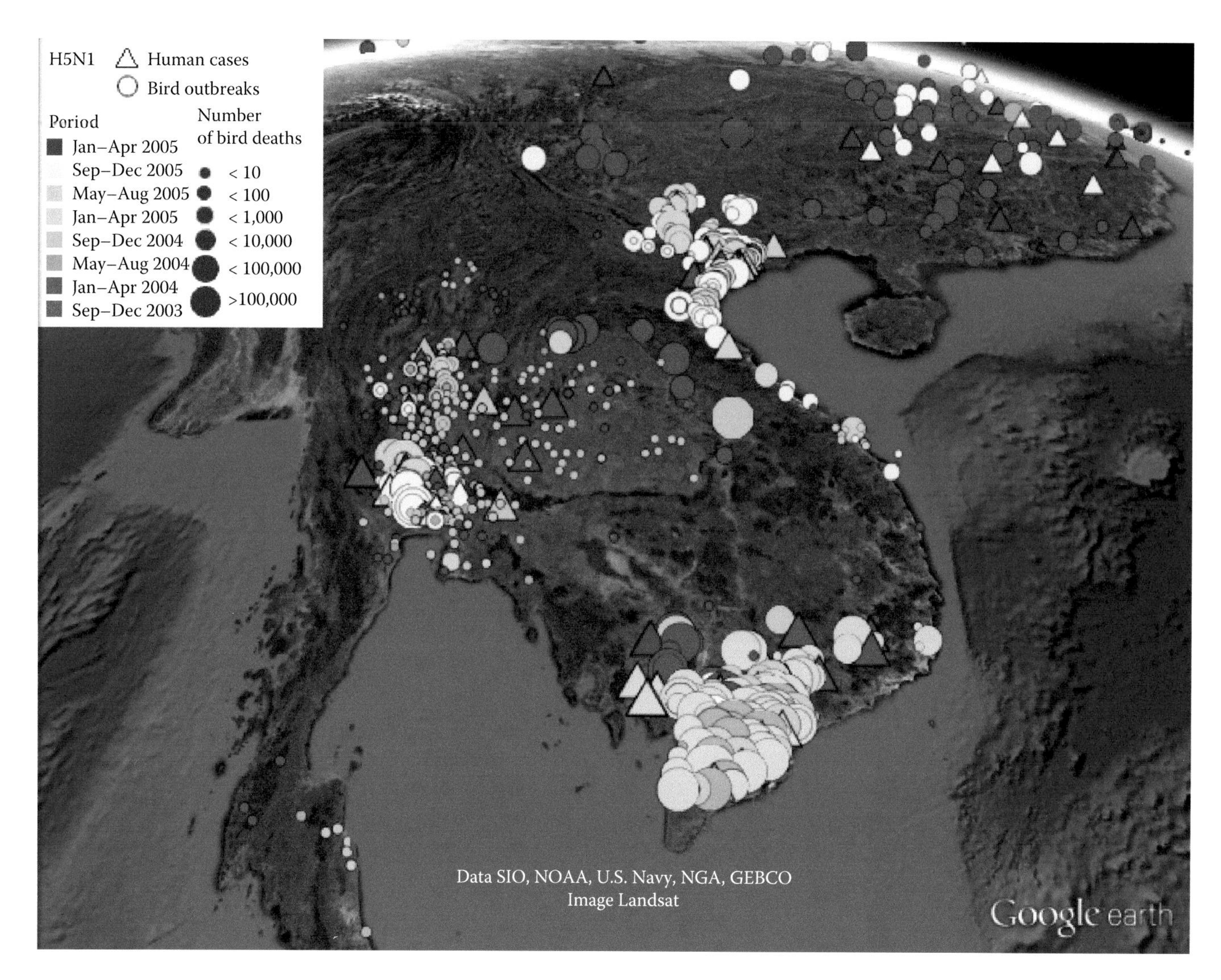

FIGURE 28.5 Visualizing the spread of the N5H1 virus (aka avian flu) using Google Earth and KML (Butler, 2006b,c).

28.3.3 Early Science Application: Volcanology

One of the first fields to demonstrate GE's capabilities to dynamically visualize Earth observations was volcanology (Figure 28.6). The Smithsonian Global Volcanology Program worked with the GE team to develop a volcano layer (Venzke et al. 2006; Figure 28.6a), which became one of the most frequently accessed datasets in the application (Rebecca Moore, personal communication). Meanwhile researchers in the Alaska Volcano Observatory remote sensing group developed the use of KML for modeling volcanic ash clouds (Webley et al. 2009a,b; Figure 28.6b) and observing real-time thermal satellite imagery (Bailey and Dehn 2006; Figure 28.6c and d).

28.3.4 Evolution in a Multitude of Science Applications

The use of GE for volcanology is just one illustration of how it has become an important remote sensing tool. Examples showing how GE has become integral to Earth observation activities could be cited from a multitude of disciplines from structural geology (Lisle 2006; De Paor 2008; De Paor and Whitmeyer 2011) or cryospheric studies (Ballagh et al. 2007, 2011; Gergely et al. 2008), to land use assessment (Sheppard and Cizek 2008; van Lammeren et al. 2010) or water resources management (Silberbauer and Geldenhuys 2008; USGS 2014), and everything in between. Ultimately all these fields (and others) are using GE because they want to use Earth imagery to tell a story.

28.4 Google Earth for Telling Geospatial Stories of Societal Importance

From simple fly-throughs of the 3D landscape (Lang et al. 2012) or KML-enhanced tours (Rueger and Beck 2012; Treves and Bailey 2012) to complex interactive experiences using the Earth API (Dordevic and Wild 2012), an obvious development of GE was for use as a platform for Virtual Field trips. Teachers embraced

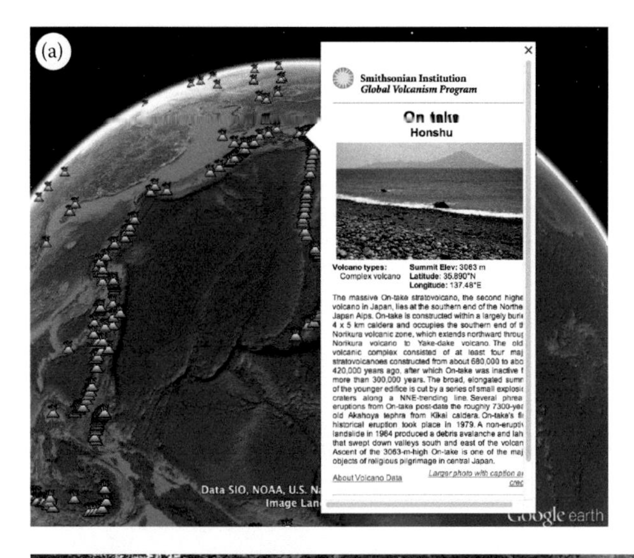

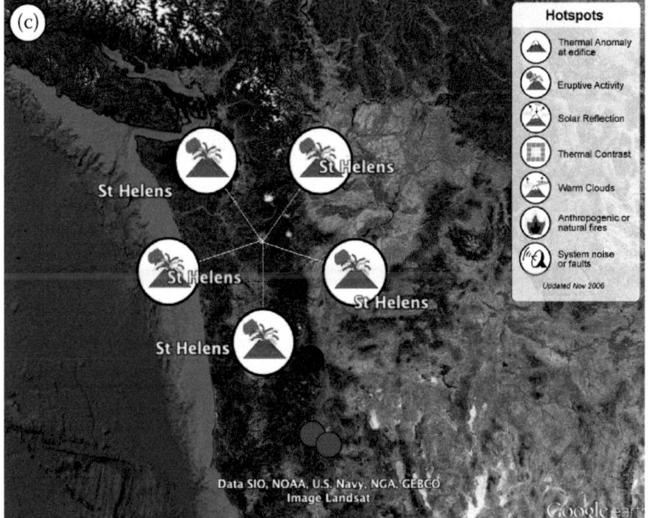

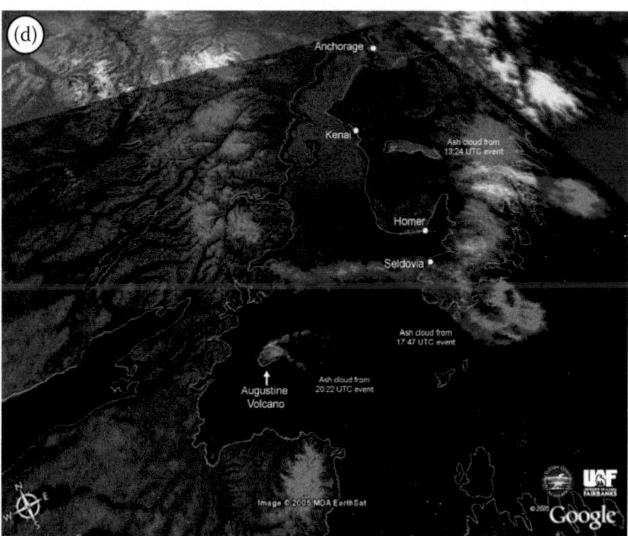

FIGURE 28.6 Examples of early uses of Google Earth by volcanologists: (a) Smithsonian GVP's Google Earth layer of Holocene volcanoes, (b) 3D visualization of an ash plume that erupted from Cleveland Volcano in the Aleutian Islands, (c) locations of thermal hotspots shown using KML derived from thermal IR images geolocated in GE, and (d) ash cloud density identified using AVHRR thermal IR imagery and visualized as translucent overlays in GE.

this ability (Lamb and Johnson 2010) as way of taking students to places they otherwise couldn't hope to go. However, nonprofits and conservation saw an opportunity to add a narrative to those tours and explorations. The catalyst for this development was a citizen action group's response to a logging plan for their neighborhood in Los Gatos, CA. The Neighbors Against Irresponsible Logging (NAIL) was a nonprofit organization formed in 2005 by Rebecca Moore, a software engineer who had recently joined the GE team. After receiving a low-quality black and white leaflet in her mail outlining a company's logging plan for the Santa Cruz forest she lived in, Moore used GE to create powerful visualizations of the true scale of the plan that helped rally her neighbors into action (Figure 28.7). The press picked up on these visualizations and the subsequent coverage led to the proposed logging plan being ruled ineligible by the California Department of Forestry (Google Earth Outreach 2014).

While not the original intent, the success of NAIL led to conservation and nonprofit groups contacting Moore for help with telling their stories. NGOs such as the Sierra Club, the Jane Goodall Institute, and Appalachian Voices began to use GE as an educational platform for their messages (Google Earth Outreach 2014; Figure 28.8). The interest in GE as an outreach tool led Rebecca Moore to form the GE Outreach team, which focuses on working with nonprofit and educational groups to use "Google Geo for Good."

28.5 Strengths and Limitations of Google Earth

The strengths in the choice of GE applications over applications using other remote sensing data or imagery are due to three primary reasons:

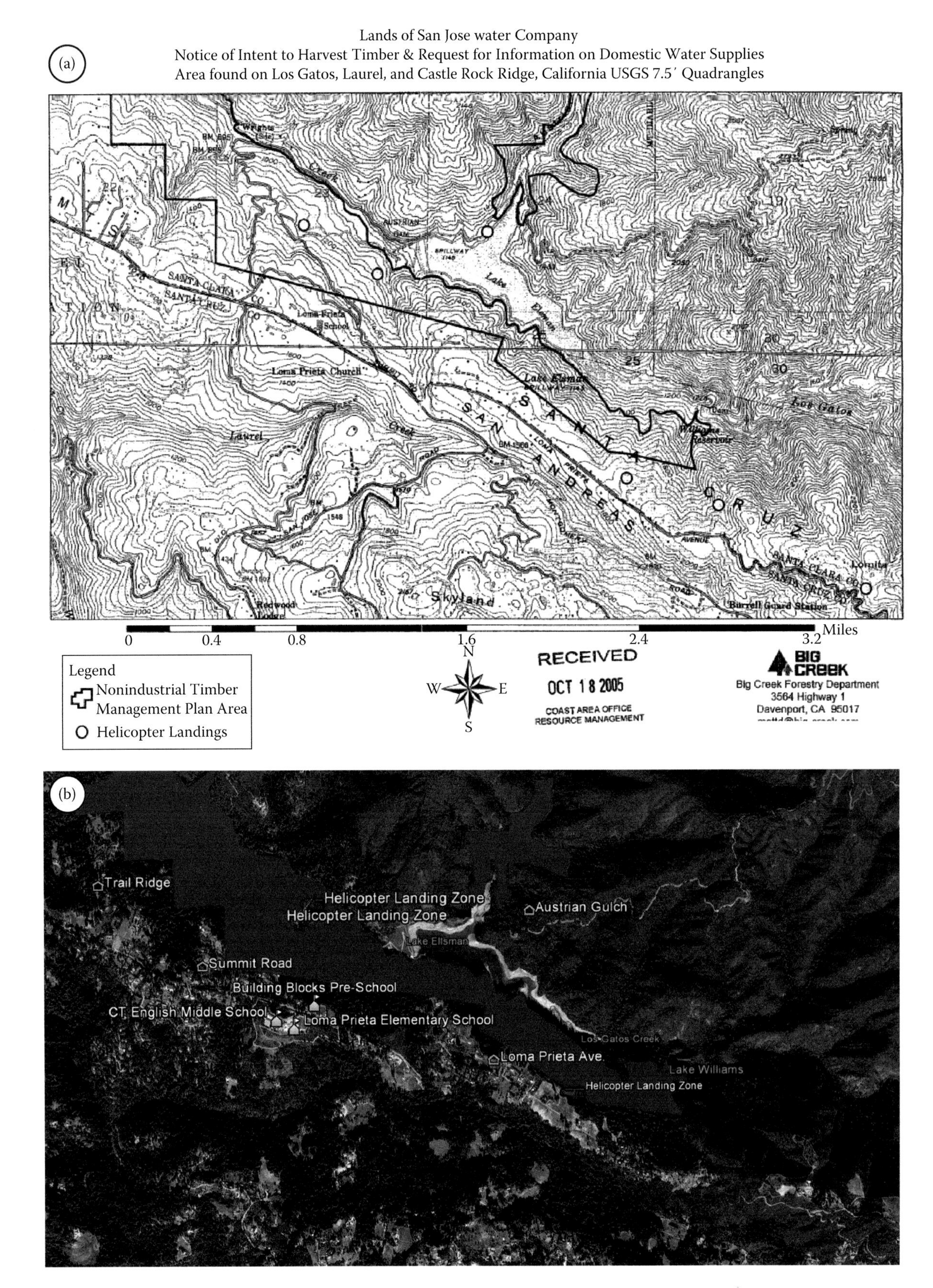

FIGURE 28.7 Information on proposed logging given to a neighborhood in Los Gatos, CA: (a) the original map circulated to neighbors by the logging company and (b) a Google Earth visualization of the same data created by a Googler who lived in the neighborhood.

1. Cost: Google Earth, and the more importantly access to the imagery, is free.
2. Accessibility: The process of continually making imagery and terrain data available at a global scale, for a variety of platforms and operating systems, requires considerable computing infrastructure. Google's resources allow them to do this at a level few others are capable of.
3. Usability: The user-friendly interface and associated what you see is what you get (WYSIWYG) graphical user interfaces (GUIs) allows anyone to explore imagery and create annotations using KML.

However, some have viewed GE as a *fun way* to look at imagery, rather than an actual remote sensing or GIS platform. They certainly do not consider it a *real* remote sensing or GIS (Avraam

FIGURE 28.8 Examples of NGO's uses of Google Earth: (a) The Sierra Club visualized the impacts of sea level rise on Vancouver, Canada (Sheppard and Cizek 2008). (b) The Jane Goodall Institute created a geolocated blog about chimpanzees' activity (Jane Goodall Institute 2010). *(Continued)*

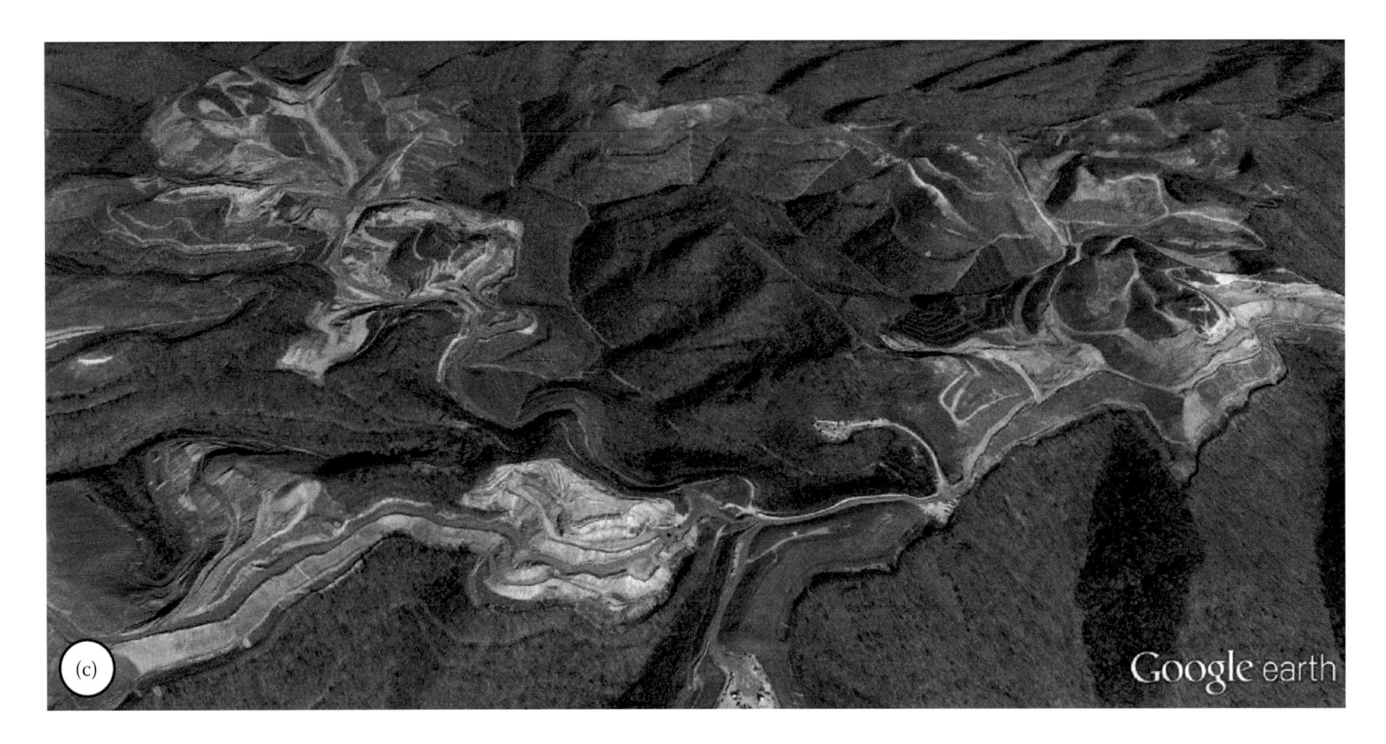

FIGURE 28.8 (*Continued*) Examples of NGO's uses of Google Earth: (c) Appalachian Voices used Google Earth imagery and KML overlays to show mountaintop removal due to open-pit mining (Appalachian Voices 2014).

2009), though others disagree (Turner 2008). For others the idea that GE was a type of *GIS for the masses* (Bader and Glennon 2007) was acceptable as it became the go-to technology for presentations and illustrations using Earth imagery.

The primary argument of the *not remote sensing or GIS* crowd is that GE is a qualitative, not quantitative, tool. It is an excellent platform for viewing imagery, but little to no analysis is possible directly within the application. But the success of GE—and Google Maps—led to continuous expansion of Google's development of Geo tools and the development of an application that now meets the needs of remote sensing analysis.

28.6 Leveraging Google Earth Imagery and Databases

Google Earth Engine (EE) (Google Inc. 2014) is a cloud computing platform for hosting and processing satellite imagery and other Earth observation data. It leverages Google's large distributed server capabilities to analyze imagery.

Despite the name, EE does not use the GE technology (the interface displays its output using the Maps API), but it does use the same imagery and GIS databases created to support Earth, and it was developed to meet the needs of the users using that imagery to observe changes on our planet's surface. EE brings together 40 years of satellite imagery and makes it available to detect changes, map trends, and quantify differences on the Earth's surface. Applications include detecting deforestation, classifying land cover, estimating forest biomass and carbon, and mapping the world's roadless areas (Hansen et al. 2013; Giri et al. 2014; Patel et al. in press).

An example of the power of EE is the Timelapse project, which leverages the archive of Landsat data, a program managed by the USGS that has been acquiring images of the Earth's surface since 1972 (see Chapter 1). Using EE a global mosaic was constructed using 29 years of Landsat satellite data, covering the globe from 1984 to 2012. Each frame of the Timelapse map is constructed from a year of Landsat imagery, creating an annual 1.7 terapixel snapshot of the Earth at 30 m resolution. In total the project processed 2,068,467 scenes, a total of 909 TB of data. This required 2 million CPU (central processing unit) hours and processing was distributed over 66,000 CPUs allowing the mosaics to be computed in 1.5 days. Through a combination of image stacking and pixel matching, the composite images were created free from clouds and without the data loss created by the scan line corrector failure on Landsat 7. River system changes (Figure 28.9), urban growth, receding glaciers, and an expansion of mining operations are a few examples of Earth surface changes that are clearly highlighted by Timelapse (Klugger and Walsh 2013).

28.7 Next-Generation Google Earth

This ability to map on a global scale is powerful enough by itself, given the availability of Landsat, moderate resolution imaging spectroradiometer (MODIS), advanced spaceborne thermal emission and reflection radiometer (ASTER), and other large archives of other land observing satellites. However, the future offers even more potential with the recent acquisition by Google of Skybox Imaging, a company building and launching its own satellites to acquire high-resolution satellite imagery and

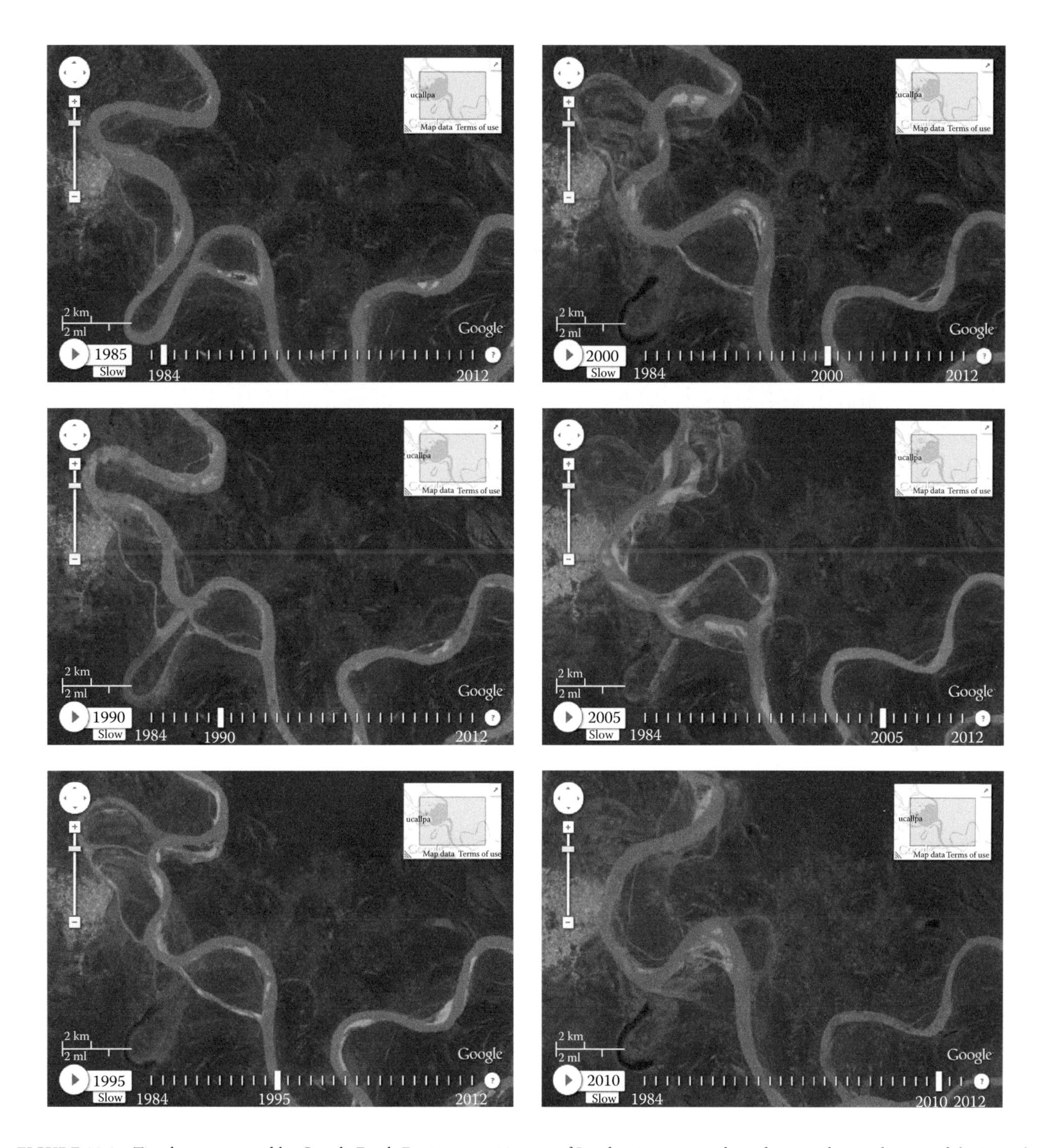

FIGURE 28.9 Timelapse, powered by Google Earth Engine, uses 29 years of Landsat imagery to show the meandering changes of the Ucayali River, near Pucallpa, Peru.

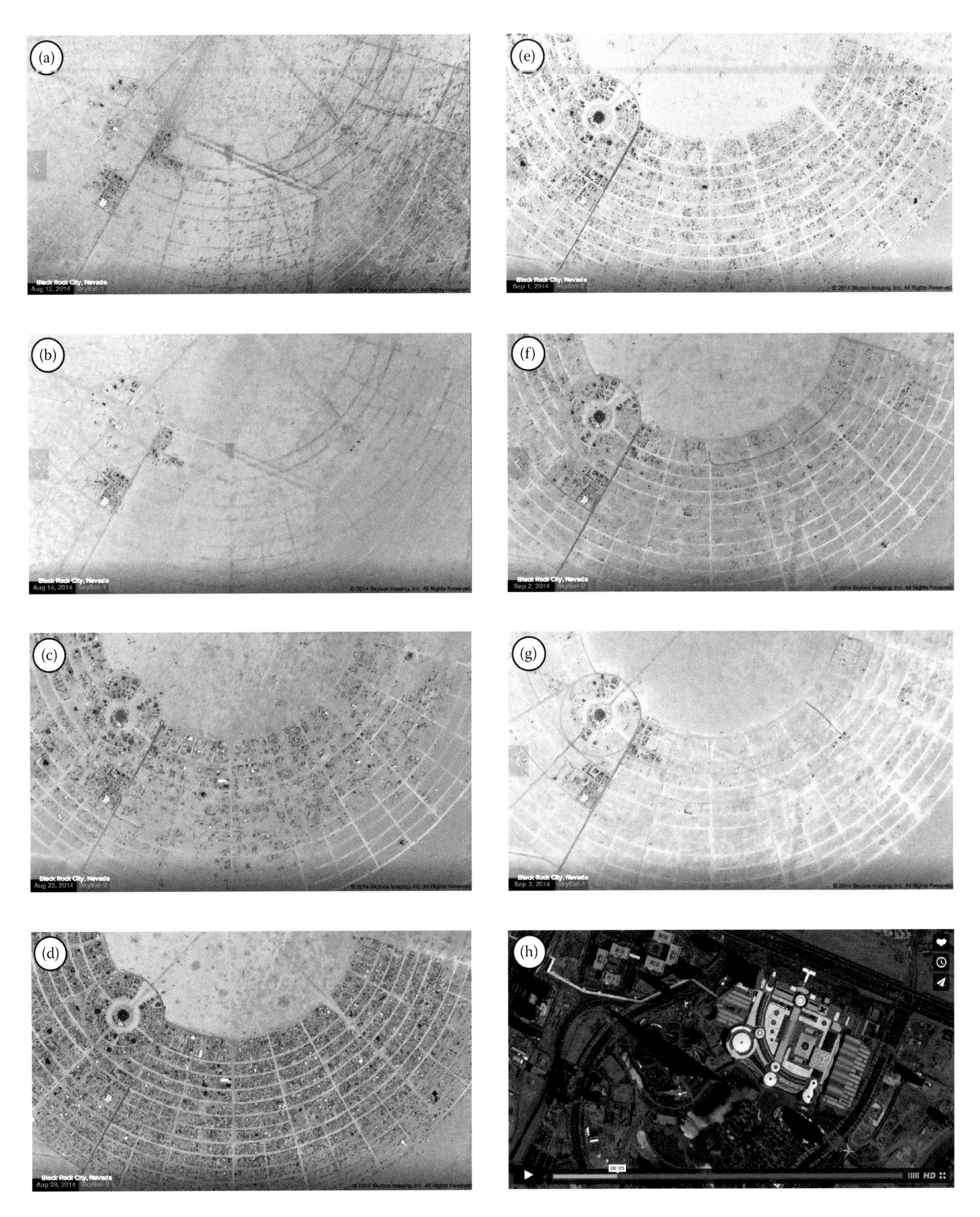

FIGURE 28.10 Imagery and video from SkySat-1 and -2. (a to g) An example of how these satellites can be used to see the growth and decline of a city in the Black Rock Desert, Nevada, during the week of Burning Man. (h) A still from Skybox satellite video captured over the Burj Khalifa, Dubai. Note the airplane at the bottom right.

TABLE 28.1 Technical Specifications for Sensors on SkySat-1 and SkySat-2 Satellites

Image Band	Spatial Resolution (at Nadir) (m)	Temporal Resolution	Swath Width/FoV	File Format
Blue (450–515 nm)	2	Repeat coverage in 3 days to 1 week	8 km	16-bit GeoTIFF
Green (515–595 nm)	2	Repeat coverage in 3 days to 1 week	8 km	16-bit GeoTIFF
Red (605–695 nm)	2	Repeat coverage in 3 days to 1 week	8 km	16-bit GeoTIFF
Near-IR (740–900 nm)	2	Repeat coverage in 3 days to 1 week	8 km	16-bit GeoTIFF
Panchromatic (450–900 nm)—images	0.9	Repeat coverage in 3 days to 1 week	8 km	16-bit GeoTIFF
Panchromatic (450–900 nm)—video	1.1	Up to 90 s at 30 frames per second	2 km by 1.1 km	MPEG-4

high-definition video (Figure 28.10). Skybox's innovative methods allow them to build smaller and cheaper satellites (<$50 million, Truong 2013), while still maintaining performance and an ability to capture high-resolution imagery on par with capabilities of DigitalGlobe's and other high-resolution (<5 m) satellites (Table 28.1). The combination of Skybox-acquired imagery, the processing power of EE, and GE as an interface to view and explore offers exciting possibilities for the future of remote sensing.

28.8 Conclusions

As GE turns a decade old, it seems appropriate to reflect on the role it has played in terrestrial remote sensing. While its analytical potential is still being unlocked, the role it has played in democratizing satellite imagery cannot be denied. The idea that anyone, anywhere can take out their phone and in a matter of seconds see a relatively recent view from space of their current location seemed like the stuff of science fiction as recently as the end of the last millennium. Yet now GE is an accepted everyday tool for scientists, educators, government, and the general public alike and has fundamentally changed how we view our planet (and house).

Acknowledgments

True appreciation of the evolution of Google Earth involves acknowledging countless Keyhole and Google software engineers and product and program managers, who developed and continue to transform the technology. However, the role of researchers, teachers, bloggers, and citizen scientists cannot be understated either. It is fair to say that GE would not have become what it has without the active Google Earth Community (GEC—a user forum formerly known as the Keyhole BBS) and blogging community. GE has impacted the lives of countless groups and individuals, including this author who discovered it while in postdoctoral researcher and ended up on a career path that ultimately saw him become part of the Google Earth Outreach team.

References

Appalachian Voices. 2014. Mountaintop removal in Google Earth. http://ilovemountains.org/google_earth_tutorial (accessed September 26, 2014).

Avraam, M. 2009. Google Maps and GIS. *Intersecting Space and Time Through GIS Endeavors*, http://michalisavraam.org/2009/10/google-maps-and-gis, October 2, 2009 (accessed September 27, 2014).

Bader, J. and A. Glennon. 2007. Virtual Globes: GIS for the Masses? *AAG Annual Meeting*, Paper Session 4510, April 17–21, San Francisco, CA.

Bailey, J.E. 2010. Entry for "Virtual Globe". In Warf, B. ed., *Encyclopedia of Geog*, Sage Publications, Los Angeles, CA, 3528p.

Bailey, J.E. and A. Chen. 2011. The role of Virtual Globes in geoscience. *Comp Geosci* 37(1):1–2.

Bailey, J.E. and J. Dehn. 2006. Volcano monitoring using Google Earth. Geoinformatics 2006—Abstracts, *USGS Scientific Investigations* Report 2006-5201, p. 25., May 10–12, Reston, VA.

Bailey, J.E., S.J. Whitmeyer, and D.G. De Paor. 2012. Introduction: The application of Google Geo Tools to geoscience education and research. *Geol Soc Am Special Paper* 492:vii–xix.

Ballagh, L.M., M.A. Parsons, and R. Swick. 2007. Visualising cryospheric images in a virtual environment: present challenges and future implications. *Polar Record* 43(4):305–310. doi: http://dx.doi.org/10.1017/S0032247407006523.

Ballagh, L.M., B.H. Raup, R.E. Duerr, S.J.S. Khalsa, C. Helm, D. Flower, and A. Gupte. 2011. Representing scientific data sets in KML: Methods and challenges. *Comp Geosci* 37(1):57–64. doi:10.1016/j.cageo.2010.05.004.

Butler, D. 2006a. The Web-Wide World. *Nature* 439:776–778.

Butler, D. 2006b. Avian flu maps in Google Earth. http://declanbutler.info/blog/?p=16 (accessed September 22, 2014).

Butler, D. 2006c. The spread of avian flu with time; new maps exploiting Google Earth's time series function. http://declanbutler.info/blog/?p=58 (accessed September 22, 2014)

De Paor, D.G. 2008. Enhanced visualization of seismic focal mechanisms and centroid moment tensors using solid models, surface bump-outs, and Google Earth. *J Virtual Explorer*, 29:100.

De Paor, D.G. and S.J. Whitmeyer 2011. Geological and geophysical modeling on virtual globes using KML, COLLADA, and *Javascript*. *Comp Geosci* 37(1):100–110. doi:10.1016/j.cageo.2010.05.003.

Dordevic, M.M. and S.C. Wild. 2012. Avatars and multi-student interactions in Google Earth–based virtual field experiences. In Whitmeyer, S.J., J.E. Bailey, D.G. De Paor, and T. Ornduff, eds., *Google Earth and Virtual Visualizations in Geoscience Education and Research. Geol Soc of Am Special Paper* 492:315–321. doi:10.1130/2012.2492(22).

Eames, C. and R. Eames 1977. *Powers of Ten: A Film Dealing with the Relative Size of Things in the Universe and the Effect of Adding Another Zero*. IBM.

Fuller, B. 1962. *Education Automation, Freeing the Scholar to Return to His Studies*. Southern Illinois University Press, Carbondale and Edwardsville, Feffer & Simons Inc., London, U.K.

Geens, S. 2009. Media stupidity Watch: No, It's not Atlantis, *Ogle Earth Blog*. http://ogleearth.com/2009/02/media-stupidity-watch-no-its-not-atlantis/ (accessed September 23, 2014).

Gergely, K.L., T.M. Haran, and B. Billingsley. 2008. Virtual Globe Visualizations of Cryospheric Data at the National Snow and Ice Data Center. *Eos Trans. AGU*, 89(53), Fall Meet. Suppl., Abstract IN41A-1119, December 15–19, San Francisco, CA.

Giri, C., J. Long, S. Abbas, R.M. Murali, F.M. Qamer, and D. Thau, D. 2014. Current status and dynamics of mangrove forests of South Asia. *J Environ Management* doi:10.1016/j.jenvman.2014.01.020.

Google Earth Outreach. 2014. Get inspired by organizations who have used Google mapping tools for good, http://www.google.com/earth/outreach/stories/index.html (accessed September 26, 2014).

Google Inc. 2005. Google Launches Free 3D Mapping and Search Product, *Google Blog*, http://googlepress.blogspot.com/2005/06/google-launches-free-3d-mapping-and_28.html, (accessed September 21, 2014).

Google Inc. 2009. Atlantis? No, it Atlant-isn't, *Google Blog*. http://googleblog.blogspot.com/2009/02/atlantis-no-it-atlant-isnt.html (accessed September 27, 2014).

Google Inc. 2014a. Google Earth. http://www.google.com/earth (accessed September 30, 2014)

Google Inc. 2014b. Google Earth Engine. https://earthengine.google.org (accessed September 30, 2014)

Gore, A. 1998. The Digital Earth: Understanding our planet in the 21st Century. Speech given at the California Science Center, Los Angeles, CA, on January 31, 1998, http://www.digitalearth.gov/VP19980131.html (accessed September 30, 2014).

Gramling, C. February 2007. Google Planet—With virtual globes, Earth scientists see a new world. *Geotimes*, p. 38–40.

Hansen, M.C., P.V. Potapov, R. Moore, M. Hancher, S.A. Turubanova, A. Tyukavina, D. Thau et al. 2013. High-resolution global maps of 21st-century forest cover change. *Science* 342:850–853.

Jane Goodall Institute. 2010. Video Highlights Partnership with Amazon Tribe, Google Outreach. http://www.janegoodall.org/media/news/google-earth-video-tour-highlights-jgi-partnership-amazon-tribe (accessed September 22, 2014).

Klugger, J. and B. Walsh. 2013. Timelapse, *TIME*. http://world.time.com/timelapse/ (accessed September 5, 2014).

Knabb, R.D, J.R. Rhome, and D.P. Brown. 2005. Hurricane Katrina: August 23–30, 2005 (Tropical Cyclone Report). *US National Oceanic and Atmospheric Administration's National Weather Service* (accessed September 30, 2014).

Lamb, A. and L. Johnson 2010. Virtual expeditions: Google Earth, GIS, and geovisualization technologies in teaching and learning. *Teacher Librarian* 37(3):81–85.

Lang, N.P., K.T. Lang, and B.M. Camodeca. 2012. A geology-focused virtual field trip to Tenerife, Spain. In Whitmeyer, S.J., J.E. Bailey, D.G. De Paor, and T. Ornduff, eds., *Google Earth and Virtual Visualizations in Geoscience Education and Research. Geol Soc of Am Special Paper* 492:323–334. doi:10.1130/2012.2492(23).

Lisle, R.J. 2006. Google Earth: A new geological resource. *Geology* 22(1):29–32

Lubick, N. 2005. Spinning around the globe online. *Geotimes*, December 2005.

National Oceanic and Atmospheric Association. 2005. Hurricane Katrina Images. http://ngs.woc.noaa.gov/katrina/ (accessed September 18, 2014).

Nourbakhsh I., R. Sargent, A. Wright, K. Cramer, B. McClendon, and M. Jones. 2006. Mapping disaster zones. *Nature* 439:787–788, doi :10.1038/439787a.

Patel, N.N., Angiuli, E., Gamba, P., Gaughan, A., Lisini, G., Stevens, F.R., Tatem, A.J., and Trianni, G. Multitemporal settlement and population mapping from Landsat using Google Earth Engine. *J. Appl. Geophys.*

Rueger, B.F. and E.N. Beck. 2012. Benedict Arnold's march to Quebec in 1775: An historical characterization using Google Earth. In Whitmeyer, S.J., J.E. Bailey, D.G. De Paor, and T. Ornduff, eds., *Google Earth and Virtual Visualizations in Geoscience Education and Research. Geol Soc of Am Special Paper* 492:347–354. doi:10.1130/2012.2492(25).

Sheppard, R.J. and P. Cizek. 2008. The ethics of Google Earth: Crossing thresholds from spatial data to landscape visualisation. *J Environ Management* 90(6):2102–2117. doi:10.1016/j.jenvman.2007.09.012.

Silberbauer, M.J. and W. Geldenhuys. 2008. Using Keyhole Markup Language to create a spatial interface to South African water resource data through Google Earth. In *FOSS4G 2008 Free and Open Source Software for Geospatial*, OSGeo, GISSA.

Simonite, T. 2007. Virtual Earths let researchers "mash up" data. *New Scientist*, www.newscientist.com/article/dn11773 (accessed September 15, 2014)

Stephenson, N. 1992. Snow Crash. Bantam Dell, Random House Inc., New York, 470p.

Tewksbury, B.J., A.A.K. Dokmak, E.A. Tarabees, and A.S. Mansour. 2012. Google Earth and geologic research in remote regions of the developing world: An example from the Western Desert of Egypt. In Whitmeyer, S.J., J.E. Bailey, D.G. De Paor, and T. Ornduff, eds., *Google Earth and Virtual Visualizations in Geoscience Education and Research. Geol Soc of Am Special Paper* 492:23–36, doi:10.1130/2012.2492(02).

Treves, R. and J.E. Bailey. 2012. Best practices on how to design Google Earth tours for education. In Whitmeyer, S.J., J.E. Bailey, D.G. De Paor, and T. Ornduff, eds., *Google Earth and Virtual Visualizations in Geoscience Education and Research. Geol Soc of Am Special Paper* 492:383–394. doi:10.1130/2012.2492(28).

Truong A. 2013. Proof that cheaper satellites still can take incredibly detailed photos of Earth, Fast Company, http://www.fastcompany.com/3023325/fast-feed/proof-that-cheaper-satellites-still-can-take-incredibly-detailed-photos-of-earth, (accessed September 6, 2014)

Turner, A. 2008. Is GoogleMaps GIS? *High Earth Orbit*, http://highearthorbit.com/is-googlemaps-gis/, June 12, 2008 (accessed September 17, 2014).

USGS. 2014. WaterWatch, http://waterwatch.usgs.gov/kml.html, (accessed September 22, 2014).

van Lammeren R., J. Houtkamp, S. Colijn, M. Hilferink, and A. Bouwman. 2010. Affective appraisal of 3D land use visualization. *Comp Environ Urban Syst* 34:465–475.

Venzke, E., L. Siebert, J.F. Luhr. 2006. Smithsonian volcano data on Google Earth. *Eos Trans. AGU*, 87(52), Fall Meet. Suppl., Abstract N43A-0900, December 11–15, San Francisco, CA.

Webley, P.W., J. Dehn, J. Lovick, K.G. Dean, J.E. Bailey, and L. Valcic. 2009a. Near-real-time volcanic ash cloud detection: Experiences from the Alaska volcano observatory. In Mastin L., and P. Webley. eds. *Volcanic Ash Clouds, J. Volcanol. Geotherm. Res.* Special Issue 186:79–90, doi: 10.1016/j.jvolgeores.2009.02.010

Webley P.W., K. Dean, J.E. Bailey, J. Dehn, and R. Peterson. 2009b. Automated forecasting of volcanic ash dispersion utilizing Virtual Globes. *Natural Hazards* 51(2):345–361. doi:10.1007/s11069-008-9246-2.

Wernecke, J. 2009. *The KML Handbook: Geographic Visualization for the Web*. Addison-Wesley, Upper Saddle River, NJ, 368p.

Whitmeyer, S.J., J.E. Bailey, D.G. De Paor, and T. Ornduff. 2012. Google earth and virtual visualizations in geoscience education and research. *Geol Soc of Am Special Paper* 492, 468p.

Yu, L. and P. Gong. 2012. Google Earth as a virtual globe tool for Earth science applications at the global scale: progress and perspectives. *Int J Rem Sens* 33(12):3966–3986. doi:10.1080/01431161.2011.636081

Accuracies, Errors, and Uncertainties of Remote Sensing–Derived Products

29

Assessing Positional and Thematic Accuracies of Maps Generated from Remotely Sensed Data

Russell G. Congalton
University of New Hampshire

Acronyms and Definitions

ASPRS	American Society for Photogrammetry and Remote Sensing
AVHRR	Advanced very-high-resolution radiometer
CMAS	Circular Map Accuracy Standards
FEMA	Federal Emergency Management Agency
FGDC	Federal Geographic Data Committee
GPS	Global positioning system
KHAT	Estimate of the Kappa statistic
LiDAR	Light detection and ranging
MAS	Map Accuracy Standards
MMU	Minimum mapping unit
MODIS	Moderate resolution imaging spectroradiometer
MSS	Multispectral scanner
NMAS	National Map Accuracy Standards
NSSDA	National Standard for Spatial Data Accuracy
OBIA	Object-based image analysis
RMSE	Root mean square error

29.1 Introduction

This chapter is devoted to assessing the accuracy of maps created from remotely sensed data. In the ideal handbook, it would be great to list the 10 steps that the reader must follow in order to conduct such an accuracy assessment. Unfortunately, map accuracy assessment does not follow such simple procedures. Instead, there are a considerable number of important factors that must be carefully considered from the beginning of the project as well as several statistical and practical methodologies that must be balanced to achieve a successful and valid assessment. Therefore, instead of a few simple steps, this chapter presents these considerations and methodologies in a flowchart (Figure 29.1) to help the reader begin to see all the components that must be thought out and planned for to conduct a valid accuracy assessment. The rest of this chapter deals with each of the parts of this flowchart.

29.2 Assessing Map Accuracy

Today, assessing the accuracy of a map generated from remotely sensed imagery is a routine component of most mapping projects. However, this was not always the case. By the end of World War II, the use of aerial photography and photo interpretation was a well-established means of learning about the earth's resources. Maps generated from aerial photos for such uses as agricultural monitoring, forest inventory, geologic exploration, and many others became commonplace. A key component of every photo interpretation project included the necessary field visits to train the interpreter to recognize the objects of interest on the ground and on the aerial photos. It was generally recognized that the human interpreter drew lines on the photo (i.e., created polygons) around areas that seemed to be distinct from each other (i.e., had more variation between polygons than within a polygon) and then did their best to appropriately label the polygons. Those areas that were difficult to label were checked in the field during another field visit. Very little thought was given to any

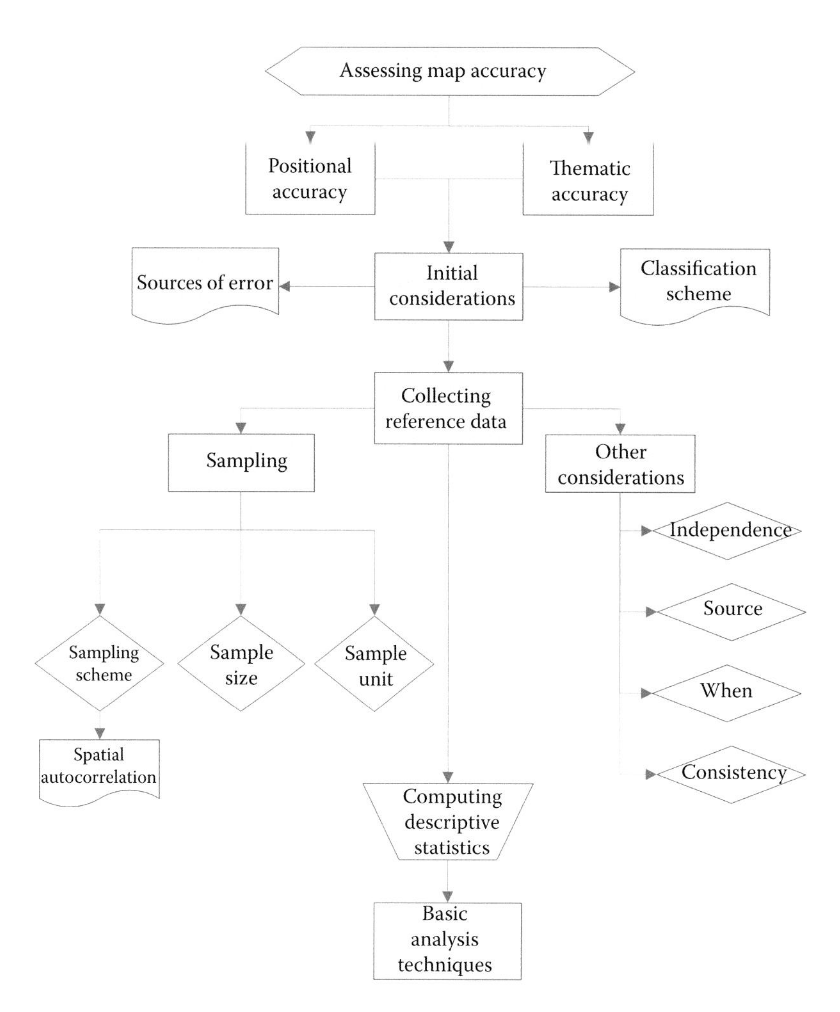

FIGURE 29.1 Flowchart of accuracy assessment considerations.

type of quantitative evaluation or accuracy assessment of the resulting photo-interpreted maps.

A notable exception to this lack of interest in photo interpretation accuracies occurred in the early 1950s. Some researchers wishing to promote the field of photo interpretation as a science recognized the need to evaluate the accuracy of their work and published a number of papers on the topic (Sammi, 1950; Katz, 1952; Colwell, 1955; Young, 1955). A panel discussion entitled, "Reliability of Measured Values" was held at the 18th Annual Meeting of the American Society of Photogrammetry in 1952. At this discussion, Mr. Amrom Katz (1952), the panel chair, made a strong case for the use of statistics in photogrammetry. The results of all these papers and discussions culminated with a paper by Young and Stoeckler (1956). In this paper, these authors proposed techniques for a quantitative evaluation of photo interpretation, including the use of an error matrix to compare field and photo classifications, and went on to present a discussion of the boundary error problem when labeling polygons. Unfortunately, nothing really became of these forward-thinking ideas for about the next 30 years.

In 1972, the first Landsat was launched thus strongly promoting the field of digital image processing. That first Landsat multispectral scanner had only four bands and sensed in only three wavelengths (green, red, and two NIR), but it completely changed the way that we looked at creating maps from remotely sensed imagery. Instead of using the human being as the interpreter, a computer could now be used to manipulate the digital values in the imagery and produce a map. Since the computer, and more specifically some mathematical/statistical algorithm implemented by the computer, was now the entity producing the map from the digital remotely sensed imagery, questions about how good the map was soon followed.

Actually, in any new technology, there is an initial exuberance of use of that technology without much thought to the quality of what is being produced. This is quite natural as the excitement of the new technology dominates and many new uses of the technology are investigated. However, as the technology begins to mature, questions about quality and limitations are inevitable. Such was the case for digital image processing. During the first 5–10 years after the launch of Landsat, many exaggerated claims about the quality and detail that could be produced from the imagery prevailed such that there was a great overselling of the technology. Fortunately, by the early 1980s, a number of researchers began to question some of these claims and started the development of methods for assessing the accuracy of maps derived from remotely sensed data.

It must be noted at this point that map accuracy really has two components: positional accuracy and thematic accuracy.

Positional accuracy deals with the accuracy of the location of map features and measures how far a spatial feature on a map is from its true or reference location on the ground (Bolstad, 2005). Thematic accuracy deals with the labels or attributes of the features of a map and measures whether the mapped feature labels are different from the true or reference feature label. This chapter deals with both types of accuracy, and the flowchart in Figure 29.1 is appropriate for both types also.

29.3 Positional Map Accuracy Assessment

In order to make a map, you must know where you are. Therefore, it is critical in any mapping project that the determination of the exact same location on both the image or map and the reference data (often the ground) be assessed (see Figure 29.2). If this correspondence is not attained, then being in the wrong location may result in a thematic error. For example, it is possible to be in the right place and mislabel (incorrectly measure, observe, or label) the attribute. This error would be a thematic error. However, it is also possible to correctly label the attribute, but be in the wrong place. This locational error could then also lead to a thematic error. Positional error/accuracy and thematic error/accuracy are not independent of each, and it is critical that every possible effort be made to control both of them in order to effectively assess the accuracy of any map.

The history of assessing the positional accuracy of maps begins with the National Map Accuracy Standards (NMAS) from the U.S. Bureau of the Budget in 1947. NMAS is a very simple and straightforward standard that states the following:

- "For horizontal accuracy, not more than 10% of the points tested may be in error by more than 1/30th of an inch (at map scale) for maps larger than 1:20,000 scale, or by more than 1/50th of an inch for maps of 1:20,000 scale or smaller," and
- "For vertical accuracy, not more than 10% of the elevation tested may be in error by more than one half the contour interval."

While easy to apply, NMAS provides no information about determining statistical bounds around the error, but rather just uses the *percentile method* for accepting or rejecting the map as accurate. Therefore, NMAS is not commonly used today.

The next step in developing methods for positional accuracy assessment resulted from the work of Greenwalt and Schultz (1962 and 1968). Their report (The *Principles of Error Theory and Cartographic Applications*, Greenwalt and Schultz 1962 and 1968) proposed equations by which to estimate the maximum error interval for a given probability. They computed a one-dimensional map accuracy standard (MAS) used for elevation/vertical (z) data assessment and a two-dimensional circular map accuracy standard (CMAS) for horizontal (x and y) data assessment. Greenwalt and Schultz assumed that map errors are normally distributed and used the equations for MAS to estimate the interval around the mean vertical error and the equations for CMAS to estimate the interval around the mean horizontal error within which 90% of the error should occur. These equations can also be used to estimate the distribution of

FIGURE 29.2 Example of positional accuracy showing that the road intersections as indicated by the yellow dots (reference points) and the roads are not in the same location.

errors at probabilities other than 90% and offers more flexibility than the previous NMAS.

In 1989, the American Society for Photogrammetry and Remote Sensing (ASPRS) released the ASPRS Interim Accuracy Standards for Large-Scale Maps (ASPRS, 1989). Instead of stipulating that no more than 10% of the errors may exceed a given maximum value (the NMAS approach), the ASPRS standard computes a mean error from a set of samples and then provides a threshold value that this mean error cannot exceed. This standard is expressed in ground units and not map units as NMAS had used. The document also reviews and confirms the work of Greenwalt and Schultz, but does not make any recommendations about the use of these equations.

The National Standard for Spatial Data Accuracy (NSSDA) was released by the U.S. Federal Geographic Data Committee (FGDC) in 1998. This standard represents the currently approved method for assessing positional accuracy and establishes some much needed guidelines for measuring, analyzing, and reporting positional accuracy for maps as well as other geo-referenced imagery (FGDC, 1998). These guidelines have been widely adopted by those in the federal, state, and local governments and also in the private sector. The full version of the NSSDA standards (FGDC, 1998) can be downloaded at http://www.fgdc.gov/standards/projects/FGDC-standards-projects/accuracy/part3/chapter3.

NSSDA does not use a maximum allowable error as the decision point if a map has acceptable positional error at any scale. Instead, it recommends determining an allowable error threshold as needed such that accuracy is then reported in ground distances at the 95% confidence level. This 95% level is stricter than previous standards that tended to use the 90% level. NSSDA incorporates the work of Greenwalt and Schultz and used these equations to determine accuracy as a maximum threshold error at a given or specified probability. However, there are issues with the calculations of the NSSDA as there is a mistake in the calculations that will be discussed later in this section.

Most recently, three new documents/guidelines have been developed that deal with positional accuracy, but especially with regard to LiDAR data. These documents include (1) Guidelines and Specifications for Flood Hazard Mapping Partners (FEMA, 2003), which specifies that there must be 20 samples taken in each of the major vegetation cover types (at least three cover types); (2) the ASPRS Guidelines for Reporting Vertical Accuracy of LiDAR Data (ASPRS, 2004), which confirms the FEMA standard, but uses a slightly different vegetation cover type classification; and (3) Guidelines for Digital Elevation Data from the National Digital Elevation Program (NDEP, 2004), which agrees with the various accuracy measures specified in the ASPRS (2004) guidelines.

29.3.1 Initial Considerations

Important initial considerations when assessing positional accuracy include both sources of error and appropriate classification scheme. In fact, in positional accuracy assessment, these two considerations are intimately linked together. Historically, guidelines such as NMAS and even NSSDA did not specify anything about where the samples should be taken to conduct the assessment. However, with the advent of LiDAR data, it was quickly recognized that land cover or vegetation type at the location the samples were collected significantly impacts the accuracy of the results. Therefore, vegetation or land cover type is now considered an important source of error in positional accuracy assessment and is mitigated by taking a minimum number of samples (usually 20) in each of the major vegetation or land cover types regardless of the size of the area (FEMA, 2003; ASPRS, 2004). As previously mentioned, the exact vegetation/land cover classes (i.e., the classification scheme) varies between the FEMA and ASPRS guidelines, but both agree that samples must be acquired in a variety of cover types in order to have a valid assessment.

29.3.2 Collecting Reference Data

There are many factors that must be considered when collecting the reference data used to assess the positional accuracy of the map (see Figure 29.1). If the reference data are not collected properly, then the entire assessment may be invalid. Many of these considerations involve sampling while others deal with issues such as independence, source, timing of collection, and consistency.

Independence is a key component of the assessment process. The data used must be independent from any data used in the registration of the map or other spatial data. This fact is commonly understood in statistical analysis, but there seems to still be many examples in positional accuracy where statements about the goodness of the position are still expressed using the data that were employed to register the map to the ground. These data are clearly not an independent data set and represent a most optimistic and invalid estimate of the actual positional accuracy.

Obviously, the source of the reference data must be more accurate than the map that is being assessed. Sometimes, a map of larger scale is sufficient. In other situations, a survey using GPS or other equipment is required. NSSDA recommends data be "of the highest accuracy that is feasible and practical" (FGDC, 1998). Others have suggested that the positional reference data be from one to three times more accurate than the anticipated accuracy of the map being tested (e.g., Ager, 2004; ASPRS, 2004; NDEP, 2004).

When and who collects the reference data can also be important although typically these factors are more important in thematic accuracy assessment. It is important that major changes have not taken place due to earthquakes or other natural phenomena that could alter the positions of objects (again, this is a rare situation). Also, the collection of the data must be consistent between multiple collectors. In other words, if more than one individual is collecting the data, it is important to establish objective procedures to ensure that the data are collecting similarly by all collectors.

Since the assessment of the positional accuracy of a map is performed not at every place on the map, but rather at a series of points, it is critical that the sampling be valid to achieve an appropriate assessment. Sampling involves determining the number of samples, the sample unit (identification), and the sampling scheme or distribution. Again, failure to properly plan for these factors will result in an invalid assessment. The NSSDA (FGDC, 1998), the current standard, states that a minimum of 20 samples must be used in the assessment. Some of the new guidelines that require selecting samples from a number of vegetation/land cover classes require 20 samples per class, while some have increased that number to 30 per class (ASPRS, 2004; NDEP, 2004). From a statistical standpoint, taking 30 samples is often recommended. Unfortunately, many assessments have been conducted with either nonindependent samples or too few samples resulting in statistically invalid results.

The sample unit in positional accuracy assessment is actually a point. In fact, it is critical that these points be well defined and "represent a feature for which the horizontal position is known to a high degree of accuracy and position with respect to the geodetic datum" (FGDC, 1998). Any ambiguity about where a point is located will disqualify it as a sample point for use in the positional accuracy assessment.

Finally, the sampling scheme determines how the samples are collected throughout the map (i.e., the sample distribution). It is important that the samples are distributed throughout the map so that the entire map is assessed. The full range of variation in the map should be considered including topography, vegetation/land cover, and important features. Figure 29.3 demonstrates a stratified approach proposed by ASPRS (1989) that guarantees that the samples are distributed throughout the map by dividing the map into quadrants and forcing a minimum of 20% of the samples into each quadrant. This method also minimizes spatial autocorrelation in the sampling by setting a minimum spacing between sample points such that no two points can be closer together than *d*/10 where *d* is the diagonal dimension of the map. This method seems to be a very effective way to ensure that the sampling considerations for positional accuracy assessment are not ignored.

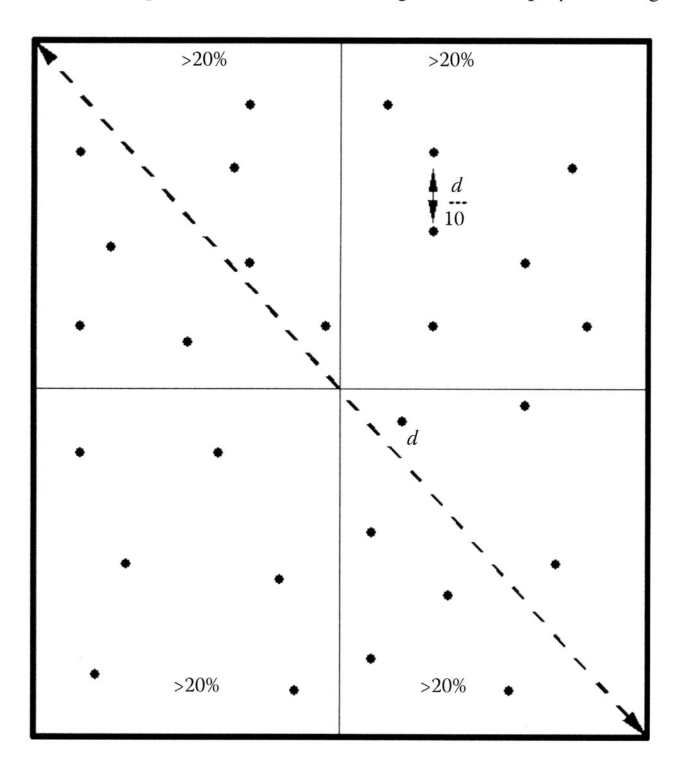

FIGURE 29.3 Sampling scheme showing samples distributed throughout the map.

29.3.3 Computing Descriptive Statistics

As previously discussed, assessing positional accuracy is performed using a sample of data to estimate the agreement (i.e., fit) between the map or other geospatial data and a reference data set that is assumed to be correct. This analysis is done through the computation of a number of statistics. The key statistic that is computed in positional accuracy assessment is the root mean square error (RMSE). The RMSE is simply the square root of the mean of squared differences between the samples on the map and those same samples on the reference data. The differences between the map and reference sample points are squared to ensure positive values since the simple arithmetic difference between the map and reference sample points can be either negative or positive.

As alluded to previously, positional accuracy can be measured in the vertical dimension (z) or in the horizontal dimension, which involves both x and y. The analysis is similar in each case, but it is little more complicated for the horizontal dimension. The equation for RMSE for vertical accuracy is

$$\text{RMSE}_v = \sqrt{\frac{\sum_i^n (e_{vi})^2}{n}} \tag{29.1}$$

where

$$e_{vi} = v_{ri} - v_{mi} \tag{29.2}$$

v_{ri} equals the reference elevation at the ith sample point
v_{mi} equals the map elevation at the ith sample point
n is the number of samples

The equation for RMSE for horizontal accuracy is

$$\text{RMSE}_h = \sqrt{\frac{\sum_i^n e_{hi}^2}{n}} \tag{29.3}$$

where

$$e_{hi}^2 = (x_{ri} - x_{mi})^2 + (y_{ri} - y_{mi})^2 \tag{29.4}$$

x_{ri} and y_{ri} are the reference coordinates
x_{mi} and y_{mi} are the map coordinates for the ith sample point
n is the number of samples

The NSSDA states that positional accuracy be reported at the 95% level defined as "95% of the locations in the data set will have an error with respect to the reference position that is equal to or less than the computed statistic" (FGDC, 1998). The equation for computing the NSSDA for vertical accuracy is given in the guidelines as

$$\text{NSSDA Vertical Accuracy}_v = 1.96(\text{RMSE}_v) \quad (29.5)$$

The equation for computing the NSSDA for horizontal accuracy is a little more complicated because of the two dimensions (x and y). It is possible to have a different distribution of errors in the x direction than in the y direction causing the errors around the position to be oblong in shape rather than circular as one would expect if the errors were equally distributed. Most use the simplified equation ignoring the distribution of errors and compute NSSDA for horizontal accuracy as

$$\text{NSSDA Horizontal Accuracy} = 1.7308 * \text{RMSE}_h \quad (29.6)$$

Table 29.1 shows the computation of NSSDA and RMSE for a small horizontal data set. Careful study of this table shows why the differences between the reference position and map position are squared to eliminate the positive and negative values. These computations can easily be executed in an Excel spreadsheet or other software package to quickly compute the currently required statistics for assessing the positional accuracy of any map or geospatial data set.

It should be noted here that there has been significant confusion in the mapping community over some of the computations used in assessing positional accuracy. This confusion results because the term RMSE as used by mapping professionals differs from the term RMSE used in statistics. In addition, the term *standard error* is unfortunately used to depict different parameters in different professions. While most statistics textbooks define the standard error as the square root of the variance of the population of means, $\sigma_{\bar{X}}$, many mapping texts define the standard error as the square root of the variance of the population signified by σ. Statisticians call this term the standard deviation. As a result, some of the equations that are correct in the work by Greenwalt and Schultz have been misapplied or at least misinterpreted by the FGDC for use in developing the NSSDA. While it is not possible in this chapter to provide a full explanation of this confusion and the statistical ramifications that have resulted from this issue, a full discussion is provided in Congalton and Green (2009). Currently, the NSSDA is the accepted standard, and most positional accuracy assessments report this statistic along with the RMSE. Newer guidelines as reviewed previously have suggested refinements including sampling in the major vegetation/land cover types as well as other ways to assess positional accuracy.

TABLE 29.1 Example of Computing Positional Accuracy (RMSE and NSSDA)

Sample No.	x (ref)	x (map)	x diff	(x diff)^2	y (ref)	y (map)	y diff	(y diff)^2	Σ (x diff)^2 + (y diff)^2
1	36.5	37.2	−0.7	0.49	842.5	843.7	−1.2	1.44	1.93
2	12.0	11.1	0.9	0.81	900.0	896.2	3.8	14.44	15.25
3	66.8	66.7	0.1	0.01	1010.1	1009.1	1.0	1.00	1.01
4	4.1	4.2	−0.1	0.01	786.5	782.4	4.1	16.81	16.82
5	9.1	8.5	0.6	0.36	655.8	658.2	−2.4	5.76	6.12
6	77.0	79.0	−2.0	4.00	676.0	672.1	3.9	15.21	19.21
7	112.1	112.1	0.0	0.00	655.1	666.2	−11.1	123.21	123.21
8	99.0	99.8	−0.8	0.64	688.1	689.2	−1.1	1.21	1.85
9	55.5	54.9	0.6	0.36	744.3	737.1	7.2	51.84	52.20
10	102.1	110.5	−8.4	70.56	810.7	811.2	−0.5	0.25	70.81
11	6.8	6.7	0.1	0.01	845.6	847.2	−1.6	2.56	2.57
12	200.5	198.2	2.3	5.29	902.1	902.4	−0.3	0.09	5.38
13	252.1	250.9	1.2	1.44	937.4	939.2	−1.8	3.24	4.68
14	260.8	262.0	−1.2	1.44	945.5	940.1	5.4	29.16	30.60
15	300.9	299.5	1.4	1.96	960.8	962.0	−1.2	1.44	3.40
16	266.8	267.0	1.8	3.24	971.2	971.1	0.1	0.01	3.25
17	142.1	135.8	6.3	39.69	1000.3	999.1	1.2	1.44	41.13
18	96.7	96.5	0.2	0.04	1006.8	1008.2	−1.4	1.96	2.00
19	42.0	36.9	5.1	26.01	1101.3	1110.0	−8.7	75.69	101.70
20	12.2	15.1	−2.9	8.41	999.9	1002.0	−2.1	4.41	12.82
								Sum	515.94
								Ave.	25.78
								RMSE	5.08
								NSSDA	8.79

29.4 Thematic Map Accuracy Assessment

The history of assessing the thematic accuracy of maps derived from remotely sensed data is shorter than that of positional accuracy. Except for a brief effort to quantitatively assess the accuracy of photo interpretation in the 1950s, the history of thematic accuracy assessment began shortly after the advent of digital imagery (i.e., the launch of the first Landsat). Researchers, notably Hord and Brooner (1976), van Genderen and Lock (1977), and Ginevan (1979), proposed criteria and basic techniques for testing overall thematic map accuracy. In the early 1980s, more in-depth studies were conducted and new techniques proposed (Rosenfield et al., 1982; Congalton et al., 1983; Aronoff, 1985). Finally, from the late 1980s up to the present time, a great deal of work has been conducted on thematic accuracy assessment. As we develop increasingly high spatial and spectral resolution sensors, more complex classification algorithms including object-based image analysis (OBIA), and faster computer hardware with easier-to-use software, it is vital that our methods for effectively assessing thematic maps continue to improve as well.

The history of thematic accuracy assessment has progressed through a number of stages as it developed and matured. As already discussed, the initial excitement of any new technology typically results in a time of intensive use of the technology without much, if any, attention paid to quality of the results. This stage quickly moved into the next stage in which the map *looks good.* There are a number of ways to qualitatively assess that a map looks good including visual comparison of the thematic map to the original imagery (Congalton, 2001) or perhaps to other image sources (e.g., Google Earth or the like). The map could be shown to those especially familiar with the area or the map could be visually compared to other thematic maps of the area. Each of these methods could convince the analyst that the map looks good. It is important that a thematic map look good. If the analyst is not convinced the map is of sufficient quality, there really is no sense in conducting a quantitative accuracy assessment. Therefore, that it looks good is a necessary but not a sufficient characteristic of any thematic map. Unfortunately, some are still stuck in the *it looks good* stage of thematic map accuracy assessment and never go any further to effectively assess the map.

Quickly following the *it looks good* stage of thematic map assessment was the stage called non-site-specific assessment. In this stage, some effort was made to obtain quantitative estimates of the accuracy of the thematic map. However, these estimates represent results of the map as a whole but do not say anything about the accuracy at any specific location on the map. Instead, estimates such as the total hectares of forest on the map are compared to some reference estimate of the total forest hectares in that same area. While non-site-specific assessments do begin to introduce some quantitative assessment into the process, the results are left lacking in that nothing can be said about any specific location on the map. Therefore, this method rapidly gave way to the current stage of thematic map accuracy assessment called site-specific accuracy assessment.

Site-specific thematic accuracy assessment is represented in the form of an error matrix (Figure 29.4), also called a contingency table in statistics. The process of creating an error matrix is demonstrated in Figure 29.5. An error matrix is a square array of numbers set out in rows and columns that expresses the number of sample units assigned to a particular category (i.e., land cover class) in one classification as compared to the number of sample units assigned to a particular category (i.e., land cover class) in another classification. Typically, one of the classifications is considered to be correct (or at least of significantly higher accuracy than the other) and is called the reference data. Unfortunately, it has become common practice to call these reference data the *ground truth.* While the reference data should have high accuracy, there is no guarantee that it is 100% correct and using the term *ground truth* is not appropriate. This author hopes that the geospatial community will abandon this term for ones that better represent the actual data such as reference data or ground data or field data or the like. The columns of the error matrix typically represent the reference data, while the rows indicate the map derived from remotely sensed data or other geospatial data set. It is possible to switch the headings for the row and columns so the analyst must take care to notice how the axes of the matrix are labeled before any further analysis is undertaken.

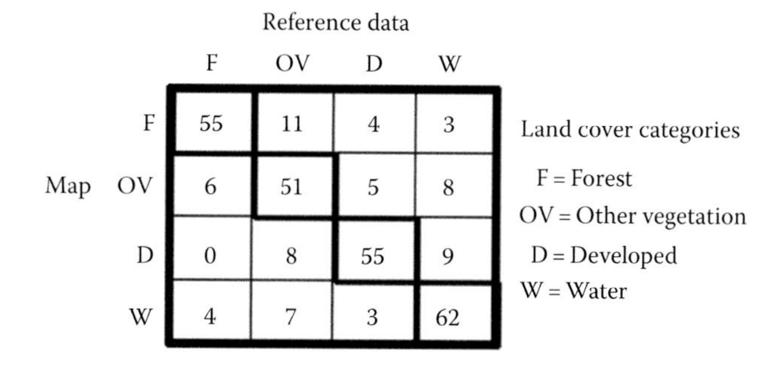

Map \ Reference data	F	OV	D	W
F	55	11	4	3
OV	6	51	5	8
D	0	8	55	9
W	4	7	3	62

FIGURE 29.4 An example error matrix.

29.4.1 Initial Considerations

There are two very important initial considerations that must be taken into account before beginning a thematic accuracy assessment. These are classification scheme and sources of error.

29.4.1.1 Classification Scheme

Thematic maps are an abstraction of the real world and as such use some type of classification scheme to group together and simplify reality. Determining which classification scheme to use or developing your own must be completed at the beginning of the mapping project. Failure to select an efficient and effective scheme from as early in the project as possible dooms the project to significant inefficiencies and potential failure. Whether the thematic map is of land cover or vegetation or soil types or some other thematic information, it is critical that the scheme represents the information that is necessary for the end user. However, it is also important that this information be discernible with the remotely sensed data selected for the mapping. Therefore, selecting the appropriate scheme at the beginning of the project facilitates choosing the

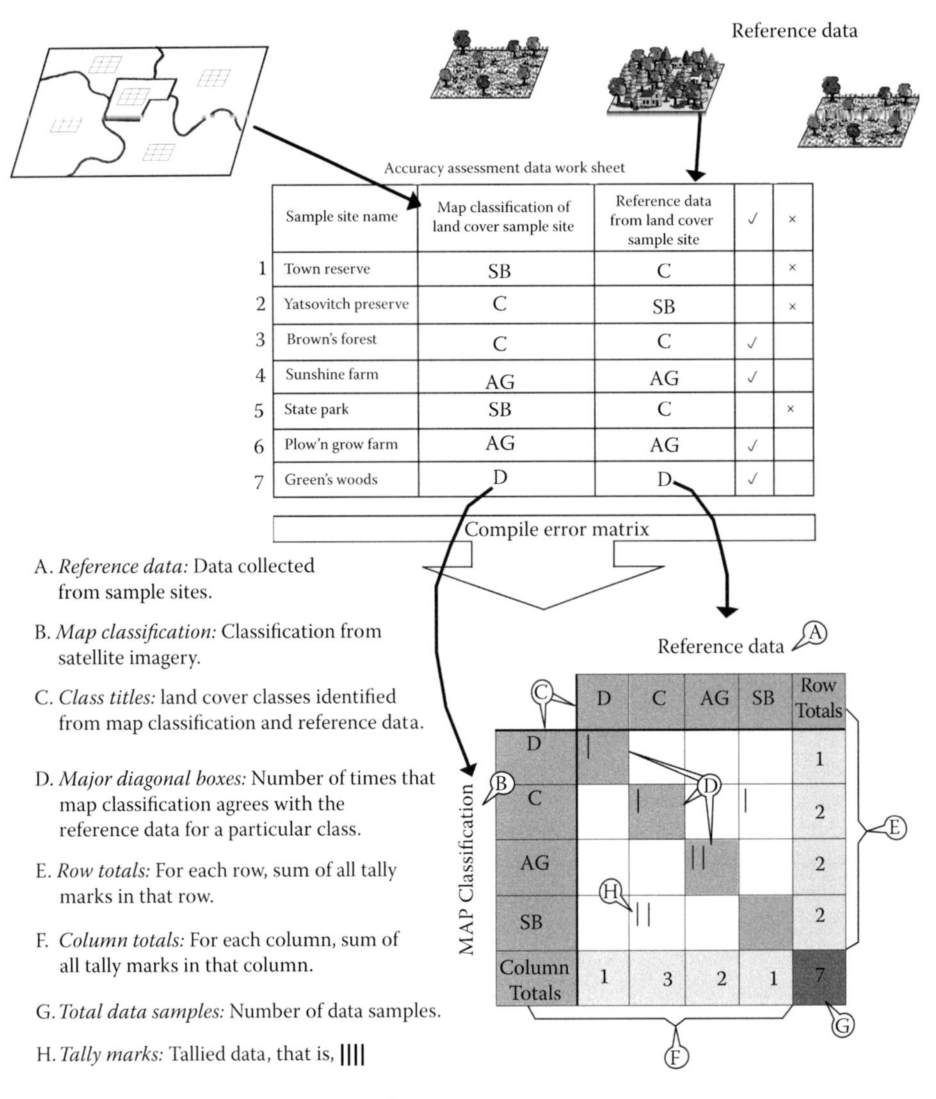

FIGURE 29.5 A diagram showing how an error matrix is generated from a series of samples. (From Landcover Protocols of the GLOBE Teacher's Manual, www.globe.gov.)

suitable imagery and allows reference data to be collected for use in assessing the accuracy of the resulting thematic map.

Any classification scheme selected should have the following four characteristics: (1) *Definition*: The scheme must contain rules or definitions that explicitly define each of the map classes. It is not sufficient to simply list the map classes and assume that everyone agrees on exactly what each class is. For example, a forest may be defined as trees greater than 5 m tall. Using this definition, a grouping of trees that are only 3 m tall would not be defined as a forest. (2) *Mutually exclusive*: The scheme must be mutually exclusive. In other words, every effort must be made using the class definitions to clearly eliminate any overlap between classes so that every area on the map or ground falls into one and only one class in the scheme. (3) *Totally exhaustive*: The scheme must be totally exhaustive. In other words, every area on the map or ground must fall into one of the classes in the classification scheme. An effective means of ensuring that your scheme is totally exhaustive is to include an *other* class. However, if a large portion of the thematic map ends up in the other class, then it may be appropriate to rethink the classification scheme as perhaps some important information on the map has been neglected. (4) *Hierarchical*: Finally, it is useful if the classification scheme is hierarchical. That is having multiple levels of detail (a hierarchy) such that the map can be generated and especially assessed for accuracy at different levels of detail. For example, the forest class may be further divided in conifer and deciduous classes that add more detail. The conifer class may be further divided into a pine class and a non-pine class. As we will see shortly, it may not be possible to assess the accuracy of the map at the most detailed level because of costs of collecting the reference data. Therefore, it may be possible with a hierarchical classification scheme to map to one level of the hierarchy but to assess the accuracy to a lesser level of detail.

In determining the appropriate classification scheme for a thematic mapping project, the analyst must also consider the concept of *minimum mapping unit* (mmu). The mmu is the smallest area that is uniquely delineated on a thematic map. This concept is widely known and used in photo interpretation where the maps are created from aerial photographs of a given scale. However, the concept has not been as widely adopted in digital imagery as the imagery can be rendered at any scale, and pixel size seems to be more of the limiting factor. As will be seen in the discussion of sample unit, the pixel should not be selected as the minimum area to be mapped on digital data, and therefore, consideration of an mmu even on digital mapping projects is important.

29.4.1.2 Sources of Error

Creating thematic maps from remotely sensed imagery is possible because there is a strong linkage between what can be sensed on the imagery and what is actually happening on the ground. However, this correlation is not perfect, and every situation where this correlation breaks down is a source of potential error between the map and the ground. Lunetta et al. (1991) presented a discussion of these sources of error, which included errors from image acquisition, errors from data processing, errors that occur within the analysis including data conversion, errors that can occur in the accuracy assessment process itself, and finally error in decision making and implementation.

While it is clear that error can enter a thematic mapping project from a great variety of sources, little has been done to evaluate these errors and prioritize methods for dealing with them. Congalton and Brennan (1999) and Congalton (2009) proposed an error budget analysis method that not only clearly lists the

TABLE 29.2 Error Budget Analysis for a Remote Sensing Project

Error Source	Error Contribution Potential	Implementation Difficulty	Implementation Priority	Error Assessment Technique
Systematic				
Sensor	Low	5	21	Instrumentation and analysis
Natural				
Atmosphere	Medium	4	20	Instrumentation and analysis
Preprocessing				
Geometric registration	Low	2		Positional accuracy assessment
Image masking	Medium	3	7	Single date error matrix
Derivative data				
Band ratios	Low	1	18	Data exploration
NDVI	Low	1	18	Data exploration
PCA	Low	1		Data exploration
TCA	Low	1	8	Data exploration
Other ancillary data	Medium	3	9	Data exploration
Classification				
Classification scheme	Medium	2	14	Single date error matrix
Training data collection	Medium	3	15	Single date error matrix
Classification algorithm	Medium	3	11	Single date error matrix
Post-processing				
Data conversion	High	2	12	Single date error matrix
Accuracy assessment				
Sample unit	Low	2	13	Single date error matrix
Sample size	Low	2	16	Single date error matrix
Sampling scheme	Medium	3	17	Single date error matrix
Spatial autocorrelation	Low	1	3	GeoStatistical analysis
Positional accuracy	High	2	2	RMSE/NSSDA
Final product				
Decision making	Medium	2	1	Sensitivity analysis
Implementation	Medium	2	6	Sensitivity analysis

Note: Error Contribution Potential, ranked from low to medium to high; *Implementation Difficulty*, ranked from 1: not very difficult to 5: extremely difficult; *Implementation Priority*, ranked from 1 to *n* showing order in which to implement improvements.

sources of error in any mapping project, but also documents a method for evaluating the error contribution, implementation difficulty, and implementation priority for each error. Table 29.2 shows an example of this error budgeting.

29.4.2 Collecting Reference Data

There are many factors that must be considered when collecting the reference data used to assess the thematic accuracy of the map. Collection of reference data is more complicated for thematic accuracy assessment than positional accuracy assessment, and therefore, there is even more risk that the entire assessment may be invalid if the reference data are poorly collected. Many of the factors that must be considered involve sampling, while others deal with issues such as independence, source, timing of collection, and consistency.

29.4.2.1 Independence

It is absolutely critical that the reference data used to assess the thematic map accuracy be independent of any other data used for training or any other purpose during the mapping project. This independence can be achieved in two ways. First, the collection of the reference data can be done at a separate time and/or by different personnel from the collection of the training data. While effective, this method is inefficient. A second approach involves collecting the training and reference data simultaneously and then splitting the data into two sets whereby the reference data set is put aside and not viewed by any project personnel until it is time to conduct the accuracy assessment.

29.4.2.2 Sources of Reference Data

The source of the reference data in any thematic map accuracy assessment is a function of the complexity of the classification scheme (i.e., level of map detail) and the budget of the project. It is fortuitous if existing maps (e.g., USDA Cropland data layer) or ground data (e.g., U.S. Forest Service Inventory Data) can be used as the reference data. However, differences in classification scheme and most often size of the sample unit typically limit using existing reference data, no matter how tempting they are. Often, the existing data are too old or the sampling plot size (sample unit) is too small to be used as reference data. Therefore, more commonly, the reference data are newly collected information assumed to be more accurate than the map that is being assessed. Aerial photo interpretation has often been used to assess thematic maps generated from moderate resolution digital imagery, ground/field visits are used to assess maps made from high-resolution airborne imagery, and manual image interpretation has been used to assess maps created from automated classification techniques. No matter what source is selected as the reference data, it is imperative that the data are effectively and appropriately collected so as to provide a viable data set from which to validly assess the accuracy of the thematic map. Therefore, explanation and justification of the data used for reference data must be part of the accuracy assessment report.

29.4.2.3 When Should the Reference Data Be Collected

Timing for collecting reference data for assessing the accuracy of thematic maps clearly depends on the type of information being mapped. Some information is just more timely than others. For example, if mapping agricultural crops, it is very important that the reference data be collected as near to the day the imagery was collected as possible (in some cases, same day collection is important). In other examples where change occurs on a much slower time scale, collection of the reference data within a year or 2 or 5 may be sufficient. Clearly, if a change occurs between the date of the remotely sensed image capture and the date of the reference data collection, then the reference data will be of little or no use in effectively assessing the accuracy of the map.

29.4.2.4 Consistency in Reference Data Collection

A good classification scheme consisting of labels with clear rules or definitions goes a long way to ensuring consistency in reference data collection. A guidebook documenting the procedures used to collect the data is also important. Finally, a field form, whether on paper or using computer input, completes the process. Having these three components facilitates training of all field personnel whether it is a single collector or a large group being disbursed throughout the study area. Objectivity in the data collection process is key, and having a good reference data collection form will help enforce this. Regardless of the classification scheme, the form will have certain common components such as the names of the collectors, date of collection, location (GPS) information, a key to the map classes (dichotomous keys are excellent), the actual map class determined by the collection, and a place to describe any issues, special findings, or problems that occurred at the collection site.

29.4.2.5 Sampling

As in positional accuracy assessment, thematic accuracy assessment relies on a statistical valid sample of reference data to compare to the map to compute the required statistics. It is not possible to assess the entire map (a total enumeration) because of time, money, and effort constraints. Therefore, sampling (a subset of the map) is used to facilitate the accuracy assessment. Sampling for collecting reference data for assessing the accuracy of thematic maps derived from remotely sensed data can be divided into three components: the sample unit, the sample size, and the sampling scheme including spatial autocorrelation. Each of these is very important to ensure proper and valid reference data collection.

29.4.2.5.1 Sample Unit

Historically, a single pixel has often been used as the sample unit for assessing thematic map accuracy. Unfortunately, a single pixel is a very poor choice for the sample unit because a pixel is arbitrary, difficult to locate, and typically smaller than the minimum mapping unit defined by the project. Even with all the latest advances in GPS, terrain correction, and geometric registration, all reference data sample units have some positional errors. Mid-resolution sensors such as Landsat consider positional errors of one half a pixel (10–15 m) to be reasonable. Higher spatial resolution sensors have smaller pixel sizes but increased positional errors if expressed in terms of pixels (i.e., a sensor with 1 m pixels could have a positional error of 10–15 m—the same amount as Landsat, but what is half of a Landsat pixel represents a 10–15 pixel error with this sensor). If a Landsat pixel is registered to the ground with 10–15 m accuracy and the GPS unit used to locate the center of the pixel on the ground has an accuracy of 10–15 m, then it is totally impossible to use that single pixel as the sample unit for assessing the thematic accuracy of the map because the pixel simply cannot be located accurately. The problem is only exacerbated more with increased spatial resolution imagery.

Therefore, the appropriate sample units are either a cluster of pixels or a polygon. Considering positional accuracy limitations, a cluster of pixels, typically 3 × 3 pixels for moderate resolution imagery, has been a good choice. A homogeneous cluster of pixels minimizes positional error problems while maintaining a constant-sized sample unit. If assessing the accuracy of a thematic map made from course resolution imagery such as MODIS or AVHRR, it is unlikely to find many homogeneous single pixels let alone a 3 × 3 cluster. In this case, it may be necessary to document the proportions of the sample unit in each land cover class and set rules for labeling the reference data accordingly. If assessing the accuracy of a thematic map made from fine-resolution imagery such as GeoEye or Digital Globe data, it is important to select appropriate sampling units that account for the positional error. For example, if the imagery has 1 m pixels, then a 3 × 3 cluster will not compensate for positional error. Instead, a sample unit of perhaps 20 × 20 pixels is needed to compensate for a 10 m registration error.

It is important to remember that if a cluster of pixels is selected as the sample unit, then the cluster (sample unit) represents a single sample and is tallied as such in the error matrix. There are far too many examples in the literature where an analyst has selected a cluster of pixels and then considered each pixel in the cluster as a separate sample unit. Collecting reference data in this way is incorrect and results in an invalid assessment. It should also be noted that typically the clusters selected for sampling are homogeneous (this may not be possible with coarse-resolution imagery). Selecting homogenous clusters has the great advantage of minimizing positional error, but may result in a biased or inflated assessment as borders between classes (i.e., heterogeneous areas) are avoided, and therefore, these edge errors may not be assessed. Finally, remember that while a 3 × 3 pixel cluster should work well for moderate resolution imagery, much larger clusters (in terms of number of pixels, not meters) must be collected for higher spatial resolution imagery in order to account for positional error.

Thematic maps generated from aerial photo interpretation and more recently from OBIA are vector maps where polygons are delineated on the imagery instead of the typical rasters (i.e., pixels). In these situations, the polygon should be used as the sample unit for collecting the reference data. There are a few new issues that arise when a polygon is used for accuracy assessment. First, the polygon may be much larger that a 3 × 3 pixel cluster, and therefore, more effort may need to go into accurately labeling that polygon on the reference data (MacLean et al., 2013). Second, the polygons are no longer of the same area as was the case for a 3 × 3 pixel cluster. Therefore, a more area-based error matrix may be appropriate instead of just tallying the sample units (Maclean and Congalton, 2012, 2013). This topic will be covered more in the section on advanced analysis techniques later in this chapter.

29.4.2.5.2 Sample Size

Collecting enough samples to obtain a statistically valid reference data set is perhaps the most challenging component of assessing the accuracy of a thematic map. In almost every situation, there is a balance between what is required to be statistically valid and what is affordable within the given budget. Early on, many researchers, including Hord and Brooner (1976), van Genderen and Lock (1977), Hay (1979), Ginevan (1979), Rosenfield et al. (1982), and Congalton (1988b), published equations and guidelines for choosing the appropriate sample size. Initial efforts used the binomial equation to estimate the number of samples required to report the overall accuracy of a map. However, sampling to complete an error matrix is not a binomial situation, but rather a multinomial one in which there is one correct answer and $n - 1$ wrong answers (where n is the number of thematic classes) for each sample (Congalton, 1988b; Congalton and Green, 2009). Tortora (1978) presents the equations needed to compute the required sample size using the multinomial sampling approach. The number of samples depends on the number of thematic classes and the allowable error and the desired confidence level. Congalton (1988b) determined, using an extensive series of Monte Carlo simulations, that most maps could be assessed using 50 samples per thematic map class (maps less than about 500,000 ha and 12 or less thematic map classes). More complex maps with more than 12 thematic map classes and covering larger areas should use 75–100 samples per thematic map class. Assessing the thematic accuracy of global and/or continental maps requires partitioning the area into some type of ecological zones with the requisite number of samples per zone. Therefore, assessing these maps requires some minimum number of samples per zone times the number of zones.

It should be noted that practical considerations are a key component for determining sample size. There is a trade-off between the budget and the number of samples needed. There is a point that is reached when the samples per thematic class are under 30 per class that the assessment loses statistical validity completely, and it may be better not to even attempt the accuracy assessment.

29.4.2.5.3 Sampling Scheme

A number of sampling schemes (how to collect the samples) have been suggested for collecting the reference data including simple random sampling, systematic sampling, stratified sampling, stratified random sampling, and cluster sampling. While random sampling has very nice statistical properties, it is often impractical to implement because of the costs of getting to every random location and because thematic classes consisting of small areas are left under-sampled. Systematic sampling ensures that samples are distributed over the entire study area and is very effective if higher resolution imagery is the source used to obtain the reference data. Systematic sampling has similar problems of access when collecting ground reference data. Cluster sampling, that is, collecting a number of sample units within close proximity to each other, offers certain efficiencies, but care must be taken to not take too many samples too close together because of spatial autocorrelation (samples are not independent of each other). Therefore, most reference data collection efforts use stratified sampling in order to ensure some minimum number of samples in each stratum (thematic map class).

Many accuracy assessments use the rule expressed earlier and include 50 or so samples in each thematic map class. In some situations, it may be appropriate to have more samples in certain thematic classes than in others. These situations include when some classes are more important or when some classes occupy more of the map area than the others. Care is necessary here to ensure some minimum number of samples in each thematic class so as to be able to determine the accuracy of every thematic map class. For example, imagine an area that is 95% sand and water (see Figure 29.6) and the 5% divided among commercial, industrial, and residential areas. If a stratified sampling approach was not employed where a minimum number of samples were taken from each map class and instead a random sample of 200 samples were selected, it is possible to achieve a high map accuracy and yet poorly map 3 of the 5 map classes. Probability tells us that if 200 samples were randomly selected from a map that was 95% sand and water, then 95% or 190 of these samples would fall in the sand and water. Only 10 samples would fall within the other 3 map classes (commercial, industrial, and residential).

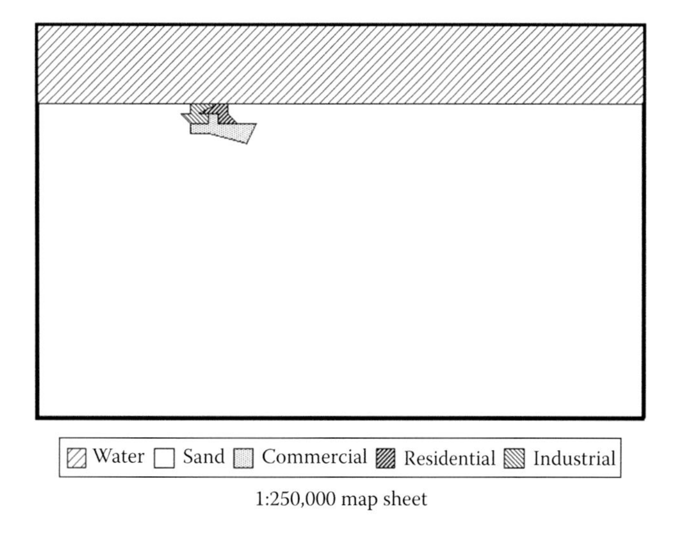

FIGURE 29.6 A map showing an area of 95% sand and water with a small portion (5%) of the area commercial, residential, and industrial land cover types.

If the sand and water were correctly mapped, but the other classes were incorrect all the time, we would still determine that this map was 95% correct based on this sampling scheme. This is a statistically valid and yet completely misleading result of our random sampling strategy. Therefore, selection of the appropriate sampling scheme is as important as selecting the correct sample unit and number of samples.

A final concept that influences the choice of sampling scheme and how the samples are selected is called spatial autocorrelation. Spatial autocorrelation occurs when the presence, absence, or degree of a certain characteristic affects the presence, absence, or degree of that same characteristic in neighboring units (Cliff and Ord, 1973). In other words, the concept involves independence between neighboring samples. If an error is made at a certain location and spatial autocorrelation exists, then it is more likely that the same error will occur in nearby locations. Congalton (1988a) and Pugh and Congalton (2001) have demonstrated that spatial autocorrelation is an important consideration in the collection of reference data, and therefore, samples should be taken so as to maximize the separation between samples as practical. Additionally, if cluster sampling is necessary, the number of clusters taken at a location should be as few as possible, and they should be as far apart as possible.

Given the complexities, considerations, and choices that are involved in collecting proper reference data to assess the thematic accuracy of a map generated from remotely sensed data, it is absolutely required that a report be produced to document the process. Simply reporting an error matrix or, even worse, an overall accuracy measure without describing the choices made and the justification for these choices results in an incomplete assessment. Such a document is necessary for the map user to thoroughly understand the accuracy assessment process.

29.4.3 Computing Descriptive Statistics

Assuming that the error matrix has been created using a valid reference data collection approach as described earlier, the matrix is then the starting point for a number of descriptive statistics including overall accuracy, producer's accuracy, and user's accuracy. Figure 29.7 contains the same error matrix as was presented in Figure 29.4, but now shows the computation of some descriptive statistics. Remember that an error matrix is a very effective way to represent map accuracy because the individual accuracies of each category are easily discerned along with both the errors of inclusion (commission errors) and the errors of exclusion (omission errors) present in the map. Just like a coin has two sides (heads and tails), map errors have two components (omission error and commission error). A commission error can be defined as including an area into a thematic class when it doesn't belong to that class, while an omission error is excluding that area from the thematic class in which it truly does belong. Each and every error is an omission from the correct thematic map class and a commission to a wrong thematic map class.

Figure 29.7 shows that six areas (sample units) were mapped as other vegetation, while the reference data show that they were

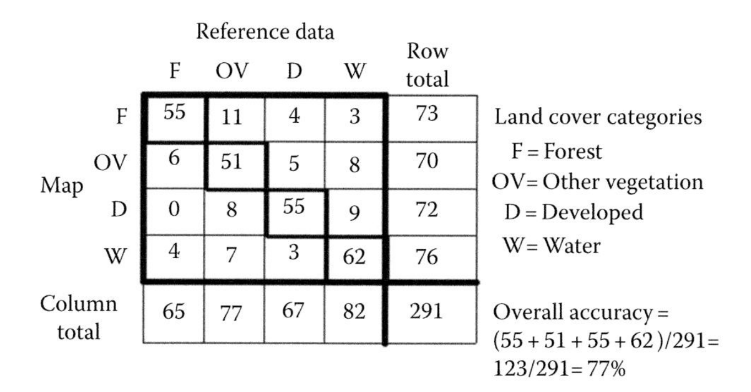

Map \ Reference data	F	OV	D	W	Row total
F	55	11	4	3	73
OV	6	51	5	8	70
D	0	8	55	9	72
W	4	7	3	62	76
Column total	65	77	67	82	291

Land cover categories
F = Forest
OV= Other vegetation
D = Developed
W= Water

Overall accuracy =
(55 + 51 + 55 + 62)/291=
123/291= 77%

Producer's accuracy	User's accuracy
F = 55/65 = 85%	F = 55/73 = 75%
OV = 51/77 = 66%	OV = 51/70 = 73%
D = 55/67 = 82%	D = 55/72 = 76%
W = 62/82 = 75%	W = 62/76 = 82%

FIGURE 29.7 Error matrix showing the calculations for overall, producer's, and user's accuracies. *Note*: (1) As the map producer, 85% of the forest sample units (in the reference data) are correctly classified as forest on the map. (2) As a map user, 75% of the sample units classified as forest on the map are indeed forests.

actually forest. In other words, six areas were omitted from the forest class and committed to the incorrect other vegetation class. In addition to showing the omission and commission errors, the error matrix in Figure 29.7 also shows the computation of three other accuracy measures: (1) overall accuracy, (2) producer's accuracy, and (3) user's accuracy (Story and Congalton, 1986). Overall accuracy is computed by summing up the major diagonal of the error matrix (i.e., the correctly mapped sample units) and dividing by the total number of sample units in the matrix. While this statistic is perhaps the most commonly reported measure of thematic map accuracy, this value alone is not sufficient. It is vital that the error matrix be reported so that other measures of accuracy can be computed and that the errors in the map can be clearly seen and understood.

Overall accuracy represents the accuracy of the entire map. However, in addition, we may want to know the accuracy of an individual thematic map class. Computation of individual class accuracies is a little more complicated than overall accuracy and requires two measures: producer's accuracy and user's accuracy (Story and Congalton, 1986). The producer of the map may want to know how well they mapped a certain thematic map class, the producer's accuracy. This value is computed by dividing the value from the major diagonal (the agreement) for that class by the total number of samples in that map class as indicated by the sum of the reference data for that class. Looking at Figure 29.7 shows that the map producer called 55 areas forest, while the reference data indicate that there were a total of 65 forest areas. Ten areas were omitted from the forest, and of these, six were committed to other vegetation and four were committed to water. So, 55/65 samples were correctly called forest for a forest producer's accuracy of 85%. However, this is only half the story. If you now observe the user's perspective of the map, you see once again that 55 samples were called forest on the map that were actually forest, but in addition, the map called 11 samples forest that were actually other vegetation, 4 samples forest that were actually developed, and 3 samples forest that were actually water. The map, therefore, called 73 samples forest, but only 55 are actually forest. There was a commission error of 18 samples into the forest that were not forest. The forest user's accuracy is then computed by dividing the major diagonal value for the forest class by the total number of samples mapped as forest, 55/73 = 75%. In evaluating the accuracy of an individual map class, it is important to consider both the producer's and the user's accuracies.

Overall, producer's, and user's accuracies can also be explained with the following equations. Begin with n samples that are distributed into k^2 cells where each sample is assigned to one of k thematic map classes in the map (usually the rows) and, independently, to one of the same k thematic map classes in the reference data set (usually the columns). Then, let n_{ij} denote the number of samples mapped into thematic map class i ($i = 1, 2,\ldots, k$) in the map and thematic map class j ($j = 1, 2,\ldots, k$) in the reference data set.

Let

$$n_{i+} = \sum_{j=1}^{k} n_{ij} \tag{29.7}$$

be the number of samples classified into thematic map class "i" in the map, and

$$n_{+j} = \sum_{i=1}^{k} n_{ij} \tag{29.8}$$

be the number of samples classified into thematic map class "j" in the reference data set.

Then, the overall accuracy between map and the reference data can then be computed as follows:

$$\text{Overall accuracy} = \frac{\sum_{i=1}^{k} n_{ii}}{n} \tag{29.9}$$

Producer's accuracy can be computed by

$$\text{Producer's accuracy}_j = \frac{n_{jj}}{n_{+j}} \tag{29.10}$$

and the user's accuracy can be computed by

$$\text{User's accuracy}_i = \frac{n_{ii}}{n_{i+}} \tag{29.11}$$

29.4.4 Basic Analysis Techniques

After computing the descriptive statistics from the error matrix, some additional analysis techniques can also be implemented to learn even more about the results of the accuracy assessment.

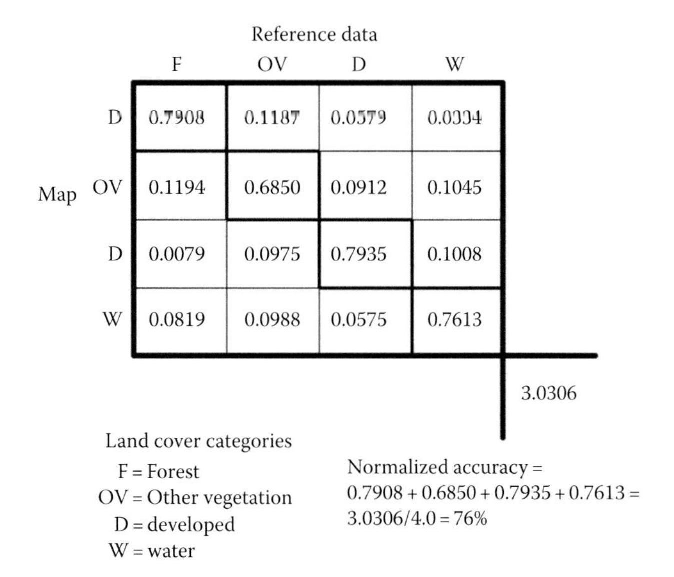

		Reference data			
		F	OV	D	W
Map	D	0.7908	0.1187	0.0579	0.0334
	OV	0.1194	0.6850	0.0912	0.1045
	D	0.0079	0.0975	0.7935	0.1008
	W	0.0819	0.0988	0.0575	0.7613
					3.0306

Land cover categories
F = Forest
OV = Other vegetation
D = developed
W = water

Normalized accuracy =
0.7908 + 0.6850 + 0.7935 + 0.7613 =
3.0306/4.0 = 76%

FIGURE 29.8 Error matrix showing the results of the Margfit analysis in Figure 29.7.

The first of these techniques is called "Margfit" and is used to normalize the error matrix to remove the effect of sample size for comparison with other error matrices. Margfit uses a standard iterative proportional fitting algorithm that normalizes the matrix by forcing each row and column total (i.e., the marginal) to sum to 1. The result is a normalized matrix in which differences in sample sizes used to generate different matrices are eliminated and the individual cell values in the matrix are directly comparable with another matrix. Figure 29.8 shows the results of the Margfit analysis on Figure 29.7. Notice that each value in the matrix is now a percentage of 1, and multiplying each value by 100 would result in the accuracy for each cell. The major diagonal values represent the accuracy of each thematic map class and are a combination of the individual producer's and user's accuracies for each class. For example, the accuracy of the forest class is shown as 0.7908 or 78%. Finally, as in computing overall accuracy, the normalized accuracy of the error matrix can be computed by summing the values in the major diagonal and dividing by 4 (the number of map classes and also the sum of all values in the matrix). The normalized accuracy of Figure 29.8 is 76%, while the overall accuracy is 77% (see Figure 29.7).

In order to fully appreciate the usefulness of the Margfit analysis, it is necessary to have a second error matrix to compare to the first. Figure 29.9 presents an error matrix for the same study area as was assessed in Figure 29.7. However, the analyst collected more accuracy assessment sample units from which to assess the quality of the map. Overall, producer's, and user's accuracies were computed as described and shown in Figure 29.9. Note that 356 sample units were used for this assessment, while only 291 sample units were selected for the assessment in Figure 29.7. The overall accuracy of the error matrix in Figure 29.9 is 84%.

The results of the Margfit analysis on Figure 29.9 are shown in Figure 29.10. Again, the individual cell values represent thematic class accuracies, and the normalized accuracy is 83%.

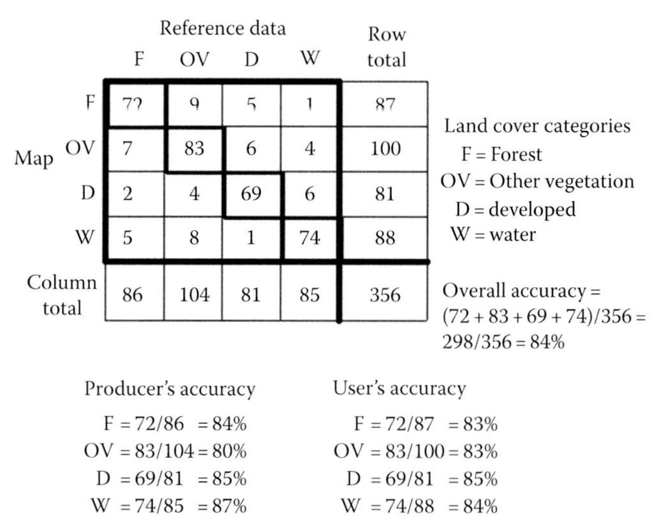

		Reference data				Row total
		F	OV	D	W	
Map	F	72	9	5	1	87
	OV	7	83	6	4	100
	D	2	4	69	6	81
	W	5	8	1	74	88
Column total		86	104	81	85	356

Land cover categories
F = Forest
OV = Other vegetation
D = developed
W = water

Overall accuracy =
(72 + 83 + 69 + 74)/356 =
298/356 = 84%

Producer's accuracy
F = 72/86 = 84%
OV = 83/104 = 80%
D = 69/81 = 85%
W = 74/85 = 87%

User's accuracy
F = 72/87 = 83%
OV = 83/100 = 83%
D = 69/81 = 85%
W = 74/88 = 84%

FIGURE 29.9 An error matrix from the same study area as Figure 29.7, but the analyst selected more sampling units for accuracy assessment.

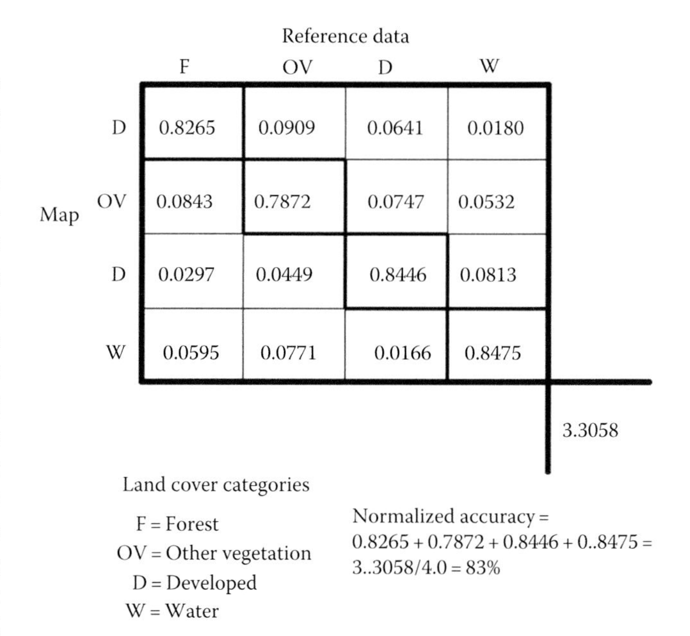

		Reference data			
		F	OV	D	W
Map	D	0.8265	0.0909	0.0641	0.0180
	OV	0.0843	0.7872	0.0747	0.0532
	D	0.0297	0.0449	0.8446	0.0813
	W	0.0595	0.0771	0.0166	0.8475
					3.3058

Land cover categories
F = Forest
OV = Other vegetation
D = Developed
W = Water

Normalized accuracy =
0.8265 + 0.7872 + 0.8446 + 0..8475 =
3..3058/4.0 = 83%

FIGURE 29.10 Error matrix showing the results of the Margfit analysis in Figure 29.9.

Because the original error matrices (Figures 29.7 and 29.9) have been normalized (Figures 29.8 and 29.10), the values in the normalized error matrices can be directly compared. For example, comparing the 55 correct sample units in Figure 29.7 for the forest class with the 72 correct sample units in Figure 29.9 for the forest class would not have much meaning since the matrices were generated using different total sample sizes (291 and 356, respectively). However, with Margfit, the matrices are normalized so that these values are now directly comparable with Figure 29.8 having a forest accuracy of 79% and Figure 29.10 having a forest accuracy of 83%. It seems that the analyst who created the map assessed using the error matrix in Figure 29.9 used a method that produced better forest class accuracy. This normalization process produces a single accuracy value for each of the map classes, which is a combination of the producer's and user's accuracies for that class. In this way, a single number rather than two separate values can be used to represent a class accuracy.

A second technique that has often been used in accuracy assessment is called "Kappa" (Cohen, 1960). The results of performing a Kappa analysis is a KHAT statistic (an estimate of Kappa), which like overall accuracy and normalized accuracy can be used as another measure of agreement or accuracy. The KHAT statistic is computed as

$$\hat{K} = \frac{N\sum_{i=1}^{r} x_{ii} - \sum_{i=1}^{r}(x_{i+} * x_{+1})}{N^2 - \sum_{i=1}^{r}(x_{i+} * x_{+1})} \quad (29.12)$$

where

r is the number of rows in the matrix

x_{ii} is the number of observations in row i and column i

x_{i+} and x_{+i} are the marginal totals of row i and column i, respectively

N is the total number of observations (Bishop et al., 1975)

The equations necessary for computing the variance of the KHAT statistic and the standard normal deviate can be found in many publications including Congalton et al. (1983), Rosenfield and Fitzpatrick-Lins (1986), Hudson and Ramm (1987), Congalton (1991), and Congalton and Green (1999, 2009). Table 29.3 shows three accuracy measures that can be computed easily from the error matrix. These measures include overall accuracy computed from the original matrix, normalized accuracy computed from the normalized matrix, and kappa. Which measure or measures to use for expressing thematic map accuracy is open for discussion. Each brings a different amount of information into the calculation. Overall accuracy simply sums the major diagonal and divides by the total number of sampling units. Normalized accuracy directly incorporates the off-diagonal elements of the matrix (i.e., the errors) through the normalization (i.e., iterative proportional fitting) process. Kappa indirectly incorporates the errors by using the sums of the row and column totals in the computation of the statistic. Therefore, each measure represents different information and should be evaluated accordingly. Landis and Koch (1977) proposed certain range values of the Kappa statistic to represent different levels of agreement. A value greater than 80% represents high agreement, while a range between 60% and 80% represents moderate agreement and values below 40% represent weak agreement. Presenting the full error matrix allows the analyst to compute any or all of these three measures of thematic accuracy.

TABLE 29.3 Three Measures of Thematic Map Accuracy

	Matrix #1	Matrix #2
Overall accuracy (%)	77	84
Normalized accuracy (%)	76	83
Kappa (%)	69	78

While the Kappa statistic (KHAT) has been used extensively as a measure of accuracy, a number of researchers have pointed out the shortcomings of using it as such (Zhenkui and Redmond, 1995; Naesset, 1996; Pontius and Millones, 2011). However, the major contribution of the Kappa analysis is not as a measure of accuracy, but rather as a technique used to determine if one error matrix is statistically significant from another. Such a test using a matrix derived from one classification approach could then determine if that matrix is statistically significantly different than a matrix derived from a second classification approach. In other words, is technique A statistically significantly better than technique B. Actually, two tests are possible with the Kappa analysis. The first is to test if a single error matrix is significantly better than random. In other words, is the map better than a random assignment of pixels (hopefully it is)? The KHAT statistic is asymptotically normally distributed, and therefore, confidence intervals can be calculated using the approximate large sample variance (Congalton and Green, 1999). The test to determine if a classification is better than random uses the Z test as follows:

$$Z = \frac{\text{KHAT}}{\sqrt{\hat{\text{var}}(\text{KHAT})}} \quad (29.13)$$

where Z is standardized and normally distributed.

The test to determine if two error matrices are significantly different from one another uses the following Z test:

$$Z = \frac{|\text{KHAT}_1 - \text{KHAT}_2|}{\sqrt{\hat{\text{var}}(\text{KHAT}_1) + \hat{\text{var}}(\text{KHAT}_2)}} \quad (29.14)$$

where

KHAT_1 represents the KHAT statistic from one error matrix

KHAT_2 is the statistic from the other matrix

Running the Kappa test of significance on the matrix in Figure 29.7 produces a Z value of 20.8. At the 95% confidence level, at Z value above 1.96 is considered significant so it can be concluded that the process that went into making this map is significantly better than a random distribution of thematic labels to the pixels. The same can be said for the matrix in Figure 29.9 with a Z value of 29.9. However, the most valuable question is whether these two matrices are statistically significantly different than the other. The resulting test statistic for this Kappa analysis is 2.2 again indicating statistical significance. Therefore, the matrices are significantly different and looking at the accuracy values from Table 29.3 shows that matrix 2 (Figure 29.9) is the better of the two. If matrix 2 (i.e., the map assessed in Figure 29.9) was produced with a different classification algorithm than matrix 1, then it can be said that the algorithm is better. This ability to determine if one matrix, and therefore map, is statistically significantly better than another is the real power of the Kappa analysis.

29.4.5 Advanced Analysis Techniques

In addition to the basic analysis techniques for thematic map accuracy assessment described earlier, there are two other more advanced techniques that must be seriously considered depending on the methodologies used to make the map. These techniques include fuzzy accuracy assessment and area-based error matrix generation.

29.4.5.1 Fuzzy Accuracy Assessment

One of the four characteristics of a good classification scheme is that it is mutually exclusive. That is, any area on the map should fall into one and only one map class. While this is a valid goal for any scheme, in reality, there are always issues with achieving this goal. Gopal and Woodcock (1994) proposed the use of fuzzy set theory in thematic accuracy assessment to recognize the possibility that ambiguity exists in determining the appropriate map class. They further state that it is rare to find the situation where one map class is exactly right and all the other map classes are equally wrong.

There are many places that fuzziness can enter into a thematic map assessment. The classification scheme itself is a large source as many map classes are more a continuum that has been grouped into a single class. For example, mapping hardwood forest vs. conifer forest is easy if the forest is 100% hardwoods or 100% conifer. However, as the percentages of each get closer to 50%, it becomes harder to decide which type of forest it is. Human interpretation also is an important source of variation. In many cases, training data used in the classification process and accuracy assessment data used in the validation process are generated from visual interpretation of high-resolution imagery. Even visiting field sample sites on the ground is not without variation as one observer may determine the boundary between classes is in one place and another observer put it in another. Therefore, depending on the classification scheme used and the methods employed to collect the training and reference data, some acknowledgment of the potential fuzziness in the map is justified.

Green and Congalton (2004) and more fully described in Congalton and Green (2009) have proposed the modification of the standard error matrix to incorporate fuzziness into the assessment process. In this case, the matrix consists of the correct values, the acceptable values, and the unacceptable values. Use of the fuzzy error matrix is extremely powerful as it incorporates all the benefits of the standard error matrix while allowing for situations where classification scheme breaks are artificial points along a continuum and/or where interpreter variability is difficult to control. The best way to understand the fuzzy error matrix is to provide an example as shown in Figure 29.11.

A quick examination of Figure 29.11 reveals that it is the same error matrix as in Figure 29.7 with the exception that there are now two values in the off-diagonal elements of the matrix. The values in the major diagonal still represent the best determination of the correct classification. For example, there are 55 sample units that the map said were the forest class and the reference data agreed. However, the off-diagonal elements now have two values: the first value represents acceptable matches while the second value represents the errors. Looking at Figure 29.11 shows that the map labeled an area as forest when it was really other vegetation a total of 11 times. However, four times this label is considered acceptable because the area, when visited to collect the reference data, was with actually young trees that had not yet grown tall enough to meet the definition of forest, which stated that the trees must be 3 m tall to be considered a forest. The other seven times, the other vegetation was actually other vegetation (i.e., not trees), and the map is in error. Fuzziness comes into play in the matrix only where appropriate. It is not an excuse to accept errors, but rather a mechanism to account for variability and is especially helpful with the reference data. Note that no confusion between the forest class and the developed

		Reference data				User's accuracies			
		F	OV	D	W	Deterministic totals	Deterministic accuracies	Fuzzy totals	Fuzzy accuracies
Map	F	55	4, 7	0,4	0,3	55/73	75%	59/73	81%
	OV	2,4	51	0,5	2,6	51/70	73%	55/70	79%
	D	0,0	0,8	55	0,9	55/72	76%	55/72	76%
	W	1,3	0,7	0,3	62	62/76	82%	63/76	83%

Producer's accuracies				
Deterministic totals	55/65	51/77	55/67	62/82
Deterministic accuracies	85%	66%	82%	75%
Fuzzy totals	58/65	55/77	55/67	64/82
Fuzzy accuracies	89%	71%	82%	78%

Overall accuracies

Deterministic 223/291 = 77%

Fuzzy 232/291 = 80%

Land cover categories F = Forest OV= Other vegetation D= developed W= water

FIGURE 29.11 Deterministic and fuzzy error matrix example.

class is considered acceptable. Instead, the first value in the off-diagonal is zero, and the second value shows the errors (4). So, what about the other vegetation/water confusion? Again, looking at the fuzzy error matrix shows that twice the map said the sample unit was other vegetation when the reference data said that it was water. The fuzziness here can be explained in that there were lily pads and other green vegetation growing on the surface of a pond. The pond is clearly water, but it is acceptable in this case to call it other vegetation because of what was floating on the surface.

This new error matrix (Figure 29.11) is really two error matrices in one. It is the original error matrix (now called the deterministic error matrix) and it is also a fuzzy error matrix. The difference is simply how one handles the two values in the off-diagonals. In the case of the deterministic error matrix, only the values in the cells of the major diagonal are considered correct. Therefore, the deterministic overall accuracy, the deterministic row and column totals, and the deterministic producer's and user's accuracies are identical in every way to the calculations shown in Figure 29.7. However, the fuzzy error matrix calculations are a little bit different as they include the acceptable values (represented by the first value in the off-diagonal cells) along with the correct values (represented by the cells on the major diagonal). Therefore, the fuzzy producer's accuracy for the forest class is 58/65 = 89% where the 58 is equal to the 55 correct values plus 2 acceptables from the other vegetation class and 1 acceptable from the water class. The fuzzy overall accuracy is 232/291 = 80%, which includes summing the values on the major diagonal plus all the acceptables (first values in the off-diagonal cells). That is adding 55 + 51 + 55 + 62 + 2 + 1 + 4 + 2 = 232.

In some cases, the difference between the deterministic error matrix and the fuzzy error matrix can be quite small. This situation would occur when the classification scheme is rather general, and there is little variation between thematic map classes. In other situations, the differences in the matrices can be quite large especially if the classification scheme is quite complex with great variability among the classes. Another situation in which the fuzzy error matrix can be very important is when the only reference data available are difficult to interpret causing great variation in the reference data. For example, using medium-scale aerial photographs (i.e., 1:25,000) as the reference data source for assessing a map generated from Landsat Thematic Mapper imagery using a classification scheme that has over 20 land cover types. In this case and because only the photos could be used without any ground visits, the interpreters would likely select a best class label and then perhaps one to three acceptable labels for each of the reference data sample units.

An effective way to implement this fuzzy accuracy assessment method is through the use of a reference data collection form. All the thematic map classes can be listed on the form, and the reference data collector can then indicate which map class they believe is the correct one. The collector would also have the option then of indicating if any of the other map classes would be an acceptable label for that particular sampling unit and also which would be not acceptable. This approach would work equally well if the collector was in the field making measurements or observations from the center of the sampling unit or if they were interpreting some imagery to label the reference data. Clearly, making measurements in the field to label a certain sample unit should result in less fuzziness (i.e., fewer acceptable labels included) than reference data collected by the interpretation of some medium-scale imagery. Regardless of the reference data collection approach, producing a fuzzy error matrix with its counterpart, the deterministic error matrix does provide an effective way of compensating for fuzziness in the classification process while also giving the information contained within the traditional error matrix.

29.4.5.2 Area-Based Error Matrix Generation

Historically, most land cover/vegetation/thematic maps created from digital imagery have analyzed the individual image pixels. More recently, a new method called object-based image analysis (OBIA) has become increasingly popular. In OBIA, pixels are grouped together into segments or objects and classified together instead of as individual pixels. This approach more closely mimics human interpretation and incorporates significant additional information about the object than was possible for the individual pixel (Blaschke, 2010). With a pixel-based map, it is very appropriate to take a cluster of pixels as the sample unit for the reference data collection used in the error matrix generation. As previously described in this chapter, for medium-resolution imagery such as Landsat, a 3 × 3 cluster of pixels is used as a single reference sample unit. Therefore, since all the sample units are exactly the same size, the error matrix is generated by simply tallying each sample unit as either correct (on the major diagonal of the error matrix) or incorrect (in the appropriate off-diagonal cell). However, for assessing the accuracy of maps created using OBIA, objects of various sizes have been created to replace the equal-area pixels. It is possible then to use these objects as the reference sampling unit (i.e., polygons). If this is done, the sampling units are no longer equal areas and therefore cannot simply be tallied in the error matrix (Radoux et al., 2010; MacLean and Congalton, 2012). Instead, the actual area of each of the reference sampling units must be incorporated into the error matrix, and the matrix becomes an area-based error matrix. In the traditional error matrix, each reference sample unit has equal weight since each has the same area. In the area-based error matrix, each reference sample unit is weighted by the area of the sample. Creating this area-based error matrix is no more difficult than the tally method used to create the traditional matrix accept that the areas are input into the matrix instead of just a tally. MacLean and Congalton (2012) present a comparison of these methods. Radoux et al. (2010) not only document these methods, but also show that some sampling efficiencies can be gained by knowing the sizes of all the objects in the map and using this information to reduce the number of sample units needed to create the matrix.

It should be noted that area-weighted error matrices are possible even with pixel-based classifications if different-sized clusters of pixels are selected for the reference sampling units. Also, it is possible to use a cluster-based reference sampling unit for OBIA-based maps. However, the unequal area of the objects in an OBIA-based map more readily leads one to conduct an area-based accuracy assessment and, therefore, this technique is described in this part of the chapter.

29.5 Conclusions

This chapter has presented a discussion of the many considerations necessary to conduct a valid accuracy assessment. Both positional error and thematic error must be carefully considered as an error if one can result in an error in the other. While there is no definitive step-by-step method for conducting a valid assessment, it is clear that the assessment must be carefully thought out and planned from the very beginning of the mapping project. All decisions regarding how the assessment is conducted must be thoroughly documented so that the process can be easily understood and reproduced by anyone interested in the map.

Accuracy assessment is an essential component of any mapping project. However, accuracy assessment is also an extremely expensive component. Therefore, every effort must be made to balance statistical validity with what is practically attainable. Improper assessments can cost the entire budget for the project. Poorly designed assessments can turn out to be statistically invalid causing the assessment to not be worth the effort. However, with care, proper consideration, and wise planning, accuracy assessment can be an extremely satisfying part of the project that not only shows the accuracy of the map, but also indicates where the map can be improved in the future.

References

Ager, T. March, 2004. *An Analysis of Metric Accuracy Definitions and Methods of Computation.* Unpublished memo prepared for the National Geospatial-Intelligence Agency. InnoVision, Washington, DC.

Aronoff, S. 1985. The minimum accuracy value as an index of classification accuracy. *Photogrammetric Engineering and Remote Sensing*, 51(1), 99–111.

ASPRS. 1989. ASPRS interim accuracy standards for large-scale maps. *Photogrammetric Engineering and Remote Sensing*, 54(7), 1038–1041.

ASPRS. May 24, 2004. *ASPRS Guidelines, Vertical Accuracy Reporting for LiDAR Data.* American Society for Photogrammetry and Remote Sensing, Bethesda, MD.

Bishop, Y., S. Fienberg, and P. Holland. 1975. *Discrete Multivariate Analysis: Theory and Practice.* MIT Press, Cambridge, MA, 575pp.

Blaschke, T. 2010. Object based image analysis for remote sensing. *ISPRS Journal of Photogrammetry and Remote Sensing*, 65, 2–16.

Bolstad, P. 2005. *GIS Fundamentals*, 2nd edn. Eider Press, White Bear Lake, MN, 543pp.

Cliff, A.D. and J.K. Ord. 1973. *Spatial Autocorrelation.* Pion Limited, London, U.K., 178pp.

Cohen, J. 1960. A coefficient of agreement for nominal scales. *Educational and Psychological Measurement*, 20(1), 37–40.

Colwell, R. N. 1955. The PI picture in 1955. *Photogrammetric Engineering*, 21(5), 720–724.

Congalton, R. G. 1988a. Using spatial autocorrelation analysis to explore errors in maps generated from remotely sensed data. *Photogrammetric Engineering and Remote Sensing*, 54(5), 587–592.

Congalton, R. G. 1988b. A comparison of sampling schemes used in generating error matrices for assessing the accuracy of maps generated from remotely sensed data. *Photogrammetric Engineering and Remote Sensing*, 54(5), 593–600.

Congalton, R. 1991. A review of assessing the accuracy of classifications of remotely sensed data. *Remote Sensing of Environment*, 37, 35–46.

Congalton, R. 2001. Accuracy assessment and validation of remotely sensed and other spatial information. *The International Journal of Wildland Fire*, 10, 321–328.

Congalton, R. 2009. Accuracy and error analysis of global and local maps: Lessons learned and future considerations. In: *Remote Sensing of Global Croplands for Food Security*, P. Thenkabail, J. Lyon, H. Turral, and C. Biradar (eds.). CRC/Taylor & Francis, Boca Raton, FL, pp. 441–458.

Congalton, R. and M. Brennan. 1999. Error in remotely sensed data analysis: Evaluation and reduction. *Proceedings of the Sixty Fifth Annual Meeting of the American Society of Photogrammetry and Remote Sensing.* Portland, OR, pp. 729–732 (CD-ROM).

Congalton, R. and K. Green. 1999. *Assessing the Accuracy of Remotely Sensed Data: Principles and Practices.* CRC/Lewis Press, Boca Raton, FL, 137pp.

Congalton, R. and K. Green. 2009. *Assessing the Accuracy of Remotely Sensed Data: Principles and Practices*, 2nd edn. CRC/Taylor & Francis, Boca Raton, FL, 183pp.

Congalton, R. G., R. G. Oderwald, and R. A. Mead. 1983. Assessing Landsat classification accuracy using discrete multivariate statistical techniques. *Photogrammetric Engineering and Remote Sensing*, 49(12), 1671–1678.

Federal Emergency Management Agency (FEMA). 2003. *Guidelines and Specifications for Flood Hazard Mapping Partners.* http://www.fema.gov/fhm/dl_cgs.shtm. Accessed 3/25/15.

Federal Geographic Data Committee, Subcommittee for Base Cartographic Data. 1998. Geospatial Positioning Accuracy Standards. Part 3: National Standard for Spatial Data Accuracy. FGDC-STD-007.3-1998, Federal Geographic Data Committee, Washington, DC, 24pp.

Ginevan, M. E. 1979. Testing land-use map accuracy: Another look. *Photogrammetric Engineering and Remote Sensing*, 45(10), 1371–1377.

Gopal, S. and C. Woodcock. 1994. Theory and methods for accuracy assessment of thematic maps using fuzzy sets. *Photogrammetric Engineering and Remote Sensing*, 60(2), 181–188.

Green, K. and R. Congalton. 2004. An error matrix approach to fuzzy accuracy assessment: The NIMA Geocover project. A peer-reviewed chapter. In: *Remote Sensing and GIS Accuracy Assessment*, R. S. Lunetta and J. G. Lyon (eds.). CRC Press, Boca Raton, FL, 304pp.

Greenwalt, C. and S. Melvin. 1962 and 1968. *Principles of Error Theory and Cartographic Applications*, ACIC Technical Report Number 96. United States Air Force, Aeronautical Chart and Information Center, St. Louis, MO, 60pp. Plus appendices. This report is cited in the ASPRS standards as ACIC, 1962.

Hay, A. M. 1979. Sampling designs to test land-use map accuracy. *Photogrammetric Engineering and Remote Sensing*, 45(4), 529–533.

Hord, R. M. and W. Brooner. 1976. Land-use map accuracy criteria. *Photogrammetric Engineering and Remote Sensing*, 42(5), 671–677.

Hudson, W. and C. Ramm. 1987. Correct formulation of the kappa coefficient of agreement. *Photogrammetric Engineering and Remote Sensing*, 53(4), 421–422.

Katz, A. H. 1952. Photogrammetry needs statistics. *Photogrammetric Engineering*, 18(3), 536–542.

Landis, J. and G. Koch. 1977. The measurement of observer agreement for categorical data. *Biometrics*, 33, 159–174.

Lunetta, R., R. Congalton, L. Fenstermaker, J. Jensen, K. McGwire, and L. Tinney. 1991. Remote sensing and geographic information system data integration: Error sources and research issues. *Photogrammetric Engineering and Remote Sensing*, 57(6), 677–687.

MacLean, M., M. Campbell, D. Maynard, M. Ducey, and R. Congalton. 2013. Requirements for labeling forest polygons in an object-based image analysis classification. *International Journal of Remote Sensing*, 34(7), 2531–2547.

MacLean, M. and R. Congalton. 2012. Map accuracy assessment issues when using an object-oriented approach. *Proceedings of the Annual Meeting of the American Society of Photogrammetry and Remote Sensing*. Sacramento, CA, 5pp. (CD-ROM).

MacLean, M. and R. Congalton. 2013. Applicability of multi-date land cover mapping using Landsat 5 TM imagery in the Northeastern US. *Photogrammetric Engineering and Remote Sensing*, 79(4), 359–368.

Naesset, E. 1996. Conditional tau coefficient for assessment of producer's accuracy of classified remotely sensed data. *ISPRS Journal of Photogrammetry and Remote Sensing*, 51(2), 91–98.

NDEP. May 10, 2004. *Guidelines for Digital Elevation Data*. Vertion 1.0. National Digital Elevation Program, Washington, DC.

Pontius, G. and M. Millones. 2011. Death to kappa: Birth of quantity disagreement and allocation disagreement for accuracy assessment. *International Journal of Remote Sensing*, 32(15), 4407–4429.

Pugh, S. and R. Congalton. 2001. Applying spatial autocorrelation analysis to evaluate error in New England forest cover type maps derived from Landsat Thematic Mapper Data. *Photogrammetric Engineering and Remote Sensing*, 67(5), 613–620.

Radoux, J., R. Bogaert, D. Fasbender, and P. Defourny. 2010. Thematic accuracy assessment of geographic object-based image classification. *International Journal of Geographical Information Science*, 25(6), 895–911.

Rosenfield, G. H. and K. Fitzpatrick-Lins. 1986. A coefficient of agreement as a measure of thematic classification accuracy. *Photogrammetric Engineering and Remote Sensing*, 52(2), 223–227.

Rosenfield, G. H., K. Fitzpatrick-Lins, and H. Ling. 1982. Sampling for thematic map accuracy testing. *Photogrammetric Engineering and Remote Sensing*, 48(1), 131–137.

Sammi, J. C. 1950. The application of statistics to photogrammetry. *Photogrammetric Engineering*, 16(5), 681–685.

Story, M. and Congalton, R. 1986. Accuracy assessment: A user's perspective. *Photogrammetric Engineering and Remote Sensing*, 52(3), 397–399.

Tortora, R. 1978. A note on sample size estimation for multinomial populations. *The American Statistician*, 32(3), 100–102.

United States Bureau of the Budget. 1941/1947. National Map Accuracy Standards. Washington, DC. http://nationalmap.gov/standards/pdf/NMAS647.PDF. Accessed March 25, 2015.

van Genderen, J. L. and B. F. Lock. 1977. Testing land use map accuracy. *Photogrammetric Engineering and Remote Sensing*, 43(9), 1135–1137.

Young, H. E. 1955. The need for quantitative evaluation of the photo interpretation system. *Photogrammetric Engineering*, 21(5), 712–714.

Young, H. E. and E. G. Stoeckler. 1956. Quantitative evaluation of photo interpretation mapping. *Photogrammetric Engineering*, 22(1), 137–143.

Zhenkui, M. and R. Redmond. 1995. Tau coefficients for accuracy assessment of remote sensing data. *Photogrammetric Engineering and Remote Sensing*, 61(4), 435–439.

Space Law and Remote Sensing

30

Remote Sensing Law: An Overview of Its Development and Its Trajectory in the Global Context

P.J. Blount
University of Mississippi School of Law

Acronyms and Definitions

CD	Conference on Disarmament
EO	Earth observation
GNSS	Global Navigation Satellite System
INPE	National Institute for Space Research (Brazil)
NTM	National Technical Means
UNCOPUOS	United Nations Committee on the Peaceful Uses of Outer Space
UNGA	United Nations General Assembly

30.1 Introduction

The law relating to remote sensing is a complex mix of international and national laws, regulations, policies, and agreements. This body of remote sensing law has numerous facets that regulate a number of uses of remote sensing satellites, related technology, and collected data. This chapter will serve as a survey of the laws and regulations that governs remote sensing activities across a variety of applications and a variety of legal spaces.

This overview of remote sensing law will assert that there are essentially four bodies of law that govern remote sensing and define its regulatory regime. These are not distinct bodies; they overlap and intersect with each other. The typology is not intended to obscure these interactions but instead to serve as an explanatory framework. This chapter will address each of these bodies of law chronologically in the order in which each emerged, though it will not attempt to identify the specific temporal locations due to the fluidity and subjectivity of such assignments. The chapter begins with a brief examination of relevant international space law as developed in the space treaty regime. Next, it will turn to remote sensing law as developed in relation to military uses with a particular focus on disarmament treaty verification. Then, it will address international remote sensing law applicable to civil remote sensing satellites as articulated in the United Nations Principles on Remote Sensing (UNGA Res. 41/65, 1986). Finally, it will argue that remote sensing law is increasingly being subsumed into the developing field of geospatial law, which is resulting from the processes of commercialization, globalization, and technological convergence. The purpose of this exercise will be to trace a trend of increasing adoption of domestic law regimes that intersect with international governance systems (Gabrynowicz, 2007, 7). These laws are in a constant state of flux as a result of rapid advancements in technology and different perceptions on how we manage space data gathering that is agreeable to all.

The thrust of this chapter will be to address laws and regulations specific to remote sensing activities. Many of the provisions of the Outer Space Treaty apply to all space activities including remote sensing activities, and reference will be made to these provisions as a whole, but in this chapter, we will endeavor to focus our attention on legal provisions that are specific to remote sensing activities. The goal will be to highlight the regulations that are most relevant to remote sensing activities and the end uses of remotely sensed data. Finally, it should be noted that while this survey of remote sensing law seeks to give a broad overview of the law, it is not by any means comprehensive, and the reader's attention is drawn to the footnotes for further readings and elaboration.

30.2 Four Pillars of Remote Sensing Law

Remote sensing law is a multifaceted legal regime that draws on a number of regulatory systems. This chapter will attempt to address the four core regulatory systems, namely, traditional space law, disarmament law, international remote sensing law, and geospatial law. These regimes should be understood as separate regulatory spaces that interact with each other in a variety of ways. Indeed, depending on the issue being addressed, they can reorient themselves to the others; it would be folly to suggest that they are always oriented in a specified way. These regimes shift from discrete and separate to hierarchical to overlapping as the context of their application changes. The best description would be that their relationship is discursive in nature. It is system of regulatory feedback loops across different governance spaces (see Figure 30.1). For example, international law creates obligations for states, but state implementation of these obligations often leads to influencing the content of the obligation (Blount, 2012).

This chapter will proceed by addressing these varying regimes as discrete, but will endeavor to highlight the areas wherein we can observe their interactions. It should not be assumed that at any given moment, one regime maintains primacy over another or that the delineations among them are clear by any means. Instead, this serves as a framework for understanding the overall governance system for remote sensing technologies.

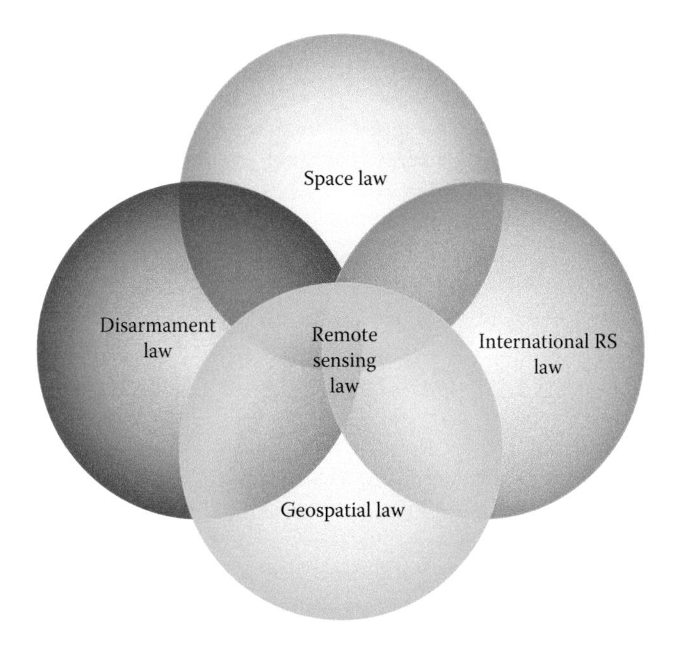

FIGURE 30.1 Remote sensing law is the confluence of multiple legal regimes.

30.2.1 Space Treaty Regime

The core body of international space law is found in four multilateral treaties (Treaty on Principles Governing the Activities of States, 1967; Agreement on the Rescue of Astronauts, 1968; Convention on International Liability, 1972; Convention on Registration, 1976; on the space-law-making process, see Dembling and Arons, 1967; Marchisio, 2005). These treaties set the parameters for lawful conduct by states when engaging in outer space activities. There are several introductions to these core treaties; therefore, this chapter will refrain from a full evaluation of these and instead focus on provisions of special concern to remote sensing satellites.*

Article I of the Outer Space Treaty grants all states free access to outer space, without interference by other states. By coupling freedom of access with the principle of nonappropriation found in Article II, which places space outside the sovereign territory of all states, the negotiators of the Outer Space Treaty enshrined the principle of freedom of overflight by space objects over other state's territories (Vásquez, 1965; Jahku, 2003, 73–77). This is of critical importance to remote sensing activities, since it allows states to observe other states from the nonsovereign vantage point of space. Since no state has ever protested the overflight of its territory by other states space object, overflight is a customary norm that preexists the Outer Space Treaty (Jahku, 2003, 73–74).

Article VI of the Outer Space Treaty places a duty on states to "authorize and supervise" nongovernmental actors and gives states "international responsibility" for nongovernmental actors. This means that many states have adopted full regulatory regimes in order to manage the risk of commercial actors. These regimes, which take the form of licensing regimes, are meant to ensure government oversight over nongovernmental actors. A primary concern with many of these licensing regimes is to ensure that the nongovernmental satellite operators comply with minimum standards of conduct in its activities, so that the state avoids situations of international responsibility. These regimes emphasize "cradle to grave" operations plans that are meant to reduce liability issues as well mitigate space debris.† These licenses are often attached to the act of launching.‡ Some states, however, have licensing regimes specifically for remote sensing activities that ensure compliance with national security and global data policies.§ These regimes will be addressed in Section 30.2.3.1.

* For a variety of approaches, see Diederiks-Verschoor (1999), Lyall and Larsen (2013), Reynolds and Merges (1997), and Lachs (1972, 2010).

† For instance, one of the primary goals of the Canadian licensing regime is to regulate "the operation of the satellite itself" (Gillon, 2008, 26–27). See also, Mann (2008, 69–71).

‡ For instance, the Chinese regime is based on launch licenses (Chinese Law, 2007).

§ Canada, Remote Sensing Space Systems Act, Gillon, "Regulating Remote Sensing," 26 (noting a second concern of the Canadian licensing regime is "the distribution of raw data and remote sensing products produced by satellites"); Germany, Satellite Data Security Act (2008); and the United States, Licensing of Private Remote Sensing Systems (2013).

There are numerous information-sharing clauses in the Outer Space Treaty, which read together lead to an overall duty to warn other states of situations of impending disaster. Most relevant is Article IX's requirement that states engaging in space activities be guided by "the principles of cooperation and mutual assistance."* While the information-sharing provisions are not directly related to remote sensing, they build strong evidence that there is a customary norm of information sharing in a wide variety of contexts as a way to increase cooperation in the use space "for the benefit of all mankind." This norm is further articulated in the Remote Sensing Principles (Section 30.2.3, *infra*).

Finally, reference should be made to the liability provision found in Article VII of the Outer Space Treaty and further articulated in the Liability Convention. Whether damage caused by remote sensing data might represent "damage caused by [a state's] space object" is a contested interpretation of the treaty.† However, this author would argue that it is an increasingly irrelevant debate. While parsing the language of the treaty is an excellent exercise, state practice is a better touchstone for determining the practicality of the argument. State reluctance to apply the Liability Convention in clear cases such as *Cosmos 954* and the *Cosmos-Iridium* collision points to the likelihood that damage as a result of data presents a weak case for the application of the Liability Convention (see generally, Listner, 2012). In this case, state practice indicates that such damage would most likely sit outside the scope of damage defined in the Outer Space Treaty and the Liability Convention, though this is admittedly an argument based less on strict legal analysis and more on a constructivist view of norm creation and definition, in which practice identifies the content of the norm.‡

30.2.2 Disarmament and Verification Law

The space race and the resulting international law of outer space are a product of the Cold War; the law's early development was linked to both technological diplomacy and military uses (see Gabrynowicz, 2004). The launch systems represented ballistic missile technology; the communications applications could support the emerging concept of a globally projected military; and remote sensing could open up intelligence secrets about the other sides weapons.§ While one of the core guiding principles and central legal obligations in space law is the use of space for peaceful purposes, it is well established that this allows states to use space technologies in accordance with *peaceful* as set out in the UN Charter as a *de minimis* threshold (Petras, 2005, 88–90; Blount, 2009). As a result, states are considered to be engaging in peaceful activities when they deploy and use defensive technologies in a nonaggressive way; this is supported by the fact that until the launch of *Landsat-1*, "remote sensing was exclusively developed and used for military purposes" (Jahku, 2003, 70).

This use of remote sensing was critical in the negotiation of disarmament and arms control agreements between the United States and the Soviet Union. "[F]reedom of space" was "perhaps the principal aim of U.S. space policy throughout the first ten years of the space age" (Petras, 2005, 86). The technology allowed the two states to overcome the problem of verification. The Soviets rejected verification by on-site visits or overflights, and the United States rejected any agreement without reliable verification measures (Vásquez, 1965, 163–164). Once it was clear that both sides were using satellites for remote sensing of the other's territory, they were able to agree on satellite technology as the core method of verification.¶ While some states have challenged the military and intelligence remote sensing regimes as nonpeaceful, the "international acceptance of the legitimacy of intelligence gathering from outer space was effectively secured with the enshrinement of space-based photoreconnaissance in the Cold War strategic arms control regime" (Petras, 2005, 89–90). This was made possible by a legal structure, articulated in the Outer Space Treaty and embodied in customary international law, that ensured a nearly complete freedom of access and use outside of the earth's atmosphere. As a result, there was no way either state could object to the use of satellites to sense its territory, because such activity was "consistent with international law" (Petras, 2005, 87). These technologies were indicated in disarmament agreements under the term National Technical Means (NTM),** a phrase that would remain officially undefined until its meaning was revealed in a speech by president Carter (1978).††

The use of satellites as the central method of verification has implications for space security, since treaties that use NTM often require that parties to the treaty not interfere with the other's NTM (Diederiks-Verschoor, 1999, 85; Petras, 2005, 91). Arguably, this helped to solidify space security by creating a *de facto* drawdown of space weaponry development, due to the risk that any interference with space objects could interfere with NTM and thereby offset the strategic stability these agreements provided (Harrison, 9). It can be argued that this limitation is becoming less relevant as space is populated by an increasingly diverse set of actors.

While there has been scholarship on expanding the verification possibilities of remote sensing satellites to other areas, little movement has been made. France made a proposal at the Conference on Disarmament (CD) in 1978 for an "international satellite body for monitoring disarmament treaties"

* Outer Space Treaty, Art. IX; see also Articles III, V, VIII, X, XI, and XII.

† Ito notes that the Outer Space Treaty and the Liability Convention "are silent as to the types of damage associated with satellite remote sensing" (Ito, 2008, 51).

‡ The author has developed and applied this type of analysis to other areas of space law. See generally, Blount (2009, 2011).

§ Legally these opportunities were facilitated by a "regime of complete freedom" (John et al., 2011).

¶ For a history of such a system, see generally, Richelson (1999).

** For example, SALT I & SALT II, Diederiks-Verschoor (1999, 84–85); ABM Treaty, Petras (2005, 90); CFE Treaty, Petras (2005, 92–93); and Treaty between the United States of America and the Russian Federation on Measures for the Further Reduction and Limitation of Strategic Offensive Arms (signed April 10, 2010) Art. X(b) and (c).

†† Importantly, Petras notes that NTM includes a wider range of technologies than just remote sensing satellites (Petras, 2005, 91).

(Diederiks-Verschoor, 1999, 84; Achilleas, 2008, 3). This measure was rejected by the CD, and there has been no progress in the CD on such a system. It is unlikely, in the near term, under current political conditions, that there is sufficient political will for such a system to be implemented. There is, however, a great deal of potential for remote sensing technologies to play a role in the international lawmaking toolkit for its verification and monitoring capabilities in a variety of areas outside disarmament such as human rights and the environment (see, for example, Onoda, 2005; Alvarado, 2010).

30.2.3 Remote Sensing Principles

The core of international legal principles dealing with remote sensing activities is contained in UN General Assembly (UNGA) Resolution 41/65, Principles Relating to Remote Sensing of the Earth from Outer Space. These principles are the "first and foundational source of policy guidance for remote sensing activities" (Uhlir et al., 2009, 209).

The Principles are contained in a UNGA resolution, and as such, The Principles themselves are not binding as legal principles.* However, to the extent that they represent customary international law, these principles are binding. The existence of customary international law is evinced by two criteria: *opinio juris* and state practice. *Opinio juris* represents a subjective belief by states that they are legally bound by a rule. In the case of the Remote Sensing Principles, *opinio juris* is found in the consensus process of negotiation at the United Nations Committee on the Peaceful Uses of Outer Space (UNCOPUOS) and by the unanimous adoption by the General Assembly.† In light of the fact that there are no denunciations of The Principles, that states often appeal to The Principles in their domestic laws, and that states make claims of compliance with The Principles, a sufficient level of acceptance seems to have been achieved to constitute *opinio juris*.‡

The second evidentiary parameter that a rule constitutes a binding customary norm requires that states indeed follow the rule they believe to be binding. State practice, essentially, helps define the content of the norm. If states are found to be claiming compliance and are doing so in good faith, then their compliance becomes an epistemic unit informing that norm's interpretation. In other words, when states think they are complying with a norm, their actions construct the meaning and content of that norm.§ Since states may disagree as to the exact content of a norm's meaning, customary norms are often better defined along a spectrum of acceptable state practice.¶ In this analysis, the key question is not if states are complying, but instead, how states are complying. This is because "[c]ustom as a source of international law, leads to the recognition of the legality of the existing practice if there is general consent … to the observable rule of conduct" (Vereschetin and Danilenko, 1985, 30).

The following inquiry will address the Remote Sensing Principles in two different groupings. First, it will identify principles that are substantively based in preexisting binding legal principles. It will then address the *legal innovations* of the UN Principles, which articulated new content into the international law governing remote sensing. These innovations will be evaluated by examining the spectrum of state practice to illustrate acceptable state conduct that defines the content of these norms.

30.2.3.1 Principles Based on Preexisting Legal Obligations

Numerous of The Principles appeal to preexisting legal obligations that can be found in general international law or in the *lex specialis* of international space law. Essentially, these principles makes explicit the application of more general legal standards explicit that arguably would apply regardless of the acknowledgment of such in The Principles. There are three different groupings of these: those that implement general international law, those that implement international space law, and those that seek to increase international cooperation.

30.2.3.1.1 General International Law: Principles III, IV, and XV

Principle III makes the application of general international law to remote sensing activities explicit.** This principle makes specific reference to three treaty instruments: the UN Charter, the Outer Space Treaty, and International Telecommunications Union Instruments.†† This is explicit acknowledgment that these treaties represent preexisting law and that the principles should not be read in such a way as to conflict with those instruments. Principle III also acknowledges that other general international

* On the legal status of the UN Principles, see generally, Terekov (1997, 100–101) and Kopal (1988, 14–20).

† See UNGA Meeting Record no. A/41/PV.95 (1986). Some commentators argue that this rises to the level of "instant custom." For instance, Jackson (1997, 873). Additionally, Hosenball notes that the language evinces that states made a "commitment to live by these *Principles*" (Gabrynowicz, 2002, 35).

‡ For instance, the UN Principles are the "guiding principle for the practice of Earth remote sensing and was, accordingly, considered in the preparation" of the German satellite data law (Schmidt-Tedd and Kroyman, 2008, 105). For the United States and Japan as evidence of custom, see Gabrynowicz (2002, 42). The nondiscriminatory access principle has repeatedly been incorporated in the U.S. law and in international agreements (Gabrynowicz, 2007, 6). Jahku (2003, 86–88) notes that there is a divide on the issue, but that Principle XII at least creates a legal obligation.

§ So, for instance, nondiscriminatory access is being "more narrowly construed" as a norm due to state practice that has expanded national security exceptions over data (Gabrynowicz, 2007, 11).

¶ For instance, while all jurisdictions recognize a crime of "murder," the elements of the crime changes as jurisdictional borders are crossed. As a result, we can say that the prohibition against murder is a general principle, but that the specific content of that principle exists in a spectrum of meaning. Customary norms suffer from a similar lack of clarity and require a spectrum analysis (if you'll allow me the pun). Furthermore, this type of analysis is particularly apt in high-technology cases where there is unequal distribution of access coupled with wide acceptance of negotiated rules.

** Compare to the Outer Space Treaty, Article III.

†† UN Principles, III. The Outer Space Treaty is also referenced in Principle IV.

law principles apply, so, for instance, the law of armed conflict, private international law principles, and environmental law principles could all be applied to space in the proper context. While this acknowledgment is important, it is hardly necessary, since The Principles are embodied in a document defined under international law, and the form reifies the notion that its content exists within those parameters.

While referencing a preexisting right to the freedom of use of outer space, Principle IV specifically notes the that remote sensing activities shall be

> conducted on the basis of respect for the principle of full and permanent sovereignty of all States and peoples over their own wealth and natural resources, with due regard to the rights and interests, in accordance with international law, of other States and entities under their jurisdiction. Such activities shall not be conducted in a manner detrimental to the legitimate rights and interests of the sensed State.*

This principle seeks to reaffirm "the freedom data collection and distribution," while at the same time ensuring "respect of the principle of full and permanent sovereignty of all States and peoples over their own wealth and natural resources" (Achilleas, 2008, 2). The content of this norm is consistent with the principles of nonintervention and sovereign equality that sit at the heart of the international legal system.† The formulation in The Principles reflects the concern of developing nations that developed nations would use satellite technology to exploit the resources of the developing world and advance economic imperialism (Sekhula, 2011, 233).‡ However, as addressed earlier, the right (Ling, 2010, 436–437) to overflight had long been established, and "such protestations have never adversely affected operational progress" (Diederiks-Verschoor, 1999, 83). Once deployed, technology is difficult to withdraw, and as a result, the developing nations settled for a compromise in the requirements for data distribution (see Section 30.2.3.2). This principle cannot be read as extending or elaborating on the concept of sovereignty as already established in international law and, in no way, requires a state to "seek consent or authorization" from a state it intends on sensing (Jahku, 2003, 79).

It is worth noting that the intersection of remote sensing and sovereignty could become more relevant in the future if the principle of responsibility to protect gains traction as a justification for military intervention for gross human rights violation (UNGA Res. 60/1, 2005).§ In such scenarios, the earth observation data could be used to expose the violations that give rise to a legitimate international action. In this capacity, remote sensing as a global technology can continue to play a role in the way in which the international community constructs the idea of sovereignty. Indeed, there are current efforts to leverage remote sensing data to expose gross human rights violations such as Sudan (Satellite Sentinel Project, 2014), Darfur (Eyes on Darfur, 2014), and North Korea.¶

Finally, Principle XV states that disputes will be settled in accordance with the norm of the peaceful settlement of disputes, which is previously enshrined in Article 1 of the UN Charter.**

30.2.3.1.2 General Space Law: Principles IX and XIV

As noted earlier, the Remote Sensing Principles acknowledge the applicability of the Outer Space Treaty to remote sensing activities, and The Principles reference the treaty throughout. Two provisions, though, are concerned explicitly with general space law provisions.

Principle IX affirms the registration obligation found in the Registration Convention†† as well as the Article IX obligation to provide information on request of affected states.‡‡ It lowers the threshold for requesting such information from the Article IX standard of "harmful interference" to "affected" states. However, the scope of the obligation is likely the same in practice, where relevant information will likely be readily available for civil systems and scarce for military systems (see Section 30.2.3.2, *infra*).

Principle XIV incorporates Article VI of the Outer Space Treaty, which requires supervision and authorization of remote sensing activities, which has already been discussed earlier. It also acknowledges that in no way does the Article VI obligation preclude the application of the law of state responsibility to remote sensing activities (see generally, International Law Commission, 2001). Though not explicitly stated in the Outer Space Treaty, the customary law of state responsibility is certainly applicable to states as they engage in space activities, so this can hardly be seen as an innovation.

30.2.3.1.3 International Cooperation: Principles II, V, VI, VII, VIII, and XIII

A large portion of international space law is dedicated to promoting the soft obligation of international cooperation. It is soft because the obligation is always defined by terms to be negotiated by the states engaging in the cooperation. Additionally, while it can be said that there is an obligation to engage in international cooperation, there is no corresponding affirmative right

* UN Principles, IV.

† UN Charter, Art. 1 and 2.

‡ Such contentions reflect current critiques of neoliberalism, for instance, see Harvey (2005) ("Information technology is the privileged technology of neoliberalism").

§ R2P is an emerging norm that supports broader ability of the international community to intervene when states are committing gross human rights violations against their own citizens, such as genocide. *Id.* para. 138–140.

¶ United Nations High Commission on Human RIghts (2014) (relying on satellite data as evidence).

** UN Charter, Art. 1.

†† Outer Space Treaty, Art. VII; Registration Convention, Art. II; and UNGA Res. 62/102 (2007). See generally, Schrogl and Hedman (2008). It should not be assumed that all states necessarily comply with registration requirements, for a remote sensing specific case, see Mayence (2008, 92–94).

‡‡ Outer Space Treaty, Art. IX.

to international cooperation. The following provisions are continuations of this cooperation framework.

Principle II states that remote sensing should be done for the benefit of all nations and "taking into particular consideration the needs of the developing countries."* This principle is an extension of Article I of the Outer Space Treaty that states that space activities "shall be carried out for the benefit and in the interests of all countries, irrespective of their degree of economic or scientific development."† The Remote Sensing Principles add an additional obligation to take into consideration the special needs of developing countries. At best though, this obligation is weak. The Outer Space Law regime puts strong emphasis on cooperative activities in space, but the terms and extent of the cooperation are, by design, entirely up to the states involved. As a result, the addition of the additional clause on developing countries represents the extension of an obligation that is more akin to an international policy statement. However, it is an important statement, since "[i]n developing nations… these technologies have proven essential for developing public policies on issues such as deforestation assessment and management, urban planning, agricultural production, and environmental assessment" (Ferreira and Camara, 2008).

That is not to say that states willfully ignore this obligation. On the contrary, there is abundant evidence that states actively engage in such cooperation.‡ The need to extend an extra obligation to developing countries, which are burdened by technological lag, has been adopted as a set of principles in UNGA Resolution 51/122, The Declaration on International Cooperation in the Exploration and Use of Outer Space for the Benefit and in the Interest of All States, Taking into Particular Account the Needs of Developing Countries (UNGA Res. 51/122, 1996; see also, Carpanelli and Cohen, 2012). This resolution represents international consensus that developing countries should have access to space technologies, but the principles enunciated give states broad leeway to determine the extent to which they will engage in such cooperation. This certainly makes the terms of this principle very aspirational.

Despite the aspirational nature, such principles of conduct have led to an expansion of space technologies to developing nations, and "remote sensing is no longer a technology, which only superpowers can enjoy; it is now getting diffused among states around the globe, including emerging and developing ones" (Kuriyama, 2010). Specifically, in the field of remote sensing, states have sought to assist developing nations through their data policies, which will be addressed in depth later (Section 30.2.3.2.3). Japan, for instance, seeks to make a "safe, secure, and affluent society through the establishment of an effective space-based data use policy" (Aoki, 2010, 337). Another mode of assistance is allowing direct access to the satellite data via ground stations. A great example of this is the highly successful *Landsat* program that the United States implemented, which allowed any state to build and receive the raw data from the *Landsat* system.§

Principles V, VI, VII, VIII, and XIII all incorporate the general obligation for international cooperation in space activities and can arguably be read more as a list of possible cooperative activities. Principle V makes a general appeal for cooperation, which echoes the "principle of cooperation" found in the Outer Space Treaty's Article IX.¶ Principle VI puts forth a specific example of a type of cooperation in the form of data receiving and processing facilities within the framework of regional agreements,** and Principle VII encourages states to exchange technical assistance on "mutually agreed" terms. Principle VIII requires the United Nations to promote international cooperation. Article XIII gives sensed states a right to request consultations to seek cooperative opportunities with the sensing state.†† Each of these principles sets forth cooperation in relatively soft terms under which states can pursue international cooperation as they see fit.

30.2.3.2 Legal Innovations

The following principles represent legal innovations, the content of which is defined by state practice. These principles added new substance to the legal framework governing state remote sensing activities and as such represent legal innovations.

30.2.3.2.1 Definitions: Principle I

The Principles begin with a definitions section that serves the scope of the application of the principles. The principles begin by defining remote sensing as "the sensing of the Earth's surface from space by making use of the properties of electromagnetic waves emitted, reflected or diffracted by the sensed objects, for the purpose of improving natural resources management, land use and the protection of the environment."‡‡ This is a twofold

* UN Principles, II.

† Outer Space Treaty, Art. I.

‡ Diederiks-Verschoor (1999, 83) (noting a trend toward increased cooperation).

§ For instance, Iran maintains a receiving station for Landsat data (Tarikhi, 2008).

¶ Outer Space Treaty, Art. IX; Diederiks-Vershoor (1999, 82) ("Principle V strengthens the other provisions of the Space Treaty's Article IX").

** An example of such cooperation would be the Landsat Ground Station Operators' Working Group, which "attempt[s] to address … how data policies should be formulated" (Graham and Gabrynowicz, 2002a, 294–295). On ground station policy and operations, see generally, Gabrynowicz (1997, 229).

†† It should be noted that this consultation request is different in scope than the consultations afforded by Article IX of the Outer Space Treaty. The Outer Space Treaty's consultation provision is triggered by harmful interference, whereas Principle XIII is triggered by remote sensing activities (as defined in the principles) and an opportunity for cooperation between the two states. While arguably, this expansion could have been included in the following innovations section, it is included here due to both its softness as an obligation and its strong connection to the use of space for international cooperation. Additionally, tracking compliance and use of this principle is difficult due to the informal nature of "consultations."

‡‡ UN Principles, I(a). This definition is legal in nature and is lacking as a technical definition. Vasquez and Lara suggest—as a definition to help bridge the gap between law and policy makers and scientists—that remote sensing is "a way of obtaining information about an object by analyzing the data acquired through a device that is not in physical contact with said object" (Vásquez and Lara, 2010, 367).

definition. It incorporates both a technical and intent parameters. The first element describes in a technical sense what remote sensing activities will be governed by The Principles (so, for instance, GNSS is excluded). The second element, though, limits the scope of application by making the principles applicable only to activities of a specific nature, which is the political dimension of the definition. Specifically, this definition purposely places military remote sensing beyond the pale of The Principle's applicability.*

Next, The Principles define three types of information, making distinction among primary data, processed data, and analyzed information. These definitions are consistent with technical usage. Primary data are the "raw data that are acquired by the remote sensors."† Processed data are value-added data that have been processed into a usable form, and analyzed information is the knowledge garnered when remote sensing data are interpreted along with other inputs (Diederiks-Verschoor, 1999).‡

Finally, Principle I defines remote sensing activities as "the operation of remote sensing space systems, primary data collection and storage stations, and activities in processing, interpreting and disseminating the processed data."§ The significance of this definition is its expansiveness in including not only ground-link stations, but also the data analysis and dissemination. This holistic definition is important as it scopes these activities sufficiently broad to allow The Principles to articulate obligations in regard to data access.

30.2.3.2.2 Environmental and Disaster Provisions: Principles X and XI

Principles X and XI are environmental provisions. Principle X states that remote sensing should be used to protect the earth's environment and that data collected that can help to avert environmental degradation. This acknowledges that "remote sensing data can be used to assess and locate damage, monitor the progression and effect of corrective measures, verify the application of environmental treaties and assist in the response to man-made and natural disasters" (Carpanelli and Force, 2011). Article XI on the other hand seeks to encourage states to exchange data in the wake of natural disasters as a way to alleviate suffering occurring in affected areas. Both of these are data-sharing provisions and reflect core values found in Article IX of the Outer Space Treaty.¶ These principles support a "*general duty to inform*" other states of environmental damage (Carpanelli and Force, 2011, 33).**

Principle X finds support in the "general principles of environmental and human rights law" (Carpanelli and Force, 2011). It first requires that states use remote sensing to *promote* the protection of the environment, which is a fairly broad, yet soft, requirement.†† Then, it requires states to share "information... that is capable of averting any phenomena harmful to the Earth's natural environment." The second obligation is surprisingly narrow in application. It requires only that these data should be shared when the information is "*capable of averting*" the environmental harm. Post-incident information as well as cumulative data that might give indications as to environmental trends is not included in this sharing regime.‡‡ While compliance with this provision is difficult to monitor, it seems that the narrowness is likely unimportant. With the amount of available data and expanding access, states are increasingly able to get the information they need to manage the domestic environmental concerns.

Indeed, state practice indicates that states are actively engaging in environmental monitoring through remote sensing and increasingly making that data openly available. Japan has adopted the policy that remote sensing satellites should be used as "The Guardian of the Environment" (Aoki, 2010, 345, 348).§§ Korea uses remote sensing for "environmental protection purposes" (Lee, 2010, 419, 423–424), and China uses satellites for "environmental monitoring" (Ling, 2010, 439, 451–452). An EU directive mandates that "all information held by public authorities relating to imminent threats to human health or the environment is immediately disseminated to the public likely to be affected" (EU Directive 2003/4/EC, 2003; see Harris, 2008, 37). State practice will continue to evolve in relation to this principle as the international community seeks to deal with global environmental problems.¶¶ In particular, as the climate change debate intensifies, remote sensing technologies will likely be critical in both engineering and monitoring responses to climate change. These technologies will gather the evidence that will underlay the political debates surrounding the existence of, scope of, and responses to the problem. Additionally, it is likely that environmental treaties negotiated in the future will likely rely on satellite technology for verification purposes (see, for example, Onoda, 2005; Vásquez and Lara, 2010, 371–377; Froehlich, 2011, 221–222).

Principle XI is worded similarly to Principle X and also has a two-pronged obligation. First, it invokes a broad humanitarian obligation to *promote* the use of remote sensing technology to protect humans from natural disasters.*** The second prong is that states should promptly share *processed data and analyzed information* with states that have been or will soon be affected by

* See Graham and Gabrynowicz (2002a, 18–19) (quoting Neil Hosenball as stating that the UN Principles "do not apply to domestic military systems" and that the issue was never raised during the negotiations), Achilleas (2008, 2), Diederiks-Verschoor (1999, 86).

† Remote Sensing Principles, I.

‡ For more on the different definitions of data, see Jahku (2003, 66–67).

§ Remote Sensing Principles, I(e).

¶ Outer Space Treaty, Art. IX.

** Carpanelli and Force (2011, 34) argue that this particular provision is part of a dynamic process of customary international law development and that the extent to which it represents custom is still being defined. Onoda (2005, 341) also argues that there is "general obligation of international environmental law for States to cooperate in promotion of global Earth observation to protect the environment."

†† Remote Sensing Principles, X.

‡‡ It should be noted that meteorological data are "considered a public good used for the benefit of all" (Jahku, 2003, 84).

§§ For example, see Japan's *Greenhouse Gases Observation Satellite* (*GOSAT*) (Aoki, 2010, 338–340).

¶¶ "Global cooperation in environmental data is on the rise" (Oprong and Rwehumbiza, 2011, 252).

*** Remote Sensing Principles, XI.

a natural disaster. This is a much broader secondary obligation than that articulated under Principle X and is certainly geared toward the immediacy of the threat to life that disasters cause.

Principle XI has seen a great deal of state action. Several international systems have been established to facilitate the sharing of satellite data for disaster response, including the Disasters Charter (Charter on Cooperation, 2000), UN-SPIDER,* and Asia Sentinel (Sentinel Asia, 2014). One of the most prominent of these initiatives is The Charter on Space and Disaster Cooperation, the purpose of which is to "provide EO data at times of natural or technological disasters... to all authorised users free of charge" (Beets, 2010; Harris, 2013, 47). This agreement was established in 2000 and includes "a broad range of participants beyond Nation-States to enable pragmatic responses to a disaster" (Uhlir et al., 2009, 211). China has launched satellites for disaster management and is involved in both UN-SPIDER and Asia Sentinel (Ling, 2010, 439, 453). While there have been a number of successful programs, Ito argues that specific shortcomings in the principles themselves inhibit effective use of remote sensing technologies for disaster warning, response, and management (see generally, Ito, 2008).†

30.2.3.2.3 Nondiscriminatory Access: Principle XII

The most significant legal innovation in The Principles is nondiscriminatory access. Principle XII requires that sensed states be given "primary data and the processed data concerning the territory under its jurisdiction ... on a non-discriminatory basis and on reasonable cost terms."‡ Additionally, sensed states are given a right of access, on the same terms, to analyzed data that any state engaged in remote sensing activities may have. There are two keys to this obligation: "nondiscriminatory" and "reasonable cost terms." Nondiscriminatory has a "common interpretation ... that the sensing States have an obligation to provide the data to the sensed States under the same conditions as other States that wish to access the data" (Ito, 2008, 49–50). Reasonable cost basis, on the other hand, is "ambiguous and open to different interpretations," and "it in no way serves as a general guideline for price settings" (Ito, 2008, 50; see also, Gabrynowicz, 2002, 31; Oprong and Rwehumbiza, 2011, 257). Principle XII is often seen as a compromise between the developed and developing nations in the negotiating process, wherein data access is given in exchange for freedom to sense (Harris, 2008, 45; Schmidt-Tedd and Kroyman, 2008, 105).

Nondiscriminatory access has been implemented by a critical mass of states, and "[g]enerally speaking, all national data policies and laws contain the same fundamental principles" (Gabrynowicz, 2007, 3). Gabrynowicz notes that these principles are "making data available for scientific, social, and economic benefit and restricting access to some data for national security reasons," and that "[d]ifferences occur in variables" (Gabrynowicz, 2007).§ Significantly, data provisions (both the nondiscriminatory access mandates and the national security exceptions) are found in the licenses and agreements between states and private or commercial actors.¶

The policy of nondiscriminatory access was "driven by Cold War foreign policy" and finds its roots in the U.S. national policy, which had the "goal of influencing allies and non-aligned Nations by demonstrating technological superiority and encouraging them to use the data" (Gabrynowicz, 2007, 5).** The U.S. law and policy grants nondiscriminatory access to all data from government-funded civil satellites and provides that commercial satellite systems can use "reasonable commercial terms and conditions" (Jahku, 2003, 88–89).†† It does give the governments of sensed states increased rights to data from commercial actors, but that access is still on "reasonable terms and conditions" (Jahku, 2003, 89; 15 CFR § 960.11(b)(10); 15 CFR 960.12). Nondiscriminatory access policies include Japan's *GOSAT* data policy (Aoki, 2010, 339); Brazil's National Institute for Space Research's (INPE) policy, which is "to give out free on the Internet all remote sensing data received by INPE, the resulting maps, and the software for image processing and GIS" (Ferreira and Camara, 2008, 15); *Landsat* data, which are available free on the Internet‡‡; and similarly, GeoScience Australia's policy to make "*data free on the Internet*" (Gabrynowicz, 2007, 16). Korea and Japan both have graduated system wherein data for different types of users are priced at different points (Gabrynowicz, 2007, 18–19; Lee, 2010, 429–430). India grants nondiscriminatory access for data above a certain resolution (Gabrynowicz, 2007, 20). The Canadian law provides that the government of a sensed state has a right to obtain data from a commercial actor on "*reasonable terms*."§§ The French civilian remote sensing policy is "based on the promotion of a space imagery global market where data could be acquired on a nondiscriminatory basis" (Achilleas, 2008, 2).¶¶ The German Act on Satellite Data Security "favourable to commercial dissemination, create[s] de facto a wide database accessible to all

* UNOOSA (2014) (UN-SPIDER is a "United Nations programme, with the following mission statement: 'Ensure that all countries and international and regional organizations have access to and develop the capacity to use all types of space-based information to support the full disaster management cycle'").

† These critiques are echoed by Jackson (1997, 872–873).

‡ Remote Sensing Principles, XII.

§ There is at least one exception to this rule found in Israel's ImageSat, which "openly promotes exclusivity and secrecy" (Gabrynowicz, 2007, 21). For more on ImageSat, see Crook (2007).

¶ For instance, France in the SPOT Image commercial policy (Achilleas, 2008, 5). Gabrynowicz (2007, 7–10) notes that policies are still more common than formal legal regulation.

** For more on law and policy connected to *Landsat*, see generally Graham and Gabrynowicz (2002a) and Gabrynowicz (2005).

†† For an extensive, though dated, overview of data availability under U.S. law, see Gabrynowicz (1999). *Landsat* data are now available free on the Internet (NASA Landsat Science, 2014).

‡‡ USGS, Landsat, http://landsat.usgs.gov/ (accessed May 15, 2014).

§§ Remote Sensing Space Systems Act (S.C. 2005, c. 45), 8(4)(c). See also Mann (2008, 78).

¶¶ Achilleas (2008) further notes that in order to "secure the market from a legal point of view, France has supported the adoption" of the Remote Sensing Principles.

third persons on a non-discriminatory basis" (Schmidt-Tedd and Kroyman, 2008, 105).* Other examples include Argentina, Malaysia, South Africa, and Thailand (Gabrynowizc, 2007, i–xvii).

In addition to actions at the state level, the Group on Earth Observations has developed data sharing principles that endorse "the full and open access to data and … Information" (Uhlir et al., 2009, 220). This is based on an underlying philosophy that these data are a public good, and data should be distributed as such (Uhlir et al., 2009, 220). The European Union has adopted a directive to "ensure[] that environmental information is systematically available and disseminated to the public" (Harris, 2008, 37),¶ as well as the INSPIRE Directive, which "promote[s] unrestricted access to spatial data" (Doldirina, 2013, 303). China and Brazil distribute *CBERS* data free of charge to many users and have established a specific program to distribute these data to African states (Ferreira and Camara, 2008, 15; Ling, 2010, 446–447).†

Full and open access is limited by national security concerns, but *"[d]ata denial is the exception, not the rule"* (Gabrynowicz, 2007, 11; Uhlir et al., 2009, 230–231).‡ Not only do The Principles not apply to military activities, but also they do not apply to civil or commercial imagery with national security implications.§ As a result, when a state determines that a civil or commercial remote sensing satellite has taken imagery that affects national security concerns, that state does not have to distribute that data on a nondiscriminatory basis. Proper analysis of such provisions begs the question of "where on the spectrum is the point passed that … required data access moves into the reasonable national security constraints" (Gabrynowicz, 2007, 22). National security exceptions are also often implemented via the licensing process that states have implemented in accordance with Principle XIV and Article VI of the Outer Space Treaty.¶ National security restrictions come in two different categories: transactional restrictions on the distribution of data and the use of "shutter control," which are regulations that allow a government to restrict, for national security reasons, a nongovernmental satellite company's "collection or distribution of data" (Petras, 2005, 95).**

The current trend is for states to restrict data on a transactional basis in which the *"specifics of each request"* are examined (Gabrynowicz, 2007, 3). One of the most developed of these regimes is in the German satellite data law, which implements "a two phase procedure: the sensitivity check undertaken by the data provider and the granting of the permit by the responsibility authority" (Schmidt-Tedd and Kroyman, 2008, 109). Under this regime, data dissemination is undertaken only after that data have been cleared through a process that determines "any potential endangerment of security" (Schmidt-Tedd and Kroyman, 2008).†† France's 2003–2008 Military Program Law seeks to "[ensure] that certain civil programs comply with defense requirements" (Achilleas, 2008, 3).‡‡ While *CBERS* has some open access, Zhao argues that remote sensing "data are strictly controlled by the Chinese government" on the grounds of protecting 'state secrets'" (Zhao, 2010, 563).§§ Additionally, Italy, India, Israel, Japan, and Korea have all implemented national security exceptions (Gabrynowicz, 2007, 11–12; Aoki, 2010, 338; Lee, 2010, 431–432).

Shutter control provisions are found in a handful of state regulations, but likely exists *de facto* in many nations. Canada (Gillon, 2008, 29–30; Mann, 2008, 80–81) and the United States (15 CFR 960, 2013) have shutter control provisions that they can use to interrupt a satellite operator from distributing or collecting data that may harm national security. Canada's legal regime was adopted explicitly to ensure that remote sensing providers would comply with national security issues (Gillon, 2008, 25).¶¶ The regime allows the government to "interrupt normal service of a satellite" in cases of "the most serious of national security, national defence, and foreign policy/international obligations

* Nondiscriminatory in this context means that "it is impossible for a customer to prevent a third person from accessing data about a specific region" (Schmidt-Tedd and Kroyman, 2008).

† Harris (2008) notes that the term "reasonable cost" is similar to, but less definite than "cost of fulfilling a user request."

‡ Jahku (2003, 89–90) argues that such laws and policies are inconsistent with Principle XII. However, it should be noted that these exceptions have become prevalent enough that they can be considered part of the acceptable state practice under the norm. This is especially so in light of the prevalence of national security exceptions across a wide range of international law regimes. See also Sekhula (2011, 229) (arguing that the nondiscriminatory access rule to "have proved unworkable in practice and [to] require new interpretation consistent with contemporary practice")

§ Gabrynowicz (2007, 13) ("regardless of actual practice, no Nation or data supplier wants to appear to denounce the nondiscriminatory access policy").

¶ But see Jahku (2003, 83–84) (noting a contractual agreement between India and Space Imaging restricting imagery distributed in India).

** Gabrynowicz (2007, 13) notes "[a]ll current and pending national legislation and policy provide for some sort of 'shutter control,'" which correctly classifies all national security exceptions to nondiscriminatory access as forms of shutter control. This is a classification made on the outcome. The distinction made in this paper is at the point of government intervention. The transactional controls are made at the data level, whereas shutter control, though including distribution limitations, also gives a state authority over collection activities as well. This division is meant to be explicatory, and it hardly represents two distinct categories.

†† This sensitivity check takes into account all aspects of the transfer including the identity of the customer (Schmidt-Tedd and Kroyman, 2008).

‡‡ According to Achilleas, though France lacks a specific remote sensing law "governmental control is imposed on the SPOT Image commercial policy" (Achilleas, 2008, 5). The state can restrict "data when hostile entities might use data representing protected and sensitive French areas …, location of French troops abroad …, or the location of allied troops abroad" (Achilleas, 2008b, 6).

§§ Kuriyama delineates between a U.S. approach to data distribution and a Chinese approach (Kuriyama, 2010, 569).

¶¶ Mann adds three additional "public interest: justifications for the law," "the environment," "public health," and "safety of persons and property" (Mann, 2008, 69).

concerns" (Gillon, 2008, 29–30).* The U.S. provisions require licensees to agree to "[s]pecific limitations on operational performance, including, but not limited to, limitations on data collection and dissemination" in their licenses on a case-by-case basis (15 CFR 960.11, 2013).

30.2.4 Geospatial Law

Remote sensing data, primarily as a result of the Internet, has become increasingly available to individuals and more prominent in society. Products such as Google Earth and extensive media usage of satellite imagery have brought remote sensing data into mainstream consciousness. Along with other technological advances, such as GPS, an array of geo-location and mapping applications are increasingly available to individuals globally through devices like smartphones and tablets, and geotechnologies are being embedded directly into these devices. This means that an array of legal issues have arisen, primarily in the domestic law context, which situates remote sensing activities within a broader complex of geo-location technologies. This has given rise to the area of geospatial law, which is loosely defined as the body of law that governs the collection, use, and distribution of geospatial information.

This new trend is directly related to three phenomena: commercialization, globalization, and technological convergence. The first is commercialization. This process occurred in a number of countries and drove the development of the domestic law in those countries.† Canada adopted The Remote Sensing Space Systems Act of 2005 after "advances in satellite remote sensing technology in the private sector started to drive the development of commercial space systems" (Gillon, 2008, 19). Similarly, Germany's data security law was triggered by the development of public–private partnerships and commercial use of data (Schmidt-Tedd and Kroyman, 2008, 100–102). Commercialization does not just refer to the operation of the satellite; it also cuts across issues such as "whether or not data should be distributed by the public sector or the private sector" (Graham and Gabrynowicz, 2002a, 291). For instance, France's SPOT Image is a commercial data distributor of public data (Jackson, 1997, 859), and India has made data from government remote sensing satellites commercially available (Gabrynowicz, 2007, 6). Commercialization creates legal issues because the UN principles "were agreed [to] at a time when the commercialization of remote sensing activities were still not envisaged" (John et al., 2011, 261). Commercialization is not a completed process (and will most likely remain that way indefinitely), and an important lesson of its limitations can be found in the failed *Landsat* commercialization in the United States (Gabrynowicz, 2005, 50–59; Gabrynowicz, 2007, 5–6; and, generally, Graham and Gabrynowicz, 2002a).

Next is the process of globalization,‡ the modern version of which has cultural and conceptual roots in the space age and the reimagination of the world as a global space.§ Held argues that globalization is the "widening, deepening and speeding up of worldwide interconnectedness in all aspects of contemporary social life, from the cultural to the criminal, the financial to the spiritual."¶ It leads to the combining of the "local and the global" to create "distant proximities" (Ferguson and Mansbach, 2012). Essentially, globalization changes the space that society exists in, and importantly, "satellite imagery, digital maps, and associated information have transformed our ability for understanding the forces that shape geographical space" (Ferreira and Camara, 2008). As such, globalization is an information-driven process, and one of the effects of globalization has been increased access to global technologies.** This is certainly true in the field of Remote Sensing with numerous states now engaging in remote sensing activities including new entrants such as Nigeria, Algeria, Morocco, Egypt, Kenya, South Africa, Colombia, and Turkey (Jahku, 2003, 68–69; Gabrynowicz, 2007, 7; Oprong and Rwehumbiza, 2011, 255). More importantly, though, is the fact that remote sensing is a global technology, and as such, it facilitates the conceptualization of the world as a global space.††

Finally, the process of technological convergence has changed the way in which we interact with this type of data. Technological convergence is the process through which technologies that

* Shutter control restrictions in Canada can apply to spatial, temporal, or resolution elements of the remote sensing activities (Gillon, 2008, 30). The regime also gives the government priority access to data in "cases of emergency response …, in support of requests for aid of a civil power, or in support of Canadian Forces," which moves such requests "to the front of the order queue" (Gillon, 2008, 30–31; Mann, 2008, 81).

† Gabrynowicz notes that "[t]he distinction between 'public' and 'private' in the remote sensing space segment is disappearing worldwide. What constitutes 'commercial' operations varies among nations" (Gabrynowicz, 2007, 3, 15–16). While these definitional elements are important, the purpose here is to acknowledge the general trend as opposed to engaging with the competing definitions. In the context of this paper, commercialization represents the processes along this spectrum that result in an increased number of private, nongovernmental actors in the field. Additionally, Gabrynowicz notes that there is a rising trend of new government entities "organized like private corporations" (Gabrynowicz, 2007, 16–18). See generally, Jackson (1997) (arguing that the UN Principles Regime is inadequate to cope with increased commercialization). For an early view of the remote sensing industry, see generally, Graham and Gabrynowicz (2002b).

‡ This author ascribes to the view that globalization is a process that can retract and recede. For a fuller account of different theories of globalization and its meaning, see generally, Ferguson and Mansbach (2012).

§ Ferguson and Mansbach (2012, 21) note that one of the features of globalization is the nonexclusiveness of "territoriality" and note that outer space is one of the environments that have allowed territoriality to be reimagined as nonexclusive. For a more in-depth argument that the space age is an important historical factor in modern globalization, see Blount and Fussell (2014).

¶ David Held referenced in Ferguson and Mansbach (2012, 17).

** Ferguson and Mansbach (2012, 22, 167) ("Globalization advocates also reject critics' concerns about modern technology … More information creates an informed citizenry… And, although a digital divide exists new technologies are reaching more and more people and accelerating economic development in poor countries."). In relation to space technologies, see generally Gabrynowicz (2005, 30).

†† For example, Gabrynowicz argues that Landsat is a "national program with an inherently global function" (Gabrynowicz, 2005, 47).

were once significantly distinct (e.g. a map and a phone) are integrated and become indistinguishable. Current convergence trends are connected to the way in which information has been redefined as data in a technical sense. This is not to imply that information is data, but instead that, technologically, they are indistinguishable. For example, when a user downloads a remote sensing image, s/he receives information when s/he look at the image. However, as that image was transferred over the Internet, it was simply data broken down into manageable packets, and the application on the end users' device reassembled it into information.* This means that data are now highly integrable (depending on creativity and innovation in applications) and highly portable (depending on bandwidth). This can cause an array of issues since integrated data packages pull from a variety of sources. For example, mapping software on mobile devices may use geo-location information provided by the device, map databases, remote sensing data, and crowd-sourced information, among other sources. When this information becomes data, it becomes indistinguishable as separate sets of knowledge and becomes a single information point for the user. This is one of the features that has allowed "ICT technologies [to reduce] geography and physical distance in politics, economics, and war … render[ing] territory less important" (Ferguson and Mansbach, 2012, 112). While territory has become less important, location has become increasingly important,† meaning that "[g]lobal politics no longer reflects … neat, exclusive territorial boxes," creating a need to "rethink … and design a wide variety of maps that take into account the many forms of political activity" (Ferguson and Mansbach, 2012, 112). Remote sensing data have become part of technological convergence, and there is a trend toward "increas[ing] the use of data" (Gabrynowicz, 2007, 14).‡ Additionally, convergence has made the Internet the core tool for improving transparency, and geospatial technologies play an important role in transparency (Gabrynowicz, 2007, 11).

These three processes have significantly changed the place of remote sensing data in society. This naturally results in legal tensions as the law is recalibrated to cope with emerging technologies. Commercialization challenges remote sensing law by introducing new types of stakeholders that do not fit into the international law regime. Globalization challenges the regime because "the overlap of authority and governance among subnational, national, transnational, and international polities is fraught with difficulties" (Ferguson and Mansbach, 2012, 134). And technological convergence challenges the regime by reorienting how and why remote sensing data are used.§ The resulting complexity situates remote sensing law within the bounds of geospatial law, at least as far as data and information issues are involved. This section will not attempt to give a full account of geospatial law, which is a broad category encompassing a variety of legal issues; instead, it will attempt to focus on prominent issues that are directly related to remote sensing technologies.

* This is the result of the network design of the Internet (Clark and Landau, 2010, 27). This design is critical to the functionality of the Internet as we know it.

† Location is identity (Clark and Landau, 2010, 25).

‡ As an example of convergence specific to remote sensing, India has seen moves to regulate "web-based image suppliers" (Gabrynowicz, 2007, 21).

§ Ferreira and Camara (2008, 17) link the Internet to a change in how remote sensing data are used.

30.2.4.1 Privacy and Security

The 9/11 terrorist attacks shattered the notion that globalization was necessarily a sign of positive progress. Not only did globalization enable global public goods and mass communication, but it also enabled global threats like terrorism. In this context, privacy has become an increasingly important issue. This results from both of the faces of globalization. The increasing access to and use of remote sensing imagery by the general population have generated significant privacy issues based on global connectivity and content coupled with global threats. The UN Principles "offer only fairly minimal guidance on the topic of privacy of individual persons and entities," and as a result, privacy protections "are consequently to be found only at national level" (von der Dunk, 2013, 257), but at the national level, "[p]rivacy is a broad area of law, and relevant legislation is unclear in concept and application" (Cho, 2013, 261).

In the wake of the 9/11 attack, there were "*sweeping organizational changes*" in the U.S. defense and intelligence organizations, which "g[ave] rise to myriad potential statutory and regulatory impediments" (Petras, 2005, 81).¶ This creates specific issue about the use of U.S. military satellites for use of surveillance of the U.S. citizens, which raises "particularly vexing and controversial" issues (Petras, 2005, 81).** While the freedom of use of remote sensing satellites is established at the international law level, these technologies are often treated differently under domestic law (Petras, 2005, 94). In the American context, this represents a cultural value that has been embedded in the law (Petras, 2005, 102). The Reconstruction era Posse Commitatus Act restricts the use of the military in law enforcement, and the values found in the Posse Commitatus Act have been furthered formalized through laws, regulations, policies, and procedures.†† The application of Posse Commitatus and its progeny to remote sensing technologies is less clear in practice, though there are clear exceptions (Petras, 2005, 110–113).‡‡ However, the core value of separating these institutions from domestic law enforcement can be found in the law and policies governing remote sensing technologies.

¶ Petras notes specifically that 9/11 caused the U.S. national security regime to turn inward (Petras, 2005, 82–83).

** The securitization phenomenon will be addressed in the American context since it is the locus of this trend.

†† While this is a value, it should be noted it is not a constitutional value, though it reflects core American Values. Congress has the ability to waive the prohibition in certain circumstances, which in a securitization context means that political will can shift this value. See Petras (2005, 104–105) (noting the Congressional waiver for military participation in antidrug trafficking activities near the U.S. Borders).

‡‡ For a common line of cases addressing the Posse Commitatus Act, see *U.S. v. Yunis* (1991), *Wrynn v. U.S.* (1961), *U.S. v. Red Feather* (1975), *Bissonette v. Haig* (1985), *U.S. v. Rasheed* (1992), *U.S. v. Kahn* (1994).

The U.S. law requires defense collection of information on U.S. persons to be gathered using the "least intrusive means," which requires the use of a four-step methodology (Petras, 2005, 97). Petras points out that satellite data present an issue since it is unclear whether that data represent a "publicly available" source of information, in light of commercial imagery available on the Internet (the first tier of the methodology), or whether such information retrieval requires "a warrant or approval by the Attorney General" (the fourth tier of the methodology) (Petras, 2005, 97–98). National Geospatial-Intelligence Agency procedures require that imagery gathered by domestic satellites must be for a "foreign intelligence mission" and only if commercial imagery is "unsuitable or unavailable" (Petras, 2005, 100, 108–109). Any other military request must be for "authorized activities of the armed forces" (Petras, 2005, 100–101). Consistent with this theme, there are further procedures to limit the use of such technologies in the American context. The collection cannot be "focused on a specific U.S. person" and must be in connection with activities that are not "domestic" in nature (Petras, 2005, 101).*

Since satellite technology is not entirely out of the picture for law enforcement, the restrictions on government observation of citizens are primarily contained in search and seizure law, and as remote sensing technology continues to "*become a viable law enforcement tool*," issues of proper use by governments will become more important (Gunasekara, 2010). In the United States, this is governed by a Constitutional restriction found in the Fourth Amendment, which forbids the government from engaging in unreasonable searches without a proper warrant.† The standards involved come from a line of cases that hold that searches that breach a "reasonable expectation of privacy" violate Constitutional protections (Gunasekara, 2010, 118–119). However, courts have been challenged by technologies that make observations of individuals by remotely sensing them or their property. While the Supreme Court has never ruled directly on the use of satellite remote sensing data, there is a line of cases that can be traced, which gives some indication of how satellites might be treated under the U.S. law.

Several cases address the use of aerial surveillance and hold that the government has fairly wide discretion to use aerial surveillance without a warrant.‡ Of particular note, though, is the dictum in *Dow Chemical* in which the Court posits that certain technologies not available to the general public, like satellite technology, might violate the Fourth Amendment (Petras, 2005, 99). This theme is taken up in the *Kyllo v. United States* case. In this case, police used a thermal imager to determine whether grow lights were being used inside a residence to grow marijuana (Gunasekara, 2010, 124–126). The Court determined that the technology that was used was not available to the general public, and as a result, it violated a reasonable expectation of privacy (Gunasekara, 2010). Of course, remote sensing technology is now arguably available to the general public as a result of technological convergence (Petras, 2005, 99; Gunasekara, 2010, 135–136). A final and more recent case is *Jones v. United States*, which addressed the use of a GPS device on a car. The opinion was split, but one of the concurring opinions posited that such searches created a mosaic of the life of an individual and that such mosaics breach a reasonable expectation of privacy (U.S. *v.* Jones, 2012). If such a standard were to be adopted by the Court in the future, it would certainly have implications for uses of remote sensing data in law enforcement.

In general, states have a variety of approaches to privacy issues. Korea, in Article 17 of its Space Development Promotion Act, has legislated that the "Government shall endeavor to avoid any invasion of privacy during the course of utilizing satellite information" (Lee, 2010, 425). Lee notes that privacy is a "fundamental right" provided for in the Korean Constitution. Legal regimes create a variety of contexts in which privacy must be considered; for example, remote sensing data issues have been identified in the Health Insurance Portability and Accounting Act in the United States (Secunda, 2004, 251). Other states do not have fully formed privacy protections. Zhao notes that Hong Kong has "no comprehensive data protection law" (Zhao, 2010, 551). In the Canadian remote sensing regime, "there are no provisions … dealing with privacy, and no privacy conditions have been incorporated in the first remote sensing satellite system license"; therefore, privacy concerns must be addressed under Canada's Constitution and Privacy Act (Mann, 2008, 85–87).§ Additionally, commercial use of data has caused numerous concerns. In particular, the Google suite of geo-technologies has come under extensive fire for privacy violations (Foresman, 2011; Google Faces Lawsuits over Gmail, 2013), and as content providers increasingly incorporate geospatial data, privacy breaches are becoming more common (e.g., Chen, 2011; Frizell, 2014). This creates cross-jurisdictional problems since content providers operate in a multiplicity of jurisdictions at any given time.

30.2.4.2 Use of Remote Sensing Data as Evidence

Data derived from remote sensing activities has become increasingly important in judicial proceedings; this is related to both the integration of geo-location technologies into everyday lives and large-scale problems such as environmental damage. The ability to introduce this type of data is governed by rules of evidence for the particular jurisdiction meaning that geospatial and remote sensing data must

* Domestic activities are those "that take place within the United States that do not involve a significant connection with a foreign power, organization, or person" (Petras, 2005). Petras specifically argues that international terrorist organizations operations within the borders of the United States would sit outside this definition (Petras, 2005, 110).

† U.S. Constitution. 4th Amendment.

‡ Gunasekara identifies *Ciraolo, Dow Chemical,* and *Riley* (Gunasekara, 2010, 120–124).

§ Mann (2008, 87) argues that in Canada "[p]rivacy rights are adequately protected." It should also be noted that the Canadian Supreme Court in a case very similar to the United States' *Kyllo* case addressed earlier split from the U.S. rational (Mann, 2008, 86–87; Gunasekara, 2010, 130–134).

be fit into a preexisting evidentiary categories.* One of the core issues has been authentication of remote sensing evidence in courts due to its digital nature (Shipman, 2013, 359–377). This problem has been addressed through technical means such as "*digital signatures*" (Croi et al., 2013, 379–398) and complete authentication procedures, as in the case of USGS/EROS data (Rychlak et al., 2007, 212–218).

Korea (Lee, 2010, 423–424), Queensland, Australia (Goulevitch, 2013), Singapore (Escolar, 2013, 108–110), and the United States (Rychlak et al., 2007, 206–210; Dighe et al., 2013) use satellite technology for investigating and prosecuting breaches of environmental laws. Satellite data have also found its way into evidence in a variety of international Institutions (see generally Williams, 2013, 195–216). The International Court of Justice has used satellite data in numerous cases, which often involve boundary disputes (see generally Froehlich, 2011, 222–225). The Permanent Court of Arbitration and the International Criminal Court have both made use of earth observation data as evidence.† The use of these data in both civil and criminal courts will increase across a number of areas as data become increasingly embedded into society.

30.2.4.3 Torts

Misuse of geospatial data could result in torts. Navigation systems providers have been repeatedly sued for incorrect data resulting in incorrect directions (e.g., Hadhazy, 2010; Gannes, 2012). A U.S. court held a map provider liable for incorrect data that it purchased from a third party (Brocklesby *v.* U.S., 1985), and Nigerian law can hold "GIS professionals ... legally accountable for the accuracy and reliability of information stored in their databases, sold, or issued to the public" (John et al., 2011, 263). Ito states that "[d]ata suppliers bear liability risks in cases where the population is affected by disasters or aid workers are injured as a result of inappropriate instructions" (Ito, 2008, 57). While Ito is arguing for a more complete international legal regime, data torts will likely continue to be handled *ad hoc* at the domestic judicial level.

30.2.4.4 Intellectual Property

The UN Principles are "completely silent" as to intellectual property rights in data generators (Ito, 2008, 50). This is likely because, at the time of the negotiations, these activities had not yet seen a sufficient amount of commercial or private activity to warrant addressing the issue. Commercialization has resulted in a variety of users making intellectual property rights a growing concern for remote sensing data providers and users. Generally, "current practice is that the ownership of data stays with data generators and the copyright is claimed by the majority of data generators," but it is not clear to what extent these claims can be maintained for raw data especially in jurisdictions where a "certain degree of human intellectual intervention" is required for information to be copyright protected (Ito, 2008, 55).‡ The nature of remote sensing data means that "[c]opyright may not be the most appropriate regime to protect EO data" (Doldirina, 2013, 298).§

State regulations vary significantly. Goh notes a Singaporean case in which a data provider was held to violate copyright protections by distributing maps created with unlicensed data (Escolar, 2013, 111–112). The European Union protects intellectual property rights in "databases to secure remuneration to the maker of the database," which Harris argues could have implications for remote sensing technologies (Harris, 2008, 37–38). He notes that problems could arise when database protections are involved with public data (Harris, 2008, 38). Nigeria explicitly grants intellectual property rights to data producers via its National Geoinformation Policy (John et al., 2011, 262; see also Birisbe, 2006, 241–245), and JAXA maintains copyright to data it creates in Japan (Gabrynowicz, 2007, 18). The Korean regime can protect remote sensing imagery via copyright or database protections (Lee, 2010, 426–428). Additionally, Germany has denied copyright of *Landsat* data, while France has allowed it (Mejia-Kaiser, 2006).

30.3 Conclusion

Remote sensing will remain an important technology in the foreseeable future. Based on the international principles examined in this chapter, remote sensing is regulated within a stable international legal framework that is permissive in nature. Emerging challenges will come from the increasing use of remote sensing data in a data-driven society. National regulations will need to balance among competing interests of national security, individual liberty, and commercial growth in an increasingly complex technological infrastructure.

References

Achilleas, P., French remote sensing law, *The Journal of Space Law*, 34 (2008) 1.

Agreement on the Rescue of Astronauts, the Return of Astronauts and the Return of Objects Launched into Outer Space (the "Rescue Agreement") (entered into force on December 3, 1968).

Alvarado, S.C., The analysis of existing international space cooperation initiatives for UNESCO's World Heritage Sites, IAC-10.E3.1A.14 (2010).

Aoki, S., Japanese law and regulations concerning remote sensing activities, *The Journal of Space Law*, 36 (2010) 335.

* For analysis of these rules in different jurisdictions, see, England and Wales, Germany, the Netherlands, Belgium, Mosteshar (2013, 147–175); Nigeria, John et al. (2011, 262–263); Singapore, Escolar (2013, 98–108); United States, Dighe et al. (2013, 81–90) and Wright (2013, 313–320).

† PCA, Macauley (2013, 224–225); ICC, Macauley (2013, 226–238).

‡ For an overview of copyright implications for EO data, see Doldirina (2013, 293–310). For an example of such a regime, see Zhao on Hong Kong, Zhao (2010, 552–554).

§ Cromer presents a full analysis of the conflicts between the copyright regime and the space law regime Cromer (2006).

Beets, J., The international charter on space and major disasters and international disaster law: The need for collaboration and coordination, *Air & Space Lawyer*, 22/4 (2010) 12–15.

Birisbe, T., Outer space activities and intellectual property protection in Nigeria, *The Journal of Space Law*, 32 (2006) 229.

Bissonette *v.* Haig, 776 F.2d 1384 (1985).

Blount, P.J., Limits on space weapons: Incorporating the law of war into the Corpus Juris Spatialis, in *Proceedings of the 51st Colloquium on the Law of Outer Space* (AIAA 2009).

Blount, P.J., Developments in space security and their legal implications, *Law/Technology*, 44/2 (2011) 18.

Blount, P.J., Renovating space: The future of international space law, *Denver Journal of International Law and Policy* 40 (2012) 515–686.

Blount, P.J. and J. Fussell, Musical counter narratives: Space, skepticism, and religion in American music, in *AIAA 52nd Aerospace Sciences Meeting* (January 2014), National Harbor, MD.

Brocklesby *v.* U.S. 767 F.2d 1288 (9th Cir. 1985).

Carpanelli, E. and B. Cohen, A legal assessment of the 1996 declaration on space benefits on the occasion of its fifteenth anniversary, *The Journal of Space Law*, 38 (2012) 1.

Carpanelli, E. and M.K. Force, The protection of the earth natural environment *through* space activities: A general overview of some legal issues, in *Proceedings of the International Institute of Space Law*, The Hague, the Netherlands: Eleven (2011), p. 31.

Carter, J., Remarks at the congressional space medal of honor awards ceremony, Kennedy space Center, FL, *Weekly Compilation of Presidential Documents*, 14(40) (October 8, 1978) 1684.

CBERS Data Policy, *The Journal of Space Law*, 31 (2005) 281.

Charter on Cooperation to Achieve the Coordinated Use of Space Facilities in the Event of Natural or Technological Disasters Rev. 3 (April 25, 2000), http://www.disasterscharter.org/web/charter/charter. Accessed February 22, 2014.

Chen, B.X., iPhone tracks your every move, and there's a map for that, *WIred* (April 20, 2011), http://www.wired.com/gadgetlab/2011/04/iphone-tracks/. Accessed February 22, 2014.

Chinese Law: Registration, launching and licensing space objects, *The Journal of Space Law*, 33 (2007) 437.

Cho, G., Privacy and EO: An overview of legal issues, in R. Purdy and D. Leung (eds.), *Evidence from Earth Observation Satellites*, Leiden, the Netherlands: Martinus Nijhoff Publishers (2013), p. 259.

Clark, D.C. and S. Landau, Untangling attribution, in *Committee on Deterring Cyberattacks, Proceedings of a Workshop on Deterring Cyberattacks: Informing Strategies and Developing Options for U.S. Policy*, Washington, DC: National Academies Press (2010), p. 25.

Convention on International Liability for Damage Caused by Space Objects (the "Liability Convention") (entered into force on September 1, 1972).

Convention on Registration of Objects Launched into Outer Space (the "Registration Convention") (entered into force on September 15, 1976).

Croi, W., F.-M. Foeteler, and H. Linke, Introducing digital signatures and time-stamps in the EO data processing chain, in R. Purdy and D. Leung (eds.), *Evidence from Earth Observation Satellites*, Leiden, the Netherlands: Martinus Nijhoff Publishers (2013), p. 379.

Cromer, J.D., How on earth terrestrial laws can protect geospatial data, *The Journal of Space Law*, 32 (2006) 253.

Crook, J., Corporate Sovereign symbiosis: *Wilson v. ImageSat International*, Shareholders' actions, and the dualistic nature of state-owned corporations, *The Journal of Space Law*, 33 (2007) 411.

Dembling, P.G. and D.M. Arons, The evolution of the outer space treaty, *Journal of Air Law & Commerce*, 33 (1967) 419–456.

Diederiks-Verschoor, I.H.Ph., *An Introduction to Space Law*, The Hague, the Netherlands: Kluwer (1999).

Dighe, K., T. Mikolop, R.W. Mushal, and D. O'Connell, The use of satellite imagery in environmental crimes prosecutions in the United States: A developing area, in R. Purdy and D. Leung (eds.), *Evidence from Earth Observation Satellites*, Leiden, the Netherlands: Martinus Nijhoff Publishers (2013), p. 65.

Doldirina, C., The impact of copyright protection and public sector information regulations on the availability of remote sensing data, in R. Purdy and D. Leung (eds.), *Evidence from Earth Observation Satellites*, Leiden, the Netherlands: Martinus Nijhoff Publishers (2013), p. 293.

Escolar, G.G., The use of EO data as evidence in the courts of Singapore, in R. Purdy and D. Leung (eds.), *Evidence from Earth Observation Satellites*, Leiden, the Netherlands: Martinus Nijhoff Publishers (2013), p. 93.

EU Directive 2003/4/EC, On public access to environmental information and repealing Council Directive 90/313/EEC (January 28, 2003).

Eyes on Darfur, http://www.eyesondarfur.org/ (last visited February 22, 2014).

Ferguson, Y.H. and R.W. Mansbach, *Globalization: The Return of Borders to a Borderless World?*, London, U.K.: Routledge (2012).

Ferreira, H.S. and G. Camara, Current status and recent developments in Brazilian remote sensing law, *The Journal of Space Law*, 34 (2008) 11.

Filho, J.M. and A.F. dos Santos, Chinese–Brazilian accord on distribution of CBERS products, *The Journal of Space Law*, 31 (2005) 271.

Foresman, S., Google faces $50 million lawsuit over Android location tracking, *Ars Technica* (April 30, 2011), http://arstechnica.com/tech-policy/2011/04/google-faces-50-million-lawsuit-over-android-location-tracking/. Accessed February 22, 2014.

Frizell, S., Tinder security flaw exposed users' locations, *Time* (February 19, 2014), http://techland.time.com/2014/02/19/tinder-app-user-location-security-flaw/. Accessed February 22, 2014.

Froehlich, A., Space related data: From justice to development, in *Proceedings of the International Institute of Space Law*, The Hague, the Netherlands: Eleven (2011), p. 221.

Gabrynowicz, J.I., Earth observation: The view from the ground up, *Space Policy* (August 1997), 229.

Gabrynowicz, J.I., Defining data availability for commercial remote sensing systems under United States federal law, *Annals of Air and Space Law*, 23 (1999) 93.

Gabrynowicz, J.I. (ed.), *The UN Principles Relating to Remote Sensing of the Earth from Space: A Legislative History—Interviews of Members of the United States Delegation*, University, MS: National Center for Remote Sensing and Space Law (2002).

Gabrynowicz, J.I., Space law: Its cold war origins and challenges in the era of globalization, *Suffolk University Law Review*, 37 (2004) 1043.

Gabrynowicz, J.I., The perils of *Landsat* from grassroots to globalization: A comprehensive review of US remote sensing law with a few thoughts for the future, *Chicago Journal of International Law*, 6/1 (2005) 45.

Gabrynowicz, J.I., The International space treaty regime in the era of globalization, *Ad Astra* (Fall 2005), 30.

Gabrynowicz, J.I., *The Land Remote Sensing Laws and Policies of National Governments: A Global Survey*, University, MS: National Center for Remote Sensing, Air, and Space Law (2007).

Gannes, L., GPS App Strava sued over cyclist's death, All Things D (June 19, 2012), http://allthingsd.com/20120619/gps-app-strava-sued-over-cyclists-death/.

Gillon, T., Regulating remote sensing space systems in Canada—New Legislation for a New Era, *The Journal of Space Law*, 34 (2008) 19.

Google Faces Lawsuits over Gmail, Street View privacy, *CBC News* (October 2, 2013), http://www.cbc.ca/news/business/google-faces-lawsuits-over-gmail-street-view-privacy-1.1876594.

Goulevitch, B., Ten years of using earth observation data in support of Queensland's vegetation management framework, in R. Purdy and D. Leung (eds.), *Evidence from Earth Observation Satellites*, Leiden, the Netherlands: Martinus Nijhoff Publishers (2013), p. 113.

Graham, J.F. and J.I. Gabrynowicz (eds.), *Landsat 7: Past Present and Future*, University, MS: National Remote Sensing and Space Law Center (2002a).

Graham, J.F. and J.I. Gabrynowicz, *The Remote Sensing Industry: A CEO Forum*, University, MS: National Remote Sensing and Space Law Center (2002b).

Gunasekara, S.G., The March of science: Fourth Amendment implications on remote sensing in criminal law, *The Journal of Space Law*, 36 (2010) 115.

Hadhazy, A., Bad directions from Google maps lead to lawsuit, *TechNewsDaily* (June 3, 2010), http://www.technewsdaily.com/556-bad-directions-from-google-maps-lead-to-lawsuit.html. Accessed February 22, 2014.

Harris, R., Current status and recent development in UK and European remote sensing law and policy, *The Journal of Space Law*, 34 (2008) 33.

Harris, R., Science, policy, and evidence in EO, in R. Purdy and D. Leung (eds.), *Evidence from Earth Observation Satellites*, Leiden, the Netherlands: Martinus Nijhoff Publishers (2013), p. 43.

Harrison, R.G., *Space and Verification, Vol. 1: Policy Implications*, USAFA, CO: Eisenhower Center for Space and Defense Studies (2011).

Harvey, D., *A Brief History of Neoliberalism*, Oxford, U.K.: Oxford University Press (2005).

International Law Commission, Draft articles on Responsibility of States for Internationally Wrongful Acts, with commentaries (2001).

Ito, A., Improvement to the legal regime for the effective use of satellite remote sensing data for disaster management and protection of the environment, *The Journal of Space Law*, 34 (2008) 45.

Jackson, S.M., Cultural lag and the international law of remote sensing, *Brooklyn Journal of International Law*, 23 (1997) 853.

Jahku, R., International law governing the acquisition and dissemination of satellite imagery, *Journal of Space Law*, 29 (2003) 65.

John, O.N., E. Ezekiel, and S.O. Mohammed, Legal regime of remote sensing and geographic information systems in Nigeria, in *Proceedings of the International Institute of Space Law*, The Hague, the Netherlands: Eleven (2011), p. 260.

Kopal, V., The role of United Nations declarations of principles in the progressive development of space law, *The Journal of Space Law*, 16 (1988) 5.

Kuriyama, I., Environmental monitoring cooperation paves the way for common rules on remote sensing, *The Journal of Space Law*, 36 (2010) 567.

Lachs, M., *The Law of Outer Space: An Experience in Contemporary Law-Making*, Leiden, the Netherlands: Martinus Nijhoff (1972, 2010).

Lee, J.G., Remote sensing issues as they relate to Korea, *The Journal of Space Law*, 36 (2010) 415.

Licensing of Private Remote Sensing Systems, 15 CFR 960 (2013).

Ling, Y., Remote sensing data distribution and application in the environmental protection, disaster prevention, and urban planning in China, *The Journal of Space Law*, 36 (2010) 435.

Listner, M., Iridium 33 and Cosmos 2251 three years later: Where are we now? *The Space Review* (February 13, 2012), http://www.thespacereview.com/article/2023/1. Accessed February 22, 2014.

Lyall, F. and P.B Larsen, *Space Law: A Treatise*, Aldershot, U.K.: Ashgate (2013).

Macauley, E.D., The use of EO technologies in court by the office of the prosecutor of the international criminal court, in R. Purdy and D. Leung (eds.), *Evidence from Earth Observation Satellites*, Leiden, the Netherlands: Martinus Nijhoff Publishers (2013), p. 217.

Mann, B., First license issued under Canada's remote sensing satellite legislation, *The Journal of Space Law*, 34 (2008) 67.

Marchisio, S., The evolutionary stages of the legal subcommittee of the United Nations Committee on the peaceful uses of outer space (COPUOS), *The Journal of Space Law*, 31 (2005) 219.

Mayence, J.-F., Belgian legal framework for earth observation activities, *The Journal of Space Law*, 34 (2008) 89.

Mejia-Kaiser, M., Copyright claims for Meteosat and Landsat images under court challenge, *The Journal of Space Law*, 32 (2006) 293.

Mosteshar, S., EO in the European Union: Legal considerations, in R. Purdy and D. Leung (eds.), *Evidence from Earth Observation Satellites*, Leiden, the Netherlands: Martinus Nijhoff Publishers (2013), p. 147.

NASA Landsat Science, The numbers behind Landsat, http://landsat.gsfc.nasa.gov/?page_id=9 (last visited February 22, 2014).

Onoda, M., Satellite Earth Observation As 'Systematic Observation' in multilateral environmental treaties, *The Journal of Space Law*, 31 (2005) 339.

Oprong, A.A. and V. Rwehumbiza, A glance at the earth observation policies and regulations and the impact on developing countries: Focusing on the African continent, in *Proceedings of the International Institute of Space Law* (2011), p. 251.

Petras, C.M., 'Eyes' on freedom—A view of the law governing military uses of satellite reconnaissance in U.S. Homeland defense, *The Journal of Space Law*, 31 (2005) 81.

Remote Sensing Space Systcms Act (S.C. 2005, c. 45).

Reynolds, G.H. and R.P. Merges, *Outer Space: Problems of Law and Policy*, Boulder, CO: Westview Press (1997).

Richelson, J.T., *America's Space Sentinels: DSP Satellites and National Security*, Lawrence, KA: University Press of Kansas (1999).

Rychlak, R.J., J.I. Gabrynowicz, and R. Crowsey, Legal certification of digital data: The earth resources observation and science data center project, *The Journal of Space Law*, 33 (2007) 195.

Satellite Data Security Act, *The Journal of Space Law*, 34 (2008) 115.

Satellite Sentinel Project, http://www.satsentinel.org/ (last visited February 22, 2014).

Schmidt-Tedd, B. and M. Kroyman, Current status and recent developments in German remote sensing law, *The Journal of Space Law*, 34 (2008) 97.

Schrogl, K.-U. and N. Hedman, The U.N. General Assembly Resolution 62/102 of 17 December 2007 on 'Recommendations on Enhancing the Practice of States and International Intergovernmental Organizations in Registering Space Objects,' *The Journal of Space Law*, 34 (2008) 141.

Secunda, P.M., A mosquito in the ointment: Adverse HIPPAA implications for health-related remote sensing research and a 'Reasonable' Solution, *The Journal of Space Law*, 30 (2004) 251.

Sekhula, P.P., The right to remote sense data: Impact of multilateral cooperation on international space law [sic], in *Proceedings of the International Institute of Space Law*, The Hague, the Netherlands: Eleven (2011), p. 228.

Sentinel Asia, Sentinel Asia: Disaster management support system in the Asia-Pacific Region, http://www.aprsaf.org/initiatives/sentinel_asia/pdf/Sentinel-Asia.pdf (last visited February 22, 2014).

Shipman, A., Authentification of images, in R. Purdy and D. Leung (eds.), *Evidence from Earth Observation Satellites*, Leiden, the Netherlands: Martinus Nijhoff Publishers (2013), p. 359.

Tarikhi, P., Mahdasht satellite receiving station verging into a space center, *Res Communis* (October 13, 2008), http://rescommunis.olemiss.edu/2008/10/13/guest-blogger-parviz-tarikhi-mahdasht-satellite-receiving-station-verging-into-a-space-center/. Accessed February 22, 2014.

Terekov, A., UN general assembly resolutions and outer space law, in *Proceedings of the International Institute of Space Law* (1997), p. 97.

Treaty between the United States of America and the Russian Federation on Measures for the Further Reduction and Limitation of Strategic Offensive Arms (signed April 10, 2010).

Treaty on Principles Governing the Activities of States in the Exploration and Use of Outer Space, including the Moon and Other Celestial Bodies (the "Outer Space Treaty") (entered into force on October 10, 1967).

Uhlir, P.F., R.S. Chen, J.I. Gabrynowicz, and K. Janssen, Toward implementation of the global earth observation system of systems data sharing principles, *The Journal of Space Law*, 35 (2009) 201.

UN Charter (1945).

UNGA Meeting Record no. A/41/PV.95 (December 3, 1986).

UNGA Res. 41/65, Principles relating to remote sensing of the earth from outer space (December 3, 1986).

UNGA Res. 51/122, The Declaration on International Cooperation in the Exploration and Use of Outer Space for the Benefit and in the Interest of All States, Taking into Particular Account the Needs of Developing Countries (December 13, 1996).

UNGA Res. 60/1, 2005 World Summit Outcome (October 24, 2005).

UNGA Res. 62/102, Recommendations on Enhancing the Practice of States and International Intergovernmental Organizations in Registering Space Objects (December 17, 2007).

UNOOSA, About UN-SPIDER, http://www.unoosa.org/oosa/en/unspider/index.html (last visited February 22, 2014).

United Nations High Commission on Human RIghts, Report of the commission of inquiry on human rights in the Democratic People's Republic of Korea, A/HRC/25/63 (February 17, 2014).

U.S. Constitution.

U.S. *v.* Jones, No. 10-1259, Concurring Opinion of Justice Sotomayor (January 23, 2012).

U.S. *v.* Kahn, 35 F.3d 426 (1994).

U.S. *v.* Rasheed, 802 F.Supp. 312 (1992).

U.S. *v.* Red Feather, 392 F.Supp. (1975)

U.S. *v.* Yunis, 924 F.2d 1086 (1991).

Vásquez, F.R. and S.C. Lara, What lawyers need to know about science to effectively make and address laws for remote sensing and environmental monitoring, *The Journal of Space Law*, 36 (2010) 365.

Vásquez, M.S., *Cosmic International Law*, Detroit, MI: Wayne State University Press (1965), p. 164.

Vereschetin, V.S. and G.M. Danilenko, Custom as a source of international law of outer space, *The Journal of Space Law*, 13 (1985) 22.

von der Dunk, F.G., Outer space law principles and privacy, in R. Purdy and D. Leung (eds.), *Evidence from Earth Observation Satellites*, Leiden, the Netherlands: Martinus Nijhoff Publishers (2013), p. 243.

Williams, M., Satellite evidence in international institutions, in R. Purdy and D. Leung (eds.), *Evidence from Earth Observation Satellites*, Leiden, the Netherlands: Martinus Nijhoff Publishers (2013), p. 195.

Wright, M., The use of remote sensing evidence at trial in the United States—One State Court Judge's observations, in R. Purdy and D. Leung (eds.), *Evidence from Earth Observation Satellites*, Leiden, the Netherlands: Martinus Nijhoff Publishers (2013), p. 313.

Wrynn *v.* U.S., 200 F.Supp. 457 (1961).

Zhao, Y., Regulation of remote sensing activities in Hong Kong: Privacy, access, security, copyright and the case of Google, *The Journal of Space Law*, 36 (2010) 547.

XIV

Summary

31

Remote Sensing Data Characterization, Classification, and Accuracies: Advances of the Last 50 Years and a Vision for the Future

Prasad S. Thenkabail
United States Geological Survey (USGS)

Acronyms and Definitions

ALI	Advanced Land Imager
ALOS	Advanced Land Observing Satellite
ANN	Artificial Neural Networks
ASPRS	American Society for Photogrammetry and Remote Sensing
AVHRR	Advanced Very High Resolution Radiometer
CDMA	Code Division Multiple Access
CEOS	Committee on Earth Observation Satellites
CHRIS	Compact High Resolution Imaging Spectrometer
CYGNSS	Cyclone Global Navigation Satellite System
DEM	Digital Elevation Model
DESDynI	Deformation, Ecosystem Structure and Dynamics of Ice
DSM	Digital Surface Model
EO	Earth Observation
ERTS	Earth Resources Technology Sateelites
ESRI	Environmental Systems Research Institute
ETM+	Enhanced Thematic Mapper+
EVI	Enhanced Vegetation Index
FGDC	Federal Geographic Data Committee
GEOBIA	Geographic Object-based Image Analysis
GIMMS	Global Inventory Modeling and Mapping Studies
GIS	Geographic Information System
GLONASS	Global Orbiting Navigation Satellite System
GNSS	Global Navigation Satellite Systems
GOES	Geostationary Operational Environmental Satellite
GEOSS	Global Earth Observing System of Systems
GLAS	Geoscience Laser Altimeter System
GNSS	Global Navigation Satellite System
GPS	Global Positioning System
ICESat	Instrument aboard the Ice, Cloud, and land Elevation
ISODATA	Iterative Self-organizing Data Analysis Technique
LiDAR	Light Detection and Ranging
LST	Land Surface Temperature
MODIS	Moderate Resolution Imaging Spectroradiometer
NDVI	Normalized Difference Vegetation Index
NOAA	National Oceanic and Atmospheric Administration
NPOESS	National Polar-orbiting Operational Environmental Satellite System
NPP	NPOESS Preparatory Project
OBIA	Object Oriented Image Analysis
OMI	Ozone Monitoring Instrument
PALSAR	Phased Array type L-band Synthetic Aperture Radar
PROBA	Project for On Board Autonomy
RS	Remote Sensing
RADAR	Radio Detection and Ranging
SAR	Synthetic Aperture Radar
SMTs	Spectral Matching Techniques
SST	Sea Surface Temperature
SODAR	Sonic Detection and Ranging
SONAR	Sound Navigation and Ranging
SPOT	Système Pour l'Observation de la Terre
SRTM	Shuttle Radar Topographic Mission
SVM	Support Vector Machines
TIN	Triangular Irregular Network
TIR	Thermal Infrared
TM	Thematic Mapper
TOA	Top of Atmosphere
UAS	Unmanned Aircraft Systems
UAV	Unmanned Aerial Vehicles
VGI	Volunteered Geographic Vegetation
VHRI	Very High Resolution Imagery
VI	Vegetation Index

This chapter provides a summary of each of the 30 chapters in *Remotely Sensed Data Characterization, Classification, and Accuracies* of the Remote-Sensing Handbook. The topics covered in the chapters of *Remotely Sensed Data Characterization, Classification, and Accuracies* include: (a) satellites and sensors; (b) remote-sensing fundamentals; (c) data normalization, harmonization, and standardization; (d) vegetation indices (VIs) and their within and across sensor calibration; (e) image classification methods and approaches; (f) change detection; (g) integrating remote sensing with other spatial data; (h) Global Navigation Satellite Systems (GNSS); (i) crowdsourcing; (j) cloud computing; (k) Google Earth™ remote sensing; (l) accuracy assessments; and (m) remote-sensing law. Under each of the aforementioned broad topics, there are one or more chapters. For example, there are nine chapters under image classification methods and approaches. In a nutshell, these chapters provide a complete and comprehensive overview of these critical topics, capture the advances over the last 50 years, and provide a vision for further development in the years ahead. By reading this summary chapter, a reader can have a quick understanding of what is in each of the chapters of this book, see how the chapters interconnect and intermingle, and get an overview on the importance of various chapters in developing complete and comprehensive knowledge of the remote-sensing data characterization, classification, and accuracies. These chapters, together, capture the advances of the last 50 years and provide a vision for the future.

31.1 Remote-Sensing Satellites and Sensors

Today, remote-sensing satellites and sensors are ubiquitous. Spatial data from these sensors are used to understand, query, measure, model, monitor, store, and map a wide array of land, water, and atmospheric features and resources. Sensors continuously and repeatedly collect data in various spectral, spatial, radiometric, and temporal resolutions. Indeed, for many applications, remote sensing has become indispensable and often the most important data source. Its strengths are obvious: repeated coverage of the planet at different scales; ability to collect data in wide array of spatial, spectral, radiometric, and temporal resolutions; collecting consistent data without human subjectivity; and ability to collect data from inaccessible parts of the world without any political boundary restrictions. However, the complexities of data collection from satellites and sensors are many. They vary due to orbital distances, Sun angles, view angles, spectral bands, bandwidths, spatial resolutions, sensor platforms, type of sensors (e.g., active or passive) (e.g., Figure 31.1), radiometry, data download links, ground stations, calibration of data collected, normalization of data collected, geometric registration, data delivery platforms, and a host of other factors. As a scientist involved in the study of planet Earth, apart from the full understanding of the remotely sensed (RS) data, it is equally important to understand the process involved in design, launch, and operation of satellites and sensors. Especially, given many countries operate their own satellites and sensors, and there are many private players.

Chapter 1 by Dr. Sudhanshu Panda et al. is focused on providing a comprehensive overview of wide array of satellites and sensors operated by various governments and private entities. The idea of the chapter is to provide readers with a full understanding of the variety of remote-sensing data characteristics as well as their sources. The chapter provides a progressive development of remote sensing from the early days of airborne remote sensing, mainly developed, perfected, and used during the World Wars I and II, to the more modern hyperspectral, light detection and ranging (LiDAR), and Unmanned Aircraft Systems (UAS) sensors. The early era of remote sensing was mainly for military usage. Subsequently, owing to its strengths in mapping and survey, civilian applications of remote sensing grew mainly through airborne, ground-based, and platform-mounted sensors. With the launch of Sputnik and the National Oceanic and Atmospheric Administration's Advanced Very-High-Resolution Radiometer (NOAA-AVHRR), spaceborne remote sensing began its initial steps in late 1950s and early 1960s. With the launch of Earth resources technology satellites (ERTS-1 later renamed Landsat) in the year 1972, remote sensing of the Earth system science truly evolved. More recently or currently, data on planet Earth are routinely gathered by numerous satellite sensors that include very high spatial resolution (e.g., IKONOS, QuickBird, GeoEye, RapidEye) and multispectral moderate to coarse resolution (e.g., Landsat, Sentinel, Resourcesat, Moderate-Resolution Imaging Spectroradiometer [MODIS] in Terra/Aqua, AVHRR, Geostationary Operational Environmental Satellite (GOES), The National Polar-orbiting Operational Environmental Satellite System (NPOESS) Preparatory Project or NPP, Radarsat, Advanced Land Observing Satellite (ALOS) Phased Array type L-band Synthetic Aperture Radar (PALSAR), Japanese Earth Resources Satellite (JERS) Synthetic Aperture Radar(SAR)). Less frequently collected data come from hyperspectral (e.g., Earth Observation [EO]-1 Hyperion, Compact High Resolution Imaging Spectrometer (CHRIS) Project for On Board Autonomy (PROBA)). These are either Sun synchronous or geostationary, and collected data are wide array of spatial, spectral, radiometric, and temporal resolutions that allow routine, consistent, and repeated study of the planet's land, water, and atmosphere. Chapter 1 captures a wide range of these satellites and sensors detailing their characteristics. The United States, France, India, China, Germany, the United Kingdom, Brazil, and several others (see Chapter 1) now routinely launch, operate, and share remote-sensing data. With the 1992 Land Remote Sensing Policy Act of United States, which permitted private companies to enter the satellite imaging business, private companies like WorldView Imaging Corporation participated in launching and operating commercial Earth-observing satellites, and very recently, it has picked up tremendously. These initiatives have helped in obtaining very high spatial and spectral resolution satellite imageries. Increasingly, UASs are also being used for carrying sensors of various kinds.

In immediate years ahead, as we enter the second decade of the twenty-first century, we are likely to have numerous remote-sensing satellites and sensors from government and several private operators. There is also likely greater role for UASs and UAVs. Compared to conventional remote sensing that maps and analyzes the Earth, water, and atmospheric phenomena, importance of RAdio Detection And Ranging (RADAR), and SOund NAvigation Ranging (SONAR), Light Detection and Ranging (LiDAR), remote-sensing technology is gaining ground in the twenty-first century due to their unconventional application of real-world problem solving like accurate ground topography mapping and tree heigh measurement, or other Earth object identification and measurements, accurate weather forecasting (e.g., Next-Generation Radar (NEXRAD or Nexrad) and Doppler SOnic Detection And Ranging (SODAR)), and deep-sea exploration. The private enterprise is also dwelling into newer, more revolutionary, approaches and techniques of capturing remote-sensing data. These initiatives come from companies such as the Planet Labs Inc., the Skybox Imaging Inc. (recently purchased by Google), and the Boeing company recently receiving its first commercial order for the 502 Phoenix small satellite from HySpecIQ of Washington, DC. The satellites planned by these initiatives will allow for gathering spectral signatures, or submeter data on a continuous basis, or even videos, and many other innovative ways.

Further, the chapter also provides the details of the principles of electromagnetic spectrum in remote sensing. It also describes the future of remote sensing, that is, transformation of computer/lab imaging spectrometry technology to field-imaging spectrometry to determine crop, vegetation, fruit, and other physical object quality.

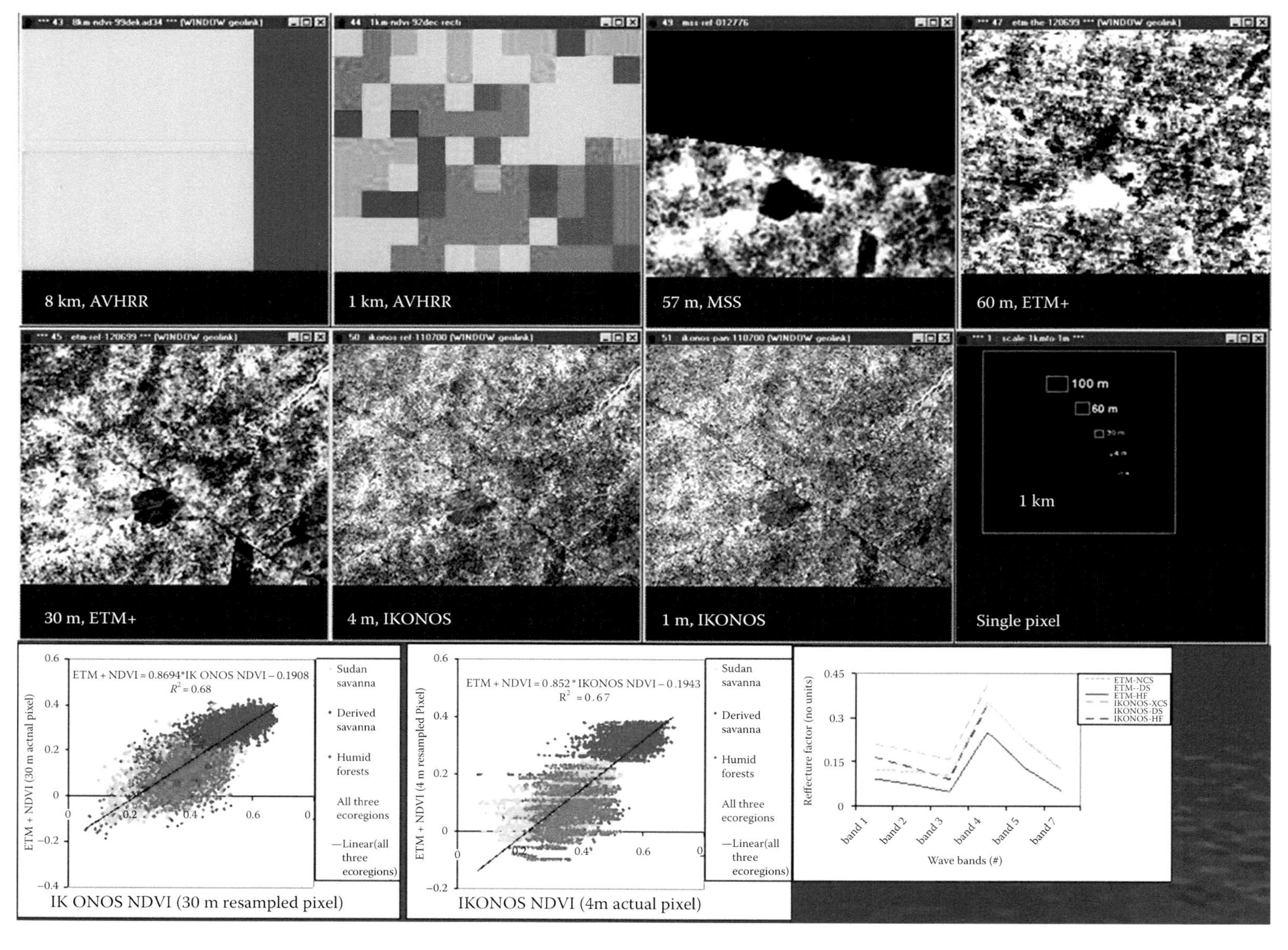

FIGURE 31.1 The top two rows of images depict the impact of spatial resolution on information content. An area in National Oceanic and Atmospheric Administration's Advanced Very-High-Resolution Radiometer (NOAA-AVHRR) 8 km resolution (top left) is geo-linked to images acquired over the same areas from various other resolutions: 1 km NOAA-AVHRR, 57 m Landsat MSS, Landsat ETM+ thermal image of 60 m, Landsat ETM+ multispectral image of 30 m, IKONOS multispectral of 4 m, IKONOS panchromatic of 1 m. The leftmost bottom row plot shows the intersensor relationships between the Landsat ETM+ 30 m normalized difference vegetation index (NDVI) and IKONOS NDVI (resampled to 30 m) derived from images acquired about the same dates. In the last row, the left and the center plots show the intersensor relationships between the Landsat ETM+ NDVI (resampled to 4 m) and IKONOS 4 m NDVI derived from images acquired about the same dates. The line chart shows the normalized reflectance derived from various sensors that collected data over various African sites.

31.2 Fundamentals of Remote Sensing for Terrestrial Applications

Early remote sensing was primarily based on images and data gathered from aircrafts. This had limited scope, given the complexities of covering large areas and the cost factor. Thereby, land remote sensing from space commenced in earnest once the first Landsat was launched in 1972. Even though NOAA AVHRR and Sputnik preceded Landsat and the data from these satellites were used for land applications, it is with the launch of Landsat that the real momentum in terrestrial applications of spaceborne remote sensing began. Oppelt et al. in Chapter 2 provide a brief overview on the remote-sensing platforms as well as the electromagnetic spectrum from where these data are captured. Limitations of early visual interpretations are pointed out. Then, the chapter provides a focus on typical applications at local, regional, and global levels as well as specialized thematic applications. What is clear in this chapter is the following:

1. Specific applications (e.g., urban, vegetation, water, terrain) have specific data needs (e.g., thermal, normalized difference vegetation index [NDVI], near-infrared, microwave). For example, one cannot study species using coarse spatial and/or coarse spectral resolutions. Similarly, a particular sensor performs better than another in certain applications, for example, radar sensors are suitable to derive digital elevation models. So knowledge of sensor performance for specific terrestrial applications is required.
2. Multisensor approaches (e.g., combine Landsat with Sentinel or optical with microwave sensors) become crucial for improved accuracies in mapping, modeling, and monitoring complex terrestrial themes.
3. For a value adding beyond the sole physical measurement of remote-sensing data, multidisciplinary approaches using remote-sensing data in combination with other data sources became increasingly important. Figure 31.2 presents an example for such an approach to monitor coastal vegetation combining different analysis techniques, biological mapping, hydrography, and a geographic information system.
4. Best results are obtained when studies incorporate various dimensions (e.g., spectral, spatial, temporal, radiometric) of RS data.
5. Data policy is changing; free and open data policy supports existing trends to provide remote-sensing data (e.g., Landsat, MODIS) and standardized products (e.g., global land cover maps) free of charge.
6. Global networks have formed to develop robust methodologies and services for decision support tools and community portals to meet specific needs of stakeholders, scientists, and user communities (e.g., Global Earth Observation System of Systems [GEOSS], Global Forest Watch).
7. Continuous and comparable data are of high importance to develop standardized measures, especially for global monitoring. To increase acceptance of standardized products and to ensure their widespread application, standards for accuracy assessment will have to be set.

Developments and trends mentioned earlier require overlapping missions and well- (inter-) calibrated sensors. To fulfill these requirements, international cooperation and data sharing will be increasingly important. Joint commitments between intergovernmental, international, and regional organizations therefore are a powerful means of stabilizing new space initiatives and hence data availability and continuity. Even though great advances have been made as a result of use of data from multiple sensors using advanced methods, techniques, and modern computing power, there is still a long way to go in reducing uncertainties of various terrestrial themes studied using remote sensing.

31.3 Overview of Satellite Image Radiometry

A primary goal of remote-sensing data acquired from the wide array of EO satellites is to establish protocols and mechanisms for delivering calibrated and normalized data over the entire planet. This allows analyses of data over space and time with minimal and\or known uncertainties.

In the early years, remote-sensing applications of EO satellites were often done using raw digital counts (Level-0 data) without physical units. Great advances have been made over the years. Currently, remote-sensing data are analyzed routinely after calibration to their physical units (e.g., radiance expressed in W m^{-2} sr^{-1} μm^{-1}) using appropriate radiometric, geometric, spectral, and atmospheric corrections, as explored in detail in Chapters 3 through 7 of this book. These calibrations lead to data being converted to

1. At-sensor or top-of-atmosphere (TOA) radiance and/or reflectance (Level-1)
2. Surface radiance or reflectance (Level-2)

Currently, almost all EO data are provided in one of these units, typically scaled for digital processing. Yet calibrations remain challenging and complex given that satellite sensors are launched by various space agencies of different nations as well as private initiatives, which have widely varying geometric, radiometric, and spectral performance characteristics. Chapter 3 by Teillet provides a comprehensive summary of the wide array of radiometric factors involved in EO data. However, not all these parameters are used in developing standard products, produced through various algorithms, due to difficulty and/or uncertainty in obtaining some or many of these parameters.

In addition, intersensor calibrations (cf. Chapter 4) further facilitate the use of data from multiple sensors from multiple sources.

In order to overcome these difficulties, the Committee on Earth Observation Satellites (CEOS; http://www.ceos.org/), under the umbrella of the Group on Earth Observations (GEO; http://www.earthobservations.org/index.shtml), established an international calibration/validation (Cal/Val) working group (http://calvalportal.ceos.org/), where many of these issues are discussed and implementation recommendations are provided. The ultimate goal of these efforts is to create an effective Global Earth Observation System of Systems (GEOSS) that will deliver

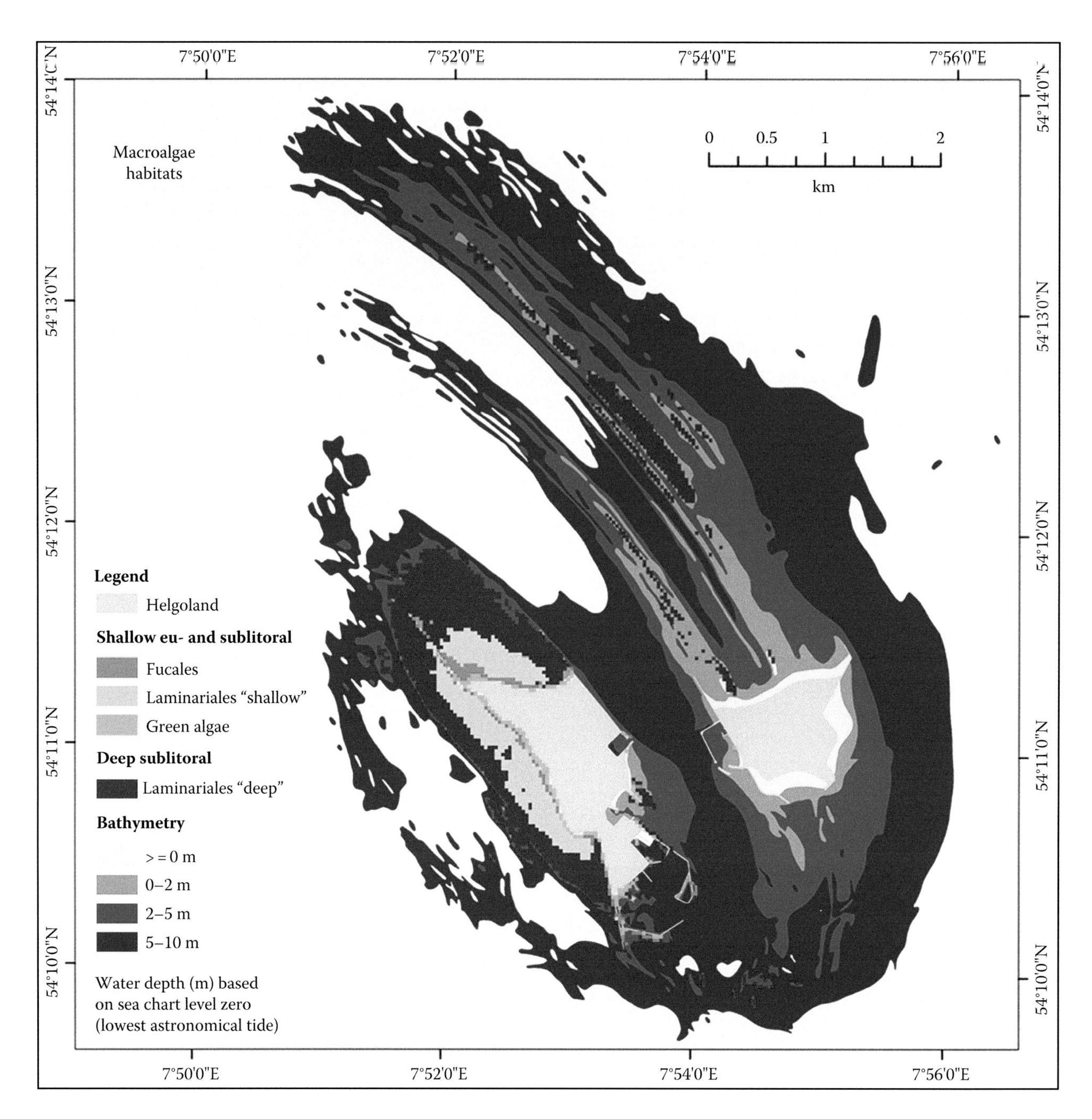

FIGURE 31.2 Map of macroalgae habitats in the coastal waters of Helgoland (North Sea, Germany); the map was generated in the scope of an interdisciplinary research project of geographers, biologists, and remote-sensing experts. The data sources include airborne hyperspectral data ($AISA_{eagle}$), floristic mappings, diving mappings, and bathymetry information from the German Federal Maritime and Hydrographic Agency. The remote-sensing data have been analyzed separately for algae cover in the eu- and sublitoral (the methods are described in Oppelt et al. (2012) [eulitoral] and Uhl et al. (2013) [sublitoral]). Since analysis of optical remote-sensing data is limited by visibility depth (approximately 5 m during airborne data acquisition), the analysis is valid up to a depth of ~4 m. All data have been combined in a GIS, which enables validation of remote-sensing results and mappings. The resulting user accuracy is 97.5% (number of validation points = 45), and GIS analysis also revealed that large sublitoral brown algae (kelp) cover 248.31 ha of the area with water depth ≤ 5 m. (From Uhl, F. et al., KelpMap—Development of an EnMAP approach to monitor sublitoral marine macrophytes [KelpMap—Entwicklung eines EnMAP Verfahrens zur Bestimmung von sublitoralen marinen Makrophyten], Final report of research project FKZ: 50EE1020, funded by the German Federal Ministry of Economy and Technology (BMWi), 36pp., 2014.)

EO data that are well calibrated, reliable, and consistent and have comprehensive coverage over space and time.

Current approaches to data delivery by many space agencies and commercial data providers are already moving in the direction of surface reflectance products, such as for MODIS (http://modis.gsfc.nasa.gov/data/dataprod/dataproducts.php?MOD_NUMBER=09) and Landsat 8 (http://earthexplorer.usgs.gov/). The approaches and methods are discussed in detail in Chapters 3 through 7. Some difficulties yet to be overcome include when

1. Space agencies and commercial data providers do not yet provide calibrated data
2. Calibration coefficients provided are incomplete (e.g., do not allow for all the factors listed in Chapter 3)
3. Gaps in intersensor calibration exist (e.g., the Système Pour l'Observation de la Terre [SPOT] High Resolution Visible (HRV) series of satellite data are not adequately intercalibrated with the Landsat series of satellites; Goward et al., 2012, Price, 1987) (Figure 31.3)

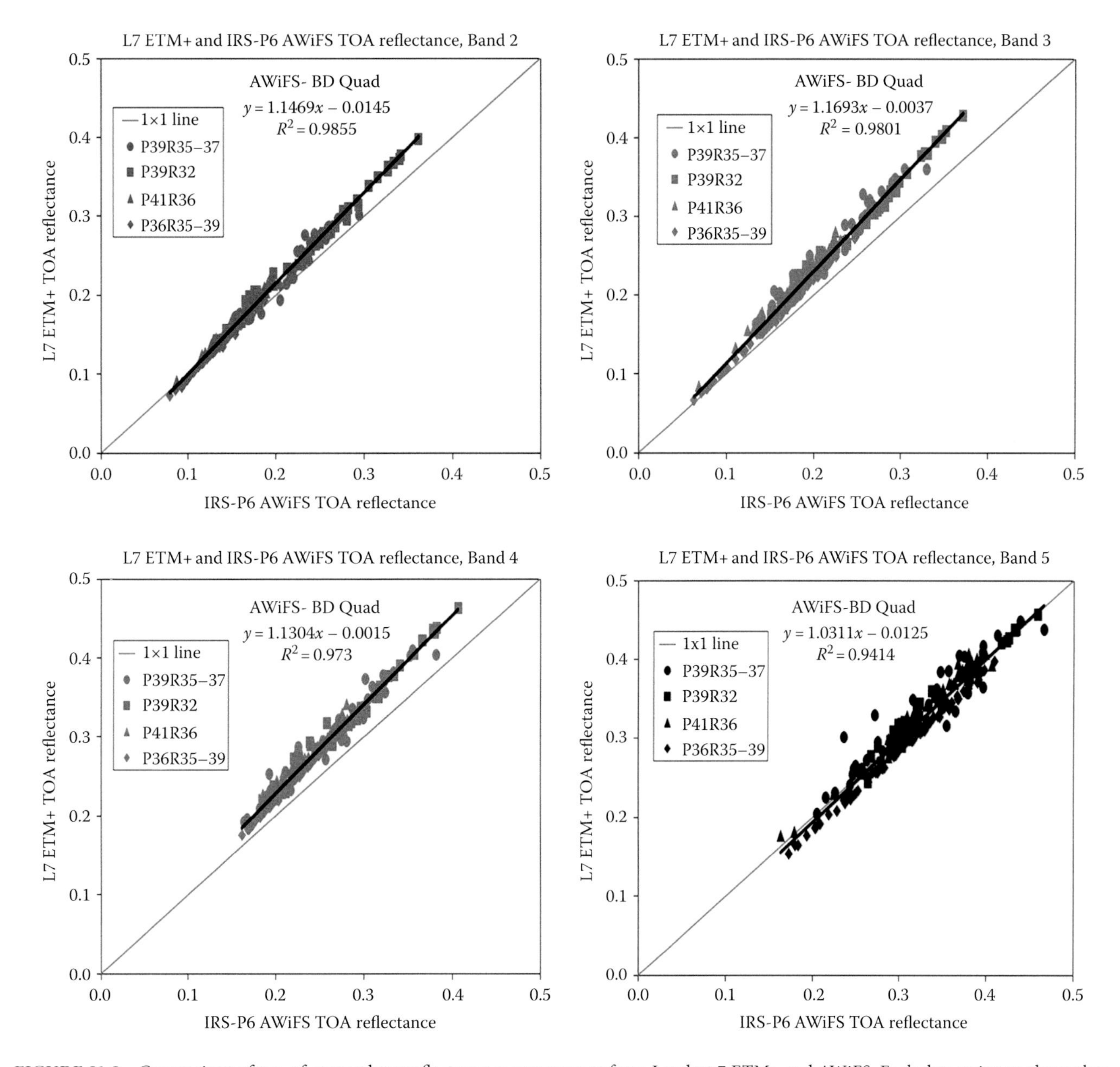

FIGURE 31.3 Comparison of top-of-atmosphere reflectance measurements from Landsat 7 ETM+ and AWiFS. Each data point on these plots represents an ensemble average of all pixels in a defined ROI for a given day and spectral band. (From Goward, S.N. et al., *Remote Sens. Environ.*, 123, 41, August 2012, ISSN: 0034-4257, http://dx.doi.org/10.1016/j.rse.2012.03.002. Accessed June–October, 2014.)

31.4 Post-launch Radiometric Calibration of Optical Satellite Sensors

The scientific integrity of satellite sensor data can only be maintained if the sensors are well calibrated. Once a sensor is launched into orbit, postlaunch calibrations need to come into effect and be reiterated for the entire life of the sensor, which is often from a few years to even decades. Chapter 4 by Teillet and Chander provides an exhaustive state-of-the-art review of postlaunch radiometric calibration of optical sensors. The standard and well-established approach adopted by various agencies and researchers for post-launch calibration is to use built-for-purpose onboard systems initially and adjust calibration coefficients over time using suitable reference targets vicariously. The international community has adopted and recommends the use of approximately 20 such targets, selected with specific criteria in mind (cf. Chapter 4).

Currently, the best practice is for optical sensors to be calibrated vicariously using

1. Eight CEOS-endorsed reference instrumented standard sites based in the deserts and playas of the world
2. Six CEOS reference pseudoinvariant calibration sites located in various parts of the Sahara desert

The multigeneration Landsat series of sensors from 1972 to the present are recalibrated postlaunch using vicarious targets and are known to have uncertainties of within 5%–10% across sensors and spectral bands. The calibration efforts in the contexts of GEO, GEOSS, and CEOS are important and are discussed in Chapter 4 by Teillet and Chander. Well-calibrated sensors allow multisensor data continuity, facilitating long-term studies of Earth and environment (Figure 31.4).

31.5 Normalization of Remotely Sensed Data

Modern remote sensing involves using data/imagery from multiple dates and from multiple sensors. The only way such data can be used in scientific investigations is through appropriate calibration and normalization. Sensors, once launched, go through degradation over time. They also operate in different seasons and climates. Various sensor characteristics such as push broom (or along track) versus whisk broom (or across track), solar zenith angles, field of view, and angle of the sensor also add to the need for standardization and normalization of the sensors. In Chapter 5, Dr. Rudiger Gens and Dr. Jordi Cristóbal Rosselló address these issues and provide a clear approach to normalization of sensor characteristics. They recommend

1. Absolute normalization based on radiative transfer models that account for atmospheric and sensor characteristics
2. Relative normalization where one or more images are corrected to a standard master image

They illustrate these two normalization methods separately for optical, thermal, and radar imagery.

Instruments are always calibrated prelaunch. Then, vicarious calibration takes place-throughout the life of the sensor on board either using an onboard calibration reference unit or by focusing the sensor to time-invariant sites like the full moon. Factors that need to be considered during calibration include (Figure 31.5)

1. Satellites
   a. Height of acquisition (e.g., 500, 700, 36,000 km above Earth)
   b. Orbital parameters

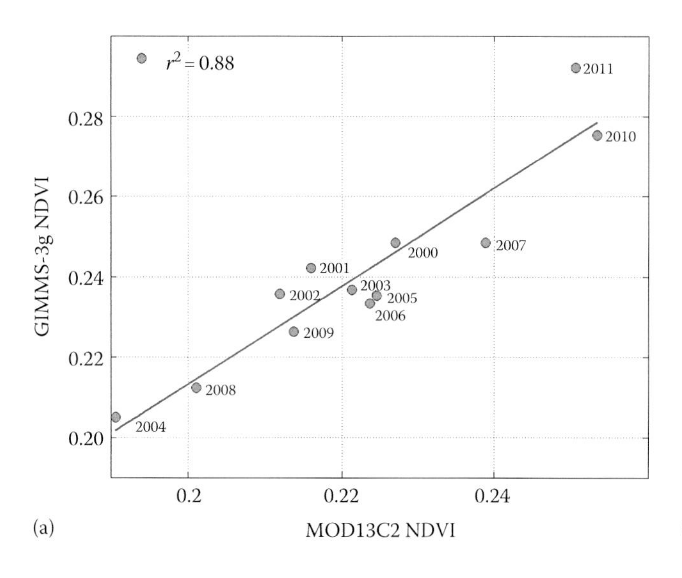

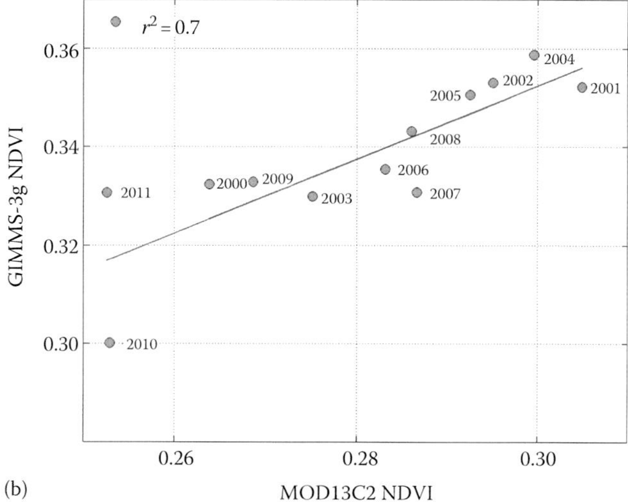

FIGURE 31.4 Linear regression between third-generation Global Inventory Modeling and Mapping Studies 8 km and Moderate-Resolution Imaging Spectroradiometer Terra normalized difference vegetation index (MOD13C2) at 5.6 km resolution over (a) the Gourma (Mali) region and (b) the Fakara (Niger) region. (From Dardel, C. et al., *Remote Sens. Environ.*, 140, 350, January 2014, ISSN: 0034-4257, http://dx.doi.org/10.1016/j.rse.2013.09.011. Accessed June–October, 2014.)

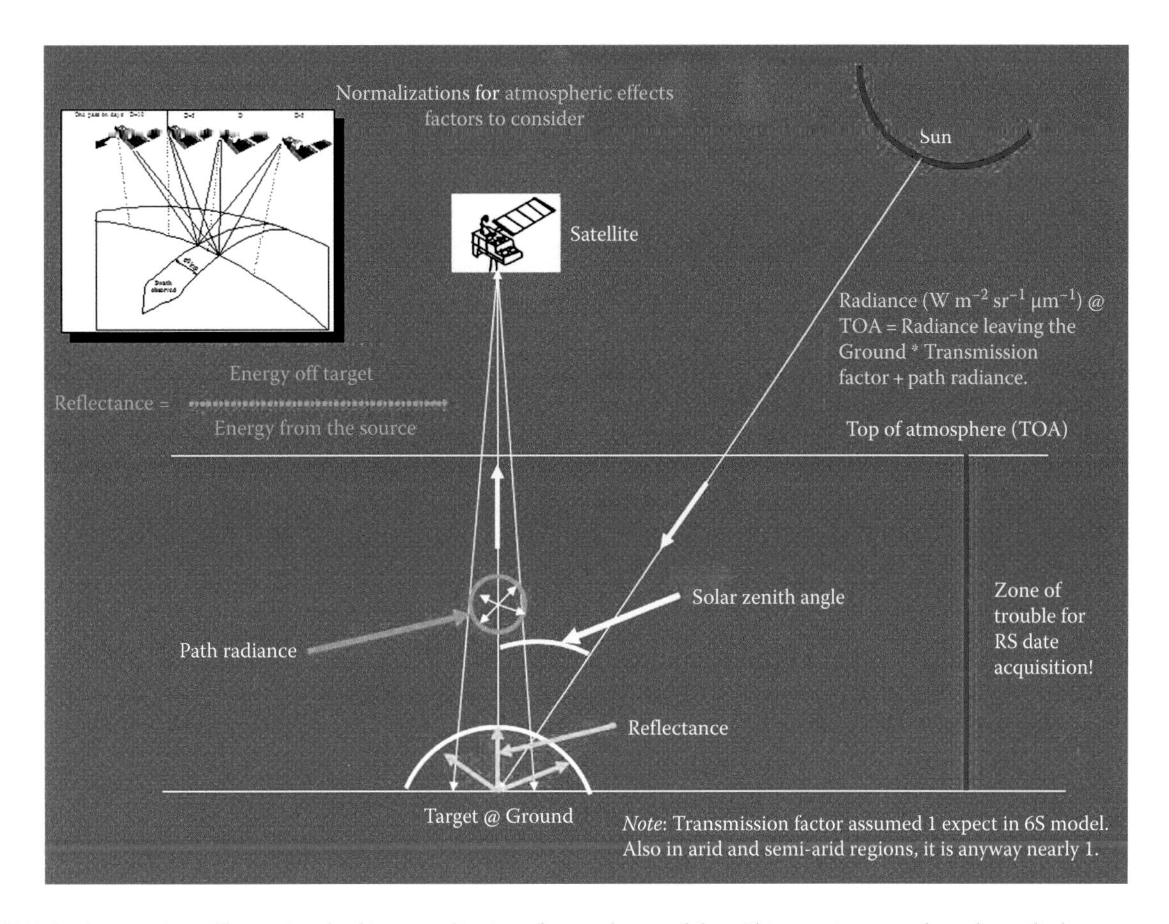

FIGURE 31.5 An overview of factors involved in normalization of remotely sensed data. This generic approach can be applied across any satellite sensor system.

2. Sensors
   a. Radiometry
   b. Bandwidth
   c. Optics/design
   d. Degradation over time
   e. Nadir, off-nadir viewing
3. Solar flux or irradiance
   a. Function of wavelength
4. Sun
   a. Sun elevation at time of acquisition
5. Sun–Earth
   a. Distance between Earth and Sun
6. Stratosphere or atmosphere
   a. Ozone, water vapor, haze, aerosol
   b. Path radiance
7. Surface of Earth
   a. Topography
8. Seasons
   a. Earth–Sun distance

Normalization steps involve three major steps:

- Digital number (unit less) to radiance (W m^{-2} sr^{-1} µm^{-1})
- Radiance to top of the atmosphere reflectance (also referred to as at-satellite exoatmospheric (or apparent or) reflectance (%)
- TOA reflectance to surface reflectance (%)

The data are first normalized to TOA reflectance using the following common approach, irrespective of the sensors involved:

$$\text{Reflectance }(\%) = \frac{\text{Energy off Target}}{\text{Energy from the Source}}$$

$$= \frac{\text{Radiance}\left(\text{W m}^{-2}\,\text{sr}^{-1}\,\mu\text{m}^{-1}\right)}{\text{Irradiance }\left(\text{W m}^{-2}\,\text{sr}^{-1}\,\mu\text{m}^{-1}\right)} \times 100$$

$$\text{Reflectance }(\%) = \frac{\pi L_\lambda d^2}{\text{ESUN}_\lambda \cos\theta_s}$$

where

TOA reflectance (%) is the at-satellite exoatmospheric reflectance
L_λ is the radiance (W m^{-2} sr^{-1} µm^{-1})
d (dimensionless) is the Earth-to-Sun distance in astronomic units at the acquisition date (see Markham and Barker, 1987)
ESUN_λ is the irradiance (W m^{-2} sr^{-1} µm^{-1}) or solar flux (Neckel and Labs, 1984)
θ_s (degrees) is the solar zenith angle

Note: θ_s is the solar zenith angle in degrees (i.e., 90° minus the Sun elevation or Sun angle when the scene was recorded as given in the image header file).

The aforementioned data are then converted to ground reflectance through atmospheric correction as described in Chapter 5. Atmospheric corrections are part of normalization and eliminate or reduce path radiance resulting from haze (e.g., thin clouds, dust, harmattan haze, aerosols, ozone, and water vapor). Atmospheric correction methods include (a) dark object subtraction technique (Chavez, 1988); (b) improved dark object subtraction technique (Chavez, 1989); (c) radiometric normalization technique, bright and dark object regression (Elvidge et al., 1995); and (d) 6S model (Vermote et al., 2002). As pointed out by Dr. Rudiger Gens and Dr. Jordi Cristóbal Rosselló in Chapter 5, the main difficulty in atmospheric radiosondes are usually not available at the time of satellite pass, and data from a single atmospheric radiosonde or aerosol robotic network are inadequate for large areas covered by single swath (e.g., AVHRR, MODIS, Landsat) of most of the satellites. Even when the swath are not big (e.g., IKONOS, QuickBird, GeoEye), local weather data may not be representative since these networks are often quite away from the images. A classic paper on intersensor calibration involving multiple sensors is provided by Chander et al. (2009). So most uncertainty in atmospheric correction of imagery is due to lack of detailed, reliable input data for the models.

31.6 Satellite data degradations and their impacts on high level products

One of the greatest challenges for terrestrial remote sensing has always been to ensure well-calibrated data from within and between sensors throughout its data acquisition. All sensors are precalibrated before launch and continue to be calibrated during their life span in space vicariously and/or on board. Chapter 6 by Dongdong Wang discusses three issues affecting stability of calibration:

1. Radiometric calibration with onboard calibration systems
2. Radiometric calibrations without onboard systems
3. Data degradation due to orbit drift of Sun-synchronous satellites

There are very many factors influencing sensor degradation such as orbital drift, solar zenith angle, and sensor degradation in space due to adverse conditions, clouds, haze, shadows, and surface changes even on so-called time-invariant sites. Some of these key factors are summarized in Table 6.1. What is important for a terrestrial scientist to ensure is that the measurements made by satellites are reliable and temporally stable and, if there are uncertainties, those are well characterized. This important and fundamental first step in using satellite remote-sensing data in Earth sciences is well illustrated in this chapter. For example, the chapter summarized some significant factors of different sensors such as the following:

1. Difference in reflectances from early AVHRR sensor derived with different versions of calibration coefficients can be as large as 20%.
2. There are systematic differences between MODIS Terra- and Aqua C5-derived information at certain places. This lead to recommendation that for interannual variability studies, it is better to use MODIS Aqua data rather than MODIS Terra data before new release with improved calibration stability is available.
3. Annual degradation of Landsat Enhanced Thematic Mapper Plus (ETM+) are less than 0.4% and are well characterized.
4. Landsat Thematic Mapper (TM) showed trends in sensor response over time. So new method that involved desert sites and cross calibration with ETM+ was preferred for better understanding Landsat TM calibration.

These inferences, drawn from Chapter 6, only depict some of the issues. These issues clearly indicate the complexities as well the critical need for well-established calibration of various orbiting sensors. Images are also normalized taking time-invariant sites into consideration (e.g., Figure 31.6). In such a case, an image is considered reference, and all other images of the same area are normalized to the reference image by taking time-invariant sites. For example, in Figure 31.6, Wei et al. (2014) selected the features that emit temporally stable, and invariant nighttime lights that are used as pseudoinvariant features (PIFs) for the Defense Meteorological Satellite Program Operational Linescan System (DMSP-OLS) image normalization. The fully developed and stable urban districts, where no significant further development and changes occur, satisfy the stable nighttime light requirement and can be used as PIFs (Wei et al., 2014).

31.7 Inter- and Intrasensor Spectral Compatibility and Calibration of the Enhanced Vegetation Indices

Long-term consistent global studies from satellite sensors are only feasible if there is within and between sensor calibration leading to clear, well-understood relationships between these sensors. Establishing relationships is required for both within a family of sensors such as Landsat series and AVHRR series and across family of sensors such as Landsat versus MODIS versus AVHRR versus IKONOS. This is because of a host of issues such as (a) sensor degradation over time; (b) their characteristic differences in spatial (e.g., Figure 31.1), spectral, radiometric, and temporal resolutions; (c) atmospheric conditions under which data are acquired; and (d) different processing algorithms applied over time.

Chapter 7 by Dr. Tomoaki Miura et al. provides the latest understanding of intersensor spectral compatibility across multiple sensors, including visible infrared imaging radiometer suite, MODIS, second-generation global imager, VEGETATION (VGT), AVHRR/3, AVHRR/2, and ETM+, for two well-known vegetation indices (VIs): enhanced vegetation index (EVI) and EVI2. EVI uses three bands (red, near-infrared, and blue), whereas EVI2 does not use the blue band.

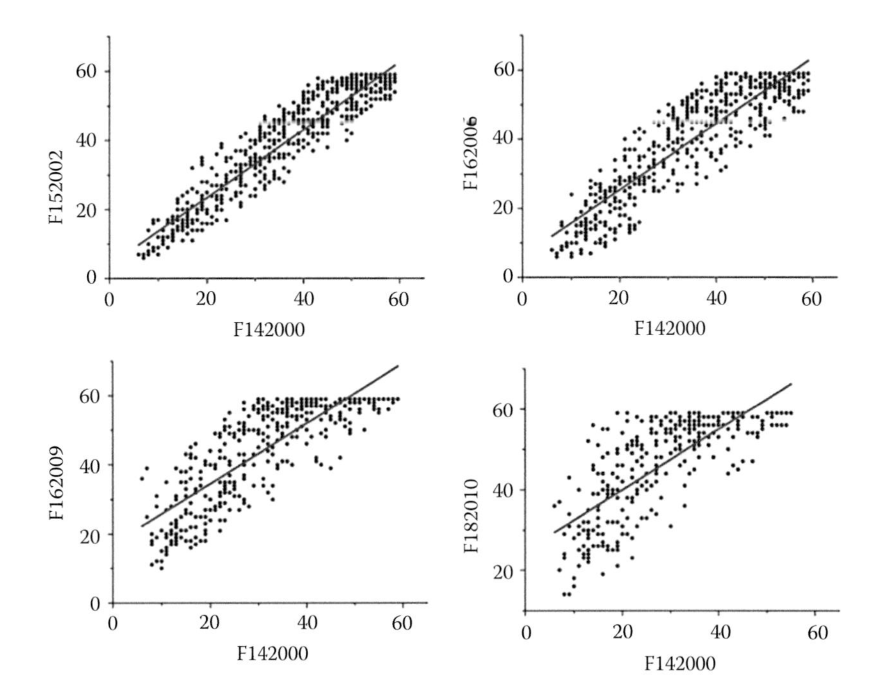

FIGURE 31.6 Scatter plots of the digital numbers (DN) values of the target years against the DN values of the reference year (2000) for pixels within desaturated pseudoinvariant features for the U.S. Defense Meteorological Satellite Program Operational Linescan System satellites. The solid line stands for the trend line of regression equation. (From Wei, Y. et al., *Landscape Urban Plan.*, 128, 1, August 2014, ISSN: 0169-2046, http://dx.doi.org/10.1016/j.landurbplan.2014.04.015. Accessed June–October, 2014.)

It is clear from this chapter that correlation between the same families of sensors (e.g., AVHRR/2 vs. AVHRR/3) is typically high and can be reasonably well established. However, correlation between dissimilar sensors (e.g., MODIS vs. AVHRR) can be poor. Nevertheless, it is feasible to establish intersensor calibrations (e.g., Figure 31.7a) that will help studies beyond a particular sensors' life span (e.g., Figure 31.7b). Chapter 7 results introduce potential methodologies that allow for the derivation of robust intersensor relationships. Even though Chapter 7 uses EVI and EVI2 to illustrate these intersensor relationships, other indices like NDVI can also be used with equal effect.

31.8 Toward Standardization of Vegetation Indices

Monitoring tools such as the NDVI have been part of remote-sensing studies ever since pioneering work reported by Tucker (1979). Simple reflectance from a fixed object will vary over time, even if the object has not changed, due to changes in the measurement conditions, so that, for example, two plants of identical type, biomass, and health will reflect differently if they are observed under different conditions. A number of such factors influence reflectivity making it difficult to conduct scientific studies over space and time using raw RS data. This difficulty is overcome by using a VI: either a simple ratio derived by dividing reflectivity in the near-infrared waveband by reflectivity in the red waveband, or its equivalent, the NDVI, or a more complex index involving adjustment coefficients for one or more confounding environmental factors. Over the years, many satellites and sensors have been launched, and their characteristics can differ in many ways: spectral band placement (e.g., differences between broad and narrow bands and their positions within a given spectral range), viewing characteristics (e.g., nadir or off-nadir viewing), conditions under which data are acquired (e.g., Sun angle, atmospheric conditions, Earth–Sun distance), sensor degradation over time, processing methods, and many others (see Chapter 8 by Michael D. Steven, Timothy J. Malthus, and Frédéric Baret). Consistency in the integration of data from different sources and continuity in long-term studies of vegetation change on planet Earth requires us to standardize VIs to adjust for intersystem differences such as, say, Landsat, with MODIS or SPOT or IRS, or SPOT VGT with AVHRR Global Inventory Modeling and Mapping Studies (GIMMS) as well as intrasystem changes such as changes of the Landsat sensor from 1972 to the present day (e.g., Figure 31.8).

Such standardization will involve normalization for various factors (e.g., as mentioned earlier), deriving VIs, and building relationships that will allow for consistent and stable monitoring of vegetation over time and space. For example, VIs of certain time-invariant areas (e.g., certain areas of Sahara desert) should remain the same over time, and extraneous factors such as view angle, atmosphere, and solar elevation should have minimal influence on measurements at such sites. Therefore, measurements over these invariant sites remain the same over time, and one can use such time-invariant sites to normalize and correct for system variations.

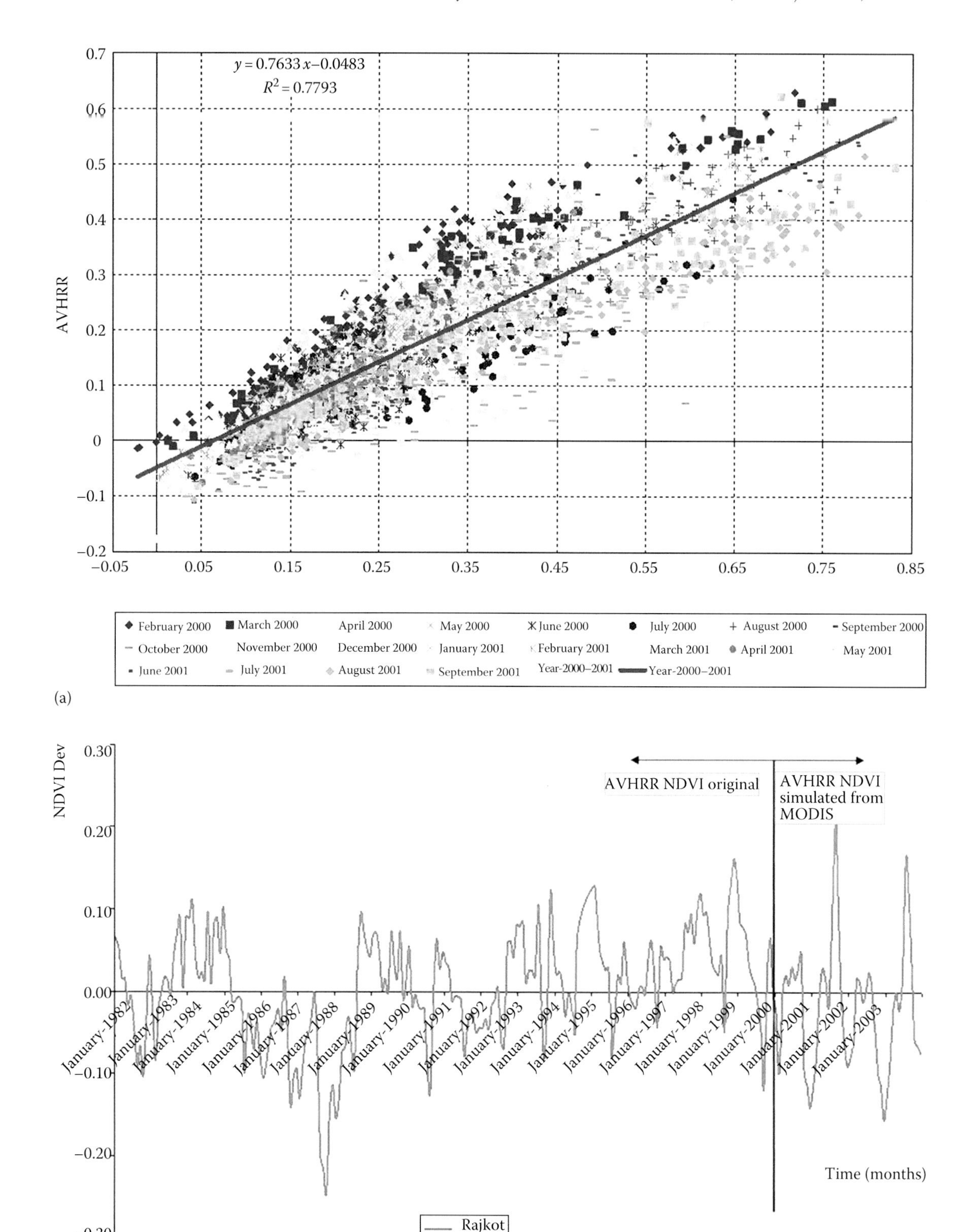

FIGURE 31.7 (a) Intersensor relationships between Moderate-Resolution Imaging Spectroradiometer (MODIS) 500 m derived normalized difference vegetation index (NDVI) versus Advanced Very-High-Resolution Radiometer 8 km NDVI. Relationship was developed taking data from various years to ensure robustness. (b) The intersensor relationship developed in (a), enabled continuation of drought studies using NDVI deviation from long-term mean ($NDVI_{dev}$). From 1982 to 2000, the study was conducted using AVHRR 8 km data. Beyond the year 2000, MODIS 500 m data were used. (From Thenkabail, P.S. et al., IWMI Research report # 85. Pp. 25. IWMI, Colombo, Sri Lanka, 2004. http://www.iwmi.cgiar.org/Publications/IWMI_Research_Reports/PDF/pub085/RR85.pdf. Accessed June–October, 2014.)

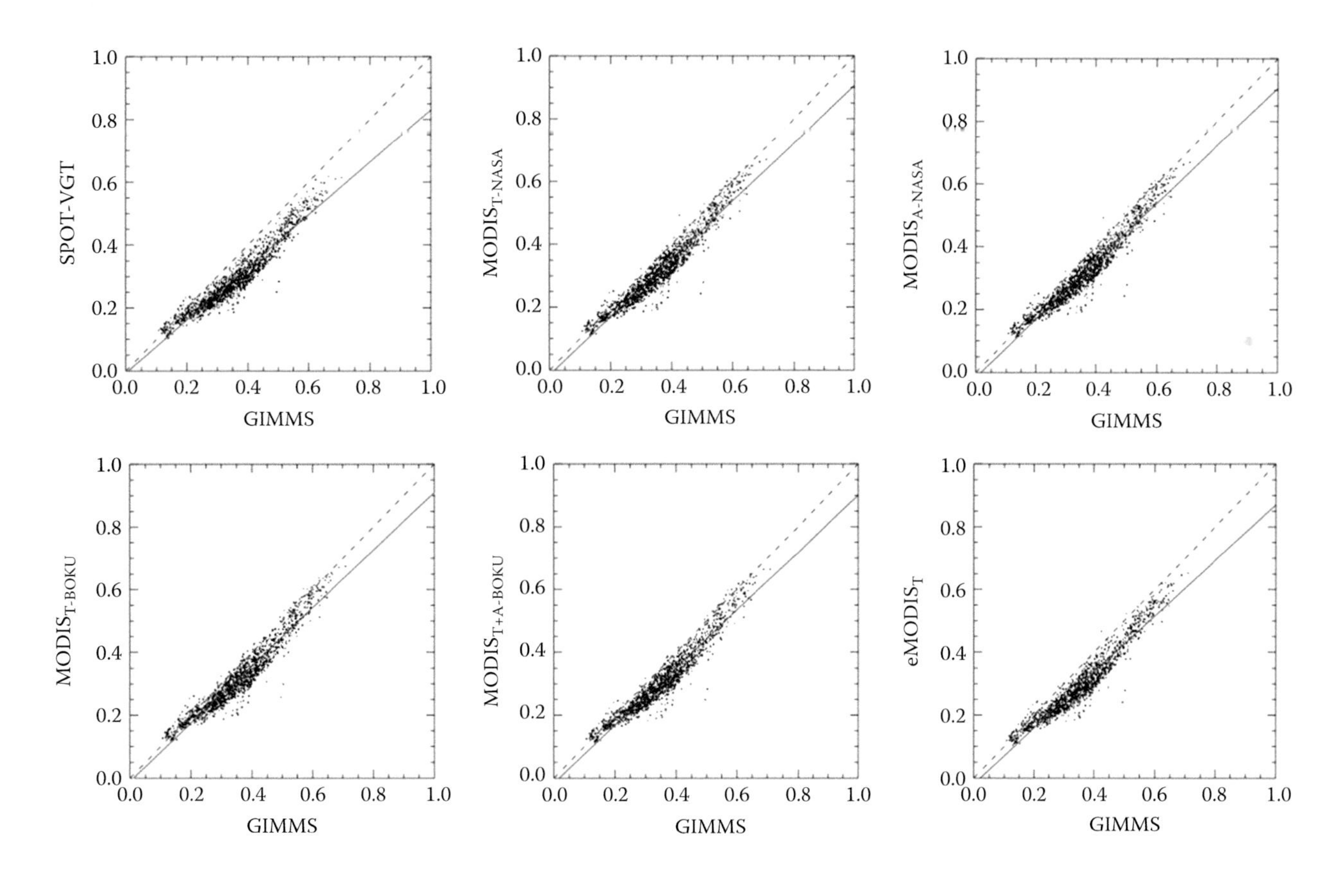

FIGURE 31.8 Example of empirical scatterplots showing intercalibration between seasonally averaged and regionally aggregated normalized difference vegetation indices (NDVIs) from various systems over Kenya. Scatterplots showing intercalibration between NDVI* (seasonally averaged NDVI and aggregated per division) derived from Global Inventory Monitoring and Modeling System (*x*-axis) against NDVI* from each of the other NDVI products such as the Système Pour l'Observation de la Terre–VEGETATION and various Moderate-Resolution Imaging Spectroradiometer data. Each plot contains a total of 1512 data points (84 divisions × 9 years × 2 seasons/year). (From Vrieling, A. et al., *Int. J. Appl. Earth Observ. Geoinform.*, 28, 238, May 2014, ISSN: 0303-2434, http://dx.doi.org/10.1016/j.jag.2013.12.010. Accessed June–October, 2014.)

In Chapter 8, Drs. Steven, Malthus, and Baret start with how VIs are computed, what factors influence variation in VIs, and VI intercalibration approaches. The chapter then provides equations for cross-sensor intercalibration, based on precise simulations of system spectral responses from a common dataset of field observations. Then, the chapter lays out certain universal principles for VI standardizations. The principles include that the standardization should be (a) applicable to VIs from all systems, (b) fully traceable as standards evolve and develop over time, and (c) modular. Finally, the chapter outlines specific proposals for VI standards and discusses the precision of the adjustments required.

31.9 Digital Image Processing: Methods and Techniques

Chapter 9 by Dr. Sunil Narumalani and Paul Merani provides a step-by-step approach to understanding the fundamental methods and techniques involved in digital image processing. These steps are summarized in Table 31.1 and are broadly grouped into image: (1) quality assessment, (2) preprocessing, (3) enhancement, (4) registration and reprojections, (5) mosaicking and megafile creation, (6) analysis, (7) classifications, and (8) accuracy assessments. Fundamental methods and approaches (see Jensen, 1996; Lillesand et al., 2008; Richards and Xiuping, 2006) have remained firm over the years and are a must for anyone wanting to master the remote-sensing data-processing methods and techniques. In recent years, numerous advances have been made in digital image processing (Nagamalai et al., 2011) that include data fusion, object-based image analysis (OBIA) or geographic OBIA (GEOBIA), support vector machines (SVMs), decision tress, spectral unmixing, artificial neural networks (ANN), phenological matrices, Fourier transforms, and spectral matching techniques (SMTs). In data composition, multidate, multispectral data are time composited to long time periods into single megafile data cubes, akin to hyperspectral data cubes. This has facilitated applying hyperspectral data analysis techniques like SMTs for multidate multispectral data composited over long time periods. Such data enable analysis of time series phenological matrices. All aspects of image analysis (Table 31.1) have become efficient as a result of speed in computing power as well as smarter and better algorithms.

TABLE 31.1 Standard Digital Image Processing Steps

1. Image quality evaluation
Histograms evaluation (e.g., frequency distribution)
Basic statistical evaluation (e.g., mean, median, mode)
Visual evaluation (e.g., for cloud, haze)
Digital number (brightness value) evaluation (e.g., in various bands indication conditions)
2. Image preprocessing
Radiometric correction (digital numbers to at-sensor reflectance)
Atmospheric correction (deriving ground reflectance)
Geometric correction (corrections for various geometric distortions)
3. Image enhancement
Contrast enhancement (e.g., histogram stretch)
Edge enhancement (e.g., filtering)
Color enhancement (e.g., true and false color)
Differencing (e.g., from one date with another)
Ratioing (e.g., vegetation indices)
Transformations (e.g., PCA, vegetation indices)
4. Image registration and reprojections
Image to image (e.g., all images of a area to a master reference image)
Image to map (e.g., image to a reference map or GPS points)
Image to coordinate systems (e.g., assigning projections, datum)
5. Image mosaicking, fusion, and megafile creations
Mosaicking in required coordinate systems (e.g., entire country, continent, or globe)
Data fusion (e.g., multispectral with panchromatic)
Time compositing (e.g., monthly maximum value composites)
Megafile data cubes (e.g., single file from multiple images of same year over several years)
6. Image analysis
Thresholding (e.g., NDVI thresholds)
Edge detection (e.g., to highlight lineaments)
Texture analysis (e.g., to segment an image into distinct objects)
Change analysis (e.g., drought year vs. non-drought year)
7. Image classifications
Supervised classification (e.g., maximum likelihood, minimum distance to mean)
Unsupervised classification (e.g., ISOCLASS clustering)
Decision trees (e.g., rule base)
Object based image analysis (e.g., segment an image into objects and then classify)
Spectral unmixing (e.g., linear spectral unmixing)
Support vector machines
Artificial neural networks
Fourier transforms
Spectral matching techniques (e.g., spectral similarity value, spectral correlation similarity)
8. Image accuracy assessment
Error matrices (e.g., remote sensing versus reference data, leading to user's producer's and overall accuracies)
Fuzzy classification accuracies (e.g., accuracy measures in terms such as absolutely correct to absolutely wrong)
Comparison of statistics with other reference data (e.g., areas computed from remote sensing with national statistics)

31.10 Urban Image Classification Methods and Approaches

Chapter 10, by Dr. Soe W. Myint et al., discussed four distinct methods of image classification for urban mapping. These four methods, discussed in Chapter 10, are summarized as follows:

1. Per-pixel spectrally based methods (hard classifications, e.g., Figure 31.9a) are used in all types of classification (e.g., land use/land cover, urban, or forests categories) using algorithms such as maximum likelihood, iterative self-organizing data analysis technique (ISODATA), and minimum distance to mean. In per-pixel classifiers, each pixel falls into a class or a group and forms minimum mapping unit. It is noted that the per-pixel classifiers have significant limitations in detecting and accurately mapping individual urban features, but they can produce simple and convenient thematic maps with reasonable accuracies (often in realm of 70%) in producing few urban classes.
2. Subpixel methods (soft classifications, e.g., Figure 31.9b) quantify fractions of multiple classes within a pixel. Due to complexity of landscape, the presence of multiple classes within a pixel is common. Methods used to resolve subpixel composition of pixels include linear methods (e.g., linear spectral mixture analysis, regression trees, and regression analysis) and nonlinear methods (e.g., ANN). Success of linear spectral mixture analysis will depend on the purity of collecting endmember (preferably through exact ground knowledge). Typically, the number of endmembers does not exceed the number of bands. However, multiple endmember spectral mixture analysis can overcome this. Nevertheless, it is complex to identify all the possible endmembers, and the many unidentified endmembers lead to uncertainties.
3. With the advent of very-high-resolution (submeter to 5 m) imagery around year 2000, OBIA or GEOBIA (Blaschke, 2010) became a standard approach to discern numerous urban features like individual buildings, pools, roads, sidewalks, urban trees, and golf courses. This involves use of spectral and textural characteristics of the image data to first segment the image into distinct groups and then apply classification algorithms separately on individual segments to identify detailed urban classes accurately. Yet computing challenges of large number of segments that need to be carried out based on various scale parameters in object-oriented algorithms (e.g., Definiens or eCognition) remains challenging. Nevertheless, numerous studies that have shown an increase of anywhere between 10% and 20% increase in accuracies (reaching the realm of 90% accuracy) using OBIA or GEOBIA have been reported when compared with pixel-based approaches (e.g., Figure 31.9).
4. Some geospatial methods such as spatial co-occurrence matrix, spatial autocorrelation, and fractal and lacunarity approaches capture the local texture similarities into distinct group of pixels, before a classification algorithm

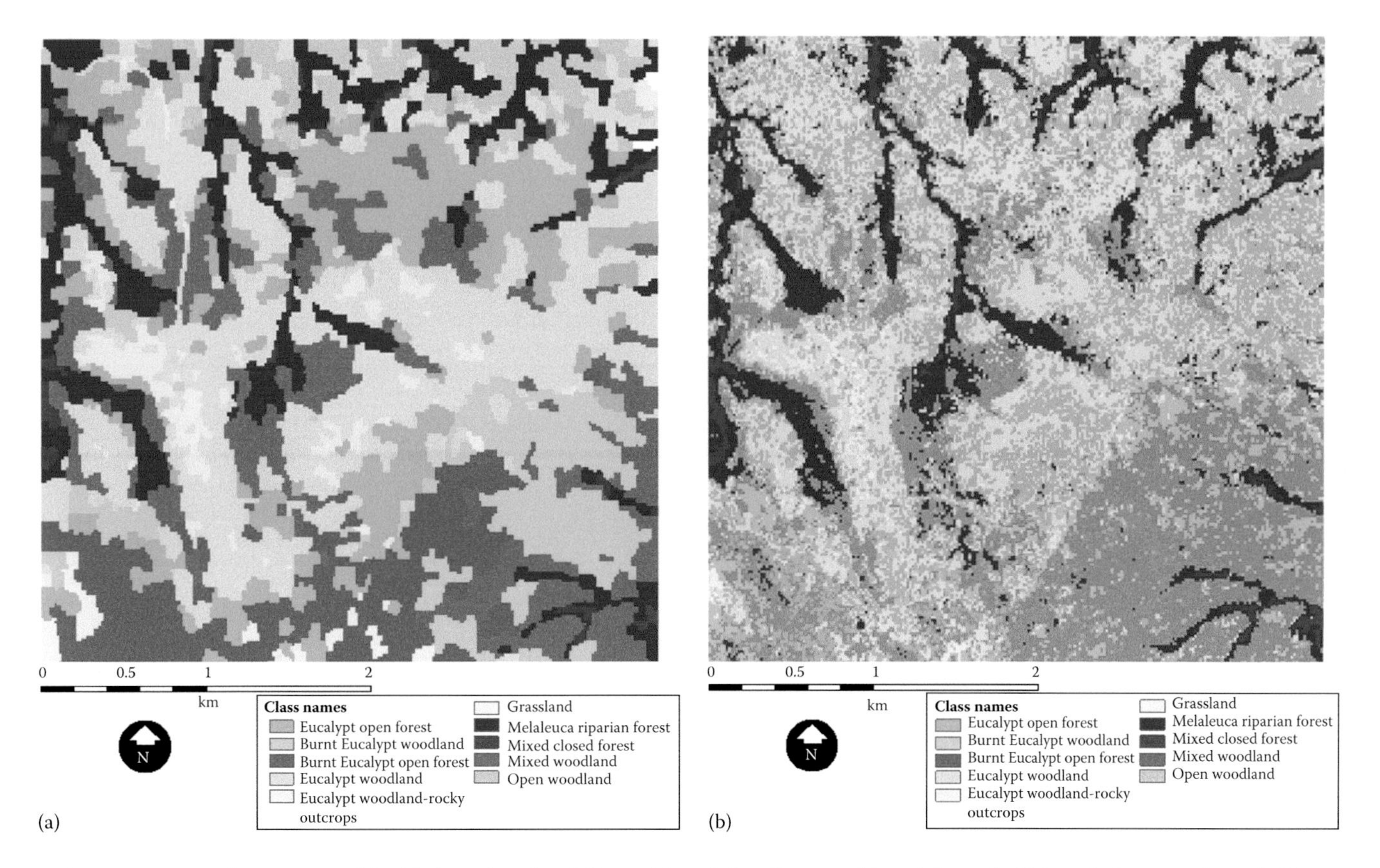

FIGURE 31.9 Resultant images of the (a) object-based and (b) pixel-based classifications. (From Whiteside, T.G. et al., *Int. J. Appl. Earth Observ. Geoinform.*, 13(6), 884, December 2011, ISSN: 0303-2434, http://dx.doi.org/10.1016/j.jag.2011.06.008. Accessed June–October, 2014.)

is applied to improve classification accuracy. This can be done by local moving windows (e.g., 2 × 2, 3 × 3, 5 × 5). In contrast to the aforementioned advanced geospatial approaches, a new spatial frequency-based algorithm called wavelets characterizes spatial features in different directions at multiple scales. A window-based approach employing wavelet theory and dyadic decomposition procedures to measure spatial arrangements of features at multiple scales has been determined to be more effective than the aforementioned advanced geospatial approaches in directly identifying urban classes (Xu et al., 2013). While this approach outperformed other spatial approaches, it is limited by a finite decomposition procedure that requires a large geographic area (or a large window size) to perform a classification. An overcomplete wavelet analysis using an infinite-scale decomposition procedure is more powerful in measuring the spatial complexity of geographic features and identifying detailed urban classes more effectively.

31.11 Image Classification Methods in Land Cover and Land Use and Cropland Studies

Chapter 11 by Dr. Mutlu Özdoğan discussed methods of classification of land cover and land use (e.g., Figure 31.10), vegetation, croplands, and other land themes. The classification methods are broadly classified into two categories:

1. Parametric
2. Nonparametric

Parametric classification algorithms assume a known (often normal) statistical distribution of data. Parametric classification methods include minimum distance to mean, Gaussian maximum likelihood, Mahalanobis distance, and parallelepiped.

Nonparametric methods are not limited by any statistical distribution assumption of data and are based on class statistics that include mean vectors and covariance matrices. This is specifically suitable in conditions where there is great heterogeneity in landscape and variances between classes are high. Nonparametric classification methods include classification and regression trees, ANN, *k*-nearest neighbor, and SVMs.

These classification algorithms are applied to satellite data through either supervised or unsupervised logic. In a supervised approach, classes are trained through available knowledge (e.g., ground data, accurate secondary maps, expert knowledge). Widely used supervised methods include Gaussian maximum likelihood and minimum distance to mean. Decision tree algorithms, random forest (RF), SVMs, ANN, and nearest neighbor classifier are also supervised classifications but fall into the nonparametric category. Gaussian maximum likelihood, *k*-means

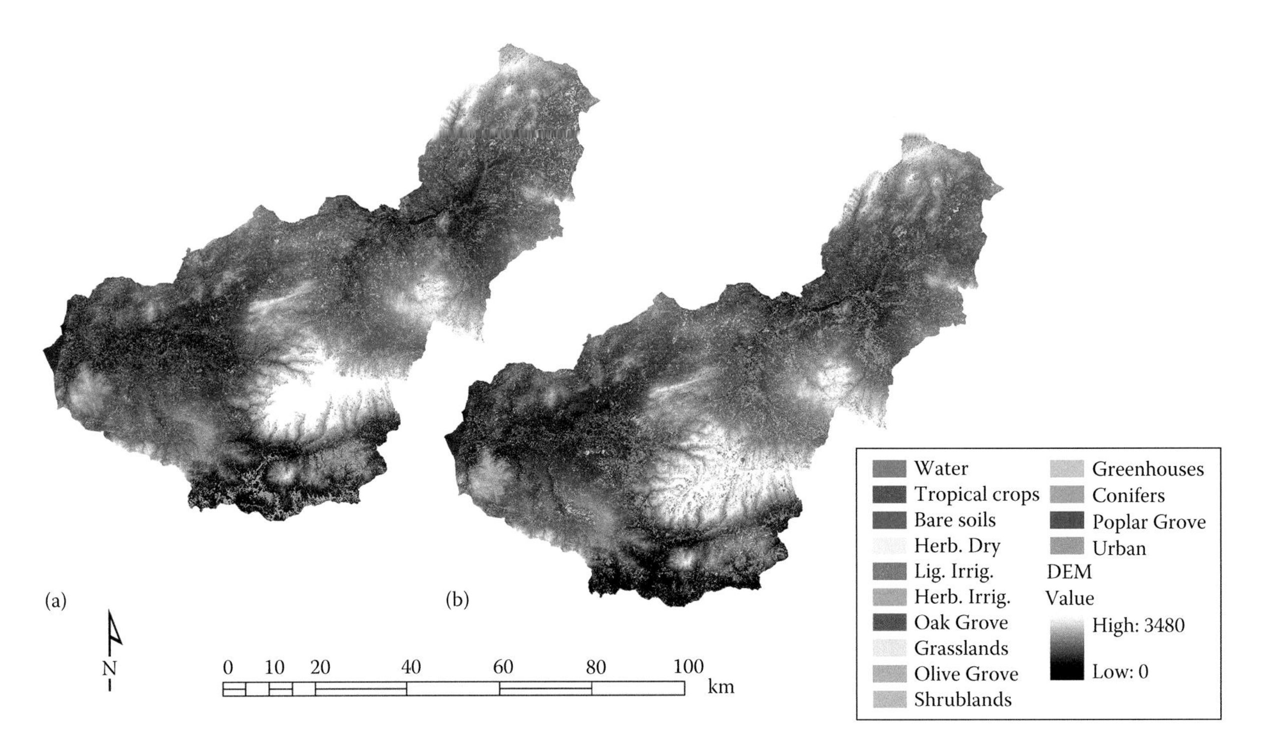

FIGURE 31.10 Land cover change maps between classification trees (CT) and random forest (RF) for Granada Province: (a) RF and (b) CT. (From Rodriguez-Galiano, V.F. et al., *ISPRS J. Photogram. Remote Sens.*, 67, 93, January 2012, ISSN: 0924-2716, http://dx.doi.org/10.1016/j.isprsjprs.2011.11.002. Accessed June–October, 2014.)

nearest neighbor, and SVMs, on the other hand, are widely used supervised parametric methods.

In unsupervised classification, the algorithm assigns classes based on spectral characteristics (e.g., mean, standard deviation, matrices such as covariance, and correlations) of a pixel. Classes are then identified based on available knowledge (e.g., ground data, reference data), which will then turn the information classes generated by the algorithm into land cover classes. Unsupervised classification is also known as clustering and includes ISODATA clustering, *k*-means clustering, and Narendra–Goldberg clustering. Unsupervised classification approach is recommended when a priori knowledge of the landscape (e.g., ground data) is unavailable, limited, or inadequate. *k*-Means is an unsupervised nonparametric classification algorithm. Note that there are times that unsupervised classification is used as a first step, before applying supervised classification. This is known as a hybrid classification.

These image classification methods can be performed based on (a) per pixel, (b) subpixel, (c) per field, and (d) per object. These can also be soft classification (e.g., mixed classes such as vegetation that may include croplands, rangelands, forests) meaning that the class boundaries can intermingle to some extent or hard classifications (e.g., croplands, soils, rangelands) when the class boundaries are distinct without any intermingling.

Chapter 11 also discusses other classification approaches that are becoming more popular. For example, object-oriented classifications or a combination of object-oriented segmentation followed by pixel-based classification is propagated by some (e.g., Chapters 14 through 16).

31.12 Hyperspectral Image Processing Methods and Approaches

Hyperspectral data are rich in information content but bring with it many challenges of data handling and processing. In Chapter 12, Dr. Jun Li and Dr. Antonio Plaza explore some current trends in hyperspectral image processing. The classification approaches discussed in the chapter include the following:

1. Supervised classification consisting discriminant analysis techniques. In these discriminant classifiers, discriminant functions such as nearest neighbor, decision trees, linear functions, and nonlinear functions are applied.
2. Kernel methods for supervised classification including the well-known SVM classifier.
3. Spectral–spatial classification approaches, which provide significantly improved classification results compared to spectral-based classifiers. Spatial features are extracted through advanced morphological techniques such as morphological profiles and Markov random fields as described in the chapter.
4. Probabilistic classification methods such as the multinomial logistic regression.

The main problem with the supervised classifiers is addressing the huge data dimensionality of hyperspectral data in comparison with the limited availability of training samples. Specifically, supervised approaches require sample data from pure locations both for classification and validation. The number of samples required increases with data dimensionality, and as a result, the challenge of using a large number of training samples to address the high dimensionality of hyperspectral data is costly and often prohibitive. This issue, known as Hughes phenomenon, is hard to address and is often the bottleneck in the effective use of supervised classification approaches. So, often, the uncertainty in classification results from the supervised hyperspectral image analysis is high.

The concept of semisupervised classification, as discussed by Dr. Jun Li and Dr. Antonio Plaza, is to use not only the labeled samples for training but also unlabeled data. Semisupervised learning is performed, for example, through transductive (or machine learning) SVMs, among other strategies discussed in the chapter. However, the unlabeled data may not be confident enough and hence can also lead to uncertainties.

The aforementioned difficulties related with the Hughes phenomenon in supervised classification methods can be effectively overcome through kernel methods such as the SVM. This is also discussed extensively in Chapter 7. Authors use a well-known and widely used hyperspectral dataset, collected by the reflected optics spectrographic imaging system over the University of Pavia, Italy, to provide a series of results comparing the most commonly used SVM classifier with (1) several discriminant classifiers, (2) composite SVM obtained using summation kernel, (3) combination of the morphological EMP for feature extraction followed by SVM for classification (EMP/SVM), (4) pixel-wise SVM classifier with the morphological watershed, (5) SVM classifier combined with the segmentation result provided by the unsupervised recursive hierarchical segmentation (RHSEG) algorithm (Tilton et al.), and (6) a subspace-based multinomial logistic regression classifier followed by spatial postprocessing using a multilevel logistic prior (MLRsubMLL). For this purpose, they use nine land use classes and show how SVM performs relative to other methods. In general, the performance of classification increases progressively as we move from step 1 to step 6. The chapter does not discuss hyperspectral image analysis using unsupervised classification, even though RHSEG is discussed briefly.

Some of the well-known hyperspectral data dimensionality reduction and classification methods are summarized in Figure 31.11 by Damodaran and Nidamanuri (2014). They concluded that the choice of dimensionality reduction method and classifier significantly influences the classification results obtained in hyperspectral image analysis.

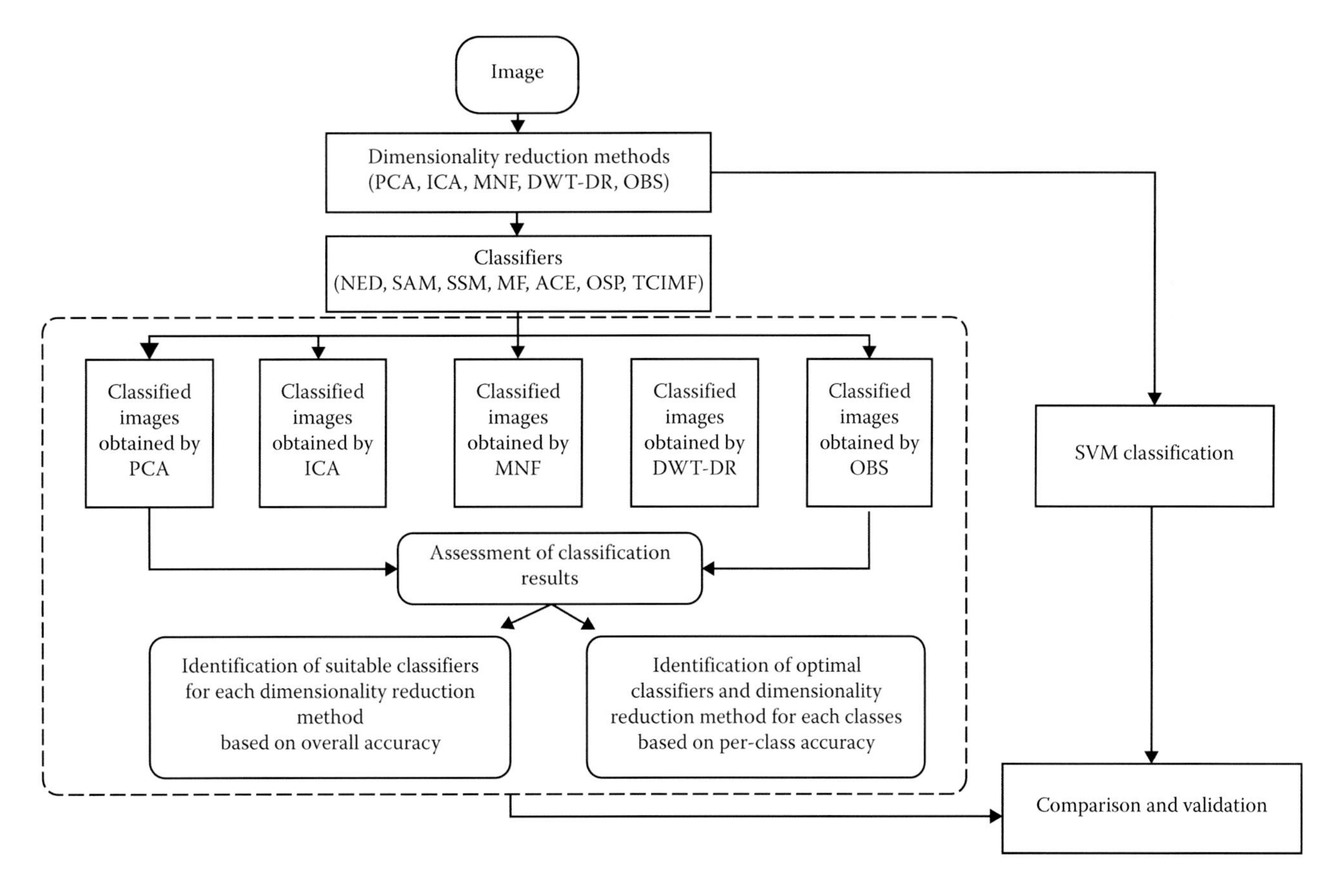

FIGURE 31.11 Flowchart showing hyperspectral image data reduction and classification methods. Dimensionality reduction methods include principal component analysis, independent component analysis, minimum noise fraction, discrete wavelet transform–based dimensionality reduction, and optimal band selection. Multiple classifier system includes normalized Euclidian distance, spectral angle mapper, spectral similarity measure, matched filter, adaptive coherence estimation, orthogonal subspace projection, and target constraint minimum filter. (From Damodaran, B.B. and Nidamanuri, R.R., *Adv. Space Res.*, 53(12), 1720, 2014, ISSN: 0273-1177, http://dx.doi.org/10.1016/j.asr.2013.11.027. Accessed June–October, 2014.)

31.13 Thermal Infrared Remote Sensing: Principles and Theoretical Background

Different authors define the wavelength range of thermal infrared (TIR) domain differently, but most sensor measurements are undertaken in the 8–14 μm wavelength range. In general, TIR is the field of remote sensing, which utilizes emitted radiation in the wavelength domain, where our planet has its emission maximum. A large example of TIR data is its independence of an illumination source: TIR sensitive sensors can collect TIR data during the day as well as during the night. The principles and theoretical background of thermal data and thermal data analyses are presented in Chapter 13 by Kuenzer et al. This chapter not only provides a comprehensive overview of the theoretical background including all the laws applicable for the TIR domain but also elucidates numerous application examples.

TIR data usage is exemplarily demonstrated for the analyses, the general derivation of land and sea surface temperature (LST), the derivation of urban heat islands, and the detection of hot spots, including thermal anomalies resulting from forest fires, coal fires, or gas flaring activities. Also, examples of sea surface temperature (SST) products and thermal water pollution are presented. The authors furthermore explain the valuable principle of diurnal temperature changes (ΔT), which makes use of the fact that different surface types exhibit a different thermal behavior over the day. Some heat up faster than others in the morning, some cool down slower than others, and the extent of the diurnal temperature change (ΔT) contains valuable clues about the physical characteristics of a surface. For example, water bodies warm up very slowly during the day and cool down slowly during the night—overall, their diurnal temperature change (ΔT) of water is very low and usually lies within the range of only few degrees Celsius. ΔT is much more accentuated for, for example, a dark asphalt surface, which heats up very fast during a sunny day and also cools down relatively fast after dawn. Here, diurnal temperature change is much more accentuated and can reach several tenths of degrees Celsius. This principle can be exploited when diurnal data are available, for example, daytime and nighttime data of the Landsat thermal band or diurnal data of the MODIS sensor thermal bands, and can support the TIR-based mapping of different surface types. Furthermore, statements on the moisture content of the surface can be derived as, for example, a wet soil will exhibit a lower diurnal temperature change (ΔT) than a dry soil.

However, TIR data can also be used for many other applications, which could not be presented within the chapter. Airborne TIR data can support the monitoring of irrigated areas and agricultural monitoring of crop water stress, can support with the detection of technical accidents and leakages (e.g., pipe bursts, nuclear accidents), and are also frequently employed in medical imaging.

One major limitation of spaceborne thermal data is their relatively low spatial resolution. Whereas optical and multispectral data in the visible and near infrared domains can be collected with resolution of better than 1 m pixel size, spaceborne thermal sensor data are collected at between 60 m and 1 km spatial resolution. Whereas Landsat 7 ETM+ has a thermal band acquiring the data at 60 m, the novel Landsat 8 (formerly LDCM or Landsat Data Continuity Mission) only offers 100 m spatial resolution in the thermal bands, and sensors such as AVHRR or MODIS only allow for the monitoring and mapping at 1 km resolution. In the TIR domain, bandwidths and spatial resolution usually need to be wider and lower to ensure that a proper amount of incoming energy is collected. Furthermore, higher-resolution TIR sensors are very costly to build. However, as demonstrated in Chapter 13 by Kuenzer et al., the number of nations launching satellites including TIR sensors is growing rapidly. We can therefore be confident that remote sensing in the TIR domain will continue to strive catering to unique applications as well supplementing\complimenting optical remote sensing (e.g., Figure 31.12).

31.14 Object Based Image Analysis (OBIA): Evolution and State of the Art

OBIA or GEOBIA—the latter is used when emphasizing scales and applications in remote sensing and geographic information science—evolved with the advent of very high-spatial-resolution satellite imagery (submeter to 5 m) around the year 2000. However, the underlying concepts of OBIA have been around far longer as enumerated in Chapter 14 by Dr. Thomas Blaschke et al. OBIA involves segmenting an image into distinct objects and then performing the classification on distinct objects separately by utilizing the spatial, spectral, radiometric, and temporal characteristics of image data and eventually auxiliary data. OBIA addresses object properties such as shape, size, pattern, tone, texture, shadows, and association (see Blaschke, 2010; Blaschke et al., 2014). An overview of GEOBIA approach to classification is shown in Figure 31.13 (Ke et al., 2010) in classifying forest species using very-high-spatial-resolution QuickBird multispectral imagery and LiDAR data. QuickBird imagery is used for spectral-based segmentation and LiDAR data used for LiDAR-based segmentations (such as height and intensity Figure 31.13).

Chapter 14 discusses the evolution of OBIA, concepts of image segmentation and fusion, classification and synthesis, and various applications of GEOBIA. As Chapter 14 shows us, there is substantial increase in accuracies using OBIA as opposed to traditional per-pixel classification results. For example, GEOBIA is now increasingly used for various integration tasks including new technologies such as mobile applications (e.g., locating where you make a call), geo-intelligence, and volunteered geographic information (VGI).

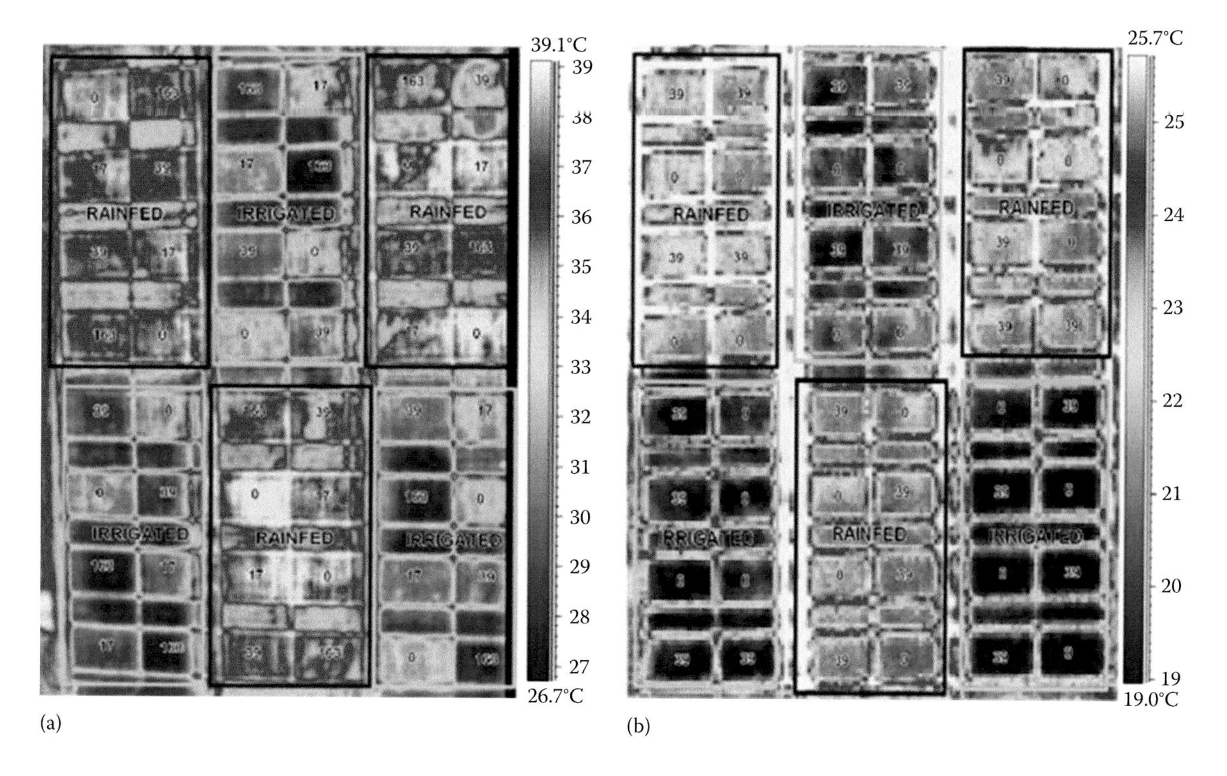

FIGURE 31.12 Thermal images gathered using an airborne ThermaCAM P40 (FLIR systems AB, Danderyd, Sweden) infrared digital imager measuring emitted energy in the range of 7.5–13 μm for an experiment site located near Horsham, Australia (36°44′S, 142°06′E; elevation 133 m). The study areas comprised 48 plots (18 m × 12 m) planted to wheat arranged in a randomized block design with three replications for (a) October 2004 and (b) October 2005. Rainfed or irrigated blocks are indicated. Numbers denote kg/ha N applied to each plot. Lighter shades correspond to higher temperatures. However, readers should note that 90% of Chapter 14 (Section 13.13) focuses on spaceborne. Readers should note thermal airborne thermal infrared data are really rare and rarely used. But this is a good application example that one can easily replicate with spaceborne sensors as well. (From Tilling, A.K. et al., *Field Crops Res.*, 104(1–3), 77, October–December 2007, ISSN: 0378-4290, http://dx.doi.org/10.1016/j.fcr.2007.03.023. Accessed June–October, 2014.)

31.15 Geospatial Data Integration in OBIA and Implications on Accuracy and Validity

OBIA is now extensively supporting geographic information needs. The type of information derived from OBIA is wide ranging and could include such features as forest species, buildings, road networks, and farm field boundaries. Availability of very-high-spatial-resolution imagery (submeter to 5 m) from various satellite sensors (e.g., IKONOS, QuickBird, GeoEye), LiDAR, and UAVs has made deriving objects over large areas feasible. However, remote-sensing data are acquired in a number of platforms, in number of resolutions (spatial, spectral, radiometric, and temporal). It is handled by different users in different way (e.g., different atmospheric correction models, with and without atmospheric correction). Methods and approaches in analyzing the data are often different.

In Chapter 15, Dr. Stefan Lang and Dr. Dirk Tiede discuss strategies and approaches of geospatial data integration in OBIA. The chapter introduces various concepts of object validation that include multistage validation, object fate analysis, object-based accuracy assessment, object-based change detection, and object linking. Object identification means able to delineate objects accurately through such means as color, texture, shape, size, scale, or form. Recent advances in OBIA have resulted in increasing accuracies in the objects delineated such as the building features and their areas (e.g., Figure 31.14). Beyond that, OBIA enables to assess the way how objects are delineated, in terms of appropriate scale, complexity, shape, and the fitness to existing geospatial datasets.

31.16 Image Segmentation Algorithms for Land Categorization

Early remote sensing heavily depended on land categorization and classification algorithms that are purely spectral based. These methods include simple thresholding (e.g., using NDVI or band reflectivity), single or multiband classifications using supervised or unsupervised approaches as *k*-means algorithms, maximum likelihood, and minimum distance to mean. But, over the years, remote-sensing scientists have been looking to improve our understanding of the land categorization with improved accuracies and reduced uncertainties. One of the powerful approaches has been to include image segmentation in the analysis. Chapter 16 by Dr. Tilton et al. provides a comprehensive overview of various segmentation algorithms and illustrates their strengths and limitations. The chapter

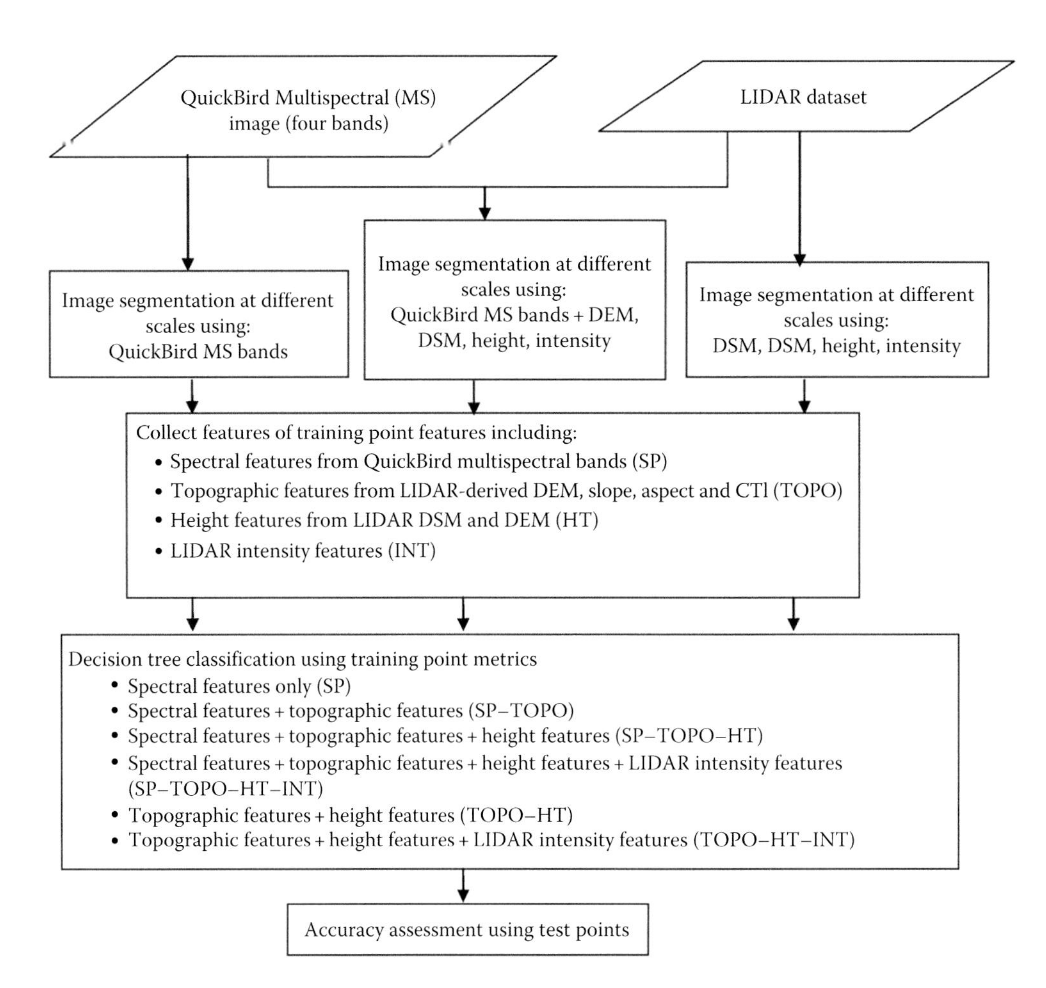

FIGURE 31.13 An object-based image analysis protocol for forest species classification through synergistic use of QuickBird multispectral imagery and LiDAR data. (From Ke, Y. et al., *Remote Sens. Environ.*, 114(6), 1141, 2010, ISSN: 0034-4257, http://dx.doi.org/10.1016/j.rse.2010.01.002. Accessed June–October, 2014.)

also summarizes in a table various commercial and noncommercial segmentation software packages available and discusses on where and how they have been used. Chapter 16 shows that segmentation algorithms involve using spectral, textural, spatial, and other secondary (e.g., elevation, slope) information of the landscape obtained from remote sensing to "segment" the land into unique units or segments (e.g., uplands from lowlands or agricultural lands from natural vegetation) and then classify them separately. This reduces the complexities within each segment and also enables shape-based as well as contextual modeling of segments that are not possible using pixel-based approaches. Furthermore, when classification algorithms are applied separately on each segment, we have more meaningful and accurate classes. In addition, we can also get more unique classes. For example, lowland vegetation is often distinct from upland vegetation. When we classify the image that included both upland and lowland segments, we will be often not able to distinguish upland and lowland vegetation with great degree of accuracy due to spectral limitations of data. But, when we look at them separately, we will be able to categorize upland vegetation and lowland vegetation with greater degree of certainty. There are many other segmentation algorithms evolving, for instance, region-based image segmentation algorithm based on *k*-means clustering (Wang et al., 2010; e.g., Figure 31.15). Chapter 16 concludes that the existing and evolving wide range of image segmentation approaches provide a rich menu for the users to choose the method, which is the best adapted for each particular application.

31.17 LiDAR Data Processing and Applications

LiDAR data are collected as "point clouds" with each point having 3D coordinates: horizontal (x, y) and vertical (z). LiDAR data are acquired with a laser scanning system in very high point rate. For each point measurement, a laser pulse is transmitted and the reflected energy is caught. The return signal can be fully recorded as a waveform or interpreted to be a single or multiple returns to form point clouds. The features embedded in LiDAR "point clouds" include location (x, y, z), intensity, echo, and waveform. The data can be used to model, map, and study various Earth surface features such as tress, buildings, and terrain in 3D. There are four broad categories of LiDAR systems: (1) terrestrial or ground-based laser scanning with footprint of 1–10 cm and point accuracy of 1–5 cm, (2) airborne laser scanning (ALS) with footprint of 5–50 cm and

(a)

(b)

FIGURE 31.14 Assessing accuracy of geographic object-based image analysis (GEOBIA) extracted building features and their areas (a) relative to reference data (b) for Lisbon, Portugal. Data used for GEOBIA include QuickBird pan sharpened 0.6 m, QuickBird multispectral 2.4 m, QuickBird image-derived normalized difference vegetation index 2.4 m, normalized digital surface model (nDSM) 1 m, and reference map of bridges. GEOBIA extracted 317 of the 330 features found in reference data. (From Freire, S. et al., *ISPRS J. Photogram. Remote Sens.*, 90, 1, 2014, ISSN: 0924-2716, http://dx.doi.org/10.1016/j.isprsjprs.2013.12.009. Accessed June–October, 2014.)

point accuracy of 10–100 cm, (3) mobile laser scanning with footprint of 1–10 cm and point accuracy of 1–30 cm, and (4) spaceborne LiDAR with footprint of 30–100 m and vertical accuracy of 25 cm–10 m (e.g., GLAS, DESDYnI, ICESatII). Waveform LiDAR is recent advance in ALS technology. The ability to capture 3D structure of plant and objects is a major advantage of LiDAR data, unlike other remote sensing. In 3D modeling, it has big advantages compared to photogrammetry or land surveying due to its relatively fast acquisitions and automated processing. However, LiDAR data often have small coverage of data collection and large processing volume of data and are affected by cloud and haze.

Chapter 17 by Dr. Yi-Hsing Tseng and others focuses on LiDAR data processing and applications. The LiDAR data processing involves noise filtering, processing 3D (x, y, z) points, independent variables such as laser use time, scanner position, and scanner orientation. The LiDAR data are often delivered in the American Society for Photogrammetry and Remote Sensing (ASPRS) laser (LAS) format. Data quality is required to meet accuracies of ASPRS large-scale mapping standard and/or the national spatial *data* accuracy standard. The chapter extensively discusses LiDAR data:

1. Quality assessment and control (e.g., system components, error budgets, quality assessment and control)
2. Management (e.g., storage of LiDAR point cloud, spatial indexing, hierarchical representation)
3. Point cloud feature extraction (e.g., spatial feature in LiDAR data, methods of extraction of spatial features that include extraction of line and plane features)
4. Full-waveform airborne processing (e.g., shape of received waveform, echo detection, feature extraction)

The chapter enumerates various applications of LiDAR data such as in the following:

1. *Forestry studies*: Structure, diameter at breast height, crown size, basal area, tree height, aboveground biomass (e.g., Kronseder et al., 2012; Figure 31.16). LiDAR ability to penetrate vegetation and look through it and capture information is a huge benefit that is infeasible using other remote-sensing techniques.
2. digital elevation model (DEM), digital terrain model (DTM), digital surface model (DSM), and triangular irregular network (TIN) generation.
3. City modeling (e.g., building detection/construction, road extraction).
4. Structures and objects (e.g., dams, bridges).
5. Agriculture (e.g., farm topography, plant height).
6. Hydrology and geomorphology (e.g., stream network, slopes derived off DEM).
7. Land cover classification.
8. Flood mapping and assessment.
9. Disasters (e.g., earthquakes).

LiDAR data are often used by integrating with other remote-sensing data to achieve higher accuracies and discern features that is otherwise infeasible.

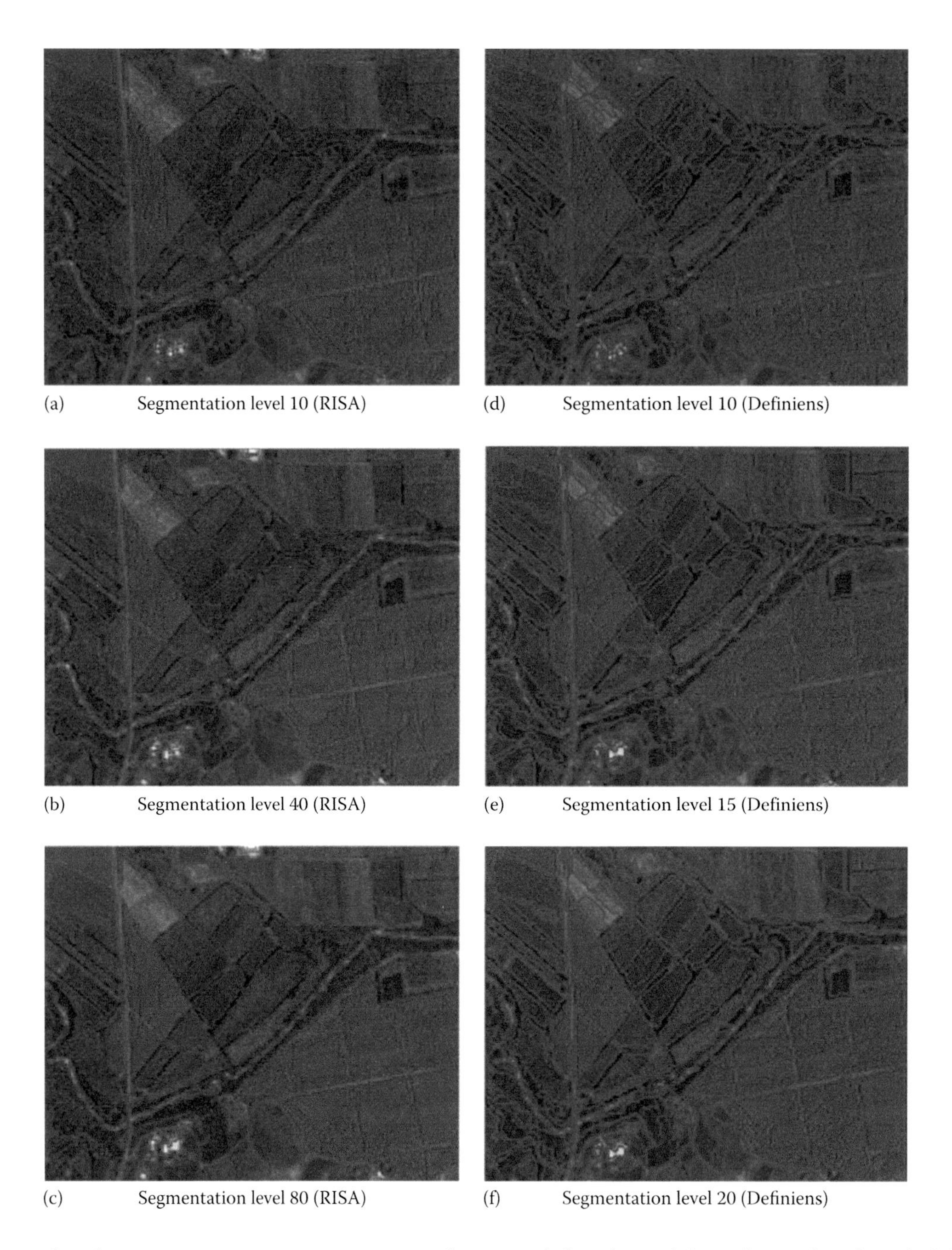

FIGURE 31.15 Multiscale image segmentation comparison using the proposed algorithm and the Definiens algorithm. The left three images (a, b, and c) were segmented using region-based image segmentation algorithm based on *k*-means clustering (RISA) with the default parameter values and segmentation scales of 10, 40, and 80, respectively. The right three images (d, e, and f) were segmented using the Definiens software with the color criterion of 0.9, shape criterion of 0.1 (compactness 0.2 and smoothness 0.8), and segmentation scales of 10, 15, and 20, respectively. (From Wang, Z., et al., *Environ. Modell. Softw.*, 25(10), 1149, October 2010, ISSN: 1364-8152, http://dx.doi.org/10.1016/j.envsoft.2010.03.019. Accessed June–October, 2014.)

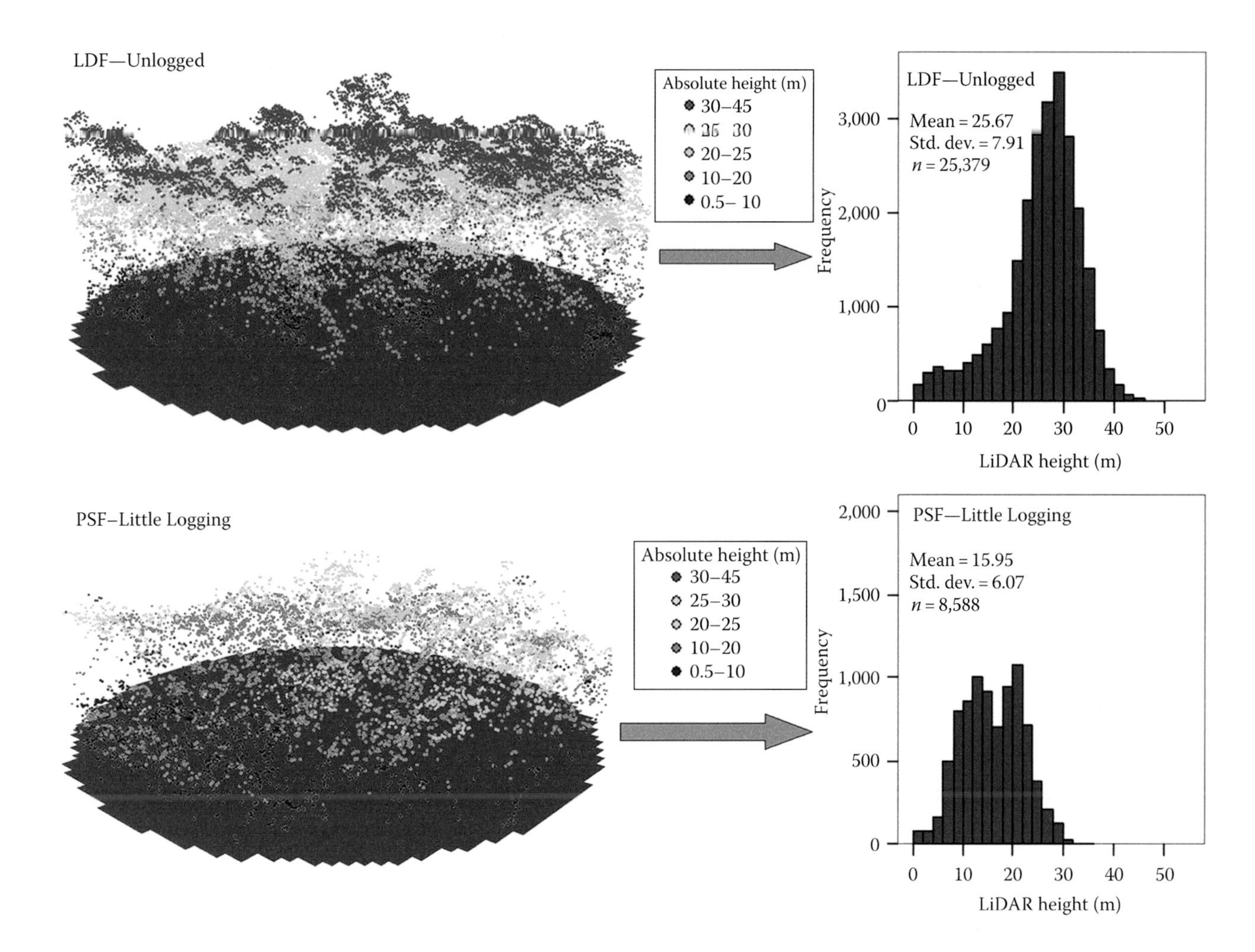

FIGURE 31.16 Distribution of LiDAR point heights within 1 ha plots: comparison of lowland dipterocarp forest and mixed peat swamp forest. Mean tree height, its standard deviation (std. dev.), and the total number of points higher than 0.5 m (n) are given in the histograms. (From Kronseder, K. et al., *Int. J. Appl. Earth Observ. Geoinform.*, 18, 37, August 2012, http://dx.doi.org/10.1016/j.jag.2012.01.010, ISSN: 0303-2434. Accessed June–October, 2014.)

31.18 Change Detection

Remote sensing by its inherent characteristics such as repeat coverage of any location- and consistency of data offers the best data type for change detection anywhere on planet Earth. Therefore, remote sensing, particularly spaceborne remote sensing, is widely used for change detection of a wide range of factors such as the land cover (physical footprint), land use (anthropogenic), forests, agriculture, glaciers, and biomass. Daniela Anjos, Dengsheng Lu, and others in Chapter 18 provide an outline of change detection methods, reason out why remote sensing is central, identify various approaches for change detection, and present an example of land cover/land use change detection in the Brazilian Amazon. There are four broad approaches to change detection studies:

1. *Conversion*: Detailed land cover change trajectories
2. *Binary change*: Change versus no change
3. *Specific changes*: Deforestation, urbanization
4. *Continuous change*: Forest disturbance

Chapter 18 presents a complex case of performing land cover/land use change detection using multisensor data. They use for 1 year radar remote-sensing data from Radarsat and for another date use optical remote-sensing data from Earth Observing-1 Advanced Land Imager (EO-1 ALI). This is not an ideal situation to get best results since using characteristically different data sources for change detection leads to higher degree of uncertainties. Nevertheless, using distinctly different remote-sensing data is necessary, especially in areas of high cloud cover such as Amazon or Congo rainforests. The ability of radar to penetrate the clouds becomes useful. So, in this regard, the case study presented for change detection in Chapter 18 is important. Further, Chapter 18 also shows us the utility of using the direct classification approach of SVMs as a preferred advanced method for change detection. Whereas

FIGURE 31.17 Detailed excerpts from the classifications (right column) and their respective areas in false color composite images in the left. The six rows show the following areas in descending order: Beijing 1990, Beijing 2010, Shanghai 1990, Shanghai 2010, Shenzhen 1990, and Shenzhen 2010. Images used are Landsat TM and HJ-1A/B. Images were studied using random forest classifier, tassel cap transformations, urban indices, and landscape matrices. (From Haas, J. and Ban, Y., *Int. J. Appl. Earth Observ. Geoinform.*, 30, 42, August 2014, ISSN: 0303-2434, http://dx.doi.org/10.1016/j.jag.2013.12.012. Accessed June–October, 2014.)

simple and direct change detection approaches such as the difference imaging (e.g., NDVI differences) can be used, methods such as SVMs allow us to use the power of multiband data and from multiple sensors leading to relatively greater accuracies. Many other methods such as RF classifiers (e.g., Figure 31.17; Haas and Ban, 2014) can be used in change detection studies.

31.19 Geoprocessing, Workflows, and Provenance

Today remote-sensing data go through complex chains of events from acquisition, curation, preprocessing, and applying analytical methods and approaches to derive and synthesize information, leading to postprocessing and dissemination to users (e.g., Figure 31.18). Different laboratories or individuals may apply different models, workflows, inferences, and processing chains toward achieving the same goal. For example, a forest change map may be produced using at-sensor surface reflectance from similar but distinct sensor systems (e.g., Landsat 5 TM vs. Landsat 8 operational land imager). Another forest change map may be produced using a combination of optical, radar, and LiDAR data. Yet another study may also use the same combination of input data to produce a forest change map, but using a different analytical change detection algorithm. Unless a user has a full understanding on how the maps are produced, it is difficult to put real value to them or to even replicate another scientist's remote-sensing workflow. This is where careful geospatial provenance (or lineage) becomes very valuable and effective. Provenance in this case entails keeping an exact historical and machine-readable record of the geospatial processes whereby a remote-sensing product originated and was developed.

In Chapter 19, Dr. Jason A. Tullis et al. provide a historical context of geospatial provenance with a focus on David P. Lanter's (1992b) Geolineus project in the early 1990s; show us how remote-sensing product quality, reproducibility, and trust depend on provenance; and provide a number of examples of recent provenance-aware remote-sensing geoprocessing systems. They discuss current international standards, specifications, and recommendations pertaining to implementation of shared remote-sensing provenance information. These include, for example, International Standards Organization's (ISO) 19115-2 metadata standard, endorsed by the Federal Geographic Data Committee (FGDC) (ISO, 2009), and World Wide Web Consortium's PROV Data Model (Moreau and Missier, 2013) and accompanying recommendations of keen interest in the provenance research community. They identify questions pertaining to remote-sensing data products that only provenance can help answer such as the exact processing steps applied to develop a remote-sensing product, processing order, parameters, execution time, and how errors were expressed and propagated. Finally, they discuss provenance-aware systems applied to remote sensing such as MODIS Adaptive Data Processing System and OMI Data Processing System designed for use by NASA to manage satellite imagery and provenance from MODIS sensors on the Terra and Aqua satellites and the OMI sensor on Aura, respectively (Tilmes and Fleig, 2008).

31.20 Toward Democratization of Geographic Information

Chapter 20 by Dr. Gaurav Sinha et al. discusses the democratization of geo-information and geospatial information communication technologies (Geo-ICTs). Early use of geographic

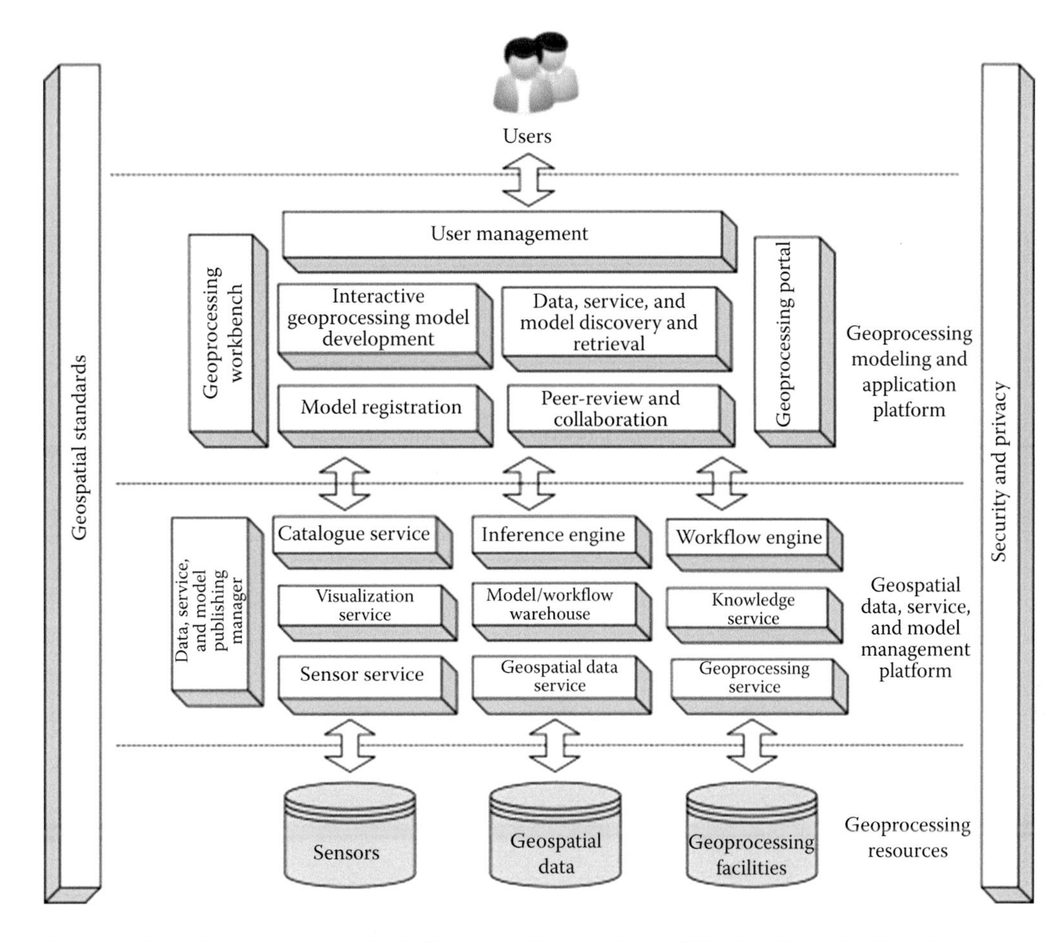

FIGURE 31.18 Framework for the Geoprocessing Web. (From Yue, P. et al., *Trans. GIS*, 14(6), 755, 2010; Zhao, P. et al., *Comput. Geosci.*, 47, 3, October 2012, ISSN: 0098-3004, http://dx.doi.org/10.1016/j.cageo.2012.04.021. Accessed June–October, 2014.)

information technologies, much like computing, was an elitist intellectual activity where general public had hardly any participation. The information was processed, analyzed, and disseminated by very few highly specialized experts knowledgeable in remote sensing, geographic information system (GIS), global positioning systems (GPS)/GNSS, and spatial modeling. This top–down approach was criticized as a "hegemonic system perpetuating the grip of powerful agencies and trained professionals on geographic information" as viewed by several thinkers. This has been discussed in detail in Chapter 20. This was also a period where GIS, remote sensing, and GNSS were mainly tools of well-funded national and international specialized institutes and military establishments.

Chapter 20 discusses how geo-information has become ubiquitous, linking web mapping, wireless delivery, crowdsourcing, and other digital technologies through expert collaboration and user participation into a seamless platform for powerful delivery of data and knowledge. One of the earliest influences of this democratization and merger of technologies was U.S. Vice President Al Gore's "Digital Earth" vision. As a result, access to digital geo-information through virtual globes (e.g., Google Earth, Microsoft Bing, ESRI ArcGIS Explorer™) is now integral to billions of people's daily lives. In addition to the accessibility of Geo-ICTs, neogeography (or new geography), participatory GIS, crowdsourcing, VGI, and social networking have further integrated geo-information into people's lives. For example, there exist several crowdsourcing initiatives, especially for managing disasters like earthquakes, floods, and tsunamis. One such example are flood databases, such as the International Disaster Database (EM-DAT), ReliefWeb (launched by the United Nations Office for the Coordination of Humanitarian Affairs), the International Flood Network, and the Global Active Archive of Large Flood Events (created by the Dartmouth Flood Observatory) (Wan et al., 2014; Figure 31.19). Another global initiative called Geo-wiki (Fritz et al., 2012) helps improve global land cover assessment and mapping. Readers can expand on these ideas after reading Chapter 20. Long back, Prof. Duane Marble, one of the pioneers of geographic information science, predicted that "all information will be geospatial," meaning that the information will be tied to precise geographic location. Today, an enormous amount of information is captured location specific and delivered with location specificity. However, this ubiquitous nature of geo-information, and the ease with which it can now be captured, stored, shared, and sold, has also raised many questions about individual rights and freedom. There also arise issues related to the quality of information generated

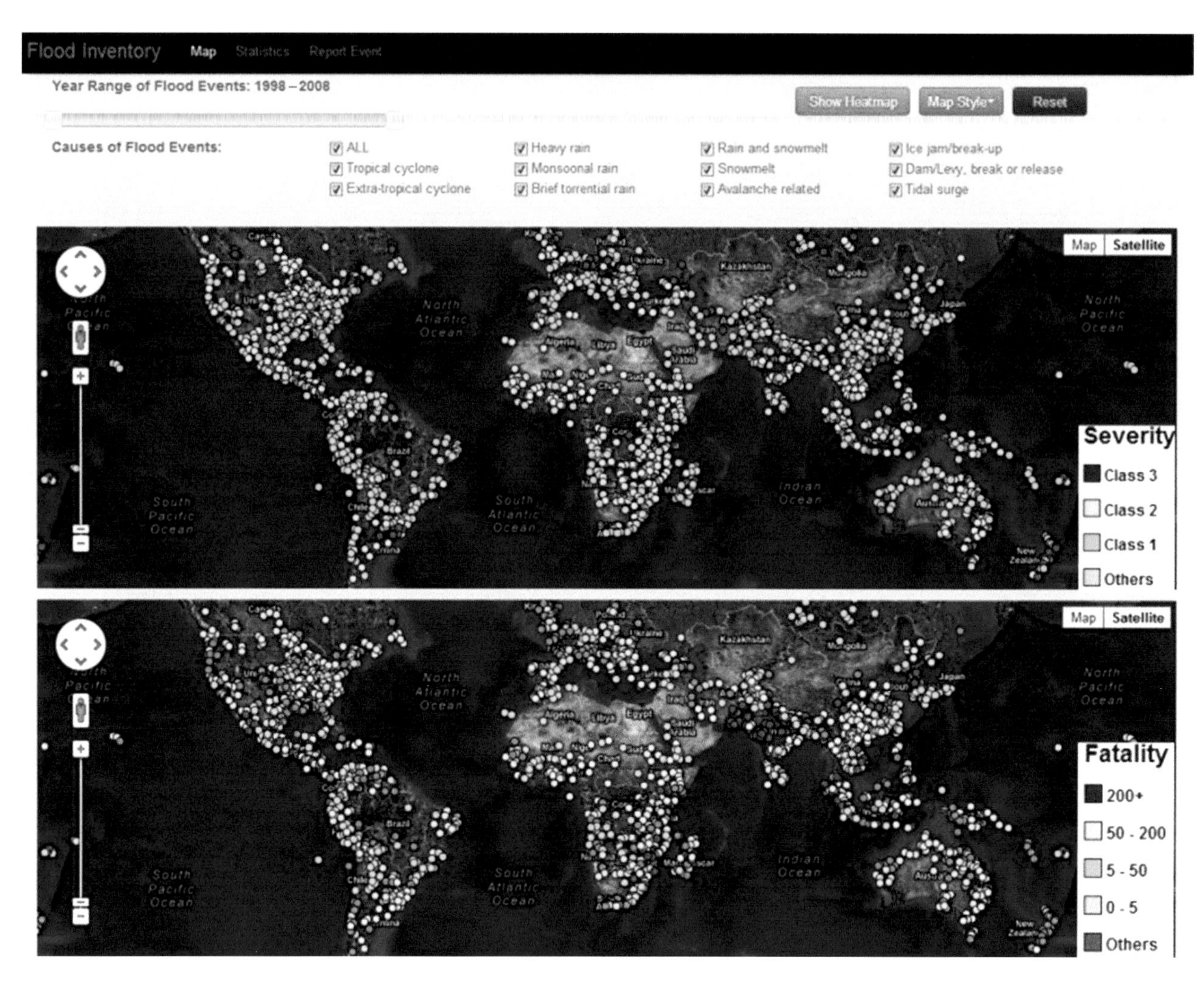

FIGURE 31.19 The map visualization of global flood cyber infrastructure. The top and bottom maps are color-coded by severity and fatalities, respectively. (From Wan, Z. et al., *Environ. Modell. Softw.*, 58, 86, August 2014, ISSN: 1364-8152, http://dx.doi.org/10.1016/j.envsoft.2014.04.007. Accessed June–October, 2014.)

and processed by nonprofessionals. For example, when data are crowdsourced, experts as well as nonexperts share data. Thus, quality of such data can vary widely depending on the source, which could even be detrimental to the progress of precise geoinformation that is reliable and trusted.

31.21 GIScience

GIScience is the science of geographic information and analysis that substrates the development of GIS and a host of related technologies (cartography, geodesy, surveying, photogrammetry, GPS, remote sensing, spatial modeling) for solving problems (e.g., detecting a landfill site, Figure 31.20) related to a wide spectrum of themes such as agriculture, water, forestry, geography, geology, geophysics, oceanography, ecology, environmental science, and social sciences. GIScience is contributing spatial thinking and spatial computing that propels many disciplines to "take a spatial turn," such as spatial ecology, spatial epidemiology, spatial social sciences, and spatial humanities. Soon after the advent of computers, early pioneers (see Chapter 21) started the development of GIS technology in 1960s and 1970s. The maturing of GIS technology has led to a wide adoption of mapping and spatial analysis in domain applications with extensive data from multiple sources (e.g., GIS, GPS, remote sensing) and sophisticated algorithms (e.g., image-processing algorithms, map algebra, spatial econometrics) for spatial decision support. GIScience grounds GIS technologies with theories of spatial representation, spatial relationships, and a range of computational methods that transform the underlying conceptualization and problem-solving approaches. Today, GIScience research expands into data uncertainty, space–time analytics, and social implications on web mapping, crowdsourcing, and cloud computing as well as mobile delivery. In Chapter 21, Dr. May Yuan presents an outline and history of GIScience and highlights its core epistemology: abstraction, algorithms, and assimilation. Also see Chapter 26 (crowdsourcing), Chapter 27 (cloud computing), and Chapter 28 (web mapping) and how GIScience is intricately linked to these techniques, especially going forward as discussed by Dr. Yuan in this chapter.

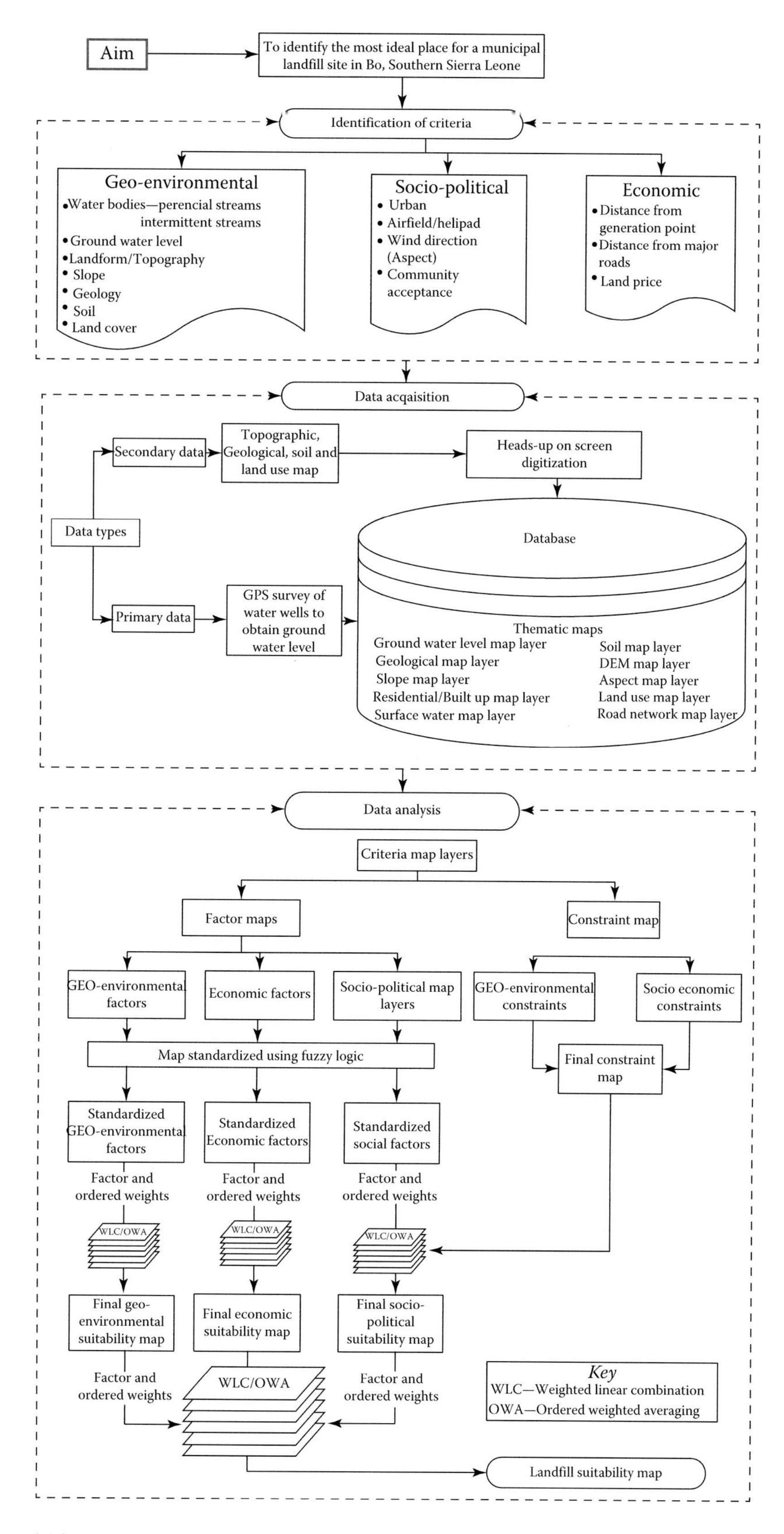

FIGURE 31.20 Spatial model for the identification of a potential landfill site by integrating spatial data from various sources (e.g., remote sensing, GIS, GPS). (From Gbanie, S.P. et al., *Appl. Geogr.*, 36, 3, January 2013, ISSN: 0143-6228, http://dx.doi.org/10.1016/j.apgeog.2012.06.013. Accessed June–October, 2014.)

31.22 Object-Based Regionalization for Policy-Oriented Partitioning of Space

One of the biggest challenges facing policy makers and others using remote sensing and other geospatial data–derived information is to ensure its integrity, robustness, and reliability. Inherently, remote-sensing data are consistent and robust (with proper calibration and normalization). But methods and approaches in analyzing the data are often different. Further, in order to derive new information or enhance existing geospatial information, remote-sensing data are used as a data source to be integrated with existing geospatial data in different ways. Resultant products are often dissimilar, depending on the algorithms used, or the weights assigned to each of the information source (e.g., Figure 31.21; Branger et al., 2013). These issues are discussed in Chapter 22 by Dr. Stefan Lang et al., introducing the geon approach as a strategy to integrate multiple sets of geospatial data.

Chapter 22 begins with definitions of latent, complex, and multidimensional phenomena and the related concept of geons and sets the platform for its spatial representation that is robust and useful to decision makers in different domains. The authors discuss the principles of regionalization, how geons are used in domain-specific regionalization, and OBIA in image segmentation and regionalizations. They take four case studies (socioeconomic vulnerability of hazards, social vulnerability to malaria, landscape sensitivity, and climate change susceptibility) to show readers how latent phenomenon is addressed, what indicators and datasets are used, and the results.

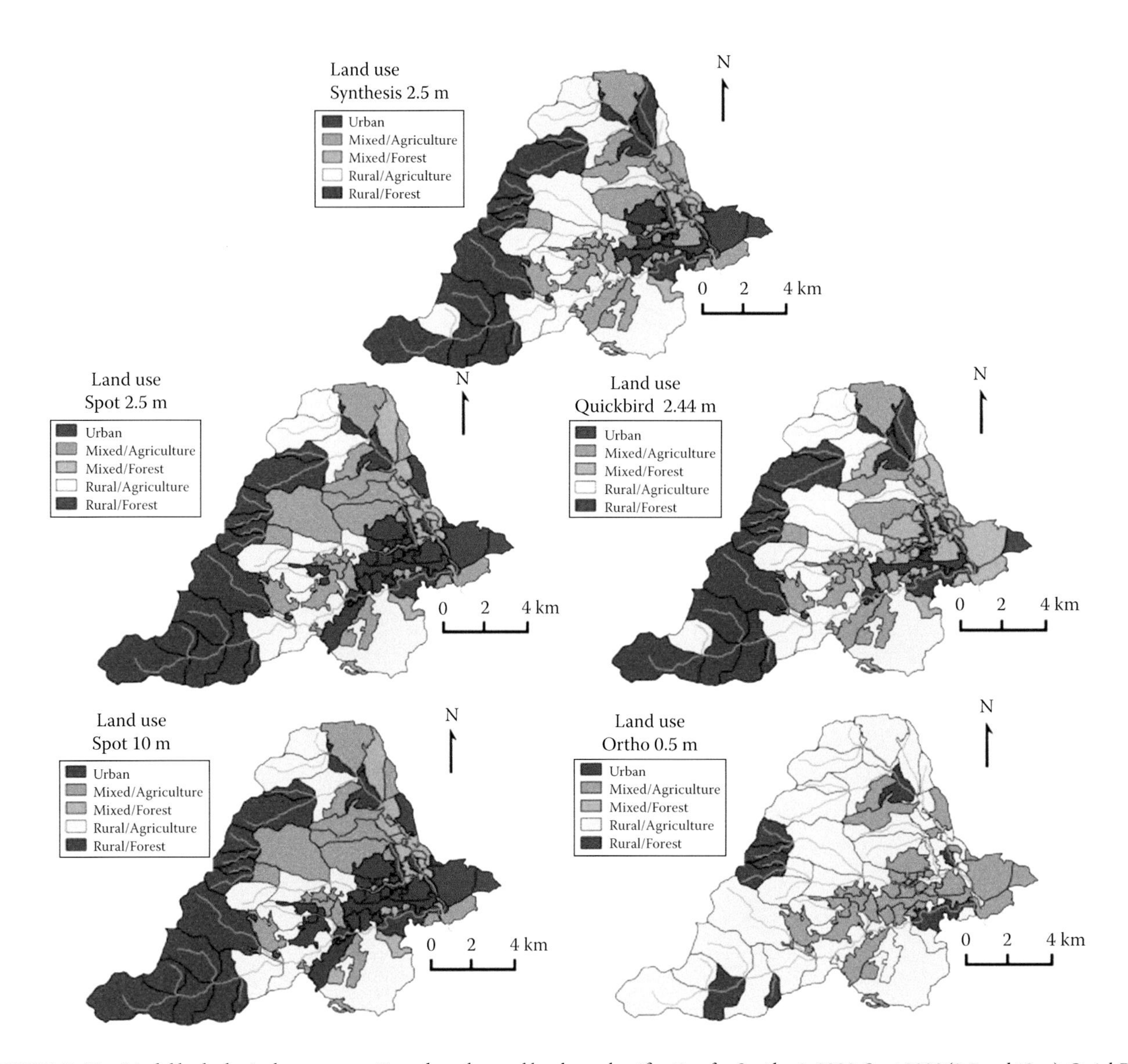

FIGURE 31.21 Model hydrological response units and reaches and land use classification for Synthesis 2008, Spot 2008 (2.5 and 10 m), QuickBird 2008, and Ortho 2008 land use maps. (From Branger, F. et al., *J. Hydrol.*, 505, 312, November 15, 2013, ISSN: 0022-1694, http://dx.doi.org/10.1016/j.jhydrol.2013.09.055. Accessed June–October, 2014.)

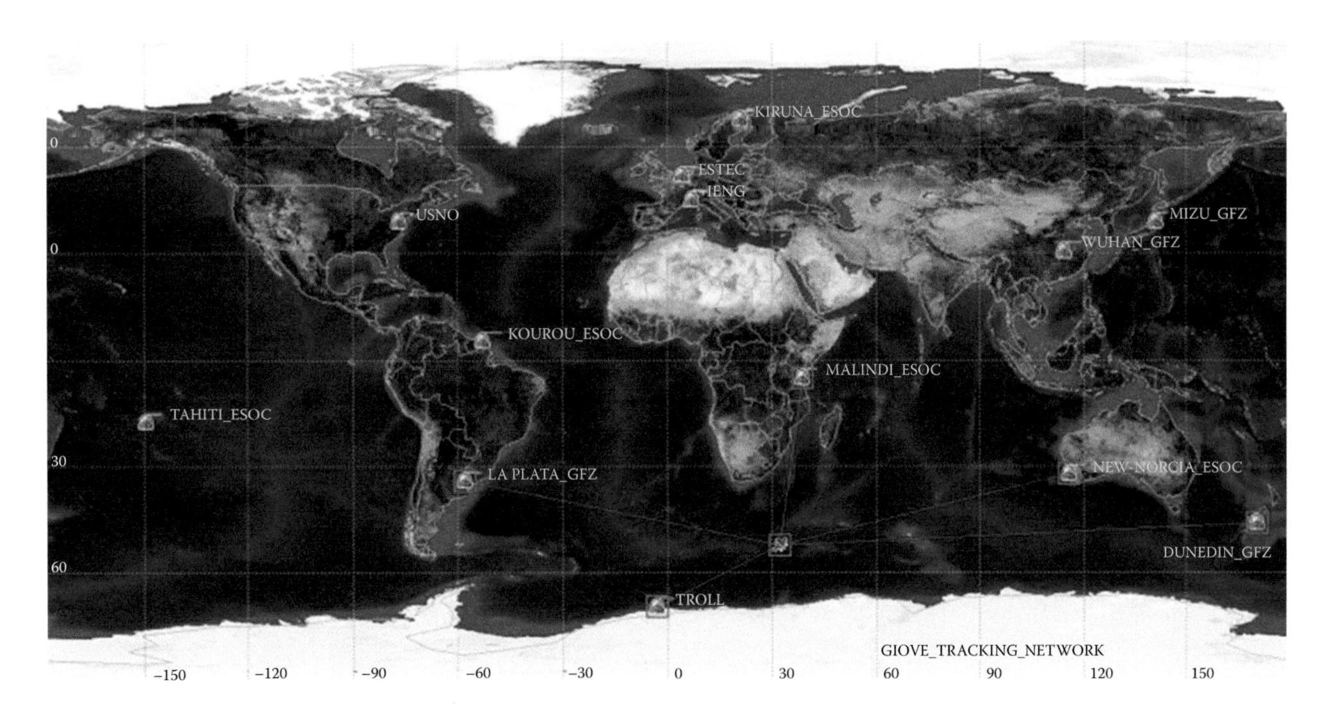

FIGURE 31.22 Galileo System Test Bed, Version 2 (GSTB-V2) station network. (From Dow, J.M. et al., *Adv. Space Res.*, 39(10), 1545, 2007, ISSN: 0273-1177, http://dx.doi.org/10.1016/j.asr.2007.04.064. Accessed June–October, 2014.)

31.23 Global Navigation Satellite Systems

In the modern-day world, GNSSs have become ubiquitous. They guide us from place A to place B; pinpoint any target in the world for gathering information of various natures; control air, road, and sea traffic; precisely locate an agricultural crop or forest species, or a water body as tiny as a well; and numerous other applications. Currently, there are five distinct operational GNSS: (1) GPS controlled by the United States; (2) Global Orbiting Navigation Satellite System (GLONASS) controlled by Russia; (3) Galileo controlled by Europe (e.g., Galileo System Test Bed, Version 2 [GSTB-V2, Figure 31.22]); (4) BeiDou (Compass) controlled by China, and 5. Indian Regional Navigational Satellite System (IRNSS). Typically, GNSSs have upto to 36 satellites, of which a minimum of four satellites always accessible for signals from any part of the world. These systems can provide positional accuracy at any location on Earth to within a few centimeters or even millimeters. Chapter 23 by Grewal discusses in detail GNSS characteristics, including signal spectrums, differences between civil and military signals (for GPS), and the nature and characteristics of the different signals, for example, frequency division multiple access, code division multiple access (CDMA), and time division multiple access.

31.24 GNSS Reflectometry for Ocean and Land Applications

Chapter 24 by Kegen Yu et al. discusses the increasing interest and usage of GNSS reflectometry (GNSS-R) in ocean and land applications. The chapter provides a solid foundation on the theoretical concepts of GNSS-R. NASA and ESA are launching series of satellites focused on GNSS-R such as the Cyclone Global Navigation Satellite System (CYGNSS). Even though the use of GNSS-R data in remote sensing is still a novelty, with increasing number of low-Earth-orbit satellites equipped with GNSS receivers, the situation is going to change soon. The chapter highlights the key land applications of GNSS-R such as soil moisture and forest change detection. In many other applications like that for tsunamis (e.g., Figure 31.23; Stosius et al., 2011), GNSS-R is very powerful. Similarly, it highlights the key ocean applications of GNSS-R such as sea surface undulation and sea surface height. Just as the UAVs are becoming increasingly important in land remote sensing, GNSS-R is also becoming very important. They add to conventional remote sensing and bring in unique capabilities that only GNSS-R can provide such as CYGNSS.

31.25 GNSS and Wide Array of Applications

GNSS include satellite constellations (typically, 20–30) from NAVSTAR GPS of the United States, GLONASS of Russia, Galileo of European Union, BeiDou navigation systems of China, and IRNSS of India that provide instant, freely available position data at high accuracies (few centimeters to few meters) for any place in the world (see Figure 31.24; Jin et al., 2011). As a result of global international GNSS service constellation, a dense network of GNSS satellites and sites are available. The first and the most reliable GNSS acquires position data and altitude (x, y, z) through a baseline constellation of 24 satellites positioned about 20,000 km from Earth. To get accurate position for any

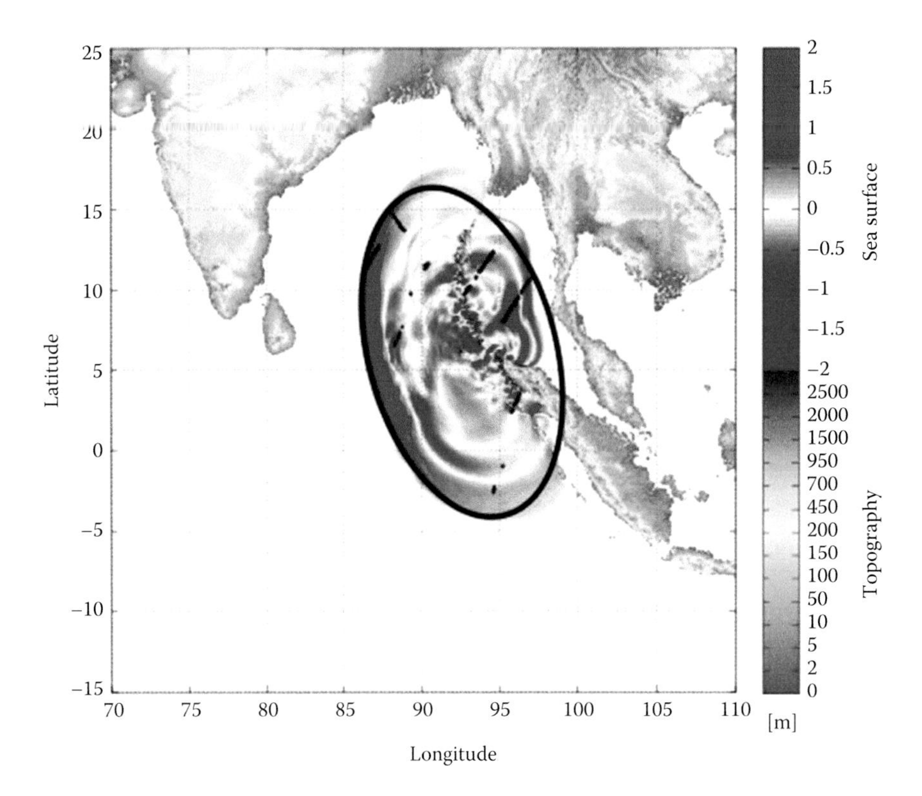

FIGURE 31.23 Simulation of the Sumatra tsunami 1 h after the earthquake. Ground tracks (black) show Global Navigation Satellite System reflectometry detections within 1 min of observation. The ellipse encloses all detections made up to this moment and show the current tsunami expansion. (From Stosius, S. et al., *Adv. Space Res.*, 47(5), 843, March 2011, ISSN: 0273-1177, http://dx.doi.org/10.1016/j.asr.2010.09.022. Accessed June–October, 2014.)

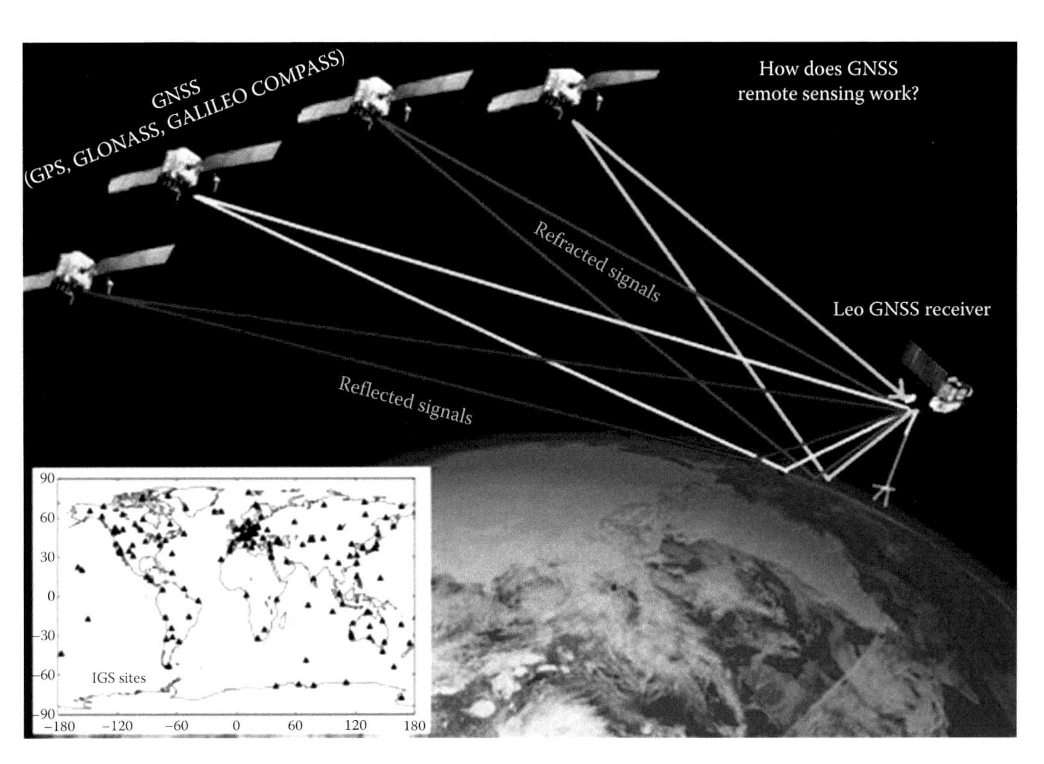

FIGURE 31.24 Remote sensing using future denser Global Navigation Satellite System (GNSS) network and more satellite constellations. The left lower corner shows the global International GNSS Service (IGS) sites distribution. (From Jin, S. et al., *Adv. Space Res.*, 47(10), 1645, May 17, 2011, ISSN: 0273-1177, http://dx.doi.org/10.1016/j.asr.2011.01.036. Accessed June–October, 2014.)

location on Earth's surface, one needs data from at least three satellites. Typically, at least four satellites are available from any spot on Earth. Of the 24 GPS satellite constellation at any given time, 21 are in operation continuously, and 3 are used as spare to replace any failing satellite. The signals (e.g., CDMA) transmitted by them are either for military (encrypted) or civilian (open).

Dr. Myszor et al. in Chapter 25 discuss various applications of GNSS. These are broadly categorized into civilian, military, and aeronautic. The applications discussed include the following:

1. Consumer applications such as the interactive maps to locate any place or location of interest (e.g., house, restaurant, airports, bus stations). These are ubiquitous today in our daily life, in cars and smart phones.
2. Industrial applications such as in agriculture (e.g., detecting location of farms, crops grown in farms), stores, and city services (e.g., location of severs, pipelines).
3. Transportation applications such as route from point A to B.
4. Surveying applications such as mapping large areas and isolated areas.
5. Disaster applications where disaster maps that provide information such as people injured, property lost, and damage areas are instantly mapped. GNSS data are used in wide array of disasters like earthquakes, floods, droughts, famine, and fire.
6. Health applications (e.g., monitoring and managing disease spread).
7. Tracking applications (e.g., animal grazing paths, wildlife tracking).
8. Military applications (e.g., locating and monitoring troop movements, military installations).
9. Aeronautic applications (e.g., tracking aircraft movements).

Chapter 25 also provides the GNSS satellite characteristics as well as errors in GNSS and ways and means to control them and keep the errors to a minimum.

31.26 Crowdsourcing in Remote Sensing and Spatial Technologies to Study Planet Earth

Remote sensing made it possible to gather data from any spot on our planet Earth: remotely, consistently, repeatedly, and in various levels of detail. Nevertheless, the need "to be there on the spot physically" to decipher or verify the actual conditions on the ground remains. For example, satellite remote sensing can help establish important information such as location, size, and orientation of buildings in a residential area through processing of multispectral, high-resolution images. Still, the most important proxies for building vulnerability to natural disasters, such as presence of steel frames or seismic retrofitting, are not set to become visible in spaceborne data anytime soon. Or even when remote sensing establishes these facts, there is a need to verify, to be absolutely sure or nearly so. Naturally, gathering such ground-based data extensively across the planet timely, routinely, and cost-effectively is extremely difficult if not infeasible. This is overcome to significant extent through crowdsourcing. In Chapter 26, Dr. Fabio Dell'Acqua illustrates some of the major advances made in crowdsourcing such as the following:

1. OpenStreetMap, which is widely used.
2. "Did you feel it service," a pioneering earthquake intensity mapping based on crowdsourced information gathered from people on the ground. This leads to Community Internet Intensity Maps.
3. Geo-wiki for land cover mapping.

There are, of course, numerous other applications of crowdsourcing for spatial information (e.g., Figure 31.25; Heipke, 2010). This can be, for example, used for conducting election pool surveys, product surveys, and even mapping field boundaries or host of other services. Crowdsourcing assumes that there are sufficient numbers of volunteers who will provide such data either for free (personal satisfaction) or for a fee. Chapter 26 discusses the evolution of crowdsourcing and provides many practical examples of crowdsourcing.

In an increasingly sophisticated, Internet-dominated world of today, crowdsourcing is both attractive and desirable. Nevertheless, one needs to be cautious of many aspects of crowdsourced data. It is not scientifically sound data for many applications. For example, when millions of users feed data on where agricultural croplands are or say something about their productivity situation (e.g., drought or stress condition), uncertainties in such data are likely to be high. Even when such data are collected by experts, there are differences as a result of the factors such as the depth of one's understanding and definition issues. So, when untrained volunteers provide such data, the likelihood of these uncertainties increases. Yet, besides statistical approaches to data cleaning, it is expected that the technology itself will develop increasingly smart solutions to reduce these uncertainties and make the crowdsourced data more accurate and reliable. For example, when one takes a photo of a particular crop and uploads, the technology may identify the crop type automatically along with its precise location, removing the uncertainties that may creep from a volunteer input. Given these facts, crowdsourcing will become an important and integrated component of the future remote sensing and spatial data analysis practices.

31.27 Cloud Computing and Remote-Sensing Data

Cloud computing can generate and involve a massive amount of data, especially in GeoSciences and specifically when remote sensing is involved. Today, data are gathered in increasing quantities by both public and private entities, and in every dimension: spatial, spectral, radiometric, and temporal, covering the entire planet. The challenges of processing these gigabytes, terabytes, and petabytes of "big data" over and over again and converting this data into meaningful information for the study of our planet are real today. How this can be done is presented and discussed in

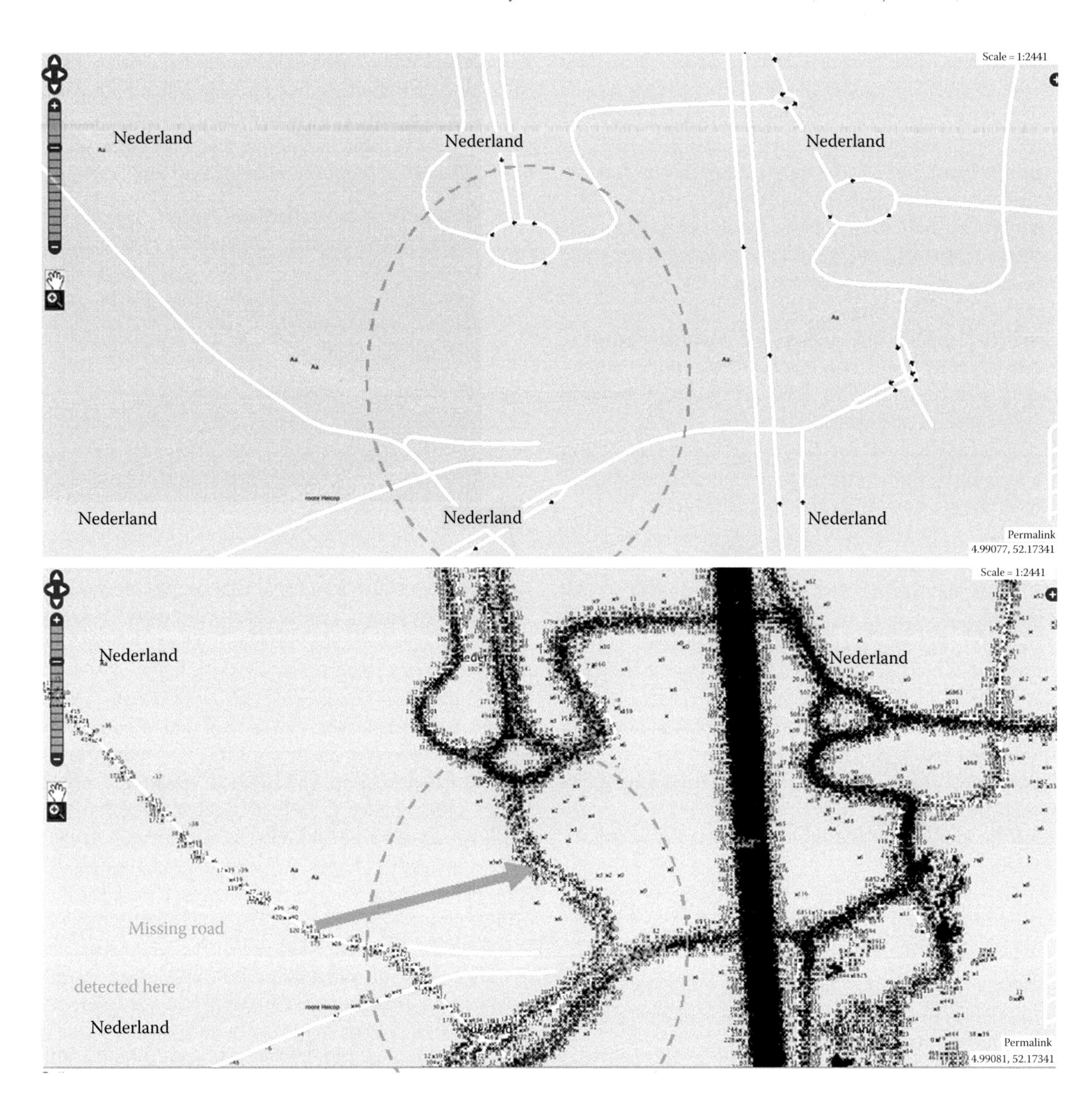

FIGURE 31.25 Example of a result from passive mapping. Upper, existing map data; lower, superimposition with data from global positioning systems tracks derived from cars. The missing road in the center is clearly visible; however, the individual tracks need to be aggregated into an attributed graph structure for use in routing applications (Tele Atlas©). (From Heipke, C., *ISPRS J. Photogram. Remote Sens.*, 65(6), 550, November 2010, ISSN: 0924-2716, http://dx.doi.org/10.1016/j.isprsjprs.2010.06.005. Accessed June–October, 2014.)

Chapter 27 on cloud computing and remote-sensing data by Dr. Sugumaran et al. They define cloud computing and contrast it with cluster, grid, high-performance computing, and high-throughput computing. The chapter discusses cloud-computing paradigms such as the public cloud, private cloud, and hybrid cloud. Cloud-computing services such as Amazon's elastic compute cloud allow researchers to use paradigms such as infrastructure as a service, platform as a service, and software as a service and make possible the processing of massive amounts of data for nominal costs. They illustrate the ability to process one large LiDAR dataset (~6 TB) covering the entire state of Iowa. An open system architecture for geospatial cloud computing is shown in Figure 31.26 (Evangelidis et al., 2014). In spite of these advances, cloud computing is still in its nascent stages as far as remote-sensing data processing is concerned. Some of the biggest challenges are

1. Ability to process very high-resolution spatial (e.g., a few meters), spectral (e.g., hundreds of bands), radiometric (e.g., 16 bit or higher), and temporal (e.g., daily) images of the world
2. Development of image-processing algorithms in the cloud environment
3. Data security
4. Storage and backup of "big data" over long time periods

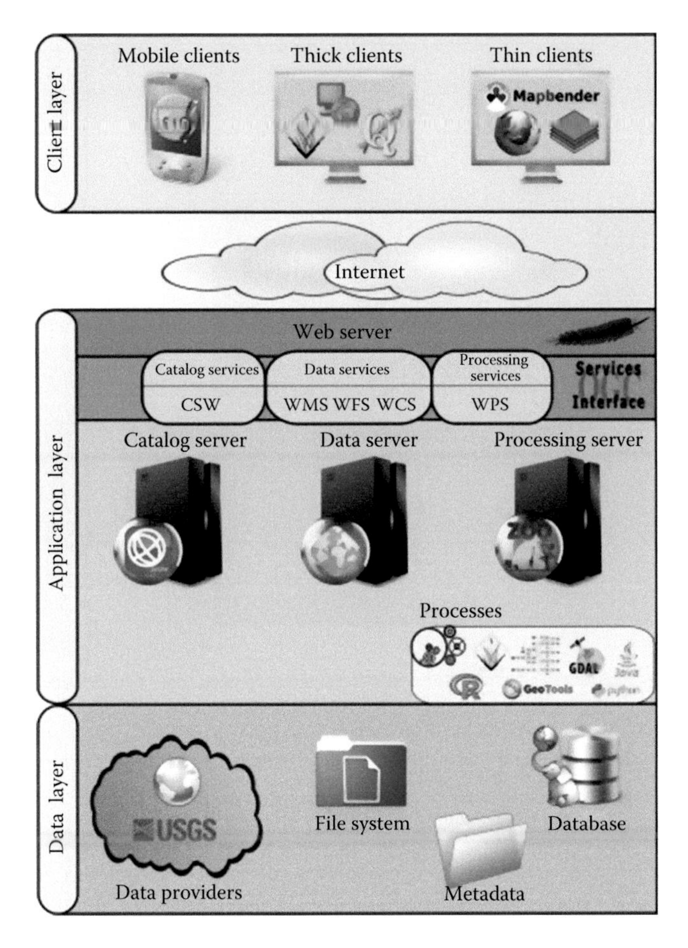

FIGURE 31.26 Open system architecture for Geospatial Cloud Computing. (From Evangelidis, K. et al., *Comput. Geosci.*, 63, 116, February 2014, ISSN: 0098-3004, http://dx.doi.org/10.1016/j.cageo.2013.10.007. Accessed June–October, 2014.)

Nevertheless, in due time, these challenges will be overcome and allow ubiquitous processing of "big remote-sensing data" in various cloud-computing environments.

31.28 Google Earth for Earth Sciences

In Chapter 28, Dr. John E. Bailey outlines how Google Earth, launched in the year 2005, brought complex remote-sensing technology to the doors of common person. Based on the philosophy of "Digital Earth," a concept popularized by former Vice President of the United States Mr. Al Gore in 1998, Google Earth enabled some powerful possibilities. These included

1. Visualization of the entire Earth from your desktop, laptop, or mobile seamlessly
2. Ability to zoom in and zoom out of any part of the planet and visualize a number of details such as cities, forests, agricultural lands, houses, trees, and networks (e.g., roads, rivers)
3. Precisely locate annotations
4. Providing the "power of information" to common person who may not have any idea of remote sensing or geoscience
5. Enabling people to tell stories or inform changes through images

Google Earth involves use of geocoded imagery of any various resolutions along with the 3D elevation derived from Shuttle Radar Topographic Mission (SRTM) and other (e.g., LiDAR) high-resolution data. The baseline imagery for the Google Earth is Landsat 30 m imagery, but for large swaths of the world, very high-spatial-resolution imagery (VHRI; submeter to 5 m) from providers such as DigitalGlobe are also available. The system is supported and revolutionized by Keyhole Markup Language (KML).

Further, the ease of use and the power of data and information available in Google Earth for anyplace and anywhere and to anyone truly democratized remote sensing and GIS sciences.

Chapter 28 also shows us how several scientific applications and applications of societal importance of Google Earth soon became apparent. They show us example of volcanoes of the world mapped by the Smithsonian. Others like Thenkabail et al. (2009a,b) have extensively made use of VHRI as ground data to identify and label croplands of the world as well as assess cropland accuracies. These days, VHRIs in Google Earth are extensively used by remote-sensing scientists as data for developing or testing their algorithms in certain applications. Nongovernmental organizations and bloggers have used Google Earth to illustrate and demonstrate issues of environmental and societal importance that often influence public opinion. Crowdsourcing of information gathered from remote-sensing imagery is depicted on Google Earth during such events as earthquakes, floods, hurricanes, and tsunamis (e.g., Figure 31.27).

Even though Google Earth has tremendous strengths, it cannot be used for all scientific studies that use remote sensing. This is because the imagery available is used only "visually," without the underlying values and metadata available, which are required for quantifying variables and studying changes. For example, Google Earth imagery cannot be used for quantifying drought, deriving drought indices, and determining drought conditions over time. However, in order to overcome these limitations, Google Earth Engine (EE) has been introduced. EE stores real images. For example, the entire Landsat imagery archive (from 1972 to present) is currently available in EE. This imagery can be used for scientific studies by building algorithms directly in the EE platform to analyze the imagery. This also allows for time-lapse analysis. EE offers a cloud-computing platform to quickly and instantly analyze images at global scale even at 30 m spatial or higher resolutions. This platform is still being developed, but already some interesting global applications are becoming available (e.g., forest cover and wetland changes using Landsat mentioned in Chapter 28). The recent purchase of Skybox Imaging by Google also sets the stage for more frequent and innovative acquisition of satellite imagery for Google EE applications.

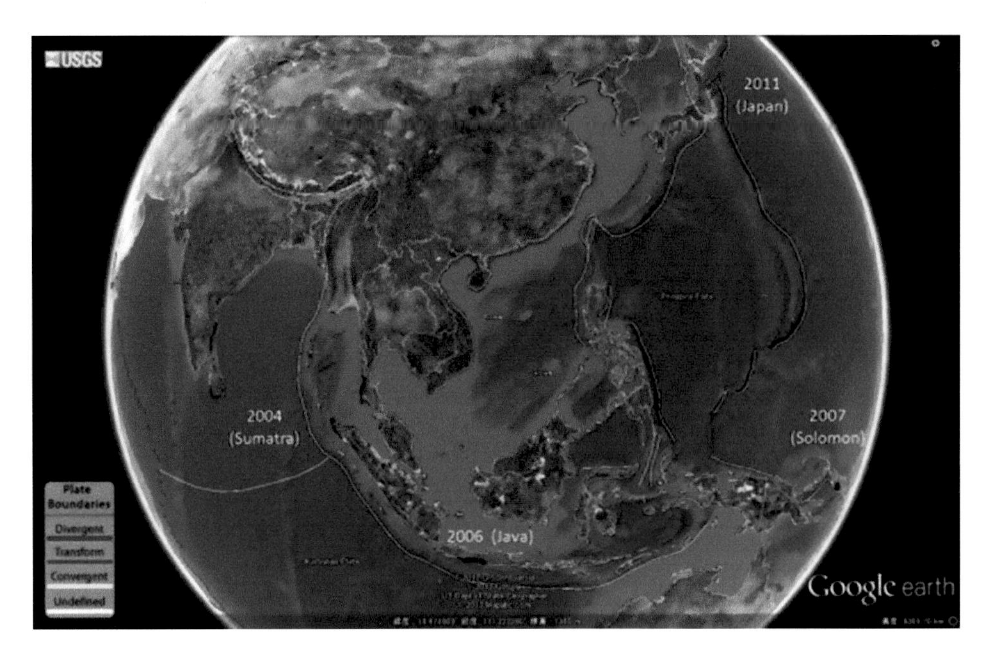

FIGURE 31.27 Tsunami field survey data available in Google Earth ".kml" format since 2004 and plate tectonics in the Indian and southwest Pacific Oceans. A collection of the Google Earth (http://www.google.com/earth/index.html) ".kml" files of tsunami measurement data from field surveys after the 2004 event including the 2004 Indian Ocean, 2006 Java, 2007 Solomon, and 2010 Chile are summarized (Fujima, 2011; Mori and Takahashi, 2012) and shown with plate tectonics (USGS, 2012) in this figure. (From Suppasri, A. et al., *Int. J. Disaster Risk Reduct.*, 1, 62, October 2012, ISSN: 2212-4209, http://dx.doi.org/10.1016/j.ijdrr.2012.05.003. Accessed June–October, 2014.)

31.29 Map Accuracies

Chapter 29 by Dr. Russell G. Congalton provides an in-depth guideline on assessing thematic and positional accuracies of maps. When RS data are processed, classified, and information deciphered through digital image processing and interpretations (or through manual interpretations as in the past), a thematic map is produced. Examples of thematic maps include products such as land cover types, species types, biomass categories, and leaf area index. Scientific use of any thematic map requires assessing it's thematic as well as positional accuracies. Thematic accuracy addresses how well classes in thematic map correspond to what is observed or measured in the field or ground. Positional accuracy refers to quantifiable differences between two map locations or a map location and a location on ground. Dr. Congalton addresses these issues systematically and comprehensively in Chapter 29, and these are summarized as follows:

1. *Error sources*: Errors in thematic maps can occur as a result of any number of causes—systematic (e.g., sensor calibration errors), natural (e.g., atmospheric), preprocessing (e.g., geometric and radiometric correction), derivative (e.g., NDVI), classification schemes and methods (e.g., digital image processing), class definitions, reference data collection design (e.g., sample design, position inaccuracies), and user interpretation of classes and class labeling.
2. *Classification*: Thematic maps are produced by digital image analysis. The four characteristics of classification are identified as (a) complete definition, (b) mutually exclusive, (c) totally exhaustive, and (d) hierarchical.
3. *Reference data*: Typically, reference data are collected through ground visits. A rule of thumb is to collect 50 samples for each land cover category when total numbers of classes are 12 or less. When classes are more than 12, the recommendation is to gather data from 75 or 100 samples per class. However, the number of samples can go up for large areas. Reference data are collected using well-designed sampling strategies (e.g., random or stratified random). Further homogeneous areas are selected for collecting ground data. For example, collecting data from a single 30 m pixel could lead to errors due to misregistration (locational error). Therefore, it is better to select homogeneous areas for a land cover class such that a 3×3 cluster of 30 m pixels are chosen to reduce the positional error.

Dr. Congalton then provides a series of error matrices to evaluate map accuracies. These error matrices compute the following:

4. *Three basic accuracies*: An error matrix computes three basic accuracies—producer's, users', and overall. Accuracies are computed by comparing the thematic map classes with reference data. Producer's accuracy refers to percentage of class X in reference map that is correctly classified as class X in a thematic map. A user's accuracy is the percentage of the pixels classified as a class X in a thematic map is indeed class X. Overall accuracy is the percentage of pixels correctly classified from all classes to total number of pixels from all classes. If sample sizes vary, so does the producer's, user's, and overall accuracies. The larger the sample size, the more robust these accuracies are.

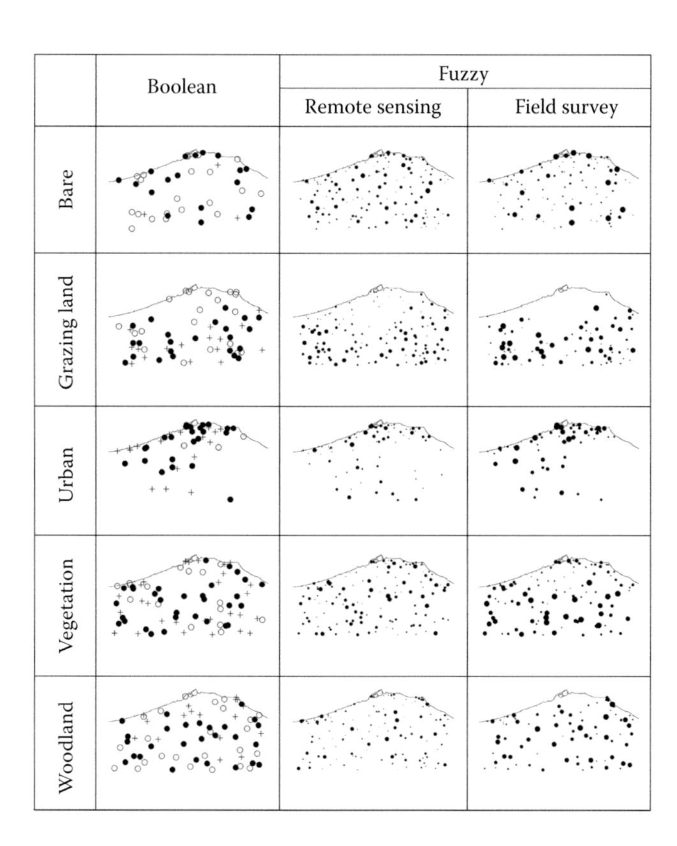

FIGURE 31.28 The data collected in the field and classified from remotely sensed data. In the Boolean maps, solid symbols (●) indicate where both field and RS agree, hollow circles (○) where only the RS indicates the class, and crosses (+) where only field survey indicates the class. In the maps of fuzzy classes, the size of the plot characters indicates the degree of membership to the class. (From Comber, A., et al., *Remote Sens. Environ.*, 127, 237, December 2012, ISSN: 0034-4257, http://dx.doi.org/10.1016/j.rse.2012.09.005. Accessed June–October, 2014.)

5. *Normalized accuracies*: "Margfit" normalized the error matrix by removing the effect of sample size. "Kappa" KHAT statistic is another measure of accuracy and is effectively used as a way to test if one map is statistically better than another.
6. *Fuzzy accuracy assessment*: Instead of deterministic accuracy assessment, like the one's mentioned earlier, fuzzy classification provides measure of accuracy that considers possibilities such as correct, mostly correct, somewhat correct, mostly incorrect, somewhat incorrect, and incorrect (e.g., Figure 31.28; Comber et al., 2012).

31.30 Remote-Sensing Law or Space Law

Remote-sensing law is part of international space law as well as the quickly developing field of geospatial law. The early evolution of space law was based primarily on the military and strategic requirement of nations. However, wide and ubiquitous use of remote sensing has resulted in broad set of legal principles governing remote sensing for civilian uses. The evolution of technology has contributed to remote sensing becoming an essential component of many other technological delivery systems (e.g., Google Earth, web maps, and mobile delivery of location specific data). This means that the legal principles governing remote-sensing technologies are increasingly being developed in the context of the broader field of geospatial law. This is especially so in light of the increasing number of EO satellites launched and operated by various nations (e.g., Figure 31.29). Dr. P.J. Blount in Chapter 30 not only provides an excellent discussion on these matters but refers to an extensive set of international and national treaties, regulations, agreements, and policies that when combined form the complex web of regulations governing these activities.

Broadly, the four pillars of remote-sensing law are

1. International space treaty regime
2. International remote-sensing law for military era or disarmament and verification law
3. International remote-sensing principles for civilian purposes
4. National geospatial law principles

First, space treaty regime defines laws pertaining to

1. Free access to outer space for any nation
2. Observation and collection of data for any part of the world, cutting across sovereignty
3. Governments to "authorize and supervise" nongovernmental or other players (e.g., commercial) satellites

Second, disarmament and verification law was for military purposes involving control of ballistic missiles, freedom to monitor military activities, and intelligence gathering.

Third, remote sensing for civilian purposes is guided by the United Nations General Assembly Resolution 41/65 and the United Nations Committee on the Peaceful Uses of Outer Space, which outlines the core legal principles dealing with remote-sensing activities for civilian purposes. These principles can be divided into three groups:

1. Preexisting binding legal principles
2. Innovations involving international law governing remote sensing
3. Promoting the soft obligation of international cooperation

The centerpiece of the international law principles are "the freedom of data collection and distribution," while at the same time ensuring "respect of the principle of full and permanent sovereignty of all states and peoples over their own wealth and natural resources." Through these measures, a large number of cooperation of internal remote-sensing data collection and access is determined for human welfare. These efforts are further solidified through efforts through GEO and GEOSS where web-enabled (free) remote-sensing data sharing and distribution is encouraged for a wide range for peaceful applications (e.g., the nine societal beneficial areas such as forests, agriculture, water, biodiversity, disasters).

Fourth, remote-sensing law is increasingly perceived to be part and parcel of geospatial law. This as a result of the use of

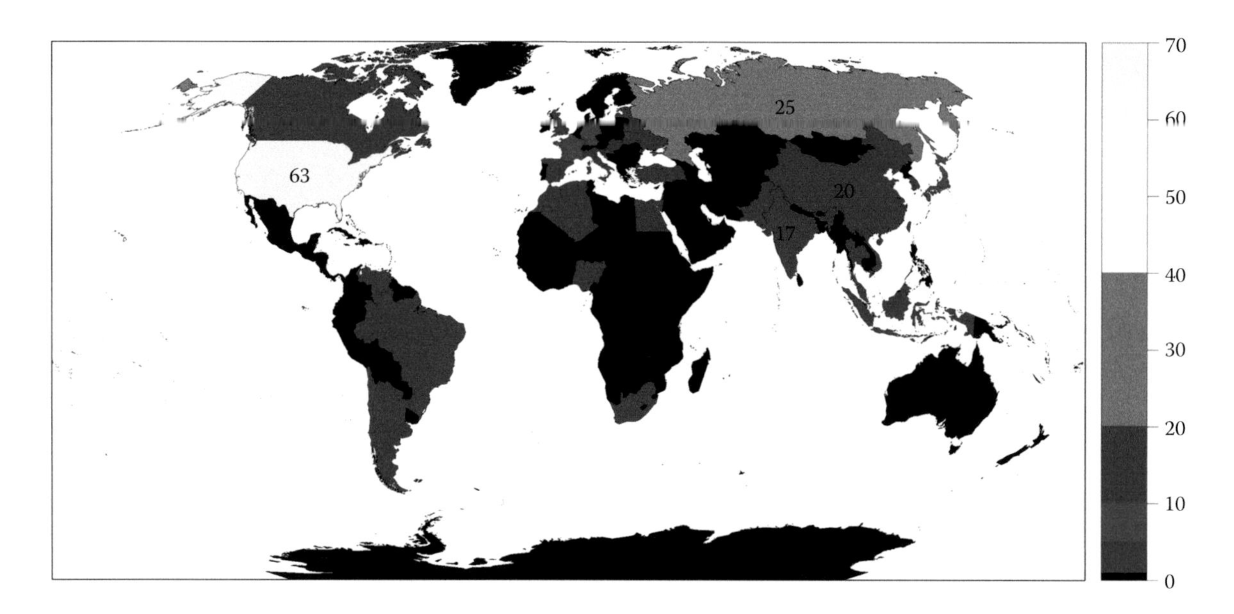

FIGURE 31.29 Map showing the total number of near-polar orbiting, land imaging civilian satellites launched by (or on behalf of) different geographical regions between July 23, 1972, and December 31, 2013. The legend to the right of the map shows the number according to seven groups: 0, 1–5, 6–10, 11–20, 21–40, 41–60, and 60–70. Note that no region falls into the 41–60 category. The numbers of launches made by the top four individual countries (India, China, Russia, and the United States) are specifically cited—note that collectively Europe has launched 30. (From Belward, A.S. and Skøien, J.O., *ISPRS J. Photogram. Remote Sens.*, 2014, ISSN: 0924-2716, http://dx.doi.org/10.1016/j.isprsjprs.2014.03.009, available online: April 28, 2014.)

many interconnected technologies such as remote sensing, GIS, GPS, Internet delivery of data, mobile delivery of data, and number of other technological and commercial convergence of spatial and nonspatial data and their delivery systems. Remote sensing has the ability to gather information at great detail, repeatedly, and over any area of the globe. This means that remote-sensing activities and data have become implicated in a wide variety of legal fields such as evidence, torts, and privacy, among others. These issues are most often domestic in nature and operate outside the international regimes

Acknowledgments

I thank the lead authors and coauthors of each of the chapters for providing their insights and edits of my chapter summaries.

References

Belward, A.S., Skøien, J.O. 2014. Who launched what, when and why; trends in global land-cover observation capacity from civilian earth observation satellites. *ISPRS Journal of Photogrammetry and Remote Sensing*, ISSN: 0924-2716, http://dx.doi.org/10.1016/j.isprsjprs.2014.03.009, available online: April 28, 2014.

Blaschke, T. 2010. Object-based image analysis for remote sensing. *ISPRS International Journal of Photogrammetry and Remote Sensing*, 65(1), 2–16.

Blaschke, T., Hay, G.J., Kelly, M., Lang, S., Hofmann, P., Addink, E., Feitosa, R., van der Meer, F., van der Werff, H., Van Coillie, F., Tiede, D. 2014. Geographic object-based image analysis: A new paradigm in remote sensing and geographic information science. *ISPRS International Journal of Photogrammetry and Remote Sensing*, 87(1), 180–191.

Branger, F., Kermadi, S., Jacqueminet, C., Michel, K., Labbas, M., Krause, P., Kralisch, S., Braud, I. November 15, 2013. Assessment of the influence of land use data on the water balance components of a peri-urban catchment using a distributed modelling approach. *Journal of Hydrology*, 505, 312–325, ISSN: 0022-1694, http://dx.doi.org/10.1016/j.jhydrol.2013.09.055.

Chander, G., Markham, B.L., Helder, D.L. 2009. Summary of current radiometric calibration coefficients for Landsat MSS, TM, ETM+, and EO-1 ALI sensors. *Remote Sensing of Environment*, 113, 893–903.

Chavez, P.S. 1988. An improved dark-object subtraction technique for atmospheric scattering correction of multispectral data. *Remote Sensing of Environment*, 24, 459–479.

Chavez, P.S. 1989. Radiometric calibration of Landsat thematic mapper multispectral images. *Photogrammetric Engineering and Remote Sensing*, 55, 1285–1294.

Comber, A., Fisher, P., Brunsdon, C., Khmag, A. December 2012. Spatial analysis of remote sensing image classification accuracy. *Remote Sensing of Environment*, 127, 237–246, ISSN: 0034-4257, http://dx.doi.org/10.1016/j.rse.2012.09.005.

Damodaran, B.B., Nidamanuri, R.R. 2014. Assessment of the impact of dimensionality reduction methods on information classes and classifiers for hyperspectral image classification by multiple classifier system. *Advances in Space Research*, 53(12), 1720–1734, ISSN: 0273-1177, http://dx.doi.org/10.1016/j.asr.2013.11.027.

Dardel, C., Kergoat, L., Hiernaux, P., Mougin, E., Grippa, M., Tucker, C.J. January 2014. Re-greening Sahel: 30 years of remote sensing data and field observations (Mali, Niger). *Remote Sensing of Environment*, 140, 350–364, ISSN: 0034-4257, http://dx.doi.org/10.1016/j.rse.2013.09.011.

Dow, J.M., Neilan, R.E., Weber, R., Gendt, G. 2007. Galileo and the IGS: Taking advantage of multiple GNSS constellations. *Advances in Space Research*, 39(10), 1545–1551, ISSN: 0273-1177, http://dx.doi.org/10.1016/j.asr.2007.04.064.

Elvidge, C.D., Yuan, D., Weerackoon, R.D., Lunetta, R.S. 1995. Relative radiometric normalization of Landsat multispectral scanner (MSS) data using an automatic scattergram controlled regression. *Photogrammetric Engineering and Remote Sensing*, 61, 1255–1260.

Evangelidis, K., Ntouros, K., Makridis, N., Papatheodorou, C. February 2014. Geospatial services in the cloud. *Computers & Geosciences*, 63, 116–122, ISSN: 0098-3004, http://dx.doi.org/10.1016/j.cageo.2013.10.007.

Freire, S., Santos, T., Navarro, A., Soares, F., Silva, J.D., Afonso, N., Fonseca, A., Tenedório, J. 2014. Introducing mapping standards in the quality assessment of buildings extracted from very high resolution satellite imagery. *ISPRS Journal of Photogrammetry and Remote Sensing*, 90, 1–9, ISSN: 0924-2716, http://dx.doi.org/10.1016/j.isprsjprs.2013.12.009.

Fritz, S., McCallum, I., Schill, C., Perger, C., See, L., Schepaschenko, D., van der Velde, M., Kraxner, F., Obersteiner, M. May 2012. Geo-Wiki: An online platform for improving global land cover. *Environmental Modelling & Software*, 31, 110–123, ISSN: 1364-8152, http://dx.doi.org/10.1016/j.envsoft.2011.11.015.

Fujima, K. 2011. Tsunami measurement data compiled by IUGG tsunami commission, http://www.nda.ac.jp/cc/users/fujima/TMD/index.html, accessed December 2, 2011.

Gbanie, S.P., Tengbe, P.B., Momoh, J.S., Medo, J., Kabba, V.T.S. January 2013. Modelling landfill location using Geographic Information Systems (GIS) and Multi-Criteria Decision Analysis (MCDA): Case study Bo, Southern Sierra Leone. *Applied Geography*, 36, 3–12, ISSN: 0143-6228, http://dx.doi.org/10.1016/j.apgeog.2012.06.013.

Goward, S.N., Ghander, G., Pagnutti, M., Marx, A., Ryan, R., Thomas, N., Tetrault, R. August 2012. Complementarity of ResourceSat-1 AWiFS and Landsat TM/ETM+ sensors. *Remote Sensing of Environment*, 123, 41–56, ISSN: 0034-4257, http://dx.doi.org/10.1016/j.rse.2012.03.002.

Haas, J., Ban, Y. August 2014. Urban growth and environmental impacts in Jing-Jin-Ji, the Yangtze, River Delta and the Pearl River Delta. *International Journal of Applied Earth Observation and Geoinformation*, 30, 42–55, ISSN: 0303-2434, http://dx.doi.org/10.1016/j.jag.2013.12.012.

Heipke, C. November 2010. Crowdsourcing geospatial data. *ISPRS Journal of Photogrammetry and Remote Sensing*, 65(6), 550–557, ISSN: 0924-2716, http://dx.doi.org/10.1016/j.isprsjprs.2010.06.005.

ISO, 2009. ISO 19115-2:2009 Geographic Information-Metadata-Part 2: Extensions for imagery and gridded data Workbook (2.99 MB) - Guide to Implementing ISO 19115-2:2009(E), the North American Profile (NAP), and ISO 19110 Feature Catalogue.

Jensen, J.R. 1996. *Introductory Digital Image Processing: A Remote Sensing Perspective*, 3rd edn. Prentice Hall, p. 318, ISBN: 0-13-145361-0. Publisher: Prentice Hall, NJ.

Jin, S., Feng, G.P., Gleason, S. May 17, 2011. Remote sensing using GNSS signals: Current status and future directions. *Advances in Space Research*, 47(10), 1645–1653, ISSN: 0273-1177, http://dx.doi.org/10.1016/j.asr.2011.01.036.

Ke, Y., Quackenbush, L.J., Im, J. 2010. Synergistic use of QuickBird multispectral imagery and LIDAR data for object-based forest species classification. *Remote Sensing of Environment*, 114(6), 1141–1154, ISSN: 0034-4257, http://dx.doi.org/10.1016/j.rse.2010.01.002.

Kronseder, K., Ballhorn, U., Böhm, V., Siegert, F. August 2012. Above ground biomass estimation across forest types at different degradation levels in Central Kalimantan using LiDAR data. *International Journal of Applied Earth Observation and Geoinformation*, 18, 37–48, ISSN: 0303-2434, http://dx.doi.org/10.1016/j.jag.2012.01.010.

Lillesand, T.M., Kiefer, R.W., Chipman, J.W. 2008. *Remote Sensing and Image Interpretation*, 6th edn., p. 768, ISBN: 978-0-470-46555-4. Publisher: Wiley.

Markham, B.L., Barker, J.L. 1987. Radiometric properties of U.S. Processed Landsat MSS data. *Remote Sensing of the Environment*, 22, 39–71.

Luc Moreau; Paolo Missier; eds. PROV-N: The Provenance Notation. 30 April 2013, W3C Recommendation. URL: http://www.w3.org/TR/2013/REC-prov-n-20130430/.

Mori, N., Takahashi, T. 2012. The 2011 Tohoku Earthquake Tsunami Joint Survey Group Nationwide post event survey and analysis of the 2011 Tohoku Earthquake Tsunami. *Coastal Engineering Journal*, 54, 1250001.

Nagamalai, D., Renaulat, E., Dhanuskodi, M. 2011. *Advances in Digital Image Processing and Information Technology: First International Conference on Digital Image Processing and Pattern Recognition*, Tirunelveli, India. Series: Communications in Computer and Information Science. Paperback. Springer, Tirunelveli, Tamil Nadu, India p. 478, ISBN-10: 3642240542.

Neckel and D. Labs. 1984. The solar irradiance between 3300 and 12500a.° *Sol. Phys.*, 90:205–258.

Oppelt, N., Schulze, F., Bartsch, I., Doernhoefer, K., Eisenhardt, I. 2012. Hyperspectral classification approaches for intertidal macroalgae habitat mapping: A case study in Helgoland. *Optical Engineering*, 51, 111703, doi: 10/1117/1.OE.51.11.111703.

Price, J.C. 1987. Calibration of satellite radiometers and the comparison of vegetation indices. *Remote Sensing of the Environment*, 21, 15–27.

Richards, J.A., Xiuping, J. 2006. *Remote Sensing Digital Image Analysis*, Vol. XXV, 4th edn. Springer-Verlag, p. 439 Publisher: Berlin: Springer.

Rodriguez-Galiano, V.F., Ghimire, B., Rogan, J., Chica-Olmo, M., Rigol-Sanchez, J.P. January 2012. An assessment of the effectiveness of a random forest classifier for land-cover classification. *ISPRS Journal of Photogrammetry and Remote Sensing*, 67, 93–104, ISSN: 0924-2716, http://dx.doi.org/10.1016/j.isprsjprs.2011.11.002.

Stosius, S., Beyerle, G., Hoechner, A., Wickert, J., Lauterjung, J. March 1, 2011. The impact on tsunami detection from space using GNSS-reflectometry when combining GPS with GLONASS and Galileo. *Advances in Space Research*, 47(5), 843–853, ISSN: 0273-1177, http://dx.doi.org/10.1016/j.asr.2010.09.022.

Suppasri, A., Futami, T., Tabuchi, S., Imamura, F. October 2012. Mapping of historical tsunamis in the Indian and Southwest Pacific Oceans. *International Journal of Disaster Risk Reduction*, 1, 62–71, ISSN: 2212-4209, http://dx.doi.org/10.1016/j.ijdrr.2012.05.003.

Thenkabail, P., Lyon, G.J., Turral, H., Biradar, C.M. 2009a. *Remote Sensing of Global Croplands for Food Security*. Boca Raton, FL: CRC Press, Taylor & Francis Group, p. 556 (48 in color). Published in June 2009.

Thenkabail, P.S., Biradar, C.M., Noojipady, P., Dheeravath, V., Li, Y.J., Velpuri, M., Gumma, M. et al. July 20, 2009b. Global irrigated area map (GIAM), derived from remote sensing, for the end of the last millennium. *International Journal of Remote Sensing*, 30(14), 3679–3733.

Thenkabail, P.S., Gamage, N., and Smakhin, V. 2004. The use of remote sensing data for drought assessment and monitoring in south west Asia. IWMI Research report # 85. Pp. 25. IWMI, Colombo, Sri Lanka. http://www.iwmi.cgiar.org/Publications/IWMI_Research_Reports/PDF/pub085/RR85.pdf.

Thenkabail, P.S., Smith, R.B., De-Pauw, E. 2002. Evaluation of narrowband and broadband vegetation indices for determining optimal hyperspectral wavebands for agricultural crop characterization. *Photogrammetric Engineering and Remote Sensing* 68(6), 607–621.

Tilling, A.K., O'Leary, G.J., Ferwerda, J.G., Jones, S.D., Fitzgerald, G.J., Rodriguez, D., Belford, R. October–December 2007. Remote sensing of nitrogen and water stress in wheat. *Field Crops Research*, 104(1–3), 77–85, ISSN: 0378-4290, http://dx.doi.org/10.1016/j.fcr.2007.03.023.

Tilmes, C., Fleig, A.J. 2008. Provenance tracking in an earth science data processing system. In: J. Freire, D. Koop (eds.), *Provenance and Annotation of Data and Processes*. Berlin, Germany: Springer-Verlag, pp. 221–228. http://ebiquity.umbc.edu/_file_directory_/papers/445.pdf.

Tucker, C.J. 1979. Red and photographic infrared linear combinations for monitoring vegetation. *Remote Sensing of the Environment*, 8, 127–150.

U.S. Geological Survey (USGS). 2012. Earth's Tectonic Plates—USGS, http://earthquake.usgs.gov/regional/nca/…/kml/Earths_Tectonic_Plates.kmz, accessed May 25, 2012.

Uhl, F, Oppelt, N., Bartsch, I. 2013. Mapping marine macroalgae in case 2 waters using CHRIS PROBA. *Proceedings of the ESA Living Planet Symposium*, September 9–13, Edinburgh, U.K., *ESA Special Proceedings SP-722* (CD-ROM).

Uhl, F., Oppelt, N., Bartsch, I., Geisler, T., Heege, T., Nehring, F. 2014. KelpMap—Development of an EnMAP approach to monitor sublitoral marine macrophytes (KelpMap—Entwicklung eines EnMAP Verfahrens zur Bestimmung von sublitoralen marinen Makrophyten). Final report of research project FKZ: 50EE1020, funded by the German Federal Ministry of Economy and Technology (BMWi), 36pp.

Vermote, E.F., El Saleous, N.Z., and Justice, C.O. 2002. Atmospheric correction of MODIS data in the visible to middle infrared: First results. *Remote Sensing of Environment*, 83(1–2), 97–111.

Vrieling, A., Meroni, M., Shee, A., Mude, A.G., Woodard, J., (Kees) de Bie, C.A.J.M., Rembold, F. May 2014. Historical extension of operational NDVI products for livestock insurance in Kenya. *International Journal of Applied Earth Observation and Geoinformation*, 28, 238–251, ISSN: 0303-2434, http://dx.doi.org/10.1016/j.jag.2013.12.010.

Wan, Z., Hong, Y., Khan, S., Gourley, J., Flamig, Z., Kirschbaum, D., Tang, G. August 2014. A cloud-based global flood disaster community cyber-infrastructure: Development and demonstration. *Environmental Modelling & Software*, 58, 86–94, ISSN: 1364-8152, http://dx.doi.org/10.1016/j.envsoft.2014.04.007.

Wang, Z., Jensen, J.R., Im, J. October 2010. An automatic region-based image segmentation algorithm for remote sensing applications. *Environmental Modelling & Software*, 25(10), 1149–1165, ISSN: 1364-8152, http://dx.doi.org/10.1016/j.envsoft.2010.03.019.

Wei, Y., Liu, H., Song, W., Yu, B., Xiu, C. August 2014. Normalization of time series DMSP-OLS nighttime light images for urban growth analysis with Pseudo Invariant Features. *Landscape and Urban Planning*, 128, 1–13, ISSN: 0169-2046, http://dx.doi.org/10.1016/j.landurbplan.2014.04.015.

Whiteside, T.G., Boggs, G.S., Maier, S.W. December 2011. Comparing object-based and pixel-based classifications for mapping savannas. *International Journal of Applied Earth Observation and Geoinformation*, 13(6), 884–893, ISSN: 0303-2434, http://dx.doi.org/10.1016/j.jag.2011.06.008.

Xu, H., Huang, S., Zhang, T. October 15, 2013. Built-up land mapping capabilities of the ASTER and Landsat ETM+ sensors in coastal areas of southeastern China. *Advances in Space Research*, 52(8), 1437–1449, ISSN: 0273-1177, http://dx.doi.org/10.1016/j.asr.2013.07.026.

Yue, P., Gong, J., Di, L., Yuan, J., Sun, L., Sun, Z., Wang, Q. 2010. GeoPW: Laying blocks for geospatial processing web. *Transactions in GIS*, 14(6), 755–772.

Zhao, P., Foerster, T., Yue, P. October 2012. The geoprocessing web. *Computers & Geosciences*, 47, 3–12, ISSN: 0098-3004, http://dx.doi.org/10.1016/j.cageo.2012.04.021.

Index

D

E

H

P

T